BRINGING FOSSILS TO LIFE

An Introduction to Paleobiology

Donald R. Prothero

Occidental College

Boston, Massachusetts Burr Ridge, Illinois Dubuque, Iowa
Madison, Wisconsin New York, New York San Francisco, California St. Louis, Missouri

This book is dedicated to the great paleontologists who have instructed and inspired me:

Michael Woodburne
Michael Murphy
Malcolm McKenna
Niles Eldredge
Steve Stanley
and
Stephen Jay Gould

WCB/McGraw-Hill

A Division of The ***McGraw-Hill*** *Companies*

Bringing Fossils to Life: An Introduction to Paleobiology

This book is printed on recycled, acid-free paper containing 10% postconsumer waste.

6 7 8 9 0 QPD/QPD 9 0

ISBN 0-07-052197-2

Publisher: *Edward E. Bartell*
Sponsoring Editors: *Lynne M. Meyers/ Anne Duffy*
Marketing Manager: *Lisa L. Gottschalk*
Project Manager: *Donna Nemmers*
Production Supervisor: *Sandra Hahn*
Designer: *K. Wayne Harms*
Cover painting: *Pat Linse*
Typeface: *9/10 Palatino and 9 Helvetica*
Printer: *Quebecor Printing Book Group/Dubuque*

Library of Congress Cataloging-in-Publication Data
Prothero, Donald R.
Bringing fossils to life : an introduction to paleobiology / Donald R. Prothero. – 1st ed.
p. cm.
Includes bibliographical references and index.
ISBN 0-07-052197-2
1. Paleobiology. 2. Fossils. I. Title.
QE719.8.P76 1998
560—dc21 97-30208
CIP

www.mhhe.com

CONTENTS

LIFE OF THE PAST AND PRESENT

Preface

Scientists in general might be excused for assuming that most geologists are paleontologists and most paleontologists have staked out a square mile as their life's work. A revamping of the geologist's image is badly needed.

Editorial, *Nature*, 1969

As well as being lumps of patterned stone, fossils are also historical documents. History per se has had a bit of bad press recently. . . There is a tension between the documentation of history (famously referred to as "one bloody thing after another"), and the search for universal principles that are ahistoric and possibly timeless. After a period in the doldrums, the bearers of the historical tidings, the paleontologists, are making tentative movements toward the legendary High Table where, just visible through the clouds of incense (and rhetoric), the high priests of evolutionary theory smile benignly.

S. Conway Morris, "Early Metazoan radiations: what the fossil record can and cannot tell us," 1996

Paleontology is at a crossroads. For decades, introductory paleontology classes focused on the taxonomy and anatomy of the major phyla of fossil invertebrates. This traditional approach dates back at least a century, and has long given paleontology the bad reputation of drudgery. Paleontology was often equated with "stamp collecting." Paleontologists were caricatured as narrow specialists who were interested only in the trivial details of the names, ages, and shapes of their fossils.

In the 1970s, however, paleontology experienced a revolution. A new generation of "Young Turks" who earned their doctorates in the 1960s challenged the old obsession with trivial details and asked broader theoretical questions of the fossil record. How good is the fossil record, and what can (and can't) we read from it? What does the fossil record say about the tempo and mode of evolution? What do the patterns of mass extinction reveal about fragility of life on this planet? How do the movements of continents explain the distribution of fossils in time and space? How do we determine the relationships and evolutionary histories of extinct organisms? In the preface to the pioneering book *Models in Paleobiology* (1972), Tom Schopf described a revealing encounter with a young Ph.D. student who could not decide on his dissertation topic. Should he describe a collection of fossils his adviser had recently assembled? It never occurred to this student to ask *what problems he should be solving* or *what hypotheses he should be testing* using the fossil record.

This purely descriptive approach to paleontology became outmoded as the "Young Turks" changed the way we look at fossils. In 1972, Niles Eldredge and Stephen Jay Gould published their punctuated equilibrium model (in the Schopf book just mentioned), and that debate has been raging ever since. In 1975, Tom Schopf and Ralph Johnson founded *Paleobiology,* and it became the premier theoretical journal in the profession. In 1975, Steve Stanley first suggested the model of species selection, and evolutionary theory has since begun to rethink the issue of macroevolution. In 1980, the asteroid hypothesis for the extinction of dinosaurs was published, and the last 17 years have seen a tremendous bandwagon of interest in mass extinctions and their implications. For the last 20 years, the Paleontological Society meetings have featured papers of broad theoretical interest, with few talks that are purely local or descriptive in nature.

Ironically, this revolution in the practice of paleontology has barely worked its way into the teaching of paleontology. Most textbooks are highly detailed reviews of the anatomy and systematics of the major groups of fossil invertebrates, with only short chapters on theoretical topics. Having taught from several of these books, I have found that most of today's undergraduates are not prepared for this level of detail when they don't see the point in learning it all. Such an approach is more appropriate to advanced level or graduate courses in paleontology, after the student has become hooked on fossils and is motivated to master all those names. First, however, the instructor should give the student a sense of the excitement of the field, and explore the reasons why so many people find fossils fascinating, instead of killing their interest with excessive memorization. Other textbooks stress theoretical topics at a very high level but do not give the necessary background in the taxonomy and systematics so that the students can place the theory in context.

This book is an attempt to bridge the gap between purely theoretical paleobiology and purely descriptive invertebrate paleontology books. It is written for the general geology or biology majors who take paleontology for fun or as part of their requirements, not as an encyclopedia for professionals. The theoretical topics are covered in Part I, stressing the newer developments in paleobiology that should excite the student. Part II outlines the major features of the fossil record of vertebrates and invertebrates. Here, I have deliberately reduced the excessive memorization of taxa and anatomy to a bare minimum of terms that are essential in talking about each group. In my experience, most of this memorization is quickly forgotten unless there is a point to it, or a problem to be solved with it. Only students who plan to follow paleontology as a career should be expected to know the quantity of information that some other textbooks demand.

In both parts, I have tried to write at a level of detail appropriate to undergraduate geology majors with a previous course in historical geology, but with limited backgrounds in biology. Throughout the book, I have tried to capture a sense of the excitement of paleobiology and have tried to convey *why* we try to do what we do. Students today are very results-oriented, so in the back of my mind their persistent question "What's the point?" guided my writing. As I wrote, I kept in mind what 20 years of experience with some of the best undergraduates in the country (at Columbia, Vassar, Knox, and Occidental Colleges) has taught me about what students retain and what ideas excite them and keep their interest. If we can interest them at this level, then they will come back for more and be willing to learn more taxonomy and morphology. Then they will see how exciting paleobiology has become and possibly consider it to be their future as well.

Acknowledgments: I thank my McGraw-Hill editor, Anne Duffy, for encouraging me to write this book, and Lynne Meyers, Wayne Harms, Donna Nemmers, and Tracie Kammerude at WCB/McGraw-Hill for all their help. James W. Bradley copyedited the text. The cover art was painted by Pat Linse. I thank Linda Ivany and Norman MacLeod for carefully critiquing Part I, and Marie-Pierre Aubry, Bill Ausich, Loren Babcock, John Barron, Bill Berggren, Dave Bottjer, Gilbert Brenner, Carl Brett, Sandy Carlson, Alan Cheetham, Mary Droser, Tony Ekdale, George Engelmann, James Firby, Tom Holtz, Jim Hopson, Conrad LaBandeira, Neil Landman, Dave Lazarus, Jere Lipps, Bill Orr, Kevin Padian, Roy Plotnick, John Pojeta Jr., J. Keith Rigby, Dave Schwimmer, William Siesser, Jim Sprinkle, George D. Stanley, Jr., and Karen Whittlesey for their helpful comments on various parts of the book. Many colleagues were very helpful in providing photographs or illustrations, and their contributions are acknowledged in the figure captions.

This book was written, laid out, and produced electronically direct to film by the author using a PowerMac 6500, an Agfa StudioStar scanner with Adobe Photoshop, Microsoft Word 6.0.1, and QuarkXpress.

Donald R. Prothero
Los Angeles, California
July, 1997

To the student: Why study fossils?

Fossil hunting is by far the most fascinating of all sports. It has some danger, enough to give it zest and probably about as much as in the average modern engineered big-game hunt, and danger is wholly to the hunter. It has uncertainty and excitement and all the thrills of gambling with none of its vicious features. The hunter never knows what his bag may be, perhaps nothing, perhaps a creature never before seen by human eyes. It requires knowledge, skill, and some degree of hardihood. And its results are so much more important, more worthwhile, and more enduring than those of any other sport! The fossil hunter does not kill, he resurrects. And the result of this sport is to add to the sum of human pleasure and to the treasure of human knowledge.

George Gaylord Simpson, *Attending Marvels*, 1934

At the beginning of any course, students often ask, "Why study this subject? What will I get out of this class?" Many subjects seem very abstract and intangible to students, and the relevance to their careers, or importance in their daily lives, is not readily apparent. Like many other topics in the curriculum, paleontology has had to fight to justify its existence. However, the benefits of paleontology are easy to enumerate:

1. **Biostratigraphy**—*Fossils are the only practical means of telling time in geology.* Radioisotopic decay methods, such as potassium-argon or uranium/lead dating, work only in rocks that have cooled down from a very hot state, such as igneous or metamorphic rocks. Most of geological history is contained in sedimentary rocks, which cannot be dated by radioisotopes. Consequently, fossils are the only practical method of determining the age of rocks in most geological settings. For a long time, the largest employers of paleontologists were oil companies, who relied heavily on biostratigraphers to tell them where to find oil. Modern civilization would have been impossible without them. No matter what fads come and go in geology, there will always be demand for paleontologists who can answer the basic question, "How old is it?"

2. **Evolution**—*Fossils are the only direct record of the history of life.* Although evolutionary biology has made enormous strides studying living organisms such as bacteria, fruit flies, and lab rats, these studies see evolution only in the thin slice of time known as the Recent. Fossils provide the only direct evidence of 3.5 billion years of the history of life, and in many cases, they suggest processes that might not be explainable by what is known from living organisms. Fossils provide a fourth dimension (time) to the biology of many living organisms. Many groups of organisms, such as conodonts and graptolites, are extinct and are known only from the fossil record.

3. **Paleoecology**—*Fossils can provide direct evidence of ancient environments.* Although many sedimentary rocks deposited in different environments look very similar, the fossils and trace fossils found within them are often their most diagnostic feature. They can be used to pinpoint the depositional environment more precisely than any other property of the sedimentary rock.

4. **Paleogeography**—*Fossils can be critical to determining ancient continental positions and connections.* Some of the earliest evidence for continental drift came from the similarities of fossils on different continents, and paleontological evidence is critical to any understanding of biogeography.

5. **Simple fascination**—*Fossils are interesting, in and of themselves.* Although the reasons listed above are important, most people become paleontologists because they love fossils and are fascinated by extinct organisms. In what other discipline but paleontology could a discovery only slightly larger than any previously known (a new gigantic dinosaur specimen) make the front page of the *New York Times*? Why is it that anything with dinosaurs in it (including the highest-grossing movie of all time, *Jurassic Park*) is immensely popular? Although dinosaurs are only a small part of paleontology, the popular fascination with things prehistoric can be extended even to the invertebrates of the Cambrian (witness the enormous popularity of trilobites, or the response to Stephen J. Gould's book *Wonderful Life*).

In short, there are many reasons, both practical and fun, for studying fossils. The eminent early twentieth-century paleontologist William Diller Matthew (writing in 1925 during the height of anti- evolutionism coming from the Scopes "Monkey" Trial) said it eloquently (and with the traditionally masculine language of his time, when there were few female paleontologists):

A few men, a mere handful scattered among the millions of civilization, devote their lives to collecting and studying fossils. A larger number take a more casual interest in the results of these studies. The vast majority have never heard of fossils or ask indifferently, "Why should anyone waste his time upon such useless studies?" This is the answer:

In the first place, what are fossils? They are teeth or bones or shell of animals found buried in solid rock. These fossils, remains of animals that have lived in past ages but no longer survive, are the records of the history of life. We collect and study and compare them so that we can reconstruct these extinct animals, so that we can find out, as far as may be, just how they differed from those that preceded and from those that succeeded them, how they lived and to what environment they were adapted. Our aim is to reconstruct the history of life during the vast periods of time that have elapsed since the first rocks were formed.

Current history is but a passing phase, a stage in the march of events, past, present, and future. We cannot follow present events without becoming keenly interested in the past, which explains the present, and in the future, which we can predict more certainly if we have an adequate knowledge of the past and the present. The more we know of these, the more clearly can we discern the laws that govern the orderly progress of events, the more definitely and positively can we perceive what is to come, at least in its broader outlines. Herein lies the great fascination of historical studies: in the attempt to synthesize and arrange the infinite multiplicity of events great and small, to find the underlying causes to whose interaction they may be ascribed, to test and prove the soundness of our theories by bringing them to bear upon other groups of events, above all to apply the acid test of fulfilment to our prophecies, the confirmation of our theories by new discoveries and forthcoming events. To read the future is the dearest wish of man and it can be done in so far as his knowledge and understanding of the past show whence we have come and whither we are going.

But history in the ordinary sense of the word deals with but a limited portion of the past of man. The world in which we live has a far wider scope and its history extends backward through enormous periods of time in comparison with the few thousand years covered by recorded human history. From these fossil records, the "documents" of earth history, it has been possible to build up a great and splendid science, secure and fixed in its massive foundations and its broader lines of structure, more doubtful and speculative in some of its lighter tracery and ornamentation. Those who, through field work and study, have been able to add brick by brick to extend and amplify its solid basis, who have learned the laws of its architecture and aided in building up its superstructure, who have at times been privileged to add some bright pinnacle or favorite cornice to its glittering towers—these men have come to love their science beyond all else. It is their home which they have helped to build, and its beauty and symmetry, its noble and appropriate proportions, and its perfection fill them with an ever-growing admiration and affection.

Do you wonder that the paleontologist, absorbed in contemplation of his splendid edifice, walks a little apart from the ways of men; that the little personal affairs and interests of the fleeting present which make up the world of his fellows, seem to him but gewgaws and trifles of no importance? His field of vision embraces the whole of life. His time scale is so gigantic that it dwarfs to insignificance the centuries of human endeavor. And the laws and principles which he studies are those which control the whole great stream of life, upon which the happenings of our daily existence appear but as little surface ripples.

Pre-eminent among the laws which govern the architecture of our world of life is evolution. To the zoologist the law of evolution appears as a theory, an explanation of the world of nature that lies about him. It is the only theory that really explains it, and it fits all the marvelously complex details of adaptation, the perfections and likewise the imperfections of structure of every animal and plant so perfectly and accurately that few or no zoologists can question the theory, however they may dispute about the precise method of its action. To the paleontologist, however, evolution appears not as a theory but as a fact of the record. He does not and cannot doubt the gradual development of diversely specialized races from a common ancestral stock through a long series of intermediate gradations, for he has before him all these stages in the evolution of the race preserved as fossils, each in its appropriate place in the successive strata of a geologic period. It is not a matter of deduction but of observation, at least in those races of animals whose fossil record has been discovered; for the rest it is a matter of obvious inference. Concerning the causes and methods of this evolutionary process he finds wide room for discussion; but of the fact, of the actuality of it he can have no doubt. Evolution is no more a theory to the man who has collected and studied fossils than the city of New York is a theory to the man who lives in it.

But, in truth, evolution is only one aspect of the order of nature, of the relations of cause and effect, of continuity of space and time, which pervade the universe and enable us to comprehend its simplicity of plan, its complexity of detail. The paleontologist, engaged in adding year by year to the mass of documents which record the history of life, in deciphering their meaning and interpreting their significance, has no more occasion to doubt the its continuity and orderly development than the historian has to doubt the continuity and consecutive evolution of human history, or the student of current affairs to doubt that the events of today will result in the conditions of tomorrow.

Such is the value of palæontology. It provides an essential part of the evidence for scientific study of the rocks, which has made possible the huge expansion of the mining industries upon which our modern material civilization is so largely based. Its higher value lies in adding to our knowledge and aiding in our comprehension of the world we live in, in tracing the past history of life and finding in it the explanation of the present, in observing the ordered progress of evolution under natural law from the beginnings of the world down to the present day, in helping us to discern through a better knowledge of the past what may be the course of future events.

W.D. Matthew, 1925, *Natural History* 25(2): 166-168.

BRINGING FOSSILS TO LIFE
An Introduction to Paleobiology

PART I
The Fossil Record:
A Window on the Past

Figure 1.1. Seventeenth-century illustrations of fossil ammonoids (*Cornua ammonis*, also know as "snake stones") prepared by the versatile scientist Robert Hooke, the father of microscopy and paleontology in Britain. (From the posthumous works of Robert Hooke, 1703.)

Chapter 1

The Fossil Record

In the mountains of Parma and Piacenza multitudes of rotten shells and corals are to be seen, still attached to the rocks. . . And if you were to say that such shells were created, and continued to be created in similar places by the nature of the site and of the heavens, which had some influence there—such an opinion is impossible for the brain capable of thinking, because the years of their growth can be counted on the shells, and both smaller and larger shells may be seen, which could not have grown without food, and could not have fed without motion, but there they could not move.

And if you wish to say that it was the Deluge which carried these shells hundreds of miles from the sea, that cannot have happened, since the Deluge was caused by rain, and rain naturally urges rivers on towards the sea, together with everything carried by them, and does not bear dead objects from sea shores towards the mountains. And if you would say that the waters of the Deluge afterwards rose above the mountains, the movement of the sea against the course of the rivers must have been so slow that it could not have floated up anything heavier than itself.

Leonardo da Vinci, c. 1500

WHAT IS A FOSSIL?

When we pick up fossils in a roadcut, or see a dinosaur skeleton in a museum, we have no problem connecting it to some sort of extinct organism. We have been conditioned since our early education to interpret fossils as remains of extinct organisms, and it is hard for us to imagine any other explanation.

Centuries ago, however, such an interpretation was not automatic or even easy to make. The ancient Greeks interpreted the giant bones of mammoths as the remains of mythical giants, but were puzzled with sea shells found hundreds of feet above sea level and miles inland. Had the sea once covered the land, or had these objects grown within the rocks like crystals do? In the sixth century B.C., Xenophanes of Colophon saw the seashells high in a cliff on the island of Malta and suggested that the land had once been covered by the sea. The oldest recorded statement that fossils are the remains of once-living animals that have been entombed in rocks was made by Xanthos of Sardis around 500 B.C. Aristotle (born 384 B.C.) suggested that the fossils of fish were remains of sea animals that had swam into cracks in rocks and were stranded there, and his ideas were influential for the next 2000 years.

From the late days of the Roman Empire, almost all people in Western society were raised to believe in the literal interpretation of the book of Genesis, and the stories of the six days of Creation and Noah's flood colored their view of rocks and fossils. For those of us in the end of the twentieth century, a fossil snail shell looks so similar to its living descendants that we cannot imagine any other explanation. We forget that most people of that time (other than fishermen) had limited familiarity with life on the bottom of the ocean. In fact, many fossils bear no resemblance to anything that fifteenth-century Europeans could have seen. Until the living chambered nautilus was discovered in 1829, who could imagine that the coiled objects known as *Cornua ammonis* ("Horns of Ammon"), "serpent stones" were relatives of the squid and octopus with a coiled, chambered shell (Fig. 1.1)? Who could imagine that the strange bullet-shaped objects known as a belemnites were also related to squid? Even today, most people who pick up the odd cylindrical objects known as crinoid columnals (Fig. 1.2) do not recognize them as relatives of the sea star or sea urchin, because only a few people have seen the rare stalked crinoids that still live on the seafloor. For centuries, scholars were impressed with the star-shaped patterns in the centers of the columnals (and the radial patterns in fossil corals), and thought they had been produced by thunderbolts or fallen from the sky; they were known as "star stones" (*Lapis stellaris* or *Astroites stellis*).

During the Middle Ages and Renaissance, learned men began to speculate on the meaning of fossils, producing a wide range of interpretations. Originally, the word "fossil" (from the Latin *fossilis*, "dug up") applied to any strange object found within a rock. These included not only the organic remains that we call fossils, but also crystals and concretions and many other structures that were not organic in origin. Most scholars thought that fossils had formed spon-

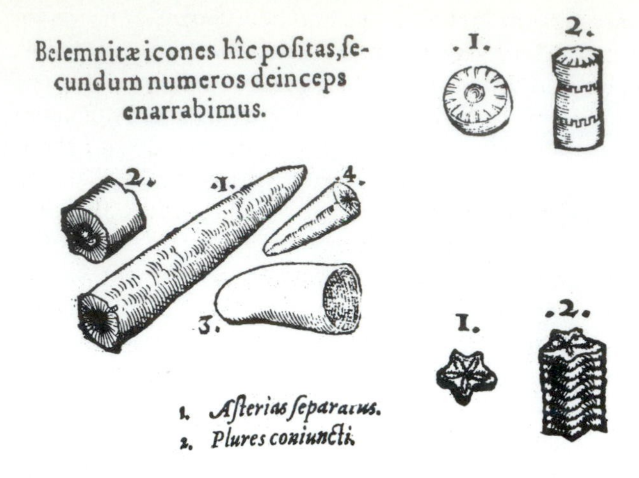

Figure 1.2. (above) Conrad Gesner's 1565 illustrations of bullet-shaped belemnites and crinoid columnals, neither of which looked like marine organisms familiar to Renaissance Europeans. To many of them, including Gesner, the star-like pattern of some crinoid stems suggested that they might be produced by falling stars.

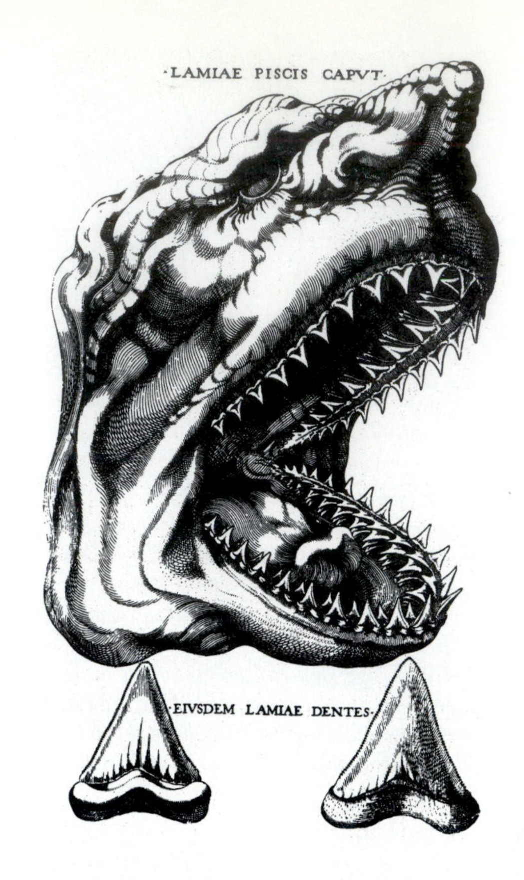

Figure 1.3. (right) Steno's 1669 illustration of the head of a shark, showing that the "tongue stones" or *glossopetrae* are extremely similar to modern shark teeth.

taneously within the rock; those that resembled living organisms were thought to have crept or fallen into cracks and then converted to stone. Others thought they were grown in rocks from seeds, or were grown from fish spawn washed into cracks during Noah's flood. Many scholars thought they were supernatural, "pranks of nature" (*lusus naturae*), or "figured stones" produced by mysterious "plastic forces." Still others considered them to be works of the Devil, placed in the rocks to shake our faith. As quaint and comical as these ideas seem to us today, in their own time they were perfectly rational for people who believed in a literal interpretation of Genesis, and thought the Earth had been created just as we see it about 6000 years ago, with little or no change since then except for decay and degradation due to Adam's sin.

Some Renaissance men, however, were ahead of their time. Around 1500, Leonardo da Vinci (1452-1519) recognized that the fossil shells in the Apennine Mountains of northern Italy represented ancient marine life, even though they were miles from the seashore. Unlike his contemporaries who thought they had been washed there by the Flood, da Vinci realized that they could not have washed that far in 40 days, and many shells were too fragile to have traveled that far. Many shells were intact and in living position and resembled modern communities found near the seashore; clearly they were not transported. In some places, there were many shell beds separated by unfossiliferous strata, so they clearly were not due to a single Flood. However, most of da Vinci's ideas remained in his unpublished notebooks, and even if he had tried to publicize them, they would not have been accepted at that early date.

In 1565, the Swiss physician Conrad Gesner (1516-1565) published *De rerum fossilium* ("On the nature of fossils"), the first work that actually illustrated fossils. With this step, the vague verbal descriptions of earlier authors could be made more precise (Fig. 1.2). Gesner based his descriptions on both his own collections and those of his friends, beginning the modern tradition of scientific exchange, analysis, and comparison. Gesner was correct in comparing most fossils to their living relatives, but he thought that some objects (crinoid columnals, belemnites) were formed by mineral precipitation. Like most of his contemporaries, Gesner interpreted fossils as supernatural representations of Neoplatonic "ideal forms" and did not explore most of the implications that would seem obvious to us today.

Through all of these early writings, four main questions about fossils were disputed:

1. Are fossils really organic remains?
2. How did they get into the rocks?
3. When did they get there—as the rock was being formed, or long after?
4. How did they become petrified?

Essentially modern answers to all these questions were first proposed by a Dane named Niels Stensen,

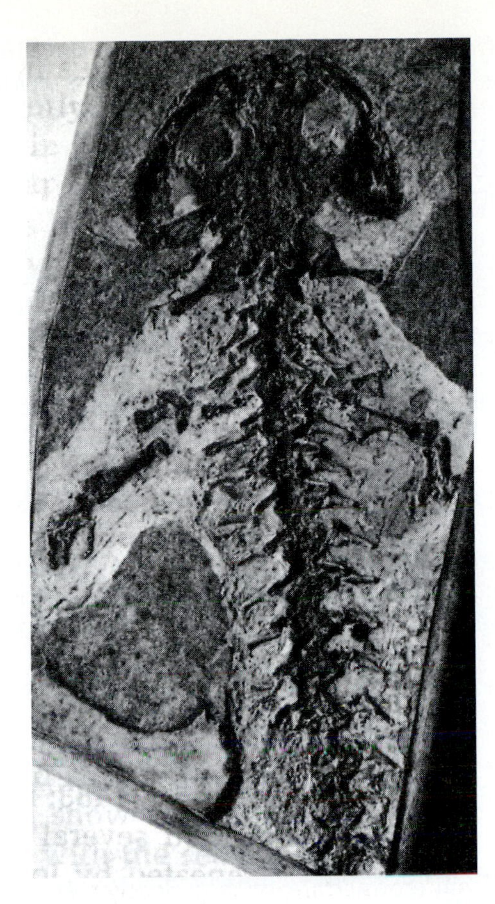

Figure 1.4. Johann Scheuchzer's *Homo diluvii testis*, or "Man, a witness of the Flood," described in 1726. Unfortunately, Scheuchzer's anatomical skills were not up to his Biblical knowledge, since it is actually the fossil of a giant salamander.

known to later generations by the Latinized version of his name, Nicholaus Steno (1638-1686). Steno was the court physician to the Grand Duke of Tuscany, so he had ample opportunity to see the shells in the rocks of the Apennine Mountains above Florence. In 1666 he had a chance to dissect a large shark caught near the port town of Livorno. A close look at the mouth of the shark showed that its teeth closely resembled fossils known as "tongue stones" (*glossopetrae*), which had been considered the petrified tongues of snakes or dragons (Fig. 1.3). Steno realized that tongue stones were actually ancient shark teeth, and that fossil shells were produced by once-living organisms.

In 1669, Steno published *De solido intra solidum naturaliter contento dissertationis prodromus* ("Forerunner to a dissertation on a solid naturally contained within a solid"). The title may seem peculiar at first until you realize the central problem that Steno faced: how did these solid objects get inside solid rock? Steno realized that the enclosing sandstone must have once been loose sand and was later petrified into sandstone. With this idea, he overturned the longstanding assumption that all rocks had been formed exactly as we see them during the first days of Creation. Steno extended this insight into a general understanding of relative age of geological features. Fossils that were enclosed in rock that had been molded around them must be older than the rock in which they lie. On the other hand, crystals that clearly cut across the preexisting fabric of a rock must have grown within the rock after it formed. From this, Steno generalized the principles of superposition, original horizontality, and original continuity that are the fundamental principles of historical geology and stratigraphy.

As the *Prodromus* was being published, Steno converted to Catholicism and gave up his scientific interests, so the "forerunner" was never followed by the promised dissertation. He was eventually made the Bishop of Titiopolis, a region in eastern Europe that had not converted to Catholicism, so he never lived there or ministered to its people. Instead, he returned to Denmark to serve the Church for the rest of his life.

About the time that Steno's writings appeared, a pioneering British scientist came to similar conclusions. Robert Hooke (1635-1703) is better known as the "Father of Microscopy," because he built one of the first microscopes and made the first drawings of microorganisms and the details of cellular structure. In 1665, Hooke made observations of many natural objects, including the first accurate drawings of fossils, which were published posthumously in 1705 (Fig. 1.1). Hooke even suggested that fossils might be useful for making chronological comparisons of rocks of similar age, much as Roman coins were used to date ancient historical events in Europe. He speculated that species had a fixed "life span," for many of the fossils he studied had no living counterparts. This was one of the first hints of the extinction of species, because few people at that time doubted that all the species on earth had been created 6000 years ago and were still alive.

However, most of Hooke's and Steno's ideas would not be accepted for another century. In the early 1700s, ideas about fossils were still heavily influenced by the Bible. For example, in 1726 the Swiss naturalist Johann Scheuchzer (1672-1733) described a large fossil as "the bony skeleton of one of those infamous men whose sins brought upon the world the dire misfortune of the Deluge." He named it *Homo diluvii testis* ("Man, a witness of the Flood"). Unfortunately, since comparative anatomy was not very sophisticated at this time, his specimen turned out to be giant fossil salamander (Fig. 1.4). Another scholar, Dr. Johann Beringer (1667-1740), dean of the medical school of Wurzburg, Germany, was fascinated with the "petrifactions" that collectors had brought him from the local hills. Some bore resemblance to frogs, shells, and many other natural objects; others had stars and many other curious shapes and patterns. In 1726, as Beringer was about to publish a massive monograph of his "figured stones," two colleagues whom Beringer had offended confessed to the prank. They had carved the figured stones, correctly guessing that he was gullible enough to accept them, but their warning came too late to stop publication of the hoax. Beringer died a ruined man, spending his last pfennig trying to buy back all the copies of his book.

By the mid 1700s, however, naturalistic concepts of fossils began to prevail. When Linnaeus published his first edition of his landmark classification of all life, *Systema Naturae*, in 1735, fossils were treated and named as if they were living animals. Around 1800, Baron Georges Cuvier (1769-1832) made great strides in comparative anatomy, skillfully showing that certain anatomical features, such as claws and sharp teeth, or hooves and grinding teeth, were correlated. He became so adept at this knowledge that he started the paleontological tradition of predicting unknown parts of the animal by comparison to the known anatomy of close relatives. Cuvier also showed that the bones of mastodonts and mammoths were the remains of elephant-like beasts that clearly had to be extinct, because the explorers had not found them on even the most remote continents. Prior to that time, most people could not accept the fact of extinction, because it went against their notion of Divine Providence. After all, if God watched after the little sparrow, surely He would not allow any of his creations to go extinct? Cuvier went on to become the founder of comparative anatomy and vertebrate paleontology, and brought much of paleontology out of Biblical supernaturalism and into a firm comparative basis.

Just before 1800, British engineer William Smith (1769-1839) was surveying England for the great canal excavations prompted by the Industrial Revolution. From these fresh canal exposures and regular visits to mines, Smith began to realize that fossils showed a regular pattern—each formation had a different assemblage of fossils. As he wrote in 1796, he was struck by "the wonderful order and regularity with which nature has disposed of these singular productions [fossils] and assigned each to its own class and peculiar Stratum." Smith was so good at recognizing the fossils of each formation that he amazed private collectors by correctly identifying the layers from which their specimens had come. He used this understanding of faunal succession to map the strata of England and Wales, culminating in the first modern geological map, which was finally published in 1815. At about the same time, Cuvier and his colleague Alexandre Brongniart were mapping fossils and strata in the Paris Basin and also began to realize that there was a regular succession of fossils that differed from formation to formation. In two different regions (apparently independently) each made the discovery that eventually led to our modern concepts of biostratigraphy as a tool for unraveling earth history.

By the time of the publication of Darwin's *On the Origin of Species* in 1859, the realization of the complexity of the fossil record had reached the point where few scholars took Noah's flood literally. However, the notions about what the fossil record tells us about the history and evolution of life has continued to change, as we shall see in later chapters.

HOW DOES AN ORGANISM BECOME A FOSSIL?

Being a paleontologist is like being a coroner except that all the witnesses are dead and all the evidence has been left out in the rain for 65 million years.

Mike Brett-Surman, 1994

There are over 1.5 million named and described species of plants and animals on Earth at this moment, and probably many more that have never been named or described; some estimates place the total number at about 4.5 million species. Yet the fossil record preserves only a small fraction of this total and does so in a very selective manner. Some groups of organisms with hard parts (such as shells, skeletons, wood) tend to fossilize readily and much is known about their past. Many others are soft-bodied and rarely if ever fossilize, so paleontology has little to say about their history. The study of how living organisms become fossilized is known as **taphonomy** (Greek for "laws of burial").

There are several ways to get a sense of just how unlikely fossilization can be. For example, modern biological studies show that the typical sea bottom is often dense with shells. One-quarter of a square meter of seafloor off Japan (Thorson, 1957) yielded 25 individuals of a large bivalve (*Macoma incongrua*), 160 of a smaller cockle shell (*Cardium hungerfordi*), and 12 of the tusk shell (*Dentalium octangulatum*). The average age of these mollusks is 2 years. At this rate, there would be 1000 shells in just 10 years, or one hundred million in a million years—over one-quarter of a square meter! Extrapolated over the whole seafloor and over geological time, this suggests that a staggering number of shells could have been fossilized. In fact, that tiny area of seafloor near Japan could produce more fossilizable shells than is actually known from the entire fossil record! Clearly, most organisms do *not* become fossils.

The study of taphonomy has become very popular in the last 20 years for one simple reason: to under-

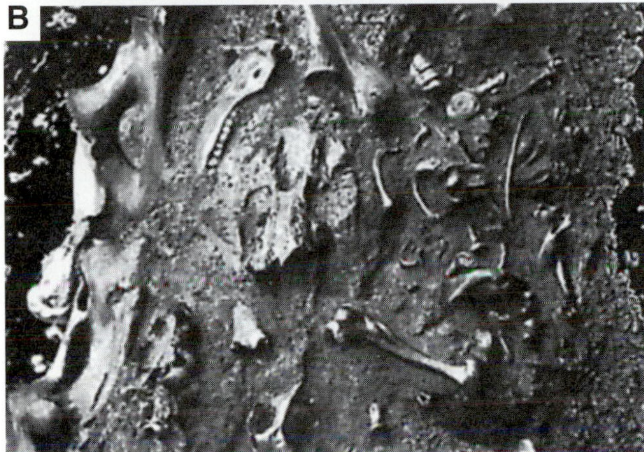

Figure 1.5. (A) A frozen mammoth discovered in Beresovka, Siberia, in 1901 was essentially unaltered except for freezing. It still had its last meal in its mouth and a full stomach. Most of the skin and flesh of the head and trunk had been eaten by wolves, even though it was 30,000 years old. (Courtesy of the American Museum of Natural History.) (B) Typical 40,000-year-old fossil bones from the Rancho La Brea tar pits in Los Angeles, California. They still retain their original bone material, but have been pickled in tar. (Courtesy of the Natural History Museum of Los Angeles County.)

Figure 1.6. (A) This fly trapped in amber preserves even the finest details of wings and bristles, but most of the original organic material has degraded. (From Briggs and Crowther, 1990). (B) Petrified wood from the Triassic Petrified Forest of Arizona has been completely permineralized with silica, preserving the delicate plant tissues and cell structure. (Photo by the author.)

stand and interpret the preserved fossil record, you must first determine how taphonomic processes have biased your sample. From the moment an organism dies, there is a tremendous loss of information as it decays and is trampled, tumbled, and broken before it is buried. The more of that lost information that we can reconstruct, the more reliable our scientific hypotheses are likely to be. In this sense, every paleontologist has to act as a coroner/forensic pathologist/detective, determining what killed the victim and trying to reconstruct the events at the "scene of the crime."

The first step is to determine just what type of fossilization has taken place. Most fossils have been dramatically altered from the original composition of the specimen, and often their original shape and texture are hard to determine unless one has some idea of what took place. The major types of preservation processes are discussed next.

Unaltered Remains

In a few exceptional cases, organisms are preserved with most of their original tissues intact. Ice Age woolly mammoths have been found thawing out of the Siberian tundra with all their soft tissues essentially freeze-dried, including their last meals in their digestive tracts (Fig. 1.5A). Some were so fresh that humans and animals could eat their 30,000-year-old meat with no ill effects. An Ice Age woolly rhinoceros was found intact in a Polish oil seep; the petroleum pickled the specimen and prevented decay. These examples are extremely rare, but when they occur,

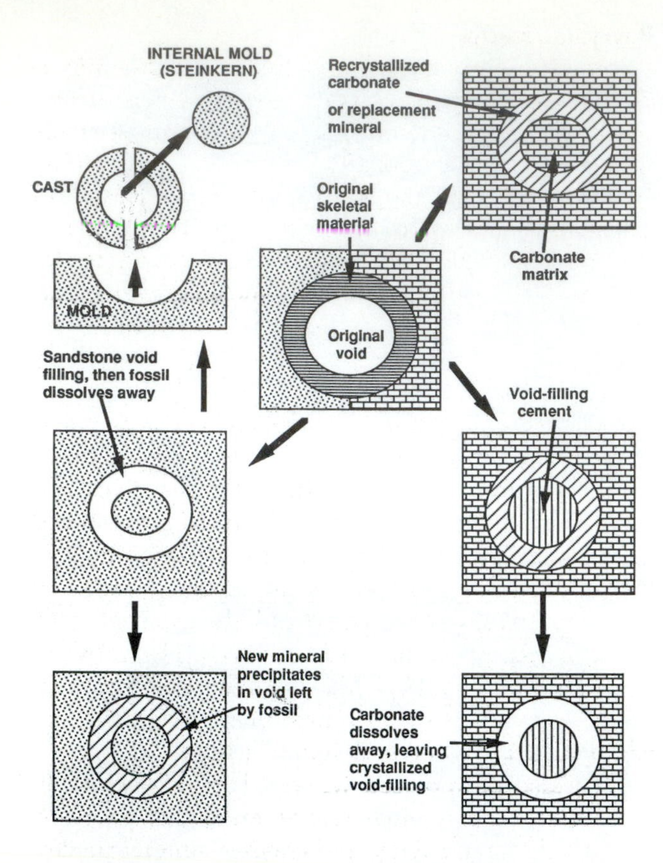

Figure 1.7. Many different diagenetic processes can change the original material into a fossil. If it is preserved in sandstone (stippled pattern), the sand may fill the original cavity in the fossil, leaving a cast. If the original skeletal material dissolves away, a void may be left which can be filled by another mineral. If it is preserved in limestone (brick pattern), fine-grained carbonate mud may fill the original void, or it may be filled by precipitation of cement (which can be calcite, but also silica or other minerals). If the original skeletal material was carbonate, it usually recrystallizes to coarsely crystalline calcite, but it may also dissolve away, leaving a void.

A

B

Figure 1.8. (A) Internal molds formed by filling the cavity left by a dissolved fossil are know as steinkerns. In this specimen, the shell of these pentamerid brachiopods has been dissolved away, leaving a steinkern with clefts where the internal partitions once were. (B) All articulate brachiopod shells are originally made of calcite, but these specimens have been replaced by pyrite (top) and silica (bottom). In the latter case, the siliceous replacement allows the specimen to be etched from limestone using acid, and preserves the delicate spines. (Photos by the author.)

they give us detailed insight into the color, diet, muscles, hair texture, and other anatomical features that paleontologists seldom see.

Some fossils have decayed so that the soft tissues are gone, but their hard parts are unaltered. The famous tar pits of Rancho La Brea in Los Angeles are full of 40,000-year-old bones that retain their original composition, but are pickled in tar, so they are black and smell like petroleum (Fig. 1.5B). However, enough of their original chemistry remains that scientists have been able to extract their DNA and compare it to that of living relatives. Similarly, many shells (particularly those of Pleistocene age) still have their original shell material unaltered, including the iridescent "mother of pearl" aragonitic layer that frequently lines their interior. In a few places, Cretaceous ammonites have been found with their original aragonite intact, but there are few fossils much older than this that retain their original aragonite.

Preservation in amber is a special case (Fig. 1.6A). These specimens are fossilized when tree resin oozes downward and entraps insects, spiders, and even frogs and lizards. The resin then hardens and forms a tight seal around the organism. Most specimens are only carbonized films of insects and spiders, but some are so well preserved that their original biochemicals are still intact. Some of these molecules have been successfully extracted and sequenced, as depicted in the movie *Jurassic Park*, but the genetic material is so incomplete that we will never be able to reconstruct a complete organism.

Figure 1.9. (A) Carbonized films are all that remain of most graptolite fossils, although they may still preserve enough detail to be identifiable. These graptolites have also been aligned by currents. (Photo courtesy R. B. Rickards.) (B) The body outline of this ichthyosaur, a marine reptile, is preserved in a carbonized film around the skeleton. (Photo courtesy R. Wild.)

Permineralization

Many biological tissues are full of pores and canals. The bones of animals are highly porous, especially in their marrow cavity, and most wood is full of canals and pores. After the soft parts decay, these hard parts are buried and then permeated with groundwater that flows through them. In the groundwater are dissolved calcium carbonate or silica, which precipitate out and fill up the pores, completely cementing the bone or wood into a solid rock. Unlike replacement (discussed below), new material comes in but none of the original material is removed. The famous multicolored fossil logs of Petrified Forest in Arizona (Fig. 1.6B) are permineralized by silica, and many other examples of petrified wood and bone are permineralized by carbonate. Permineralization can be so complete that even the details of the cell structure are preserved.

Recrystallization

Some shells are made of relatively unstable minerals, such as aragonite. Once the shells leave surface conditions, most aragonite reverts to the more stable form of calcium carbonate, the mineral calcite (Fig. 1.7). In other cases, shells made out of tiny crystals of calcite recrystallize into larger crystals. In these cases, the original shape of the fossil is preserved, but under the microscope, the difference in the texture is apparent.

Dissolution and Replacement

As water seeps through sediments filled with shells or bone, there is also a tendency for the original material to dissolve (Fig. 1.7). If the fossil dissolves and leaves a void, then the shape of the fossil is preserved in the surrounding sediments. The internal filling of this specimen is known as an internal mold, or *steinkern* ("stone cast" in German); the external mold of the specimen is often also preserved (Fig. 1.8A). In other cases, the void is filled with sediment and a natural cast of the fossil is formed, mimicking the original in surprising detail. Original bone or shell material can also be replaced without leaving a void. In these cases, the original calcite, aragonite, or phosphate is dissolved away, and another mineral precipitates almost immediately in its place. This is easiest to detect when a fossil is made of some mineral that is clearly not original, such as calcitic brachiopods now made of silica or pyrite (Fig. 1.8B).

Carbonization

Many fossils are preserved as thin films of carbon on the bedding planes of sandstones and shales (Fig. 1.9). When the organism dies, most of the volatile organic materials disperse and leave a residue of coal-like carbon, in the form of a black film that preserves the outline and sometimes the detailed structures of the organism. This kind of preservation is typical of most plant fossils; indeed, coal is the accumulated carbonized films of countless plants. There are many examples of carbonized animal fossils, especially among the graptolites (Fig. 1.9A), that are virtually always preserved as carbonized films. There are also Eocene fossil insects preserved in extraordinary quality in places such as Florissant, Colorado, and the body outlines of ichthyosaurs from the Jurassic Holzmaden Shale in Germany (Fig. 1.9B).

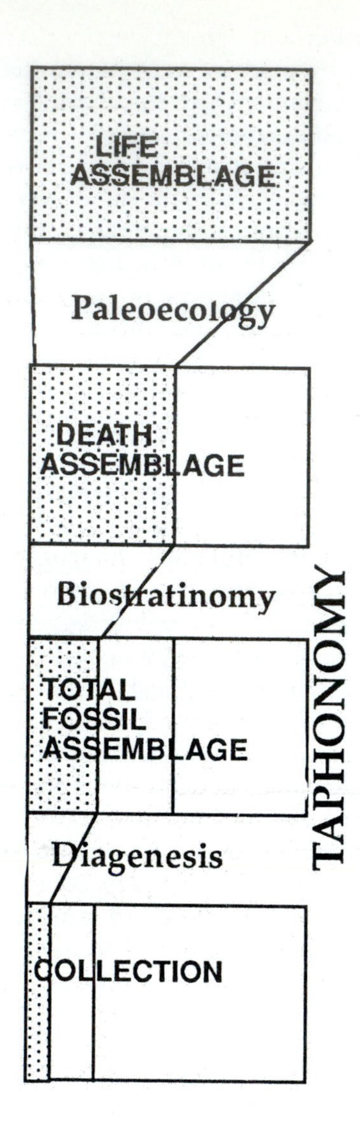

Figure 1.10. Diagrammatic summary of the processes of fossilization. At each step, a larger percentage of the original assemblage is lost, until only a small fraction remains (stippled area).

WHAT FACTORS AFFECT THE FOSSILIZATION POTENTIAL OF AN ORGANISM?

> *Now let us turn to our richest geological museums and what a paltry display we behold! That our collections are imperfect is admitted by everyone. Many fossil species are known from single and often broken specimens. Only a small portion of the earth has been geologically explored, and no part with sufficient care. Shells and bones decay and disappear when left of the bottom of the sea where sediment is not accumulating. We err when we assume that sediment is being deposited over the whole bed of the sea sufficiently quickly to embed fossil remains.*
>
> Charles Darwin, *On the Origin of Species*, 1859

As the list of modes of preservation suggest, there are many factors that affect the preservation potential of a dead organism. These factors operate at different stages in the "life history of a fossil." Figure 1.10 shows a diagrammatic summary of the sequence of such processes. From the original complete assemblage of living organisms, known as a "life assemblage" or **biocenosis**, many events can occur that screen out certain organisms, leaving a much smaller death assemblage, or **thanatocenosis**.

The process of breakup and decay of organisms immediately after death is known as **necrolysis** (literally, "death breakup"). For example, the thanatocenosis on a modern seafloor will consist mostly of durable mollusc shells; all the soft-bodied invertebrates in the living biocenosis, such as jellyfish and worms, have decayed. Numerous studies have documented the biases inherent in the processes of death and decay and estimated the **preservation potential** of various marine invertebrates. For example, Johnson (1964), Stanton (1976), and Schopf (1978) censused the marine invertebrate fauna in three different regions (Tomales Bay, California; southern California shelf; and Friday Harbor, Washington, respectively), and then estimated which organisms had the highest preservation potential. All three studies concluded that 25% to 30% of the fauna is likely to be preserved in the fossil record, with snails and clams having the best potential, and the soft-bodied groups such as flatworms, segmented polychaete worms (which may make up 40% of the species in modern shallow marine habitats), and other wormlike organisms (sipunculids, phoronids, and echiurids) having very little chance of fossilization. Some arthropods (such as heavy-shelled crabs and barnacles) may fossilize, but other thin-shelled crustaceans such as shrimp rarely do. A few thick-shelled echinoderms such as sea urchins fossilize, but sea stars and brittle stars have little chance of becoming fossils.

In addition, Schopf (1978) found that there were differences based on substrate and ecology as well. Organisms living on mud, sand, or rocky substrates appeared to fossilize almost equally well, although muddy bottoms have the best potential (because they represent quiet water with few energetic currents) and rocky habitats the worst (because they are seldom buried). Even more striking was the difference between ecological types. About 67% of the sedentary organisms (herbivores and filter feeders) are fossilized, but carnivores and mobile detritus feeders are much less frequently preserved (16% to 27% of the taxa). Schopf pointed out that herbivores and filter feeders tend to have heavier skeletons, or may even be solid masses of calcite (such as corals), whereas detritus feeders and carnivores have much lighter and less durable skeletons because they must move

Figure 1.11. The processes of death, decay, scavenging, and burial act upon organisms in the African savanna, such as on this antelope. Only a tiny fraction of its bones ever have a chance of making it into the fossil record. (From Shipman, 1981.)

around to find their food. An estimate of the ratio of herbivores to carnivores from such data would be distorted beyond recognition.

After a death assemblage accumulates, many other factors operate on the hard parts to break them up and scatter them around, so an even smaller percentage ends up buried for future fossilization. These processes occur after necrolysis and are known as **biostratinomy** (loosely, "the laws by which living things become stratified"). These agents of destruction can be biological, mechanical, or chemical.

Biological Agents

Biological agents are the most important factor in most environments, both marine and terrestrial. Both predators and scavengers are very active in breaking up shells and bones to extract almost all the useful nutrition out of them. On the seafloor, a variety of organisms (especially fish, crabs, and lobsters) are effective in cracking shells to extract their food content. In an interesting set of experiments, Plotnick (1986) buried dead shrimp in marine sediment and then monitored their breakup. He found that scavengers (probably crabs) broke most of them up very quickly, even if they were buried as deep as 10 cm. Burrowing organisms were also very important in disturbing, breaking up, and consuming the carcasses. Only if they were buried deeper than 10 cm in relatively anoxic waters did the shrimp carcasses last, but eventually even these decayed due to bacterial action. A great variety of similar studies in experimental taphonomy were summarized by Briggs (1995).

In addition to predators and scavengers breaking shells to extract soft parts for food, the shells themselves are subject to other biological agents of destruction. The most important of these are organisms that use the shell as a substrate or as a source of food or nutrients. A variety of organisms, including boring algae, boring sponges, worms, and bryozoans, erode holes and canals in dead shells and eventually weaken them so that they fall apart.

On land, a variety of predators and scavengers work very quickly to break up carcasses or vegetation. Once a tree falls in the forest, a wide variety of organisms, from termites, ants, beetles, and worms, to fungi and bacteria of various kinds reduce it to organic material that can be recycled back into the food chain. Studies of animals on the African savanna have shown that there is a distinct pecking order among predators and scavengers (Fig. 1.11). After lions bring down their prey, they consume not only the best meat, but may even break some bone. Soon thereafter, the hyenas and jackals move in to scavenge the remaining meat and break up the bones for their marrow. The vultures hover nearby, getting whatever scraps then can. Finally, the ants and dermestid beetles strip the bones of the very last scraps of soft tissue, including tendons and cartilage. But these scavengers are not the only important biological agents of destruction. If the bones remain exposed, they are likely to be trampled by herds of antelope or zebra and quickly scattered around and reduced to splinters.

In summary, the key factor that prevents biological destruction is *rapid burial.*

Mechanical Agents

Mechanical agents of destruction such as wind, waves, and currents can be very important. These processes are most effective in shallow waters, where both waves and storms have their highest energies (Norris, 1986). In the marine environment, a number

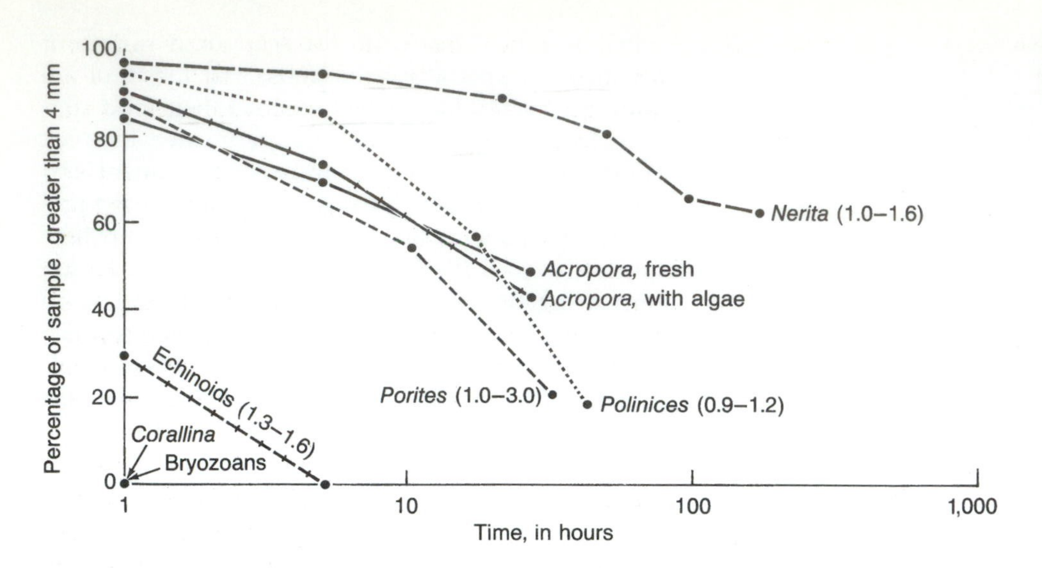

Figure 1.12. Tumbling experiments simulate the breakdown of various shells and skeletons and show which are the most delicate and which are the most resistant to the processes of mechanical abrasion. (After Chave, 1964.)

of studies (Driscoll, 1967, 1970; Driscoll and Weltin, 1973; Stanton, 1976; Warme et al., 1976) have simulated the processes of mechanical breakup of skeletons. One of the earliest, simplest, and most meaningful studies of these processes was performed by Chave (1964). He placed the skeletal parts of a number of marine invertebrates in a tumbling barrel with chert pebbles and then tumbled them for over 100 hours (Fig. 1.12). As expected, thin-shelled organisms such as coralline algae, bryozoans, echinoids, and jingle clams and razor clams broke up in minutes or at most a few hours. Thicker corals, bivalves such as mussels and oysters, and some snails were more durable, lasting up to 50 to 70 hours. The most durable were thick-shelled snails and clams, such as the round snail *Nerita* and the clam *Spisula*. Shell texture was also important. The densest, most fine-grained shells were the most durable, but skeletons with coarsely crystalline structure (such as oysters) or porous structure (such as corals) were less durable, even if they were relatively thick.

Similar studies have been performed for terrestrial environments (Behrensmeyer, 1975; Boaz and Behrensmeyer, 1976; Korth, 1979; Hanson, 1980; Shipman, 1981). A classic study by Voorhies (1969) took a series of mammal bones and tumbled them in a flume to simulate the processes that occur in a stream. They clustered into three well-defined groups (now known as "Voorhies groups") of increasing durability. The least durable were thin, delicate, elongate Group I bones, such as ribs, vertebrae, and the breastbone (Table 1.1). Slightly more durable were shoulder blades and toe bones. Group II consisted of thicker and less delicate bones such as most of the limb bones, and parts of the hip. The most durable elements (Group III) were the skull and jaws, which are dense and heavily sutured together. Even more durable are the individual teeth, which are coated in enamel, the hardest substance in the skeleton. Not surprisingly, most mammal fossils consist of teeth and jaws, or less commonly skulls. Limb bones for fossil mammals are known less often, and identifiable ribs and vertebrae are so seldom preserved in association with diagnostic skull or tooth material that they are seldom even collected or described.

In summary, the *shape, density, and thickness* of the bone or shell are the most important factors in determining its survival under mechanical transport of waves, storms, or river currents.

Diagenesis and Discovery

After burial, a variety of diagenetic changes in the rock (especially metamorphism) can easily destroy the shells and prevent their preservation. As discussed earlier, aragonitic fossils are much more prone to dissolution than calcitic fossils, so fossils that are made primarily of aragonite are discriminated against in the fossil record. For example, there are a number of calcareous plankton in the fossil record, including the foraminifera, the ostracodes, and a group of tiny aragonitic planktonic snails known as pteropods. All three groups can be very abundant in tropical surface waters, but as they die and sink to the bottom, the pteropods dissolve first because they are made of aragonite, not calcite. Many calcareous oozes on the seafloor are not a good reflection of the original plankton in the surface waters, because the pteropods have been selectively dissolved away.

In another study, Voight (1979) censused a living oyster bank and found over 300 species (mostly soft-bodied worms). Of the shelly invertebrates, 16% (nearly all the snails and many of the bivalves besides oysters) had aragonitic shells. Once these had dissolved, the remaining "oyster community" appears to be of low diversity, but this is a false conclusion—the only species fossilized are those with calcitic shells.

Table 1.1. Durability of the bones in a typical mammalian skeleton. (After Voorhies, 1969.)

Group I: immediately removed by low-velocity currents; high surface-area/volume ratio
Ribs
Vertebrae
Hip bone
Breastbone

(*slightly more durable*)
Shoulder blade
Finger bones

Group II: removed gradually by moderate currents; low surface area/volume ratio
Thigh bone
Shin bone
Upper and lower arm bones
Ankle and wrist bones
Hip bone

Group III: lag deposit, moved only by high-velocity currents; low surface area/volume ratio
Lower jaw
Skull

Many apparently unfossiliferous rocks probably had fossils at one time, but diagenesis has removed them. This is usually hard to document, but McCarthy (1977) described a Permian beach sand that had no apparent fossils until he examined scattered concretions. Inside the concretions were abundant fossils, showing the original texture of the rock before groundwater flushed through to dissolve everything not protected by a concretion.

The diagenetic history of a fossiliferous deposit can be very complex. The Miocene Leitha Limestone of southern Austria (Dullo, 1983) can be subdivided into several biofacies based on relative abundance of mollusks, corals, red algae, bryozoans, and foraminifera. However, a closer examination shows that these relative abundances are largely a diagenetic artifact rather than original ecological abundances. Calcitic molluscs such as oysters and scallops are well preserved in all facies, so they dominate the heavily dissolved facies. Aragonitic mollusks are preserved only in basinal muddy facies. Aragonitic fossils represented by molds, or replaced by calcite, are found in the chalk and well-cemented limestone facies. Aragonitic fossils are completely absent from the carbonate sand facies. In this case, the degree of exposure to groundwaters and marine waters is more important than the original community composition.

In summary, the *original composition* and *groundwater chemistry* are the most important factors in determining whether diagenetic changes are likely to alter or dissolve a fossil.

Finally, only a small portion of all the fossiliferous rocks in the world happen to be exposed during just the last few centuries, when people began to collect them. Even a smaller proportion of those outcrops are ever seen by a qualified collector before the fossil erodes out and is destroyed. Thus, the chances of a given animal having the extraordinary luck of being preserved and collected by a paleontologist are extraordinarily small.

Taphonomic research has come a long way since the pioneering studies of the 1960s and 1970s. Paleontologists can no longer afford to naïvely take the fossil record at face value, but must always keep in mind the taphonomic "noise" that may obscure the original biological "signal." However, not all taphonomic processes are negative. Applied properly, taphonomic information can add to our understanding of the fossil record, and in many cases we can appreciate dimensions of the past that would otherwise have escaped our notice (see Behrensmeyer and Kidwell, 1985, 1993, for a review of taphonomic processes and their important positive implications).

WHAT FACTORS ARE REQUIRED FOR EXTRAORDINARY PRESERVATION?

The processes outlined above account for most fossilization, and what we know of the living organism is usually quite incomplete. However, there are extraordinary fossil deposits around the world that preserve soft tissues and sometimes even skin texture and color patterns, giving us a much more complete picture. These are known as *Lagerstätten* (German for "mother lode"), and they have produced some of the most important fossils known. Some of them are so famous that they form an "honor roll" that most paleontologists know by heart. These are discussed next.

The Middle Cambrian Burgess Shale, British Columbia

In recent years, the most celebrated of all the *Lagerstätten* is the Middle Cambrian Burgess Shale, from the Canadian Rockies in Yoho National Park, British Columbia. A few quarries high on the slopes of Mount Field have produced over 65,000 specimens of mostly soft-bodied animals, representing at least 93 species. Most are preserved as flat impressions on the shales, although the trilobites are preserved with their calcified skeletons as well (Fig. 1.13). These organisms were apparently transported from a nearby shallow carbonate bank into these anoxic, deep-

Figure 1.13. Soft-bodied impressions from the Middle Cambrian Burgess Shale, showing the exquisite detail of the preservation of the appendages and even the bristles of these strange creatures. (Photo courtesy the Smithsonian Institution.)

water shales by submarine landslides, where they were abruptly buried in fine mud and not allowed to decay.

What is extraordinary about the Burgess Shale is not just the preservation of typical Cambrian animals, such as trilobites, but the great number of unique and bizarre soft-bodied animals that we would otherwise never know about. Although the majority can be assigned to existing phyla, many are clearly unique "evolutionary experiments" that belong to no living phylum. In his book *Wonderful Life*, Gould (1989) argues that this "window in time" of the Burgess Shale shows that there were many experimental phyla during the early Cambrian radiation that did not survive even to the later Paleozoic. Contrary to the long-held expectation that everything in the Cambrian is simple, primitive, and ancestral to living groups, the Burgess Shale shows that there was a great deal of experimentation in unusual body plans during the Cambrian, most of which did not survive.

The Burgess Shale is not unique in this regard. *Lagerstätten* from Chengjiang, Yunnan Province, China, and Sirius Passet, Greenland, have produced similar forms, showing that the explosion of experimental forms took place around the world in the Cambrian.

The Upper Jurassic Solnhofen Limestone, Southern Germany

This deposit was first quarried because it yields extremely fine-grained slabs of limestone that were perfect for the acid etching process to produce lithographic printing plates. Apparently, these limestones were produced when fine lime mud washed from the ocean into a shallow, stagnant lagoon, which was low in organic productivity (so the rocks are not black with organic matter, but white) and completely hostile to all forms of life on the bottom. As animals died and sank to the bottom, there were no scavengers or decomposers to decay them, no currents to break them up, and apparently they were rapidly buried as storms brought in new layers of lime mud. There are

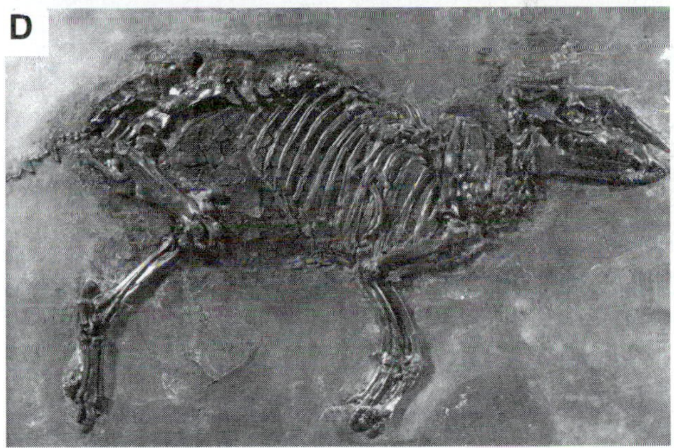

Figure 1.14. Fossils from some famous *Lagerstätten.* (A) *Pterodactylus* specimen from the Upper Jurassic Solnhofen Limestone of Germany. (Photo courtesy P. Wellnhofer.) (B) Typical siderite nodule from the Pennsylvanian Mazon Creek beds of Illinois, split in half to reveal both sides of a delicately preserved fern fossil. (Photo by the author.) (C) This Devonian trilobite from the Hunsrück Shale of Germany has been x-rayed to show the soft appendages and gills. (Photo courtesy W. Stürmer.) (D) Complete articulated specimen of the horse-like palaeothere *Propalaeotherium* from the Eocene lake deposits of Messel, Germany. Not only is every bone in place, but the body outline and even the fine structure of the vegetation in the stomach are preserved. (Photo courtesy J. L. Franzen.)

more than 600 fossil species preserved in the Solnhofen Limestone, including over 180 species of insects, and many marine arthropods (especially crustaceans) such as a well-known specimen of the horseshoe crab *Mesolimulus* preserved at the end of its final trail (Fig. 18.1). The most striking specimens are the terrestrial vertebrates, including numerous pterodactyls (Fig. 1.14A), which sometimes preserve their body outline and wing membrane, and all seven known specimens of the earliest bird, *Archaeopteryx*, complete with feather impressions (Fig. 5.1). These specimens are important as the "missing link" between birds and dinosaurs, especially because the first specimen was found just two years after Darwin's *On the Origin of Species* was published, providing dramatic evidence for his theory.

The Pennsylvanian Mazon Creek Beds, Northeastern Illinois

Exposures of the Middle Pennsylvanian Francis Creek Shale of the Carbondale Formation in the

Mazon Creek area of northeastern Illinois yield nodules of siderite (iron carbonate) that, when split open, preserve soft-bodied organisms in exquisite detail (Fig.1.14B). Over 350 species of plants (the most diverse flora known in North America), 140 species of insects, and over 100 additional taxa, including bivalves, jellyfish, worms, sea cucumbers, centipedes, millipedes, scorpions, spiders, crustaceans, fish, and amphibians are known. In addition, there are soft-bodied enigmas like the "Tully monster" (a bizarre worm-like creature) that cannot be assigned to any known phylum.

The absence of normal marine taxa and the sedimentology suggest that these fossils were quickly buried at the mouth of an estuary-delta complex, when sediment-laden floods of freshwater rapidly buried everything on the muddy bottom. The rotting tissues changed the chemistry of the mudstone so that iron carbonate could precipitate and nucleate around the fossil, forming the concretion. These concretions must have formed very early, for many of the fossils are undistorted, even though the surrounding shales have undergone much compaction.

Other *Lagerstätten*

In addition to these classic localities, there are the Lower Devonian **Hunsrück Shale** of western Germany, which produces over 400 species of animals with soft parts preserved, including segmented worms and trilobites (Fig. 1.14C) with all their soft limbs, gills, and antennae. The Middle Jurassic Posidonienschiefer of **Holzmaden**, southern Germany, yield over 100 species of Jurassic marine life, including squids and ichthyosaurs with their body outlines preserved as a dark film (Fig. 1.9B); one ichthyosaur was preserved in the process of giving birth (Fig. 8.10A). Both of these deposits are classic black shales, formed in deep, stagnant water where there were few currents or scavengers to break up the carcasses.

The middle Eocene deposit at **Messel**, in western Germany, is an oil shale deposit formed at the bottom of a stagnant lake (Fig. 1.14D). Many different plants and animals (especially insects, fish, frogs, turtles, lizards, snakes, crocodiles, birds, and about 35 species of primitive mammals) are preserved with impressions of the soft-tissue and body outlines. There are also extraordinary examples of preservation of cell walls of plants, hair of mammals, scales from the wings of moths, color patterns on many of the insects, and the stomach contents of many of the animals.

From all these examples, several general trends emerge. The best fossilization occurs when there is *rapid burial and anoxic conditions to prevent scavenging, no reworking currents, and little or no diagenetic alteration* to destroy the fossils.

HOW GOOD IS THE FOSSIL RECORD?

> *To those who believe that the geological record is in any degree perfect will undoubtedly at once reject the theory* [of evolution]. *For my part, following out Lyell's metaphor, I look at the geological record as a history of the world imperfectly kept, and written in a changing dialect; of this history, we possess the last volume alone, relating only to two or three countries. Of this volume, only here and there a short chapter has been preserved; and of each page, only here and there a few lines.*
>
> Charles Darwin, *On the Origin of Species,* 1859

All of the descriptive and anecdotal evidence we have just reviewed underlines the incompleteness of the fossil record. Can we also get a quantitative estimate of its quality? Can we answer in numerical terms the question: how good is the fossil record?

Let us start with some simple estimates. We have already estimated that there are 1.5 million described species, or as many as 4.5 million described and undescribed species of organisms alive on Earth today. How many species are known as fossils? It turns out that there are only about 250,000 described species of fossil plants and animals presently known, or only 5% of the total for species living today. But the present is only one moment in geologic time. If we multiply the present diversity by the 600 million years that multicellular life has existed on this planet, the estimate is much worse. No matter how one does this calculation, it is clear that the quarter of a million species known as fossils represents only a tiny fraction of a percent of all species that have ever lived.

But all is not lost. Almost half of the 1.5 million described species are insects, which have a poor fossil record. Let us just focus on nine well-skeletonized phyla of marine invertebrates and see if we come up with better estimates. These nine phyla are the Protista, Archaeocyatha, Porifera, Cnidaria, Bryozoa, Brachiopoda, Mollusca, Echinodermata, and Arthropoda (excluding insects). In these groups, there are about 150,000 living species, but over 180,000 fossil species (Valentine, 1970; Raup, 1976). To translate these numbers into completeness estimates, we need to know the turnover rate of species and the number of coexisting species through time. Different values have been used for each of these variables, but the results of the calculations are remarkably similar. Durham (1967) estimated that about 2.3% of all the species in these nine phyla were fossilized. Valentine (1970) gave estimates that ranged from 4.5 to 13.6%. No matter which method we use, we must conclude that *85% to 97% of all the species in these nine well-skeletonized phyla that have ever lived have never been fossilized.*

This is a very sobering estimate. It forces us to step back and reassess the limitations of almost any study based on fossil data. However, there is another consideration to keep in mind: the quality of the record depends on the level of detail we require. For a census of all the phyla or classes of invertebrates in a given sample, it would not be hard to get a complete sample. Obtaining every species is much harder. The reason for this is simple: a higher taxon like a phylum or class contains many different genera and species. If we obtain one species in each given phylum or class in a sample, we have a complete sample of phyla or classes with only a few specimens. But we need huge samples to get every species, or even every family or genus that might have lived in a given time and place.

In short, if we want to conduct large-scale studies of evolutionary trends, we must concentrate on the higher taxonomic levels (kingdom, phyla, classes, and orders); we cannot expect to work on the species level. But in local problems, we can often select faunas with the best possible preservation, which might allow us to trust the species-level data.

CONCLUSIONS

The answer to the question, "How good is the fossil record?" is a complex one. For certain types of studies, it is excellent; for others, it is terrible. A good paleontologist knows when the data are worth pursuing and when they are so bad that any analysis is "garbage in, garbage out." In the following chapters, we will review some of the excellent research that has taken advantage of the strengths of the fossil record. Naturally, we do not feature the many studies (mostly never published) that did not take these limitations into account.

For Further Reading

Barthel, K. W. 1978. *Solnhofen—Ein Blick in die Erdegeschichte*. Ott Verlag, Thun, Switzerland.

Behrensmeyer, A. K., and A. Hill, eds. 1980. *Fossils in the Making*. University of Chicago Press, Chicago.

Behrensmeyer, A. K., and S. M. Kidwell. 1985. Taphonomy's contributions to paleobiology. *Paleobiology* 11:105-119.

Behrensmeyer, A. K., and S. M. Kidwell, eds. 1993. *Taphonomic Approaches to Time Resolution in Fossil Assemblages. Short Courses in Paleontology* 6. Paleontological Society and University of Tennessee Press.

Briggs, D. E. G. 1995. Experimental taphonomy. *Palaios* 10:539-550.

Donovan, S. K., ed. 1991. *The Processes of Fossilization*. Columbia University Press, New York.

Durham, J. W. 1967. The incompleteness of our knowledge of the fossil record. *Journal of Paleontology* 41:559-565.

Faul, H., and C. Faul. 1983. *It Began with a Stone*. Wiley, New York.

Gould, S. J. 1989. *Wonderful Life: The Burgess Shale and the Nature of History*. Norton, New York.

Kidwell, S. M., and K. W. Flessa. 1995. The quality of the fossil record: populations, species, and communities. *Palaios* 26:269-299.

Nitecki, M. H. 1979. *Mazon Creek Fossils*. Academic Press, New York.

Poinar, G., and R. Poinar. 1994. *The Quest for Life in Amber*. Addison-Wesley, Reading, Mass.

Rudwick, M. J. S. 1972. *The Meaning of Fossils: Episodes in the History of Palaeontology*. Macdonald, London.

Schaal, S., and W. Ziegler. 1992. *Messel: An Insight into the History of Life and of the Earth*. Clarendon Press, Oxford.

Shipman, P. 1981. *Life History of a Fossil*. Harvard University Press, Cambridge.

Valentine, J. W. 1970. How many marine invertebrate fossil species? A new approximation. *Journal of Paleontology* 44:410-415.

Figure 2.1. Variability within a single population of the scallop *Guizhoupecten cheni willisensis* from the Permian of Texas. All six specimens are now interpreted as belonging to a single subspecies, but an earlier generation of paleontologists might have put each shell in a different species. (From Newell and Boyd, 1985.)

Chapter 2

Variation in Fossils

Individuals of the same species often present great differences of structure, independently of variation, as in the two sexes of various animals, in two or three castes of sterile females or workers amongst insects, and in the immature and larval states of many of the lower animals.

Charles Darwin, *On the Origin of Species*, 1859

THEME: VARIATION

If you make a collection of fossils from a quarry or roadcut, the first thing you will notice is that no two specimens are exactly identical (Fig. 2.1). If you sort them into clusters that seem to be the same kind, you will notice variations in size, shape, preservation, and many other features. What is the significance of this variation? How much is due to differences in age and growth between specimens? How much is due to differences between individuals in the same population? How much is due to post-mortem breakage and distortion? How much can be attributed to the differences between species? These questions are not always easy to resolve in field studies of living organisms whose behavior, soft anatomy, and breeding habits can be directly observed. They are much more difficult to answer when studying fossils, because the organisms are not only dead and decayed, but many are extinct with no close living relatives to compare them to.

The changing concept of variation among organisms has been critical to biology in the last two centuries. From the Renaissance until Darwin's time, most scientists viewed organisms as specially created after an ideal "**type**" or "blueprint" in the mind of God. All deviation from this ideal was simply imperfection and was ignored as "noise" obscuring God's handiwork. This is known the **typological concept of species**. One of Darwin's greatest insights was the realization that *natural populations are variable*, not just imperfect copies of an ideal blueprint. Such variation among closely related individuals in a population gives them their ecological flexibility, so that when natural selection weeds them out, some variants are better able to survive than others. Variation is the matrix for evolution. Darwin changed the focus of biology from ignoring variation as insignificant imperfection to what we now call the **population concept of species**, and it is central to our conceptions of evolution and speciation in modern biology.

Variation among organisms can be separated into two categories: variations that occur within an individual during its lifespan (**ontogenetic variation**); and variations between different individuals in the same population (**population variation**). In this chapter, we will examine what we know of ontogenetic and population variation in living organisms and then see how we might detect the same in fossils.

HOW DO ORGANISMS VARY DURING THEIR LIFESPANS?

Phylogenesis is the mechanical cause of ontogenesis. The connection between them is not of an external or superficial, but of a profound, intrinsic, and causal nature.

Ernst Haeckel, *Anthropogenie*, 1874

Anyone who has ever watched a baby grow up, or watched a tadpole turn into a frog, or a caterpillar into a pupa and then into a butterfly, knows that individuals can vary tremendously in their own lifetimes. If we did not have living insects to study, would we ever guess that a fossilized caterpillar could have been the same individual as a fossilized butterfly? Organisms are capable of enormous transformations during their lifespans, especially during the early embryonic history, when we all start as a fertilized, single-celled egg, and then divide into two, four, eight, sixteen cells, then a ball of cells, and on and on. All changes that occur during an individual's life-

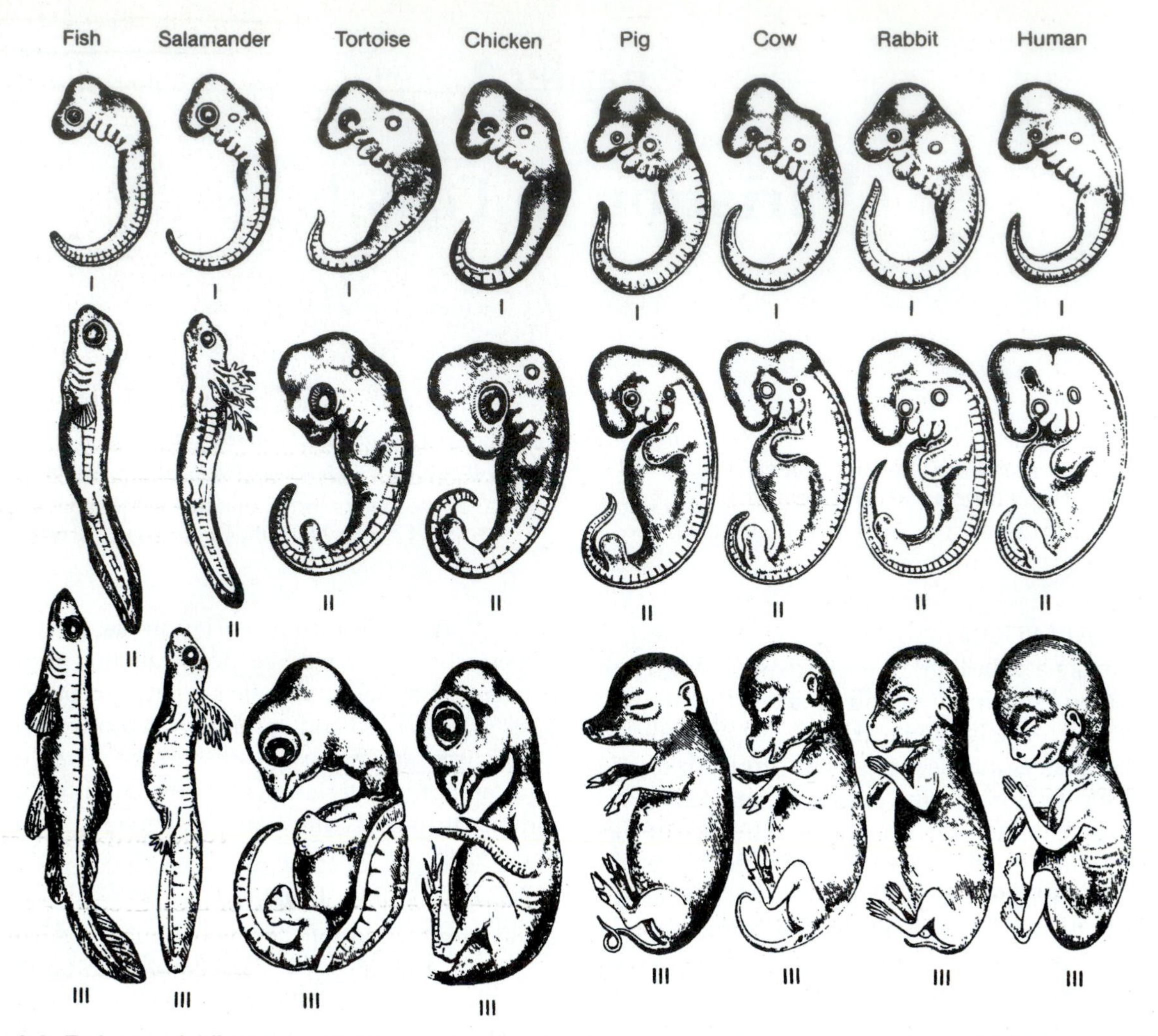

Figure 2.2. Embryos of different vertebrates at comparable stages oт development (top row) are strikingly similar in every group, regardless of their adult anatomy. Note that each embryo begins with a similar number of gill arches (pouches below the head) and similar vertebral column. In the later stages of development, these structures are modified to yield a variety of different adult forms. (From Romanes, 1910, after Haeckel.)

span, from the moment of fertilization until death, are known as **ontogenetic** (Greek, "coming into being") changes (Fig. 2.2). The embryonic and post-embryonic history of an organism is also known as its **ontogeny**.

The word "ontogeny" may sound familiar, especially if you have taken biology classes. It is most often heard as part of the phrase, "ontogeny recapitulates phylogeny." This so-called "biogenetic law" was coined by the German scientist Ernst Haeckel (1834-1919) in 1866 (along with a great many other terms in biology, such as "ecology"). Haeckel was one of Darwin's staunchest advocates in Germany after *On the Origin of Species* was published in 1859, although he pushed Darwinism further than Darwin ever intended. Haeckel was interested in showing how embryonic development revealed the evolutionary history of organisms.

Ever since the pioneering work of the German embryologist Karl Ernst von Baer (1792-1876), biologists had known that many organisms show evidence of their evolutionary past in their embryos. For example, humans go through embryonic stages that are very fish-like (complete with gill slits), and they also have a tail that is lost late in embryology. Once Darwin's theory was published, Haeckel took the embryological approach to extremes. He argued that the entire evolutionary history ("phylogeny") of an organism is repeated ("recapitulated") during its embryological history ("ontogeny"), or in his phrase, "ontogeny recapitulates phylogeny." Under this postulate, Haeckel felt that he could reconstruct the missing ancestral forms of living things just by looking at their embryos.

We now know that Haeckel took this idea a bit too far. Embryology does repeat some of the *juvenile* stages of an organism's evolutionary past, but it does not yield a sequence of *adult* ancestors, as Haeckel postulated. There were many objections to Haeckel's ideas (reviewed by Gould, 1977), but the most important were pointed out by von Baer in 1828 long before Haeckel began his research. (Haeckel simply skirted these objections in his 1866 book, even though he followed von Baer in most other ways.) One of the strongest objections is that embryos have many specializations that have nothing to do with the adult stages of an organism, such as the umbilical cord and placental membrane in mammals, the fetal circulation pattern, the lack of an immune system (to prevent rejection of the fetus by the mother's immune system), and many other features. These characteristics are critical to embryonic development, but reveal nothing about phylogeny. Von Baer also pointed out that not every feature in embryology moves in the same sequence as in the organism's evolutionary past. For example, the chick embryo at one stage has a heart and circulation like that of a fish, but otherwise lacks all the rest of the anatomy of adult fishes. In other cases, features that characterize later stages of phylogeny, such as the vertebral column, appear anomalously early in embryology. In short, ontogeny gives general indications of the sequence of juvenile forms seen in the ancestors, but it is not a complete guide to evolutionary history.

In this short section, we cannot completely summarize all of embryology and developmental biology, which are major fields of research. Instead, we will focus on aspects of ontogeny that can be recognized in fossils and discuss their implications. The first thing to recognize is that the spectrum of ontogenetic changes is enormous, from subtle continuous changes (as occur in the development of most mammals) to abrupt, major transformations we call metamorphosis (whether it be in insects with distinct larval, pupal, and adult forms, or in frogs, or in many other organisms.). As a general rule of thumb, these changes have a lot to do with the kinds of skeletons the organism possesses. Soft-bodied organisms can add tissues in a continuous fashion as they grow, requiring no major adjustments; others undergo complete metamorphosis from their larval stages into very different-looking adults. Skeletonized organisms are often constrained by the permanence of skeletal tissues. Many animals (such as most vertebrates) can allow their skeletons to grow slowly and continuously. Others (such as insects) molt their skeletons abruptly, then grow quickly before a new larger skeleton can harden. This abrupt molting event gives insects the opportunity to change their anatomy rapidly before the next skeleton hardens.

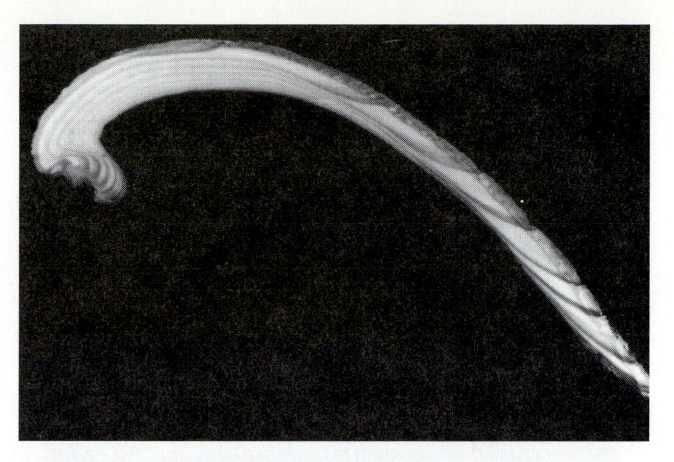

Figure 2.3. Growth lines in a 3-mm-thick section through the common clam, *Mercenaria mercenaria*. Each pair of white and gray bands forms during a single annual growth cycle. (Photo courtesy I. R. Quitmyer and D. Jones)

Growth Strategies

As if we could not count, in the shells of cockles and snails, the years and months of their life, as we do in the horns of bulls and oxen, and in the branches of plants.

Leonardo da Vinci, ca. 1500

There are a wide variety of growth strategies within the animal kingdom, many of which fossilize readily. Many animals grow by **accretion**, adding on discrete growth layers to their skeletons as they get larger (Fig. 2.3). This type of growth is common in many hard-shelled organisms, especially molluscs and corals. The growth lines themselves have often been a powerful paleobiological tool. In many cases, animals grow slowly or rapidly in response to regular fluctuations in their environment. Tree rings, for example, are routinely used to establish a chronology of dry and wet years when growth is hampered or accelerated. Many molluscs add new layers on a seasonal basis, and can almost be "read" like tree rings.

One of the most interesting results of growth ring analysis has been implications of the spacing of the growth bands in fossil corals and many other organisms. This research was first initiated by John Wells in the early 1960s, when he noticed that Devonian corals appeared to show two different kinds of growth layers: thin layers that apparently corresponded to the day/night cycle and thicker layers due to the annual change of the seasons (Fig. 2.4). By counting the number of daily bands per yearly seasonal cycle, Wells discovered that there were about 400 days in the Devonian year and about 387 in the Pennsylvanian (compared to 365.25 in our present year). This startling conclusion actually verified an

Figure 2.4. Devonian rugose coral *Heliophyllum*, showing a spectrum of growth lines. The white lines bracket the annual band, and the fine lines between them are presumed to be daily growth lines. From evidence such as this, paleontologists have inferred that there were about 400 days in a Devonian year, and the day was shorter (about 22 hours). (From Dott and Prothero, 1994.)

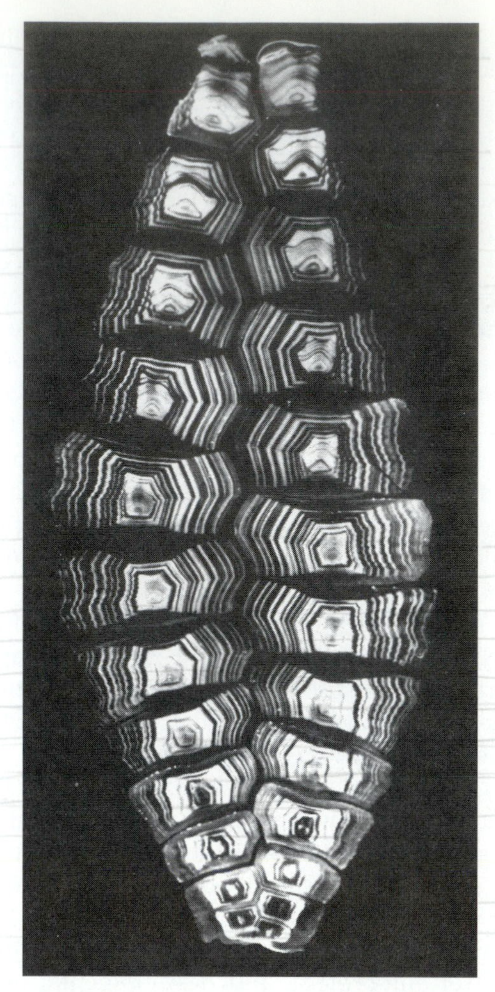

Figure 2.5. These columns of paired plates from the sea urchin *Strongylocentrotus* show both additional growth (each plate is added as a discrete unit) and accretionary growth (each plate grows by accretion after it is added). (From Raup, 1968.)

idea long held by astronomers: the Earth's rotation is slowing down. We know from astronomical calculations that the earth's rotation is decelerating by about 2 seconds per 100,000 years from an initially higher speed at the time of the formation of the solar system. This deceleration is due to the friction caused by the tides on the Earth's surface. If one extrapolates this rate back to the Cambrian, the Cambrian day would have contained 21 hours and the Cambrian year 420 days, in remarkably good agreement with the fossil data.

A second growth strategy is the **addition** of discrete new parts, which grow very little after they are formed. Throughout their lives, many invertebrates follow this strategy. Chambered organisms, such as the foraminifera, add new chambers that are slightly larger than the last. Many arthropods (such as trilobites) add new segments as they grow. Echinoderms add new plates to their skeletons as discrete elements initially, then these plates grow to their final diameter by accretion (Fig. 2.5).

A third type of growth, known as **molting**, is found almost exclusively in arthropods (Fig. 2.6). Members of this phylum have no internal skeleton, but their outer covering, or exoskeleton, provides all their structural support. However, a hard outer skeleton cannot grow, so all arthropods periodically must absorb the minerals from their old skin, split it open along the middle of the back, and then emerge as a soft-bodied but larger individual. After a few hours, they expand to a larger size, their new exoskeleton hardens, and they are no longer vulnerable.

As we saw, the molting process gives the arthropods one clear advantage—they can make radical changes in their body plan between molts. But it also has a major disadvantage. When they are in the soft-bodied stage between hard exoskeletons, they are very vulnerable. In particular, it places important

Figure 2.6. Beautifully preserved specimen of the Late Cretaceous lobster *Glyphea robusta* caught in the act of molting. (From Feldmann and McPherson, 1980; photo courtesy R. M. Feldmann.)

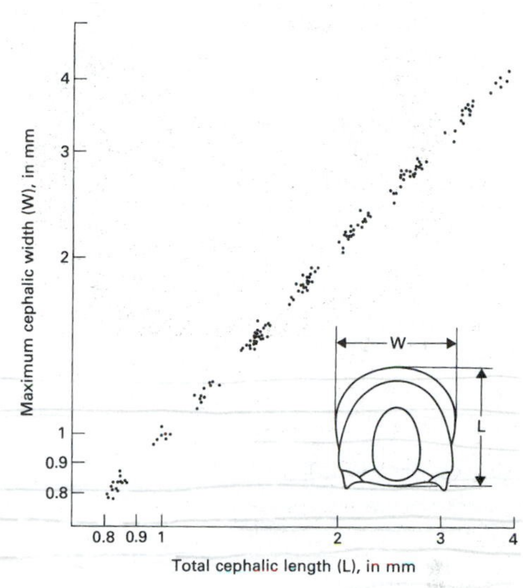

Figure 2.7. Bivariate plot of length and width of cephalon of the Ordovician agnostid trilobite *Trinodus*, showing clusters of specimens at each growth stage (instar) separated by gaps of rapid growth during molting. (Modified from Hunt, 1967.)

constraints on the maximum size of arthropods. The largest living terrestrial arthropods are goliath beetles, which may reach 15 cm in length, although extinct dragonflies had wingspans of almost a meter. King crabs, which reach about 5 m in legspan, are the largest living arthropods in the marine realm (where the density of water can support larger bodies), although some extinct "sea scorpions" were almost 3 m long.

The molting process produces some other unusual features in the arthropod fossil record. Since the various elements of their exoskelton break up after they are shed, it is typical to find arthropod fossils of only part of the animal. In many trilobite localities, for example, one finds only head segments (cephala) or tail segments (pygidia), or isolated thoracic segments, but few whole individuals. Many of these shed molts have been winnowed and concentrated by currents to form a thanatocenosis that tells us little about the original life habitats, or biocenosis.

Because arthropods grow rapidly during their soft-bodied hours between hardened skeletons, their growth is discontinous. In large samples of some arthropods, for example, the skeletal elements plot in clusters, with gaps in between (Fig. 2.7). These clusters probably represent a sequence of skeletonized growth stages, and the gaps between them represent periods in the animal's life when it grew rapidly between molts and left no hard skeleton.

A fourth growth strategy involves continuous **modification** of existing skeletal elements as the animal grows. This type of growth is typical of vertebrates, which constantly remodel and add to their bones throughout their life.

Most organisms use a mixture of these growth strategies. Although arthropods grow mainly by molting, they also grow by addition of segments and appendages. Shelled cephalopods, such as *Nautilus* and ammonites, grow by continuous accretion at the edge of the shell and also by addition when they grow a new wall, or septum, to enclose another chamber. Echinoderms grow by adding new plates or skeletal segments through addition, then add to these by accretion until they reach adult size.

Describing and Quantifying Ontogenetic Change

> *I once overheard a children's conversation in a New York playground. Two young girls were discussing the size of dogs. One asked: "Can a dog be as large as an elephant?" Her friend responded: "No, if it were as big as an elephant, it would look like an elephant." How truly she spoke.*
>
> Stephen Jay Gould, "Size and Shape," in *Ever Since Darwin*, 1977

A biologist can study ontogeny of both soft and hard tissues directly with living animals or plants. Fossils, however, are no longer living or growing, so paleontologists must try to deduce their ontogeny by one of two strategies.

1. Try to reconstruct ontogenetic history with a **growth series** of specimens that are thought to represent different stages of growth of the same organism (Fig. 2.7). This is the only practical strategy for organisms that modify their juvenile skeleton through growth (as most vertebrates do), or lose it entirely (as many arthropods do).
2. In organisms that preserve their earlier growth stages in their skeletons, we can actually see some of the ontogeny still preserved. This pattern is particularly common in organisms (such as foraminifers, snails, and cephalopods) that have spiral shells and grow by addition of new chambers. By breaking away or slicing through the larger chambers, we can see each stage of their growth from their earliest shell. Paleontologists can also use the accretionary growth lines of certain organisms (such as bivalves or trees) or to reconstruct the size and shape of earlier ontogenetic stages of a given fossil.

Once the paleontologist has some measure of the changes in size and shape of the organism through its ontogeny, the next step is to examine the growth series. In almost every case, organisms show significant shape change through their ontogenetic history. These shape changes range from the extreme transformation of metamorphosis from caterpillar to butterfly to more subtle changes in most organisms that gradually change proportions as they grow larger. How do we plot these changes in a way that allows us to interpret them?

The simplest method plots one variable (for example, width) against another (for example, length) in a bivariate graph (Fig. 2.8). Any pair of variables (such as length, width, thickness) can be plotted this way, but ideally one tries to select the variables that best capture the changes in size and shape. When two linear variables are plotted against one another, a straight line with a 45° slope is the result. This kind of plot shows that the two dimensions are changing at the same rate so that shape stays the same; this is known as **isometric** ("equal measure") growth. The formula for a line fit through these points can be easily calculated by dividing the change along the y-axis (ordinate) by the change on the x-axis (abscissa). The formula for this line is given as

$$y = mx + b$$

where the values of x and y at any point on the line are specified by the slope (m) and the y-intercept (b). If the line passes through the origin, then $b = 0$ and the formula reduces to $y = mx$.

When two linear dimensions plot as a straight line with a 45° slope, then $m = 1$. Frequently, however, the two variables do not plot with a slope of 1, but with slopes greater or lesser than that value. Under these circumstances, one variable is changing faster than another; usually this happens to maintain the same shape at larger sizes. For example, if the length of an organism begins at exactly twice the width, then the length will have to double in size for every unit increase in width to maintain the same shape (Fig. 2.8). Although these changes are not occurring at the same rate, they are still considered isometric changes as long as they maintain a constant shape.

However, not all dimensions of an organism change in a linear fashion. For example, the cross-sectional area of a specimen is measured as length squared (L^2); the volume is length cubed (L^3). In a plot of a simple linear dimension against area or volume, one axis will increase by linear increments, but the other will increase exponentially. It is no longer possible to fit a line through these points (Fig. 2.7), but instead they must be approximated by a curve. The formula for that curve can be given by

$$y = mx^c$$

where c is a constant of exponential growth. If $c = 1$, then growth is linear and isometric. However, if the value of c does not equal 1, then one dimension changes as a power of the other. Under these circumstances, we have a net change in shape, known as **allometric** ("different measure") growth, or **allometry**.

In many cases, allometric growth is required by the simple mathematics of size change. For example, if one plots a linear dimension (such as length or width) against area (which increases by powers of 2), then the value of c will be either 2 or 1/2. If one plots a linear dimension against a volume (which increases by powers of 3), then the value of c will be either 3 or 1/3. A plot of an area against a volume will give an exponent c of either 2/3 or 3/2 (0.67 or 1.5). All these examples show that some shape change is occurring,

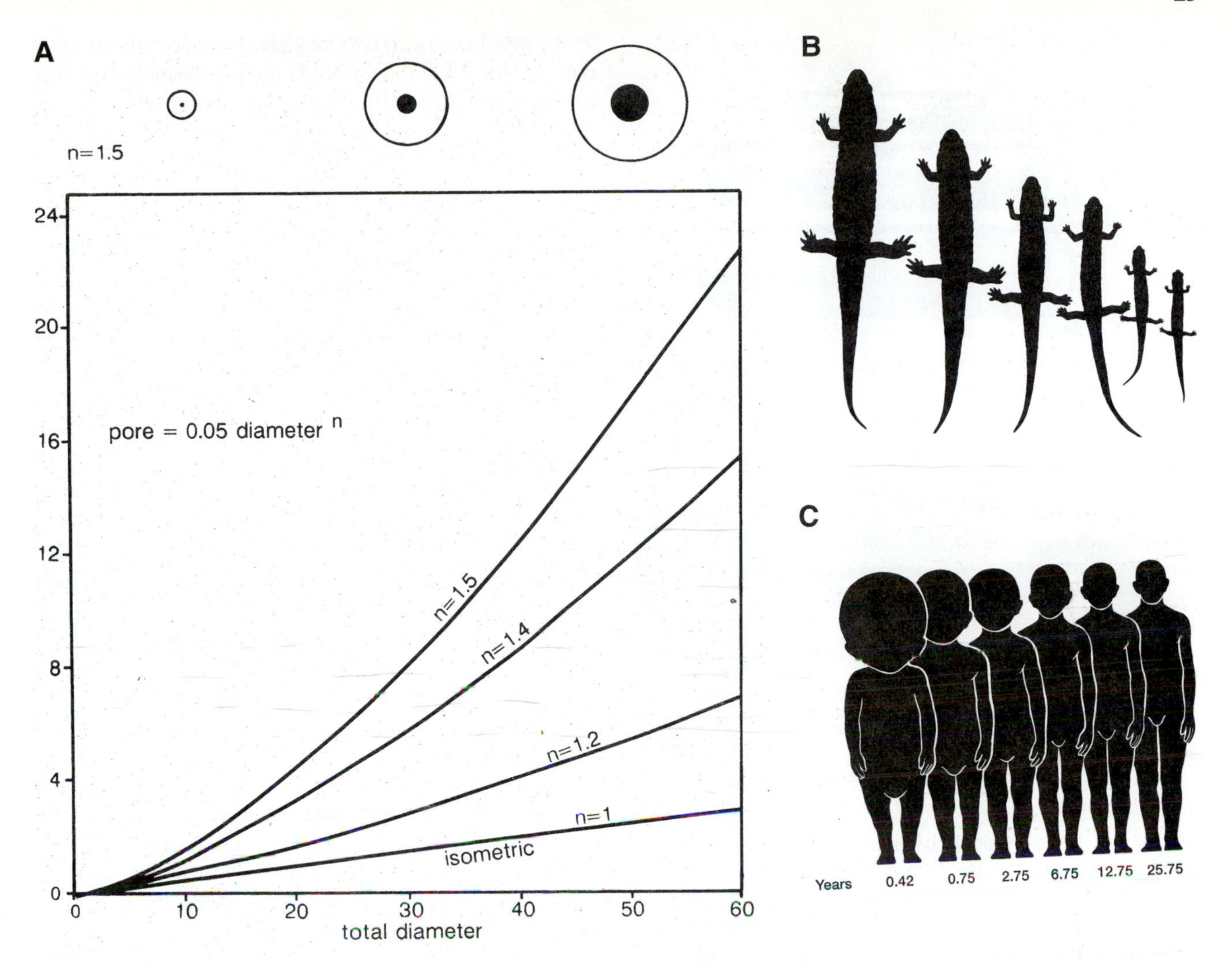

Figure 2.8. (A) Plot of growth of a hypothetical organism with a round shape and a central pore. Its growth can be expressed as an allometric growth equation, with an exponent n. If n = 1, then diameter increases as a constant ratio of pore width, and the change is isometric. If n > 1, then the pore increases faster than the diameter. (From Stearn and Carroll, 1989). (B) Isometric growth in these salamanders maintains the same shape. (C) Allometric growth in humans, with limbs growing much faster than the head. Numbers show age in years. (From Kardong, 1995.)

although this is the simplest possible example of allometric change.

What do these changes mean in biological terms? That is the most interesting question of all and is one that has fascinated many biologists and paleontologists for decades. In fact, isometric growth is very rare in nature; allometry is the rule rather than the exception. In each case, we make our plots, calculate our exponential constant c, and ask the question: why is this shape changing in the way it does?

Let us take a very simple example. We can approximate the limbs of a four-legged vertebrate as a set of cylinders. A very small animal can have thin spindly legs; as the limbs get longer, we would expect the cross-sectional area of the limbs to increase by a power of 2 relative to length. However, in the real world, the limbs of large animals are much more robust than would be expected by this simple relationship of length to area. Indeed, the legs of elephants and dinosaurs are thick pillars, much more robust than scaled-up legs of smaller animals. This fact was first recognized by Galileo in 1638 (Fig. 2.9) and was called the **principle of similitude** by the pioneering biologist D'Arcy Wentworth Thomson (1917).

Why does this occur? Remember that the limbs of a land animal are not just simple cylinders; rather, they support the body of the animal. Because the mass of the animal (a function of its volume) is increasing by a power of 3, the mass is increasing much faster than the linear dimensions or cross-sectional area. To support this rapid increase in mass, the limbs must be much more robust than expected, or else they would snap like matchsticks.

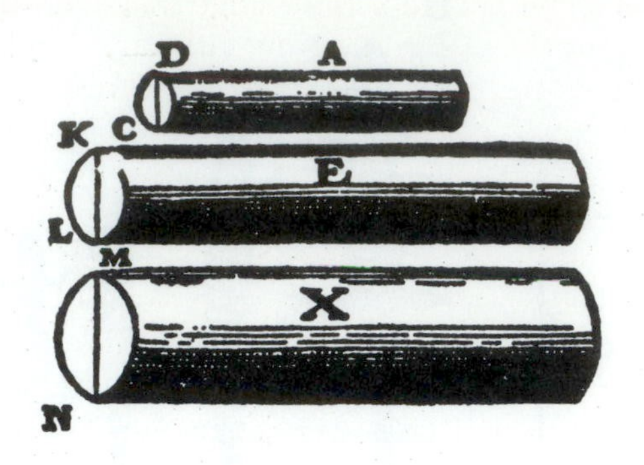

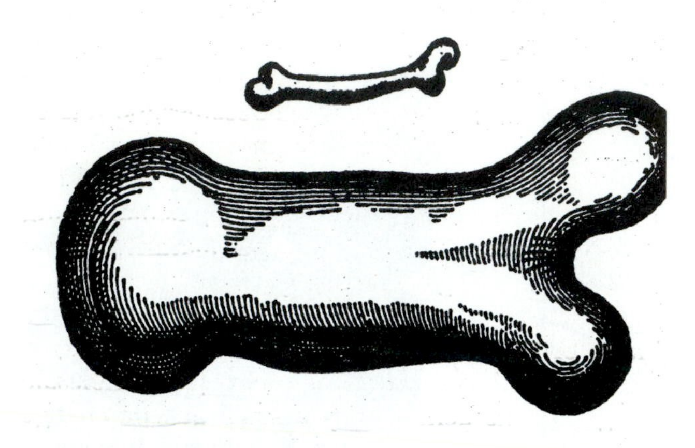

Figure 2.9. Galileo's original illustration of the relationship between size and shape. To maintain the same strength, large cylinders must be relatively thicker than small ones. For the same reason, large animals must have relatively thick leg bones.

An understanding of this principle explains why the gigantic insects or spiders featured in low-budget horror movies are biologically impossible. First of all, a giant ant or praying mantis the size and weight of an elephant would need the stocky limbs of an elephant as well. More importantly, if that giant insect molted, its huge soft body would be crushed under its own weight, because it completely lacks skeletal support during this critical phase.

There are many other examples of allometric growth in nature. For many animals, the surface area of animal bodies is critical to how fast they gain or lose heat or how fast they take in oxygen and diffuse wastes through their skins. A tapeworm can reach 20 feet in length, but it must always remain very thin, because each segment has no lungs or other organs and must exchange gases and nutrients directly with the environment. Similarly, a small organism can have a very simple digestive or respiratory tract; as it gets larger, the surface area of these systems must increase disproportionately faster to work effectively. Most large animals accomplish this by having incredibly convoluted surfaces in their lungs (with all their complex passages) or their digestive systems (such as the intricate walls of the long, coiled intestine).

The best example of this process is demonstrated by the effects of size on thermoregulation. Very small organisms (such as insects, small birds, and small mammals) have a large surface area relative to their volume. Consequently, they gain or lose heat rapidly through their skin, although birds use feathers and mammals use fur as insulation to prevent this. Nevertheless, a warm-blooded animal such as a hummingbird or shrew loses heat so fast, simply due to its size, that it must eat almost constantly to feed its metabolic furnace that keeps its body temperature constant. Indeed, hummingbirds spend almost all their time feeding and sleep very little; if they starve for more than a day or two, they die. A shrew spends almost 24 hours a day feeding voraciously and is the fiercest animal for its size anywhere in the forest; it too will die after only a day or two without food.

As size increases, however, volume increases by a power of 3 while the surface area only increases by a power of 2. This means that a large animal has much more volume relative to its surface area, so its problems are the opposite of those that small animals face. Its surface area is insufficient to dump the heat in its body without special adaptations. Elephants, for example, are constantly fighting heat stress by wallowing in mud or bathing, and their huge ears are primarily used to radiate heat from their bodies. Camels survive in the desert by allowing their body temperature to drop during the cold night, then letting their large bodies slowly warm up during the day; their size gives them thermal inertia against rapid changes in temperature. Indeed, the debate about whether dinosaurs were warm-blooded reduces to this problem of thermal inertia. Large dinosaurs had so much volume relative to their surface area that they would have automatically maintained stable body temperatures through thermal inertia without generating their own internal body heat. Indeed, if they had been "warm-blooded," they would have had insufficient surface area to dump this body heat and would have died of thermal stress.

Some organisms solve the problem of increasing body size in other ways. Individuals in colonies, for example, do not grow larger themselves; the colony grows larger by adding more chambers (as in sponges) or more individuals (as in corals and bryozoans). In this manner, each individual can retain the same small size and shape, but the entire colony can continue to grow.

There are other methods of quantifying shape besides the simple bivariate examples given above. For complex three-dimensional shapes, the bivariate method is insufficient.

Ontogenetic Change and Evolution

> Ambystoma's *a giant newt who rears in swampy waters*
> *As other newts are wont to do, a lot of fishy daughters;*
> *The axolotls, having gills, pursue a life aquatic,*
> *But when they should transform to newts, are naughty and erratic.*
>
> *They change upon compulsion, if the water grows too foul,*
> *For then they have to use their lungs, and go ashore to prowl;*
> *But when a lake's attractive, nicely aired, and full of food,*
> *They cling to youth perpetual, and rear a tadpole brood.*
>
> *And newts perennibranchiate have gone from bad to worse;*
> *They think aquatic life is bliss, terrestrial a curse.*
> *They do not even contemplate a change to suit the weather,*
> *But live as tadpoles, breed as tadpoles, tadpoles altogether!*
>
> Walter Garstang, *Larval Forms*, 1951

> *There is another possible mode of transition: some animals are capable of reproduction at a very early age, before they have acquired their perfect characters; if this power became thoroughly developed in a species, it seems probable that the adult stage of development would sooner or later be lost; and in this case, the character of the species would be greatly changed or degraded.*
>
> Charles Darwin, *On the Origin of Species*, 1859

The fact that organisms change shape through their ontogeny has important evolutionary implications. By speeding up or slowing down their growth to adult body form, a lineage can gain evolutionary flexibility and escape from the specialization of adult body forms. For example, most salamanders metamorphose from a gilled juvenile stage (that must stay in the water except for short periods of time) to a lunged adult organism (that can spend most of its life on land). However, the Mexican salamanders known to the Aztecs as axolotls (genus *Ambystoma*) rarely complete this metamorphosis in the wild; rather, they become sexually reproducing adults while still retaining their larval gills (Fig. 2.10). This allows them to take advantage of aquatic habitats throughout their lives that would otherwise be occupied only by juveniles. However, when the water becomes too foul or stagnant, they complete their metamorphosis into lunged adults so they can live on land or migrate to a fresher lake. A small change in developmental timing of metamorphosis relative to sexual maturity allows tremendous ecological flexibility.

This sort of evolution by changing developmental timing, or **heterochrony** ("different time" in Greek) is quite common, because it does not require wholesale changes in the genes—only a small genetic change that controls the timing of development. Instead of trying to evolve new anatomical features, an organ-

Figure 2.10. The Mexican salamander Ambystoma, known to the Aztecs as the axolotl. During normal conditions, it retains its larval gills into sexual maturity, allowing it to retain the aquatic lifestyle of the larva. However, if the water becomes stagnant and foul, it completes its metamorphosis into an adult salamander with lungs, allowing it to find a new lake and survive. By this heterochronic change, the axolotl has great evolutionary flexibility. (From Dumenil, 1867.)

ism takes advantage of features that are already present in its ontogeny. Gould (1977) describes numerous examples, especially in the arthropods, where such changes allow high reproductive rates. For example, aphids go through asexual cloning as wingless juveniles when the conditions are good; in this way, they can multiply rapidly to infest a choice plant. When conditions become too crowded, and the plant has little room left for aphids, some individuals will develop into asexual winged forms that can fly to a new plant. As day length decreases in the fall, some individuals complete their development to sexually reproducing males and females and lay eggs that will lie dormant until the next spring.

Several different types of heterochrony are possible (Gould, 1977; McKinney and McNamara, 1991). The retention of juvenile features into sexual maturity is called **paedomorphosis** (Greek for "child formation"). It can be caused by stopping development at an early stage (**progenesis**, as in the aphid example above) or by slowing down developmental timing (**neoteny**, as in the axolotl). The opposite of paedomorphosis is is the addition of ontogenetic stages beyond the adult reproductive stage; this is called **peramorphosis** ("overdevelopment"). This can be caused by an increased rate of growth (**acceleration**), so that more stages are added in a shorter time, or by shutting off the growth at a later stage (**hypermorphosis**, or "growth beyond"), so that growth continues past the usual endpoint. Examples of peramorphosis are less common, but the addition of ammonite body chambers beyond the ancestral number is often cited as an example. Many features of human development, such as the late enlargement of the brain after birth, are hypermorphic.

There are several ways of displaying ontogenetic variation and its relationship to phylogeny. Figure 2.11 shows examples of a range of variation (a **morphocline**) that can be due to paedomorphosis (a **paedomorphocline**) and peramorphosis (a **peramorphocline**). In each diagram, the evolutionary history of the organism can be compared to the ontogenetic history (represented by the comparison between juveniles and adults). Diagrams such as these help the paleontologist visualize how the changes in evolutionary history can be related to ontogeny.

Since the stimulus of Gould's (1977) book *Ontogeny and Phylogeny*, many paleontologists have been motivated to find examples of heterochrony throughout the fossil record (see McKinney, 1988, and McKinney and McNamara, 1991, for a recent summary). It has become clear that heterochrony is one of the most important factors in evolution, and paleontologists must be on the lookout for changes that could be explained in this manner.

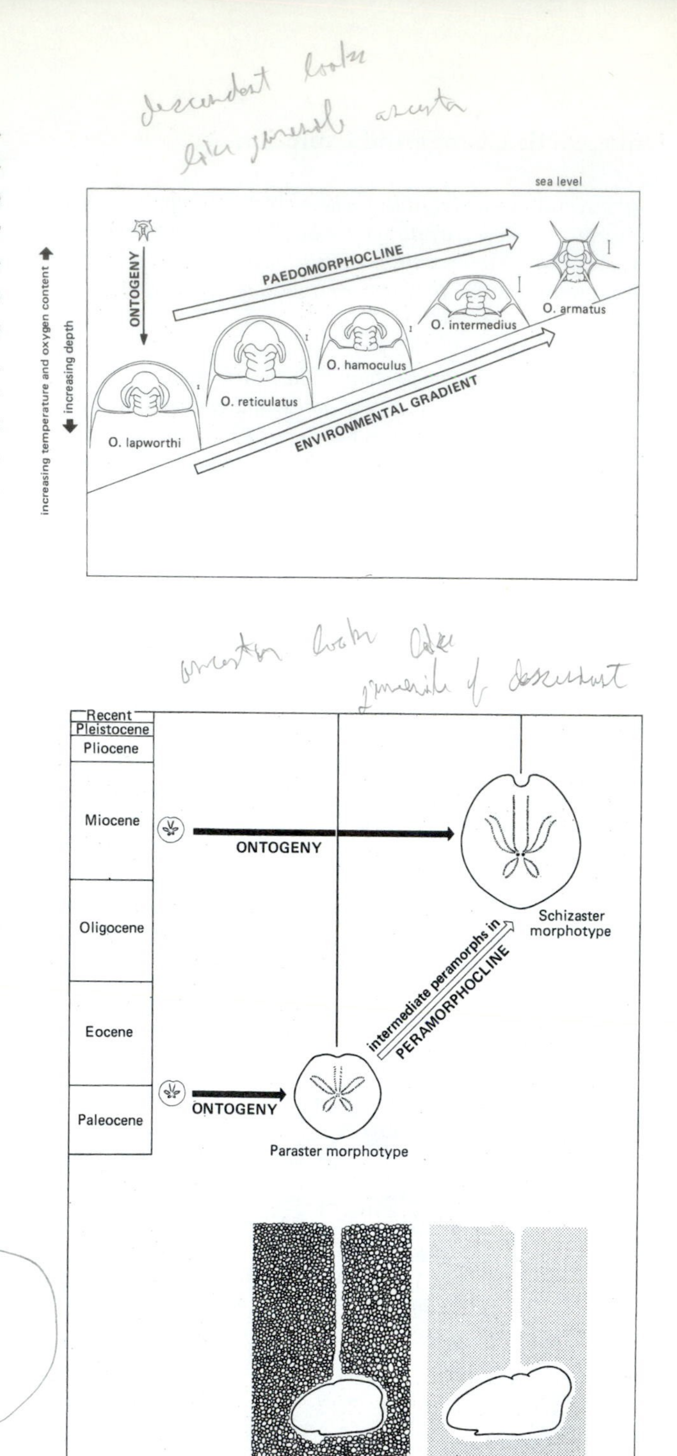

Figure 2.11. (Top) Paedomorphocline in the Early Cambrian trilobite *Olenellus*. Later species (culminating with *O. armatus*) show progressively more and more juvenile features (as shown by the ontogenetic trajectory in the upper left). (Bottom) Peramorphosis in the Cenozoic heart urchin, from the Paleocene sand-dweller *Paraster* to the later forms culminating in the mud-dwelling *Schizaster*. Through the peramorphocline, there is increased development of many traits, as well as a progressive increase in size as later species adapted for living in deeper, muddier water. (From McNamara, 1982.)

HOW DO POPULATIONS OF ORGANISMS VARY?

Shells at the southern limit, and when living in shallow water, are often more brightly coloured than those of the same species from farther north or from a greater depth; birds of the same species are more brightly coloured under a clear atmosphere than when living near the coast or on islands; residence near the sea also affects the colours of insects. Certain plants, when growing near the sea-shore, have their leaves to some degree fleshy, though not elsewhere fleshy.

Charles Darwin, *On the Origin of Species*, 1859

We have seen how the variation during an individual's lifetime (**ontogenetic** variation) can result in tremendous changes in body form. If you glance around the room, however, it should be obvious that there are also tremendous variations between individuals in the same population, as in the interbreeding population of *Homo sapiens* of which you are a part. This is **population** variation, and it can have enormous effects as well. Both types of variation occur within a single species, yet their effects can be so great that an unwary paleontologist might interpret the differences as representing different species.

What is a population? Most biologists define a population as a group of individuals living closely together so that they are capable of interbreeding. In genetics, the same concept is often called a local breeding population, or a deme. All the genes of this population are considered a common **gene pool**.

Ecophenotypic Variation

Within populations, we tend to think of the individuals that share many genes as all alike, but they are not. Even identical twins (developed from the same fertilized egg and therefore genetically identical) have significant differences because of different rates of growth or other differences in ontogeny. Any changes due to differences in the environment and ontogeny, rather than due to genetic differences, are considered **ecophenotypic variation**. There are many examples in nature where a slight change in nutrients, or light, or water temperature or chemistry, can result in changes in individuals that are genetically identical. We are all familiar with plants that grow well in ideal sunlight, temperature, soil, and moisture conditions, whereas identical seeds of the same plant produce weak and spindly plants when conditions are bad. The same population of corals (Fig. 2.12) shows wide variations of growth form, depending upon the conditions to which they are exposed. Those found in the surf zone are subjected to constant pounding of the waves, so they are low and massive. Those growing in the protected zone behind the reef tend to be delicately branching, because they experience no wave pounding that might break them down. Likewise, calcareous red algae tend to be massive and encrusting when they live in turbulent waters but delicate and branching in quieter waters. These organisms have the same genetic makeup, but different genes are expressed when triggered by different environmental stimuli.

Perhaps the most dramatic example of variation within genetically identical populations occurs in social insects (termites, ants, bees, and wasps). All the individuals in the colony are genetically identical, since they are all descended from a single queen who was fertilized only once during her mating flight. Thus, each egg is identical as well. However, by feeding and nurturing the eggs differently, the workers cause them to have different fates. Most eggs hatch to become new workers. However, if the queen dies, the workers feed a special food to certain larvae that become new queens. Other changes in growing conditions trigger certain eggs to become male, and these winged drones are driven out of the colony to perform their brief mating flight and then die.

Size can be strongly controlled by local ecological variables, such as temperature and nutrition. The best documented examples occur in human populations that experience great cultural change, and therefore changes in their nutrition. For example, Japanese populations at the beginning of the century were, on the average, composed of small individuals, largely due to their poor diets. But the postwar prosperity of Japan brought much improved diets, and the last two generations of Japanese have been noticeably bigger than their grandparents. This change is entirely ecophenotypic, because no significant genetic change could have taken place in so few generations.

Many other examples of variation are clearly ecophenotypic. The common Eastern oyster, *Crassostrea virginica*, grows best when seawater temperatures average 15°C (typical of Chesapeake Bay). The growth yields drop off significantly if the mean temperature is too warm or too cold. Alexander (1974) found that the size of scallop populations from the Pliocene rocks of the Kettleman Hills of California varied tremendously over the region. Apparently, the salinity of the waters had a strong controlling effect. Those individuals that grew in waters of normal salinity were the largest, whereas those found in lower salinity, brackish waters tended to be smaller. Salinity also controlled non-size variation as well. Specimens of the clam *Anadara* had a mean value of 28 ribs in the normal salinity regions but as few as 24 ribs in areas of low salinity. Similar trends have been observed in modern populations, such as the clam *Cerastoderma edulus* from the Netherlands, which has fewer ribs in lower salinity waters (Eisma, 1965).

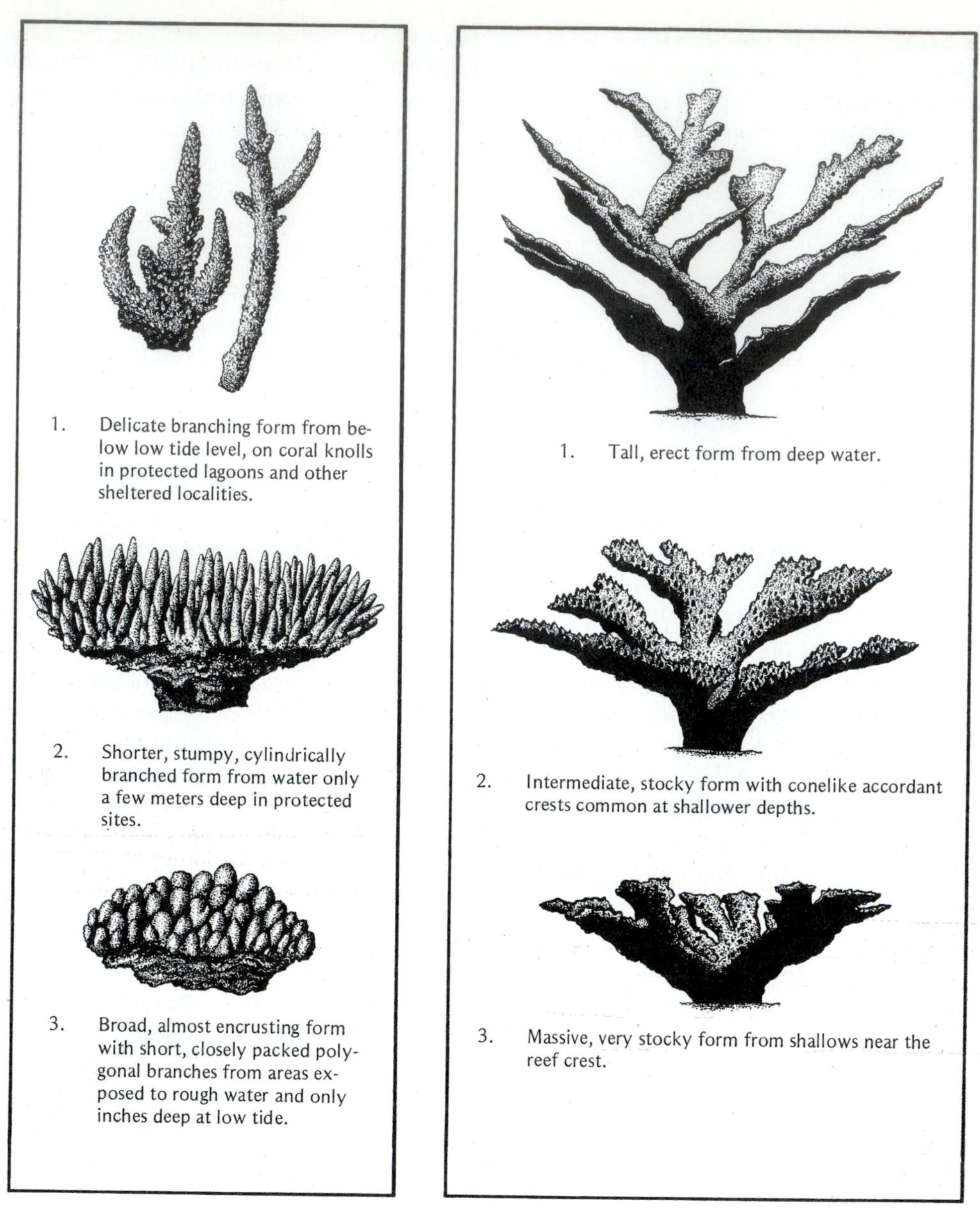

Figure 2.12. Ecophenotypic variation in growth forms of corals from the same species. (Left) Three growth forms of *Acropora humilis* from the Marshall Islands in the Pacific, modified by differences in water depth and turbulence. (Right) Three growth forms of *Acropora palmata* from Andros Island in the Bahamas modified by differences in water turbulence. (From Anstey and Chase, 1979.)

Some ecophenotypic variations can be even more extreme than simply size, number of ribs, or growth form. Newell and Boyd (1985) analyzed large samples of the Permian scallop *Guizhopecten cheni willisensis* from West Texas and found that they varied not only in size but also in their ornamentation and shape (Fig. 2.1). The shells shown in Figure 2.1 might have been split into numerous species by an earlier generation of paleontologists who were inclined to make new species of every slight variation. However, based on the continuous range of variation observed in their large population samples, Newell and Boyd saw that all of these variants were members of the same species. Some planktonic foraminifera undergo radical changes in their growth and coiling direction due to temperature. The cold-water-loving *Globigerina pachyderma* (the only species now living in the Arctic Ocean) normally coils to the left in the Arctic, but is right-coiling in the warmer North Atlantic. Another species, *Globotruncana truncatulinoides,* does just the opposite, coiling left in warmer waters and right in cooler waters. Apparently, there is some sort of critical temperature threshhold during their early embryonic development such that a small difference in temperature makes a big difference in the development of the organism.

Sexual Dimorphism

A second major source of intrapopulational variability is **sexual dimorphism** (Box 2.1). We are all familiar with examples of organisms that differ tremendously depending upon whether they are male or female but that are part of the same population with the same genes (except for their X or Y chromosomes). Lions have prominent manes, while lionesses do not; among deer, bucks have large antlers, while does have none; many male antelopes have horns, while the females do not. In most birds, the males have the showiest plumage, while the females are often drab. In some organisms, the differences between the sexes can be extreme. Bull elephant seals have a long, flexible proboscis (hence the name) and weigh up to 2200 kg (5000 pounds), almost three times as large as the females; indeed, they may crush them while rushing across the beach to drive off rivals. Although the males tend to be larger in most birds and mammals, the reverse is often the case in insects, reptiles, amphibians, and fish. In some deep-water fish that live in total darkness, the tiny males attach to the much larger female and then degenerate into a sperm sac.

Once again, recognizing sexual dimorphism in the fossil record can be a challenge. If we did not have living deer to compare to, would we ever guess that the antlered males and antlerless females belong to the same species? With a fossil, all we have is hard parts (and often only fragments of them), so there is little direct evidence of the sex of the specimen. However, we frequently find variability in fossil collections that suggests two separate sexes; some of these variants may have already been assigned to separate species by earlier paleontologists. How could we decide whether two distinct forms are really two different species, or simply males and females of the same species?

There are no ironclad criteria for distinguishing sexual dimorphism from true species differences, simply some general rules of thumb.

1. *Analogy with living relatives.* If we know enough about the biology of living relatives of the extinct organisms, we may infer whether sexual dimorphism is a reasonable hypothesis. For example, fossil deer with antlers are almost always considered male because nearly all living deer follow that rule (except for the reindeer, where both sexes have antlers).

2. *Ratios of presumed males to females.* In the majority of dimorphic organisms, there should be two (not three or more) distinct morphs; rarely are natural populations composed only of a single sex. In addition, the ratio of males to females is typically 1:1. Of course, there are many exceptions to this rule, especially among organisms where one dominant male guards a harem of females or among herds of elephants, which are all related females with no males.

Genus *Kosmoceras*

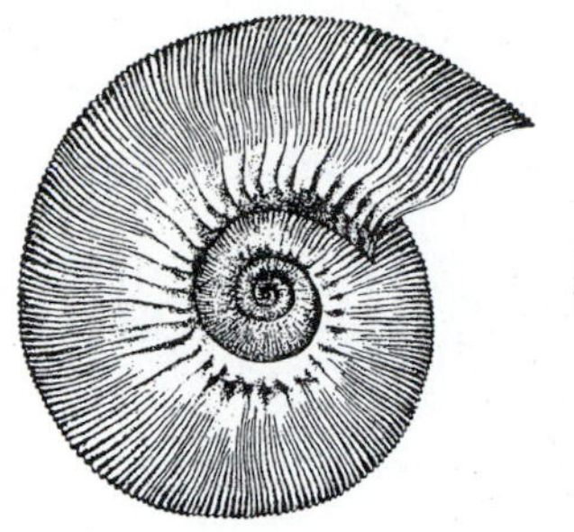

K. (Zugokosmoceras) grossouvrei

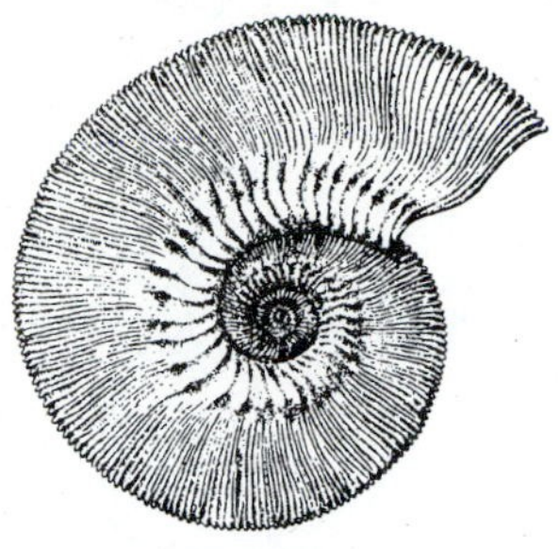

K. (Lobokosmokeras) phaeinum

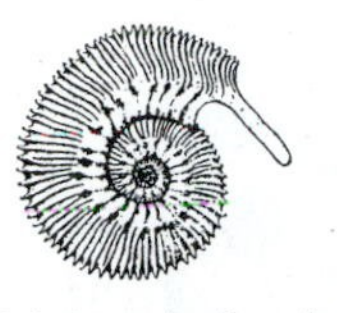

K. (Gulielmiceras) aff. *gulielmi*

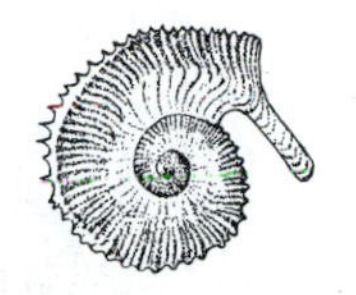

K. (Spinikosmokeras) acutistriatum

Figure 2.13. For almost a century, these different shapes of ammonites from the Middle Jurassic of England were placed in different species and subgenera (note the names under each specimen). However, Callomon (1963) amd Makowski (1963) argued that the smaller forms ("microconchs") with the long lappets around the aperture are from males, and the larger shells ("macroconchs") are from females. (From Clarkson, 1993.)

3. *The two presumed sexes should have the same distribution in space and time.* This is the most powerful line of evidence because it is very unlikely that males and females of the same species lived millions of years apart or on different continents—breeding would be very difficult, to say the least.

Using rules of thumb such as these, a number of important discoveries have been made. For example, in the 1860s, paleontologists described paired groups of the Middle Jurassic ammonites of the Family Oppeliidae (Fig. 2.13). They recognized that these "species" had identical stratigraphic ranges, that their early chambers were identical (they differed only in their outer chambers), and that they evolved in pairs with simultaneous appearance of new characters in both members of the pair. Although the possibility of sexual dimorphism was actively debated, these morphs were long regarded as different species. Almost a century later, Callomon (1963) and Makowski (1963) reviewed the evidence and came to the opposite conclusion. The larger shell of each pair is known as the macroconch; the smaller shell ("microconch") is a third to a quarter of the size of the macroconch and has small horns or "lappets" that protrude from the opening. Which conch is male and which is female is still debated.

Box 2.1

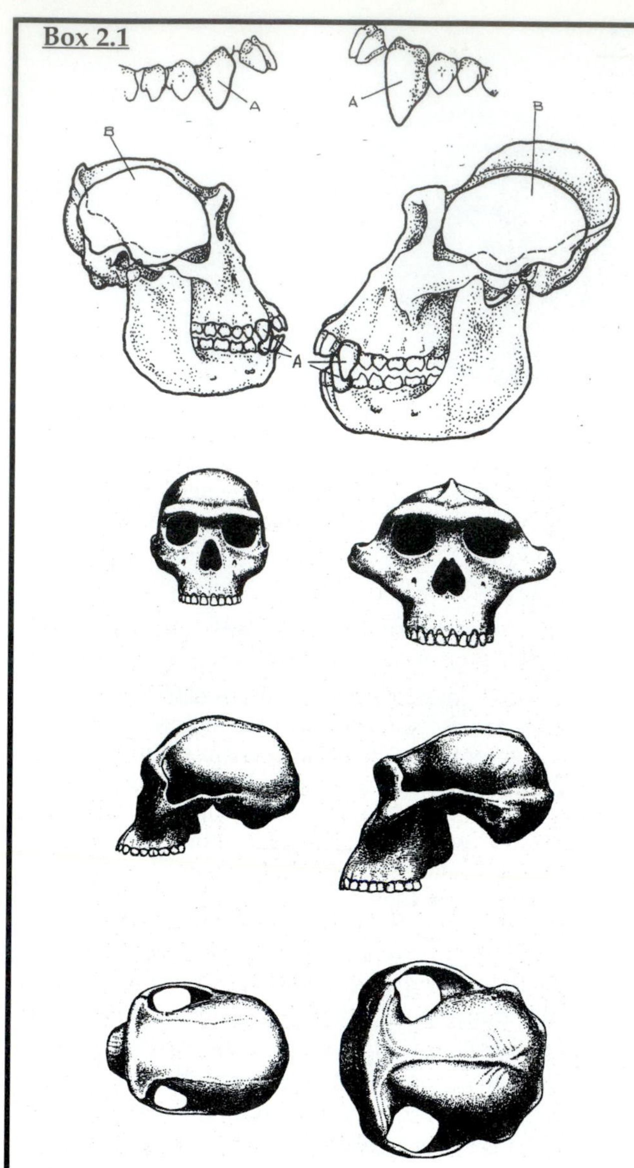

Figure 2.14. (Top) Typical dimorphism in living gorillas, in which the males (right) weigh almost twice as much as females, and have a large sagittal crest along the top of the skull, and larger canines (A) and brains (B). (After Zihlman, 1982). (Bottom) Contrast between the robust (right) and gracile (left) australopithecines. Are they males and females of the same species, or different taxa? (After Walker and Shipman, 1996.)

Primate Sexual Dimorphism and Fossil Hominids

Among the mammals, the order Primates demonstrates some of the more extreme forms of sexual dimorphism. In the Old World monkeys (but not the New World monkeys), the females are about 70% the size of the males, and have less prominent canine teeth. In some monkeys, such as the mandrills and baboons, the females are only half the size of the males, and the males have much more prominent snouts and canines (Fig. 2.14). Male gorillas are also significantly larger than females, with a much more robust skull including larger canines and a crest along the top of the braincase (sagittal crest) for their stronger jaw muscles.

Living humans show a slight dimorphism in size (males slightly larger than females in most cultures), and females also have a larger opening in their pelvis for the birth canal, but otherwise the differences are slight. This does not necessarily apply to our fossil ancestors, however. For example, early discoveries of fossil *Australopithecus* in Africa produced two basic kinds: a delicate, gracile form originally called *Australopithecus africanus*, and a more robust form with heavy teeth and a larger sagittal crest, originally called *Paranthropus robustus*. Although they were originally described as distinct species, later anthropologists tended to view human evolution as one single, highly variable species. Under this scheme, the robust individuals were interpreted as males and the gracile forms as females. The huge crests and large teeth of the robust forms seemed remarkably different, but anthropologists pointed to the robust teeth and large sagittal crests of male gorillas as support for the interpretation of sexual dimorphism. By the early 1960s, this faith in the "single species" model of human evolution was so strong that paleoanthropologists denied that the earliest species in our genus, *Homo habilis*, was anything more than a gracile australopithecine.

By the 1970s, however, there were so many different kinds of australopithecine specimens that the "single species hypothesis" could no longer be supported. The most conclusive evidence was geographic and stratigraphic distribution. It is now clear that there are at least two robust forms, *Paranthropus robustus* and *P. boisei*, and both had different distributions in time and space than did *Australopithecus africanus*. Today, opinion has swung from the extreme taxonomic splitting of the pre-1950s to the extreme lumping of the 1960s and 1970s back to a more speciose view of human evolution.

When the famous hominid specimens of *Australopithecus afarensis*, known as "Lucy" and her relatives, were found in Ethiopia in 1973, there was much discussion about how many species these specimens represented. The original describers, Don Johanson and Tim White, decided to incorporate all of the variability into a single species and attribute it to sexual dimorphism. Others, such as Richard and Mary Leakey, were more skeptical, and considered the variability too great for a single species; some specimens belonged to a different species. This debate has not yet been resolved.

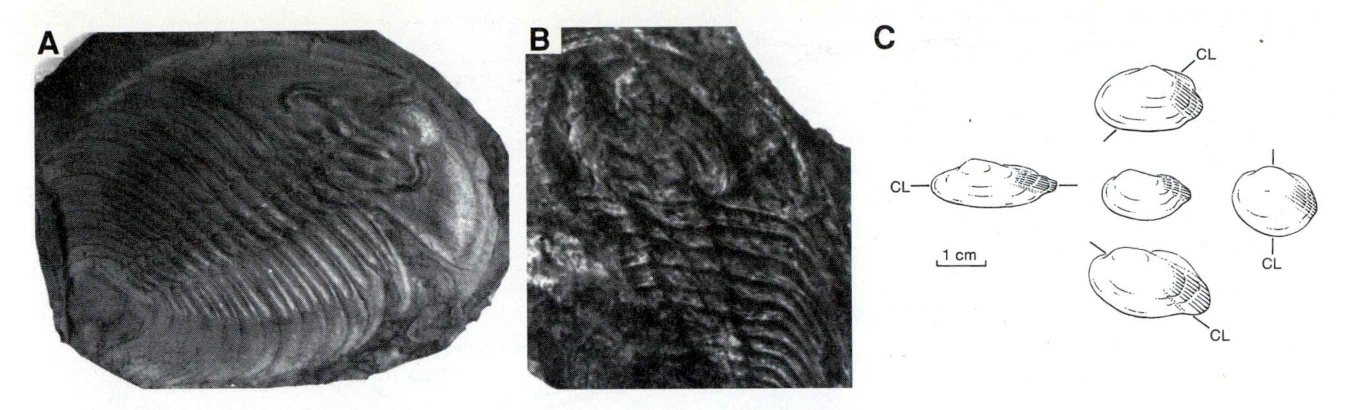

Figure 2.15. (A) and (B) Trilobites which have been distorted from their normal bilaterally symmetrical shape by tectonic forces. (Photos by the author) (C) Effects of rock deformation on the bivalve *Arisaigia postornata*. The specimen in the center is undistorted, and the others show different patterns of deformation, depending upon how they were originally oriented with respect to the stress axis. The direction marked "CL" corresponds to the rock cleavage direction and is perpendicular to the maximum direction of shortening. (After Bambach, 1973.)

Taphonomic Variability

So far, we have discussed the variability of populations due to biological causes. However, the process of fossilization can add even more variability to the specimens that was not present when the organisms were alive and members of a natural population.

The most common taphonomic effect on specimens is **post-mortem distortion**. Many fossils have been subjected to tectonic stresses, compaction, crushing, dissolution, and other processes that greatly alter the shape and composition of the fossil (Fig. 2.15A, B). For example, Bambach (1973) noted that the shapes of the Silurian bivalve *Arisaigia postornata* varied tremendously, and in peculiar, asymmetric ways. He measured the proportions of specimens and also noted the direction of cleavage in the rocks. As expected, the direction of cleavage predicted the distortion axis of the fossils (Fig. 2.15C). In fact, all the specimens were so distorted that there were none of the undistorted shape (middle of diagram) in the entire sample. This type of distortion has been used by structural geologists as a measure of rock deformation, since the original symmetry of the fossil is a good indicator of the undistorted fabric of the rock.

If paleontologists fail to account for post-mortem distortion, they can make erroneous judgments. For example, one of the commonest fossils of the Oligocene Big Badlands of South Dakota is the sheep-sized hoofed mammal known as *Miniochoerus* (Fig. 2.16). Schultz and Falkenbach (1956, 1968) split the genus into multiple genera and subfamilies based on post-mortem distortions that they failed to recognize. Skulls that were flattened top-to-bottom were christened *Platyochoerus* ("flat pig") and those that were crushed side-to-side were named *Stenopsochoerus* ("narrow pig"); undistorted specimens remained in the genus *Miniochoerus*. When Stevens and Stevens (1996) studied these specimens in detail, they found no statistical basis for separation into multiple generic and subfamilial lineages. They even simulated this crushing by casting a specimen in rubber, and distorting the rubber cast in various ways to show that the differences recognized by Schultz and Falkenbach (1956, 1968) were all due to crushing. In all characters that were not affected by post-mortem crushing, the specimens at the same stratigraphic level were identical.

Another important taphonomic effect is selective sorting by transport. In Chapter 1, we discussed the distortions introduced by transport in water and noted that more robust shells or bones tend to be the most often preserved, even if their original abundance in the population was much lower. But transport can work in surprising ways. In one study, a collection of clams from a beach in Trinidad showed a remarkable bias (Martin-Kaye, 1951). The shells on the northern beaches were almost all left valves (from 87% to 89%), those in the southern beaches were almost all right valves (87% at the southernmost sampling site), and those in the middle were about 50-50. Because each clam has one left and one right valve, clearly some sorting bias is at work here. In this case, it seems that the prevailing longshore currents pick up one valve more easily than its mirror image, so that one valve is left behind as a residue at one end, while the other valve is preferentially carried to the other end.

This kind of bias would be hard to detect if we did not already know that the ratio was originally 50:50. Similarly, trilobite fossils are seldom found complete; instead, we tend to find their molted segments. Many localities are famous for containing nothing but head segments or tail segments or disarticulated segments of the thorax. Because each trilobite had all three segments in life, some sorting by currents must have occurred.

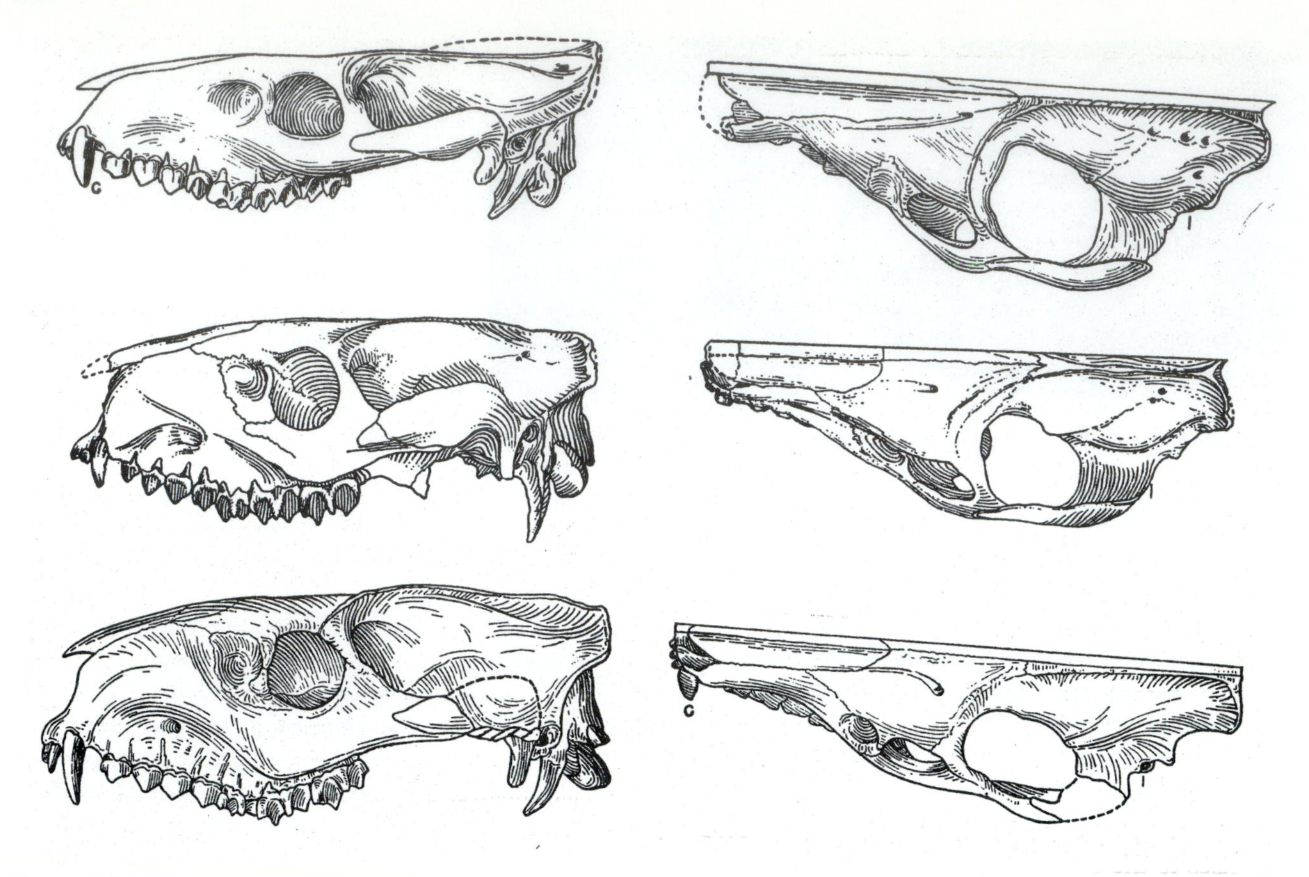

Figure 2.16. Distortion of the Eocene-Oligocene oreodont *Miniochoerus* (center row of specimens) was misinterpreted by Schultz and Falkenbach (1956, 1968) as different taxa. For example, specimens that had been flattened top-to-bottom so that the skull was flatter and wider (top row) were called "*Platyochoerus*" and those that were crushed side-to-side so that the skull was narrower and taller (bottom row) were named "*Stenopsochoerus*." (Modified from Schultz and Falkenbach, 1956.)

In some cases, post-mortem effects can have serious biases. We have seen in Chapter 1 that more delicate shells and bones tend to be more easily broken and lost from the fossil record. In many cases, this means that the delicate juvenile shells or bones are missing from a large fossil sample. In fact, for most fossil vertebrate species, juvenile specimens are unknown. An unwary paleontologist might conclude that the juveniles were rare, or lived somewhere else, or that the sample was accumulated at a season of the year when few juveniles were born. In fact, most such cases of juvenile rarity are simply due to taphonomic bias, not biological factors.

In other cases, certain fossils can undergo tremendous post-mortem transport that can give misleading notions. Fossil cephalopod shells fill up with gas from the decaying organism and become buoyant. They are capable of drifting across entire oceans before they sink to the bottom and are fossilized, giving the impression that they lived in regions where they didn't. Similarly, pollen is capable of being blown across enormous distances; it is common to find pollen in the sediments of the deepest oceans. Clearly, they do not reflect the local ecology.

In terrestrial environments, fossils can be eroded from the bedrock and redeposited into younger beds, giving the impression that the organism lived much later than it actually did. A few years ago, there was a great controversy over carnivorous dinosaur teeth found in Paleocene deposits, suggesting that these dinosaurs survived the great Cretaceous extinctions (Rigby et al., 1987). Although the teeth appeared fresh, unworn, and untransported, several critics pointed out how durable these teeth were and showed that reworked Cretaceous shark teeth with no signs of transport were very common in nonmarine Paleocene rocks (Eaton et al., 1989). Obviously, these were not "land sharks" but reworked specimens; similarly, the evidence for Paleocene dinosaurs has been discredited.

Things are not always as they seem, especially when it comes to fossil assemblages. Transport and selective fossilization can greatly bias the composition of a death assemblage (**thanatocenosis**) from the original life assemblage (**biocenosis**) that produces it. How do we decide if a fossil sample is a reliable approximation of the orginal biocenosis or is a highly biased thanatocenosis that cannot be used for biological analysis? As we showed in Chapter 1, the best criteria are those showing specimens in life position or, failing that, those showing little evidence of transport or breakage or wear. There are other tools as well. Boucot et al. (1958) looked at the brachiopods in a block of Devonian limestone and found that certain genera had both shells preserved in about equal numbers, suggesting that they had undergone little transport. Other genera were represented almost exclusively by one of the two shells, showing that they had been transported into the area and that any biological conclusions based on their abundance would be unreliable. Since that study, paleontologists have developed many other sophisticated methods of detecting taphonomic biases so that the reliability of fossil assemblage can be assessed. Whenever you see a fossil assemblage in an outcrop, the first thing to remember before you jump to conclusions is that it could be highly biased.

CONCLUSIONS

We have seen how the variability of fossil samples can be a very complex problem. Some of this variability is due to ontogenetic change within the lifespans of individuals, some is due to ecophenotypic responses to the environment, and some may be due to sexual dimorphism. Much of the variability of the fossil sample is also biased by taphonomic effects such as post-mortem distortion, selective sorting, reworking, and other nonbiological factors. All these problems must be kept in mind when a paleontologist is analyzing the biological meaning of a sample. Before a new species can be named and described, the paleontologist should attempt to demonstrate that the differences of this "new species" are not simply due to some of the factors we have just discussed.

For Further Reading

Gould, S.J. 1977. *Ontogeny and Phylogeny*. Harvard University Press, Cambridge.

Macurda, D. B., ed. 1968. *Paleobiological Aspects of Growth and Development, A Symposium*. Paleontological Society Memoir 2:1-119.

McKinney, M. L., ed. 1988. *Heterochrony in Evolution, A Multidisciplinary Approach*. Plenum Press, New York.

McKinney, M. L., and K. J. MacNamara. 1991. *Heterochrony, The Evolution of Ontogeny*. Plenum Press, New York.

Newell, N. D. 1956. Fossil populations, *in* Sylvester-Bradley, P.C., ed., *The Species Concept in Paleontology*. Systematics Association Publication 2:63-82.

Thompson, D. W. 1942. *On Growth and Form*. Cambridge University Press, Cambridge.

Figure 3.1. What is a species? In captivity, lions and tigers interbreed to form "ligers" or "tigrons," although their hybrids are infertile. In nature they normally don't interbreed because lions are now restricted to Africa and tigers to Asia. However, in the Gir Forest of India is the last remaining population of Asiatic lions, which does not interbreed with the local tigers—suggesting that they are separate species in nature. (Photo from IMSI's Master Photo Collection.)

Chapter 3

Species and Speciation

The boundaries of the species, whereby men sort them, are made by men.

John Locke, *An Essay Concerning Human Understanding*, 1689

I look at the term species as one arbitrarily given, for the sake of convenience, to a set of individuals closely resembling each other.

Charles Darwin, *On the Origin of Species*, 1859

WHAT IS A SPECIES?

In the previous chapter, we discussed the sources of variation of natural living populations and how they are reflected in fossil populations. All this variation clearly occurs within genetically related populations of individuals. Throughout this discussion, another question lurked in the background: how much variation is normal for individuals within a population, and how much variation can be attributed to individuals belonging to a different species?

This question is not simply a matter of deciding what to name a specimen. The species is the fundamental taxonomic unit in nature, the only such category that has biological reality. How we determine whether a particular group is a phylum, class, order, or genus is largely arbitrary—but species have an existence outside the minds of scientists. Most valid species were recognized not only by scientists, but even by pre-Darwinian natural historians and by non-Western native cultures. More important, animals of the same species recognize each other. Because the formation of new species is the basic mechanism of evolutionary change, species are the stuff of evolution. How we recognize species in our fossil collections has many implications.

Species Concepts

The definition of a species is not simple and straightforward. Before Darwin's time, the species was considered an arbitrary category erected by naturalists to reflect God's handiwork. Pre-Darwinian natural historians were largely **typological** in their thinking, seeing each species as a fixed group of organisms created by God, with all the variations within a species as imperfections from the ideal "type" that was God's original blueprint. Thus, the original concept of species was largely one of convenience: clusters of "individuals closely resembling each other" reflecting God's plan of nature.

Darwin fundamentally changed the assumptions about species. To Darwin, species were a part of the continuous variation of natural populations. If natural populations varied enough, sooner or later they would change enough to be recognized as new species. In his thinking, variation was not imperfection from a perfect type, but the basis for evolution. Natural selection picks and chooses among all the variants in a population and creates new species. Darwin introduced an evolutionary meaning to a pre-evolutionary concept of species. Species were dynamic, changing evolutionary units, not just fixed clusters of similar organisms delineated by God.

These two different conceptions of species should complement each other, but in practice they can conflict. To some biologists, the evolutionary implication of species is the most important aspect, and the species definition must reflect this. To others, the practical business of subdividing specimens into species is more important. These different conceptions even lead to different definitions. Many biologists use a simple **morphological species concept**, such as that described by Niles Eldredge and Joel Cracraft (1980): "a species is a diagnosable cluster of individuals within which there is a pattern of ancestry and descent, and beyond which there is not." Ernst Mayr (1942, 1954, 1963) has been the foremost proponent of the **biological species concept**, where "a species is an array of populations which are actually or potentially interbreeding, and which are reproductively isolated from other such arrays under natural conditions." This definition has been widely discussed and critiqued by biologists in the last 30 years, but it is the most widely accepted in biology today.

There are many practical problems with the biological species definition. The major one is the criterion of interbreeding. Ideally, before we decide whether any two populations are the same species or different species, we must observe them interbreeding in nature. But this is simply impossible for the vast majority of living organisms. Most named species are known only from dead specimens in jars and cabinets, and nothing is known of their field biology. Only a tiny fraction have been studied in the field in any detail, and they may not breed often. If we don't observe interbreeding, how do we know it didn't occur when we were not watching? This entire issue begs a larger question: by asking whether two clusters of organisms are interbreeding or not, haven't we already assumed a morphological species definition in order to recognize them as distinct in the first place? In addition, the biological species concept only works for sexual organisms: for asexual organisms, there is not interbreeding, only budding and cloning.

It's possible (but not always easy) to tell if two populations interbreed when they live in the same area (**sympatric**), but what about animals that live in different areas (**allopatric**)? How can we tell whether they are capable of interbreeding if they never come in contact? To get around this problem, Mayr introduced the phrase "potentially interbreeding" to the definition. The only way to tell if they are potentially interbreeding is to bring allopatric populations together in unnatural conditions, such as zoos. For example, lions are normally found in Africa and tigers in Asia, so they are allopatric. They look so different that even a child can tell them apart, so most people would regard them as distinct species. When mixed in zoos, however, they can interbreed and hybridize to form lion/tiger mixtures known as "ligers" and "tigons." This would suggest that they are the same interbreeding species. But are they? In the Gir Forest of India, the only surviving Asian population of lions lives in sympatry with tigers—and they do not interbreed (Fig. 3.1). Many other examples of distinct species in nature do hybridize in zoos, but does this make them the same species (considering the unnatural conditions in most zoos)? Unlike the lion and tiger example, we have no sympatric natural populations to decide whether they would interbreed in nature. Clearly, there are major problems with the interbreeding criterion as applied to most living organisms. (See Sokal and Crovello, 1970, for a critique of the biological species concept.)

The situation is even more complex when you consider **sibling species**. These are species that cannot be told apart by any external anatomical features and are recognized only by differences in behavior or by subtle molecular or chromosomal differences. For example, the *Anopheles maculipennis* mosquito and its relatives carry malaria, and other sibling species do not. The adult mosquitos cannot be distinguished, but the larvae are somewhat different, and the eggs are so distinctive that the officers of the U.S. Army Sanitary Corps were able to teach untrained soldiers stationed in Italy during World War II to recognize the different species by their eggs and decide if they needed to worry about malaria. In other insects, biologists could not recognize sibling species until they noticed the difference in the frequency of the hum of their wings (which is very easy for flies and mosquitoes to hear and recognize and thus find their mates). After biologists had made this behavioral distinction, they were able to find molecular differences as well and recognized two species. But if these insects were known only from dried adult specimens pinned to a board or sitting in a vial, it would have been impossible to tell the difference. Sibling species are common not only among flies and mosquitoes but also in butterflies, crickets, and other insects. They have also been documented in birds (tyrant flycatchers), fish (whitefishes), water fleas (*Daphnia*), limpets, and sponges. How many other examples of sibling species are there among our morphological species? We cannot tell.

If the biological species definition is so impractical that most biologists define species on morphological criteria alone, why do we keep using it? The reason is simple: organisms recognize each other as members of the same species, especially when they are choosing a mate. If the species differences have meaning to the organisms themselves, then biologists view them as the "real thing" that we are trying to approximate when we name species based on morphological features.

Speciation

At an even more fundamental level, biologists prefer the biological species definition because they feel that **reproductive isolation** is the key to the formation of new species, or **speciation**. Ironically, the issue of the origin of new species was never really addressed in Darwin's book, even though it was entitled *On the Origin of Species*. Just by demonstrating how organisms could transform from one shape to another, Darwin thought that the origin of species would be answered. But modern biologists view the problem in a different light, largely based on the advances in our understanding of genetics in the last century. Tranformation from one shape to another is not enough. Modern biologists want to understand how we get a new species population to split off from the original ancestral population. In other words, speciation is not about transformation but about splitting and multiplication of species.

Modern genetics views natural populations as a large reservoir of interbreeding individuals; all their combined genes can be thought of as a **gene pool**. The members of most natural populations freely interbreed among themselves, so there is continous **gene flow** throughout the gene pool. But this presents a problem. Suppose a novel mutation appeared that would confer an advantage under natural selection to a few individuals in a large population. If these mutants interbred with normal individuals, this advantageous feature would quickly hybridize out by diluting the genes that control it, and it would disappear. For an individual with novel mutations (and thus unusual gene frequencies), it is critical that they be part of a small, reproductively isolated population (or a **peripherally isolated population**). Under such circumstances, their unusual gene frequencies have a chance of becoming dominant in the population. If their descendants find these genes advantageous, and survive and flourish, they may eventually become a new species.

This problem became apparent in the 1940s and 1950s, when genetic studies of natural populations began to determine their gene frequencies. Large natural populations with their high rates of gene flow were not capable of changing rapidly when a new mutation appeared. But small populations easily changed in response to the unusual gene frequencies of a few of their members. The most dramatic demonstration of this is the **founder principle**. Oceanic islands, for example, are usually populated by organisms descended from a few individuals, or founders, that came there by chance dispersion (rafting, storms, or brought by humans). Those founders may have had unusual gene frequencies compared to their ancestral populations on the mainland. Since all their descendants will carry these same gene frequencies, populations on islands can quickly change and differentiate genetically from the mainland population due to their isolation. If some of those island inhabitants are reintroduced to mainland species, they may no longer interbreed and they have become new biological species as well.

Although islands provide the best examples, genetic isolation of small populations with unusual founders can occur in many other ways. For example, isolated religious sects, such as the Pennsylvania Amish and Dunkers, have unusual gene frequencies compared to the rest of the population of the state. The Lancaster County Amish are well known for their unusually high incidences of polydactylous (mulitple fingers and toes) dwarfism. Like island populations, these religious sects were founded by a few members (apparently with some unusual gene frequencies) over a century ago and have had relatively little gene flow with outside populations. People leave the sect to marry out, but few are recruited, so the religious sect is highly inbred, increasing the chances of expression of their unusual genes.

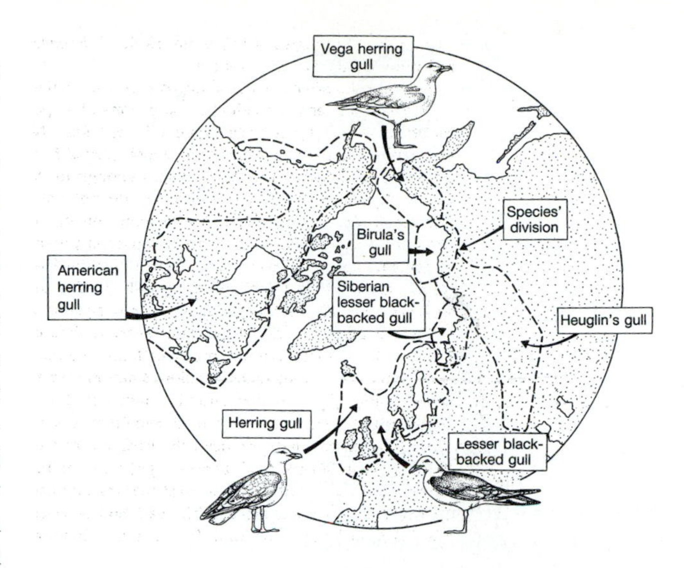

Figure 3.2. In northern Europe, the herring gull and lesser black-backed gulls are good species that do not interbreed. However, their related subspecies are distributed in a ring around the Arctic Circle, and adjacent populations around the ring can interbreed. This example of a circular cline or "ring species" suggests that the gull populations are in the process of splitting into distinct species, and those in direct sympatry in northern Europe are already distinct. (From Ridley, 1993.)

Studies by Mayr and others suggested that geographic separation and reduction of gene flow can occur in many ways. Many species are distributed over a large geographic area with very different climatic and ecological factors. For example, animals in colder climates tend to be larger in body size since it helps them conserve heat (Bergmann's rule), and they have shorter stubbier limbs and ears to prevent heat loss (Allen's rule); those from warmer climates tend to have smaller body sizes and long limbs and ears. This kind of gradient in features within a species is known as a **cline**. Slight anatomical differences in local populations over a large area may be enough so that they can be recognized as **subspecies**. Each subspecies still interbreeds with adjacent subspecies, but not with subspecies that are more distant.

Widely distributed subspecies with complex patterns of interbreeding are best demonstrated by cases of circular overlap, also known as **ring species**. The most famous example comes from the gulls that live around the Arctic Circle (Fig. 3.2). The subspecies of the herring gull (*Larus argentatus*) apparently originated in Siberia and then spread east, populating

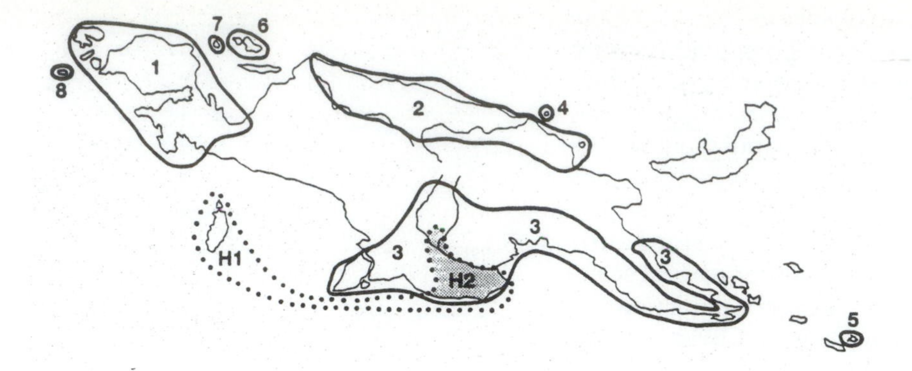

Figure 3.3. (above) Map showing the distribution of kingfishers of the *Tanysiptera hydrocharis-galatea* group in New Guinea. The three forms on mainland New Guinea (1, 2, and 3) are almost indistinguishable. The five island forms (4-8) are strikingly distinct and most were originally described as species. H1 and H2 make up the form *hydrocharis.* The Aru Islands (where H1 now lives) and southern New Guinea (H2) once formed a separate island and *hydrocharis* apparently evolved there. When southern New Guinea rejoined the mainland (dotted line), the southeastern New Guinea subspecies (3) invaded the area, but did not interbreed with *hydrocharis. Tanysiptera hydrocharis* had apparently evolved into a full species while isolated (From Mayr, 1942.)

Figure 3.4. (below) Model of the process of allopatric speciation, which transforms a single homogeneous population into different races, subspecies, and eventually species. (From Mayr, 1942.)

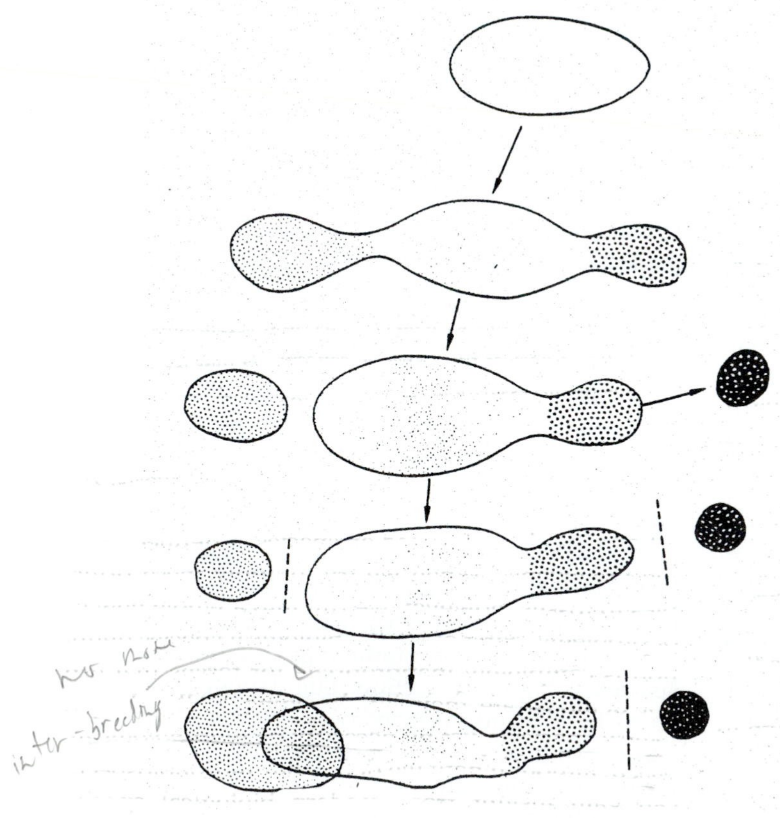

First stage.
A single population in a homogeneous environment.

Second stage.
Differentiation of environment, and migration to new environments produces racial differentiation of races and subspecies (indicated by different kinds of shading).

Third stage.
Further differentiation and migration produces geographic isolation of some races and subspecies.

Fourth stage.
Some of these isolated subspecies differentiate with respect to genic and chromosomal changes which control reproductive isolating mechanisms.

Fifth stage.
Changes in the environment permit geographically isolated populations to exist together again in the same region. They now remain distinct because of the reproductive isolating barriers which separate them, and can be recognized as good species.

northern Canada and Alaska. The subspecies of the lesser black-backed gull (*Larus fuscus*) also originated in Siberia, but spread westward and populated Russia, nothern Europe, and Iceland. All along this "ring around the Arctic," the subspecies can interbreed with adjacent subspecies. However, the herring gull eventually spread to northern Europe, where it became secondarily sympatric with the lesser black-backed gull. In the zone of overlap, the two species do not interbreed—yet if we trace their gene flow back through the circle, they do! This interesting paradox shows how crucial geographic separation is to the formation of species.

Many other examples of geographic variation of subspecies have been documented, and these have convinced most biologists that geographic separation and reduction in gene flow are the key to speciation. For example, the kingfishers of the *Tanysiptera hydrocharis-galatea* group show a striking pattern in New Guinea (Fig. 3.3). The three mainland forms (ranges 1, 2, and 3) are barely distinguishable, whereas the island species (4 to 8) were so distinct that they were originally described as different species. The Aru Islands (H_1) and South New Guinea (H_2) were originally isolated areas on which *T. hydrocharis* (H) differentiated. When the geographic separation of these areas ended, the southern New Guinea species (3) became sympatric with *T. hydrocharis*, but they do not interbreed. This suggests that the isolation of *T. hydrocharis* was sufficient for it to become a new species. Mayr (1964, 1963) visualizes this process in a series of stages (Fig 3.4). In the first stage, a large uniform population is spread over a wide area; eventually it becomes subdivided into geographical subspecies (stage 2). When some form of genetic isolation occurs (stage 3), the individual geographic variants begin to diverge genetically. If they come into contact again (secondary sympatry), they may have become so different that they cannot interbreed and are now new species (stage 4), or they still interbreed but form a hybrid zone between the subspecies (stage 5). This is known as the geographic, or **allopatric speciation,** model.

Although geographic separation can be very important for establishing new species, large mountain ranges or oceans are not necessarily required. Anything that causes reproductive isolation can result in new species, even if they are sympatric. Many organisms develop different behaviors (especially in courtship and mate recognition) that prevent interbreeding. In other cases, organisms that live in the same broad geographic area actually live in very distinct niches (such as birds that live in the crowns of trees versus their close relatives in the undergrowth, or insects that feed on completely different types of plants), so they rarely interact or interbreed. In other cases, organisms reproduce at different times of the year, so they cannot interbreed because they are reproductively out of synch. We think of geographic areas in terms of organisms that can freely move over large distances, but for attached or sedentary organisms a few meters is far enough to prevent interbreeding. After all, plants are rooted to the ground and cannot physically interbreed (their pollen or spores must do the job of interacting and dispersing). Many insects attach to a plant and never leave it, and most benthic marine invertebrates rarely move from their place on the bottom (especially sponges, corals, crinoids, barnacles, attached bivalves, and brachiopods). In some of these cases, these organisms overcome their isolation by having mobile larvae, or seeds, or juvenile stages. The most extreme cases are the internal parasites, ranging from bacteria and viruses to roundworms and tapeworms. For these creatures, their environment is their host, and they never leave their host once they have invaded. However, nearly all internal parasites have mechanisms of spreading to other hosts. Sometimes this happens when the parasite exits through the feces or when the host is eaten by a predator (such as when humans get trichinosis from eating pork).

Although the allopatric speciation model may not apply to all cases (or we may need to think of allopatry on very small scales), there is widespread agreement that reproductive isolation of small subpopulations is critical to forming new species.

THE SPECIES PROBLEM IN PALEONTOLOGY

> *The paleontologist has to deal with many types of discrete and objectively recognizable organic remains whose homology with living species is in doubt. . . In order to deal with these fossils the paleontologist must perforce call them something. In this discussion I shall call them "species" for want of another name. I shall use the word to cover any of that large class of objects of organic origin that are of sufficiently distinctive and consistent morphology so that a competent paleontologist could define them so that another competent paleontologist could recognize them. For purposes of communication he would apply a Linnean name to them. This definition is intended to be sufficiently flexible to cover any sort of fossil that would normally be named. The only essentially biologic part of this "species concept" is that fossils are necessarily derived from organisms.*

Alan Shaw, *Time in Stratigraphy*, 1964

So far, we have focused on species as defined in living organisms. When we try to apply these principles to fossils, however, we run into a whole new set of problems. First of all, the biological species definition is completely inappropriate for fossils that cannot demonstrate their interbreeding. All paleontological species must by necessity be morphologically based, although how we define our morpholospecies can gain insight from modern biological species. Similarly, we cannot recognize sibling species without behavior or molecular evidence, so it is impossible to tell how many similar fossils might have once been different sibling species. It is difficult enough deciding if two different fossils are sexual dimorphs, or different stages of ontogeny of the same organism, let alone worrying about whether they could interbreed.

A much bigger problem than this lack of information, however, is the dimension that biologists lack: time. Biologists are concerned only with the relationships of living populations and whether or not they interbreed now. Their data consist only of the modern time frame, not the geologic past. But paleontologists can see how these same populations have changed through time. Two different fossils that look so different that they are clearly different species might be connected by many transitional forms. If we find fossil populations gradually changing from one form to another (**anagenesis**), how do we decide where to subdivide the continuum? For many paleontologists, the difference between the beginning and the end of the series is so great that the lineage must be subdivided, no matter how arbitrary this makes each species definition. In practice, the boundaries are rarely a problem, since the fossil record is full of gaps that give us artificial but convenient breaks between species (Fig. 3.5). There are instances, however, where some remarkably continuous, uninterrupted transformation series have been discovered (especially among the marine microplankton). Then the problem can be acute. How do we define the species? Is it when the first specimen shows the diagnostic feature (even though the remaining 99% of the specimens do not)? Or is it when 50% of the population has the diagnostic feature? Biostratigraphers in particular have strong motivation to finely subdivide their fossils into as many species as possible. The more species they can distinguish, the more biostratigraphic events and zones they can recognize.

The arbitrariness of dividing continua into segments creates problems. When one species (usually called a "successional species" or "paleospecies") in a lineage transforms into another, the parent species "goes extinct" by definition (**pseudoextinction**), but this is not the same as extinction of a lineage (true extinction, where all the members die out). For this and many other reasons, some paleontologists have argued that anagenetic lineages should not be subdivided into species. Instead, species should be recognized only by splitting or branching events. Every member of a continuous lineage, no matter how different from its ancestors or descendants, must be the same species. This was called the **evolutionary species concept** by Simpson (1961), who defined an evolutionary species as a "lineage evolving separately from others and with its own unitary evolutionary role and tendencies." In other words, species are defined by branch points in a family tree; each branch and twig is a different species, and each branch can be considered only one species (Fig. 3.6). This concept has its own practical problems, especially in the cases of anagenetic lineages that undergo enormous transformations so that almost every paleontologist would concede that the end members must be called different species. At least it does not suffer from the criticism of arbitrarily subdividing continua, since the evolutionary species are defined by their own branching history and not by human decisions.

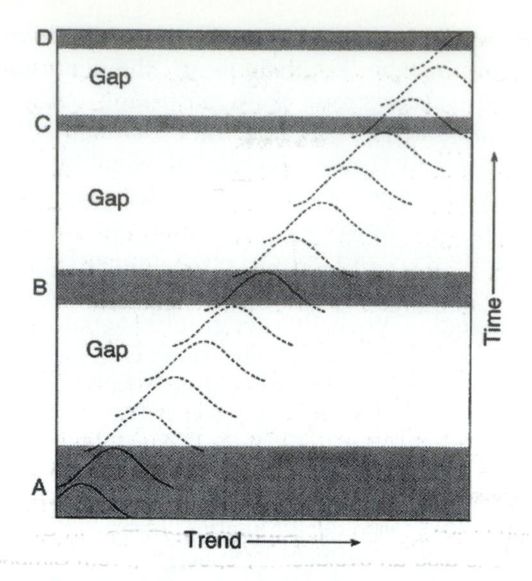

Figure 3.5. If species change gradually through time as recorded in the fossil record (as shown by the shift in the histogram of some anatomical feature), then the specimens at time A and time D would be so different that most taxonomists would call them different species. Yet how does one subdivide a continuum of variation between the two endpoints into distinct species? However, the gaps in the stratigraphic record (unshaded areas) mean that we will seldom have a complete continuum, and those gaps provide a convenient way of separating species in an anagenetic continuum. (From Newell, 1956.)

Punctuated Equilibria and the Species Problem

For almost a century, paleontologists have agonized over the problem of subdividing evolutionary continua into species. In 1972, Niles Eldredge and Stephen Jay Gould proposed an idea that radically changed the long-running "species problem in paleontology" debate. Eldredge and Gould (1972) pointed out that paleontologists had been conditioned to expect gradual, anagenetic transformation series (**phyletic gradualism**) in the fossil record and to dismiss the fossil record as imperfect when these do not appear. But in their own research, Eldredge and Gould found that fossil species remained static and unchanged through millions of years of strata and seldom showed any gradual change. As they inquired further, it became apparent that this pattern of fossils unchanged through millions of years was the norm in paleontology. Because of inherited gradualistic biases, paleontologists had long ignored these examples of stability as uninteresting and has

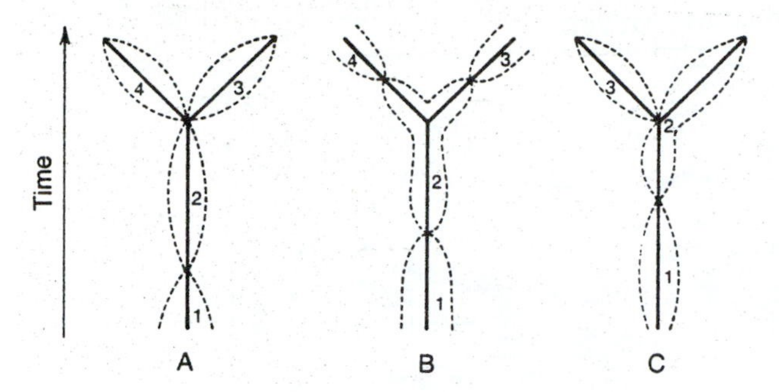

Figure 3.6. Three possible ways of subdividing a branching fossil lineage into three or four species. In all of them, the distinctions between 1 and 2 are arbitrary breaks in a continuous lineage, but in A, species 3 and 4 are non-arbitrarily defined by a splitting event. These are what Simpson (1961) called **evolutionary species**. In C, species 3 is also an evolutionary species. (From Simpson, 1961.)

focused on a few examples of apparent gradual change.

But is gradual change in fossils the right expectation? In modern allopatric speciation theory, most evolutionary change occurs in small, peripherally isolated populations in a few tens of generations (which is a geological instant). Large, mainland populations were relatively buffered against genetic change and would not be expected to evolve in any direction, let alone gradually in one direction for millions of years. If we take allopatric speciation seriously, we should expect the fossil record to show fossil populations that are in stable *equilibrium* for long periods of time, *punctuated* by the sudden introduction of new species that formed in some peripherally isolated area and then migrated back to our area of study. Eldredge and Gould (1972) called this idea **punctuated equilibrium**. Their paper generated a storm of controversy that has not subsided in 25 years (see Somit and Peterson, 1992; Prothero, 1992; Gould and Eldredge, 1993). It has many implications for evolution (discussed in later chapters), but for our present discussion it makes one important point: if gradual, anagenetic continua are extremely rare in the fossil record (as most paleontologists now concede), then the problem of subdividing them into species disappears. There are some (particularly micropaleontologists) who do have documented good examples of gradual transformation series and still must wrestle with how (or whether) to subdivide them, but for most paleontologists, the issue of subdividing continua is no longer a major debate. Most apparent "gradual anagenetic series" turn out not to be real when examined in detail, so subdividing them is not a issue.

CONCLUSIONS

The idea of species has changed from the static typological concept of pre-Darwinian natural historians, to the dynamic part of a fluctuating population proposed by Darwin, to the genetic isolation mechanisms proposed by modern evolutionary biologists. Yet through it all, the morphological criteria of species are still of paramount importance, since most biologists still rely on anatomical features to tell species apart. The paleontologist, of course, can use only the morphology; behavioral or molecular differences are lost in the mists of time. However, the paleontologist has something else that neontologists do not: the dimension of time, and how species have changed through time and are interconnected with other species. For some paleontologists, this creates a problem when two or more species appear to be connected in time, and gradually transform, so that they cannot be distinguished without some arbitrary cutoff. The flurry of studies stimulated by the punctuated equilibrium debate, however, showed that such cases of gradual transformation were extremely rare, and most fossil species can clearly be defined by branching speciation events.

For Further Reading

Bush, G. 1975. Modes of animal speciation. *Annual Reviews of Ecology and Systematics* 6:339-364.

Claridge, M. F., ed. 1997. *Species, The Units of Biodiversity*. Chapman & Hall, London.

Eldredge, N. 1985. *Time Frames*. Simon & Schuster, New York.

Eldredge, N., and S. J. Gould. 1972. Punctuated equilibria: an alternative to phyletic gradualism, pp. 82-115, *in* Schopf, T.J.M., ed. *Models in Paleobiology*. Freeman, Cooper, San Francisco.

Endler, J. A. 1977. *Geographic Variation, Speciation, and Clines*. Princeton Univ. Press, Princeton, N.J.

Gould, S. J., and N. Eldredge. 1977. Punctuated equilibria: the tempo and mode of evolution reconsidered. *Paleobiology* 3:115-151.

Gould, S. J., and N. Eldredge. 1993. Punctuated equilibrium comes of age. *Nature* 355:223-227.

Mayr, E. 1942. *Systematics and the Origin of Species*. Columbia University Press, New York.

Mayr, E. 1963. *Animal Species and Evolution*. Harvard University Press, Cambridge, Mass.

Otte, D., and J. A. Endler. 1989. *Speciation and Its Consequences*. Sinauer Associates, Sunderland, Mass.

Sokal, R. R., and T. J. Crovello. 1970. The biological species concept: a critical evaluation. *American Naturalist* 104:127-153.

White, M. J. D. 1978. *Modes of Speciation*. W.H. Freeman, San Francisco.

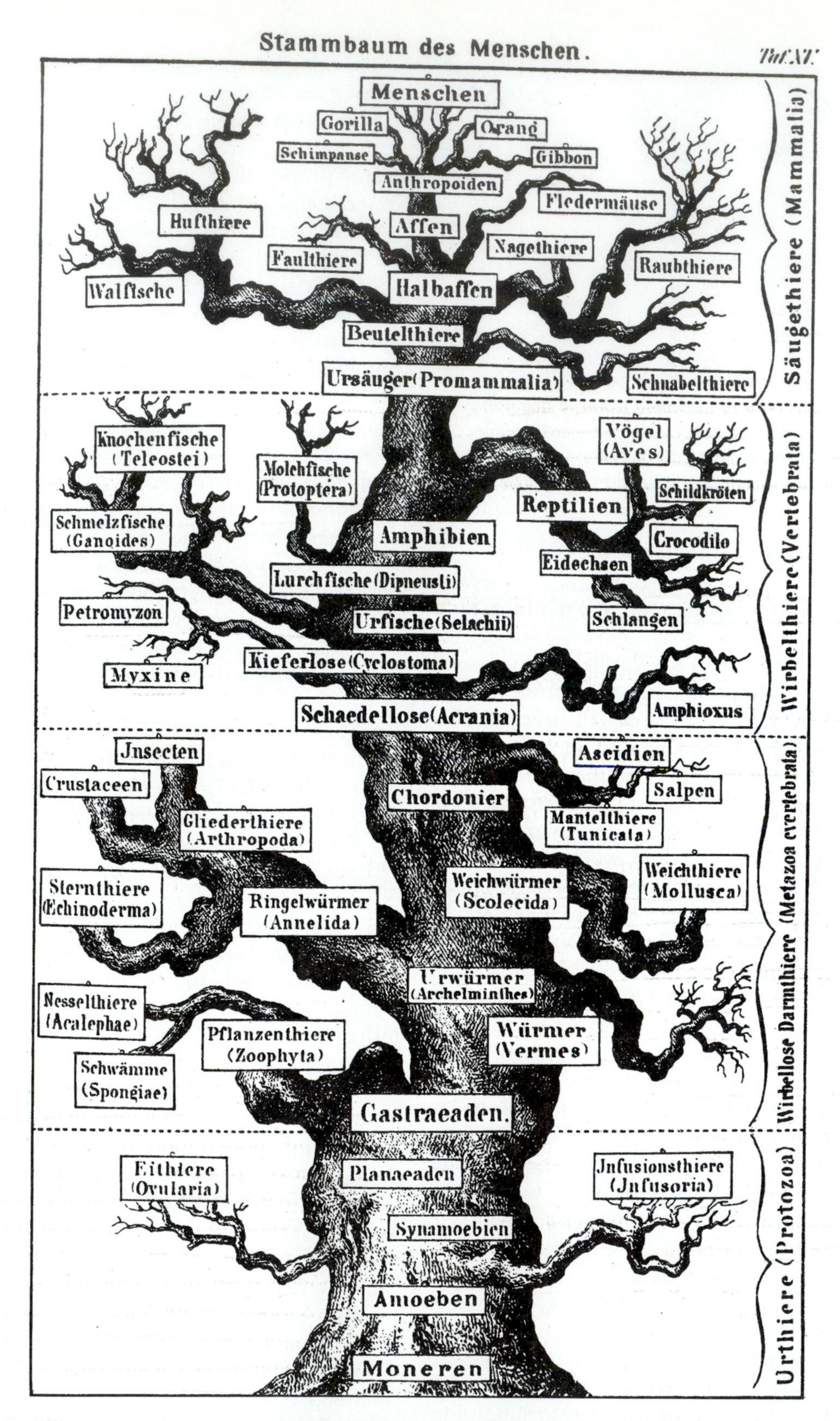

Figure 4.1. Ernst Haeckel's (1874) evolutionary tree, one of the first graphic representations of phylogeny as a "tree of life" ever to appear in print. Although many of the branches are real animals, the first five stages (monera, amoeba, synamoeba, planaea, and gastraea) are reconstructed based on ontogeny of higher forms, not based on living organisms then known to Haeckel. (From Haeckel, 1874.)

Chapter 4

Systematics

The amount of diversity in the living world is staggering. About 1 million species of animals and half a million species of plants have already been described, and estimates on the number of still undescribed species range from 3 to 10 million. . . . It would be impossible to deal with this enormous diversity if it were not ordered and classified. Systematic zoology endeavors to order the rich diversity of the animal world and to develop methods and principles to make this task possible.

Ernst Mayr, *Principles of Systematic Zoology*, 1966

The extent to which progress in ecology depends upon accurate identification and upon the existence of sound systematic groundwork for all groups of animals, cannot be too much impressed upon the beginner in ecology. This is the essential basis of the whole thing; without it the ecologist is helpless, and the whole of his work may be rendered useless.

Charles Elton, *Animal Ecology*, 1947

WHY SYSTEMATICS?

In the previous chapter we focused on the meaning of an important category in nature: the species. How are species grouped into larger categories? How are classification schemes set up, and what do they mean? The science of classifying is known as **taxonomy** (Greek, "laws of order"); any named grouping of organisms (a species, a genus, etc.) is called a **taxon** (plural, **taxa**). Deciding how to name a new species and genus may seem to be a highly specialized, legalistic dimension of biology and paleobiology, not nearly as glamorous as ecology or behavior or physiology. But taxonomy is not just naming species, because *species and higher taxa reflect evolution*. Taxonomists do much more than label dusty jars in a museum. They are interested in comparing different species and deciding how they are related and ultimately in deciphering their evolutionary history. They look at the diversity of organisms in time and space and try to understand the large-scale patterns of nature. They look at the present and past geographic distributions of organisms and try to determine how they got there. In short, they look at the total pattern of natural diversity and try to understand how it came to be. Contrary to stereotypes, they are among the most eclectic of biologists and paleobiologists.

All these various enterprises go beyond conventional taxonomy and are usually given the broader label **systematics**. Systematics has been defined as "the science of the diversity of organisms" (Mayr, 1969, p. 2) or "the scientific study of the kinds and diversity of organisms and of any and all relationships among them" (Simpson, 1961, p. 7). Its core consists of taxonomy, but it also includes determining evolutionary relationships (phylogeny) and determining geographic relationships (biogeography). The systematist uses the *comparative approach to the diversity of life to understand all patterns and relationships that explain how life came to be the way it is.* Put this way, systematics is one of the most exciting and stimulating fields in all of biology and paleobiology.

Taxonomists and systematists may not be as numerous or well funded as ecologists or physiologists or behaviorists, but their labors are essential. All other disciplines in biology and paleobiology depend upon taxonomists to give their experimental subjects a name and, more important, to give them a comparative context. If a physiologist wants to study the organism that is most like humans, it is the taxonomist who points to the chimpanzee, our closest evolutionary relative. If an ecologist wants to understand how a particular symbiotic relationship may have developed, or the ethologist wants to understand a peculiar type of animal behavior, they need to know the evolutionary relationships and phylogenetic history of each organism—these are the domain of the systematist. Systematics provides the framework of understanding and interconnection upon which all the rest of biology and paleobiology are based. Without it, each organism is a random particle in space, and what we learn about it has no relevance to anything else in the living world.

In our present age, taxonomists have become scarce as grant funding dries up and students go into more glamorous specialities that require big, expensive machines. Yet one of the most important issues on this planet today—**biodiversity**—is within the domain of systematists. Without someone to describe, name, and count all the species on this planet, how will we know whether we are wiping them out catastrophically, or whether they are holding their own, or even flourishing? Without the perspective of past diversity changes on this planet, how can we decide the severity of the human-induced mass extinction? Each time someone surveys a patch of rainforest, trying to determine how humans have impacted the life there, their first task is taxonomy. Ecologists complain that they cannot find anyone who has the right training to identify and describe all the new species of insects and birds and plants that are being destroyed even before we get to know them. Without knowing that they are there, how can we decide how important they might be? One of these species might hold the cure to some deadly disease or the solution to the control of a nasty pest, but without systematic and taxonomic research, these species go extinct before we even encounter them.

In the context of paleontology, the situation is analogous. The public may think that collecting big dinosaur specimens in exotic places is exciting, but it is just a tiny part of paleontology. Collecting and preparing fossils is a specialized task, often performed by people with little advanced scientific training. Analyzing and understanding their taxonomy, geography, and phylogenetic relationships is the domain of the systematic paleontologist. Without a properly trained paleontologist to correctly identify, name, and analyze the fossils, they are mute stones. Hours in the laboratory and museum collections spent measuring and describing specimens may not seem as glamorous as visiting exotic places, but they are equally essential. From this naming and description comes the understanding of larger problems in paleobiology, such as: how is all life interrelated? what is the past history of life? how has diversity on this planet changed? Without the foundation of systematics, these questions cannot even be approached.

Compared to most biologists, paleobiologists are much more likely to practice taxonomy as part of their research, since nearly every study requires some kind of taxonomic or phylogenetic analysis at its foundation. In some cases, one can use the work of previous taxonomists, but in most research, there is more taxonomy to be done, or the paleontologist needs to do his or her own taxonomy just to determine if past taxonomy can be trusted. A paleontologist without adequate training in systematics is severely handicapped.

EVOLUTION AND CLASSIFICATION

God created, but Linnaeus classified.

Carolus Linnaeus, 1758

From the most remote period in the history of the world, organic beings have been found to resemble each other in descending degrees, so that they can be classed in groups under groups. This classification is not arbitrary like the groups of the stars in constellations. The existence of groups would have been of simple significance if one group had been exclusively fitted to inhabit the land, and another the water; one to feed on the flesh, another on vegetable matter, and so on; but the case is widely different, for it is notorious how commonly members of even the same subgroup have different habits. Naturalists try to arrange the species, genera, and families in each class on what is called the Natural System. But what is meant by this system? Some authors look at it merely as a scheme for arranging together those living objects which are most alike, and separating those most unlike. Many naturalists think something more is meant by the Natural System; they believe it reveals the plan of the Creator. I believe that a community of descent—the one known cause of close similarity in organic beings—is the bond which is partially revealed by our classifications.

Charles Darwin, *On the Origin of Species*, 1859

There are many ways to classify things. Children often classify objects by similarity in color, or shape, or texture. As adults, we may use more subtle means of telling things apart. For example, a child may label all objects with four wheels as "cars" but as adults, we recognize the difference between cars, trucks, and vans, or between cars with diesel or gasoline engines, or between Fords, Chevys, and Toyotas. Some classification schemes attempt to have a logical basis or structure to make them easier to use. For a long time, the Dewey Decimal system was the most widely used means of cataloging books, until it was replaced in many libraries by the Library of Congress system. Both try to cluster books by **natural groups** (such as a category for science books, subdivided into physics, chemistry, biology, geology, and so on) but the Library of Congress system is apparently more flexible at handling larger numbers of books. Both natural classification schemes attempt to organize the same array of objects, but one is apparently more successful than the other.

In the realm of life, a wide variety of classification schemes were developed since the 550 kinds of animals recognized by Aristotle. Some grouped organisms on properties that humans favored ("good to eat" vs. "eat only in emergency" vs. "inedible" vs.

"poisonous") or on properties of their ecology (for example, most animals in the ocean were called "fish," including "starfish" and "shellfish" and whales). By the early 1700s, there were over 6000 recognized species of plants and 4000 of animals, organized into a great array of classification schemes proposed by natural historians. Most of these classifications were arbitrary and highly unnatural (for example, flying fish and birds were put together because they both fly, or turtles and armadillos because of their armor), and everybody had their own favorite scheme. The classification method that eventually won out was proposed by the Swedish botanist Carl von Linné, known to us by his Latinized name, Carolus Linnaeus (1707-1778). As a botanist, Linnaeus recognized that the most fundamental and diagnostic properties of plants occur in their reproductive structures, particularly their flowers. His "sexual system" for classifying plants was published as *Species Plantarum* in 1752 and created a scandal. Eventually it won out over all the competing systems, since flowers are clearly more useful than any other structure. Linnaeus tried a similar approach in animals, using fundamental structures (such as hair and mammary glands in mammals) rather than superficial ones (such as flight, or armor). His *Systema naturae, regnum animale* ("the system of nature, animal kingdom") was first published in 1735, and its tenth edition (1758) is now regarded as the starting point of modern systematic zoology.

Linnaeus' original classification became outdated as thousands of new species were described since 1758, but his fundamental system still survives. Each species is given a **binomen** (two-part name), consisting of the **genus** (plural, **genera**) name (always italicized or underlined, and always capitalized) and the trivial name indicating the species (always italicized or underlined but never capitalized). For example, our genus is *Homo* ("human" in Latin) and our trivial name is *sapiens* ("thinking" in Latin), so our species name is *Homo sapiens* (abbreviated *H. sapiens*). The trivial name can never stand by itself (since trivial names are repeated over and over in taxonomy), but must always accompany its genus. To prevent confusion, the genus name can never be used for any other organism in the animal kingdom (there are few generic names that are reused for different animals and plants). Genera are then grouped into higher categories, such as the family (always capitalized, but never underlined or italicized, and with the "-idae" ending in animals, the "-aceae" ending in plants), then orders, classes, phyla (singular, "phylum"), and kingdoms. For example, humans are members of the Kingdom Animalia (there are also kingdoms for plants, fungi, and single-celled organisms), the Phylum Chordata (including all other backboned animals), the Class Mammalia (mammals), the Order Primates (including lemurs, monkeys, apes, and ourselves), the Family Hominidae (including our own genus and the extinct *Australopithecus, Ardipithecus,* and *Paranthropus*), the genus *Homo* (including other extinct species such as *Homo habilis* and *H. erectus*), and our species *H. sapiens*.

Notice that this classification scheme is **hierarchical**. Each rank is grouped into larger ranks, so that there may be several species in a genus, several genera in a family, and so on. However, some genera have only one species, some families have only one genus, some orders have only one family, and so on; these one-member groups are called **monotypic**. The great reason for the success of Linnaeus' scheme is this flexibility created by groups hierarchically clustered within larger groups, with infinite room for expansion as new species are discovered. The Latinized binomen is also very flexible, and universally recognizable in science. Local vernacular names in a single language may vary greatly. The word "gopher" refers to both a tortoise and a burrowing rodent in English, and every other language uses completely different names for the same animals. But in all languages, the scientific name is always based on Latin or Greek (since these were the languages of scholars in Linnaeus' time), or a Latinized version of other words. A scientist can pick up a publication in some unfamiliar (to most scientists) alphabet, such as Cyrillic or Hebrew or Chinese, and not recognize a word except the scientific names; these stand out and at least communicate the essential content of the paper.

Linnaeus and his contemporary natural historians viewed their task as a religious mission. They thought that deciphering the "Natural System" of life would reveal the workings of the mind of the Creator that set up this "Natural System." But the obvious clusters of organisms into groups within groups suggested something else to Darwin. This hierarchical, nested, branching structure of life only made sense if life had descended from common ancestry in a branching fashion (Fig. 4.1). Although Linnaeus has not intended to provide evidence for evolution, a century later his classification scheme became one of Darwin's best arguments. In doing so, Darwin changed the goals of classification. It was no longer just a nice but arbitrary system of arranging things into pigeonholes. Taxonomy now had an evolutionary meaning as well, and taxonomists were trying to create natural groups that reflected evolutionary history. Although these goals are not contradictory, they do not always agree, either. Some taxonomists view organisms of similar descent and ecology, such as the fish, as a formal group, "Pisces." But in evolutionary terms, not all fish are created equal. Lungfish, for example, are more closely related to four-legged land vertebrates (tetrapods) than they are to a shark or a tuna. In other words, a lungfish and a cow are more closely related than a lungfish and a tuna. Here we see a clear tension between ecological groupings, such as "fish," and evolutionary groups, such as the lungfish-tetrapod group (known as the Sarcopterygii). Which is better? The different priorities and goals of taxonomists has led to much debate over the proper methods of classification.

COMPETING SYSTEMATIC PHILOSOPHIES

> *Taxonomists have always had the reputation of being difficult. Intransigence may be rooted in the necessity of defending prolonged self-immersion in a taxon that others find a total bore; it is frustrating to have one's life work greeted with a yawn. Numerical taxonomists have proved to be just as prickly as conventional taxonomists, possibly more so because some of the brightest people in systematics are involved in the current taxonomic battles. The political maneuvering and character assassination that characterize certain taxonomists today may not be atypical for science; they certainly provide a fine example of its seamier side. If Feyerabend is correct, it may even be a requirement of human nature that scientific progress occur in this manner.*
>
> W. W. Moss, "Taxa, taxonomists, and taxonomy," 1983

> *If contemporary philosophers of science agree on anything, it is that scientific classification cannot be theoretically neutral. Nor can there be any prescribed order in which theoretical combinations are introduced into a classification. One cannot begin by producing a theoretically neutral classification, and then only later add theoretical interpretations.*
>
> David Hull, "The principles of biological classification: the use and abuse of philosophy," 1981

What is the proper way to classify organisms? That question had been the center of a very intense scientific debate since the 1960s. As the historian of science David Hull (1988) pointed out, the debate reveals almost as much about the sociology of science as it does about the science itself. In the late 1950s, there was relatively little argument, since the majority of taxonomists practiced a vaguely formulated method later called "evolutionary taxonomy," exemplified by Simpson's (1961) *Principles of Animal Taxonomy* or Mayr's (1966) *Principles of Systematic Zoology*. This mainstream, orthodox school of taxonomy was challenged by two upstarts in the 1960s and 1970s. Both schools of thought followed very different basic assumptions and used new jargon to distinguish themselves from the amorphous orthodoxy. Sometimes they took very extreme positions so that they could be seen as different and not be absorbed into the mainstream as a minor variant. Those extremes may have been moderated by later practicioners as the controversies died down, but they were important in the early phases of the movements.

Hull points out that both these movements had great similarities to religious movements in trying to establish themselves, attract converts and grow, and make themselves distinct and recognizable. They both had "prophets," a "bible," a "high priest," a "Mecca," "acolytes," and a central philosophical dogma. The first movement, **numerical taxonomy** (later known as **phenetics**), was introduced by "prophets" Robert Sokal and Peter Sneath in several papers in the late 1950s, culminating with their 1963 "bible" *The Principles of Numerical Taxonomy*. Their "high priest" was bee taxonomist Charles Michener, and phenetics was promoted by important "acolytes," such as James Rohlf, Steve Farris, and Paul Ehrlich. The movement was spread from the "Mecca" of the University of Kansas at Lawrence. Their central philosophical tenets revolved around statistics and objectivity and the idea that computers could do taxonomy better than humans.

By the mid-1970s, numerical taxonomy had been eclipsed by the second movement, **cladistics**, which began with its "bible" *Phylogenetic Systematics*, by "prophet" Willi Hennig, a German entomologist. First published in German in 1950, *Phylogenetic Systematics* had little impact outside of entomology until it was translated into English in 1966 and promoted by its "high priest," the American ichthyologist Gareth Nelson in the late 1960s and 1970s. Other important "acolytes" included ichthyologists Donn Rosen and Colin Patterson, and arachnologist Norm Platnick. The central tenets of cladistics were that phylogenetic hypotheses must be testable and that classification should reflect evolutionary history and nothing else. Nelson, Rosen, Platnick, and their colleagues at the "Mecca" of the American Museum of Natural History in New York were at first regarded as too radical and extreme, but by the mid-1980s cladistics itself had become mainstream and it is now used by most taxonomists.

Phenetics

Let us take a closer look at these two movements. Numerical taxonomy was precipitated by several factors: the availability of the first practical computers; an increase in interest in statistical methods; and a widespread dissatisfaction with conventional taxonomy as being an intuitive, arbitrary "art" that was only valid and reproducible in the mind of the taxonomist. To get away from this element of subjectivity, the numerical taxonomists argued that classification should be a purely objective, statistical exercise that can be coded and deciphered by a computer. Numerical taxonomists concluded that since classifications cannot reflect both evolutionary history and degree of overall similarity, we should give up trying to make our classifications phylogenetic and instead base them on objective statistical similarities and differences, or overall phenetic similarity. To them, a "natural" classification is judged by how successfully it clusters groups with the most in common and how well it creates stable classification schemes that are maximally useful to scientists. Typically, this is accomplished by measuring and coding numerous anatomical features, or **characters**, in each specimen or taxon (called **OTUs,** or **operational taxonomic units**) to create a large **data matrix** of OTUs versus characters. Next, a computer program sorts the data and finds clusters of OTUs that have the most characters in common. When the computer analysis is finished, a branching diagram of similarity is produced.

The response to numerical taxonomy was predictable: most systematists did not like computers and statistics intruding into their arcane domain. According to Hull (1988, p. 120), "one systematist volunteered that he hoped they would never succeed in making taxonomic judgments sufficiently quantitative so that a computer could make them, because, if they did, it would take all the fun out of systematics." In response to this hostility and rejection, the more outspoken numerical taxonomists, such as Sokal and Ehrlich, did not hesitate to step on toes. When Ehrlich was "asked indignantly, 'You mean to tell me that taxonomists can be replaced by computers?' Ehrlich responded, 'No, some of you can be replaced by an abacus.' Thereafter, Ehrlich did not consider the give-and-take after a paper truly successful unless he brought at least one taxonomist to the point of tears" (Hull, 1988, p. 121). At first, the establishment fought back by preventing numerical taxonomic papers from being published. The editor of the journal *Systematic Zoology* in 1961 was Libbie Hyman, who reportedly said, "One paper with numbers is enough." Shortly thereafter, she was replaced as editor and the journal moved to Lawrence, Kansas, where it became almost the unofficial house organ of numerical taxonomy. *Systematic Zoology* went from a staid, obscure journal to a "must-read," almost doubling its page count and circulation with the excitement of the debates.

A few years after numerical taxonomy gained this status, it went into decline. According to Hull (1988), several factors were responsible. The original concentration of pheneticists in Lawrence, Kansas, broke up as Sokal, Rohlf, and Farris all went to the State University of New York at Stony Brook in 1969. Another problem was that the numerical taxonomists tried to apply their methods so widely that their efforts in systematics were dissipated. More important, the majority of systematists never accepted the fundamental goals of phenetics. Most still wanted classification to reflect evolutionary relationships in some way, even if this was a difficult task. Many were alienated by the great emphasis on computers and statistics (this was when computers were huge, slow, and cumbersome, and only a few scientists had access to one). Taxonomists preferred to study their favorite organisms and were less interested in math or statistics for its own sake.

The most serious blows came when a number of studies showed that the "objectivity" of phenetics was a myth. Coding and weighing the importance of the characters in the data matrix cannot be done objectively. When one systematist decides that a wing represents a single character state and another subdivides it into numerous character states, which is the correct approach? Once again, the "art" of systematic judgment comes into play. Taxonomists ultimately must decide what is a character, and that decision is filtered by their own prejudices. Even more serious were studies that showed that the same data matrix gave different results with different computer programs, and occasionally even with the same computer program! If the methods were not truly objective and reproducible, and gave up on the whole idea of evolutionary classification, then what was the advantage? If a purely phenetic classification placed unrelated animals such as whales and fish together, then what good was it?

Cladistics

The other primary reason for the decline of phenetics was competition from an even more radical school of systematics, cladistics. In many ways, cladistic methods are the opposite extreme from phenetics. Rather than abandoning the evolutionary meaning behind taxonomy, cladists argued that classification should reflect *only* evolutionary history and ignore overall phenetic similarity. Rather than throwing all the characters, unweighted, into a computer, Hennig argued that not all characters are created equal. The only characters that are useful in a given problem are those that are shared evolutionary novelties, or **shared derived characters (synapomorphies** in Hennig's terminology). For example, the presence of hair and mammary glands is a shared specialization of all mammals, unique to them and not found elsewhere in the animal kingdom. Those characters are synapomorphies that help define the taxon Mammalia. Characters such as the presence of four limbs, or a backbone, would not be very useful in dis-

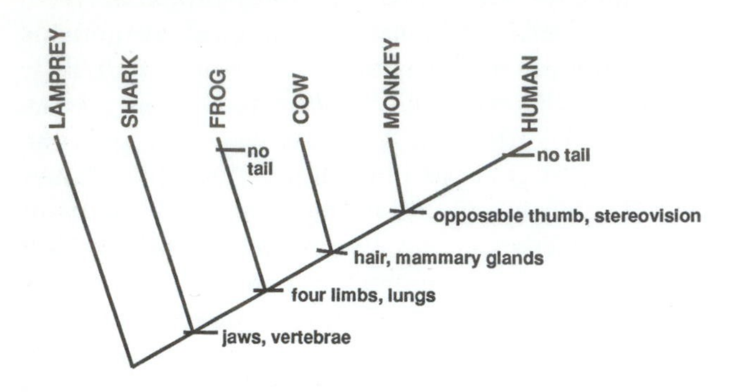

Figure 4.2. Evolutionary relationships of some vertebrates, showing the shared specializations that support each branching point (node) on the cladogram.

tinguishing mammals from other vertebrates, because they are found in reptiles, birds, and amphibians; they are **shared primitive characters**, or **symplesiomorphies**. Taxa are recognized and defined only by their unique evolutionary novelties. Characters already found in their near relatives and ancestors are considered primitive and irrelevant to the problem of the evolutionary relationships within their group.

Whether a character is primitive or derived depends upon how it is used. For example, within the Mammalia, hair and mammary glands are the primitive state, and cannot be used to decide the relationships of different groups of mammals. If a systematist wanted to find shared derived characters uniting monkeys, apes, and humans, he or she would use unique specializations such as the opposable thumb and stereovision, not primitive features found in all other mammals, such as hair and mammary glands. This means that hair and mammary glands are derived characters at the level of Class Mammalia, but primitive at the level of the orders and families within the Mammalia.

To analyze these characters, the systematist draws a branching diagram known as a **cladogram** (Fig. 4.2). At the tip of each branch are the taxa, and the branching geometry shows which taxa are more closely related to each other. In Figure 4.2, the cladogram shows that monkeys and humans are more closely related to each other than either is to anything else in the diagram. Monkeys, humans, and cows, as mammals, are more closely related to each other than they are to any non-mammal. Frogs, cows, monkeys, and humans, as four-legged tetrapods, are more closely related to each other than they are to any non-tetrapod, such as a shark. Notice that the fork of each branch (the **node**) represents the common ancestor of the organisms at the branch tips, and the node is defined by the shared derived characters that unite those taxa. For example, opposable thumbs and stereovision are derived characters that define the node uniting monkeys and humans; hair and mammary glands are derived characters that define the node Mammalia; the presence of derived characters such as four legs and lungs define the tetrapods and distinguish them from other vertebrates.

How is a cladogram constructed? Typically, the systematist sets up a data matrix of characters and taxa, just as in phenetics. However, instead of letting the computer cluster these characters for overall similarity, the cladist evaluates each character and decides which state is primitive and which is derived. The best way to do this is to examine more primitive relatives, known as **outgroups**, and see what character state they possess. For example, the absence of hair and mammary glands in outgroups such as the frog, shark, or lamprey establishes that these features are derived for the mammals; the absence of four limbs, lungs, or hands and feet instead of fins (as found in sharks and lamprey) shows that those characters are derived for tetrapods. Once these decisions are made about the derived character states, then the systematist (or a computer) clusters only the derived character states to see which taxa have the most shared specializations in common and which have fewer. Ideally, this generates a branching cladogram.

Although a cladogram looks superficially like an evolutionary tree (such as in Fig. 4.1), it is not. A cladogram shows a nested pattern of evolutionary specializations within diverse taxa. It works with any three or more taxa, no matter how distant their relationships. (A family tree works only with immediate ancestors and descendants.) The nodes in a cladogram summarize the derived character states of the hypothetical common ancestor, but cladists avoid putting real ancestors at these nodes. There are several reasons for this. Because the fossil record is so incomplete, it is very unlikely that we will actually find a true ancestor. A more fundamental problem is that ancestors have nothing but primitive character states compared to their descendants. It is impossible to tell if a potentially ancestral fossil was truly an ancestor, because it has no unique evolutionary specializations that link it with another taxon. At a more fundamental level, cladists avoid searching for ancestors because this search emphasizes primitive characters and may ignore important derived characters. In many cases, troubling phylogenetic problems were solved when the systematists stopped looking for ancestors, and instead discovered a remarkable pattern of shared derived similarities that had long been neglected. Instead of ancestors and descendants, cladograms show only that two taxa are closely related **sister groups**. For example, monkeys and humans are sister groups compared to cows or other animals that are not primates.

A cladogram is simply a branching diagram of relationships, supported by unique shared derived characters states, and makes no statements about ancestry. This minimalist approach to systematics may seem dull at first, but it has a major advantage: it is **testable**. Each cladogram is a scientific hypothesis, and to test it, the systematist can look at additional character states or additional taxa (especially more outgroups). Cladists find this very appealing, because one of their central philosophical tenets is that *all science must be testable*. Family trees with ancestors may be more interesting, but there is no way of testing their more complicated hypotheses and assumptions, so they fall outside this narrow definition of science.

One of the great advantages of the cladistic method is that it provides a simple, straightforward set of rules that any systematist can follow, and each phylogenetic hypothesis can be immediately tested and shot down if better data emerge. In the old school of "evolutionary taxonomy," phylogenetic trees had no characters at the nodes, so they could not be tested. Trees came out of the intuition and experience of the systematist, and since there was no way of evaluating them or seeing how they were constructed or supported, there was no way to criticize them. By contrast, a cladogram "lets it all hang out." If there are few or no derived characters to support a node, it is immediately apparent; if the characters have been incorrectly coded or evaluated, that too is clear. Systematists could no longer hide behind the foggy obscurity of phylogenetic trees, but had to suffer immediate criticism if their work did not hold up. A lot of early cladograms were overturned, but eventually, as fewer and fewer problems were found, the cladograms began to converge upon consistent and often surprising answers to longstanding phylogenetic problems.

If nature were ideally cooperative, all character data matrices would give a single, unique cladogram and there would be no doubts. But the real world is much more complex than this. Although life has had a single evolutionary pathway, character states have changed more than once, and in confusing ways, and sometimes they have reverted back to the ancestral state. In Figure 4.2, for example, humans and frogs could be united by the loss of a tail. If we were basing our cladogram on just that one character, then humans and frogs would be more closely related than humans and monkeys. However, there are far more characters that support human-monkey relationships, so it is simpler (or more **parsimonious**) to suggest that humans and frogs have independently and secondarily lost their tails (especially since both have tails in their embryonic state). The loss of a tail is considered an evolutionary **convergence**. The criterion of simplicity, or parsimony, may not work in every case, but most problems are resolved without too many conflicts of this type. They are especially useful when a knotty problem appears. For example, a superficial analysis might suggest that whales, ichthyosaurs, and fish share many phenetic similarities, such as fins, a paddle tail, a streamlined body, and other features related to swimming in water. But a more detailed analysis incorporating non-aquatic animals would find many more shared derived features (mostly concerning the internal anatomy not related to swimming) that overwhelmingly show that whales are mammals, ichthyosaurs are reptiles, and fish are more primitive than either—all of their phenetic similarities are due to convergence.

When cladistics burst upon the scene in the early 1970s, paleontologists were among the harshest critics. The denial of ancestors certainly alienated them, as did another extreme claim: stratigraphic order is of no relevance to deciding whether a character is primitive or derived. Most cladists felt that the stratigraphic record is too full of gaps, and that fossils are too incomplete, to ever use them reliably in determining derived character states (Schaeffer et al., 1972; Patterson, 1981). This extreme position was partly a response to generations of paleontologists who stacked fossils in stratigraphic order and "connected the dots" without any independent anatomical analysis of the character states. It is true that there are many gaps in the fossil record, and species may appear earlier than their presumed ancestor, and most fossils are much less complete than a living specimen for the purposes of phylogenetic analysis. Since these initial debates, a number of paleontologists (Fortey and Jefferies, 1982; Lazarus and Prothero, 1984; Paul, 1992; Huelsenbeck, 1994; Fisher, 1994; Smith, 1994; Clyde and Fisher, 1997; Hitchin and Benton, 1997; Huelsenbeck and Rannala, 1997) have developed methods of rigorously analyzing stratigraphic data so they can be incorporated into a cladistic analysis with some degree of testability.

Throughout the 1970s, the battles over cladistics were fought fiercely within the walls of the American Museum in New York, in journals, and at scientific meetings around the world. (See Hull, 1988, for a blow-by-blow description.) Scientific seminars frequently became shouting matches as the different camps became more polarized, and many vicious words were said and even printed as the debates spread across the country. Unlike phenetics, however, cladistics did not vanish after a few years and leave everyone in peace. Although determined and skillful advocacy by people such as Gary Nelson certainly promoted the cause, cladistics became the dominant method *because it works*. Many systematists who were alienated by the combat and by the arcane terminology nevertheless tried it on their own organisms and found that this simple, rigorous method of analyzing taxa and characters cut through decades of confusion. As the years went by, the shouting died down, and systematists quietly began doing their job using cladistics, without many of the extreme positions taken by the early cladists. A transformation is apparent in the programs of scientific meetings. Very few cladistic talks were given in the late 1970s or early 1980s, except when a session revolved around debating cladistics. By the late 1980s, however, cladistic systematic papers were common, and during the 1990s, virtually every systematic paper delivered at the meetings of the Paleontological Society or the Society of Vertebrate Paleontology uses at least some cladistics. There are still many areas where cladistics does not work well, and still many unresolved problems, and still many who resist the idea, but by and large, the war is over.

Cladistics and Classification

Many traditional systematists have no problem with cladistics for inferring phylogenetic relationships, but they draw the line at another issue: **cladistic classification**. As we mentioned earlier, some cladists argue that classification should be a strict reflection of phylogeny and nothing else. Once the branching sequence has been determined, it dictates the ranks and clustering of higher taxa. Although traditional evolutionary systematics also tries to reflect

Table 4.1. Differences between traditional classification and cladistic classification of hominoids.

TRADITIONAL CLASSIFICATION

Superfamily Hominoidea
- Family Pongidae (apes)
- Family Hominidae (humans)

A CLADISTIC CLASSIFICATION

Superfamily Hominoidea
- Family Hylobatidae (gibbons)
    - Genus *Hylobates*
    - Genus *Symphalangus*
- Family Hominidae
    - Tribe Pongini (orangutans)
        - Genus *Pongo*
    - Tribe Gorillini (gorillas)
        - Genus *Gorilla*
    - Tribe Hominini (chimps and humans)
        - Subtribe Panini (chimps)
            - Genus *Pan*
        - Subtribe Homininae
            - Genus *Homo*

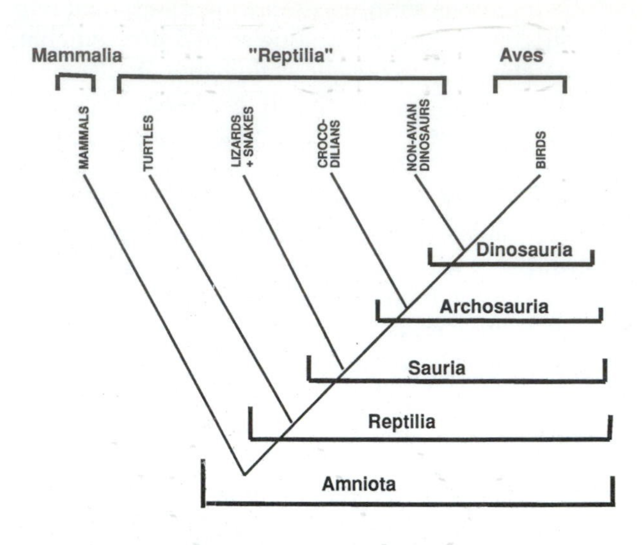

Figure 4.3. Different ways of classifying the same group of organisms. Traditional classifications (top) prefer to emphasize the tremendous evolutionary radiation of birds and mammals and place them in their own classes, equal in rank to the rest of the amniotes, lumped in the paraphyletic "Reptilia." A cladistic classification (bottom) does not permit mixing of phylogeny and other factors such as evolutionary divergence. Instead, every group is monophyletic and defined strictly by evolutionary branching. In this view, birds are a subgroup of dinosaurs, archosaurs, saurians, and reptiles.

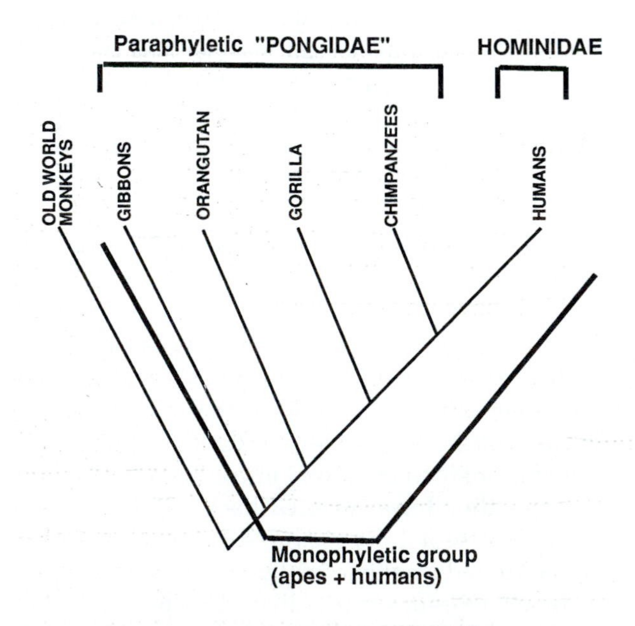

Figure 4.4. Traditional classifications emphasize the huge difference between humans and the rest of the apes by placing ourselves in a separate family Hominidae, and all the rest of the apes in the paraphyletic wastebasket taxon "Pongidae." To a cladist, humans must be included within the group that includes the apes (whether this means expanding the Hominidae to include the apes, or broadening the definition of the Pongidae to include the humans).

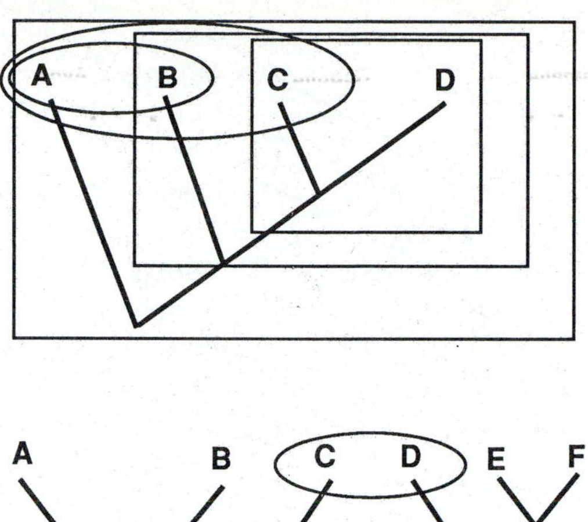

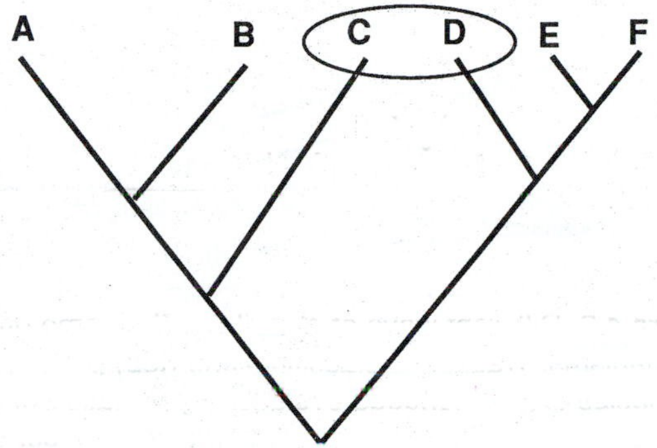

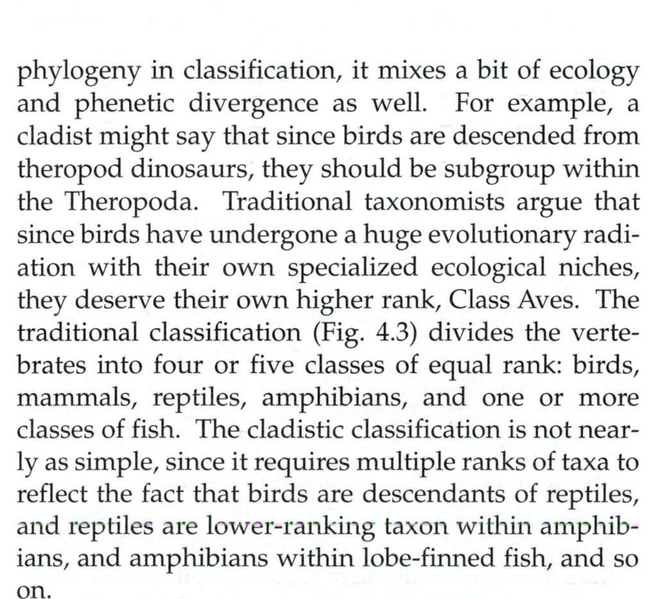

Figure 4.5. A monophyletic group includes all the descendants of a common ancestor, while a paraphyletic group excludes some of these descendants. In the top figure, the groupings outlined by rectangles are monophyletic, since they include all descendants of the common ancestor, while those in ovals are paraphyletic, because they exclude at least one descendant (group D). A polyphyletic group is composed of taxa from two or more unrelated lineages. In the lower figure, the grouping of C + D is polyphyletic, since they are from very different lineages.

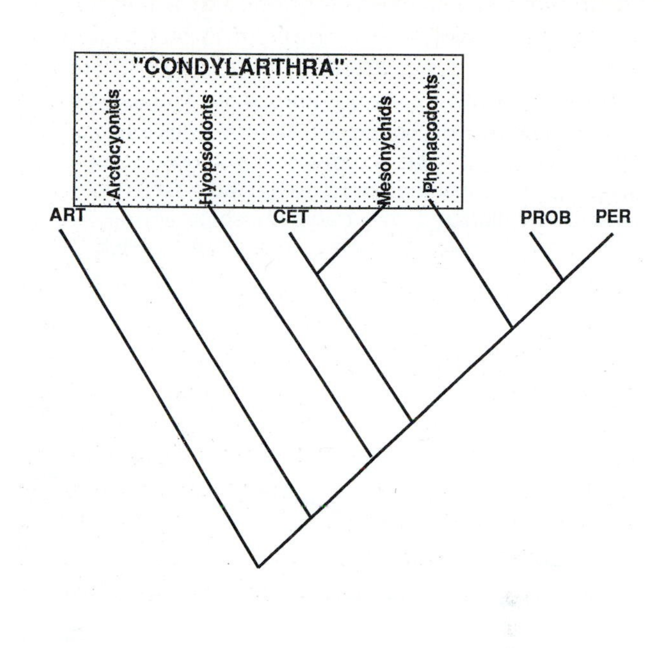

Figure 4.6. The wastebasket taxon "Condylarthra" (shaded box) is an example of a polyphyetic group, since it includes many different groups of animals with widely different phylogenetic origins. These include the primitive arctocyonids and hyopsodonts, the mesonychids (which are closer to whales), and the phenacodonts (which are closer to horses and elephants). ART = artiodactyls (even-toed hoofed mammals); CET = cetaceans (whales); PROB = proboscideans (elephants); PER = perissodactyls (odd-toed hoofed mammals).

phylogeny in classification, it mixes a bit of ecology and phenetic divergence as well. For example, a cladist might say that since birds are descended from theropod dinosaurs, they should be subgroup within the Theropoda. Traditional taxonomists argue that since birds have undergone a huge evolutionary radiation with their own specialized ecological niches, they deserve their own higher rank, Class Aves. The traditional classification (Fig. 4.3) divides the vertebrates into four or five classes of equal rank: birds, mammals, reptiles, amphibians, and one or more classes of fish. The cladistic classification is not nearly as simple, since it requires multiple ranks of taxa to reflect the fact that birds are descendants of reptiles, and reptiles are lower-ranking taxon within amphibians, and amphibians within lobe-finned fish, and so on.

Let's try another example. Figure 4.4 shows the cladogram of higher primates, and there is little disagreement over the geometry of their relationships. Traditionally, the human-ape clade was divided into two families: the Hominidae for ourselves and the Pongidae for all the non-human apes (Table 4.1). This classification scheme reflected both the huge divergence between ourselves and the rest of the apes and also our own egotism and anthropocentrism. This is unacceptable to a strict cladist. The Hominidae are a natural group with shared derived characters that support it (a **monophyletic** group), but the Pongidae becomes a "wastebasket" taxon of all the apes that don't happen to be human (Fig. 4.5). The Pongidae have some shared derived characters (such as the loss of a tail) that define them (but so do hominids). As long as Hominidae is a separate but equal family, the Pongidae is partially defined by the *lack* of the characters that define Hominidae. In cladistic terms, a group that does not include all its descendants is a **paraphyletic** group, defined by the absence of synapomorphies.

There are many such unnatural, wastebasket groups in classification schemes, such as the "invertebrates" (defined by the absence of a derived character, the vertebral column). If Reptilia does not include birds, then Reptilia is paraphyletic. If Amphibia does not include all higher vertebrates descended from them, then Amphibia becomes para-

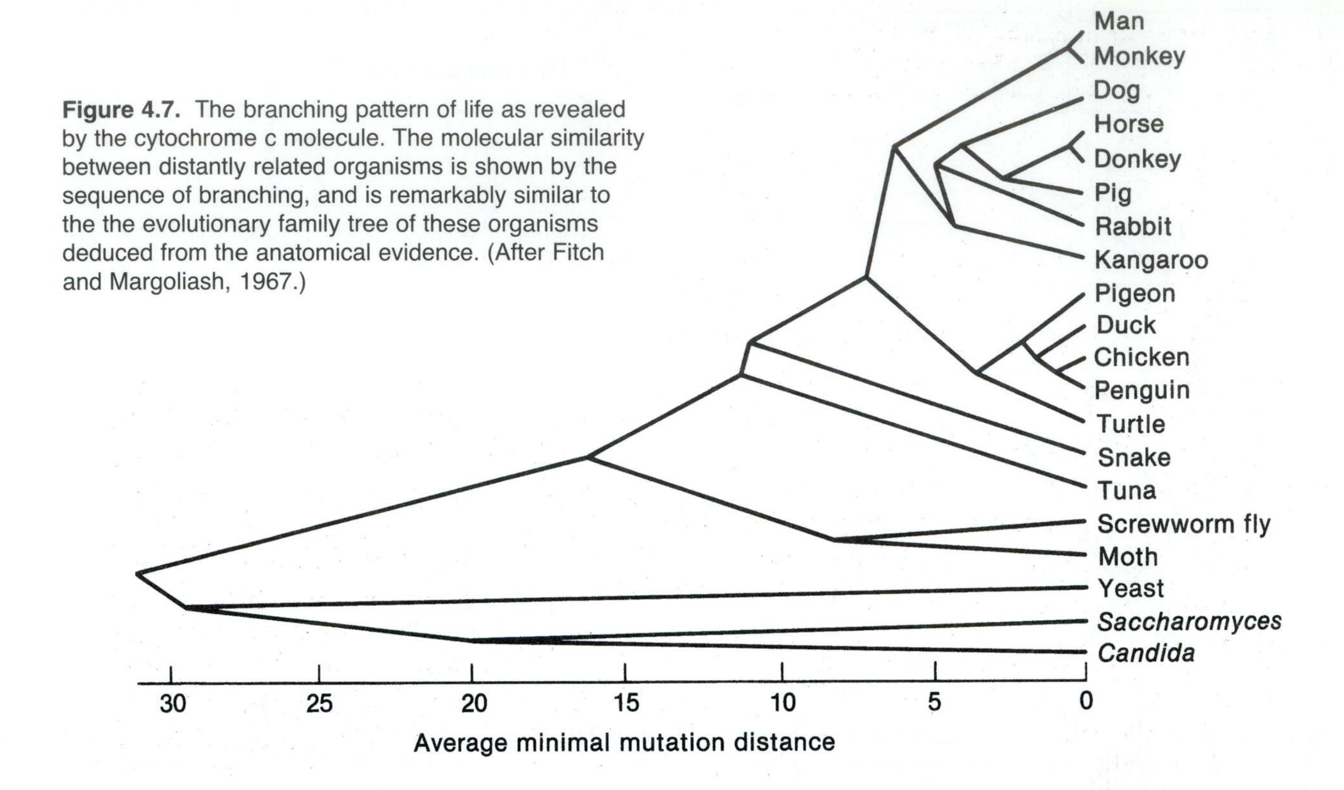

Figure 4.7. The branching pattern of life as revealed by the cytochrome c molecule. The molecular similarity between distantly related organisms is shown by the sequence of branching, and is remarkably similar to the the evolutionary family tree of these organisms deduced from the anatomical evidence. (After Fitch and Margoliash, 1967.)

phyletic—and so on. In rare cases, taxa converge from very different parts of the cladogram and might be put in a group together, forming a **polyphyletic group** (Figs. 4.5, 4.6). These are clearly unnatural, and few systematists would defend them once their polyphyly is revealed.

Once again, this debate is not merely an argument over semantics, but represents two fundamentally different ways of looking at the world. Traditional systematists don't mind mixing a little bit of ecology or phenetics into the classification, such as when they place Class Aves in an equal rank with Class Reptilia, so the evolutionary relationship is no longer apparent in the classification. Cladists say that mixing two or more data types in the classification is confusing. How can the student or general reader recognize which groups are strictly based on phylogeny (monophyletic) and which ones mix phylogeny with ecology (paraphyletic or polyphyletic)? The larger, more cumbersome classification schemes demanded by cladistics also might seem harder for students to learn than the simple, parallel taxa of equal rank in the old system. But cladists would argue, if the old system was a confusing and inconsistent mixture of natural and unnatural groups, why are we teaching it to our students in the first place?

Here, we run into established tradition. Old textbook conventions die hard and often do not reflect what the scientists at the cutting edge of the field might think. Much of the phylogenetic analysis in the taxa described later in this book has not yet caught up with cladistic methods, and not every group will be consistently monophyletic. As the years go by, however, more and more textbooks are catching up with the cladistic revolution, and students are no longer being asked to memorize classification schemes that the majority of active systematists no longer believe in.

MOLECULAR SYSTEMATICS

Traditionally, systematics has focused on analyzing anatomical characters that are visible to the naked eye or microscope. In the last decade, however, another approach has begun to revolutionize our understanding of phylogenetic relationships: **molecular systematics**. Instead of traditional anatomical characters, molecular biologists look at the similarity or differences of molecules. For example, such important molecules as hemoglobin, myoglobin, or cytochrome *c* (Fig. 4.7) have different sequences of amino acids in different organisms. By analyzing the similarities in amino acid sequences in many different organisms, a branching diagram can be produced. In most cases, the organisms that were thought to be most closely related by traditional anatomical comparisons also turn out to have the most similar molecular sequences.

Methods

A wide variety of molecules and techniques have been used over the years (see Hillis and Moritz, 1990, for a more complete discussion). A partial list might include:

1. *Gel electrophoresis*—One of the simplest, earliest, and least expensive techniques in molecular biology has long been used to detect the presence of certain proteins. A concentrate of proteins is placed in a number of wells at the end of a thin sheet made of gel, and then an electrical field is applied across the gel. Different amino acids move at different rates in an electrical field (small molecules move faster than larger ones), so they "race" at different speeds across the gel as the field is applied. One the field is turned off, the gel is stained, and the final position of each amino acid shows up as a dark band in its individual track. (Trial watchers might be familiar with DNA fingerprinting from the transparencies with multiple dark bands in a trial.)

Gel electophoresis gives an idea of the amino acids and proteins present in a sample, although the method is not as precise for determining sequence as some of those described below. Its main advantage is that it is relatively cheap, so it is very widely used. When Lewontin and Hubby (1966) published their classic studies that showed organisms to be much more genetically variable than previously supposed (discussed in the next chapter), they used gel electrophoresis. It has been widely used to determine enzyme efficiency, genetic variability of natural populations, and gene flow and hybridization, and to recognize species boundaries and determine phylogenetic relationships.

2. *Amino acid sequencing*— Through electrophoresis and even more sophisticated methods, the sequence of amino acids in a given protein can be determined (Goodman et al., 1987). This technique has been widely used over the years to determine the molecular similarities between many organisms (such as the hemoglobin, myoglobin, and cytochrome *c* examples just mentioned). However, it has largely been replaced by direct DNA sequencing, which can reconstruct not only the amino acids and proteins but even the genetic code of each amino acid in the sequence.

3. *Immunological methods*—Most animals produce **antibodies** that react to foreign substances (**antigens**) as part of their immune protection against disease and infection. When the antibodies from two organisms are mixed, the stronger the immunological reaction observed, the more similar two proteins are in their genetic sequence. Through a technique called microcomplement fixation, small amounts of antibodies of several different animals are placed in wells in a gel (Maxson and Maxson, 1990). The reaction between the antibodies in two different gels can be observed, and this reaction gives a semiquantitative measure of the degree of similarity.

4. *Chromosomal methods*—Ever since the double-stranded bands of DNA known as chromosomes were first seen under the microscope, their visible appearance has been important in determining genetic properties (Sessions, 1990). The methods can be as simple as counting how many chromosomes are present and looking for visible differences (known as a **karyotype**), to staining them to look for bands that reveal the presence of certain genes, to various methods that make them react to each other and even hybridize with other chromosomes.

5. *DNA-DNA hybridization*—This method was the first that directly measured the differences and similarities between DNA from two different kinds of organisms (Werman et al., 1990). When DNA is heated to 100°C in boiling water, the two strands of the molecule separate. As the solution cools, the individual strands seek to recombine with their exact matching strands. However, in a mixture of DNA from two different organisms, some strands from one organism will combine with strands from the other, forming a hybrid. Then the mixture is reheated, breaking the DNA apart once again. The more similar the two strands are, the more tightly they will bond together, and a higher temperature is required to break them. DNA strands from two less closely related organisms, however, are less similar and less tightly bonded and will dissociate at lower temperatures.

This method has been extensively used over the last decade to get a percentage similarity between the DNA of two different organisms. It has revealed some startling things, such as the fact that humans and chimps share about 98% of their DNA (discussed further in the next chapter). Sibley and Ahlquist (1983) have used it to completely reshuffle bird phylogeny and classification. However, it is not as informative as direct DNA sequencing, since it gives only a semiquantitative estimate of similarity and not the actual base-by-base sequence.

6. *DNA sequencing*—This is the cutting edge of molecular biology, since it allows the scientist to directly determine each base pair along the strand of DNA, and look at the gene sequence directly (Hillis et al., 1990). Until recently, it was too difficult and expensive to get enough DNA to make this procedure possible. Then the discovery of the **polymerase chain reaction** (**PCR**) amplification method revolutionized molecular genetics by making it possible to generate hundreds of copies of a DNA strand very cheaply and quickly. Many of the new developments in genetics that are reported in the news come from DNA sequencing, and the multimillion-dollar Human Genome Project is compiling the complete code for every gene in the human body.

Results

Although molecular methods have provided tremendous support for most anatomically based phylogenies, there are some striking disagreements between molecular and morphological methods; then both the systematists and the molecular biologists must go back to the drawing board and see if they can resolve their differences. Sometimes the conflict lies in the fact that there are relatively few molecular differences between organisms, so each branch point is supported by only one change in amino acids. These weakly supported phylogenies are unstable, since a few new data points can change them radically. In other cases, the conflict may lie in the fact that it is very easy for a molecule to switch back and forth from the primitive to the derived state and back again. For example, there are only four bases—adenine, guanine, cytosine, and thymine—that make up the genetic code in DNA. (Uracil replaces thymine in RNA). If a gene is going to change, it can very easily revert back to the primitive base pair condition, since that may be the only alternative. This molecular "noise" can often obscure the phylogenetic "signal," and molecular biologists have developed sophisticated methods to analyze and deal with it.

In a few cases, however, molecular systematics showed that traditional systematics was wrong. A classic case involved an argument over our earliest human ancestors. Anthropologists had long promoted fossils known as *Ramapithecus*, from beds about 12 Ma (million years before present) in Pakistan, as the earliest member of our human lineage. Based on these fossils, the great apes diverged from our lineage before 12 Ma. But molecular biologists looked at one system after another and found that they all consistently showed that our divergence from the common ancestor with chimpanzees must have occurred about 5 to 6 Ma. For a decade, the two sides argued, with neither giving any ground. Then, in the late 1980s, more complete specimens of *Ramapithecus* (now called *Sivapithecus*) were found which showed that the original fossils were misinterpreted. They were actually more similar to orangutans, and there was no longer any fossil evidence to deny the idea that the ape/human divergence took place about 5 to 6 Ma.

Molecular systematics has become one of the hottest new fields in paleontology as more and more paleontologists find ways to use it, even though their organisms are extinct. Unfortunately, it is not possible to review it in detail in this textbook. The reader is referred to the excellent book edited by Hillis and Moritz (1990), or the papers in the symposia edited by Runnegar and Schopf (1988) or Patterson (1987) to explore this topic further (see "For Further Reading.")

CODES OF SYSTEMATIC NOMENCLATURE

When Linneaus and other early natural historians developed different schemes of classification, there was no general agreement on how it should be done. Linnaeus' system became so successful that it soon became the standard in most parts of the world, but still there were no official rules, and chaos reigned. If one systematist didn't like a particular name for an organism, he might rename it for no good reason. Another systematist might use a name that had already been used for some other animal. Still another might name the species in his native language or name the species after himself. Some taxa were given more than one name. Systematics became a battleground of natural historians squabbling over proper names, and there were no referees to break up the fights.

To bring order out of this chaos, rules were needed. In 1813 de Candolle proposed the first code of nomenclature for botany, and in 1842 Strickland proposed the first code for zoology. Over the years these codes have evolved, and the first international code of zoological nomenclature was published in 1905. The current International Code of Zoological Nomenclature (ICZN) was published in 1985, and the International Code of Botanical Nomenclature was last revised in 1983. There is also a code for bacterial nomenclature.

Each code has a similar purpose: *to enhance stability and improve communication when creating or using taxonomic names and making taxonomic decisions*. All three have very similar basic principles, with slight variations in practice due to the differences between animals, plants, and bacteria. Systematists around the world are bound to follow these rules if they want their taxonomy to be recognized by other systematists. If they fail to do so, their work may not be published (since editors follow the codes strictly), or if it is published, it may be corrected by someone else who does follow the rules. At times, it seems that systematics becomes bogged down in legalistic trivia (see Box 4.1), but the rules are essential if taxonomists want to avoid unnecessary squabbles and wasted or duplicated effort. It is comparable to knowing the rules of the road before you take your driver's test to get your license. The Department of Motor Vehicles doesn't want you on the streets if you don't know the rules that everyone else is following. Similarly, the international community of systematists avoids "collisions" and "mistakes" by following their own internal set of "traffic rules."

Because all three codes are similar, we will concentrate on the ICZN as an example. Bound in bright

(*continued on p. 58*)

Box 4.1
An Example of the Formal Description of a New Species (from Prothero, 1996)

SYSTEMATIC PALEONTOLOGY

Class MAMMALIA Linnaeus, 1758
Order ARTIODACTYLA Owen, 1848
Suborder TYLOPODA Illiger, 1811
Superfamily CAMELOIDEA Gill, 1872
Family CAMELIDAE Gray, 1821

Hierarchy of taxa to which this genus belongs, with original authors and dates

Miotylopus Schlaikjer, 1935

Oxydactylus Loomis, 1911 (in part)
Protomeryx Loomis, 1911 (in part)
Miotylopus Schlaikjer, 1935
Dyseotylopus Stock, 1935
Gentilicamelus Loomis, 1936 (in part)

List of other invalid genera that are now synonyms of* Miotylopus*, with original authors and dates

Type Species—*Miotylopus gibbi* (Loomis, 1911). ***Each genus must have a type species.***

Included Species—*Miotylopus leonardi* (Loomis, 1911), new combination; *Miotylopus taylori*, new species. ***Parentheses around "Loomis, 1911" indicate the original author (Loomis) did not place these species in* Miotylopus*, but in other genera.***

Range—Arikareean (Gering to Harrison formations), Wyoming, Nebraska, and South Dakota; early Arikareean, southern California. ***Geographic and stratigraphic range must be indicated.***

Revised Diagnosis—Stenomyline camels with highly reduced premolars and elongate rostra. Differs from *Pseudolabis* in these two features and in the lack of a pseudolabine flexure on P4 lingual selene. Differs from the Stenomylini in having lower-crowned teeth; P1/1-C1/1 diastemata still present; and no diastemata between P2/2 and P3/3. Differs from all other camels in having high-crowned teeth with very weak or no mesostyles and dorsal premaxilla extended posterior to the level of P1. Size quite variable. P2-M3 length ranges from 65 to 84 mm. ***Diagnosis gives characters critical to distinguishing* Miotylopus *from other closely related camels.***

Discussion— **[*History of original naming and ideas about specimens now referred to* Miotylopus].** Two small, primitive oxydactyline-like camel jaws were described by Loomis (1911). The larger jaw was named *Oxydactylus gibbi* (p. 67) and the smaller jaw was called *Protomeryx leonardi* (p. 68). Loomis thought that these specimens were from the "Upper Harrison Formation" in the Muddy Creek area, Goshen County, Wyoming, but McKenna and Love (1972, p. 26) believed that they were from much lower in the Arikaree Group. In the Frick Collection, camels from the Muddy Creek area are from sediments equivalent to the Gering and Monroe Creek formations, and occasionally equivalent to the Harrison Formation—but none are equivalent to the "Upper Harrison" rocks of Nebraska. Apparently, Loomis' Muddy Creek specimens are early or middle Arikareean in age.

In May, 1935 ***[Date is critical to priority because these two taxa are synonyms but were named in different months of the same year]***, Schlaikjer described the skull and jaws of a camel from the "Lower Harrison" on the south side of 66 Mountain, Goshen County, Wyoming. According to McKenna and Love (1972, p. 28), this specimen was actually from Monroe Creek Formation or older rocks. Schlaikjer named this form *Miotylopus bathygnathus* and compared it only to "*Paratylopus*" *sternbergi*, *Poebrotherium*, and "*Protomeryx*." He recognized that the mesodont molars without a mesostyle were a new combination for camels, and this was part of his diagnosis for the genus. In July of the same year, Stock (1935) described a camel, *Dyseotylopus migrans*, from the early Arikareean Kew Quarry in the Sespe Formation, Las Posas Hills, Ventura County, California. It also had mesodont teeth and weak mesostyles on the upper molars. Schlaikjer (1935, p. 176) briefly discussed this specimen and considered it distinct from his *Miotylopus*, or possibly intermediate between *Miotylopus* and *Poebrotherium*.

Loomis (1936) placed the enigmatic *Protomeryx leonardi* in his new genus *Gentilicamelus*. As discussed above, nothing united the different species that Loomis assigned to *Gentilicamelus* except their Whitneyan-Arikareean age. Most of the contents of *Gentilicamelus* are here referred to *Paralabis*, *Oxydactylus*, and *Miotylopus*, except for the type species, *G. sternbergi*.

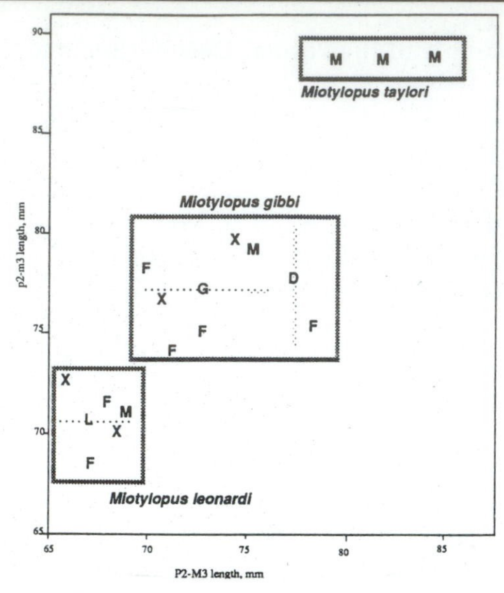

Figure 4.8. Plot of p2-m3 vs. P2-M3 of selected specimens of *Miotylopus*, showing size clusters here recognized as *M. leonardi* (small species), *M. gibbi* (medium species), and *M. taylori* (large species). "M" and "F" indicate specimens that can be sexed as male or female; "X" indicates specimens whose sex is indeterminate. "L" = type specimen of *"Protomeryx" leonardi* (a lower jaw, so P2-M3 length is unknown); "G" = type of *"Oxydactylus" gibbi* (a lower jaw, so P2-M3 length is unknown; "D" = type of *"Dyseotylopus migrans*" (a partial skull, so p2-m3 length is unknown.) (After Prothero, 1996.)

McKenna (1966) straightened out much of the confusion regarding small Arikareean camels with weak mesostyles. He transferred *"Oxydactylus" gibbi* to *Miotylopus*, although he did not formally synonymize it with *M. bathygnathus* (consistently misspelled *"brachygnathus"* by McKenna, 1966, and McKenna and Love, 1972). McKenna (1966) also noted the similarities of *Miotylopus* with *Dyseotylopus*, although he felt they represented separate lineages. In his 1966 paper, McKenna suggested that *"Protomeryx" leonardi* might be referable to *Dyseotylopus*. But in 1972, McKenna and Love formally synonymized *leonardi*, *"brachygnathus"* [*sic*], and some other specimens with *M. gibbi* (new combination). They also described a new specimen from the Arikareean of Darton's Bluff, Johnson County, Wyoming, which was considerably smaller than other specimens of *M. gibbi*. McKenna and Love (1972) referred specimens from the Castolon l.f. of Texas described by Stevens et al. (1969) to *M. gibbi*. Stevens (1977, p. 48) has since referred these camels to *Michenia*, based on more complete material, including upper molars with strong mesostyles. Frick and Taylor (1971) suggested that *Dyseotylopus* might be related to *Michenia* and the protolabidines, but Taylor (cited in Stevens, 1977, p. 51) later rejected this idea.

In my studies of the large samples of Arikareean camelids with weak mesostyles in the Frick Collection, I found considerable variation in size. Specimens with relatively broad rostra, unreduced premolars, and the typical *Pseudolabis* P4 were all referred to *Pseudolabis*. The remaining specimens were all united by the derived condition of reduced premolars and relatively slender rostra. All had weak or absent mesostyles, and their dorsal premaxillae are always extended posterior to the level of P1, so they are clearly stenomylines. Most specimens have the deep, elongate premaxillary fossa characteristic of stenomylines, although there is some variation in this feature (see *M. taylori* below). After sorting specimens by size and canine development (which seem to be the best indicator of sexual dimorphism in camels), I found no strong size dimorphism in most *Miotylopus* (unlike in *Poebrotherium* or *Pseudolabis*). For example, one of the smallest jaws, F:AM 36427, has male-shaped canines, yet is the same size as a jaw with presumed female canines, F:AM 36806, from the same deposits. In sorting the sample, there seemed to be three distinct size clusters (Fig. 4.7) with both males and females represented. The small form includes the type of *"Protomeryx" leonardi*. The medium-sized form includes the types of *"Oxydactylus" gibbi*, *Miotylopus bathygnathus*, and *Dyseotylopus migrans*. A very large form was also found that has never been named or described, and must be a new species. These size clusters are much more apparent in overall proportions of skull and mandible than they are in tooth dimensions (Fig. 4.8).

The largest forms are a new species, described as *Miotylopus taylori*. The medium- and small-sized forms are more difficult to separate. McKenna and Love (1972) lumped them together as *Miotylopus gibbi*. However, I am not comfortable with such a wide difference in size and morphology (not attributable to sexual dimorphism) in a single species. The differences are clearly not due to ontogeny, either. F:AM 36441 (a small male) and F:AM 36446 (a medium-sized female) are strikingly different in size, yet their M3's are fully erupted and show comparable wear. I find that the specimens can easily be sorted by size into small and medium-sized *Miotylopus*. Therefore, I recognize two species: the medium-sized *M. gibbi* and the smaller *M. leonardi* (new combination).

Miotylopus is the first valid generic name for this group of camels. It is the senior synonym of *Dyseotylopus* by two months. The name *Miotylopus* was originally chosen because these camels are typical of the Arikareean,

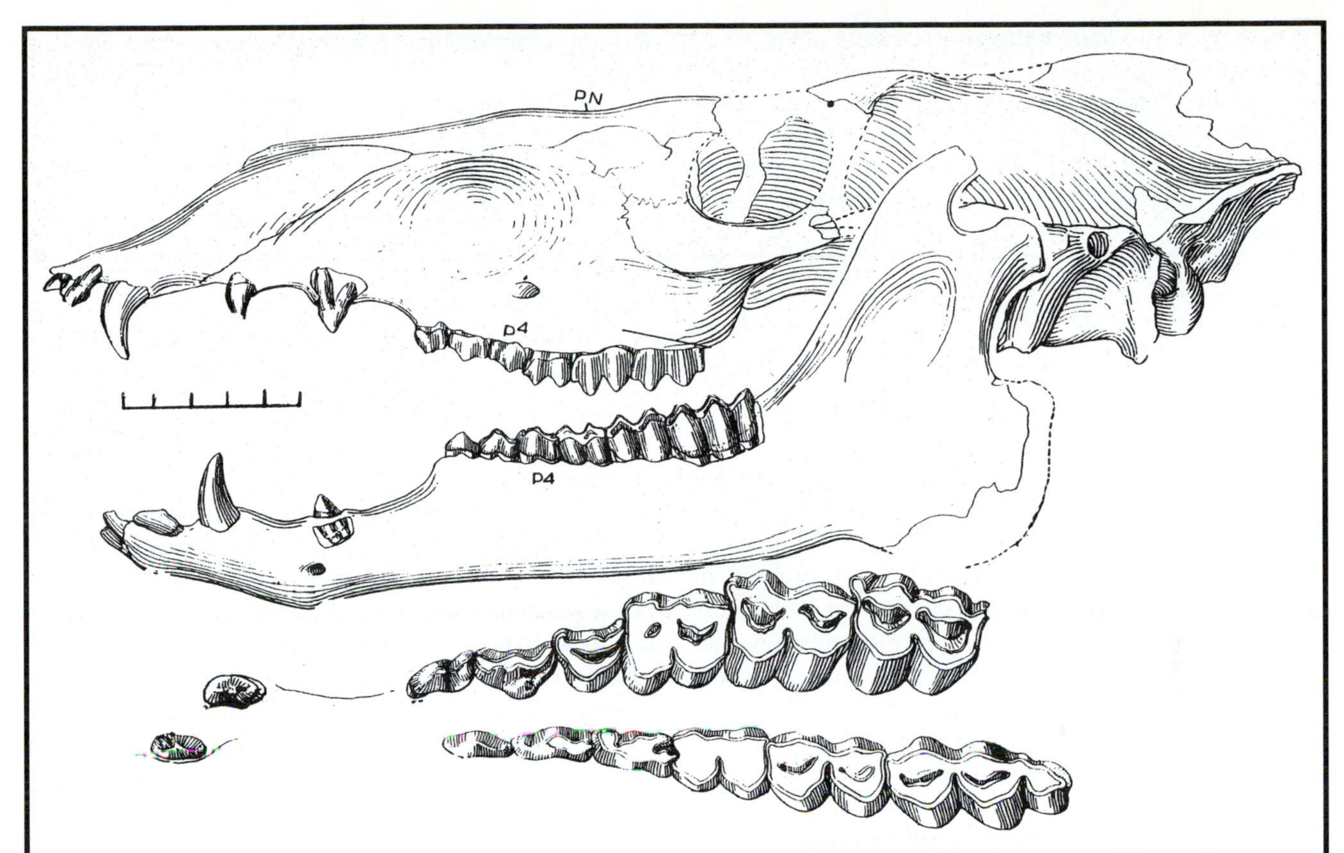

Figure 4.9. *Miotylopus taylori*. F:AM 36459, type specimen, showing left lateral view of skull and mandible, and crown views of upper and lower dentititions. Scale in cm for skull and jaws, 2 cm increments for teeth. (After Prothero, 1996.)

which was then considered early Miocene. Ironically, nearly all the early and middle Arikareean (and thus nearly all *Miotylopus*) are now considered late Oligocene (Tedford et al., 1987; Prothero and Rensberger, 1985), so the name *Miotylopus* has become a misnomer.

Miotylopus taylori, new species

Type—F:AM 36459, male skull, mandible (Fig. 4.9), atlas, and axis. From the early Arikareean (Gering Formation correlative) rocks of the Willow Creek area, Goshen County, Wyoming. ***Museum catalog number, nature of material, and original locality information must be indicated.***

Hypodigm—From the Muddy Creek area (middle Arikareean, Monroe Creek equivalent), Niobrara County, Wyoming: F:AM 36824, male skull and mandible; F:AM 41855, partial skeleton; F:AM 41829, partial skeleton; F:AM 36461, male palate, mandible, and partial skeleton; F:AM 34460, skull, mandible, partial skeleton. ***List of additonal referred specimens that the author considers part of this species, with geographic and stratigraphic range.***

Etymology—In honor of Beryl Taylor, who devoted most of his career to the study of the Frick Collection camels and whose hard work and insights made this research possible. ***Source of the name***.

Diagnosis—Largest species of *Miotylopus*. P2-M3 length = 81-85 mm. The p4 posterior crests are separate, unlike other species in the genus. ***How to distinguish this species from others in the genus.***

Description—***A detailed description is given; see Prothero (1996)***

Discussion—The striking feature of the skull and skeleton of *M. taylori* is its more giraffe-like proportions, with elongate cervical vertebrae and distal limb elements. In this respect, it converges on many of the oxydactylines and aepycamelines. However, the diagnostic *Miotylopus* skull features clearly show that these features are convergent (as happens many times in camels).

The difference in overall proportions may also explain why three closely related species of *Miotylopus* could live sympatrically through most of the Arikareean in Wyoming and Nebraska (Fig. 4.8). Apparently, they were ecologically differentiated. *M. gibbi* maintains the more primitive proportions seen in many camelids. *M. leonardi* has the gazelle-like size and build that eventually was the niche of the stenomylines. *M. taylori* had the longer neck and legs of the giraffe-camel niche, which was later occupied by *Oxydactylus*, the aepycamelines, and other later camels.

Although sparse material referable to *M. gibbi* and *M. leonardi* is known from the upper Arikareean Harrison Formation and equivalents, it appears that the heyday of *Miotylopus* was over by the late Arikareean. Instead, the Harrison and younger formations are characterized by the oxydactylines (which were also highly differentiated in size and ecological adaptations) and by the highly specialized stenomylines.

(*continued from p. 54*)

red, the 1985 ICZN runs to 338 pages covering 88 articles and 83 recommendations, with left-side pages in French and right-side pages in the equivalent English. (Except for the French, most international zoologists use English in international communication and publication.) The arbitrary starting point of the code is the 1758 tenth edition of Linnaeus' *Systema naturae*; names and taxa proposed before that date are not bound by these rules (but may not be recognized, either). The code is built around several basic principles:

1. *Binomial nomenclature* (Article 5)—These are the basic rules by which genera and species are created, named, and described. Each binomen must be based on Latin or latinized words from other languages to enhance international communication across language barriers. The latinized binomial is not always based on actual Latin words, but it must still follow the rules of Latin grammar. For example, if the species name is an adjective, it must be in the same gender (masculine, feminine, or neuter) as the genus name that it modifies. (Although few scientists know Latin these days, it is still useful in surprising ways.) The new taxon must be adequately diagnosed, described, illustrated, named, and published in a recognized scientific journal that is widely distributed and available to most systematists. This does not include unpublished dissertations and local newsletters with limited circulation. In addition to a clear definition and description, the author must also indicate the geographic or stratigraphic range of the taxon and list any relevant measurements or statistics. (See Box 4.1 for an example.) The origin and meaning, or **etymology**, of the new name, is also usually indicated (although this is not required). You can base names on any word as long as it is properly latinized, except that you cannot name a taxon after yourself. (You can, however, name it after a friend, and have your friend do the same for you with a different species.) Once a name has been used (even if it later proves to be invalid), it can never be used again for another animal.

2. *The principle of priority* (Article 23)—For the sake of stability and simplicity, the first available name proposed (after 1758) for a taxon is the valid name, except under highly unusual circumstances. This conflict usually arises when two different scientists give different names to the same organism because they were unaware of each other's work, or when more than one name is given to the same organism because some scientists name new species based on the most trivial of criteria. Once the valid name is established, all the later names become invalid **synonyms**, which cannot be used again. The synonyms can be **objective** (two scientists actually gave different names to the same specimen) or **subjective** (a later reviser thinks that two species or specimens are the same, and so one is a synonym of the other).

Normally, this synonymy is established early, so when most scientists learn a name, its priority is no longer in question. Occasionally, however, there are problems. If careful library work shows that some obscure scientist gave a different but prior name to a familiar taxon, that long-forgotten name legally has precedence over the much more familiar name. It doesn't matter that this obscure name was poorly described and poorly illustrated in a minor journal that nobody reads. As long as the name does not violate any of the rules, it has priority. As Charles Michener put it, "In other sciences the work of incompetents is merely ignored; in taxonomy, because of priority, it is preserved."

If the overthrow of a well-established name causes too much hardship for scientists, there is one last legal recourse: the International Commission of Zoological Nomenclature can suppress the obscure name through use of its plenary powers. To suppress the name, the taxonomist submits a formal application and justification to this international committee of about 30 scientists, who then publish the case, invite commentary, and decide it by majority vote. This procedure has served very well. For example, the widely studied protozoan *Tetrahymena* has been mentioned in over 1500 papers published over 27 years using that name. However, there are at least 10 technically valid but long-forgotten names that had priority. Because no purpose would be served by resurrecting these obscure names, the Commission voted unanimously to suppress them.

Sometimes the case is not so clear. Take the dinosaur that every kid knows as "*Brontosaurus*." In 1877, Yale paleontologist O. C. Marsh published two paragraphs without illustrations on a juvenile specimen of a sauropod he called *Apatosaurus ajax*. Two years later, he described another slightly larger, more complete, and more mature specimen from the same beds as *Brontosaurus*. Like most paleontologists of his time, Marsh was a taxonomic "splitter" who created a new taxon on every slightly different variant he found. By 1903, Elmer Riggs realized they were the same dinosaur, and without fanfare sank the name *Brontosaurus* as a junior synonym of *Apatosaurus*. As far as scientists are concerned, the case is closed—and the name "*Brontosaurus* "cannot be used, except in an informal sense.

Unfortunately, Marsh's "*Brontosaurus*" was the most complete sauropod specimen then known, and it became a famous museum display. The reconstructions of this mounted skeleton were then copied and were the basis of hundreds of drawings, paintings, book illustrations, and movie monsters—all bearing the scientifically invalid name "*Brontosaurus*." Since children's books and popular movies seldom check the scientific accuracy of their content with scientists, but shamelessly copy older books and movies, the name was perpetuated, even though no paleontologist has taken the name seriously since 1903. In 1989, the U.S. Postal Service made the news when they issued a "*Brontosaurus*" stamp and then got criticism from paleontologists for using an invalid name. Some think that the name *Apatosaurus* should be suppressed, since "*Brontosaurus*" is much more familiar (see Steven Jay Gould's essay, "Bully for *Brontosaurus*.") However, the Commission is unlikely to agree, since the synonymy was established over 90 years ago and professional paleontologists haven't used the invalid name since. It may be obscure to the general public (although more and more children's books and popular books now have it right), but that doesn't matter—it's not obscure as far as scientists are concerned.

There are a number of taxonomic names that are widely known but scientifically invalid. For example, most books still call the earliest horse "eohippus" even though the first name proposed was *Hyracotherium*. However, Hooker (1989) showed that the original material of *Hyracotherium* is not a horse after all, but a member of a closely related group called palaeotheres. Does that allow us to use "eohippus" again? Unfortunately, the original material of "eohippus" is not a horse, either. The problem is still under study, but it may turn out that the oldest valid name for the first horse is *Protorohippus* (see Prothero and Schoch, 1989).

3. *Principle of first reviser* (Article 24)—When it's not clear which name has priority (for example, when two names are published on the same date or with different original spellings), then the Code says that the first person to revise these names decides which name has priority. In earlier codes, when two names were published on the same date (usually in the same publication) and then later shown to be synonyms, the one mentioned first (page priority) was the valid name; this is no longer true in the 1985 Code.

4. *Principle of coordination* (Articles 36, 43, and 46)—The names of families, subfamilies, and some higher taxa are based on the name of a genus within that family. It doesn't matter if anyone formally establishes this family name—its proper name is automatically established by the naming of that genus. For example, in the case of the extinct rhinoceros family Hyracodontidae, the genus *Hyracodon* was described in 1856 as the first named of all the genera in the family. Even though no one actually published the name Hyracodontidae in 1856, it is considered established. This is despite the fact that names for the same family of rhinos, such as "Triplopodidae," were published before the name "Hyracodontidae" finally appeared in print.

5. *Homonymy* (Article 52)—Names that have already been used cannot be used again, even if they are spelled differently but sound the same (a **homonym**). This helps prevent confusion when the names are spoken rather than written; scientific names are hard enough to keep track of without having two different names that sound the same! Cases of homonymy are rare, but they do crop up now and then. For example, the extinct camel *Protomeryx campester* was named in 1904, and another extinct camel was named *Oxydactylus campestris* in 1909. Later it was shown that the name "*Protomeryx*" was invalid, and some scientists thought that the species "*P.*" *campester* should be included in *Oxydactylus*. But that creates a problem: the genus *Oxydactylus* already had a species *campestris* (both trivial names would be spelled the same to agree with the gender of the genus). If this homonymy had actually occurred, a different name would be required for one of these camels. Fortunately, a determination of priority by the principle of first reviser showed that "*campester*" was not valid. The correct species name for the specimens originally called "*P.*" *campester* was *cedrensis*—so the dilemma was avoided (Prothero, 1996).

6. *Type specimens* (Article 61)—When a taxonomist names a species, he or she must also designate one specimen as a standard of reference to represent his or her concept of that species. That specimen is know as the **type specimen,** or **holotype**. Typically, it is the best available specimen of the species, and in the original publication it must be clearly illustrated and

indicated by museum catalog number. Normally, type specimens are deposited in a major museum or other reference collection with scientific access, so other scientists can examine them. Museums put a special label on their type specimens and often store them in a special place.

Although this idea recalls the typological concept of species prevalent before Darwin, it is really a practical matter. Telling similar species apart is hard enough without having vague definitions. If the original author picks a type specimen, it is possible for later scientists to determine his or her concept of the species, even if the original descriptions or diagnoses or illustrations are inadequate. In addition to the type specimen, most taxonomists indicate a number of additional specimens that are considered referable to that species; these are known as the referred material, or **hypodigm**. This gives later scientists a chance to see the original author's concept of the range of variation within the species. Some scientists prefer to name several **syntypes**, including the holotype and several **paratypes**, to give an idea of this spectrum of variation.

Sometimes, problems emerge with the original type specimen. If the original author did not designate which specimen among the syntypes is the holotype, then a later scientist can do so. This specimen is a later designated type, or **lectotype**. If the type specimen is lost, a later reviser is obligated to name a new type specimen, or **neotype**. If a taxonomist decides that the type specimen is inadequate to tell if the species is really distinct, he or she may sink the species name as a ***nomen dubium*** ("doubtful name"). If a taxonomist describes a species without following the rules properly (usually because they left out a diagnosis, a description, a type specimen, or an illustration), the invalid name is dropped as a ***nomen nudum*** ("naked name").

By now, most students' heads are reeling with all this jargon and legalese. After all, most people enter biology or paleontology because they like organisms or fossils, not law books. If you were going to be a lawyer, you wouldn't be taking paleontology! It is true that there are systematists who act like lawyers, spending all their time ferreting out obscure names and correcting other people's mistakes in print. However, most systematists regard the Code as a necessary skill to be mastered, just as most drivers learn the rules of the road before they drive. Although relatively few biologists will ever describe a new species or need to know the Code well, this is not true of paleontologists. Most find it necessary to do at least some of their own systematics, and many describe more than one new species during their careers. For this reason, serious students of paleontology should get a copy of the Code and read it through before they reach the point where they need to describe new species. Even if you never describe a new species, a good paleontologist knows the Code well enough to understand its requirements and to realize when another author has made a mistake.

CONCLUSIONS

Although some people are alienated by the noisy controversies in systematic theory, or by the legalisms of taxonomic nomenclature, most paleontologists find that systematics is central to their field. Without proper names, identifications, and classifications, we cannot understand the specimens we work on. Systematics is the key to some of the most interesting problems in comparative biology and paleontology, especially such central issues as the reconstruction of evolutionary history and the determination of phyletic relationships as well as the reconstruction of biogeographic history. More than any other group of bioscientists, paleontologists find themselves doing their own systematics in order to solve any other kind of paleobiological problem. A good paleontologist has to have a solid understanding of systematics and taxonomic nomenclature. Besides, part of the fun of deciphering the fossil record is discovering, naming, and describing new species. It represents a permanent legacy of a scientific career that may long outlive the paleontologist!

For Further Reading

Eldredge, N., and J. Cracraft. 1980. *Phylogenetic Patterns and the Evolutionary Process*. Columbia University Press, New York.

Gould, S.J. 1991. *Bully for Brontosaurus*. Norton, New York.

Hennig, W. 1966. *Phylogenetic Systematics*. Univ. Illinois Press, Urbana, Illinois.

Hillis, D. M., and C. Moritz, eds. 1990. *Molecular Systematics*. Sinauer Associates, Sunderland, Mass.

Hull, D. 1988. *Science as a Process*. University of Chicago Press, Chicago.

International Commission on Zoological Nomenclature. 1985. *International Code of Zoological Nomenclature* (3rd ed.). International Trust for Zoological Nomenclature, London.

Krishtalka, L. 1989. "The naming of the shrew,"pp. 28-37, *in* Krishtalka, L., *Dinosaur Plots*. William Morrow, New York.

Mayr, E. 1966. *Principles of Systematic Zoology*. McGraw-Hill, New York.

Nelson, G., and N. Platnick. 1981. *Systematics and Biogeography: Cladistics and Vicariance*. Columbia University Press, New York.

Patterson, C. 1981. Significance of fossils in determining evolutionary relationships. *Annual*

Review of Ecology and Systematics 12:195-223.

Patterson, C., ed. 1987. *Molecules and Morphology in Evolution: Conflict or Compromise?* Cambridge University Press, Cambridge.

Runnegar, B. and J. W. Schopf. 1988. *Molecular Evolution and the Fossil Record*. Paleontological Society Short Course Notes 1.

Schaeffer, B., M. K. Hecht, and N. Eldredge. 1972. Phylogeny and paleontology. *Evolutionary Biology* 6: 31-46.

Schoch, R. M. 1986. *Phylogeny Reconstruction in Paleontology*. Van Nostrand Reinhold, New York.

Simpson, G.G. 1961. *Principles of Animal Taxonomy*. Columbia University Press, New York.

Sokal, R. R., and P. H. A. Sneath. 1963. *Principles of Numerical Taxonomy*. W.H. Freeman, San Francisco.

Voss, E. G., and others. 1983. *International Code of Botanical Nomenclature*. W. Junk, The Hague.

Wiley, E. O. 1981. *Phylogenetics: The Theory and Practice of Phylogenetic Systematics*. Wiley Interscience, New York.

Wiley, E. O., D. Siegel-Causey, D. R. Brooks, and V. A. Funk. 1991. *The Compleat Cladist, A Primer of Phylogenetic Procedures.* University of Kansas Museum of Natural History Special Publication 19.

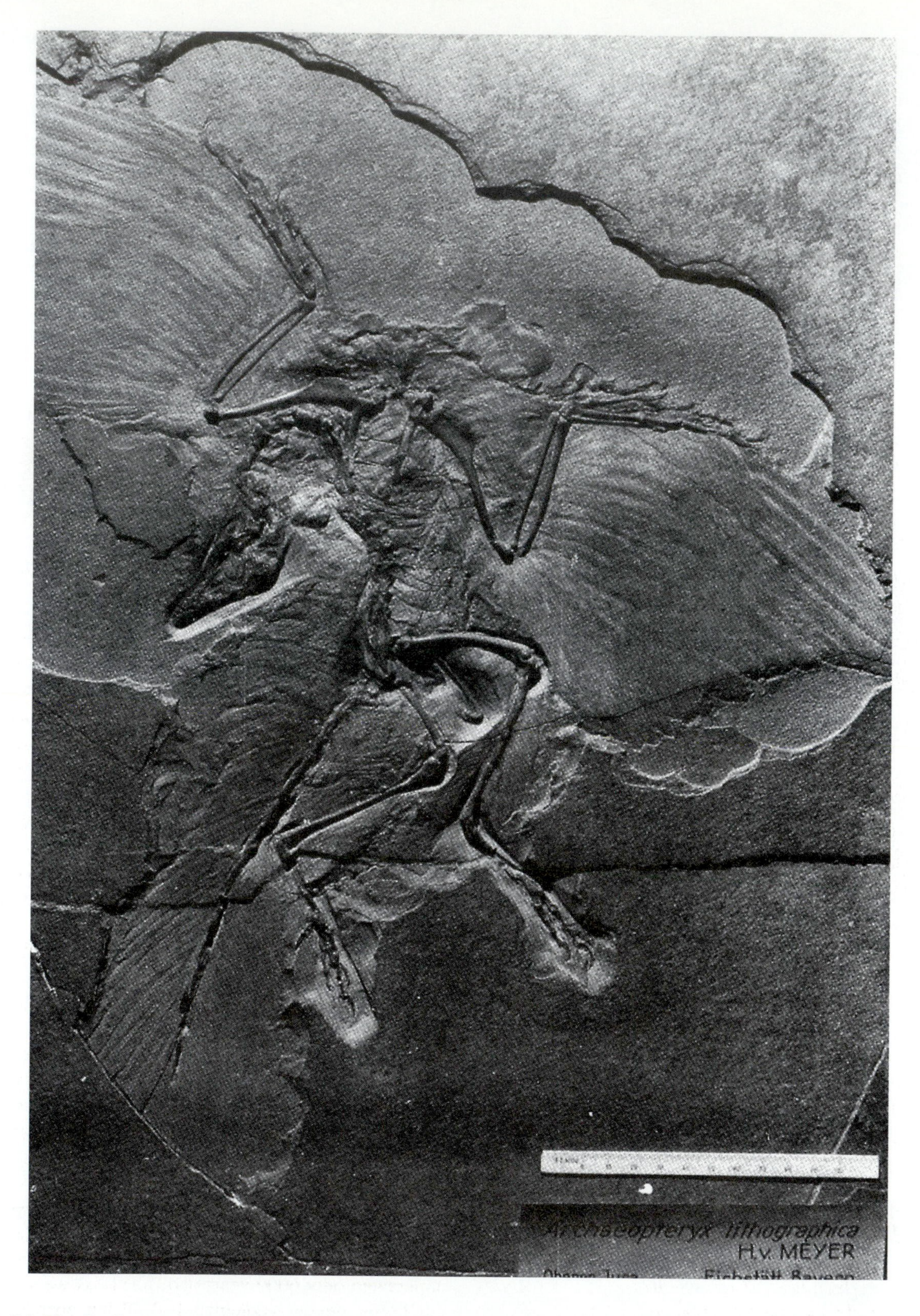

Figure 5.1. The Berlin specimen of *Archaeopteryx* is the best preserved of the seven known specimens from the Jurassic Solnhofen Limestone of Germany. It was discovered in 1877, and provided crucial evidence as a "missing link" between reptiles and birds. (Photo courtesy J. H. Ostrom.)

Chapter 5

Evolution

Natural selection is daily and hourly scrutinizing throughout the world every variation, even the slightest; rejecting all that which is bad, preserving and adding up all that is good; silently and insensibly working . . . We see nothing of these slow changes in progress until the hand of time has marked the long lapse of ages.

Charles Darwin, *On the Origin of Species*, 1859

How exceedingly stupid of me not to have thought of that!

Thomas Henry Huxley, 1859, after reading *On the Origin of Species*

Nothing in biology makes sense except in the light of evolution.

Theodosius Dobzhansky, 1973

THE EVOLUTION OF EVOLUTION

When Charles Darwin published *On the Origin of Species* in 1859, he started a scientific revolution that transformed natural history into modern biology and forever changed how we view ourselves as well. Prior to 1859, it was still possible (although increasingly difficult) to interpret life, fossils, and geology as evidence of the book of Genesis. After 1859, not only the scientific community, but virtually all educated people had to come to terms with the fact that all life is interrelated, and humans are descended from other forms of life, not supernaturally created.

Yet Darwin's book did not spring out of a vacuum. The idea that one species might be transformed into another is an old one that continually crops up in western thought. The Greek philosophers of the fifth century B.C. promoted the idea that "everything changes," as did the Roman philosopher Lucretius, whose poem *De rerum naturae* ("On the nature of things") postulated the existence of atoms and was centered around the idea that nature is constantly in flux. However, the dominance of the Church in western civilization put an end to this kind of free thought for over a millennium. Since scholars took the book of Genesis literally, they believed that all life, and indeed, the entire earth, had been created exactly as we see it. No changes had occurred since the Creation some 6000 years ago, except for decay and corruption due to the sins of Adam and Eve.

During the 1700s, however, it became increasingly difficult to take Genesis literally. It was possible for the ancient Hebrews to believe that every animal they knew had once fit onto Noah's Ark and then migrated from Mt. Ararat to their present homes. But by the mid-1700s, the work of taxonomists like Linnaeus and his successors had already recognized over 6000 species of animals, far too many for the Ark. A century of exploration in exotic places—Africa, South America, Australia, southeast Asia—produced more and more new creatures previously unknown to Europeans, and the Ark story became ludicrous. Even more impossible to believe was that they had all migrated from Mt. Ararat (now in Turkey). If so, then why were the pouched marsupials found almost entirely on Australia and placental mammals elsewhere?

Literal belief in Genesis was also being challenged by the fossil record. By the early 1800s, the discovery of faunal succession made it impossible to treat all fossils as if they had been victims of a single Noah's flood, or even a series of floods. The complexity of the fossil record showed that life had undergone many changes over time, and no stretching of the biblical accounts explained it. In addition, Hutton, Lyell, and other geologists made it clear that the earth was immensely old and constantly changing, not created only 6000 years ago exactly as we see it. By the 1840s, devout geologists had all but abandoned the Bible as having anything to do with the fossil record as they knew it. This directly contradicts the creationist assertion that scientists use circular reasoning with the fossil record. Allegedly scientists arrange the fossil record to prove evolution, and then evolutionists point to it as evidence of evolution. Nothing could be further from the truth! The sequence of fossils was established by creationist geologists more than 40 years before Darwin published his book.

In the late 1700s, the idea of evolution became more and more popular. In 1749, the great French philosopher George-Louis Leclerc, the Count of Buffon (1707-1788), began a 34-volume work entitled *Histoire naturelle*, which included a highly speculative account of evolution and even suggested that the earth may be 75,000 years old—highly unorthodox, even for the French radicals of his time. The foremost French evolutionist of the next generation was Jean Baptiste Antoine de Monet, the Chevalier de Lamarck (1744-1829). Lamarck began as a botanist, but after the French Revolution, he was assigned to the least desirable job in the Museé d'Histoire Naturelle—keeper of insects, shells, and worms. Lamarck used this position to completely reorganize invertebrate zoology, laying the foundation for our modern understanding of invertebrates. Lamarck saw all life as part of a single whole, and even coined the word "biology" to emphasize the unity of zoology and botany. His experiences also taught him a tremendous amount about the diversity of life, and from this understanding, he developed his own ideas about evolution, published as *Philosophie Zoologique* in 1809.

Lamarck saw the interconnectedness and variability of life and realized that species were not fixed. He arranged animals in the "scale of nature" that was popular in his time. This "ladder of creation" placed corals and worms and other invertebrates at the bottom, "lower" vertebrates in the middle, mammals higher, and humans at the top of the earthly ladder. Various ranks of cherubim, seraphim, angels, and archangels were on the higher rungs, culminating with God. Like many of his contemporaries, Lamarck believed that life was constantly being spontaneously generated out of the mud. From these ideas, Lamarck developed his own concept of evolution. Living things were constantly being transformed and moving up the ladder to the top. Those that had arisen recently from the mud were still worms or molluscs. Those that had been evolving for a long time were mammals or even humans.

Although Lamarck's ideas were evolutionary, they were very different from our modern concepts. Instead of a bushy, branching family tree of life that we now visualize, Lamarck saw life as a series of individual blades of grass, independently growing out of the mud. Although life was constantly changing, Lamarck did not see it as interrelated. Typical of the philosophers of his time, Lamarck's work was highly philosophical and speculative, with little hard evidence or experimental data. One minor idea in Lamarck's long book was that organisms could inherit the characters that were acquired in their parent's lifetimes. For example, the blacksmith would pass on his large muscles to his children. According to Lamarck, the ancestral giraffe that stretched its neck would pass its longer neck on to its descendants, until giraffes reached their current neck length. These ideas of the **inheritance of acquired characters**, and the effects of use and disuse, were held by nearly every evolutionist of that generation, and even by Darwin himself.

Unfortunately for Lamarck, he became the enemy of the most powerful man in French science, Baron Georges Cuvier. Cuvier made sure that Lamarck's ideas were ridiculed while he was alive and distorted or forgotten after he died. In later years, the term "Lamarckism" was attached only the inheritance of acquired characters, an idea that was only a minor part of Lamarck's thought and widely believed by nearly every biologist until early in this century. However unfair it is to Lamarck's memory, we still use the term "Lamarckian inheritance" since it is commonly understood in this context.

The Evolution of Darwin

Charles Darwin (1809-1882) was born on the same day and year as Abraham Lincoln and the same year that Lamarck published his evolutionary theory. His grandfather, Erasmus Darwin, was the King's own physician and had published a controversial poem that promoted evolutionary ideas, so evolution was in the Darwin family inheritance. Charles' father was also a doctor, and he planned for Charles to follow in the family calling. However, Darwin did not have the stomach for the sickening sights of medical school in Edinburgh in the 1820s. Instead, he developed a passion for natural history and was very influenced by the zoologist and sponge taxonomist Robert Grant, who taught in Edinburgh but was a follower of the radical French evolutionists.

After dropping out of medical school, Darwin was sent to Cambridge to become a theologian. His father hoped that Charles could retire to a country parsonage and indulge his mania for natural history without disgracing the family. At Cambridge Charles became even more interested in natural history and ignored theology entirely. He received most of his training from the botanist John Stevens Henslow and the first professor of geology, Adam Sedgwick. After more wasted years, Charles was again about to drop out, when he heard of an opportunity to take part in a round the world oceanographic voyage on the H.M.S. *Beagle*. Originally hired to be the "gentlemanly dining companion" for the aristocratic ship captain (who could not socialize with lower-class sailors), Darwin eventually became the ship's naturalist when the original naturalist dropped out early in the voyage. The voyage took five years to go around the world, spending most of its time off the coast of South America. While the ship was at sea surveying, Darwin spent weeks on land, collecting animals and fossils. His most important stop was on the Galapagos Islands off the coast of Ecuador. There he found that each island had its own unique species of tortoises, and the finches had been modified to take over many of the roles taken by other birds on the mainland. Yet the importance of this discovery did not occur to Darwin until after he returned.

The voyage continued to Australia and around the southern tip of Africa, returning to the Argentinian coast for more surveying before finally reaching England in 1836. When Darwin returned home, his father said, "Why even the shape of his head has changed!" The contents of Darwin's head had changed even more. While on the *Beagle*, he had read each of the three volumes of Lyell's *Principles of Geology* as soon as they were published and began to see not only geology but also biology in the Lyellian uniformitarian way. After he returned, he settled in London to publish the results of his voyages and also to do his own research. Here he encountered the writings of Thomas Malthus, who argued that humans reproduce at a far higher rate than their resources can support. The only thing keeping human populations in check, said Malthus, was death and disease. Darwin also took an active interest in breeding domesticated pigeons, which gave him an important insight into the variability of species and what artificial selection by breeders can accomplish.

By 1842, Darwin had a rough sketch of his evolutionary theory, but he sealed it and put it away with instructions for his wife to publish it after he died. Darwin was justifiably afraid of the reaction it might provoke, because evolution was a very dangerous notion in those days. It was popular in the radical medical schools in London, who were heavily influenced by French biology and evolutionism. For that reason, the conservative upper classes and the scholars from Oxford and Cambridge did their best to suppress dangerous notions such as evolution as a threat to their society and to their religion. In 1844, a Scottish publisher, Robert Chambers, anonymously wrote a tract on evolution entitled *Vestiges of the Natural History of Creation* that caused a national furor. Although its biology was amateurish and easy to discredit, the implications struck fear in the hearts of the upper classes.

It is no wonder then that Darwin, a wealthy Cambridge-educated gentleman, was fearful of speaking out in favor of this dangerous idea. After writing his ideas on evolution down, he put them aside for 15 years while he worked on other research, particularly the systematics of barnacles. He may never have published if it were not for an extraordinary historical accident. A young naturalist, Alfred Russel Wallace (1823-1913), had developed his own concept of natural selection. He came up with his inspiration while delirious from malarial fever, contracted while he was collecting animals in what is now Indonesia. Wallace wrote his ideas down and sent them in a letter to Darwin, not realizing that Darwin had been working on the problem for decades. Darwin was aghast when he received the letter. Not wishing to rob Wallace of due credit for the idea but also not wanting to lose credit for his own prior discovery of natural selection, Darwin appealed to his friends, the geologist Charles Lyell and the botanist Joseph Hooker. They suggested that both Darwin's 1842 outline and Wallace's letter be read to the 1858 meeting of the Linnean society so that they would get equal credit. Yet the papers received little notice. In his annual summary for 1858, the President of the Linnean Society wrote that there were no important discoveries to report that year. Nevertheless, Darwin realized he could no longer procrastinate. He quickly abandoned his planned gigantic work on the species and summarized his ideas in a 155,000-word (brief by Victorian standards) book, *On the Origin of Species by Means of Natural Selection*. It sold out all 1250 copies on the day it was published and went through six editions while Darwin was alive.

Darwin's argument was very simple, yet very clever. First, he drew an analogy to the artificial selection practiced by the breeders of domesticated animals. If they could modify the original wild dog into beasts as different as a Dachshund and a Great Dane, then species were much more flexible than naturalists admitted. Then he drew an argument from Malthus. Animals are capable of exponential growth of populations, yet in nature most animal populations remain constant. His first deduction: more young are born than can survive. Next, Darwin described how natural populations are also highly variable, and pointed to the experiments on domesticated animals that showed how these variations were heritable. His major conclusion: *organisms that inherit favorable variations for their immediate environment will tend to survive more often than others*. Darwin called this idea **natural selection**.

Darwin was not the originator of the concept of evolution. Why then does he get most of the credit for changing western civilization with his idea? First of all, Darwin was in the right place at the right time. In 1844, Chambers' *Vestiges* was too early and was written by an amateur biologist, so it was easily discounted. By 1859, scientists were primed to accept evolution (as Wallace's letter shows), and Darwin had spent his entire life building his scientific reputation, as well as an overwhelming case for evolution. In addition, Darwin was of the wealthy upper class, so he could not be attacked like one of the radicals from the working class London medical schools. If you read *The Origin*, you will find that it is argued very carefully, from the analogy to artificial selection, to the intricate argument about natural selection, and then case after case to pound his point home. Darwin provided two things that no other previous work had provided: overwhelming evidence that life had evolved, and a convincing mechanism for how evolution took place, namely, natural selection.

Although controversies raged for decades after 1859, most of the scientific community and the educated people in the western world soon came to terms with evolution. By the time Darwin died in 1882, he was hailed as one of Britain's greatest scientists and was buried with honor in Westminster Abbey near the tomb of Isaac Newton. Yet there was one problem that Darwin never solved while he was alive: the nature of inheritance. One of Darwin's critics pointed out that if an organism inherited a favorable variation, that advantage would be blended out when that organism bred with normal strains. Ironically, the solution to this problem of "blending inheritance" had been discovered in 1865 by an obscure Czech monk by the name of Gregor Mendel (1822-1884). Mendel worked with hybrid pea plants, which happened to have very simple inheritance patterns. From his experiments, Mendel found that cross-breeding produced very characteristic ratios of different features that could be explained by very simple genetic mechanisms. More important, Mendel showed that rare characters are not blended out, but can reappear after many generations (due to what we now call recessive genes). The importance of this type of **discrete inheritance** was not appreciated until 1900, when three different geneticists independently rediscovered Mendel's work 16 years after he died. From the turn of the century onward, genetics dominated evolutionary biology, since the fundamental problems Darwin did not solve were all genetic in nature. In 1953, the discovery of the DNA molecule and its role in inheritance put molecular genetics in the position to actually decipher the code for the evolution of genes. Much of the work in genetics since then has been focused on the fine-scale details of evolution at the molecular level.

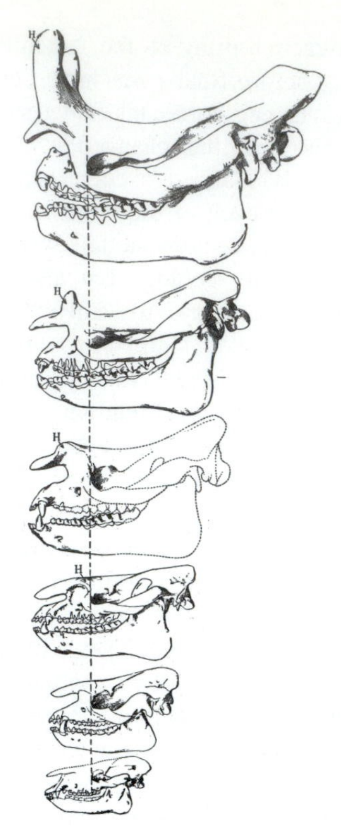

Figure 5.2. Orthogenetic sequence of Eocene brontothere skulls as portrayed by Henry Fairfield Osborn (1929). Osborn believed that the brontothere horns grew larger and larger through internal forces not under the control of natural selection. His ideas were not widely accepted in his time, and today even this linear sequence of brontotheres has been replaced by mulitple branches. (From Osborn, 1929.)

THE "EVOLUTIONARY SYNTHESIS"

Evolution is a change in gene frequencies through time.

Theodosius Dobzhansky, *Genetics and the Origin of Species*, 1937

Evolution is merely a reflection of changed sequences of bases in nucleic acid molecules.

John Maynard Smith, *The Theory of Evolution*, 1958

We are survival machines—robot vehicles blindly programmed to preserve the selfish molecules known as genes. . . They swarm in huge colonies, safe inside gigantic lumbering robots . . . they are in you and me; they created us, body and mind; and their preservation is the ultimate rationale for our existence.

Richard Dawkins, *The Selfish Gene*, 1976

When Darwin died in 1882, the world had come to terms with the fact that life has evolved, but there were few people who agreed with his mechanism, natural selection. By 1900, natural selection was out of favor since many different geneticists put forth more radical notions of how evolution takes place. And where were paleontologists throughout this entire period? After all, it would seem that the interpreters of the fossil record would lead the parade in discoveries about evolution. On the contrary, paleontologists were among the lesser contributors to evolutionary biology. Britain's foremost paleontologist was Richard Owen, the man who named the dinosaurs and described *Archaeopteryx* (Fig. 5.1) and the extinct South American mammals from Darwin's *Beagle* voyage. He was an ardent antievolutionist until the day he died. Darwin himself spent an entire chapter apologizing for the imperfection of the fossil record, since in 1859 it offered little direct support of his ideas. In the late 1800s, prominent American scientists such as Edward Drinker Cope, a vertebrate paleontologist, and Alpheus Hyatt, an ammonite specialist, were both believers in Lamarckian inheritance. As late as 1938, the Harvard paleontologist Percy Raymond said of his colleagues, "Probably most are Lamarckians of some shade; to the uncharitable critic it may seem that some out-Lamarck Lamarck."

The early twentieth century saw even less of a contribution from paleontology. While genetics was breaking new ground, most paleontologists avoided evolutionary theory altogether and remained skeptical about whether any of their discoveries could be applied to the fossil record. Other paleontologists held their own views, like the arrogant Henry Fairfield Osborn, who had his own peculiar view that evolution was driven by a mysterious internal force he called "aristogenesis," causing organisms to strive for greater and higher achievements. (His concepts were heavily influenced by his high position in the American aristocracy and his belief in his own superiority.) Osborn also believed that these internal forces could drive animals to develop structures (such as the gigantic antlers of the Irish elk, or the large canines of the sabertooth cat) that went beyond the control of natural selection (Fig. 5.2). In his opinion, animals could evolve out of control until they were actually maladaptive. Needless to say, very few other scientists took Osborn's ideas seriously.

By the early 1930s, anarchy ruled evolutionary biology. Geneticists were in one corner, doing their own research, and virtually ignoring natural selection. Paleontologists had almost completely given up reconciling the biological aspects of evolution with their knowledge of the fossil record. Systematists were busy describing new species, but few saw the evolutionary implications of their work. Through all this, Darwinian natural selection was supported by only a minority of scientists, even though the reality of evolution had become even more convincing than it was in 1859. What was most needed was a way of showing that natural selection was compatible with the understanding of genetics that had developed by then.

Into this breach stepped three mathematicians and geneticists—Sir Ronald Fisher, J.B.S. Haldane, and Sewall Wright—who introduced mathematical rigor to the problem. In a series of papers, they developed the foundations of **population genetics**, which uses mathematical models to determine how factors such as selection and mutation affect the changes in gene frequencies within populations. Through these mathematical experiments, they showed that natural selection could make major changes in the gene pool, even if it worked very slowly with very low rates of mutation. By the late 1930s, several scientists wrote books that pulled all these threads together, although the publication of several of these books was delayed by World War II. In 1937, the geneticist Theodosius Dobzhansky published *Genetics and the Origin of Species*, which synthesized all that was known of genetics at the time and showed that it was consistent with Darwinism. In 1942, Ernst Mayr published *Systematics and the Origin of Species*, which addressed the problem of natural selection in wild populations and laid the groundwork for his allopatric speciation model (see Chapter 3). And in 1944, paleontologist George Gaylord Simpson published *Tempo and Mode in Evolution*, which attempted to reconcile the fossil record with Darwinian natural selection. In the late 1940s and 1950s, many other scientists joined this emerging consensus, so that Julian Huxley described their ideas as the "the modern synthesis," or as others labeled it, **Neo-Darwinism**. By the time of the centennial celebration of *The Origin of Species* in 1959, the Neo-Darwinian synthesis dominated evolutionary biology, and most evolutionists thought that all the major problems had been solved. Most current evolution textbooks still reflect a strict Neo-Darwinian approach.

What are the central ideas of the Neo-Darwinian synthesis? As we have seen, it is basically the synthesis of modern genetics with Darwin's idea of natural selection as the primary mechanism of evolutionary change. Decades of genetics experiments (especially with fruit flies) has revealed an enormous amount about how selection can work on populations and cause them to change. Each organism has its own **genotype** (genetic code) that results in a **phenotype** (the physical characteristics of the organism that develop from those genes). Neo-Darwinism views populations as gene pools, and the variability in gene frequencies results in different phenotypes in the population. Natural selection then acts on those phenotypes to weed out the less fit, and this selection results in changing gene frequencies in the population. In its more extreme forms, Neo-Darwinism defined evolution simply as the gradual substitution of gene frequencies through time. What happens after fertilization or during embryology was almost irrelevant, because the focus was entirely on how gene frequencies can be affected by natural selection. Some biologists, like Richard Dawkins, take it even further. In *The Selfish Gene*, he argues that the body is simply a device for genes to make more copies of themselves.

From population genetics and fruit fly studies, Neo-Darwinians argued that the primary source of variability was random substitutions of base pairs in the genotype plus recombination when two different strands of DNA get together during sexual reproduction. Evolution had no internal or inherent directionality, but was simply a process of weeding out random variants. Rates of phenotypic change directly reflect the strength of selection; organisms are constantly being changed by the selective forces in their environment. In some versions, natural selection was treated as an all-powerful, all-pervasive force that, in Darwin's words, "is daily and hourly scrutinizing throughout the world every variation, even the slightest; rejecting all that which is bad, preserving and adding up all that is good; silently and insensibly working." More extreme Neo-Darwinians asserted that *all changes are adaptive* in some way, even if we can't detect them; there are no features in the organism that are not affected by natural selection. Such a belief in the all-powerful force of natural selection is called **panselectionism**.

Coupled with this reductionist attitude that organisms can be reduced to their genes was an attitude that the small genetic changes observed in fruit flies and lab rats could be extrapolated indefinitely upward to explain all evolutionary changes. All of evolution was essentially defined as **microevolution**, or the small-scale phenotypic changes (such as the change in the wing veins of a fruit fly) that occur in a few generations by natural selection. Gradual changes in phenotype, added up over hundreds of generations, produced new anatomical features and eventually new species and new fundamental body plans. These four "isms"—reductionism, panselectionism, extrapolationism, and gradualism—were central to orthodox Neo-Darwinism of the 1950s and 1960s and are still followed by a majority of evolutionary biologists today.

Neo-Darwinism had a great many successes that demonstrated the strengths of their approach. Decades of genetic experiments had shown how new characters appear and how populations respond to selection. Natural selection has been documented in the wild many times. The most famous example was the case of the peppered moth, *Biston betularia*. Normal peppered moths, as their name implies, are a speckled black and white color, but there is also a dark mutant. The normal moth is well camouflaged when it sits on the speckled background of a tree trunk. During the Industrial Revolution, the tree trunks in central England became black with soot, and the normally peppered moths were suddenly conspicuous against this black background, while the dark forms were well concealed from bird predators. As expected, the frequency of normal varieties decreased, and the black varieties came to dominate the population. After environmental laws helped reduce air pollution and the sooty tree trunks disappeared, the normal variety returned and the black variant became rare again.

The case of sickle-cell anemia demonstrates another aspect of Neo-Darwinism. Population geneticists were long puzzled how a lethal gene, even if it were a rare recessive, could still manage to occur within a population. This question is important, for it goes back to Darwin's dilemma of how a rare variety (or in Neo-Darwinian terms, a rare gene) can persist for generations when selection works against it and then reappear when selection changes to favor it. Sickle-cell anemia is coded by a fairly simple dominant-recessive genetic system. Most individuals have no copies of the sickle-cell gene. In genetic terms, their genotype has two copies of the normal dominant gene, symbolized "AA." Since they have two copies of the same gene from their two parents, they are considered **homozygous**. Individuals with the sickle-cell condition are homozygous for the recessive gene, symbolized "aa." This homozygous recessive condition leads to sickle-cell anemia, which is eventually lethal. If these individuals mate with normal individuals, half their offspring will be **heterozygous**, carrying a copy of both the dominant and recessive gene (symbolized "Aa"). Since the normal gene ("A") is dominant, these individuals will not have sickle-cell anemia.

In tropical regions with a high incidence of malaria, there is an unusually high frequency of the sickle-cell gene (Fig. 5.3). Why might this be? It turns out that heterozygous individuals ("Aa") have a greater resistance to malaria than normal individuals. Thus, when "AA" individuals are selected against by malaria and "aa" individuals die from sickle-cell anemia, the healthiest individuals are the heterozygotes (this is known as "heterozygote advantage"). When two heterozygous individuals mate, one-quarter of their offspring will have the lethal "aa" combination and die. But as long as strong selection by malaria maintains this advantage, the heterozygotes will continue to dominate the population. Heterozygote advantage is one mechanism that allows rare, even lethal, genes to persist in a population through many generations.

Another classic study demonstrates how a variety of phenotypes can be maintained in a population, even in the face of natural selection. A common European snail, *Cepaea nemoralis*, was long known for the enormous variation of shell colors it exhibits; this is known as a color **polymorphism**. Why didn't selection just pick the "best" color variety and weed out the rest? It turns out that different color varieties are better camouflaged in different habitats. In the grass, the light-colored and banded snails were more difficult to see; in the woods, the solid dark shells are less conspicuous. Since the snails live in all these habitats, they must maintain a high diversity of shell colors so that some individuals will survive, no matter what their background. How do we know that natural selection is causing this? Cain and Sheppard (1954) looked at the "thrush anvils," rocks that birds had used to break snail shells and eat their contents. In the wooded areas, there were many light-colored and banded shell fragments, but few dark shells; in the grasslands, the opposite was the case. Clearly, the more conspicuous shells had been selected against in each habitat, and the variety of habitats was responsible for maintaining the high polymorphism in these snails.

There are many other classic studies that could be described here, but other textbooks that focus on Neo-Darwinism give these cases in detail. In a paleontology textbook, however, we want to know how well Neo-Darwinism has held up to the scrutiny of the fossil record. After all, fossils are the only direct record of how evolution actually did take place, and this record is not dependent upon extrapolation from a few generations of living organisms. Experiments with fruit flies are fine, but what do the fossils say?

CHALLENGES TO THE NEO-DARWINIAN SYNTHESIS

> *The variations detected by electrophoresis may be completely indifferent to the action of natural selection. From the standpoint of natural selection they are neutral mutations.*
>
> Richard Lewontin, *The Genetic Basis of Evolutionary Change*, 1974

> *You have loaded yourself with an unnecessary difficulty in adopting* Natura non facit saltum *so unreservedly.*
>
> Thomas Henry Huxley, in a letter to Charles Darwin, 1859

> *The characters of subspecies are of a gradient type, the species limit is characterized by a gap, an unbridged difference in many characters. This gap cannot be bridged by theoretically continuing the subspecific gradient or cline beyond its actually existing limits. . . Subspecies are actually, therefore, neither incipient species nor models for the origin of species. They are more or less diversified blind alleys within the species. The decisive step in evolution, the first step towards macroevolution, the step from one species to another, requires another evolutionary method than that of sheer accumulation of micromutations.*
>
> Richard Goldschmidt, *The Material Basis of Evolution*, 1940

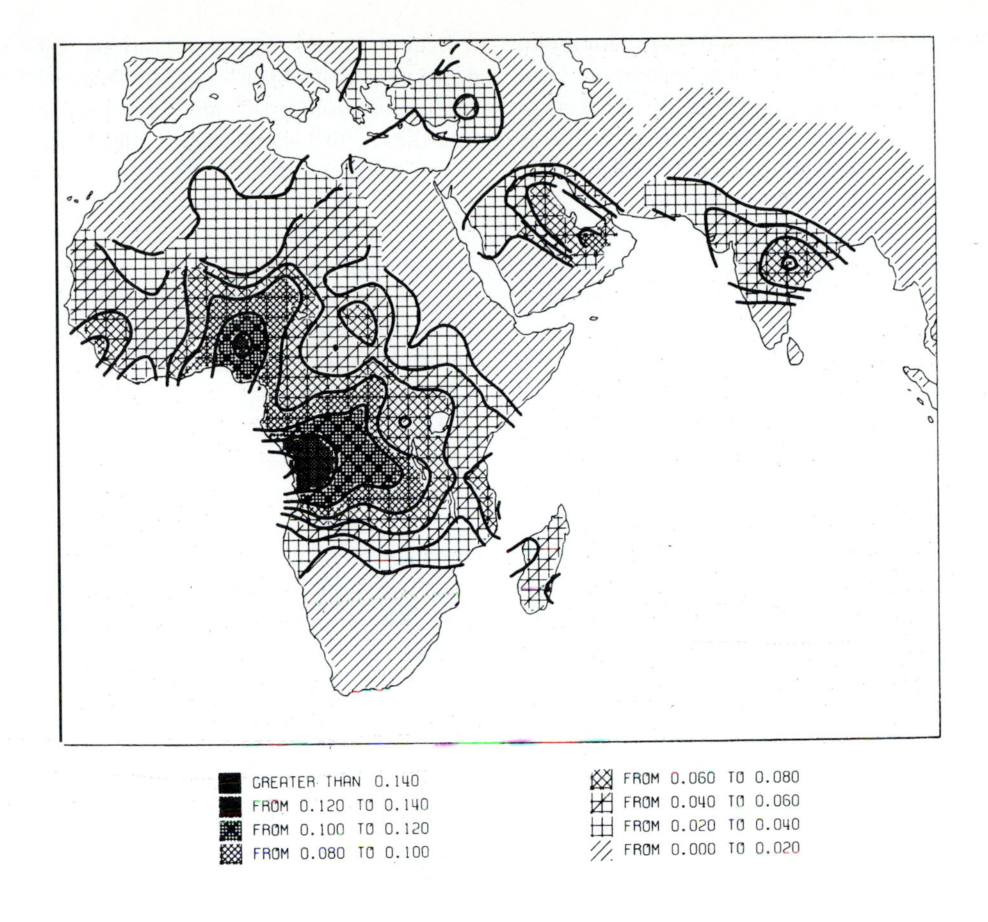

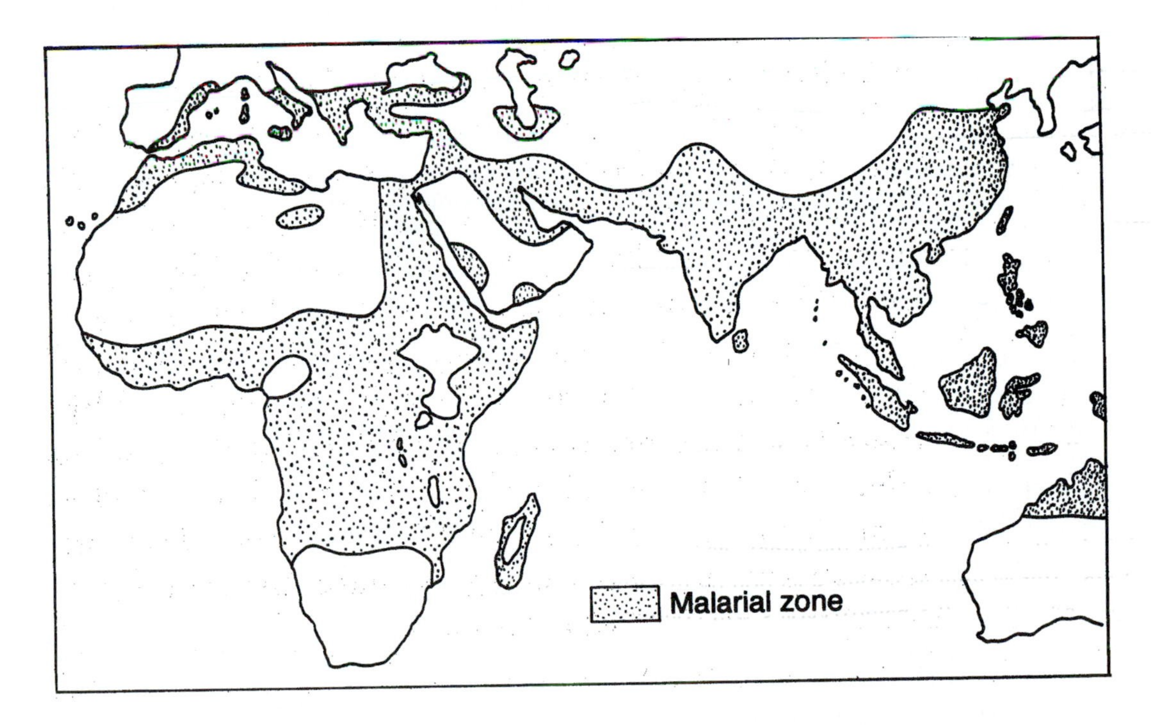

Figure 5.3. (Top) Frequency of the gene for sickle-cell anemia in Africa, southern Asia, and Europe. This distribution closely matches the incidence of malaria (bottom), suggesting that malaria selects for sickle-cell gene through heterozygote advantage. (After Ridley, 1993.)

Neo-Darwinism swept through evolutionary biology in the 1940s and 1950s and achieved near unanimity. But whenever a scientific field seems to have all the answers, it is not necessarily a good thing. New unsolved problems and controversies are essential to scientific progress. Unchallenged orthodoxy and dogma can become stale, since the practitioners are no longer asking tough questions that might challenge their view of the world and are no longer submitting their core hypotheses to critical testing. Contrary to popular belief, science does not reach final truth, but is always being revised as new discoveries overturn old hypotheses. If scientists stop questioning things, then a field becomes more dogma than science.

Throughout the last four decades, there have been numerous scientific ideas and discoveries that have challenged the dogmas of classical Neo-Darwinism. Some have been accommodated into a more expanded, inclusive version of the synthesis. Others, however, are still very controversial, and evolutionary biologists have fought hard against these ideas because they challenge the very foundations of the orthodoxy. In the following discussion, remember one essential point: none of these controversies disputes the *fact* that evolution has occurred. They are only about whether Neo-Darwinism is the only *mechanism* by which it has occurred.

The Inheritance of Acquired Characters Revisited

As we saw earlier, the idea that acquired characters can be passed on from parent to child was widespread in the days before modern genetics. It has been mislabeled "Lamarckism" but it was believed even by Darwin. The appeal of this idea is obvious. It allows an advantage to be acquired in just one generation in direct response to the environment so that organisms can adapt rapidly on demand. Classic Darwinian selection, on the other hand, is very slow and wasteful. Many offspring are born so that only a few will survive with the favorable variations, and many generations are required for a whole new population to become established as a new species.

After Darwin's ideas transformed biology, however, geneticists began to test the idea of acquired inheritance more rigorously. The German biologist August Weismann (1834-1914) ran a series of experiments in the 1880s that seemed to discredit acquired inheritance once and for all. Weismann cut off the tails of 20 generations of mice, but each succeeding generation of mice had normal tails, not shorter tails in response to this extreme environmental pressure. Weismann concluded that changes that occur in our phenotype ("*soma*" in Weismann's terminology) cannot ever get back into the genotype (what Weismann called the "*germ line*"). In other words, the flow of information is strictly one way. Changes in the genotype dictate changes in the phenotype, but not the other way around. This came to be known as the "central dogma" of molecular genetics. Later, DNA discoverer James Watson redefined the "central dogma" to mean the one-way path from DNA to RNA to protein.

Over the years, however, various scientists have proposed ideas that appear to violate the central dogma. In the 1950s, embryologist Conrad Waddington subjected larval fruit flies to heat shock and produced a mutation where the wings lacked a cross-vein. This procedure was carried on generation after generation, until after 14 generations, cross-veinless flies appeared without administering heat shock. Had the environmental stress somehow changed the genotype directly, rather than by selection? Waddington called this phenomenon **genetic assimilation**, and Neo-Darwinists still argue over how to explain it without Neo-Lamarckism.

More recently, immunologists have conducted experiments that seem to show acquired inheritance. Whenever an organism is exposed to a disease, its immune system develops antibodies that kill the foreign infection. This immunity is acquired during one's lifetime and should not be able to work its way back into the genome. However, experiments have shown that laboratory mice could pass on their immunity to their offspring (Gorczynski and Steele, 1980; Steele, 1981). Although Neo-Darwinists are still arguing that this can be explained by non-Lamarckian means, it raises serious questions about the inviolability of the germ line.

In molecular biology, more and more examples have been documented where the genes can be changed after the organism is born. Barbara McClintock won the Nobel Prize for her discovery of "jumping genes" that move from one spot on the DNA strand to another, changing the gene code. Other experiments have shown that external DNA can be incorporated into a cell and possibly into the host DNA. In one case, different bacteria appeared to exchange bits of genetic material, a switch that allowed them to all evolve a new mutation rapidly. A recently discovered group of viruses called retroviruses (such as the HIV virus that causes AIDS) copy their own genetic information into the DNA of the host. Could this be the mechanism that allows environmental changes to be translated directly back into the genetic code?

John Campbell (1982) has summarized an entire range of studies that suggest some form of environmental sensitivity of the genome, if not true acquired inheritance. At the lowest level of these "structurally dynamic" genes are those that have specific areas that respond to external circumstances by turning on or off a certain enzyme. At a higher level are cases of genes that apparently sense their environment, and change their structure in response to the external conditions. Even more sophisticated are automodulating genes, which change their future responsiveness to stimuli when stimulated. Finally, the most Lamarckian of all are called "experiential genes," which transmit specific modifications induced in the phenotype to their descendants. The example from immunology is one of several such cases that may fit this description.

Although hard-core Neo-Darwinians are still debating and disputing the implications of these studies, it is now clear that the genome is far more complicated and flexible

First base in the codon	Second base in the codon				Third base in the codon
	U	C	A	G	
U	Phenylalanine	Serine	Tyrosine	Cysteine	U
	Phenylalanine	Serine	Tyrosine	Cysteine	C
	Leucine	Serine	Stop	Stop	A
	Leucine	Serine	Stop	Tryptophan	G
C	Leucine	Proline	Histidine	Arginine	U
	Leucine	Proline	Histidine	Arginine	C
	Leucine	Proline	Glutamine	Arginine	A
	Leucine	Proline	Glutamine	Arginine	G
A	Isoleucine	Threonine	Asparagine	Serine	U
	Isoleucine	Threonine	Asparagine	Serine	C
	Isoleucine	Threonine	Lysine	Arginine	A
	Methionine	Threonine	Lysine	Arginine	G
G	Valine	Alanine	Aspartic acid	Glycine	U
	Valine	Alanine	Aspartic acid	Glycine	C
	Valine	Alanine	Glutamic acid	Glycine	A
	Valine	Alanine	Glutamic acid	Glycine	G

Table 5.1. The genetic code, which specifies by three letters in the genome (A = adenine; C = cytosine; G = guanine; U = uracil) any one of 20 amino acids, or a stop or start command.

than the original static entity visualized by Weissman. As molecular biology and immunology find more and more exceptions to the central dogma, the once disreputable idea that organisms can respond to environmental stresses by changing their genomes directly may no longer be so outrageous.

Neutralism

One of the central dogmas of evolutionary biology is panselectionism, the idea that every feature, no matter how minute, is always under the scrutiny of natural selection. By definition, this view denies that any feature can be selectively neutral. Before the 1960s, geneticists thought that every gene coded for a single protein and that each protein (and the structures built from them) was under the influence of natural selection. This was known as the "one gene, one protein" dogma.

In 1966, however, a series of discoveries led to a new school of thought known as **neutralism**, or the idea that some changes are adaptively neutral. The first research was published in 1966, when Lewontin and Hubby used the new technique of electrophoresis (see Chapter 4) and discovered that organisms were much more genetically variable than ever imagined. Study after study showed that organisms have far more genes than they actually use or that can be expressed. If these genes are silent and not expressed, then how can selection act upon them?

At one level, there is no question that most genetic changes are adaptively neutral. For example, the 64 possible combinations of the genetic code (Table 5.1) specify only 20 amino acids plus a few "stop" codes. In most cases, the first two nucleotides of the triplet determine which amino acid is produced, and a change in the third nucleotide makes no difference. For example, if the first nucleotide is a uracil (U) and the second is a cytosine (C), the amino acid serine will be produced, no matter what the third nucleotide is. Clearly, the great majority of changes in nucleotide sequences make no difference if they do not change the amino acid that is produced.

If neutralism is correct and most genetic changes occur without interference from natural selection, then random mutations can accumulate without being weeded out. They might eventually accumulate to the point where they may code for a new phenotypic feature. At this point, they are no longer neutral, and selection may occur to favor or suppress this random variant. However, it is clear that new features can arise by random mutations without the action of selection.

Traditional Neo-Darwinians concede that a large part of the genome may be adaptively neutral, but they argue that much of the supposedly "neutral" genetic information may provide subtle selective advantages that we cannot yet detect. Experiments have found unexpected selective advantages for variations that were previously thought to be useless. Nevertheless, the old dogma of "one gene, one protein" has clearly been discredited. Not only do the majority of genes code for no protein at all, but other genes code for more than one protein (**pleiotropy**). If selection maintains a certain pleiotropic gene because it codes for one very important feature, then all the other features it determines may be passively "carried along" even if they are selectively neutral or slightly harmful. In this case, natural selection might even allow harmful features to persist simply because they are tied to more important features with strong positive selective advantage.

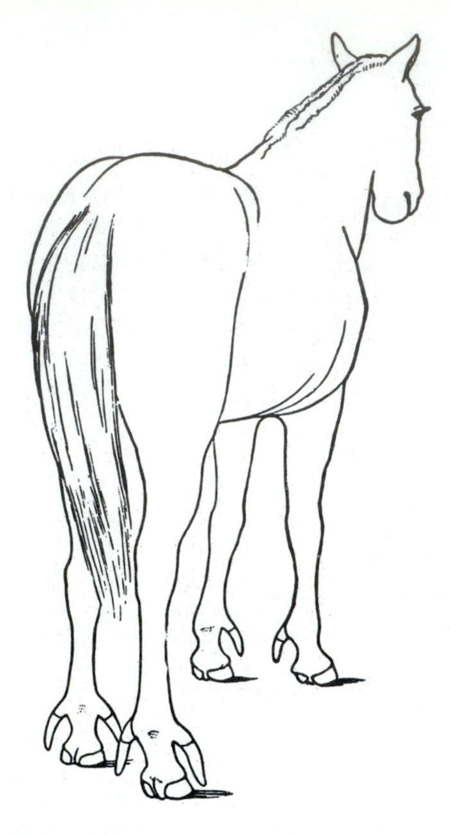

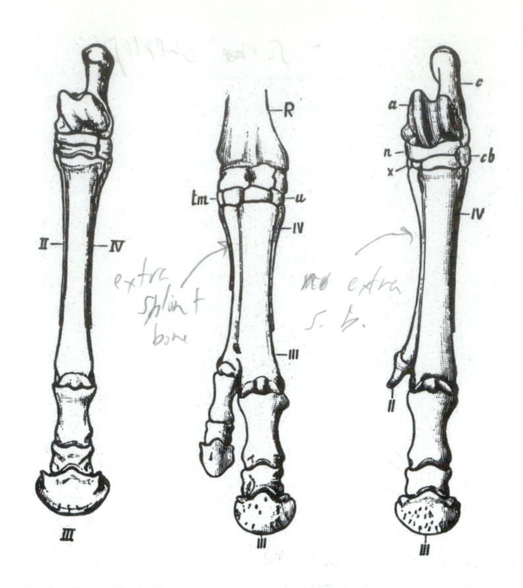

Figure 5.4. (Left) A famous example of rare mutant horse with three toes, rather than one. (Above) Bony structure of the foot of these mutant horses. On the left is a normal foot; in the middle is an extra toe formed by duplicating the central toe; on the right is an extra toe formed by enlarging the reduced side splint bones (which were once the functioning side toes in the ancestral horses). (From Marsh, 1892.)

Structural and Regulatory Genes

The discovery that most of the genome apparently codes for nothing has led to some startling changes in our understanding of genetics. Most of the genes in your chromosomes are considered **structural genes**. These code for each individual protein and structure in the body, and many apparently no longer code for anything (these are often called "junk DNA"). Thirty to 70% of the mammalian genome consists of repetitive sequences, some repeated hundreds of times, that apparently yield nothing. Whether or not a given structural gene will be expressed is determined by **regulatory genes**, which are the "switches" that turn on or off the expression of a given gene. A small mutation in a regulatory gene can make a radical change in the expression or silence of a whole string of structural genes that it controls. For example, humans still have the genes for the monkey-like tail, and every once in a while the gene regulation fails and a human is born with a tail. Horses still carry the structural genes of their three-toed ancestors, since occasional mutant horses are born with three toes on their feet (Fig. 5.4). Although no living bird has teeth, they still have the genes of the ancestral tooth-bearing birds. When embryonic tissues from a chick were grafted onto the mouth area of a developing mouse, the mouse developed teeth—not those of a mouse, but peglike reptilian teeth found in the earliest birds.

One of the most striking demonstrations of the difference between structural and regulatory genes can be seen in ourselves. When DNA-DNA hybridization studies of humans and chimpanzees were done, they were found to be 98% identical. This means that no more than 2% of our DNA is regulatory genes that are different from the chimp, and they must be responsible for all our phenotypic differences. The remaining 98% are structural genes, some of which code for a wide variety of anatomical features we no longer exhibit (such as the monkey's tail and many other aspects of our primate relatives). According to Jared Diamond, we are so genetically similar to the other two living species of chimpanzee that if an alien biologist knew us only from our DNA sample, it would class us as a "third chimpanzee."

The most important breakthroughs in genetics have come from the study of **homeotic genes**, which make huge and abrupt transformations by producing normal anatomical parts in odd places. For example, most flies have two sets of wings and two balancing organs called halteres where the second pair of wings would be. A homeotic mutation changes these halteres back into the ancestral wings, producing four-winged flies. Another homeotic mutation produces the "antennipedia" condition, where a normal fly grows a leg out of its head in place of an antenna (Fig. 5.5). More recently, it has been discovered that a series of homeotic genes known as the *Hox* complex control the basic segmentation of the body, not only of arthropods but also of vertebrates. Clearly, homeotic genes are fundamental to the body plans of almost all animals, and a small change in the homeotic genes can have huge effects on the phenotype, producing new body plans, extra limbs, or extra segments in only a single generation. This directly contradicts the old Neo-Darwinian assertion that novel features arose only by gradual selection over many generations.

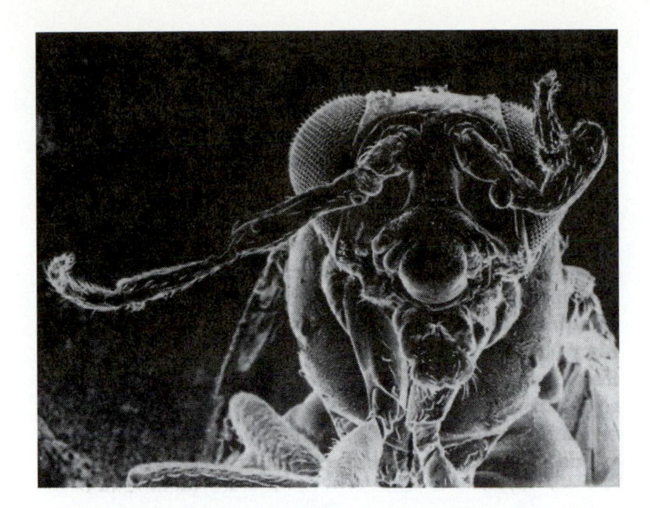

Figure 5.5. The homeotic antennipedia mutation changes the embryonic development of appendages, so that the antenna of this fly has developed into a limb instead. (SEM photo courtesy F. M. Turner.)

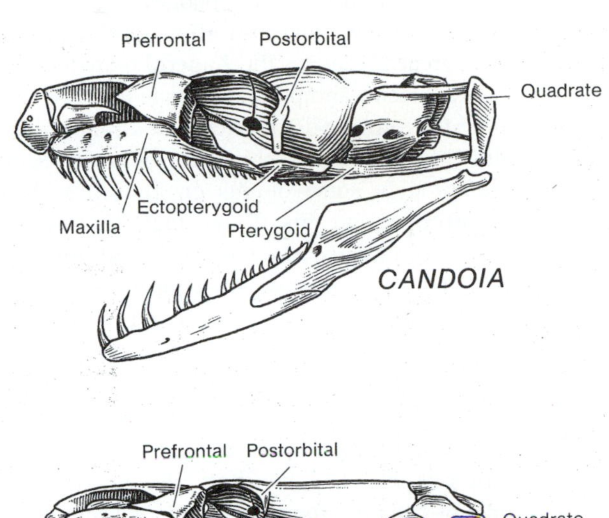

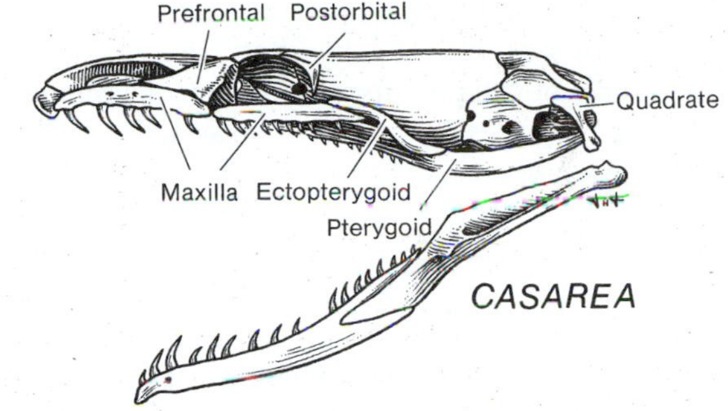

Figure 5.6. At the top is a normal boa constrictor of the genus *Candoia,* and at the bottom is one of the unique bolyerine snakes of Madagascar, *Casarea.* The bolyerine has a hinge between the two maxillary bones of the upper jaw that allows it to flex its skull around the prey when it is swallowing. Frazzetta (1975) argued that some complex adaptations cannot have evolved from gradual transformation, but require an abrupt change from one functional state to another. In this case, if a normal snake thinned its maxillary bone more and more, it would eventually break, not evolve the hinge mechanism found in bolyerines. Thus, the change must have been abrupt. (From Frazzetta, 1970.)

Hopeful Monsters and Macroevolution

A number of Darwin's critics were in favor of evolution by natural selection but did not like his emphasis on gradual change. Darwin firmly believed that evolution must occur in "infinitely numerous transitional links" forming "the finest graduated steps." Darwin's friend and ally, Thomas Henry Huxley, told him, "You have loaded yourself with an unnecessary difficulty in adopting *Natura non facit saltum* [nature does not make jumps] so unreservedly." Nevertheless, the Neo-Darwinian school of thought inherited Darwin's gradualistic bias in their idea that evolution was caused by gradual gene substitutions, and no more. Speciation was caused by the continual gradual accumulation of variation, extrapolated on and on until we recognize the descendant as a new species.

Over the years, a number of biologists have agreed with Huxley. In their view, the origin of major adaptive features, such as the origin of a bird wing (**macroevolution**), is a different process from the small-scale changes in the wing veins or bristles on a fruit fly (microevolution). There are many examples of evolutionary transformations that are difficult to imagine as a gradual stepwise process. At some point, an abrupt change is required. One of the best cases was described by Thomas Frazzetta in his book *Complex Adaptations in Evolving Populations.* On the tiny island of Mauritius, east of Madagascar, live snakes known as the bolyerines (Fig. 5.6). They are descended from the boa contrictors, but they have a unique feature: the upper jawbone is hinged in the middle. Frazzetta points out that it is impossible to imagine a gradual transition from the solid boid jawbone to the hinged bolyerine jaws. If the boid jaws became thinner and thinner, they would break and the snake would die; it wouldn't evolve a hinge. Frazzetta argued that only a rapid change between the solid jaw and the hinged jaw would keep the snakes functional and alive.

One of the major dissidents from the synthesis during the 1940s and 1950s was a geneticist named Richard Goldschmidt (1878-1958), who fled Nazi persecution and spend the rest of his career at Berkeley, where he wrote *The Material Basis of Evolution* (1940). Goldschmidt was an expert on the genetics and development of the gypsy moth caterpillar, famous for its destruction of forests all over the eastern United States. In his studies, he found that the variants of these insects appeared quite abruptly and discontinuously, and he suggested that these changes might be due to "controlling genes" (now called regulatory genes). In Goldschmidt's view, speciation was a discontinuous, rapid process that was caused by small changes in these "controlling genes." If a new **macromutation** appeared to give an organism a great advantage, this **hopeful monster** could establish a new species abruptly rather than gradually.

Naturally, the gradualistic Neo-Darwinians hated Goldschmidt's ideas and subjected them to scorn. Goldschmidt brought some of this on himself with his combative attitude and his claim that his form of macroevolution discredited Darwinism (it only discredits gradualism). To the orthodoxy, the central question was always, "How does the hopeful monster find a mate?" Without one, the "hopeful monster" could not breed or establish a population, and therefore no new species can arise.

Ironically, the last few years of genetics have shown that Goldschmidt was ahead of his time. Regulatory genes have become of central importance, yet Goldschmidt postulated them decades before they were actually discovered. In addition, embryology has come to Goldschmidt's aid. A number of studies have shown that an environmental influence on early embryology, such as a heat shock, can cause many organisms to go through the same new path in embryonic development. If they mature, they will all be "hopeful monsters," and there will be no problem finding a mate with a similar macromutation.

This is why the studies of heterochrony discussed in Chapter 2 are so important to evolution. As we saw with the example of the axolotl, a small change in the regulatory gene that controls developmental timing can make a big change in the phenotype of the organism. All it takes is stagnant water, and the gilled axolotl becomes a lunged salamander, occupying a completely different ecological niche. No great rearrangement of the genome is required to make a new body plan through a major change in development.

A number of evolutionists are now convinced that macroevolutionary change is more than just microevolution extrapolated upward. In their view, the establishment of new body plans and macromutations to form new species is fundamentally different from the processes that control microevolution. Gould (1980) called this the "Goldschmidt break," in honor of the much-maligned scientist who first proposed it. If such macroevolutionary ideas are correct, then decades worth of experiments on fruit flies have little relevance to the interesting part of evolution, namely, the origin of new species and body plans. Most of what is now studied by evolutionary biologists is just "noise" in the long run. Naturally, the orthodoxy does not take well to the idea of decades of their work becoming irrelevant, so there is much resistance to these ideas in conventional evolutionary literature. The best place to resolve this issue is to see whether the fossil record fits a macroevolutionary viewpoint or not. That brings us to the ultimate goal of this chapter: what *does* the fossil record say about evolution?

EVOLUTION AND THE FOSSIL RECORD

Not long ago paleontologists felt that a geneticist was a person who shut himself in a room, pulled down the shades, watched small flies disporting themselves in milk bottles, and thought that he was studying nature. A pursuit so removed from the realities of life, they said, had no significance for the true biologist. On the other hand, the geneticists said that paleontology had no further contributions to make to biology, that its only point had been the completed demonstration of the truth of evolution, and that it was a subject too purely descriptive to merit the name "science." The paleontologist, they believed, is like a man who undertakes to study the principles of the internal combustion engine by standing on a street corner and watching the motor cars whizz by.

George Gaylord Simpson, *Tempo and Mode in Evolution*, 1944

Experimental biology . . . may reveal what happens to a hundred rats in the course of ten years under fixed and simple conditions, but not what happened to a billion rats in the course of ten million years under the fluctuating conditions of earth history. Obviously, the latter problem is more important.

George Gaylord Simpson, *Tempo and Mode in Evolution*, 1944

Why be a paleontologist if we are condemned only to verify what students of living organisms can propose directly?

Stephen Jay Gould and Niles Eldredge,
"Punctuated equilibria: the tempo
and mode of evolution reconsidered," 1977

As mentioned earlier in this chapter, paleontologists have always been uneasy with their role in the community of evolutionary biologists. Originally, they actively opposed Darwinian natural selection (Owen), or were Lamarckians (Cope and Hyatt), or proposed their own ideas (Osborn), or were agnostic about the mechanism of evolution (for example, William Berryman Scott), even as they found more and more evidence that evolution had occurred. Gould (1983) describes this as the period of "irrelevance," when paleontologists contributed very little to evolutionary theory.

This situation changed completely in 1944 with the publication of George Gaylord Simpson's *Tempo and Mode in Evolution*. Simpson was in contact with Dobzhansky and Mayr and was aware of the new developments in genetics and evolution. With *Tempo and Mode in Evolution*, Simpson set out to undo all the confusion about the nature of the fossil record caused by the speculations of people like Cope and Osborn. Using his vast first-hand knowledge of fossils, Simpson tried to show that nothing in paleontology was inconsistent with the concepts of Neo-Darwinism, as they were just emerging. In some respects, he was *too* successful. Simpson gave the evolutionary biology community no reason to believe that the fossil record contained anything new or unusual that they could not see better with their rats or fruit flies. Even though he protested that the scale of geologic time was far beyond the level of experimental biology (see the second Simpson quote on p. 74), he succeeded in putting paleontology in a subservient role (as Gould, 1983, describes it).

Only a few scientists registered any protest to this artificial unaninimity. One of these was the distinguished vertebrate paleontologist Everett C. Olson. During the 1959 Darwin centennial that proclaimed the victory of Neo-Darwinism, Olson (1960, pp. 527-531) wrote:

> *The statement is made, in effect, that those who do* not *agree with the synthetic theory do* not *understand evolution and are incapable of so doing, in most cases, because they think typologically . . . Some avid proponents of the synthetic theory would appear to . . . eliminate as competent students of evolution, because of their inability to understand* the *theory, those who may disagree. . The situation proposes a frustrating dilemma for the sincere student who feels from his observations that there is more to evolution than can be studied, tested, and integrated under the synthetic theory, who is confident that real problems exist but also sees no way of making progress toward an understanding by means of the materials that raise the questions in his mind. Few feel that the genetic-selection theory is invalid, but rather consider that there is much evidence that is not adequate.*

Unfortunately, few other paleontologists voiced opinions like those of Olson, and most followed Simpson's lead instead. During the 1960s, paleontologists had very little to contribute to evolutionary biology except for studies that were cast in strict Neo-Darwinian terms.

If the first two stages of paleontology's role in evolution theory might be described as irrelevance and subservience, then the third stage began with the paleobiological revolution in the early 1970s. In Chapter 3, we described how the idea of punctuated equilibrium began with a classic Neo-Darwinian concept, allopatric speciation theory, that was belatedly applied to our recognition of species in the fossil record. The prevalence of the belief in phyletic gradualism among paleontologists prior to 1972 is a testament to how far they were out of touch with the current ideas in evolutionary biology. But punctuated equilibrium had far more implications than the simple idea that speciation is geologically rapid. The prevalence of stasis in species over millions of years was something that was not expected by Neo-Darwinists. Even though paleontologists had known for years that most fossil species are static through long periods of geologic time, they never emphasized this, since they were taught to look for gradual evolution. As Gould and Eldredge (1977) put it, "Stasis is data." When paleontology's "dirty little secret" of the prevalence of stasis in the fossil record finally got out, it caused great problems for evolutionary theory.

Conventional Neo-Darwinism had always treated species as infinitely flexible and responsive to the environment. But fossil species showed no change across long periods of geologic time when there were clearly many changes in the environment. In some cases, well-documented and sometimes extreme climatic changes led to no changes in the fossils (Jackson, 1992; Prothero, 1992; Prothero and Heaton, 1996). Some Neo-Darwinists attempted to dismiss this evidence as an example of stabilizing selection (Charlesworth et al., 1983; Levinton, 1983; Lande, 1985). But such conventional models are appropriate only on scales of a few generations, or at most a few thousand years. No environment is so constant that stabilizing selection can operate for millions of years. From this evidence, biologists have had to reconsider their concept of species and organisms. More and more, they are treating organisms as integrated wholes, with complex interactions between their various parts, so that they cannot change one part without disrupting the whole organism. The old idea that selection could act on one character at a time just by changing its gene frequencies has been discredited. Evolution is more than a "change in gene frequencies through time."

The existence of static species, resistant to changes through millions of years, implies that species have a reality beyond the fact that they are composed of individuals in populations. Instead of being arbitrary slices of a continuum, species are distinct entities with their own history of "birth" (speciation) and "death" (extinction). Species are a hierarchical level above that of the levels of genes, individuals, and populations. Many evolutionary paleontologists are now saying that this requires a whole new spectrum of evolutionary theories to deal with processes operating at the species level. Gould (1980) coined the term "Wright break" for the gap between processes that operate on populations and those that operate on species, since this concept first emerged from the work of Sewall Wright.

Although glimmerings of this idea were present in the original punctuated equilibrium paper in 1972, it first emerged explicitly in a brief paper by Stanley (1975), followed by his provocative and controversial book *Macroevolution* (1979). Stanley called this concept "species selection," and it became the basis for a whole new round of debates about the mode of evolution. Since the original proposal, Vrba and Gould (1980) argued that it should be called **species sorting**, since the process is not really analogous to natural selection on the level of individuals and populations.

In a nutshell, the argument postulates that species are real entities that have characteristics that are more than just the sum of the characteristics of their component populations. When two or more species come into competition, the differential survival that sorts out the "winners" and "losers" may be due to intrinsic species properties rather than to natural selection on individuals or populations. The causes of survival of a given species cannot be reduced directly to the survival of its component populations, but seem to be due to properties that are inherent to the species.

For example, the tendency of a group to speciate rapidly or slowly is not a property of its component individuals. Organisms do not speciate, species do. Vrba (1980, 1985) has suggested that the evolution of African antelopes is an example of this. The relatively conservative impala clade seems to have an intrinsically low rate of speciation. Only three very similar species in one lineage are known for the last 5 million years. By contrast, the wildebeest tribe has speciated profusely during the same period of time, with multiple episodes of evolutionary branching and extinction. Hansen (1978, 1982) argued that marine snails without planktonic larvae speciated more rapidly than those who disperse their larvae as plankton all over the ocean. The less mobile nonplanktotrophic snails are more likely to be genetically isolated than species whose planktonic larvae spread their genes all over the ocean. Since the larval condition is a property of the species, not merely of its component individuals, it might represent an example of species sorting.

Two groups of South American burrowing rodents, the tuco-tucos (*Ctenomys*) and the coruros (*Spalacopus*) both have evolved adaptations for a burrowing, gopher-like existence (Vrba and Gould, 1986). Tuco-tucos are far more speciose than coruros, even though they have the same ecology and home range. The difference lies in the fact that tuco-tucos have very low gene flow, so they can speciate rapidly, while coruros are genetically homogeneous. Other possible examples of species sorting were reviewed by Gilinsky (1986).

Traditional Neo-Darwinists fail to see any difference between natural selection and species sorting (Mayr, 1992; Hecht and Hoffman, 1986; Hoffman, 1982, 1984, 1989, 1992). It is clear that the debaters are talking past each other, since each has fundamentally different perceptions of the world. Traditional Neo-Darwinists come from a reductionist viewpoint that cannot see species as entities, even after all the evidence has accumulated. Some paleontologists see the world as hierarchically ordered, with each level having its own reality. As long as this difference in world view underlies the argument, neither side will convince the other.

The primary point here is that if species are real entities with their own properties, then macroevolutionary processes operating on the level of species are not necessarily the same as those microevolutionary processes operating on the level of populations. Macroevolution is not just microevolution scaled up. Once again, the simple reductionism of the orthodoxy has been challenged.

CONCLUSION

The debates about the mechanisms of evolution are more than just trivial disputes among specialists. The stakes are enormous. Now that the Neo-Darwinian side has invested decades worth of research into the genetics of fruit flies and rats, it may turn out that the most interesting evolutionary phenomena can be studied only in the fossil record, or in the embryology lab. With publications, prestige, and grant money on the line, the conventional Neo-Darwinists do not want to become irrelevant to the most interesting issues in macroevolution. Paleontologists, in turn, have begun to shed their subservience to geneticists (Gould, 1983) and assert the importance of the fossil record for detecting phenomena that are too large in scale for biologists to observe. Clearly, all of evolutionary biology is undergoing ferment and change after the stagnation of the Neo-Darwininan orthodoxy of the middle part of this century. Gould (1980) asked, "Is a new and general theory of evolution emerging?" Many paleontologists and biologists would say, "Yes!" To paraphrase a Chinese curse, we indeed live in interesting times.

For Further Reading

Campbell, J. H. 1982. Autonomy in evolution, pp. 190-200 *in* Milkman, R., ed. *Perspectives on Evolution*. Sinauer Associates, Sunderland, Mass.

Darwin, C. R. 1859. *On the Origin of Species*. John Murray, London.

Dawkins, R. 1976. *The Selfish Gene*. Oxford University Press, New York.

Dawkins, R. 1982. *The Extended Phenotype*. Oxford University Press, New York.

Dawkins, R. 1987. *The Blind Watchmaker*. Norton, New York.

Desmond, A., and J. Moore. 1991. *Darwin: The Life of a Tormented Evolutionist*. Warner, New York.

Dobzhansky, T. 1937. *Genetics and the Origin of Species*. Columbia University Press, New York.

Eldredge, N. 1985. *Unfinished Synthesis*. Oxford University Press, New York.

Eldredge, N. 1989. *Macroevolutionary Dynamics: Species, Niches, and Adaptive Peaks*. McGraw-Hill, New York.

Eldredge, N. 1995. *Reinventing Darwin*. Wiley, New York.

Frazzetta, T. H. 1975. *Complex Adaptations in Evolving Populations*. Sinauer Associates, Sunderland, Mass.

Gilinsky, N. 1986. Species selection as a causal process. *Evolutionary Biology* 20:249-273.

Goldschmidt, R. 1940. *The Material Basis of Evolution*. Yale University Press, New Haven, Conn.

Gould, S. J. 1980. Is a new and general theory of evolution emerging? *Paleobiology* 6:119-130.

Gould, S. J. 1982a. The meaning of punctuated equilibrium and its role in validating a hierarchical approach to macroevolution, pp. 83-104, *in* Milkman, R., ed., *Perspectives on Evolution*. Sinauer Associates, Sunderland, Mass.

Gould, S. J. 1982b. Darwinism and the expansion of evolutionary theory. *Science* 216:380-387.

Gould, S. J., 1983. Irrelevance, submission, and partnership: the changing role of paleontology in Darwin's three centennials, and a modest proposal for macroevolution, pp. 347-366, *in* Bendall, D. S., ed., *Evolution from Molecules to Men*. Cambridge University Press, Cambridge.

Gould, S. J. 1992. Punctuated equilibria in fact and theory, pp. 54-84, *in* Somit, A., and S. A. Peterson, eds., *The Dynamics of Evolution*. Cornell University Press, Ithaca, New York.

Gould, S. J., and N. Eldredge. 1977. Punctuated equilibria: the tempo and mode of evolution reconsidered. *Paleobiology* 3:115-151.

Hoffman, A. 1989. *Arguments on Evolution*. Oxford University Press, New York.

Levinton, J. S. 1983. Stasis in progress: the empirical basis of macroevolution. *Annual Reviews of Ecology and Systematics* 14:103-137.

Lewontin, R. C. 1974. *The Genetic Basis of Evolutionary Change*. Columbia University Press, New York.

Mayr, E. 1942. *Systematics and the Origin of Species*. Columbia University Press, New York.

Mayr, E. 1982. *The Growth of Biological Thought*. Harvard University Press, Cambridge, Mass.

Milner, R. 1990. *The Encyclopedia of Evolution*. Facts on File Publications, New York.

Minkoff, E. C. 1983. *Evolutionary Biology*. Addison-Wesley, Reading, Mass.

Prothero, D.R. 1992. Punctuated equilibria at twenty: a paleontological perspective. *Skeptic* 1(3): 38-47.

Prothero, D. R., and T. H. Heaton. 1996. Faunal stability during the early Oligocene climatic crash. *Palaeogeography, Palaeoclimatology, Palaeoecology* 127:257-284.

Ridley, M. 1993. *Evolution*. Blackwell, Cambridge.

Simpson, G. G. 1944. *Tempo and Mode in Evolution*. Columbia University Press, New York.

Somit, A., and S. A. Peterson, eds., *The Dynamics of Evolution*. Cornell University Press, Ithaca, New York.

Stanley, S. M. 1975. A theory of evolution above the species level. *Proceedings of the National Academy of Sciences* 72:646-650.

Stanley, S. M. 1979. *Macroevolution: Patterns and Process*. W.H. Freeman, New York.

Strickberger, M. W. 1996. *Evolution* (2d ed.). Jones and Bartlett, Sudbury, Mass.

Vrba, E. S. 1980. Evolution, species, and fossils: how does life evolve? *South African Journal of Science* 76: 61-84.

Vrba, E. S., and S. J. Gould. 1986. The hierarchical expansion of sorting and selection: sorting and selection cannot be equaled. *Paleobiology* 12:217-228.

Wesson, R. 1991. *Beyond Natural Selection*. MIT Press, Cambridge, Mass.

Wills, C. 1989. *The Wisdom of the Genes, New Pathways in Evolution*. Basic Books, New York.

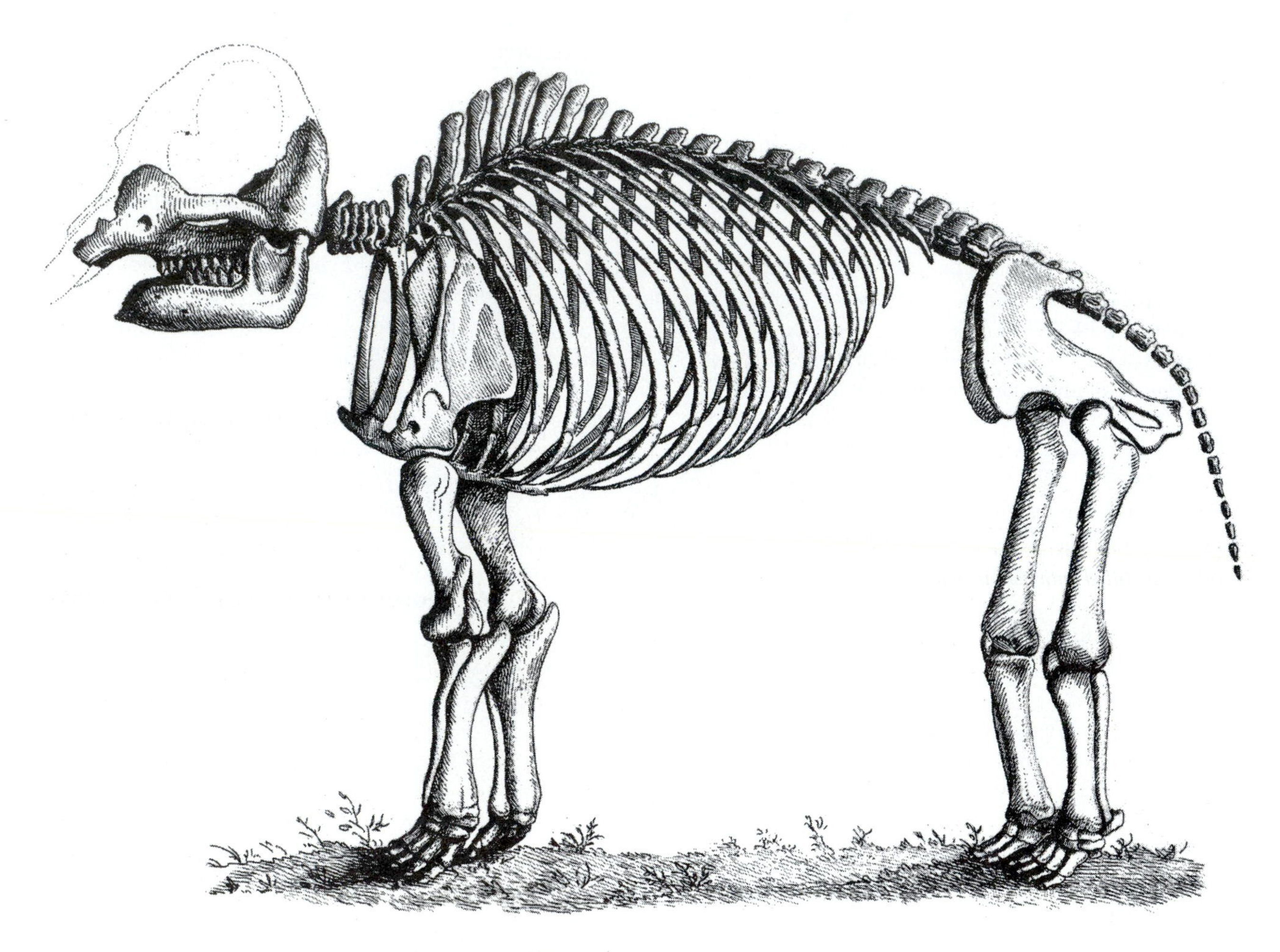

Figure 6.1. Cuvier's (1806) reconstruction of the American mastodon. Not only was this one of the first successful reconstructions of an extinct animal, but it was also clear to Cuvier that it was no longer on this planet, and must be extinct. (From Cuvier, 1806.)

Chapter 6

Extinction

There are as many species as the infinite being created diverse forms in the beginning, which, following the laws of generation, produced as many others but always similar to them. Therefore, there are as many species as we have different structures before us today.

Linnaeus, *Philosophia Botanica*, 1751

Who sees with equal eye, as God of all,
A hero perish, a sparrow fall . . .
Where, one step broken, the great scale's destroy'd:
From Nature's chain whatever link you strike,
Ten or ten thousandth, breaks the chain alike.

Alexander Pope, *Essay on Man*, 1735

The extinction of species has been involved in the most gratuitous mystery. No one can have marvelled more than I have done at the extinction of species. When I found in La Plata the tooth of a horse embedded with the remains of Mastodon, Megatherium, Toxodon, *and other extinct monsters, which all co-existed with still living shells at a very late geological period, I was filled with astonishment.*

Charles Darwin, *On the Origin of Species*, 1859

THE FACT OF EXTINCTION

As naturalistic interpretations of fossils became more prevalent after the work of Steno and Hooke, natural historians worried that many of these "petrifactions" bore no resemblance to any living organism. In the 1600s, no one knew of a living relative of the strange ammonites or the odd cylinders we call crinoid stems. Yet the idea that these strange fossils might represent animals that were now extinct was rejected by most natural historians. The whole concept of extinction went against several deeply held Christian dogmas, such as the notion of Divine Providence. An omnipotent, benevolent God would never allow any creature to become extinct. If, as Alexander Pope wrote, God cared for the sparrow, certainly he would not allow any of his creations to become extinct. Another deeply held concept was that of a "Great Chain of Being" linking the animals to man to the angels to God. Breaking any link in that chain implied the destruction of the whole chain. Finally, the idea of extinction went against the Christian dogma of plenitude, or the "fullness of nature." Anything that God had created to complete and perfect nature could not possibly go extinct and render God's creation incomplete and imperfect.

By the mid 1700s, however, it was becoming obvious that the recently discovered fossils of mammoths found in Siberia had no living counterparts. The remote corners of the world were being explored, and although many new and surprising beasts were discovered, clearly the gigantic mammoths were not hiding in South America or Africa or the East Indies. Many strange fossils, such as the bones of hippopotami unearthed in Paris, were clearly related to tropical animals, but it was assumed that the bones had been washed out of the tropics during the Great Flood. The notion of extinction was still considered blasphemous.

The New World soon provided unequivocal evidence that these large bones were not simply gigantic humans. In 1739, Charles le Moyne, the second Baron de Longueil, left Montreal with French and Indian troops to fight the Chickasaw Indians along the Ohio River. Somewhere along the Ohio, he found the remains of what appeared to be three elephants. When the war ended in 1740 le Moyne collected the bones and shipped them to New Orleans and ultimately to Paris, where they came to the attention of French naturalists. In the 1740s and 1750s, English settlers in the region sent more of these bones from Big Bone Lick, Kentucky, off to England and also to Benjamin Franklin in America. Most of the bones (especially the tusks) were clearly like those of elephants and mammoths, but the teeth were puzzling. They were clearly unlike any living elephant, yet they were part of an animal of elephantine size. (We now know that these were specimens of the American mastodon.) Franklin speculated that the teeth were reminiscent of a carnivorous animal, although he and others later decided it was a vegetarian. In 1769, the famous British anatomist William Hunter took the carnivory suggestion seriously and suggested this was not a true elephant, but a "pseudelephant" or "American *incognitum*" (Latin for "unknown") that had independently developed ivory tusks. "This monster, with the agility and ferocity of a tiger. . . cruel as the bloody panther, swift as the descending eagle, terrible as the angel of night . . . . And if this animal was indeed carnivorous, which I believe cannot be doubted, though we may as philosophers regret it, as men we cannot but thank Heaven that its whole generation is probably extinct."

This very precocious suggestion was still not accepted by naturalists of the time. Nevertheless, remains of the "*incognitum*" and also of the Siberian mammoth (including frozen carcasses with hair and skin) were turning up again and again. Georges Louis Leclerc, the Comte de Buffon (1707-1788), concluded in his *Théorie de la terre* in 1749 that although most of the supposedly extinct animals were hiding somewhere in an unknown region, it was likely that the large terrestrial mammals such as the mammoth and "*incognitum*" had actually perished. By 1778, Buffon was relating their disappearance to his ideas of violent cataclysms in Earth's early history. During this time, the climate was warmer, so that polar regions had once been tropical and elephants (meaning mammoths) could live in Siberia. This implied a non-Biblical Earth of much greater antiquity. Buffon suggested it was as much as 75,000 to 3,000,000 years old, rather than the 6000 years demanded by most literalist Biblical scholars. Naturally, such revolutionary ideas were not popular with the theologians in the Sorbonne. Buffon was protected by the King, however, so he was not persecuted for his heresy, although his ideas were not widely accepted, either.

Thomas Jefferson was a firm believer in the Great Chain of Being. As he wrote in 1799,

> *The bones exist: therefore the animal has existed. The movements of nature are in a never-ending circle. The animal species which has once been put into a train of motion, is probably still moving in that train. For if one link in nature's chain might be lost, another and another might be lost, till this whole system of things should vanish by piece-meal . . . If this animal had once existed, it is probable on this general view of the movement of nature that it still exists.*

He was convinced the *incognitum* was still living out in the unexplored Northwest. As he wrote in 1781,

> *It may be asked why I* [list] *the mammoth as if it still existed? I ask in return, why should I omit it, as if it did not exist? Such is the economy of nature, that no instance can be produced of her having permitted any one race of her animals to become extinct; of her having formed any link in her great work so weak as to be broken. To add to this, the traditional testimony of the Indians that this animal still exists in the northern and western parts of America, would be adding the light of a taper to that of the meridian sun.*

When delays prevented Lewis and Clark from leaving on their expedition to the great Northwest until 1804, President Jefferson instructed them to go to Big Bone Lick and collect some more of the mysterious beast. When he received some gigantic claws from some cave deposits, he instructed Lewis and Clark to look for a gigantic lion in the American West. (They turned out to be claws of the extinct ground sloth now known as *Megalonyx jeffersoni*, not those of a lion.)

The fact of extinction was finally proved by one of the greatest biologists of all time, the Baron Georges Cuvier (1769-1832). He was one of the outstanding figures in French science, surviving the reign of Louis XVI, the French Revolution and Reign of Terror, Napoleon, and subsequent French kings without loss of status or position. Cuvier was the founder of comparative anatomy and of vertebrate paleontology, developing a tremendous skill in describing and recognizing bones of vertebrates.

In 1796, Cuvier read a paper before the French Institute on living and fossil elephants. He was the first scientist to recognize the difference between the Asian and African elephant. Then he showed that the mammoth and the American *incognitum*, although related to elephants, were clearly not the same as living elephants and did not require that Earth be much warmer in the past (Fig. 6.1). Since they were different animals, it was possible that they lived in cold climates, unlike modern elephants. He pointed to the Siberian woolly rhinoceroses, the giant cave bear, the strange marine reptile from Maastricht (now known as a mosasaur, a giant marine monitor lizard from the Cretaceous), or the giant ground sloth from Patagonia he had just described. Where could such great animals be hiding today if they were still alive? They "prove the existence of a world before ours, and destroyed by some catastrophe." Cuvier went on to develop his ideas of great catastrophes that had preceded our world and were not mentioned in Genesis. This was the "antediluvian" world before the flood, a time of darkness, great monsters, and cataclysmic changes. Over his long career, Cuvier described many other important fossils, such as the palaeotheres from the Eocene gypsum deposits of the Paris Basin and the first pterodactyls. Although his notion of catastrophist geology was gradually replaced by the uniformitarian ideas of James Hutton and Charles Lyell in the 1830s and 1840s, his proof of extinction was never challenged. Indeed, more and more bizarre beasts, including the first dinosaurs, were described in the 1840s, driving home the reality of extinction beyond any doubt.

Lyell was one of the last scientists to resist the notion of

Figure 6.2. Satirical cartoon by Henry de la Beche, one of Lyell's geologist colleagues, showing Professor Ichthyosaurus lecturing on the strange creature from the previous creation (the human skull). Lyell was one of the last geologists to deny the reality of extinction, and to believe that earth history was cyclic, with extinct species returning in a later incarnation. Eventually, Lyell had to concede that the fossil record shows directional change, and extinct species never return.

extinction. In his early works, he embraced an extreme form of uniformitarianism that viewed the earth and life as proceding in cycles rather than changing progressively through time. In this concept, organisms never vanished from the planet completely, but would come back at some later time. Lyell's colleagues found this view hard to accept, since the unidirectional change in life's history had long been clear from Smith's discovery of faunal succession. In one famous cartoon (Fig. 6.2) by British geologist Henry de la Beche, Lyell's idea is ridiculed as a "Professor Ichthyosaurus," holding a human skull, lectures his students about a strange beast from the previous creation. Eventually, even Lyell was forced to admit that the fossil record did show change through time, and that there was no reason to believe that extinct species would ever return.

THE CAUSES OF EXTINCTION: BAD GENES OR BAD LUCK?

The inhabitants of each successive period in world's history have beaten their predecessors in the race for life, and are, insofar, higher on the scale of nature. If the Eocene inhabitants were put into competition with the existing inhabitants, the Eocene fauna or flora would certainly be beaten and exterminated, as would a secondary [Mesozoic] *fauna by an Eocene, and a Palaeozoic fauna by a secondary fauna.*

Charles Darwin, *On the Origin of Species*, 1859

All species that have ever lived are, to a first approximation, dead.

David Raup, *The Nemesis Affair*, 1986

If 5 to 50 billion species have lived on this planet, but only about fifty million are alive today, then 99.9% of all species that have ever lived are now extinct. As the statistician might say, to a first approximation, all species are extinct. Extinction is a hard fact of nature, not a mark of obsolescence. We laugh at the dinosaurs because their extinction seems to prove their inadequacy, but they dominated this planet for over 150 million years. Our species has been on this planet for about 100,000 years, so we are in no position to laugh (especially when nuclear weapons and our own destructive consumption habits make our probability of long-term survival much less likely). Most species in the fossil record last a few million years and then become extinct. We will be lucky to last as long as the average species, let alone try to match the record of some living fossils that have been around for tens to hundreds of millions of years.

Yet despite its importance to life's history, there has been surprisingly little research on extinction. Biologists have spent decades and written hundreds of books on how species evolve, but not on how they disappear. Until 1980, there was relatively little interest in extinction from paleontologists either, even though they were confronted with it constantly. Most paleontologists took extinction for granted, but there was little impetus to study the phenomenon in and of itself. Although this has changed since 1980 (due to the asteroid extinction hypothesis and our concern for species that we humans are destroying), extinction theory is still not a major part of evolutionary biology. As Raup (1991) puts it, extinction is still a "cottage industry" without the trappings or funding of Big Science, such as the Supercollider or Human Genome Project or Hubble Space Telescope. Yet it is comparable "to a demographer trying to study population growth without considering death rates. Or an accountant interested in credits but not debits" (Raup, 1991, p. 12).

Many discussions of extinction have little explanatory content, but amount to saying "organisms become extinct because they become extinct." As Raup (1991, p. 12) describes it, "Textbooks of evolutionary biology contain little about extinction beyond a few platitudes and tautologies like 'species go extinct when they are unable to cope with change' or 'extinction is likely when population size approaches zero.' The *Encyclopaedia Brittanica* (1987) says, 'extinction occurs when a species can no longer reproduce at replacement levels.' These statements are almost free of content."

When extinction is discussed at all, it is usually in strict Darwinian terms. Organisms become extinct because they fail to adapt to new circumstances. Although this is probably true for the majority of extinctions, it is a very hard hypothesis to test in the fossil record, except by the fact that the organism is extinct. Rarely can we find geologic evidence (independent of the last appearance of the organism) that establishes a cause for that extinction. As Raup (1991, p. 17) wrote, "Sadly, the only evidence we have for the inferiority of victims of extinction is the fact of their extinction—a circular argument. . . The disturbing reality is that for none of the thousands of well-documented extinctions in the geologic past do we have a solid explanation of why extinction occurred."

Many of the classic stories of competition and extinction in the fossil record do not hold up under closer scrutiny. One of the most frequently cited is the case of the Great American Interchange (reviewed in Stehli and Webb, 1985; Webb, 1991). According to the traditional account, South America was an island continent until the Panamanian land bridge rose up and connected it with North America about 3.5 million years ago. At that time, there was a tremendous migration of land mammals across the new land bridge, with some South American natives, such as the opossums, porcupines, capybaras, sloths, anteaters, armadillos, and huge armadillo-like glyptodonts, migrating north. A much larger percentage of the North American mammalian fauna migrated south and became established in South America, including mammoths, mastodonts, horses, camels (now represented by llamas, guanacos, and vicuñas in South America), deer, tapirs, peccaries, dogs, saber-toothed cats, pumas, bears, and a variety of rabbits and rodents migrating south. When this wave of North American immigrants came south, they allegedly drove many of the South American natives that occupied similar niches to extinction. However, recent discoveries have shown that this story is not so clear-cut. Raccoons, mastodonts, and other North American immigrants have been found in Miocene deposits in the Amazon Basin that are at least 6 to 7 million years old (Campbell, 1995; Campbell and Frailey, 1996). If many of these immigrants crossed the water barrier before the land bridge was formed, and coexisted with South American natives for millions of years, then clearly the competition from North American immigrants is not the direct cause of their extinction.

We know much more about extinctions that have happened in historic times due to human influence. Such cases include the passenger pigeon, which once darkened the skies with their flocks of billions of birds, but were decimated since they were so abundant and easy to hunt. In 1878, one hunter shipped some 3 million birds from Michigan, its last stronghold, and the passenger pigeon was extinct in that state 11 years later. The last passenger pigeon, a bird named Martha, died in the Cincinnati Zoo in 1914.

A similar story can be told about the heath hen. These birds were a common and easily caught food source in colonial America, so they were hunted mercilessly. By 1840, they were limited to just a few places, and by 1870, they were found only on the island of Martha's Vineyard, off the coast of Massachusetts. They continued to decline until 1908, when a refuge was established to protect the last 50 birds. Their populations recovered to about 2000 by 1915, but then a series of natural events spelled their doom. In 1916, a fire destroyed most of their breeding grounds, followed by a hard winter and an unusual invasion of predatory goshawks that severely depleted their numbers. Heath hens were also weakened by severe inbreeding caused by the reduced population size, by distorted sex ratios, and also by a poultry disease contracted from domestic turkeys. In 1927, only 11 males and 2 females were left, and the last bird disappeared in 1932.

These cases, and many other examples, show how both human interference and natural causes have reduced species numbers until they vanish. Unfortunately, they are not that helpful in understanding extinctions that happened before humans could have caused them. For these, we need to turn to other lines of evidence and hope that a hypothesis can be proposed and tested by data other than the last appearance of the fossil. In particular, we want to answer the question: are most extinctions caused by the failure to adapt ("bad genes" as Raup puts it)? Or are many extinctions caused by random accidents ("bad luck"), such as the fires and diseases that wiped out the heath hen or the impact of asteroids? If the latter is the case, then the Darwinian fitness of the species may be irrelevant. Even very successful, well-adapted organisms can do nothing about major environmental catastrophes that change all the ground rules and make their success in normal times irrelevant. As Gould (1984, p. 16) puts it, "If mass extinctions . . . [are caused by agents] . . . so utterly beyond the power of organisms to anticipate, then life's history either has an irreducible randomness or operates by new and undiscovered rules for perturbations, not (as we always thought) by laws that regulate predictable competition during normal times."

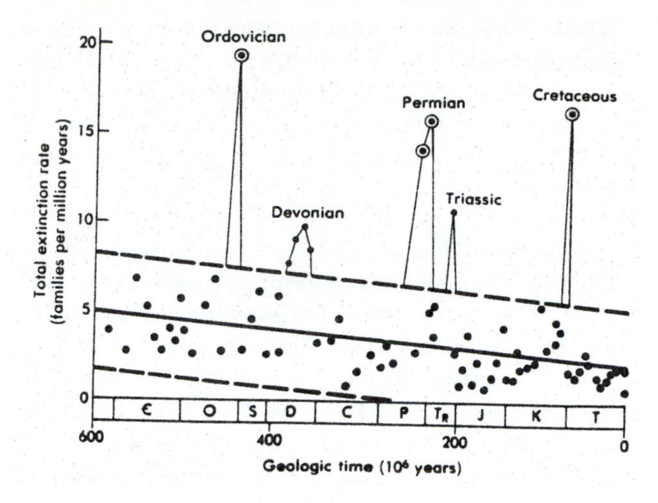

Figure 6.3. Raup and Sepkoski's (1982) plot of extinction rates through the Phanerozoic. Most extinction rates fall with a broad band called "background" rate, but five stand out as peaks above the background. (From Raup and Sepkoski, 1982.)

BACKGROUND EXTINCTIONS AND THE RED QUEEN HYPOTHESIS

Here, you see, it takes all the running you can do, to keep in the same place.

The Red Queen, from Lewis Carroll's *Alice Through the Looking Glass*, 1872

Since we cannot pinpoint the cause of each prehistoric extinction on a case-by-case basis, a number of studies have tried to analyze the record of extinctions in the geologic past by looking at them statistically. A pioneering study undertaken by Raup and Sepkoski (1982) plotted the number of extinctions in marine invertebrate families per million years (Fig. 6.3). They found that there was a steady background extinction rate averaging between 2.0 and 4.6 families per million years, declining slightly over the duration of the Phanerozoic. However, five intervals stood out against this steady "background," with between 10 and 20 families dying out each million years. These are the so-called "Big Five" **mass extinctions**, and we will discuss each in detail in the next section. However, first let us consider the nature of these "background" extinctions.

Are mass extinctions just amplified versions of normal background extinctions, or are they fundamentally different? A number of studies have been done since Raup and Sepkoski (1982). Some paleontologists view the mass extinctions as just a part of the continuous spectrum of extinction. Jablonski (1986) attempted to address this question by analyzing properties of organisms that would make them extinction-resistant during normal "background" times. Species with widespread geographic ranges better resist extinction, because a local event that

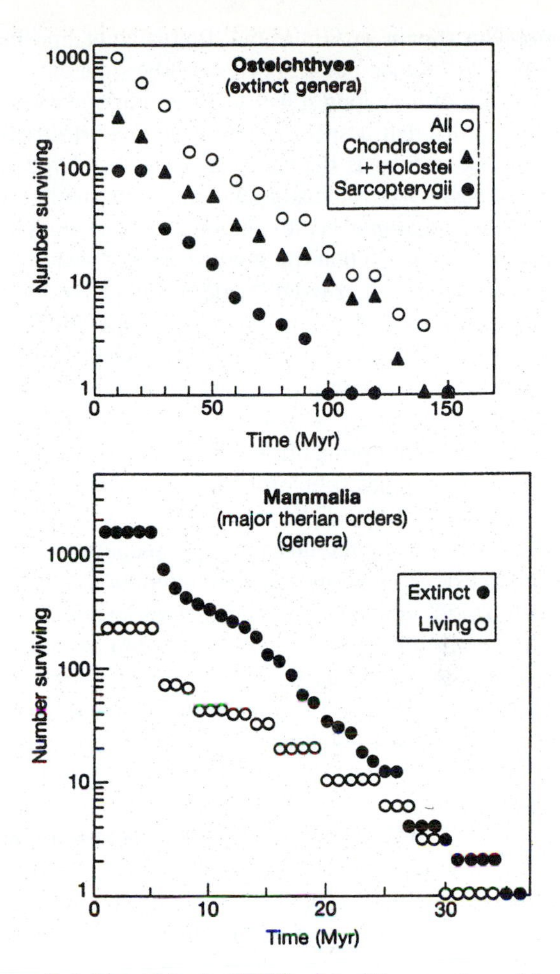

Figure 6.4. Van Valen's (1973) plot of the number of taxa surviving for a given length of time. Note that the lines, with this logarithmic y-axis, are approximately straight, rather than concave upward. This suggests that long-lived species have no greater probability of survival than short-lived species. (Adapted from Van Valen, 1973.)

wipes out part of the population will not wipe out the entire species (as the heath hen example demonstrates). Similarly, species with planktonic larvae that can disperse widely should also be extinction-resistant. Instead, when Jablonski looked at the molluscan victims and the survivors of the terminal Cretaceous extinctions, he found that all the rules had changed. Extinction-resistant taxa were just as likely to vanish as those without these properties. Mass extinctions are not simply background extinctions writ large.

Granted that they are different from mass extinctions, what can we say about background extinctions? In 1973, Leigh Van Valen published a study based on the measurements of the durations of 24,000 taxa (Fig. 6.4). He plotted the durations of these taxa against the number that survived for each duration, producing what is known in demography as a **survivorship curve**. As expected, the majority of taxa have very short durations, and there are fewer and fewer with long life spans, so the plot slopes down to the right. What was surprising was the linear relationship shown in Figure 6.4; this straight line indicates that the chance that a species will become extinct is independent of how long it has survived. In other words, the probability that a species with a very short duration will become extinct is the same as for a long-lived survivor species. One can easily imagine an alternative scenario. For example, if evolution was progressively improving species so that those who survived were better adapted, then the longer-lived species should have a lower probability of extinction. Or, if one views evolution as producing more and more specialized species over long durations of time, one might expect the long-lived species to be more vulnerable to extinction. Neither of these predictions is true: the probability of extinction is the same whether a species has been around a long time or a short time.

How do we interpret this result? According to Van Valen (1973), it suggests that evolution is a zero-sum game: species compete for limited resources; natural selection constantly improves organisms to keep up with competing species; and each species' environment deteriorates as competitors evolve new adaptations. In other words, species must constantly improve to avoid extinction. Van Valen called this the "Red Queen Hypothesis," after the Red Queen chess piece in *Alice Through the Looking Glass*, who told Alice that she must keep running and running to stay in the same place.

Although there has been much criticism and discussion of Van Valen's methods and results, the last 24 years have generally vindicated the Red Queen hypothesis. It is clear that species do not grow better or worse at avoiding extinction as they persist in time; old species have the same chance of extinction as young ones. Thus, even "background" extinctions are not trivial, inevitable consequences of natural selection, but tell us a lot about the nature of species and their resistance to extinction.

THE MAJOR MASS EXTINCTIONS

> *The Age of Reptiles ended because it had gone on long enough and it was all a mistake in the first place.*
>
> Will Cuppy, *How to Become Extinct*, 1941

> *Mass extinction is box office, a darling of the popular press, the subject of cover stories and television documentaries, many books, even a rock song . . . At the end of 1989, the Associated Press designated mass extinction as one of the "Top 10 Scientific Advances of the Decade." Everybody has weighed in, from the* Economist *to* National Geographic.
>
> David Raup, *Extinction: Bad Genes or Bad Luck?* 1991

By the 1830s, geologists were aware that there were big differences between fossils found in the Cambrian through Permian, those found in the Triassic through Cretaceous, as well as those found in the Cretaceous and younger rocks. These great differences led to the naming of the "Palaeozoic" by Adam Sedgwick in 1838 and "Mesozoic" and "Cainozoic" by John Phillips in 1840. These great eras of geologic time were not picked arbitrarily, but are sharply subdivided by two of the biggest mass extinctions in earth history: the Permo-Triassic extinction that separates the Paleozoic from the Mesozoic, and the Cretaceous-Tertiary (K/T) extinction that separates the Mesozoic from the Cenozoic. Mass extinctions clearly stand out from the background of normal extinction, and some are truly massive. Raup estimated that the Permo-Triassic event wiped out 96% of marine species and 80% of the marine genera on earth, and the K/T extinction may have terminated 60 to 75% of marine species and 50% of the genera.

The topic of mass extinction has been discussed several times (including a classic paper by Newell, 1967), but the subject really got rolling in 1980 with the announcement of the asteroid extinction hypothesis for the K/T extinction. Suddenly, mass extinctions were front-page news and the subject of many meetings and symposia. Numerous books and thousands of scientific papers were published, mostly on the K/T event. The debate became very heated, both in public and in print, with several violent clashes and personal insults that reveal as much about the sociology of science as they do about the actual data. Although the intense momentum of the mass extinction debate has slowed down somewhat since the 1980s, there are still many who are attempting to answer the many unsolved questions and understudied aspects of the problem.

Through all the discussion of each individual mass extinction event discussed next, we will keep in mind one important question: do the mass extinctions have any common themes or causes that might lead to a general theory of extinction?

Permo-Triassic Extinctions

By far the biggest of all extinction events was the Permo-Triassic extinction, which may have wiped out 96% of the species in the marine realm. Erwin (1993) calls it "the mother of all mass extinctions" in reference to Saddam Hussein's threat that the land war in Iraq would be the "mother of all battles" (it was the "mother of all routs" instead). The extinction was so severe that it radically changed the basic taxonomic composition of marine life. The crinoid-coral-bryozoan-brachiopod-dominated "Paleozoic fauna" described by Sepkoski (1981) was replaced by the bivalve-gastropod-echinoid "Modern fauna" that still prevails on the seafloor today.

The list of Permian victims is long and surprising. Some groups had been in decline well before the Permian, including the trilobites, blastoids, tabulate and rugose corals, orthid brachiopods, and archaic molluscs. Other groups were thriving in the Permian, but were still strongly affected. Fusulinid foraminiferans and productid brachiopods covered the seafloor through much of the late Paleozoic but were completely wiped out. Both Paleozoic groups of bryozoans disappeared, and only one family of crinoids survived. Only two genera of ammonoids survived to found another great radiation in the Mesozoic. Bivalves and gastropods were less severely affected, with about 30% of their genera going extinct. In addition, 75% of the families of land vertebrates went extinct over this interval.

The first thing to note is that this extinction was not only severe, but protracted, spanning about 8 million years of the final two stages of the Permian. Thus, any model (such as a meteorite impact) that attempts to explain it by a sudden, single event does not account for the protracted pattern. Several attempts have been made to find impact evidence at the end of the Permian, but so far without success. Many other hypotheses have been proposed for the Permo-Triassic event. Schopf (1974) noted that the Permian was the time of coalescence of Pangaea, which drastically reduced most of the area of shallow marine seaway that used to lie between the separate continents. Coupled with a marine regression, these conditions would have greatly reduced the available area of the shallow marine seafloor and could have crowded a lot of marine species to extinction. Unfortunately for this hypothesis, Pangaea had already assembled by the middle Permian, well before the main wave of extinctions.

Stanley (1984, 1987) attributed the Permian extinctions to global cooling, caused by the presence of glaciers not only on the South Pole in Gondwanaland but also on the North Pole in Siberia. Although it is true that a large percentage of the victims were tropical, warm-adapted species, once again the timing is all wrong. The bipolar glaciers were in place by the middle Carboniferous, and the latest evidence suggests that they melted back by the end of the Permian. In fact, the end of the Permian is marked by an abrupt warming event.

Although we can rule out extraterrestrial impacts, reduced marine shelf habitat, and global cooling as causes, a great number of other environmental stresses were clearly operating at the end of the Permian. Chief among these was a huge marine regression, which greatly reduced the shallow marine shelf area and exposed it to erosion, and also allowed the climates on Pangaea to become more extreme. The rapid warming at the end of the Permian is characterized by climatic extremes, including large episodes of evaporite deposition in many parts of the world. In addition to climatic instability, the carbon isotopes in the ocean suggest that the marine regression oxidized a great deal of organic matter, which may have in turn depleted atmospheric oxygen levels from about 30% to less than 15% in the Late Permian. Finally, the end of the Permian was marked by one of the biggest eruptions of flood basalts, the Siberian traps, which are in excess of 1.5 million cubic kilometers in volume. These eruptions would have released a tremendous quantity of carbon dioxide and sulfates into the atmosphere and greatly warmed it. Recently, Chinese scientists have found evidence of an immense explosive silicic volcanic eruption right at the Permo-Triassic boundary, which may have formed global ash clouds and blocked out sunlight for weeks, further stressing an already tenuous situation. Knoll et al. (1996) pointed to the high concentration of anomalous carbonate precipitates, and the carbon isotopic anomalies at the Permo-Triassic boundary, and argued that there was a massive overturn of the oceans which poisoned the surface waters with high concentrations of carbon dioxide. Visscher et al. (1996) found evidence of fungal spores at the terrestrial Permo-Triassic boundary all over the world, suggesting a major collapse of the land plants consistent with the climatic and atmospheric stresses just outlined. As Erwin (1993) sees it, there are so many disasters occurring at the end of the Permian that the cause of the extinction might be "all of the above." Erwin calls this the "Murder on the Orient Express" model, after the Agatha Christie story where the victim is killed by all 12 suspects on the train who had reason to want him dead.

Cretaceous-Tertiary Extinctions

By far the most interest and attention has been focused on the great extinction that ended the Mesozoic. The obvious reason is that this event wiped out the dinosaurs, and everyone wants to know why they disappeared. (Dinosaurs can make the cover of *Time* magazine, but not productids or fusulinids.) A second reason is the evidence of an extraterrestrial impact, first proposed in 1980, that brought astronomers, geochemists, geophysicists, and atmospheric scientists into an area of research that had once been the exclusive domain of paleontologists. Consequently, thousands of papers have been published on this event, and much (sometimes unpleasant) controversy has occurred because so many scientists with different research methods and different agendas have become involved.

Prior to 1980, there were a number of ideas for the extinction of the dinosaurs (cooling, warming, disease, inability to digest angiosperms, mammals ate their eggs), but they all missed the central point: the K/T extinctions wiped out not only the dinosaurs, but also the ammonites, the marine reptiles, most marine plankton, and many other marine invertebrates. Any explanation that focuses on only the dinosaurs misses the global nature of this extinction, and fails to show why plants and animals from every level on the food chain died out. The dinosaurs are really an afterthought. If the proposed cause can wipe out plankton, ammonites, many other marine invertebrates, and most land plants, then it makes sense that the top of the food pyramid—the dinosaurs—would also disappear.

The breakthrough came out through sheer serendipity. In 1978, Berkeley geologist Walter Alvarez was studying deep marine sections that spanned the K/T boundary near the town of Gubbio in the Apennine Mountains of Italy. Between the Cretaceous limestones and Paleocene limestones was a thin clay layer that marked their mutual boundary. Alvarez was trying to find a way to estimate how much time was represented by this clay layer. His physicist father Luis Alvarez suggested that the clay layer might record the steady rain of cosmic dust that constantly falls to the earth from space. A low content of cosmic dust would suggest that the clay was accumulated rapidly, but a high content of cosmic material might indicate a long-term accumulation. The best marker of cosmic input seemed to be a rare platinum-group metal, iridium, that is found in trace amounts in extraterrestrial rocks and in the earth's mantle and core, but not in the crust. The Alvarezes sent their samples to nuclear chemists Frank Asaro and Helen Michel, who found far more iridium than even the longest accumulation could explain. Luis Alvarez wracked his brain for an explanation, and the only one that made sense was an impact by an asteroid about 10 km in diameter.

At first, the iridium anomaly and the impact hypothesis were questioned, because marine clays are notorious for accumulating all sorts of trace elements. But when the high levels of iridium were found not only in other marine sections, but also in coal layers that marked the K/T boundary on land, there was little doubt that it was a global event. (Since then, it has been found in over 100 places around the world.) Unlike previous ideas about the K/T extinctions, however, this model generated independent predictions that motivated many different scientists to look for further evidence. Within a few years, other supporting evidence had been found: tiny spherules that appeared to be impact droplets; grains of quartz with shock features, and a high-pressure polymorph of quartz called stishovite; apparent impact breccias and tsunami deposits in several places around the Caribbean. Glen Penfield, a petroleum geologist working in Mexico, first identified a likely impact crater called Chicxulub in the northern Yucatan Peninsula of Mexico in 1978, but his discovery was completed missed by the impactors until 1990, when Alan Hildebrand independently rediscovered the same evidence. At the time of this writing, there seems to be little doubt that some sort of impact occurred at the K/T boundary.

But is this the whole story? Ever since the impact theory hit the scene in 1980, another group of scientists (led by Charles Officer and Charles Drake of Dartmouth) have argued that the huge flood basalt eruptions known as the Deccan traps, now found in western India and Pakistan, could have been responsible as well (Officer and Page, 1996). These eruptions covered over 10,000 square kilometers, erupting a total thickness of 2400 meters of lava flows, and were apparently caused by a hotspot eruption from the Réunion hotspot. Such mantle-derived volcanism produces immense amounts of greenhouse gases that can cause global warming, or it can produce ash clouds that block the sun and cause global cooling. The Deccan eruptions could have produced high levels of iridium (as mantle-derived volcanoes still do today) and may have had an early, explosive phase that could produce something resembling shocked quartz and melt droplets. More important, the Deccan eruptions were recently dated at 65 Ma, right at the K/T boundary, although they started in the latest Cretaceous. Thus, one prediction that would distinguish the volcanic versus impact hypotheses is one of timing: the volcanic effects should be gradual and prolonged, but an impact would produce a sharp, instantaneous extinction horizon.

Here's where the controversy gets even more confusing. Some paleontologists argue that dinosaurs were dying out gradually well before the end of the Cretaceous, while others say that it is an artifact of sampling and that the gradual pattern disappears when every last scrap of dinosaur bone is collected right up to the boundary coal. At one time, paleobotanists said that the extinction of land plants was gradual, but later they noticed the abundance of fern spores in the boundary coal and concluded that it was sudden. One group of micropaleontologists state that the extinction of foraminifera was sudden at the boundary, but another has long argued that there were gradual extinctions both before and after the boundary. At one time, the cephalopod specialists were convinced that ammonites died out gradually, but later they revised their opinion to the opposite conclusion.

Even more confusing is the geographic pattern in the data. Stets et al. (1995) claim to have discovered a gradual pattern of extinction in the rocks of China, including Paleocene dinosaurs, although the dating is still disputed. Most of the evidence of abrupt extinction of plankton or ammonites comes from tropical latitudes, but the data from the polar regions (especially Seymour Island on the Antarctic Peninsula) suggest a more gradual pattern of extinction, with the last ammonite occurring about 100 feet above the iridium anomaly (Zinsmeister et al., 1989). A similar gradual pattern was found with the plants of Seymour Island (Askin et al., 1995) and in the foraminifera recovered from cores in the Antarctic Ocean (Keller, 1993). Marincovich (1993) reports that many bivalves and gastropods that supposedly died out at the end of the Cretaceous survived on into the Tertiary in the Arctic.

Some data are not disputed. Many Cretaceous groups, such as the inoceramid and rudistid bivalves, most of the calcareous nannoplankton, and the marine reptiles, were clearly in decline during the Late Cretaceous and were wiped out by something that preceded the K/T impact. Other microplankton, such as diatoms and dinoflagellates, show no effect whatsoever. On land, the pattern is even more puzzling (Archibald, 1996). The tiny rat-sized mammals went right through the boundary with no obvious effects, except that placentals replaced the opossum-like marsupials in abundance. Even more striking is the lack of change in terrestrial vertebrates, such as turtles, crocodilians, and salamanders. Most of the impact scenarios postulate a long period of cold and darkness caused by the clouds of impact debris and dust, and some also suggest global wildfires and acid rain as well. Yet turtles or crocodilians cannot hide from these conditions any more than the dinosaurs could. It is true that some of them can hibernate, but they require a long preparation time before they go into hibernation; the impact would have happened without warning. Besides, the Cretaceous greenhouse world was warm all the way to the poles, unlike today, so that they may not have hibernated at all. The most sensitive of all were the salamanders, which cannot tolerate acid waters. In fact, their populations are in serious trouble today due to human-induced acid rain. The heavy acid rain postulated by the impact model would have wiped them off the face of the earth.

A third factor that is usually ignored is the massive regression that occurred at the end of the Cretaceous. Archibald (1996) points out that a regression explains the pattern of extinction of terrestrial vertebrates much better than either an impact or volcanism.

From the present evidence, it is clear that both massive volcanic eruptions, a major regression, and an impact occurred at the end of the Cretaceous. It is also clear that many groups were extinct before the Cretaceous ended, probably due to a variety of climatic stresses that may have been triggered by the atmospheric effects of mantle-derived volcanism from the Deccan eruptions. Whether dinosaurs, land plants, ammonites, and planktonic foraminifera were decimated by the impact is less clear, since specialists disagree on the pattern of their decline. However, the impact-generated cooling and darkness could not have been so severe that turtles, crocodiles, and salamanders could not survive.

But this 17-year controversy goes far beyond just a dispute over the causes of the K/T extinction. Ever since the evidence brought in astronomers and other non-geologists, the debate has been highly emotional and polarized, with the paleontologists resisting the impact hypothesis based on their first-hand familiarity with the nature and limitations of the fossil record, and astronomers supporting it (even though they tend to not know much about how these rocks accumulated or what the fossils say). Each meeting quickly developed into polarized camps, with much hostility and even unseemly name calling. For example, Luis Alvarez told a reporter in 1986, "I don't like to say bad things about paleontologists, but they're really not very good scientists. They're more like stamp collectors." This kind of name-calling did not reflect well on the Nobel-Prize-winning physicist, and he was taken to task for it. The other camp has been equally outspoken at times. The dinosaur paleontologist Robert Bakker was quoted in the same article saying, "The arrogance of those people [the impactors] is simply unbelievable. They know next to nothing about how real animals evolve, live and become extinct. But despite their ignorance, the geochemists feel that all you have to do is crank up some fancy machine and you've revolutionized science. The real reasons for the dinosaur extinctions have to do with temperature and sea level changes, the spread of diseases by migration and other complex events. But the catastrophe people don't seem to think such things matter. In effect, they're saying this: 'We high-tech people have all the answers, and you paleontologists are just primitive rock hounds'" (quoted in Raup, 1986, p. 104).

Many scientific papers have been rejected, grant proposals sabotaged, and careers ruined in the bare-knuckle fisticuffs over the K/T boundary controversy. The debate has become so angry and polarized that almost no evidence will change the minds of the major players, because they are so committed to the positions they have argued for 17 years that they cannot afford to change positions and lose face as well as funding. Thus, the impactors will ignore the evidence of land reptiles and amphibians, no matter how compelling it is. The gradualists have also been stubborn and hard to convince when the evidence such as stishovite and Chicxulub crater finally proved the impact had occurred. It seems likely that the argument will not be ended by some conclusive piece of evidence, but only after all the key participants are either exhausted or extinct.

Late Devonian Extinctions

Although it has received much less publicity than the Permo-Triassic or K/T events, the Late Devonian extinction was almost as impressive, wiping out perhaps 75% of the species and 50% of the genera in the marine realm (McGhee, 1989, 1990, 1995). Unlike most of the other mass extinctions that mark the boundary *between* two geologic periods, this event occurred *within* the Late Devonian (between Frasnian and Famennian stages of the Late Devonian), and spanned about 4 million years of the late Frasnian. Once again, the warm-adapted shallow marine invertebrate community was particularly hard-hit. Pentamerid brachiopods, which dominated the Silurian, disappeared, and only 15% of the Frasnian brachiopod fauna survived. The early goniatitic ammonoids were devastated, and trilobites and graptolites also nearly vanished. The most dramatic extinctions, however, happened in the giant reef communities that dominated the Frasnian. Tabulate and rugose corals, and stromatoporoid sponges were the main reef-builders, and all three were so reduced that they rarely appear in the later Paleozoic. The planktonic acritarchs that had persisted since the Proterozoic were nearly wiped out, and there was also a crisis in the conodonts. Many of the groups of archaic fish (especially the armored jawless fishes and jawed placoderms) were eliminated.

Major anomalies in carbon and oxygen isotopes are found in the latest Frasnian, suggesting significant oceanographic events. One scenario points to the appearance of glaciers on Gondwanaland for the first time since the Ordovician, which suggests the beginning of the global "icehouse" phase that dominated the late Paleozoic. A severe cooling event would wipe out the tropical reefs and other warm-adapted taxa and triggered a massive overturn of the ocean. This in turn would bring cold, oxygen-poor waters up from the nutrient-rich bottom and cause the enrichment of organic carbon formerly trapped in the deep ocean that can be seen in the carbon isotope record. Further evidence for this cooling effect is shown by the fact that the heaviest extinction took place in tropical taxa, while some of the high-latitude faunas (such as the cold-adapted faunas of Siberia and Gondwanaland) expanded their ranges toward the tropics.

The evidence of possible extraterrestrial impacts has been more controversial. Digby McLaren, an early impact advocate, first proposed the idea in 1970, but was not taken seriously. Several attempts have been made to find impact evidence, but iridium anomalies are absent at the boundary, or too late in the Famennian (McGhee, 1989). Recently, however, some evidence has been found that there were several impacts at the boundary after all (McGhee, 1995). Like the Permian, however, the protracted nature of the extinctions in the late Frasnian argues against a single impact being the sole cause (McGee, 1995, postulates several impacts), and the dramatic cooling and overturn of oceanic carbon are hard to explain by impact.

Late Ordovician Extinctions

Compared to the three events just discussed, relatively little attention has been paid to the Late Ordovician event (Brenchley, 1989, 1990). Yet it was just as severe as the others, wiping out about 57% of the marine genera (Sepkoski, 1989). The once-flourishing trilobites were decimated, never again to recover the abundance they enjoyed in the Cambrian and Ordovician. Some of the archaic groups of echinoderms, such as paracrinoids, eocrinoids, and parablastoids, also died out. Planktonic acritarchs and conodonts were strongly affected, and graptolites were nearly wiped out, with only a few genera surviving into the Silurian. More than half of the species of brachiopods (especially among the orthids and strophomenids) and bryozoans died out, and the reef community of tabulate and rugose corals (50 out of 70 genera extinct) and receptaculitid algae was severely hit. Even predators such as the orthoceratid nautiloids that dominated the Ordovician were severely affected.

As in the Devonian event, warm-adapted taxa seem to be the chief victims, suggesting that a global cooling event might have been responsible. In fact, there was a brief pulse of glaciation during the Late Ordovician, centered over the North African part of Gondwanaland, which was then situated over the South Pole. The cooling and glacial regression were especially hard on the sensitive shallow marine benthic community, which had developed during a long period of stable, warm, high Ordovician sea levels. These climatic changes may also have triggered oceanic changes that brought biologically toxic waters to the surface (Wilde and Berry, 1984). Like the Permian and Devonian extinctions, the Ordovician extinctions were very protracted (spanning about 2 million years of the Rawtheyan and Hirnantian stages), arguing against a sudden impact. No clear evidence of a single extraterrestrial iridium anomaly has yet been documented for the Ordovician (Orth, 1989).

Late Triassic Extinctions

The fifth event of the "Big Five" was the extinction at the end of the Triassic, which eliminated 48% of the marine genera (Sepkoski, 1989), making it only slightly less severe than the previous four. It has been getting more attention in recent years (Hallam, 1981, 1990; Benton, 1986, 1991, 1993, 1994; Johnson and Simms, 1989; Simms and Ruffell, 1989, 1990), although not nearly as much as the K/T event. This "event" is actually a composite of several extinctions spanning the last two stages of the Triassic, the Carnian and Norian (the last half of the Norian used to be considered a separate stage, the Rhaetian), which lasted over 17 million years (Benton, 1986, 1991). Over 90% of the Late Triassic bivalve species and 80% of the brachiopod species (especially the spirifers, which dominated the mid-Paleozoic and survived the Permian event) became extinct by the Jurassic. The abundant ammonoids were nearly wiped out, going through another bottleneck of a handful of genera (as they did at the end of the Permian) before radiating profusely in the Jurassic. The hardy conodonts, which had weathered all the Paleozoic extinction events, finally gave up the ghost. There were also significant extinctions in the marine gastropods and crinoids as well. Reef communities were also affected, and the only marine reptiles that survived were the fish-like ichthyosaurs. Land vertebrates (especially the archaic amphibians, synapsids, and non-dinosaurian reptiles) were hit very hard, although in two separate pulses in the Carnian and Norian (Benton, 1986).

Like the Permian, Devonian, and Ordovician, these extinctions were very protracted, spanning three stages of the Late Triassic. Some extinctions (especially the ammonoids, bivalves, and the terrestrial and marine vertebrates) were concentrated in Carnian, and others (especially the brachiopods and conodonts) in the Norian, and some groups showed a gradual decline throughout the interval (Johnson and Simms, 1989). The severe effect on Triassic reefs, especially in the Tethys, suggest that cooling was a significant factor. In addition, the abundance of black shales and the geochemical anomalies suggest that major oceanic changes were important. The presence of a major impact crater in Quebec known as Manicouagan that may date to the Late Triassic has inspired the possibility that impacts may play a role (Olsen et al., 1987), although the date on this impact is closer to the Carnian/Norian boundary than the end of the Norian. "Shocked quartz" has been reported from the Triassic/Jurassic boundary in Italy (Bice et al., 1992), but it does not have the lamellae that distinguish impact-derived grains from tectonically shattered grains (Hallam, 1990), and several layers of "shocked quartz" also occur

1 to 3 m below the boundary. Many attempts to find an iridium anomaly have also failed (Hallam, 1990). The protracted nature of the Carnian and Norian extinctions also argue against impact being the sole cause.

Eocene-Oligocene Extinctions

Although not among the "Big Five" in terms of extinction of genera and species, the extinctions at the end of the Eocene have received attention because they are recent enough to have a detailed record and may be associated with impacts. At one time, these events were treated as a single "Terminal Eocene Event" (TEE), and iridium anomalies were reported *near* the Eocene/Oligocene boundary (Asaro et al., 1982; Alvarez et al., 1982). However, as the interval was studied in greater detail and the dating improved, a completely different picture emerged (Berggren and Prothero, 1992; Prothero, 1994a, 1994b). The major pulse of extinction occurred at the end of the middle Eocene (37 Ma), in response to global cooling and oceanographic overturn. Major extinctions affected the bivalves and gastropods and especially the warm-adapted benthic foraminifera, such as the coin-shaped nummulitids. There were also major changes in the marine plankton, especially in the foraminifera, diatoms, and calcareous nannoplankton. Four impact horizons are known, but they occur in the middle of the late Eocene (36.0 to 35.5 Ma) associated with no extinctions of consequence (Miller et al., 1991). The Eocene/Oligocene boundary (33.7 Ma) turns out to be a very minor episode, with little extinction outside the planktonic foraminifera. The final pulse occurred in the earliest Oligocene (33 Ma), when major Antarctic glaciation began and triggered global cooling. Bivalves, gastropods, and echinoids all suffered another wave of extinction at this time, but only minor extinctions occurred in land vertebrates.

The clear signal from all these data is that the planet underwent global cooling from the greenhouse world of the early Eocene to the icehouse world of the Oligocene. The extinctions seem to be clearly correlated with multiple episodes of cooling and climatic change, especially when glaciers covered Antarctica in the earliest Oligocene. Although there is some doubt about how to explain the middle-late Eocene cooling, most specialists agree that the earliest Oligocene glaciation of Antarctica was triggered by its separation from Antarctica and the development of the circum-Antarctic current. This flow traps cold water in the Southern Ocean and prevents exchange with the temperate and tropical regions. These tectonic changes, in turn, triggered the beginning of modern oceanic circulation, especially the Antarctic Bottom Waters that sink from the South Pole and flow along the bottom of the ocean basins to the tropics. The beginning of the North Atlantic Deep Water may have also coincided with the earliest Oligocene event.

Massive flood basalt eruptions in Ethiopia have also been blamed (Rampino and Stothers, 1988), but their dating is imprecise and the eruptions do not peak during the middle Eocene or early Oligocene extinctions. Besides, volcanic eruptions do not explain the glacial and oceanographic changes of the Eocene.

Extinction of the Pleistocene Megafauna

This extinction event is restricted to land mammals, mostly to larger land mammals—the so-called "Pleistocene megafauna" (reviewed in the book edited by Martin and Klein, 1984). These included not only the mammoths and mastodonts in most of the world (except for some dwarf mammoths which survived until 6000 years ago on an island in the Bering Strait) but also horses, camels, ground sloths, glyptodonts, saber-toothed cats, giant bears, and dire wolves in North America. In South America, the last of the native South American hoofed mammals died out, as did the ground sloths and glyptodonts, along with immigrants such as the mammoths, mastodonts, horses, and sabertoothed cats. Eurasia and Africa suffered fewer losses, the major ones being sabertoothed cats and woolly rhinos. Australia's giant marsupial fauna was wiped out. There were few extinctions in the smaller mammals (particularly rodents), and no effect in the marine realm.

In the Americas, most of these extinctions were concentrated around 11,000 years ago, leading Paul Martin (1984) and others to propose the "overkill hypothesis." Martin points out that extinctions peaked at the time when humans with Folsom points on their spears migrated into the Americas over the Bering land bridge. However, only a few mammal species actually show evidence of being hunted, and this model does not explain why humans could hunt horses and camels to extinction, but not bison (which have been found with Folsom points). Beck (1996) tested the overkill hypothesis in North America by examining the date of the last occurrence of each extinct species. According to the prediction, they should have gone extinct first in areas near the Bering land bridge, and last in areas most remote from it. In fact, the opposite pattern was found, with extinctions occurrng first in areas most remote from the Bering land bridge. Other peculiarities of the extinction pattern also did not fit the predictions of the overkill model.

Humans evolved alongside these animals in Africa and Eurasia, yet these mammals experienced mass extinctions between 20,000 and 11,000 years ago. Klein (1984) argue that African and Eurasian hunters must have developed more sophisticated hunting techniques. The Australian extinctions occur between 24,000 and 11,000 year ago, too early for the first human hunters. In addition, many of the late Pleistocene giant marsupials exhibit a dwarfing trend toward the Holocene, presumably in response to late Pleistocene climatic deterioration.

The opposing school of thought (Lundelius, 1983; Graham and Lundelius, 1984) blames the extinctions on the rapid transition from the last glacial maximum to the present interglacial about 11,000 years ago. These scientists point out that there are many unusual biogeographic patterns that suggest that the most recent glacial/interglacial transition was completely different from any that occurred earlier in the Pleistocene. Changes in pollen suggest that the climate went from mild and equable to harshly drier and more seasonal, and many plant communities (and presumably their dependent mammals) died out or greatly reduced their range. Critics of this hypothesis are not satisfied that the last glacial/interglacial transition was that different from the earlier ones in the Pleistocene, which did not cause a mass extinction.

Of course, it is possible that both factors may have been important, since nature is complex, and effects can have multiple causes (Marshall, 1984). Perhaps some climatically stressed species were driven to final extinction by the human blitzkrieg. The explanation may even be more complex than our simple models suggest. As Owen-Smith (1987) points out, in East Africa today the diversity of savanna mammals in maintained by certain "keystone" species such as the elephant, which is constantly breaking down trees and keeping the vegetation open so that other species can thrive. When poaching wiped out the elephants in certain regions, dense brush grew and the mixed grassland habitat (and all its mammals) disappeared. Owen-Smith argues that a similar effect may have occurred at the end of the Pleistocene. If climate and/or humans wiped out the mammoths and mastodonts, their extinction may have had a cascading effect and indirectly caused the extinction of many other mammals that depended on them for their habitat.

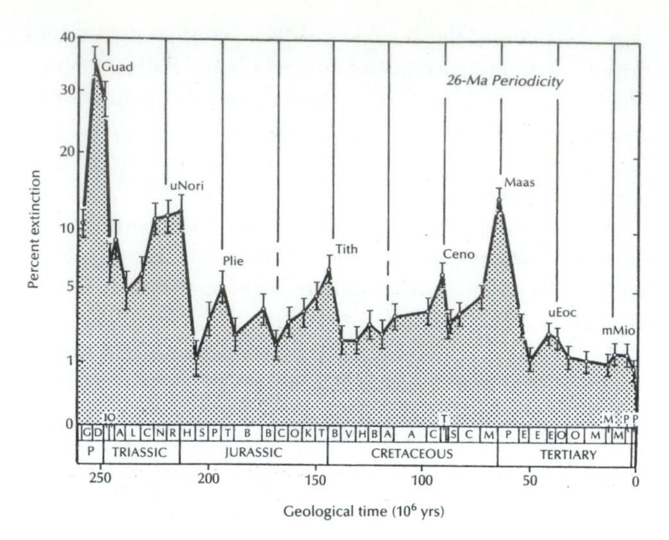

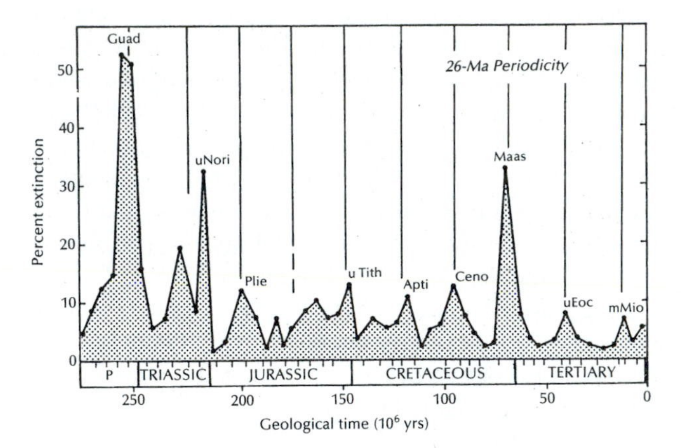

Figure 6.5. Percentage of extinction for marine families (top) and genera (bottom) during the last 250 million years. In this plot, the apparent 26-million-year periodicity shows up, as do the major extinction peaks. (From Raup and Sepkoski, 1986.)

IS THERE A COMMON CAUSE?

The great tragedy of science—the slaying of a beautiful hypothesis by an ugly fact.

Thomas Henry Huxley, *Biogenesis and Abiogenesis*, 1894

Facts are simple and facts are straight
Facts are early and facts are late
Facts all come with points of view
Facts don't do what I want them to

Talking Heads, *Crosseyed and Painless*, 1980

For every problem, there is a solution that is simple, neat, and wrong.

H.L. Mencken

Behind all the frenzy of research on mass extinctions is a broader question: is there a general theory of mass extinctions? Do these events have a common cause, or are they each individually caused by different factors? The impact advocates argue that extraterrestrial signals (such as iridium, shocked quartz, and impact droplets) occur at many of the major mass extinctions, and by extrapolation, that many mass extinctions are impact-induced. In 1989, Digby McLaren told a stunned audience at the International Geologic Congress in Washington, D.C., that *all* mass extinctions were caused by impacts, *whether or not there was evidence of impact in the fossil record!* Raup (1991) has written that all extinctions (even normal background extinctions) might be caused by impacts. With statements such as these, why bother gathering data any more? Extinctions occurred, and impacts occurred—therefore impacts caused all the extinctions.

A hard look at the evidence shows that this is far from the case. The evidence for the K/T impact is pretty strong, but it is not clear that it caused many of the late Cretaceous extinctions. The late Eocene impact evidence is also well established but irrelevant—it caused no extinctions of consequence, it was too late for the middle Eocene extinction, and it was too early for the early Oligocene event. The possible impact evidence for the late Triassic and late Devonian extinctions has been controversial, and so far there is no clear evidence that impacts had anything to do with the Permo-Triassic or late Ordovician extinctions. The protracted nature of most of these extinctions also argues against a single, sudden impact causing them. Hut et al. (1987) suggested pulses of comet showers as an extinction mechanism, but that idea died quickly because it does not explain the details of the major extinctions. Finally, the stratigraphic record is full of major impacts that caused *no* extinctions. These include not only the late Eocene impacts but also the early Eocene Montagnais impact off Nova Scotia (Bottomley and York, 1988; Aubry et al., 1990), and the Miocene Ries impact in Germany (Heissig, 1985). If the late Triassic Manicouagan impact is correctly dated, then it too is too late for the Carnian extinctions and too early for the Norian extinctions.

Rampino and Stothers (1988) suggested that massive eruptions of flood basalts, which release greenhouse gases and sulfur compounds into the atmosphere, were correlated with major mass extinctions. Although the dates on the Siberian traps support this for the Permian extinctions, and the Deccan traps for the Cretaceous extinctions, the rest of the eruptions are too poorly dated to establish a connection. In addition, the flood basalt model does little to explain the frequent evidence of cooling and oceanographic changes that accompany several of the mass extinctions.

Stanley (1984, 1987) argued that climatic cooling seems to be the common signal of most of the mass extinctions, many of which are also correlated with pulses of glaciation. This is true of the late Ordovician, late Devonian, and late Eocene extinctions, but not of the Permian or Cretaceous extinctions. In addition, physiological studies (Clarke, 1993) suggest that marine organisms are much more tolerant of temperature change than we give them credit for. Instead of cooling directly causing extinctions, Clarke (1993) points out that changes in temperature are usually associated with changes in many other oceanographic conditions (changes in oxygen content and other water properties) as well as restrictions on biogeographic range and ability to reproduce.

Hallam (1989) pointed out that many mass extinctions are correlated with marine regressions and that regression and loss of shelf area are critical to mass extinction. Although this is true of the late Ordovician, late Permian, late Triassic, and late Cretaceous extinctions, it is not true of the late Eocene, and is questionable in the others. As Stanley (1987) and Paulay (1990) showed, historical examples of regressions on many tropical islands seemed to have little effect on the biota, so it is hard to see how regressions are a direct cause of extinction. They may be an effect of global cooling and glaciation, however, and so are closely correlated with extinction horizons.

If our general review of mass extinctions has shown anything, it is that each extinction is individual and idiosyncratic, and each has a different pattern of environmental perturbations and different durations and patterns of extinction pulses. This is the strongest evidence that the search for a common cause is probably futile.

Are Mass Extinctions Periodic?

The search for common causes was triggered by a controversial idea first published in 1984. Sepkoski (1982) had compiled a data base of the ranges of families of marine organisms and found that mass extinctions were truly distinct from background extinctions (Fig. 6.3). As he refined this data base for the Mesozoic and Cenozoic, Sepkoski noticed that there appeared to be an even spacing of the extinction peaks, suggesting a periodicity of about 26 million years (Fig. 6.5). After subjecting these data to statistical analysis, Raup and Sepkoski (1984) published their ideas about periodic extinctions. While the paper was still circulating as a preprint, a number of astronomers jumped on the bandwagon, proposing causes for this alleged periodicity. They postulated periodic comet showers (Davis et al., 1984), the oscillation of the solar system through the galactic plane (Rampino and Stothers, 1984; Schwartz and James, 1984), an unknown Planet X (Whitmire and Jackson, 1985), and even an undetected companion star to the sun named Nemesis (Whitmire and Jackson, 1984). Loper and McCartney (1986) and Loper et al. (1988) suggested that there was a 26-million-year periodicity in mantle overturn within the earth, triggering pulses of volcanism and global climate change that then caused extinctions.

Unfortunately for the pro-impact stampede, several ugly little facts killed their beautiful hypotheses. No evidence for Nemesis or Planet X has ever been found, nor has any evidence tied comet showers or the motion through the galactic plane to mass extinctions (Shoemaker and Wolfe, 1986; Tremaine, 1986; Sepkoski, 1989). Similarly, the mantle periodicity model has been discredited. In fact, the very existence of the 26-million-year extinction cycle has been challenged on statistical grounds (Kitchell and Pena, 1984; Kitchell and Easterbrook, 1986; Hoffman and Ghiold, 1986; Harper, 1987; Stigler and Wagner, 1987; Noma and Glass, 1987; Quinn, 1987). Cladistic taxonomists have criticized Sepkoski's database because it is full of paraphyletic and monotypic taxa that are not real, monophyletic groups, as well as bad taxonomy and misidentifications. When echinoid specialist Andrew Smith and paleoichthyologist Colin Patterson examined the data base in their taxa of specialization and eliminated the mistakes and non-monophyletic groups, the periodicity disappeared (Patterson and Smith, 1987; Smith and Patterson, 1988).

Another problem with the data is the way they are compiled. Since the quality of the data is highly variable, Sepkoski lumped all the data by stages of 3 to 5 million years in duration. This means that all extinctions that happened in a given stage are treated as if they occurred exactly at the end of the stage, even if they were evenly spaced through the duration of the stage. Such a method artificially bunches all the extinctions at stage boundaries and makes a gradual extinction pattern appear catastrophic. The dating is not very reliable either. The time scales have changed so much in recent years that the 26-million-year prediction can succeed or fail depending upon which time scale is used. For example, Raup and Sepkoski (1984,

1986) predicted a late Eocene extinction peak at 39 Ma, and at the time, the Eocene/Oligocene boundary date was disputed, ranging from 36.5 to 32 Ma. Even with time scales in use in 1984, it appeared that their prediction was off. Today, we place the middle/late Eocene extinction at 37 Ma, the Eocene/Oligocene boundary (not much of an extinction) at 33.7 Ma, and the earliest Oligocene extinction at 33 Ma, so none of these match Raup and Sepkoski's (1984, 1986) prediction.

Finally, there is a serious question whether many of the extinction "peaks" are real. The middle Miocene "extinction peak" at 13 million years is based on a few species of molluscs and does not show up in the excellent record of land mammals (Webb, 1977; Barry, 1995). The early Jurassic peak was barely above background noise levels, and Sepkoski (1989) abandoned the mid-Jurassic extinction peak. Some extinction peaks (the late Triassic, the mid-Jurassic, the early Cretaceous, and Pliocene) fall well outside the predicted time interval (Sepkoski, 1989). If only half of the "peaks" appear to be real and occur on schedule, and there are long gaps with no extinction at the predicted 26-million-year interval, what does this imply about the "periodicity"?

Stanley (1990) suggested a much simpler explanation for this apparent periodicity. Major mass extinctions tend to kill the highly specialized taxa, leaving only extinction-resistant generalists known as "survivor" taxa. In the aftermath, it takes many millions of years for diversity to recover and for a wide variety of extinction-prone specialists to evolve and fill the vacant ecological niches. If some extreme event occurred soon after a major mass extinction, there would be no further extinctions, because the only organisms alive would be the extinction-resistant "survivor" taxa. Only after 10 to 20 million years does diversity return with specialized taxa that would be vulnerable to a major environmental perturbation. (Sepkoski, 1990, does not agree with this argument, suggesting that faunas recover much faster than that.) The 26-million-year "periodicity" may simply be a reflection of the time it takes for a fauna to recover before it can feel the effects of the next climate change or impact. This would also explain why the "cycles" are not precisely 26 million years, but vary in duration. An astronomically caused cycle would be much more regular.

CONCLUSIONS

Extinction theory has come a long way from the denial of extinction prior to 1800 to the hottest topic in paleobiology in the 1980s. The enormous volume of research triggered by the asteroid extinction hypothesis has generated much important new data, and many new insights, even if many of the original ideas eventually proved wrong. Extinction theory is no longer the neglected flip side of speciation and evolution, but it has become almost as intensely studied as evolutionary theory. From this research, we have learned a lot about both background and mass extinctions. More important, we have discovered how little we know and how much more there is to learn. We still cannot pinpoint the direct cause of most extinctions in the fossil record, and repeated attempts to find a common cause for mass extinctions have failed. Clearly, there is much more research to be done with extinctions, and paleontologists will be exploring this topic for decades to come.

For Further Reading

Alvarez, L. W., W. Alvarez, F. Asaro, and H. V. Michel. 1980. Extraterrestrial causes for Cretaceous-Tertiary extinctions. *Science* 208:1095-1108.

Archibald, J. D. 1996. *Dinosaur Extinction and the End of an Era: What the Fossils Say*. Columbia University Press, New York.

Donovan, S. K., ed. 1979. *Mass Extinctions: Processes and Evidence*. Columbia University Press, New York.

Ehrlich, P., and A. Ehrlich. 1981. *Extinction: The Causes and Consequences of the Disappearance of Species*. Random House, New York.

Erwin, D. 1993. *The Great Paleozoic Crisis: Life and Death in the Permian*. Columbia University Press, New York.

Glen, W., ed. 1994. *Mass-Extinction Debates: How Science Works in a Crisis*. Stanford University Press, Stanford.

Leakey, R., and R. Lewin. 1995. *The Sixth Extinction*. Doubleday, New York.

MacLeod, N., and G. Keller, eds. 1995. *Cretaceous-Tertiary Mass Extinctions, Biotic and Environmental Changes*. Norton, New York.

Martin, P. S., and R. G. Klein, eds. 1984. *Quaternary Extinctions: A Prehistoric Revolution*. University of Arizona Press, Tucson.

McGhee, G. R., Jr. 1995. *The Late Devonian Mass Extinction: The Frasnian/Famennian Crisis*. Columbia University Press, New York.

Newell, N. D. 1967. Revolutions in the history of life. *Geological Society of America Special Papers* 89:63-91.

Nitecki, M. H., ed. *Extinctions*. University Chicago Press, Chicago.

Officer, C., and J. Page. 1996. *The Great Dinosaur Extinction Controversy*. Addison Wesley, New York.

Prothero, D. R. 1994. *The Eocene-Oligocene Transition: Paradise Lost*. Columbia University Press, New York.

Raup, D. M. 1986. *The Nemesis Affair: A story of the Death of Dinosaurs and the Ways of Science*. Norton, New York.

Raup, D. M. 1991. *Extinction: Bad Genes or Bad Luck?* Norton, New York.

Raup, D. M., and J. J. Sepkoski, Jr. 1982. Mass extinctions in the marine fossil record. *Science* 215:1501-1503.

Raup, D.M., and J. J. Sepkoski, Jr. 1984. Periodicity of extinctions in the geologic past. *Proceedings of the National Academy of Sciences* 81:801-805.

Raup, D.M., and J. J. Sepkoski, Jr. 1986. Periodicity of

extinctions of families and genera. *Science* 231:833-836.

Sepkoski, J. J., Jr. 1989. Periodicity in extinction and the problem of catastrophism in the history of life. *Journal of the Geological Society of London* 146:7-19.

Sharpton, V., and P. Ward, eds. 1990. *Global Catastrophes in Earth History, An Interdisciplinary Conference on Impacts, Volcanism, and Mass Mortality*. Geological Society of America Special Paper 247.

Silver, L. T., and P. H. Schultz, eds. 1982. *Geological Implications of Impacts of Large Asteroids and Comets on the Earth*. Geological Society of America Special Paper 190.

Stanley, S. M. 1987. *Extinction*. Scientific American Books, New York.

Ward, P. 1994. *The End of Evolution: On Mass Extinctions and the Preservation of Biodiversity*. Bantam, New York.

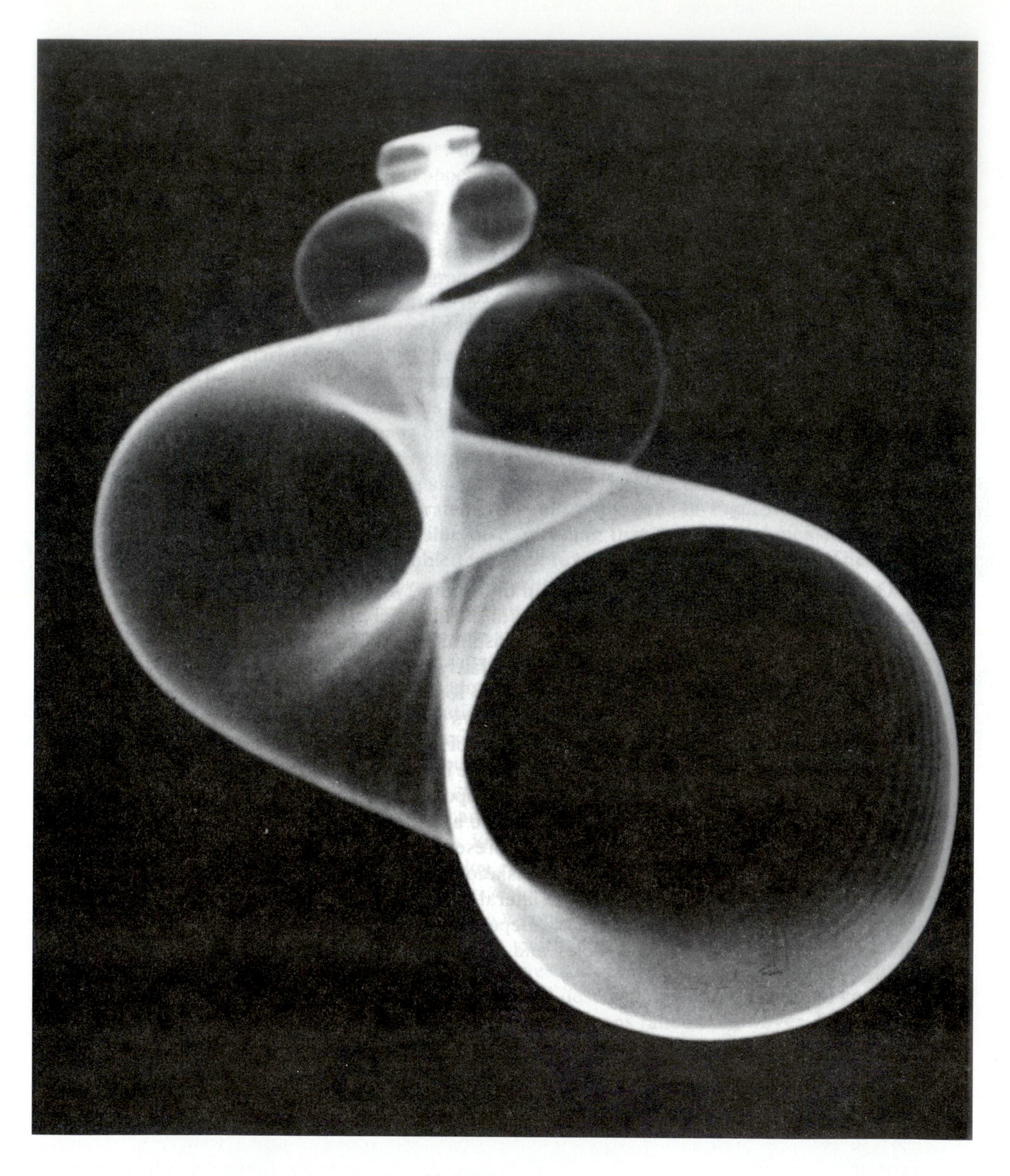

Figure 7.1. Computer-generated trochospiral form that closely resembles the spiral forms of coiled gastropods. Such computer simulations allow us to generate theoretical morphologies that can be compared to real animals. (From Raup and Michelson, 1966.)

Chapter 7

Functional Morphology

The web foot determines, you say, the duck to swim: but what would that avail, if there were no water to swim in? The strong, hooked bill, and sharp talons, of one species of bird, determine it to prey upon animals; the soft straight bill, and weak claws, of another species, determine it to pick up seeds: but neither determination could take effect in providing for the sustenance of the birds, if animal bodies and vegetable seeds did not lie within their reach. . . . Faculties thrown down upon animals at random, and without reference to the objects amidst which they are placed, would not produce to them the services and benefits which we see.

William Paley, *Natural Theology*, 1802

It is on this mutual dependence of functions and the assistance which they lend one to another that are founded the laws that determine the relations of their organs; these laws are as inevitable as the laws of metaphysics and mathematics, for it is evident that a proper harmony between organs that act one upon another is a necessary condition of the existence of the being to which they belong.

Georges Cuvier, *Lessons in Comparative Anatomy*, 1800

It was equally evident that neither the action of the surrounding conditions, nor the will of the organisms, could account for the innumerable cases in which organisms of every kind are beautifully adapted to their habits of life—for instance, a woodpecker or a tree-frog to climb trees, or a seed for dispersal by hooks or plumes. I had always been struck by such adaptations, and until these could be explained it seemed to me almost useless to endeavor to prove by indirect evidence that species had been modified.

Charles Darwin, *Autobiography*, 1887

FORM AND FUNCTION

Since the time of Aristotle, one of the central problems of comparative biology has been determining the function of a particular anatomical form. Another related problem was determining how organisms reached their present forms. Both of these aspects of the problem of form and function have puzzled the greatest minds in biology for many centuries (Russell, 1916). In 1802 the Reverend William Paley wrote an entire book, *Natural Theology*, that marveled on the perfection of design in nature and attributed it all to a divine designer. This explanation was extremely popular among devout natural historians, but as a scientific explanation it was deficient. In Paley's view, whatever exists must do so for a reason, no matter how badly designed it may appear. Under such circumstances, everything can be twisted to fit this worldview. Thus it explains nothing.

Baron Georges Cuvier, the founder of comparative anatomy, was less interested in the theological implications of biological design. Instead, he developed the laws of form and how various shapes and forms were strongly correlated with life habit and function. Cuvier became famous for his ability to predict the unknown aspects of an extinct animal from the parts that were preserved. Hoofed animals typically had grinding teeth for eating plants; carnivorous animals had both sharp, pointed teeth and sharp claws. Knowing one, Cuvier could often predict the other. This was known as the **law of correlation of parts**. In one apocryphal story, a prankster bursts into Cuvier's bedchamber at night, dressed as the Devil and shouting that he would eat Cuvier alive. Cuvier allegedly sat up and coolly replied, "You have horns and hooves, so you must eat plants."

In his student days, Darwin was an avid follower of Paley, so he was well aware of the marvelous designs in nature. Later he became more interested in a non-supernatural way to explain *how* life achieved such perfection of form. When *On the Origin of Species* was published in 1859, the old supernaturalistic explanations of form were replaced by naturalistic efforts to find out how adaptation produced such intricately constructed beings. Since the time of Darwin, this has been the main task of functional morphology: observation of organisms to see how they live and function and determination of the principles that allow us to predict function from bones and shells and whatever else happens to be preserved. In the 1960s and 1970s, functional morphology was one of the hottest areas of

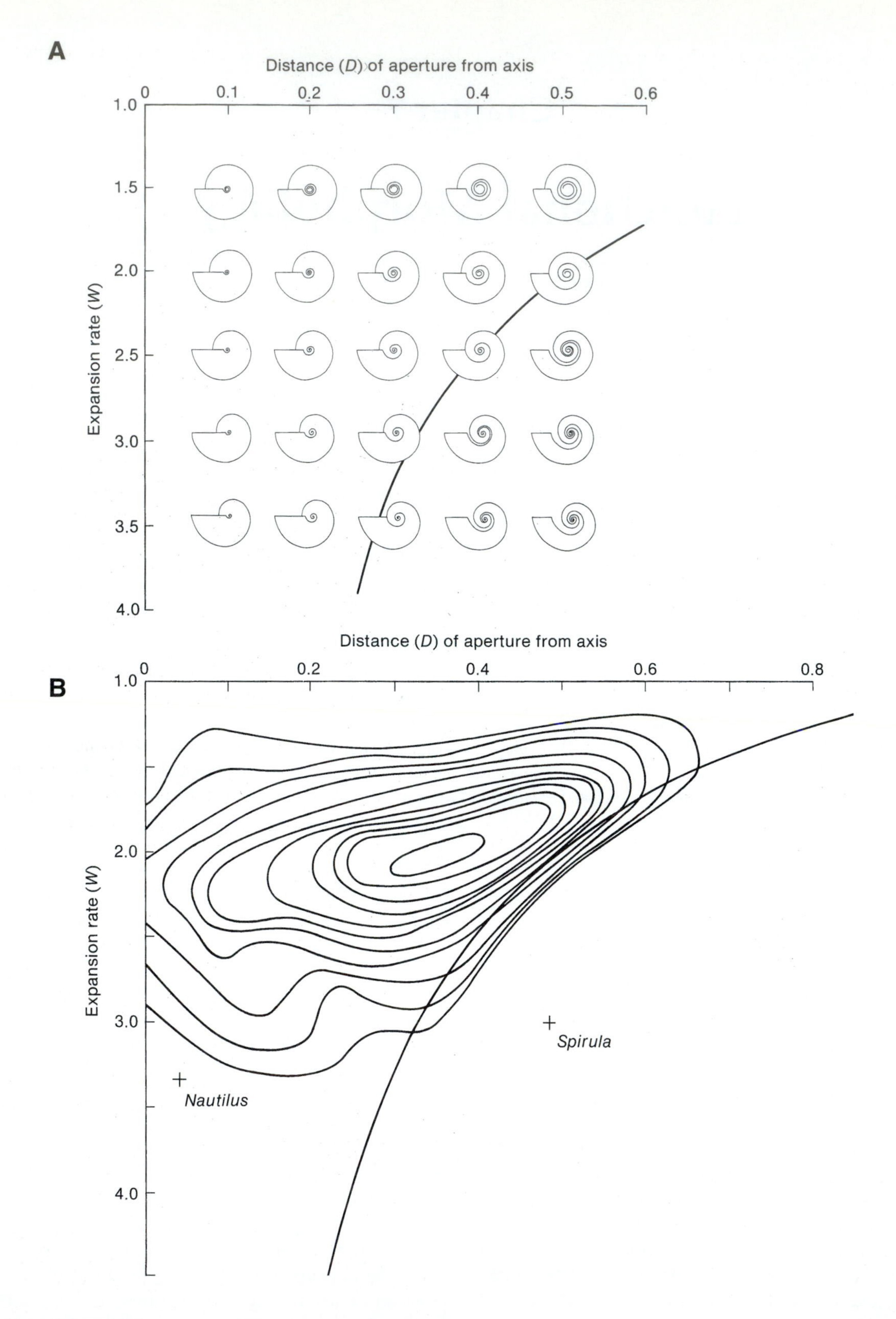

Figure 7.2. The variation in the form of planispirally coiled cephalopods can be summarized by a two-dimensional morphospace, simply by varying expansion rate (W) and distance of aperture from axi (D). At the top is a computer simulation of all possible forms of planispiral shells. The curved line (W = 1/D) separate shells with overlapping whorls from those whose whorls do not touch. At the bottom is a contoured distribution of 405 genera of Paleozoic and Mesozoic ammonoids. Ninety percent of the sample lies within the outermost contour, and none lie outside the curved line. However, the living *Nautilus* has a much higher W than any ammonoid, and the living *Spirula* has a shell whose whorls do not touch (but the shell is internal, so the constraints of the external shell do not apply.) (From Raup, 1967.)

research, both in biology and in paleobiology, and a number of fascinating studies were published, in addition to books that analyzed the biomechanics of organisms (see, for example, Alexander, 1968; Schmidt-Nielsen, 1972; Dullemeijer, 1974; Gans, 1974; Vogel, 1981, 1988; Hildebrand et al., 1985; Rayner and Wootton, 1991; Thomason, 1995, cited at the end of this chapter).

Paleontologists have been no less fascinated than biologists with how extinct organisms functioned. However, it is not possible to directly observe how extinct animals behaved, so the paleontologist must resort to indirect evidence. Before we look at studies about how specific organisms might have functioned, however, we need to ask about the general principles behind evolutionary morphology.

THEORETICAL MORPHOLOGY

The world is so full of a number of things
I'm sure we should all be as happy as kings.

Robert Louis Stevenson, *A Child's Garden of Verses*, 1885

Morphospaces and Adaptive Landscapes

Evolution has produced an enormous variety of living organisms, but not everything that *could* exist *does* exist. Certain shapes occur again and again in nature, such as the fish-like body of swimming vertebrates in the fishes, the ichthyosaurs, and dolphins, or the conical and cylindrical shapes of filter-feeding colonial organisms, such as archaeocyathids, many sponges and corals, richthofenid brachiopods, and rudistid bivalves. Other shapes simply do not occur. A snail shell with no opening would not work very well, nor would a horse with only three legs. These kinds of examples seem trivial, but can we devise a more sophisticated way of analyzing the problem? Is there a theoretical foundation behind functional morphology?

Raup (1966, 1967; Raup and Michelson, 1965) pioneered the study of theoretical morphology, using computers to simulate biological shapes (Fig. 7.1). With the computer, it is possible to generate not only shapes that are based on real organisms but also shapes that do not occur in nature. Figure 7.2 shows that only two variables are needed to simulate the variety of possible planispirally coiled shells. One is the expansion rate (W), which varies from no expansion (essentially a coiled cylinder, as in the top row in Fig. 7.2A) to rapid expansion (a coiled cone, such as in the bottom row of Fig. 7.2B). The other is distance of the aperture from the axis (D), which measures how rapidly the coil spirals outward. From these two variables it is possible to draw the full variety of planispirally-coiled shapes, from snake-like shells with a low expansion rate, to rapidly expanding *Nautilus*-like shells. The curved line (W = 1/D) represents a threshold within the two-dimensional space. Those shells to the left of the line have each successive whorl in contact with the whorl inside them; those to the right are open spirals that do not have each whorl in contact.

Raup (1967) examined the shapes of 405 genera of Paleozoic and Mesozoic ammonoids to see which ones actually occur and with what frequency. These results are contoured in Figure 7.2B. From this diagram, it is clear that ammonoids with W = 2 and D = 0.35 are the most common, and most other shapes occur, but with lesser frequencies. Notice that there is a sharp break at the threshold between open and closed spirals, and no ammonoids used shells shaped like open spirals. There are good functional reasons why this might be so. When the whorls of very thin shells contact each other in a closed spiral, the shells are reinforced. In addition, a volume of dead space between the whorls of the shell creates turbulence and drag when the shell is propelled through the water. Note, however, that the two living shelled cephalopods both fall outside the contoured area for the ammonoids (Fig. 7.2B). The paper nautilus, *Spirula*, has a thin shell that is an open spiral, but the shell lies within its squid-like body, so the hydrodynamic and reinforcement aspects are no longer important. The chambered nautilus, on the other had, has an external shell, yet it has a higher expansion rate than any ammonoid. We do not understand why this is so, but since most nautiloids fall in this region, there must be more complex factors operating here than just the two used in this model.

The three-dimensional version of this model uses not only the expansion rate (W) and the distance from the generating curve (D), but also a third variable, translation rate (T). This is a measure of how rapidly the spiral moves up an axis. Shells with zero translation do not move up an axis, but remain in a plane; they are planispiral, as in Figure 7.2. Those with a high translation rate move rapidly down the axis of coiling, and are trochospiral (Fig. 7.3). With these three variables, we can now describe a three-dimensional **morphospace** of possible coiled shell forms. The expansion rate (W) increases downward from the top surface of the cubic morphospace; the distance of the generating curve from the axis (D) increases from the front of the curve toward the back; the translation rate (T) is zero on the right edge and increases to the left. Arrayed around this cubic morphospace are computer-generated images of the different possible shapes represented by a given combination of variables.

Nearly all gastropod shells (area A in Fig. 7.3) cluster near the top front edge, with a gradual expansion rate, variable values of D, and a variety of translation rates from planispiral to high-spired trochospiral. Ammonoids occupy area B, with relatively low expansion rates and low translation rates (they are all planispiral) but variability in the values of D. By contrast, bivalve shells (area C) are mostly characterized by high expansion rates, but very small values of D, and relatively small translation rates. Finally, brachiopod shells (area D) also have high values of expansion rate, but zero translation (they are always planispiral) and relatively low values of D.

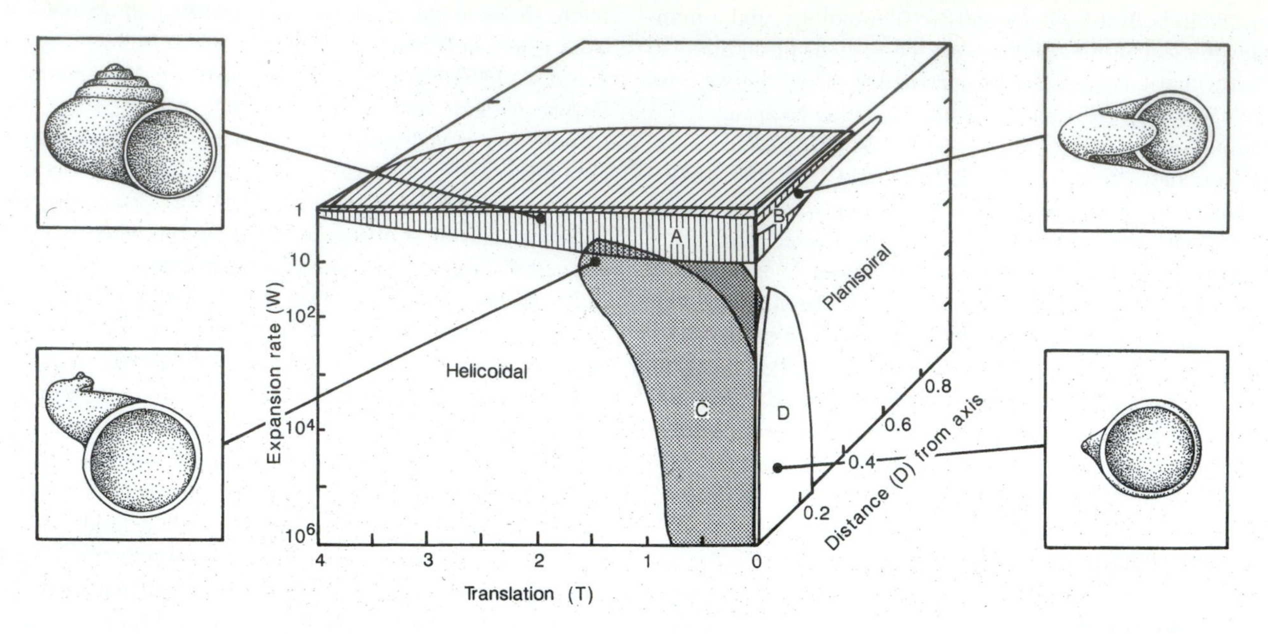

Figure 7.3. Three-dimensional morphospace, which varies not only W and D, but also the rate of translation (T) up an axis. Gastropods occupy the region (A) of slow expansion rate and small to medium D, but their T varies with their spire height. Cephalopods (B) occupy the region of no translation, low expansion rate, and variable D. Bivalves (C) occupy the region of limited D and T, but highly variable expansion rate. Brachiopods occupy the region (D) of no translation and limited D, but variable expansion rate. Most of the morphospace of theoretically possible shells is not occupied by a real shell in nature, so clearly there are constraints on what types of shell is practical versus those that are possible. (Modified from Raup, 1966.)

Since Raup's initial studies, several paleontologists have tried to apply them to ammonites (Ward, 1980; Saunders and Swan, 1984) and biconvex brachiopods (McGhee, 1980), but not yet to gastropods. Schindel (1990) suggests that the method is too difficult to use for gastropods and proposed another, easier way of measuring the coiling of gastropods. Using his methods, it was possible to see more variation within gastropod shell form than in Raup's (1966) original plot.

The striking thing to note about this diagram is that the actual area occupied by real shapes is only a small part of the total possible geometrical shapes represented by this cubic morphospace diagram. As shown in Figure 7.3, almost all real organisms cluster near the upper right corner, and few occupy the morphologies represented by the remaining faces or by the internal volume of the morphospace. Diagrams such as this, and many similar analyses, suggest that real organisms rarely show the full variety of forms that are mathematically possible. Why is this so?

Conventional Neo-Darwinian theory proposed an explanation for this paradox over 60 years ago. In 1932, the geneticist Sewall Wright proposed the **adaptive landscape** model of evolution (Fig. 7.4). In this concept, the various combinations of alleles could be represented as a topographic map of an "adaptive surface." The peaks of the surface were particularly useful combinations of alleles, that might be favored by selection, and the valleys and low spots were less useful combinations. Wright's original formulation of the idea suggested that harmonious combinations of alleles could be achieved by selection, or by random genetic drift. However, later evolutionary biologists took the concept far beyond Wright's original meaning (see Eldredge, 1989). Dobzhansky (1937) and Simpson (1944)

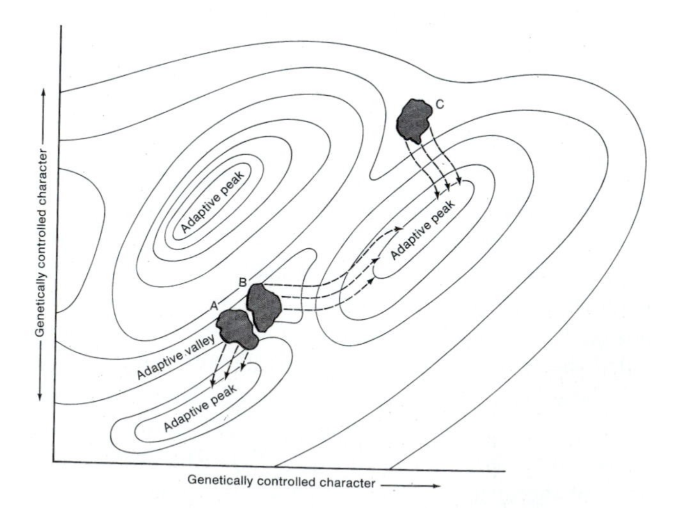

Figure 7.4. Contoured representation of an adaptive landscape, with adaptive peaks of optimality, and valleys where organisms are less than optimally adapted. Populations A and B are climbing different adaptive peaks and undergo divergence, while B and C climb the same adaptive peaks and become convergent. (From Raup and Stanley, 1978.)

treated the peaks not as just combinations of alleles but as *species* driven to the tops of peaks by natural selection. The landscape was constantly changing and shifting, just like the surface of the sea, so species could not "stay put" on an adaptive peak but had to constantly change to track these shifting peaks of optimal adaptation. Those species that could not reach the peaks, or that got stuck in the valleys and did not change, became extinct. For a new species to form, it might have to cross an indaptive valley, which might result in its extinction before the new species could become established. Wright's original concept hardened into a strict Neo-Darwinian explanation that emphasized selection and ignored random genetic drift.

Can this adaptive landscape model be applied to our morphospace concepts just discussed? Figure 7.2 does resemble a landscape in some ways, with a distinct "peak" of the commonest ammonite shell forms, analogous to the "adaptive peaks" of the model. The absence of forms that should occupy the empty morphospace could be explained in several ways. The non-existent forms could be adaptive valleys that are functionally inferior (as the unreinforced open shells seem to be in Fig. 7.2), or they could be biologically impossible. There is also a third alternative: a particular shape might be adaptive, but no organism has found a way to reach this unoccupied peak given its original starting point morphology and its available range of variation. This is an interesting suggestion, but it is hard to find an example that supports it.

In recent years, the entire idea of an adaptive landscape has been under attack. After all, it is purely an abstract concept, not something based on empirical observation. The clumping of organisms in morphospace may be due to adaptive peaks, but it is hard to test such a hypothesis (Lewontin, 1978). Moreover, the idea that species are infinitely flexible and shift constantly in response to changes to their environment has also been challenged (see Chapter 5). If species are static over millions of years, and do not change even in the face of enormous environmental changes (Prothero and Heaton, 1996), then it does not seem likely that the adaptive landscape actually exists. Some have suggested that species are affected more by biotic interactions with other species, which may be stable over longer periods of time than the physical environment, and even survive climatic changes.

Morphospaces and Ecological Niches

Van Valkenburgh (1985) provided another example of a morphospace analysis, using carnivorous mammals from modern and extinct mammalian faunas. She chose several variables that best distinguished the different predatory behaviors of carnivorans. Three characteristics of the skeleton were used to capture the variety of locomotory adaptations: body weight (LBW in Fig. 7.5); the ratio of hand bones to finger bones (metacarpal/phalangeal ratio, or MCP), with open-habitat runners having large MCP values, and forest-dwelling ambush predators having lower MCP; and the ratio of the thigh bone to the toe bones (femur/metatarsal ratio, or FMT), with positive FMT values correlated with short toes, and with slow-moving, ambush predators, and negative FMT values correlated with longer toes and with running.

The modern Serengeti savanna in Africa (Fig. 7.5A) had 13 carnivore species, all of which were tightly clustered toward running (except for the ratel, #12, a wolverine-like mongoose relative that is partly adapted for digging), not surprising considering that they live in an open grassland. The temperate forests of Yellowstone National Park (Fig. 7.5B), on the other hand, had only 10 species of predators, but they were widely spread out, with bears (#15 and 16) occupying a very different morphospace (short-limbed, large body weight) from the more running-adapted cats and dogs. Once again, the digging badger (#22) occurs near the front of the morphospace in the most negative MCP values. The tropical rainforests (Fig. 7.5C and D) of Malaysia and Chitawan in Nepal tended to have very tight clusters of carnivores, which as a group were adapted more for ambush (lower MCP and higher FMT values) than for running.

With these modern analogues for comparison, it is interesting to see how the Oligocene carnivores of the Big Badlands of South Dakota compared (Fig. 7.5E). Most of them clustered in the same area as the tropical rainforest ambush carnivores, even though the evidence from ancient soils seems to indicate a mixed forest-open habitat (Retallack, 1983). There were none with the body form of bears, so that morphospace (shaded area to left) was unoccupied. Three species fell completely outside the range of modern animals: the saber-toothed nimravid *Hoplophoneus* (#1), and two large-bodied, hyena-like species of the creodont *Hyaenodon* (#2 and 3). This suggests that sabertooths are a type of ambush predator completely unlike anything alive today. *Hyaenodon* is more difficult to interpret. None are as fast runners as living hyenas or other living predators, and their skulls and teeth are a mixture of features for slicing and bone crushing, unlike living hyenas. Their position outside the modern morphospace suggests that they have no modern analogue.

Diversity and Disparity

The concept of theoretical morphospace has been applied to test another set of hypotheses. In his provocative book *Wonderful Life*, Gould (1989) proposed that the great range of taxa and shapes seen in the Cambrian Burgess Shale fauna contradicted the usual notion that life started out simply and became more complex. Instead, life was a complex, branching bush at the very beginning that has been successively pruned by extinction. Gould (1989) suggested that this morphological stability since the Cambrian radiation might be due to decreasing developmental flexibility, as the major phyla became stereotyped in certain body plans and could not make radical changes once they had reached this embryological rut. Briggs et al. (1992) and Ridley (1990) criticized this hypothesis, largely on the grounds that the apparent diversity of Cambrian

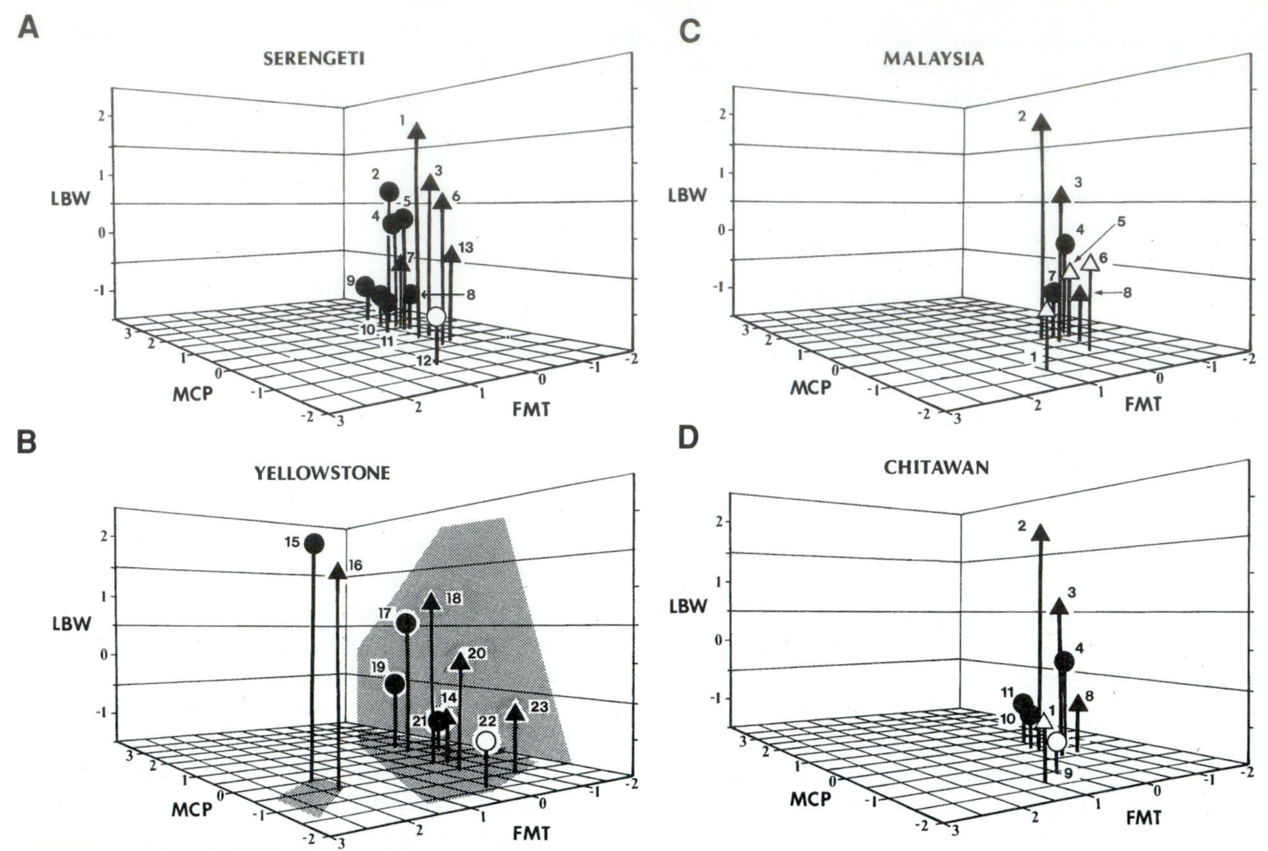

Figure 7.5. A three-dimensional morphospace plot of carnivorous mammals shows their clustering by ecological specializations. The vertical axis is body weight (LBW), and the horizontal axes are metacarpal/phalangeal ratio (MCP) and femur/metatarsal ratio (FMT). Arboreal carnivores are shown by open triangles, scansorial (limited climbing) forms by solid triangles, terrestrial carnivores by solid circles, and burrowers by open circles. Species are as follows: 1, lion; 2, spotted hyena; 3, cheetah; 4, wild dog; 5, striped hyena; 6, leopard; 7, serval; 8-10, jackals; 11, civet; 12, ratel; 13, caracal; 14, bobcat; 15, grizzly bear; 16, black bear; 17, wolf; 18, puma; 19, coyote; 20, wolverine; 21, red fox; 22, badger; 23, lynx. The Yellowstone carnivores occupy a greater variety of guilds than do the Serengeti carnivores. In the Malaysian and Chitawan predator guilds (upper right), an even smaller range of locomotor types are present. Species are: 1, binturong; 2, tiger; 3, leopard; 4, dhole; 5, Temminck's cat; 6, clouded leopard; 7, civet; 8, fishing cat; 9, ratel; 10, jackal; 11, civet. In the early Oligocene predatory guilds (below), an even smaller range of locomotor types are represented, and some have no modern analogs. Species are: 1, 6, *Hoplophoneus* (saber-toothed nimravid); 2-3, *Hyaenodon* (creodont); 4, 7, *Daphoenus* (beardog); 5, *Dinictis* (dirk-toothed nimravid).

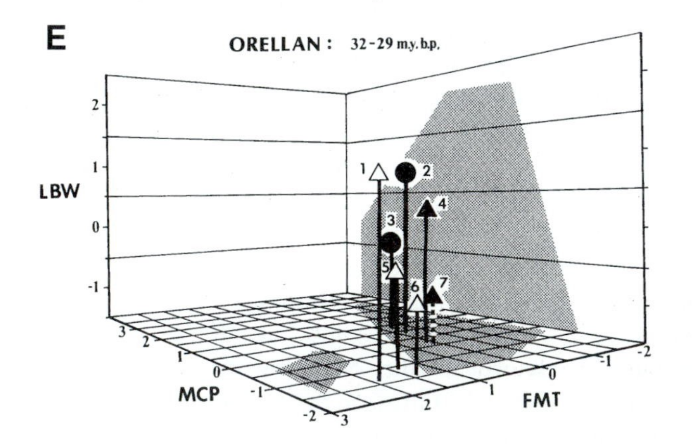

life forms might be exaggerated by taxonomic splitting of each unusual specimen into a new phylum or class. According to Briggs et al. (1992), the actual variation of *shapes* (**disparity**) is much less than the variation in number of *taxa* (**diversity**). Gould (1991) and Foote and Gould (1992) answered these criticisms and then a second round of argument followed (Ridley, 1993; Gould, 1993; Wills et al., 1994). As McShea (1993) points out, the argument reached an impasse, since there was no method of measuring how morphological disparity had changed through time.

Wills et al. (1994) attempted to answer this question by generating a multivariate morphospace for the arthropods of the Burgess Shale and the living world. They found that there was no significant difference in the morphological disparity between the Cambrian and the Recent. However, their study looks at only two time frames and does not answer the question of how disparity has changed during the time between the Cambrian and Recent. In addition, the Burgess Shale sample did not include some of the truly weird organisms, so they did not get the maximum possible disparity, and the modern sample included only one member from each living class of arthropods. Nevertheless, the fact that they found no real difference in disparity between the Cambrian and Recent suggests that the arthropods reached their maximum disparity early in their history, which agrees with Gould's suggestions.

Foote (1992, 1993a, 1993b, 1995) provided the first empirical data that might test these hypotheses. Using first the blastoids and trilobites (Foote, 1992, 1993a, 1993b) and then the crinoids (Foote, 1995), he coded their shapes by multivariate methods, reducing a large number of shape variables to just two principle components (PC1 and PC2 in Figs. 7.6 and 7.7). These simple two-dimensional morphospaces for all the trilobites or blastoids in a given time period could then be stacked in stratigraphic order, and the position of the specimens in this time-space continuum could be tracked. The trilobites (Fig. 7.6) reach their maximum disparity early in their history, and then disparity gradually decreases, agreeing with Gould's prediction (although reaching a maximum in the Ordovician, not in the Cambrian as Gould would have suggested). The blastoids (Fig. 7.7), on the other hand, reach their maximum disparity at the end of their history, contrary to what Gould predicted. It will be interesting to see whether such analyses of other major groups produce more cases that show disparity was highest in the beginning or has increased through time.

Another interesting comparison can be made with Foote's (1993b) data. How do diversity and disparity track each other? After all, the original debate focused on the problem of using taxonomic diversity as a means of measuring the range of shapes present, but now we have a more direct measure of morphological variability. In the case of the trilobites (Fig. 7.6), both diversity and disparity peak early, although their peaks do not exactly coincide. In the blastoids (Fig. 7.7), however, diversity peaks in the Early Carboniferous, but disparity is higher in the Late Devonian and Permian, and low in the Early Carboniferous. Clearly, the two are not tracking one another in this taxon. This raises serious questions about using taxonomic data such as diversity to analyze questions about how morphology has changed through time. It will be interesting to see how this line of research continues in the next few years.

FUNCTIONAL HYPOTHESES AS TESTABLE SCIENCE

> *Pangloss proved admirably that there is no effect without a cause and that, in this best of all possible worlds, My Lord the Baron's castle was the finest of castles. "Things cannot possibly be otherwise, for everything being made for an end, everything is necessarily for the best end. . . . Everything is made for the best purpose. Our noses were made to carry spectacles, so we have spectacles. Legs were clearly intended for breeches, and we wear them."*
>
> Voltaire, *Candide*, 1759

Panselectionism and the Adaptationist Programme

The fossil record is full of strange and weird animals that puzzle and fascinate us. Probably the first questions we want to answer are, "How did this animal live? Why is it shaped the way it is?" A 40-ton *Brachiosaurus* practically cries out for explanations about how it fed, how it supported its massive weight, how it pumped blood all the way up to its head five stories above the ground, and whether it was warm-blooded or cold-blooded. The existence of a pterosaur the size of a small airplane demands answers to questions such as: How did it fly? Could it merely soar, or did it use active flight? How did it feed? Other fossils may be less spectacular, but paleontologists have been no less intrigued by them and ingenious in trying to explain how they lived.

In doing so, however, scientists have often let their imaginations run way ahead of the data. Until recently, it was common to see stories in scientific meetings and popular books that were more like the "just-so stories" of Rudyard Kipling. Kipling told charming tales about how the elephant got its trunk by having its nose seized by a crocodile and pulled out, or how the camel got its hump by saying "harumph" too much. Similarly, scientists would propose an adaptive explanation for a particular feature in an animal, and that was all—the story stands or falls on its own plausibility and on how well the author has told it. Some of these sound positively silly in retrospect. For example, in *The Naked Ape* (1967, p. 75), Desmond Morris suggests that breasts are unusually large and rounded in human females (compared to most other primates) because males were used to being sexually stimulated by large rounded buttocks when they copulated from behind. According to Morris, when humans shifted to upright posture and face-to-face copulation, females needed large breasts to attract males to copulate from the front. In *The Aquatic Ape* (1982), Elaine Morgan proposed that humans are hairless because they were adapted for swimming. Even the beginning student can think of problems with both of these stories, but the fundamental one is that they are not testable, scientific hypotheses. These ideas were just tossed out to explain the available facts, and no attempt was made by their authors to evaluate them further (if there is any way to test them, which is debatable).

Some attractive, popular scenarios have not only been resistant to testing, but have even hampered investigation. Since the discovery of the earliest human fossils, anthropologists were convinced that the brain was the salient future that made us human, and therefore humans must have acquired their large brains early in their evolution. Therefore, any fossil that had a small brain, but other human features (such as upright posture) was simply another ape. When Raymond Dart discovered the first specimen of *Australopithecus africanus* in 1924 and argued that it was an early hominid, European anthropologists scoffed because his specimen had a small brain (even though it had an upright posture). At the same time, the Piltdown hoax was hailed as our ancestor, because it had a

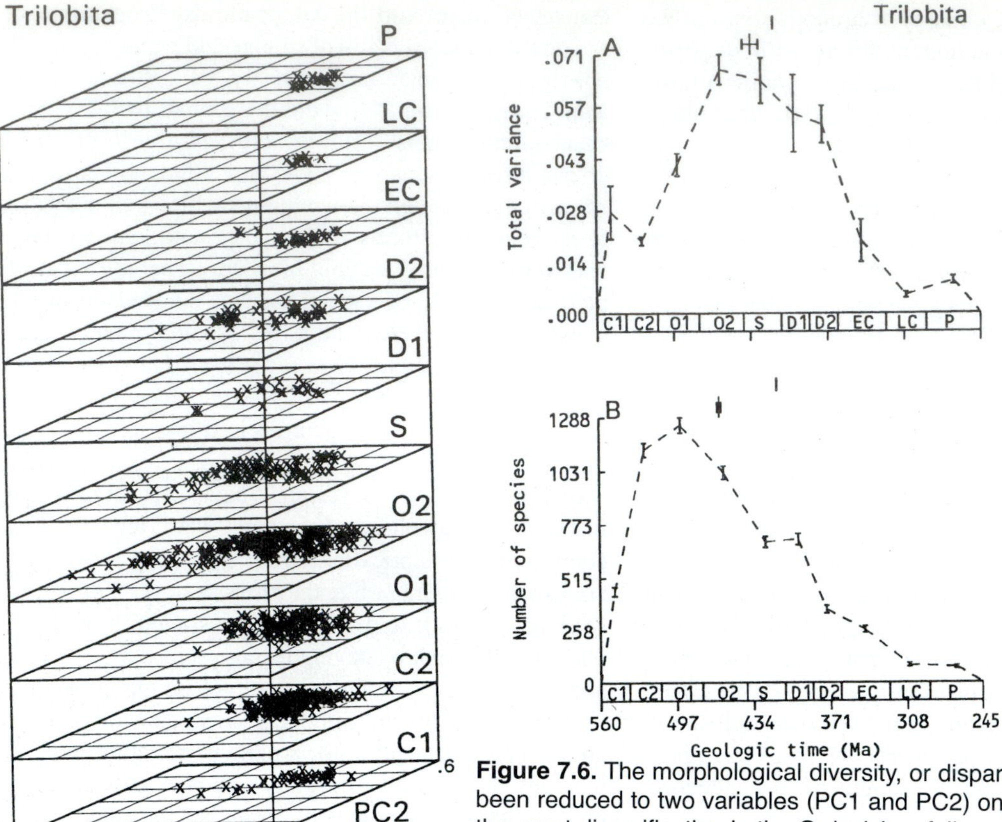

Figure 7.6. The morphological diversity, or disparity, of the trilobites has been reduced to two variables (PC1 and PC2) on the plot on the left. Notice the great diversification in the Ordovician, followed by a decline. In the plots above, the disparity peaks in the Ordovician, but the taxonomic diversity peaks in the Cambrian and declines sooner. (After Foote, 1993.)

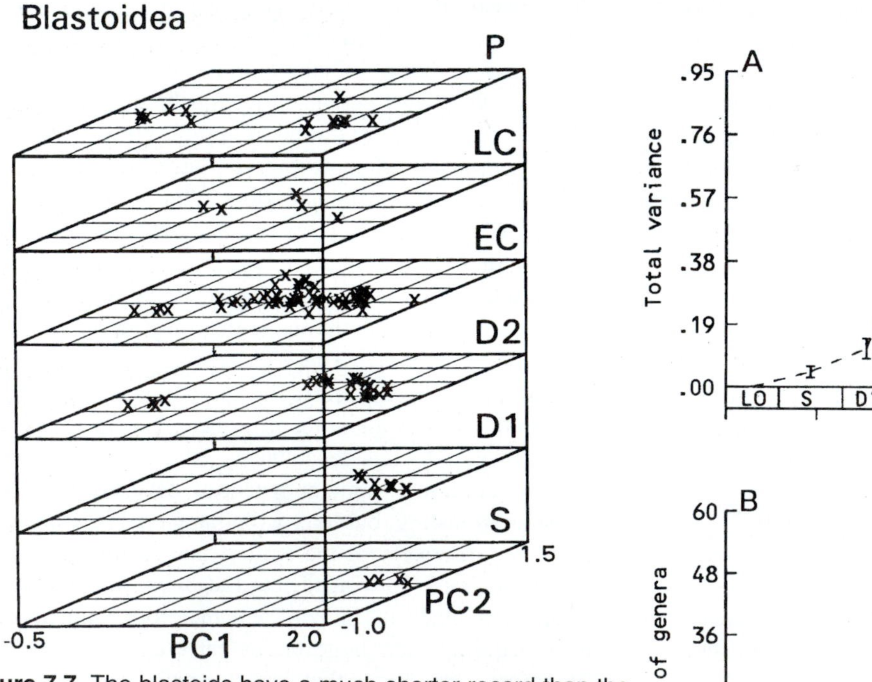

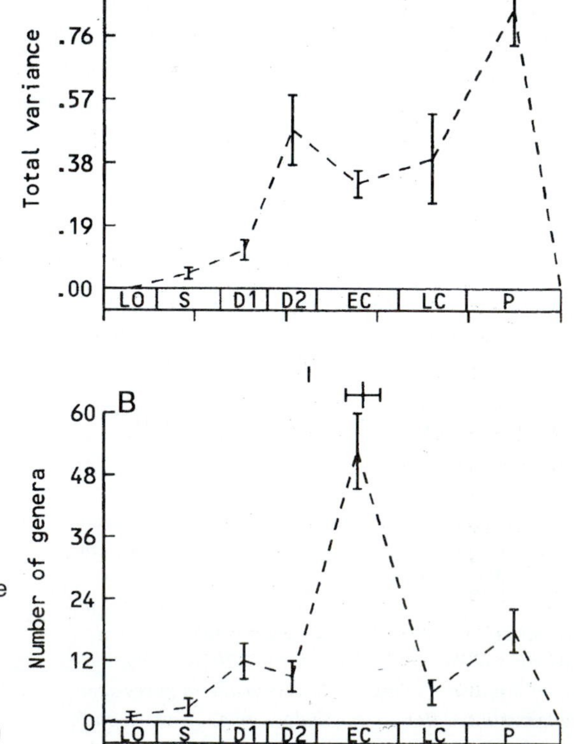

Figure 7.7. The blastoids have a much shorter record than the trilobites, but appear to show their maximum disparity in the Permian in the plot above and in the plot in the upper right (even though this disparity is based on far fewer taxa). Their taxonomic diversity, however, peaks in the Early Carboniferous. Clearly, there is no direct association between diversity and disparity in blastoids, and disparity does not peak at the beginning of their evolution. (After Foote, 1993.)

large brain but an ape-like jaw (as the hoaxer knew his peers would expect). As time passed, however, it became more and more clear that upright posture occurred much earlier in human evolution than larger brain size. More and more australopithecine specimens proved this in the 1940s and 1950s. When the Piltdown hoax was exposed in the 1950s, it forced anthropologists to admit that their expectations had colored their perceptions of the human fossil record (see Lewin, 1987, for an excellent account of other cases where anthropologists have been misled by their expectations).

The most extreme form of story-telling has been labeled the **adaptationist programme** by Gould and Lewontin (1978). In classic panselectionist Neo-Darwinism (see Chapter 5), selection is an all-powerful force that is constantly acting upon every part of the anatomy. No feature escapes the scrutiny of selection. Organisms are typically atomized into individual parts, and an adaptive explanation is suggested for each part. However, when a particular adaptive explanation fails, the Neo-Darwinian quickly substitutes another. "The notion that suboptimality might represent anything other than the immediate work of natural selection is usually not entertained" (Gould and Lewontin, 1978). They compare this kind of explanation to the ideas of Dr. Pangloss, the character in Voltaire's *Candide* who constantly voices the Leibnizian view (also seen in Paley's *Natural Theology*) that in the best of all possible worlds, everything is created for a purpose. Extreme panselectionism is truly a "Panglossian paradigm," with selection (instead of God) acting so that everything has a purpose.

Functional Constraints and Some Caveats

Instead of this approach, a number of paleontologists (Schindewolf, 1950; Remane, 1977; Seilacher, 1970, 1972; Dullemeijer, 1974; Grassé, 1977; Gould, 1980) have advocated a more pluralistic approach to functional morphology. In their view, there are more factors acting upon a structure than just the direct force of natural selection. Some of the things to keep in mind include:

1. Structural constraints—Many features of an organism are simply a product of fundamental design constraints. If certain stresses are placed on a shell or bone, other features are required just as a fact of engineering. To illustrate this concept, Gould and Lewontin (1978) use an analogy of the "spandrels of San Marco." In this famous Venetian cathedral, the spandrels (the tapering triangular spaces between two arches at right angles) are decorated with religious art. A panselectionist (or Dr. Pangloss) might argue that the spandrels were created to place the art on, but this misses the point. The spandrels are a necessary architectural element required by engineering to support the arches; without them the building would fall. The art is purely decorative, and has no bearing on the function. Similarly, many Neo-Darwinist "just-so stories" are actually explanations for the "decoration" on a fundamental structure that must be there for engineering reasons. For example, many people have marveled at the beauty of the hexagonal pattern seen in the lens of an insect's compound eye, or the structure of a honeycomb, or the hexagonal pattern of some colonial rugose corals, such as *Hexagonaria*. As D'Arcy Thompson (1942) pointed out, however, such a regular hexagonal pattern is the necessary result of the closest possible packing of spheres or cylinders (such as a honeycomb cell) that then form flat, compromise boundaries. There is no necessary special story for such a structure; it simply results from closely packed cylindrical units. In another example, an organism can start only with its fundamental body plan (frequently called *Bauplan*, since the Germans are the foremost advocates of the concept). Molluscs have a basic body plan that ensure that none will ever fly. Arthropods have a different body plan that gives them flexibility to modify the segments and appendages, but the fact that they are soft and vulnerable when they molt means they can never become gigantic.

2. Evolutionary heritage—An organism can build a new anatomical feature only out of the raw materials that were furnished by its ancestor. If the elements were already present in the system, then they require no special adaptive explanation. Gould (1980) describes a case of an adaptationist functional morphologist asserting that land vertebrates have four legs because that is the optimal number for land locomotion. This person forgot that tetrapods have four limbs because they happened to have evolved from fish that had four primary fins, not because four limbs had some special significance. After all, ostriches run even better on two legs, insects do very well on land with six legs, spiders with eight, and millipedes and centipedes with dozens to hundreds.

3. Pleiotropy—Many features are linked to others genetically by a process known as **pleiotropy**, where a single gene has multiple effects. If the selection is strong enough on a given feature, the others that are linked to it may be "carried along" passively, even if they have no positive effect, or are even slightly disadvantageous. For example, the adaptationist might speculate about the selective forces that give us a big toe and might argue that it is favored to make our bipedal locomotion more efficient (a podiatrist would disagree). But in our genome the big toe is linked to the thumb, and there are strong selective forces that act to give us our large, opposable thumb, passively giving us large big toes as well. Actually, our large big toe is a hindrance to locomotion. Witness how we have reduced it and lost our opposability compared to less bipedal apes, such as the gorilla and chimpanzee. An ideally designed bipedal foot would be much more symmetrical (as they are in hoofed mammals and carnivores that run).

4. There may be no selective advantage whatsoever—Some features may simply have no selective advantage or may not even be true anatomical "features" in the sense that selection could act upon them. For example, many scientists have speculated about the advantages of the human chin, arguing that it makes male beards more prominent for sexual selection, and other absurd ideas. But the chin is not a trait, rather it is passively affected by

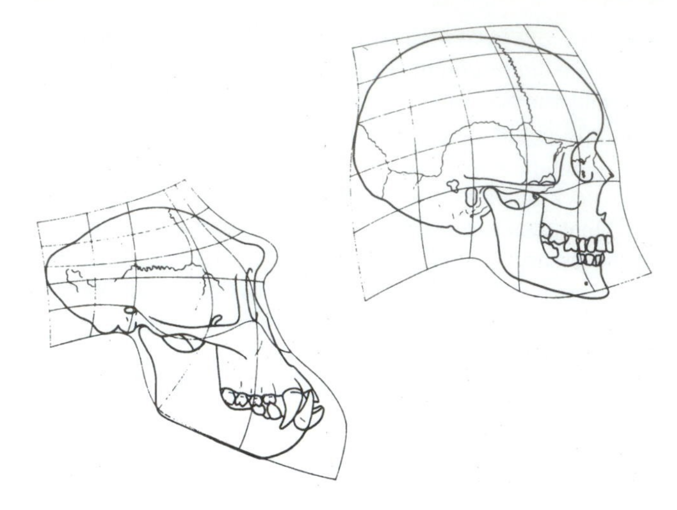

Figure 7.8. Many scientists have proposed adaptive explanations for the function of the human chin, but the chin may simply be an artifact of the way in which humans have reduced their front teeth (notice how the grid is much less deformed in the box around the snout) compared to the ape) Retraction of the tooth row simply left the chin "hanging out." (Modified from Gould, 1977.)

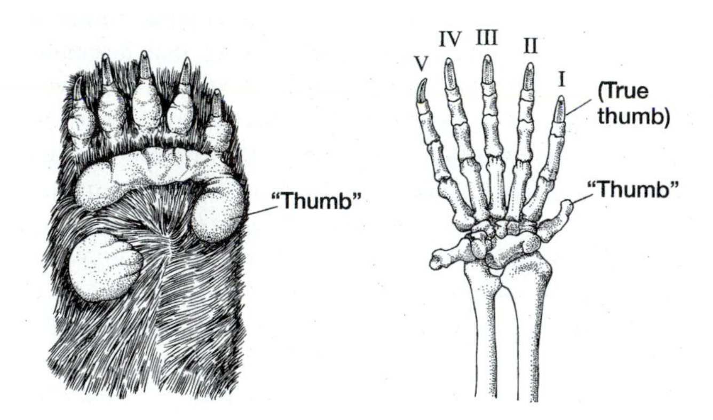

Figure 7.9. Not ever feature is optimally adapted for a purpose. The hand of the panda is built like other carnivorans with all five fingers (including the true thumb) fused into a paw. Yet the panda lives on bamboo rather than meat, and has modified a wrist bone (the radial sesamoid) as a clumsy, jury-rigged "thumb" which it can use to grasp and strip leaves off bamboo. It is not as effective as a real thumb, but it works just well enough for the panda to survive. (From Kardong, 1995.)

another trait—the retraction of the tooth row as humans have reduced their front teeth, particularly the large stabbing canines (Fig. 7.8). In our change in diet, there has been strong selection to reduce the front teeth, but selection has ignored the chin, which "hangs out" since there is no reason for selection to act upon it (Gould, 1977; Lewontin, 1978).

5. Not every feature is optimally designed—Neo-Darwinian panselectionism emphasizes the perfection of the design of organisms to prove that selection must be involved in producing them. But as Gould (1977, 1980b) points out, this is the same evidence that Paley (1802) used to prove the existence of a Divine Designer. It is nature's *imperfection* that falsifies the notion of a perfect designer. Unlike God, natural selection works with whatever raw materials are available to produce adequate, if suboptimal, organisms. All it needs to be is good enough to permit survival, not optimal. Gould (1977, 1980b) cites a number of examples, but one of the best is the panda's thumb. Pandas are true carnivorans (related to bears) that have become secondarily herbivorous, feeding almost exclusively on bamboo shoots. Their original fingers and thumb form a solid paw, so they cannot oppose their true thumb to grasp. Yet they strip the leaves off the bamboo as they feed by running it between their paw and a sixth digit, a false "thumb" made out of a wrist bone known as a radial sesamoid (Fig. 7.9). This "thumb" is not very flexible or strong, but it works just well enough for the panda to survive. Gould (1977, 1980b) points to many other examples of jury-rigged devices—a spine on an angler fish that looks enough like a small fish to lure its prey; a brood pouch in the freshwater clam *Lampsilis* that looks just enough like a fish to attract bigger fish who will bite the pouch and then carry the clam's larvae on their gills; the crude facsimiles of insects that orchid flowers use to lure their pollinators—that are not ideally designed, but are adequate to enhance the survival of the organism. Similarly, in functional studies, we should not assume that every feature is optimally adapted, but may just work well enough to get by.

6. The correlation between structure and function is not perfect— Although muscles usually leave muscle scars on the bone or shell, not every feature is obvious. Some muscles leave no scars; in other cases, large roughened areas on a bone that appear to be muscle scars have no muscle attached to them. These facts give good reason to doubt the elaborate reconstructions of extinct animals with all their muscles shown in detail. Likewise, there are many animals whose external body form shows no obvious adaptation to their current existence. Darwin (1859) mentioned the water ouzel (also known as the dipper), a small bird that walks along the bottoms of streambeds finding its insect prey. As he described it, "the acutest observer by examining its dead body would never have suspected its sub-aquatic habits; yet this bird, which is allied to the thrush family, subsists by diving—using its wings under water, and grasping stones with its feet" (Darwin, 1859). Darwin also pointed to examples of wasps that swim with their wings, birds with webbed feet that never swim, and woodpeckers that feed on fruit and never use their bills to dig in wood. None of these animals has changed their morphology to reflect their new behaviors. Cases such as these should make the paleontologist a bit more cautious when asserting that "this extinct animal had this feature, therefore it must have done this activity."

Recently, biologists have developed methods of actually observing the structure as it is used by the animal, and recording the firing of nerve impulses that control the muscles in the structure. The most surprising conclusion is that structures do not always work as their shape and muscles would have predicted. In many cases, the muscles do not act as expected, and the nerve impulses are not recorded when they should be. Lauder (1995) concludes that the control and modification of nerve networks is much more flexible than the change in the structure, so that new nerve pathways can be generated while the structure remains the same, preserving the vestigial shape of a function it no longer performs.

7. Structures may have more than one function—To conduct a functional analysis, the scientist tries to hypothesize a simple cause-and-effect relationship between a single function that might explain the structure. However, the more scientists look at the actual function of structures, the more they find that they serve multiple functions, some of which are entirely unexpected; some structures that appear to be designed for one function are used for something completely different (Lauder, 1995). In other words, real structures are much more complex than the simple hypotheses we use to explain them. "Given this track record, it is perhaps unwise to place much faith in detailed predictions of function from structure, especially those predictions involving taxa and structures in which no experimental test can be conducted to test function independently" (Lauder, 1995, p. 15).

Testing Functional Hypotheses

We have seen some of the possible pitfalls in trying to explain the functional anatomy of fossils. A number of authors (Fisher, 1985; Greene, 1986; Padian, 1987, 1995; Hickman, 1988; Lauder and Liem, 1991) have suggested methods of setting up functional hypotheses in a rigorous, testable framework. Most of them agree on the following:

1. *Define and diagnose* the adaptation that is being analyzed, and determine which taxa have this feature or adaptation.

2. *Conduct a phylogenetic analysis to determine the state from which the feature evolved*. This allows the scientist to examine the evolutionary heritage of the feature and to decide how much can be attributed to the basic building materials that the organism inherited from its forbears.

3. *Examine the possibility that the character is a structural or engineering artifact*. This is comparable to the "null hypothesis" in statistics, or the "hypothesis of no difference." In this case, if the feature is simply an engineering necessity, then there is nothing unusual or extraordinary to be explained, and no further analysis is warranted.

4. *Propose a hypothesis to explain how the structure once functioned*, after historical, phylogenetic, and structural constraints have been considered. Most often these hypotheses come from comparing the structure to those seen in its living relatives, which is a classic uniformitarian approach. However, in some organisms, there are no living relatives, so then we must seek unrelated organisms that have similar structures (what are known as **analogues**). For example, we can use dolphins or fish as analogues for the swimming behavior and life habits of ichthyosaurs, since they are so similar in many ways, although not closely related. For some truly bizarre structures, however, there are no living relatives or good analogues. Then we must find whatever idea seems reasonable and work from there. Most scientists simply use their knowledge of anatomy and their (sometimes *too* vivid) imaginations to generate hypotheses.

A good example of how novel hypotheses can be generated was called the **paradigm method** by Rudwick (1964). Rudwick (1961) attempted to understand how the peculiar cone-shaped productid brachiopods known as the richthofenids might function (Fig. 7.10). He hypothesized that they used their tiny lid-like brachial valves (which serve as a "cap" in the open mouth of the cone-shaped pedicle valves) as a flapping device to pump water past

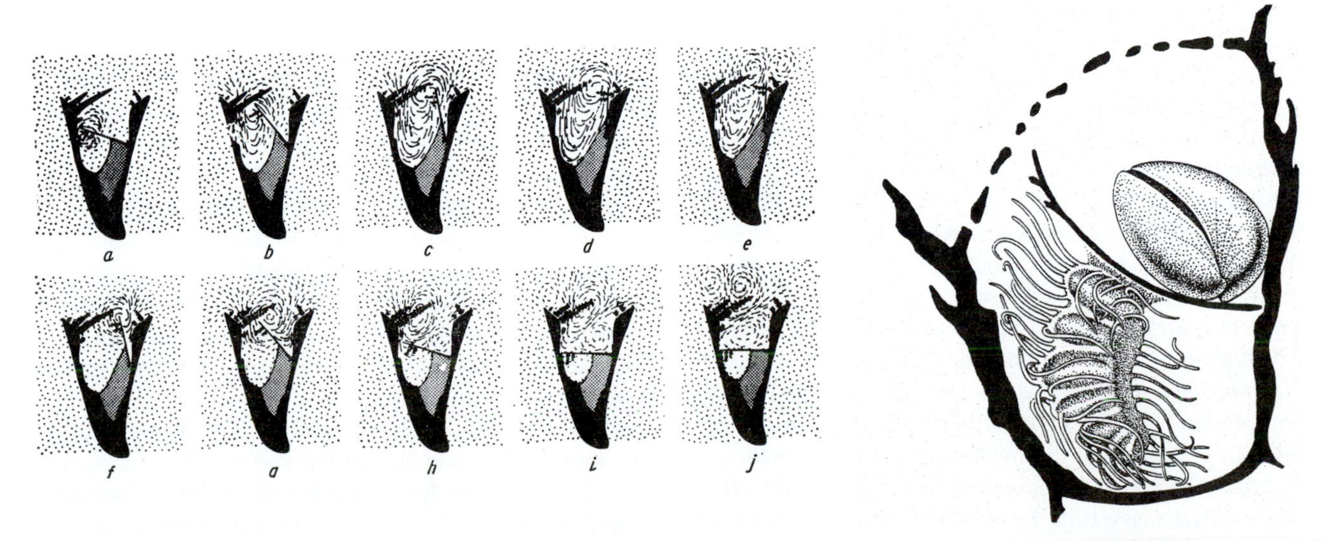

Fig. 7.10. (Left) Reconstruction of the feeding mechanism of Prorichthofenia, showing the currents generated by flapping the dorsal valve. (After Rudwick, 1961.) (Right) Grant (1975) illustrated this specimen, which has another brachiopod cemented in life position on top of the dorsal valve, showing that it could not have flapped up and down as a pump.

their lophophore and aid in filter feeding. This is an unusual proposal, because most brachiopods pump the water through their shells by ciliary action. He constructed a mechanical model, a sort of "robot richthofenid" in an aquarium, made its brachial valve flap and pump, and watched the flow of currents using colored dyes in the water. As expected, the flapping mechanism moved the water through very efficiently, and so seemed to explain the peculiar shape of these animals. Rudwick (1964) called such a model a **paradigm**; in his words, it is "a structure that would be capable of fulfilling the function with the maximal efficiency attainable under the limitations imposed by the nature of the materials" (Rudwick, 1964, p. 36).

The paradigm approach has been criticized because it assumes optimality for the structure, which is not necessarily always the case in a suboptimal, jury-rigged world, and because it assumes only one function for a structure. It also focuses on the mechanical aspects of the structure, ignoring the possibilities of architectural constraints and the importance of soft tissues (Hickman, 1988). Signor (1982) argued that although a paradigm can show an optimal design for a form, there are no rigorous criteria for the minimum resemblance between the model and the organism needed to justify its function. But Fisher (1985) pointed out that the problem can be framed in terms of probability and suggested a test of a minimally sufficient or threshold conditions for achieving a particular functional effect.

The paradigm method has also been criticized (Grant, 1972, 1975) and defended (Cowen, 1975; Paul, 1975) as a method for explaining richthofenid feeding. Grant (1972) argued that no other brachiopods seemed to require such an unusual method, so it made no sense that only richthofenids required more than normal ciliary pumping. In addition, the lophophore would have been attached to the flapping brachial valve, weighing it down. Finally, the flapping valve mechanism was not a very good pumping mechanism, and Grant (1972) suggested that there were no actual examples of such a pump. Cowen (1975) pointed to a Chinese blast furnace that had a flapping door that pumped air to stoke the fire. But Grant (1975) clinched his case against the flapping valve mechanism by describing a specimen (Fig. 7.10) that had another smaller brachiopod trapped in life position on top of the brachial valve. Apparently it had landed there as a larva and grown in that space, causing the brachial valve of the richthofenid to warp as it grew larger. Clearly, it could not have flapped its valve with another brachiopod wedged in the space.

4. *Test the hypothesis*. Go beyond the "just so story" level of analysis and look for evidence that can critically rule out part or all of the hypothesis. If the hypothesis is falsified, then look for another and test it as well. If the original hypothesis fails the test, do not try to rescue it with special pleading. Once this happens, the hypothesis has left the world of testable science, and becomes nothing more than belief.

CASE STUDIES IN FUNCTIONAL MORPHOLOGY

The flowering of functional morphology has yielded a panoply of elegant individual examples and few principles beyond the unenlightening conclusion that animals work well. . . Newtonian procedures yield Newtonian answers, and who doubts that animals tend to be well designed?

Stephen Jay Gould, "The promise of paleobiology as a nomethetic, evolutionary discipline," 1980

The literature of paleobiology is full of functional analyses of many different extinct organisms. Some of these amount to little more than "just-so stories" with no testable science. Too many can be described (in the words of Fisher, 1985) as "analogous, ad hoc, and anecdotal" hypotheses that are based on little more than comparisons with analogues or a single anecdotal story. When these fail, they are rescued by ad hoc (Latin, "for this [purpose]"), stopgap, special pleading explanations rather than abandoned. More recently, there have been some quite elegant case studies that use sophisticated methods to test their hypothesis. We will look at just a few selected examples to show how the method can be properly applied. In most of these cases, the authors looked at more than just a single interpretation of the organism, and often constructed mechanical models of different alternatives to allow rigorous testing. Sometimes, they came up with unexpected insights that allowed them to learn more than "the unenlightening conclusion that animals work well."

Archaeocyathids as Cylindrical Passive Filter Feeders

Many different types of organisms filter feed while attached to the sea bottom. Corals build their skeleton out of massive calcite, and trap food particles with their tentacles. Brachiopods and clams encase their bodies in shells, and filter feed by pumping water past a screening device (lophophore in brachiopods, gills in bivalves). The sponges and archaeocyathids, however, make skeletons that are not much more than tubes or cones with porous walls and an open top. The cells of living sponges live along the pores and canals in the skeleton, where they use tiny flagella to propel the water past them.

Cylindrical filter feeders use an additional factor that passively pumps water through their structure (Vogel, 1981). **Bernoulli's principle** states that the sum of the velocity and pressure on an object in flow must be constant. If the velocity increases, the pressure decreases, and vice versa. When a flow speeds up, it exerts less pressure than a slower flow. The most famous example of this principle is the airfoil cross section of a wing (Fig. 7.11A). The air moving over the curved top surface of the wing must travel farther than the air moving along the flat bottom, so it must move faster to keep up with the rest of the flow. By Bernoulli's principle, since the air moves faster, it exerts less pressure than the flow at the bottom of the wing, and there is a net force upward.

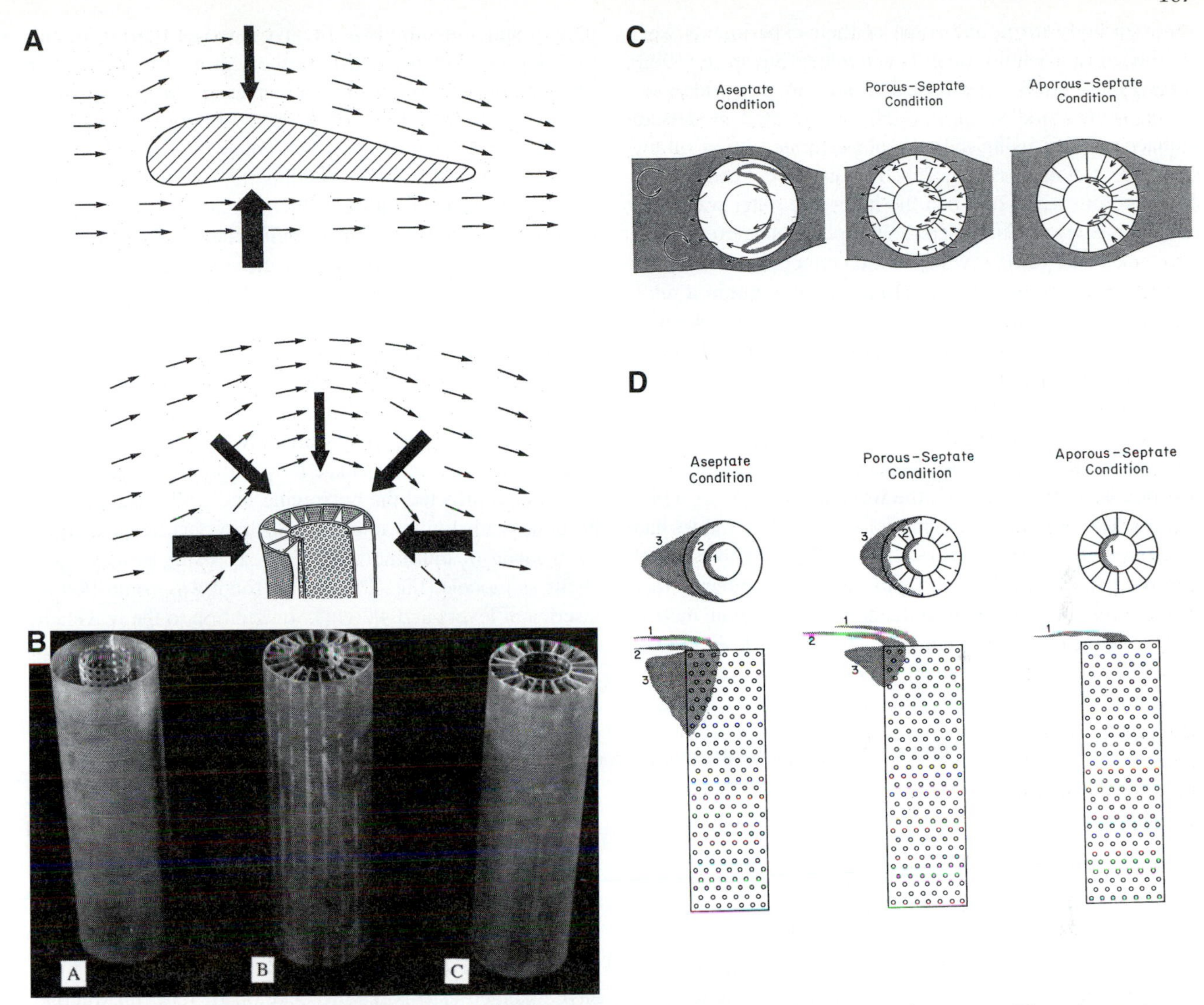

Fig. 7.11. (A) Bernoulli's principle states that if the velocity of fluid increases, its pressure decreases. The air passing over the top of an airplane wing (top) must move faster since it is diverted further, so it has less pressure than the air on the bottom of the wing. Similarly, the fluid passing over the a protruding object (such as a chimney or archaeocyathid) has less pressure than the fluid on the sides (bottom), so there is a tendency for fluid to flow in the direction of least pressure (out the top). (B) Photos of three brass models of archaeocyathids. A lacks septae between the cylinders; B has septae with pores in them, and C has septae without pores. (C) Flow of currents through the three model archaeocyathids. The porous-septate model had a smooth flow through the cylinder, while the other two models tended to produce irregular turbulent currents, or trap water in pockets. (D) Lateral view of the flow of water (marked with dye) through the archaeocyathid models. The porous-septate model had the largest current flowing between the outer and inner walls (labeled 2), while the aporous septate model allowed only a weak upward flow through the central cavity (labeled 1), and the aseptate model showed extensive leakage (3) out of the side walls. (B, C, and D from Savarese, 1992.)

A cylindrical filter feeder works on a similar principle. The top of the sponge sticks up into the flow along the bottom of the seabed and deflects the currents up and around its opening (Fig. 7.11A). Consequently, the flow moving over the sponge is moving faster, and there is decreased pressure at the top opening of the sponge. Since the water can flow through the pores in the walls of the sponge, it is drawn by the suction at the top opening of the sponge through the pores and out the top. A chimney works on the same principle, only it sucks the air and smoke up and out of the room. Otherwise, the smoke would go into the room.

Archaeocyathids are an enigmatic extinct group of Early Cambrian reef builders whose biological affinities are controversial (see Chapter 12). Whatever their biological affinities, however, it seems likely that they worked on the same principle as sponges, since they were cylindrical or conical with porous walls and an open top that protruded above the base of the seafloor. Balsam and Vogel (1973) conducted some simple experiments on models of archaeocyathids that showed that in a unidirectional current they could produce a passive flow out the top. Unfortunately, their models were not very similar to a real archaeocyathids in body form, and some of their experiments were conducted in a wind tunnel, not in water. Savarese (1992) carried these studies much further, looking at a wide variety of shapes and variables. He constructed models of archaeocyathids using cylindrical perforated brass tubing of different diameters (Fig. 7.11B). One model had a series of nonporous septa between the inner and outer walls like real archaeocyathids; others had septa that were porous, allowing water to flow within the chambers in the wall, or had no septa at all, but were simply a tube within a tube. The modes were then placed in a flume, and the velocities of the water and their flow patterns were carefully measured and documented.

Using a dye to track the movement of the water currents, Savarese (1992) showed that the nonporous septate model (closest to the real thing) produced the most efficient movement of water through the walls and out the top (Fig. 7.11D). The porous septate model and aseptate model had much more turbulent and less efficient flows, with much of the dye leaking through one wall and out another, rather than entirely out the top as in the nonporous septate model. Thus, it appears that the real archaeocyathid design is the most efficient of those tested.

Savarese's study is an elegant example of using a mechanical model (a paradigm in the sense of Rudwick, 1964) to test hypotheses of mechanical function. Unlike earlier studies, Savarese did not simply take one model and show that it worked. Instead, he considered several mechanical alternatives and tested all of them under the most rigorous possible conditions. Savarese looked at the variety of actual archaeocyathid shapes known from the fossil record to see if the model made sense in terms of real diversity, and also examined the changes due to growth and ontogeny. It was not possible to make phylogenetic comparisons, since we do not know the biological affinities of archaeocyathids. Nevertheless, he discussed the possibility that archaeocyathids were algae with a solid core, and showed how this interpretation was less plausible than one in which they are interpreted as having a hollow cavity like sponges. He took into account nearly all the relevant factors and made a strong case that archaeocyathids functioned like sponges.

Robot Clams and Bivalve Burrowing

Most bivalves burrow into the sand or mud at the sea bottom and then filter feed the water brought in from the surface by their tubular siphon. Consequently, bivalve shell shape is strongly affected by how they burrow, and much of their shell form can be directly traced to their burrowing mechanics. Paleontologists have been studying the shell form of bivalves for over a century, and biologists have worked with the habits of living bivalves for an equal amount of time. Yet, until recently very little was known about the actual mechanisms of bivalve burrowing, nor was there much observational behavioral data on how each species lived.

Stanley (1970) published a classic study in which he photographed hundreds of bivalves as they burrowed, and took x-rays of their position in the sediment to show their life position. He followed the pioneering studies of bivalve burrowing of Trueman (1966) by looking at the detailed motions of dozens of species. Trueman had shown that bivalves burrow by first protruding their muscular foot into the sediment, then bulging its tip to anchor it, and then contracting the foot so that it pulls the shell downward. However, it is very difficult to pull a wedge-shaped clam shell directly into the sand, so the bivalve rocks its shell back and forth to loosen the sand in front of it and to ease its movement through the sediment. In addition, the rocking must be asymmetrical, so that the shell slices forward and downward; a symmetrical shell would go nowhere.

Observing a living clam burrow was just the first step. Stanley (1975) wanted to see how variations in the details of the shell affected the burrowing rate, and this could not be done with living animals. So he constructed a robot clam made of aluminum-filled epoxy using casts of real shells as models (Fig. 7.12). This robot was controlled by a series of levers and weights. In addition to the real shell, he generated alternative shapes such as a discus-like form that did not have the depressed lunule in the front. He then conducted a series of experiments with the different models, varying the grain size of the sediment and the nature and rate of the movement. Although the depressed lunule in the front was not actively involved in the burrowing itself, it was important in a different way—it takes weight away from the front end (compared to the discus-shaped shell), shifting the center of gravity backward and making the asymmetric rocking motion more effective.

Stanley (1975) also looked at primitive bivalves without this depressed lunule and showed how the function must have changed as this feature developed. By constructing alternative models, he avoided the "just-so story" aspects of previous studies, which looked only at a single living form and tried to explain it. With the unnatural discus-like form, he was able to test its efficiency against a real shell shape. In addition, by looking at the ancestral bivalve shape, he was able to address the phylogenetic aspects of their form.

The Function of Horseshoe Crab Spines

Horseshoe crabs, like many arthropods, have a series of spines around the front and sides of their bodies as well as in the tail region. Various scientists have interpreted their function as protection, support or stabilization on the substrate, control of orientation during swimming, and slowing of settling in water (Fisher, 1977). Although many of these functions may be valid, it is difficult to test hypotheses about each.

Fisher (1977) chose to test one of these hypotheses—the slowing of settling in water—using models of the Pennsylvanian horseshoe crab *Euproops danae*, which had a long spine in front and two long spines on the corners of the body (Fig. 7.13). Based on his underwater observations of the swimming of the living horseshoe crab (Fisher,

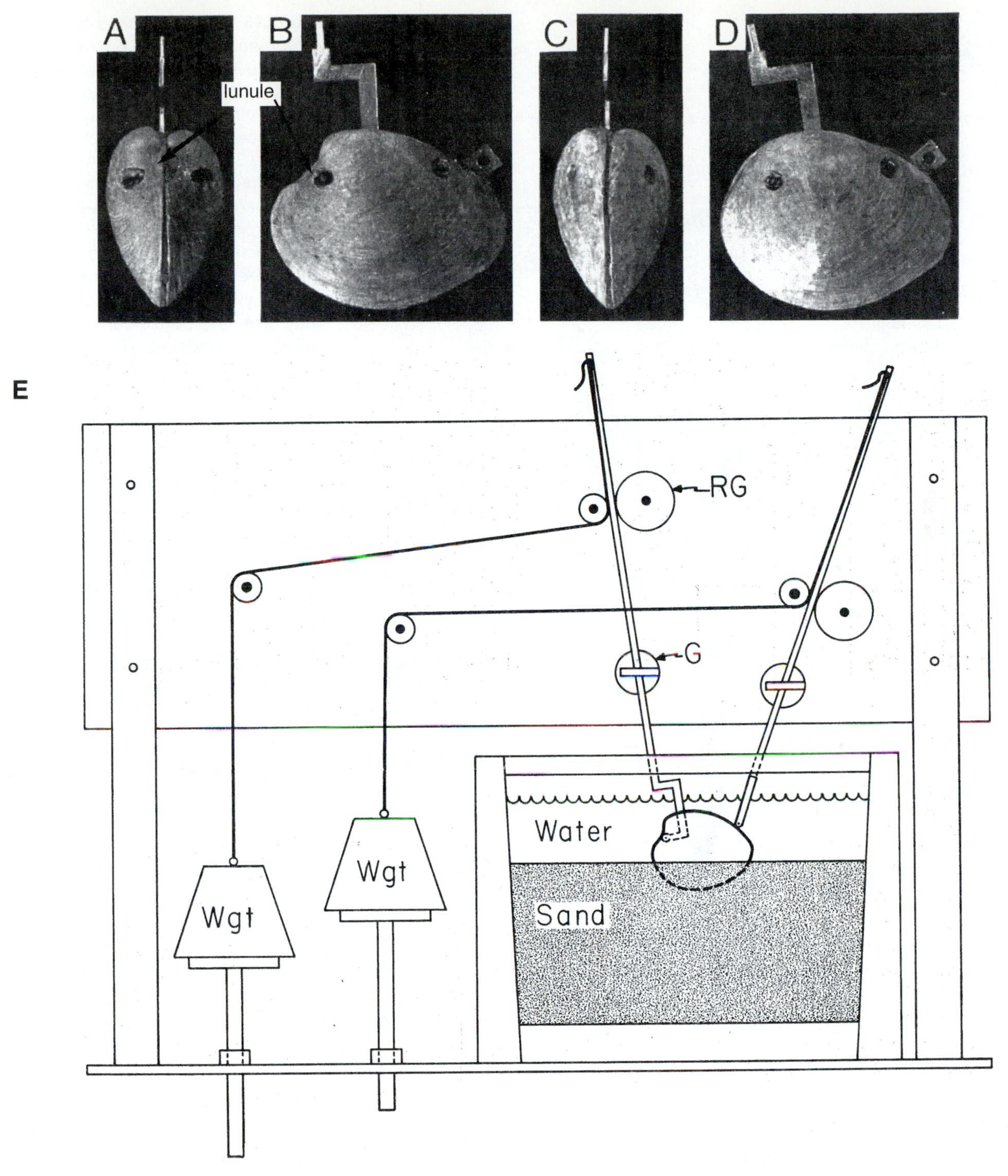

Figure 7.12. (A-D) Models used in burrowing experiments. (A) anterior and (B) left-lateral view of robot based on living clam; (C-D) robot with rounded false front attached. (E) Apparatus used to simulate burrowing in robot clams. The robot is pushed into the sand in an aquarium by two rods pulled downward by cables, to which weights are attached by pulleys. Roller guides (RG) and clasp guides (G) hold the rods in place. (From Stanley, 1975.)

1975), he predicted that the spines would help when *Euproops* rolled into a tight ball and tried to sink out of sight of predators. To test alternative hypotheses, he constructed models that varied the proportions of the spines and other parts of the skeleton (Fig. 7.13). He tried out each model in a large aquarium, filming their downward motion and constructing a time-lapse sequence of their movements (Fig. 7.13). Models with no spines (such as A-1) tended to somersault in a random fashion and spin around an axis as well, giving them no control as they sank. Most of the models with spines (including the models most resembling the real animal) tended to oscillate back and forth at first, but eventually settled in a stable position. However, if the spines were too long (as in A-3), the model

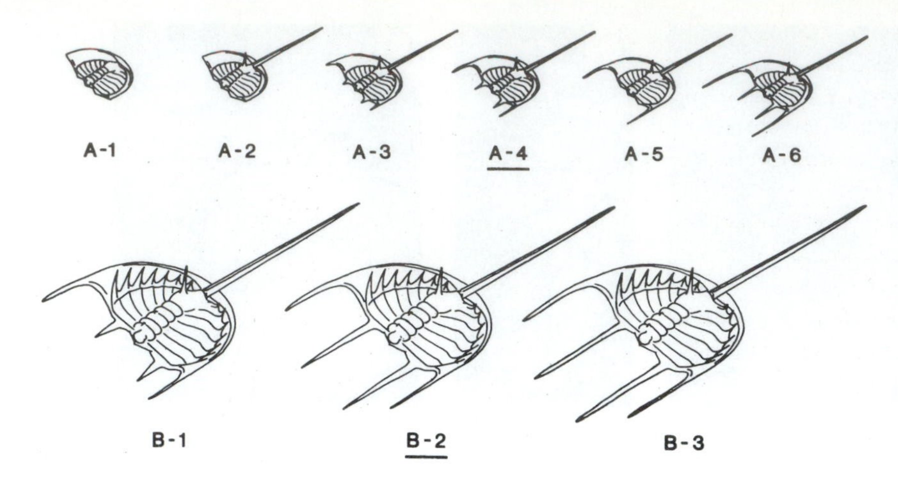

Fig. 7.13. (Left). Sketches of the different models used in Fisher's (1977) settling experiments. A-4 and B-2 are closest to the actual fossils of *Euproops.* (Bottom) Time-lapse sequences of the settling of various models. Spineless forms (A-1) tended to slowly oscillate as they sank, while long-spined forms (A-6) oscillated rapidly as they sank. Models like A-3 and A-4 (closest to the fossil) oscillated initially, but then sank smoothly in a sustained fashion, so they have less horizontal motion to attract fish predators. (From Fisher, 1977.)

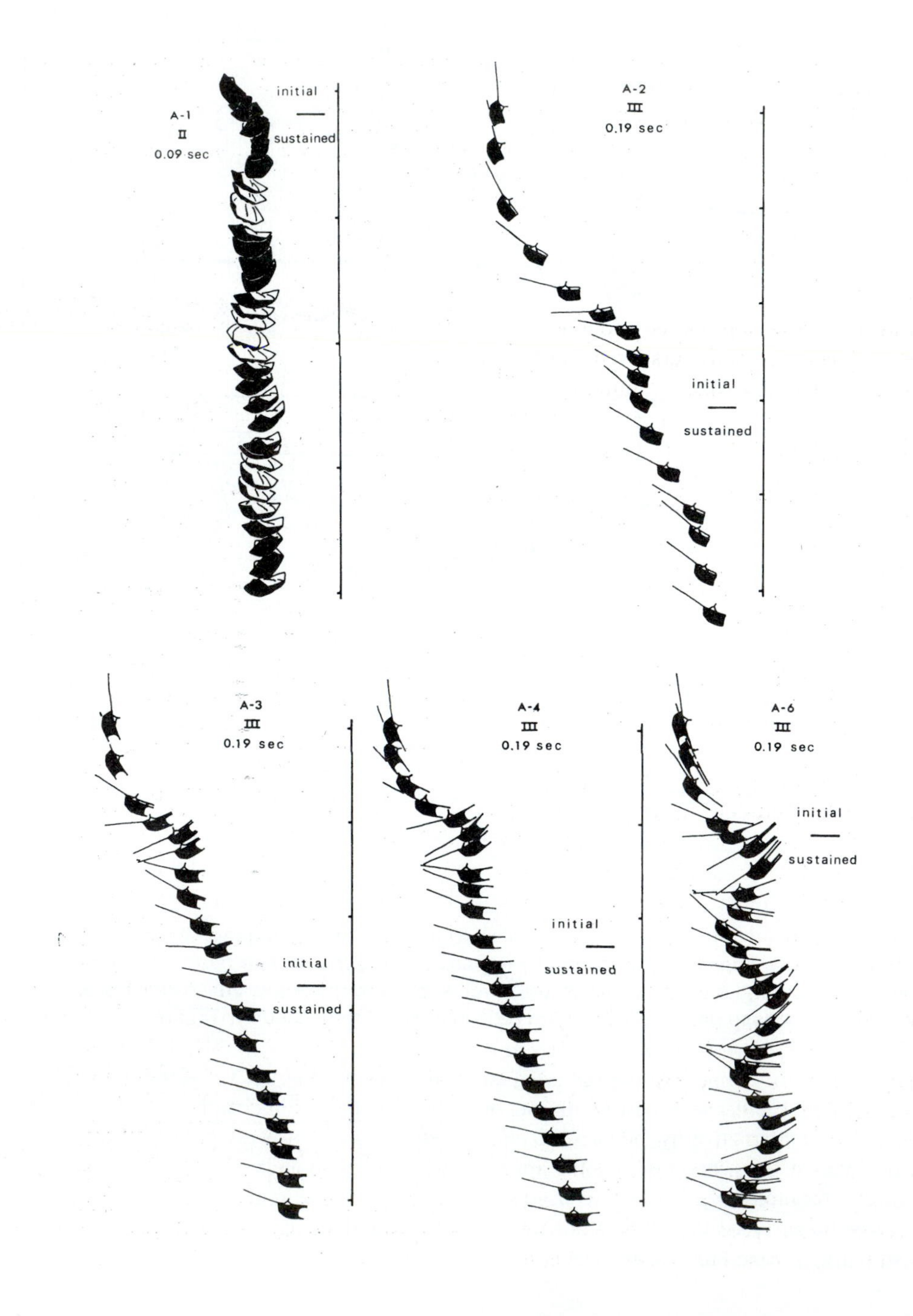

tended to oscillate continuously and never stabilized.

From these experiments, Fisher concluded that steady settling was the primary motion employed by *Euproops* during its escape. Since the dominant predators of *Euproops* during the Carboniferous were probably freshwater fish, Fisher suggested that this was the best mechanism to escape danger without attracting the attention of the predator. Studies on living fish perception (sight and motion sensing) show that they are very sensitive to horizontal motions but less so to vertical motions (a principle that has been employed in many successful fishing lures). It appears that the slow vertical settling of the actual *Euproops* specimen is the motion least likely to catch the attention of predatory fish.

Fisher (1977) also considered many other aspects of the shape and function of *Euproops* that might modify or compromise his conclusions, including ontogenetic development, structural constraints, and the condition found in near relatives. His study is a classic example of a well thought-out, rigorous, carefully executed functional analysis that considers several alternatives, tests them experimentally, and then draws conclusions only as far as the data allow.

Pterosaur Flight

Many different ideas have been proposed about how the extinct flying reptiles known as pterosaurs lived and flew. Some were small, sparrow-sized animals, but *Quetzalcoatlus* from the Cretaceous of Texas was the size of a small airplane. Although they are superficially similar to birds and bats, pterosaur wings were constructed in a very different fashion from either of these living flyers. Birds have limited bony structure in their wing and support the airfoil with their stiff feathers. Bats stretch a web of skin between all five of their finger bones and also between their fifth ("pinky") finger and their legs. But pterosaurs supported a membrane using an extremely elongate fourth ("ring") finger and had no bony supports within the wing itself, only stiffened fibers that show up in well preserved specimens.

Hankin and Watson (1914) made the first attempt to model the aerodynamics of pterosaurs, and a later series of studies (Kripp, 1943; Bramwell, 1971; Heptonstall, 1971; Bramwell and Whitfield, 1974; Stein, 1975; Brower, 1980, 1983; McMasters, 1984; Pennycuick, 1988) focused on the larger species, such as the well-known Cretaceous pterosaur *Pteranodon*, which had a 7-meter (23-foot) wingspan. The most influential study was published by Bramwell and Whitfield (1974), who used the insights from aerodynamics to interpret the flying abilities of *Pteranodon*. Using estimates of weight divided by the wing area based on the fossils, they were able to calculate wing loading. This in turn allowed them to calculate the curves of aerodynamic behavior for an optimal glider. A typical range of aerodynamic behaviors is shown by a plot of sinking speed versus flying speed (Fig. 7.14A). All flying structures try to balance these two forces, so that they

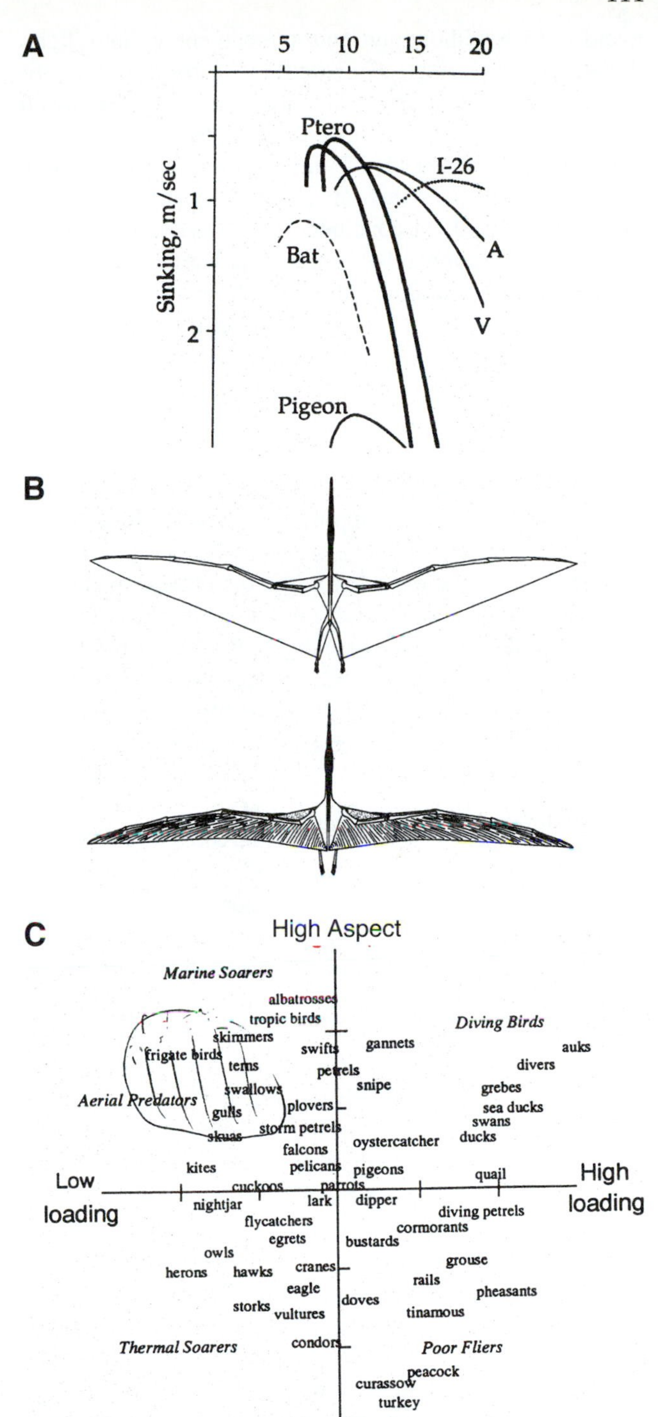

Fig. 7.14. (A) Aerodynamic data for several types of gliders, including an albatross (A), vulture (V), and a sailplane (I-26), and the calculated performance of two pterosaurs (Ptero). Pterosaurs were apparently better low-speed gliders than anything alive today, or anything we have ever built. (Modified from Brower, 1983.) (B) Comparision of conventional pterosaur reconstruction (top) with a more modern reconstruction (bottom) which does not connect the legs to the wing membrange. (C) Plot of aspect vs. loading ratios of various living birds. The pterosaurs (oval area) plot among the marine soarers and aerial predators. (Modified from Hazlehurst and Rayner, 1992.)

avoid sinking without putting too much energy into flight. Above optimal flying speed, the flier wastes energy inefficiently. Below that speed, it stalls and sinks. Bramwell and Whitfield (1974) estimated that *Pteranodon* had an optimal flying speed of 6.7 m/s and was a considerably more efficient glider than an albatross or a human-made glider. These authors concluded that *Pteranodon* was an efficient glider but barely capable of powered flight. They suggested that it hung upside down from cliffs like a bat and required winds or a drop from a cliff to become airborne. Stein (1975) carried out wind tunnel studies of a model *Pteranodon* and concluded that it had an optimal flying speed of only 4.9 m/s. Contrary to previous studies that argued that *Pteranodon* was a good soarer but a poor flier, Stein (1975) concluded that *Pteranodon* was primarily adapted to slow, flapping flight with long flight endurance.

All these calculations were based on a reconstruction of pterosaurs with a wing that connected to their legs, as occurs in bats. However Padian (1979, 1983, 1985, 1987) and Padian and Rayner (1993) looked at the specimens with preserved wing impressions and found only one that suggests this connection. Most of the nearly 80 specimens of *Rhamphorhynchus* and *Pterodactylus* from the Jurassic Solnhofen Limestone (see Fig. 1. 14A) that preserve wing impressions seemed to show that the wing was a narrow structure that did not connect to the legs (Fig. 7.14B). However, Wellnhofer (1987) has disputed this claim, arguing that most pterosaurs still had a connection between the wings and the legs (see reply by Padian and Rayner, 1993). In addition, Padian (1983) re-examined the skeletons of these pterosaurs and found further evidence that discredits the bat-winged model. For example, the arm bones are articulated for the forward and downward motion of a powered flapping stroke, and the breastbone is large for the attachment of powerful flight muscles. The bones themselves are hollow, like those of birds. In addition, the hind limb was very bird-like, capable of walking, but not in the right plane to attach to the wing membrane or to hang upside down like a bat. (Again, Wellnhofer, 1988, disagrees, pointing out specimens with pelves that suggested a sprawling posture and and four-legged stance.) The ultimate vindication of this reconstruction came when engineer Paul Macready constructed flying models of large pterosaurs and showed they were capable of flapping flight as well as gliding (Macready, 1985). From these considerations, Padian and Rayner argued that large pterosaurs like *Pteranodon* were constructed more like gulls or albatrosses than like bats. They were capable of powered takeoff and some flapping flight, although they were even more efficient gliders.

Brower (1980, 1983) used the smaller wing area estimates to re-examine the problem of the aerodynamics of *Pteranodon* and the very similar (but smaller) *Nyctosaurus*, using recently developed computer models of aerodynamics and information from a Sailvane hang glider, which has very similar aerodynamic properties. Even with the reduced wing area, he still concluded that *Pteranodon* was a soarer, but *Nyctosaurus* probably had enough sustained power output for continuous level flapping. Both pterosaurs were capable of slow takeoff speeds (4 m/s or less), and slow landing speeds, and were well suited for soaring along a low-relief coastline and for convection-current soaring. Since they are known from the Cretaceous Interior Seaway, Brower concluded that they could soar far from the shore, using only the gentle updrafts generated by the warm, calm climates of the Cretaceous. He compared their feeding behavior to certain marine birds, which skim the surface of the water with their bills to catch fish. However, these pterosaurs were much more specialized for lower power requirements and a narrow range of horizontal flight speeds.

Small pterosaurs, however, are a different matter. Hazlehurst and Rayner (1992) looked at the wing shape of a variety of the smaller Triassic and Jurassic pterosaurs and compared them to birds in terms of wing loading and wing aspect (the ratio of the square of the wingspan divided by wing area). High aspect ratio denotes narrow wings (typical of efficient fliers, such as albatrosses and swifts), and low aspect ratio denotes broad wings essential for rapid flapping flight (typical of fowl and doves, which must accelerate rapidly to escape but are poor fliers over distance). When a variety of birds were plotted in terms of wing loading and wing aspect (Fig. 7.14C), a clear pattern emerges. The poor fliers (like fowl and doves) are all high loading and low aspect. Diving birds (auks, grebes, ducks, swans) are high loading and high aspect. Marine soarers and aerial predators (gulls, albatrosses, petrels, frigate birds, swifts, and swallows) tend to be high aspect and low loading. Thermal soarers (hawks, owls, eagles, vultures, storks, cranes, and condors) are low loading and low aspect. And where do the smaller pterosaurs fit? If they are reconstructed with gull-like narrow wings, they fall largely in the area of high aspect, low loading marine soarers. If, however, they are reconstructed with bat-like wings extending to the feet, then they fall along the axis of lowest loading and intermediate aspect, along with kites, flycatchers, and cuckoos. From this analysis, Hazlehurst and Rayner concluded that smaller pterosaurs were capable of relatively slow and highly efficient flight, with high maneuverability.

Hazlehurst and Rayner (1992) were also struck by the lack of variability in pterosaur wing shapes. Unlike birds, which occupy the full spectrum of possible shapes in the aspect/loading morphospace, known pterosaurs occupy only a small portion of it. Perhaps it is because most pterosaur specimens are known from marine rocks and were apparently nearshore fliers, but this does not explain why no pterosaurs had wings shaped like those of ducks or auks. It may also be due to a structural constraint. Bird wings can be shaped in a number of ways since they can vary the length and proportion of the feathers. But the wings of pterosaurs were not composed of individual feathers, only a single wing membrane with limited support. This kind of wing may have been mechanically incapable of the entire potential range of wing shapes.

Unlike the functional constraints in the studies we've seen earlier, the aerodynamics of powered flight or gliding is such a limiting factor that we can be quite specific about which conditions pterosaurs must have met in order to fly. Nevertheless, we have seen how much the aerodynamic predictions can be wrong if they are based on the wrong reconstruction of the wing area. This is a common problem in functional studies. Soft tissues, which are rarely preserved, may be critical to the model that best explains the data. This study also illustrates another principle. Macready's flying pterosaur model is the acid test of the hypothesis—if it doesn't fly, we know that something is wrong with our interpretations.

The Bite of a Saber-Toothed Cat

Ever since they were first described in 1853, the spectacular canine teeth of saber-toothed carnivores have invited all sorts of speculations. Some have suggested they functioned as can openers to pierce the armor of glyptodonts or were used for slicing carrion, for grubbing for marine molluscs as walruses do, as tree-climbing aids, or for stabbing with the mouth closed. Some paleontologists could not imagine that the saber teeth were very functional and argued that they were an example of evolution driven by internal forces and no longer under the constraint of natural selection. However, this last argument can be easily discredited, because saber-toothed carnivorous mammals evolved at least four times independently in four different groups of mammals: the Pliocene South American marsupial *Thylacosmilus*, the Eocene creodont *Apataelurus,* and two groups of true carnivorans: false cats, or Nimravidae (which are cat-like, but not closely related to cats); and the true cats, or Felidae (Fig. 7.15A). Clearly, such a feature was a successful adaptation, and the question should not be whether it functioned successfully, but why don't we have any sabertooths living today, since that ecological niche was usually occupied in the geologic past.

Since the analyses of Matthew (1910), Bohlin (1940), Simpson (1941), Hough (1949), and Kurtén (1952), most scientists have accepted the idea that the sabers were for killing prey, but there has still been much argument and speculation. Some suggested that the sabers were specializations for cutting through the thick skin of large, pachydermous mammals. Most of these scientists were puzzled by how these animals could have opened their mouths wide enough to get a bite on anything, and how strong their bite could be if their mouth was opened that wide. Typically, sabertooths were reconstructed jumping on the backs of prey and stabbing downward with their heads. However, the thick skins of large hoofed mammals are hard to pierce in this fashion, and it is difficult to imagine the predator getting much force behind its head with a short, downward head movement. Besides, the necks and backs of most prey species are bony and heavily muscled, making this a particularly difficult region to slash open without breaking a saber.

Emerson and Radinsky (1980) conducted a rigorous functional analysis of the problem, observing living cats and manipulating casts of skulls. First, they photographed the maximum gape angle that living cats in zoos are capable of, including one panther that tried to get its mouth around a bowling ball. Most living cats can gape about 65 to 70°. Next, they looked at the maximum gape of a variety of sabertooths and found that they could gape as much as 95° (Fig. 7.15B). This allowed just as much clearance between upper and lower canines as a living cat has, so sabertooths could get their mouth around similar-sized objects. Emerson and Radinsky (1980) also analyzed the bite force exerted by the muscles at maximum gape and found that the sabertooth bite was no weaker than that of living cats.

From these considerations, and also from the strongly retractile claws, Emerson and Radinsky (1980) concluded that saber-toothed predators killed their prey with a shallow bite to the throat, avoiding the bones and strong muscles of the upper neck. They argued that such a bite would be more effective in killing large prey than the method employed by living large cats, such as lions, which grab their prey at the back of the skull and crush its neck or at the throat or muzzle and suffocate it. Sabertooths would not have been able to do this, but they would have been able to kill a larger prey species quickly by slashing its throat.

Akersten (1985) agreed with most of these conclusions in his analysis of the Rancho La Brea saber-tooth cat, *Smilodon*. However, he noted that the sabers are relatively dull and would not be very good at slashing thick hide. In addition, the upper and lower canine teeth clearly were adapted for biting and shearing against one another, so the upper canines were not used for stabbing or slashing by themselves. Akersten proposed an analogy with the giant monitor lizard, the Komodo dragon, which ambushes its prey along a game trail and then slashes its underbelly with its saber-like, serrated teeth (Fig. 7.15C). From these and other considerations, Akersten (1985) concluded that saber-toothed carnivores used their strong build to pull down a prey animal, then they used their teeth to grab a fold of skin on the abdomen of the prey and slice it open, quickly disemboweling the hapless creature.

The controversies over the killing habits of saber-toothed predators shows the range of interpretations possible, even with animals that have close living analogues. Most of these studies have relied on the full spectrum of evidence from the skull, gape, bite force, retractile claws, and short, stocky body of the animal to make their cases. Observations of the gapes of living carnivores, and the killing behavior of living cats and Komodo dragons provided analogues for the extinct forms. Still, these studies could not test several alternative models in a definitive fashion, so their conclusions are still more tentative and controversial than we have seen in other such analyses.

A

A. *Dinictis*

5 cm.

B. *Hoplophoneus*

C. *Eusmilus*

D. *Pseudaelurus*

5 cm.

E. *Homotherium*

F. *Smilodon*

G. *Barbourofelis*

H. *Apataelurus*

I. *Thylacosmilus*

B

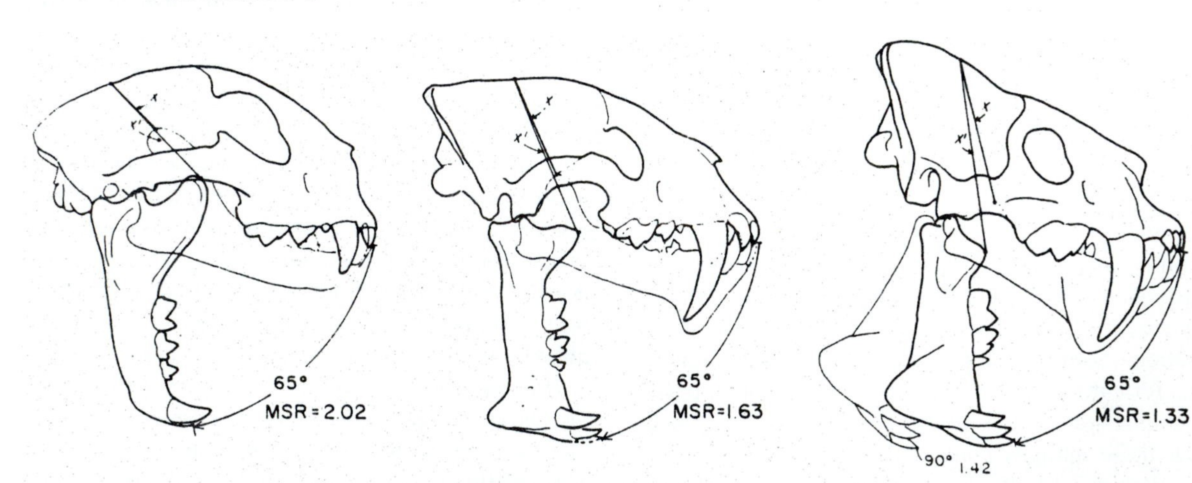

Figure 7.15. (A) Diversity of saber-toothed mammals. A-C are nimravids, while D-G are true cats. H is a creodont (not a true carnivoran) and I is a pouched marsupial (not even a placental mammal). (B) Estimates of the stretch capabilities of the jaw muscles in various cats. Despite the differences in canine length, each cat is capable of opening its mouth about 65° as measured from the points of the upper and lower incisors. Saber-toothed carnivores accomplished this by lowering the coronoid process on the upper back part of the jaw, so that the jaw muscles could pull a much shorter distance when the mouth is fully opened. (After Emerson and Radinsky, 1980.) (C) Akersten's reconstruction of a saber-toothed cat using its sabers to bite through a fold of flesh in the underbelly of its prey. (After Akersten, 1985.)

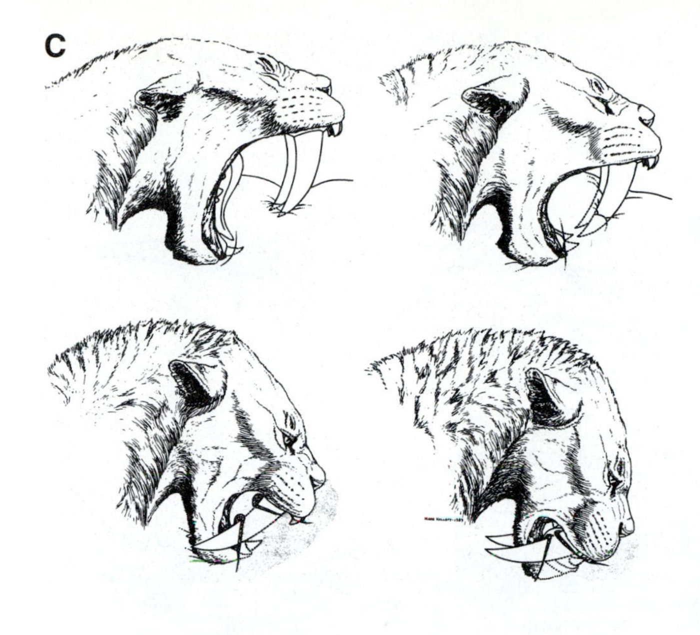

CONCLUSIONS

Paleontologists have long been fascinated with the behavior of extinct organisms and seldom lack ingenuity in trying to explain their function. In many cases, this meant untestable speculation and scenarios, which were limited only by the imaginations of the scientist. Fortunately, the field of functional has moved beyond the "analogous, ad hoc, and anecdotal" to the realm of testable hypotheses. We have seen a number of examples in which mechanical models (paradigms in Rudwick's sense) were constructed and used to test predictions, rejecting models that were less efficient. Still, the functional morphologist must always keep several things in mind: the influence of structural and phylogenetic constraints, pleiotropy, suboptimality, and especially the fact that structures do not always mirror their presumed function. Ultimately, the promise of functional morphology is that we get beyond proving that individual animals are well designed, and discover more general "laws of form" that have implications beyond individual case studies. That promise has not yet been achieved, but progress has been made.

For Further Reading

Alexander, R. M. 1968 (2d ed., 1983). *Animal Mechanics*. Blackwell Scientific Publications, Oxford.

Dullemeijer, P. 1974. *Concepts and Approaches in Animal Morphology*. Van Gorcum, the Netherlands.

Fisher, D. C. 1985. Evolutionary morphology: beyond the analogous, ad hoc, and anecdotal. *Paleobiology* 11:120-138.

Gans, C. 1974. *Biomechanics, An Approach to Vertebrate Biology*. J. B. Lippincott, Philadelphia.

Gould, S. J. 1970. Evolutionary paleontology and the science of form. *Earth-Science Reviews* 6:77-110.

Gould, S. J. 1980. The evolutionary biology of constraint. *Daedalus* 109:39-52.

Gould, S. J., and R. C. Lewontin. 1970. The spandrels of San Marco and the Panglossian paradigm: a critique of the adaptationist programme. *Proceedings of the Royal Society of London* (B) 205:581-598.

Hickman, C. S. 1988. Analysis of form and function in fossils. *American Zoologist* 28:775-793.

Hildebrand, M., D. M. Bramble, K. F. Liem, and D. B. Wake, eds. *Functional Vertebrate Morphology*. Harvard University Press, Cambridge.

Lauder, G. V. 1981. Form and function: structural analysis in evolutionary morphology. *Paleobiology* 7:430-443.

Lewontin, R. C. 1978. Adaptation. *Scientific American* 239:212-230.

Raup, D. M. 1966. Geometric analysis of shell coiling: general problems. *Journal of Paleontology* 40:1178-1191.

Rayner, J. M. V., and R. J. Wootton, eds. 1991. *Biomechanics in Evolution*. Cambridge University Press, Cambridge.

Rudwick, M. J. S. 1964. The inference of structure from function in fossils. *British Journal for the Philosophy of Science* 13:27-40.

Russell, E. S. 1916. *Form and Function, A Contribution to the History of Animal Morphology*. University of Chicago Press, Chicago.

Schmidt-Nielsen, K. 1972. *How Animals Work*. Cambridge University Press, Cambridge.

Seilacher, A. 1973. Fabricational noise in adaptive morphology. *Systematic Zoology* 22:451-465.

Thomason, J. J., ed. 1995. *Functional Morphology in Vertebrate Paleontology*. Cambridge University Press, Cambridge.

Thompson, D'Arcy W. 1942. *On Growth and Form*. Cambridge University Press, Cambridge.

Vogel, S. 1981. *Life in Moving Fluids*. Princeton University Press, Princeton, N.J.

Vogel, S. 1988. *Life's Devices*. Princeton University Press, Princeton, N.J.

Wainwright, S.A., W. D. Biggs, J. D. Currey, and J. M. Gosline, eds. 1976. *Mechanical Design in Organisms*. Edward Arnold, London.

Figure 8.1. Artistic reconstruction of a typical Silurian reef assemblage built of a variety of colonial organisms: tabulate corals like *Halysites* (c) and *Favosites* (b), rugose corals like *Heliolites* (a) and *Streptelasma* (e), branching bryozoans like *Hallopora* (d). In addition, there were crinoids (g), brachiopods such as *Leptaena* (h) and *Atrypa* (f), trilobites such as *Dalmanites* (i), and straight-shelled nautiloids (j). (After McKerrow, 1978.)

Chapter 8

Paleoecology

Our task, then, is to identify the remains that lived together, reconstruct the community structure and infer its ecological and evolutionary significance.

James Valentine, *Evolutionary Paleoecology of the Marine Biosphere*, 1973

ECOLOGY AND PALEOECOLOGY

In the 1960s and 1970s, ecology was one of the hottest ideas in western culture. Everyone was aware of conserving resources and recycling, and many important steps were taken and laws were passed to help our environment. The word "ecology" became such a popular buzzword that it was used far beyond its original biological meaning of the interaction of organisms and their environment. Anything that was good was "ecological," and advertisers managed to attach the word to all sorts of products that were no friend of the environment. Paleontologists are no less a product of their culture, so they too jumped on this bandwagon. Along with the explosive growth of other paleobiological ideas in the 1970s, the field of paleoecology grew by leaps and bounds as paleontologists tried to extend ecological principles to the fossil record and interpret the paleoecology of fossil assemblages. The pinnacle of this optimistic agenda for paleobiology was epitomized by Valentine's (1973) classic book *Evolutionary Paleoecology of the Marine Biosphere*, which embraced most of paleontology into the context of paleoecology. Other important books of this period were Ager's (1963) *Principles of Paleoecology*, Imbrie and Newell's (1964) *Approaches to Paleoecology*, Laporte's (1968) *Ancient Environments*, culminating with the encyclopedic book, *The Ecology of Fossils*, edited by McKerrow (1978), along with Dodd and Stanton's (1981) *Paleoecology, Concepts and Applications*, and Boucot's (1981) *Principles of Benthic Marine Paleoecology*.

As happened with functional morphology, however, much of what was called "paleoecology" was not very original, nor was it testable science. Many studies simply consisted of reconstructing extinct communities as if they were living ecosystems, ignoring many of the real differences between the organisms and their analogues—what Everett Olson called "me-too ecology." This approach was typified by the book edited by McKerrow (1978), which consists of page after page of pretty pictures of extinct organisms in life position (Fig. 8.1), with few general principles that can be deduced. In his review of this book, Surlyk (1979, p. 446) wrote: "Paleoecology as a science still has a long way to go. . . It is immensely clear how little we know within the fields of morphological adaptation and modes of life of the great majority of fossil taxa. . . Paleoecology, to be accepted as a modern science, has to shift away from cartoons to much more detailed and quantitative studies." As Gould (1980, p. 101) put it, "The reconstruction of communities sounds like the right thing to do. . . But where does it all lead, and why is it being done? Suppose that we could proceed unambiguously, that we could enumerate taxa, determine relative abundances, assign trophic roles, and calculate biomass. At the end, we decide that ancient communities worked much like modern ones. Did we ever doubt it (and if we did doubt it, would this be the way to nurture suspicion)? Community reconstruction will gain theoretical interest when it addresses unresolved questions, but not while it measures success by the fit of individual solutions to modern analogues, and proceeds by enumerating more and more individual solutions." A number of scientists have leveled similar criticisms (Hoffman, 1979; Hill, 1981; Lewontin, 1982; Paine, 1983; Kitchell, 1985; Benton, 1987; Peters, 1991).

Consequently, the interest in the small-scale aspects of paleoecology has receded since the 1970s and 1980s. Instead, paleontologists have realized that the fossil record is better for examining the large-scale and long-term processes that no biologist can observe. These concepts are now being called "evolutionary paleoecology," and we will examine some examples at the end of this chapter. Like macroevolution, this large-scale view of ecology is beginning to change the way both neontologists and paleontologists look at the living world, as well as at the fossil record.

Paleoecology is often subdivided into **autecology** (the behavior of individual organisms and their relation to the environment) and **synecology** (the ecology of communities of organisms and their relationship to the environment). The former is essentially the same as functional morphology, discussed in the previous chapter, so we will not consider it further here. This chapter is chiefly concerned with what is called synecology.

As in functional morphology, the fundamental approach of paleoecology is uniformitarian in nature. We use our understanding of modern processes and modern organisms and communities to attempt to decipher past ecologies. In functional morphology, if the organism has close living relatives, this is easy to do; so too in paleoecology, especially with relatively recent organisms and communities that have many living relatives. For example, a study of Miocene mammalian communities of Africa (Van Couvering, 1980) can largely rely on the modern African savanna as a model, because so many of the organisms are very similar. The use of Neogene microfossils to interpret ancient ocean currents and climates is also straightforward, since most of the species have living relatives with apparently similar tolerances of temperature, depth, and salinity.

But when there are no close living relatives, then the problem becomes more difficult. As in functional morphology, we look for analogues. But is a community of modern bivalves really that good an analogue for an extinct community of Paleozoic brachiopods? Are modern coral reefs really that similar to Cambrian archaeocyathid reefs? In some cases, there simply are no modern analogues. There is nothing on this planet today that approaches the harsh, low-oxygen world of the Proterozoic, with its lack of an ozone layer to screen ultraviolet and its vast sheets of stromatolites with no advanced animals to feed upon them. There is no modern analogue for the early Paleozoic land surface, lacking both vascular land plants to change the pattern of weathering and erosion, and land animals, with their own complex relationships. In these cases, we must use whatever principles of chemistry, physics, and biology seem applicable, but clearly we are in a brave new world where the rules are much less clear-cut.

Still, there are many lines of evidence that can be used in paleoecology, even when modern analogues fail. We can observe the positions of animals with respect to each other, and with the substrate, when they are extraordinarily well preserved. We can look for evidence of associations of organisms. We can look for biogeochemical lines of evidence, especially with respect to the chemical composition of the skeleton and what it indicates. Some of these approaches have been very successful, and we will see their applications in this chapter.

ECOLOGICAL RELATIONSHIPS

The term "animal community" is really a very elastic one, since we can use it to describe on the one hand the fauna of the equatorial forest and on the other hand the fauna of a mouse's caecum.

Charles Elton, *Animal Ecology*, 1927

Ecological communities are merely an epiphenomenon of the overlap in distributional patterns of various organisms controlled primarily by the environmental framework.

Antoni Hoffman, "Community paleoecology as an epiphenomenal science," 1979

The Ecological Hierarchy

Many terms used in ecology are also widely used in common language, so that their precise meaning has been blurred. To clarify this confusion, we will try to define these terms as ecologists mean them. The basic structure of the ecological world is hierarchical in nature, with smaller-scale units clustered into larger, more inclusive units.

The broadest of all categories is the **biosphere**, the region of the earth's atmosphere and surface that is inhabited by life. It is a highly abstract term for the biotic envelope of the earth. The biosphere is divided into **ecosystems**, which are the sum of all the physical and biological characteristics in a given area. Ecosystems are usually very large-scale units, such as the shallow marine ecosystem or the terrestrial ecosystem. Ecosystems are in turn divided into **communities**, which are local associations of organisms, such as the intertidal community or the savanna community. Each organism in a community has its own **habitat**, which is the actual physical environment in which the organism lives. An abstract extension of this concept is the **niche**, which is the sum of all the physical, chemical, and biological limits on the organism, its way of life, and the role it plays in the ecosystem. For example, the habitat of squirrels in the park is the local trees and the ground below them, but their niche is that of a tree-dwelling, nut-gathering small mammalian herbivore. Its community might be the North American temperate forest, which is a subdivision of the temperate terrestrial ecosystem. Even with an example, it is obvious that most of these concepts are not so clear-cut and discrete, since they vary in scale and content. What might be considered an ecosystem in one case might be a community in another context. Ecologists have been battling about the precise meaning of these terms for decades, and some have recommended abandoning them altogether (Peters, 1991). Most ecologists find that, although their absolute definitions are vague, they are useful concepts within the confines of a particular study and so continue to use them.

There are different ways to classify ecosystems, communities, habitats, and niches. One of the simplest is a classification based on environmental factors. For example, in the marine realm (Fig. 8.2), depth of water and the nature of the substrate are frequently used to subdivide the marine ecosystem into smaller parts. The bottom-dwellers (**benthic** organisms) may live above the range of the tides (**supratidal**), between them (**intertidal**, also called **littoral**) or below the tides but on the shallow shelf (**subti-**

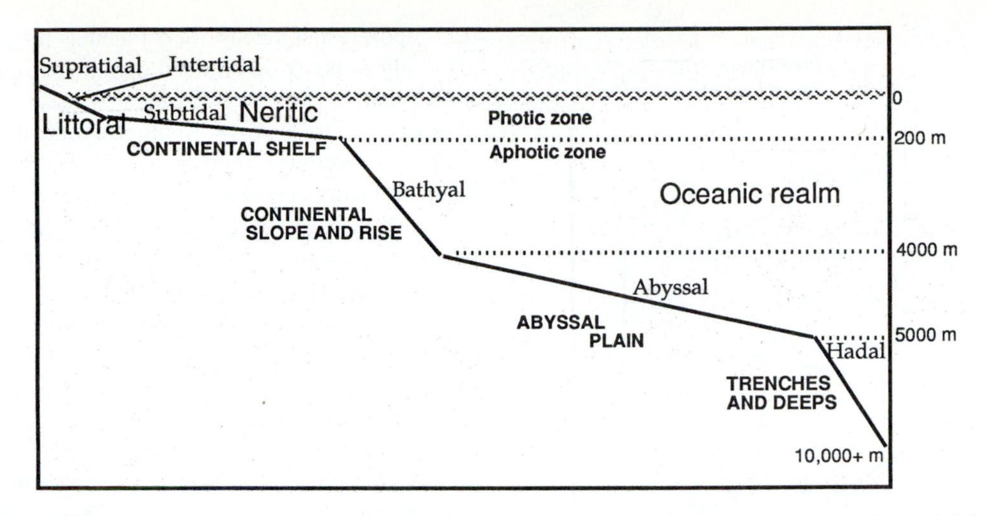

Figure 8.2. Definitions of the major marine environments. Although the correlation is not exact, the subtidal, bathyal, abyssal, and hadal environments roughly correspond to the continental shelf, slope-rise, abyssal plain, and deep ocean trenches. Within the littoral, neritic, and oceanic realms, there can be planktonic, nektonic, and benthic organisms.

dal). The depths of the continental slope are called **bathyal**, and the broad open ocean seafloor is called the **abyssal** plain. A few organisms live in deep oceanic trenches, or the **hadal** depths. Benthic organisms can also be classified by how they live with respect to the substrate: **infaunal** burrowers, or **epifaunal** organisms that live directly on the sea bottom. Some epifauna are attached to the seafloor, but other erect large structures (such as coral reefs or crinoid stems) that let them filter feed high above the seafloor.

This hints at another way of subdividing marine life: how they feed. Some infaunal organisms are detritus feeders, living on the food extracted from the mud, while others simply live in the mud as protection but filter feed by sucking in seawater. Still others live on or in the sea bottom but crawl around catching prey. Most epifaunal animals are filter feeders (also called suspension feeders) but many are predators. Marine plants cannot live below a certain depth because they need light, so they are usually attached to the bottom, or free-floating; there is no real infaunal plant life.

Organisms that live in the open waters of the ocean (**pelagic** organisms) can inhabit waters above the shallow subtidal shelf (**neritic** realm) or above the deep ocean (**oceanic** realm). Pelagic organisms can also be classified by their way of living in the water. Passive floaters (especially microfossils and jellyfish) are **planktonic**, whereas active swimmers (especially fish, squids, and whales) are **nektonic**.

Of all the problematic concepts in ecology, one of the most heavily criticized is the niche. By its very nature, it is a vague, abstract concept. Lewontin (1978) points out that it is impossible to recognize the niche without the organism that fills it. If the niche exists only when an organism fills it, is it really a discrete concept, or are the definition of the niche and its organism inherently circular? One can easily imagine an arbitrary "niche" that is unfilled (such as herbivorous snakes, which do not exist), but without an organism to fill it, is there any real meaning or explanatory value to the concept? Still, there does seem to be some value to the idea of niche. For example, the saber-toothed carnivores discussed in the previous chapter appear to occupy a distinct niche that has been occupied four different times by four different groups of mammals. At present (and through most of geologic history), that niche is unfilled, but the strong tendency for different groups of animals to fill it suggests that the concept has some merit.

Similarly, the concept of community has been questioned. Traditionally, communities were viewed as tightly integrated entities, with many strong interactions between the members and long-term stability as a unit. But the recent ecological literature seems to show that communities are much more dynamic and ephemeral than traditionally conceived, and that the interactions between the members are much weaker than previously supposed (Hoffman, 1979, 1983; McIntosh, 1980; Kitchell, 1985; Benton, 1987). Hoffman (1979, p. 370) argues that "ecological communities are merely an epiphenomenon of the overlap in distributional patterns of various organisms controlled primarily by the environmental framework . . . There is no intrinsic, biotic mechanism inducing community dynamics that is an inherent trend to maximize a selection value in either ecological or evolutionary time." Today, many paleoecologists are less enthusiastic about the meaning of community reconstructions as exemplified by McKerrow (1978). Kaesler (1982) wrote that "in the early seventies, community paleoecologists were convinced that their science was indispensable to marine ecologists. What could be more important . . . than knowing the history of marine communities through evolutionary time? The answer was loud and clear: 'almost anything.'"

Such negative assessments were a bit premature. Although many of the concepts of ecology based on short

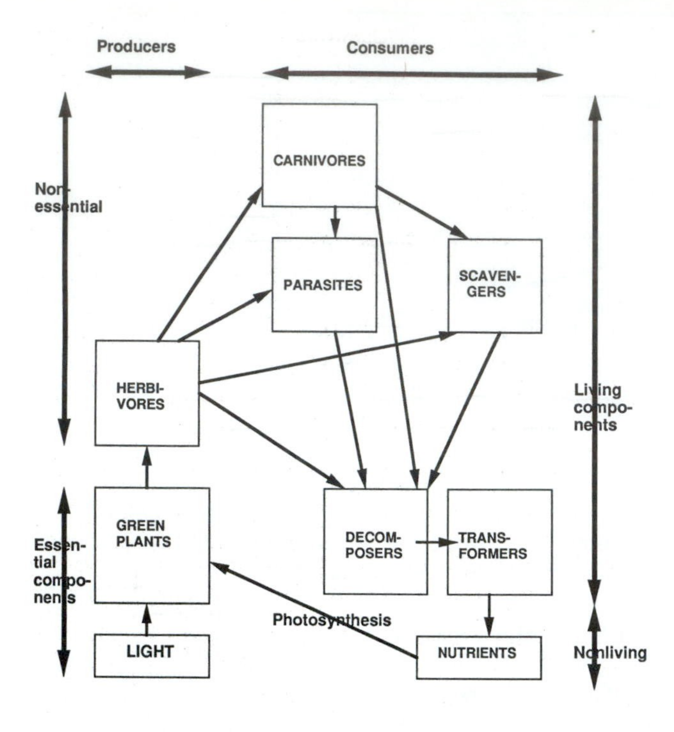

Figure 8.3. The flow of materials through a typical food web.

time scales are difficult to apply to paleoecology, the fossil record has its own unique contribution to make to ecological theory that could not be discovered from the living biota alone. Such insights are now labeled "evolutionary paleoecology" and will be discussed later in this chapter.

Trophic Dynamics

Another fundamental aspect of the interactions of organisms is energy flow, or "who eats whom." All living systems ultimately depend on the sun for all their energy, but it is transferred from organism to organism in complex ways. Green plants are the fundamental producers at the base of the **food chain**, converting the sun's energy and carbon dioxide into organic matter by means of photosynthesis. Free oxygen is the waste product of this reaction. Most animals are consumers, eating other organisms to get their energy and nutrition and consuming free oxygen to metabolize this food; their waste product is carbon dioxide, which returns to the plants. The primary level of consumers is herbivores, which eat the plants directly. They, in turn, are eaten by at least one secondary level of predators (either carnivorous meat-eaters or insectivores), and in many food chains, there are multiple levels of predatory animals, each larger than the last and feeding on smaller prey.

Plants, herbivores, and predators are not the only elements in an ecosystem. A variety of organisms feed on others in secondary ways. For example, parasites live off other organisms without killing them (at least not immediately). Scavengers feed on the remains of organisms, and decomposers and transformers (such as fungi and bacteria) also help to break down dead organisms into free organic matter that can be recycled as nutrients, which help living plants and animals grow. In other words, the food chain is better described as a **food web**, with every organism intricately linked in one way or another with every other organism in the ecosystem (Fig. 8.3). The feeding relationships between organisms is known as **trophic** structure, or trophic dynamics (from the Greek, "trephein," to feed).

Until recently, it was thought that photosynthesis and plants were the fundamental basis of the food chain for all life. In 1977, however, a discovery was made which rocked biology (Corliss et al., 1979). Deep-sea submarines diving along the rift valley in the mid-ocean ridges found a previously unsuspected realm of life that does not depend on light, photosynthesis, or plants. These organisms live in the springs of hot water seeping up from the lava-filled vents in the rift valley, and derive their energy from breaking down hydrogen sulfide that seeps up from below. In the place of plants converting carbon dioxide, water, and sunlight into biomass by photosynthesis, these vents are inhabited by bacteria that reduce methane or hydrogen sulfide by **chemosynthesis** (Cavanaugh, 1985; Fisher, 1990). These bacteria are the base of a food chain that includes foot-long clams, meter-long tube worms, odd crustaceans, and bizarre creatures not seen in any other part of the marine realm. The clams and tube worms lack ordinary feeding structures and digestive systems, since they obtain their nutrition directly from a symbiotic relationship with the chemosynthetic bacteria that live inside them. Detailed analysis of the animals (Tunnicliffe, 1992) has shown that they are unique to this environment (95% of the species and 22% of the families are endemic to vents), and most are evolutionary holdovers from the Mesozoic that have not interacted with the rest of the marine ecosystem in tens of millions of years.

Since these discoveries, paleontologists have discovered a variety of fossil localities that seem to be deep-sea vent faunas, some as old as the Paleozoic. In addition to the hot sulfide-rich vents of mid-ocean ridges, a number of cold, methane-rich vents have been discovered in other regions of the seafloor. These "cold seeps" also have a unique fauna that depends on bacterial chemosynthesis of methane, producing mound-like bodies of carbonate in deep-sea strata. These mound-like carbonate deposits with their peculiar bivalves had long been a puzzle, because they were found in such deep-water deposits, but a chemosynthetic cold seep explanation appears to solve this mystery (Campbell and Bottjer, 1993). [For further information, see the special issue on chemosynthesis in *Palaios*, vol. 7(4), August, 1992].

Paleontologists have often tried to reconstruct ancient food webs (Fig. 8.4). Since the Ordovician, the marine seafloor has harbored organisms that appear to occupy many of the same ecological niches in the food web, even if they are unrelated to their living analogue. For example, fish today perform the main roles as swimming (**nektonic**)

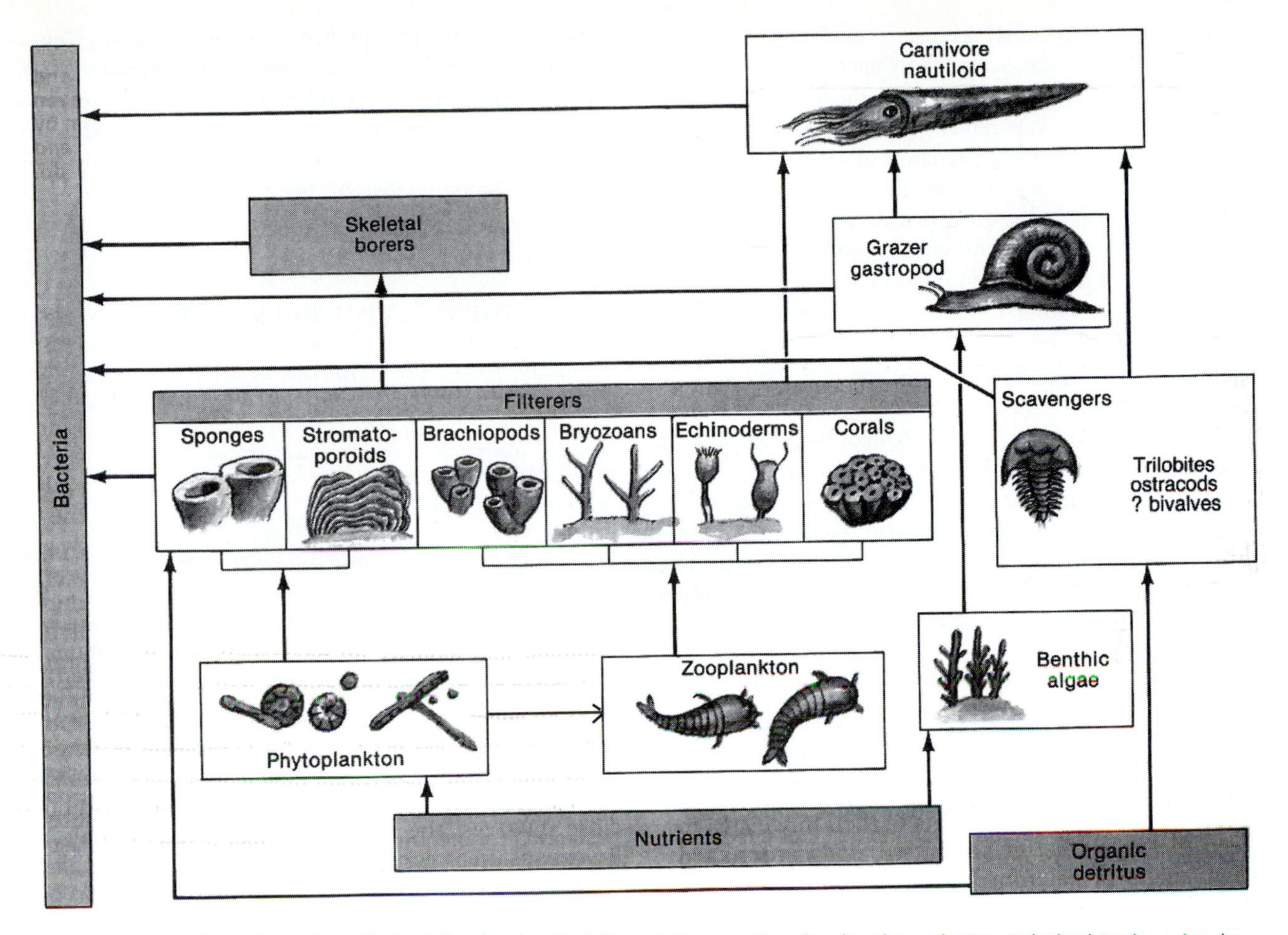

Figure 8.4. Reconstruction of an Ordovician food web. Like modern marine food webs, plants and planktonic animals are the food for a variety of filter feeders (although these are mostly extinct Paleozoic groups.) Unlike modern ecosystems, there are relatively few deposit feeders (mainly trilobites) and very few predators (mainly straight-shelled cephalopods). The muliple levels of crustacean and fish predators that now characterize the ocean was a later development. (From Dott and Prothero, 1994.)

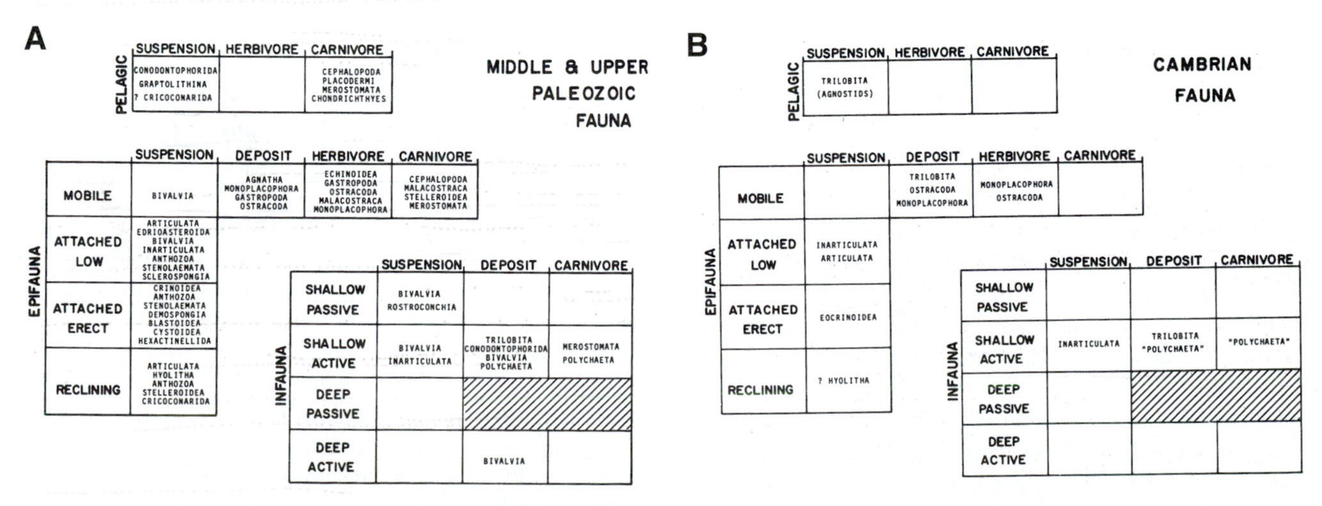

Figure 8.5. The various organisms of the marine realm can be partitioned into ecological niches. In modern faunas, all of the boxes are occupied. In the middle and upper Paleozoic (A), most of the niches had occupants, except for many of the burrowing (infaunal) niches. By contrast, in the Cambrian (B), only a few of the possible ecological niches had occupants. Most animals were suspension or deposit feeders, with almost no burrowers, pelagic forms, or large carnivores. (After Bambach, 1983.)

predators, but before fish with jaws evolved in the Late Silurian, that niche was occupied by nautiloids and eventually ammonoids. The nautiloids were the top carnivores of the Ordovician. But apparently there were not as many different levels of swimming predators among the nautiloids as there are with modern fish, and there were no swimming herbivores until fish occupied that niche much later (Fig. 8.5A). Unlike today, Ordovician corals (from extinct groups, such as rugosids and tabulates) were a minor part of the attached filter-feeding community, with brachiopods, bryozoans, crinoids, and stromatoporoid sponges playing a much larger role than they do today. A number of other niches were occupied by a different cast of players in the Paleozoic (Fig. 8.5A), but the most striking difference occurs in the burrowing, infaunal organisms. Until bivalves (and to a lesser extent polychaete worms and crustaceans) successfully invaded the deeper burrowing niches in the Mesozoic, there were no shallow passive deposit feeders or carnivores, no deep passive suspension feeders, and almost no active deep infaunal feeders of any kind.

If the differences between the Ordovician and modern seafloor are striking, this is even more so prior to the Ordovician radiation. For example, the Cambrian seafloor (Fig. 8.5B) was very different in ecological structure, with very few of the modern niches having occupants. Except for the floating agnostid trilobites and a few soft-bodied forms, there were almost no swimming organisms; the top predator appears to have been the soft-bodied *Anomalocaris* known from the Burgess Shale, which was only about a meter long. There were abundant shallow deposit-feeding trilobites, but no mobile predators or carnivores on the sea bottom, and very few attached suspension feeders; there was certainly no complex reef community as would appear in the later Paleozoic. The infaunal niches were even more vacant than they were in the rest of the Paleozoic. Thus, the uniformitarian approach breaks down in the Cambrian, when the food web is so much simpler because nothing had evolved yet to fill many of the niches. Community uniformitarianism is even harder to apply in the Proterozoic world, before there were even hard-shelled organisms such as trilobites or especially grazing gastropods. The only common organisms were the cyanobacterial mats that formed huge stromatolites, with no organisms that could feed upon them, let alone predators or scavengers. Essentially none of the boxes in Figure 8.5B (except for the possibility that there were plankton that fed on other plankton) had occupants prior to the late Proterozoic.

The complex food web (Fig. 8.3) that we take for granted today is actually a relatively recent (since the last 500 Ma) invention, which did not exist through most of life's history. Even more important, its complexity changed dramatically in the Late Proterozoic, and again in the Late Cambrian, before stabilizing into its present guise in the Mesozoic. This kind of large-scale view of ecology would never have been possible without the fossil record, and is one of several examples where paleontology has given us insights that ecologists would never have known about otherwise. We must be careful when we generalize about modern phenomena, such as a complex food web, and assume that it is the obvious or inevitable product of living systems. Through most of life's history, the food web looked nothing like it does today.

ENVIRONMENTAL LIMITING FACTORS

> *In solving ecological problems we are concerned with what animals do . . . as whole, living animals, not as dead animals or a series of parts of animals. We study the circumstances under which they do things and . . . the limiting factors which prevent them from doing certain other things.*
>
> Charles Elton, *Animal Ecology*, 1927

Although it is not easy to specify the detailed behavior of an extinct organism, there are certain constraints in the environment that almost certainly applied to the past as they do the present, since these are largely based on invariant laws of physics and chemistry. In many cases, a uniformitarian approach to paleoecology is easily justified, and we can clearly identify which conditions an ancient community must have tolerated and which ones they could not.

Temperature

Temperature is one of the most obvious limiting factors for a number of reasons. Many different organisms tolerate only a limited range of temperature fluctuation. Others can tolerate extreme temperatures and thus have little competition in habitats such as the freezing polar regions. Its most direct effect, however, is on the physiology and metabolism of organisms. Most biological reaction rates vary with temperature, so that metabolism, development, and reproduction typically operate best at an optimal temperature and are dramatically less efficient (or shut down altogether) as temperatures deviate further from the optimum. In many organisms, physiology is governed by **van't Hoff's rule**, which states that for every 10°C of temperature increase up to the optimum, biological reaction rates increase by a factor of 1 to 6 (depending upon species). Although there are exceptions, this increase in biological reaction rates with temperature is a widespread phenomenon. Once the optimal temperature has been exceeded, however, the reaction rate declines rapidly until temperatures are so hot that they become lethal.

Although it is difficult to determine physiological variables directly in fossils, the growth rate recorded in the skeleton can be measured in many different fossil groups (as indicated by growth rings and other characteristics discussed in Chapter 2). Data from living organisms (Fig. 8.6) show that higher growth rates occur with warmer temperatures, so tropical species tend to reach breeding age

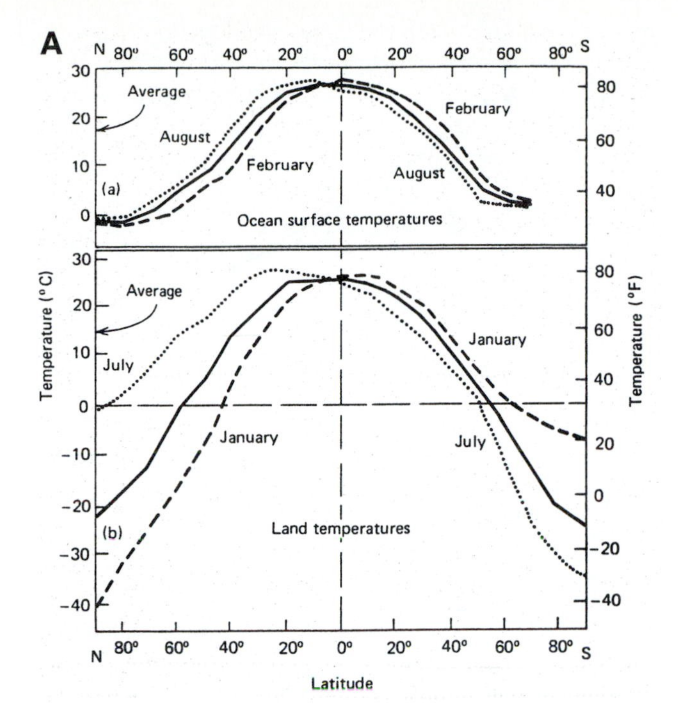

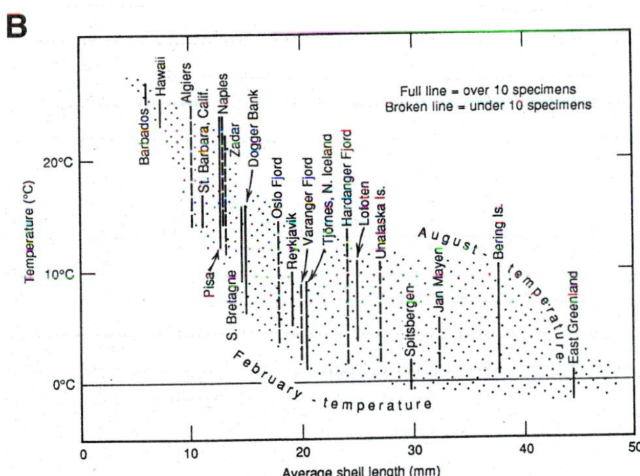

Figure 8.6. (A) Average and temperature ranges at the earth's surface as a function of latitude and season. In both the land and sea, temperatures are warmest and least variable in the tropics, and progressively cooler and more variable as one moves away from the tropics. In the oceans (top), the maximum variation occurs in the middle latitudes, while on land (bottom), variation increases toward the poles. In addition, the range of annual temperature is much higher on land than in the sea, due to the buffering effects of seawater. (After Wüst et al., 1954.) (B) Size variation of the bivalve Hiatella arctica with annual temperature range. As the temperature decreases and becomes more variable toward the poles, the shell size increases. (From Strauch, 1968.)

earlier and are smaller in body size (Fig. 8.6B). In addition, warm water is more likely to be supersaturated with calcium carbonate, further enhancing the growth of calcified tropical marine life. By contrast, organisms that grow slowly in cold waters breed much later and thus reach larger body sizes. As mentioned in Chapter 3, body size also increases with higher latitude in land animals because larger body size relative to surface area helps conserve heat. In other words, rabbits have larger body size near the poles than they do near the equator.

In local marine communities, temperature is one of the most stable variables, because water has tremendous heat capacity and thermal inertia. Marine seawater in a given area seldom fluctuates more than a few degrees centigrade over the course of the year, so most marine organisms are intolerant of temperature extremes (Fig. 8.6A, top). In the tropics, the fluctuation is even less, but in the temperate latitudes, there may be as much as 5 to 8°C fluctuation due to seasonal changes. When long-term extremes persist for months or years, it can result in a crisis for marine organisms. For example, the unusually warm waters along the Pacific Coast due to the El Niño current decimated the marine life over the last few years, as any Pacific beachcomber or fisherman can attest.

Over the entire world ocean, temperature varies according to latitude, with the warmest temperatures (about 28°C) in the tropics (Fig. 8.6A) and extreme subzero temperatures in the polar regions (since salty seawater freezes at -1.4°C, lower than the freezing point of freshwater, 0°C). Many organisms live in distinct biogeographic provinces based on temperature, particularly in the large water masses of the open ocean. Temperature also varies with depth in the ocean, with surface waters the warmest and deep waters (which typically arise in polar regions, sink to the bottom, and the flow toward the equator) being very cold. For example, most of the deep oceans are bathed in the Antarctic Bottom Water, which forms around the Antarctic continent and then sinks and flows northward along the bottom to as far north as 50°N in the Pacific and 45°N in the Atlantic; this one current makes up about 59% of the ocean's water. In the North Atlantic, the North Atlantic Deep Water forms in the Arctic and then sinks and flows south around Greenland, forming much of the bottom water in this region.

Although temperature is a major controlling variable in the marine realm, it is closely related to other variables, such as oxygen content and salinity. For example, the solubility of many minerals (especially calcium carbonate) varies with temperature. Warm water carries a higher concentration of calcium carbonate than cold water, as we mentioned earlier, affecting carbonate skeleton precipitation rates. Conversely, increased temperature lowers the solubility of gases such as oxygen and carbon dioxide, making conditions more difficult for some organisms at high temperature. It is not always easy to tease these variables apart when examining living systems, let alone in extinct organisms.

In the terrestrial realm, temperature is also one of the most important environmental parameters, along with precipitation. Together, they determine the major plant biomes (tundra, taiga, grassland, deciduous forest, desert, and rainforest, and so on), which are largely distributed by the latitudinal gradient in temperature. Unlike the marine realm, however, terrestrial temperatures are much more variable (Fig. 8.6A, bottom graph). In the continental interiors, where the buffering effect of the ocean is not felt, temperatures can change from scorching to freezing in a matter of hours, such as when an Arctic cold front rolls through the Plains on a hot day. (On one summer day in 1996 in Minnesota, the temperature changed 100°F in a few hours). In addition to being more variable, temperatures on the land can be much more extreme, with subfreezing temperatures covering most of the high latitudes through most of the year, while the seawater in the Arctic or Antarctic remains mostly unfrozen.

Paleoclimatologists seek highly sensitive organisms that are good indicators of paleotemperature. There are many species with very narrow temperature tolerances, so when we find their near relatives in the fossil record, we can make very precise predictions about ancient temperatures. For example, benthic molluscs are arrayed into distinct biogeographic provinces (see Chapter 9) based on temperature and latitude, so a tropical species found much farther north in the fossil record than they occur today indicates that ancient temperatures were higher. Some marine organisms are extraordinarily sensitive to temperature. Coral reefs are restricted to a warm tropical belt between 25° north and south latitude, and individual hermatypic corals live no farther from the equator than 35°. Micropaleontologists have long used the restriction of Cenozoic plankton to water masses of certain temperatures to interpret the distribution of temperatures in the past (Kennett, 1976). For example, the planktonic foraminifer *Globorotalia menardii* is an indicator of tropical conditions, while the thick-shelled *Globigerina pachyderma* indicates cold conditions. In some cases, they respond in even more dramatic ways. Several species (including *G. menardii, G. pachyderma,* and *G. bulloides*) switch from right-coiling during warmer interglacial periods to left-coiling during cold glacials (see Fig. 10.4). *Globotruncana truncatulinoides* changes from highly conical forms in tropical waters to more compressed forms in cooler waters.

In terrestrial ecosystems, there are many species that can be used as paleotemperature indicators. For example, reptiles and amphibians cannot survive where it gets too cold for them to metabolize, so there are definite latitudinal limits on the distribution of such common fossils as crocodilians and turtles. Today, very few reptiles lives north of 70° latitude, and crocodilians are restricted to a belt between 35° north and south latitude. Paleontologists have used this fact to determine terrestrial paleotemperatures and to infer that the polar regions were much warmer in the Cretaceous and Eocene, since they supported a diverse fauna of cold-intolerant reptiles (Colbert, 1964; Hutchison, 1982; Estes and Hutchison, 1980). Perhaps the most sensitive paleothermometers, however, are the land plants. Paleobotanists have long used their distribution and physical characteristics to infer ancient climates, especially in locating the ancient tropics by the distribution of thick coal deposits. One of the most sensitive techniques, however, uses the shapes of leaves as paleothermometers (Fig. 8.7). Tropical leaves tend to be larger, thicker, with smooth ("entire") margins and a drip tip, while leaves from cooler climates are smaller, thinner (and possibly deciduous), with jagged margins. The percentage of entire versus jagged margined leaves directly tracks mean annual temperature (Fig. 8.7A), and using that one parameter alone, Wolfe (1978) was able to plot a paleotemperature curve for floras from the Gulf Coast to Alaska (Fig. 8.7B). Recently, Wolfe (1990, 1994) has refined his methods using multiple variables of leaf shape, so that his ability to predict mean annual temperature and the seasonal variation is even more robust and precise.

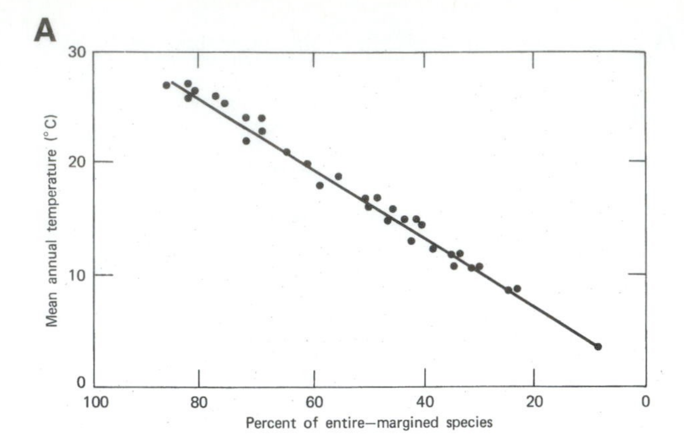

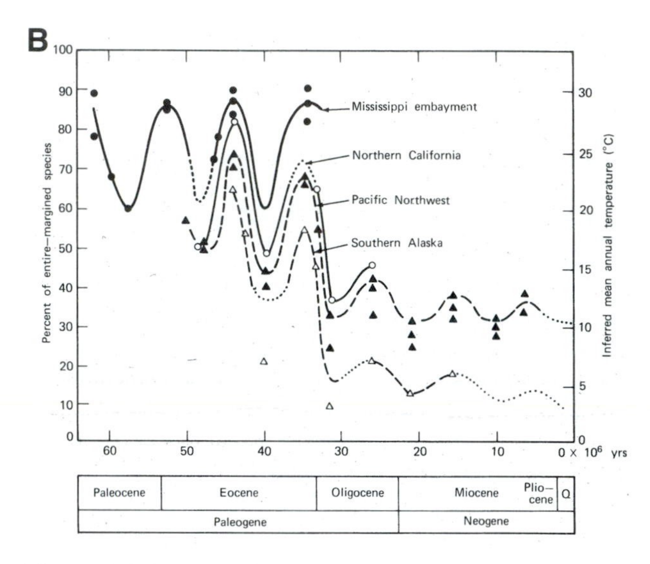

Figure 8.7. Land plants are very sensitive indicators of climate, particularly temperature. (A) Wolfe (1978) has shown that leaves with smooth ("entire") margins tend to be correlated with warm climates, while those with jagged margins are found in colder regions. (B) Using this correlation, Wolfe (1978) was able to predict the mean annual temperature curve for various regions in North America during the Cenozoic. (Modified from Wolfe, 1978.)

Oxygen

Oxygen is essential for the cellular respiration in animals, so its abundance is another critical variable in marine

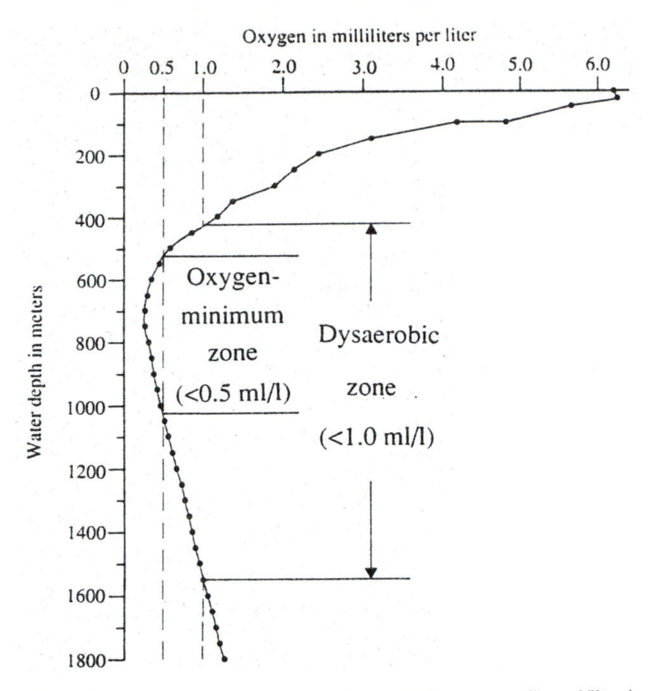

Figure 8.8. Concentration of dissolved oxygen (in ml/liter) versus water depth for the central California margin. The surface waters are highly oxygenated, but the oxygen content decreases rapidly in the dysaerobic zone due to the activity of organisms, and due to oxygen comsumption from organic decay. Deeper waters tend to have slightly more oxygen, since few organisms live at this depth to consume it. (Modified from Thompson et al., 1985.)

and freshwater systems. (Clearly, it is rarely a problem in the terrestrial realm.) Near the surface, the actions of wind and waves constantly mixes oxygen in the water, and plants produce oxygen by photosynthesis, so the surface waters are highly oxygenated (as much as 6 ml O_2/liter of water); these conditions are called **aerobic**. Dissolved oxygen levels drop off rapidly with depth (Fig. 8.8), however, until they reach a threshold of 1.0 ml/l, typically encountered at depths of 400 m. Below this value, conditions are **dysaerobic**, and typically there is an **oxygen-minimum zone** in the ocean at depths of about 600 to 1000 m. This low oxygen is caused by the heavy respiration by marine life in near-surface waters, and by the fact that most decay of sinking organic matter occurs at this depth, severely depleting the available oxygen. Below the oxygen-minimum zone, deep ocean waters tend to be well oxygenated, because their flow is mostly derived from the Antarctic and Arctic, where the cold bottom waters formed at the surface and then sank, carrying their high oxygen levels with them.

In extremely stagnant, stratified bodies of water, where little exchange occurs between the surface and depths (such as in the Black Sea), bottom conditions can become **anaerobic** (less than 0.1 ml/l). Such conditions may be devoid of animal life (except for anaerobic bacteria, which cannot tolerate much oxygen), and the extreme reducing conditions produce black shales with pyrite. They also tend to produce stagnant conditions with few or no currents or scavengers, so that extraordinary preservation is possible (see Chapter 1).

Biologically, the oxygen level is critical to many kinds of organisms. Most marine life cannot tolerate low oxygen levels, and lives in the near-surface waters. Corals and cephalopods in particular are known to be intolerant of low oxygen. Some annelid worms, on the other hand, thrive under low oxygen conditions, and thus they tend to be the most numerous animals in the abyssal depths (along with brittle stars and sea cucumbers, which are also tolerant of those conditions).

Lakes can also have depth and oxygen stratification, particularly if they are very deep and have a highly restricted flow. In addition, the bottoms of lakes can have high concentrations of organic matter from decaying plants, so they can quickly become reducing and anaerobic even at shallow depths (as in the typical coal swamp). It is common to find highly stratified lake deposits, with light-colored oxygenated sediments overlying, or even alternating with, anaerobic black muds. Rapid alternation of oxygen levels is particularly characteristic of lakes, since they can fluctuate wildly in these variables (unlike the ocean).

Salinity

Salinity is the third most important parameter in marine environments. Measured as the total dissolved salts (especially sodium chloride) per volume of seawater, it is commonly expressed in parts per thousand, or parts per mil (‰). Freshwater ranges from 0 to 0.5‰, while brackish water varies from 0.5 to as much as 30‰. Normal seawater salinity is very stable at between 30 and 40‰, whereas conditions between 40 and 80‰ are hypersaline, and greater than 80‰ is a brine. Most marine waters fluctuate very little in salinity around a mean of 35‰. It is slightly higher (up to 36.5‰) in the centers of oceanic circulation gyres in the "horse latitudes" (about 30° north and south latitude), since evaporation is highest there. Conversely, it is slightly lower (as low as 32‰) in polar regions, since the evaporation rate is very low and there is a lot of freshwater runoff. Nearshore regions, especially estuaries and lagoons, are characterized by very large fluctuations in salinity. At certain times there are great flushes of freshwater from runoff, but at other times the runoff is insufficient to keep normal marine waters out of the lagoon or estuary. Hypersaline conditions are usually found in lakes, lagoons or inland seas that have restricted flow of water into them, especially if they are located in the great high-pressure belts of evaporation (between 10 and 30° north and south latitudes), where most of the world's deserts also occur.

Salinity variations chiefly affect the osmotic regulation of organisms. Some animals and plants have mechanisms regulating the content of body salts when they live in waters that are more or less saline than their internal body fluids; these organisms are known as **euryhaline** (*eury* means "broad" and *halos* means "salt" in Greek). Others cannot do so, and will literally explode or shrivel up if they are in waters that are dramatically different in salinity from their internal fluids; such narrow tolerances characterize **stenohaline** organisms (*steno* is "narrow" in Greek). Most marine organisms are stenohaline. The few examples of euryhaline organisms (such as oysters, mussels, certain crustaceans, certain foraminifera) are so distinctive that a fauna dominated by them almost certainly inhabited brackish water. Diversity is also strongly related to salinity. The highest diversities are found in normal salinities and drops off dramatically in regions of fluctuating fresh- and saltwater, and even more so in hypersaline waters. Most hypersaline environments can support only extremely salt-tolerant organisms, such as algae, bacteria, ostracodes, and brine shrimp, and have very low diversity.

Depth and Light

A fourth important variable in aquatic environments is the depth of the water. Water depth has several effects. The most important is that light penetration decreases with depth, so that photosynthetic organisms (especially algae, phytoplankton, and organisms that are symbiotic with plants, such as hermatypic corals, larger foraminifera, and giant clams) can live only in the upper waters (the **photic zone**). The intensity of light drops off so rapidly below the surface that the base of the photic zone is rarely deeper than 200 m in the clearest tropical waters, and it is usually much shallower (less than 50 m) because particles in the water absorb light. In most marine environments, 80% of the light is absorbed in the upper 10 m of the water column. There is a concentration of marine life near the surface, and the diversity of organisms drops off rapidly with depth.

Below the photic zone is a region of perennial darkness, and thus no photosynthesis or primary productivity takes place. Here, the trophic relationships are much simpler. Nearly all animals (no plants, obviously) feed on detritus, or scavenge dead organisms that sink down, or prey upon these detritus feeders and scavengers. The sparse concentration of suspended food makes filter feeding unprofitable. Although there is less diversity than in the photic zone, recent oceanographic surveys have shown that there is a surprising diversity of detritus feeders.

As depth increases, the loss of light is not the only major change. Temperatures and dissolved oxygen decrease, and other variables (such as salinity and carbonate content) also change. Most ocean waters below the **carbonate compensation depth** (the CCD, or the depth at which carbonate dissolves, about 3000 to 4000 m) are undersaturated with calcite, so that calcareous organisms cannot be preserved. At bathyal and abyssal depths, the pressure of thousands of meters of seawater is so great that most organisms must be adapted to these extreme pressures as well. This is the realm of the bizarre deep-sea fish with their own light sources and the amazing gaping mouths with long, sharp teeth. On the sea bottom, worms, brittle stars, and sea cucumbers tend to predominate (Fig. 16.2).

Paleobathymetry is a very difficult parameter to estimate in the stratigraphic record. Usually the best clue is the fossils themselves, but one must be careful of circular reasoning. One cannot argue that strata are shallow marine deposits because of their fossils and then use that inference to argue that the fossils are shallow marine. Depth can also fluctuate rapidly with changes in sea level, so that a stratigraphic sequence may have been formed at one depth but then overprinted with effects from another depth. For example, an offshore marine deposit may suddenly become shallower or emergent, causing the sediment to be cemented into a **hardground**. Sedimentologists have traditionally used the fining of grain size in deeper and deeper water as a first line of evidence of paleobathymetry, although the presence of coarse sandy turbidites in abyssal depths violates this rule of thumb.

The best fossils for determining paleobathymetry are those that are diagnostic of certain depths. Clearly, tidepool organisms like limpets and mussels indicate intertidal environments, and a high diversity of shelly organisms in fine sands are indicative of nearshore, neritic environments. Finer-grained silts and muds that are low in diversity but have abundant burrows are usually interpreted as deeper and more offshore. One of the most useful groups are the foraminifera, which are strongly depth-controlled. Benthic species are well known to have preferences for certain depths of water (see Chapter 11). Some are associated with shallow lagoons, others with nearshore environments, and still others with offshore shelf, slope, or abyssal depths. The planktonic species that float in the water column above also rain down to the bottom, but they are rare in nearshore waters and progressively tend to dominate the faunas preserved in deeper waters. As depths approach the CCD, the calcareous foraminifera become scarcer, and most of the benthics are made of agglutinated sand grains, which require little calcite. No calcareous species are found far below the CCD, and foraminifera are scarce at abyssal depths (as are most animals).

Substrate

Benthic organisms can also be restricted by the nature of the bottom, or substrate, on which they live. For example, most tidepool animals (limpets, periwinkles, barnacles, rock-boring clams) require hard rocky substrates on which to attach, and cannot live anywhere else. Other animals require a soft, soupy, muddy substrate in which to burrow or wriggle along, feeding on detritus. Still others are adapted to the rapidly shifting sands of the surf zone, and thrive by burrowing rapidly once the waves have deposited them on the bottom. Burrowing organisms that filter feed (especially bivalves) do better in coarser sands and silts, because loose muds tend to clog their gills; consequently, there is a much lower diversity of filter feeders on the muddy bottom. Other organisms, such as productid brachiopods, evolved stilt-like spines to prevent their sinking into the soft muddy bottom. Many organisms can live on loose substrates as long is there is one solid piece of rock or shell on which to attach or anchor. In some environments, these attachment sites become a scarce resource, and every one will be heavily covered and encrusted by colonizers who took advantage. This is especially true of the larvae of marine invertebrates that require a hard surface for attachment, as corals do. Although millions will be released and most eaten by predators, those that survive are largely determined by which ones were lucky enough to land on a hard substrate that wasn't already crowded with previous tenants.

Unlike temperature, salinity, oxygen, or depth, which are hard to infer from the fossil record, substrate is the only parameter that is directly preservable, since fossils are often entombed in the substrate where they lived. This generalization applies only to organisms that are found in their life habitat; once they have been transported, the sedimentary matrix may have no relation to the substrate originally inhabited by the organism.

In terrestrial environments, substrate (in the form of soils) is one of the most critical biological factors determining what type of vegetation will live (and thus which animal communities can exist). Certain soils, such as the rich windblown proglacial loess clays of the American Plains, central Europe, or China, are extremely rich soils; they are known as the "breadbaskets" of the world. Other soils are extremely poor. For example, the tropical lateritic soils found beneath most of the world's rainforests are extremely thin and lacking in nutrients. When the rainforest is clear-cut, all the nutrients leave with the trees, exposing a poor soil that quickly turns brick red and rock hard in the sun. Within a few years, it is incapable of supporting much vegetation, let alone another rainforest. Some soils have an unusual chemistry that dictates which vegetation will grow. For example, soils developed on ultramafic rocks, such as gabbros and peridotites, have unusually large concentrations of magnesium (from Mg-rich minerals such as olivine and pyroxene) and calcium (making them alkaline). Consequently, they support a specialized plant community that is tolerant of high magnesium and calcium concentrations in the soil.

Summary

Much is known about how various environmental parameters in living systems constrain the distribution of organisms and ecological communities. In many cases, we can make the uniformitarian assumption and extend this understanding to their fossil counterparts, leading to reliable inferences about the environmental parameters constraining a fossil community. Except for substrate, however, many of these inferences are based on indirect evidence, since there is little direct evidence of temperature, salinity, oxygen content, or depth in a given fossil assemblage. Wherever it is possible to obtain a second line of evidence to test these paleoenvironmental hypotheses (as with stable isotopic ratios discussed in the next section), the reliability of the paleoecological hypothesis can be greatly enhanced.

DIRECT PALEOECOLOGICAL EVIDENCE

> *Surprisingly detailed interpretations of past environments are possible, and the fossils themselves acquire much added significance whenever they are studied in the full context of their geologic setting.*
>
> Norman Newell, "The nature of the fossil record," 1959

Although many ecological processes are hard to document in the fossil record, there are also many positive things that we can determine about ancient ecology. In many cases, behavioral patterns and ecological relationships can be reliably inferred from a number of lines of evidence.

Fossilized Behavior

The strongest evidence comes from extraordinary fossilization that actually preserves a behavior. The best source in this topic is Boucot's (1990) *Evolutionary Paleoecology of Behavior and Coevolution*, a 725-page book full of examples of fossilized behavior of nearly every imaginable kind. In a short chapter like this, it is impossible to give a sense of the variety of behaviors that have been fossilized, but we can mention a few exceptional cases. There are numerous specimens where symbiosis or commensalism (mutual living together for each other's benefit) are preserved in fossils, such as the association between the giant Cretaceous inoceramid bivalve *Platyceramus* and the schools of fish that apparently sought shelter inside (Fig. 8.9A). Many different organisms attached to others for a variety of purposes. For example, juvenile productid brachiopods apparently started out life by attaching to a crinoid stem and stayed there until they grew so large (Fig. 8.9B) that they broke free and settled to the bottom, where they rested on the muddy bottom using their spines as stilts. Many organisms attach to others, especially when there is no other substrate available. In addition to familiar cases such as barnacles and encrusting bryozoa, there are more unusual cases, such as the edrioasteroids, which are habitually found attached to large strophomenid brachiopods (Fig. 8.9C). There are many specimens where a parasite is fossilized with its host or a parasite leaves a distinctive scar on the host that can be seen in the fossil.

Even sex can be fossilized. Among the insects found in amber are many cases of individuals caught in the act of copulation (Fig. 8.9D). The cap-shaped gastropod known as the slipper shell, *Crepidula fornicata*, is frequently found stacked one upon the other (Fig. 8.9E). Living *Crepidula* have a very unusual sexual system—the large ones at the bottom of the stack are females and the smaller, younger ones at the top are males. The males have a very long penis that can reach down the stack and fertilize a female far below. When larvae settle, they are attracted to the top of the stack, where they grow up to be male. As the stack becomes larger, the individuals in the middle change sex from male to female. If the top of the stack should be

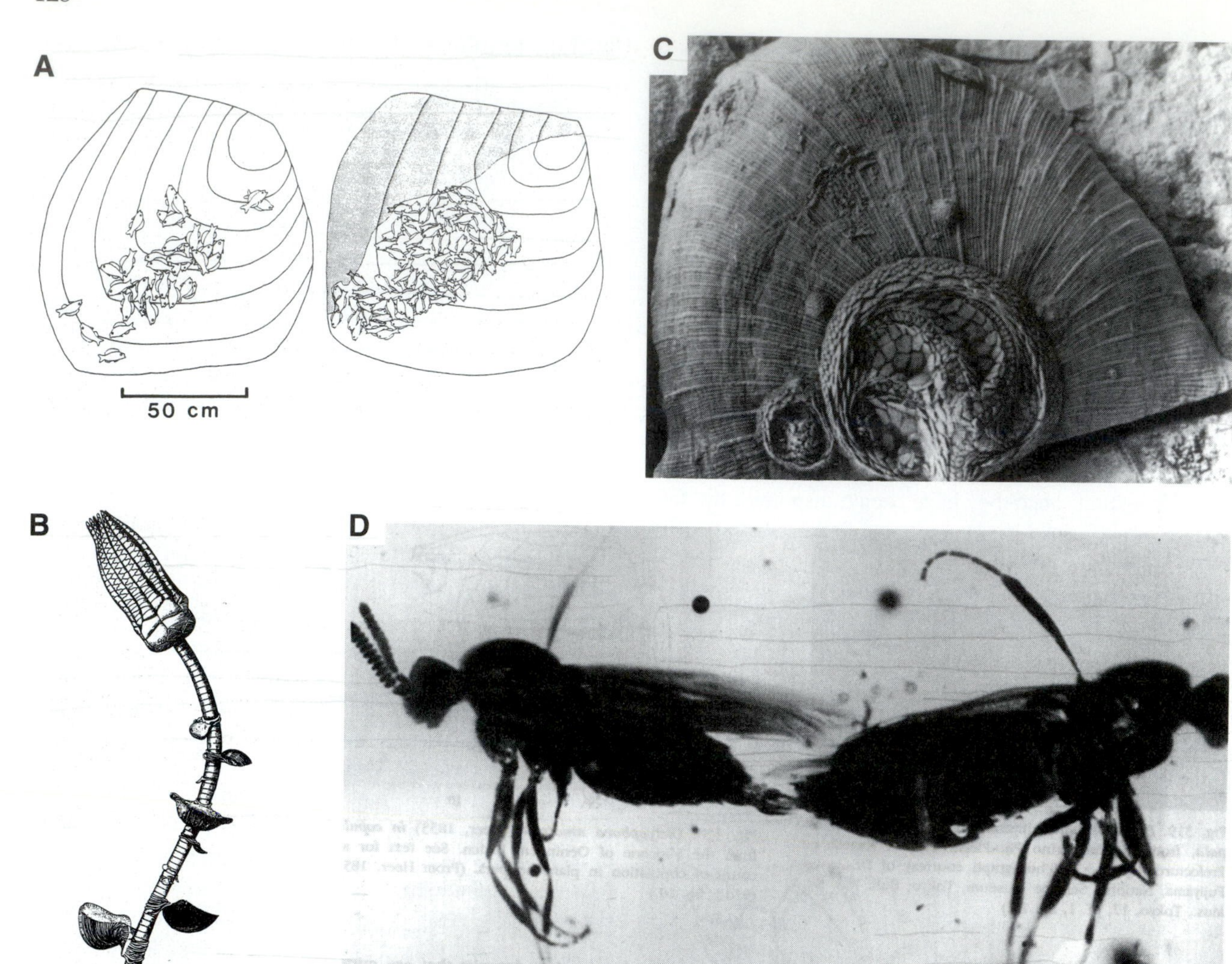

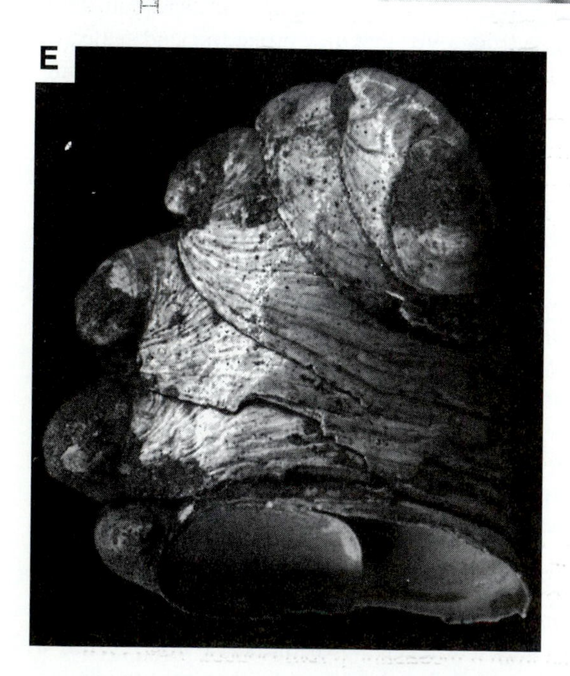

Figure 8.9. Examples of fossilized behavior. (A) The meter-wide flat inoceramid bivalves of the muddy Cretaceous seafloor were frequently inhabited by schools of fish. (From Boucot, 1990.) (B) Juvenile productids of *Linoproductus angustus* apparently attached to crinoid stems until they reached a certain size, then dropped off and settled on the bottom. (From Grant, 1963.) (C) The disk-shaped edrioasteroid *Isorophus* is typically found encrusting a larger shell. (Photo courtesy D. Meyer.) (D) Male (left) and female flies (genus *Rhegoclemina)* caught in amber during the act of copulation. (From Boucot, 1980.) (E) This stack of fossil slipper shells (*Crepidula*) consisted of males on the top, and females on the bottom. (Photo by D. R. Prothero.)

knocked off, the females remaining on the top can switch back to being male.

Other aspects of reproduction have also been revealed in the fossil record. A Jurassic ichthyosaur from the Holzmaden shales was preserved with a juvenile emerging from the birth canal (Fig. 8.10A), evidence that ichthyosaurs (like dolphins) gave birth to live young (since they could not crawl out on land to lay eggs). A number of examples of dinosaur eggs and nests are known. In the case of Jack Horner's (1986) famous dinosaur nests from the Cretaceous Two Medicine Formation in Montana,

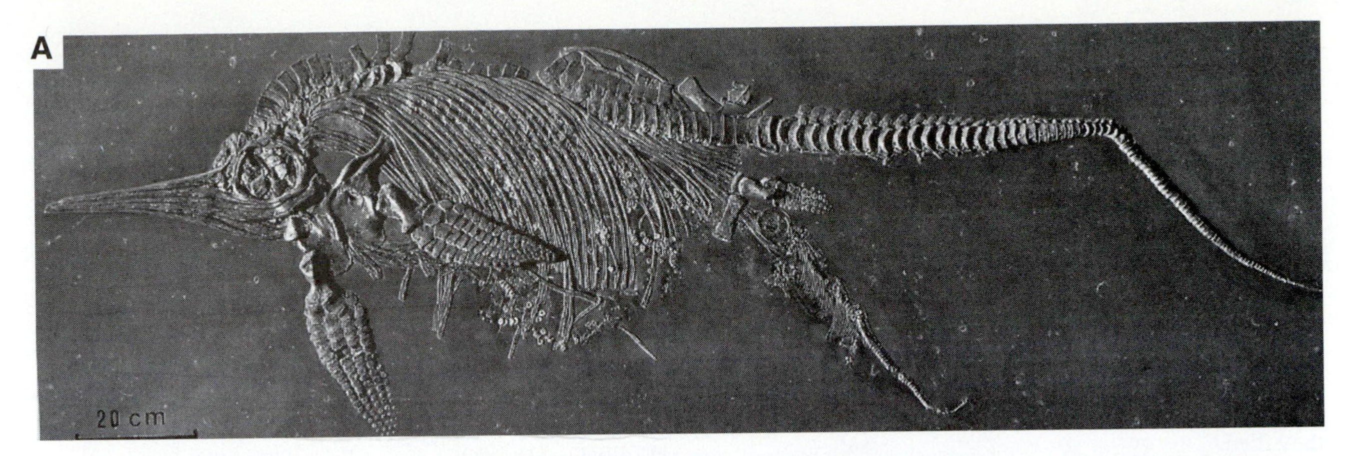

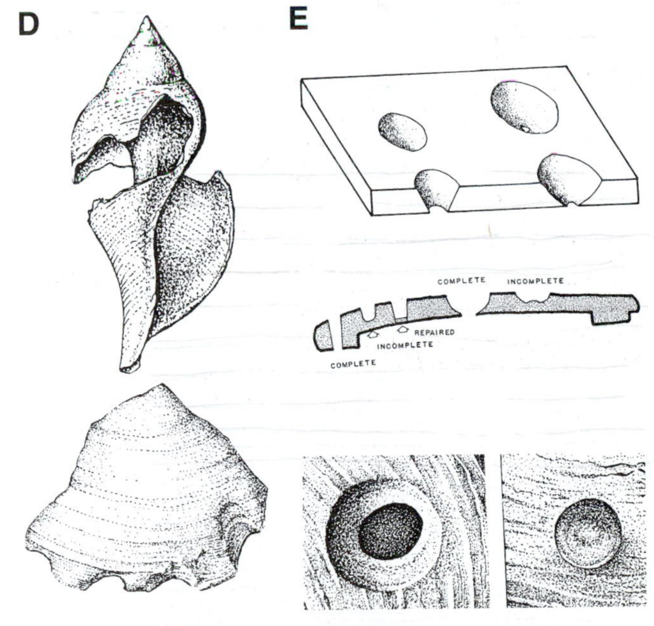

Figure 8.10. (A) This famous specimen of a Jurassic ichthyosaur from the Jurassic Holdmaden shale was preserved with a young individual just emerging from the birth canal. (Photo courtesy R. Wild.) (B) An Eocene fish, *Mioplosus,* died attempting to swallow a *Knightia,* as preserved in the Green River Shale. (Photo courtesy L. Grande.) (C). This ammonite *Placenticeras* from the Cretaceous Pierre Shale has a V-shaped row of conical tooth marks from a mosasaur. (From Boucot, 1980.) (D) Two examples of gastropod shells that have been broken or peeled open by predators (probably crabs.) (From Briggs and Crowther, 1990.) (E). Naticid and muricid snails drill beveled holes in their prey. (From Boucot,

some nests show that the young were capable of leaving the nest and feeding immediately after birth, while others (those of the duckbill dinosaur *Maiasaura*, the "good mother lizard") have been interpreted as evidence of parental care, with young that could not leave immediately and were fed by the parents. In the famous Ashfall Fossil Bed State Park in eastern Nebraska, a large herd of the hippo-like rhinoceros *Teleoceras* were entombed in volcanic ash, giving it the nickname "Rhino Pompeii" (Voorhies, 1981). They are preserved exactly as they died, with stomach contents and their last meals in their throats. Perhaps the most touching scene are the baby rhinos, which lie against their mother's bellies in nursing position, where they died.

Predation is commonly fossilized, both as specimens that have been preyed upon and as predators with stomach contents preserved. Many of the extraordinary fossils in the *Lagerstätten* discussed in Chapter 1 preserve stomach contents. In the famous Eocene lake deposits of Messel, we can analyze the last meals of many of the animals, because the details of the plant cuticle are delicately preserved. Mammoths that died and were frozen in the Arctic typically have preserved stomach contents (buttercups seem to be favored), and sometimes a mouthful of partially chewed food as well. Many places that are famous for excellent preservation of fossil fish preserve one fish inside another, showing the last meal of the larger fish (Fig. 8.10B). Several ammonite specimens are known that have a distinctive set of conical bite marks arranged in a "V" pattern, suggesting that they were bitten by mosasaurs (Fig. 8.10C).

Specimens such as these, however, tend to be rare exceptions, and not enough examples are known to make broad generalizations or do statistical analysis. Certain kinds of predatory behavior, however, have been studied many times, since there are large samples. Crabs and lobsters can crush mollusc shells with their large claws or break open the aperture and peel it away (Fig. 8.10D). A number of gastropods, especially the murexes (Family Muricidae) and the moon snails (Family Naticidae) prey upon other molluscs by drilling a hole through their shells and then eating their prey inside (Fig. 8.10E). Naticids leave a distinctive beveled drill hole in their prey, and these are very common in the rich shell beds of the Mesozoic and Cenozoic. Numerous studies on naticid boreholes (Kitchell et al., 1981; Kitchell, 1986; Anderson, 1992; Kelley, 1988, 1989, 1991) show that these snails are very stereotyped in their behavior, preferring one valve over the other and certain species over others, and always drilling near the center of the shell. Such data have also been used to assess the long-term changes in predatory behavior over time and the spatial variation in predatory behavior as well (Kelley and Hansen, 1993; Hansen and Kelley, 1995).

Although body fossils are the best direct evidence of behavior, tracks and traces are also valuable clues. We will discuss trace fossils in greater detail in Chapter 18.

Paleobiogeochemistry

Although biological evidence of ancient temperatures or salinity conditions are valuable, it is much better when we can find independent evidence to test our estimates of these parameters. Such evidence can come from the chemistry of the shell or bone, and often it tells us much more than we would have expected based on the fossils alone.

Oxygen—The most widely used chemical system is **oxygen isotopes**. Most (99.756%) of the Earth's oxygen is the light isotope ^{16}O, which has 8 protons and 8 neutrons. The slightly heavier isotope ^{18}O, which has 8 protons and 10 neutrons, is normally present in the ocean in rare amounts (0.205%). The two isotopes are compared using a standard formula:

$$\delta^{18}O = \frac{1000 \times [(^{18}O/^{16}O)\text{ sample} - (^{18}O/^{16}O)\text{ standard}]}{(^{18}O/^{16}O)\text{ standard}}$$

The ratio is expressed as $\delta^{18}O$, with more positive ("heavier") values indicating increases in ^{18}O and more negative ("lighter") values indicating increased ^{16}O. Like the salinity discussed above, oxygen isotopes are measured in parts per thousand (parts per mil, or ‰) against a standard material, usually calcite in belemnites from the Cretaceous Pee Dee Formation in South Carolina (abbreviated "PDB").

In 1947, Harold Urey and Cesare Emiliani found that these two isotopes fractionated with changes in temperature. A change of 1 per mil corresponded to an apparent change of temperature of 4.5°C. Water with less ^{18}O is lighter, so it evaporates more easily than ^{18}O-rich water; consequently, the clouds are enriched in ^{16}O and depleted in ^{18}O. When temperatures are warmer, there is more evaporation and removal of ^{16}O, and the oceanic waters are thus enriched in ^{18}O; the carbonate of marine organisms should reflect this change in the ratio. Initially, most studies assumed that the relationship between oxygen isotopes was straightforward. In addition, the carbonate precipitated from seawater during colder temperatures is more enriched in ^{18}O, while warm water the carbonate is richer in ^{16}O.

However, there is another complicating factor. When the Earth has polar glaciers (as it did in the late Paleozoic, and has since the beginning of the Oligocene), it turns out that ice volume has an even larger effect than temperature (Fig. 8.11). During non-glaciated times, clouds rain their ^{16}O-rich water on the land, but it immediately runs off and returns to the ocean, keeping the $^{18}O/^{16}O$ ratio in the oceans relatively negative. During glacial times, however, much of this ^{16}O-rich (–30‰) water is locked up in the ice caps, making the oceans ^{18}O-rich by about +1.6‰. Thus, the $\delta^{18}O$ signal trapped in the skeletons of marine organisms is primarily (about two-thirds) an ice volume signal during glacial times, but more directly measures the global temperature during non-glaciated times.

Although the relationship is complicated, oxygen isotopes have become the primary tool of paleoclimatologists

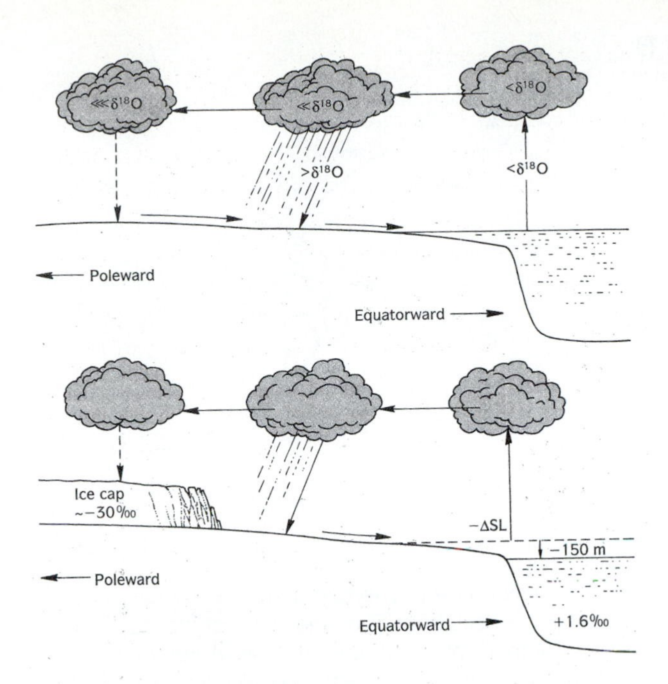

Figure 8.11. Rain falling on the land (top) is depleted in oxygen-18. When it is locked up in an icecap (bottom), this enriches the ocean in oxygen-18. (From Matthews, 1984.)

and paleoceanographers for determining ancient temperatures, especially in the oceans. Micropaleontologists and isotope geochemists have collaborated on the analysis of thousands of specimens of benthic and planktonic foraminifera, and have produced a detailed record of climatic change during the Mesozoic and Cenozoic (Fig. 8.12A). When the Deep-Sea Drilling Project began routinely coring Cenozoic sediments on the deep ocean floor, oxygen isotopic analysis became a standard tool in all carbonate sediments. During the glacial-interglacial cycles of the Plio-Pleistocene, the fluctuations of oxygen isotopes are so regular and predictable that oxygen-isotopic "stages" were numbered and used for correlation all over the world oceans (Fig. 8.12B). Paleoclimatologists used oxygen isotopes to estimate the temperature of water masses, and produce oceanic temperature maps of the peak of the last glacial, about 18,000 years ago.

Where temperatures can be assumed to be relatively constant, oxygen isotopes can be used to estimate ancient salinities (Rye and Sommer, 1980; Talbot, 1990). Since clouds are isotopically lighter (more negative $\delta^{18}O$) than seawater, so is the freshwater and ice on the land that is precipitated from them. Thus, when isotopically light

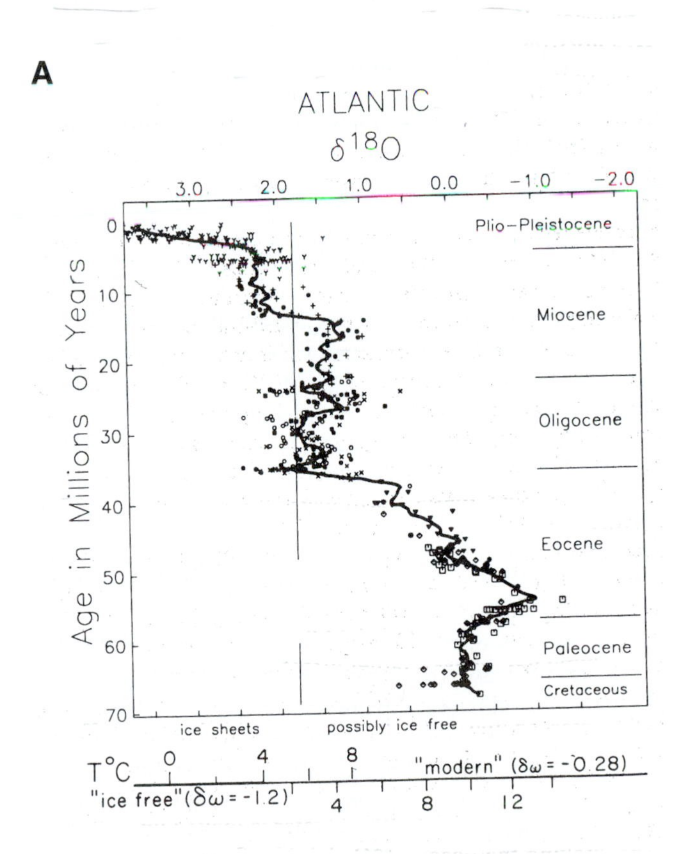

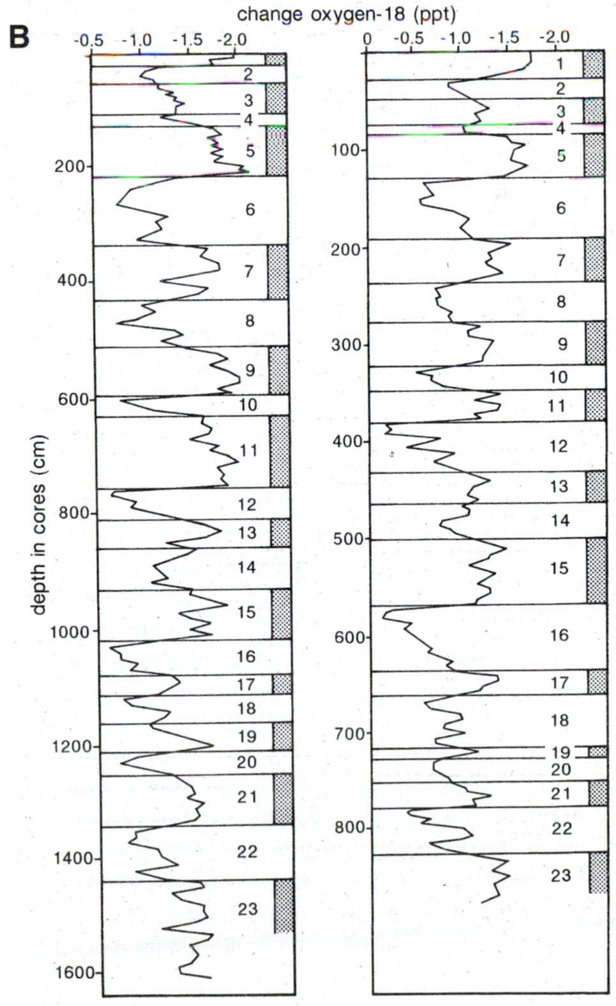

Figure 8.12. (A) The oxygen-isotope ratios of the Cenozoic are a good indicator of global temperature and ice volume. (From Miller et al., 1987.) (B) The oxygen-isotopic cycles of the ice ages can be matched up precisely from one oceanic core to another all over the world. (From Shackleton, 1976.)

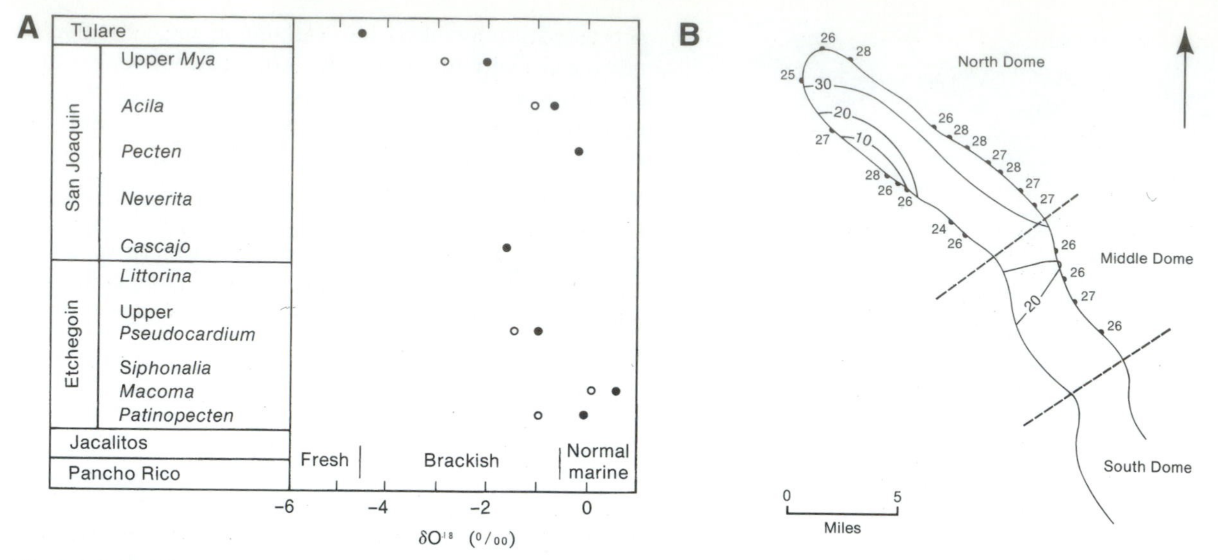

Figure 8.13. Oxygen isotopes can also be used to detect variations in salinity as well as temperature. (A) Oxygen isotope values increase (move to the right) with an increase in salinity in these fossils from Pliocene rocks of the Kettleman Hills in central California. (After Stanton and Dodd, 1970.) (B) The mean number of ribs in the ark shell *Anadara* (shown by the numbers next to the data points) highly correlates with the salinity gradient (shown by the contour lines) in the ancient marine basin that once covered the Kettleman Hills region. (From Alexander, 1974.)

freshwater mixes with isotopically heavy water of normal salinity, the resulting $\delta^{18}O$ will be depressed by as much as -6‰. However, it takes an extreme change in salinity to show up in the carbon and oxygen isotopes, and other factors, such as humidity, can complicate the interpretation (Matyas et al., 1996). Stanton and Dodd (1970) and Dodd and Stanton (1975) were able to use oxygen isotopes to interpret the salinity changes that occurred in the Pliocene bay flooding the San Joaquin Valley of California. Salinity was lowest at the north end of the basin where the freshwater influx was highest, and increased to almost normal marine values of 35‰ toward the south end of the basin (Fig. 8.13). This interpretation correlated with the changes in ribs in the bivalve *Anadara*, which seemed to vary depending on salinity (Alexander, 1974) (see Chapter 2).

Geochemists continue to find paleontological problems that might be solved with oxygen isotopes. For example, Barrick and Showers (1994) argued that this method could test whether dinosaurs were warm- or cold-blooded. The latter have poor temperature regulation, and thus a great difference in temperature between the core of the body and the extremities, while warm-blooded animals keep the temperature nearly constant throughout the body. Since $\delta^{18}O$ strongly responds to temperature, the difference in oxygen isotopic ratios in bones found in the core of the body versus bones from the extremities should be able to test this hypothesis. Barrick and Showers (1994) examined the $\delta^{18}O$ of *Tyrannosaurus* and other dinosaurs and found less than 4°C difference, suggesting they were warm-blooded. However, if dinosaurs were cold-blooded animals that maintained constant body temperature due to their size and reduced surface area, this would have produced the same result.

Carbon—The second most commonly used geochemical system is **carbon isotopes**. Most of the carbon (98.89%) in the Earth is in the form of ^{12}C, which has six protons and six neutrons. The heavier stable isotope, ^{13}C, which has an extra neutron, constitutes about 1.11% of the Earth's carbon. Like oxygen isotopes, these two carbon isotopes cycle through the atmospheric, oceanic, and terrestrial environments, and are found in atmospheric CO^2, in carbonate rocks, and in organic materials. Any material with carbon in it—shells, bone, wood—will contain one or both of these isotopes and can be sampled and measured as oxygen isotopes are. The formula for measuring $\delta^{13}C$ is similar to the oxygen isotope formula given above, with appropriate substitutions of ^{13}C for ^{18}O and ^{12}C for ^{16}O.

A number of different processes control the ratios of ^{13}C to ^{12}C, so there is no simple relationship with temperature and ice volume, as there is for oxygen isotopes. Nevertheless, several factors are known to affect carbon isotope ratios in the oceans, as recorded in the carbonate of marine fossils. Organic materials tend to be low in ^{13}C, so when they decay they release a lot of ^{12}C, causing the value of $\delta^{13}C$ to decrease (become more negative, or lighter). Deep ocean waters are a major source of ^{12}C-rich organic materials, so that when deep waters are brought to the surface during upwelling, it causes the $\delta^{13}C$ to shift to more negative values. Thus, changes in $\delta^{13}C$ in the oceans usually reflect changes in oceanic circulation and, by extension, oceanic productivity. Combined with oxygen isotopes, these two systems can be valuable tools in interpreting paleoclimates and paleoceanography.

Because carbon is the most abundant element in most living systems, carbon isotopes have been used in many other contexts besides paleoceanography. For example, the carbonate in terrestrial soils also contains a distinctive ratio of $\delta^{13}C$, which can change due to a number of effects (Cerling, 1984; Cerling et al., 1989). Carbon isotope studies have shown that grasslands and savannas did not appear until about 7.5 Ma (Cerling, 1992; Cerling et al., 1993; Wang et al., 1994; Quade et al., 1994; Quade and Cerling, 1994). The shift in $\delta^{13}C$ is apparent not only in the soil carbonate, but even in the teeth of the fossil mammals from these beds. This has led to an interesting paradox: horses, camels, and other mammals with high-crowned teeth for eating gritty grasses appear as early as 15 Ma, but the geochemical evidence shows quite clearly that there were no extensive modern-type grasslands until 7 million years later. If there were no true grasslands at 15 Ma, what were these animals eating that required such high-crowned teeth? This puzzle has still not been resolved.

Another important discovery derived from carbon isotopes is the great oceanographic change at the Paleocene/Eocene boundary. At that time, a dramatic decrease of 3‰ in marine carbon isotopes shows that a major circulation change took place; oxygen isotopes and other paleoclimatic signals show that it was a peak of warming as well (Kennett and Stott, 1990, 1991; Stott, 1992). The carbon isotope shift even shows up in the soil carbonate of the Bighorn Basin of Wyoming (Rea et al., 1990; Koch et al., 1992), as well as in the teeth of the fossil mammals found in these beds. The exact timing of the abrupt climatic warming at the Paleocene/Eocene boundary can be seen in the terrestrial record, and its effect on the mammals can be precisely calibrated.

Scientists continue to find new paleoecological uses for carbon isotopes. For example, chemosynthetic organisms that live in hydrothermal vents and cold seeps have a $\delta^{13}C$ that is considerably lighter (more negative) than shallow marine invertebrates that depend on a normal food chain (Rio et al., 1992); their ratios of strontium to calcium and magnesium are also distinctive. If one suspects that a peculiar assemblage of animals might be a chemosynthetic vent fauna, the isotopes provide a test.

In recent years, paleobiogeochemistry has become one of the most powerful tools in paleoecology and paleoclimatology, so a modern paleontologist needs to know more geochemistry than we have room to discuss in this chapter. Within a few years, there will be other novel applications of geochemical techniques to solve paleoecological problems. One of the best aspects of paleobiogeochemistry is that it provides an independent test of ecological predictions based on behavioral biology or morphology, allowing much greater scientific rigor in paleoecology. In some cases (such as the paradox of horses with high-crowned teeth before there were modern grasslands), it has led to some surprises, so our classic hypotheses need to be rethought.

SOME ECOLOGICAL IDEAS THAT HAVE BEEN APPLIED (AND MISAPPLIED) TO FOSSILS

We see, in the vastness of geologic time, events that bear superficial similarity to phenomena of local populations—and we assign a similar cause without realizing that the extended time itself precludes such an application.

S. J. Gould, "The promise of paleobiology as a nomothetic evolutionary disclipline," 1980

One of the biggest problems in applying ecological principles to ancient deposits is the matter of scaling. Most phenomena in ecology operate on a time scale of minutes to years, and rarely on a scale of decades to centuries. To the paleontologist, this is an unresolvable instant between bedding planes (Schindel, 1980). Very few examples from the fossil record preserve processes that operate on such rapid time scales. When we try to fit most fossil assemblages into modern ecological models, we have completely ignored the implications of "millions and millions of years." Much of what has been written about paleoecology based on modern ecology is interesting, but utterly inapplicable to the fossil record for this reason. In this section we will look at some ecological concepts that seemed to apply to the fossil record, only to be undermined by problems of scaling or taphonomy or other aspects of fossilization that biologists seldom encounter.

Food Pyramids

An important aspect of trophic dynamics is the rate at which energy is transferred between organisms. In the process of feeding on other organisms, animals must use some of that energy for their own growth and metabolism. Consequently, for a given total weight (**biomass**) of plants, there must be a smaller biomass of herbivores and even a smaller biomass of predators. This is known as the food pyramid (Fig. 8.14A). The slope of the pyramid reflects the fact that the transfer of energy from level to level is relatively inefficient. Some portion is unavailable as food to the next level, since it is used for growth, metabolism, and other phenomena and is not incorporated into the tissues of the organisms at that level (and thus not available to be eaten by the next level). This means that a given biomass of plants necessarily supports a much smaller biomass of herbivores, and predators are necessarily fewer than their prey. If there are multiple levels of predators, the loss of energy at each level becomes more and more attenuated, so that a top predator (like a hawk, or a lion) is necessarily much more rare than its prey items. Typically, such top predators are not only scarce, but must range over large areas in order to find enough food, which further prevents them from reaching large population sizes.

But the marine food pyramid (Fig. 8.14B) is inverted. It appears that there is less plant biomass than there are consumers. How could this be? It turns out that generalizations about food pyramids work well as long as the *rate of*

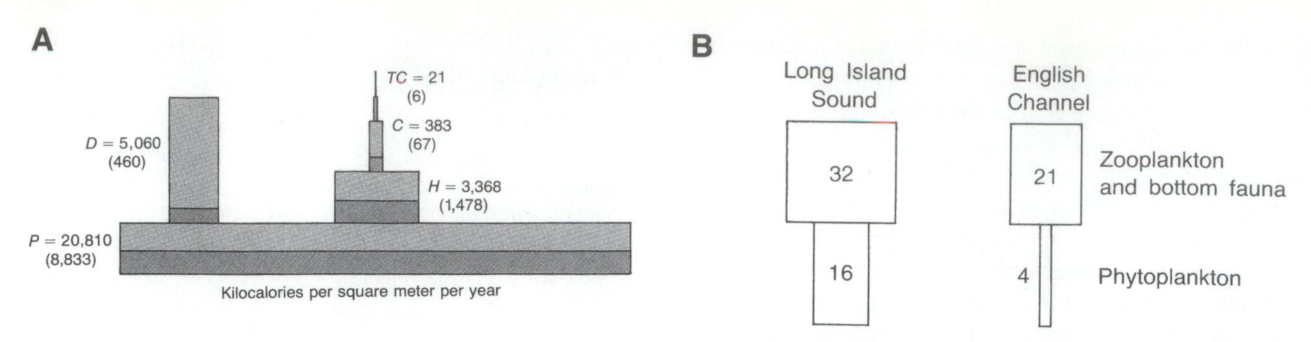

Figure 8.14. (A) The energy pyramid for a normal ecosystem has much more producer biomass (P) than herbivores (H), and even less carnivore biomass (C; TC = top carnivore) and decomposer biomass (D). Each bar represents the total energy flow through a given trophic level. The darker portion of each bar represents energy locked up in the biomass in the area studied, while the lighter portion represents energy that is lost through respiration or movement downstream and out of the study area. (B) In a marine ecosystem, the energy pyramid appears to be inverted, because the plants at the base (phytoplankton) have such short generation times compared to their consumers (zooplankton) that they are replaced much faster than the consumers. The numbers are biomass (in g/square m) for the water column and seafloor beneath one square meter at the sea surface. (From Raup and Stanley, 1978.)

overturn at each level is similar. Marine systems, on the other hand, are based on planktonic algae that have a much higher rate of overturn than the planktonic and nektonic animals that feed upon them. At any given time, the plant biomass is small, but it is replaced so rapidly that it more than keeps up with the consumers, which have a much longer generation time. By contrast, the lifespans of terrestrial plants (especially trees and shrubs) are as long as or longer than the animals that feed upon them, so their biomass is much more consistent with the normal pyramid shown in Figure 8.14A. Turnover rate and generation times are important considerations when comparing the biomass at each level of a food pyramid.

The concept of food pyramids has important implications that paleontologists have tried to apply to understanding extinct organisms. For example, if the predator is warm-blooded, or **endothermic** (like a bird or mammal), it uses most of its food (about 90%) just to keep warm, burning its fuel to produce metabolic heat. Endothermic predators (like birds or mammals) are much less efficient than cold-blooded (**ectothermic**) predators (like crocodiles), which use very little of their food for metabolism and can feed infrequently (think of how long a snake takes to digest a single meal). This means that endothermic predators must eat more often, and eat much more biomass per predator, than ectothermic predators. Thus, a larger biomass of prey species is required for each predator (Fig. 8.15). In terms of food pyramids, the slope of a system with ectothermic predators is much steeper than one with endothermic predators, since a given biomass of prey supports fewer of the former.

Bakker (1972, 1977) used this relationship as an argument that dinosaurs were endotherms. He pointed out that modern communities with endothermic predators have a very low predator/prey ratio, typically 1:10, or 10%, so the food pyramid has a very shallow slope. By contrast, fossil communities with ectothermic predators will have many more of them, so the predator/prey ratio will be close to 4:10, or 40%, and the slope will be steep (Fig. 8.15). Bakker then looked at the predator/prey ratios in fossil terrestrial vertebrate communities. Early Permian communities, for example, had many predators (chiefly the big finbacked mammalian relative, *Dimetrodon*), but by the late Permian, their predator/prey ratios were very low. Mesozoic dinosaur-dominated communities had very low predator/prey ratios (always less than 5%, and typically even lower), as did Cenozoic communities with mammalian predators (which were undoubtedly endothermic).

However, the story is not so simple. As a number of critics (Thulborn, 1973; Charig, 1976; Tracy, 1976; Farlow, 1976, 1980; Beland and Russell, 1980) have pointed out, the distinction between endothermic and ectothermic predator/prey ratios is not so clear cut. The more data are considered, the more the two ranges of ratios overlap, so that there are ectothermic predator communities with low ratios and endothermic predator communities with high ones. More important, it is very difficult to find any suitable modern analogue for an ecosystem with a large ectothermic predator (and no endothermic predators) on which to base this comparison. Practically the only examples are communities where lizards are the top carnivores, and they don't begin to approach the dinosaurs in size. (Some of Bakker's examples used spiders as top predators, and they're not even vertebrates.) For it turns out that body size is an important variable that needs to be considered. A huge predator (such as a large carnivorous dinosaur) loses body heat very slowly, simply due to the fact that it has a large body mass relative to its surface area (discussed in detail in Chapter 2). Consequently, it may have been warm-blooded with a stable body temperature

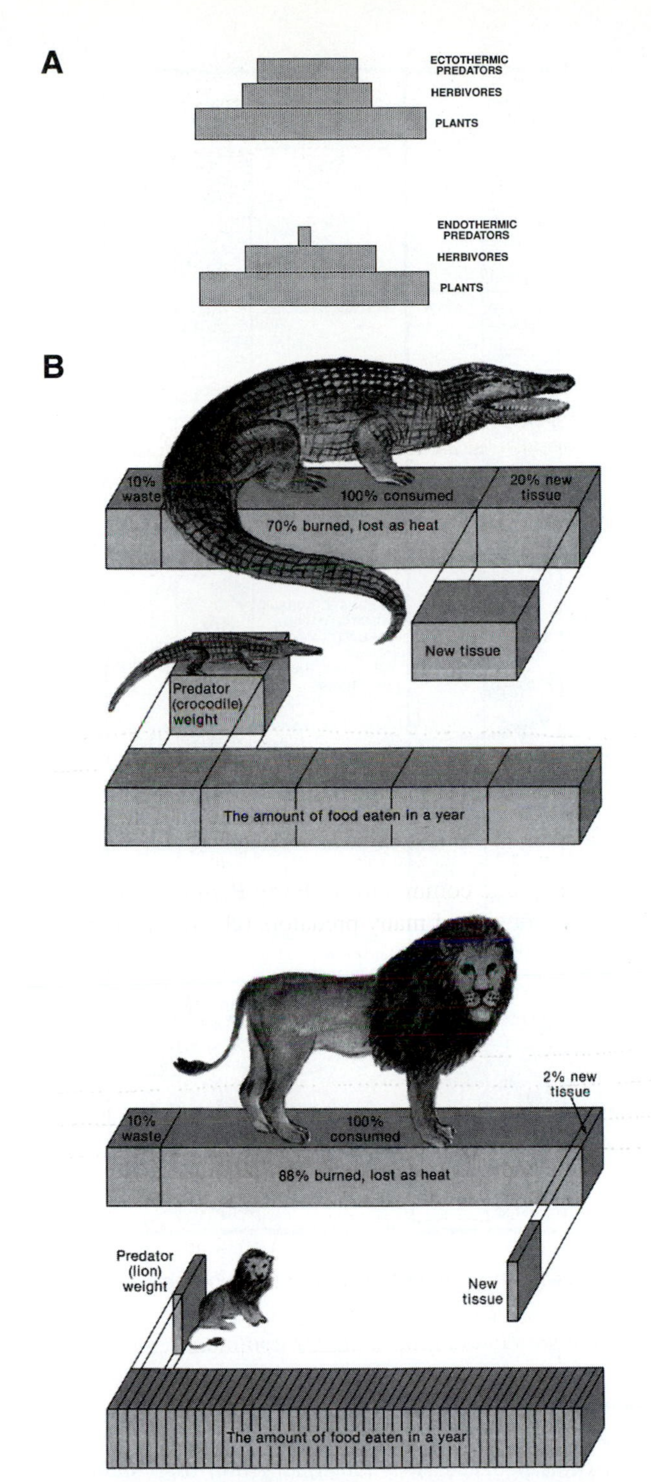

Figure 8.15. (A) If the predator is endothermic and expends most of its food for metabolism, then there can be fewer of them for a given biomass of prey (and a shallow-sloping food pyramid. By contrast, there can be almost as much biomass of ectothermic predators as prey. (B) This is demonstrated by the different amount of food needed to keep an endothermic lion and an ectothermic crocodile alive. About 20% of the crocodile's food goes to new tissue, so it only eats about 5 times its body weight in a year. By contrast, 88% of a lion's food goes for metabolism, and only 2% to new tissue, so it must consume 50 times its body weight in a year. (From Dott and Prothero, 1994.)

simply due to its size, and may not have needed endothermy to keep active and catch prey. A tyrannosaur may have gotten by with a few kills a month (as does a large crocodile) and still have been an active predator.

Another important consideration is taphonomy. In Chapter 1, we pointed out that the process of fossilization introduces all sorts of biases and distortions to the death assemblage that is actually preserved and fossilized. If the predator is much more or less likely to be fossilized than its prey, then the predator/prey ratio in the fossil record would be meaningless. Bakker (1977) argues that both the predatory and prey dinosaurs are of similar body sizes, so their bones should be preserved with the same probability. However, there are plenty of examples that force us to question this assumption. The famous Cleveland-Lloyd Dinosaur Quarry, in the Upper Jurassic Morrison Formation near Price, Utah, yields almost entirely predatory *Allosaurus*, with few prey specimens. Either the predators were cannibals, or this quarry is not yielding a reliable predator/prey ratio. Cleveland-Lloyd may have been a predator death-trap, like the famous Pleistocene La Brea tar pits in Los Angeles (Fig. 1.5B), which preserve far more predators (saber-toothed cats, dire wolves, bears, lions, coyotes) than they do prey (mammoths, mastodonts, horses, camels, ground sloths). Apparently a single trapped prey animal, struggling for freedom, attracted many predators and scavengers, who all became trapped and died. Further examples like this cause us to doubt whether the ratios preserved in museum collections can be used at all. One could easily explain the low predator/prey ratio in dinosaurs by arguing that the predators may have lived in upland areas, away from floodplains where fossilization takes place, or were usually too clever to get caught. By contrast, the stupid, lowland-dwelling sauropods and duckbills were much more likely to get caught in floods, and when they died, their bones were much more likely to be buried in the floodplain. Since we cannot rule these possibilities out, the entire data base of predator/prey ratios has to be regarded with suspicion as evidence of past biology.

Community Succession

A very popular concept in ecology is known as **succession**, or the regular changes that take place in a community as it becomes established and matures to a stable endpoint. Perhaps the most familiar example, found in every introductory science book, is the forest succession. It starts when a wet habitat (such as a lake) is gradually invaded by **pioneer** marsh plants and filled in with sediment. Once the ground becomes drier, the marsh plants are replaced by **intermediate** plants, such as grasses or shrubs, and ultimately by larger trees, forming the stable **climax** community. Unless fire or other disturbance disrupts this sequence, successions will naturally tend to the stable climax community. In other cases, succession occurs when a habitat is disturbed—by fire, volcanic ash, flash flood, or the retreat of an ice sheet—and the bare ground is taken

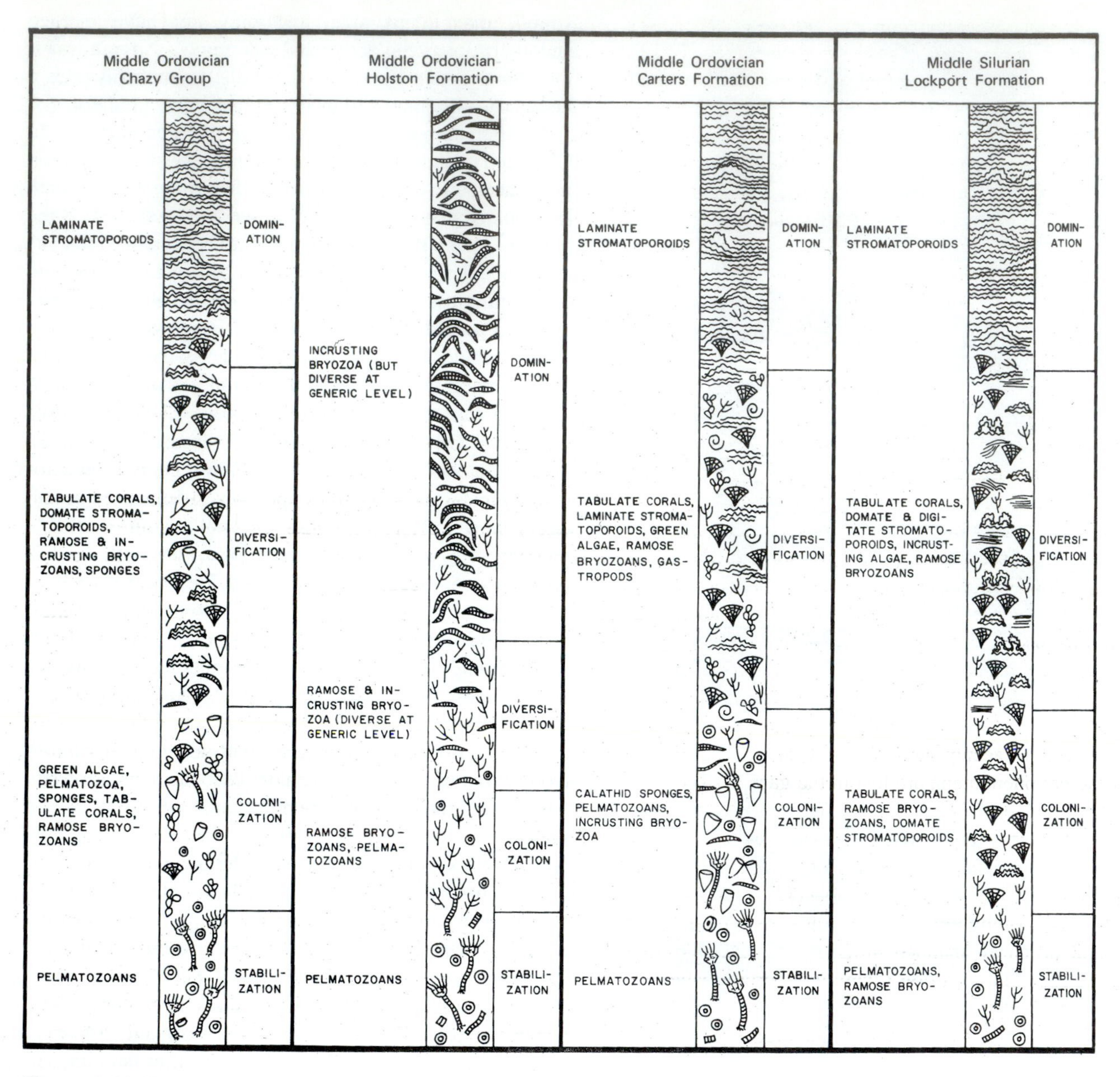

Figure 8.16. Examples of the successional stages of four ancient reefs, showing the vertical succession of the reef biota. (From Walker and Alberstadt, 1975.)

over by pioneering weeds, and then (if there is no further disturbance) by larger plants that can root more deeply and withstand environmental fluctuations better than the opportunistic weeds. All these cases are examples of **autogenic**, or internal, self-driven, successions, where the activity of the early members of the community alters the environment and allows the later members to dominate. There are also cases of allogenic succession, or **replacement**, where the sequence of organisms and/or communities is due to changing external factors, such as climatic changes, salinity changes, sea level changes, or other nonbiological factors.

Paleontologists have long sought examples of succession in the fossil record. Raymond (1988) described the sequence of ancient plant communities that occurred as a Pennsylvanian deltaic complex progrades from brackish marine to freshwater to terrestrial environments. Johnson (1977) described a case where one group of pioneering brachiopods, which were tolerant of muddy bottoms, provided a hard substrate on which other brachiopods could attach. Stanton and Dodd (1981, p. 436) show an example of a succession of a coral attached to a bryozoan attached to a scallop shell. Apparently, corals do not attach to scallop shells directly, but must have a bryozoan substrate.

The best-documented examples of succession are ancient reef communities (Fig. 8.16). Walker and Alberstadt (1975) illustrated a number of cases where the pioneers are soft-substrate tolerant organisms (often crinoids, brachiopods, or bryozoans during the Paleozoic) that stabilized the substrate, followed by a colonization stage, where larger colonial animals take over once the substrate is hard and stable. These are then replaced in the diversification stage by a complex community consisting of not only the colonial organisms (corals, sponges, archaeocyathids, or whatever is important at the time) but also a variety of organisms that live in the sheltered back-reef environment (including delicate bryozoans, brachiopods, crinoids, and other lagoon dwellers). Most Paleozoic reefs are capped by a domination stage, where a single organism (typically stromatoporoid sponges) completely takes over the reef as a stable climax community. Walker and Alberstadt (1975) show seven examples of Ordovician, Silurian, or Devonian reefs that seem to follow this pattern, as well as the rudistid bivalve reefs of the Cretaceous, which have some similarities (Fig. 8.16).

These examples can legitimately be called true autogenic successions, because the stages are clearly very short in duration. Reefs in particular are good examples of succession because they trap the early stages of their history in their layers (but see Copper, 1988, for some problems and caveats). However, most stratigraphic sequences represent much longer time scales than the days to years that are required for normal ecological successions. Each bedding plane may be separated by gaps that represent hundreds to thousands of years, so a series of communities through a thick stratigraphic sequence represents thousands to millions of years of accumulation. The years to tens of years of ecological phenomena are lost in the gaps between beds (Schindel, 1980). Gould (1980) points to examples of "successions" in the Ordovician studied by Bretsky and Bretsky (1975), or in the Devonian by Walker and Alberstadt (1975), where the "pioneer," "intermediate," and "climax" communities are separated by tens of feet of strata and probably spanned millions of years. As Gould (1980, p. 103) put it, "I don't deny that certain patterns of faunal change, observed over millions of years in geological sequences, offer some interesting analogies to classical succession. But the scale is all wrong; they cannot represent the process itself, and attempts to force such sequences into a successional mold obscure a phenomonon [*sic*] that may be new and revealing." Bretzky and Bretzky (1975) talk about "opportunists" that persist through many feet of strata, but Gould (1980) points out, "opportunists cannot persist for millennia if the concept of self-induced change has any meaning." Such "successions" are more likely caused by allogenic factors, such as repetitive systematic changes in sea level, than they are by internal autogenic forces that cause true biological succession. In the terminology introduced earlier, these are examples of community replacement and have little to do with succession as recognized by most biologists. Such patterns of replacement may be interesting in their own right, but they are not equivalent to the biological processes involved in autogenic succession. Most paleontologists and ecologists (e.g., Gould, 1980; McCall and Tevesz, 1983) agree that "only in exceptional circumstances does the fossil record preserve the evidence of ecological succession. There is also agreement that most large-scale evolutionary patterns previously attributed to ecological succession are not. Both global extinction and speciation are embedded in the pattern, and the time scale is too long" (Kitchell, 1985, p. 98).

Competition

Another important principle of ecology is **competition**, or the interaction between two organisms striving for the same thing (food resources, space to live, mates, light, or whatever resource is limiting). Many examples of competition have been documented among living organisms. One of the central concepts that has come from these observations is **Gause's competitive exclusion principle**, which states that whenever two organisms try to occupy the same niche, they will tend to subdivide the niche, or one will drive the other out completely. For example, if two species of birds inhabit the same trees, they will subdivide the tree canopy into different habitats or seek different food resources. In other cases, competition is so severe that only one species will occupy the niche without the intervention of outside forces. For example, the exposed rocks of the intertidal zone are a limited resource exploited by a variety of barnacles, algae, limpets, periwinkles, and mussels (Paine, 1966). However, this balance between species would not exist were it not for predators, since mussels are capable of growing much more aggressively and crowding the others out. Only when predators keep the population of mussels in check can there be a higher diversity of intertidal life.

In recent years, the entire concept of competitive exclusion has been challenged by ecologists (Heck, 1980). Nevertheless, paleontologists have tried to explain aspects of the fossil record by competition and competitive exclusion. Once again, we run into the problems of scaling and duration. Most competitive interactions in biology take place in a matter of hours to years at the most, a time frame too short to see in the fossil record. At best, we would see one group replace another within the span of one bedding plane, and not a protracted replacement that stretches over many feet of strata and millions of years.

One of the classic cases of competition was the alleged replacement of brachiopods by bivalves during the Mesozoic. As early as 1857, Louis Agassiz wrote, "Every zoologist acknowledges the inferiority of the Bryozoa and Brachiopoda when compared with the Lamellibranchiata [bivalves] . . . Now if any fact is well established in Paleontology it is the earlier appearance and prevalence of Bryozoa and Brachiopoda in the oldest geological formations, . . . until Lamellibranchiata assume the ascendancy which they maintain to the fullest extent at present." Simpson (1953, pp. 115-116) wrote, "A major feature of the fossil record, and correspondingly of the history of life, is that of succession and replacement of one group of organisms by another . . . Some mollusks and some brachiopods are enough alike so that there is a certain relation between the expansion of one phylum and contraction of the other." In their classic paleontology textbook, Shrock and Twenhofel (1953, p. 352) attributed the decline of the brachiopods to their "constitutional inaptitude to compete successfully with the Mollusca under changing conditions."

Gould and Calloway (1980) looked at the problem in detail and found no evidence of competitive displacement. First of all, clams and brachiopods do not compete in any real sense. Nearly all brachiopods (except the inarticulates) live epifaunally on the seafloor, and the majority of bivalves are infaunal, burrowing beneath it. Although these two groups may both filter the same seawater for food, they are not competing for space. In fact, during the Paleozoic, the brachiopods dominated the offshore shelf environments, while clams were restricted to nearshore habitats. Through most of the Paleozoic, the two groups lived without either one increasing at the expense of the other; brachiopods maintained a high level of diversity, and bivalves a lower one. Then the great Permian extinction decimated brachiopods (especially productids, orthids, and strophomenids, which became extinct) much more severely than bivalves. In the Triassic aftermath, the bivalves recovered, while the brachiopods did not, for reasons we do not yet fully understand. The spiriferid brachiopods survived into the Triassic but then disappeared in the end-Triassic event, leaving only rhynchonellid and terebratulid brachiopods to survive until the present, as well as the "living fossil" inarticulates. If Vermeij (1977, 1987) is right, the increased predation on epifaunal shelled organisms in the Mesozoic made the exposed brachiopods (such as spirifers) more vulnerable than the burrowing bivalves, or the rhynchonellid and terebratulid brachiopods, which today hide in crevices in rocks. Clearly, the idea that the brachiopods and bivalves could compete with each other for tens of millions of years before one pushed the other out is an unwarranted extrapolation from ecological processes happening in a geological instant, and is not supported by a close analysis of the data. In Gould and Calloway's (1980, p. 393) words, "Instead of acting as competitive antagonists, continually pressing upon each other during hundreds of millions of years, brachiopods and clams may have behaved more like Longfellow's 'ships that pass in the night'— 'only a signal shown and a distant voice in the darkness.'"

Perhaps the strongest case for competitive exclusion in the fossil record involves the barnacles. The chthamaloid barnacles are commonly known to be outcompeted by the balanoid barnacles in modern intertidal habitats. Stanley and Newman (1980) suggested that the replacement of chthamaloids by balanoids in the Cenozoic might also be due to competitive exclusion. However, Paine (1981) challenged this conclusion on several grounds: (1) the smaller chthamaloids would be less susceptible to predation and can live in the refuge of the higher intertidal area; (2) their supposed "competitive exclusion" is largely maintained by differential predation in the tropics, and in temperate regions, where disturbance by bulldozing limpets is more important than predation, the chthamaloids live much lower down in the intertidal zone, and outcompete the balanoids; (3) there is no real evidence in the fossil record that such competition took place in the past. Newman and Stanley (1981) acknowledged many of Paine's criticisms, but still argued that competition was the primary factor. Palmer (1982) and Branch (1984) also argued that predation and environmental factors were more important than competition. Kitchell (1985) regarded the fossil record of the two groups as too poor to provide a conclusive test.

Another classic example of alleged competitive exclusion is the displacement of the archaic mammals known as multituberculates, and the primitive primates known as the plesiadapids, by the invasion of the rodents from Asia to North America. All three groups were small herbivorous mammals with chisel-like incisors, adapted for living in trees and gnawing fruits and nuts. Multituberculates and plesiadapids flourished in North America in the Paleocene (multituberculates even go back to the Jurassic), and when rodents arrived in the early Eocene (55 Ma), the other two groups went into gradual decline. The last plesiadapids occurred in the middle Eocene (47 Ma), and the multituberculates finally expired in the latest Eocene (35 Ma). Krause (1986) found that the evidence was consistent, if not conclusive, for the rodents displacing the multituberculates, and Maas et al. (1988) came to a similar conclusion for the plesiadapids being replaced by rodents. Nevertheless, the three groups overlapped by about 20 million years. If this was true competitive exclusion, it operated at an excruciatingly slow pace.

Other classic examples of competitive exclusion have not fared as well. For example, the synapsid ancestors of mammals were allegedly driven to extinction in the Late Triassic by competition from the early dinosaurs. However, Benton (1983, 1987) argued that the data were inconsistent with competition, especially over the long span of the Late Triassic. The odd-toed perissodactyls (horses, rhinos, tapirs, and their relatives) were supposedly outcompeted by the even-toed artiodactyls (pigs, camels, cattle, deer, and their relatives) during the Cenozoic, but Cifelli (1981) found no evidence to support this. In summary, competitive exclusion is an important ecological process, but in most cases it operates too rapidly to be seen in the fossil record.

EVOLUTIONARY PALEOECOLOGY

Evolution cannot be understood except in the frame of ecosystems.

Ramón Margalef, *Perspectives in Ecological Theory*, 1968

The history of organisms runs parallel with, is environmentally contained in, and continuously interacts with the physical history of the earth.

George Gaylord Simpson, "Historical Science," 1963

If we cannot observe many important ecological processes in the fossil record, what is left for paleoecology? Paleontologists have responded by looking at phenomena that operate *only* on the large scale of millions of years and are invisible to ecologists of the living biota. This approach was first labeled "evolutionary paleoecology" by Valentine (1973) and has become one of the most exciting areas in paleobiology. We have already discussed one example of the process—the marine community structure and food web before the Ordovician was much simpler than anything living today. Such a discovery forces us to re-examine the assumptions about ecology based on the modern world, because through most of life's history, food webs were not as complex as they are today. What other large-scale ecological trends can we find in the fossil record, and what do they tell us about our ecological theories based on the modern world?

Phanerozoic Diversity and the Three "Evolutionary Faunas"

One of the first questions that Valentine (1969, 1970, 1973) posed was, "What is the pattern of diversity of life?" Collecting the data then available on the total number of fossil species throughout the Phanerozoic, Valentine produced a curve showing that diversity was low through most of the Paleozoic, but accelerated rapidly in the Mesozoic and Cenozoic, with the highest levels of diversity occurring at present (Fig. 8.17A). Valentine viewed this increase as a result of greater specialization and ecospace utilization as life became more sophisticated.

Raup (1972, 1976a, 1976b) was not convinced by Valentine's analysis and suggested a number of non-ecological reasons why the curve might have the shape it does. For example, the shape of the diversity curve closely correlates with the curve for total rock area, suggesting that the diversity at each time horizon might be an artifact of the total rock available to be sampled for fossils (Fig. 8.18B, C). The younger the rock, the more likely it will be exposed at the surface, and the less likely it has been destroyed by erosion. All of these might contribute to its fossils being more abundant and diverse.

Another possible bias is the fact that well-exposed rocks tend not only to yield more fossils but also to attract more attention from paleontologists (Sheehan, 1977). This suggests that the apparent diversity curve might actually be a curve of paleontological interest and activity (Fig.8.18D). Such a suggestion is supported by "monographic bursts" that occur when the publication of major monographs appear and increase the number of published species. For example, publication of major monographs on Permian brachiopods by Cooper and Grant suddenly increased the apparent diversity in the Permian.

A third possible bias is the "pull of the Recent." Diversity might rise so sharply to its modern-day maximum because the modern biota is much better preserved and contains soft-bodied animals that seldom fossilize. If those animals have just one previous occurrence in the fossil record, their range is automatically extended backward into the Cenozoic and occasionally into the Mesozoic. Thus, there may not be many more fossils at each level in the Mesozoic or Cenozoic, but these soft-bodied animals are counted as present at time intervals when they are not fossilized, simply by the fact that they occur at least once in the fossil record and are still living.

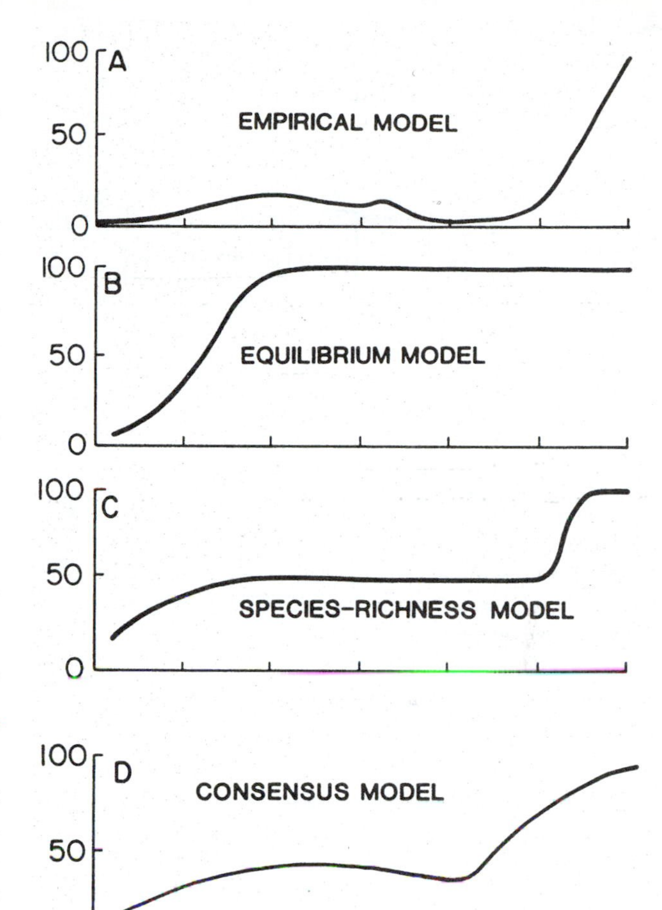

Figure 8.17. Comparison of four different interpretations of the history of Phanerozoic marine faunal diversity. (A) Valentine (1970) collected data that showed empirically a rapid increase in the Mesozoic-Cenozoic. (B) Raup (1972) argued that this increase might be an artifact of numerous biases, and that if they were filtered out, diversity might increase rapidly in the Cambrian and Ordovician and then reach an equilibrium. (C) Bambach (1977) tested these two hypotheses by looking at the species richness in well-preserved faunas. He found that diversity was indeed lower in the Paleozoic, but not as low as Valentine suggested. (D) Consensus of Sepkoski et al. (1981), based on a newer tabulation of the data. (From Signor, 1985.)

Since all these biases tend to produce curves that correlate with the apparent diversity curve of Valentine, Raup (1972, 1976a, 1976b) suggested that the true shape of diversity may not be slow acceleration to a modern-day

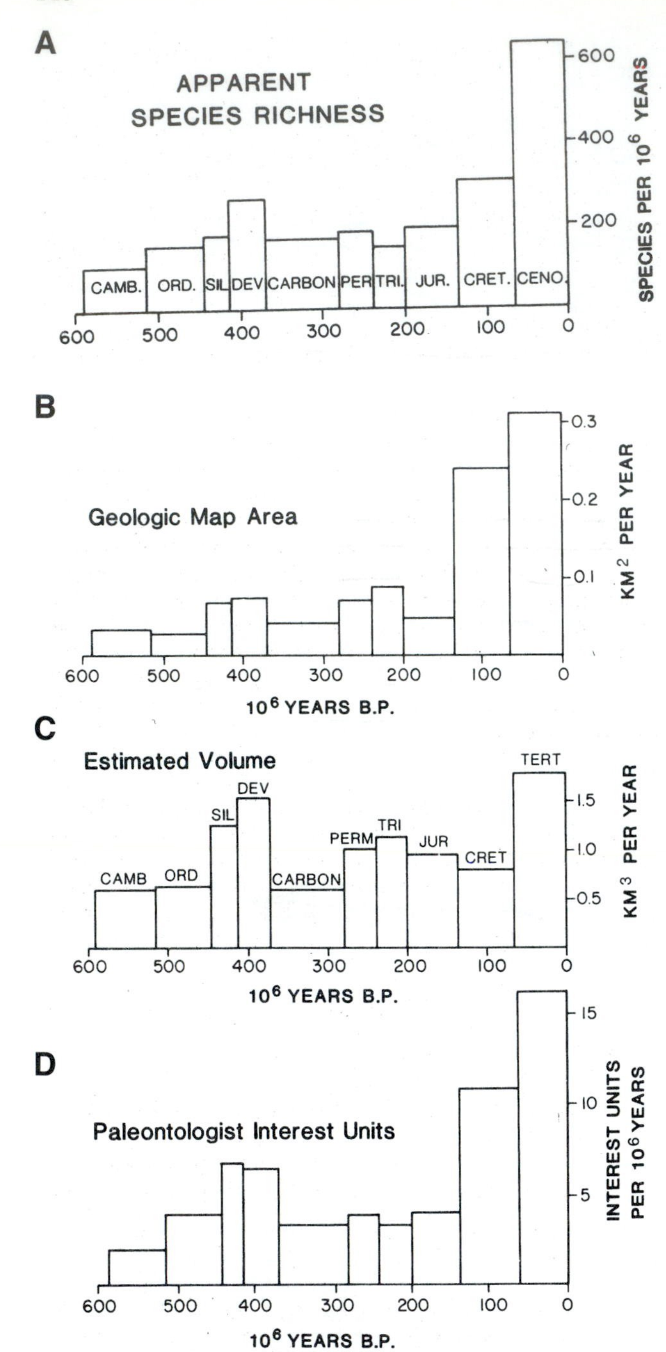

Figure 8.18. (A) The empirical data of marine invertebrate diversity through time shows a Siluro-Devonian peak, and a rapid increase in the Cretaceous and Cenozoic. But the exposed area (B) and volume (C) of rocks is also biased toward the Siluro-Devonian, Cretaceous and Cenozoic, as is the research interest of paleontologists (D). (After Signor, 1985.)

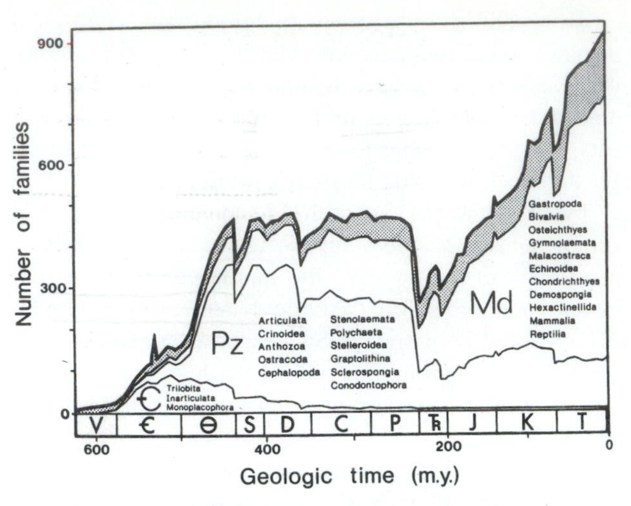

Figure 8.19. Sepkoski's (1977, 1978) factor analysis of Phanerozoic marine diversity. He found that the marine fauna was composed of three main components: a "Cambrian fauna" that dominated the Cambrian and declined after the Ordovician; a "Paleozoic fauna" (Pz) that dominated the Ordovician through Permian; and a "Modern fauna" (Md) that was present in the Paleozoic, but dominated the Mesozoic and Cenozoic. (From Sepkoski, 1981.)

maximum. Instead, he argued that life diversified rapidly in the Cambrian and Ordovician, filling most of the niches and then reaching an equilibrium state (Fig. 8.17B). This curve is almost the diametric opposite of Valentine's (and also of the empirical data). How could we test which of these is closer to the truth? Just counting the species all over again doesn't solve the theoretical dispute behind the data. Is there any other way of measuring diversity besides species counting?

Bambach (1977) proposed a novel way to test these alternatives. He studied 386 well-sampled, well-preserved fossil communities throughout the Phanerozoic and tabulated their species diversity. In this manner, he could address the question of whether diversity of a typical Cambrian or Ordovician community is actually less than a Cenozoic community, or whether this is an artifact of the compilation of the data (Fig. 8.17C). Bambach found that there was an increase in diversity through the Phanerozoic (contrary to Raup), but not nearly as extreme (only 4 times as large, not 10 times as large as suggested by Valentine).

While these disputes were raging in the literature, Jack Sepkoski was working on a new, much more accurate tabulation of the total number of families in the fossil record (finally published in 1982). His tabulations (Sepkoski et al., 1981) suggested that the truth was between the two extremes proposed by Raup and by Valentine (Figs. 8.17D, 8.19). Diversity had indeed increased rapidly in the Cambrian and Ordovician and then reached a plateau through most of the Paleozoic (as suggested by Raup). However, after the great Permian extinction caused a dip in diversity, it began to increase in the Mesozoic and

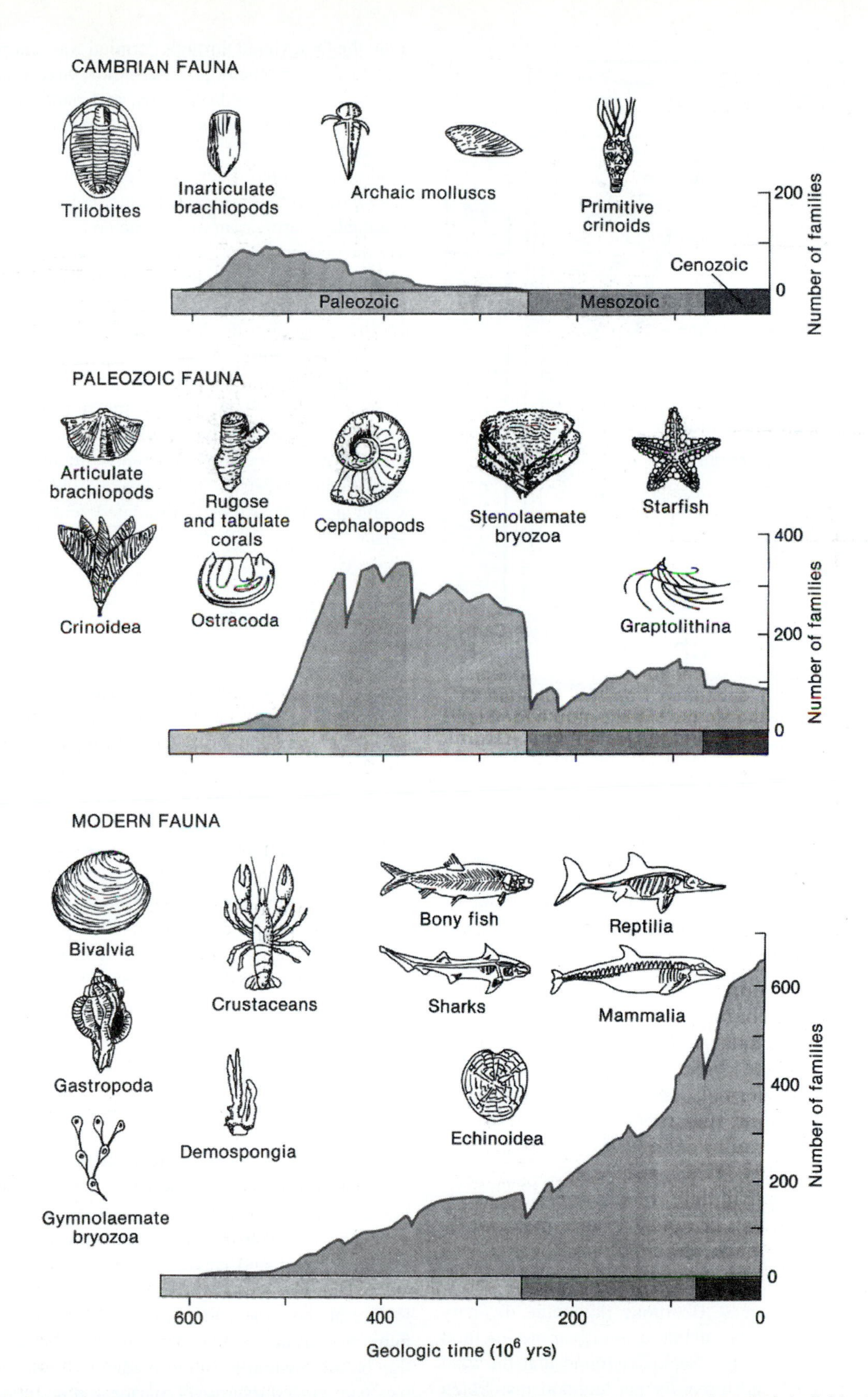

Figure 8.20. Sepkoski's "three evolutionary faunas" (see Fig. 8.19) are shown here separated so their individual diversity histories are clear. Also shown are cartoons of the major groups that dominate each "evolutionary fauna." (Modified from Sepkoski, 1981.)

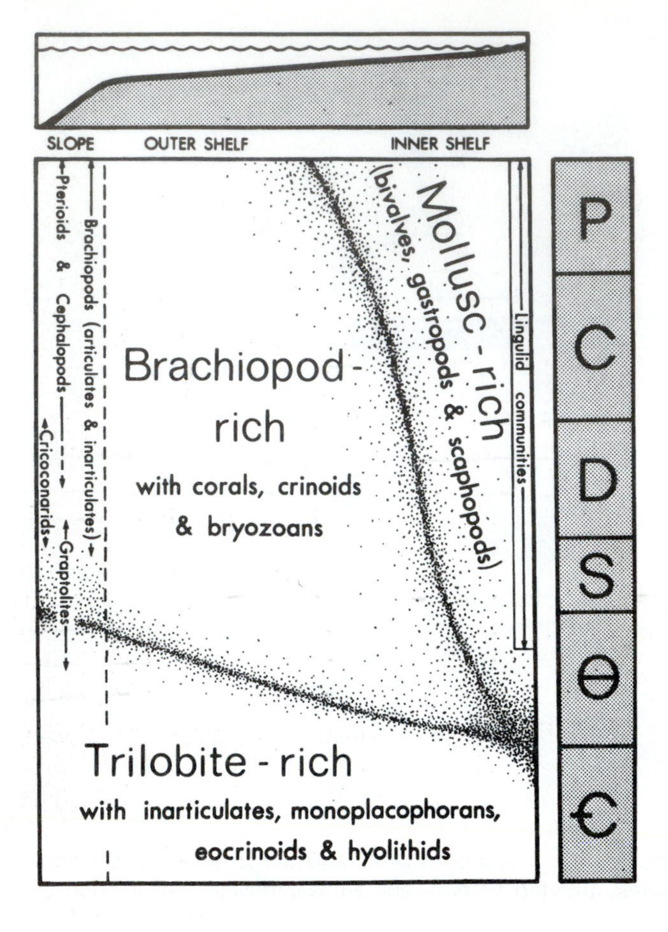

Figure 8.21. By looking at the taxonomic composition and apparent depths of various Paleozoic marine communities, Sepkoski and Miller (1985) found definite onshore-offshore trends through the Paleozoic. During the Cambrian, a trilobite-rich fauna was found at all depths. The Ordovician radiation of the Paleozoic fauna took over most of the middle and outer shelf habitats, displacing the trilobite-rich fauna offshore or replacing it entirely (except for the lingulid community, which still inhabits the nearshore-intertidal region.) As the Paleozoic fauna diversified, a mollusc-rich fauna of bivalves and gastropods (essentially the "Modern fauna" came to dominate the nearshore habitats, while the brachiopod-rich fauna still flourished offshore. (From Sepkoski and Miller, 1985.)

Cenozoic well beyond Paleozoic levels (as Valentine had suggested, but more in line with the orders of magnitude suggested by Bambach).

Sepkoski (1977, 1978, 1979) went beyond looking at the total diversity, however. He broke down the diversity curve into components, using a multivariate method known as factor analysis. Sepkoski found that the total diversity separated into three distinct components, which he called "evolutionary faunas" (Fig. 8.20). His "Cambrian fauna" was a distinct assemblage dominated by trilobites, inarticulate brachiopods, and primitive echinoderms and molluscs. The "Paleozoic fauna" that dominated in the Ordovician through Permian was composed mostly of articulate brachiopods, crinoids, rugose and tabulate corals, stenolaemate bryozoans, and nautiloids and gonitatitic ammonoids. This faunal dominance was completely rearranged by the Permian extinction, and during the Triassic a "Modern Fauna" arose, dominated by bivalves, gastropods, echinoids, crustaceans, bony fish, sharks, and eventually marine mammals—the creatures that still populate the oceans today.

Paleontologists have long known that there are fundamental differences between the Cambrian world, the rest of the Paleozoic, and the Mesozoic-Cenozoic, but could not describe this difference in precise quantitative terms. What originally seemed to be an insoluble dispute over the overall shape of diversity has turned into an insight into the fine structure of the last 540 million years of evolution. It would not have been as apparent without the large-scale tabulation of diversity of Sepkoski (1982), which in turn would never have been stimulated without the original debate between Valentine and Raup over Phanerozoic diversity.

Onshore-Offshore Trends

In the process of analyzing the total diversity of the Phanerozoic and breaking it down into evolutionary faunas, Sepkoski found that his data base could produce other interesting results as well. For example, the elements of the modern faunas (especially bivalves and gastropods) were present in the Paleozoic, but did not dominate in most environments. Sepkoski and Miller (1985) looked at the depth and substrate information associated with these faunas and found that there was an ecological separation between these two evolutionary faunas. During the Ordovician, the brachiopod-crinoid-bryozoan-coral-rich "Paleozoic fauna" was dominant in the more offshore habitats, but bivalves and gastropods tended to dominate the inner shelf and nearshore regions (Fig. 8.21). In essence, the Permian extinction did not simply wipe out the Paleozoic fauna—it also meant that the offshore shelves were stripped of brachiopods and other "Paleozoic fauna" groups, and when life returned in the Triassic, those offshore habitats were taken over by bivalves and gastropods.

Marine ecologists have long argued whether the nearshore environment, with its high variability due to tides and storms, or the deep ocean, with its long-term stability, would be better places for new taxa to originate. The fossil record is the ideal place to test this hypothesis. Dave Jablonski and Dave Bottjer (Jablonski and Bottjer, 1983; Jablonski et al., 1983; Bottjer and Jablonski, 1988) looked at the time and place of the first occurrence of numerous groups in the fossil record, plotting their position on the offshore-onshore gradient. A number of scientists had noticed that novel marine communities tended to occur in nearshore habitats first, while archaic forms often persisted in deeper waters, and Jablonski and Bottjer found that this was indeed true: new communities did indeed arise in the nearshore, and tended to displace older com-

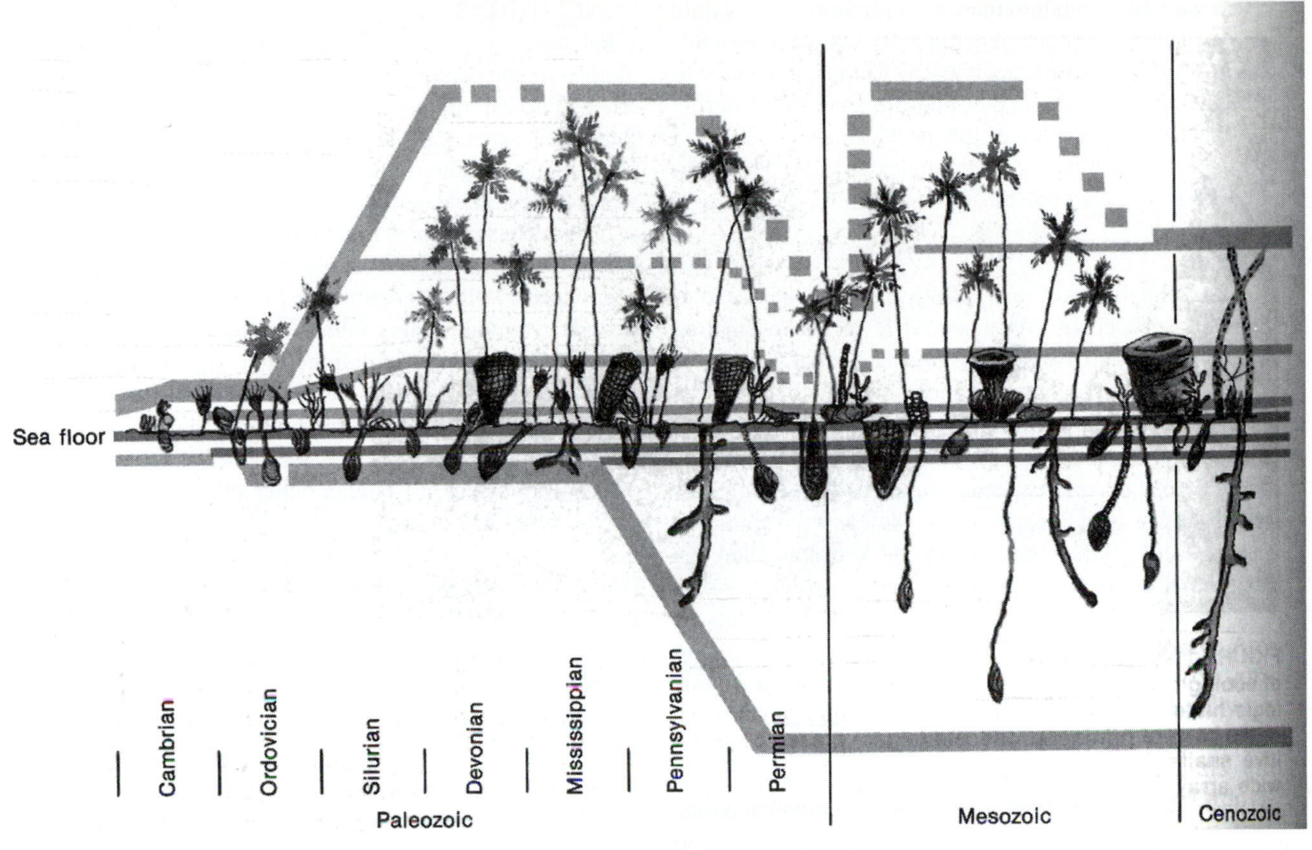

Figure 8.22. Another long-term change in marine faunas is the number of levels, or tiers, above and below the seafloor that are occupied. In the Cambrian, the limited fauna of trilobites, archaeocyathids, and archaic molluscs, echinoderms, and brachiopods lived on the surface, or only a few centimeters above or below it. The Ordovician radiation provided long-stemmed crinoids which fed as much as a meter above the seafloor, and through the Paleozoic, these various tiers became taller and more finely subdivided by various tall filter-feeding organisms, such as corals, blastoids, and delicate bryozoans. In the late Paleozoic and especially the Mesozoic, the deep burrow habitat began to be exploited, not only by various worm-like organisms, but also by burrowing bivalves and gastropods in the early Mesozoic. Thus, there has been a net increase in tiering through time. (Modified from Ausich and Bottjer, 1991.)

munities as they expanded to the more offshore habitats. However, the reasons *why* this is so are less obvious (Jablonski and Bottjer, 1988). It may be that nearshore habitats, with their high variability, are stressful and drive organisms to adapt to new conditions and speciate. Finally, onshore communities may just be more resistant to extinction, so that even if novelties arise with equal probability at any depth, those that occur nearshore have a better chance of surviving and spreading out.

Tiering

Total diversity and community composition are not the only fundamental differences between the Cambrian, Paleozoic, and Modern faunas. Ausich and Bottjer (1982, 1985; Bottjer and Ausich, 1986) looked at another aspect of ancient communities: how many different levels, or **tiers**, above and below the sea bottom did they occupy? They found that during the Cambrian, the trilobites, brachiopods, and archaeocyathids occupied only the surface of the seafloor (or dug shallow burrows), and few were tall enough to filter feed more than a few centimeters off the bottom (Fig. 8.22). By the Ordovician, however, crinoids were feeding over a meter above the sea bottom, exploiting an ecological niche that was unavailable until organisms grew that tall. Other stalked echinoderms and tall bryozoans, as well as corals, were capable of filter feeding at intermediate levels, allowing for a fine-scale subdivision of the water column several meters above the bottom. Through most of the Paleozoic, however, there were no deep burrowers. In the late Paleozoic and Mesozoic, bivalves with long siphons, as well as a variety of worms and crustaceans, began to dig deep burrows as much as meter below the seafloor, exploiting yet another ecological niche that was unoccupied until organisms with deep burrowing adaptations came along. This multiple tiering was temporarily disrupted by the Permian extinction, but returned in the Triassic. Giant crinoids declined after the Jurassic, so the highest tiers of the Paleozoic have not been duplicated in the Cenozoic. The deep burrowers, however, have maintained their Permian levels of tiering.

When we look at modern marine communities, we would never suspect that ecological complexity was ever very different in the past. But the analyses by Bambach mentioned in the early part of the chapter and the analysis of tiering show that we cannot think of the pre-Permian world in the same terms as the present—the complexity of modern communities had not yet evolved.

Escalation

Another phenomenon that is invisible to modern ecologists was proposed by Vermeij (1987) in his fascinating book *Evolution and Escalation, an Ecological History of Life*. Vermeij has long studied phenomena such as the history of predation on molluscs by other species that peel them open, or drill their shells, and noticed that when new predators arose, the prey species tended to evolve defenses (thicker shells, smaller apertures, defensive spines, or swimming ability) that protected them. Vermeij called this phenomenon **escalation**. It is analogous to the arms race during the Cold War—each development in armaments on one side tended to provoke a defensive reaction and stronger armaments by the other. The best example of this phenomenon is Vermeij's (1977) "Mesozoic marine revolution" discussed earlier in the chapter. During the Triassic, a number of new predators—marine reptiles, bony fish, crustaceans, and starfish—appeared that had superior methods of crushing shells. None of these animals existed in great abundance in the Paleozoic, so there was no pressure on bivalves or brachiopods back then to protect themselves. But when the seafloor suddenly became more dangerous for shellfish in the Triassic, the bivalves, gastropods, and echinoids that could burrow, or protect themselves with spines, or swim away, survived. Epifaunal brachiopods and bivalves and unprotected gastropods and echinoids went into decline. In short, the arms race between predators and prey escalated during the early Mesozoic, and the world changed in such fundamental ways that the old "Paleozoic fauna" could never return to its former glory (even if it had done better in the Permian crisis).

The phenomena of escalation and tiering together suggest reasons why Sepkoski found that diversity increased since the Paleozoic and why the modern fauna is so different from the Paleozoic fauna. The presence of new ecological niches in the Mesozoic (especially the deep burrowing tiers) allowed for greater diversity than in the Paleozoic, and the escalation of predators required that the prey become more diverse in their adaptations to escape new forms of predation. These ecological phenomena, which are only visible on the scale of millions of years, would never have been suspected by an ecologist looking only at the living world. These are the types of insights that "evolutionary paleoecology" has produced, and they give us a whole new way of thinking about ecology.

CONCLUSIONS

Paleoecology flourished in the 1970s as paleontologists jumped on the ecology bandwagon and tried to apply modern ecological principles to the fossil record. Then it foundered when it became apparent that many ecological phenomena operated on scales that were invisible at the scale of geological events, and when "me-too" paleoecological reconstructions led to few original insights. In recent years, however, "evolutionary paleoecology" has begun to show that there are many phenomena—community diversity, evolutionary faunas, onshore-offshore origination trends, tiering, escalation—that are invisible to the ecologist working only in the modern time frame. These insights promise not only to revitalize paleoecology but also to challenge many of the assumptions of modern ecology as well, leading to a much richer and more realistic picture of how life works.

FOR FURTHER READING

Ager, D. V. 1963. *Principles of Paleoecology*. McGraw-Hill, New York.

Behrensmeyer, A. K., J. D. Damuth, W. A. DiMichele, R. Potts, H.-D. Sues, and S. L. Wing, eds. 1992. *Terrestrial Ecosystems through Time*. University of Chicago Press, Chicago.

Boucot, A. J. 1981. *Principles of Benthic Marine Paleoecology*. Academic Press, New York.

Boucot, A. J. 1990. *Evolutionary Paleoecology of Behavior and Coevolution*. Elsevier, Amsterdam.

Dodd, J. R., and R. J. Stanton, Jr. 1981 (2nd ed., 1990). *Paleoecology, Concepts and Applications*. Wiley, New York.

Hoffman, A. 1979. Community paleoecology as an epiphenomenal science. *Paleobiology* 5:357-379.

Imbrie, J., and N. D. Newell, eds. 1964. *Approaches to Paleoecology*. Wiley, New York.

Jablonski, D., D. E. Erwin, and J. H. Lipps, eds. 1996. *Evolutionary Paleobiology*. University of Chicago Press, Chicago.

Kaesler, R. L. 1982. Paleoecology and paleoenvironments. *Journal of Geological Education* 30:204-214.

Kitchell, J. A. 1985. Evolutionary paleoecology: recent contributions to evolutionary theory. *Paleobiology* 11: 91-105.

Laporte, L. F. 1968. (3rd ed., C. R. Newton and L. F. Laporte, 1989). *Ancient Environments*. Prentice-Hall, Englewood Cliffs, N.J.

McCall, P. L., and M. J. S. Tevesz, eds. 1983. *Biotic Interactions in Recent and Fossil Communities*. Plenum, New York.

McKerrow, W. S., ed. 1978. *The Ecology of Fossils*. M.I.T. Press, Cambridge, Mass.

Raup, D. M. 1972. Taxonomic diversity during the Phanerozoic. *Science* 177:1065-1071.

Sepkoski, J. J., Jr. 1981. A factor analytic description of the Phanerozoic marine fossil record. *Paleobiology* 7:36-53.

Sepkoski, J. J., Jr. 1982. *A Compendium of Fossil Marine Families*. Milwaukee Public Museum Contributions to Biology and Geology 51. Milwaukee Public Museum, Milwaukee.

Sepkoski, J. J., Jr., R. K. Bambach, D. M. Raup, and J. W. Valentine. 1981. Phanerozoic marine diversity and the fossil record. *Nature* 293:435-437.

Valentine, J. W. 1969. Patterns of taxonomic and ecological structure of the shelf benthos during Phanerozoic time. *Palaeontology* 12:684-709.

Valentine, J. W. 1973. *Evolutionary Paleoecology of the Marine Biosphere*. Prentice-Hall, Englewood Cliffs, N.J.

Valentine, J. W., ed. 1985. *Phanerozoic Diversity Patterns: Profiles in Macroevolution*. Princeton University Press, Princeton, N.J.

Vermeij, G. J. 1987. *Evolution and Escalation, An Ecological History of Life*. Princeton University Press, Princeton, N.J.

Figure 9.1. Different biogeographic regions have different native animals, sometimes demonstrating striking ecological convergence. For example, the native marsupials, or pouched mammals, of Australia, have converged on the body forms of placental mammals in many other parts of the world. Even though they look very much like their placental counterparts, they are not closely related, since their common ancestors diverged over 100 million years ago. (Modified from Simpson and Beck, 1965.)

Chapter 9

Biogeography

In considering the distribution of organic beings over the face of the globe, the first great fact which strikes us is that neither the similarity nor the dissimilarity of the inhabitants of various regions can be wholly accounted for by climatal or other physical conditions.

Charles Darwin, *On the Origin of Species*, 1859

ORGANISMS IN SPACE AND TIME

Biogeography is the study of the geographic distribution of organisms and how they got to be where they are found today. As a scientific discipline, it has some very unusual characteristics. Very few biology departments have a position for a biogeographer (although they sometimes find work in geography departments), and few institutions teach courses in biogeography. There is no specialized journal, or professional society, or separate scientific meeting for biogeography. Instead, biogeography tends to be a subspeciality of biologists and paleontologists who combine it with another primary area of research. Most biogeographers tend to be systematists, since deciphering the evolutionary relationships of a group of organisms naturally leads to questions of why they have their present distribution as well. Consequently, there has not been much rigorous study or analysis of the methods of biogeography until recently. In the past, most biogeography was done is a very casual, informal basis—the distribution of the organisms was mapped, and an explanation was formulated to explain this distribution. In recent years, the theory and methodology of biogeography has developed greatly, so now there are competing schools of theoretical biogeography as there are in systematics and other fields.

Biogeography also had unusual origins (see Nelson and Platnick, 1981, and Browne, 1983, for a historical overview). Before the late 1700s, natural historians such as Linnaeus did not know much of animals outside Europe and had little reason to doubt that they were similar to familiar plants and animals. The strong hold of religious dogma made the Noah's Ark story the only allowable explanation of their geographic distribution. According to Genesis 6, all animals migrated from Mt. Ararat (now in Turkey) to their present locations. But scientific expeditions in the years after 1730 led to many discoveries that demolished this view of biogeography. Far-off places such as South America, Africa, Madagascar, southeast Asia, Australia, and the Pacific islands yielded strange animals and plants that completely changed the Eurocentric view of nature. Tropical regions in particular had much richer and more unusual faunas and floras than the depauperate wildlife of northern Europe, with its large human populations, widespread agriculture, and deforestation. Natural historians were overwhelmed trying to describe and catalogue it all. More importantly, the geographic distributions could not be explained by the Noah's Ark story. Why would continents such as Australia have a fauna dominated by pouched mammals (marsupials), with no native placental mammals? Did the marsupials run straight from Mt. Ararat toward Australia, while placentals didn't even try? Even more striking was the fact that many of the marsupials had body forms that mimicked the placentals occupying a similar ecological niche on other continents (Fig. 9.1). There were marsupial equivalents of wolves, badgers, cats, flying squirrels, groundhogs, anteaters, moles, rabbits, and mice—yet all were pouched mammals unrelated to their placental counterparts on other continents. We now view this as an outstanding example of convergent evolution, but these major discoveries were made a century before evolution explained them.

As he was with evolution, extinction, and many other scientific ideas, Georges Leclerc, the Count de Buffon, was one of the pioneers in zoogeography, first formulating the general principles of the differences between the faunas of southern continents (especially Africa and South America) in 1776. The great explorer and scientist Alexander von Humboldt is considered the founder of plant geography, publishing most of his ideas around 1805. Among his many radical notions, he proposed that continents were once connected by land bridges and that others had drifted apart (this was a hundred years before Wegener proposed continental drift). Von Humboldt also realized the importance of climate on plant distributions as well. By 1805, Von Humboldt and Bonpland were articulating many of the

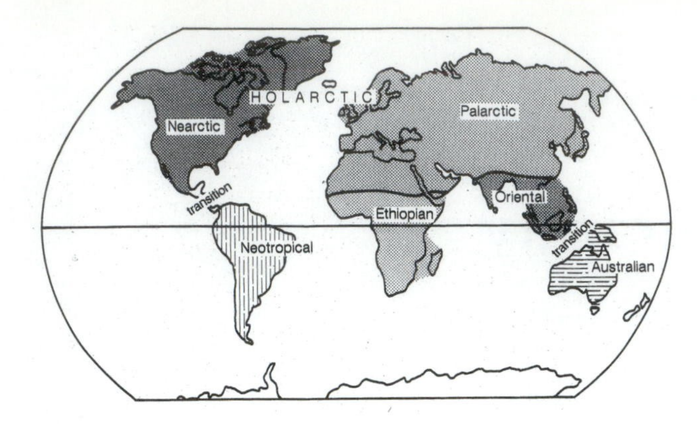

Figure 9.2. Sclater's original division of land faunas into biogeographic realms. The transition in the Indonesian archipelago is known as Wallace's line. (From Stearn and Carroll, 1989.)

central ideas of biogeography: that organisms arose from a "center of origin" (a relict idea from the Mt. Ararat center of origin in the Noah's Ark story); that changes in the Earth's climate could explain distributions (long before the idea of an Ice Age came along); and that present-day patterns were largely determined by climate.

In 1820, the Swiss botanist Augustin de Candolle divided the world into 20 biogeographic regions, and in 1858, the British ornithologist Philip Sclater proposed a classification of the land faunas into the familiar regions or **realms**: Neartic (North America), Neotropical (Central and South America), Palearctic (northern Eurasia and northern Africa), Ethiopian (sub-Saharan Africa), Oriental (southeast Asia), and Australian (Fig. 9.2). By the time of the publication of *On the Origin of Species* in 1859, biogeography had advanced far enough that it was a powerful line of evidence for evolution. Darwin devoted two chapters to it. The evidence of the Galápagos finches was particularly important in showing that evolution had occurred, since their geographic distribution made no sense without evolution.

After years of collecting animals in South America and southeast Asia, Alfred Russel Wallace, the codiscoverer of natural selection, synthesized these trends in biogeography with his publication of *The Geographical Distribution of Animals* (1876). In that work, he brought together all the new information on faunal provinces and Sclater's realms and formalized many of the ideas of biogeography that persist today. One of his most famous contributions is the discovery of the boundary in the Indonesian archipelago between islands that have mostly Asian-influenced faunas and those that are dominated by Australian marsupials. This boundary came to be known as "Wallace's line" and today it is a striking example of how two distinct faunas can mingle once a continent (Australia) drifts into the influence of another continent (Asia).

Today, many biologists try to explain the present distribution of organisms from modern climatic and ecological factors. This approach is known as **ecological biogeography**, and is primarily concerned with present-day processes and conditions or those of the recent past. By contrast, some biologists and most paleontologists are more concerned with past distributions of organisms, and how those past distributions explain many of the present-day occurrences (especially those that cannot be explained by modern climate and ecology). This is known as **historical biogeography**. The methods, data, and approaches of these two subdisciplines are so different that we will review them separately in the next two sections.

ECOLOGICAL BIOGEOGRAPHY

So far I have tried to prove that habitations, considered in their totality, appear to be determined by temperature. Without doubt, it is necessary to combine with temperature the considerations deduced from the study of stations; it is clear, for example, that the more a certain country is sandy, the more one will find plants of the sand growing there, etc. But even when one gives to these causes all of the latitude possible, does one do justice to the best-known facts? It is this that I doubt; it is this that requires a new discussion.

Augustin de Candolle, *Géographie Botanique*, 1820

The third great fact is the affinity of the productions of the same continent or sea. The naturalist, in travelling from north to south, never fails to be struck by the manner in which successive groups of beings specifically distinct, though nearly related, replace each other.

Charles Darwin, *On the Origin of Species*, 1859

Everyone knows that the same animals and plants are not found everywhere . . . but that they are distributed so as to be gathered together in distinct zoological and botanical provinces, of greater or less extent, according to their degree of limitation by physical conditions, whether features of the earth's outline, or climate.

Edward Forbes and R. Godwin-Austen,
The Natural History of the European Seas, 1859

Provinces, Biomes, and Realms

Ecological biogeography attempts to explain the distribution of animals and plants in terms of present-day climatic and ecological conditions. Such research begins with maps of biogeographic realms or provinces (Figs. 9.2 and 9.3). Some provinces, such as the realms recognized for terrestrial animals (Fig. 9.2), correspond roughly to the boundaries of continents. The exceptions are areas of marked ecological change, such as the Sahara Desert that separates the sub-Saharan Ethiopian Realm from North Africa (which is climatically more like the rest of the Mediterranean), or the tropical rainforests of southeast Asia or Central America, which are distinct from the temperate regions of North America or Eurasia.

In the case of plant **biomes**, however, there are much finer-scale subdivisions (Fig. 9.3). Plants are much more

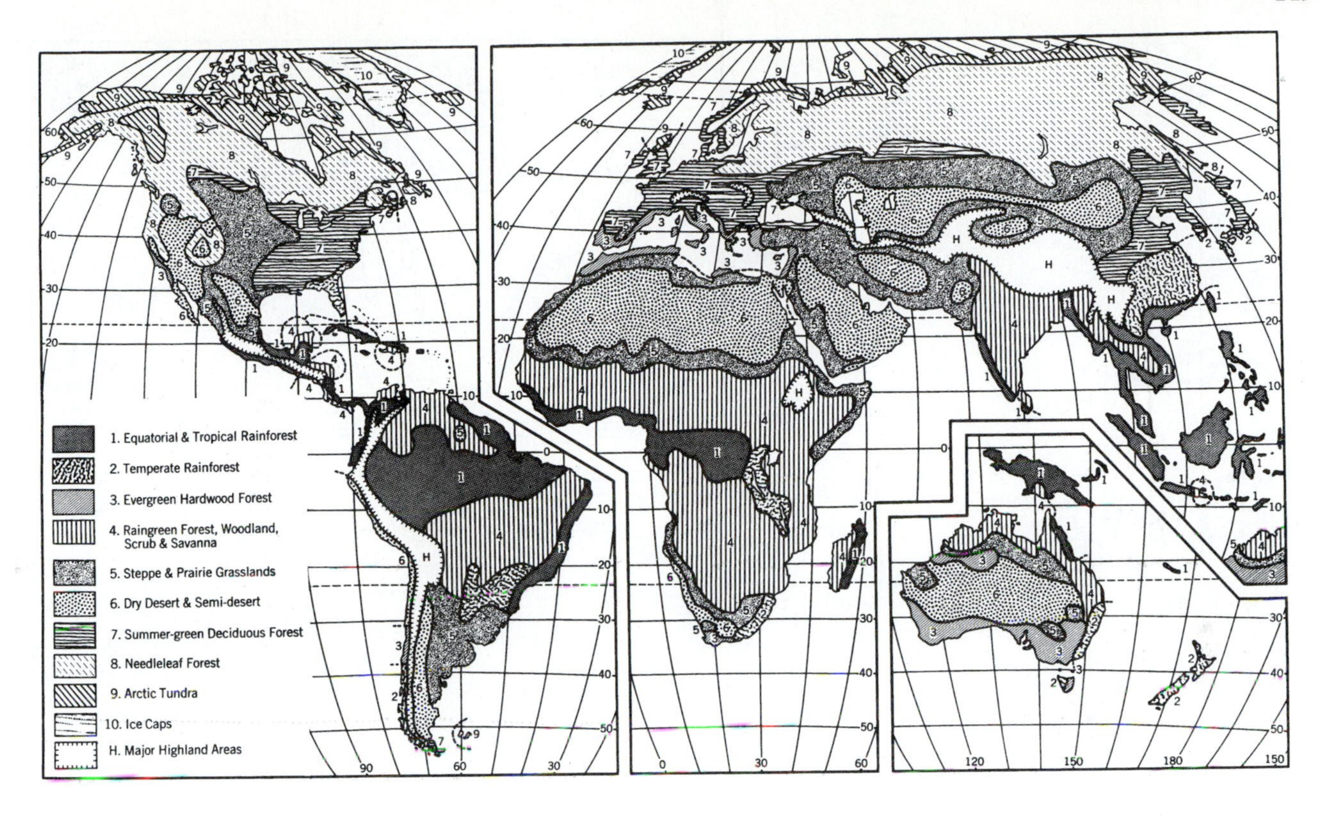

Figure 9.3. The pattern of distribution of vegetational provinces around the world. Most of the major plant provinces are controlled by the latitudinal climatic belts, with other effects due to elevation and rain shadow effects. (From Strahler, 1971.)

sensitive to variations in temperature and precipitation, so their biomes are subdivided into latitudinally defined climatic belts. For example, there are distinct regions of tundra, evergreen coniferous forest, deciduous forest, grasslands, deserts, and rainforest that correspond to certain latitudes, regardless of which continent they are found on. Clearly, the primary controlling factor is temperature, with the highest temperatures found in the tropics and the lowest found near the poles. Another important factor is precipitation, which is also latitudinally controlled. As you may recall from a physical geology course, atmospheric circulation determines the location of rainforests and deserts. Since the greatest amount of solar energy is received near the equator, the oceans and air are constantly circulating in huge cells that transport this excess energy from the equator to the poles (Fig. 9.4). In the tropics, the air is constantly being heated, so the air rises, producing the tropical low-pressure belts. As the warm air rises, it cools and condenses, raining its moisture back to the land; this is responsible for the great tropical rainforests. The rising tropical air then circulates away from the equator and sinks toward the earth in the subtropical high-pressure belts (between 10 and 30° north and south). In this region, the drier air from the tropics begins to cool, so it sinks toward the earth. As this cool dry air sinks, it begins to warm up as it absorbs heat from the land, enabling it to absorb more moisture. The subtropical high-pressure belts are regions of drier climate, and most of the world's deserts

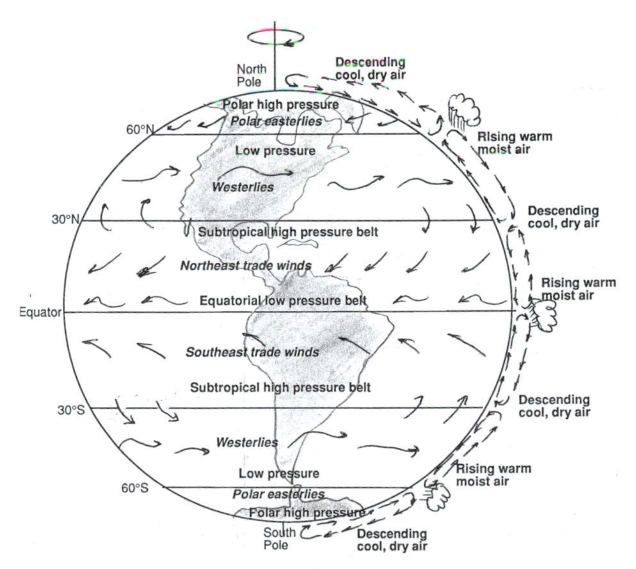

Figure 9.4. The circulation of the earth's atmosphere is divided into cells of high and low pressure controlled by latitude, by the differences in solar radiation between the pole and equator, and by the spin of the earth beneath the atmosphere. Along the equator, the rising columns of air at the edge of the pressure cells are full of moisture. As they rise, they cool and condense, forming the tropical rainforest belt. By contrast, in the subtropics (between 10° and 30° north and south latitude) the descending column of air is cool and dry, and warms as it nears the ground, sucking up moisture and producing most of the world's deserts. Similar rainy and dry regions are found near the poles.

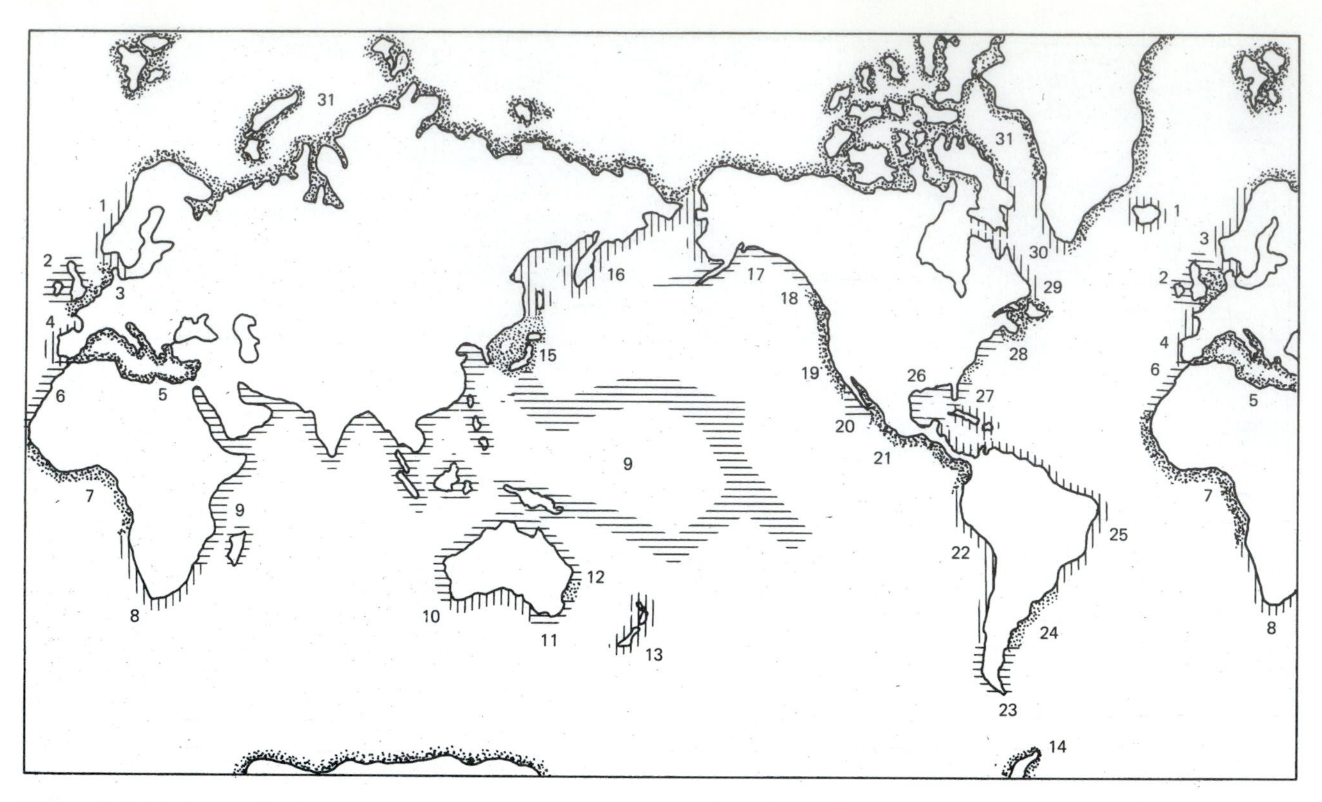

Figure 9.5. Marine molluscan provinces of the shallow continental shelves of the world. 1, Norwegian; 2, Caledonian; 3, Celtic; 4, Lusitanian; 5, Mediterranean; 6, Mauritanian; 7, Guinean; 8, South African; 9, Indo-Pacific; 10, South Australian; 11, Maugean; 12, Peronian; 13, Zeolandian; 14, Antarctic; 15, Japonic; 16, Bering; 17, Aleutian; 18, Oregonian; 19, Californian; 20, Surian; 21, Panamic; 22, Peruvian; 23, Magellanic; 24, Patagonian; 25, Caribbean; 26, Gulf; 27, Carolinian; 28, Virginian; 29, Nova Scotian; 30, Labradorian; 31, Arctic. (From Valentine, 1973.)

are found in these latitudes. Other atmospheric circulation cells go all the way to the poles (Fig. 9.4), but their biotic effects are less obvious than the example above.

In addition to the latitudinal belts determined by atmospheric circulation, other factors contribute to non-latitudinal variations in temperature and precipitation. For example, in the temperate regions, the prevailing winds come out of the west due to the circulation patterns caused by the spin of the earth beneath its atmosphere (Fig. 9.4). Thus, much of the moisture in North America or South America comes from Pacific Ocean clouds that blow east. If a high mountain range stands in the path of these prevailing westerlies and their rain clouds, it will force the air to rise, cool, and drop its moisture on the west side of the mountains in the form of rain or snow. As the cool dry air then descends down the eastern flank of these mountains, it receives heat from the land and warms up, allowing it to absorb moisture from the area to the east of the mountains, thus turning it into a desert. This is known as the **rain shadow effect**, and it is responsible for coastal Oregon and Washington being wet and lush, while the eastern part of those states is dry. Similarly, the Sierra Nevada Mountains cause moisture to drop in coastal California, while the dry air passing over them produces the Great Basin desert of Nevada and Utah. The mighty Himalayas create a rain shadow in Mongolia and China known as the Gobi Desert.

A third important factor controlling moisture is the presence of cold oceanic currents. These cause the clouds over the ocean to cool and condense before they reach land, so they are stripped of their moisture. The cold Humboldt Current (named after Alexander von Humboldt), which flows north from Antarctica up the coast of Chile and Peru, makes this region a desert. The Atacama Desert of northern Chile is the driest place on Earth, having had no measurable rainfall in decades. Similarly, the cold California Current flowing south from the northern Pacific causes deserts in southern California, and the Kalahari Desert in Namibia is caused by the cold Benguela Current flowing up the west coast of Africa.

In the marine realm, temperature and oceanic circulation are major controlling factors, analogous to temperature and atmospheric circulation on land. Valentine (1973) divided the shallow marine shelf into 31 molluscan provinces based on discrete assemblages of living species (Fig. 9.5). A coral specialist, on the other hand, would recognize only three, since corals do not occur outside tropical and subtropical waters. Valentine et al. (1978) found that the number of provinces in the Phanerozoic ranged from 2 (and usually less than 5) to our present-day high of 31, although Schopf (1979) recognized a range from 8 to 18 provinces. Valentine's (1973) molluscan provinces are also latitudinally controlled, so the Indo-Pacific province (#9) covers

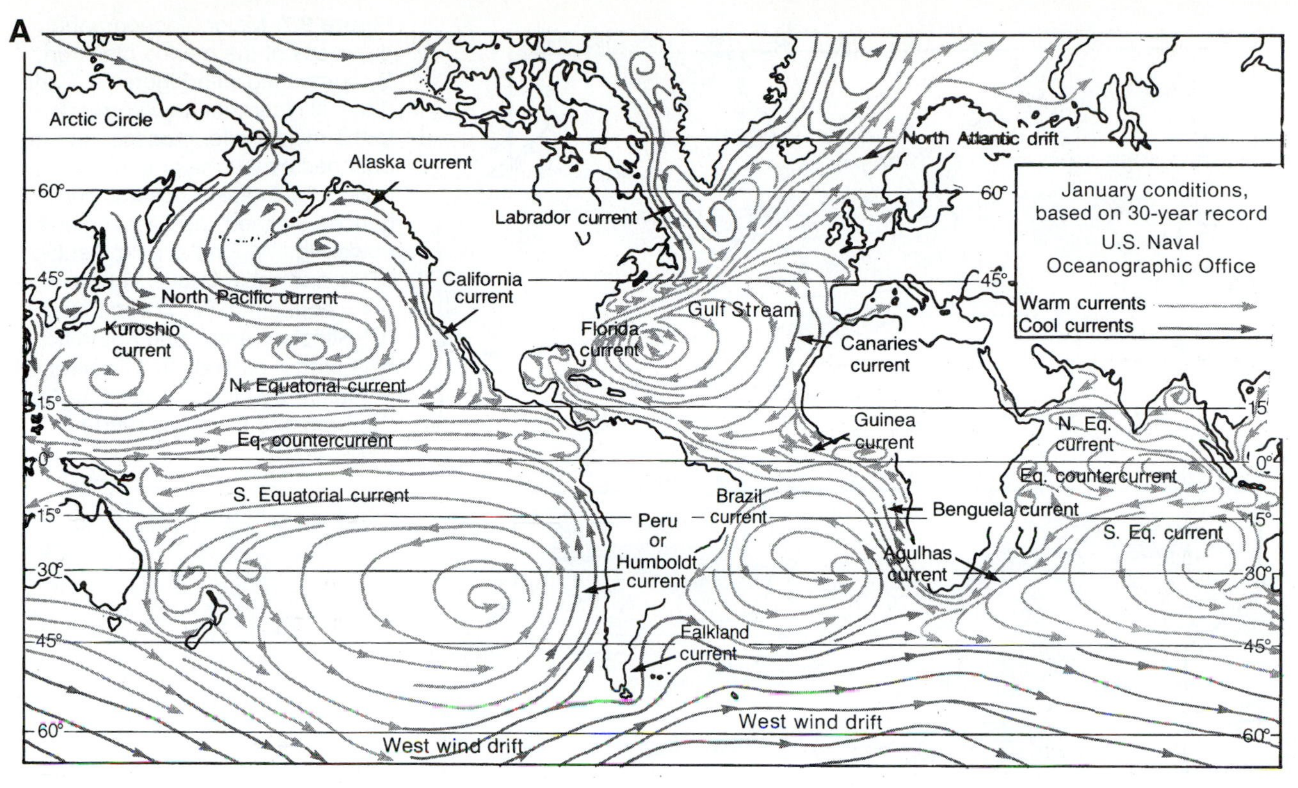

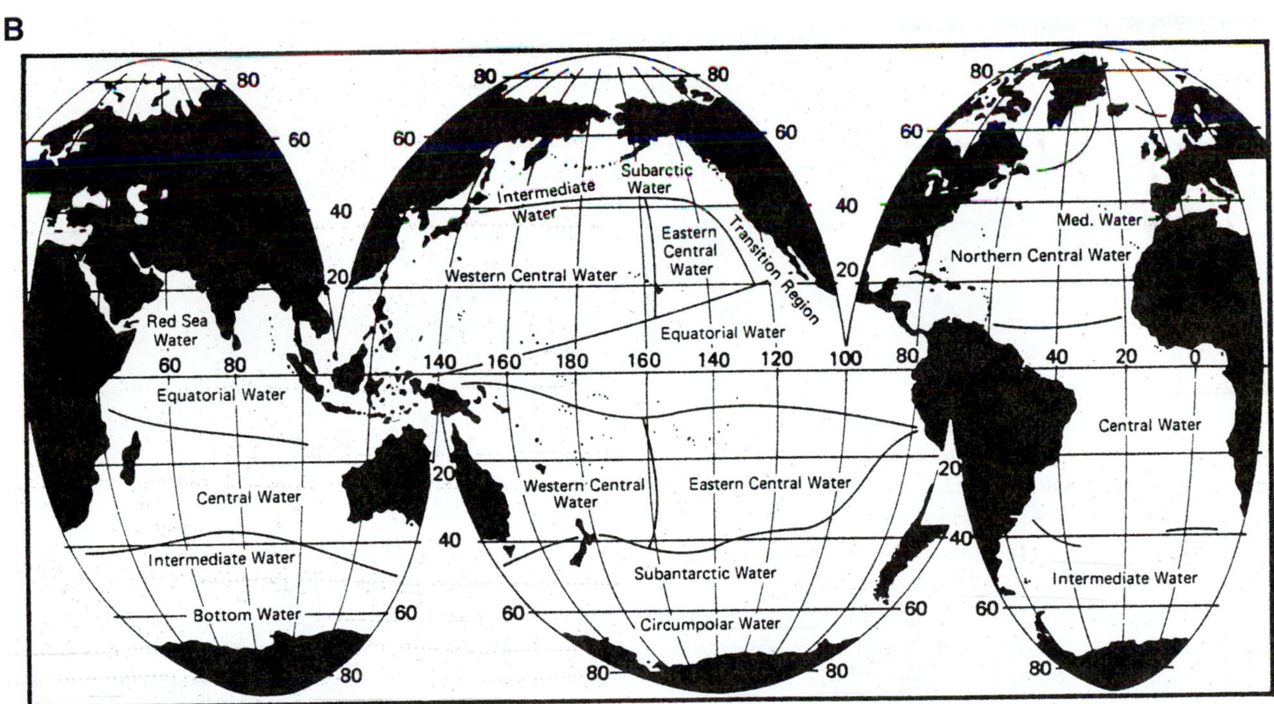

Figure 9.6. (A) Major surface currents of the modern ocean. (B) Boundaries of major surface and intermediate depth water masses in the oceans. Note that the boundaries match many of the boundaries between major oceanic currents. (From Prothero and Lazarus, 1980.)

the tropical waters of the Indian Ocean from Madagascar to the central Pacific Ocean. Along north-south-trending coastlines in the temperate regions, the rapid change in temperature with latitude yields a series of provinces. For example, along the Pacific Coast of North America, there are at least five: the tropics have a distinct Panamic fauna (#21); northern Mexico, Baja, and southern California have a Surian molluscan fauna (#20); central and northern California constitute the Californian province (#19);, the coasts of Oregon, Washington, and southern British Columbia are part of the Oregonian province (#18); and the Gulf of Alaska and Aleutian Islands have an Aleutian

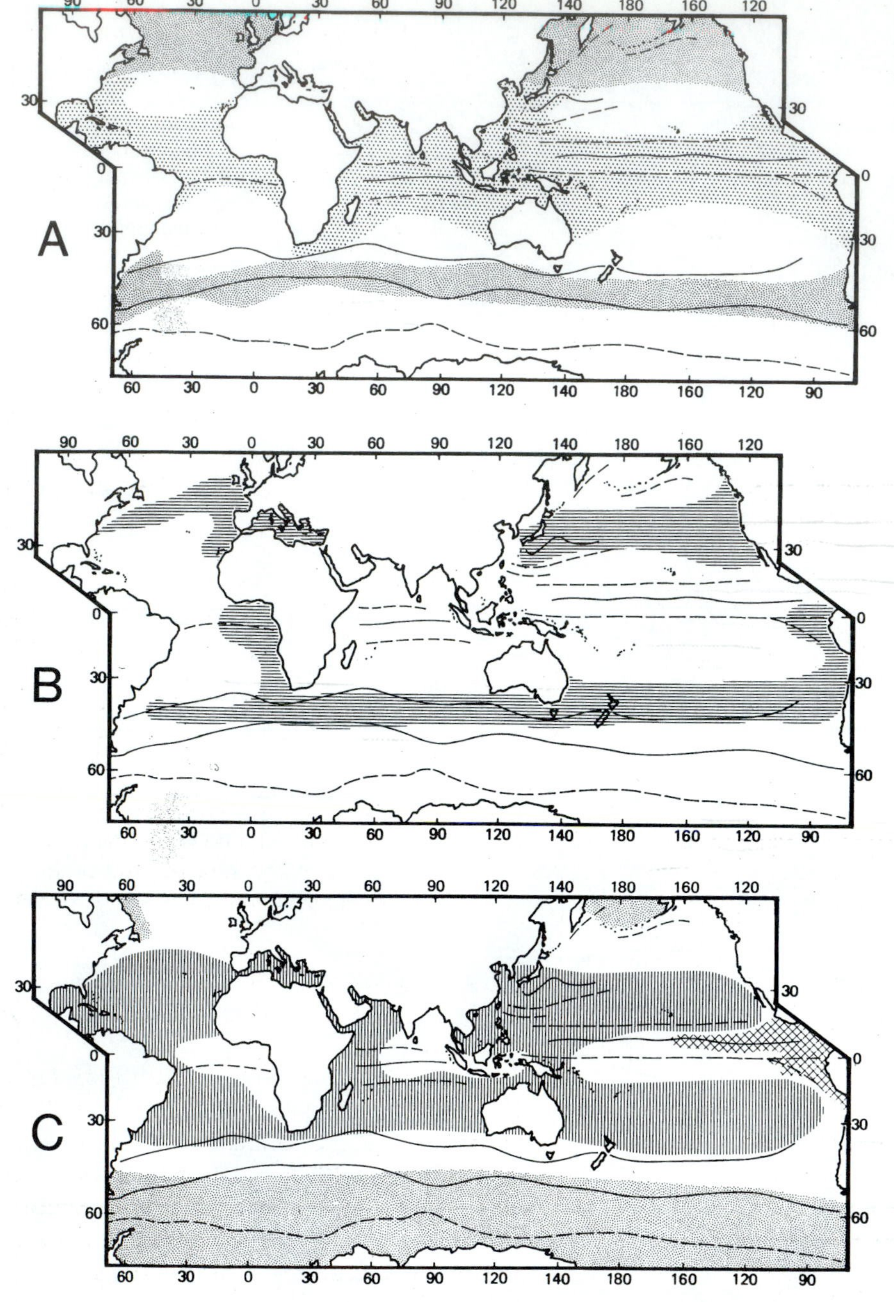

Figure 9.7. Major biogeographic provinces of the marine plankton. By comparing with Figures 9.5 and 9.6, it is clear that they correspond to the major water masses, and are bounded by the major currents. Solid lines indicate oceanic convergences; dashed lines are regions of upwelling. (A) Fine stipple = subarctic and subantarctic; coarse stipple = equatorial. (B) Transitional. (C) Stipple = Arctic and Antarctic; hachured = central; cross-hachured = eastern equatorial Pacific. (After Prothero and Lazarus, 1980.)

assemblage (#17). There is also a distinct assemblage in the Bering Sea (#16). The Gulf-Atlantic-Arctic Coast of North America has six provinces: Gulf (#26), Carolinian (#27), Virginian (#28), Nova Scotian (#29), Labradorian (#30), and Arctic (#31). Western Europe also has five provinces. There are fewer provinces around the equatorial coasts of Africa, South America, and Australia, because the equatorial temperature gradient is repeated both north and south of the equator.

In the open ocean, the marine planktonic realm can also be divided into biogeographic provinces. These, too, are largely controlled by latitude and temperature, which in turn drive oceanic circulation patterns (Fig. 9.6). The Atlantic and Pacific are both divided into large **gyres** of water circulating clockwise in the North Atlantic and Pacific and counterclockwise in the South Atlantic and Pacific (Fig. 9.6A). Lesser currents circulate around the Antarctic, Subantarctic, Arctic, and the equatorial regions. These currents divide the ocean into **water masses** that have distinctive temperature and salinity characteristics (Fig. 9.6B). Since plankton are very sensitive to these conditions, the boundaries of their biogeographic provinces correspond quite closely to the water mass boundaries (Fig. 9.7). For example, the Subarctic and Subantarctic planktonic provinces (Fig. 9.7A) are found chiefly in the Subarctic and Subantarctic waters (Fig. 9.6B); the Equatorial province is confined to the Equatorial water mass. The planktonic biogeographic provinces cover huge areas (as large as or larger than the terrestrial realms of Sclater and Wallace) and are much more homogeneous. In

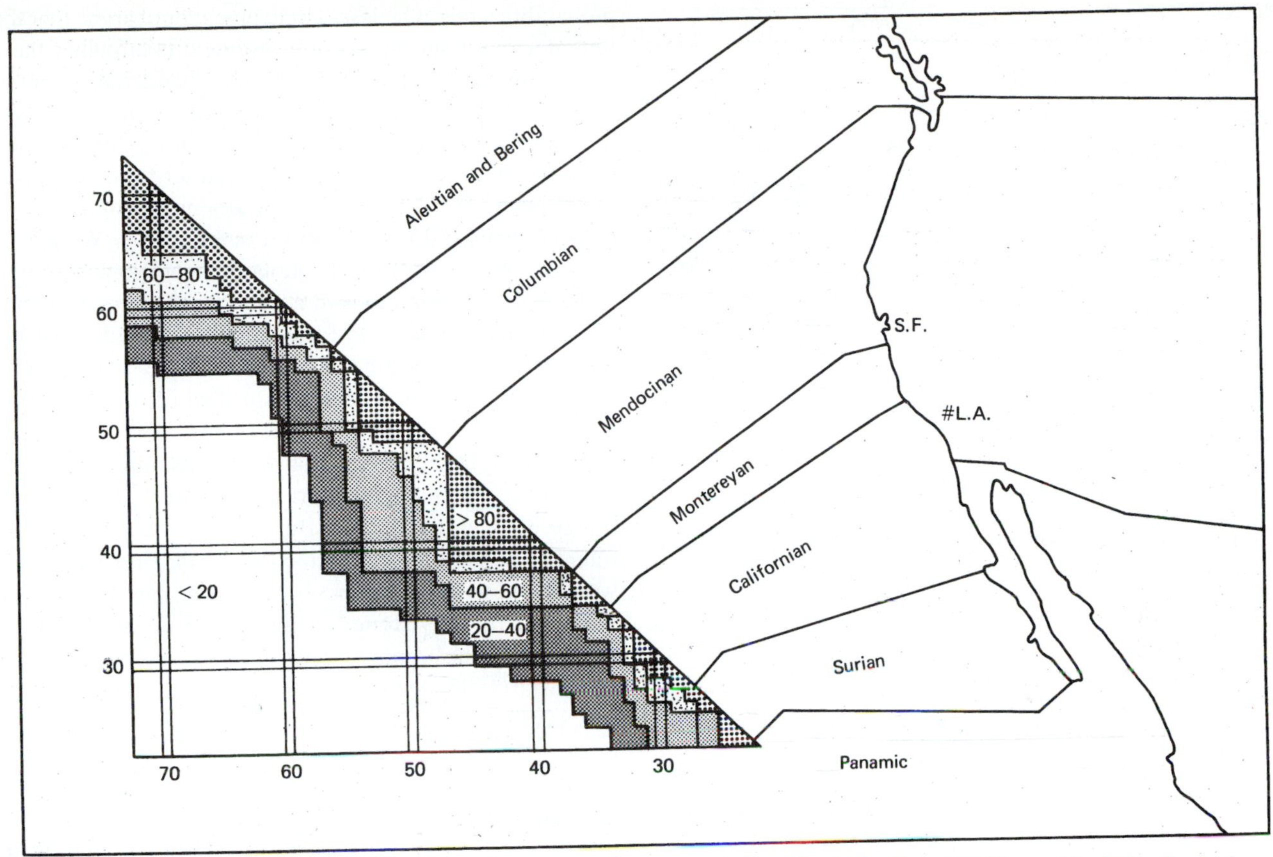

Figure 9.8. The major northeastern Pacific molluscan provinces, as defined by Valentine (1966), and their similarity (as measured by Jaccard coefficients) in 1° latitude increments, indicated by the various shades and stipples. Notice that some provinces (for example, the Mendocino) share 80% or more similarity within the province, but the similarity drops sharply at the boundaries with adjacent provinces. (From Valentine, 1966.)

this regard, planktonic biogeography is the simplest of all (Prothero and Lazarus, 1980).

How are the boundaries between provinces drawn? In some cases, there are natural barriers (such as the oceans that surround the terrestrial realms in Figure 9.2 or the Sahara Desert, which separates North Africa from Sub-Saharan Africa). In other cases, the distinctions are more subtle. For example, the molluscan ranges of the Pacific Coast of North America (Fig. 9.8) form a continuous intergradation of taxa (Valentine, 1966). Some molluscs range through several provinces, while others are found in only one. A number of different quantitative methods can be used to estimate the similarity between molluscan assemblages scattered along the coast line. In Figure 9.8, the coarsely stippled areas have more that 80% similarity of their species and form the natural clusters (see the large triangular stippled areas found in the Mendocinan or Columbian province, for example). Where that similarity breaks down is a major change between provinces, and these are the obvious places to draw provincial boundaries. Notice that this scheme of Pacific Coast molluscan provinces of Valentine (1966) has different subdivisions from those shown in Figure 9.5 (from Valentine, 1973). The Montereyan, Mendocinan, and Columbian of Valentine (1966) in Figure 9.8 were combined into the Californian and Oregonian of Valentine (1973) in Figure 9.5. This discrepancy underlines the difficulty in subdividing a continuum of molluscan distributions into provincial units.

A number of different methods of calculating faunal similarity have been generated (Hagmeier and Stults, 1964; Hagmeier, 1966; Cheetham and Hazel, 1969; Valentine, 1973; Fallaw, 1979; Flessa et al., 1979; Flynn, 1986). The simplest and most widely used is the **Simpson coefficient** (Simpson, 1936, 1960). Its formula is written as C/N x 100, where C is the number of taxa (at a specified taxonomic level) in common between two faunas or samples and N is the total number of taxa (at the same taxonomic level) present in the smaller of the two samples. Another popular metric is the **Jaccard coefficient** of faunal similarity. Its formula can be written as $C/(A + B - C)$, where C is the number of taxa in common between the two samples, and A and B are the number of taxa occurring only in sample a and sample b, respectively. (The percentage similarity in Figure 9.8 is calculated as a Jaccard coefficient, for example.) Of these metrics, the Simpson coefficient is still the most popular. Not only is it simpler to calculate and understand (it is a simple percentage of shared taxa), but it is also more sensitive to a variety of faunal changes than the Jaccard coefficient (Cheetham and Hazel, 1969; Fallaw, 1979; Flynn, 1986). The Jaccard coefficient works well unless samples a and b are of remarkably

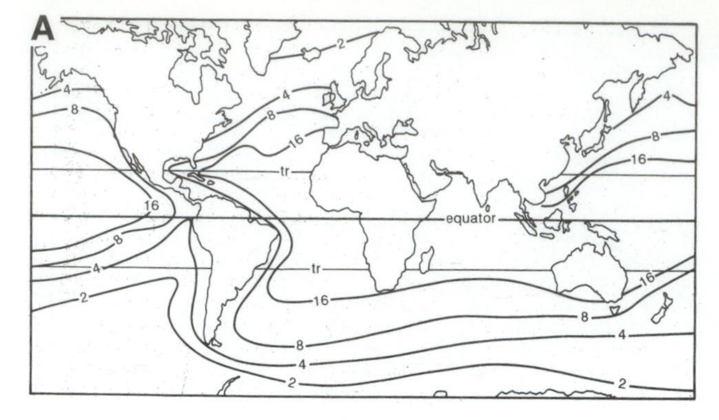

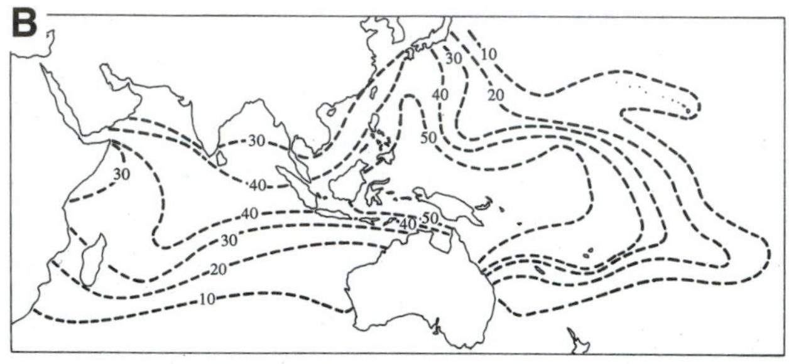

Figure 9.9. Most organisms show their highest diversity in the tropics, and their diversity declines toward the poles. This is known as the latitudinal diversity gradient. (A) Species diversity of recent planktonic foraminifera. (Modified from Stearn and Carroll, 1989.) (B) Generic diversity of hermatypic corals in the Indo-Pacific region. (Modified from Wells, 1954.)

unequal size. In these cases, the much larger sample distorts the ratio and gives a lower indication of shared taxa than is actually the case. As Hazel (1970) and Valentine (1973) pointed out, the Simpson coefficient emphasizes similarity, while the Jaccard coefficient emphasizes differences. (See Cheetham and Hazel, 1969, and Valentine, 1973, for a detailed review of the different coefficients.)

The Tropics: Cradle or Museum?

This subtitle describes in a nutshell one of the most interesting and controversial facts of biogeography: the highest diversity of organisms is found in the tropics, and diversity declines as one reaches higher latitudes. In ecological parlance, this is known as the **latitudinal diversity gradient**. This generalization holds not only for the tropical rainforests that we hear so much about (which cover only a sixteenth of the land surface, but harbor more than half its species) but for the shallow marine world as well (Fig. 9.9) (Stehli, 1968; Stehli et al., 1969; Stehli and Wells, 1971), and even the deep-sea benthos (Rex et al., 1993). The debate goes back at least as far as Matthew (1915), who argued that organisms originated in the temperate zone but could maintain higher diversity in the richer tropics (the "museum"). Conversely, Darlington (1957) argued that the tropics were the ideal place for new taxa to originate (the "cradle"). Either hypothesis seems reasonable. If the tropics are more stable and less variable, the extinction probabilities might be less—but some would argue that the high degree of specialization in tropical taxa makes them *more* prone to extinction. On the other hand, the increased specialization and niche subdivision of the tropics might enhance the potential of speciation—or it might mean that there are more competitors and it is harder for new species to originate. The debate still rages among ecologists, without a consensus (Ricklefs, 1973; Pianka, 1983). What does the fossil record have to say about this question?

Stehli et al. (1969) looked at the record of Permian brachiopods and Cretaceous foraminifera and found that the latitudinal diversity gradient persists all the way back to the Paleozoic. Thus, it is not an artifact of the instability of climate in the higher latitudes due to the Pleistocene glacial-interglacial cycles, but a long-standing feature of the biosphere, lending support the idea of the tropics as a refuge of long-lived taxa. But Stehli et al. (1969) also found a higher rate of extinction in the tropics, which would tend to argue against it being a refuge. In addition, Stehli et al. (1969) found a higher rate of origination of new taxa in the tropics. This has been corroborated by Jablonski (1993), who found that the first appearance of new taxa happened more often in the tropics than in temperate regions, supporting the idea that the tropics are a cradle of diversity. Although the debate goes on, it seems that the "cradle" effect of the tropics is more important than the "museum" effect in maintaining their high diversity.

However, these limited results do not resolve the issue entirely. As Gould (in Luria et al., 1981, p. 735) put it, "No single-factor theory can completely explain latitudinal gradients, and few ecologists are foolish enough to think that a simple answer will be found. Proper natural experiments require that one factor vary while all others remain constant. Nature in her complexity has not been kind to us in this respect. The tropics are older, and they are more stable, more predictable, richer in resources, and so forth. All these factors probably play a role in the maintenance of latitudinal gradients. But some factors must be more important than others, and we would love to know whether the historical effect of time or the equilibrial effects of stability and richness predominate. We can identify these effects, but, unfortunately, we do not know how to assess their quantitative importance."

Species Diversity and Geographic Area

Another important ecological discovery is the relationship between diversity and geographic area. This insight was first articulated by Robert MacArthur and Edward O. Wilson (1967) in their classic book, *The Theory of Island Biogeography*. MacArthur and Wilson noticed a relationship between the land area of an island and the diversity of life it supported; the larger the island, the more species were present. Small islands of less than 100 square miles may support less than a dozen species, while large islands, like Cuba or Hispanola, support hundreds of species (Fig. 9.10). In mathematical terms, the species number, S =

CA^z, where A is the island area, C is a constant, and z is a number less than 1 (usually about 0.20 to 0.35). Wilson and Bossert (1971) constructed mathematical models to explain how this phenomenon might work (Fig. 9.10B). They predicted that as a new island is colonized and begins to fill up, the rate of new arrivals should drop. Wilson and Bossert (1971) modeled this declining rate of immigration (λ_S) as a declining straight-line curve. The rate of immigration will drop to zero when the island reaches saturation (P). Another factor is also at work. As the population of the island increases, there will also be a higher rate of local extinction (μ_S), which will increase with the increasing number of species. At some point, these two lines will intersect, producing an equilibrium number of species (S). If we know the rate of immigration and the rate of extinction, we can predict the equilibrium species number.

Although these models are simplistic, they still produce useful implications and predictions. For example, let us imagine that there are two islands, one large and one small, that are equidistant from the mainland, so the immigration rate is constant (Fig. 9.10C). If this is the case, then the difference between the diversities of the two islands must be due to local extinction rates—extinction is much easier on a small island, with its limited habitats, than on a larger island, so the equilibrium species number (S) will be lower on the smaller island. Alternatively, if we imagine two islands of the same size but different distances from the mainland, then the constant island size keeps the extinction rate constant, but the difference in S is due to different immigration rates—near islands have a higher immigration rate than do far islands, so they can hold a larger diversity.

Wilson and his student Dan Simberloff (1969, 1970) tested the hypothesis with a series of famous experiments. They covered some small mangrove islands off the Florida Keys with fumigation tents and then gassed them to exterminate the entire native fauna. After the tents were removed, the scientists revisited the islands over the course of several years, and censused the return of life to them. Wilson and Simberloff found that diversity returned much more slowly to the smaller islands than it did to the bigger ones. Simberloff (1974) found that the MacArthur-Wilson models were very effective in predicting the bird diversity of islands, and showed that diversity is largely a function of area and isolation, as the model suggests. In recent years, the theory of island biogeography has been applied to many other areas, such as the tropical rain forest. Logging chops the rain forest into tiny remnants that have many of the same properties of oceanic islands—isolated patches of natural diversity surrounded by a barren wasteland of stumps and pastures (rather than ocean). In a set of crucial experiments, Thomas Lovejoy (1986) examined rainforest remnants of various sizes and determined their species diversity. As with the islands, the larger the area of rainforest, the greater the diversity. This kind of research has important implications for efforts to stem the destruction of the rainforest, for it shows that only the largest con-

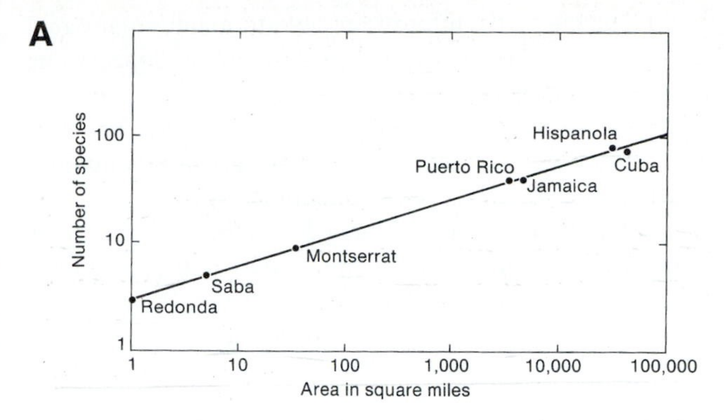

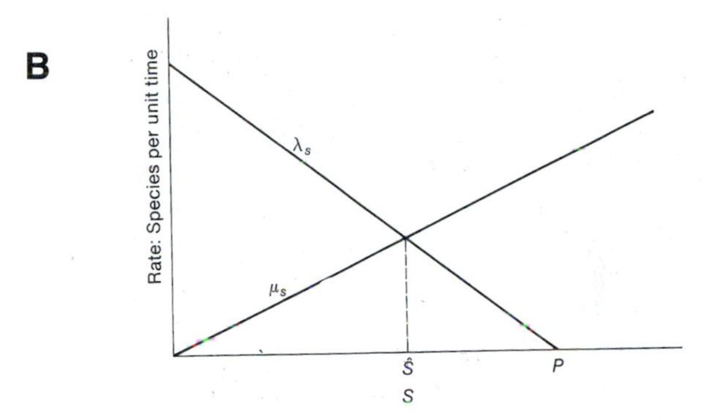

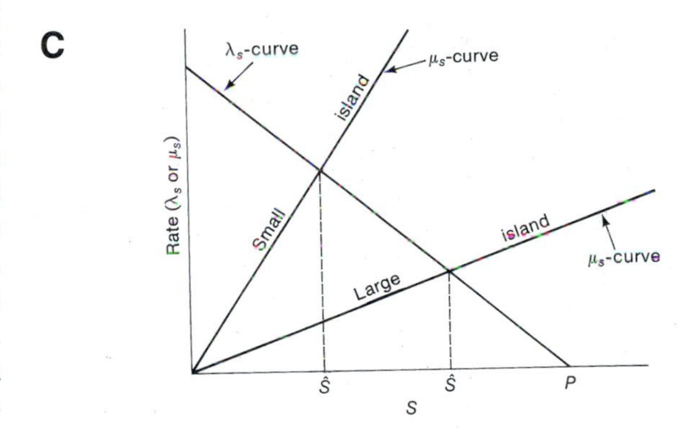

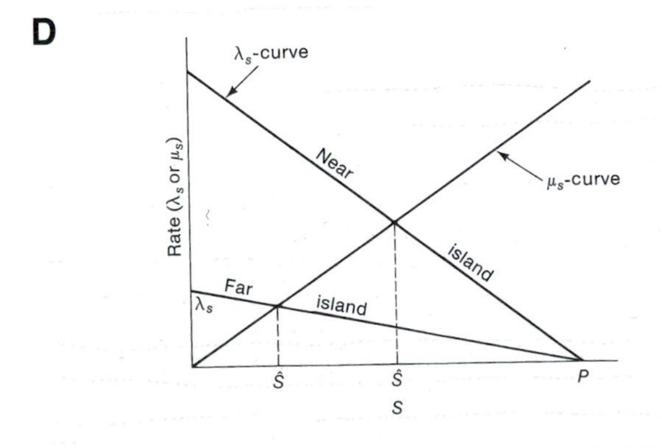

Figure 9.10. (A) The species-area curve for West Indian amphibians and reptiles. (From MacArthur and Wilson, 1967.) (B-D) Graphs illustrating the theory of island biogeography. (B) The basic model, contrasting immigration rate and extinction rate, which intersect at the equilibrium point, S. (C) If the island area is changed, the intersection with the rate of extinction line changes. (D) If the distance from the source of immigrants changes, the immigration rate lines shift in slope and produce a new equilibrium. (From Wilson and Bossert, 1971.)

tinuous patches of rainforest are able to maintain any real diversity. Small remnants, such as those left as concessions by loggers, are useless.

Paleobiologists have applied this relationship between species diversity and area to the fossil record in a number of ways. Schopf (1974) and Simberloff (1974) argued that the reduction in marine shelf area due to the coalescence of Pangaea caused the reduction in diversity at the end of the Permian. As we saw in Chapter 6, however, Pangaea was formed long before the end of the Permian, so shelf area had already been reduced before the mass extinction took place. The causes of the great Permian extinction are apparently more complex. Besides, the relationship between shelf area and diversity is not as straightforward as first supposed. Stanley (1979, 1984) pointed out that sea level drops and shelf reduction occurred repeatedly in the Pleistocene, yet many marine faunas were unaffected by these reductions in area. Several paleontologists (Valentine, 1967, 1973; Valentine and Moores, 1970; Boucot, 1975; Schopf, 1979) have noticed the relationship between the diversity of provinces and the amalgamation and fragmentation of continental plates, with the highest diversity (as in the early-middle Paleozoic and the late Mesozoic-present) occurring when the plates are separated, and the lowest diversity (especially in the Permo-Triassic) occurring during supercontinent formation. This relationship seems reasonable, but as we saw in Chapter 8, there are several things that correlate with Phanerozoic diversity (Fig. 8.18), including total outcrop exposure, paleontological interest, and other factors, so it is not clear how much the correlation of diversity with provinciality is a direct cause and effect and how much is coincidence.

Paleobiology and Ecological Biogeography

The role of the geographic range of taxa in paleobiology has proved to be important. Allopatric speciation theory (see Chapter 4) suggests that widespread species would have the greatest chance to fragment into peripheral isolates and form new species. However, several paleontological studies (Scheltema, 1977, 1978, 1979; Hansen, 1980; Jablonski, 1980, 1982, 1985) showed that gastropods with the greatest geographic ranges were mostly species with easily dispersed planktonic larvae, and they tended to have low speciation rates and long species durations. Apparently, the gene flow in these widespread populations is great enough that they do not easily fragment into genetically isolated populations. By contrast, gastropod species with small areas (mainly those that do not have planktonic larvae) exhibited higher speciation rates, possibly because they disperse so slowly that their populations have low gene flow and can become fragmented easily.

Flessa and Thomas (1985) looked at the relationship between evolutionary rate and biogeographic distribution in the fossil record. They found that most genera existed in only a few regions, while a handful of cosmopolitan genera were extremely widespread. This suggests that taxa with the highest probability of range expansion are those that already have large geographic ranges, or in their words, "like the rich getting richer, the cosmopolitans become more cosmopolitan."

Extinction patterns also seem to be affected by species-area effects and provinciality. For example, Stanley (1984) found that tropical taxa are much more extinction-prone than those from higher latitudes, presumably since the warm, equable tropics are much more sensitive to environmental change than are the temperate regions. However, as we saw in Chapter 6, temperature decline is not the only explanation for mass extinctions. Stanley and Campbell (1981) found that the Pliocene extinction of molluscs in the North Atlantic was much more severe than in the Pacific, possibly because there were more faunal provinces in the North Atlantic to be affected, but also because the cooling was more severe in the North Atlantic.

A number of paleontologists (Jackson, 1974; Hansen, 1980; Jablonski, 1980, 1982) have shown that taxa with larger geographic ranges tend to be longer-lived. This is probably because if they spread out over a larger area, then their extinction in one local area will not mean the extinction of the entire species. In particular, widespread species seem to be more resistant to mass extinction (Bretsky, 1973; Boucot, 1975; Hallam, 1981; Sheehan, 1982; Jablonski, 1985). This is one of the only properties of a species that reliably predicts its survival during mass extinction. Other properties of the species that had been important during normal, "background" extinction times (such as species richness, speciation rate, and the mode of development) conferred no special advantage during mass extinctions. In other words, the conditions during mass extinctions are completely different from those during normal times, so the normal Darwinian fitness of a species may be irrelevant during the unusual conditions of mass extinction.

If widespread taxa are extinction-resistant, then we can extend the MacArthur-Wilson models to a higher level. As Lovejoy used MacArthur-Wilson models to predict the rate of extinction within forest remants of various sizes, so could we predict the rate of extinction during mass extermination events of different sizes. Raup (1982) calculated that today's terrestrial vertebrates are too widespread (despite greater provinciality) for an extinction of K/T magnitude to be produced by local perturbations. Similarly, Jablonski (1985) argued that marine regression and shelf area was insufficient to cause the Permo-Triassic extinction, since even a complete extermination of today's marine shelf biota would still leave 87% of the living marine families on islands, where they would survive lowered sea level.

The relationships between geographic range, provinciality, environmental variation, and the paleobiological properties of organisms are very complex and are still being actively explored. Clearly, however, paleontologists have moved beyond passively mapping faunal provinces and are now proposing testable paleobiological hypotheses based on our understanding of ecological biogeography.

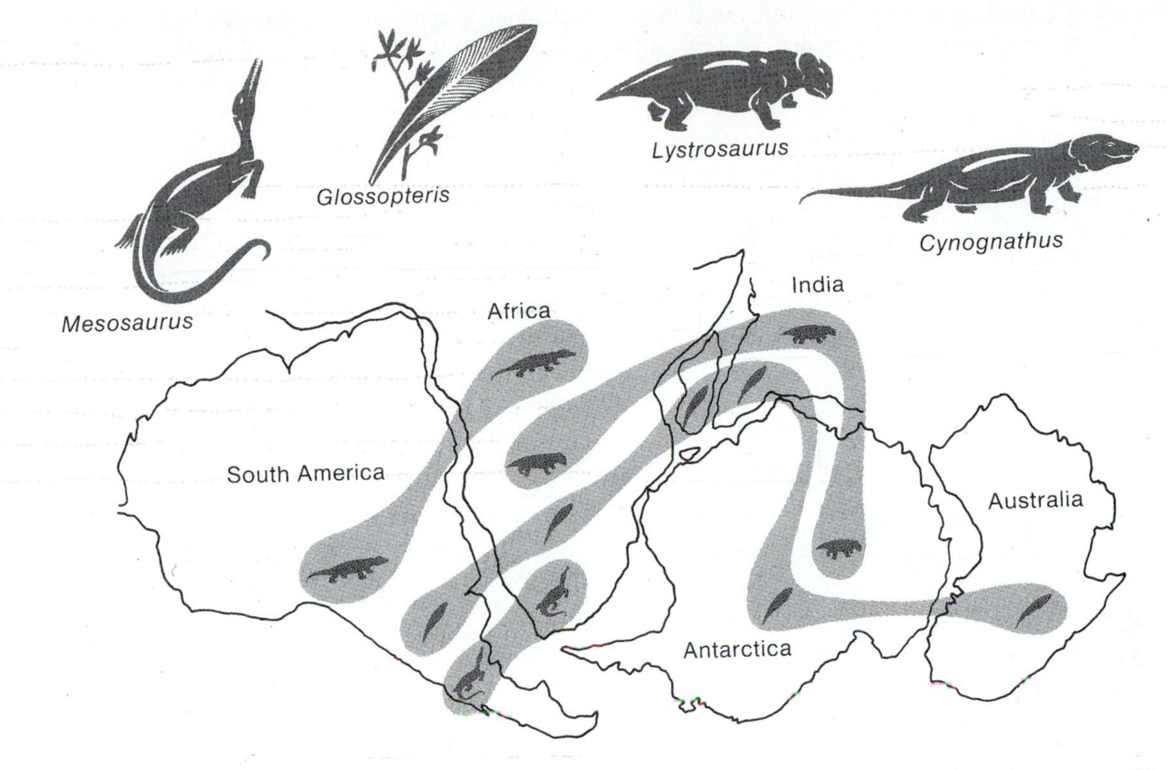

Figure 9.11. The distribution of certain animals and plants from the Permian and Triassic only makes sense if these continents were united as a Gondwanaland supercontinent during that time. For example, the aquatic reptile *Mesosaurus* and the bear-like protomammal *Cynognathus* are only known from South Africa and South America, while the herbivorous protomammal *Lystrosaurus* is found in Triassic beds of India, South Africa, and Antarctica. The distinctive seed fern *Glossopteris* is found on all the Gondwana continents, including Madagascar, India, and Australia. (From Colbert, 1973.)

HISTORICAL BIOGEOGRAPHY

In order to come to a decision as to the existence in ancient times of a connection between neighboring continents, geology bases itself on the analogous structure of coastlines, on the similarity of animals inhabiting them, and on ocean surroundings. Plant geography furnishes most important material for this kind of research. It can, up to a certain point, determine the islands which, at one time united, have become separated from one another; it finds the separation of Africa and South America occurred before the development of living organisms.

Alexander von Humboldt and A.J.A. von Bonpland, *Essai sur la geographie des plantes*, 1805

Paleobiogeography and Plate Tectonics

The concepts of ecological biogeography, such as faunal provinces, can also be recognized in the fossil record. For example, the Pacific Coast molluscan provinces of Valentine (1966, 1973) can be recognized in the Pleistocene faunas of the Pacific Coast, and are useful for determining how the coastal waters had cooled during the ice age cycles. Stanton and Dodd (1970) examined the provincial affinities of Pliocene molluscs from the San Joaquin embayment in California, and found that most could be assigned to the modern Montereyan or Californian province of Valentine (1966). The distinctiveness of some of the terrestrial realms (Fig. 9.2) was even greater in the past. For example, South America was an island continent through most of the Cenozoic, with its own unique fauna that had little to do with other continents. Australia's unique marsupial fauna was even more distinctive in the Tertiary when there were no dingos, rabbits, or other recent placental immigrants.

As we look at older and older fossil distributions, we begin to notice discrepancies with the modern geography. Prior to plate tectonics, some of these discrepancies were a mystery. For example, Wegener (1915), DuToit (1937), and other early advocates of continental drift used the peculiar distribution of distinctive Permo-Triassic plants (such as the seed fern *Glossopteris*) and animals (the aquatic reptile *Mesosaurus*, the synapsids *Lystrosaurus* and *Cynognathus*) as some of their best evidence for the former existence of the Gondwana continent (Fig. 9.11). In the late Paleozoic, there were three distinct floral provinces: a southern flora, dominated by *Glossopteris*, found on most of the Gondwana continents, which were at temperate and polar latitudes; a northern flora found in the great coal swamps of the late Carboniferous, which was dominated by lycopsids (scale trees and club mosses) and located in the ancient tropics; and a third, distinctive Asiatic floral province. Note that these provinces look oddly scattered on a modern map of the world, but if they were placed in their late Paleozoic continental positions, all the Gondwana fragments (Africa, South America, India, Australia, Antarctica) would come together, as would Laurasia (North America plus western Eurasia). The fragments of Asia had not yet collided with Siberia to form Pangaea, explaining their distinctiveness.

Paleontologists had long been puzzled by the fact that Cambrian trilobite provinces did not follow the present-day distribution of continents. The European faunal province, characterized by the trilobite *Paradoxides*, was found not only in Europe but also in the southeastern United States, Boston, Rhode Island, Nova Scotia, and eastern Newfoundland (Fig. 9.12). Most of North America was one distinct American province, the *Ogygopsis* fauna, but so was Scotland and western Newfoundland. Paleontologists tried to explain this odd pattern by suggesting differences in water depths or land barriers, but nothing made sense. Only after plate tectonics came along did an explanation present itself. Apparently the southeastern United States, eastern New England, Nova Scotia, and eastern Newfoundland were on the European side of the Proto-Atlantic Ocean during the Cambrian, while Scotland was on the North American side. The Proto-Atlantic then closed in the late Paleozoic, forming the supercontinent of Laurasia. When the present-day Atlantic was formed by splitting up this supercontinent in the Triassic and Jurassic, it did not follow the line of closure of the old ocean, but split open along a new line, and left parts of ancient Africa (southeastern United States) and Europe (New England, Nova Scotia, and eastern Newfoundland) attached to present-day North America and parts of ancient North America (Scotland) attached to Europe.

Paleobiogeographic evidence not only suggested that continents had drifted apart, but even opened the door to our understanding of how continents are put together. Decades ago, paleontologists (Thompson et al., 1950) found Permian fusulinid foraminifera in the central British Columbia Rockies that had affinities with the Tethyan seaway that once stretched from the Straits of Gibraltar to Indonesia. How did these peculiar Permian foraminifera get there? Was there an arm of the warm tropical Tethys that reached up to British Columbia? Or were they part of a moving continental fragment that used to be in the Tethys but crossed the Pacific and became sutured to British Columbia? Although the latter idea seems outrageous at first, as more and more evidence accumulated, it became the best explanation (Hamilton, 1969; Monger and Ross, 1971; Ross and Ross, 1981; Jones et al., 1977). It turns out that the western Cordillera of North America is made up of numerous exotic terranes that have rafted from across the Pacific and collided with North America at various times during the Mesozoic and Cenozoic (Fig. 9.13A). Several of these terranes (Wrangellia, Alexander, Cache Creek) not only have peculiar Tethyan faunas, but the temperature preferences of their fossils suggest that in the Permian the Cache Creek terrane was part of the tropical Tethys, the Alexander terrane was at subtropical to warm-temperate latitudes, and Wrangellia came from the cool southern latitudes (Fig. 9.13B). Today, Wrangellia is the bedrock for southeastern Alaska and Vancouver Island, and the Cache Creek and Alexander terranes are located deep in the British Columbia Rockies—quite a distance for that much land to travel in about 250 million years! In addition to fusulinids, the Triassic ammonoids (Tozer, 1982) show the previous position of these terranes, and the Jurassic ammonites (Tipper, 1981) have been used as evidence of dramatic northward movement of many of these terranes (for a review of these data, see Hallam, 1986).

Dispersalist Biogeography

Considering that these and other means of transport have been in action year after year for tens of thousands of years, it would be a marvellous fact if many plants had not thus become widely transported. It should be observed that scarcely any means would carry seeds for very great distances.

Charles Darwin, *On the Origin of Species*, 1859

Before the advent of plate tectonics, biogeographers looked at faunal and floral provinces as static entities on static continents. The chief problems were defining the province and its boundaries, and determining how adjacent provinces had interacted and how organisms could disperse from one to another. Whichever method of measuring similarity one chooses, the first task is to evaluate the similarity and/or differences between regions. Simpson (1940, 1947, 1953) measured the effect of barriers of dispersal on faunas (particularly land mammals) found in different regions. For example, the mammals of North America are fairly similar across the continent, allowing for differences in climate and vegetation. At the ordinal level, 100% of the mammals in New York also occur in Oregon, and 83% of the mammals in Oregon also occur in New York. The coefficient of similarity decreases slightly at the level of families, genera, and species (which are more restricted in geographic range), but it is clear that as a whole, temperate North America is faunally homogeneous, with few barriers to dispersal (see also Flynn, 1986). Simpson called this situation a **corridor**, since there are few or no barriers to dispersal, and organisms can move about freely within suitable habitats.

When one compares regions, however, the barriers to dispersal, such as oceans or mountain ranges, rapidly decrease the faunal similarity. For example, there are striking differences between the mammals of Eurasia and those of North America (separated by the Bering Sea) or between those of North and South America (connected only by the Isthmus of Panama). As we look at this pattern in the geologic past, the contrast is even more striking. For example, during the Ice Ages, the Bering Strait was a major land bridge, allowing many Old World mammals (such as mammoths and bison) to migrate from Eurasia to North

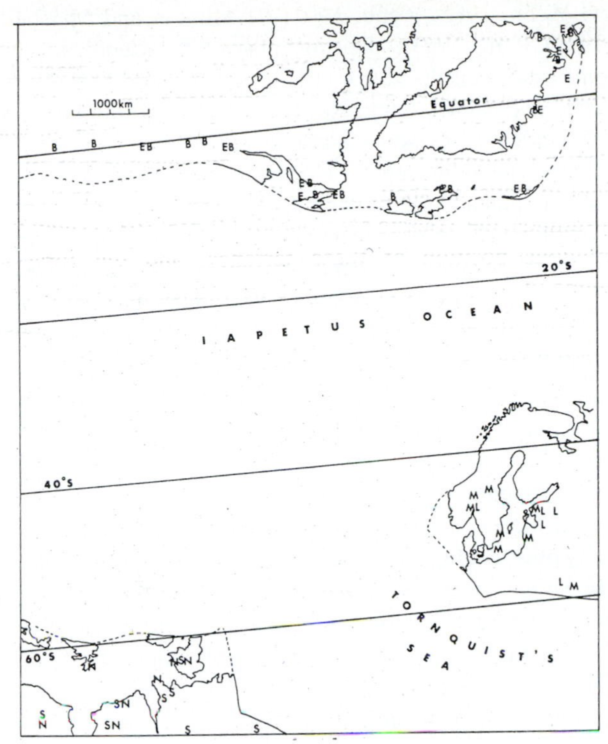

Figure 9.12. Cambrian and Ordovician shallow marine faunas show a peculiar trans-Atlantic distribution that long puzzled geologists. For example, most of North America, as well as northern Scotland and Ireland, have a distinctive assemblage of bathyurid trilobites (B) and the mollusc *Euchasma* (E). But southeastern New England, Nova Scotia, and Newfoundland, and most of western Europe had a different fauna, dominated by trilobites such as *Selenopeltis* (S) and *Neseuretus* (N). The Baltic region had yet another assemblage, dominated by megitaspid trilobites (M) and a brachiopod assemblage characterized by *Lycophoria, Clitambonites,* and *Antigonambonites* (L). These distributions only make sense if these three regions were widely separated in the Cambrian and Ordovician (here shown in their Early Ordovician paleolatitudes), and were sutured together when these continents collided as the Iapetus Ocean closed in the late Paleozoic. When the Atlantic Ocean opened in the Mesozoic, some parts of North America (e.g., northern Scotland) were carried off to Europe, while parts of Europe (e.g., eastern New England, Nova Scotia, and Newfoundland) were left with North America. (From Cocks and Fortey, 1982.)

America, and New World mammals (such as horses and camels) to migrate to Eurasia. However, the Bering route was not a freeway. For some reason, several Old World natives (such as woolly rhinos and pigs) never migrated to North America, even though woolly rhinos were just as cold-adapted as woolly mammoths. Likewise, some North American native animals (such as pronghorns and peccaries) never migrated to the Old World, although pronghorns live side by side with bison and other Old World immigrants. Such a selective route was called a **filter bridge** by Simpson, since it allows some animals to pass, while others do not cross.

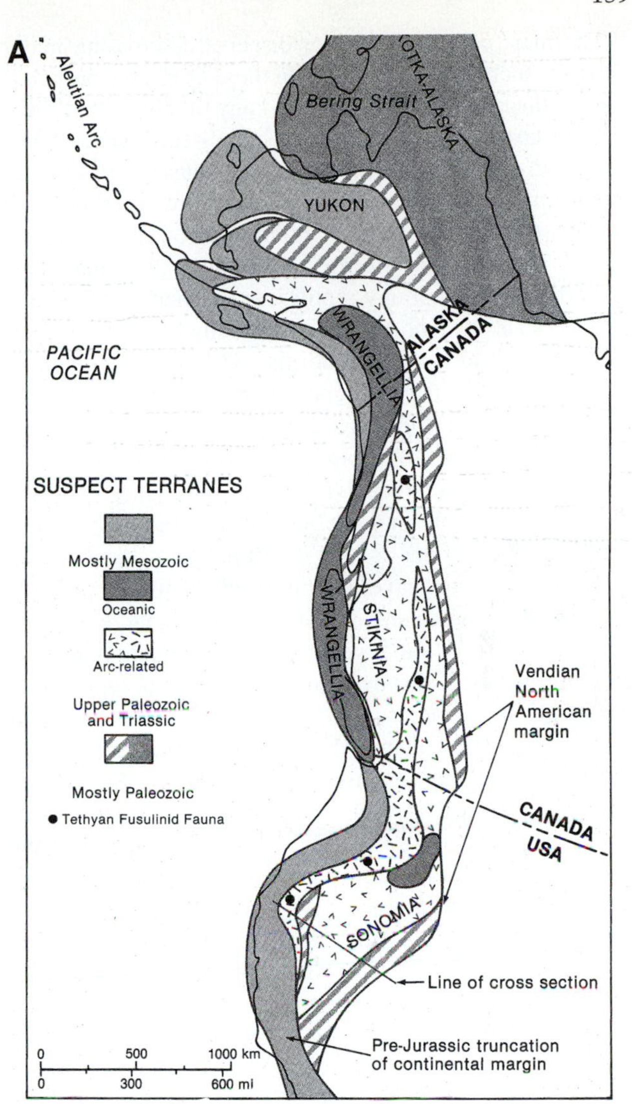

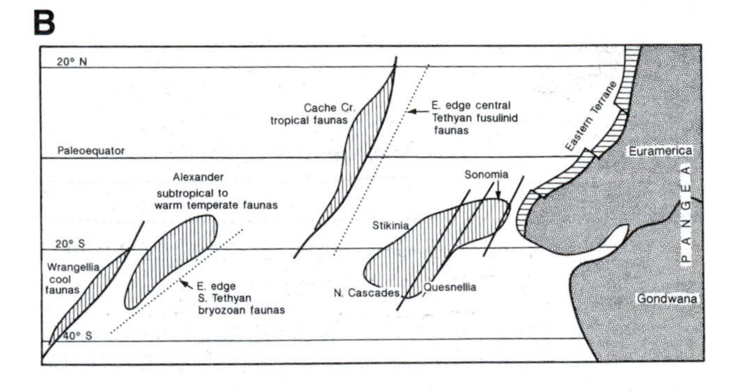

Figure 9.13. (A) Most of western North America is a collage of exotic terranes from around the Pacific, assembled by collisions during the Mesozoic and Cenozoic. (From Dott and Prothero, 1994.) (B) Reconstructed late Paleozoic positions of some of the exotic terranes that were accreted to North America in the Mesozoic. Many terranes carry distinctive fossils, such as fusulinid foraminiferans or bryozoans, that place them in latitude, or within the old Tethys seaway (now running from southeast Asia to the Mediterranean). Although their latitudinal position is well constrained, their horizontal longitude is less certain. (After Stearn and Carroll, 1989.)

The most difficult route for dispersal is crossing major barriers, such as large oceans. In these cases, the probabilities of dispersal are very low, and any organism that manages to cross does so under unusual circumstances. The islands in the middle of the ocean are thousands of kilometers from land, so their only inhabitants are organisms that blew in on a major storm or rafted there on floating vegetation. Hawaii, for example, has no native mammals (except for bats) nor any reptiles, amphibians, or freshwater fish, so all its land animals (especially birds, snails, and insects) were blown in by storms and have since evolved into a distinctive native fauna that is **endemic** (restricted) to Hawaii. Such a low-probability transport was called a **sweepstakes** route by Simpson, since its odds of success are about the same as winning a sweepstakes or the lottery—it is highly improbable, but over the long span of geologic time, sooner or later it will happen.

Biogeographers have documented many remarkable ways in which organisms can disperse across large barriers. Large rafts of floating vegetation with small mammals clinging to them have been found at sea, apparently launched when a major river flooded. These are capable of transporting small mammals across oceans, although the odds are decidedly against their making it. Small land invertebrates, such as snails, have been known to be carried long distances in the mud on birds' feet. Land plants have remarkable ways of getting across oceans. Many different seeds float long distances or fly in the wind for thousands of kilometers. The coconut's hard seed is adapted for floating long distances to a new island, where the pounding of the surf finally cracks the shell and allows it to germinate near the beach. In addition to birds and bats, many other land animals also can cross enormous distances in the wind. Young spiders disperse from their mother's web by releasing strands of silk as a parachute or balloon and letting the wind carry them wherever it will.

In the marine realm, the problem of dispersal is the mirror image of that on land. What are considered bridges to land animals are barriers to the dispersal of marine life. For example, when the Panamanian land bridge closed in the Pliocene, it allowed mammals to migrate between the Americas, but shut off the connection between Caribbean and Pacific marine faunas, which have since diverged. In addition, the ocean is not homogeneous, but divided into provinces and water masses defined by temperature, salinity, and other properties. Nevertheless, most marine organisms have mechanisms of spreading over much of the ocean. Most have planktonic larvae that are released by the millions, allowing them to float wherever the currents take them and settle in any unoccupied spot. Others, such as jellyfish, sea turtles, and bony fish, can migrate or float huge distances across the ocean under the right circumstances. The principles of dispersal in the marine realm are similar to those on land, although there are also important differences.

South America provides an excellent example of many of these concepts (Simpson, 1980; McKenna, 1980; Stehli and Webb, 1985; Webb, 1991; de Muizon and Marshall, 1992; Cifelli, 1993). During most of the Cenozoic, this continent was isolated from the rest of the world. It had no direct land connection to North America until the Pliocene and had lost its Gondwana connection to Antarctica sometime in the Oligocene. Consequently, it has a highly endemic fauna of land mammals and large predatory birds, most of which are only distantly related to animals in other continents. Some time in the Cretaceous it began to lose its faunal connection to the Gondwana continents, so it ended up with only a few native groups: the edentates (sloths, armadillos, anteaters, and their kin); marsupials (which evolved into the main carnivorous groups in the absence of placental carnivores, including remarkable hyena-like, wolf-like, and saber-toothed marsupials); and archaic hoofed mammals (which evolved in parallel to resemble horses, hippos, camels, giraffes, antelopes, mastodonts, and many other ecological niches found on other continents). A few South American natives (such as the arctostylopids) managed to get to North America and China, and a few North American taxa (such as the pantodonts) seemed to get from Asia and North America to South America in the Paleocene (de Muizon and Marshall, 1992), so there was the possibility of some sweepstakes dispersal, or island hopping, during the Paleocene, but through most of the early Cenozoic, the continent was isolated.

In the Oligocene, two more mammalian groups appeared in South America: the New World monkeys and the caviomorph rodents (today represented by chinchillas, guinea pigs, capybaras, agoutis, cavies, and their kin). Both groups have their nearest relatives in Africa, so they must have rafted across the South Atlantic by some kind of sweepstakes route. South America's isolation continued until the late Miocene, when mastodonts, raccoons, and some other North American groups apparently island-hopped across the Central American archipelago (Campbell, 1995; Campbell and Frailey, 1996). Finally, the isolation ended in the mid-Pliocene with the completion of the Panamanian isthmus, which became a filter bridge. Many North American natives (mastodonts, horses, camels, deer, and placental carnivores, such as bears, lions, and saber-toothed cats) moved south, but others (such as pronghorns) did not. Only a few of the South American natives (armadillos, ground sloths, and capybaras) managed to invade North America, while the majority stayed at home and may have died out in the face of the competition from North American invaders (see Chapter 6).

Most of the concepts of dispersalist biogeography were developed under the assumption that the continents were fixed. When plate tectonics came along, it was possible for organisms (or their fossils) to get from point A to point B without doing the walking themselves. McKenna (1973, 1983) suggested several other possible mechanisms where the plates do most of the traveling. If a landmass were to rift away from one area, travel across the ocean, and then collide into another, it could transport its inhabitants to a new land. McKenna (1973) called this a "**Noah's Ark**"

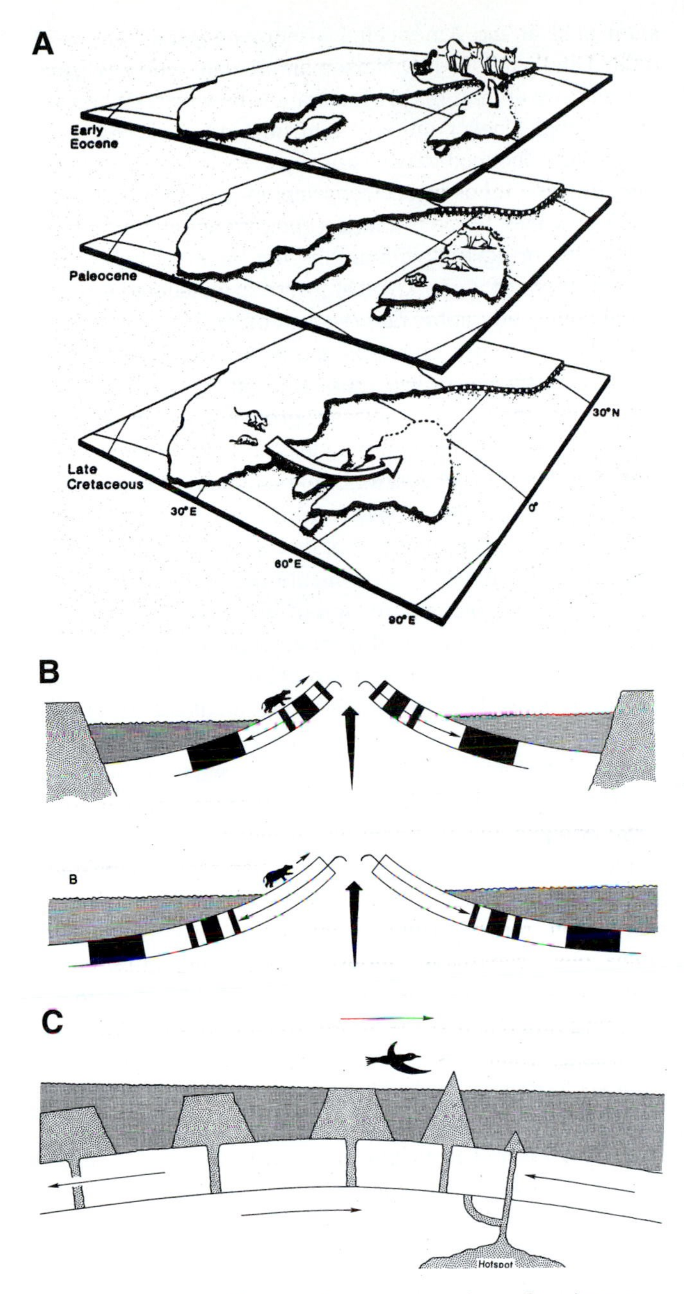

Figure 9.14. (A) When India tore away from Gondwanaland in the Cretaceous and slammed into Asia during the Eocene, it may have acted as a "Noah's ark," carrying an fauna of Cretaceous mammals which evolved in isolation until the "docking event." The sudden appearance in Asia during the Eocene of artiodactyls, perissodactyls, euprimates, and several other groups with no ancestors in the rest of world may support this hypothesis. (From Krause and Maas, 1990). (B) An island on a mid-ocean ridge (like Iceland) continually gets younger as it rifts apart and sinks below sea level. Yet an inhabitant of that island could survive there for millions of years longer than the apparent age of the island by walking "up the down escalator." (C) Similarly, hotspot islands (like the Hawaiian chain) continually sink away from the hotspot, but earlier colonists could just hop to the next island erupting over the hotspot, and so their distribution in the region is older than the islands they inhabit. (B and C from McKenna, 1983.)

since it bears similarities to the transport of animals by Noah's boat. Although it is hard to document examples of Noah's Ark transport in the fossil record, several authors (McKenna, 1983; Krause and Maas, 1990) have suggested that the transport of India from Gondwanaland to the southern part of Asia may be such a case. India rifted away from Gondwanaland in the Cretaceous and may have had its own endemic archaic mammals (Fig. 9.14A). India began its collision with Asia, forming the Himalayas, in the early Eocene. This is when several advanced groups of mammals (both even-toed artiodactyls and odd-toed perissodactyls, as well as advanced primates) suddenly appeared in Asia and around the world. If those groups had evolved in isolation in India and then spread during the early Eocene after India docked, it would explain why we do not see them, or their ancestors, anywhere in the world during the early Paleocene. Unfortunately, we do not yet have much of a fossil record for the Cretaceous or Paleocene in India to test this hypothesis.

But the organisms don't even have to be alive for the plates to move them around. They can just as easily be incorporated as fossils in the rocks, and then be found in another area when continents collide. McKenna (1973) called these "**beached Viking funeral ships**," in reference to the way the Vikings used to send their dead warriors to Valhalla on their longboats, filled with their weapons and wealth and then set ablaze. The Cambrian European trilobites in North America and American trilobites in Scotland mentioned previously (Fig. 9.12) are examples of this phenomenon, as are the Tethyan fusulinids now found in British Columbia (Fig. 9.13).

It is even possible that the geographic range of the organisms is older than the land they are living on. McKenna (1983) suggested two mechanisms that could produce this interesting paradox. Mid-ocean ridge islands, such as Iceland, are continually changing as their spreading ridge rifts apart and sinks down into the ocean. The oldest rocks in Iceland are only 13 million years old, yet it is likely that there has been an island in Iceland's position since the North Atlantic first opened in the Jurassic (Fig. 9.14B). If so, then an organism could stay on the island for millions of years while it slowly spread apart and sank away underneath it, analogous to staying in one place by walking up the down escalator (or **escalator counterflow** in McKenna's terminology).

A similar mechanism might be called **escalator hopscotch** (Fig. 9.14C). The Hawaiian Islands, for example, are part of a long chain of seamounts that extend to the submerged Emperor Seamount chain in the western Pacific. Each Hawaiian island erupted and formed as it sat over the mid-Pacific mantle hotspot, then sank away as its plate moved away from the hotspot. The hotspot is currently under the big island of Hawaii (producing the active Kilauea volcano), and the rest of the extinct volcanoes of the other Hawaiian Islands are progressively older as you move northwest along the chain. Thus, an organism could hopscotch from a dead, sinking island that is about to become a submerged seamount to an active volcanic island, and the faunas of such an island chain could be older than the islands themselves. So far, no clear-cut cases of either of these mechanisms have been documented, but they are clearly plausible.

Vicariance Biogeography

> *In summary, I judge that Darwin's "geographic distribution" and the zoogeography and phytogeography that it sired, are worthless as instruments of learning. The text expounding the stuff attracts those unaccustomed to exactitude in thought because of its lavish display of more or less harmoniously related details. It repels those persons who, satisfied that 2 and 5 are true as numbers, do not yet accept that 2 + 2 = 5. In summary, the line is by now sharply and finally drawn: either the Darwinians bury me, or I them.*
>
> León Croizat, "Biogeography: Past, Present, and Future," 1981

When plate tectonics came along in 1960s and 1970s, some predicted that it would revolutionize biogeography. After all, what could be more fundamental than the overturning of the assumption that geographic regions were fixed? Paleontologists eagerly replotted their fossils on the new plate tectonic maps, and novel transport mechanisms (such as those suggested by McKenna) were proposed. But some scientists expected more. As Gould (1980, p. 109) put it, "The building of a new earth as a framework for the facts of geographic distribution has forced us to rework hundreds of particular examples, but I cannot see that it has been a source of new paleobiological ideas. By appropriate twisting, juggling and special pleading, these facts had formerly been grafted upon stable continents; now we read the same facts in a different and more satisfactory light. The new earth has provided a foundation for important synthesis—but it has been a synthesis of groups and areas, not a production of new and general theory."

One group of scientists, however, looked at biogeography in a wholly new light and found that plate tectonics fit their predictions nicely. This school of thought is now known as **vicariance biogeography**, and in the 1970s it was embraced by the same scientists (mostly at the American Museum of Natural History in New York) who had fought the wars over cladistics a few years earlier (see Chapter 4). They argued that vicariance methods changed biogeography from the ad hoc, storytelling, untestable approach of dispersalist theory to something rigorous and testable. The battle between the vicariance and traditional biogeographers raged for years along many of the same battle lines that had been drawn between cladistic and traditional systematists: cladograms of relationships (in this case, of landmasses as well as organisms); formulation of testable hypotheses; rejection of scenarios, storytelling, or anything considered untestable (see Hull, 1988, for a detailed account of the battles over biogeography). Vicariance biogeography reached its peak with a symposium held at the American Museum on May 2-4, 1979 (edited and published by Nelson and Rosen, 1981). Unlike the cladistic revolution in systematics, however, vicariance biogeography never quite caught on, and in recent years interest has declined to occasional papers in a few journals. Nevertheless, it has forced all scientists to be more careful about how they propose and test their biogeographic ideas.

The idea originated with an obscure Venezuelan botanist by the name of León Croizat, who spent decades documenting the geographic ranges of many different species of plants and animals. His approach was so unconventional, his writing was so idiosyncratic, difficult to read, and quarrelsome (the quote to the left is representative) that he was forced to privately publish his ideas at his own expense in three long monographs spanning more than 10,000 pages (Croizat, 1952, 1958, 1964). Thus, he was not widely read in the scientific community. For decades, his influence was so minor that Simpson never mentioned his work in print except in a footnote, a sure sign of obscurity. Eventually, some similar ideas emerged from the work of entomologist Lars Brundin (1966). Then when cladistics came along in the 1970s, Gary Nelson and Donn Rosen at the American Museum saw how vicariance methods could be combined with cladistics to make biogeographic hypotheses that were more rigorous and testable. Suddenly, Croizat was rescued from obscurity to the center of a noisy controversy, although he was unhappy with the way his ideas had been transformed (Croizat, 1982). Before he died in 1982, he was at odds not only with his critics, but even his supporters like Nelson (see Hull, 1988). He even hated the term "vicariance biogeography" for a concept that he had originally christened "panbiogeography."

What is vicariance biogeography, and why is it so controversial? Croizat rejected the central assumption of dispersalist biogeography: that organisms have a center of origin and *expand* from that point to their present range. Instead, Croizat argued that biogeographic ranges are initially large (in other words, he viewed dispersal as instantaneous rather than gradual), and that biogeographic patterns are determined more by the *fragmentation* (whether by plate movements, sea level changes, mountain ranges) of formerly large ranges into smaller areas. In each of these areas, a *vicar* species represents a fragment of a formerly much larger range (hence the name "vicariance"). (A *vicar* in Latin is a representative or replacement for something else, so we get *vicarious* thrills when we experience them for someone else; a *vicar* in the clergy is Christ's representative.) Croizat analyzed these fragmented ranges by drawing lines (**tracks**) between the remnants of a geographic distribution to show their original connectedness. When this was done for a variety of organisms, the tracks began to overlap and many had the same pattern, producing **generalized tracks** that show the biogeographic affinities of major regions. For example, Croizat (1958) was struck by the large number of tracks that connected South America and Africa, yet geologists insisted that there was no way the continents could have been connected. His ideas were developed before plate tectonics was proposed, so they were at least a decade ahead of the geological discoveries that made them plausible.

When Croizat's ideas were embraced by Nelson, Rosen, and others, they adapted it to cladistic methods and hypothesis testing. Instead of generalized tracks, they generated a cladogram of landmasses, based on their time of divergence (Fig. 9.15A). They would then analyze the cladistic relationships of a variety of different organisms found on those landmasses and see what patterns emerged. If the cladograms of landmasses and taxa were congruent, it suggested that the present distribution of those taxa was due to the fragmentation of the region on which they originated, not due to recent dispersal from one place to another. For example, a number of animals have cladistic relationships that suggest that their distribution resulted from fragmentation of Gondwanaland (Fig. 9.15B). The flightless ratite birds—the ostrich in Africa, rhea in South America, emu and cassowary in Australia, kiwi and extinct moa in New Zealand, and extinct elephant bird *Aepyornis* in Madagascar—are all found on Gondwana fragments (but see Chapter 17—the ratite birds were once distributed worldwide, and today they just happen to survive on Gondwana continents). The same is true of marsupial mammals, osteoglossine fishes, galliform birds, hylid frogs, and a variety of insects (Fig. 9.15B). When many different cladograms are congruent, it is no accident—it is statistically significant. Rosen (1978) did a similar analysis of the freshwater fish of Central America and found that their cladograms were congruent. He calculated the probability that this pattern arose by chance is less than one percent (but see Simberloff et al., 1981).

The primary appeal of vicariance biogeography, however, lies in its testability. If the cladograms of areas and taxa are congruent, then they provide a positive test of the hypothesis. If they are not, then the vicariance interpretation is falsified. Such cannot be said for the old school of dispersalist biogeography, which was dominated by storytelling rather than hypothesis testing. If the oldest fossil of a given taxon was found in area A, then the dispersalist said that it migrated from area A to area B; if another, older fossil was found in area B, then the direction of dispersal was arbitrarily reversed. There was no rigorous method of deciding where a taxon originated and when and in what direction dispersal had occurred. Just about any scenario could be made to fit the distributional data, so there was no criterion for deciding whether a given hypothesis had been falsified or not.

Vicariance biogeographers do not deny that dispersal occurs, only that hypotheses based on dispersalist ideas cannot be rigorously tested. Instead, they advocate doing the necessary systematic and cladistic analyses first to see if a vicariant pattern emerges, in which case there is no need to postulate dispersal at all. For too long, biogeographers had been content to invoke easy dispersalist explanations without looking harder at systematic relationships to see if vicariant explanations fit better. Of course, if there is no vicariant pattern, then one can fall back to dispersalism by default, but it should not be the first step in the analysis (in their view).

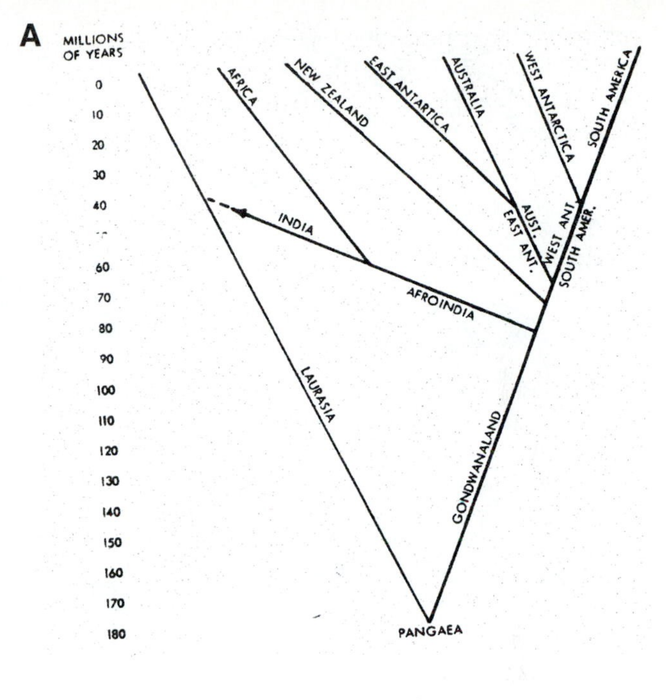

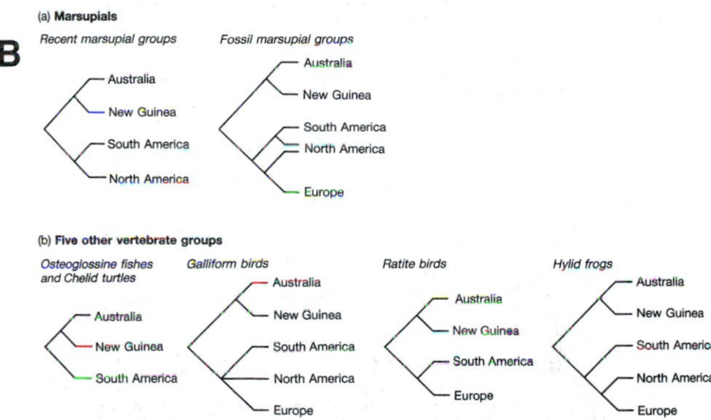

Figure 9.15. (A) Cladistic representation of the history of the breakup of Pangea (greatly simplified), with the approximate time scale of the fragmentation events and one conjunction event (India and Laurasia). (B) Area cladograms of recent and fossil marsupials, and five other vertebrate groups with congruent distributions. (From Patterson, 1981.)

One would think that vicariance explanations would be unnecessary in groups with good dispersal ability. Yet even in these cases, there are surprises. For example, young spiders can disperse hundreds of miles by ballooning with threads from their spinnerets. Yet Platnick (1976) analyzed the Laroniine spiders, and found their distribution could be explained entirely by vicariance. The first branch point on the cladogram separated Gondwana and Laurasian taxa, and then among the Laurasian taxa, there was a clear split between Asian and North American clades, corresponding to the breakup of Laurasia.

During the 1979 vicariance symposium, the dispute over

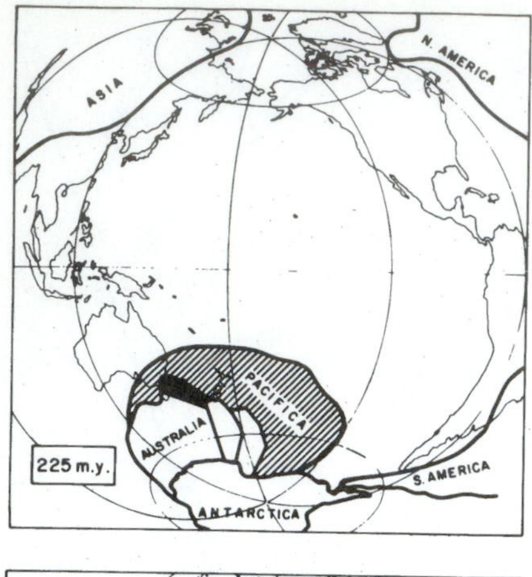

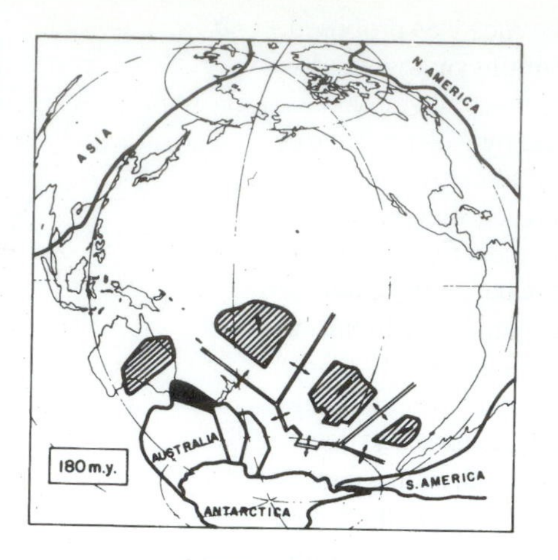

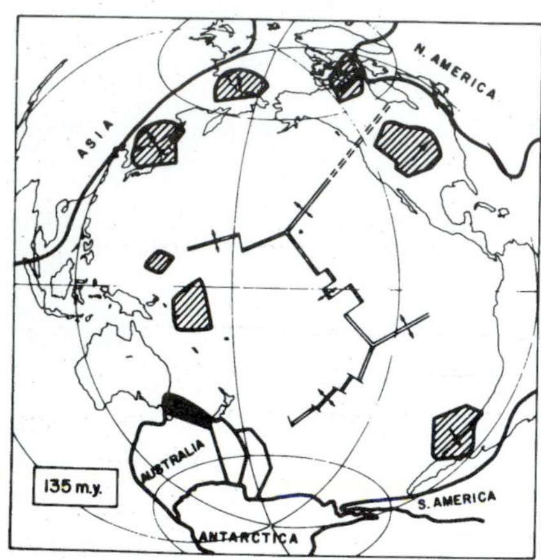

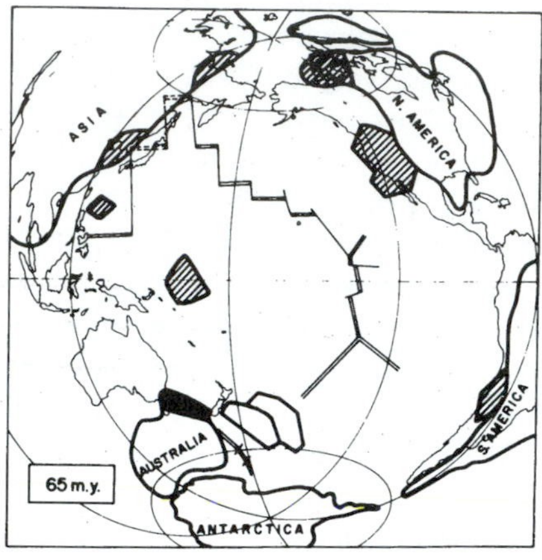

Figure 9.16. Nur and Ben-Avraham's (1981) speculative model of the "lost Pacifica continent." According to their interpretation, the Pacifica continent was attached to the Australian portion of Gondwanaland in the Triassic, and broke up during the Mesozoic. Its fragments rafted to various parts of the Pacific rim as spreading ridges developed between them, and each part eventually became accreted to different continents as exotic terranes in the late Cretaceous and Cenozoic. Some vicariance biogeographers see this as a mechanism for the circum-Pacific distribution of many plants and animals. Unfortunately, most of the fragments of the "lost Pacifica continent" would have sunk below sea level during their long journey, so terrestrial organisms would not have survived the trip. (From Nur and Ben-Avraham, 1981.)

the relative importance of vicariance versus dispersal grew very loud and angry (Ferris, 1980). Each speaker was criticized by several discussants, and then various members of the audience took the microphone and further raised the decibel level of the debate. One scientist in the audience proved so persistent in commenting on everything that the moderator stopped acknowledging him. Then he began hiding in the shadows of the back of the auditorium where he was hard to recognize until he was acknowledged, at which point he would jump to the floor microphone and resume his tirades. Again and again, the dispersalists pointed to the cattle egret as an example of how easily dispersal might overwhelm vicariant patterns. This bird is a classic example of dispersal, since it has migrated from Africa to South America to North America in the last century in response to the spread of cattle herds. The cattle egret was mentioned so often that Ferris (1980, p. 74) suggested it should be registered for the symposium! In the end, both sides came away unconvinced, but it was clear that biogeography had changed.

So how does vicariance biogeography stand today? Clearly, for groups that have long histories (at least back to the Mesozoic, when Pangaea broke up), it is appropriate to look for vicariant distributions to see if dispersalism is unnecessary to explain them. However, it is just as clear that the distribution of organisms with very recent histories (especially late Cenozoic) cannot be explained by the slow pace of continental fragmentation and collision, and require dispersalist explanations (however untestable they are). No one argues that the Great American Interchange in the Pliocene, or the migrations of Ice Age mammals, are anything but dispersal. Similarly, most of the migrations of mammals during the Cenozoic are simply too rapid to be anything but dispersal (Woodburne and Swisher, 1995).

Still, the vicariance advocates refuse to give up easily, and keep finding surprising new evidence in their favor. For example, biogeographers had long argued that the circum-Pacific distribution of many plants and animals required dispersal around the Pacific Rim. Unlike the Atlantic, the Pacific was not formed by continental fragmentation. However, Nur and Ben-Avraham (1981) have postulated the existence of a "Lost Pacifica continent" that once lay in the southwest Pacific (Fig. 9.16). In their view, its fragments have since migrated across the Pacific and attached to various continents as exotic terranes (Fig. 9.13). If this is true, then many circum-Pacific organisms may have gotten to their present location by floating on "Noah's Arks" rather than by dispersal. However, this suggestion is still very controversial (Schweickert, 1981; McKenna, 1981). It is not clear which if any of the exotic terranes on the Pacific Rim were once part of a "Lost Pacifica continent." More importantly, if these fragments had drifted away from a former supercontinent, they would have sunk down the subsidence curve of the spreading seafloor and eventually become submerged, making it very hard for land animals to survive above water until they docked on some other continent (McKenna, 1981). Animals can't hold their breath for millions of years!

It is hard to tell how important vicariance biogeography will be in the future. However, it generated a storm of controversy that had a positive effect in challenging many of the stale assumptions of traditional biogeography and forcing scientists to be more rigorous (Ball, 1976). No matter what the future of biogeography, that legacy was extremely important in the history of this science.

CONCLUSIONS

The geographic distribution of organisms presents many fascinating puzzles to the biologist and paleontologist. Some aspects of distribution can be explained by the modern-day properties of the Earth, as studied by ecological biogeographers. Others are relicts of the history of life and plate tectonics and are largely the domain of historical biogeographers and paleontologists. The advent of plate tectonics had a major effect on biogeography, especially in challenging the assumptions of continental stability and in the importance of dispersal over vicariance. It is safe to assume that biogeography will continue to present exciting new puzzles for future generations of paleontologists and biologists to solve, and that it will continue to be an exciting and dynamic field of research.

For Further Reading

Brown, J. H., and A. C. Gibson. 1983. *Biogeography*. Mosby, St. Louis, Mo.

Browne, J. 1983. *The Secular Ark: Studies in the History of Biogeography*. Yale University Press, New Haven, Connecticut.

Cox, C. B., and P. D. Moore. 1985. *Biogeography: An Ecological and Evolutionary Approach* (4th ed.). Blackwell Scientific Publications, Cambridge, Mass.

Croizat, L. 1964. *Space, Time, and Form, the Biological Synthesis*. Published by the author, Caracas, Venezuela.

Darlington, P. D. 1957. *Zoogeography*. Wiley, New York.

Gray, J. and A. J. Boucot, eds. 1979. *Historical Biogeography: Plate Tectonics and the Changing Environment*. Oregon State University Press, Corvallis, Oregon.

Hallam, A. 1973. *Atlas of Paleobiogeography*. Elsevier, Amsterdam.

Hallam, A. 1986. Evidence of displaced terranes from Permian to Jurassic faunas around the Pacific margins. *Journal of the Geological Society of London* 143:209-216.

Hughes, N. F., ed. 1973. *Organisms and Continents through Time*. Palaeontological Association Special Paper 12.

Humphries, C. J. and L. R. Parenti. 1986. *Cladistic Biogeography*. Oxford University Press, Oxford.

Jablonski, D., K. W. Flessa, and J. W. Valentine. 1986. Biogeography and paleobiology. *Paleobiology* 11:75-90.

MacArthur, R. H. 1972. *Geographical Ecology*. Harper & Row, New York.

MacArthur, R. H., and E. O. Wilson. 1967. *The Theory of Island Biogeography*. Princeton University Press, Princeton, N.J.

Middlemiss, F.A., P. F. Rawson, and G. Newell, eds. 1969. *Faunal Provinces in Space and Time*. Seel House Press, Liverpool.

Myers, A. A., and P. S. Giller, eds. 1988. *Analytical Biogeography*. Chapman and Hall, New York.

Nelson, G., and N. Platnick. 1981. *Systematics and Geography, Cladistics and Vicariance*. Columbia University Press, New York.

Nelson, G., and D.E. Rosen, eds. 1981. *Vicariance Biogeography: A Critique*. Columbia University Press, New York.

Pielou, E. C. 1979. *Biogeography*. Wiley-Interscience, New York.

Simpson, G. G. 1962. *Evolution and Geography*. Oregon State System of Higher Education, Eugene, Oregon.

Stehli, F. G., R. G. Douglas, and N. D. Newell. 1969. Generation and maintenance of gradients in taxonomic diversity. *Science* 164:947-949.

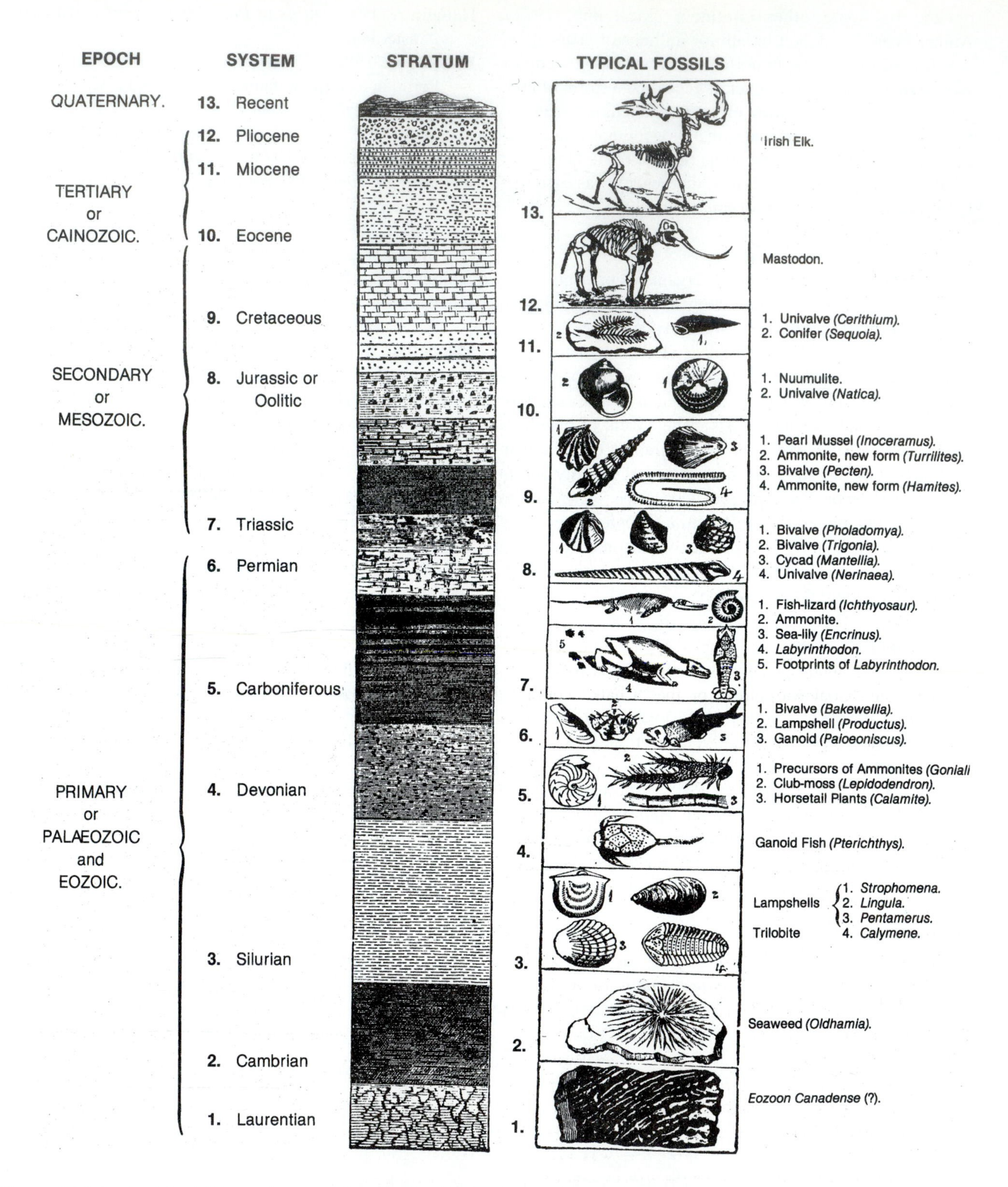

Figure 10.1. In the late 1700s and early 1800s, Smith, Cuvier, Brongniart, and others established that the fossil record shows a succession of forms, which could be used as a method for correlating strata and determining their relative age. By the time of this diagram in 1888, the sequence of strata and their characteristic fossils were well known. Notice that at this date, the recently proposed (1878) Ordovician System had not yet been accepted, nor the Paleocene (proposed 1874) or Oligocene (proposed 1854) epochs of the Cenozoic. Instead of "Precambrian" or "Primary", this time scale uses the term "Laurentian," since the studies of Precambrian rocks had made the most progress in the Laurentian region of the Canadian Shield.

Chapter 10

Biostratigraphy

Geologists, in their classifications permit themselves to be influenced by mineralogic composition of the beds, wherease I take for my point of study the annihilation of an assemblage of organisms and replacement by another. I proceed solely on the identity of faunal composition

Alcide d'Orbigny, *Terrains Jurassiques*, 1842

Every year tends to fill up the blanks between the Tertiary stages, and to make proportion between the lost and existing forms more gradual. In some of the more recent beds only one or two species are extinct, and only one or two are new. Yet if we compare any but the most closely related formations, all the species will be found to have undergone some change.

Charles Darwin, *On the Origin of Species*, 1859

FAUNAL SUCCESSION

In the late 1700s, the founders of geology were just beginning to decipher the pattern of strata in the rock record. Certain widespread, easily recognized rock units, such as the "Coal Measures" or the "Chalk" could be traced over much of northern Europe and the British Isles, and the superpositional sequence of the rock units in each area could be described and worked out (Fig. 10.1). Fossils were collected and described from these strata when they were available, but the description of rock units was the primary focus. Most natural historians were still operating under the assumption that these strata were all deposits from Noah's flood, and so the fossils were organisms that had drowned in the flood and were subsequently buried as the flood waters receded.

Closer observation, however, showed that the fossils in each stratum were not the same, as if they had all died in the same event. Instead, each layer contained its own distinct assemblage of fossils (Fig. 10.1). In other words, there was a sequence of faunas in the fossil record, or **faunal succession**. Although suggestions of this idea can be found in the work of Robert Hooke, the man who first developed this discovery was the British surveyor William Smith (1769-1839). In the late 1700s, he was given the assignment of surveying some of the great canal excavations across England, since the Industrial Revolution required a cheap form of transportation for all the coal and other commodities that had to be carried to the great industrial cities. Smith had the unusual opportunity to see the fresh bedrock exposures through much of England, which is normally covered by vegetation and difficult to map. As he saw the stratigraphic sequence of Britain repeated over and over again, he got to know not only the formations but also the characteristic fossils found within each unit. More importantly, he realized that similar-looking rock units could best be distinguished by their fossils. Eventually, he realized he could tell where he was in the sequence by the fossils alone, without even seeing the rocks from which they came. The gentlemen geologists who consulted him were amazed with his ability to accurately guess where the fossils in their collections had come from, or to arrange their collections in stratigraphic order.

By 1799, Smith's lists of fossils characteristic of each formation were widely circulated among British geologists, but Smith was too busy working on the first geologic map of England to publish his discovery until 1813. Meanwhile, other geologists were taking his discovery and making their own reputations from it, since Smith had not published and established his priority. In addition, Smith suffered from the prejudice of the wealthy, class-conscious gentlemen geologists of the time, who viewed him as a lowly working man (as engineers and surveyors were then considered). Most of them considered geology a hobby, not a "vulgar" way of making a living. Smith was one of the few who might be considered a "professional" geologist in that he worked on geological problems for an income. Eventually, however, Smith published the first geologic map of England (1815), which is so good at its scale that it can still be used today. In 1831, toward the end of his life, Smith was finally given credit for all his ground-breaking work and acknowledged by the Geological Society of London as the "father of English geology."

Smith was not the only person who noticed the sequence of faunas. In 1779, the Abbé Jean Louis Giraud-Soulavie (1752-1813) described the sequence of fossils in the Vivarais region of France and noticed there was a pattern of succession. His ideas were amplified by two other French naturalists, the Baron Georges Cuvier (whom we have discussed in several previous chapters) and the mollusc specialist Alexandre Brongniart (1770-1847). Cuvier and Brongniart worked on the Cenozoic strata of the Paris Basin, and found there was distinct succession not only of formations but also of fossils; they published their work between 1808 and 1812. Some have argued that this is a remarkable example of independent scientific discovery. When the time is ripe and an idea is in the air, several people make the same discovery simultaneously. However, Eyles (1985) has argued that Brongniart may have heard of Smith's ideas in London in 1802, when he traveled abroad during one of the brief interludes in the wars between England and France. Whether or not these discoveries were truly independent, Smith deserves credit for developing the practical applications of faunal succession. Cuvier, on the other hand, was more concerned with reconciling the changes in the fossil record with Genesis, and so postulated a series of "revolutions" in earth history that repeatedly wiped out life, followed by another creation.

The implications of faunal succession were to have a profound effect on geology. In the late 1700s, natural historians thought that rock sequences of each continent were the same, since they had been deposited by the receding Noah's floodwaters. As geologists got to know the rock record of each continent better, it was clear that the local rock sequences varied tremendously and that rock units could not be correlated across great distances. Smith and Cuvier and Brongniart had shown, however, that the fossils were distinctive, no matter what area they came from, and could be used as the basis for correlation.

A classic example of this principle comes from the establishment of the Cambrian and Silurian systems. The rocks below the Devonian Old Red Sandstone had long been given the temporary label "Transition" since they were thought to be transitional between the "Primary" rocks (the igneous and metamorphic basement rocks formed during the Creation) and the overlying "Secondary" strata deposited by Noah's flood. In 1831, Adam Sedgwick (1785-1873), the first professor of geology at Cambridge University (and the first to bear that title anywhere), and Roderick Impey Murchison (1792-1871), a wealthy gentleman geologist who gave up his fox hunting for geology, set out to decipher the "Transitional" strata in western England and Wales. Murchison started at the top of the sequence and worked down, while Sedgwick started at the bottom and worked up. Unfortunately, they used different criteria for their mapping and correlation. Sedgwick, working in the complexly deformed and sparsely fossiliferous base of the sequence, defined his "Cambrian System" on local rock units, so it could not be recognized outside of western Wales. Murchison, on the other hand, defined his "Silurian System" on both rocks and fossils, so it was useful outside Wales and western England. Eventually, they met in the middle, and Sedgwick found that some of his Cambrian strata had been assigned to Murchison's Silurian. Since Murchison's Silurian could be recognized elsewhere, but Sedgwick's Cambrian could not, the former became widely accepted, and soon began to swallow up the latter. This led to a bitter feud between Sedgwick and Murchison, which soon divided British geology into warring camps (see the excellent account in Secord, 1986). The controversy was not resolved until after both men died in the early 1870s. A later generation of geologists, led by Charles Lapworth (1842-1920), proposed an intermediate unit, the Ordovician, in 1879 as a compromise between the Cambrian and Silurian advocates. The main reason for the success of the Ordovician was that it was also defined on its fossil content, making it recognizable outside its type area in Scotland (which has a better record of the Ordovician than Wales). Eventually, the Cambrian too was redefined by its characteristic fossils, resulting in our modern concept of the first three periods of the Paleozoic.

BIOSTRATIGRAPHIC ZONATIONS

Comparison has often been made between whole groups of beds, but it has not been shown that each horizon, identifiable in any place by a number of peculiar and constant species, is to be recognized with the same degree of certainty in distant regions. This task is admittedly a hard one, but it is only by carrying it out that an accurate correlation of a whole system can be assured. It necessarily involves exploring the vertical range of each separate species in the most diverse localities, while ignoring the lithologic development of the beds; by this means will be brought into prominence those zones which, through the constant and exclusive occurrence of certain species, mark themselves off from their neighbors as distinct horizons.

Albert Oppel, *Die Juraformation*, 1856

Each objectively definable extinct fossil taxon divides geologic time into three segments—the time before it appeared, the time during which it existed, and the time since its disappearance.

Alan Shaw, *Time in Stratigraphy*, 1964

By the mid-1800s, the principle of faunal succession was well established, and geologists were using it to map and correlate strata over long distances. In 1842, the French geologist Alcide d'Orbigny published his comprehensive study of the Jurassic fossils and strata of southern France, and showed that no matter the rock unit, the fossil **assemblage** was the key to correlation. D'Orbigny used these assemblages to divide his strata into a set of **stages** (*étages*), which became biostratigraphic units independent of the lithology. D'Orbigny's original 10 stages were each thought to represent a separate creation and flood-caused extinction, followed by another creation, along the lines of Cuvier's attempts to reconcile the fossil record with Genesis. As the original 10 stages were further subdivided into as many as 27 separate creations, the multiplicity of stages had increased to the point that they were unacceptable both to uniformitarian geologists, and also to theologians. As Berry (1987, p. 125) put it, "The naming of new stages became more a sport than a scientific endeavor."

In Germany, Friedrich Quenstedt set up his own system for the Jurassic. He had measured sections and recorded the position of each fossil in minute detail, and found that d'Orbigny's stages were too broadly defined to be useful. Quenstedt's methods were developed and popularized by his student Albert Oppel (1831-1865), who traveled all over France, Switzerland, and England collecting detailed stratigraphic data on the fossils of the Jurassic. After Oppel had plotted hundreds of stratigraphic ranges, he realized that their pattern was repeated all over western Europe. These patterns could be broken up into discrete aggregates, bounded at the bottom by the appearance of certain fossils and at the top by the appearance of others. Between 1856 and 1858, Oppel described the diagnostic aggregates, or "congregations," of fossils in this scheme of **overlapping range zones**. Oppel's zonation scheme was the first that produced a clear-cut, non-arbitrary way of dividing strata into correlatable time units.

Oppel pioneered the methods that are adopted by modern biostratigraphers. The first step is the collection of a data base. Every time a geologist or paleontologist picks up a fossil in the field, he or she must record the exact stratigraphic position in a measured section at the time of collection. If the collector walks away without recording this information, it may be impossible to reconstruct it later, and the specimen is useless for biostratigraphy. For a thorough biostratigraphic study, a large collection of fossils is required, each with detailed stratigraphic data as to exactly where every specimen was collected in a local section.

Once the fossils have been collected, they have to be cleaned up and identified. This is not as trivial a task as it sounds. It is often very difficult to correctly identify a fossil to the species level. Sometimes the distinctions between one fossil species and another are so subtle that only a specialist can tell them apart. In other cases, the group may not have been studied in years, so there is no clear understanding of which species are valid or during which geologic time interval they occurred. In many cases, identification is hampered by poor preservation. Incorrect identification can lead to incorrect correlations that may be difficult to detect by non-specialists for years.

Next, the ranges are compiled into a zonation. All the strata that actually contain a fossil species are said to be in its **range zone**. In a local section, the observed range of a fossil is its *partial range zone*, or **teilzone**. Teilzones are the empirical data base from which all biostratigraphy is derived. No teilzone represents the total range of a species in time and space. To obtain the total range of the fossil species in time and space (its *total range zone*, or **biozone**), one needs to compile many teilzones until their patterns begin to converge. We shall explore methods of doing this later in the chapter.

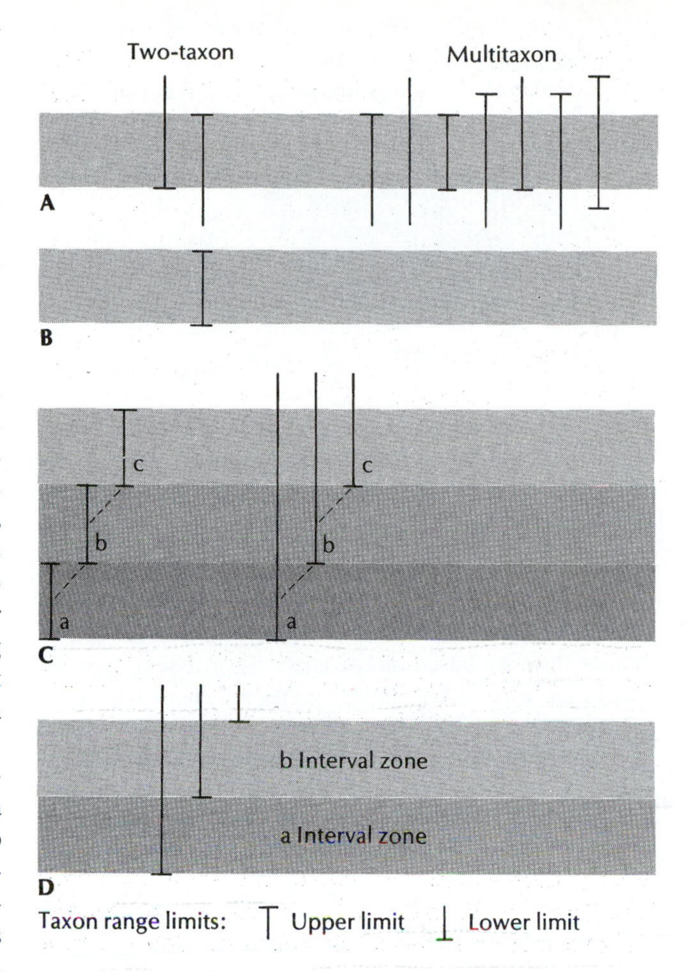

Figure 10.2. Classes of interval zones. (A) Concurrent range zones, defined by the first and last appearance of two or more taxa with overlapping ranges. (B) Taxon range zone, defined by the first and last appearance of a single taxon. (C) Lineage zones, or phylozones, defined by the evolutionary first appearance of successive taxa in a lineage. (D) Interval zones, defined by the successive first or last occurrences of partially overlapping ranges. (Modified from Hedberg, 1976.)

A number of different schemes for utilizing biostratigraphic data have been proposed over the years. In recent stratigraphic codes (Hedberg, 1976; North American Code of Stratigraphic Nomenclature, 1983; Salvador, 1994), there is general agreement about the terminology of biostratigraphic zones. The first class of zones is known as **interval zones**, since they are based on two first or last occurrences (Fig. 10.2). The most important type of interval zone is the **concurrent range zone** (Fig. 10.2A) defined by the overlap of two or more taxa. A **taxon range zone** (Fig. 10.2B) is based on the first or last occurrence of a single taxon. A **lineage zone** (Fig. 10.2C) is based on the successive evolutionary first occurrences within a single evolving lineage. An **interval zone** (Fig. 10.2D) (not to be confused with interval zone as a class of zones) is defined by two successive first or last occurrences of unrelated taxa.

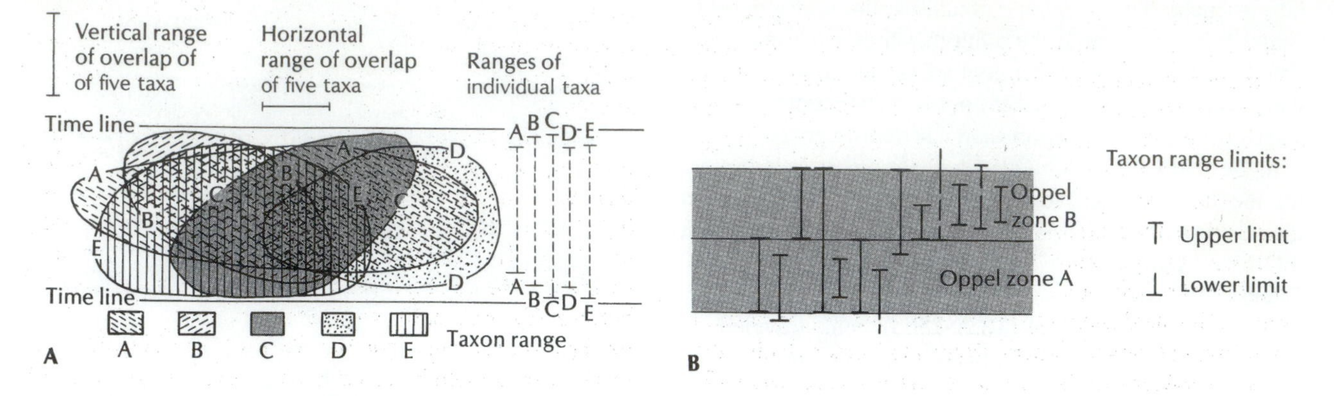

Figure 10.3. Classes of assemblage zones. (A) Typical assemblage zones are defined by a suite of taxa, so the upper and lower boundaries can be vague, depending on how many taxa are used to define the zone. The actual zone of overlap of all five taxa in this example is considerably less than the total assemblage zone. (B) Oppel zones, defined by the overlap of several taxa. (Modified from Hedberg, 1976.)

A second class of zones is known as **assemblage zones,** because they are based on associations or assemblages of three or more taxa. Assemblage zones are characterized by numerous first and last occurrences (Fig. 10.3A), so the exact top and bottom of the zone can be a bit vague since there are many different criteria for its recognition. Murphy (1977) recommended that these zones be *defined* by the first or last appearance of a single taxon, but *characterized* by additional taxa that aid in its recognition when the defining taxon is absent. An assemblage zone that is more precisely defined in this fashion can be called an **Oppel zone** (Fig. 10.3B), since it seems most similar to Oppel's original usage (although he used several different types of zones; see Hancock, 1977, and Berry, 1987).

A third class of zones is the **abundance zone**, also known as the "peak" or "**acme**" **zone**. Abundance zones are based on the sudden increase in abundance of certain fossils. Geologists have long been impressed by large concentrations of certain fossils, and used them to correlate from section to section. However, it is more likely that an increase in abundance may be due to local ecological factors, such as ideal environmental conditions, or the concentration of shells above an erosional surface. These may have little time significance. Only if abundance peaks are due to time-significant changes, such as global climatic fluctuations, will abundance zones have any time value. For example, in deep-sea cores there are chalks composed of the opportunistic coccolith *Braarudosphaera*, which bloomed in tremendous numbers in times of climatic stress. Because these chalks were triggered by global oceanographic changes, they can be correlated over long distances in the deep sea.

Certain changes in fossil lineages (other than their first or last occurrences) can also have biostratigraphic value. For example, the coiling direction in the planktonic foraminifer *Globotruncana truncatulinoides* shifts from right-coiling in warm waters to left-coiling in cooler waters (Fig. 10.4). The rapid temperature shifts in the world ocean during the Pleistocene show up in the changing coiling ratios of this foraminifer. Because these temperature changes are controlled by global climate, the coiling ratios make good time markers that can be correlated from core to core.

Beyond the theoretical aspects of biostratigraphy, there are many practical problems. All formal biostratigraphic zonations should conform to the stratigraphic procedures outlined by the the various codes of stratigraphic nomenclature. A zone must be formally named, usually by the species names of the two taxa that define its upper and lower limits or the name of one particularly abundant species that first appears in it. There must also be a description of the taxonomic content of the zone (including both

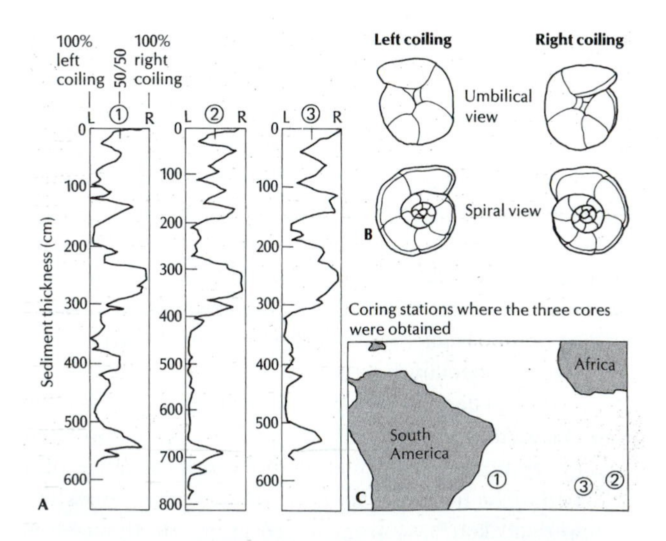

Figure 10.4. Correlation based on climatically controlled changes in coiling direction in the foraminifer *Globotruncana truncatulinoides* in the South Atlantic. This organism is predominantly right-coiling during interglacial periods, but switches to left-coiling during glacial periods. (From Eicher, 1976.)

the defining and the characterizing taxa). The describer must designate a type section and give the known geographic extent of the zonation.

FACTORS CONTROLLING FOSSIL DISTRIBUTIONS

We have no reliable chronological scale in geology but such as is afforded by the relative magnitude of zoological change—in other words, that the geological duration and importance of any system is in strict proportion to the comparative magnitude and distinctness of its collective fauna.

Charles Lapworth, *On the tripartite classification of the Lower Palaeozoic rocks*, 1879

The discovery of faunal succession had provided a method of correlation that worked better than lithostratigraphy, but it was not perfect. For example, Oppel's zonation scheme was based on Jurassic ammonites that were not worldwide in distribution. In Europe, there were two distinct faunal provinces, one in northwest Europe (Oppel's area) and another around the Mediterranean (the Tethys seaway). When Oppel's zonation was applied to the Jurassic rocks of the Tethys, it failed, because so few ammonites were found in both provinces. Clearly, zonation schemes must be worked out first in a local region, and then correlated with each other to see how they correspond.

The failure of Oppel's scheme pinpoints the major variables that control the stratigraphic distribution of fossils. Two of the most important controlling factors are *evolution* and *paleoecology.* **Evolution** is the enabling factor, providing the change in species that is the basis for biostratigraphy. Unlike almost any other means of correlation, biostratigraphy is based on this unique, sequential, non-repeating appearance of fossils, a vector of change through time. The occurrence of a single diagnostic fossil is often sufficient to pinpoint the age of a unit. By comparison, most other means of stratigraphic correlation are based on repetitive, cyclical changes: the alternation of sandstones, shales, and limestones used in lithostratigraphic correlation; the flip-flops of normal and reversed polarity used in magnetic stratigraphy; the variations in acoustic impedance used in seismic stratigraphy; the fluctuations in stable isotopic composition used in carbon or oxygen isotope stratigraphy. Besides biostratigraphy, the only other directional, non-repeating process used for dating is radioactive decay. Except for carbon-14, however, radioactive dating applies only to igneous or metamorphic rocks, so it rarely can be used in fossiliferous sedimentary sequences. For this reason, biostratigraphy will always be one of the most important disciplines in geology, because *fossils are the only practical means of dating most sedimentary rocks.* No matter how the job market changes and which fads come and go, there will always be a need for biostratigraphers as long as geologists care about the age of their strata.

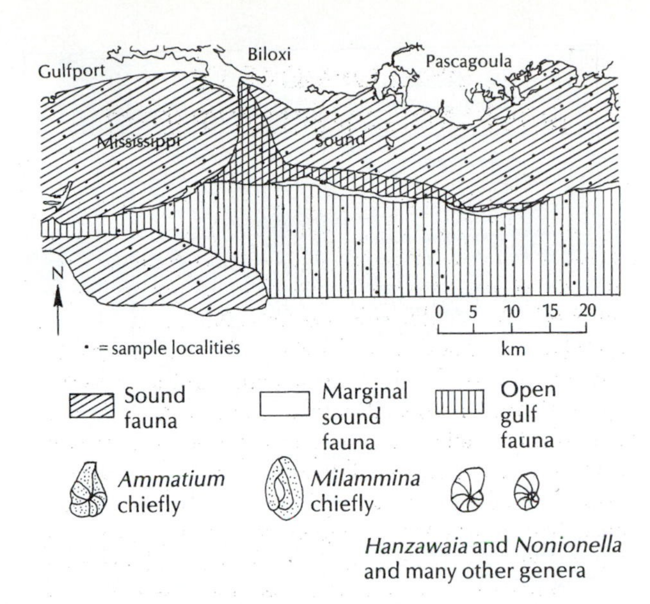

Figure 10.5. Present distribution of benthic foraminiferans in the northern Gulf of Mexico, showing their strong ecological separation. This type of facies-controlled distribution is likely to create problems for the biostratigrapher. (From Eicher, 1976.)

The directional change in the sequence of fossils through time is limited by the other important variable, **paleoecology**. No organism inhabits every environment on earth at a given time, but instead lives in a specific habitat that may have a very limited geographic distribution. Consequently, its fossils are not found everywhere on earth during the time they lived, but only in sediments that represent its preferred habitat. Some fossils are notorious for being restricted to a very narrow habitat, and their appearance and disappearance from a stratigraphic column is controlled by the changes in sedimentary facies; these are often called **facies fossils**. For example, a distinct ecological community dominated by the inarticulate brachiopod *Lingula* has persisted almost unchanged since the Cambrian. The *Lingula* community appears to track the changes in its preferred environment, and avoids much evolutionary change. When *Lingula* community fossils appear in a local section, it signifies only that the conditions were right for these organisms, not that they suddenly evolved at that point in time. Their appearance in another area might be dictated by the time that those ideal conditions appeared in that region. Accordingly, any attempt to correlate the first appearance of the *Lingula* community in the two sections would have no time significance.

Many biostratigraphic zonation schemes have come to grief when paleontologists later found that the key taxa were largely facies controlled (Fig. 10.5). For example, some robust, heavy-walled agglutinated foraminifera had long been used to provide a zonation of the Oligocene and Miocene sediments of the Gulf Coast and Caribbean. They do not occur in the Gulf of Mexico today, so biostratigra-

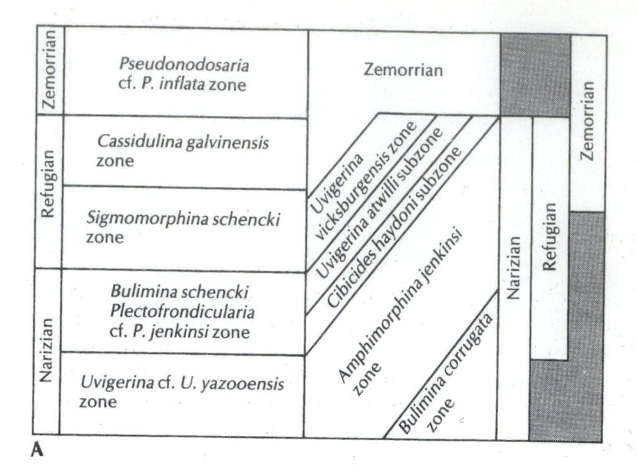

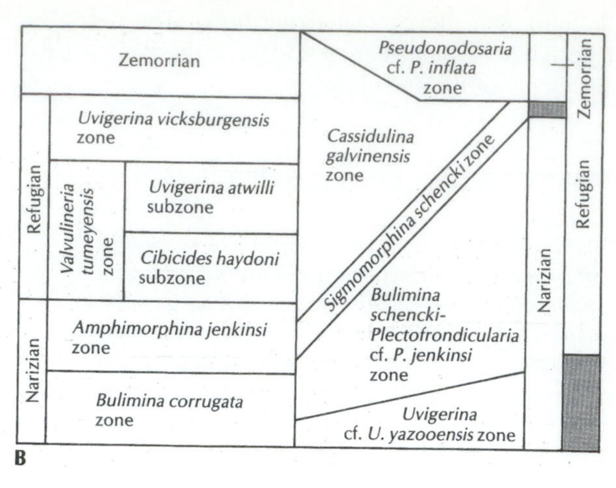

Figure 10.6. Correlation of the middle-late Eocene benthic foraminiferal zonations between California and Washington, showing the problems caused by the strong facies control on these organisms. (A) Correlation scheme assuming that the Washington zonation is valid. Under this arrangement, the California zonation is time-transgressive, and the stages that are based on California forms overlap. (B) Correlation assuming that the California zonation is valid. In this case, the Washington zonation is time-transgressive, and the stages based on Washington foraminiferans overlap. The discrepancy between these zonations occurred because most of these taxa were strongly facies-controlled; they appear and disappear from an area simply because they migrated north-south with the facies changes and varying water depth. For example, the *Sigmomorphina schencki* zone appears in the late Narizian of California, but migrates north to Washington by the early Refugian. (From McDougall, 1980.)

phers assumed they were good zonal indicators that are now extinct. However, these species were also known elsewhere in sediments from bathyal and abyssal water depths, and when offshore drilling began to recover deep-water sediments in the Gulf, their ranges were extended upward into the Pleistocene. Clearly, the occurrence of these species was controlled by conditions in the Oligocene and Miocene that yielded a high percentage of deeper, cold-water foraminifera. These did not go extinct at the end of their appearance in the shallow marine sediments of the Gulf, but simply returned to deeper waters.

A similar problem has been documented in the Pacific Coast. Since the 1930s, micropaleontologists have used benthic foraminifera to zone Pacific Cenozoic marine strata. In California, Kleinpell (1938), Kleinpell and Weaver (1936), and Mallory (1959) created a benthic foraminiferal zonation for the marine Cenozoic, and similiarly named stages were set up in Washington by Rau (1955, 1958). For its time, this zonation scheme was decades ahead of the rest in its precision and careful definition, and it has proved useful for correlating strata and finding oil all over California. However, when McDougall (1980) examined the foraminifera and their depth preferences in detail, she found that the fossils that marked one "time" in Washington marked a completely different "time" in California. For example, the *Sigmomorphina schencki* zone, which was used to mark the early Refugian in Washington (Fig. 10.6A), was based on deep-water taxa that occurred throughout the Narizian and Refugian in California (Fig. 10.6B). Similarly, the *Cibicides haydoni* subzone, which marked the early Refugian in California, was based on shallow-water forms that spanned the late Narizian and entire Refugian in Washington. When these strata were analyzed by magnetostratigraphic methods (Prothero and Armentrout, 1985; Prothero and Thompson, 1998; Prothero et al., 1998), it became clear that these stages spanned different parts of the middle and late Eocene and early Oligocene. The Narizian, Refugian, and Zemorrian benthic foraminiferal stages are not sequential time units, as had long been thought, but partially overlapped or duplicated each other between California and Washington, depending upon which criteria were used to define them.

We have seen how evolution provides the basis for biostratigraphic change, and how paleoecology restricts whether or not the key fossils will occur in a local section. The presence or absence of a fossil from a local section can be due to many reasons. The **first occurrence** of a fossil could be due to its *evolution* for the first time on this planet, but in many local sections, it first occurs sometime later than it evolves because it takes time to *immigrate* from its point of origin, or because the suitable habitat was available only at a later time. Likewise, the **last occurrence** of a fossil can be due to its final *extinction* on this planet, but its last occurrence can also be due to *emigration* when the environment is no longer suitable. Biostratigraphers document the **first appearance datum (FAD)** and the **last appearance datum (LAD)** of each fossil in the zonation.

Other factors can enter into the problem as well (Fig. 10.7). As we just saw, fossils will not be found in sedimentary environments that were unsuitable for that

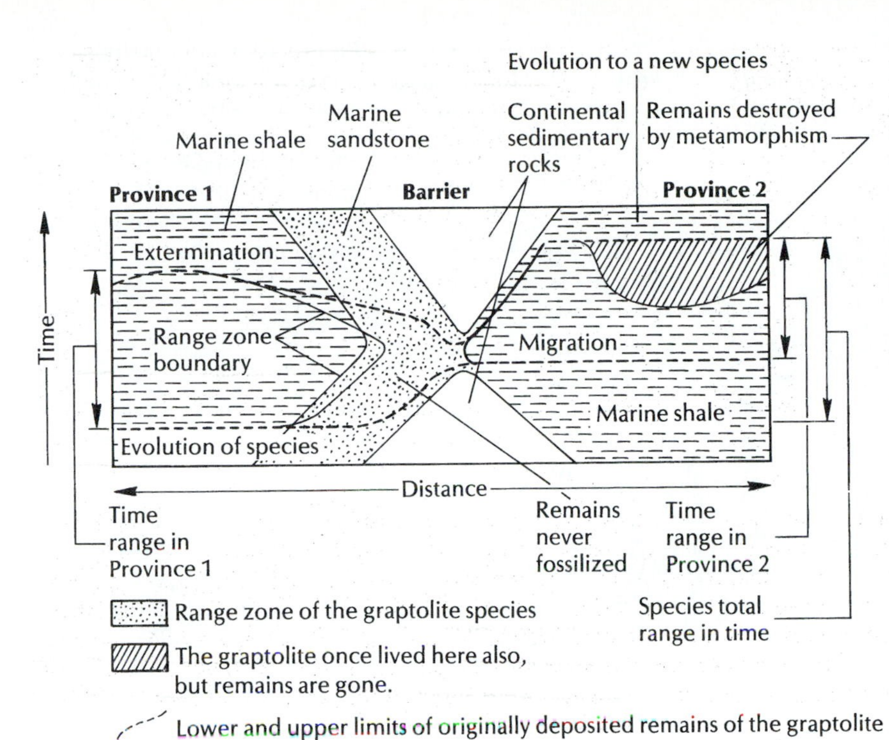

Figure 10.7. Time-space relationships of the range zone of a fossil. The species evolves first in province 1, but a barrier to province 2 prevents its migration there until long after its evolutionary first appearance. It dies off first in province 1 but survives in province 2, where it eventually disappears by evolving into another species. In addition to these factors, which prevent synchronous first and last appearances in different provinces, the range zone can be shortened by non-preservation or by erosion or metamorphism. The total species range in time is longer than its local range in any one section. (From Eicher, 1976.)

species to live in (ecological restrictions). Likewise, fossils will not be found in environments that don't preserve them, even if the organism actually lived there (preservational biases). In other cases, the organism may have lived in a particular spot and the fossils were preserved, but were later destroyed by erosion, or heated and deformed by a metamorphic event so that the fossils are unrecognizable. The final, observed range of the fossil is thus controlled by many factors that have nothing to do with its biology. Much of biostratigraphic analysis involves trying to rescue the original "signal" from the "noise" of sampling, facies restrictions, non-preservation, and destruction. The techniques described next were designed to filter biostratigraphic data and recover as much of the signal as possible.

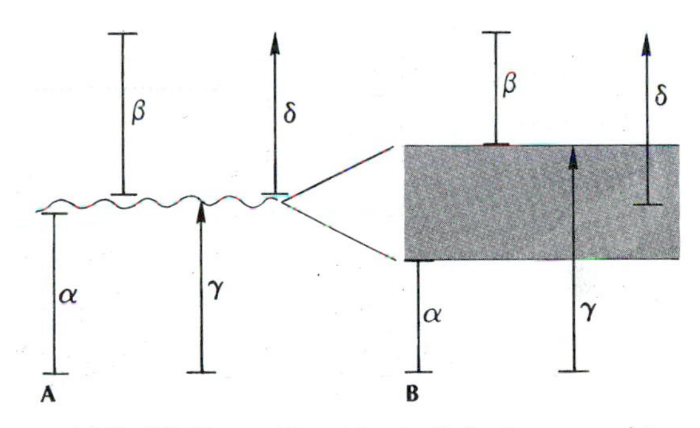

Figure 10.8. (A) Zone dfined by both bottoms and tops of ranges (α and β) and "topless" zones (γ and δ). (B) Restoration of the section removed by an unconformity in A results in a gap between α and β, but no gap results from topless zones. (From Prothero, 1990.)

BIOSTRATIGRAPHIC SAMPLING

> *Preoccupation with the unattainable is a stultifying approach to any problem. Practical paleontology cannot be concerned with any of the fossils we* cannot *find. Geologically, we can only be interested in finding the total stratigraphic range through which a species* is preserved. *While the life and death of millions of unrepresented individuals is of theoretical interest, we cannot gain practically useful information from them.*
>
> Alan Shaw, *Time in Stratigraphy*, 1964

Tops and Bottoms

As stratigraphers have become more careful in documenting ranges and describing type sections, inevitable disputes over boundaries have arisen. The type section or areas of two successive biostratigraphic units are usually in two separated areas, and there may be no overlap. Often, neither section preserves the boundary between the two units, so biostratigraphers must search for a third area where the transition is recorded. Ideally this section should be as continuous and as fossiliferous as possible, with several taxonomic groups to compare.

After the detailed local biostratigraphy of the available groups has been worked out, the stratigrapher must decide which biostratigraphic event(s) should serve as boundaries between units. At this point, philosophical disputes become acute. Some workers argue that boundaries should be drawn at mass extinctions and other faunal breaks because the high turnover of taxa would make the boundary easy to recognize. Others argue that boundaries should never be based on faunal breaks because the abrupt truncation of many ranges of taxa along a narrow zone probably indicates a major hiatus or unconformity.

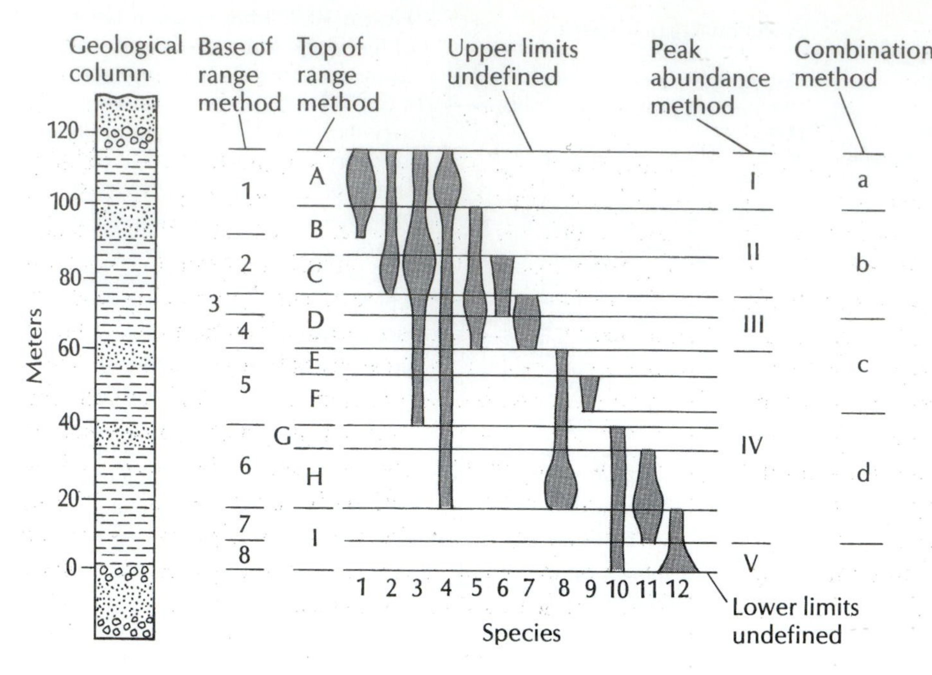

Figure 10.9. Four methds of recognizing concurrent range zones in a single section. Widths of lines indicate the relative abundances of specimens. Note that each method gives a slightly different zonation. (From Eicher, 1976.)

Drawing the boundary between two time-stratigraphic units along a major gap is indeed risky. If that gap should be filled later by some other section located somewhere else, the boundary may become blurred. Instead, some stratigraphers advocate drawing the boundary at the position of the evolutionary first occurrence of a single, distinctive taxon. If the evolutionary sequence of that taxon is relatively complete, it is less likely that a section is missing. Once such a boundary is agreed upon, Ager (1964, 1973, 1981) suggests that some physical marker (metaphorically called "the Golden Spike") should be driven into the outcrop at the precise level of the boundary. Although this procedure is more arbitrary than natural breaks, there should be no more argument once it is done. To date, a number of major boundaries have been internationally defined in this manner (Bassett, 1985; Cowie et al., 1989): the Silurian/Devonian boundary, based on the first appearance of the graptolite *Monograptus uniformis* in Klonk, Czech Republic; the base of the Cambrian (base of the *Phycodes pedum* trace fossil zone near the town of Fortune, southeastern Newfoundland, Canada); the Ordovician/Silurian boundary (base of the *Parakidograptus acuminatus* graptolite zone at Dob's Linn, Scotland); the Devonian/Carboniferous boundary (base of the *Siphonodella sulcata* conodont zone in a trench in La Serre, southern France); the Cretaceous/Tertiary boundary (placed at the iridium-bearing boundary clay at El Kef, Tunisia); and the Eocene/Oligocene boundary (the last appearance of hantkeninid foraminifera at a quarry in Massignano, near Ancona, Italy). Many other biostratigraphic boundaries are currently under discussion (reviewed in *Episodes*, vol. 8, no. 2, June 1985, and Cowie et al., 1989).

The choice of which boundary types to recognize can also vary with the method chosen. In outcrops, many stratigraphers mark zones from the bottom (first occurrence) upward because they measure a section from the base. These "topless" zones have the advantage that if a boundary in the stratotype happens to be drawn on a local hiatus, and geologists find a new section that spans this hiatus elsewhere, the new section can automatically be assigned to the zone below (Fig. 10.8). Geologists working with subsurface data tend to use tops of range zones. Not only are they accustomed to drilling downward, but the drilling mud may continue to bring up fossils from the borehole wall long after the base of a particular biostratigraphic zone has been passed. Other geologists have tried to use one of these methods in combination with abundance data, which is particularly striking in some sections (Fig. 10.9).

Stratigraphic Sampling and Mass Extinctions

The reliability of stratigraphic ranges of fossils has implications for many problems beyond biostratigraphic correlation. In recent years, one of the most important areas of research has been the pattern of mass extinctions. As we saw in Chapter 6, several major mass extinctions have been attributed to catastrophic events, such as the impacts of extraterrestrial objects. If a given extinction were catastrophic, then the ranges of taxa should terminate abruptly at the horizon of the impact. If the ranges gradually end before the impact horizon, it suggests that the extinctions were due to some long-term, protracted cause rather than to an abrupt cause. This objection has been raised numerous times by paleontologists who have argued that the K/T extinctions were gradual, and therefore could not be due to asteroid impact.

But is it that simple? In 1982, Phil Signor and Jere Lipps argued that even if there were an abrupt extinction, the effects of incomplete sampling would yield a

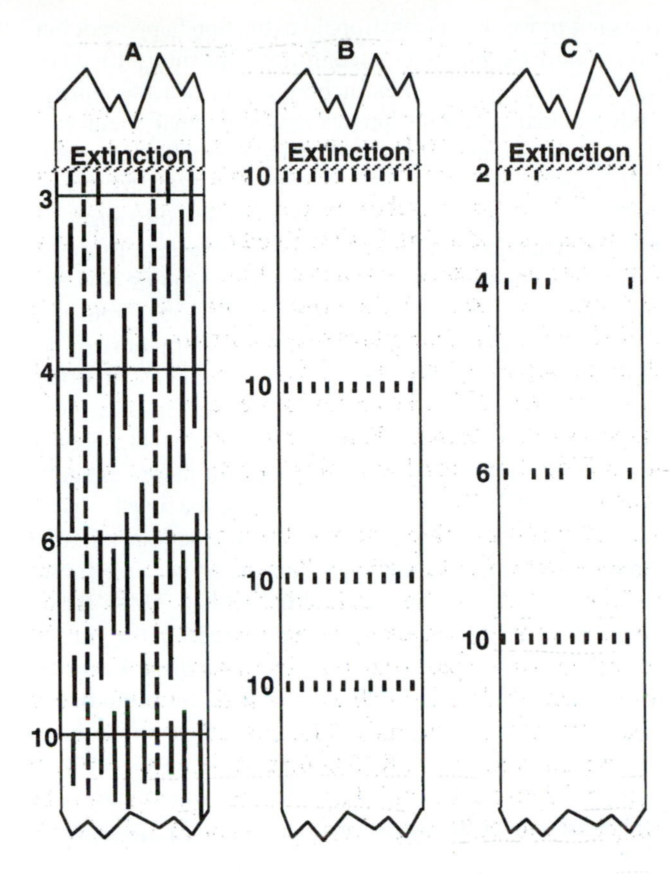

Figure 10.10. Signor and Lipps (1982) argued that an abrupt extinction in the fossil record could appear gradual just by errors in sampling. The geologic section in A shows ten species, each with a distinct and regular pattern of preservation. If we take samples at just four levels (horizontal lines), we end up with 10, 6 , 4, and 3 species at successive levels, and an abrupt extinction appears to be gradual. (B) In this section, the sampling detects no decline in species diversity through four levels, so we can be confident that this is truly an abrupt extinction. (C) This section appears to have a gradual decline in diversity, which could be real, or just a Signor-Lipps effect. (From Archibald, 1996.)

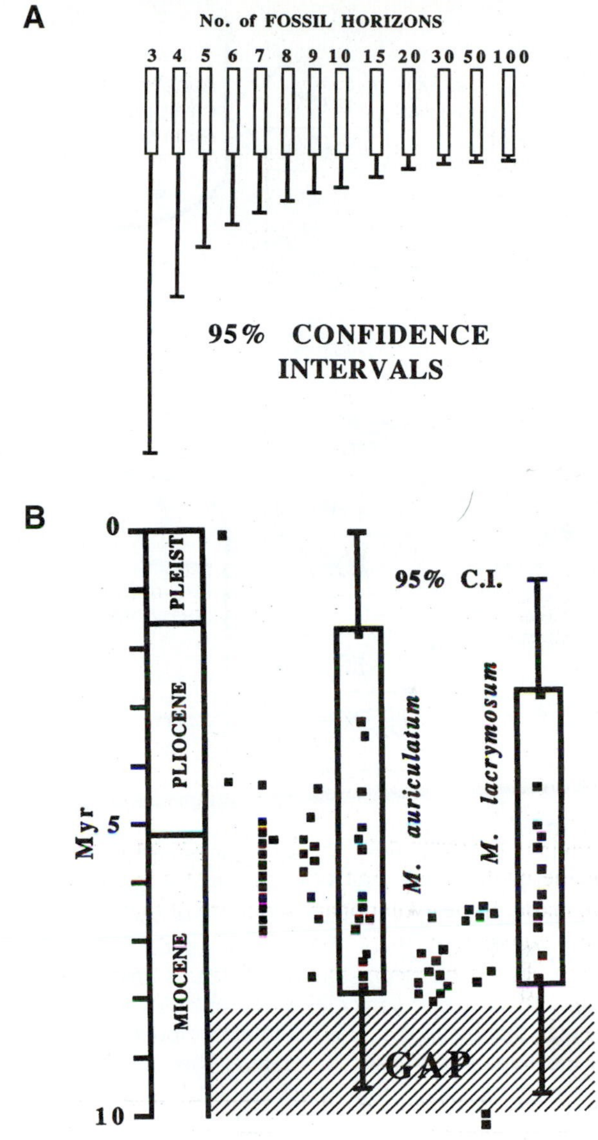

Figure 10.11. Marshall (1990) proposed a method of statistically estimating the true total range of a fossil. (A) In each of these examples, the apparent range of the fossil is the same. Within this interval, however, there can be just two or three levels, or dozens and hundreds. The more levels represented, the more confident we can be that the absence of this fossil from an interval is real. Thus, the 95% confidence levels on the range with only three levels is very large, but it gets progressively smaller with increased sampling. (B) The actual ranges (showing sample density) and 95% confidence level around those ranges for two Neogene Caribbean bryozoan species. (From Marshall, 1990.)

set of last occurrences that gradually tapered out before the impact horizon (Fig. 10.10). In other words, the vagaries of sampling means that sudden and gradual extinction patterns would be so poorly recorded that they both looked gradual, and could not be distinguished in the fossil record. This has come to be known as the **Signor-Lipps effect**, and it is usually the first problem a paleontologist must address when arguing about the reality of an apparent abrupt or gradual sequence of last occurrences (or first occurrences in the interval following the mass extinction).

Marshall (1990, 1991, 1995) has proposed one method of evaluating whether a gradual extinction pattern can be considered statistically valid. If we take the total range of a given species (Fig. 10.11), we can use the distribution of gaps between fossils within that range to determine a confidence interval of the top or bottom of the range. For example, if a total range is based on samples at only three levels (the FAD, LAD, and one in between), then statistically it is very likely that additional levels will recover more fossils that will extend the range either upward or downward, or both. In other words, the confidence interval around the actual recorded range is very large, implying a

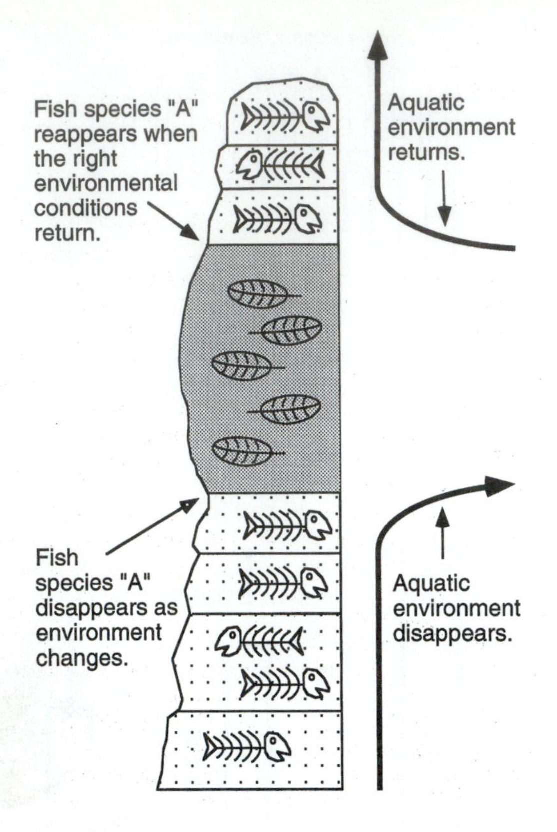

Figure 10.12. The Lazarus effect. When a taxon disappears from the record, it is not necessarily its final extinction. It may have simply migrated out of the local region, or may be restricted to a refugium where it is not fossilized. It can then reappear higher in the section when the appropriate conditions return, like Lazarus rising from the dead. This is particularly critical for studies that focus on the taxa that disappear at a mass extinction level, and do not look higher in the section to see if the "extinct" taxon later reappears. (From Archibald, 1996.)

great deal of uncertainty in both the FAD and LAD. As the number of sampling levels within the total range increases, the probabilities that the range will be extended by new discoveries decreases, and so does the confidence interval beyond the FAD or LAD. With a large number of levels, the confidence interval is very small, suggesting that the FAD or LAD is very close to the actual first occurrence or last occurrence in time. Using this method, Marshall (1990) argued that the sampling of fossils below the K/T boundary is not dense enough to conclusively prove that the gradual pattern implied gradual causes; its gradual pattern could be a statistical artifact of poor sampling, and could actual represent an incompletely recorded catastrophic extinction.

Of course, the argument can go both ways. Biostratigraphers are suspicious of large numbers of last occurrences at a single level. More often than not, these are due to a large gap in the record, which truncates many biostratigraphic ranges that do not actually coincide in more complete sections. An apparent instantaneous last occurrence pattern does not prove that a catastrophic extinction happened, but may simply be due to an unconformity. In short, it is very difficult to argue that an extinction pattern, whether apparently gradual or abrupt, proves that the actual event was gradual or abrupt, unless very dense sampling has been conducted.

There are other problems with determining the ranges of taxa at extinction horizons and using these data to make assertions about causes. Frequently, such studies concentrate on just the groups that lived before the extinction, and then look at the strata just above the extinction horizon to see if they are still around. With such counts, it often appears that a very high percentage of the species have gone extinct. However, if the biostratigrapher looks higher in the section, or in other regions, some of these "extinct" taxa may still turn up. They just happened not to have occurred in the strata right above the extinction horizon, either because it was not their habitat, or because those strata have poor preservation (Fig. 10.12). Thus, these apparently "dead" organisms seem to "rise from the grave" at some later time. Jablonski (1986) called these **Lazarus taxa**, after the man in the Bible who rose from the dead. The frequent occurrence of Lazarus taxa means that mass extinction studies cannot focus too closely on the extinction interval before counting a species as extinct, but must look widely over the post-extinction interval to prove that the organism is truly no longer around.

Some apparent Lazarus taxa may not be what they seem. After mass extinctions, organisms can evolve to a body

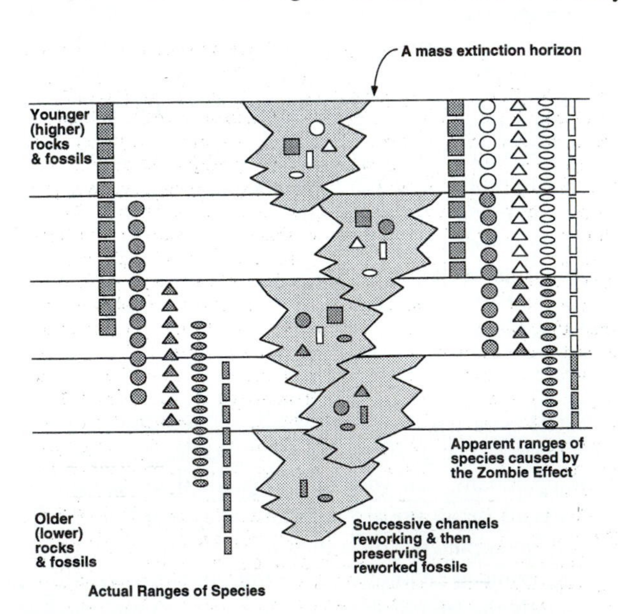

Figure 10.13. The Zombie effect. Some fossils are remarkable durable when eroded out of older deposits, so that they can be redeposited into younger beds long after the original organism is extinct. This gives the false impression that they survived longer than they actually did. Fossils deposited during the time the organism lived are shown with the filled shapes in the channels; those reworked from older beds are indicated by unfilled shapes. (From Archibald, 1996.)

form by convergent evolution that is remarkably similar to the shape that occurred before the extinction. In other words, they are remarkable facsimiles of the extinct predecessor, but if they are mistaken for the real thing, they will be counted as persistent Lazarus taxa, rather than as evidence for extinction and then repeated evolution of the same body form. Erwin and Droser (1993) called these **Elvis taxa**, in recognition of the Elvis impersonators who have proliferated long after the death of The King. Thus, Lazarus taxa give the erroneous impression that an extinction has occurred and Elvis taxa give the erroneous impression that it has not.

A third distortion of the true biostratigraphic distribution of fossils comes from reworking. It is well known that certain durable fossils can be eroded out of older strata and then reburied along with much younger fossils. These displaced fossils falsely suggest that the organism they represent lived much later than it actually did (Fig. 10.13). For example, scientists working on the K/T boundary in Montana discovered Cretaceous dinosaur teeth in lower Paleocene strata, and jumped to the conclusion that dinosaurs survived the K/T event (Rigby et al., 1987). However, Eaton et al. (1989) showed that not only were these dinosaur teeth reworked, but that such reworking was common, especially with marine shark teeth from the Cretaceous occurring in non-marine Paleocene strata. Archibald (1996) referred to this problem as the **Zombie effect**, since these fossils seem to rise from the grave like zombies and "lurk in later sediments like the living dead" (Archibald, 1996, p. 68).

As we saw in Chapter 6, the intense controversy about mass extinctions has generated a lot of excitement and anger, but who ever thought that Lazarus, Elvis, and zombies would be strolling around the debates to make them even more bizarre?

THE TIME SIGNIFICANCE OF BIOSTRATIGRAPHIC EVENTS

For anything that geology or palaeontology are able to show to the contrary, a Devonian fauna or flora in the British Islands may have been contemporaneous with Silurian life in North America, and with a Carboniferous fauna and flora in Africa. Geographical provinces and zones may have been as distinctly marked in the Palaeozoic epoch as in the present, and those seemingly sudden appearances of new genera and species, which we ascribe to new creation, may be simple results of migration.

Thomas Henry Huxley, *Anniversary Address to the Geological Society of London*, 1862

Throughout this discussion, it has been assumed that biostratigraphic datums are close approximations of time planes. How good is that assumption? What types of biostratigraphic events make the best boundaries? Many authors (e.g., Murphy, 1977) consider the evolutionary first occurrence (lineage zone) the best, because the evolution of a taxon is an unrepeatable event that happens at a unique time and place. However, there are drawbacks to this method. Evolutionary first occurrences are rare, and they may not occur in the sections of interest. If Eldredge and Gould (1972) are right in their theory of punctuated equilibrium, there are relatively few examples of gradualistic change in the fossil record. Gould and Eldredge (1977) claimed that most "gradualistic" evolutionary changes in taxa do not hold up under close scrutiny. Many cases of apparent gradualism may be due to repeated immigration events rather than to the evolutionary change of a lineage *in situ*. If gradual change is real, and is used for determining boundaries, then problems of definition arise. Does one define the evolutionary first occurrence of a species as the first appearance of the character that defines the species, or when 50 percent of the population has developed this character? Here, there is a conflict between species defined for biostratigraphic purposes and the biological meaning of species. The definition of species by the 50 percent criterion, for example, would be considered unnatural by a biologist.

Some geologists (e.g., Repenning, 1967) have argued that immigrational or emigrational events should be preferred as boundaries. Immigrational first occurences in particular can be abrupt and unambiguous. The dispersal of most marine organisms, particularly those with planktonic larval stages, is very rapid by geologic standards (Scheltema, 1977). This is also true of land vertebrates (Lazarus and Prothero, 1984; Flynn et al., 1984). The main problem with this marker is that organisms immigrate to different places at different times. The notorious "*Hipparion* datum," based on the first appearance of the three-toed horse *Hipparion* in the Old World, was once used as the base of the Pliocene. Recent work (Woodburne, 1989, 1996) has shown that the *Hipparion* immigration "event" was at least two different immigrations of different hipparionine horses at different times during the Miocene.

Extinctions are probably less reliable boundaries because it is well known that taxa can linger in a refuge long after they have disappeared elsewhere. Yet extinctions can be attractive boundaries because they often cluster at horizons that represent mass extinctions, which are major episodes in the history of life.

Stratigraphers have long assumed that biostratigraphic events are reasonably synchronous within the resolving power of the stratigraphic record. This assumption was criticized by Thomas Henry Huxley (1862), who pointed out that biostratigraphers have demonstrated only similiarity in order of occurrence (**homotaxis**), not synchrony. In recent years, other independent techniques have been employed to test the time significance of biostratigraphic events. Hays and Shackleton (1976) used oxygen isotopes to show that the extinction of the radiolarian *Stylatractus universus* was

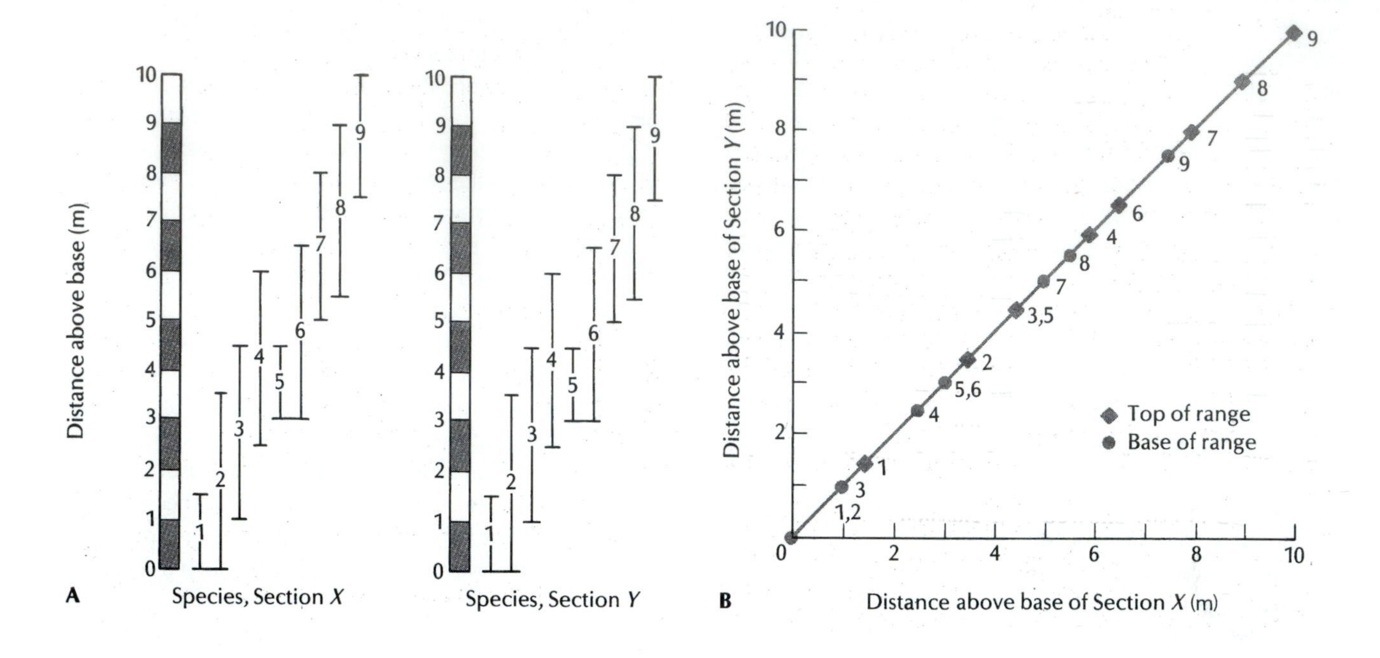

Figure 10.14. Shaw's graphic correlation method. (A) Two sections with the stratigraphic distribution of nine taxa. (B) Correlation between sections X and Y. Since the range in both sections are identical, the line of correlation between the two sections is straight and has a 45° slope. (From Shaw, 1964.)

globally synchronous. Haq and others (1980) conducted a similar test using carbon isotopes. Berggren and Van Couvering (1978) and Spencer-Cervato et al. (1995) argue that asynchrony of biostratigraphic datum planes is rare in the marine realm. Yet Srinivasan and Kennett (1981) and Dowsett (1988) showed that the last appearances of some marine microfossils can be time-transgressive. Some marine microfossils, such as the benthic foraminifers, are notorious for being facies controlled, as we saw in the example of the Pacific Coast Eocene benthic foraminifers (McDougall, 1980).

In terrestrial sections, Prothero (1982) and Flynn et al. (1984) tested mammalian biostratigraphy against magnetostratigraphy and found that mammal *assemblages* do not show significant time transgression as a rule, though the stratigraphic occurrence of *individual taxa* can be significantly time transgressive. These studies demonstrated that the basic assumptions of biostratigraphy must be constantly re-examined when new tools of correlation are developed. Most of the methods and assumptions of biostratigraphy have proved to be quite robust. This is hardly surprising because biostratigraphy has been used successfully since the days of William Smith. For these reasons, biostratigraphy will probably remain the primary tool for correlation and determining relative age.

NORTH AMERICAN LAND MAMMAL "AGES" AND BIOCHRONOLOGY

The fossil-mammal worker accepts that many mammals, marine or nonmarine, contributed fossils which are admirable tools for paleontologic stratigraphy and geochronology, and especially for age-magnitude correlations from continent to continent.

D. E. Savage, "Aspects of vertebrate paleontological stratigraphy and geochronology," 1977

Land mammal fossils have long been used as stratigraphic tools in Cenozoic terrestrial deposits. However, most fossil mammals occur in isolated quarries and pockets and are seldom distributed evenly through thick stratigraphic sections, as are marine invertebrates. As a result, mammalian biostratigraphers have not always followed classical Oppelian biostratigraphic procedures. A few thick sections that show superposition of faunas are known, but as a whole, the sequence of mammalian faunas must be worked out indirectly. The result has been called **biochronology**. Williams (1901) defined a **biochron** as a unit of time during which an association of taxa is interpreted to have lived (Woodburne, 1977). A biochron is equivalent to the biozone of an assemblage of taxa, except that it is not directly tied to any actual stratigraphic sections.

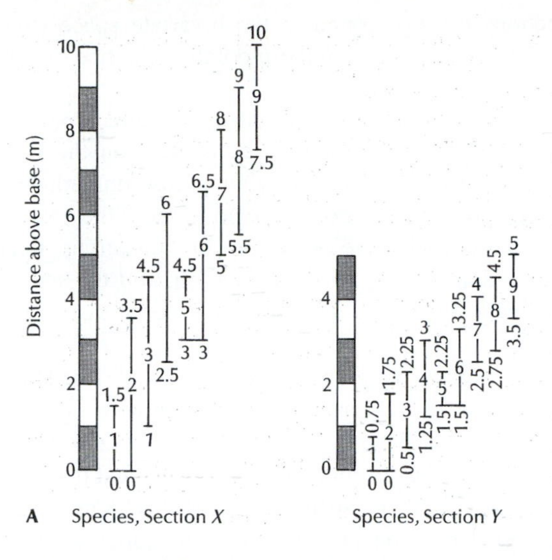

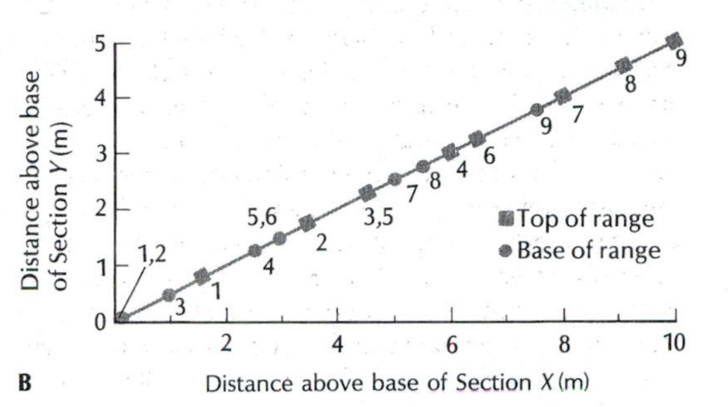

Figure 10.15. (A) Two sections that have the same relative spacing of ranges but different rates of rock accumlation. (B) Graphic correlation of these two sections. The points form a straight line, but the slope is deflected toward the axis of the section with the higher accumulation rate. (From Shaw, 1964.)

Tedford (1970) showed that classical biostratigraphic procedures were used by early North American mammalian paleontologists (e.g., the "life zones" of Osborn and Matthew, 1909). These procedures fell into disuse, and a biochronological sequence was built up by piecing together a sequence of faunas based on their stages of evolution. In 1941, the sequence of mammalian faunas was formally codified by a committee chaired by Horace E. Wood II (Wood et al., 1941). The Wood Committee set up a series of provincial land mammal "ages" that were based on the classic sequence of mammal fauna in North America. These came to be known as North American Land Mammal "Ages." However, they are not true "ages" in the formal time-stratigraphic sense because they are not built from biostratigraphic zones based on actual rock sections.

The Wood Committee attempted to erect unambiguous definitions for each unit by listing multiple criteria for each unit: index fossils, first and last occurrences, characteristic fossils, and typical and correlative areas. These multiple definitions have since led to conflicts that would not have occurred if classical biostratigraphic methods had been followed originally. For example, the late Eocene Chadronian Land Mammal "Age" is defined by two criteria: (1) the cooccurrence of the horse *Mesohippus* and titanotheres; and (2) the limits of the Chadron Formation. At the time of the Wood Committee's work, the last occurrence of titanotheres was thought to coincide exactly with the top of the Chadron Formation, so there was no conflict. Since then, titanotheres have been found in rocks above the Chadron Formation (Prothero, 1982; Prothero and Whittlesey, 1998). Now we must choose between conflicting criteria. Is the end of the Chadronian marked by the last appearance of titanotheres or by the top of the Chadron Formation?

In recent years, mammalian paleontologists have made efforts to return to classical biostratigraphic principles (reviewed by Savage, 1977; Woodburne, 1977, 1987; Woodburne and Swisher, 1995; Prothero, 1995). Much of the mammalian time scale is being restructured in terms of biostratigraphic range zones tied to local sections. There are parts of the Tertiary, however, that may never be zoned, so biochronology is still widely used. Its robustness was demonstrated when radiometric dates first became available for the land mammal record and showed that the biochronological sequence was in the correct order (Evernden et al., 1964). European paleomammalogists, who have fewer good stratigraphic sections to work from, use an explicitly biochronological approach. Every fauna is placed in order according to its stage of evolution. *Niveaux répères*, or "reference levels," of classical faunas are used in place of type sections (Jaeger and Hartenberger, 1975; Thaler, 1972).

QUANTITATIVE BIOSTRATIGRAPHY

The fossil record can normally be made as "adequate" as desired, and "adequacy" can be tested by standard quantitative analyses.

Alan Shaw, *Time in Stratigraphy*, 1964

A biostratigraphic data base can become so large that it becomes impossible for a biostratigrapher to find the pattern among so many stratigraphic sections with so many taxonomic ranges. In recent years, techniques have been introduced to quantify and handle large data bases, mostly by computer.

The most popular method, **graphic correlation**, or "Shaw plots," was introduced by Alan Shaw in his book *Time in Stratigraphy* (Shaw, 1964). Much of the work can be done on graph paper with the aid of a pocket calculator, though the entire procedure is usually done on a computer. In essence, it uses the statistician's meaning of the word *correlation*. When two variables are correlated, they form a line of points on a bivariate data plot. The statistician then attempts to fit a continuous line of correlation to the cluster. First and last occurrences of fossil taxa become data points along an axis that represents the stratigraphic section. If two stratigraphic sections are placed along the two bivariate axes (abscissa and ordinate), the data points in

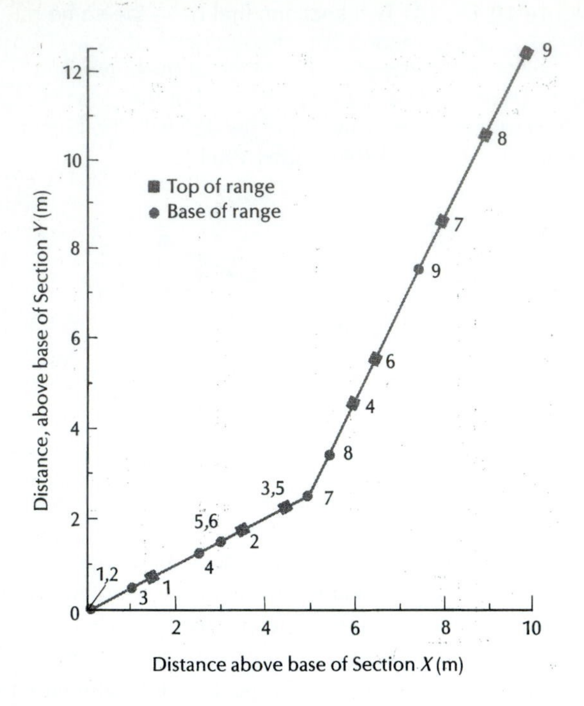

Figure 10.16. Graphic correlation of two sections that show a change in relative rates of rock accumulation. As one section increases in rate relative to the other, the line of correlation changes slope, forming the characteristic "dogleg" kink in the slope of the line. (From Shaw, 1964.)

both sections can be plotted in the bivariate space (Fig. 10.14A, B). The line through these points is the line of correlation.

The line of correlation is the most powerful aspect of Shaw's method. If all the points fit exactly on the line, there is perfect correlation; outlying points immediately become apparent and can be examined to see if they represent artificial or real range extensions or truncations. Real data sets seldom show perfect correlation, and the scatter of points near the line is a measure of the scatter of the data.

If the two sections are plotted to the same scale, the slope of the line is particularly informative. A line with a perfect 45° slope (Fig. 10.14B) indicates that the two sections have identical rates of rock accumulation and identical distributions of fossils. (Shaw speaks of "rock accumulation" because post-depositional compaction can shorten sections and distort true rates of sedimentation.) Commonly, the rates of accumulation in two sections are not identical, in which case the ranges in one section will be proportionately shorter (Fig. 10.15A). The plot of these two sections will not result in a 45° slope but will be a line inclined toward the axis that represents the section with the higher rate of accumulation (Fig. 10.15B). If the rate of accumulation changes during deposition in one section relative to the other, the slope will change, giving a "dogleg" pattern (Fig. 10.16). If there is a hiatus in one of the sections, the ranges will tend to truncate at that level (Fig. 10.17A); the result-

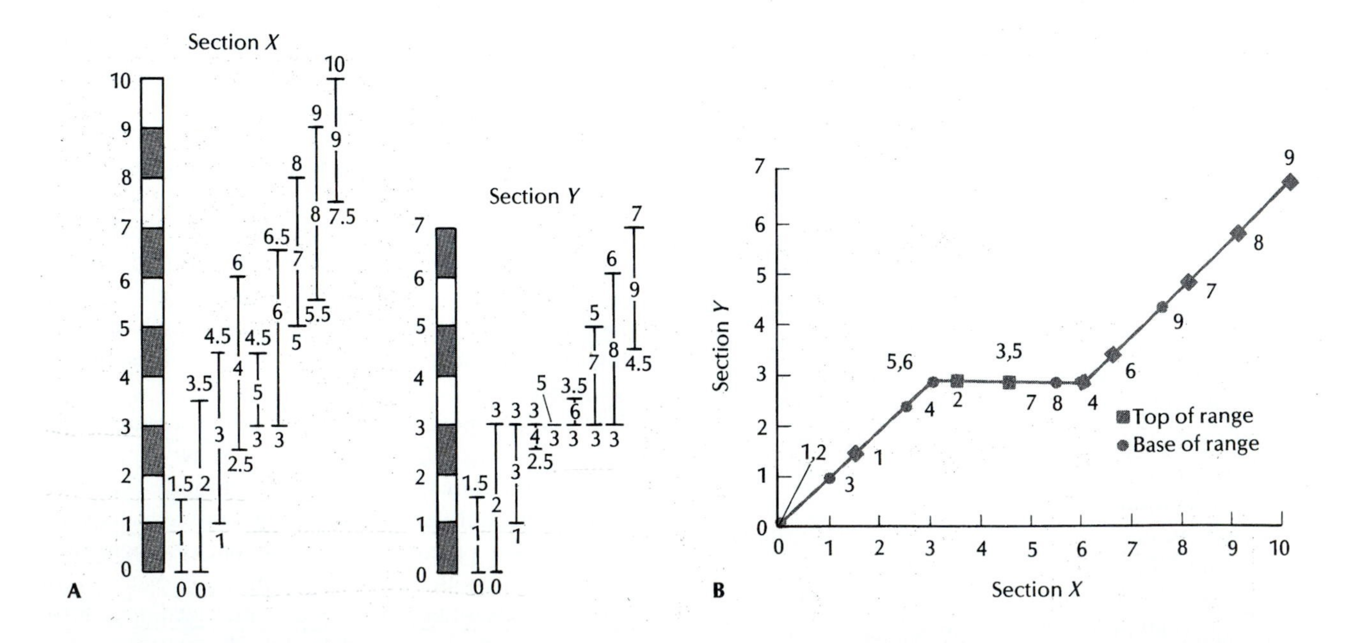

Figure 10.17. (A) Two sections with the same taxa, but one section has an unconformity that truncates the ranges of some taxa. (B) Graphic correlation plot of these two sections. The unconformity causes a "step" or "plateau" in the line. The section with the unconformity (Y) shows no change, while the complete section (X) continues to spread out the data points, producing the plateau. If the axes were flipped, the "step" would become a vertical "wall." (From Shaw, 1964.)

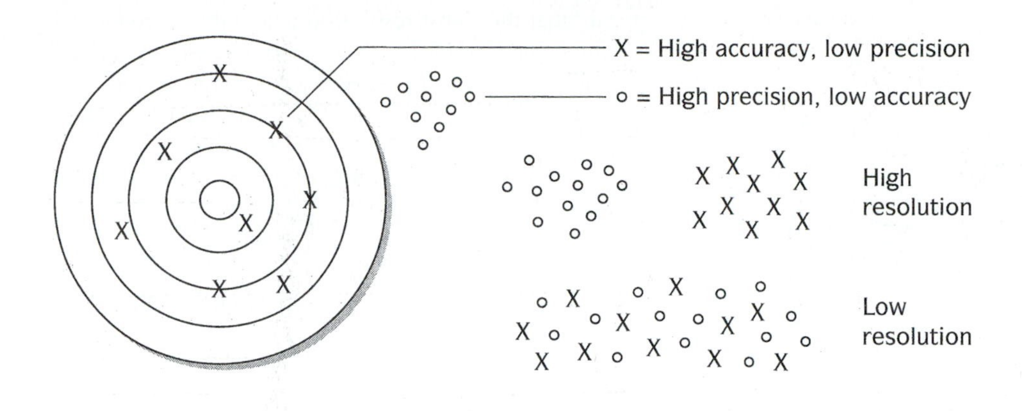

Figure 10.18. The relationships among precision, accuracy and resolution can be analogized with a target. One cluster of shots has high precision, but is not close to the mark (not accurate), while another is more accurate but not precise (poorly clustered). If two clustered are easily distinguished, it is possible to get high resolution. (From Prothero and Schwab, 1994).

ing plot will have an obvious horizontal "terrace" or vertical "wall" representing the missing section (Fig. 10.17B). It should be apparent that the graphic correlation method is a powerful tool for spotting bad range data, unconformities, and missing sections, and for interpreting rates of rock accumulation.

Shaw's method has applications beyond the correlation of two sections. It can be used as the basis for a large-scale correlation scheme among multiple sections. Shaw recommends that the stratigrapher begin by correlating the two best sections in the study area. Any outliers on the line that represent true range extensions are then added to the best reference section, producing a **composite standard**. The composite standard can then be correlated with each additional section, one at a time, and any further range extensions can be added to it. The result is a composite standard that is not a real section but a synthesis of all the information from a group of sections. The composite standard ranges are not teilzones but are the maximum ranges of each taxon in the area. If this were done for every section in the world, the total range zone would be approximated. The subdivisions of the composite standard are no longer measured in thicknesses of section but are abstract units that approximate geologic time planes. These can be used to correlate and align sections, producing results that show patterns more clearly than do correlations based on lithologic boundaries or on individual biostratigraphic planes.

Because Shaw's methods are easily learned and are applicable to a variety of biostratigraphic problems, they have had widespread use in marine geology and paleoceanography, and are even used in the oil industry (Miller, 1977). Simple graphic correlation plots can be routinely used when plotting together any two sets of stratigraphic data, including magnetostratigraphic data and isotopic data.

Shaw's methods are the most widely used among biostratigraphers, but they are not the only quantitative methods available. Further information about the many different options can be found in Prothero (1990), Cubitt and Reyment (1982), Kaufmann and Hazel (1977), and Gradstein et al. (1985). An excellent summary and appraisal of each of these methods was published by Edwards (1982).

RESOLUTION, PRECISION, AND ACCURACY

Total stratigraphic range can be established as precisely as is necessary for most problems. Precision is limited only by usual tests of geologic credibility and simple statistical tests of confidence.

Alan Shaw, *Time in Stratigraphy*, 1964

Biostratigraphy is the standard tool of correlation of strata around the world. Ideally, we would like to correlate strata in the greatest detail possible. The best way to achieve this is to collect fossils as densely as possible, and to subdivide the biostratigraphic zonation as finely as the changes in the fossils allow. The more finely divided and shorter each biostratigraphic zone, the easier it is to discriminate between two intervals of time. In other words, the finer the scale of biostratigraphic zonation, the finer the **resolution** between events. If the biostratigraphic zones are very long (for example, on the order of 2 to 5 million years), it is much more difficult to discriminate closely spaced events than if the zones are a fraction of a million years in duration.

Another related issue is the reliability, or **precision** of a given data point. Precision is a measurement of repeatability of an observed value, its "plus or minus" error. In biostratigraphy, precision is usually a factor of upward or downward mixing of fossils, improper identification or poor preservation of fossils, or unfossiliferous intervals. In the best cases, however, the precision is extremely good and is limited only by the rate of sediment accumulation. For example, some deep-sea cores have a nearly continuous record of microfossils, with no large gaps, so the first appearance of a key microfossil can be very precisely located if the preservation and identification of the fossils is good. If the core also happens to represent a time of rapid sedimentation, a precision of thousands to tens of thousands of years is possible. If there are annual climatic varves or some similar time-related feature in the strata, a precision of hundreds of years to less than a year is possible.

Ideally, the precision of a data set is a measure of its

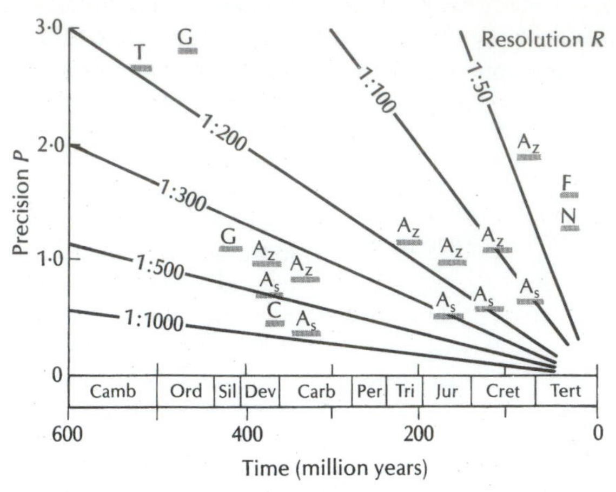

Figure 10.19. Comparative resolution and precison of some fossils. A, Ammonoidea; C, conodonts; G, graptolites; N, nannoplankton; T, trilobites; z, zones; s, subzones. Precision P (vertical axis) refers to discrimination of zonal units. Resolution R is resolving power for a given value of P for the numerical time span of the zone. Radiating lines of equal resolution are shown.

accuracy, or closeness to what actually happened. Most scientists think that the "truth" is philosophically unattainable, so the the precision is a real measure of how accurate the data are. Although accuracy and precision are usually correlated, we can imagine circumstances where they are not (Fig. 10.18). For example, if a marksman shoots at a target, the shots may be tightly clustered but far from the center (high precision but low accuracy), or poorly clustered but on the target (high accuracy but low precision).

Many groups of organisms yield abundant fossils that have evolved rapidly, so they offer the potential of very high resolution. In the right strata with dense sampling, they also offer high precision as well. For example, there are 117 ammonite zones and subzones currently recognized for the 70-million-year span of the Jurassic (House, 1985), so each zone or subzone averages only 600,000 years in length. Ramsbottom (1979) reported a zonation of the Early Carboniferous down to 25,000-year increments using ammonoids. As can be seen in Fig. 10.19, not all fossils have resolution this fine, but many have zones that shorter than 3 million years in average duration. Because the error estimate of radiometric dates is a fixed percentage plus or minus of the numerical age, the precision of numerical ages decreases with increasing age. Relative resolution of the biostratigraphy can be compared to the resolution of numerical radiometric dates to give a rough sense of the equivalence of biostratigraphic zonations. Cambrian trilobites have zones averaging about 3 million years in duration, but the radiometric dates on Cambrian rocks have error estimates of ±5 million to ±10 million years. Cretaceous zones are only a million years or less in duration, but their radiometric dates typically have error estimates of ±2 million years to ±4 million years. Scaling the zonal lengths to the error in numerical age (Fig. 10.19) puts Cretaceous ammonite subzones on the same 1:200 line on which Cambrian trilobites fall. From Figure 10.19, it appears that the finest resolution available is produced by Devonian and Carboniferous ammonoids, whose zones are extremely short despite the great age of the system. As is apparent from these examples, biostratigraphic resolution is much better than radiometric error estimates in the Paleozoic or Mesozoic, and even for much of the Cenozoic.

INDEX FOSSILS AND THE GLOBAL BIOSTRATIGRAPHIC STANDARD

It would be difficult to estimate how many nascent geologists have been turned aside from paleontology by being forced during the course of some dismal semester to learn hundreds of index fossils and the formations of which they are the index. *Many geologists' sole memory of the whole discipline of paleontology is the unerasable fact that* "Spirifer grimesi *is the index fossil of the Burlington Limestone" or some such tidbit.*

Alan Shaw, *Time in Stratigraphy*, 1964

Some fossils are so abundant and characteristic of key formations that they are known as *index fossils*. Many formations in North America can be recognized by index fossils. This information was catalogued in Shimer and Shrock's (1944) *Index Fossils of North America*. Most fossil groups (including index fossils) that make good biostratigraphic indicators are *distinctive, geographically widespread, abundant, independent of facies, rapidly evolving*, and *temporally short ranging*. From these properties, one can see that the best biostratigraphic indicators are pelagic organisms (both the floating plankton and the swimming nekton) that evolve rapidly. Pelagic organisms live in the surface waters of the oceans, so they can quickly disperse around the world. They are unaffected by the various bottom facies that control benthic organisms. Practical experience has shown that ammonoids, graptolites, conodonts, foraminifers, and other planktonic or nektonic fossils give the best results, though some benthic groups are useful.

There are also some problems with the index fossil approach to correlation. Too often the index fossil is equated with the formation, and no attempt is made to document the *actual range* of the fossil *within* the formation. This results in a loss of stratigraphic resolution. The stratigraphic range of the fossil is often reported to be the same as the total thickness of the formation, which may artificially extend the range. In addition, the use of index fossils can cause confusion of rock units with time-rock units and imply that rock units are time equivalent. Too much reliance on one index fossil may cripple a biostratigrapher who works in areas where the index fossil does not occur. Absence of a key index fossil does not necessarily mean that the rocks are not of the age of that index fossil. Finally,

Shaw (1964, p. 99) pointed out that reliance on index fossils can lead to other abuses. For example, rote memorization of index fossils and their formations *("Spirifer grimesi* is the index fossil of the Burlington Limestone") can lead a geologist to recognize formations by their fossils rather than by their lithologies.

Once this problem has been circumvented by detailed biostratigraphic range data, it is possible to create biostratigraphic zonations based on global index fossils that have both high resolution and high precision. These have been recognized as the global standard for telling time in their respective parts of the geologic column. Each part of the Phanerozoic has its own ideal groups of index fossils. In the Cenozoic, almost all marine strata in the temperate and tropical regions are correlated by planktonic microfossils, especially foraminifera and calcareous nannofossils (Berggren and Miller, 1988; Berggren et al., 1995). The standard planktonic foraminiferal zones are all numbered in sequence, with the prefix "P" for Paleogene and "M" and "PL" for Miocene and Plio-Pleistocene, respectively. Thus, P18 is an abbreviation for the *Turborotalia cerroazulensis-Pseudohastigerina* Interval Zone (early Oligocene, 33.8 to 32.0 Ma in the time scale of Berggren et al., 1995). The occurrence of just a few key foraminiferal species in temperate or tropical marine strata is sufficient to identify this zone, and pin down the age of the strata to within this 1.8-million-year interval. Similarly, Martini (1971) and Bukry (1971, 1973; Okada and Bukry, 1980) proposed numbered zonation schemes for the calcareous nannoplankton of the Cenozoic. The Martini zonation uses the prefix "NP" for Paleogene and "NN" for Neogene nannofossils. The Okada and Bukry (1980) zonation uses "CP" for Paleogene and "CN" for Neogene calcareous nannoplankton. In the Berggren et al. (1995) time scale, the P18 planktonic foraminiferal zone correlates with parts of the NP21 and NP22 nannofossil zones of Martini (1971), and with the CP16 zone of Okada and Bukry (1980). Together, these three planktonic zonations can be combined with magnetostratigraphy to produce what Berggren and Miller (1988) called **magnetobiogeochronology**, an integrated chronostratigraphy utilizing magnetics, biostratigraphy, and stable isotopes to give very high-resolution dating on a global basis.

In the Mesozoic, the standard biostratigraphic zonations are based primarily on ammonites (Kennedy et al., 1992; Hoedemaker et al., 1993; Gradstein et al., 1995) and to a lesser extent on planktonic microfossils in the Cretaceous (Bralower et al., 1995). Several different groups are used in the Paleozoic. Ammonoids are very important in the Devonian through Permian (House, 1985), and conodonts used from the Ordovician through the Triassic (Sweet and Bergström, 1971). Trilobites are the main tool of correlation in the Cambrian and early Ordovician (Taylor, 1977), and graptolites are the main basis for zonation of the Ordovician to early Devonian (Berry, 1977). Fusulinid foraminifera are the primary tool for Permo-Carboniferous zonations (Douglass, 1977). Other organisms, such as brachiopods, bivalves, gastropods, and corals, are used locally, but since they are bottom-dwellers, their biostratigraphy tends to be less global than those of pelagic organisms, such as ammonoids, conodonts, and graptolites.

CONCLUSIONS

Biostratigraphy is the only practical method of dating and correlating most sedimentary rocks, and is the fundamental tool of determining geologic history in most regions. As long as geologists need to know the age of their rocks, there will always be a need for paleontologists. Smith's discovery of faunal succession caused geologists to realize that the earth has a history too complex to describe with the Noachian deluge. Later, Darwin used the change in fossils through time as one of the best arguments for evolution. Biostratigraphy is one of the most important tools not only for determining geologic history, but also for correlating strata in the search for oil, gas, coal, and many other important resources. However, there are many factors that affect the distribution of fossils in a local section, so a good biostratigrapher must be wary of the possible problems and pitfalls in the data. Properly applied, however, biostratigraphic data can offer very high resolution and high precision in dating and correlation. Combined with other dating techniques, magnetobiostratigraphic methods can give numerical age estimates to fractions of a million years, even in ancient strata.

For Further Reading

Ager, D. V. 1981. *The Nature of the Stratigraphical Record.* (2nd ed.) Wiley, New York.

Berggren, W. A., D. V. Kent, M.-P. Aubry, and J. Hardenbol. 1995. *Geochronology, Time Scales, and Global Stratigraphic Correlation*. SEPM Special Publication 54.

Berry, W. B. N. 1987. *Growth of a Prehistoric Time Scale* (2nd ed.) Blackwell Scientific Publications, New York.

Cubitt, J. M., and R. A. Reyment, eds. 1982, *Quantitative Stratigraphy*. John Wiley and Sons, New York.

Eicher, D. L. 1976. *Geologic Time*. (2nd ed.). Prentice-Hall, Englewood Cliffs, N.J.

Gradstein, F. M., J. P. Agterberg, J. C. Brower, and W. J. Schwarzacher, eds. 1985. *Quantitative Stratigraphy*. D. Reidel, Dordrecht, Netherlands.

Kauffman, E. G., and J. E. Hazel, eds. 1977. *Concepts and Methods of Biostratigraphy*. Dowden, Hutchinson, and Ross, Stroudsburg, Penn.

Mann, K. O., and H. R. Lane, eds. 1995. *Graphic Correlation*. SEPM Special Publication 53.

Prothero, D. R. 1990. *Interpreting the Stratigraphic Record*. Freeman, New York.

Salvador, A. 1994. *International Stratigraphic Guide* (2nd ed.). Geological Society of America, Boulder, CO.

Schoch, R. M. 1989. *Stratigraphy, Principles and Methods*. Van Nostrand Reinhold, New York.

Shaw, A. B. 1964. *Time in Stratigraphy*. McGraw-Hill, New York.

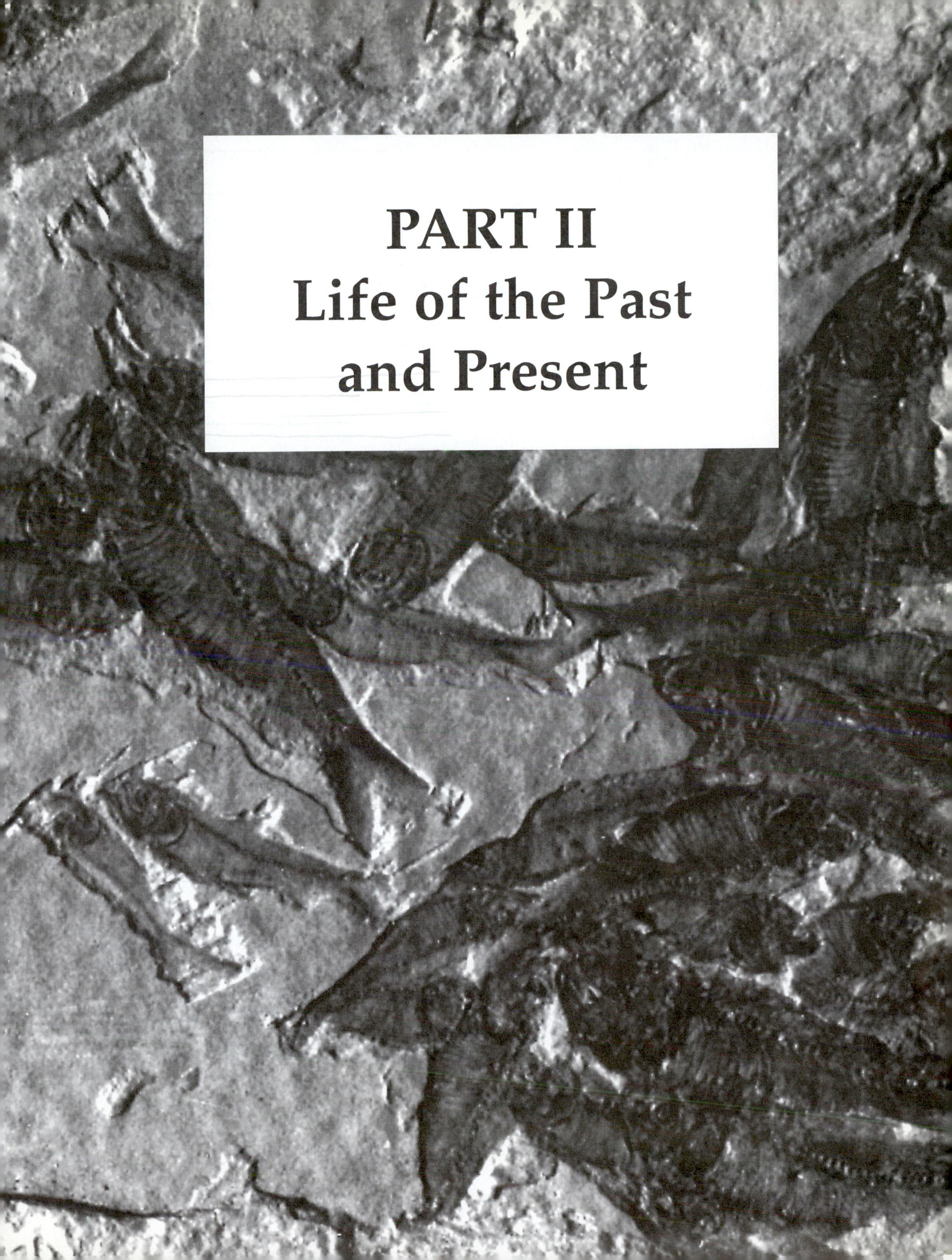

PART II
Life of the Past and Present

Figure 11.1. Scanning electron micrograph of a typical assemblage of middle Eocene foraminifera (with the bubble-shaped chambers), radiolaria (with the coarse mesh), and sponge spicules, magnified 250 times. (Photo courtesy Scripps Institution of Oceanography.)

Chapter 11

Micropaleontology
Fossil Protistans

Micropaleontology is a strange subject. It is not easily defined, its history is fairly dull, and it seems to focus only on geologic topics. Biologists by and large ignore those organisms that when fossilized become microfossils. Evolutionary biologists disdain them. Paleobiologists snub them, and "micropaleontologists" seem not to know what to do with them as once-living animals or plants. Most "micropaleontologists" are not trained in biology, and the literature of micropaleontology is an enormous edifice testifying to that fact. Although not a mere flunky of geology, micropaleontology is nevertheless largely a servant of geology, albeit an extremely powerful one.

Jere Lipps, "What, if anything, is micropaleontology?"1981

INTRODUCTION

We begin our review of the fossil record by considering the smallest and simplest of fossils, the microfossils. As Berggren (1978, p. 1) put it, "by definition micropaleontology, the study of microscopic fossils, cuts across many classificatory lines. It includes within its domain the study of large numbers of taxonomically unrelated groups united solely by the fact that they must be examined with a microscope." Micropaleontology includes some fossil plants (pollen grains), a variety of animals (including the ostracode crustaceans, the planktonic gastropod molluscs known as pteropods, as well as the conodonts, now thought to be related to chordates), prokaryotic organisms (such as bacteria), and many single-celled eukaryotic organisms that are now placed in the Kingdom Protista (including the amoeba-like foraminifera and radiolaria, and the plant-like diatoms and calcareous nannoplankton). In fact, some organisms (such as the acritarchs) have been studied by micropaleontologists for years, even though we still don't know what they are!

In most other groups of fossils, such an artificial approach based on size alone, with little understanding of the biological affinities or paleobiology of the organism, would be a serious liability. For one thing, even microfossils vary enormously in size, from foraminifera, conodonts, and ostracodes, which may reach several centimeters (but normally 0.5 to 1 mm in size), down to coccoliths, which are only a few microns to tens of microns in size (Figs. 11.1, 11.2). However, micropaleontology had a very different tradition and set of goals than the rest of paleontology. Because microfossils are abundant in most sediments and are produced by species that evolve rapidly, their primary use has been biostratigraphic. Micropaleontologists have traditionally been biostratigraphers with limited interest in paleobiology. During the heyday of the oil industry, there was a tremendous demand for micropaleontologists, and the majority of working paleontologists are still stratigraphic micropaleontologists, even with the oil industry layoffs. In such circumstances, it really doesn't matter whether we know anything about the biology of these fossils. They could be non-biological objects, like nuts, screws, bolts, and washers—as long as they give a consistent pattern of first and last occurrences, they are useful to biostratigraphers.

Another major focus of micropaleontology has been marine geology, where micropaleontologists have long used microfossils as a tool for interpreting paleoclimatic and paleoceanographic changes, in spite of their limited understanding of the biology of the organisms. In his article, "What, if anything, is micropaleontology?" Lipps (1981) bemoaned the lack of biological interest among his micropaleontologist colleagues. Paleobiological studies were long neglected by micropaleontologists for several good reasons. Many marine microorganisms are extremely hard to observe and keep alive in the laboratory, so it is difficult to study their biology (and clearly impossible with extinct groups, such as conodonts or acritarchs). This still didn't hamper the biostratigraphers, who were very successful using microfossils for correlation, even though they knew little about their biology. Oil companies paid them to date and correlate beds, not to indulge idle curiosity about paleobiological problems (unless they can show that such understanding is critical to better dating and correlation).

Despite this tradition, micropaleontologists have become increasingly interested in the paleobiology of their fossils. A number of recent books (e.g., Haq and Boersma, 1978; Lipps et al., 1979; Helmleben et al., 1989; Lipps, 1993) have summarized a wide variety of studies on the biological activity of organisms such as foraminifera, ostracodes, diatoms, and radiolaria, which were simply useful (and

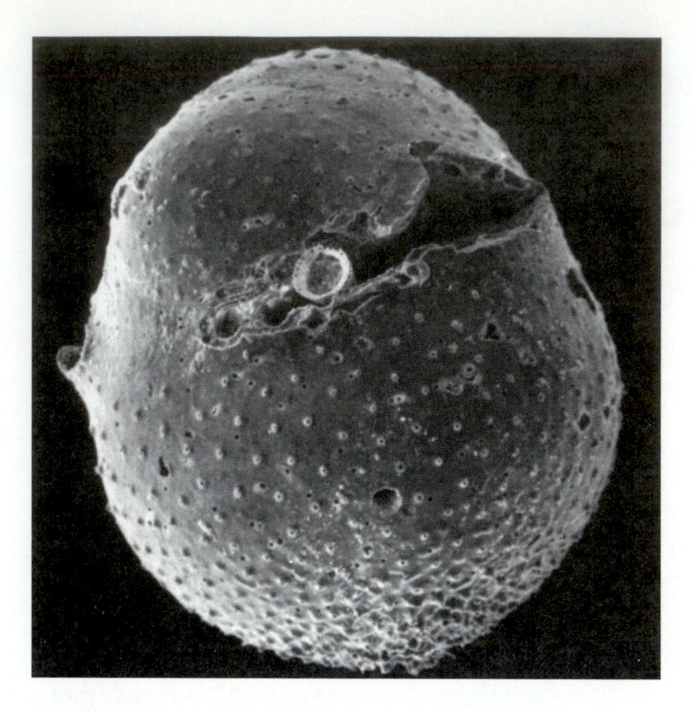

Figure 11.2. Microfossils exhibit an enormous range of sizes. The disk-shaped coccoliths (only tens of microns in diameter) are so tiny that then can fit within the holes in a foraminiferan (almost a millimeter in diameter). (Photo courtesy W. Siesser.)

sometimes even pretty) objects only decades ago. In this chapter, we will discuss such studies where they are available, although for many groups, we still know relatively little about their biology compared to most other fossil groups.

The major groups of protistan microfossils are listed in Table 11.1. We will discuss only the most important single-celled protistan microfossils, leaving the ostracodes, pteropods, and conodonts for their appropriate chapters among other phyla later in this book.

THE KINGDOMS OF LIFE

When classification first began, Linnaeus recognized only two kingdoms: plants (which produce their own food by photosynthesis; they are "self-feeding," or **autotrophs**) and animals (which derive all their food by consuming other organisms, or "feeding on others"—**heterotrophs**). As our understanding of life improved, biologists recognized that fungi (which derive their energy by decomposing other living tissues) represented a third kingdom. All other organisms, which happened to be mostly single-celled, were lumped onto a wastebasket called the "Protozoa" or later the **Protista**. One reason for this convenient but unnatural category was that many single-celled organisms (especially algae) are photosynthetic like plants, while others feed like animals, and some are both photosynthetic and have the ability to consume other organisms, so they are neither plant nor animal.

Table 11.1
Major Groups of Protistan Microfossils

Fossil	Description	Size	Composition	Habitat	Geologic Range
Foraminifera	Bubble-shaped chambers	0.01-100 mm	Organic, agglutinated, calcareous	Marine	Cambrian to Recent
Radiolaria	Spherical or conical mesh	0.03-1.5 mm	Silica	Marine	Cambrian to Recent
Diatoms	Circular, elongate, or irregular valves	<200 microns	Silica	Aquatic	Jurassic to Recent
Calcareous nannofossils	Plates, rods, stars	<50 microns	Calcite	Marine	Triassic to Recent
Dinoflagellates	Cysts	5-150 microns	Organic	Aquatic	Silurian to Recent
Acritarchs	Cysts of algae	<100 microns	Organic	Marine	Proterozoic to Recent
Silicoflagellates	Mesh of hollow rods	<100 microns	Silica	Marine	Cretaceous to Recent

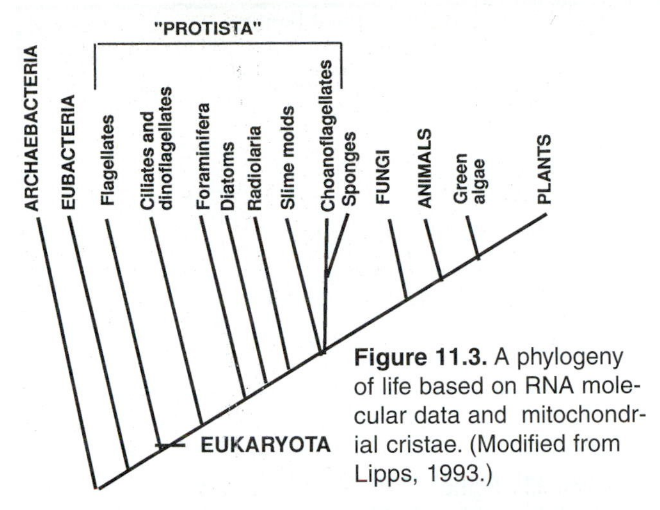

Figure 11.3. A phylogeny of life based on RNA molecular data and mitochondrial cristae. (Modified from Lipps, 1993.)

As microbiology progressed, a more fundamental distinction than single-celled vs. multicellular became apparent. The simplest single-celled organisms do not have their genetic material (nucleic acids, whether RNA or DNA) enclosed in a discrete nucleus—these are **prokaryotes**. Prokaryotes (called the Kingdom Monera in some classifications) include a wide spectrum of organisms, mostly known to us as bacteria and "blue-green algae" (properly called cyanobacteria). Organisms whose genetic material is contained within a nucleus are **eukaryotes**. Most familiar single-celled animals (the "Protista"), as well as plants, animals, and fungi, are eukaryotes. In the "five kingdom" classification scheme of Whittaker (1969), life is divided into plants, animals, fungi, protists, and prokaryotes (see Margulis and Schwartz, 1982).

Recent analyses of the genetic material of prokaryotes and protists has further rearranged the fundamental classification of life (Fig. 11.3). Woese (1987) has shown that the organisms we lump into "bacteria" are actually genetically very different. The most primitive forms, which he calls the **Archaebacteria**, include a variety of prokaryotes that are adapted for extreme temperatures, pressures, and salinities. A number of microbiologists and paleontologists regard these as relicts of the earliest forms of life, which may have been sulfur-reducing bacteria living under extreme pressure in submarine volcanic vents. The remaining prokaryotes (including most of the familiar bacteria and the "blue-green algae" or cyanobacteria) are known as **Eubacteria**. The third branch of life (including all plants, animals, fungi, and protists) are the Eukaryota. If the kingdoms were based strictly on phylogeny (see Chapter 4), then the animal, plant, and fungal kingdoms would simply be subdivisions of the Kingdom Eukaryota. However, since the ancestors of the eukaryotes were prokaryotes, and the protistans include the ancestors of plants, animals, and fungi, groups such as the prokaryotes, eukaryotes, Archaebacteria, Eubacteria, and Protista are paraphyletic. Other considerations are clearly more important than phylogenetic classification in these traditional ways of looking at life.

Even though the protistans are not a natural group, but simply eukaryotes that are single-celled yet not clearly animals, plants, or fungi, we will continue to follow tradition by considering them together in this chapter. Nonetheless, we will see that some protistans (such as diatoms and coccolithophorids) are photosynthetic like algae and are sometimes classified as plants, while others (such as the foraminifera or radiolaria) are amoeba-like forms that have many properties of animals.

SYSTEMATICS

The paraphyletic Kingdom Protista contains many different groups familiar from any microscopic examination of a sample of pond water, but only a few have skeletons and are thus readily fossilized. Of these, just four groups—foraminiferans, radiolarians, diatoms, and coccolithophores—are the dominant protistans in most micropaleontological samples. The groups conventionally placed within the Kingdom Protista that are important in the fossil record, and their most commonly encountered subgroups, include:

Kingdom Protista
Phylum Sarcomastigophora
Subphylum Sarcodina (amoebas and their relatives, including foraminifera and radiolaria; from the Greek *sarcos*, "fleshy"; they are characterized by lobe-like extensions of the cell surface, known as *pseudopodia*, which are used for locomotion and capturing food)
Class Granuloreticulosa
Subclass Rhizopoda (amoebas with thin pseudopods known as rhizopods)
Order Foraminiferida (from the Latin, *foramina*, "windows," and *ferre*, "to bear," in reference to the many pores in their shells). Over a dozen suborders are recognized, including the following:
Suborder Allogromiina (single-chambered, tubular, round, or flask-shaped test made of organic material)
Suborder Textulariina (shell made of sand grains glued together, or *agglutinated*)
Suborder Lagenina (mostly uniserial and sac-shaped foraminiferans)
Suborder Fusulinina (mostly spindle-shaped skeleton; very abundant in late Paleozoic rocks)
Suborder Miliolina (shell made of shiny, *porcelaneous* calcite, with no pores, or *imperforate*)

Suborder Rotaliina (multichambered, spiral shell; hyaline tests; includes most benthic genera)
Suborder Globigeriina (shell composed of bubble-shaped chambers arranged in an expanding spiral; all planktonic foraminifera are globigerines)
(plus five minor suborders with a limited fossil record)
Subclass Radiolaria (in reference to the radial symmetry of some taxa)
Order Spumellaria (spherical symmetry)
Order Nasellaria (conical symmetry)
Division (= Phylum) Chrysophyta ("golden-brown" algae)
Class Bacillariophyceae (diatoms)
Class Coccolithophyceae (coccolithophorids)

FORAMINIFERA

In most parts of the ocean, a typical liter of seawater from the surface will be filled with hundreds of cells, and a small tablespoon of sediment can yield thousands of shells of foraminifera (typically abbreviated to "forams" in the vernacular and some publications). The majority of these are planktonic taxa (globigerinines) that live in the surface waters and then sink to the bottom as part of the fecal matter of organisms that ate them. The rest of the specimens in the sediment are benthic foraminifera (mostly rotaliines, miliolines, and textulariines) that live on and in the sediment on the seafloor. Their density can exceed a million specimens per square meter of sediment, and weigh up to 10 grams per square meter. Indeed, much of the ocean floor (less than 4000 m deep) is covered by a calcareous ooze composed primarily of the shells of calcareous microfossils. The sand of many tropical beaches is composed entirely of the skeletons of benthic foraminifera. In a typical sample of tropical marine sediment, there may be 60 to 70 species. Over 3600 described genera and perhaps 60,000 species of foraminifera are currently recognized, making them more diverse than any other group of marine animals or plants.

Despite their small size, foraminifera are one of the most important groups in the marine biosphere, since they are a key part of the food chain. They are opportunistic feeders, capturing smaller microorganisms and detritus, and becoming food for many larger organisms.

In addition to their biological importance, foraminifera have become one of the most useful groups of fossils for solving many kinds of geological problems. Planktonic foraminifera are the global standard for correlation of marine strata during the late Mesozoic and Cenozoic, and fusulinids are the main index fossils for the late Paleozoic. Foraminifera are also very sensitive to changes in the temperature and chemistry of the open ocean, so they are the chief tool of paleoceanographers and paleoclimatologists. Paleoclimatologists use the oxygen and carbon isotopes in the shell carbonate of foraminifera to read the changes of climate and oceanic circulation. Benthic foraminifera have been widely used in statigraphic correlation, but they are also sensitive to environmental parameters that are controlled by water depth, so they are powerful indicators of paleobathymetry and depositional environment. For decades, oil company micropaleontologists have used small samples of foraminifera taken from well cuttings and cores to correlate strata, determine depositional environment, and predict paleobathymetry, greatly aiding in the search for oil. Consequently, more paleontologists (at least 2000, according to recent directories) specialize in the foraminifera than in any other group of fossils, macroscopic or microscopic.

Foraminifera were recognized and described long before most other groups of microfossils. In the fifth century B.C., the Greek historian Herodotus noticed the coin-sized spiral disks known as *Nummulites*, which are abundant in the Eocene limestones making up the Pyramids, and thought they were the petrified lentils from the meals of the slaves who built these great monuments. In 1826, Alcide d'Orbigny, one of the founders of biostratigraphy (see Chapter 10), based his early biostratigraphic zonations partially on the foraminifera, although he thought they were related to ammonites. (He also coined the word *Foraminifères*, which is the basis of the modern ordinal name.) In 1835, Felix du Jardin concluded that foraminifera were too primitive in their cellular organization to be ammonites, and realized that they were single-celled organisms. He erected much of the basis for the modern classification, and pioneered the description of their anatomy and biology. In 1884, H. B. Brady published a classic monograph from the pioneering *H.M.S. Challenger* expedition that documented which foraminifera were dredged up from different parts of the ocean floor.

The modern study of foraminifera began after World War I, when micropaleontology became important to the oil industry. The most prolific and influential foraminifer paleontologist of that period was Joseph Cushman, who published over 700 papers out of his private laboratory in Sharon, Massachusetts, and trained most of the next generation of foraminifer paleontologists. Similar pioneering work on the West Coast was stimulated by R. M. Kleinpell, who trained many micropaleontologists in the western United States during his long career at Berkeley. From the 1930s to the 1960s, M. L. Natland, Fred Phleger, Orville Bandy, and others documented the bathymetric preferences of benthic foraminifera, and established their paleoecolog-

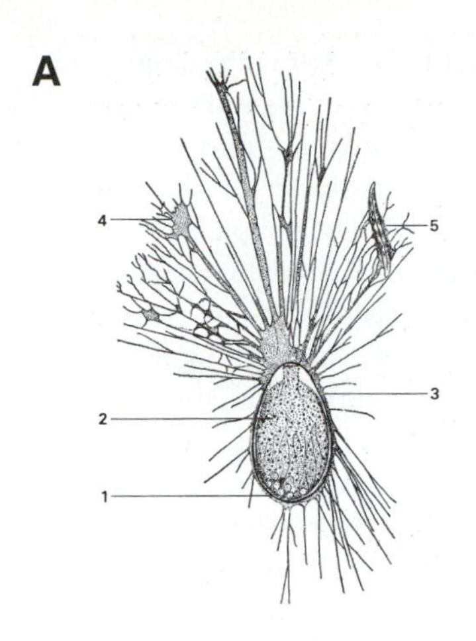

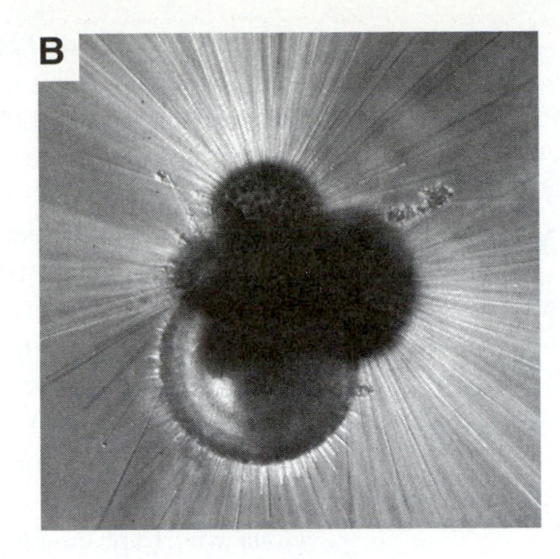

Figure 11.4. The living foraminiferan. (A) Diagram of the simple, unilocular living genus *Allogromia*. Features are: 1, test; 2 and 3, cytoplasm; 4, pseudopodia; 5, entrapped diatom. (From Boardman et al., 1987.) (B) Photograph of a living planktonic globigerinid, showing the delicate pseudopodia radiating away from the test. (From Dott and Prothero, 1994.)

ical value. At the end of World War II, the search for oil in Trinidad (where the faunas are almost entirely planktonic foraminifers) by H. M. Bolli, W. H. Blow, A. R. Loeblich, and others showed that planktonic foraminifera were not so long-ranging and difficult to use for biostratigraphic purposes (as had long been thought), but instead were one of the best organisms for marine biostratigraphy. In the 1960s and 1970s, work by W.A. Berggren and others established planktonic foraminifera as the global biostratigraphic standard for the late Mesozoic and Cenozoic. Most recently published time scales use the standard planktonic foraminiferan zones as the key that links all other biostratigraphies together.

Foraminifera are similar to amoebas except that they have an internal shell, which is usually made of calcite or aragonite. Like all other protistans, their living material (called **protoplasm**) is divided into one or more nuclei and extranuclear **cytoplasm**. Most of the protoplasm is contained within the shell, or **test**, but the foraminifer extrudes long lobes or strands of cytoplasm called **pseudopodia** from the pores in its shell to capture prey and aid in movement (Fig. 11.4). The pseudopodia can be several times as long as the test, and there can be many of them when the foraminiferan is active, and none when it is dormant. Under the microscope, the pseudopodia appear to be flowing (as in an amoeba) because of numerous granules within them that are streaming both out to the tip and back to the test. By changing the length and shape of the pseudopodia, the foraminifer can take on various shapes for locomotion and feeding.

The pseudopodia capture a variety of prey items, including diatoms, coccolithophores and other algae, radiolaria, silicoflagellates, and even crustaceans like copepods, which can be larger than the foraminiferan. Once the food is captured, it is broken down and absorbed by the surfaces of the pseudopodia, then streamed back into the aperture of the test and stored in the final, largest chamber. Meanwhile, waste products are agglomerated into small brown particles that stream outward to the tips of the pseudopodia and are excreted. In culture, foraminiferans feed several times a day. A few foraminifera are actually internal parasites, and many of the larger benthic foraminifera (probably including the extinct nummulitids and fusulinids) have symbiotic algae within their tissues that allow their rapid growth to large body size.

Although they can be sessile, foraminiferans are capable of moving as fast as 1 cm/hr, an amazing speed for an organism less than a millimeter in diameter. They can creep along surfaces, such as algal fronds or other substrates, or between sand grains in the bottom sediment, and may live as deep as 10 cm below the surface. Active foraminiferans are difficult to keep track of in a Petri dish, since they have the annoying habit of creeping up and out of the dish before the investigator is aware of it. Planktonic foraminiferans also sink and rise up and down the water column in the open ocean, probably by changing their density slightly through chemical reactions or changing the gas content in the protoplasm.

Benthic foraminifera can reproduce both sexually and asexually, usually by alternation of sexual and asexual generations (Fig. 11.5). The adult asexual form (**microspheric** form) splits its nuclei into several parts prior to asexual cloning, then apportions out its cytoplasm among the nuclei and expels them from the test. For several hours to a day or so, the young are planktonic, until (in benthic foraminiferans) they settle on the bottom and grow a new test, eventually becoming a sexual adult (**megaspheric** form). Eventually, the nucleus of the megaspheric form disintegrates and forms millions of gametes (reproductive cells), which are ejected into the water, and collides with another gamete to form a zygote, which eventually grows into a microspheric form. The entire life cycle takes from a few weeks to about a year; many are monthly, coordinated with the lunar cycles. Although the life cycle of only about 20 or 30 species has been studied, most show this basic pattern, although they may expand their sexual or asexual portions of the cycle; many reproduce asexually for many generations before reproducting sexually. In some taxa, a sexual phase has never been observed. Planktonic foraminifera have a very different life cycle, which is discussed in detail by Helmleben et al. (1989).

Although a fair amount is now known about the soft tissues of foraminifers, most taxonomy and classification is still based on the test morphology, which can be observed

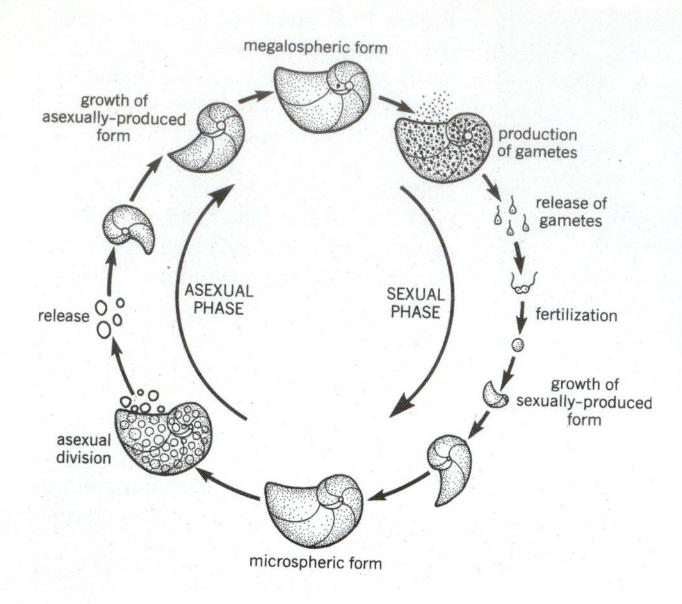

Figure 11.5. Life cycle of a benthic foraminiferan, showing the alternation between sexual and asexual generations. (From Stearn and Carroll, 1989.)

in both the living and extinct taxa. A few foraminiferans, such as the allogromiines (Fig. 11.4A), are **unilocular**, i.e., have only a single chamber. Most foraminiferans secrete calcareous tests made of successively larger chambers in different arrangements (Fig. 11.6). The simplest of these is a single row of chambers (**uniserial**), or a double row (**biserial**), or a triple row (**triserial**). By far the most common arrangement, however, is a coil of chambers increasing in size around the spiral. Some coil in a single plane (**planispiral**), but others coil up an axis like a the shell of a snail (**trochospiral**). The raised area in the center of the spire is called the **umbo**, and the depression on the other side is known as the **umbilicus**. If the chambers of the previous whorl are enveloped by later chambers, the test is **involute**; if they are visible, it is **evolute**. Some taxa are evolute on one side, involute on the other. In other forms, the chambers are added in several planes, giving a characteristic triloculine (three-chambered) or quinqueloculine (five-chambered) asymmetric shape. All foram tests are made of chambers, and the largest chamber has an opening called an **aperture**. The aperture may have a number of features, such as one or more teeth, or a variety of grillworks (known as cribrate structures) across them. In some

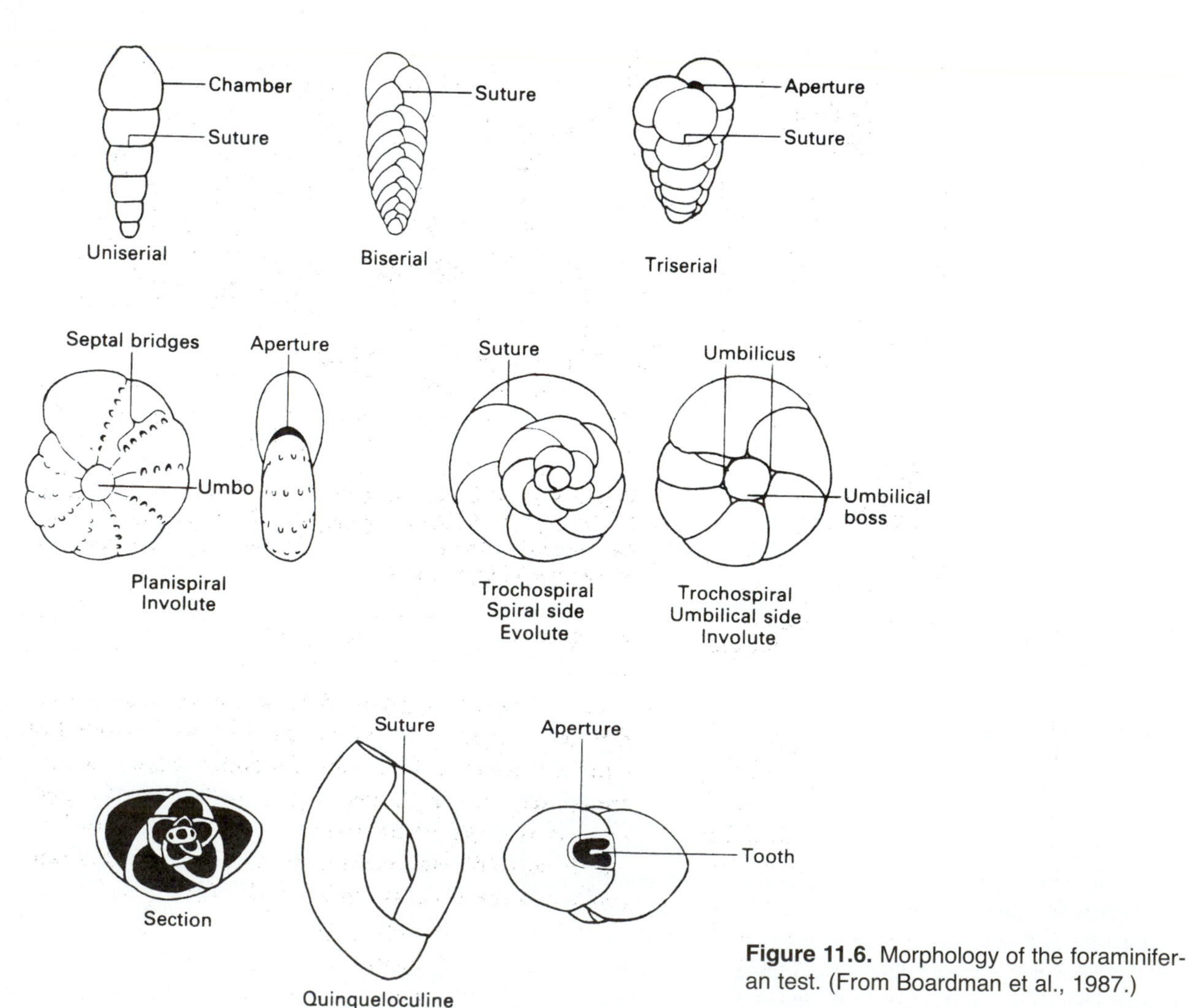

Figure 11.6. Morphology of the foraminiferan test. (From Boardman et al., 1987.)

foraminiferans, there are **secondary apertures** in various locations along the test. The lines between the chambers are known as **sutures**, and many have elaborate features (such as **septal bridges**) on their sutures. The test can also be decorated with a variety of combinations of spines, ribs, tubercles, and keels along the edge.

The wall structure of the test has long been important in foraminiferal classification. Some foraminiferans have organic-walled tests (such as the allogromiines), although these are seldom fossilized, and often destroyed by the sample processing of living assemblages. The **agglutinated** foraminiferans (textulariines) make skeletons out of cemented sand grains and other foreign particles, including tests of smaller foraminifera and sponge spicules. Fusulinids secrete tests made out of microgranular calcite, which has a sugary appearance under thin section. Many also have pores in their tests (**perforate**), although some are completely solid (**imperforate**). Some foraminiferans are imperforate on one side and perforate on the other. A few imperforate groups (especially among the miliolines) arrange their calcite crystals in a fashion that gives them a smooth, shiny appearance called **porcelanous** (because it resembles porcelain); these taxa also have a high concentration of magnesium in their calcite. Most foraminiferans have a wall structure known as **hyaline**, because it has a glassy appearance caused by the secretion of many layers, with very tiny pores 1 to 5 microns in diameter.

Ecology

Perhaps one of the most important developments in the last few decades in the study of foraminifera has been the realization that they are very powerful tools for paleoecology. Some benthic foraminifera are restricted to a narrow range of depths or substrates, or have a characteristic tolerance of variations in temperature and salinity. The majority of benthic foraminifera tolerate only a little variation from the normal marine salinity of 35‰, but *Discorbinopsis* tolerates salinities as high as 57‰ and a variety of forms (such as *Ammonia, Rosalina, Ammobaculites, Trochammina*, and others that are characteristic of marshes, estuaries, and lagoons) can tolerate salinities that fluctuate from normal marine to almost freshwater. Most foraminiferans are found in areas of normal oxygen concentration, although deep-water forms (such as *Bulimina, Uvigerina* and *Bolivina*) tolerate very low oxygen levels. Some foraminiferans (such as *Bulimina* and *Bolivina*) are even tolerant of areas of high nutrient concentrations, such as sewer outfalls, probably because these areas tend to be low in oxygen as well. In these circumstances, there are many individuals of just a few species.

A number of benthic foraminifera are known to have very specific depth and substrate preferences, and these are good indicators of paleobathymetry. Marshes are characterized by low diversities of agglutinated forms such as *Ammobaculites* and *Trochammina*, while lagoons and estuaries typically shelter perforate calcareous taxa such as *Ammonia, Elphidium,* and *Rosalina* (Fig. 11.7). The nearshore environment (0 to 50 m water depth) contains not only the lagoonal forms such as *Ammonia, Elphidium,* and *Rosalina*, but also porcelanous miliolines such as *Quinqueloculina* and *Triloculina*. Large foraminifera that have symbiotic algae (including the extinct nummulitids and fusulinids) also are found in very shallow waters because their symbionts require light. In all these shallow environments, planktonic foraminifers are rare or absent. In the outer shelf (50 to 200 m), the faunas are dominated by perforate rotaliine taxa such as *Cassidulina, Amphistegina, Nonion*, and *Nonionella*; planktonic foraminiferans are much more common. Bathyal depths (200 to 2500 m) tend to be dominated by biserial and triserial foraminiferans, such as *Bolivina, Bulimina*, and *Uvigerina*, along with *Cassidulina*, and imperforate porcelanous miliolines such as *Pyrgo*. In sediments at this depth, planktonic foraminiferans are even more abundant, since the waters above are rich in them and they rain down continuously. Abyssal depths (2500 to 7000 m) tend to be dominated by agglutinated forms because the water may be below the **carbonate compensation depth** (**CCD**), the depth below which calcite dissolves (typically 4000 to 5000 m in most parts of the ocean). Above the CCD, however, the abyssal sediment is dominated by planktonics raining down from the surface. Of course, these generalizations are oversimplified, because water temperature and substrate also play an important role. For example, the cold-tolerant foraminifera that are found in deeper waters in the tropics are found in much shallower depths in the polar regions, where the surface and intermediate waters are so cold.

Most planktonic foraminifera prefer to live in a particular part of the water column, so they are good indicators of the temperature and salinity of the water masses in which they lived. The majority live in the upper 1000 m of the water column, although their density is greatest in the 100 m near the surface (the photic zone), where the water is warmest and there is some light. The thin-walled, spherical, symbiont-bearing spiny globigerinids prefer the surface waters where there is light, while those with thicker walls and keeled shells (and lacking in photosymbionts) tend to be found at slightly greater depths (below the photic zone). Latitudinal distribution along temperature gradients and water masses is even more critical to planktonic foraminiferans. Most species are restricted to a single region, such as the tropics, the subtropics, the temperate latitudes, or the high latitudes, and can be used as an indicator of water temperature. For example, the thick-shelled *Neogloboquadrina pachyderma* is a polar indicator species, but *Globigerina bulloides* prefers subpolar waters. *Globotruncana truncatulinoides* is characteristic of temperate latitudes. *Globorotalia menardii* and the pink *Globigerinoides ruber* are found in the subtropics and tropics, while *Globigerinoides sacculifer* is a strictly tropical indicator. *Globorotalia menardii* switches from left coiling during cold conditions (see Fig. 10.4) to right coiling in warmer times (as do *Globotruncana truncatulinoides* and *Neogloboquadrina pachyderma*).

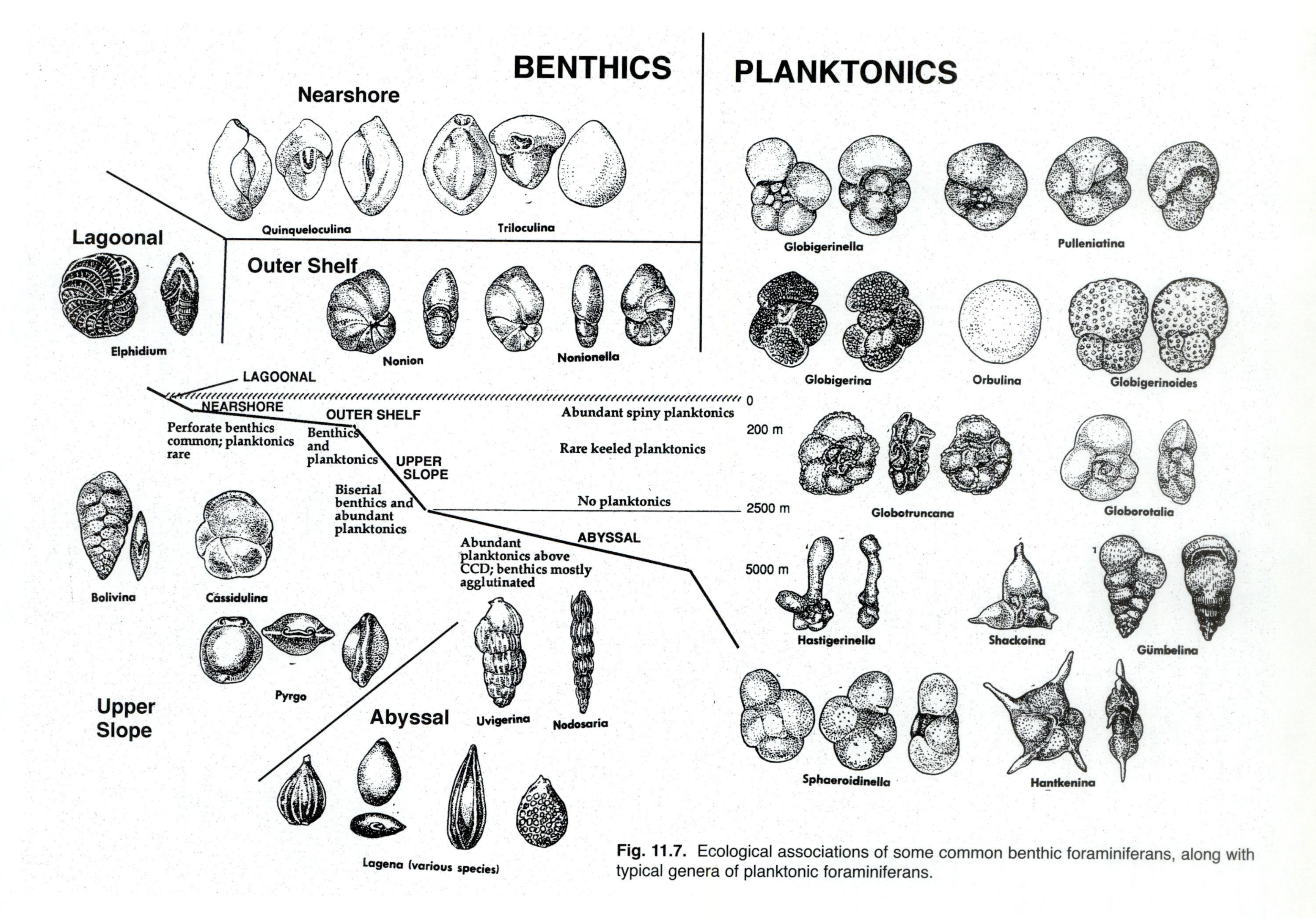

Fig. 11.7. Ecological associations of some common benthic foraminiferans, along with typical genera of planktonic foraminiferans.

Evolution

Origins—The earliest known foraminifera are cylindrical agglutinated tubes from the earliest Cambrian, so foraminiferans have a fossil record as old as any other phylum of eukaryotic organism. Foraminiferans remained simple through the early Paleozoic, with enrolled tubular forms appearing in the Silurian, and multichambered agglutinated forms appearing in the Devonian. By the Carboniferous, there were abundant biserial, triserial, and trochospiral foraminifera. For example, the genus *Endothyra* is so abundant in some Mississippian limestones that it resembles an "oolitic" limestone. These "Bedford oolites" (from Bedford, Indiana) are among the building stones known as the "Indiana limestone."

Fusulinaceans—The most spectacular Paleozoic foraminifera were the fusulinaceans, which flourished in the late Carboniferous and Permian, especially in shallow, tropical carbonate environments, where they can make up the bulk of the sediment. In some regions, limestones are made of solid fusulinaceans, with trillions of individuals covering hundreds of square miles. From simple forms like *Millerella* (Fig. 11.8), they diversified into about 150 genera and over 6000 species, and they are so abundant and distinctive that they are the primary index fossil of the late Paleozoic. They can be correlated across all the continents with a tropical Carboniferous-Permian record, and most of the standard fusulinacean zones are only 2 to 3 million years in duration.

During their evolution, some fusulinaceans increased in size until they reached almost 1 cm in length, but a few giant fusulinaceans were as long as 10 cm (4 inches). That's a pretty impressive skeleton for a single-celled organism! Fusulinaceans are most easily identified by the distinctive structure of their walls, with its intricate fluting and convolutions. The convolutions of the body wall increased the surface area needed for single-celled metabolic exchange, and may have helped shelter algal symbionts, which were necessary if these "microfossils" were to reach such large size.

Although they were most abundant in the shallow tropical seas around Pangaea, nevertheless fusulinaceans were distributed into several discrete faunal provinces, including one in the Tethyan realm (on the eastern coast of Pangaea), and another on the western margin of Pangaea (the Midcontinent-Andean realm). Because of this provinicality, fusulinaceans have proved very useful in determining the ancient position of exotic terranes (see Fig. 9.13). For example, the presence of Tethyan fusulinaceans in terranes of central British Columbia tells us that those terranes originated in the tropics, and also from the other side of the Pacific Ocean.

Finally, the fusulinaceans began to decline during the last age of the Permian, and then disappeared completely during the great Permian crisis (see Chapter 6). They were probably the most severely affected of all victims of this extinction event, since they had been so successful during most of the Permian (unlike other victims, such as trilobites, blastoids, tabulates, and rugosids, which were already in decline long before the Permian).

Mesozoic foraminifera—Although the fusulinaceans were the dominant benthic foraminifers during the late Paleozoic, the miliolines and lagenines also arose during the Carboniferous. After the great Permian extinction, these benthic groups underwent a great evolutionary radiation. By the Triassic, all the benthic foraminiferal suborders had originated and begun to diversify. At this point benthic foraminiferans began to occupy new niches, such as marshes and lagoonal habitats, and through the Mesozoic, they also began to occupy deeper habitats. By the late Cretaceous, they exhibited the depth zonation that we see today.

Planktonic foraminifera also first appeared in the Mesozoic (Middle Jurassic) with fossils known from the shallow epicontinental seas of the Tethys and Europe. By the Late Jurassic, they had become worldwide in distribution, and during the Cretaceous they underwent a great diversification, with over 300 species recognized (the highest diversity they ever attained). Indeed, their great abundance during the Cretaceous (along with the blooming of calcareous nannoplankton) is responsible for the thick chalk deposits that gave the Cretaceous its name. Their abundance, diversity, worldwide distribution, and rapid evolution makes them excellent index fossils for the Cretaceous, especially in regions where ammonites are scarce or the standard ammonite zones are difficult to recognize.

During the great K/T extinction event, planktonic foraminifera suffered severely. Only three species, *Guembelitria cretacea* and two species of *Hedbergella*, survived the K/T event, and they were apparently the ancestors of all Cenozoic planktonic foraminifera. Benthic foraminifera, on the other hand, were not severely affected. Although some micropaleontologists have argued that the shallow tropical benthics underwent severe extinction, other authors have disagreed, and showed that many of these "extinct" forms reappear as Lazarus taxa in the Paleocene. If there was an extinction event in the tropical benthic foraminifera, all authors agree that the deep-water foraminiferans were unaffected by whatever happened to the planet 65 million years ago.

Cenozoic foraminifera—After the K/T extinction, planktonics recovered and diversified. However, their evolution was punctuated by another set of extinction events. The most severe extinction of benthic foraminifera occurred at the Paleocene/Eocene boundary, probably in response to some abrupt bottom-water change that is still poorly understood. By the warm tropical conditions of the middle Eocene, however, both benthics and planktonics were at their peak, with well over 200 species of planktonic foraminifera recorded by the middle Eocene. The benthic foraminifera also diversified into new niches. For example, the coin-shaped rotaliines known as orbitoids, including the nummulitids, discocyclinids, and lepidocyclinids, diversified into as many as 20 genera in 6 families. Some

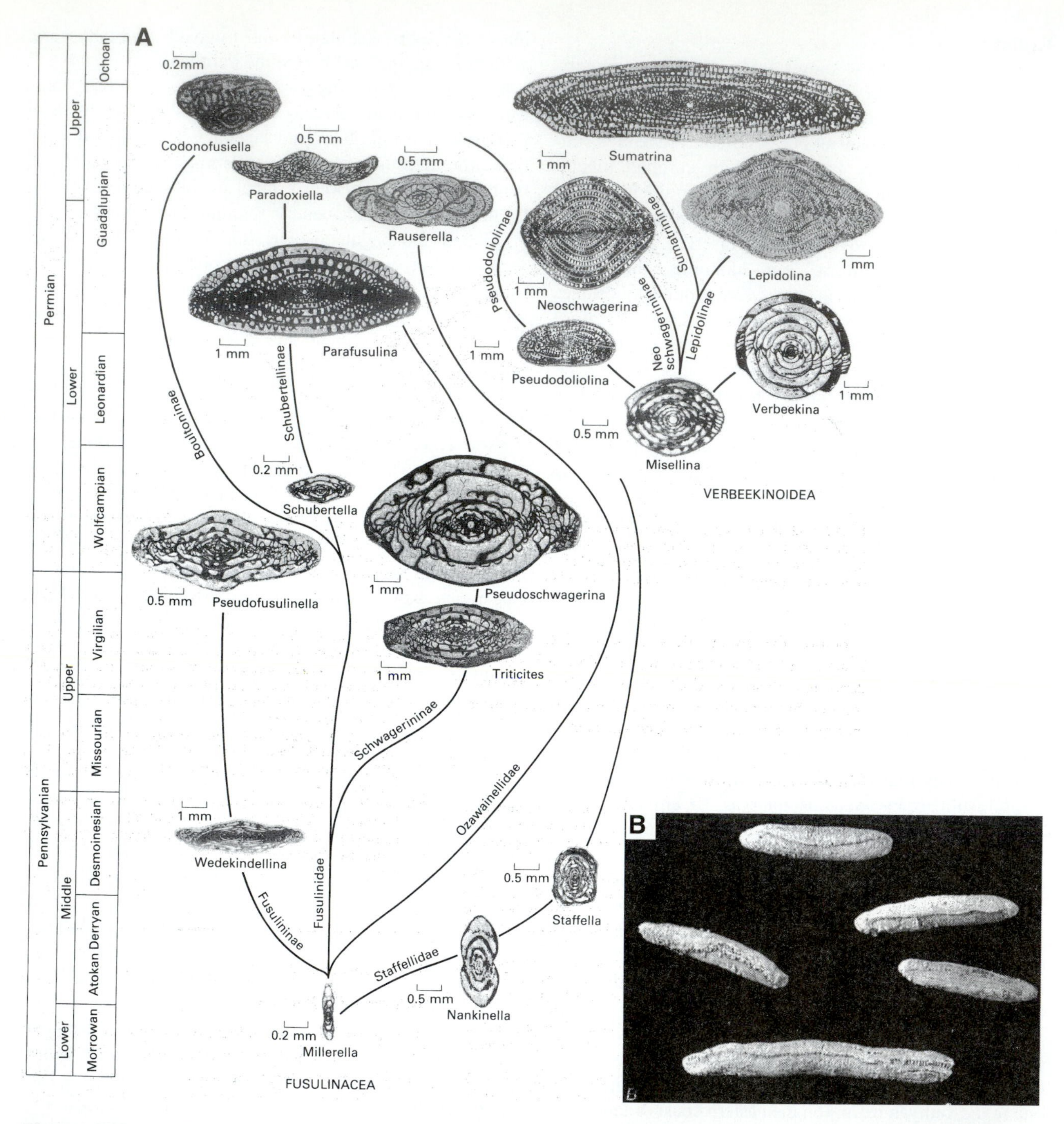

Figure. 11.8. (A) Evolution of the late Paleozoic fusulinid foraminiferans. (B) Exterior views of some large fusulinids (actual size), which were normally about the size and shape of a grain of rice. (Modified from Boardman et al., 1987.)

were true giants, reaching up to 3 cm in diameter. Again, these "microfossils" probably had endosymbiotic algae in their chambers to allow such rapid growth (especially for a single-celled amoeba-like organism). The large nummulitids and related orbitoids were most diverse in the shallow tropical seas of Tethys during the middle Eocene, where they were so abundant that they make up the limestones of the Pyramids.

The Eocene-Oligocene transition marks the second great extinction in the foraminifera, with a series of events occurring starting at the end of the middle Eocene (37 Ma), when mostly tropical, warm-water taxa were affected. As the planet became cooler and the first truly cold Antarctic bottom waters formed, there were additional extinctions, causing a great reduction in size and diversity of foraminifera by the middle Oligocene. The large tropical nummulitids and their orbitoid kin were decimated (only a few survive today in the tropical southwest Pacific), and most of the planktonic and benthic survivors of the Oligocene were small, hardy, cold-tolerant taxa with long

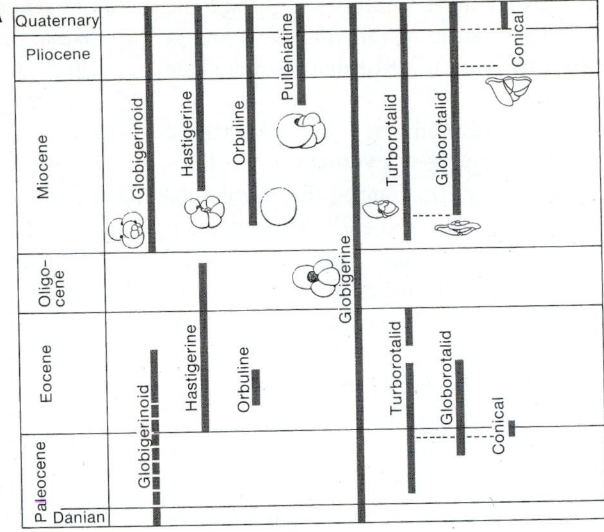

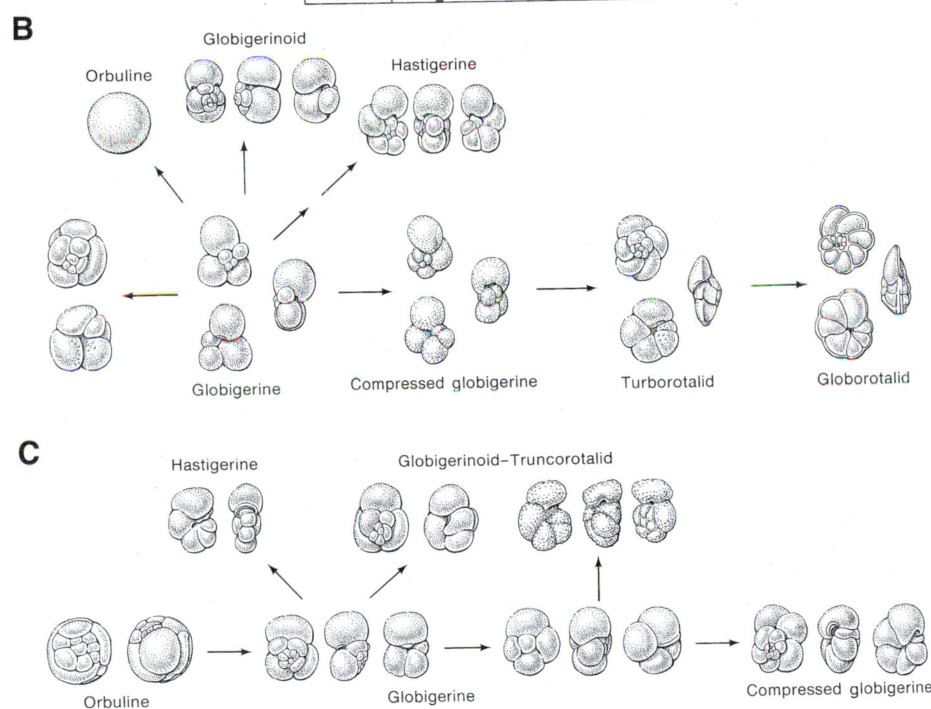

Figure 11.9. Iterative, or repeated evolution in the Cenozoic planktonic foraminifera. (A) A mass extinction of most of the foraminifer test types during the late Eocene and Oligocene left only the globigerines to repopulate the oceans in the Miocene. (B) Initial Paleogene adaptive radiation of the orbuline, globigerinoid, hastigerine, turborotalid, and globorotalid morphologies after the K/T extinctions. (C) A second Miocene radiation of foraminiferans produced the orbuline, hastigerine, globigerinoid, truncorotalid, and compressed globigerine morphologies all over again, although the detailed structure of these forms is different enough that they can be distinguished from their Eocene counterparts. (From Cifelli, 1969.)

stratigraphic ranges (Fig. 11.9). In fact, the last appearance of one of the characteristically Eocene groups, the spiny planktonic hantkeninids, is used to mark the Eocene/Oligocene boundary.

After the Oligocene cooling, the foraminifera (especially the planktonics) recovered and radiated again during the warm early Miocene (Fig. 11.9). However, the permanent refrigeration of the Antarctic in the middle Miocene (about 16 Ma) led to further extinctions, especially of benthic foraminiferans adapted to warmer conditions, and to shifts in species distribution and abundance. Since the onset of glacial conditions, foraminifera have become highly useful as indicators of oceanographic changes, especially in temperature and circulation. For that reason, they are the primary tool of most paleoceanographers.

Evolutionary patterns—Because foraminifera have such a densely continuous fossil record, they have excellent potential for studying evolutionary patterns. Unlike the discontinuous records of fossils from the shallow marine or terrestrial realms, planktonic microfossils from deep-sea cores can be sampled in enormous quantities over continuous spans of time, and we can examine their geographic variation over the entire world's oceans with very high-resolution time control (Prothero and Lazarus, 1980). Planktonic microfossils could be the "*Drosophila* of paleontology" in their potential to test evolutionary patterns. Yet when Lipps (1981) wrote his review noting the lack of paleobiological interest among micropaleontologists, little had been done. Since that time, however, micropaleontologists have documented evolutionary patterns with a vengeance, as typified by an entire symposium of the topic published in the Fall, 1983, issue of *Paleobiology*, and at least one or two papers in every issue of that journal since then.

A striking pattern observed in the planktonic foraminifera is **iterative** (repeated) **evolution**. We have seen how foraminiferans diversified during the warm conditions of the Cretaceous, Eocene, and early Miocene, and were decimated during the K/T and Eocene-Oligocene extinctions. After each extinction event, however, certain foraminiferal morphologies evolved all over again from the ancestral globigerine morphotype (Fig. 11.9). For example, the keeled globotruncanid/globorotalid morphotype independently evolved three times in the Cretaceous, Paleocene-Eocene, and early Miocene. The flattened turborotaliids, the conical shapes, the ball-shaped orbulines, the planispiral hastigerines, and the globigerinoid morphologies all reappeared in the Miocene after the extinction of the species showing these morphologies in the Eocene or Oligocene. This highly repetitive type of evolution suggests that there are specific evolutionary niches or limited evolutionary potential, so that foraminiferans continue to reoccupy the same morphological niches once the warm times have returned. In fact, the similarity between the similar convergent morphotypes (**homeomorphs**) is so great that it confused foraminiferal systematics for many years. This can be a problem for biostratigraphers, and also goes against the general tendency for evolution not to reverse itself or repeat itself.

Knoll (1987) summarized a wide spectrum of studies that had been done on patterns of evolution in the foraminifera by that date. Although numerous examples of both gradualism and punctuation had been documented (see Chapter 4), one of the most striking observations was that many lineages showed "punctuated gradualism"—periods of stasis punctuated by periods of gradual (rather than rapid) evolution, a pattern seldom observed in metazoans. As Lazarus (1983) pointed out, however, the allopatric speciation model of Mayr (1942) may not be very appropriate for protistans, and therefore they show patterns not predicted either by the punctuated equilibrium model or the gradualistic model. For one thing, many protistans are asexual clones through much or most of their life cycle, so they do not have populations or gene flow in the traditional metazoan sense. In addition, protistan populations span billions of individuals that spread across entire oceanic water masses, so there is little chance for small peripheral populations to differentiate. Finally, it is not clear how much of the change seen in microfossils is true genetic-based evolution, and how much is ecophenotypic change in response to environmental conditions. For all these reasons, the patterns of change seen in microfossils may not tell us much about metazoan evolution, even though the quality of their record is so much better.

For Further Reading

Boersma, A. 1978. Foraminifera, pp. 19-78, *in* Haq, B.U., and A. Boersma, eds. *Introduction to Marine Micropaleontology*, Elsevier, New York.

Brasier, M. D. 1980. *Microfossils*. George Allen & Unwin, London.

Broadhead, T. W., ed. 1982. *Foraminifera: Notes for a Short Course*. University of Tennessee Dept. Geological Sciences, Studies in Geology 6.

Buzas, M. A. 1987. Smaller foraminifera, pp. 72-81, *in* Boardman, R. S., A. H. Cheetham, and A. J. Rowell, eds. *Fossil Invertebrates*. Blackwell Scientific Publishers, Cambridge, Mass.

Culver, S. J. 1987. Foraminifera, pp. 169-212, *in* Lipps, J. H., ed., *Fossil Prokaryotes and Protists: Notes for a Short Course*. University of Tennessee Dept. Geological Sciences, Studies in Geology 18.

Culver, S. J. 1993. Foraminifera, pp. 203-247, *in* Lipps, J. H., ed. *Fossil Prokaryotes and Protists*. Blackwell Scientific Publishers, Cambridge, Mass.

Douglass, R. C. 1987. Larger foraminifera, pp. 81-91, *in* Boardman, R. S., A. H. Cheetham, and A. J. Rowell, eds. *Fossil Invertebrates*. Blackwell Scientific Publishers, Cambridge, Mass.

Haynes, J. R. 1981. *Foraminifera*. Macmillan, London.

Hedley, R. H., and C. G. Adams, eds. 1974. *Foraminifera*. Vols. 1-3. Academic Press, New York.

Helmleben, C., M. Spindler, and O. R. Anderson. 1989. *Modern Planktonic Foraminifera*. Springer-Verlag, Berlin.

Kennett, J. P., and S. Srinivasan. 1983. *Neogene Planktonic Foraminifera: A Phylogenetic Atlas*. Hutchinson Ross, Stroudsburg, Penn.

Lipps, J. H., W. H. Berger, M. A. Buzas, R.G. Douglas, and C. A. Ross. 1979. *Foraminiferal Ecology and Paleoecology*. SEPM Short Course 6.

Loeblich, A. R., Jr., and H. Tappan. 1988. *Foraminiferal Genera and their Classification*. Vols. 1-2. Van Nostrand Reinhold, New York.

Tappan, H., and A. R. Loeblich, Jr. 1988. Foraminiferal evolution, diversification, and extinction. *Journal of Paleontology* 62:695-714.

RADIOLARIA

Some people think of the radiolaria as exquisitely designed, delicate objects made up of a lacework of glassy bars, struts, and spines (Fig. 11.10). They have been compared to "Christmas ornaments" and "snowflakes" in their delicate, beautiful symmetry. Like foraminiferans, radiolarians consist of a blob of protoplasm with a internal skeleton, but their skeleton is made out of opaline silica, not calcite. (A closely related group, the acantharians, make their skeletons of strontium sulfate, while another, the phaeodarians, have organic skeletons; for obvious reasons, they have no significant fossil record.) Radiolarians range in size from about 60 to 200 microns, or less than half the size

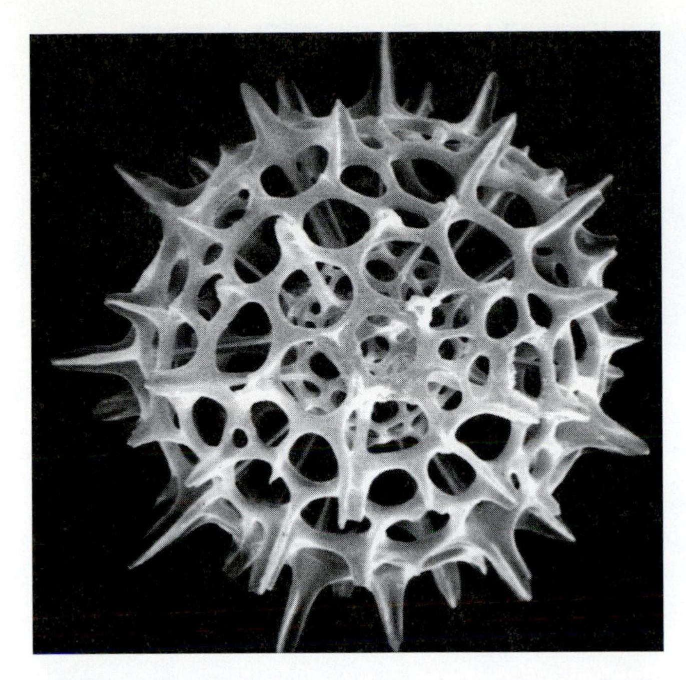

Figure 11.10. The delicate structure and symmetry of radiolarian skeletons has often been compared to snowflakes or Christmas ornaments. (From Haq and Boersma, 1978.)

of typical foraminifera. They have a continuous record from the Cambrian to the Recent, live in the marine plankton (none are benthic) from the poles to the equator, and are as diverse as foraminiferans, although not as well studied by as many micropaleontologists. They are very sensitive to differences in water temperature and chemistry, so they are excellent paleoceanographic indicators. They evolved rapidly, so their biostratigraphy is very useful, especially in Arctic and Antarctic oceanic sediments that are rich in silica but poor in calcite.

Like foraminifera, radiolaria are related to amoebas, but their protoplasm is divided into an inner **endoplasm** and an outer **ectoplasm** by a perforate organic membrane called a **central capsule** (Fig. 11.11). In the center of the ectoplasm are one or more nuclei, while the ectoplasm contains many frothy, gelatinous bubbles, known as the **calymma**, and may also contain symbiotic algae. The siliceous skeleton encloses the central capsule and endoplasm, but only the inner part of the ectoplasm and calymma, so the skeleton is completely within the the cytoplasm (unlike the external skeletons of many microplankton). Thick spines known as **axopodia**, and thin thread-like **filopodia** typically originate from the center of the cell and protrude from the outer membrane. The axopodia are sticky, and are used to trap food particles (mainly coccolithophorids, diatoms, bacteria, and other tiny plankton). Once trapped, the food is ingested by vacuoles within the calymma, and eventually passed through the central capsule to the endoplasm as it is digested.

In contrast to the foraminifera, much less is known about the biology of radiolarians. Like other plankton, they are very hard to culture or even keep alive after they have been retrieved from plankton tows. The best information comes from specimens that have a froth of bubbles around them, so they are large enough to be seen and captured in a jar by scuba divers. Radiolaria have been observed to undergo asexual reproduction through simple binary fission, although budding and multiple fission have also been reported. Typically, the daughter cells disperse from the parent skeleton, which then sinks to the bottom. Sexual

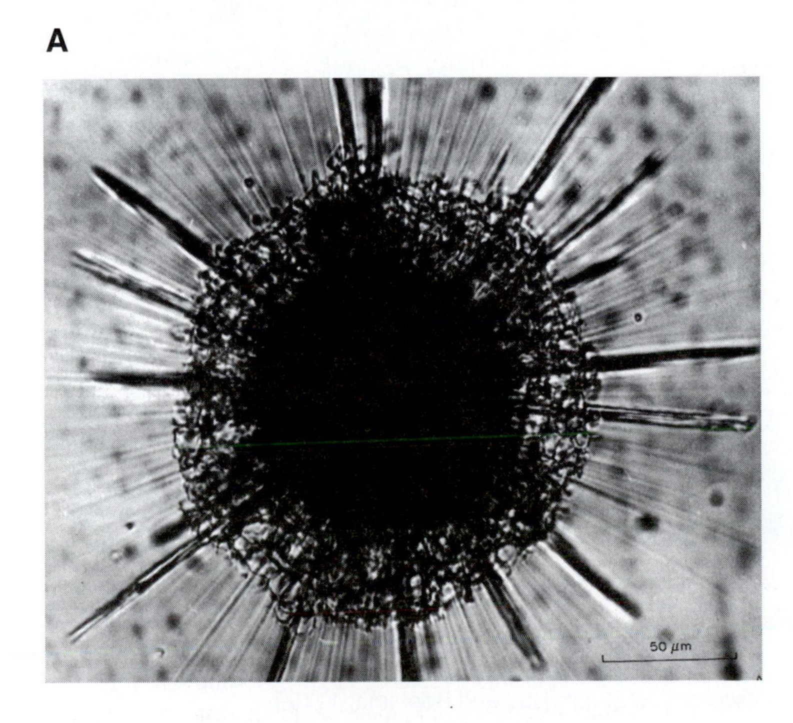

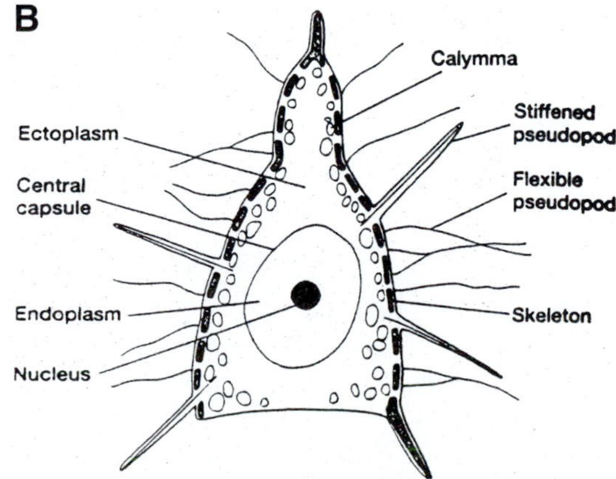

Figure 11.11. (A) Photograph of a living radiolarian. From the mass of bubbles enclosed in protoplasm protrude thick skeletal spines and thin axopodia. The dark central area is the central capsule. (B) Morphology of the radiolarian test. (After Haq and Boersma, 1978.)

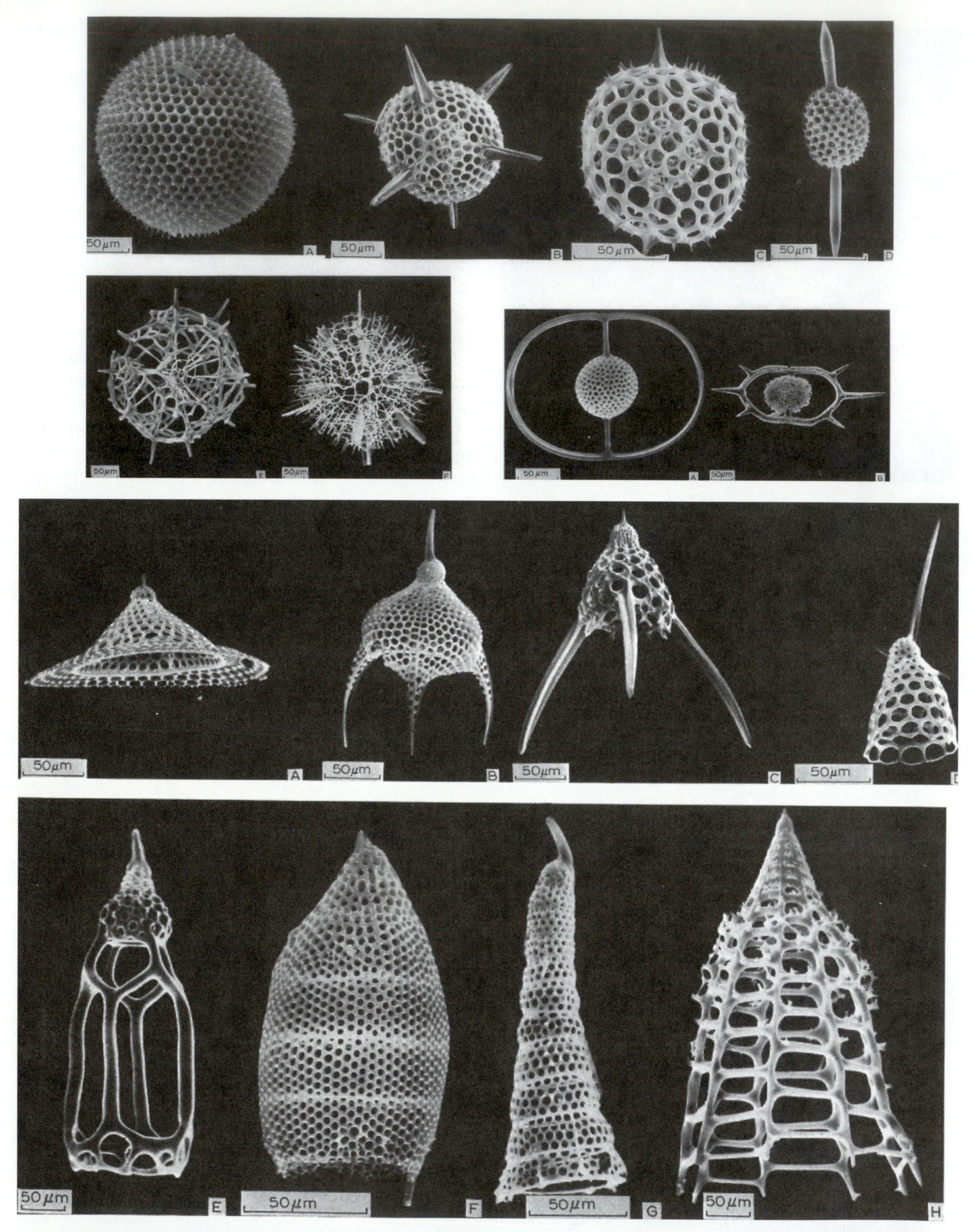

Figure 11.12. Radiolarians exhibit a great variety of test shapes. The top two rows are all spumellarians, which tend to be radially symmetrical around a central point. The bottom two rows are nassellarians, which are symmetrical around an axis, so they tend to be shaped like cones, cylinders, bells, or helmets. The top row consists of actinommid spumellarians, as are the two on the left of the second row. The two on the right of the second row are the appropriately-named saturnalins. All of the two bottom rows are theoperid nassellarians. (From Haq and Boersma, 1978.)

reproduction has been suggested, but not confirmed, by the presence of tiny flagellated cells, called swarmers, that may be the gametes. However, no one has yet proved that these swarmers unite to form a zygote in true sexual reproduction. The presence of strong dimorphism in the skeletons of some species also suggests that radiolarians have an alternation of generation, like foraminiferans and many other organisms.

The life span of radiolaria is a few months at best. In laboratory cultures, they form swarmers after about two weeks, then die. Field experiments counting specimens caught in sediment traps, and comparing them with living populations, suggest that radiolarians live from a few days to a few months, with most living about 2 weeks.

In addition to the problems of keeping captured specimens alive, the ontogeny of radiolaria is virtually unknown. Not only has it been impossible (so far) to get them to reproduce and grow in culture, but the skeletons of juveniles are seldom studied, since the incomplete juvenile shells are difficult to match to the corresponding adult forms.

Since so little is known about the living organism, all radiolarian classification must be based on the skeleton. Three major groups of living radiolaria are recognized, of which two—the **Spumellaria** and the **Nassellaria**—are common as fossils (Fig. 11.12). Spumellarians tend to have a radial symmetry around a central point, exhibiting shapes ranging from balls and spheres to disks, cylinders, stars, and even spheres with a ring around them (appropriately named the saturnalins). Nassellarians tend to be based on variations on a conical shape, with radial symmetry around a long axis. They come in many shapes, including a variety of cones, cylinders, teardrops, bells, and helmet-like forms. Other groups of radiolaria with different architectures are known from the Paleozoic and Mesozoic.

Ecology

Radiolaria are found in nearly all marine waters, sometimes in abundances as great as thousands of cells per liter of seawater. They live as deep as more than 1000 m, although most species live in the uppermost 100 m (especially those with symbiotic algae, which require light). Radiolaria show depth stratification, with certain species favoring shallow waters (especially the spumellarians), others found in intermediate or deep waters (especially the nassellarians). Of the approximately 400 to 500 species commonly found in the modern oceans, about 200 live in the shallow, warm, tropical waters, 40 to 50 in high-latitude, shallow, cool waters, 150 to 200 in waters deeper than 200 m, and about 40 to 50 live in a wide variety of depths. Many species that live in cold, deep waters in the tropics are found in the cold shallow waters of the subpolar regions. This shows how strongly they track the temperature gradients in the oceans.

To maintain their position in the water column, radiolarians regulate their buoyancy. The presence of fat globules and gas bubbles helps reduce density, while the abundant spines increase frictional resistance to sinking and allow them to be buoyed up by even the slightest current. Many radiolaria have disk-like or cup-like shapes that are effective in resisting sinking.

The distribution of radiolaria is controlled by a number of factors. The most important is the presence of nutrients, and the competition for resources. Many radiolaria living in the shallow photic zone contain symbiotic algae, which is their main food source. They are especially common in nutrient-poor regions of the ocean, where the symbiosis helps sustain them in the absence of abundant food. Others, such as the closely related phaeodarians (which have partially organic, rather than purely siliceous skeletons), live off detritus and bacteria in deep waters (1000 m or deeper). In areas of abundant resources and many competing plankton, radiolaria tend to be less common, possibly due to competition or dilution of sediment samples by the more common plankton. In addition, where large blooms of diatoms are occurring, radiolaria tend to be outcompeted, possibly because diatoms are more efficient at growth, reproduction, and silica utilization than are radiolaria. However, in silica-rich regions of the ocean with few other plankton, and below the photic zone rich in diatoms, radiolaria dominate.

The best predictor of radiolarian abundance is the pattern of oceanic circulation and upwelling. The biogeographic provinces recognized in the radiolaria correspond to the different water masses within the ocean (see Figs. 9.6, 9.7). The highest abundances occur at the meeting point between the circulating currents (or gyres) surrounding two water masses, where oceanic upwelling occurs. This upwelling brings up scarce nutrients (especially the critical ones, such as silica) from the deeper parts of the ocean, where the siliceous microfossils (especially diatoms and radiolaria) rapidly utilize it. The highest diversity of radiolaria occurs in the subtropical gyres surrounding the equatorial water masses. In such places, there are hundreds of species and thousands of individuals per cubic liter of seawater. Most of these are also symbiotic forms, which may help them dominate in this region in spite of the competition from diatoms. The temperate-subpolar regions is a transitional province, which contains a moderately diverse fauna dominated by cold-water varieties of warm-water species, and a few endemic forms. The subpolar gyres (especially the Circum-Antarctic Current) are major areas of upwelling, rich in silica, but they are dominated by diatoms, with relatively low diversities of radiolarians.

Another factor that dictates the abundance of radiolaria in the sediments is the depth of the bottom and the CCD. Because the radiolaria have an internal skeleton, they are less prone to dissolution, since the skeleton is protected by protoplasm as it sinks through the corrosive waters near the surface that are depleted in silica. In addition, many radiolaria are the food of larger organisms, and their resistant skeletons are left undigested in fecal pellets, where they remain protected until those pellets reach the bottom and break up. In tropical seafloors below the CCD, the abundant calcareous foraminifera and nannofossils have all been dissolved, leaving a residue of siliceous microfossils (called a siliceous ooze). In some silica-rich bottom waters, it is possible for enormous quantities of siliceous microfossils to accumulate. Such sediments are known as radiolarites if they are predominantly radiolarians, and can become chert when they are diagenetically altered into rock.

Because radiolaria are so strongly controlled by the temperature and chemistry of water masses, they have become powerful tools of paleoclimatology and paleoceanography. Like planktonic foraminifera, certain radiolarian species are well known to be indicators of warm or cold conditions in the ocean, and can be used to reconstruct ancient water masses and temperatures. For example, Project CLIMAP (Climate, Long-range Intepretation, Mapping And Prediction) used radiolaria as a major indicator of sea-surface temperatures during various phases of the Ice Ages. Because the abundance of silica is so important, radiolaria are excellent tools for reconstructing paleoceanographic changes that cause differences in circulation and upwelling. In other cases, certain species of radiolaria are endemic to certain water masses, so their appearance can be used to track the history of the water mass and determine its antiquity. Since siliceous microfossils are most abundant at depths below the CCD, they also help in estimating paleobathymetry. Finally, radiolaria are so resistant to dissolution that radiolarian cherts have become one of the best tools for detecting the origin of exotic terranes, especially those that have migrated long distances across the Pacific (see Chapter 10). It was radiolaria, for example, that showed that the terranes underlying the north end of the Golden Gate Bridge came from the Asian tropics in the Mesozoic.

Evolution

Evolutionary history—The earliest radiolarians are known from the Cambrian, and they are mostly simple spongy spherical forms, along with a few that are shaped like spicules or cones. By the late Silurian, both deep-water and shallow-water radiolarian faunas can be distinguished. In the late Paleozoic, the onset of glacial conditions and the appearance of the first cold, deep oceanic waters led to the evolution of radiolaria restricted to such conditions. The Permian extinctions affected the radiolaria strongly, with a number of Paleozoic groups disappearing from the fossil record.

The first unquestioned nassellarians appeared in the Triassic, and by the end of the Mesozoic, more than half of the living groups had evolved. Like other groups of plankton, radiolaria reached a peak of diversity in the Cretaceous that was not matched again until the late Cenozoic. However, radiolaria were virtually unaffected by the K/T extinction event. Most of the characteristic Cretaceous groups can be found in exceptionally well-preserved Paleocene sediments, suggesting that they were insensitive to whatever happened to the planet at the time. Perhaps this is because radiolaria tend to live in stable water masses at a variety of depths, and are not affected by short-term events that only impact shallow, mainly tropical waters. (This is also consistent with the lack of extinction in diatoms, and in benthic foraminifera during the K/T event.)

On the other hand, radiolarians were strongly influenced by major paleoceanographic changes, such as those occurring throughout the Cenozoic. The evolution of many new groups in the early Cenozoic apparently caused more extinction among Mesozoic forms than did any physical event at the end of the Cretaceous. In addition, the evolution of diatoms and their rise to dominance at the end of the Cretaceous and during the early Cenozoic also apparently triggered turnover in radiolarian faunas. The average weight and thickness of radiolarian skeletons diminished during the early Cenozoic, probably as a result of competition for silica with the diatoms. Consistent with this trend is the fact that thick, heavily silicified radiolarians remained common in deeper-water assemblages, where there are no diatoms to compete for silica.

The final major event in radiolarian history occurred during the cooling and oceanographic changes of the Oligocene and early Miocene, when modern deep, cold bottom waters developed, as well as intermediate and circumpolar water masses. Most of the Neogene groups that dominate today evolved at this time, because modern oceanographic patterns appeared during the Oligocene-Miocene interval. In particular, the modern symbiotic forms associated with nutrient-poor waters first appeared, as did those associated with nutrient-rich waters. From the Neogene onward, the radiolarian groups are so closely associated with modern water masses and oceanographic conditions that they are the best group for reconstructing those conditions in the last 20 million years.

Patterns of evolution—As we pointed out above, planktonic microfossils are potentially excellent for demonstrating evolutionary trends, since they are abundant, worldwide in distribution, lived in large biogeographic provinces, and evolved rapidly, as recorded in thick, continuous sedimentary sequences with excellent age control. The major drawback is that we know so little about their biology that it is hard to decide what caused the changes we see in long sequences. As Lazarus (1983) pointed out, we still don't know if radiolarians have any sexual stage, and apparently they reproduce by asexual cloning most of the time, so Mayr's allopatric speciation model (which is based on sexual organisms with very small, restricted, allopatric populations) and even the biological species concept itself may be irrelevant (and with it the predictions of punctuated equilibrium vs. phyletic gradualism). In addition, the possibility that many radiolarian morphotypes may just be ecophenotypic variants responding to oceanographic changes makes the significance of skeletal change through time difficult to assess. Finally, the strong possibility that different radiolarian "species" evolved by hybridizing with one another (Fig. 11.13) casts further doubt on the applicability of metazoan-based speciation models and evolutionary theory to the protistans.

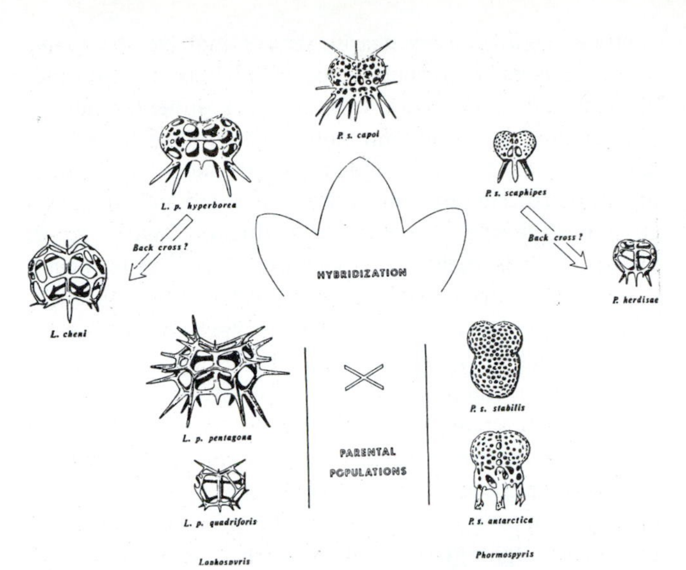

Figure 11.13. Possible example of hybridization among different radiolarian lineages. (Modified from Goll, 1976.)

Nevertheless, a number of studies have documented evolutionary patterns in the radiolaria. Both striking examples of gradualism, and also speciation, punctuation, and stasis, have been documented. Some very remarkable cases of evolutionary trends spanning extraordinary changes in morphology over millions of years can be documented in the radiolaria. For example, the *Lithocyclas-Cannartus-Ommatartus* lineage began with simple spongy spheres in the middle Eocene, then reduced to four-armed crosses in the late Eocene (Fig. 11.14). In the Oligocene, *Lithocyclas* reduced to three arms, and by the Miocene, its successor species in the genus *Cannartus* was a ball with two bipolar arms. Through the Miocene, the central spongy sphere of the skeleton reduced in size and developed a central constriction, while the bipolar arms reduced in length and become thicker. By the late Miocene, the *Ommatartus* successors of this lineage lost the arms altogether, and were made of a series of spongy, bipolar caps on the central sphere. If one were to take the segments of this lineage individually, it would be hard to imagine a spongy ball like the earliest *Lithocyclas* gradually evolving into the the multi-armed *Cannartus*, and ultimately to the bipolar *Ommatartus*, but all the transitional forms can be seen in the fossil record. Such radical rearrangements in fundamental body plan shows how loosely constrained the individual anatomical features of radiolarian skeletons must be. No anatomical feature is so fundamental that it can be considered an unchanging homologue useful for phylogenetic analysis.

For Further Reading

Anderson, O. R. 1983. *Radiolaria*. Springer-Verlag, New York.

Blome, C.D., P. A. Whalen, and K. M. Reed. 1996. *Siliceous Microfossils: Notes for a Short Course*. University of Tennessee Dept. Geological Sciences, Studies in Geology 9.

Brasier, M. D. 1980. *Microfossils*. George Allen & Unwin, London.

Casey, R. E. 1987. Radiolaria, 213-241, *in* J. H. Lipps, ed.

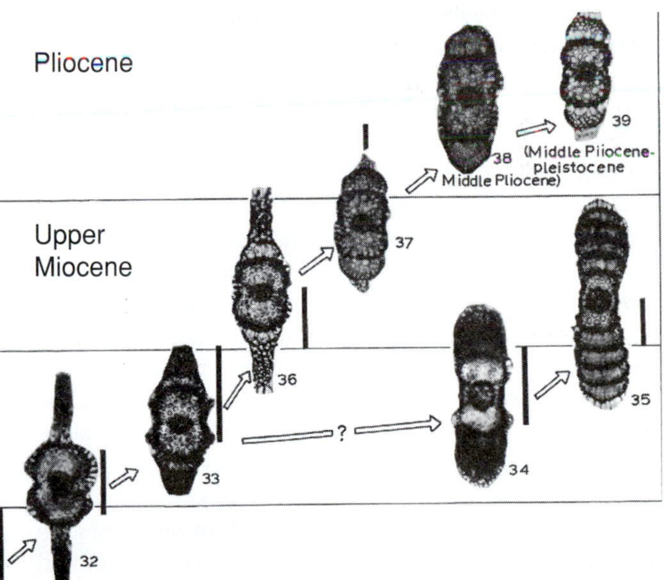

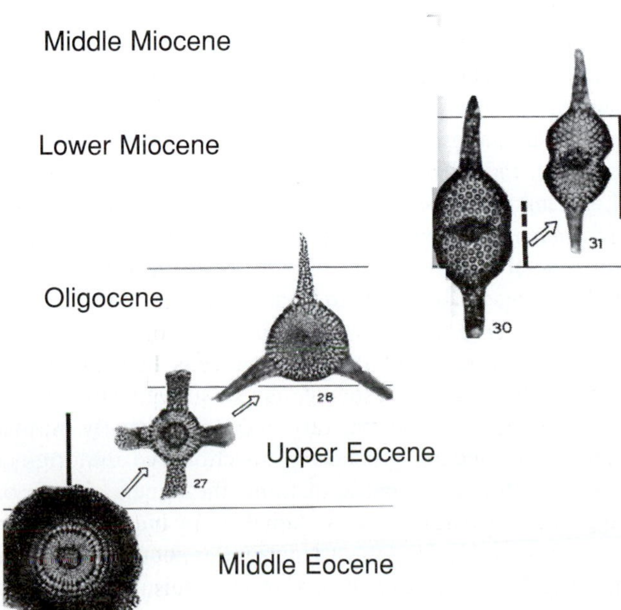

Figure 11.14. Evolution of *Lithocyclia-Cannartus-Ommatartus* lineage in the Cenozoic, from a spongy ball to a four- and then three-armed and finally two-armed bipolar structure, with variations on the spongy caps in the later Cenozoic. Taxa are as follows: 26, *Lithocyclia ocellus*; 27, *Lithocyclia aristotelis*; 28, *Lithocyclia angusta*; 30, *Cannartus tubarius*; 31, *Cannartus violina*; 32, *Cannartus mammiferus*; 33, *Cannartus laticonus*; 34, *Cannartus petterssoni*; 35, *Ommatartus hughesi*; 36, *Ommatartus antepenultimus*; 37, *Ommatartus penultimus*; 38, *Ommatartus avitus*; 39, *Ommatartus tetrathalamus*. (Modifed from Haq and Boersma, 1978.)

Fossil Prokaryotes and Protists: Notes for a Short Course. University of Tennessee Dept. Geological Sciences, Studies in Geology 18.

Casey, R. E. 1993. Radiolaria, pp. 249-284, *in* J. H. Lipps, ed. *Fossil Prokaryotes and Protists*. Blackwell Scientific Publishers, Cambridge, Mass.

Funnell, B. M., and W. R. Riedel, eds. 1971. *The Micropaleontology of Oceans*. Cambridge University Press, Cambridge.

Kling, S. L. 1978. Radiolaria, pp. 203-244, *in* Haq, B.U., and A. Boersma, eds. *Introduction to Marine Micropaleontology*, Elsevier, New York.

Lazarus, D. 1983. Speciation in pelagic Protista and its study in the planktonic microfossil record: a review. *Paleobiology* 9:327-340.

DIATOMS

Along with radiolaria, diatoms are the other major group of siliceous plankton in the fossil record. Other than their skeletal chemistry and similar size (typically 20 to 200 microns), however, diatoms have little in common with radiolarians. For one thing, diatoms are photosynthetic protists more like plants (members of the Phylum Chrysophyta, the golden-brown algae), so they require light for photosynthesis, and are therefore restricted to the surface waters (less than 100 m). Along with coccolithophores, they are important parts of the food chain for much of the marine realm. But they also occur in polar latitudes that are too cold for other phytoplankton. In these regions, they are the most important organisms at the base of the food chain.

Diatoms are not restricted to the marine plankton and benthos, but are found just about anywhere there is light and moisture. They are far more diverse and abundant in freshwaters, where they are the most common organism and make up the base of many freshwater food pyramids. They also occur in soils, ice, and attached to rocks in the splash zone of the surf. Diatoms are among the first algae to colonize any available surface in the photic zone, including rocks, shells, sand, sea grass, and the bottoms and tops of polar ice. They can even cause a yellowish coating on the skin of whales. Of the more than 600 living and fossil diatom genera, 70% are marine, 17% are freshwater, and the remaining 13% are predominantly either marine or freshwater with only a few species occurring in both habitats. In offshore waters, hundreds to thousands of cells occur per liter of seawater; in nearshore waters and estuaries, their concentrations may reach millions per liter. A square meter of shallow seafloor in the photic zone may contain over a million cells.

In addition to their ecological importance, diatoms have many other uses. Because they are abundant and evolved rapidly, they are widely used for biostratigraphy, particular in silica-rich marine sediments. In addition, they are highly sensitive to environmental conditions, so they are good paleoclimatic and paleoceanographic indicators. High concentrations of diatoms nearly always indicate an abundance of certain nutrients (especially silica, nitrate, and phosphorus), as well as little clastic sediment input to dilute them. When they bloom in huge numbers in oceans and in lakes, they can form diatomaceous oozes composed of trillions of diatoms. When these are dried and compacted into rock, they become **diatomite**, or diatomaceous earth, which is mined commercially as a filtering agent (since it is highly porous, it allows water to pass through while trapping even the tiniest impurities), an absorbant (for example, kitty litter), a mineral filler (as a paint flattening agent), and a fine abrasive. Diatomites are used for over 150 different products. Eventually, diatomite can become compacted and lithified into chert.

Diatoms are easy to grow and culture in the laboratory, so they have been studied by a wide variety of microbiologists and algalogists as well as by paleontologists. Consequently, their biology is better known than just about any other common microfossil. Each diatom cell contains yellow, olive, or golden-brown granules called **chromoplasts**, which are the sites of the photosynthesis and give the Chrysophyta their characteristic color. The cell also has a large central nucleus, and a central vacuole, but it lacks flagella or pseudopodia as seen in other protistans. Many diatoms that live on a substrate can glide over it using a coat of mucus. Others are planktonic and require low-density fat droplets, spines, and other devices to help keep them floating at a constant depth.

The diatom skeleton is composed of two valves that fit together in a nested, overlapping fashion like a Petri dish. Together, the two valves make up the entire skeleton or **frustule**. Diatom classification is based on the features of the skeleton. Those that have a radial symmetry around a structural center (mostly circular or triangular in shape) are centric forms, or the **Centrales** (Fig. 11.15). Diatoms that are bilaterally symmetrical along an axis are the pennate forms, or **Pennales**, and these are generally oval, rod-like, or triangular in shape. Since there are limited range of shapes within these two groups, the surface texture of the valve is critical for diagnosing species and genera. Diatoms show a wide variation in patterns of pores, bumps, marginal and submarginal spines, clear (hyaline) areas, longitudinal slits, and other structures that can be diagnostic.

During most of their lives, diatoms reproduce asexually. The parent cell divides in two, with each daughter cell getting one of the two valves for its large dorsal valve, or **epivalve** (Fig. 11.16). It then adds a new, slightly smaller valve (the **hypovalve**) nested within the epivalve. Each time this occurs, half the daughter cells that inherit the parental hypovalve get smaller and smaller. Eventually, the cell size becomes too small to function efficiently. At that point, they undergo sexual reproduction and the original size is restored. In centric diatoms, the female gamete or egg is nonmotile and male gametes are uniflagellate; in pennate diatoms the gametes are similar in appearance and nonflagellate. When gametes fuse, they form a specialized zygote cell called an **auxospore**, which eventually forms a

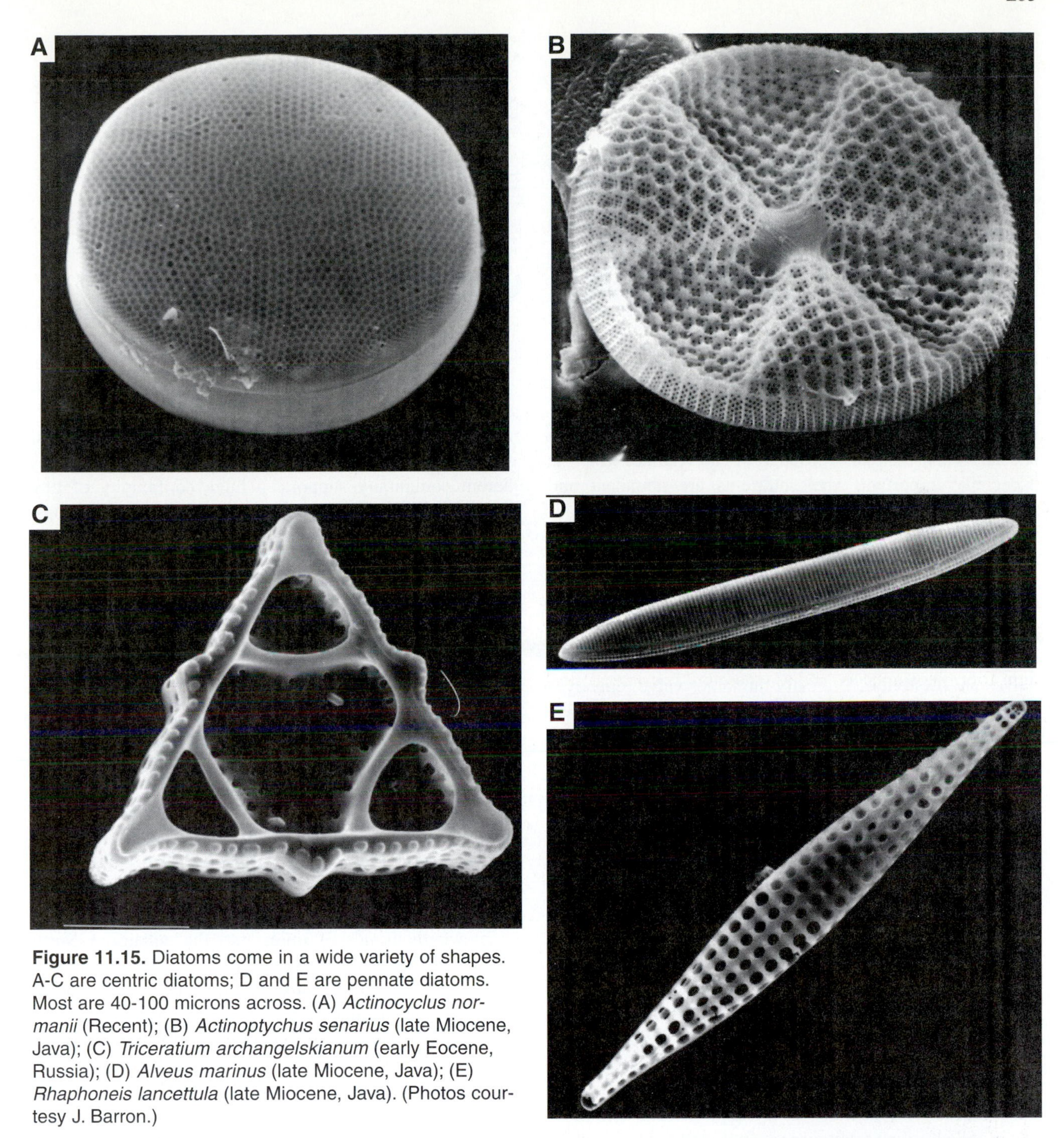

Figure 11.15. Diatoms come in a wide variety of shapes. A-C are centric diatoms; D and E are pennate diatoms. Most are 40-100 microns across. (A) *Actinocyclus normanii* (Recent); (B) *Actinoptychus senarius* (late Miocene, Java); (C) *Triceratium archangelskianum* (early Eocene, Russia); (D) *Alveus marinus* (late Miocene, Java); (E) *Rhaphoneis lancettula* (late Miocene, Java). (Photos courtesy J. Barron.)

Figure 11.16. Diatom frustules are shaped like nested petri dishes enclosing the tissue inside. During asexual reproduction, the valves split apart, and each half generates a newer, smaller valve inside it. The valves get smaller and smaller until they resume their original size by sexual reproduction. Roman numerals identify the generations; lowercase letters identify each valve in the process of reproduction. (From Barron, in Lipps, 1993.)

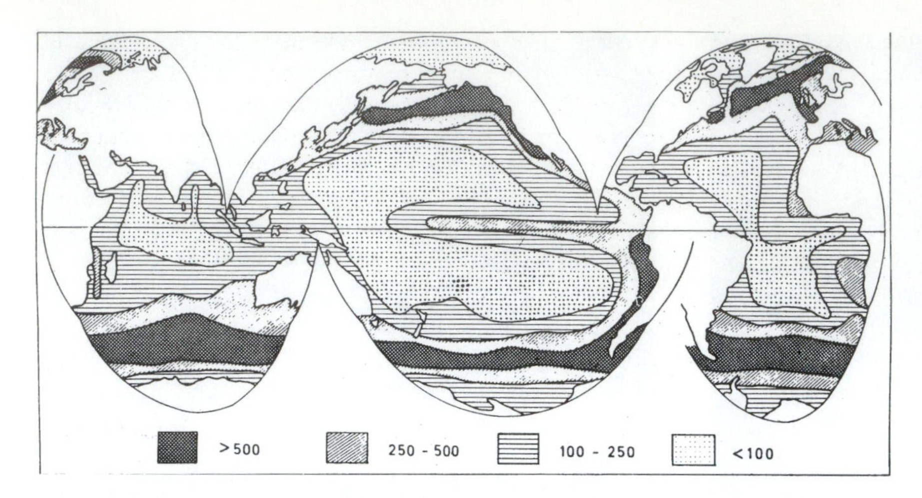

Figure 11.17. Global variation of the extraction of dissolved silica (in grams of silica per square meter per year) by phytoplankton in near-surface ocean waters. This is a close approximation of diatom production, because diatoms are overwhelmingly the dominant siliceous phytoplankton group in the oceans. (Modified from Calvert, 1974.)

new frustule around it at the maximum size for the diatom species. Then asexual reproduction can resume, which is especially important when nutrients are abundant and diatoms need to multiply as rapidly as possible.

Diatoms grow and reproduce rapidly and continuously as long as their three critical nutrients—silica, phosphorus, and nitrate—are available. Once one of these nutrients has been depleted, the diatoms stop growing, and form a thick-shelled resting spore called a **statospore**. The statospore looks similar to the normal frustule, except that its two valves are dissimilar. Typically the larger valve (epivalve) is spiny and the smaller valve (hypovalve) is smooth. Once nutrient supplies return to appropriate levels and the days are long enough to provide adequate light, the cells return to their normal vegetative state of active blooming and reproduction.

Ecology

Because diatoms depend so heavily on light and are so sensitive to a wide variety of environmental conditions, their distribution is highly predictable. They are most abundant in the oceans in areas of upwelling, where oceanic currents bring up the nutrients from the deeper waters to the photic zone. In nearshore areas of seasonal upwelling, the blooms can be phenomenal, with one to three cell divisions per diatom per day, resulting in a hundred-fold increase in cells in just a few days. These blooms last about 2 to 3 weeks until the nutrients are depleted, and then most of the cells become resting spores until the next bloom.

In the deep ocean, there are areas of permanent upwelling, especially where the oceanic gyres interact to cause deep waters to flow upward at the boundaries of water masses (Fig. 11.17). The major areas of diatom production include cold eastern boundary currents, such as the California current, the Humboldt current off the west coast of South America, and the current off southwest Africa. The other three major regions are the northern Pacific and Atlantic/Arctic oceans, the equatorial belt (which is dominated by radiolaria in volume, if not in numbers), and especially the Circum-Antarctic current. This last region is one of the most fertile regions in the world, where staggering numbers of diatoms are produced each year from the cold water circulating around the Antarctic. The enormous diatom community supports a huge community of tiny crustaceans, such as copepods, krill, and small shrimps. These in turn are fed upon by a wide variety of fish, penguins, seals and especially by most of the world's whales. It is no accident that *Moby Dick* and other whaler's tales are set in the southern ocean, because the oceanographic conditions that support its huge diatom population are ultimately reponsible for making a whale haven.

Diatoms are excellent indicators of paleoenvironmental conditions. In marine studies, they are primarily used to determine upwelling conditions. Most diatoms are also diagnostic of the temperature of the water in which they lived, so they can be used as very sensitive paleothermometers as well. Fossil diatoms have been used as indicators of paleosalinity, to recognize the glacial/interglacial stages of the Ice Ages, and to trace the history of oceanic currents (especially the Antarctic Bottom Water). Nonmarine diatoms from dry lake beds in the Sahara can be blown thousands of miles into the equatorial North Atlantic, where their presence in marine cores indicates the aridity and strength of winds over the African continent. In limnology, freshwater diatoms are the primary tool for reconstructing ancient lake conditions, especially changes in pH and fertility. They play a key role in monitoring acid rain and the pollution of the world's freshwater.

Diatoms are also excellent biostratigraphic indicators, especially in silica-rich sediments with few calcareous microfossils. In the late Cretaceous and early Cenozoic, their biostratigraphic zones are typically 1 to 3 million years in duration, but in the Neogene, the diatom zones are so short that age resolution approaches 100,000 to 300,000 years.

However, less than 1 to 5% of the living marine diatom assemblage makes it into the sediment. Surface seawater is undersaturated with respect to silica, largely because diatoms recycle and deplete it so quickly. As soon as bacterial action has dissolved the organic coating of the frustule, diatoms dissolve rapidly after they die and sink. Most of the diatom frustules that make it to the sediment are either large and robust (especially resting spores), or they settled rapidly through the water column in aggregates, or as the contents of the fecal pellets of zooplankton. Once they reach the bottom, frustules must be rapidly buried, and cannot be exposed to burial temperatures in excess of 50°C or alkaline conditions in the sediment pore water (pH > 7), or they will be dissolved. Diatoms are so aggressive in soaking up silica that when they are kept in a glass aquarium and the silica content of the water drops, they cause the inside walls of the glass aquarium to be etched as its silica dissolves away.

Evolution

Evolutionary history—The earliest records of diatoms are mostly from marine sediments. The oldest (poorly preserved) records are from the Early Jurassic, and they are poorly known until the Early Cretaceous, perhaps because they are so susceptible to diagenesis. The great diversity of shapes by the Early Cretaceous suggests that they may have had a longer evolutionary history (perhaps in freshwater), but lacked a siliceous cell wall that could be preserved. Diatom diversity increased steadily from the Cretaceous until the present. Unlike some other microfossils, diatoms showed few effects (only 23% of the genera became extinct) of the K/T event, probably because they could retreat as resting spores during the harsh conditions that may have occurred.

Although the extinctions of the Cenozoic did not reduce diatom diversity, there were major turnover events where a number of taxa were replaced by new forms when oceanographic conditions changed. The highest turnover occurred at the end of the early Eocene, in the middle Oligocene, and in the middle Miocene, with lesser turnover at the end of the middle Eocene and at the Eocene/Oligocene boundary. Each of these events was apparently triggered by changes in oceanographic circulation, shifting the position of water masses and the temperature of the world's oceans in such a way that older species were replaced by new ones. Another time of major change was the latest Miocene, when provincialism increased in the North Pacific. Finally, the onset of northern hemisphere glaciation in the mid-Pliocene (about 2.5 Ma) had a major effect on global paleoceanography, and also caused major extinctions in both high- and low-latitude diatoms. Since then, essentially modern diatom floras have persisted, changing their distributions during every glacial/interglacial cycle of the Pleistocene.

Evolutionary trends—In addition to these diversity trends, there is also a trend towards more lightly silicified frustules since the Oligocene (also seen in the radiolaria). This trend corresponds with the general increase in oceanic circulation since the growth of the Antarctic ice sheet in the earliest Oligocene. Such vigorous circulation recycles the silica and other nutrients much more quickly, which promotes more rapid growth, and presumably favors species that make their frustules out of less silica. Lightly silicified frustules are also dissolved much more easily, further enhancing the recycling of silica into new diatoms.

Patterns of evolution in diatoms have not been studied as much as they have been in foraminifera and radiolaria, but there have been a number of investigations. Most have documented long periods of stasis, although some cases of gradualism have also been reported. Since diatoms are unicellular algae, with predominantly asexual reproduction, their evolution may not be very relevant to the debate over patterns of evolution among metazoans.

For Further Reading

Barron, J. A. 1987. Diatoms, pp. 128-145, *in* Lipps, J. H., ed. *Fossil Prokaryotes and Protists: Notes for a Short Course*. University of Tennessee Dept. Geological Sciences, Studies in Geology 18.

Barron, J. A. 1993. Diatoms, pp.155-167,*in* Lipps, J. H., ed. *Fossil Prokaryotes and Protists*. Blackwell Scientific Publishers, Cambridge, Mass.

Blome, C.D., P. A. Whalen, and K. M. Reed. 1996. *Siliceous Microfossils: Notes for a Short Course*. University of Tennessee Dept. Geological Sciences, Studies in Geology 9.

Brasier, M. D. 1980. *Microfossils*. George Allen & Unwin, London.

Burckle, L. D. 1978. Marine diatoms, pp. 245-266, *in* Haq, B. U., and A. Boersma, eds. *Introduction to Marine Micropaleontology*, Elsevier, New York.

Round, F. E., R. M. Crawford, and D. G. Mann. 1990. *The Diatoms: Biology and Morphology of the Genera*. Cambridge University Press, Cambridge.

Tappan, H. 1980. *The Paleobiology of Plant Protists*. W.H. Freeman, San Francisco.

Werner, D., ed. 1977. *The Biology of Diatoms*. University of California Botanical Monographs 13.

COCCOLITHOPHORES

The tiniest of planktonic organisms are known as the **nannoplankton** (*nano* is Greek for "dwarf"; it is spelled "nanno" in Latin), and they secrete calcareous plates known as **nannofossils**. The majority of the nannofossils are made of the skeletal plates of tiny golden-brown algae known as **coccolithophores**. These algae form spherical cells about 15 to 100 microns in diameter, enclosed in a ball of calcareous plates called **coccoliths**, which are about 2 to 25 microns in diameter (Fig. 11.18). Most coccoliths are so tiny that they can fit into the pores of larger plankton, such as the foraminifera (Fig. 11.2). As a result, they are barely resolvable within the wavelengths of visible light, and require a light microscope with high-quality high-power (magnifcations of X500 to X1000) objectives, cross-polarizers and phase-contrast illumination to be studied adequately. Their structures are much better seen with the higher resolution of the SEM, although this procedure is usually too expensive and time-consuming to use for routine examination of samples.

Because of their tiny size, nannofossils were rarely studied by most micropaleontologists until about 35 years ago.

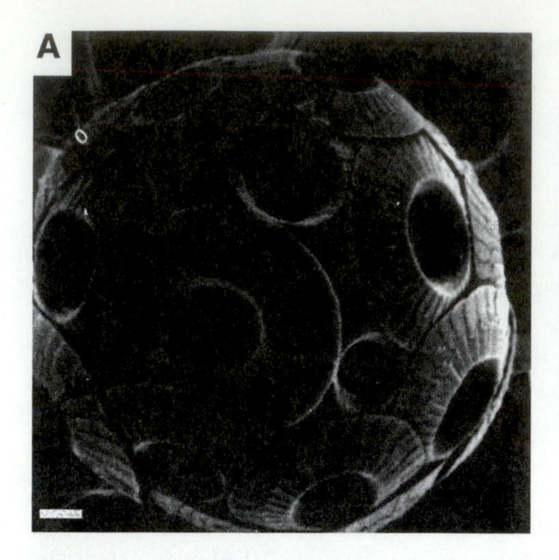

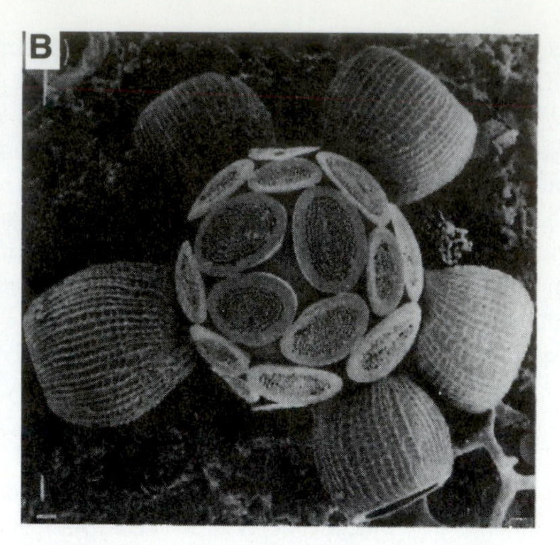

Figure 11.18. Coccoliths are secreted by a coccolithophorid, a golden- brown alga armored with calcareous plates. (A) *Umbilicosphaera sibogae.* (B) A single coccosphere which bears both the vase-shaped coccolith *Scyphosphaera apstenii*, and the flat *Pontosphaera japonica.* Scale bars = 1 micron. (Photos courtesy W. Siesser.)

Since then, their study has grown exponentially, as they have proven to be one of the best tools for Mesozoic and Cenozoic biostratigraphy. They are extremely abundant in nearly all pelagic calcareous marine sediments deposited since the Jurassic. In fact, the great Cretaceous chalks are primarily made of the clay-sized residue of coccoliths, with a few foraminifera. Even shallow marine clastic sediments, which may have few other plankton, can contain enough coccoliths to be biostratigraphically dated.

Coccoliths were long a biological mystery. When they were first mentioned by Ehrenberg in 1836 (who also described the first known radiolaria), he thought that these calcite disks were of inorganic origin. In 1857, Thomas Henry Huxley examined deep-sea sediments collected by the HMS *Cyclops*, but also concluded that coccoliths were inorganic mineral precipitates. In 1861, G. C. Wallich examined samples from the North Atlantic that still had coccoliths assembled into "coccospheres" and concluded that they were the larvae of planktonic foraminifera. In that same year, the pioneering petrographer Henry Clifton Sorby examined them under petrographic microscope, and decided that they were not foraminifera, because their calcite had different optical properties. Finally, in 1872, the great pioneering oceanographic expedition of the HMS *Challenger* allowed the first direct sampling of living coccolithophores, which John Murray found floating in the ocean water. The *Challenger* scientists discovered that coccoliths were the plates of a calcareous algae, found widely in surface waters from the tropics to the higher latitudes with water temperatures as low as 7°C.

The living coccolithophore has two golden-brown pigment spots within the cell that contain the chlorophyll and perform the photosynthesis, and a large nucleus in between them. Like many other protistans, coccolithophores have mitochondria for energy production, Golgi bodies for the production of organic scales and coccoliths, and vacuoles for waste storage. In addition, two whiplike flagella of equal length are found at one end, with a coiled flagellum-like structure called a **haptonema** between them. During the non-motile stage of most cells, the flagella are lost, although the haptonema remains. Under the stimulus of light, small vesicles form within the cell that produce the coccoliths. These eventually move to the outside of the cell, where they replace older plates from the inside, forming a multi-layered spherical envelope of calcareous plates called the coccosphere.

Coccoliths come in a great variety of shapes and patterns, although circular and oval button-like shapes are the most common (Fig. 11.19). These are distinguished by the detailed patterns within the plates, by the great variety of arrangements of the calcite crystals, and by the variations in the pores, cross-bars, and other features. In addition to circular and ovoid shapes, there are pentagonal plates (Braarudosphaeraceae—Fig. 11.19E), rhombohedral plates (Calciosoleniaceae—Fig. 11.19F), star-shaped plates (Discoasteraceae—Fig. 11.19G), and horseshoe- or wishbone-shaped plates (Ceratolithaceae). Others are shaped like crowns, mushrooms, baskets, and a variety of funnel shapes.

One of the difficulties of the taxonomy of coccoliths is that they are only disassociated parts of skeletons, and often the complete skeleton to which they belong is unknown. This can be a problem because some coccolithophores secrete more than one kind of coccolith, while other coccoliths are known from more than one species of coccolithophore. Some coccolithophores have several different layers of coccoliths, each of which has a different shape.

Although the purpose of the plates may seem obvious, there is considerable uncertainty about their function. Some suggestions include: (1) a shield to protect them against intense sunlight; (2) a by-product of carbonate detoxification by calcite fixation; (3) as another metabolic by-product; (4) as supporting and/or protective armor; (5) as a light-gathering and concentrating organ; and (6) as a ballast or stabilizing mechanism. At such a tiny body size, the plates have only limited effectiveness in preventing them from being consumed by (usually) much larger predators, so they are probably not comparable to the armor-plating of larger animals. In all likelihood, coccoliths serve several functions, including several of the metabolic and stabilizing functions listed above.

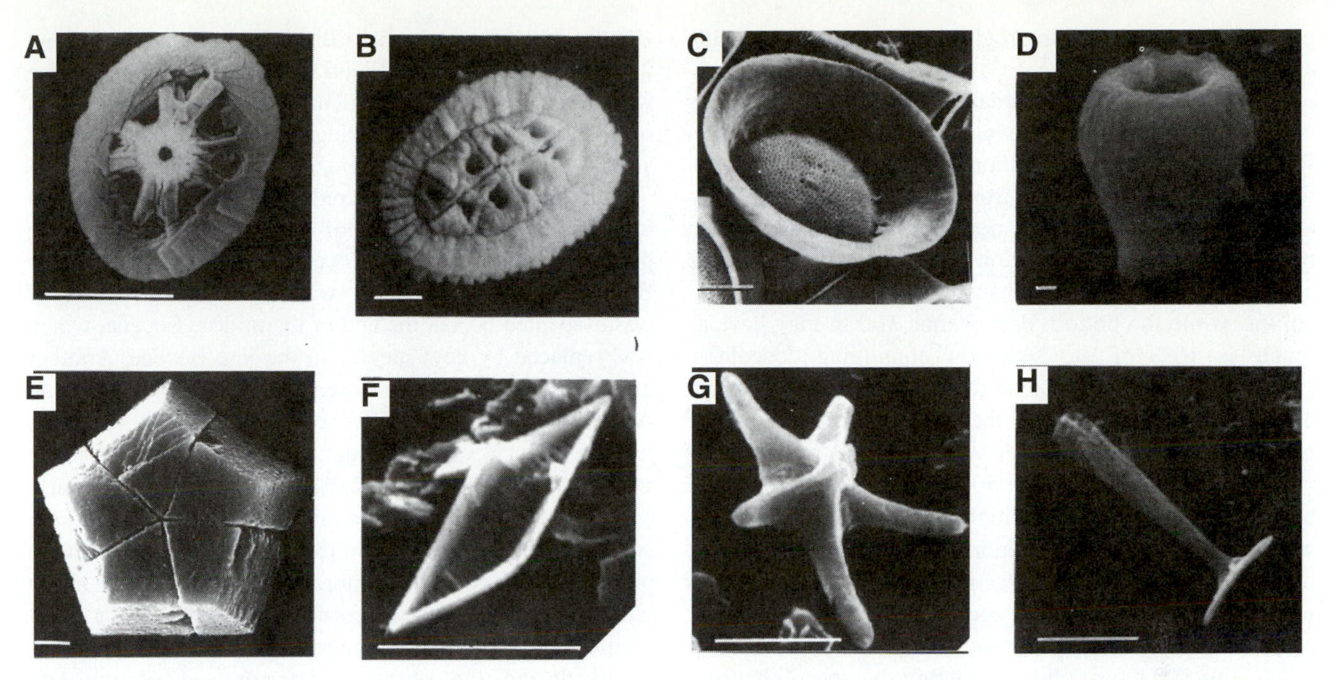

Figure 11.19. Coccoliths come in a wide variety of shapes. In each of the SEM photos, the scale bar is 4 microns. (A) *Ahmuellerella octoradiata.*(B) *Arkhangelskiella cymbiformis.* (C) *Pontosphaera syracusana.* (D) *Scyphosphaera recurvata.* (E) *Braarudosphaera bigelowii.* (F) *Anoplosolenia brasiliensis.* (G) *Discoaster quinqueramus.* (H) *Rhabdosphaera claviger.* (Photos courtesy W. Siesser.)

During most of their life cycle, coccolithophores reproduce by asexual fission into two or more daughter cells. Some coccolithophores undergo an alternation between a motile, flagellated stage and a non-motile, resting stage, enclosed in a coccosphere with different types of coccoliths than those found in the motile stage. Some authors have suggested that the non-motile phase is a zygote formed by the sexual fusion of two motile gametes, although clear proof of sexual reproduction and alternation of generations has not yet been documented in the few species whose life cycles have been studied.

Once they are undergoing asexual cell division, coccolithophores can reproduce very rapidly. In some species, each cell can divide 1 to 4 times a day, allowing them to multiply very rapidly. During coccolithophore blooms, they can quickly produce over 100,000 cells per liter, and in some places, densities of over a 30 million cells per liter have been estimated. During the active reproductive phase, they are also very rapid swimmers, capable of moving 5 to 8 mm/minute, or about 10 to 20 m/day, which is phenomenal for such a tiny cell. However, the ocean currents in which they live move faster than this, so their swimming skill is rarely needed when they can passively float much faster as plankton. It may be important, however, in maintaining a favorable position in the water column with respect to adequate light and nutrient-bearing currents.

Ecology

Although coccolithophores are algae and get their nutrition by photosynthesis, they are also active swimmers, and a few are known to ingest bacteria and small algae. Because of their dependence on photosynthesis, coccolithophores are restricted to the photic zone (usually less than 200 m in depth), with maximum concentrations in the tropics at about 50 m depth, and in temperate regions between 10 and 20 m. This is because many species cannot tolerate the intense sunlight of very shallow water, so they maintain their position in the water column at a depth where the light intensity is ideal. Most coccolithophores cannot tolerate much deviation from the normal marine salinity of 35‰, although a few species, such as *Emiliana huxleyi*, are tolerant of salinities ranging from 16 to 45‰, and are found in brackish water and in highly saline bodies, such as the Red Sea. Another species, *Braarudosphaera bigelowii* (Fig. 11.19E), thrives in low-salinity waters of the Black Sea and in nearshore, slightly brackish water, but not in the high salinity of the Red Sea. The appearance of thick chalks of the opportunistic "disaster" genus *Braarudosphaera* (with its distinctive pentagonal coccoliths) is considered an indicator of abnormal times in the marine record.

Like other microplankton, the distribution of coccolithophores is strongly controlled by the latitudinal gradient in temperature, with distinct floral provinces in the tropical, subtropical, transitional temperate, subarctic, and subantarctic latititudes. As expected, diversity is highest in the tropics (typically about 40 to 50 species, of which 10 to 15 fossilize), and lowest in the high latitudes (2 to 3 species, including the cosmopolitan cold-tolerant species *Emiliana huxleyi*, one of the commonest coccolithophores alive today). In addition, the numbers of individuals also varies with latitude, with the highest densities occurring in polar waters and upwelling regions, and density increasing away from the equator.

However, these distributional and abundance characteristics are strongly affected by taphonomic factors. Thinner-walled coccoliths are much more susceptible to dissolution than thicker-walled species, so there is a selective preservation of the flora. Shallow nearshore seafloors are much more susceptible to dissolution than deeper waters, so the continental shelf shallower than 40 m will preserve no coccoliths, while the bottom deeper than 100 m may have as much as 10% of the sediment containing coccoliths. Coccoliths are also subject to significant post-mortem transport, since they sink in the water column at a rate of only about 1 to 2 microns per second. At this rate, an individual coccolith would take 50 to 150 years to reach the bottom at 5000 m, during which time the oceanic currents would have transported it enormous distances. However, most coccoliths do not sink as individual particles, but do so inside the fecal pellets of zooplankton. Each pellet may contain as many as 100,000 coccoliths, and sinks to 5000 m in 22 to 100 days, while protecting the coccoliths from dissolution. Consequently, most micropaleontologists do not expect post-mortem transport to be a significant problem in determining the ancient biogeographic provinces preserved on the seafloor, since most coccoliths apparently got there quite rapidly in the fecal pellets of other organisms. Finally, if the calcareous ooze reaches seafloor below the CCD (typically 4000 to 5000 m deep), then the coccoliths are dissolved completely, and only siliceous microfossils and clays remain.

Using this understanding of the climatic controls on modern coccolith floral provinces, coccoliths have proved to be excellent paleoclimatic indicators. Some species are restricted to a narrow range of temperatures, and can be used as a direct paleothermometer. In other cases, entire assemblages have been characterized by their temperature preferences, and used in paleoclimatic analysis. In some cases, ratios between diagnostic taxa (such as the ratio between the warm-loving *Discoaster* versus the cold-tolerant *Chiasmolithus*) can be used as a temperature indicator. Finally, the oxygen and carbon isotopes in their calcite can be analyzed, as is done for foraminiferans, to determine paleotemperature, ice volume, or oceanic circulation changes. Unfortunately, the tiny shells of coccoliths are difficult to use since they have so little calcite, and also because they are readily diagenetically altered (unlike the much larger foraminiferal test), so most isotopic analyses have been only marginally successful.

Evolution

Evolutionary history—Although Paleozoic nannofossils have been reported, none have been confirmed. Unquestioned coccoliths first appear in the Late Triassic. There is no apparent extinction event at the Triassic/Jurassic boundary. The diversity of coccoliths increases gradually through the Jurassic and Early Cretaceous, with a rapid increase in the Santonian Stage of the Upper Cretaceous, culminating in a diversity maximum of about 240 species in the Maastrichtian (latest Cretaceous). The K/T event decimated the calcareous nannoplankton, with only 15 to 18 species surviving into the Paleocene. These survivors gave rise to all Cenozoic coccolithophores.

Diversity gradually recovered in the Paleocene and then reached another maximum of about 300 species during peak warming of the early Eocene. Through the rest of the Eocene, diversity declined, with a major extinction of warm-adapted taxa at the end of the middle Eocene, which was replaced by new species in the late Eocene. Another extinction in response to the early Oligocene cooling and Antarctic glaciation greatly reduced coccolith diversity, leaving about 45 cold-tolerant species for the rest of the Oligocene. Diversity increased slightly during the early Miocene warming, then continued to decline as the Antarctic glaciers returned in the middle Miocene, and the planet became colder and more severe climatically in the late Neogene. Present-day coccolith diversity is at a minimum for the Cenozoic, in contrast to diatoms, which diversified during the climatic deterioration of the late Cenozoic. Clearly, coccolith diversity is favored by warm, "greenhouse" conditions such as those that occurred in the Late Cretaceous and Eocene, and suppressed by the "icehouse" conditions of the Neogene.

Evolutionary patterns—Relatively few studies of evolutionary patterns have been conducted on coccoliths, since the focus is still on biostratigraphy. The studies summarized by Haq (1978) seemed to show more episodes of rapid, punctuated speciation and long-term stasis than instances of gradualism. As Haq (1978) cautions, however, such studies can be biased by a variety of problems, such as the dilemma of determining which coccoliths are actually different species (discussed above). In addition, selective dissolution of species after death and diagenetic changes in the marine sediments may bias the sample in many different ways.

Taking these problems into account, Bralower and Parrow (1996) digitized the SEM images of a Paleocene coccolith genus in an attempt to see if gradual or punctuated patterns could be distinguished. They found that most characters fluctuated randomly through time, and there were no persistent trends that could clearly be shown to be gradual.

For Further Reading

Aubry, M.-P. 1984-1990. *Handbook of Cenozoic Calcareous Nannoplankton*. Books 1-4. Micropaleontology Press, New York.

Brasier, M. D. 1980. *Microfossils*. George Allen & Unwin, London.

Bukry, D. 1978. Biostratigraphy of Cenozoic marine sediments by calcareous nannofossils. *Micropaleontology* 24:44-60.

Haq, B. U. 1978. Calcareous nannoplankton, pp. 79-107, *in* Haq, B.U., and A. Boersma, eds. *Introduction to Marine Micropaleontology*. Elsevier, New York.

Haq, B. U., ed. 1983. *Calcareous Nannoplankton. Benchmark Papers in Geology*. Hutchinson Ross, Stroudsburg, Penn.

Lord, A. R., ed. 1982. *A Stratigraphical Index of Calcareous Nannofossils*. Ellis Horwood, Chichester.

Martini, E. 1971. Standard Tertiary and Quaternary calcareous nannoplankton zonation, pp. 739-785, *in* Farinacci, A., ed., *Proceeding of the 2nd Planktonic Conference, Roma, 1970*. Edizioni Technoscienza, Roma.

Perch-Nielsen, K. 1985. Mesozoic calcareous nannofossils, pp. 329-426, *in* Bolli, H. M., J. B. Saunders, K. Perch-Nielsen, eds. *Plankton Stratigraphy.* Cambridge University Press, Cambridge.

Perch-Nielsen, K. 1985. Cenozoic calcareous nannofossils, pp. 427-534, *in* Bolli, H. M., J. B. Saunders, K. Perch-Nielsen, eds. *Plankton Stratigraphy.* Cambridge University Press, Cambridge.

Siesser, W. 1993. Calcareous nannoplankton, pp. 169-201, *in* Lipps, J. H., ed. *Fossil Prokaryotes and Protists.* Blackwell Scientific Publishers, Cambridge, Mass.

Siesser, W. G., and B. U. Haq. 1987. Calcareous nannoplankton, pp. 87-127, *in* Lipps, J. H., ed. *Fossil Prokaryotes and Protists: Notes for a Short Course.* University of Tennessee Dept. Geological Sciences, Studies in Geology 18.

Tappan, H. 1980. *The Paleobiology of Plant Protists*. W.H. Freeman, San Francisco.

Winter, A., and W.G. Siesser, eds. 1994. *Coccolithophores*. Cambridge University Press, Cambridge.

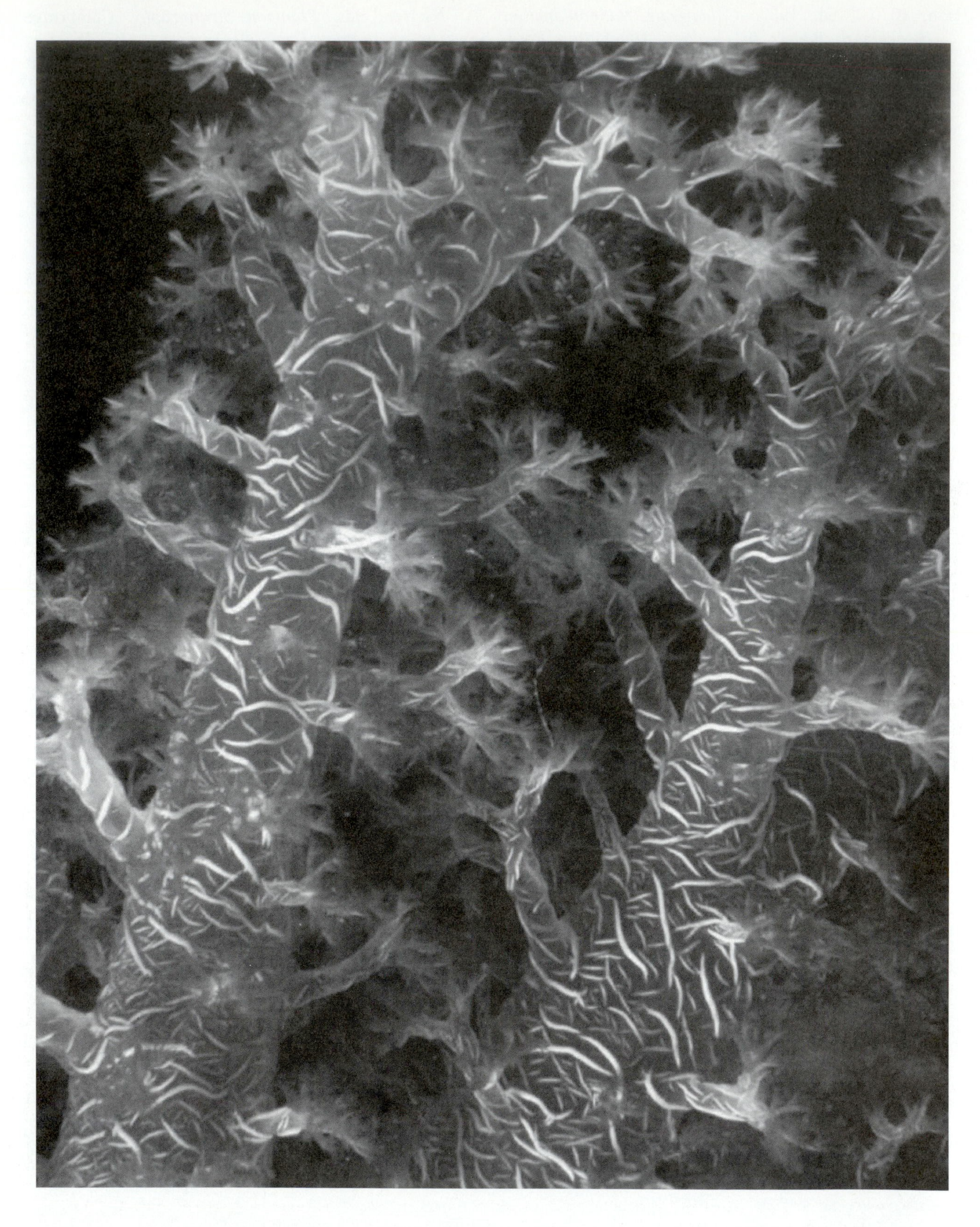

Figure 12.1. Colonial organisms, such as this branching coral, secrete enormous skeletons of calcite, but they are still the home to thousands of tiny polyps, here shown extended for feeding. (Photo from IMSI's Master Photo Collection.)

Chapter 12

Colonial Life

Sponges, Archaeocyathans, and Cnidarians

The first time you dive on a coral reef is an experience never to be forgotten. The sensation of moving freely in three dimensions in the clear sunlit water that coral favour is, in itself, a bewitching and other-wordly one. But there is nothing on land that can prepare you for the profusion of shapes and colours of the corals themselves. There are domes, branches, and fans, antlers delicately tipped with blue, organ pipes that are blood red. Some seem flower-like yet when you touch them they have the incongruous scratch of stone. But if you swim only in the day, you will hardly ever see the organisms that have created this astounding scene. At night, with a torch in your hand, you will find the coral transformed. The sharp outlines of the colonies are now misted with opalescence. Millions of tiny polyps have emerged from their limestone cells to stretch out their minuscule arms and grope for food.

Coral polyps are only a few millimetres across but working together in colonies, they have produced the greatest animal constructions the world had seen before man began his labours. The Great Barrier Reef, running parallel to the eastern coast of Australia for over a thousand miles, can be seen from the Moon.

David Attenborough, *Life on Earth*, 1979

INTRODUCTION

In the previous chapter, we studied organisms which were all single-celled. Each cell carries out all its own necessary functions (feeding, respiration, reproduction, excretion), and none were specialized for a single purpose. A few protistans, such as *Volvox*, form a colonial ball of cells, but each cell is functionally independent of the others. In this chapter, we consider the next level of organization—animals that live as multicellular colonies or organisms, but do not yet have elaborate, fully developed digestive, respiratory, reproductive, excretory, or nervous systems. They differ from the Protistans in that they are clearly heterotrophic multicellular animals (Metazoans), but they do not show the degree of integration or specialization that we shall see in the rest of the animal kingdom.

The degree of integration in these animals varies. Sponges are about as simple as a multicellular animal can be. Although their cells do have certain specializations, at any time they can switch from being one type of cell to another. Sponge cells are so independent that, if you force a sponge through a sieve, the individual cells can come together and form a new sponge. It's safe to say that *you* would not fare so well! These individual cells, however, cannot live independently very long without other sponge cells, so they are not complete animals like protistans.

The cnidarians (Fig. 12.1), on the other hand, clearly have cells permanently designated for stinging, reproduction, and for different positions on the inner, middle, or outer part of the body, but their bodies are still much simpler than those of higher metazoans. They have no one-way digestive/excretory tract, no respiratory organs, and do not have a true "head" or "tail" end. Most cnidarians spend their lives as sessile filter feeders, but even their medusa stage of reproduction (known to us as a jellyfish) are radially symmetical, with no "front" or "back." They drift at the mercy of the currents, and any direction is as good as the next. As we shall see in the following chapters, most higher metazoans (the coelomates) have bilateral symmetry with "head" and "tail" ends, a mouth and anus with a one-way digestive/excretory tract connecting them, and a fluid-filled inner body cavity (**coelom**) that gives them rigidity when they move in a purposeful direction.

SPONGES

Most people hear the word "sponge" and think of a block of synthetic foam. They are often surprised to find out that sponges are animals, and that real sponges were once used for bathing and cleaning. Even though sponges were well known to the ancients, early natural historians regarded them as plants, because they were immobile and rooted to the bottom. It wasn't until 1765 that the one-way flow of water through sponges was observed, leading to the realization that sponges are filter-feeding animals. The naturalists of the late 1700s and early 1800s, such as Linnaeus, Lamarck, and Cuvier, regarded sponges as a kind of polyp related to cnidarians. Finally, in the 1820s and 1830s, the Scottish naturalist Robert E. Grant finally determined the true nature and physiology of sponges, and recognized that they are a distinct phylum, which he named the phylum Porifera. Incidentally, Grant was Darwin's most influential

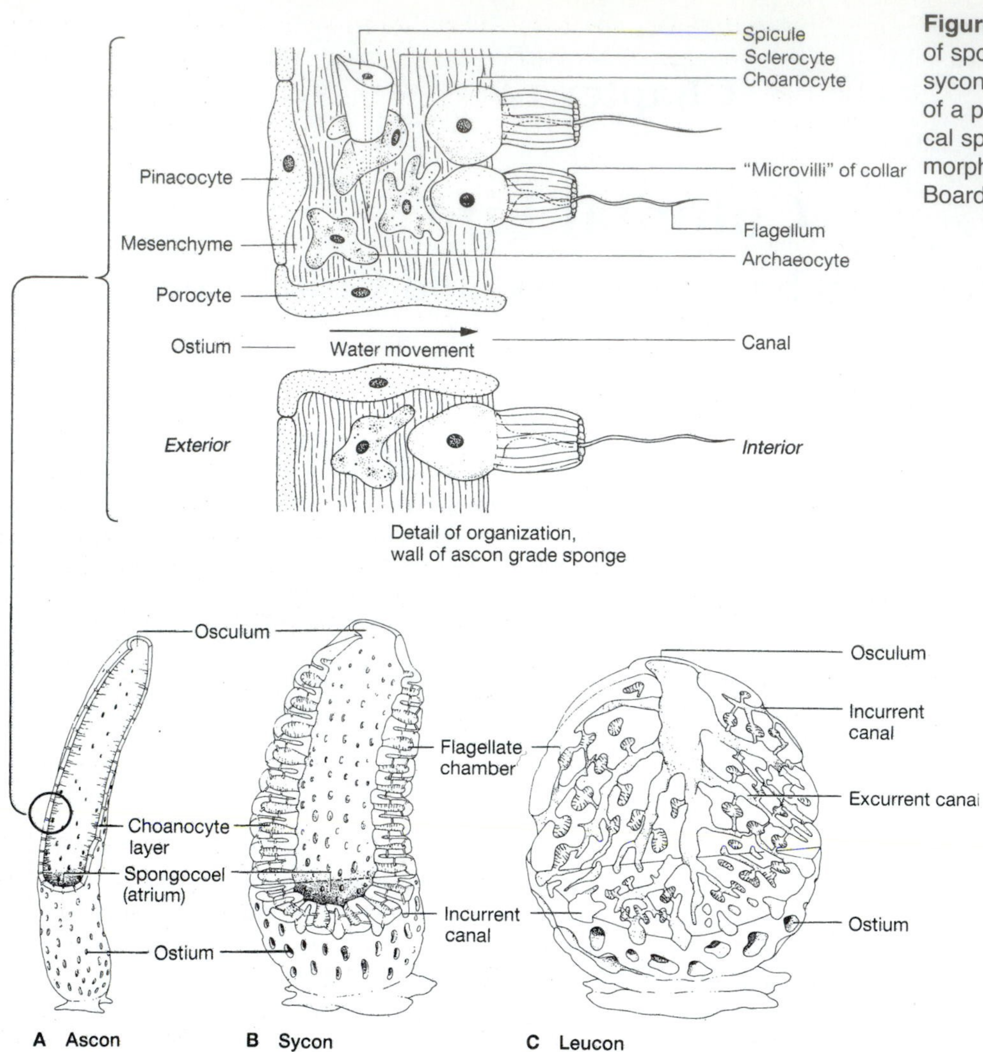

Figure 12.2. The three grades of sponges. (A) Ascon. (B) sycon. (C) leucon, with a detail of a portion of the wall of a typical sponge, showing the major morphological features. (From Boardman et al., 1987.)

teacher at medical school in Edinburgh. He filled Darwin's head with the latest ideas of the radical French evolutionists (such as Lamarck and Geoffroy), since Grant was one of Britain's earliest evolutionists.

The sponges, phylum Porifera (Latin, *porus*, "channel," and *ferre*, "to carry"), are the simplest forms of multicellular life. Their name describes them well, since they are little more than a loosely integrated assemblage of cells that secrete a shared porous skeleton. The skeleton (Fig. 12.2) is typically a simple cylinder or cone with many pores (each pore opening is called an **ostium**) and canals lining the walls, along which live the individual sponge cells. These cells create currents with a whip-like flagellum and move seawater, bringing food particles and oxygen, and eliminating their waste products. The water then flows into the hollow internal cavity of the skeleton (**spongocoel**), and out the opening at the top, the **osculum** (Latin, "little mouth" or "kiss"). Unlike any other phylum, the largest body opening in sponges is for outgoing wastes, rather than for intake of food and oxygen.

The wall structure of the sponge has some specializations. The outer part of the sponge (including the lining of the canals and the spongocoel) is covered by leathery **pinacocytes** that protect the skeleton. The inner layer has collar cells, or **choanocytes**, that gather food and pump water past using their flagella. The flagellum brings food particles close and then they are trapped by the sticky collar around the base, where they can be absorbed. Choanocytes closely resemble protistans known as Choanoflagellida, suggesting an origin for a least part of the sponge colony. In the middle of the skeleton is a gelatinous mass called the **mesohyl**, which contains **archaeocytes**, undifferentiated amoeba-like cells that digest food, transport nutrients, and develop into sex cells. The mesohyl also contains **sclerocytes** and **spongocytes**, cells that secrete the skeleton. The skeleton itself is commonly composed of fibers of an organic substance known as **spongin**, or spongin fibers plus needle-like elements known as **spicules**. These spicules may be mineralized and are the only part that is fossilized. The spicules are composed of calcite or silica, and are interconnected to support the animal. In addition, the sponge secretes a massive basal skeleton of calcite or aragonite to anchor it to the substrate.

The simplest sponges are known as **ascon** sponges (Fig. 12.2A). These are simple choanocyte-lined tubes with thin walls and very short canals. Intermediate-grade sponges are known as **sycon** sponges; they have folds and indentations (called incurrent canals) in their walls. The water passes through these canals and then into choanocyte-lined **flagellate chambers**, where it is propelled and the food captured. The highest level of complexity are **leucon** sponges, which have thick walls with many complex canals, connecting to flagellate chambers, and ultimately to a spongocoel (which may be relatively small or may be lacking altogether, but in some it is deep and wide enough for a person to stand in). Leucon sponges used to be collected for bathing before they were replaced by cheaper synthetic substitutes, but are still used extensively in art and sculpture, where there are no substitutes for the real sponge.

Sponges reproduce both sexually and asexually. During sexual reproduction, egg cells develop from the archaeocytes in the mesohyl, and sperm cells from the choanocytes. Some sponges produce both eggs and sperm, or only one kind of sex cell at a time, or alternate sexes each year. When the right season comes, sponges release great clouds of eggs or sperms from their osculum, which then float in the plankton. If they are lucky, they combine with the appropriate opposite sex cell. Millions of gametes are released, but only a few manage to form a zygote. The fertilized egg develops into a flagellated cell cluster, then into a larva. This eventually settles, attaches, and develops into a young sponge.

Sponges can also reproduce asexually by budding. The buds may eventually pinch off and attach to the substrate to form a new colony. Sponges are particularly adept at regenerating complete animals from small fragments, especially after severe storms have broken them up. Commercial sponge fishermen have found that "sowing" their fishing ground with living sponge fragments can generate a crop of new sponges, although this method hasn't proven to be economically practical.

During stressful periods, sponges form **gemmules**, which are armored, food-laden archaeocytes that can resist stress and dehydration. In some freshwater sponges, extreme temperatures, droughts, or the onset of winter will cause the parent sponge to degenerate entirely into gemmules. These cells are remarkably tough; some dried gemmules have revived and grown to adult sponges after having been out of water for 10 years.

Systematics

The classification of the phylum Porifera has been in great flux, because some taxonomists have focused on the skeleton, while others have used microscopic features of the soft tissue (which is not fossilized), and more recently, molecular biologists have tried to rearrange the classification. Traditionally, the phyum is divided into several classes:

Class Hexactinellida (glass sponges; 295 fossil genera; 130 living genera)

Hexactinellids, or glass sponges, are characterized by spicules made of silica, typically with the six rays (triaxons) of the spicules at 90° in cubic symmetry (Fig. 12.3A). Many are not attached to the hard substrate, but are anchored by ropy root tufts or mats of spicules. Hexactinellids such as *Hydnoceras* (Fig. 12.3B, C) formed extensive reef-like accumulations in the latest Devonian after the extinction of the warm-water coral-stromatoporoid reefs of the late Devonian (see Chapter 6). Hexactinellids also suffered during the Permian crisis, but recovered in the Mesozoic. They are the most common sponge in deep-marine settings, although they moved there in the middle Mesozoic. The "Venus' flower basket" sponge (*Euplectella*) is one of the best-known living hexactinellids (Fig. 12.3D).

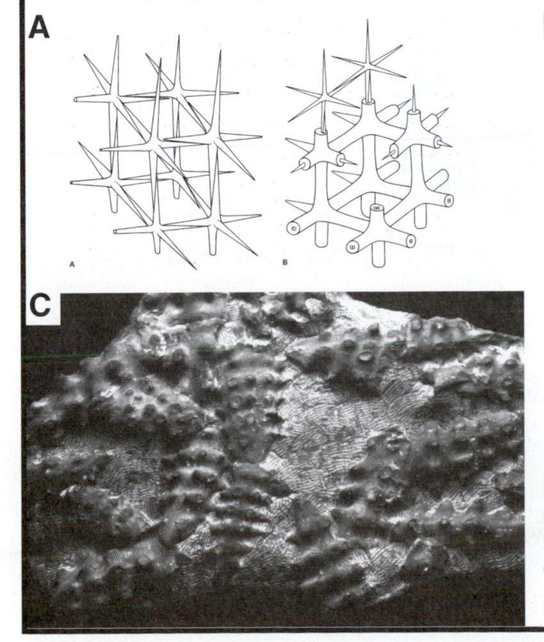

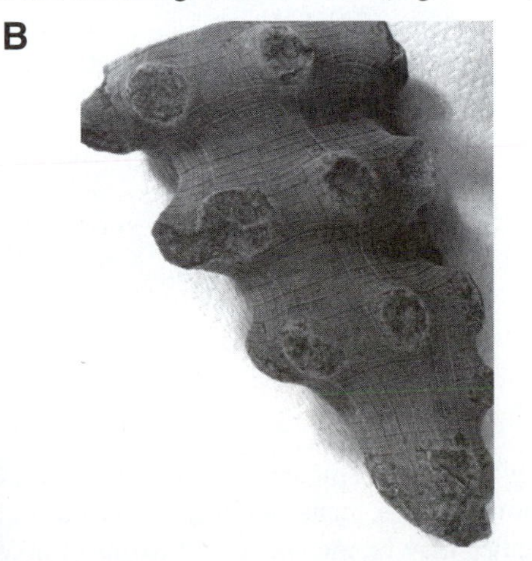

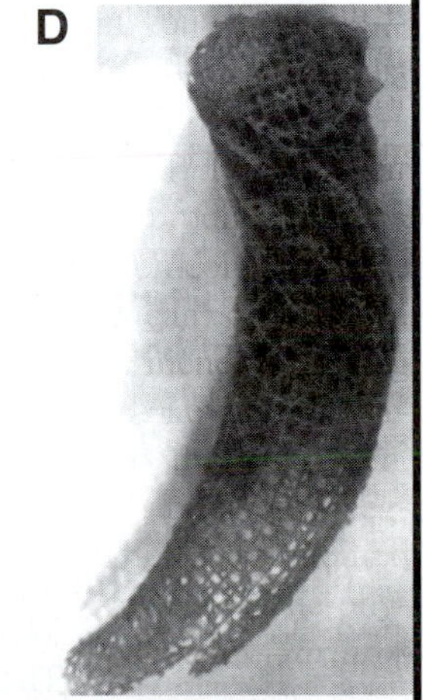

Figure 12.3. (A) Typical cubic structure of hexactinellid spicules. (B) The nodose glass sponge *Hydnoceras* built large reef-like accumulations (C) in the Late Devonian. (D) The living glass sponge *Euplectella,* or "Venus' flower basket." (Photos B, C courtesy J. K. Rigby; D by the author.)

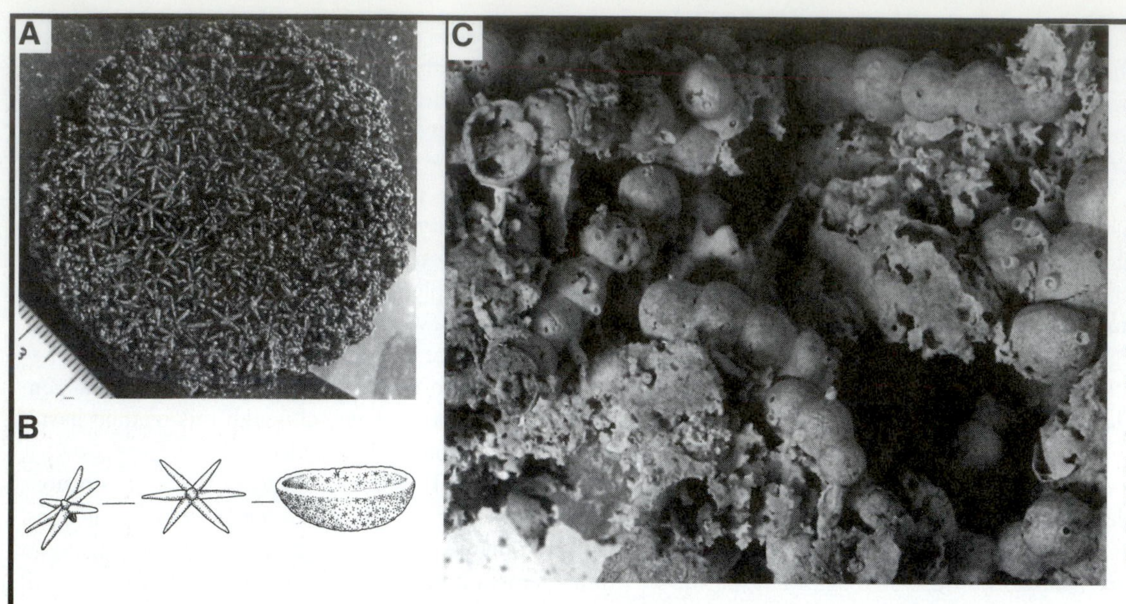

Figure 12.4. Calcareous sponges. (A, B) This saucer-shaped sponge with the star-like spicules is appropriately known as *Astraeospongia.* (C) Bead-like sphinctozoan sponges like *Girtyocoelia* were major reef-builders in the Permian. (Photos by J. K. Rigby.)

Class Calcarea (calcareous sponges; 115 fossil genera; 60 living genera)

Calcisponges have calcareous spicules (shaped like single-rayed monaxons, four-rayed tetraxons, tuning forks, stars, among other configurations) or calcareous skeletons without spicules. Four different groups are recognized, showing a great variety of shapes. The earliest group, the heteractines, were common in the Cambrian through Carboniferous, and are typified by the dish-shaped *Astraeospongia* (Fig. 12.4A, B) with its star-shaped spicules (hence the name). In the late Paleozoic, the bead-like or chambered sphinctozoan sponges, such as *Girtyocoelia* (Fig. 12.4C), were major components of the great Permian reefs and continued into the Triassic (unlike most survivors of the Permian catastrophe). Most living calcisponges are found in shallow tropical waters (less than 1000 m).

Figure 12.5. Demosponges, or common sponges, have an irregular arrangment of organic material in their skeleton. They include (A, B) the golf-ball-sized Silurian sponge *Astylospongia,* and (C) a goblet-shaped fossil demosponge. (Photo A courtesy J. K. Rigby; B from Rigby, in Boardman et al., 1987; C by the author.)

Class Demospongea (sponges with organic spicules; 390 fossil genera; 600+ living genera)

The demosponges, or common sponges, make their skeleton out of the protein spongin, although some also have siliceous spicules. Their spicules may be monaxons, tetraxons at 60°/120° angles, or knobby, irregular forms known as **desmas**, but never triaxons at 90°. Demosponges are by far the most common living group, with over 600 living genera (95% of living sponges), and are found in nearly every marine and freshwater habitat. They come in a great variety of shapes and sizes, although most are massive, globular leucon sponges like the common Silurian form *Astylospongia* (Fig. 12.5A, B).

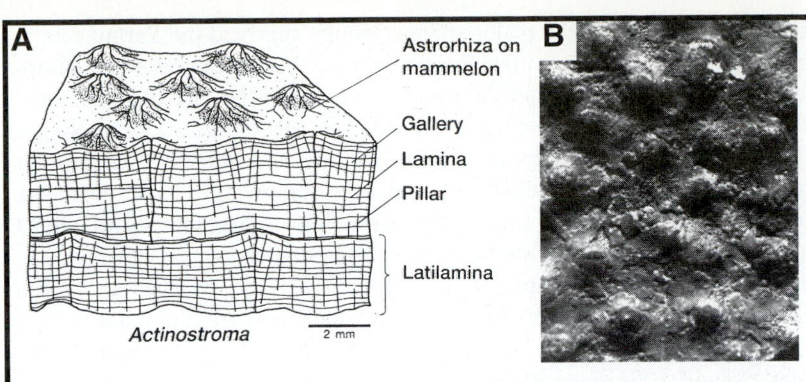

Figure 12.6. Stromatoporoids were long a zoological mystery, but most authors now consider them extinct relatives of the sclerosponges. (A) Basic anatomical features of a stromatoporoid. (B) Close-up view of a stromatoporoid, showing the bumps (or mamelons) covered with radiating canals, or astrorhizae. (A from Rigby, in Boardman et al., 1987; photo B by the author.)

Class Sclerospongea (sponges with both calcareous and siliceous spicules, or no spicules at all)

Sclerosponges were originally described by Randolph Kirkpatrick in 1911; with only 13 living species, they are considered a minor class. However, their rediscovery in **cryptic** (hidden in rock crevices) habitats in Jamaica in the 1960s solved a long-standing mystery in paleontology. These living fossils turned out to be remarkably similar to one of the most controversial extinct groups in the fossil record, the **stromatoporoids** (Fig. 12.6). Stromatoporoids are laminated calcareous fossils, with a distinct pattern of vertical pillars, horizontal laminae, and surface bumps (**mamelons**) with a star-like pattern of canals (**astrorhizae**); they rarely contain spicules, so they were usually assigned to the cnidarians, but some paleontologists had allied them with sponges, encrusting foraminiferan, cyanobacteria, or placed them in their own phylum. Hartman and Goreau (1970) argued that stromatoporoids were extinct relatives of the relict, cryptic sclerosponges, since they shared many features in common (such as the astrorhizal canals and the lack of spicules). Other paleontologists are not so convinced, although most agree that stromatoporoids were sponges (there are, of course, always a few dissenters). Regardless of the affinities of the stromatoporoids, they were one of the most important reef-builders of the Paleozoic, making up the bulk of the great Silurian and Devonian reef complexes (along with tabulate and rugose corals). Stromatoporoids then went into decline, although they survived the Permian catastrophe and straggled on until the end of the Cretaceous. (See S. J. Gould's essay, "Crazy Old Randolph Kirkpatrick," for further details of this discovery).

Ecology

Modern sponges inhabit just about any aquatic environment, from the full range of marine habitats to freshwater, and from the poles to the equator. In some warm, humid places, sponges can even grow on the trunks of trees as long as there is a regular immersion in water. Some sponges bore into calcareous substrates. The boring sponge *Cliona* is important in the degradation and breakup of shells and coral reefs. About 80% of the species of sponges, however, prefer shallow marine waters, limited primarily by the need for a firm substrate on which to attach. In clear waters with muddy bottoms, some sponges can develop a root-like base to anchor themselves, but sponges cannot tolerate muddy water because it stifles the cells. Other sponges (particularly those with siliceous spicules) prefer deeper (500 to 1000 m), colder waters, while still others are found at depths as great as 5000 m. Deep-water sponges that live in regions of weak currents tend to have long stalks, so they can stand above the muddy bottom, and bring the water in the ostia at the base and out the osculum at the top. Freshwater sponges mostly belong to the family Spongillidae, and live in lakes, rivers, and ponds. They survive the winter freeze by degenerating into gemmules, and then growing new individuals after the spring thaw.

Sponges have a variety of tolerances to turbidity. Some require clear water, while others have developed back-flushing mechanisms to prevent being clogged with sediment. Most sponges prefer areas where the currents are moderately strong; they are oriented so their "chimneys" protrude into the currents. The one-way flow of water through the walls and out the top is propelled by the flagella, but also enhanced by the fact that the top of the sponge has a weak suction drawing upward like a chimney. This is because the water flowing over the top of the sponge must flow faster than the water around it, so its increased velocity means less pressure (due to Bernoulli's principle—see p. 107). The decreased pressure over the top of the sponge relative to the rest of the surrounding water forces the water through the pores and out the top. Sponges that live in prevailing currents commonly are asymmetrical, with ostia concentrated on the upstream side and the osculum pointing downstream.

Sponges are very efficient at passing water through their canals. The entire internal volume of the sponge is replaced with new water nearly every minute. A black loggerhead sponge about 50 cm in diameter and 30 cm tall may draw about 1000 liters of water through its canals in a single day. Some sponges may flush the equivalent of 10,000 to 20,000 times their internal water volume in a single day.

Sponges have relatively few predators. In the tropics, they are fed upon by sea turtles, fish, nudibranch snails, starfish, and chitons, while their main predator in the

Antarctic are starfish. However, there is so little digestible matter compared to the bulk of indigestible skeleton that most predatory animals ignore them. Many organisms, including various worms, arthropods, fish, molluscs, and protozoans seek shelter in sponges because of their large, hollow, protective spongocoel. Some sponge predators may actually eat sponges to get at the sheltered animal inside. A single black loggerhead sponge was reported to contain over 10,000 organisms within its canals and skeleton. In some regions, sponge fishermen find so many hard-shelled molluscs in the sponges that their catches are worthless as bath sponges. The delicate glassy "Venus' flower basket" sponge, *Euplectella* (Fig. 12.3C) is prized in the Orient as a wedding present, because shrimp become trapped inside after they move in and molt enough times until they are too large to escape through the grill over the osculum. Apparently, the pair of shrimp trapped in the glassy cage is symbolic of marriage in certain cultures.

Evolution

Sponges have long been one of the more important colonial and reef-building organism on earth. Today there are about 1500 genera and 9000 species of living sponges, but far more must have existed in the geological past. Unfortunately, their fossil record is very spotty, since most of the sponge is soft and cellular, and complete sponges are rarely preserved. Spicules are far more common, but in most cases they cannot be linked to a known fossil sponge; they give only a hint of the diversity that could have been preserved. In places where sponges are well fossilized, they tend to be abundant, while most fossil localities have no sponges at all. In some cases, almost the entire diversity of sponges in a given period is known from a few excellent localities. If those are set aside, the sponge fossil record is very poor and known from only a few specimens.

Nevertheless, paleontologists have managed to piece together the general pattern of sponge evolution. All the major classes (however many are recognized) are known from the Cambrian, and they have been important parts of the reef ecosystem since that time. At times in the geologic past, they were the dominant reef builders. After the extinction of the sponge-like archaeocyathans in the Middle Cambrian (see below), sponges were among the only reef-building organisms left on earth. By the Middle Ordovician, there were significant demosponge reefs in many parts of the world. As tabulate and rugose corals continued to diversify during the Ordovician and Silurian, sponges became less important, except for the stromatoporoids, which were one of the dominant elements of Silurian and Devonian reefs. After the late Devonian extinction decimated the tropical stromatoporoid-coral reefs, hexactinellid sponges like *Hydnoceras* (Fig. 12.3B) became abundant in the latest Devonian. Sponges were relatively less important in the Carboniferous, but in the Permian, the beadlike calcareous sphinctozoan sponges were among the most important reef-builders in the great Permian reef complexes of west Texas, China, and Tunisia.

All the major sponge groups survived the Permian catastrophe, although groups like the stromatoporoids disappeared from the fossil record from the late Paleozoic and early Mesozoic. In the Jurassic, hexactinellid sponges moved into deep-water habitats, and flourished because of the expanding deep waters of the opening Atlantic Ocean. Since the Cretaceous extinction of stromatoporoids, most groups of sponges (except the demosponges) have been in decline through the Cenozoic. Today, sponges are only a minor part of most reef habitats, and are dominant only in more difficult habitats, such as the cold, deep waters where hexactinellids flourish. Some groups, like the sclerosponges, survive only as living fossils in cryptic habitats.

For Further Reading

Berguist, P. R. 1978. *Sponges*. University of California Press, Berkeley.

Gould, S. J. 1980. Crazy Old Randolph Kirkpatrick, pp. 227-235, *in* Gould, S. J., *The Panda's Thumb*. W. W. Norton, New York.

Hartman, W. D., and T. E. Goreau. 1970. Jamaican coralline sponges: their morphology, ecology, and fossil representatives. *Zoological Society of London Symposium* 25:205-243.

Hartman, W. D., J. W. Wendt, and F. Wiedenmayer. 1980. *Living and Fossil Sponges*. University of Miami Rosenstiel School of Marine and Atmospheric Sciences, Sedimenta VIII.

Rigby, J. K. 1971. Sponges and reef and related facies through time, pp. 1374-1388, *in* Rigby, J. K., ed., *Reefs through Time. Proceedings of the North American Paleontological Conference*, vol. J. Allen Press, Lawrence, Kansas.

Rigby, J. K. 1987. Phylum Porifera, pp. 116-139, *in* Boardman, R.S., A.H. Cheetham, and A. J. Rowell, eds. *Fossil Invertebrates*. Blackwell Scientific Publishers, Cambridge, Mass.

Rigby, J. K., and C. W. Stearn, eds. 1983. *Sponges and Spongiomorphs: Notes for a Short Course*. University of Tennessee Department of Geological Sciences Studies in Geology 7.

Stearn, C. W. 1975. The stromatoporoid animal. *Lethaia* 8:89-100.

ARCHAEOCYATHANS

One of the great mysteries in paleontology is the relationships of one of the earth's first reef-building organisms, the archaeocyathans. This group appeared in the Early Cambrian and quickly spread worldwide, especially in Russia, Siberia, North America, and South Australia (but they were absent from western Europe). Archaeocyathans built Early Cambrian reef complexes that occasionally reached tens of meters in thickness and covered many square kilometers. By the early Middle Cambrian, however, they were extinct, and their role was eventually filled by sponges and corals.

Since their record is only their hard parts, their biologi-

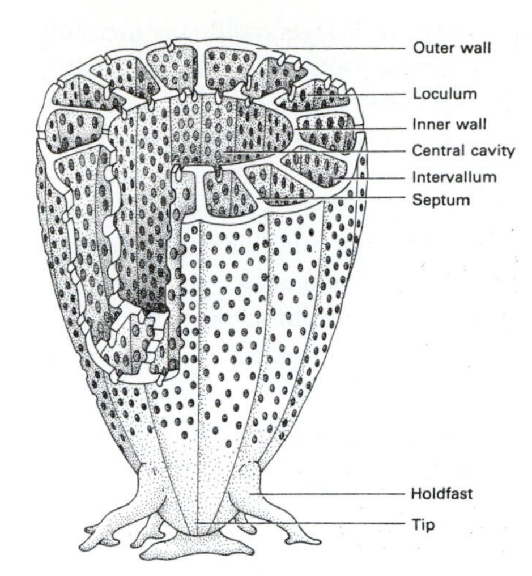

Figure 12.7. (Above) Basic morphological features of the Archaeocyatha.

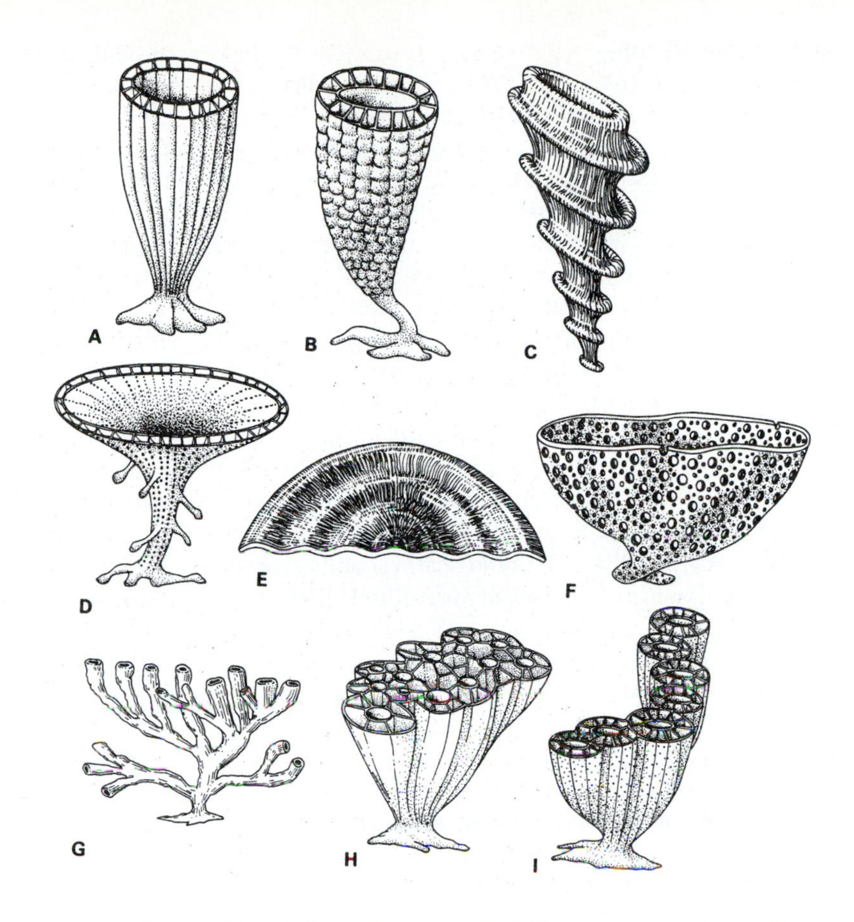

Figure 12.8 (Right) Archaeocyathans developed a surprising variety of shapes from the basic body plan. (A) *Ajacicyathus;* (B) *Kotuyicyathus;* (C) *Orbicyathus;* (D) *Paranacyathus;* (E) *Okulitchicyathus;* (F) *Cryptoporocyathus;* (G) *Archaeolynthus;* (H) Massive colonial, and (I) chain-like *Ajacicyathus.* (From Rigby, in Boardman et al., 1987.)

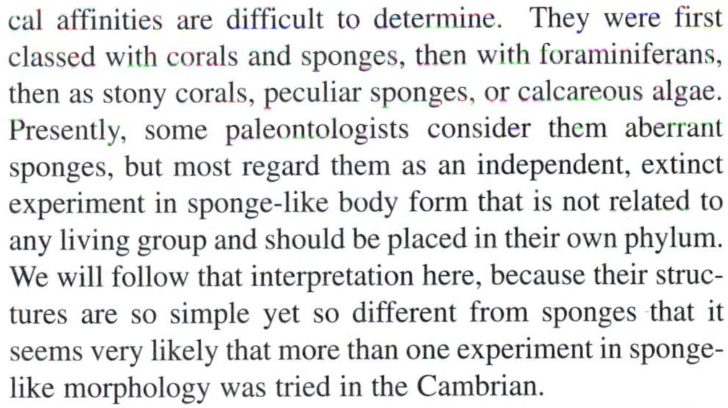

cal affinities are difficult to determine. They were first classed with corals and sponges, then with foraminiferans, then as stony corals, peculiar sponges, or calcareous algae. Presently, some paleontologists consider them aberrant sponges, but most regard them as an independent, extinct experiment in sponge-like body form that is not related to any living group and should be placed in their own phylum. We will follow that interpretation here, because their structures are so simple yet so different from sponges that it seems very likely that more than one experiment in sponge-like morphology was tried in the Cambrian.

Archaeocyathan skeletons are basically simple double-walled conical structures, with a broad, rooted base at the point of the cone, and a wide circular opening like the sponge osculum at the top of the cone (Fig. 12.7). Unlike sponges, or any other phylum, most archaeocyathans had a unique porous double-wall cone-in-cone construction, separated by perforated vertical plates (**septa**). The space between the inner and outer walls is called the **intervallum**. Some archaeocyathans also had perforated horizontal plates in the intervallum called **tabulae**, bubble-shaped structures called **dissepiments**, or curving vertical dividers called septa (similar to some corals in this respect). More than 250 genera are described, and most are distinguished by the details of the wall structure, or differences in overall shape. Although most were shaped like simple cones, there were also a variety of multibranched, fan-shaped, cup-shaped, and bowl-shaped forms (Fig. 12.8). Most individuals were about 1 to 3 cm in diameter and about 15 cm tall, although large forms reached diameters of 60 cm and were over 30 cm tall.

Although the ecology of a long-extinct group is not easy to determine, some things can be reliably inferred. Archaeocyathans preferred shallow (20 to 30 m deep) marine carbonate shelves, particularly on the shallow cratonic platform. They are less common on muddy bottoms. Archaeocyathans apparently had a high tolerance for turbidity, since some species flourished where the carbonate rocks contain up to one-third insoluble fine debris. They are not found in brackish or hypersaline deposits, or deep-water deposits. They are consistently associated with algal fossils, suggesting that archaeocyathans may have also harbored symbiotic algae and lived within the photic zone.

Although we cannot observe their soft tissues or behavior, we can use mechanical models to simulate their filter feeding (see p. 107). If these models are accurate, then archaeocyathans were simple cylindrical passive filter feeders like sponges, and their double-walled perforate construction was ideal for keeping a steady flow through their pores and up and out the top (see Fig. 7.11). Whatever archaeocyathans were doing, clearly they were very successful at it during the Early Cambrian, since they were the sole builders of the earliest reef complexes. Equally mysterious is their abrupt disappearance at the beginning of the Middle Cambrian. Although sponges eventually replaced them, no such great reef complexes as the archaeocyathan reefs appeared on earth again until the Early Ordovician.

For Further Reading

Debrenne, F. and J. Vacelet. 1984. Archaeocyatha: is the sponge model consistent with their structural organization? *Paleontographica Americana* 54:358-369.

Hill, D. 1972. Archaeocyatha, pp. 2-158, *in* Teichert, C., ed., *Treatise on Invertebrate Paleontology*. Part E. Geological Society of America and University of Kansas Press, Lawrence, Kansas.

Rigby, J. K., and R. W. Gangloff. 1987. Phylum Archaeocyatha, pp. 107-115, *in* Boardman, R.S., A.H. Cheetham, and A. J. Rowell, eds. *Fossil Invertebrates*. Blackwell Scientific Publishers, Cambridge, Mass.

Rigby, J. K., and C. W. Stearn, eds. 1983. *Sponges and Spongiomorphs: Notes for a Short Course*. University of Tennessee Dept. Geological Sciences Studies in Geology 7.

Rozanov, A. Y. 1974. Homological variability of the archaeocyathans. *Geological Magazine* 111:107-120.

Wood, R. A., A. Y. Zhuravlev, and F. Debrenne. 1992. Functional biology and ecology of the Archaeocyatha. *Palaios* 7:131-156.

CNIDARIANS

The next level of complexity is the cnidarians, familiar to most of us as corals, sea anemones, and jellyfish. In some books, you might find the old (1847) term "Coelenterata" applied to this group, but today most biologists prefer to use the term "Cnidaria," because "Coelenterata" also originally included two additional phyla, the sponges and the comb jellies (phylum Ctenophora), which were recognized as separate phyla as early as 1888. The term "Cnidaria" (from the Greek *knide*, "nettle") describes this group well, because one of their major characteristics is their stinging cells (**nematocysts**) in their tentacles that help paralyze their prey or predators. The stings of some jellyfish, such as the sea wasp, Portugese man-of-war, or Atlantic sea nettle, are capable of paralyzing or even killing a human.

The cnidarians are more complex than sponges, but only slightly more so. They have true tissues, but no discrete organs like higher animals. Unlike higher animals, they have no specific excretory, respiratory, or circulatory systems, although they do have a nervous system, muscular system, and reproductive system. Their bodies consist of specialized cells in just two cell layers with a non-cellular jelly-like material (called **mesoglea**) in between (Fig. 12.9). The outer layer of cells (**ectoderm**) encloses a sac-like body with a mouth surrounded by tentacles at one end. The inner layer of cells (**endoderm**) is primarily responsible for digesting whatever food is trapped by the tentacles and pulled through the mouth into the digestive cavity, or **enteron**. The ectoderm is responsible for secreting a skeleton in the skeletonized groups, such as corals, although most cnidarians are soft-bodied. It contains the primitive nerve network, which causes the muscular contractions of the body in response to stimuli. The tentacles are extensions of the ectoderm, loaded with stinging cells.

Unlike higher animals, cnidarians have but a single opening for the enteron, so everything they eat and digest must go out the way it came; higher animals have a separate mouth and anus that allows a one-way flow of food and waste. Along with their lack of a one-way gut, they also have radial symmetry, and lack a true "head" or "tail" end. The attached colonial forms don't care which way is front, and mobile jellyfish float any way the currents take them.

Another unique feature of cnidarians is their alternation of generation reproduction that allows them to alternate between sessile and mobile lifestyles (Fig. 12.10). Most cnidarians alternate between an attached, tentacled **polyp** stage, which reproduces asexually by budding, and a free-floating, jellyfish-like **medusa** stage, which has sexual organs and releases eggs and sperm to form a **planula** larva. This eventually grows into new polyps, and the cycle is repeated. Since many cnidarians spend most of their lives as immobile polyps, the free-floating medusa stage is necessary for sexual combinations with other individuals, and dispersal to new habitats. In many cnidarians, the two generations alternate regularly; in others, one generation may be very short, while the animal spends most of its life as a medusa (this is true of most jellyfish) or as a polyp (this is true of most hydras, corals, and sea anemones). Although polyps and medusa look very different, they are fundamentally the same structure. If you detach a polyp and turn it upside down, with its mouth and tentacles pointing downward, and add a little jelly and sexual organs, you have a medusa (Fig. 12.9).

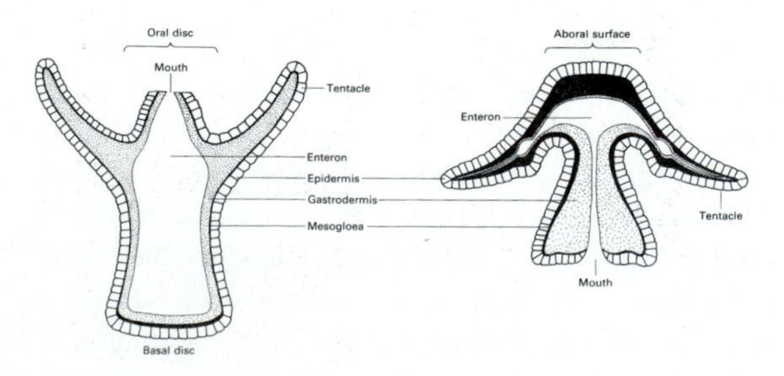

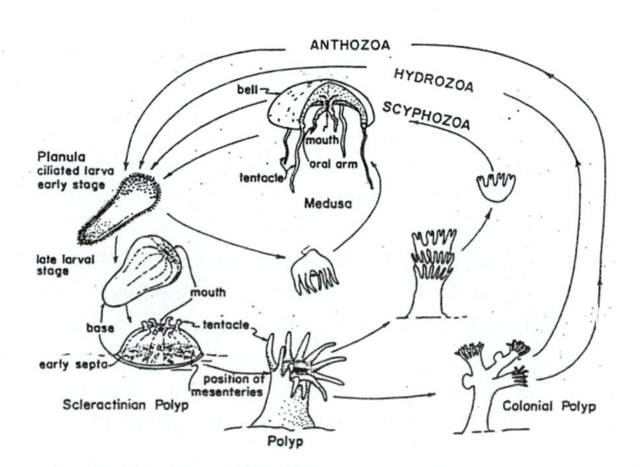

Figure 12.9. (Top) Alternation of generations between a polyp (left) and medusa (right).

Figure 12.10. (Right) Life cycle of the different groups of cnidarians.

Systematics

Although many authors have proposed different classifications of the Cnidaria, the following is the most widely accepted. In this chapter, we will emphasize only groups with a good fossil record, which neglects most cnidarians in favor of skeletonized groups.

Phylum Cnidaria

Class Hydrozoa (hydras; about 500 genera and 2700 species in 7 living orders; rarely fossilized)

Hydrozoans are mostly familiar as the polyp *Hydra*, common in any microscopic view of pond water. Unlike other classes, some hydrozoans are freshwater animals, but most are marine. Most hydrozoans spend their lives largely as polyps, although some have a medusa stage and suppress the polyp stage. The majority have no hard parts, or have a flexible chitinous skeleton that rarely fossilizes. Two orders, the "fire corals" or milleporoids, and the hydrocorals, or stylasteroids, secrete calcareous skeletons, and are important elements of the modern reef fauna. Both first appeared in the Late Cretaceous, but their fossils are much rarer than anthozoan corals.

Class Scyphozoa (jellyfish; about 90 genera and 200 living species in 4 orders; rarely fossilized)

Most familiar jellyfish are scyphozoans. They spend their lives mostly as a jelly-filled medusa, and some reach 2 m in diameter, and the tentacles of a large Portugese Man-of-War may trail many meters behind it in the water. Some scyphozoans have lost their polyp stage altogether. All are marine species.

Because they are soft-bodied, scyphozoans only rarely fossilize. Many exceptional *Lagerstätten* (see Chapter 1), such as the Vendian Ediacara localities of Australia, the Middle Cambrian Burgess Shale of British Columbia, the Pennsylvanian Mazon Creek nodules of eastern Illinois, and the Jurassic Posidonienschiefer and Solnhofen Limestone of Germany, preserve extraordinary specimens, but otherwise scyphozoan fossils are rare. A pecular group of cone-shaped fossils known as conulariids, which have a rectangular cross section and herringbone ribs down the sides, have been related to the scyphozoa based on their four-fold symmetry, but Babcock (1991) has argued that they are fossils of an unrelated group, probably an extinct phylum.

Class Anthozoa (corals and sea anemones; about 2300 genera and 6000 living species in three subclasses, with hundreds of fossil genera and species)

Anthozoans are exclusively marine, and most are colonial, sessile forms, with an elaborate hydra stage and little or no medusa stage.

Subclass Octocorallia (sea fans, sea whips, sea pens, and soft corals)

Octocorals are common members of the modern reef fauna, with a variety of shapes and habitats. They include the gorgonaceans (familiar as sea fans and sea whips), a group with a horny skeleton made of the protein **gorgonin**; the soft-bodied sea pens; and a variety of orders that have no hard parts and no fossil record. However, the fossil record of octocorals is sparse. There are soft-bodied impressions of sea pens in the Edicara fauna, and occasional fossils in other exceptional localities, but the lack of a hard calcareous skeleton limits their fossilization potential.

Subclass Zoantharia (sea anemones and stony corals)

In addition to three different orders of sea anemones, the zoantharia include three major orders of corals, one living and two extinct:

Order Tabulata (tabulate corals; about 280 genera; Early Ordovician to Permian)

The tabulate corals were colonial forms, built of many small, closely packed calcite tubes (**corallites**) with small horizontal dividers along the length of the tube (**tabulae**, Latin for "little tables") which give the group its name (Fig. 12.11). Some forms were packed into a tight, hexagonal "honeycomb" arrangement (the favositids, a common Silurian-Devonian reef builder—Fig. 12.11A). Others had their corallites arranged into "palisade" arrangements that looked like a chain in cross section (the "chain corals," or halysitids, another important Silurian reef builder—Fig. 12.11B).`The heliolitids (Fig. 12.11C) had the corallites separated by a deposit of densely packed tubes lacking tabulae (the **coenosteum**), apparently secreted colonially by more than one polyp. Tabulate corals tend to form large masses of coralline limestone which were one of the dominant reef builders during the Silurian and Devonian. They suffered near extinction during the Late Devonian event, and continued to decline in the late Paleozoic until their final extinction in the Permian.

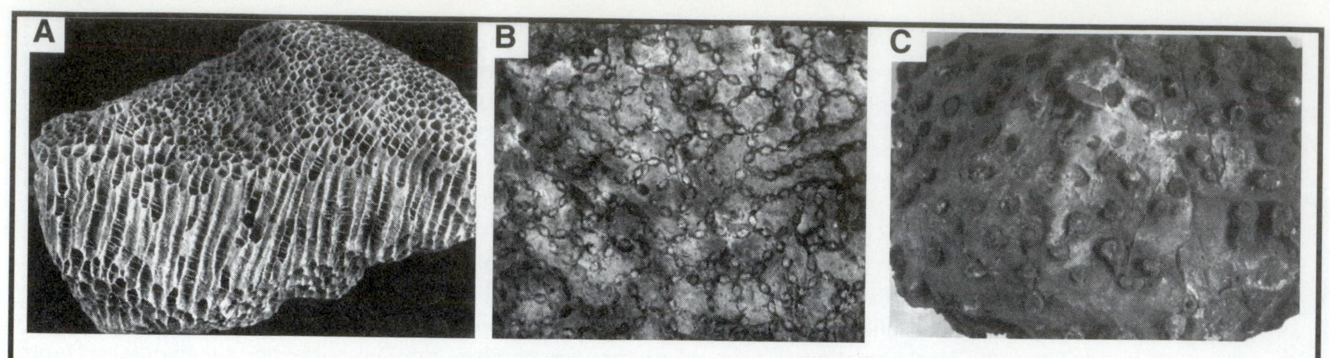

Figure 12.11. Common tabulate corals of the Paleozoic. (A) The "honeycomb coral" *Favosites,* composed of numerous vertical tubes subdivided by hundreds of tabulae (clearly visible in this photo.) (B) Top view of the "chain coral" *Halysites,* composed of numerous vertical tubes linked in chains like organ pipes. (C) The helolitids had corallites separated by large masses of skeleton known as coenosteum. (Photos by the author.)

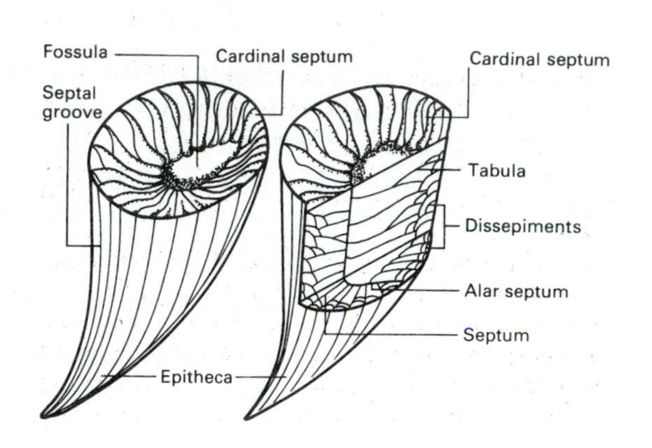

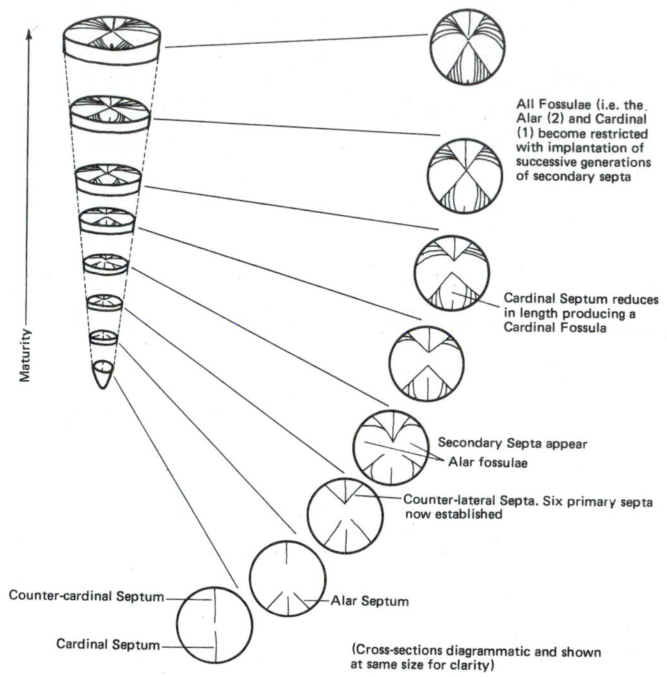

Figure 12.12 (top). Morphology of the solitary rugose corals. (From Boardman et al., 1987.)

Figure 12.13 (right) Ontogeny of a typical rugose coral, showing the sequence of insertion of the septa from the earliest ontogenetic stages at the base. (After Carruthers, 1910.)

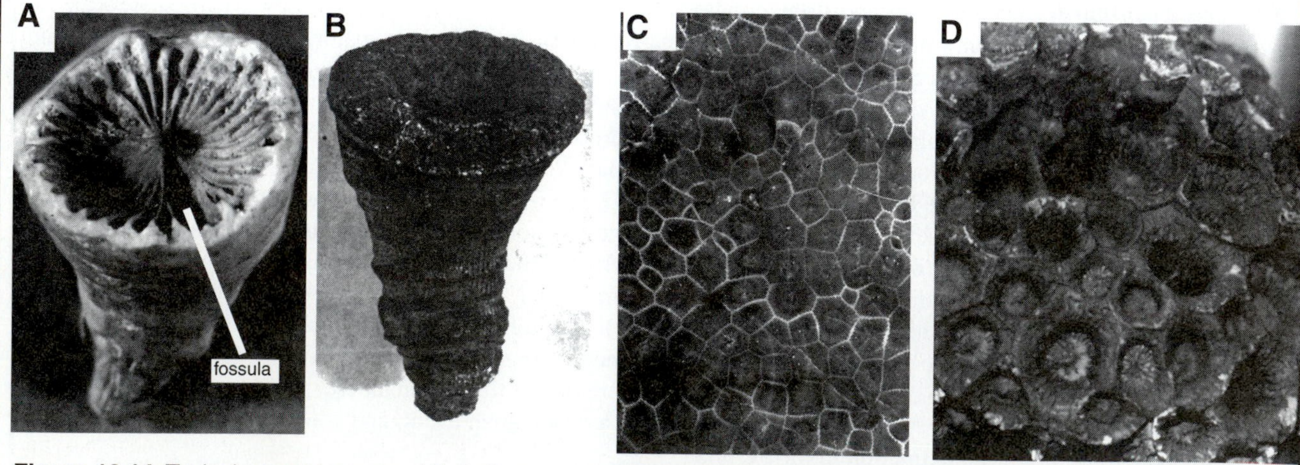

Figure 12.14. Typical rugosid genera. (A) Solitary rugosid *Zaphrentis,* showing a well-developed fossula. (B) Solitary rugosid *Heliophyllum,* with its distinctive wrinkled, irregular shape. (C) Colonial rugosid *Hexagonaria,* a major reef-builder in the Devonian. (D) Colonial rugosid *Lithostrotionella,* a Carboniferous form. (Photos by the author.)

Order Rugosa (horn corals; about 800 genera; Middle Ordovician to Permian)

The rugosids, or horn corals, were the most morphologically diverse and abundant group of Paleozoic corals. They get their name from the fact that the outer wall of the corallite, or **epitheca,** is wrinkled, or rugose, in many taxa. Some rugosids were solitary horn-shaped forms, while others were colonial. Each corallite is divided by a number of vertical walls called **septa** (singular, **septum**) that radiate outward from a central pillar called a **columella** (Fig. 12.12). Within each septal chamber could be horizontal tabulae, or curved walls known as **dissepiments**, which separate small subchambers near the **epitheca**. The cup-shaped top of the colony, or **calice** (the Greek name for a large bowl-shaped drinking cup), was the base on which an anemone-like polyp lived and grew upward, secreting a larger and larger corallite behind it.

By slicing a rugosid into numerous serial sections, it is possible to examine the ontogeny of the insertion of septa as the individual grew larger (Fig. 12.13). The first septa to insert are the cardinal septum and counter-cardinal septum, 180° from each other. Shortly thereafter, a pair of alar septa insert at an angle of about 30° from the cardinal septum, followed by counter-lateral septa at 180° from the alar septa. The four alar and counter-lateral septa tend to divide the corallite into four sections, which is reponsible for the old name of the rugosids, the "tetracorals." After these septa have inserted, a series of additional numbered septa insert in between them. The details of these septal patterns are important in taxonomic identifications, as are large gaps in the pattern of radiating septa known as **fossula** (Latin, "little trench").

Rugose corals exhibit a wide variety of shapes (Fig. 12.14), from the simple solitary forms like *Zaphrentis* or *Streptelasma* (Fig. 12.14A), to the irregular lumpy solitary forms such as *Heliophyllum* (Fig. 12.14B), to colonial forms such as the honeycomb-like *Hexagonaria* or *Lithostrotionella* (Figs. 12.14C, D). Like the tabulates, rugosids were particularly abundant and common members of the great Siluro-Devonian reef complexes, and suffered severely in the late Devonian extinctions. They straggled on through the late Paleozoic (although much more common than tabulates), and also became extinct at the end of the Permian (unless they are the ancestors of the Scleractinia).

Order Scleractinia (hexacorals; about 600 genera; Middle Triassic-Recent)

After the extinction of all calcified coral groups of the Late Permian, there are no fossils of calcified corals until the Middle Triassic, when the first scleractinians appeared. Their origins are unclear and controversial. Some think they evolved from surviving rugosids which reorganized their skeleton and secreted aragonite, not calcite. Others note the fundamental anatomical differences, and the lack of skeletonized corals in the Early Triassic, and suggest that Scleractinia independently arose by skeletonization of a previously unfossilizable anemone group. By the Late Triassic, scleractinians formed small reefs, and the group has continued to diversify until they are the primary reef builder on the planet today.

Scleractinians are similar to rugosids in that they insert the septa in their corallite in a predictable fashion. Instead of four regions, however, hexacorals have six primary septa which insert at 60° of separation from each other (hence the name "hexacorals"). Once these septa have been inserted, six additional secondary septa are inserted halfway between the primary septa, followed by additional third- and fourth-level septa. Advocates of the rugosid origin of scleractinians suggest that the alar and counter-lateral septa of rugosids (Fig. 12.13) rotated until they were at 60° angles from the cardinal and counter-cardinal septum, giving the characteristic six-fold symmetry.

Scleractinians come in an enormous variety of shapes and forms, from the massive brain corals, to the many branching varieties, as well as organ-pipes, staghorns, ruffled forms, and flattened or disk-shaped forms with large, prominent septa. In some corals, the shape is strongly influenced by the environment. For example, the common reef coral *Acropora* tends to form more massive shapes in areas of heavy wave pounding, but delicate branching shapes in protected, quiet water (see Fig. 2.12).

Ecology

The ecology of corals and other cnidarians has been studied in great detail, because corals are so important as reef builders and indicators of climate and latitude. All corals require water movement to bring in nutrients that they can trap, and to flush out waste products. Most are also intolerant of large volumes of sediment, which quickly clog them. Modern scleractinians are divided into two ecological groups: those with symbiotic algae (**zooxanthellae**) in their tissues (known as **hermatypic** corals), and those without zooxanthellae (**ahermatypic** corals). Because of the light restrictions of their endosymbionts, hermatypic corals are restricted to clear, shallow waters in the photic zone (less than 90 m of water), and they function best in the tropics at temperatures of 25 to 29°C (although they can survive in waters as cold as 16°C). With the aid of their zooxanthellae, hermatypic corals are able to grow to large size, and form enormous coral reefs with many cubic kilometers of limestone. Apparently, the rate of coral metabolism is accelerated by the presence of the zooxanthellae, which absorb their waste carbon dioxide and provide oxygen. They may also help them secrete their calcite skeletons more easily by producing by-products of photosyntheis that help calcite crystals nucleate.

Hermatypic corals are also adapted to environments of relatively poor nutrient supply, since they can rely on their zooxanthellae to provide part of their nourishment, and they are outcompeted by algae in areas of high nutrient supply (Hallock and Schlager, 1986; Wood, 1993). In regions where algae grow profusely, they will smother corals unless there are grazers (various crustaceans, gastropods, and polychaete worms) to keep them cropped.

Ahermatypic corals, by contrast, can live in much greater variety of depths, since they do not require light. Although they also thrive in shallow, warm water, they are capable of living at depths of 6000 m and surviving temperaters only 1°C above freezing. However, without zooxanthellae, they grow much slower, and never construct large reefs. They are particularly characteristic of the deeper continental shelf, where along with sponges, they are among the few common organisms.

With extinct corals, it is more difficult to determine whether a species had endosymbionts or not. Based on living hermatypic corals, most zooxanthellate corals have a smaller average corallite size, a much higher level of coloniality and integration, and their dominant shape is sheets, mounds, and branching forms with multiple series of corallites. Based on these criteria, there are good candidates for hermatypic corals in the Cretaceous (although they were not reef builders), and even some Triassic scleractinians show the appropriate shape and geochemical characteristics of hermatypic corals (Coates and Jackson, 1987; Swart and Stanley, 1989, 1995). Some paleontologists have speculated that some of the colonial rugosids and tabulates, especially the major reef builders of the Silurian and Devonian, may have been hermatypic.

Although it's hard to imagine anything eating stony corals, in fact there are numerous predators. Parrotfish have hard beaks that are excellent for crushing hard coral, and sudden population explosions of the crown-of-thorns starfish destroyed some tracts of the Great Barrier Reef. In addition, there are specialized crabs, polychaete worms, snails, and echinoids that are adapted to feeding on coral polyps, and many organisms (clionid sponges, bivalves, and certain worms) that are specialized to bore and burrow into coral skeletons.

Evolution

The overall history of the reef habitat is summarized in Figure 12.15. There are two distinct subhabitats: those with abundant nutrients, and those that are nutrient-poor. In the Early Cambrian, archaeocyathans briefly occupied the nutrient-rich zone, but there were no occupants in the nutrient-poor zone. There were no significant reef builders through the rest of the Cambrian or Early Ordovician. In the Middle Ordovician, tabulate and rugose corals originated by calcification of some soft-bodied anemone-like animal, with increasing dominance of rugosids (along with stromatoporoids and spicular sponges, and bryozoans) as the great reef complexes of the Silurian and Devonian appeared. The nutrient-poor habitat was also occupied in the Silurian and Devonian, apparently by stromatoporoids and tabulate corals. Both tabulates and rugosids were severely affected by the Late Devonian extinction, leaving glass sponge reefs in the latest Devonian. In the Carboniferous, the reef niche was essentially vacant, and it was finally occupied by calcified sponges and bryozoans in the Permian. Tabulates and rugosids straggled on into the late Paleozoic, only to die out at the end of the Permian (along with trilobites, blastoids, and fusulinaceans).

There are no fossil corals yet known from Lower Triassic rocks, but by the Middle Triassic scleractinians had appeared (by whatever origin is still controversial). Since the Late Triassic, scleractinians began to construct reefs, occupying the long-vacant nutrient-poor regions again, and they have been the major reef-builders for the rest of the Mesozoic and Cenozoic.

Paleontologists have been less successful in describing evolutionary patterns among the corals. This is due in large part to the fact that much coral morphology is ecophenotypic, responding to differences in the environment, and does not reflect genetic-based evolutionary change. Similarly, there is much iterative and convergent evolution in corals, because ecology often dictates that certain important features evolve again and again. For example, the development of dissepiments apparently occurred independently several times in rugose corals during the Paleozoic. It was probably valuable in reinforcing the walls of the corallite and allowing them to reach larger size. In a famous case study, Carruthers (1910) argued that the solitary rugosid *Zaphrentis delanouei* showed a microevolutionary pattern in the Lower Carboniferous strata of Scotland, England, and Belgium. Studying samples through the sequence, he noticed trends in the shape of the cardinal fossula and the length of the major septa. However, this study has since been criticized on a number of grounds: it is based on isolated samples from widely separated levels that may not represent a true evolutionary sequence; it has no geographic control, so some of the "trend" may be due to migration of clines; and the "trend" may simply be an ecophenotypic response to environmental changes. For these reasons, few studies of microevolutionary patterns have been attempted in corals.

For Further Reading

Babcock, L. 1991. The enigma of conulariid affinities, pp. 133-143, *in* Simonetta, A.M., and S. Conway Morris, eds., *The Early Evolution of Metazoa and the Significance of Problematic Taxa*. Cambridge University Press, Cambridge.

Bayer, F. M., et al. 1956. Coelenterata, in Moore, R. C., ed., *Treatise on Invertebrate Paleontology*. Part F. Geological Society of America and University of Kansas Press, Lawrence, Kansas.

Carruthers, R. G. 1910. On the evolution of *Zaphrentis delanouei* in the Lower Carboniferous times. *Quarterly Journal of the Geological Society of London* 66:523-536.

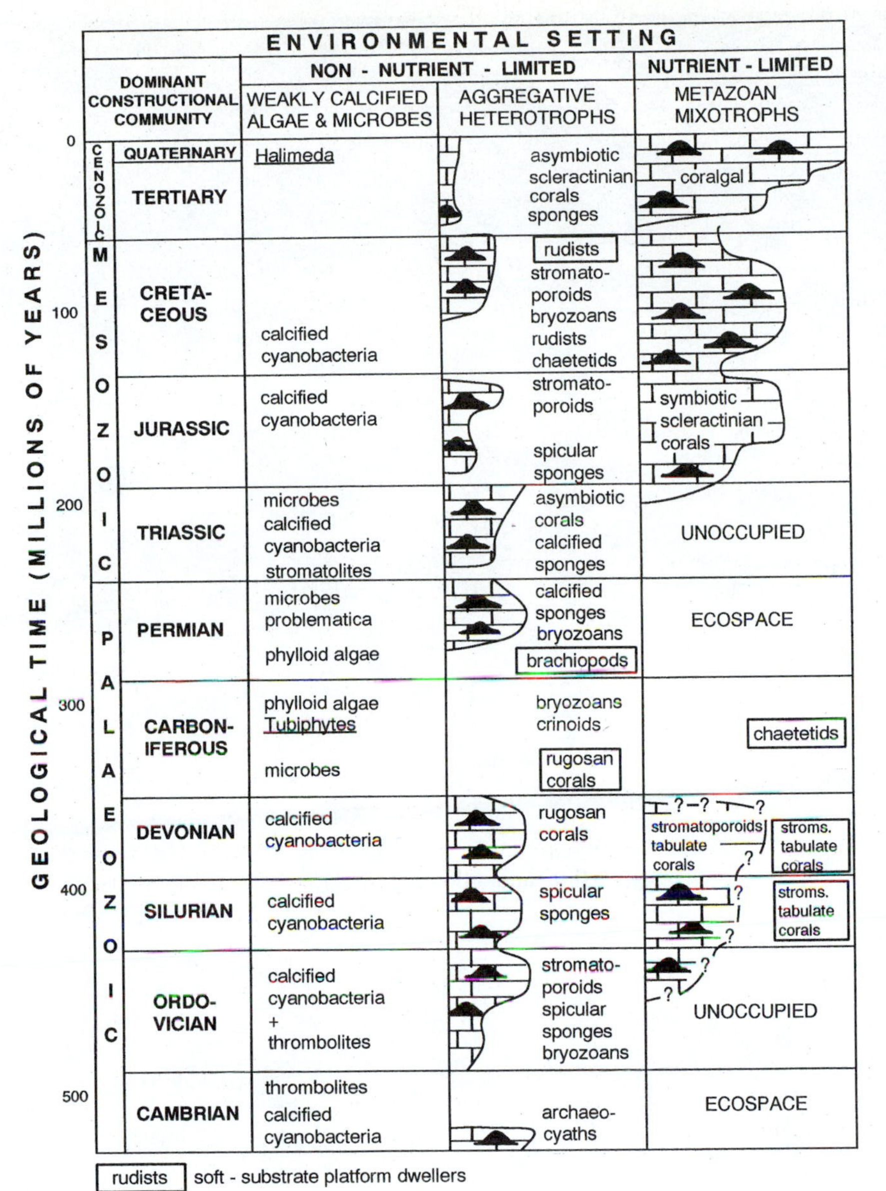

Figure 12.15. History of reef- building organisms in both nutrient-limited and non-nutrient-limited environments. Notice that hermatypic corals tend to occupy the waters with limited nutrients (since their symbiotic algae help them cope with lack of nutrients), and that ecospace was often vacant when no organisms with the appropriate adaptations were in existence. In regions where nutrients are abundant, other organisms tend to crowd the corals, and sometimes become reef-builders themselves (such as sponges, archaeocyathans, and rudistid bivalves.) (After Wood, 1995.)

Coates, A. G., and J. B. C. Jackson. 1987. Clonal growth, algal symbiosis, and reef formation by corals. *Paleobiology* 13:363-378.

Goreau, T. F., N. I. Goreau, and T. J. Goreau. 1979. Corals and coral reefs. *Scientific American* 245(5): 110-121.

Hallock, P., and W. Schlager. 1986. Nutrient excess and the demise of reefs and carbonate platforms. *Palaios* 1:389-398.

Heckel, P. H. 1974. Carbonate buildups in the geological record: a review. *SEPM Special Publication* 18:90-154.

Hill, D. 1981. Coelenterata, Supplement 1, Rugosa and Tabulata, *in* Teichert, C., ed., *Treatise on Invertebrate Paleontology*. Part F. Geological Society of America and University of Kansas Press, Lawrence, Kansas.

Muscatine, L., and H. M. Lenhoff. 1974. *Coelenterate Biology*. Academic Press, New York.

Oliver, W. A., Jr. 1980. The relationship of scleractinian corals to the rugose corals. *Paleobiology* 6:146-160.

Oliver, W. A., Jr., and A. G. Coates. 1987. Phylum Cnidaria, pp. 140-193, *in* Boardman, R.S., A.H. Cheetham, and A. J. Rowell, eds. *Fossil Invertebrates*. Blackwell Scientific Publishers, Cambridge, Mass.

Stanley, G. D., Jr. 1981. Early history of scleractinian corals and its geological consequences. *Geology* 9:507-511.

Stanley, G. D., Jr., ed. 1996. *Paleobiology and Biology of Corals: Notes for a Short Course*. University of Tennessee Dept. Geological Sciences, Studies in Geology 10.

Wood, R. A. 1993. Nutrients, predators, and the history of reefs. *Palaios* 8:526-543.

Wood, R. A. 1995. The changing biology of reef-building. *Palaios* 10:517-529.

Figure 13.1. A typical Ordovician limestone from the Platteville Group, near Dickeyville, Wisconsin. It is dominated by lophophorate fossils, including D-shaped strophomenide brachiopods, massive bryozoans (lower left) and branching bryozoans (upper center), along with kidney-bean-shaped *Leperditia* ostracodes. (Photo by the author.)

Chapter 13

The Lophophorates

Brachiopods and Bryozoans

Brachiopods are not generally familiar animals, and neither their Latin name nor the English term "lamp-shells" means very much to most people. To explain that they are "a kind of shell-fish but really quite different from other shell-fish" is still not very helpful, though it does at least convey the fact that they are aquatic animals with a hard external shell. Even among zoologists they are not well known, and most textbooks of systematic zoology give them merely a page or two in a chapter on "minor phyla." Yet on turning to any textbook on palaeontology we find that this minor phylum has become "major," and dozens of pages may be devoted to what now appears to be highly complicated objects equipped with a most formidable terminology. But at the end of this they are likely to remain no more than "objects," and the reader may still feel he has little idea what these fossils were like when they were living animals.

Martin Rudwick, *Living and Fossil Brachiopods*, 1970

INTRODUCTION

In the previous chapters, we discussed organisms with various levels of cellular organization, such as the single-celled protistans, the colonial but essentially independent sponge cells, and the cnidarians, with their specialized tissues (but no organs). The next step in organismal evolution is a worm-like organism that has a definite head and tail end (and with it, a one-way digestive tract, running from mouth to anus), which means that it also has a right and a left side (bilateral, rather than radial symmetry). Such organisms can actively seek their food and move with a purpose, rather than waiting for their food to drift by, as in sponges and cnidarians.

These primitive worms usually do not have a hard skeleton, so they seldom fossilize. However, most worms (except for the simplest, such as the flatworms), have an internal fluid-filled body cavity, called a **coelom**, which serves not only as a container for internal organs, but also as a hydrostatic skeleton. The pressure of the confined fluid inside gives them some rigidity against which their muscles can push, and allows them to move in a variety of ways, and to burrow (something a flatworm, without a coelom, cannot do). A good analogue for this system is a water balloon—it is not rigid by itself, but when pressure is placed on the confined volume of water, it resists some of the pressure, and responds by extending in any available direction. Coelomate worms are built like tiny water balloons. The fluid-filled coelom inside allows them to flex and push against the sediment, so they can move and burrow. The innovation of a hydrostatic skeleton in the coelomates is an important one, since burrowing allowed the exploitation of resources below the sea bottom that had long been unreachable. In the late Precambrian, deep burrows first appeared in the fossil record, indicating that organisms finally exploited this new niche. Burrows also provide protection, which is important for a soft-bodied animal with no skeleton or armor.

In addition to the innovation of the coelom, most coelomates have definite organs and organ systems for most of their biological functions: a digestive-excretory tract, specialized reproductive organs, a more complex musculo-nervous system, and in many groups, a discrete respiratory system as well. Once these key adaptive innovations had been achieved, the coelomates diversified into a great variety of body forms: many different worm-like phyla (such as the segmented annelid worms); segmented animals with an external skeleton (the arthropods); worm-like animals with a specialized foot, and a calcareous shell (the molluscs); calcareous animals with an internal hydraulic system (the echinoderms); and another worm-like group with a fan-like filter-feeding device, the **lophophore**. These animals are known as the lophophorates: the tiny colonial bryozoans, the shelled brachiopods, and a worm-like group called the phoronids that have no fossil record. In this chapter, we will concern ourself with only the first two phyla, which have an excellent fossil record.

BRACHIOPODS

By far the most abundant and diverse of all the skeletonized invertebrates of the Paleozoic were the brachiopods. Well over 4500 fossil genera are known (more than 900 in the Devonian alone), and in many Paleozoic localities, they are by far the most common fossil (Fig. 13.1). In some places, the ground is literally paved with brachiopods (see p. x-xi). Brachiopods have experimented with a great variety of body forms (within the contraints of

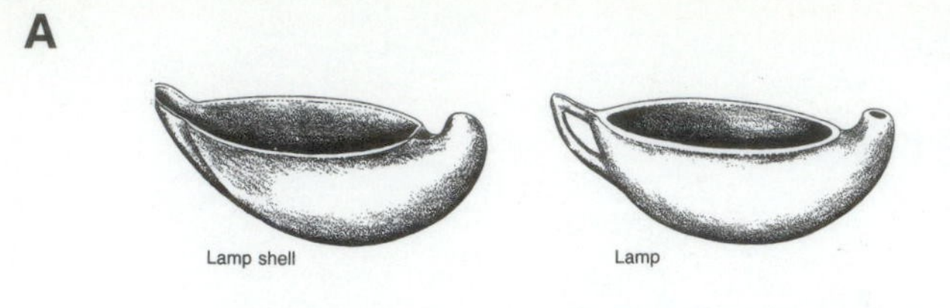

Figure 13.2. (A) The "lamp shells" got their nickname from their resemblance to a biblical oil lamp. (From Fenton and Fenton, 1958.) (B) The basic anatomical orientation and symmetry of brachiopods. (From Nield and Tucker, 1985.) (C) Internal anatomy of a brachiopod, cut along the plane of symmetry. (From Boardman et al., 1987.)

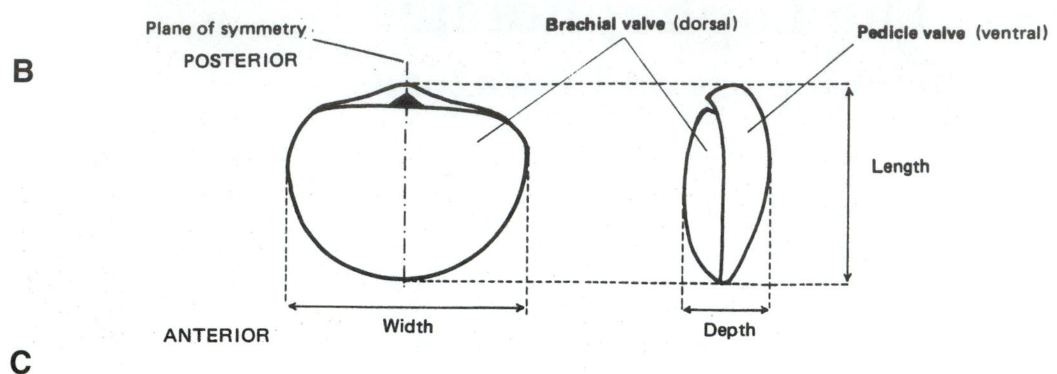

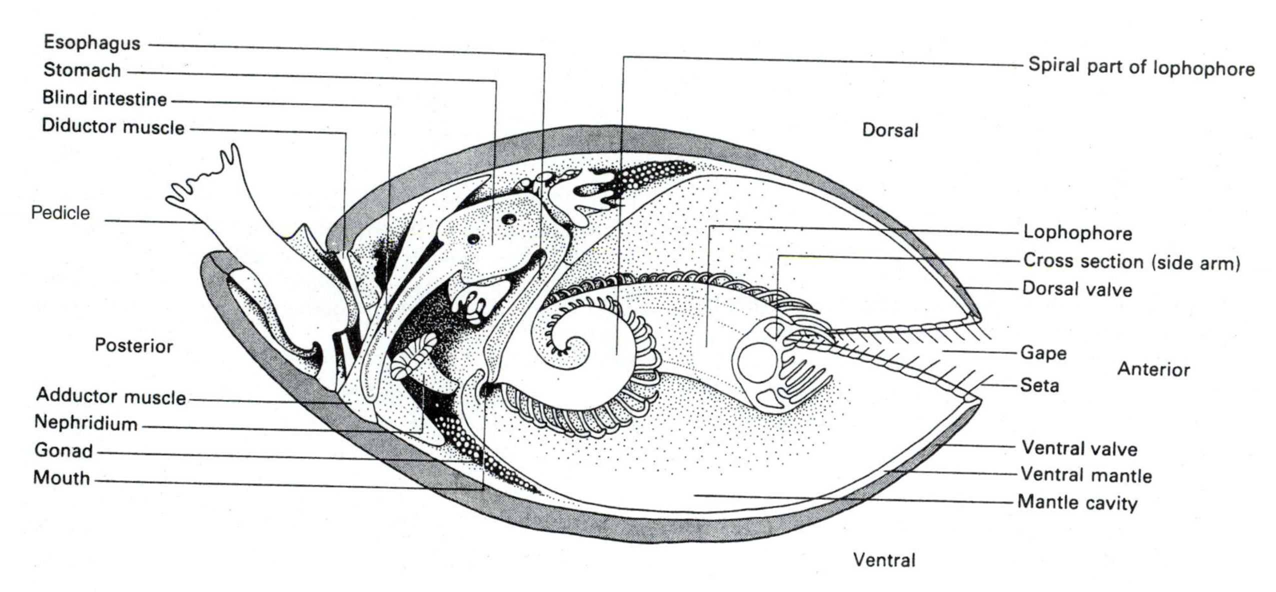

their enclosing shell, of course), and occupied a wide variety of ecological niches on the shallow seafloor. Their great abundance and diversity has made them very useful tools for paleoecology, biostratigraphy, evolutionary studies, and biogeography.

Despite their overwhelming dominance in the Paleozoic, fewer than 120 genera are alive today. Brachiopods were the most conspicuous member of the "Paleozoic fauna" to suffer from the Permian catastrophe, but failed to recover during the Mesozoic. Today the few living species are sturdy, generalist "living fossils" like *Lingula*, or relicts that live in cryptic, protected areas, or deep-water habitats. Very few beachcombers or marine biologists have even seen a living brachiopod, or collected their shells. Consequently, brachiopods are emphasized much more by paleontologists than by biologists. Most invertebrate zoology textbooks give the brachiopods only a few pages (or neglect them altogether), yet they are usually the largest single chapter in many paleontology textbooks. If they have a common name, it is "lamp shells," since the shells of one living group, the terebratulides, bears some resemblance to the oil lamps of Biblical times (Fig. 13.2A)

In contrast to the colonial bryozoans, brachiopods are solitary animals that secrete a two-valved shell around their soft bodies and lophophore. Superficially, their shell looks like that of bivalve molluscs, and they were mistakenly classified with the molluscs until the late 1800s, when scientists looked past the shell and discovered that their internal anatomy looks nothing like that of a mollusc. The quickest way to distinguish clams and brachiopods is their symmetry (Fig. 13.2B). Most brachiopods have a plane of symmetry that runs *through* both shells, so that one half of the shell is the mirror image of the other. By contrast, most clams are symmetrical *between* the valves, so the right valve is the mirror image of the left valve. Of course, there are many groups that break this rule, such as the scallops, which are nearly symmetrical in both planes, or the many oyster-like bivalves and richthofenid brachiopods, which abandoned symmetry altogether.

The typical brachiopod shell (Fig. 13.2B) has one valve

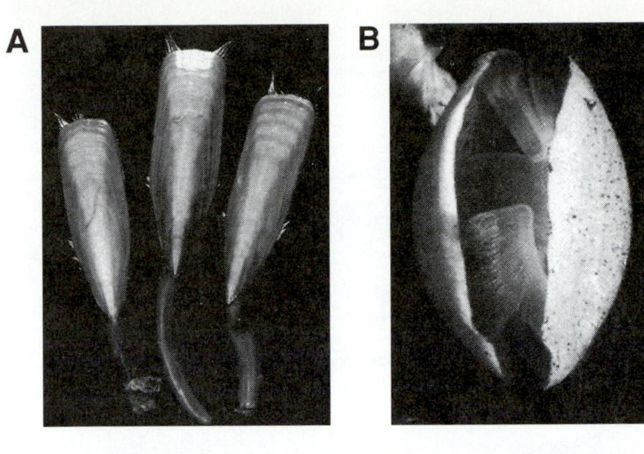

Figure 13.3. (A) The living inarticulate *Lingula*. (B) A living terebratulide brachiopod on the sea bottom, with its valves gaping slightly to show the lophophore feeding. (From Dott and Prothero, 1994.)

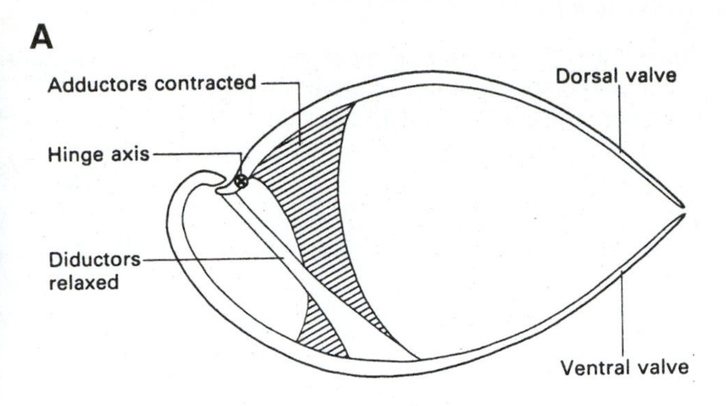

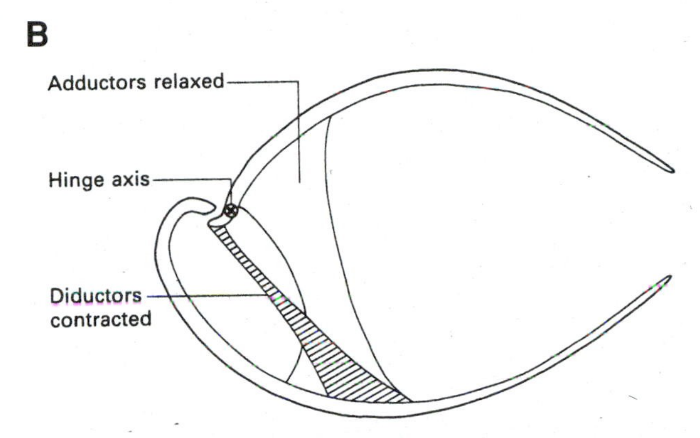

Figure 13.4. Brachiopod valves are opened and closed by pairs of opposed muscles. (A) Contraction of adductors closes the valves. (B) Contraction of diductors levers the shell around the hinge and opens the dorsal valve. (From Boardman et al., 1987.)

that is larger than the other. It is called the **ventral** (or **pedicle**) **valve**, since it usually has an opening for a fleshy stalk called the **pedicle**, with which the brachiopod attaches itself to the substrate. The opposite, smaller valve is known as the **dorsal** (or **brachial**) **valve**, because the lophophore (also called the **brachium**, or "arm" in Latin, since it was once thought of as a kind of arm) attaches to it. "Brachiopoda" means "arm-foot" in Greek, because the lophophore was once considered an appendage ("foot"). In the majority of brachiopods, known as the Class Articulata, there is a mechanical hinge that articulates the two shells on either side of the pedicle opening. The pedicle typically attaches the brachiopod to the bottom with its hinge down, and open edge sideways or up, allowing water to move through the gape so that the lophophore can filter feed (Fig. 13.3). However, brachiopods were capable of living in many different positions. For anatomical consistency, the pedicle opening is considered posterior (rear), and placed at the top in illustrations (Fig. 13.2B).

Inside this clam-like shell, however, is a body constructed very differently from that of a mollusc. The most prominent feature is the feathery lophophore (Figs. 13.2C, 13.3), which occupies most of the shell's internal volume in the **mantle cavity**. Cilia on the lophophore generate gentle currents that are drawn in from the sides of the shell and out the front. This action brings in food particles (which are trapped by the cilia and directed back to the mouth along grooves on the base of the lophophore) and oxygen, and gets rid of waste products (carbon dioxide) and gametes. Crammed into the back of the shell (Fig. 13.2C) is the coelomic cavity, which contains most of the internal organs. A U-shaped digestive tract begins with a mouth near the base of the lophophore, loops through the esophagus and stomach, and ends up in a blind intestine with no anus. When enough waste accumulates, it is regurgitated as small pellets and expelled by rapid snapping of the valves. There is also a kidney-like **nephridium** used for excretion of metabolic wastes. The coelomic cavity also contains the gonads, digestive glands, and an open circulatory system with a simple contractile heart that circulates fluids. The lophophore doubles as a respiratory system. There is a simple nervous system beginning with ganglia around the heart, and then running to the rim of the mantle. It is sensitive primarily to touch and light. Living brachiopods are particularly sensitive to light, and react strongly to avoid it, either by living at depths, or by living in sheltered overhangs in the shallows. The brachiopod has no eyes to sense light, but a set of bristles around the margin called **setae** help sense the environment. The individual cells around the mantle edge are sensitive to changes in the external environment that might cause the brachiopod to close up.

Within the shell are two sets of muscles that open and close the shell. The **adductors** (Figs. 13.2C, 13.4) run perpendicularly from the dorsal to the ventral valves, and pull the two valves together, closing the shell. The **diductors** insert on the middle of the ventral valve, and also on the **cardinal process** of the dorsal valve, so they pull the dorsal valve around its hinge line and cause it to open. The existence of two separate sets of muscles is very different from bivalves, which have only adductors to close their shells. Clam valves open automatically due to stretchy ligaments in their hinge area, so they gape easily, but must use muscles to close. This is the reason that bivalves open so easily when their muscles relax, and they are so often found disarticulated after they die. By contrast, brachiopods do not automatically, passively open, so they

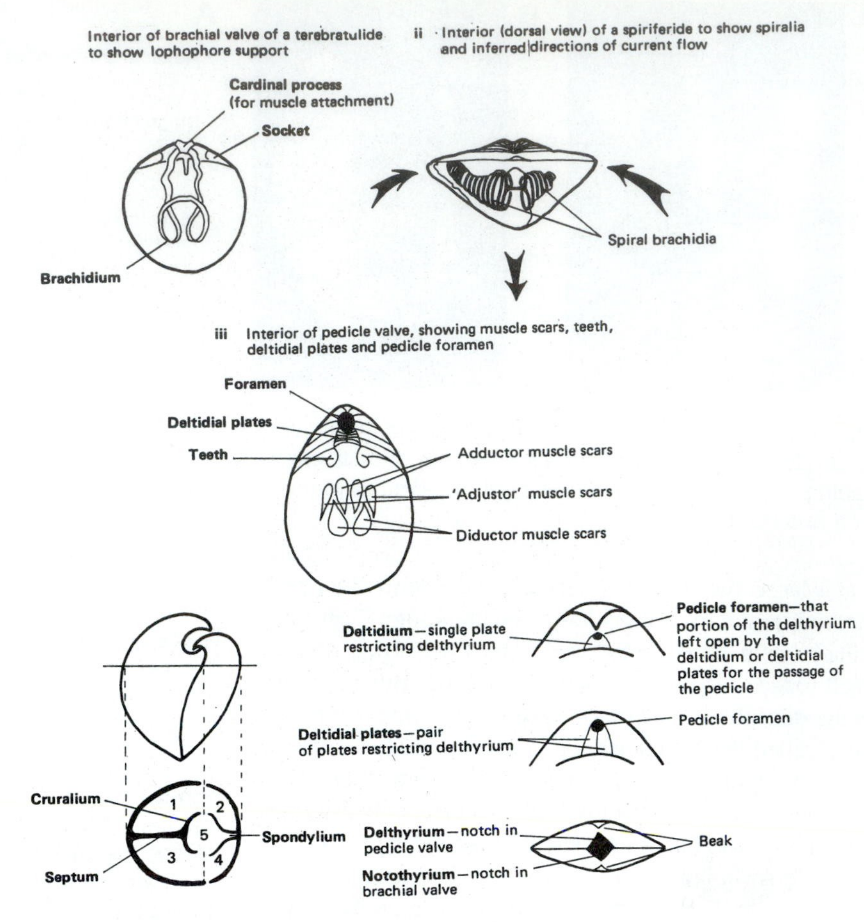

Figure 13.5. Basic terminology of the brachiopod hinge area and internal structures. (After Nield and Tucker, 1985.)

have a tendency to remain closed after they die, and are often found with both valves articulated. In most articulates, there is an additional set of **adjustor** muscles that allow the shell to move relative to the pedicle.

This describes the mechanism of the articulate brachiopods, which make up the vast majority of taxa. However, the more primitive inarticulate brachiopods do not have a mechanical hinge between the valves; they are held together by two sets of adductors, and a set of oblique muscles. Nor do they have diductor muscles to open the shell. Instead, muscles retract the body into the hinge area, forcing the shell to gape. Inarticulates have a number of other interesting differences from the articulates discussed below.

The hinge area of articulate brachiopods has long been particularly important to brachiopod classification, since it offers many intricate morphological features that are diagnostic of certain groups (Fig. 13.5). The opening for the pedicle is called the **pedicle foramen**, and it usually perforates the **beak**, or the pointed part of the ventral valve (Fig. 13.6). The pedicle foramen may be enclosed on the anterior end by a single plate, called a **deltidium**, or by a pair of **deltidial plates**. In some groups, there is no round pedicle foramen, but instead a notch called a **delthyrium**. If the delthyrium is shallow, then a similar notch on the dorsal valve called the **notothyrium** may enlarge the pedicle opening. If a plate encloses the notothyrium, it is called the **chilidium**, or there may be a pair of **chilidial plates**.

In some brachiopods, there is a large, spoon-shaped platform in the ventral valve for the hinge muscles called the **spondylium** (Fig. 13.5, lower left). This feature is particularly characteristic of the pentameride brachiopods. This group is also characterized by a **median septum**, and a corresponding spoon-shaped feature in the dorsal valve called the **cruralium**. The septum, cruralia, and spondylia of the pentamerides subdivides the internal volume of the shell into five chambers that give the group its name.

The hinge teeth and shape of the lophophore and its hard support structures are also diagnostic of many groups (Fig. 13.5). In some brachiopods, the variations in the cardinal process and the adjacent **hinge sockets** are valuable for identification. The calcareous support of the lophophore, or brachium, is called the **brachidium**, and it is attached to the the hinge area by the **crus** (plural, **crura**). It may have an elaborate **loop** to support the main part of the lophophore. The shape of the brachidia can be highly variable. In many spirifers they are arranged in a conical spiral pointing laterally (Fig. 13.5, upper right). In another group, the atrypids, the spiral brachidia lie flat in the plane of the valve and spiral upward. In other brachiopods, there is only a short, stout set of supports called **brachiophores**, rather than elaborate brachidia.

The external features of the brachiopod shell may also be useful in recognizing them (Fig. 13.6). The hinge line can be straight (**strophic**) or curved (**astrophic**). It can have a large flat or curved surface between the beak and the pos-

Figure 13.6. (right) Terminology of the external features of the brachiopod shell. (From Nield and Tucker, 1985.)

Figure 13.7. (below) Cross sections through brachiopod shells, showing curvature. b = dorsal valve; p = ventral valve. A, Biconvex, with dorsal valve less convex than ventral; B, Biconvex, with more convex dorsal valve; C, Plano-convex; D, Concavo-convex; E, Concavo-convex, but more strongly curved; F, Strongly convexi-concave; G, Gently convexi-concave; H, Resupinate, dorsal valve convex but concave near hinge line; I, Convexi-planar. (From Moore et al., 1953.)

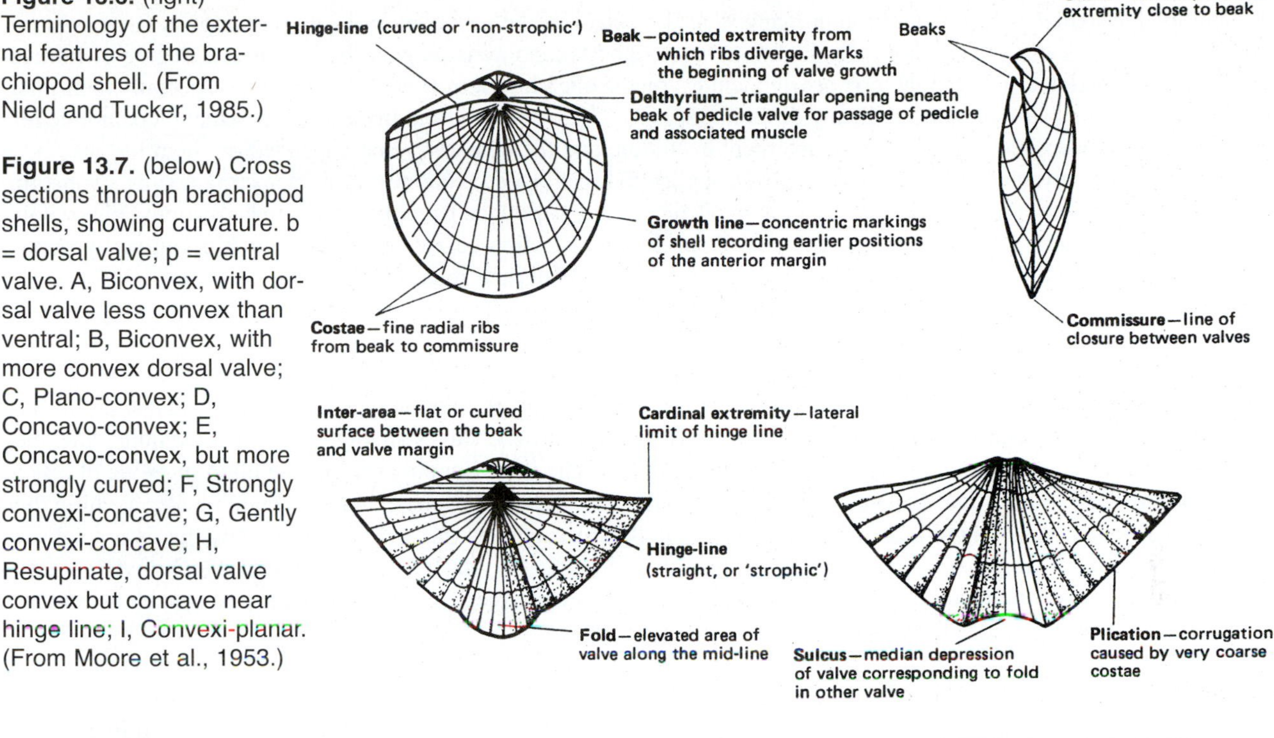

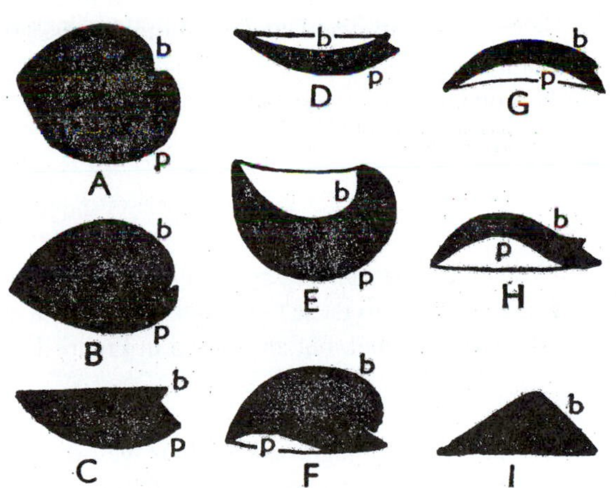

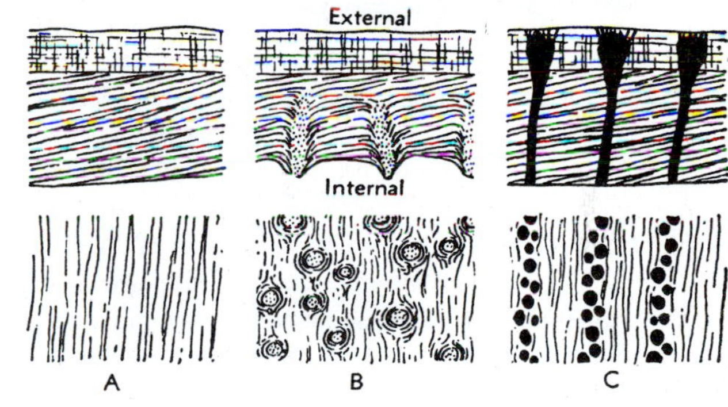

Figure 13.8. Types of shell structure in the brachopods. (A) Impunctate. (B) Pseudopunctate. (C) Punctate. (From Moore et al. 1953.)

terior margin of the other valve, which is known as an **interarea**. Since the beak often curves over, the convex posterior portion extremity of the shell is known as the **umbo**. Some hinges come to wing-like lateral points called the **cardinal extremities**. The edge of the shell along its line of closure is called the **commissure**. It may be straight, or have corrugated edges (**plication**), and it may also have a deep trough on one valve (**sulcus**) that is matched by a large elevated area (**fold**) on the opposite valve. The surface of the shell may show concentric **growth lines**, indicating the enlargement of the shell from its embryonic origin around the hinge area, as well as fine radial ribs that run from the beak to the commissure known as **costae**.

The shell can be convex outward on both sides (**biconvex**), convex on one valve (usually the ventral valve) but flat on the other (**plano-convex**), or concave on the dorsal valve and convex on the ventral valve (**concavo-convex**) (Fig. 13.7). In a few cases, the dorsal valve is convex and the ventral valve concave (**convexi-concave**). Other combinations are also possible (Fig. 13.7).

In thin section, the shell is usually made of inclined fibers of calcite in its inner layer that are overlain at a low angle by the lamellar outer layer (Fig. 13.8). If there are small tubes or pores that penetrate the shell, the shell is known as **punctate**. If there are perpendicular rods of calcite in the fibrous inner layer, it is known as **pseudopunctate**. If the shell is solid with neither of these conditions, it is **impunctate**. Although the pseudopunctate condition is useful since it is diagnostic of strophomenides, punctate and impunctate shells occur widely within the other orders of brachiopods and seem to have limited taxonomic utility. For over a century, the function of the punctae was disputed, but recent studies of living brachiopods showed that the punctae store various proteins and lipids, which helped deter external boring organisms.

Systematics

Brachiopod systematics has gone through many radical changes in the past century as paleontologists have used different key characters in their taxonomy. In addition, brachiopod taxonomy is plagued by an unusual degree of **homeomorphy**—two unrelated species that look nearly identical on the outside, but in every other feature (especially their internal anatomy and shell structure) are radically different (Fig. 13.15). For this reason, there have been many taxonomic features which have been rejected as due to convergent evolution, and the taxonomy that depended on them has also been rejected. More recently, cladistic methods have been applied to brachiopod taxonomy, with interesting results—some traditional taxa have been supported as natural, monophyletic groups, but others have turned out to be paraphyletic or polyphyletic (Carlson, 1991, 1993; Williams et al., 1996). In the classification below, we will follow a traditional grouping into classes and orders, but indicate some of the current thinking about their relationships.

Phylum Brachiopoda

Class Inarticulata

As discussed previously, these brachiopods lack a tooth-and-socket articulation between the two valves, but instead hold them together with several additional sets of muscles. Because the valves are not enclosed at the hinge line, they have room for an anus, so their digestive tract passes food in one direction only (unlike the blind intestine of articulates). The shell of most inarticulates is made of a mixture of calcium phosphate and the organic material chitin, one of the few examples of a chitinophosphatic skeleton in the invertebrates (indeed, one of the few that is not calcareous). Inarticulates were the common brachiopods of the Cambrian, but declined by the end of the Ordovician as the "Paleozoic fauna" dominated by articulate brachiopods took over, except in the marginal habitats, such as brackish mudflats, where lingulides thrive today.

Order Lingulida (Cambrian-Recent; 85 genera)

The lingulides are best known from the common living genus, *Lingula* (Figs. 13.3A, 13.9), which lives buried in mudflats using its long pedicle to dig a burrow. Lingulides are frequently considered "living fossils," since their tongue-shaped shells (*Lingula* means "little tongue" in Latin) have remained virtually unchanged since the Cambrian. Their extraordinary longevity and conservatism is probably due to the fact that they are very successful at living deep in the mud of brackish lagoons, and have tracked that common habitat through 600 million years with little pressure to change in response to predators or competitors.

Order Acrotretida (Cambrian-Recent; 120 genera)

The acrotretides are the other common group of inarticulates. They are characterized by circular or ovoid shells, rather than the tongue-shaped or rectangular shells of lingulides. In addition to these two orders, there are a number of other less common inarticulate groups that will not be discussed further, but are shown in Figure 13.9.

Class Articulata

Articulates make up about 95% of the known brachiopod genera, and they are diagnosed by their well-developed hinges with teeth and sockets between the valves, simpler musculature, and the presence of diductor muscles. They have a digestive tract with no anus, but a blind intestine instead. All articulates make their shells out of calcite, rather than chitinophosphate.

Order Orthida (Cambrian-Permian; 340 genera)

Orthides are considered to be the most primitive articulate brachiopods, first appearing in the middle Lower Cambrian, before most of the other orders differentiated (Fig. 13.9A). Consequently, paleontologists have sought the ancestry of the other orders among the orthides. Not surprisingly, phylogenetic analysis (Fig. 13.9B, C) suggests that the orthides are paraphyletic, with some placed as primitive sister-taxa to the rest of the articulates, and others as more advanced sister-taxa to various orders (Carlson, 1991, 1993; Williams et al., 1996).

Although there is a lot of variation in shell shape, most orthides are easy to recognize (Figs. 13.8, 13.10). They have a long, straight (strophic) hinge, with a wide open triangular delthyrium and notothyrium, surrounded by a distinct but narrow interarea on both valves. The majority of the taxa also have very fine costae radiating from the cardinal area, a gently biconvex shell that is circular or elliptical in outline, and fold and sulcus that are shallow or absent. Internally, they have no brachidium, but a short brachiophore instead, and relatively simple cardinalia. Primitively, they were impunctate, but several groups independently developed punctate shells.

Orthides were the dominant articulates in the Cambrian and especially the Ordovician, when genera such as *Hebertella, Resserella*, and *Dinorthis* were among the more common brachiopods (Fig. 13.10; p. x-xii). They were decimated in the Late Ordovician extinctions, but straggled on through the rest of the Paleozoic, finally succumbing in the Permian catastrophe.

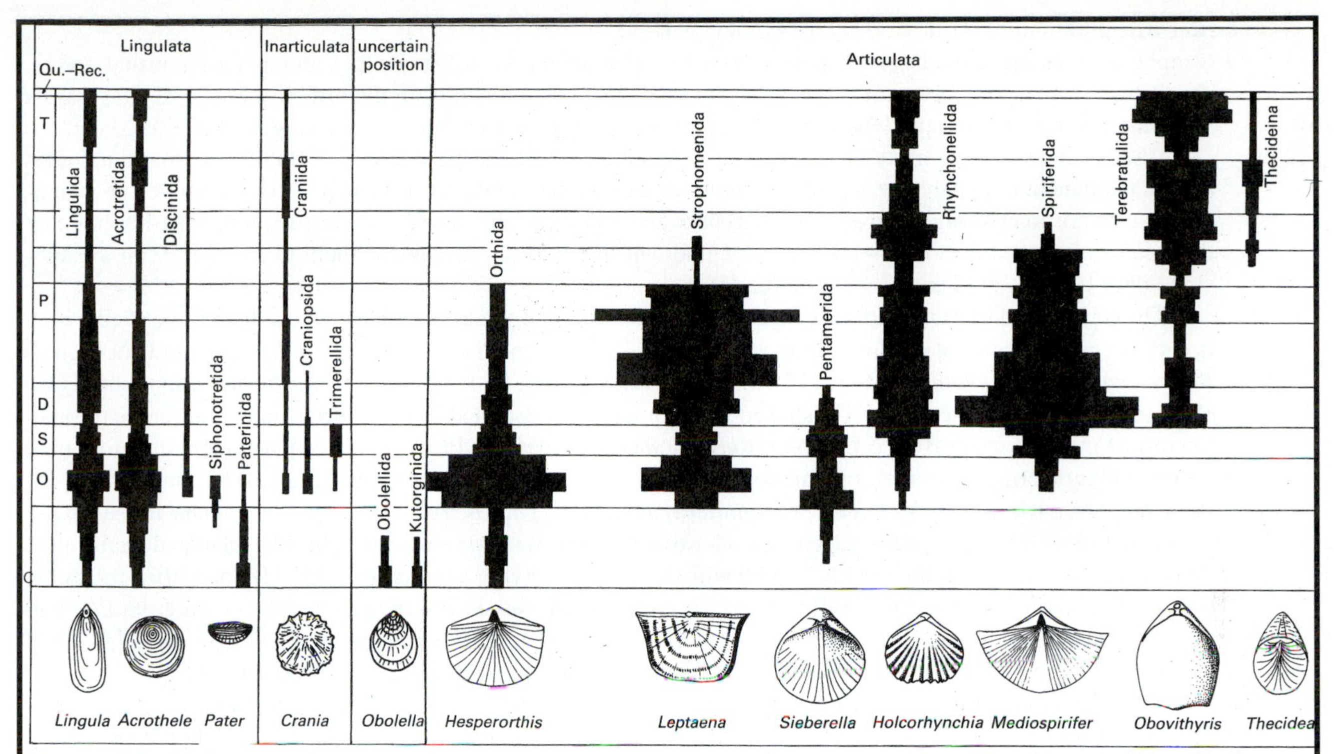

Figure 13.9. (A) Diversity, time range, and representative examples of the major orders of brachiopods (After Clarkson, 1993.)

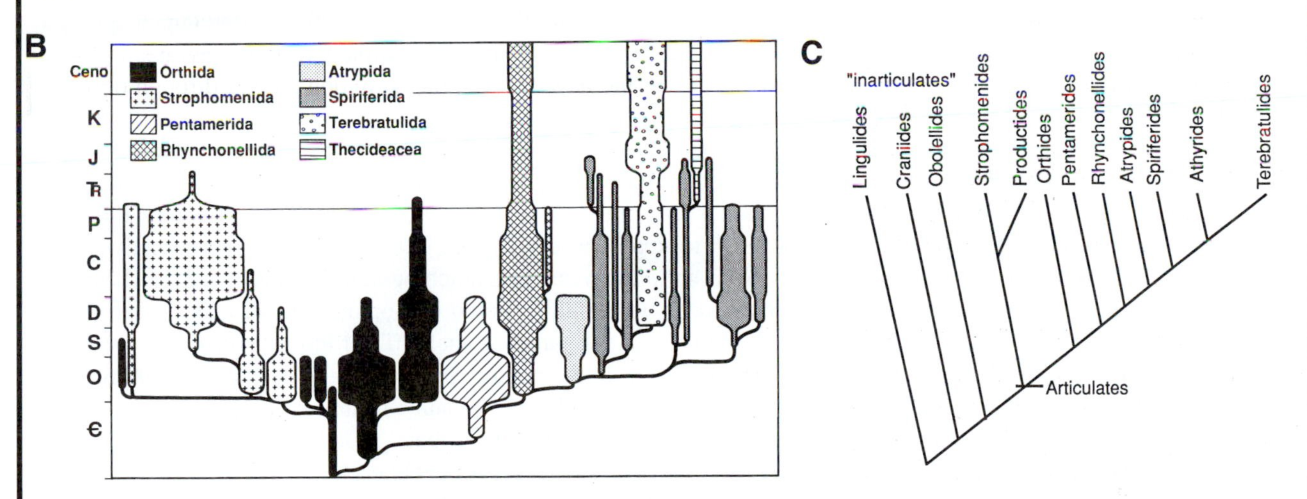

Figure 13.9. (B) Traditional phylogeny of the orders of articulate brachiopods (After Carlson, 1993.) (C) Recent cladistic analysis of the relationships of the major groups of brachiopods (Modified from Williams et al., 1996.)

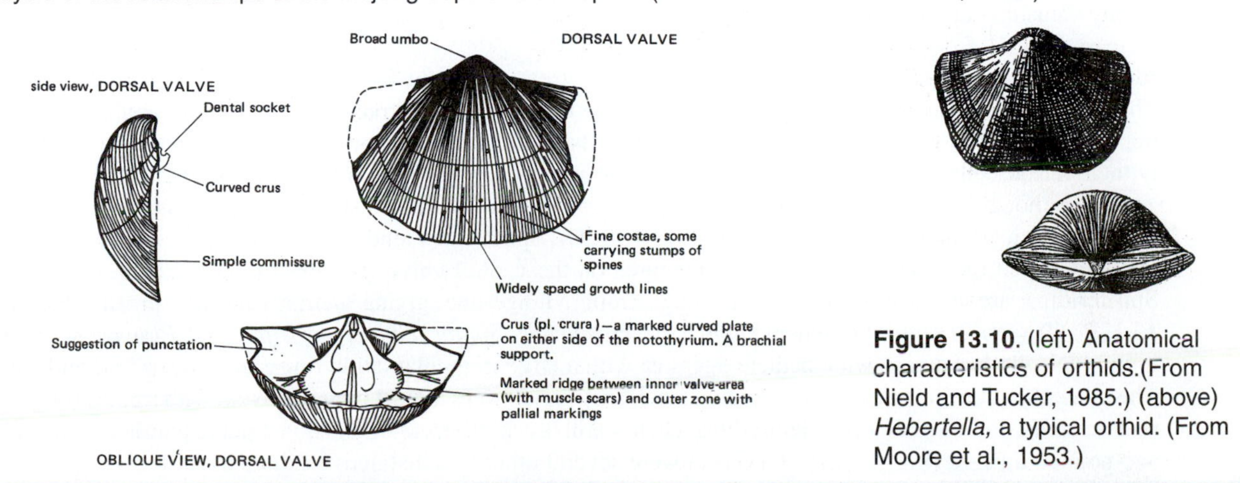

Figure 13.10. (left) Anatomical characteristics of orthids.(From Nield and Tucker, 1985.) (above) *Hebertella*, a typical orthid. (From Moore et al., 1953.)

Order Strophomenida (Ordovician-Triassic; 865 genera)

Strophomenides are the largest and most variable of brachiopod orders, with a number of unusual forms. Traditionally, they have been considered a major Paleozoic side-branch from the orthides, but cladistic analysis suggests that they are paraphyletic relatives of several clades of higher brachiopods (Fig. 13.9B, C).

With such a diverse variety of shapes, it is difficult to list diagnostic features. Virtually all strophomenides are pseudopunctate, apparently a unique feature of this group. They are typically concavo-convex to plano-convex, with a long straight (strophic) hinge line, and a completely closed pedicle foramen. This implies that their pedicle was reduced or absent, and not important in adults, so it required them to live freely on the substrate once they matured enough to lose their pedicle.

The earliest and most primitive group is the suborder Strophomenidina, which have a distinctive shell which is deeply concavo-convex and has a long strophic hinge, giving them a "D"-shaped outline seen in taxa such as *Rafinesquina* or *Strophomena* (Figs. 13.1, 13.8, 13.11A). Along with orthides, these are the commonest brachiopods of the Ordovician. The "D"-shaped strophomenidines declined in the Silurian and were scarce through the rest of the Paleozoic. By the Carboniferous, another suborder, the Productidina, soon became the dominant group, outnumbering all other brachiopods through the rest of the Paleozoic (Fig.13.11C, numbers 1-8). Productidines such as *Linoproductus, Marginifera, Juresiana, Dictyoclostus*, or *Waagenoconcha* had a highly distinctive, deeply convex cup-shaped ventral valve covered with spines, and a lid-like planar dorsal valve. Apparently, they rested on the muddy bottom with their spines serving as stilts (Fig. 13.11B), and lifted their dorsal valve to admit water so they could filter feed. The extreme version of the productidines were the Permian richthofenids, shaped like a cone with a tiny dorsal valve lid (Fig. 13.11C, numbers 20-24; see also Fig. 7.10). The most bizarre of all brachiopods were the olhaminidines (such as *Lyttonia* or *Leptodus*), with an irregularly leaf-shaped ventral valve, and the dorsal valve reduced to a comb-like support for their lophophore, so that they were incapable of closing their shell (Fig. 13.11C, number 25). These aberrant richthofenids and oldhaminidines were particularly abundant in the Permian limestones of west Texas. Although there are many diverse groups within the strophomenides, in general their history is summarized by the Ordovician radiation of "D"-shaped strophomenidines, followed by a Siluro-Devonian decline, and then a second radiation of the cup-shaped productidines, which dominated the Carboniferous and Permian, and final extinction during the Permian event.

Order Pentamerida (Cambrian-Devonian; 160 genera)

The pentamerides were a small, distinctive group that was particularly common in the Silurian, when they formed dense accumulations of thousands of individuals. Long considered a primitive offshoot of the orthides, they actually appeared in the Lower Cambrian at the same time as the earliest orthides. Traditionally, they were thought to be ancestral to the other remaining brachiopod orders (Fig. 13.9B), but recent cladistic analysis (Fig. 13.9C) places them as a monophyletic sister-group to the remaining brachiopod orders (Williams et al., 1996).

Most pentamerides had deeply biconvex shells with highly curved (astrophic) hinges, and a small uncovered delthyrium-notothyrium (Fig. 13.8, 13.12A). All had impunctate shells. Their most distinctive feature (Fig. 13.5, lower left) is the large, scoop-shaped spondylium in the ventral valve, and a large septum and cruralium complex, dividing the inside of the shell into five chambers (hence the name *penta*merides).

Although pentamerides were never as diverse as other suborders, their great numerical abundance in the Silurian makes them a useful index of that period. Many of the great pentameride "reefs" are heavily dolomitized, so the internal steinkerns are preserved but the shells are dissolved away (Fig. 1.8A). A pentameride steinkern has a distinctive cleft where the medial septum once lay. Pentamerides declined through the Devonian, finally vanishing at the end of that period.

Order Spiriferida (Ordovician-Jurassic; 720 genera)

This highly diverse and distinctive group got its name from the spiral arrangement of its lophophore (Fig. 13.5, upper right). Traditionally they were considered an offshoot of the rhynchonellides (Fig. 13.9B), but cladistically their various suborders turn out to be paraphyletic sister-taxa of several orders (Fig. 13.9C).

Even though the spiral brachidia are not visible on the outside of the shell, there are a number of external features that help diagnose spiriferides (Fig. 13.12B, C). Spiriferides tend to have highly biconvex shells, with well-developed radial costae, and a large interarea on the ventral valve. The most typical group, the suborder Spiriferidina, are characterized by a very long, strophic hinge line, giving them a wing-like profile (Fig. 13.8, 13.12B, C) seen in common genera such as *Mucrospirifer*, *Spinocyrtia, Neospirifer*, and *Platyrachella*. This group typically has a very wide pedicle interarea with a large triangular delthyrium, and a deep fold and sulcus. The other common suborder, the Atrypidina (such as *Atrypa*), were almost plano-convex, with a short hinge line, no interareas, and a small, open delthyrium. Their spiral brachidia were oriented in a plane parallel to the valves and pointed upward (Fig. 13.13A, B). There were several other less distinctive suborders as well.

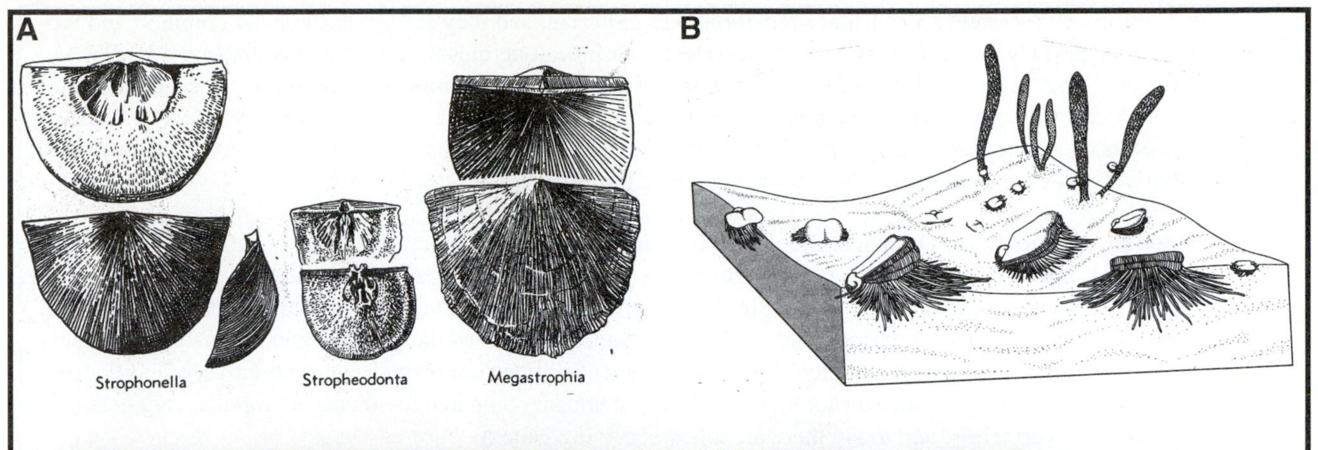

Figure 13.11. (A) Typical strophomenides showing the distinctive concavo-convex shape. (From Moore et al., 1953.) (B) Reconstruction of the life habits of the productid *Waagenoconcha* from the Permian of Pakistan. (From Grant, 1966.) (C) An assemblage of brachiopods from the Permian of west Texas, including spiny productids (numbers 1-8), cone-shaped richthofenids (numbers 20-24) and leaf-shaped *Leptodus* (25). (Photo courtesy Smithsonian Institution.)

Spirifers underwent an enormous radiation in the Silurian, and they are by far the most common and typical brachiopods of the Devonian. In some Devonian limestone quarries, *Mucrospirifer* (Fig. 13.12B) and *Atrypa* (Fig. 13.13A) litter the ground. Spiriferides were less common than productines in the later Paleozoic, but genera such as *Spirifer, Composita,* and *Neospirifer* are still important Carboniferous-Permian index taxa. Spiriferides were nearly extinguished by the Permian catastrophe, but a few taxa survived into the Triassic, only to disappear before the middle of the Jurassic.

Order Rhynchonellida (Ordovician-Recent; 520 genera)

The four suborders discussed so far were dominant groups of the Paleozoic fauna, and all suffered severely in the Permian extinction. By contrast, these last two orders, the rhynchonellides and terebratulides, were present in the background during much of the Paleozoic, but survived the Permian and Mesozoic as the only two living orders of articulate brachiopods. The rhynchonellides are a small but distinctive group that has remained little changed since the Ordovician. Never abundant compared to other brachiopods, they managed to persist nevertheless, and today there are a few surviving genera.

Their stereotyped shells are easy to spot (Fig. 13.9A, 13.13C). They have a short, bent hinge line and a pointed pedicle beak (*rhynchos* is Greek for "beak"), giving their posterior profile a distinctive pointed "V" shape. Most have a pedicle, and the delthyrium is partially closed. The majority of the genera have coarse plicated costae or ribs, giving them a highly crenulated, zig-zag commissure, and a deep fold and sulcus.

Order Terebratulida (Devonian-Recent; 540 genera)

The best known of the living articulate brachiopods are the terebratulides, which have a distinctive shape often compared to a biblical oil lamp (Figs. 13.9A, 13.2A). This is why the brachiopods have sometimes been called the "lamp shells." They were a minor part of the Paleozoic brachiopod fauna, but survived both the Permian and Cretaceous extinctions to make up the majority of living articulate brachiopods (Fig. 13.3B). Long thought to be an offshoot of the spiriferides (Fig. 13.9B), they are considered monophyletic in a cladistic analysis (Fig. 13.9C).

The shape of terebratulides is easy to recognize. A "lamp shell" is strongly biconvex, with a large pedicle foramen (and therefore a large pedicle) and beak that overhangs the short, curved hinge with no interarea. Since there is a large pedicle foramen, the delthyrium is closed by delthyrial plates. Most taxa have a relatively smooth shell with little ornamentation and no fold or sulcus. Internally, they have a loop-shaped brachidium, and all have punctate shells.

Ecology

Brachiopods are exclusively benthic marine invertebrates, living primarily on the shallow shelf and in epicontinental platform seas. All are sessile filter feeders and are incapable of moving in search of food. A few are known from deep-water graptolitic shales, and it is possible that some of the tiniest articulates may have attached to a floating substrate, but most lived in shallow, benthic habitats. A few species, such as the lingulides, can tolerate brackish salinities and burrow, but most articulate brachiopods clearly tolerate only normal marine salinities, and live on the bottom (epifaunal). Most brachiopods are incapable of burrowing because they do not have a large, fleshy mantle that can be extended into a tube-like siphon; this is the feature that allows some molluscs to burrow. Consequently, brachiopods must always stay near the surface where the currents are strong enough to bring them fresh food and oxygen, and eliminate wastes.

Although several living brachiopod species are known to be hermaphrodites (both male and female), most brachiopods are single-sex. When the proper breeding season comes, male and female brachiopods shed clouds of unfertilized eggs and sperm into the sea; if they are lucky, these gametes meet the opposite gametes. In some species, the fertilized egg is brooded in the mantle cavity, but in most cases, the zygote must swim in the plankton before quickly growing (in only a few hours) into a larva. The shell-less articulate larva swims around for only a few hours before trying to find a hard substrate on which to settle. The shortage of suitable hard substrates for attachment is probably the most crucial factor for the survival and growth of a brachiopods, although species that live on muddy substrates have found their own mechanism to keep above the bottom; they have lost the their pedicle completely. If a pediculate brachiopod does find a hard surface on which to attach, then the mantles, which used to cover the pedicle in the planktonic larvae, turn themselves inside out and begin to secrete the shell. The shell begins to grow quickly by accretion after that.

One of the most fascinating problems in brachiopod paleoecology has been deciphering their mode of life. We have already seen how the peculiar cone-shaped richthofenids have been debated over and over again (p. 105). But there are quite a range of additional morphologies that have been the subject of considerable interpretation and reinterpretation.

Figure 13.12. (A) A typical *Pentamerus* has a smooth biconvex shell; most of the diagnostic features are internal. (From Moore et al., 1953.) (B) Some typical spirifers, with long hinge line, wing-like shape, deep fold and sulcus, and prominent plication: (B) *Mucrospirifer*, and (C) *Syringothyris*. (Photos by the author.)

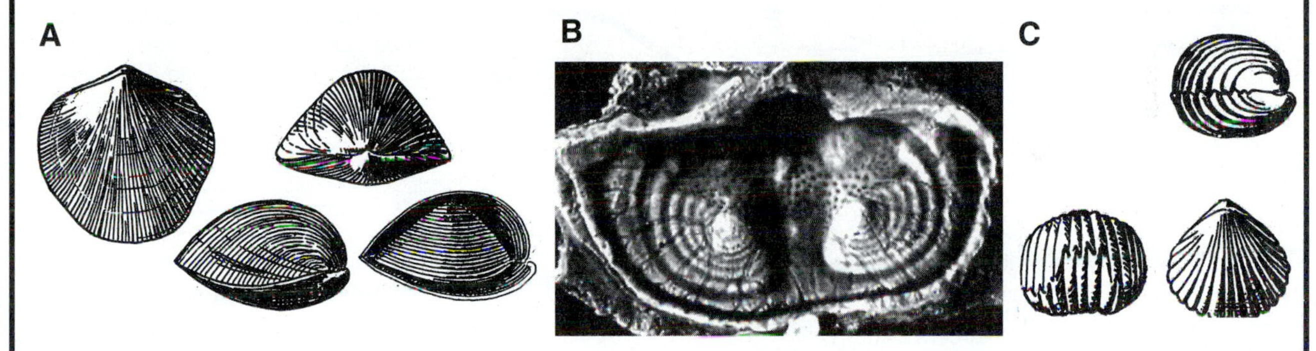

Figure 13.13. (A) Morphology of a typical *Atrypa*. (B) Inside this *Davidisonia* (an atrypid) shell are spiralia which sit in a plane parallel to the commissure. (C) Typical external morphology of a rhychonellide, *Camarotoechia*. (Photo B courtesy P. Copper; A and C from Moore et al., 1953.)

The few living species are the starting point for such studies. The long pedicles of the lingulides (Fig. 13.3A) are not used to burrow directly, but instead serve as a prop or strut to anchor them. The lingulide then rotates and rocks its shell, and twists and gapes its two valves on their muscular hinge, forcing aside the sediment with the shell going in gape first. Eventually, it digs a U-shaped burrow that collapses behind it, leaving the shell opening into the seawater at the sediment-water interface, and the pedicle pointed down below it.

Brachiopods with large pedicles (or pedicle openings in extinct groups), such as the orthides, rhynchonellides, spiriferides, and terebratulides, however, require a hard substrate. They tilt themselves up into the current on their pedicle (Fig. 13.14A), and orient their shells so that they get the best possible flow of currents through their lophophore. The fold and sulcus were apparently important in separating the incurrent water from the lateral edges of the shell from the outgoing wastewater passing out the center along the fold and sulcus. The zig-zag commissures of some brachiopods, such as the rhynchonellides, were useful because they increased the length of opening and therefore the flow rate of the valve while keeping the gape very narrow, preventing sand grains from being inhaled.

Brachiopods with no pedicle opening clearly did not attach by a fleshy pedicle, so they must have had some other method of maintaining their position on the substrate. In a number of strophomenides, there are many tiny holes near the umbo and beak that suggest their pedicle was composed of many tiny fibers for attaching to a surface (comparable to the byssal threads in bivalves). Others were clearly cemented to their substrate. However, many non-pediculate brachiopods were clearly free-living, and must have had mechanisms to maintain their orientation. The extremely concavo-convex strophomenidines (Figs. 13.11A, 13.14F), such as *Rafinesquina*, apparently sat on the substrate convex-side up, with their hinge partially buried to anchor them, and gently lifted their ventral valve to bring in currents. These animals are remarkable because they had extremely little internal volume when the shells were closed. Some flat strophomenides, such as the chonetidines, had many spines protruding from their hinge line, which are thought to serve as sensory spines for their direction of movement. It is thought that they might have clapped their valves together and jet-propelled themselves along the bottom (as scallops do). The long "snowshoe" or "stilt-like" spines of the productids (Fig. 13.11B, C) kept them upright on the muddy bottoms of the late Paleozoic,

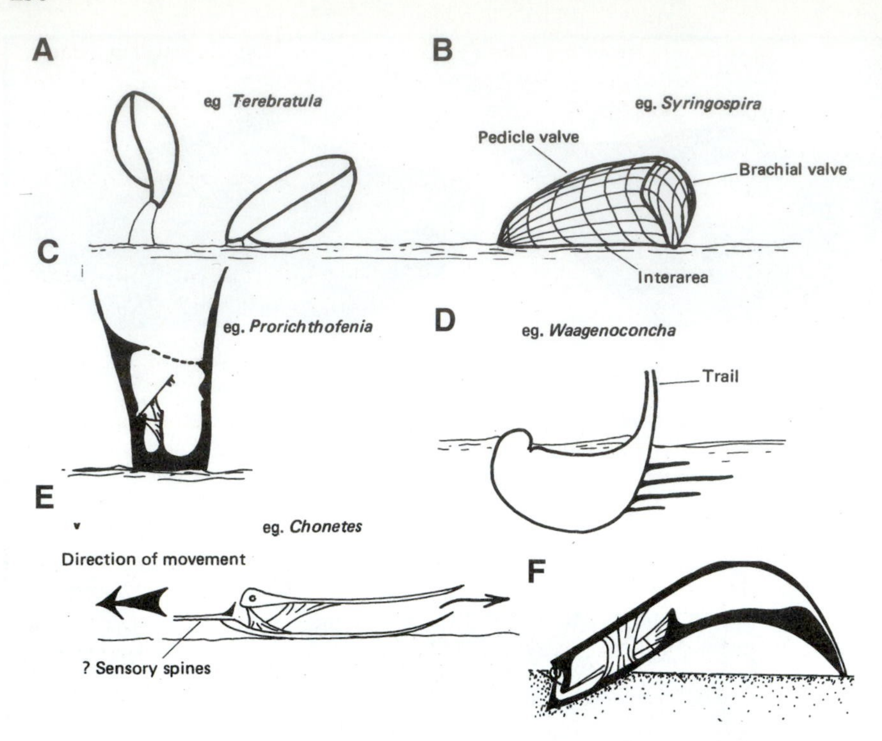

Figure 13.14. Brachiopods showed a wide variety of modes of life. (A) Most were simply attached by their pedicle to a hard substrate. (B) Others had such broad interareas in their hinge line that they could lie flat on the sea bottom. (C) Some were permanently cemented by their broad base, as in the richthofenids. (D) The cup-shaped productids tended to sit in the mud, with their spines preventing them from sinking or flipping over. (E) The spiny concavo-convex strophomenides known as chonetids may have propelled themselves along the bottom with jets of water from their mantle. (F) The concavo-convex strophomenidines rested with their concave dorsal valve upward, using their curvature to keep the shell out of the mud. (From Nield and Tucker, 1985.)

so they needed only to raise their lid-like dorsal valve to filter feed. Some even had long spines around the commissure that helped keep sediment out of the shell, or to direct currents entering the shell, or to protect against predators. In this context, the peculiar richthofenids (Fig. 7.10) can be seen as just an extreme modification of the standard productid mode of life. For further information and details, consult Rudwick (1970), an excellent book that summarizes many different aspects of brachiopod functional morphology.

In addition to functional morphology, the large assemblages of brachiopods lend themselves well to studies of community paleoecology. Many studies have delineated which brachiopods were found together, and what their apparent depth and substrate preferences were. For example, in a classic study of the Lower Silurian of Wales by Ziegler et al. (1968), there was a very clear gradient in paleoecological communities. The dry land lay to the east and south, and the deepest water was located in what is now western Wales. Nearest the shore lay a *Lingula* community dominated by burrowing lingulides and small rhynchonellides, which may have preferred brackish mudflats. Just offshore was a shallow-water community dominated by the rhynchonellide *Eocelia*. Further offshore was a community dominated by accumulations of large *Pentamerus*. Further out on the shelf was a community characterized by fine-ribbed *Costistricklandia*. Finally, the shelf edge was characterized by a community named after the small brachiopod *Clorinda*. Ziegler et al. (1968) argued that these communities were controlled primarily by depth, although Boucot (1975) suggested that the differences in temperature (with offshore assemblages considerably colder than onshore) were more important, and others arguing that substrate is a critical control as well.

Evolution

In the systematics section, we reviewed the general evolutionary trends of brachiopods. Although there are many local exceptions, knowing the dominant group of brachiopods in each period is a useful piece of paleontological information. Even without a specialist's knowledge of the phylum, any geologist can quickly recognize the age of a fossiliferous Paleozoic rock by the dominant groups. For example, a lingulid-orthid assemblage is probably Cambrian, while the first thing one sees in shallow marine Ordovician rocks is an abundance of "D"-shaped strophomenides (Fig. 13.1) and flat orthides like *Resserella* or *Hebertella*. Silurian rocks are less easy to typify, although abundant pentamerides are commonest in the Silurian. By the Devonian, spirifers are extremely abundant, and in the Mississippian, there are mixtures of spirifers and productids. The Pennsylvanian and Permian are characterized by abundant productids, with the peculiar richthofenids and oldhaminids found primarily in the Permian.

With so many diverse groups radiating and diminishing through the Paleozoic, it is hard to point to many general evolutionary trends in the brachiopods. Many features (such as the development of the punctate shell, the increase in hinge and lophophore complexity, and the development of spiralia in both the spiriferides and the atrypides) evolved independently in more than one group (Williams and Hurst, 1977). Even more striking is the high frequency of homeomorphic external shell morphology in unrelated brachiopods (Fig. 13.15). As in the example from the planktonic foraminifera (Fig. 11. 9), certain external shell morphologies are apparently either very successful or highly constrained, because they keep recurring. Carlson (1992) examined the changes in hinges in brachiopods

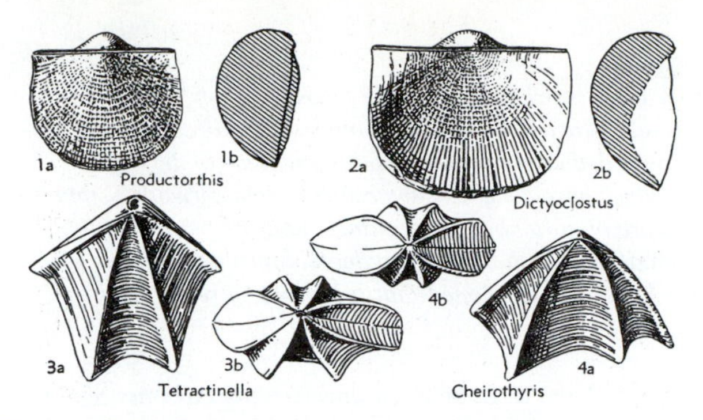

Figure 13.15. The external features of brachiopod shells show a great deal of convergence, or homeomorphy. Without knowledge of their internal anatomy, shell microstructure, and other details, these homeomorphs would be considered closely related. (1) is an orthide that resembles a productid (2); (3) is a spirifer that resembles a terebratulide (4). (From Moore et al., 1953.)

through time, and found that there is a general increase in brachiopods in several different orders that had interlocking hinge-and-socket mechanisms at the expense of those which did not.

When the punctuated equilibrium debate first appeared, a number of examples of phyletic gradualism were suggested. Johnson (1975) argued that the Devonian *Tecnocyrtina missouriensis* evolved gradually by developing plications on its fold and sulcus. However, as Gould and Eldredge (1977, p. 122) point out, there are only a few sample levels, which are insufficient to demonstrate gradualism. Makurath and Anderson (1973) attempted to demonstrate gradualism in three successive samples of Devonian *Gypidula*, but the existence of only three samples (lumped from many localities) is insufficient to prove gradualism (Gould and Eldredge, 1977, p. 123). Ziegler (1966) tried to demonstrate a progressive suppression of ribs in Ordovician *Eocelia* from Wales, but Gould and Eldredge (1977, p. 126) pointed out that the sample size and density are insufficient to prove gradualism (although it is suggestive). Hurst (1975) studied the ratio of delthyrial width to the length of the delthyrial chamber in 11 samples of *Resserella* from the Ordovician of Wales. Although the original study argued for gradualism, Gould and Eldredge (1977, p. 126) point out that the samples are again too small and not densely sampled enough to prove gradualism.

A recent study by Lieberman et al. (1995) of two Devonian brachiopod lineages from New York State showed long periods of stasis with some fluctuation around the mean, but no clear gradual trends. No outstanding examples of phyletic gradualism in brachiopods has yet been documented. On the contrary, the recent arguments about "coordinated stasis" indicates that brachiopod genera, and even their communities, are stable for millions of years. As Johnson (1975, p. 657) put it, "In subsequent years many workers have attempted to seek out and define lineages of brachiopod species and other megafossils in the lower and middle Paleozoic with little success. My conclusion, subjective in many ways, is that speciation of brachiopods in the mid-Paleozoic via the phyletic mode has been rare. Rather, it is probable that most new brachiopod species of this age originated by allopatric speciation."

For Further Reading

Carlson, S. J. 1990. The articulate brachiopod hinge mechanism: morphological and functional variation. *Paleobiology* 15:364-386.

Carlson, S. J. 1991. A phylogenetic perspective on articulate brachiopod diversity and the Permo-Triassic extinctions, 119-142, *in* Dudley, E.C., ed., *The Unity of Evolutionary Biology* 1. Dioscorides Press, Portland, Oregon.

Carlson, S. J. 1992. Evolutionary trends in the articulate brachiopod hinge mechanism. *Paleobiology* 18:344-366.

Carlson, S. J. 1993. Phylogeny and evolution of 'pentameride' brachiopods. *Palaeontology* 36:807-837.

Carlson, S. J. 1995. Phylogenetic relationships among extant brachiopods. *Cladistics* 11:131-197.

Copper, P., and J. Jin, eds. 1996. *Brachiopods:Proceedings of the Third International Brachiopod Congress, Sudbury, Ontario, Canada.* A. A. Balkema, Rotterdam.

Dutro, J. T., and R. S. Boardman. 1981. *Lophophorates: Notes for a Short Course.* University of Tennessee Dept. of Geological Sciences Studies in Geology 5.

MacKinnon, D. I., D. E. Lee, and J. D. Campbell, eds. 1991. *Brachiopods through Time: Proceedings of the Second International Brachiopod Congress, Dunedin, New Zealand.* A. A. Balkema, Rotterdam.

Rowell, A. J., and R. E. Grant. 1987. Phylum Brachiopoda, pp. 445-496, *in* Boardman, R.S., A.H. Cheetham, and A. J. Rowell, eds. *Fossil Invertebrates.* Blackwell Scientific Publishers, Cambridge, Mass.

Rudwick, M. J. S. 1970. *Living and Fossil Brachiopods.* Hutchinson and Co., Ltd., London.

Williams, A. 1968. Evolution of the shell structure in the articulate brachiopods. *Special Papers in Palaeontology* 2:1-55.

Williams, A., and J. M. Hurst. 1977. Brachiopod evolution, pp. 79-121, *in* Hallam, A., ed. *Patterns of Evolution as Illustrated in the Fossil Record.* Elsevier, Amsterdam.

Williams, A., and A. J. Rowell. 1965. Brachiopoda, *in* Moore, R. C., ed., *Treatise on Invertebrate Paleontology.* Part H. Geological Society of America and University of Kansas Press, Boulder, Colorado, and Lawrence, Kansas.

Williams, A., S. J. Carlson, C. Howard, C. Brunton, L. E. Holmer, and L. Popov. 1996. A supra-ordinal classification of the Brachiopoda. *Philosophical Transactions of the Ryoals Society of London* (B), 351:1171-1193.

Ziegler, A., L. R. M. Cocks, and R. K. Bambach. 1968. The composition and structure of Lower Silurian marine communities. *Lethaia* 1:1-27.

BRYOZOANS

Living bryozoans build tuftlike colonies which bear close resemblance to moss, whence the name "moss animals" sometimes used for the phylum. Collections of "seaweeds" made by amateurs almost invariably contain specimens of bryozoans; in fact, some of the latter make better seaweeds, so far as appearance is concerned, than the plants themselves. Study of the Bryozoa necessitates much microscopic investigation, and in the case of forms which have complex calcareous skeletal structures, thin sections must be prepared. In these studies minute structural details are of paramount importance, and the reports setting forth the results of the investigations are couched in such technical language that the average paleontologist either sends his bryozoan material to a specialist or leaves it untouched. A popular belief has developed that a lifetime of research is the essential qualification for identification of bryozoans.

W. H. Twenhofel and R. R. Schrock, *Invertebrate Paleontology*, 1935

After brachiopods and phoronids, the third major group of lophophorates are the bryozoans ("moss animals"), also known as the Phylum Ectoprocta. At first glance, most people don't recognize bryozoans as fossil lophophorates, since they are so tiny and so different from the large shelled brachiopods. However, the bryozoan animal (Fig. 13.16) is constructed in a very similar fashion, with a U-shaped digestive tract and a lophophore around the mouth. These tiny animals live in huge colonies in many marine habitats. Strands of seaweed or shells may be encrusted with thousands of individuals like mossy overgrowths, giving the phylum its name. Some colonies contains tens of millions of individuals. Bryozoans are also incredibly diverse, with over 3500 living species, and at least 15,000 fossil species.

Bryozoans are exclusively colonial and (with the exception of one freshwater family) marine organisms. Because they are colonial, they have been mistaken for corals, but there are important distinctions. The bryozoan colony usually has much tinier (typically pinhole-sized) holes for the individuals, unlike the much larger holes in corals, and they also lack the radial septa characteristic of coral skeletons. Their tiny size makes them very hard to study without a microscope and thin sections, so they tend to be neglected in undergraduate paleontology classes. They are so diverse and abundant, however, that it is worth the effort to learn to identify them, because they are useful for many different kinds of paleontological studies.

The individual bryozoan animal, or **zooid**, is typically only about 1 mm in length, but colonies can grow as large as 60 cm (2 feet) across. The body of the animal has a U-shaped digestive tract, with the mouth within a ring of tentacles of the lophophore, and the anus just outside the ring (Fig. 13.16). The name "Ectoprocta" means "outside anus" in Greek, and refers to this feature. The Entoprocta, which have an anus within their tentacle ring, were once classed with Bryozoans, but are now considered a separate phylum. The tentacles have cilia, which produce currents directed at the mouth. There are several sets of muscles for retracting and protruding the lophophore. When the lophophore is retracted into the body, it is enclosed in a tentacle sheath.

The fluid-filled body cavity (coelom) contains a small nerve ganglion in the space between the mouth and anus, but no heart or vascular system, or excretory or respiratory tract. Bryozoans are so small that they can respire and excrete their wastes by diffusion. Also within the coelom are the reproductive organs. Most individuals are bisexual, with both eggs and sperm, the former produced in the upper part of the body, and the latter lower in the zooid. The eggs and sperm are released into the ocean currents to form a zygote and a larva. The larva metamorphoses into an **ancestrula**, which (if it is lucky) settles and forms a new colony. New zooids then bud off asexually from the ancestrula, so the rest of the zooids are produced by cloning from a single individual and are genetically identical.

These soft parts are not preserved in fossil bryozoans, so the microscopic details of the calcareous colonial skeleton (**zoarium**) are studied instead. Even though all the individuals in the zoarium are genetic clones of each other, they express their genes in different ways, producing different types of zooids (**polymorphism**) that have different functions, with a corresponding variability in skeletal structure. We will examine the differences in skeletal structure in each of the classes and orders within the phylum.

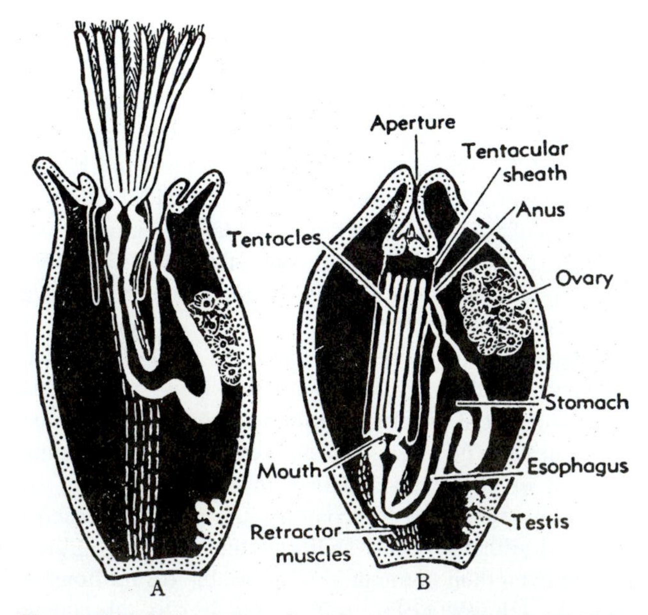

Figure 13.16. Anatomy of a living bryozoan, Alcyonidium albidium, in longitudinal section with the lophophore extended (A) and retracted (B). (From Moore et al., 1953.)

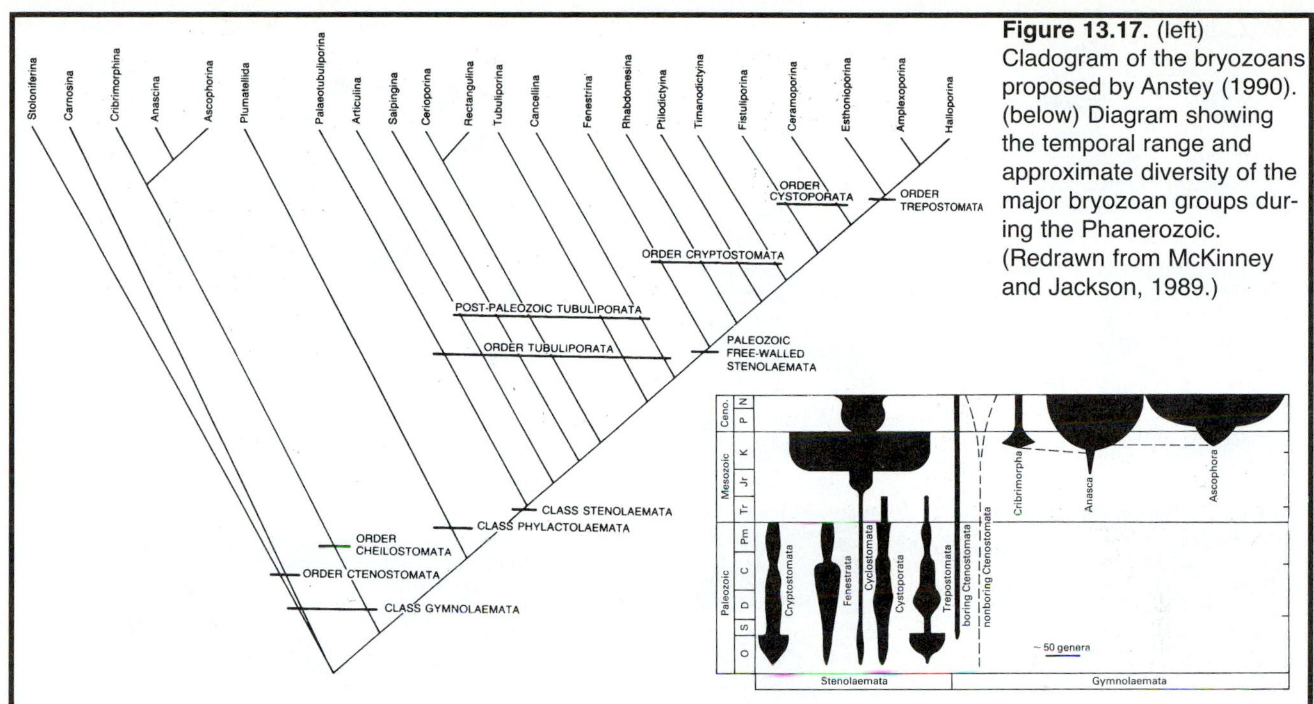

Figure 13.17. (left) Cladogram of the bryozoans proposed by Anstey (1990). (below) Diagram showing the temporal range and approximate diversity of the major bryozoan groups during the Phanerozoic. (Redrawn from McKinney and Jackson, 1989.)

Systematics

McKinney and Jackson (1989, Fig. 2.11) summarized five different schemes of relationships of bryozoan orders that had been proposed between 1953 and 1981. The geometry of relationships of the included taxa was radically different in all five, because a different combination and weighting of characters was used in each. A more thorough cladistic analysis of the Bryozoa was suggested by Anstey (1990). In this hypothesis, many of the traditional orders were paraphyletic, although the trepostomes, cheilostomes, and ctenostomes all appeared to be monophyletic (Fig. 13.17). As Anstey (1990) points out, molecular phylogenies will probably not be very helpful in resolving the problem, since the majority of bryozoan taxa have long been extinct.

Phylum Bryozoa (= Ectoprocta, Polyzoa)

Class Phylactolaemata (Recent)

This class is only known from uncalcified freshwater forms, so it has no fossil record.

Class Stenolaemata (Ordovician-Recent; about 750 genera)

Stenolaemates ("narrow gullet" in Greek) are the dominant Paleozoic class, although one order, the cyclostomates, survived and flourished in the Mesozoic and again in the Cenozoic. They have long, tubular, narrow, highly calcified zooecia, which continue to grow through the life of the colony, usually at an angle to the direction of colony growth. The tentacles are extruded by muscles that squeeze the coelomic fluid into the upper part of the zooid, forcing out the tentacles.

Order Trepostomata (Ordovician-Triassic; about 200 genera)

Trepostomes ("changing mouth" in Greek, in reference to the changes in tubes from the immature to the adult regions) were the most common early Paleozoic bryozoans, especially during the Ordovician, when they formed numerous massive and branching colonies, typified by such common genera as *Prasopora, Hallopora*, or *Dekayella*, which make up a large part of some Ordovician limestones (Fig. 13.18). They suffered severely during the Late Ordovician extinction, then straggled on through the Silurian and recovered slightly during the Devonian, declining gradually through the late Paleozoic and Permian catastrophe. Only two or three genera survived into the Triassic, when the group finally became extinct.

Trepostomes have three different kinds of tubes, or **zooecia**, within the zoarium (Fig. 13.19). The largest are known as **autopores**, and housed the individual zooids. They are recognized by their large diameter, rounded cross section, and short horizontal dividing walls known as **diaphragms**. Adjacent to these are **mesopores**, which have a smaller diameter, more angular cross section, many diaphragms, and were presumably occupied by some sort of specialized zooid unlike those living in autopores. There are also scattered, tiny, thick-walled tubes located at the junction between two or more autopores or mesopores. These are called **acanthopores**, because instead of a zooid they support a small spine (*akanthos* is Greek for "spine").

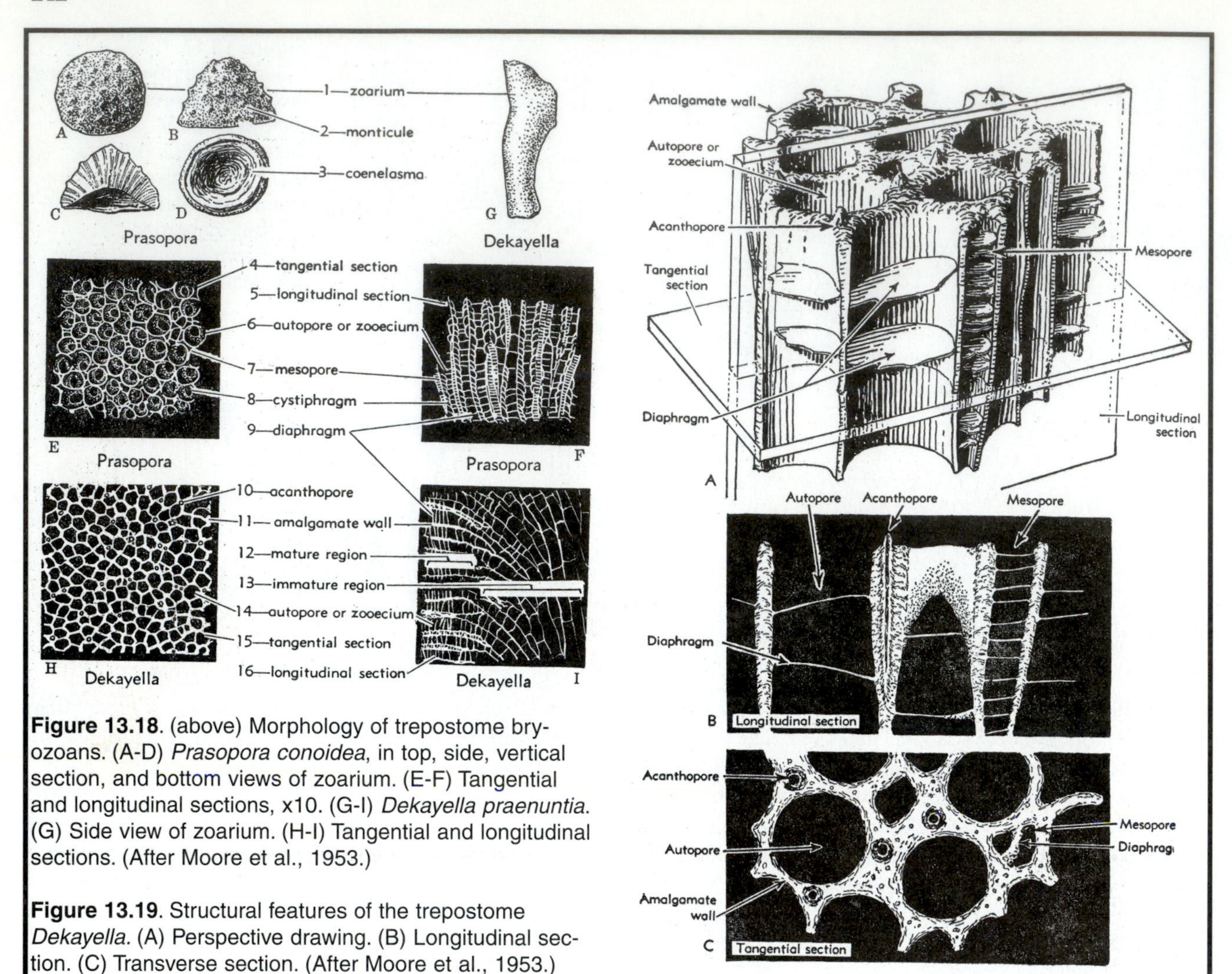

Figure 13.18. (above) Morphology of trepostome bryozoans. (A-D) *Prasopora conoidea*, in top, side, vertical section, and bottom views of zoarium. (E-F) Tangential and longitudinal sections, x10. (G-I) *Dekayella praenuntia*. (G) Side view of zoarium. (H-I) Tangential and longitudinal sections. (After Moore et al., 1953.)

Figure 13.19. Structural features of the trepostome *Dekayella.* (A) Perspective drawing. (B) Longitudinal section. (C) Transverse section. (After Moore et al., 1953.)

Order Cryptostomata (Ordovician-Permian; about 90 genera)

Cryptostomes ("hidden mouth" in Greek, in reference to how the true aperture is hidden beneath a vestibule) are the other dominant group of early Paleozoic bryozoans, radiating in huge numbers before the Late Ordovician extinctions, and then persisting through the rest of the Paleozoic before final extinction at the end of the Permian. They are distinguished by their short autopores, with a wider opening (**vestibule**) separated from the narrower inner zooecium by a **hemiseptum** (Fig. 13.20). Some have abundant acanthopores of two sizes (**megacathopores** and **micracathopores**); others do not have abundant mesopores or acanthopores, but instead have large masses of common calcified skeletal tissue (a **coenosteum**, something we have already seen in some corals).

Order Fenestrata (Early Ordovician-Permian; about 100 genera)

Fenestrates (*fenestra* is "window" in Latin) were particularly abundant in the later Paleozoic (especially the Mississippian), when their delicate lacy colonies can be found nearly everywhere. Some, like *Archimedes*, are considered Mississippian index fossils. Like the cryptostomes, fenestrates were victims of the Permian extinctions. Their zoaria are arranged in a lacy network, with short autozooids on one side of the zoarium, commonly with a hemiseptum in the base (Fig. 13.21). They are best known from the lacy sheet-like bryozoans such as *Fenestella*, or the genus *Archimedes*, whose sheet-like lacy zoarium is arranged in a spiral. This spiral arrangement reminded James Hall (who named the genus) of the screw-shaped water pump devised by the Greek mathematician and engineer Archimedes (who lived in Hellenistic Syracuse from 287 to 212 B.C.). Archimedes is famous for many advances in mathematics and science, including solving the problem of how to measure the density of an object by measuring how much water it displaces. He supposedly discovered this by watching his body displace water in a bathtub. Allegedly, he jumped out and ran naked through town shouting "Heureka!" ("I have found it!")

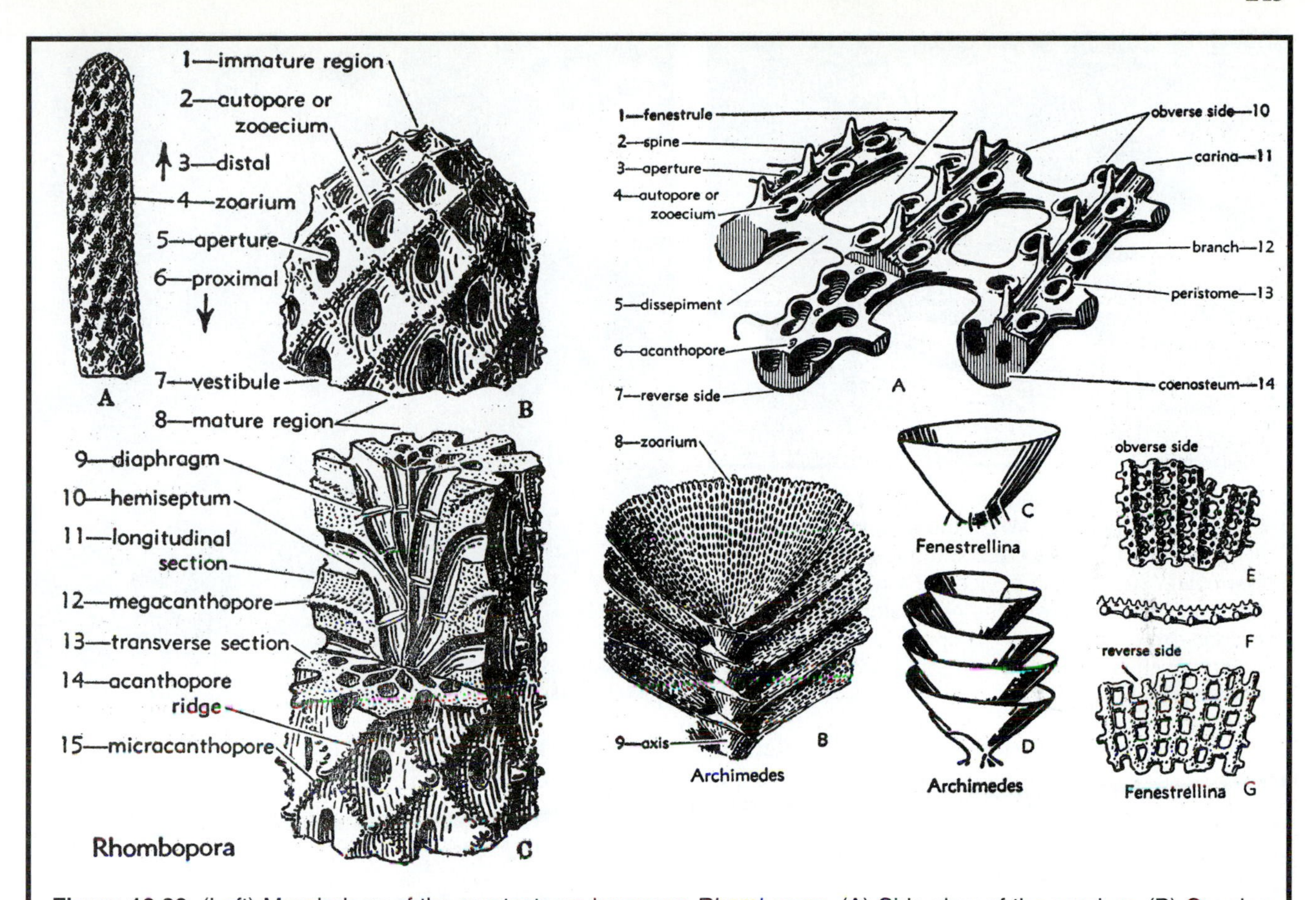

Figure 13.20. (Left) Morphology of the cryptostome bryozoan *Rhombopora.* (A) Side view of the zoarium. (B) Growing tip of the zoarium. (C) Transversely and longitudinally sectioned part of a colony, showing the internal features. (From Moore et al., 1953.)

Figure 13.21. (Right) Structural features of fenestrate bryozoans. (A) Enlarged oblique view of part of the lacy network that makes up the zoarial fronds of *Fenestrellina* and *Archimedes.* (B) Part of the zoarium of *Archimedes wortheni*, with a fan of lacy branches fanning out from the central corkscrew support. (C) Diagram of the funnel-like zoarium of *Fenestrellina.* (D) Diagram of the corkscrew fronds of *Archimedes.* (E-G) *Fenestrellina pectinis*, diagram of the obverse, edge, and reverse sides of part of a colony. (From Moore et al., 1953.)

Order Cyclostomata (= Tubuliporata) (Early Ordovician to Recent; about 250 genera)

The Cyclostomata ("round mouth" in Greek) or Tubuliporata (now preferred), were a very minor group in the Paleozoic, but were the only Stenolaemata to thrive after the Triassic. They diversified to about 175 genera in the Cretaceous, then nearly went extinct at the K/T event. They again diversified in the Cenozoic, and are one of the three common living orders, with about 50 living genera. They are constructed as erect or encrusting colonies of short, simple rounded zooecia lacking transverse partitions with a plain rounded aperture. The zooecia are usually pierced by **mural pores** that link them to adjacent autopores.

Order Cystoporata (Early Ordovician-Triassic; about 100 genera)

Cystoporates ("enclosed pores" in Greek) were a minor group in the Paleozoic, and only a few genera survived into the Triassic, when they finally went extinct. They are very similar to cyclostomes except that they have regions of curved partitions known as **cystiphragms** separating the zooecia, and may also have crescentic projections (**lunaria**) around the apertures.

Class Gymnolaemata (Ordovician-Recent; about 650 genera)

Most of the living bryozoans belong to the class Gymnolaemata ("naked gullet" in Greek). In contrast to stenolaemates, the gymnolaemates have box-like or cylindrical enclosures around the zooids. Rather than adding to the top of a tubular zooecium by accretion, each zooecium remains a fixed size, and they grow by adding additional discrete zooecia. To open their zooecium and extend their lophophore, they use their muscles to deform part of the body wall.

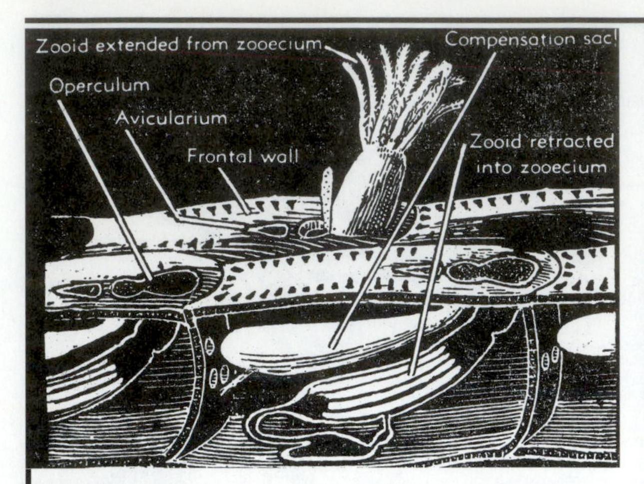

Figure 13.22. (Top) Diagram of the chambers of a cheliostome bryozoan. When the compensation sac is expanded, it pushes the zooid out through the operculum into feeding position. (From Moore et al., 1953.) (Right) Surface of the cheilostome *Hippopetraliella,* showing the opercula and several brood pouches for juveniles. (From Carroll and Stearn, 1989.)

Order Ctenostomata (Ordovician-Recent; 50 genera, of which about 40 have no fossil record)
Ctenostomes ("comb mouth" in Greek, since they have a comb-like lid over the opening) are uncalcified, with membranous or gelatinous walls, so they rarely fossilize, although they are frequently encrusting and boring, leaving traces nonetheless. Although they are known from the Ordovician, they are rare throughout their history.

Order Cheilostomata (Jurassic-Recent; about 1000 genera)
The cheilostomes ("lip mouth" in Greek, since they have a chitinous operculum on the zooecium) are by far the most diverse group of living bryozoans, undergoing an enormous radiation in the Cretaceous to three large suborders (Fig. 13.22). Their calcareous zooids are box-like with a small orifice closed by a hinged lid, or **operculum**. They have a great variety of additional unique morphological features, and they even have specialized brood cells in which their embryos are protected during early development.

Ecology

Bryozoans are found in modern oceans at all latititudes and at depths of at least 5500 m (18,000 feet), with the deepest recorded at 8500 m, although they are most abundant in shallow, tropical seas. Most require some hard substrate to attach or encrust, although several groups of bryozoans in the Paleozoic developed large, calcareous colonies that were massive or branching in structure. Bryozoans prefer clear water, since they are easily fouled by sediment at their tiny size. They can tolerate some agitation from waves and currents, but do not thrive where the bottom is composed of shifting, unstable sediment. Consequently, they are most abundant on rocky bottoms, sheltered crevices, and inside the shells of other invertebrates.

Bryozoans are exclusively filter feeders, eating tiny plankton, especially diatoms and radiolarians. Although they are not large enough to form massive reefs structures like corals, they are often important in adding to the structure of reefs, and may form a large portion of some Paleozoic limestones. At times during the Paleozoic (such as the Ordovician or Mississippian), they were the second or third most common fossil group, after brachiopods and crinoids.

Evolution

Schopf (1977), McKinney and Jackson (1989), and Anstey (1990) reviewed the wide variety of evolutionary phenomena that have been examined in bryozoans. Such evolutionary studies are complicated by the fact that these organisms are colonial, so that evolution and natural selection works on the level of the colony, rather than the individual zooid. Much effort has gone into defining and understanding the implications of coloniality for evolution. Consequently, the frequency of polymorphic zooids within a colony benefits the colonies as a whole. In stable tropical environments, about 75% of the cheilostome species are polymorphic, presumably because they have the luxury of sufficient food to allow non-feeding zooids.

There are several trends observed over the long course of bryozoan evolution, including increased calcification of zooids, increased integration of zooids (in cheilostomes), and the shift to rigidly erect species in deeper water. In the cheilostomes, the frontal wall starts out uncalcified, and then becomes increasingly calcified in all the major lineages within the cheilostomes. Anstey (1990, p. 241) demonstated that the zooids in trepostomes became progressively deeper during the Paleozoic, presumably in response to predation. A deeper zooid allows the bryozoan to retreat further into the zoarium, especially if the predators tend to nip off the surface of the colony. Since the most primitive bryozoan colonies start out simple, there is an inevitable increase in bryozoan colonies through time. McKinney (1993) suggested that cheilostome zooids are, on the average, more robust, with a greater volume of internal biomass and with a broader opening and a higher, more efficient rate than are the zooids of cyclostomes, and argued that the diversification and dominance of post-Paleozoic cheilostomes is largely an effect of their more efficient feeding. Although some authors have argued that the cheilostomes outcompeted the cyclostomes, Lidgard et al. (1993) showed that the data and the argument are very complex, and it is difficult to conclusively prove that one drove out the other.

Bryozoan morphologies are remarkably stable over long periods of geologic time. When change and variability has been documented, it is non-progressive, and apparently does not lead to new species. Cheetham (1986, 1987) and Jackson and Cheetham (1995) documented the patterns of evolution in the major lineages of cheilostomes from the Neogene of the Atlantic Coast, and found mostly stasis and rapid evolution, with no cases of gradualism. Indeed, gradualism has never been a major theme of bryozoan evolution, even before the topic became controversial.

For Further Reading

Anstey, R. S. 1990. Bryozoans, pp. 232-252, *in* McNamara, K. J., ed. *Evolutionary Trends*. University of Arizona Press, Tucson.

Boardman, R.S., et al., 1983. Bryozoa, Vol. 1, Part G, in Robinson, R. A., ed. *Treatise on Invertebrate Paleontology*. Geological Society of America and University of Kansas Press, Lawrence.

Boardman, R. S., and A. H. Cheetham. 1987. Phylum Bryozoa, pp. 497-549, *in* Boardman, R.S., A.H. Cheetham, and A. J. Rowell, eds. *Fossil Invertebrates*. Blackwell Scientific Publishers, Cambridge, Mass.

Cheetham, A. H. 1986. Tempo and mode of evolution in a Neogene bryozoan: rates of morphologic change with and across species boundaries. *Paleobiology* 12:190-202.

Cheetham, A. H. 1987. Tempo and mode of evolution in a Neogene bryozoan: are trends in a single morphologic character misleading? *Paleobiology* 13:286-296.

Dutro, J. T., and R. S. Boardman. 1981. *Lophophorates: Notes for a Short Course*. University of Tennessee Department of Geological Sciences Studies in Geology 5.

Jackson, J. B. C., and A. H. Cheetham. 1995. Phylogeny reconstruction and the tempo of speciation in cheilostome Bryozoa. *Paleobiology* 20:407-423.

Larwood, G. P., ed. 1973. *Living and Fossil Bryozoa*. Academic Press, London.

Larwood, G. P., and C. Nielsen, eds. 1981. *Recent and Fossil Bryozoa*. Olsen & Olsen, Fredensborg, Denmark.

McKinney, F. K., and J. B. C. Jackson. 1989. *Bryozoan Evolution*. Unwin Hyman, Boston, Mass.

Schopf, T. J. M. 1977. Patterns and themes of evolution among the Bryozoa, pp.159-207, *in* Hallam, A., ed. *Patterns of Evolution as Illustrated in the Fossil Record*. Elsevier, Amsterdam.

Figure 14.1. Beautifully preserved assemblage of complete specimens of the Devonian trilobite *Phacops rana*. (Photo courtesy R. Levi-Setti.)

Chapter 14

Jointed Limbs

The Arthropods

There are vastly more kinds of invertebrates than vertebrates. Recent estimates have placed the number of invertebrate species on Earth as high as 10 million and possibly more. . . . Invertebrates also rule the earth by virture of their sheer body mass. For example, in tropical rain forest near Manaus, in the Brazilian Amazon, each hectare (or 2.5 acres) contains a few dozen birds and mammals but well over a billion species of invertebrates, of which the vast majority are mites and springtails. There are about 200 kilograms by dry weight of animal tissue in a hectare, of which 93% consists of invertebrates. The ants and termites alone comprise one-third of this biomass. So when you walk through a tropical forest, or most other terrestrial habitats for that matter, vertebrates may catch your eye most of the time but you are visiting a primarily invertebrate world.

Edward O. Wilson, "The Little Things that Run the World," 1996

INTRODUCTION

Humans regard themselves as the most successful and dominant animals on the planet, but by almost any criterion, the flies, cockroaches and other arthropods rule the earth. Whether you define success by diversity, ecological variety, or just sheer numbers, this is the planet of the arthropods. This was true when they first appeared about 540 million years ago, and so they will remain long after humans are gone.

In terms of species or any other taxonomic level, arthropods are by far the most diverse phylum on the planet. Out of slightly more than a million known species of animals, there are more than 870,000 described species of insects, of which about one-third are beetles (over 340,000 species). When the great biologist J. B. S. Haldane was asked what nature told him about the Divine, he reportedly said that "God has an inordinate fondness for beetles." By contrast, there are only about 42,000 described species of vertebrates, most of which are fish, and only about 4000 mammalian species. Estimates of the number of undescribed species of animals range from 5 to 30 million, most of which are arthropods. By comparison, most vertebrates are large and conspicuous, so few new species are likely to be described in the future. Depending upon which estimates are used, arthropods (mostly insects, and mostly beetles among the insects) make up at least 85% of the species, and possibly as high as 99% of the species on earth.

Anyone who has watched a busy ant colony or beehive, or has seen the cockroaches in an infested apartment, knows that arthropods can be incredibly prolific and abundant. A single pair of cockroaches can produce 164 billion offspring in seven months if there are no restraints on population growth. In the tropics, a few acres may contain a dozen or so individual birds and mammals, but well over a billion arthropods, most of which are mites, beetles, wasps, moths, and flies. A single colony of ants may contain more than a million individuals. In the nutrient-rich parts of the ocean, the tiny planktonic copepods, krill, ostracodes, and other crustaceans number in the millions per cubic meter of seawater.

In terms of ecological variety, arthropods are as versatile and successful as any group that has ever lived. They occupy all habitats, from the the abundant crabs, lobsters, and shrimps (and formerly trilobites—Fig.14.1) of the marine realm, to the great variety of insects and other arthropods in terrestrial and freshwater habitats, to the enormous variety of flying insects. Arthropods live in the deepest oceans, the hottest deserts, the coldest ice, the highest mountains, and just about any extreme environment you can name—including a large number that are internal and external parasites on other organisms.

Why have arthropods been so successful? Part of their success can be traced to their ability to reproduce rapidly, so they can multiply in great numbers when conditions are good. With their small body size, they can subdivide a region into many tiny, highly specialized niches, allowing for a very high diversity in a small area. A key to this versatility and ability to specialize lies in their skeleton. Arthropods have a modular construction, built on a **segmented body** with pairs of jointed, segmented **appendages** on just about every body segment. The phylum gets its name from its distinctive appendages—*arthros* means "joint" (as in "arthritis") and *podos* means "foot" in Greek. Each set of appendages can be specialized for different tasks—legs, mouthparts, antennae, claws, pincers, swimming paddles, copulatory structures, and a great variety of other possible functions. With small variations in this modular construction, such as changing the number or shape of segments or appendages, they can modify their bodies for any possible habitat.

Another key to arthropod success is their hard external skeleton (**exoskeleton**) made up of the polysaccharide **chitin**, which combined with a protein makes up the **cuticle** of the exoskeleton. In some groups, the cuticle may also be mineralized with calcite. Instead of an internal skeleton of bones surrounded by muscles pulling from the outside (as vertebrates have), the arthropod skeleton is built like a hollow tube, with the muscles pulling from the inside. It serves not only as a skeleton and support for the body, but also as an external armor for protection against predators and harsh external conditions. It is not surprising that the first animals to venture out on the land in the Silurian were arthropods (millipedes, followed by scorpions, spiders, and eventually insects), since their exoskeleton protected them against drying up. No marine invertebrate with a soft body had a similar advantage for land life. The only other major groups of marine invertebrates that successfully invaded the land were the snails, which can retreat into their shells for protection against predators and desiccation, and the nematode worms, which are so tiny that they can live inside just about any moist habitat (especially animals, plants, soil, and rotting vegetation).

Another feature of the external skeleton is that it does not grow, but must be molted occasionally so the arthropod can grow larger (see Chapter 2). Molting confers several adaptive advantages, especially versatility. Each time an arthropod molts, it can make radical changes in its body plan before its new skeleton dries and hardens. For example, a caterpillar can change into a butterfly, and other arthropods can make more subtle changes each time they molt their skeletons.

The only ecological niches that arthropods do not dominate are those requiring large body size. In the ocean, where the density of the water can support them, king crabs can reach a legspan of about 5 m, but their bodies are less than 0.5 m in length. Some aquatic eurypterids reached over 2 m in length, and were much heavier than king crabs. On land, the largest arthropods were the Carboniferous dragonflies with 70 cm wingspans. The size limit is a constraint of molting their external skeleton, which leaves the soft body of the arthropod vulnerable while the new skeleton hardens. Although Hollywood loves making big spiders, scorpions, and ants the villains of their plots, the one thing arthropods will never do is grow as large as the house-sized giant ants or praying mantises of cheap science fiction movies. The first time such a large insect molted, it would dissolve into jelly under its own weight.

Given their dominance and abundance in the living world (and presumably through most of the geologic past), arthropods have a relatively poor fossil record. About 30,000 arthropod species have been described from the fossil record, a tiny fraction of their living diversity, and even a smaller fraction of a percent of all species that ever lived. Of these, about 2000 genera and thousands of species are trilobites, and another 2000 genera and many thousands of species are ostracodes, so the dominant living groups (especially insects, spiders, mites, and shrimp) are even less abundant as fossils. Such biased fossilization is largely caused by their skeletons. Most are made only of organic chitinous cuticle, which does not fossilize except in unusual circumstances, such as in amber, or in *Lagerstätten* that preserve nonmineralized tissues, such as the Burgess Shale or the Florissant fossil beds. Only those with mineralized skeletons, such as the calcified trilobites and ostracodes, have a good chance of preservation.

SYSTEMATICS

Given their tremendous diversity, it is not surprising that there are a wide variety of arthropod classifications. Some biologists, such as Manton (1973, 1977) considered the arthropods polyphyletic (derived from several independent origins) and unnatural. This conclusion was based on the old idea that such tremendous morphological disparity could not have originated from a single common origin, but required multiple origins. However, as Fortey (1990, p. 47) pointed out, these same taxonomists have no problem with including animals such as the gooseneck barnacles (which are as different from typical arthropods as could be imagined), not only within the arthropods, but even within the crustaceans. In a symposium on arthropod phylogeny (Gupta, 1979), however, opinion shifted away from the polyphyletic theory of arthropod origins. More recent analyses (Briggs and Fortey, 1989; Brusca and Brusca, 1990, p. 683) have pointed to the large number of unique specializations found only in arthropods, and argued that is simpler and more reasonable to postulate that they are monophyletic.

Relationships within the arthropods have also been controversial (see the review by Brusca and Brusca, 1990). Through all these analyses, there seems to be consensus that there are four major groups of arthropods. The interrelationships of these four groups is uncertain, but the monophyly of each seems to be well supported. Some authors class these as phyla within a superphylum Arthropoda, while others place them as subphyla of the phylum Arthropoda. In this chapter, we will not go into all the details of the relationships within the four major groups of arthropods; for further details, consult the review volumes edited by Gupta (1979) and Mickulic (1990). We will focus on groups that have the best fossil record, such as the trilobites and ostracodes, even though the insects, spiders, and other arthropods may be more numerous and ecologically important today.

Phylum Arthropoda

Subphylum Trilobitomorpha

Next to ostracodes, trilobites (Figs. 14.1, 14.2) are the most commonly fossilized of all arthropods, but they became extinct in the Permian, and their relationships to the rest of the arthropods are controversial. Their segments are fused into a discrete head plate (**cephalon**) and a tail plate (**pygidium**). In between, their thorax retains its segmentation, although the number of segments is variable. They have a pair of antennae, and branched, Y-shaped (**biramous**) limbs along the underside of their body.

Subphylum Chelicerata

Chelicerates include the arachnids (spiders, scorpions, ticks, and mites), horseshoe crabs, and an extinct group known as "sea scorpions" or eurypterids. The chelicerate body is divided into two segments, a **prosoma** combining the head and thorax, and an **opisthosoma** (abdomen). They do not have antennae, and their first appendages are modified into **chelicerae**, a specialized mouthpart. Their second appendages are called **pedipalps**, and in some groups these are modified into pincers. Behind these appendages, they have four pairs of legs, all of which are located on the head. Their appendages are unbranched (**uniramous**).

Subphylum Crustacea

The crustaceans are the dominant marine arthropod group. They include lobsters, crabs, shrimp, crayfish, barnacles, ostracodes, and many other shrimp-like animals, comprising at least 30,000 described living species. They have two pairs of antennae, and three additional pairs of appendages (maxillae and two sets of mandibles) behind the mouth. The limbs are biramous, although some groups secondarily developed uniramous limbs. Their body consists of a distinct head and a thoracic region with various degrees of specializations, and a segmented abdomen.

Subphylum Tracheata (= Uniramia)

This is the largest groups of arthropods, including the insects, centipedes, and millipedes. As the name "Uniramia" implies, their appendages are usually uniramous (although some Paleozoic insects were polyramous). The body is divided into a distinct cephalon, with four pairs of appendages in front of the mouth (antennae, mandibles, and two pairs of maxillae), which can be further modified. Centipedes and millipedes have a trunk region with many identical segments, but insects have divided the trunk region into a thorax and abdomen.

For Further Reading

Bowes, D. R., and C. D. Waterston, eds. 1985. Fossil arthropods as living animals. *Transactions of the Royal Society of Edinburgh* 76:101-399.

Briggs, D. E. G., and R. A. Fortey. 1989. The early radiation and relationships of the major arthropod groups. *Science* 256:241-243.

Brusca, R. C., and G. J. Brusca. 1990. *Invertebrates*. Sinauer Associates, Sunderland, Mass.

Clarke, K. U. 1973. *The Biology of Arthropods*. Edward Arnold, London.

Fortey, R. A., ed. 1997. *Arthropod Relationships*. Chapman & Hall, London.

Gupta, A. P., ed. 1979. *Arthropod Phylogeny*. Van Nostrand Reinhold, New York.

Manton, S. M. 1977. *The Arthropoda*. Oxford University Press, Oxford.

Mickulic, D. G., ed. 1990. *Arthropod Paleobiology: Short Courses in Paleontology* 3. Paleontological Society and University of Tennessee Press.

TRILOBITOMORPHA

Trilobites tell us of ancient marine shores teeming with budding life, when silence was only broken by the wind, the breaking of the waves, or by the thunder of storms and volcanoes. The struggle for survival already had its toll in the seas, but only natural laws and events determined the fate of evolving life forms. No footprints were to be found on those shores, as life had not yet conquered land. Genocide had not been invented as yet, and the threat to life on Earth resided only with the comets and asteroids. All fossils are, in a way, time capsules that can transport our imagination to unseen shores, lost in the sea of eons that preceded us. The time of trilobites is unimaginably far away, and yet, with relatively little effort, we can dig out these messengers of our past and hold them in our hand. And, if we learn the language, we can read the message.

Riccardo Levi-Setti, *Trilobites*, 1993

Probably the most popular of all the invertebrate fossils are the trilobites. Amateur collectors value them above all other fossils, and they have received more study than many common invertebrate groups. Such fascination is not confined to modern fossil collectors. An amulet carved out of a Silurian trilobite was found in a 15,000-year-old Paleolithic rock shelter at Arcy-sur-Cure, France. A Cambrian trilobite preserved in chert was chipped by Australian aborigines to form an implement. Both of these specimens were clearly imported from a long distance, since they do not come from the local area. Ute Indians used to make amulets out of the common Middle Cambrian trilobite *Elrathia kingi*, from the House Range of western Utah. They called it *timpe khanitza pachavee*, or "little water bug in stone house." This same trilobite is so abundant that today it is commercially mined with backhoes and found in virtually every rock shop and commercial fossil seller's catalogue.

Trilobites were by far the most common and diverse group of fossilizable invertebrates during the Cambrian, when they reached their zenith of diversity and abundance. There are over 300 genera and 65 families known in the Late Cambrian. Virtually every fossiliferous Cambrian outcrop yields trilobites, and they were so rapidly changing and abundant that they are the biostratigraphic standard for the Cambrian. Trilobites declined through the Ordovician as other groups of invertebrates came to dominate the

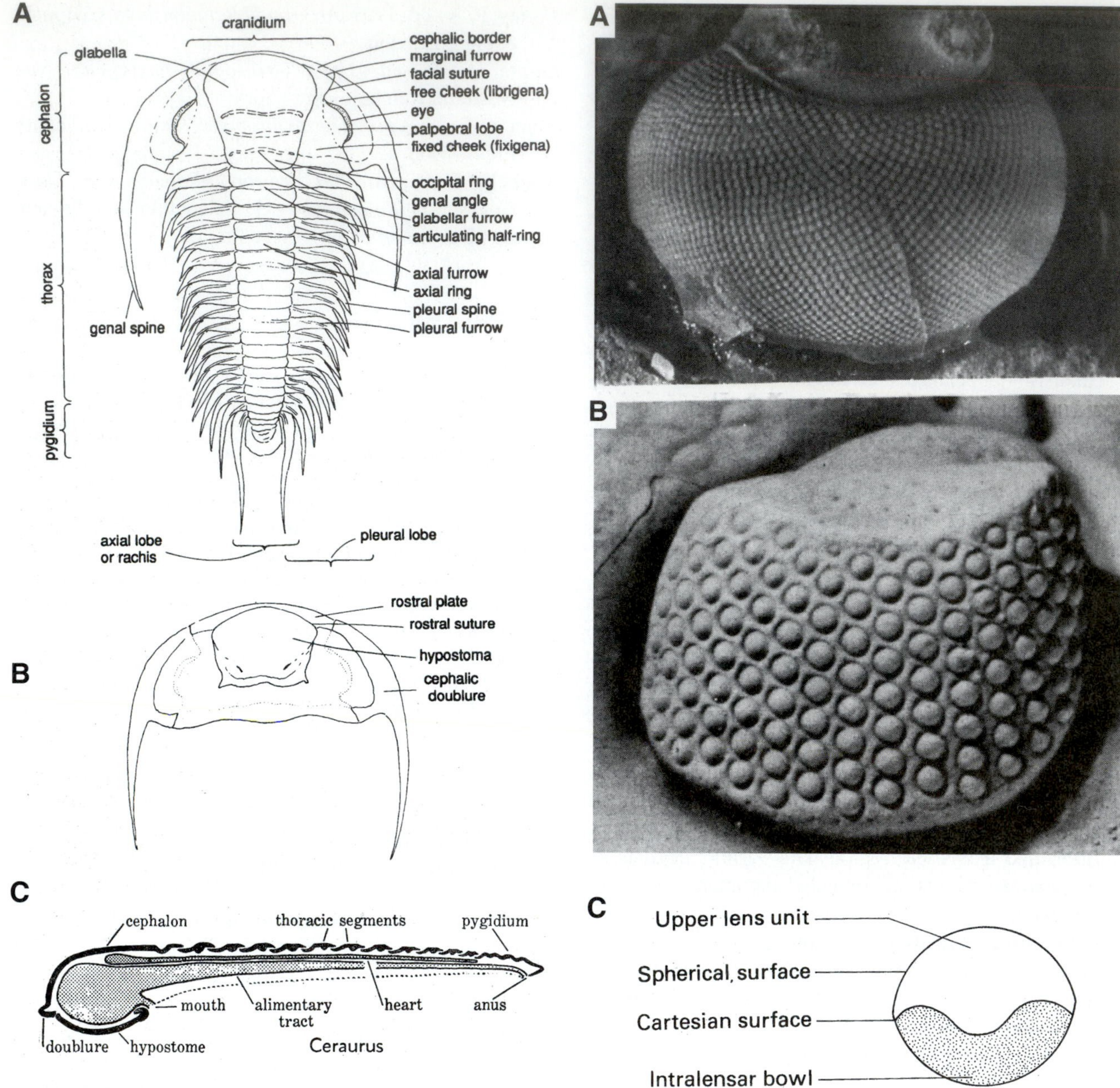

Figure 14.2. Basic anatomy of a trilobite. (A) Dorsal view of a complete exoskeleton. (B) Ventral view of the cephalon. (From Levi-Setti, 1993.) (C) Longitudinal section of a trilobite through the axis. Skeletal parts indicated by solid black. (From Moore et al., 1953.)

Figure 14.3. Anatomy of the eyes of trilobites. (A) Holochroal eye composed of closely packed lenses. (From Levi-Setti, 1993.) (B) Schizochroal eye with lenses separated by solid cuticle. (C) Cross section of the lens of a schizochroal eye, showing the division into two discrete units. (From Boardman et al., 1987.) (D) Huygens' 1690 diagram of the ray path optics of lenses, illustrating how they correct for spherical aberration. The upper lens unit of trilobite eyes solved this problem millions of years before humans did.

seafloor, and suffered severely during the Late Ordovician extinction. During the Silurian and Devonian, trilobites were a minor element of the seafloor fauna, with no more than 17 families and 60 genera. They were again hit hard by the Late Devonian extinction, leaving only a handful of taxa in a single order to straggle on through the late Paleozoic. They finally vanished during the Late Permian.

Trilobites have a calcified exoskeleton (Fig. 14.2), which is responsible for their favorable preservation compared to most other arthropods. The calcite carapace is divided into three longitudinal lobes, one on each side (**pleural lobes**)

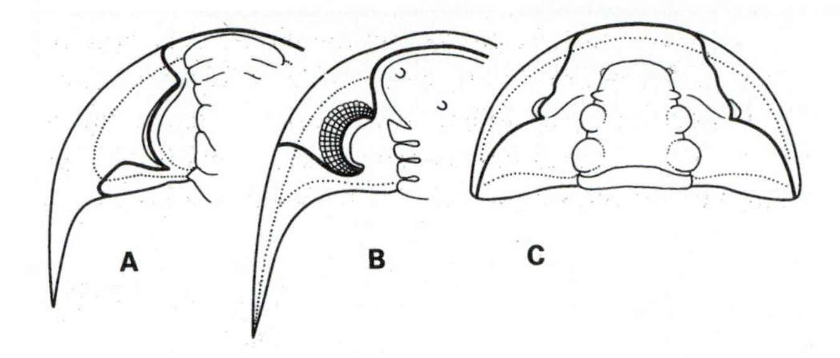

Figure 14.4. Types of facial sutures (bold lines) in trilobites. (A) Opisthoparian. (B) Proparian. (C) Gonatoparian. (From Boardman et al., 1987.)

and one down the center (**axial lobe**). The front segments are fused into a head shield called a **cephalon** ("head" in Greek), and in most taxa, the tail segments are fused into a **pygidium** ("little tail" in Greek). The dorsal (top) surface of the cephalon typically has a raised central hump or ridge called a **glabella**, which is the front end of the axial lobe. The glabella may have several grooves known as **glabellar furrows**, and the last segment of the glabella is known as the **occipital ring**, separated from the rest of the glabella by an **occipital furrow**. The edge of the cephalon may also have a groove around the edge called a **marginal furrow**, and many trilobites have a specialized broad shelf along the front (which may be ornamented in a number of ways) called a **cephalic brim**. The corners of the cephalon are known as the **genal angle**, and they may bear long **genal spines**.

On either side of the glabella are the eyes (Fig. 14.3). The eyes are connected to the glabella by the **palpebral lobe**. Trilobites were one of the first organisms on earth to have eyes, and therefore among the earliest organisms for which sight was an important sense (most marine invertebrates rely on touch and chemical clues). The earliest trilobites had compound eyes, much like those found in other arthropods. The lenses are composed of a single crystal of calcite with the c-axis oriented parallel to the incoming light direction, so it is possible for us to see what they saw. With their hundreds (some have as many as 15,000) of tiny, closely packed calcite lenses (a **holochroal** eye—Fig. 14.3A), they were able to get a composite image of the world around them, although they could not resolve tiny details. Many trilobite eyes were arranged in a broad arc for excellent peripheral vision, presumably to detect predators. Some had their eyes mounted on long stalks for a view above the seafloor, while others had eyes that were shaped like a large protruding half-dome, giving them excellent, almost 360° vision. The most sophisticated eyes are known as **schizochroal**, because each lens is surrounded by an area of interstitial opaque exoskeleton (Fig. 14.3B). Schizochroal eyes have fewer but much larger lenses that are made of two crystals of calcite, with a curved border between them (Fig. 14.3C). The design for this kind of doublet lens was invented by Huygens and Descartes in the seventeenth century as a device for correcting for spherical aberration in thick lenses (Fig. 14.3D). Trilobites developed the same system over 400 million years before humans did. Such sophisticated eyes probably allowed them to use adjacent lenses for stereoscopic vision through almost 360°, and were probably most useful for seeing in dim light. On the other hand, many different groups of trilobites lost their eyes altogether, presumably because they lived in dark or muddy waters where vision no longer mattered.

When a trilobite molted, its cephalon broke apart along a curved seam, known as a **facial suture**, that runs from the front of the cephalon, through the eyes, and then posteriorly. Several different types of facial suture are recognized (Fig. 14.4). The primitive type is known as **opisthoparian** (*opisthos* means "behind" in Greek), because the line of the suture ends behind the genal angle, on the posterior border of the cephalon. The opposite condition is called **proparian** (*pro* means "before" in Greek), because the suture emerges in front of the genal angle. A third type of suture is known as **gonatoparian** (*gonato* is "corner" in Greek), because the suture splits the genal angle. Some trilobites have a suture that runs along the margin of the cephalon. These are now called **marginal sutures**, although older classifications called them "protoparian" or "hypoparian" sutures. The facial sutures used to be very important in trilobite classification, but have since been de-emphasized. Several studies have shown that the facial suture changes position in ontogeny, and that several groups have developed the same suture type independently, so some suture-based taxa are polyphyletic. The suture is important for interpreting specimens, however, because most trilobites are known from molted partial cephalons. The two "cheek" regions outside the suture are known as the "free cheeks" (**librigenae**) and are frequently missing from the remaining core of the cheek region, known as the "fixed cheeks" (**fixigenae**). When interpreting specimens, the paleontologist must visualize complete specimens, even though the most commonly collected and identified part is the cephalon missing its free cheeks (known as the **cranidium**). In the thoracic region, the number and shape of the thoracic segments can be important, as is the presence of spines on the pleural lobes. Trilobite pygidia have many different shapes that can be used to distinguish complete specimens, but the majority of taxa are known only from isolated cephala.

On the bottom (ventral side), the cephalon curves down and under to form a **doublure**, and includes a plate along the front underside of the cephalon known as the **rostral plate**. Behind it, but in front of the mouth, is a large plate known as the **hypostome**. In exceptionally well-preserved specimens, the uncalcified biramous legs can be seen. Each pair of legs was accompanied by a branched gill, and there was also a single pair of antennae protruding from the front. In x-ray images of exceptional specimens (Fig. 1.14C), the details of the digestive tract, internal muscles, and many other features can be seen.

Systematics

In over 150 years of research, there has been no consensus on trilobite classification. There are a number of well-defined ordinal-level taxa that are easy to recognize and remember, but how they are all interrelated is still controversial. Some classification schemes relied on "key characters," such as the facial suture, number of segments, shape of pygidium, and structure of pleural lobes, but each has met with problems. Bergström's (1973) phylogeny relied on enrollment patterns, which are not known for most taxa. The first cladogram (Fig. 14.5) of trilobites was proposed by Eldredge (1977), although a comprehensive cladistic analysis has yet to be completed (see Fortey, 1990). Currently, paleontologists are trying to develop a natural, cladistic classification that is based on a number of different character states (including the morphology of the hypostome, which has proved particularly useful). In the classification given below, we will follow many of the orders recognized in the 1959 *Treatise* volume, while indicating the current thought about the status of each taxon.

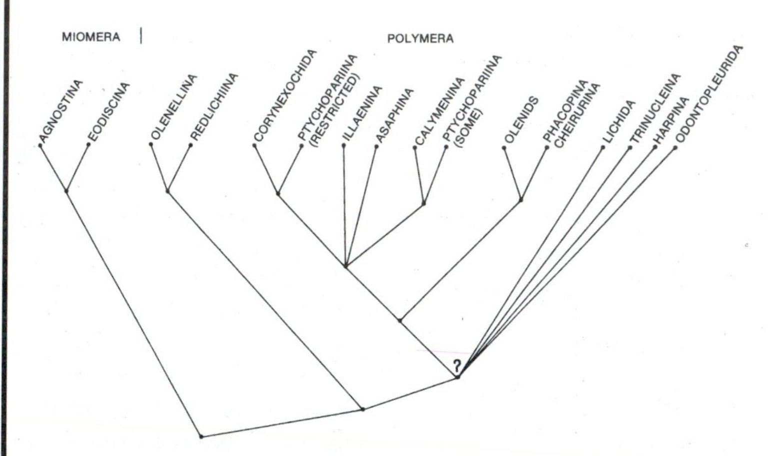

Figure 14.5. (left) Cladogram of the trilobites proposed by Eldredge (1977.)

Figure 14.6 (below) (A) Typical agnostoids, *Peronopsis interstrictus*, from the Middle Cambrian, Utah. (B) Early redlichiid, *Olenellus fremonti*, from the Lower Cambrian, Pennsylvania. (C) The large (almost two feet long) redlichiid, *Paradoxides davidis*, from the Middle Cambrian, Newfoundland. (Photos A and B courtesy L. Babcock; photo C by the author.)

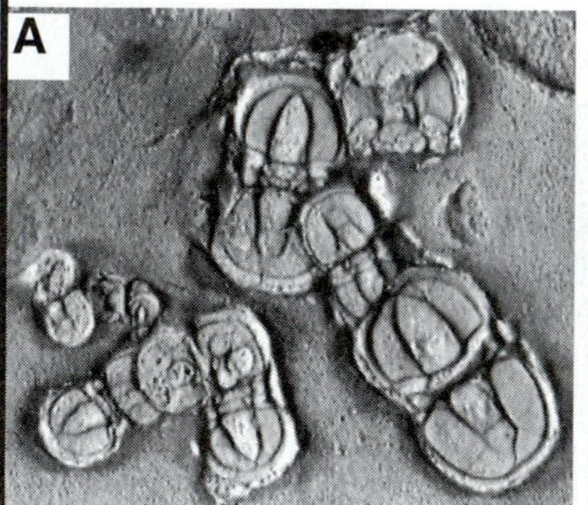

Phylum Trilobitomorpha

Class Trilobita

Order Agnostida

The tiny agnostoids (Fig. 14.6A) were common in the Cambrian but vanished by the end of the Ordovician. They are distinctive in having a button-shaped pygidium equal in size to the cephalon (that is, they are **isopygous**), and having only three (suborder Eodiscina) or two (suborder Agnostina) thoracic segments. Most were eyeless and blind, and their tiny size and widespread distribution suggests that they floated in the open ocean. This is supported by the fact that they are extremely widespread geographically, with the same species occurring in Scandinavia and North America, making them excellent for biostratigraphic correlation around the globe. Some paleontologists once questioned whether they are even trilobites at all, although the consensus now seems to place them within the trilobites (Fortey, 1990). Currently, some scientists consider the Eodiscina to be a separate order that converged on the Agnostina in their superficial features.

Order Redlichiida

The redlichiids include the most primitive trilobites known, such as the earliest Cambrian *Olenellus* (Fig. 14.6B), which is so primitive that the segments of the pygidium are not even fused together. Redlichiids tend to have a large, semicircular cephalon, strong genal spines, marginal facial suture, large half-moon-shaped eyes, and numerous thoracic segments; some have a long tail spike called a **telson**. More advanced redlichiines, such as *Redlichia* or *Paradoxides* (Fig. 14.6C), had opisthoparian sutures. As expected for such primitive trilobites, redlichiids are typical of the Early and Middle Cambrian, but were soon replaced by more advanced groups.

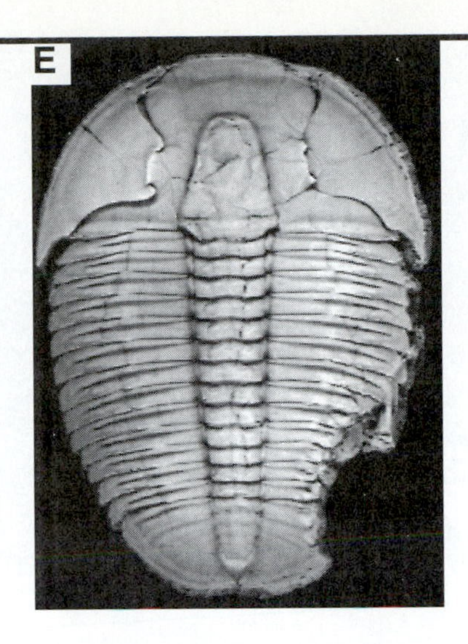

Figure 14.6 (continued) (D) Typical corynexochid, *Basidechenella lucasensis*, from the Devonian of Ohio. (E) One of the best known ptychopariids, *Elrathia kingi*, from the Middle Cambrian of Utah. These trilobites are so abundant that they are mined commercially, and sold in rock shops and as jewelry all over the world. (F) *Isotelus gigas*, a large (over a foot long) asaphid typical of the Ordovician of New York. (G) *Bumastus insignis*, a typical illaenid, from the Silurian of Ohio. (Photos courtesy L. Babcock.)

Order Corynexochida

Another common Early-Middle Cambrian group is the corynexochoids, which are characterized by a box-like glabella, and an opisthoparian facial region with subparallel sutures (Fig. 14.6D). The thorax has 7 or 8 segments, and some have large, almost isopygous pygidia. These characteristics are all typical of juveniles, and Robison (1967) argued that corynexochids are paedomorphically derived from more primitive ptychopariids.

Order Ptychopariida

The ptychopariids are a large, heterogeneous group of trilobites that is widely considered a paraphyletic wastebasket for trilobites that don't fit into other groups. Typical ptychopariines like *Elrathia* (Fig. 14.6E), *Olenus, Triarthrus*, or *Ptychoparia* have a simple glabella that tapers anteriorly, with straight glabellar furrows, and a large area in front of the glabella. They also tend to have many thoracic segments and a small pygidium. Ptychopariines are the common trilobites of the Middle-Late Cambrian, but declined soon thereafter and vanished by the Late Ordovician.

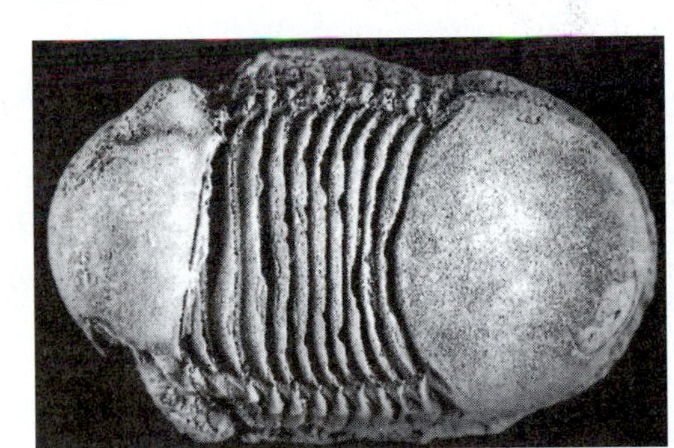

In the 1959 *Treatise*, four other highly diverse groups (asaphines, illaenines, trinucleines, and harpines) were treated as suborders of the Ptychopariida, further enhancing its status as a taxonomic wastebasket. Subsequent classifications typically promote each to ordinal status, or ally them with other groups.

Order Asaphida

Asaphids are highly distinctive trilobites. *Isotelus* (Fig. 14.6F) or *Asaphus*, with their large smooth isopygous "snowplow" cephala and pygidia, are typical. This feature presumably helped them burrow smoothly. Their free cheeks are often linked anteriorly, and some have large eyes that may even merge together. Asaphids have only 6 to 9 thoracic segments, and their pleural lobes are rounded at the ends. Although found as early as the Middle Cambrian (*Dikelocephalus* is a typical Late Cambrian taxon), they were particularly abundant in the Ordovician.

Order Illaenida

Like asaphids, illaenids are also distinctive in having a large, smooth cephalon and pygidium; frequently they are isopygous or have a pygidium larger than the cephalon. They are typified by shield-shaped forms like *Scutellum* or "snowplow" taxa like *Illaenus* or *Bumastus* (Fig. 14.6G). Unlike asaphids, they flourished through the Ordovician and survived into the Devonian.

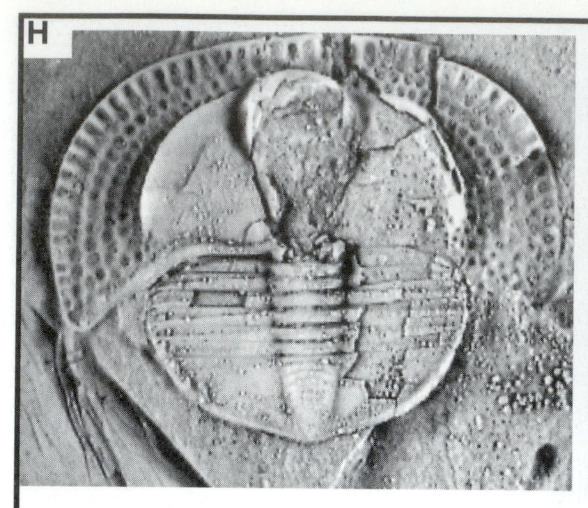

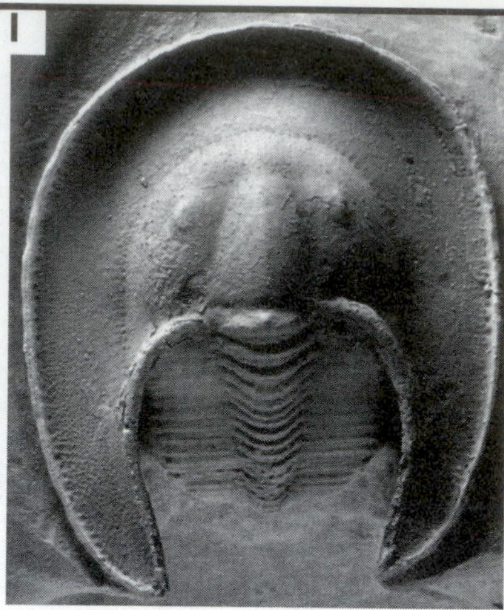

Figure 14.6 (continued) (H) A typical trinucleid, *Cryptolithus tesselatus*, from the Ordovician of Ohio. (Photo courtesy L. Babcock.) (I) The broad-brimmed harpids are easy to recognize. This *Harpes* is from the Devonian of Morocco. (From Levi-Setti, 1993.)

Order Trinucleida

The trinucleids are easy to recognize by their highly reduced pygidium and short thorax, so that their cephalon is more than half of the body. They are also typified by their broad ornamented cephalic brim and long genal spines, as seen in taxa such as *Cryptolithus* (Fig. 14.6H) or *Trinucleus*. Most taxa were blind and lacked eyes, and their tiny size and spines suggests they were adapted for shallow burrowing (although some paleontologists have speculated that they may have been pelagic). Their facial suture is marginal, and the glabella expands anteriorly. Trinucleids flourished in the Ordovician, but straggled on until the Middle Silurian.

Order Harpida

Harpids are a rather peculiar group of trilobites with an extremely broad, ornamented cephalic brim giving them a large, horseshoe-shaped head shield (Fig. 14.6I). Unlike trinucleids, they had small eyes on tubercles, a glabella that tapered anteriorly, and many thoracic segments. Despite their rarity, harpids were surprisingly long-lived, surviving from the Late Cambrian until the Late Devonian.

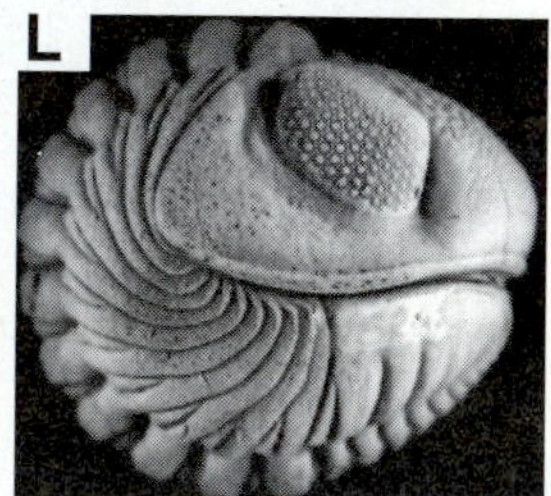

Figure 14.6. (continued) (J-K) Dorsal view (J) and side view of enrolled specimen (K) of a typical calymenid, *Flexicalymene meeki,* from the Ordovician of Ohio. (L-M) Dorsal view (M) and side view of an enrolled specimen (L) of a typical phacopid, *Phacops rana*, from the Devonian of Ohio. (Photos courtesy L. Babcock.)

Order Phacopida

The phacopids have long been separated as the only group typified by proparian facial sutures, although this is not true of all its members. They include such distinctive suborders as the calymenines (such as *Trimerus, Calymene*, or *Flexicalymene*—Fig. 14.6J-K), with their many thoracic segments, highly lobed, furrowed glabella with eyes on stalks, and their ability to roll up and lock their pygidium into the cephalon. Calymenines are particularly characteristic of the Ordovician and Silurian, although they range into the Devonian. Phacopines (such as *Dalmanites, Greenops,* and *Phacops* —Figs. 14.1, 14.6L-M) are distinguished by their broadly anteriorly expanded glabella and schizochroal eyes (Fig. 14.3B); they were particularly common in the Devonian. Cheirurids were much less common, but recognized by their four pairs of glabellar furrows, small holochroal eyes, and spiny pygidium; they too ranged from the Ordovician to Devonian.

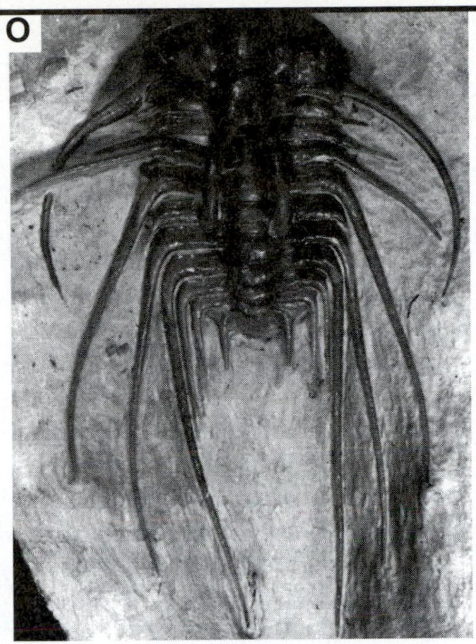

Figure 14.6. (continued) (N) A typical lichid, *Amphilichas halli*, from the Ordovician of Ohio. (Photo courtesy L. Babcock.) (O) Odontopleurids were very spiny trilobites, as shown by this specimen of *Dicranurus hamatus*, from the Devonian of Oklahoma. (Photo by the author.)

Order Lichida

Lichids are another peculiar group of trilobites with a glabella that extends to the anterior border of the cephalon, subdivided into elongate glabellar furrows, and an opisthoparian suture (Fig. 14.6N). Their most distinctive feature is the large pygidium (typically larger than the cephalon), often made up of three pairs of expanded, spiny pleurae. Lichids ranged from the Ordovician to the Devonian.

Order Odontopleurida

Odontopleurids are easily recognized by their extreme development of spines on their cephalic brim, occipital ring, genal angle, all along the thorax, and all over the small pygidium (Fig. 14.6O). Although rare, they ranged from the Ordovician to Devonian.

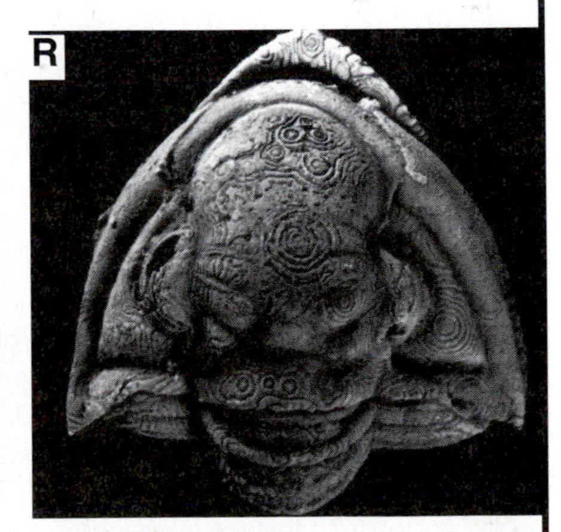

Figure 14.6 (concluded). (P) This proetid, *Kaskia chesterensis*, from the Mississippian of Indiana, was covered by a highly pustulose surface. (From Levi-Setti, 1993.) (Q-R) The last of the trilobites was this proetid *Anisopyge cooperi*, from the Upper Permian of West Texas. (Photo courtesy D. Brezinski.)

Order Proetida

The proetids were an extremely conservative but long-lived group, originating in the Late Cambrian and surviving the Late Devonian extinction that wiped out almost every other group of trilobites. They straggled on as rare elements of the Carboniferous (Fig. 14.6P) and Permian fauna, finally succumbing in the Late Permian (Fig. 14.6Q-R). Proetids were opisthoparian, with a large vaulted glabella, and many taxa are isopygous. They had large holochroal eyes, a furrowed pygidium without spines, and some also have genal spines.

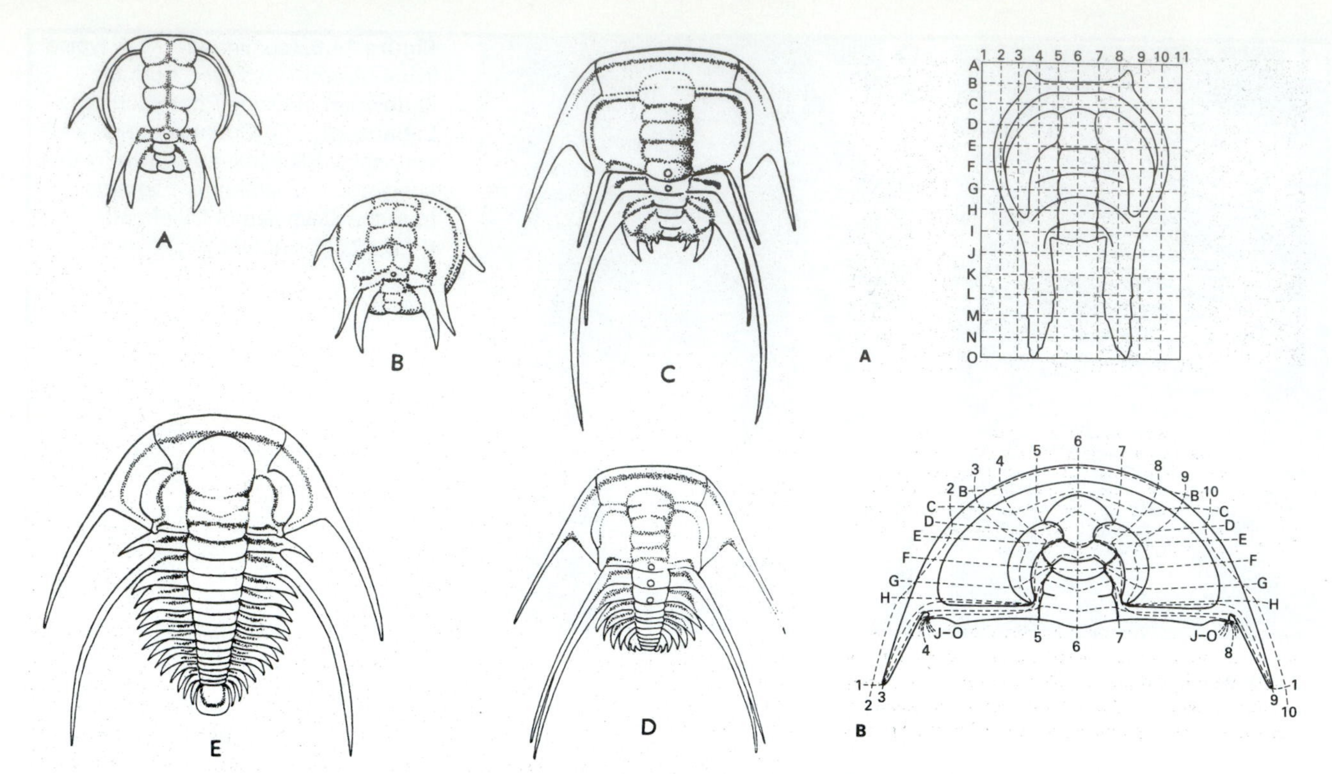

Figure 14.7. Ontogenetic changes in trilobite *Paradoxides pinus.* A and B are protaspids; C and D are meraspids; E is the adult holaspid. (Modified from Moore et al., 1959.)

Figure 14.8 The shape changes in the cephalon of *Olenellus* are illustrated by this transformation grid between juvenile and adult. (From Palmer, 1957.)

Ecology

Because they have long been extinct, it is not possible to know much about trilobite reproduction, although presumably they bred like other arthropods. However, once the juvenile trilobite was large enough to have a calcified exoskeleton, there is a good fossil record of their ontogeny. The earliest stage is known as **protaspid**, when the juvenile has only a single, undivided plate, that is usually circular or oval in outline (Fig. 14.7). When the first thoracic segments appeared, dividing the cephalon from the pygidium, it is known as the **meraspid** stage. Through the many stages of development, the original round protaspid "button" changes shape until it is like the adult cephalon, and more and more thoracic segments are added. When the final adult number of thoracic segments is finally present, it is known as the **holaspid** stage.

Studies of the changes in shape of a trilobite cephalon through ontogeny reinforce the visual impression that the overall proportions of the cephalon change radically. As shown by coordinate transformation grids (Fig. 14.8), there is a great expansion of the cheek region and a relative contraction of the back of the cephalon and the spines as the individual matures. The cheek regions probably covered important digestive and respiratory organs in the head, which needed increased surface area if they were to function efficiently in a larger animal. By contrast, the need for such large spines probably decreased as the animal changed its hydrodynamics from a free-swimming protaspid to a benthic holaspid mode of life.

These clear shape changes during ontogeny are useful when assessing the role of heterochrony in trilobite evolution (see Chapter 2). Since an expanded anterior head (especially cheek) region, reduced glabella, small pygidium, and multiple thoracic segments characterize adult individuals, this suggests that adults (holaspids) with these features (especially with additional thoracic segments) are peramorphic. Those with large cheek regions and spines, large glabella, fewer thoracic segments, and large pygidium would be paedomorphic (see Fig. 2.11).

Their tiny size and abundant spines suggest that many protaspids were pelagic, swimming in the plankton as they fed and grew (as many living arthropods do). By the time they reached adult holaspid size, however, most trilobites were apparently benthic detritus feeders. The shallow seas of the early Paleozoic were rich in trilobites, with as many as 29 different species found in the same level in some later Cambrian localities. A variety of trace fossils (see Chapter 18) are thought to be the traces of mud-grubbing trilobites, showing how they fed and moved. Using their multiple appendages they dug up and stirred the sediment, diverting organic matter and food particles found in the sediment into their mouth, where it was digested.

Although these generalizations apply to the vast majority of trilobites, there were all sorts of unusual exceptions. The tiny, blind agnostids (Fig. 14.6A) are thought to have been pelagic, swimming or floating in the open ocean. Other trilobites were extraordinarily spiny. Some apparently had spines for protection against predators, but others

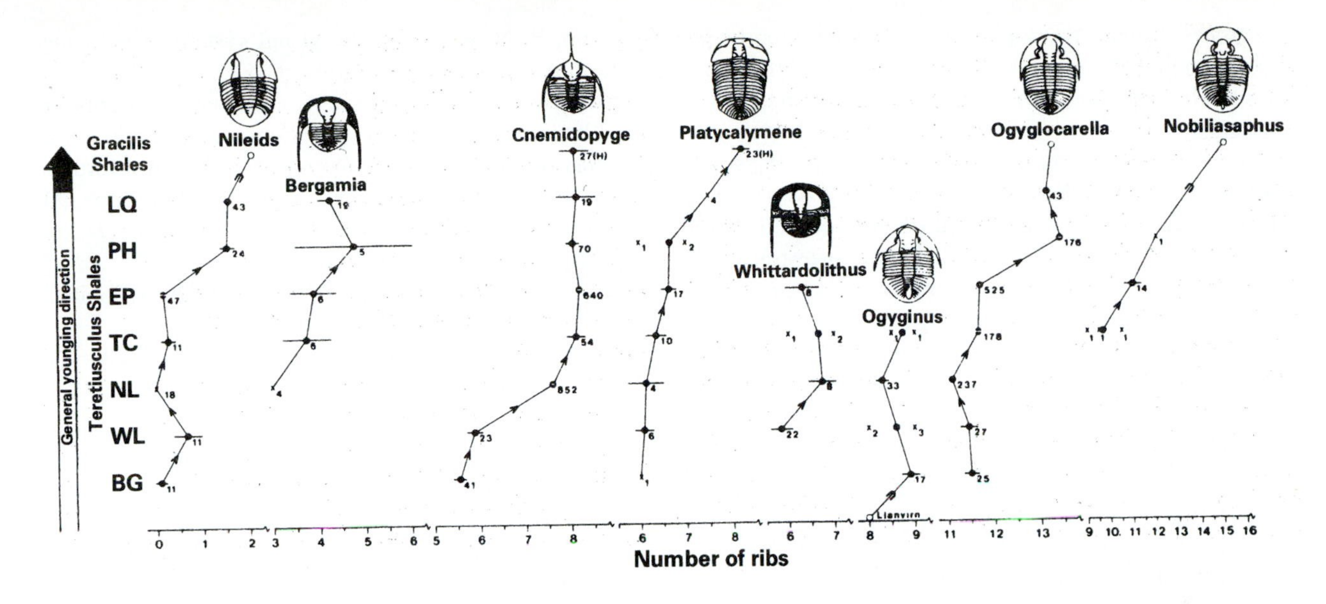

Figure 14.9. Gradual evolution of trilobites from the Ordovician of central Wales, as described by Sheldon. (From Sheldon, 1987.)

clearly used their spines for supporting themselves above the seafloor, preventing them from sinking into the mud. In many trilobites, the spines are lost and the cephalon and pygidium become large plows, presumably for burrowing in the mud and leaving only their eyes exposed. Several groups of trilobites developed the ability to roll up in a ball, locking their cephalon to the pygidium for protection, as the living isopod crustaceans (known as "sowbugs" or "pillbugs" to Americans, "wood lice" to the British) do today. A trilobite known as *Opipeuter* developed huge, hypertrophied eyes almost one-third its body length at the expense of the much reduced cheek region and tiny pleural lobes. It is thought to have swum on its back in the open ocean so that it could see everything around and above it. It didn't need large cheek regions or pleural lobes, since it was no longer a benthic mud-grubber with large digestive chambers.

Evolution

Evolutionary History—Because of their excellent, detailed fossil record, trilobites have long been a favorite group for evolutionary studies. They were among the first shelly fossils in the early Cambrian, and once they evolved, they took over the seafloor. Presumably, this was because the ecological structure of the Cambrian seafloor was very simple, so the organic matter and detritus that accumulated in the seafloor muds was an important niche to be exploited. Certainly, the lack of ecological complexity of most other Early and Middle Cambrian animals meant that few other niches were as ready for occupancy. In particular, the absence of many large predators meant that trilobites needed few anti-predatory defenses.

In the latest Cambrian, there were a series of extinction pulses that decimated most of the less specialized families (especially among the agnostids, redlichiids, corynexochids, and ptychopariids). A number of causes have been suggested, including climatic cooling and sea-level changes, although one important factor is the appearance of more efficient predators, such as the giant straight-shelled nautiloids. This suggestion is supported by the fact that the most abundant surviving groups in the Ordovician (such as the burrowing "snowplow" asaphids and illaenids, the tiny burrowing trinucleids, the enrolled calymenids, the spiny ceraurids and lichids) all have some kind of anti-predatory adaptations. Although a few unspecialized groups, such as the olenids, still persisted through the Ordovician, the majority of the groups were clearly very divergent from the primitive trilobite morphology. Another important factor may have been the general ecological diversification of the Ordovician, when many new niches became available for the first time for trilobites to diversify into and exploit.

The Late Ordovician extinctions marked another crisis in trilobite history. Their diversity was reduced to fewer than 10 families. Many of the dominant Ordovician groups, such as the trinucleids and asaphids, as well as the last of the ptychopariids, corynexochoids, and agnostoids, were wiped out. Silurian trilobites are relatively rare, with the commonest forms being large-eyed phacopids such as *Dalmanites*, and the lichids. Trilobite diversity increased slightly in the Devonian, when the typical phacopids such as *Phacops* and *Greenops* were the common forms, along with lichids and spiny odontopleurids.

The Frasnian-Famennian extinction in the Late Devonian was almost the last straw for the trilobites, with all but two orders meeting their end. The Late Devonian event wiped out the asaphids, lichids, odontopleurids, harpids, and calymenids. Only the phacopids and the conservative proetids persisted through the late Paleozoic, but always as an extremely rare element in the fauna, usually found in reef facies. Finally, even the proetids disappeared in the great Permian catastrophe, although their importance to understanding this event is limited by the fact that they were already a rare and minor element of the fauna.

Evolutionary patterns—Eldredge (1977) and Fortey and Owens (1990) have summarized a number of the evolutionary trends seen in the history of trilobites. As we have indicated, there are a number of different specializations that appear in trilobites from their simple ancestral morphotype. These include more complex eyes; improved mechanisms of enrollment; increasing size, integration, and proportions of the pygidium; and the development of various spines in several groups. Another common trend is the reduction and loss of eyes, which happened independently in several clades. Yet as Clarkson (1993, p. 374) pointed out, "trilobites as a whole remained constructed on the same archetypal plan defined in the earliest Cambrian, and, especially after the Early Ordovician, changes of real significance remained surprisingly low."

A number of paleontologists have also looked at the tempo and mode of evolution in trilobites. One of the original examples of rapid change and stasis in the punctuated equilibrium model was Eldredge's (1972) study of Devonian phacopid trilobites of the eastern and central United States. Eldredge found that most *Phacops* species were static through millions of years, with the only changes occurring in the number of files of lenses in their eyes, and these changed abruptly between species. Kaufmann (1933) documented patterns of evolution in the Upper Cambrian olenid trilobites, and found both gradual trends in their pygidial characters, and also abrupt speciation events that suggested allopatric origins. In some cases, *Olenus* showed iterative evolution, with repeated development of a character from different ancestors. Sheldon (1987, 1997) studied a 3-million-year sequence of trilobites from the Ordovician of central Wales, looking at over 15,000 specimens in 8 lineages (Fig. 14.9). Most of the lineages showed gradual increases in the number of pygidial ribs, although some also showed stasis, and a few reversed the trend by decreasing the number of ribs. The cause of these changes is not well understood, but as Sheldon (1997) pointed out, most of the gradual change took place during periods of environmental stability, while evolutionary stasis characterizes periods of environmental fluctuation. Sheldon calls this the "*plus ça change*" model, in reference to the old French adage *plus ça change, plus la même chose* ("the more things change, the more things stay the same"). A good explanation for why environmental stability allows gradual change, while environmental instability seems to prevent it and promote stasis, has been difficult to obtain, but it seems to be a common pattern of evolution in many groups.

For Further Reading

Bergström, J. 1973. Organization, life, and systematics of trilobites. *Fossils and Strata* 2:1-69.

Briggs, D. E. G., and P. D. Lane, eds. 1983. Trilobites and other arthropods. *Special Papers in Palaeontology 30.* Academic Press, London.

Clarkson, E. N. K. 1979. The visual system of trilobites. *Palaeontology* 22:1-22.

Eldredge, N. 1977. Trilobites and evolutionary patterns, pp. 305-332, *in* Hallam, A., ed. *Patterns of Evolution as Illustrated in the Fossil Record.* Elsevier, Amsterdam.

Fortey, R. A., and R. A. Owens. 1990. Trilobites, pp. 121-142, *in* McNamara, K. J., ed. *Evolutionary Trends.* University of Arizona Press, Tucson.

Fortey, R. A., and H. B. Whittington. 1989. The Trilobita as a natural group. *Historical Biology* 2:125-138.

Levi-Setti, R. 1993. *Trilobites* (2nd ed.). University of Chicago Press, Chicago.

Martinsson, A., ed. 1975. Evolution and morphology of the Trilobita, Trilobitoidea, and Merostomata. *Fossils and Strata* 4:1-467.

Mickulic, D. G., ed. 1990. *Arthropod Paleobiology: Short Courses in Paleontology* 3. Paleontological Society and University of Tennessee Press.

Moore, R. C., ed. 1959. *Treatise on Invertebrate Paleontology* Part O, Arthropoda 1. Geological Society of America and University of Kansas Press, Lawrence, Kansas.

Robison, R. L., and R. L. Kaesler. 1987. Phylum Arthropoda, pp. 205-269, *in* Boardman, R. S., A.H. Cheetham, and A. J. Rowell, eds. *Fossil Invertebrates.* Blackwell Scientific Publishers, Cambridge, Mass.

Whittington, H. B. 1957. The ontogeny of trilobites. *Biological Reviews* 32:421-469.

Whittington, H. B. 1966. Phylogeny and distribution of Ordovician trilobites. *Journal of Paleontology* 40:696-737.

Whittington, H. B. 1981. Paedomorphosis and cryptogenesis in trilobites. *Geological Magazine* 11:591-602.

CHELICERATA

> *I shall never forget when the first excellently preserved specimen of the new giant eurypterid was found. My workmen had lifted up a large slab, when they turned it over, we suddenly saw the huge animal, with its marvelously shaped feet, stretched out in natural position. There was something so life-like about it, gleaming darkly in the stone, that we almost expected to see it slowly rise from the bed where it had rested in peace for millions of years and crawl down to the lake that glittered close below us.*
>
> Johan Kiaer

The chelicerates include the arachnids (spiders, scorpions, whip scorpions, pseudoscorpions, "daddy long-legs" or harvestmen, mites, ticks, and chiggers) and the merostomes (horseshoe crabs and the extinct eurypterids). With over 70,000 named living species, chelicerates are today the largest arthropod group after the insects. The chelicerate body is divided into two segments, a **prosoma** (or cephalothorax) combining the head and thorax, and an **opisthosoma** (or abdomen). They do not have antennae, but all six segments of the prosoma bear appendages. Their first pair of appendages are modified into **chelicerae**, the structure that gives the group its name, which are typically used as mouthparts. The next pair of appendages are the **pedipalps**, which may also serve as mouthparts, or as enlarged pincers for capturing prey (in scorpions, eurypterids, and some spiders). These are followed by four pairs of walking legs. Hence, most chelicerates (not just spiders) have eight legs. Beall and Labandeira (1990) reviewed most of the recent analyses of the chelicerata.

Subclass Arachnida

Although arachnids are the vast majority of living chelicerates, none have a mineralized skeleton, so they rarely fossilize. Since they are primarily a terrestrial group, most fossil spiders, scorpions, or ticks come from amber pieces and freshwater *Lagerstätten* with unusual soft-bodied preservation, so relatively little is known about their evolutionary history. A total of about 320 extinct genera of arachnids and about 400 extinct species has been described, equivalent to less than 1% of the 70,000 described living species. Some scientists estimate that there may be more than 500,000 living species of arachnids, most of which are mites that are largely undescribed.

Possible marine arachnids are known from the Cambrian. The earliest terrestrial arachnids were the scorpions, such as *Palaeophonus* from the Late Silurian of Europe and North America, and *Waeringoscorpio* from the Lower Devonian of Germany. They looked very much like modern scorpions in their anatomy, complete with the pincer-like pedipalps and a long tail with a stinger. They were originally thought to be terrestrial airbreathers, but they are now regarded as aquatic, since they apparently had gills rather than the book lungs of terrestrial arachnids (although some authors, such as Labandeira and Beall, 1990, dispute this and claim they were fully terrestrial). Unquestioned terrestrial scorpions are known from the Pennsylvanian Mazon Creek nodules. Scorpions are true "living fossils," changing very little in morphology since the late Paleozoic.

An extinct spider-like group, the trigonotarbids, were apparently the first terrestrial arachnids, crawling out onto land in the Early Devonian. The earliest spiders, pseudoscorpions, and mites are also known from the Early Devonian Rhynie Chert of Scotland. By the Carboniferous, 12 of the 16 orders of arachnids had appeared. Spiders are particularly abundant in the Mazon Creek nodules. Like the scorpions, they are very similar to modern spiders in most of their anatomy. They even had almost complete fusion of their segments into a prosoma (=cephalothorax) and opisthoma (=abdomen). In fact, there are three living orders of spiders that have bodies with unfused segments, a condition more primitive than the earliest known fossil spiders. The earliest ticks are also known from Mazon Creek, and look essentially modern. The Mesozoic record of arachnids is particularly poor, with only four genera (two of scorpions) known from the entire era, probably due to the lack of appropriate terrestrial *Lagerstätten* preservation. Arachnids are much better known from the Cenozoic, where *Lagerstätten* such as the Oligocene Florissant lake deposits, and the Miocene Baltic amber, have yielded many important specimens. From these deposits, it is apparent that carnivorous, web-spinning spiders had radiated since the late Mesozoic, probably in response to the explosion of insect diversity in response to the diversification of flowering plants during the Cretaceous.

If the cladistic relationships of the arachnids postulated by Beall and Labandeira (1990) are correct, then the entire arachnid radiation (from the most primitive group, the scorpions, to the most derived groups, the mites) must have occurred rapidly in the Late Silurian and Early Devonian, even though many of these taxa do not have a fossil record that early. Petrunkevitch (1952) argued that the sudden diversification of arachnids, followed by their long periods of stability, required that the major arachnid body plans arose by some sort of Goldschmidtian "hopeful monster" mechanism (see Chapter 5). However, Beall and Labandeira (1990) argued that most of the body plans in the arachnids can be derived by heterochrony, especially by paedomorphosis (retention of juvenile body form).

Subclass Eurypterida

The eurypterids, or "sea scorpions," were an extinct group of merostomes that superficially resembled scorpions in their overall body proportions (Fig. 14.10). Some taxa also had pedipalps modified into pincers and a scorpion-like telson. Eurypterids were unlike scorpions in many ways. Unlike the mostly terrestrial scorpions, eurypterids were adapted for normal marine, brackish, and fresh water. Their most obvious difference is size: most eurypterids were tens of centimeters or more in length, and the giant *Pterygotus* was about 2.5 m long!

The name "eurypterid" (*eury*, "broad" and *pteros*, "wing" in Greek) refers to the final pair of legs, which may be modified into a wing-like paddle for propulsion in the water. However, they had a wide variety of body forms. Some were long-bodied with a flat, paddle-like telson and large pincers (*Pterygotus*); others had long spider-like legs, a spiky telson, and no swimming appendages (*Stylonurus*); still others had long spiny chelicherae and pedipalps, spiny walking legs, and a scorpion-like tail (*Mixopterus*). Some had large compound eyes (*Pterygotus, Hughmilleria*), while others had only small compound eyes. Most had another pair of light sensors on the top of the prosoma called **ocelli**, a feature also found in other arthropods, such as horseshoe crabs.

Eurypterids are known from about 25 genera ranging from the Ordovician to the Permian, but they are most abundant in the Silurian and Devonian. Some rock units, such as the Upper Silurian Bertie Limestone in upstate New York, are famous for the quality and quantity of their eurypterids. During the Silurian, eurypterids were among the largest known predators, only to have that role usurped by large armored fish in the Devonian. Eurypterids appar-

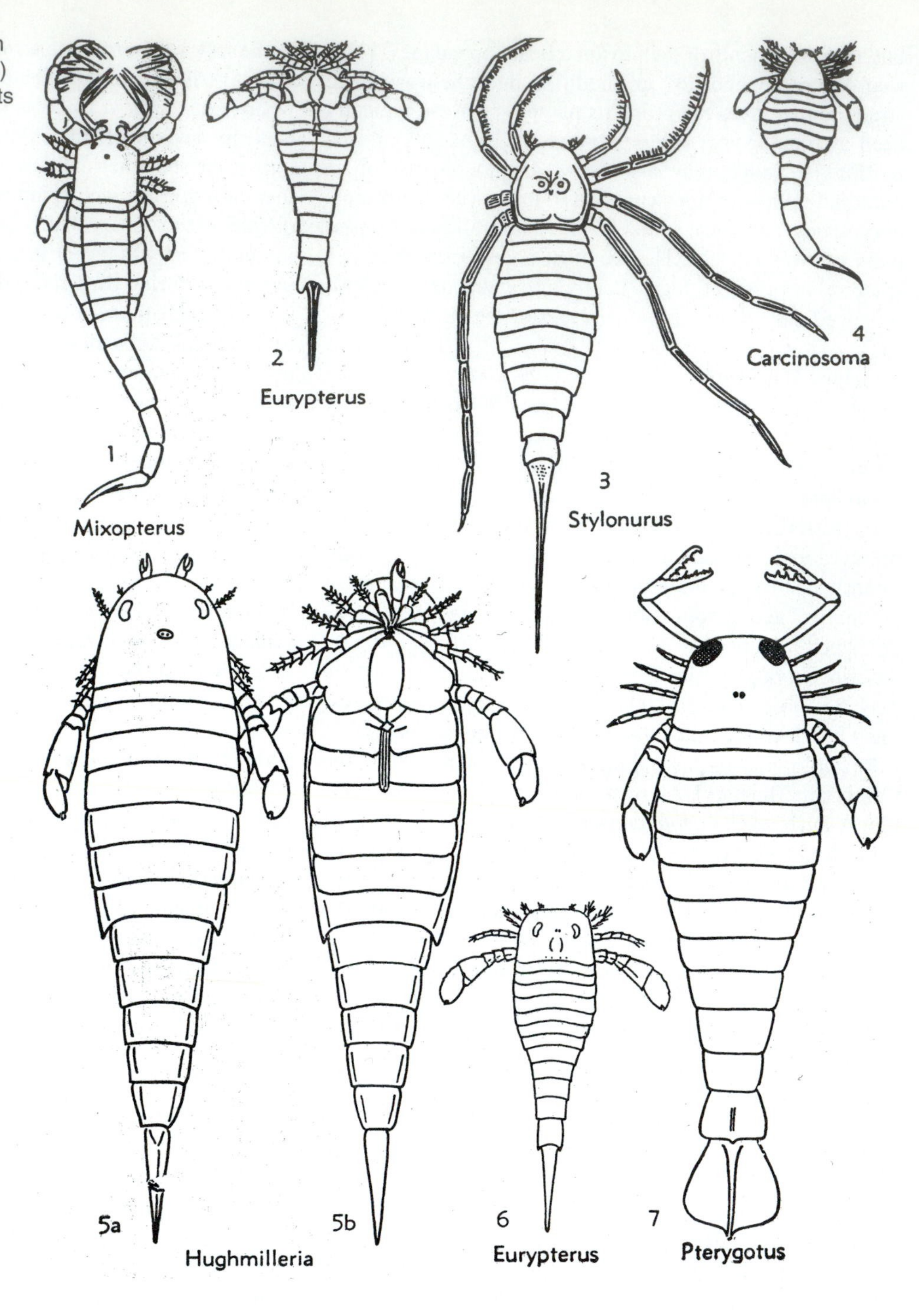

Figure 14.10. Typical Silurian and Devonian eurypterids. (1) *Mixopterus* (Ord.-Dev.) with its spiny appendages. (2, 6) *Eurypterus* (Ord.-Perm.), a very abundant small form in the Silurian. (3) *Stylonurus* (Ord.-Miss.), a form with an unusually long sharp telson and long legs. (4) *Carcinoma* (Ord.-Sil.), a genus with a wide abdomen. (5) *Hughmilleria* (Ord.-Dev.) a small eurypterid that differs from *Eurypterus* in its prosoma and appendages. (7) *Pterygotus* (Ord.-Dev.), the giant (almost 2 meters long) among the eurptyerids, characterized by its large eyes, long pincer-like chelicerae, and broad flat telson. (From Moore et al., 1953.)

ently captured a wide variety of mobile prey, such as fish and smaller eurypterids. Eurypterids apparently swam with both their paddles and legs contributing to their propulsion. Trackways of eurypterids have been described that show the motions of their appendages as they walked along the bottom. Functional studies of eurypterids, and of crabs with similar appendages, suggest that some taxa, such as the 30-cm-long, streamlined *Baltoeurypterus*, could swim rapidly at about half a meter per second. The large, flat, hydrofoil-shaped pterygotids, on the other hand, probably cruised only slowly, and only rarely used a burst of speed from their flat paddle-like telson to catch prey (Selden, 1983).

Eurypterids apparently lived in a wide variety of habitats (Kjellesvig-Waering, 1961). In the Silurian of Wales, there are taxa that lived in normal marine waters (such as the pterygotids), others that apparently preferred lagoonal or estuarine conditions (the greatest abundance and diversity of forms), and some are clearly associated with brackish or freshwater conditions (such as *Hughmilleria* and *Stylonurus*, a form with long walking legs and no paddle) (Waterston, 1979).

Although eurypterids are rare compared to other Paleozoic fossils, their role as the top predators of their world, their fearsome appearance, and their impressive pincers and size has captured the imagination of many paleontologists.

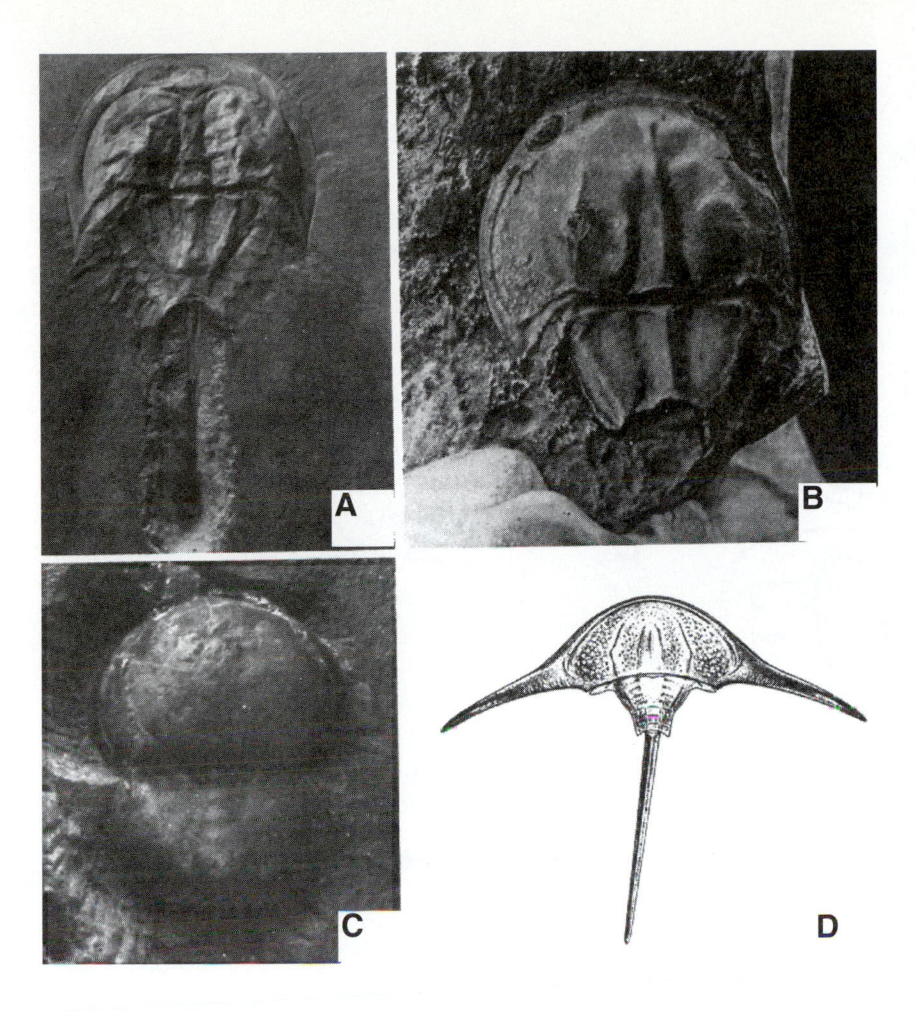

Figure 14.11. A spectrum of different types of xiphosurans. (A) *Mesolimulus walchi*, a modern-looking form from the Upper Jurassic Solnhofen Limestone (see also Fig. 18.1). (B) *"Limulus" vicensis*, from the Triassic of France. (C) The button-shaped *Pringlia birtwelli*, from the Carboniferous of Britain. (D) The boomerang-shaped *Austrolimulus fletcheri*, from the Triassic of Australia (From Fisher, 1984.)

Subclass Xiphosura

Like the misnamed "sea scorpions" (which are not true scorpions), the xiphosurans, or "horseshoe crabs," are not true crabs, either. They are merostome chelicerates, not crustaceans. The name "xiphosura" comes from the Greek *xiphos*, "sword" and *ura*, "tail" and refers to the long spike-like telson found in many taxa. Most xiphosurans have a relatively large prosoma, which makes up a large portion of the body (Fig. 14.11) in living *Limulus*, but primitive taxa still had a relatively long opisthoma. During their evolution, there was a tendency to reduce the number of opisthomal segments, shortening the body behind the prosoma. The most derived forms are the limulaceans, which include all four of the living species. They have fused all their opisthosomal segments into a single plate, from which protrudes their long telson spike.

Although xiphosurans have long been touted as examples of "living fossils" that have persisted unchanged through millions of years, this is a misconception (Fisher, 1984). It is true that the late Paleozoic *Paleolimulus*, and the Jurassic *Mesolimulus* (Fig. 14.11A) which was described from numerous well-preserved Solnhofen Limestone specimens (Fig. 18.1) looks much like the living species. During their early evolution in the Paleozoic, however, they had a wide variety of body forms. Some xiphosurans had a wide, spiky prosoma like *Euproops* (Fig. 7.13), or a broad boomerang-like shape like *Austrolimulus* (Fig. 14.11D) or the double-button shape of a taxon like *Pringlia* (Fig. 14.11C). The Late Cambrian aglaspids have long been considered the earliest, most primitive xiphosurans, but study of specimens with well-preserved appendages shows that they are not similar (Briggs et al., 1979). Briggs and Fortey (1989) considered aglaspids to be a distant sister-group of both trilobitomorphs and merostomes.

Much is known about the biology of xiphosurans, since the living horseshoe crab *Limulus* has been studied by many biologists. It is easy to keep alive in an aquarium, so it is popular for many types of biological experiments. *Limulus* lives in nearshore habitats (less than 50 m water depth), and tolerates a wide range of salinities, including brackish-water estuaries. It can even crawl on land for short distances. It swims on its back, stroking its legs in a rhythmic fashion, but crawls on its legs. It can turn itself over easily by flexing its body and digging in the long telson spike. For protection, it spends most daylight hours burrowed into the sandy bottom, with only one of its book lungs exposed. It uses its legs to dig downward, and then with legs churning and its telson whipping up and down, it stirs up a cloud of sediment that eventually settles and conceals it. *Limulus* comes out at night to feed on bivalves, polychaete worms, and other soft invertebrates, probing the

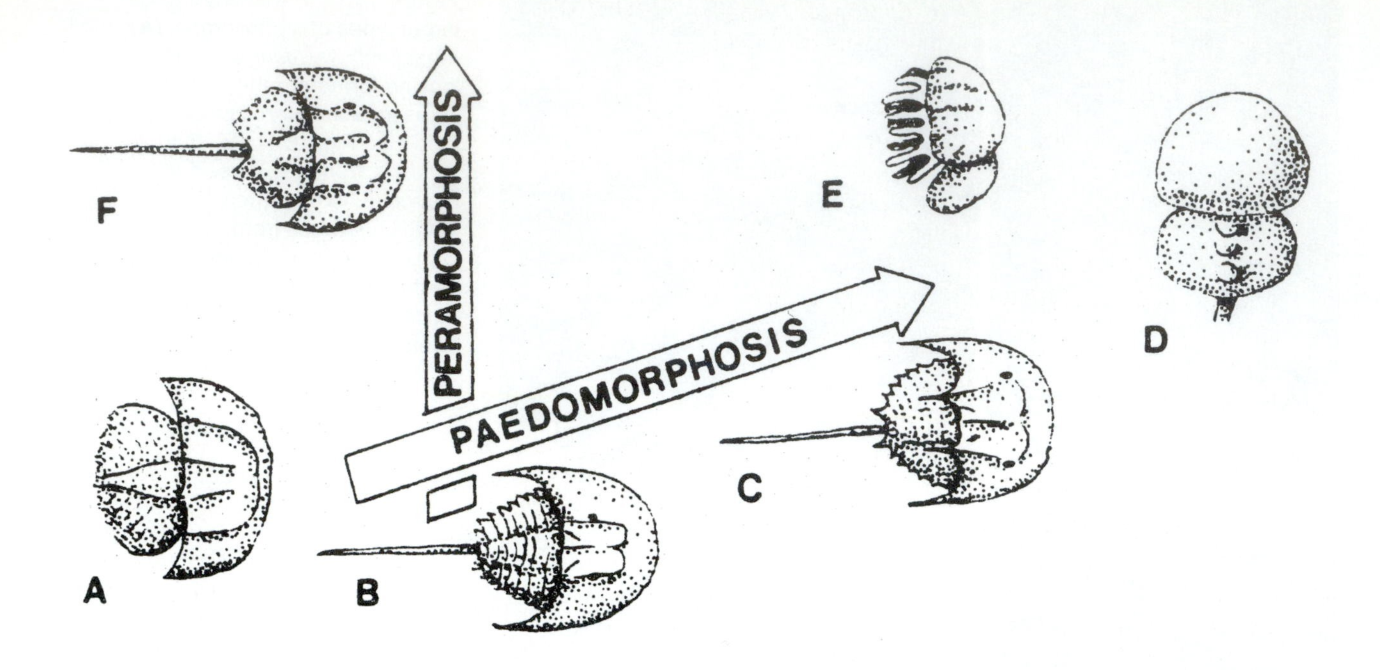

Figure 14.12. Models of heterochronic change in the Xiphosurida. (A) Using the *Euproops*-like larvae of *Tachypleus* as a starting point, a paedomorphic trend can be seen from (B) *Bellinurus* to (C) *Euproops* to (D) *Liomesaspis,* as shown by comparisons with (E) the late embryonic stage of *Limulus.* Modern limulids like (F) *Tachypleus* and the living genus, on the other hand, seemed to have developed via peramorphosis. (From Beall and Labandeira, 1990.)

sediment with its anterior appendages to locate its prey. It is known to scavenge dead fish and molluscs when they are available. In the lab, *Limulus* will eat a wide variety of food, including brine shrimp, chopped squid, dead fish, and even dog biscuits.

One of the most remarkable habits of *Limulus* is its breeding behavior. Once a year during highest spring tide, males and females swarm in the surf zone of certain beaches, where many males will pile on top of and mate with the much larger females. Once mating is completed, the females lay their eggs in the sand, where they will remain protected until a month later when the next high tide exposes them. At this point, the eggs hatch, and the larvae (which are called "trilobite" larvae because of their remarkable resemblance to trilobites) swim away. Juvenile *Limulus* live in the shallower waters and in estuaries, digging for polychaete worms. Adults live further offshore, where they prey mostly on bivalves.

Although xiphosurans have never been tremendously diverse or abundant, they do have quite a long record nonetheless, ranging from the Silurian to the present. Fisher (1982, 1984) recognizes about 30 genera throughout their history, starting with a diverse radiation of primitive forms in the Silurian, and another radiation of *Euproops*-like and *Bellinurus*-like taxa in the Carboniferous. The modern subfamily Limulacea first appeared in the Pennsylvanian and Permian with *Paleolimulus*, which closely resembles the living genus *Limulus*. From the Permian to the present, the xiphosurans have indeed been remarkably conservative "living fossils," showing little change in morphology.

Xiphosurans have been suggested as examples of heterochrony (Størmer, 1955; Beall and Labandeira, 1990). The evolution of limulaceans seems to be a peramorphic trend, where the ontogeny of the living species recapitulates the evolutionary history of the group (Fig. 14.12). Juvenile *Limulus* "trilobite" larvae (Fig. 14.12E) have many opisthosomal segments and segmentation of the prosoma, as seen in the ancestral early Paleozoic taxa. Older *Limulus* larvae closely resemble Carboniferous taxa, with fewer opisthomal segments that are not yet fused. Permian *Paleolimulus*, with its short, fused opisthoma, large prosoma, and long telson, closely resemble the modern living *Limulus*. On the other hand, xiphosurans such as *Liomesaspis* retain the proportions of the larvae of early xiphosurans, such as *Euproops* (Fig. 14.12A, C); this is apparently a paedomorphic trend.

For Further Reading

Beall, B. S., and C. C. Labandeira. 1990. Macroevolutionary patterns of the Chelicerata and Tracheata, pp. 257-284, *in* Mickulic, D. G., ed. *Arthropod Paleobiology: Short Courses in Paleontology* 3. Paleontological Society and University of Tennessee Press.

Bergström, J. 1975. Functional morphology and evolution of the xiphosurids. *Fossils and Strata* 4:291-305.

Briggs, D. E. G., D. L. Bruton, and H. B. Whittington.

1979. Appendages of the fossil arthropod *Aglaspis spinifer* (Upper Cambrian, Wisconsin) and their significance. *Palaeontology* 22:167-180.

Eldredge, N. 1970. Observations on the burrowing behavior in *Limulus polyphemus* (Chelicerata, Merostomata) with implications for the functional anatomy of trilobites. *American Museum of Natural History Novitates* 2436:1-17.

Eldredge, N. 1974. Revision of the suborder Synziphosurina (Chelicerata, Merostomata) with remarks on merostome phylogeny. *American Museum of Natural History Novitates* 2543:1-41.

Fisher, D. 1975. Swimming and burrowing in *Limulus* and *Mesolimulus*. *Fossils and Strata* 4:281-290.

Fisher, D. 1982. Phylogenetic and macroevolutionary patterns within the Xiphosurida. *Proceedings of the Third North American Paleontological Convention* 1:175-180.

Fisher, D. 1984. The Xiphosurida: archetypes of bradytely? pp. 196-213, *in* Eldredge, N., and S.M. Stanley, eds. *Living Fossils*. Springer Verlag, New York.

Kjellesvig-Waering, E. N. 1961. The Silurian eurypterids of the Welsh Borderland. *Journal of Paleontology* 35:784-835.

Kjellesvig-Waering, E. N. 1986. A restudy of the fossil Scorpionida of the world. *Palaeontographica Americana* 55:1-287.

Petrunkevich, A. 1952. Macroevolution and the fossil record of the Arachnida. *American Scientist* 40:99-122.

Plotnick, R. E. 1985. Life-based mechanisms for swimming in eurypterids and portunid crabs. *Transactions of the Royal Society of Edinburgh, Earth Sciences* 76:325-337.

Raasch, G. O. 1939. Cambrian Merostomata. *Geological Society of America Special Paper* 19.

Selden, P. A. 1983. Autecology of Silurian eurypterids. *Special Papers in Palaeontology* 32:39-54.

Selden, P. A. 1988. The arachnid fossil record. *British Journal of Entomology and Natural History* 1:15-18.

Selden, P. A., and D. J. Siveter. 1988. The origin of the limuloids. *Lethaia* 20:383-392.

Størmer, L. 1952. Phylogeny and taxonomy of fossil horseshoe crabs. *Journal of Paleontology* 26:630-639.

Størmer, L. 1955. Arthropoda 2: Merostomata, pp. 4-41, *in* Moore, R.C., ed. *Treatise on Invertebrate Paleontology, Part P*. Geological Society of America and University of Kansas Press, Lawrence.

Waterston, C.D. 1957. The Scottish Carboniferous Eurypterida. *Transactions of the Royal Society of Edinburgh* 66:265-288.

Waterston, C. D. 1979. Problems of functional morphology and classification of stylonurid eurypterids (Chelicerata, Merostomata), with observations on Scottish Silurian Stylonuroidea. *Transactions of the Royal Society of Edinburgh* 70:251-322.

CRUSTACEA

Crustaceans are, in our opinion, the most exciting of all invertebrates. They are certainly one of the most popular invertebrate groups, even among nonbiologists, for they include some of the world's most delectable gourmet fare, such as lobster, crab, and shrimp. There are more than 30,000 described living species, and probably four times that number waiting to be discovered and named. They exhibit an incredible diversity of form and habit, ranging in size from tiny interstitial and planktonic forms less than a millimeter in length to giant crabs with leg spans of 4 m and lobsters attaining weights up to 17 pounds. Crustaceans are found at all depths in every marine, brackish, and freshwater environment known; a few have become successful on land, the most notable being sow bugs, or pill bugs. Crustaceans are the most widepread and diverse group of invertebrates inhabiting the world's oceans. The range of morphological diversity among crustaceans far exceeds that of even the insects. In fact, because of their diversity of form and number, it is often said that crustaceans are the "insects of the sea." We prefer to think of insects as the "crustaceans of the land."

Richard and Gary Brusca, *Invertebrates*, 1990

Although people who specialize in other organisms might not agree with the Bruscas (both specialists in crustaceans) that they are the "most exciting of all invertebrates," no one will dispute the major points they raise: crustaceans are the most diverse group of invertebrates in aquatic habitats, and have even managed to live on land (such as the "pill bugs" or "sow bugs" in America, or "wood lice" of the British—they are not bugs but isopod crustaceans). Most crustaceans have a generalized shrimp-like body form, but there are some extreme exceptions. Barnacles are crustaceans that start out as free-swimming larvae, but then settle and attach to a substrate. Once they are attached, they secrete a series of hard calcareous plates around themselves, and their legs (no longer required for walking or swimming) rhythmically comb the water to filter out food particles. Crabs have modified the shrimp-like body form to another extreme, shortening the body and reducing their segmentation. Despite all these differences, crustaceans have several features in common. They have two pairs of antennae, and three additional pairs of appendages (maxillae and two sets of mandibles) behind the mouth. The limbs are biramous, although some groups secondarily develop uniramous limbs. Their body consists of a distinct cephalon and a thoracic region with various degrees of specialization, and a segmented abdomen. Finally, the entire group develops from a distinctive larva type known as a **nauplius**, which is composed of a simple unsegmented oval head plate with three pairs of appendages. No matter how unusual their adult body form, this larva is a unique characteristic that demonstrates the monophyly of the Crustacea.

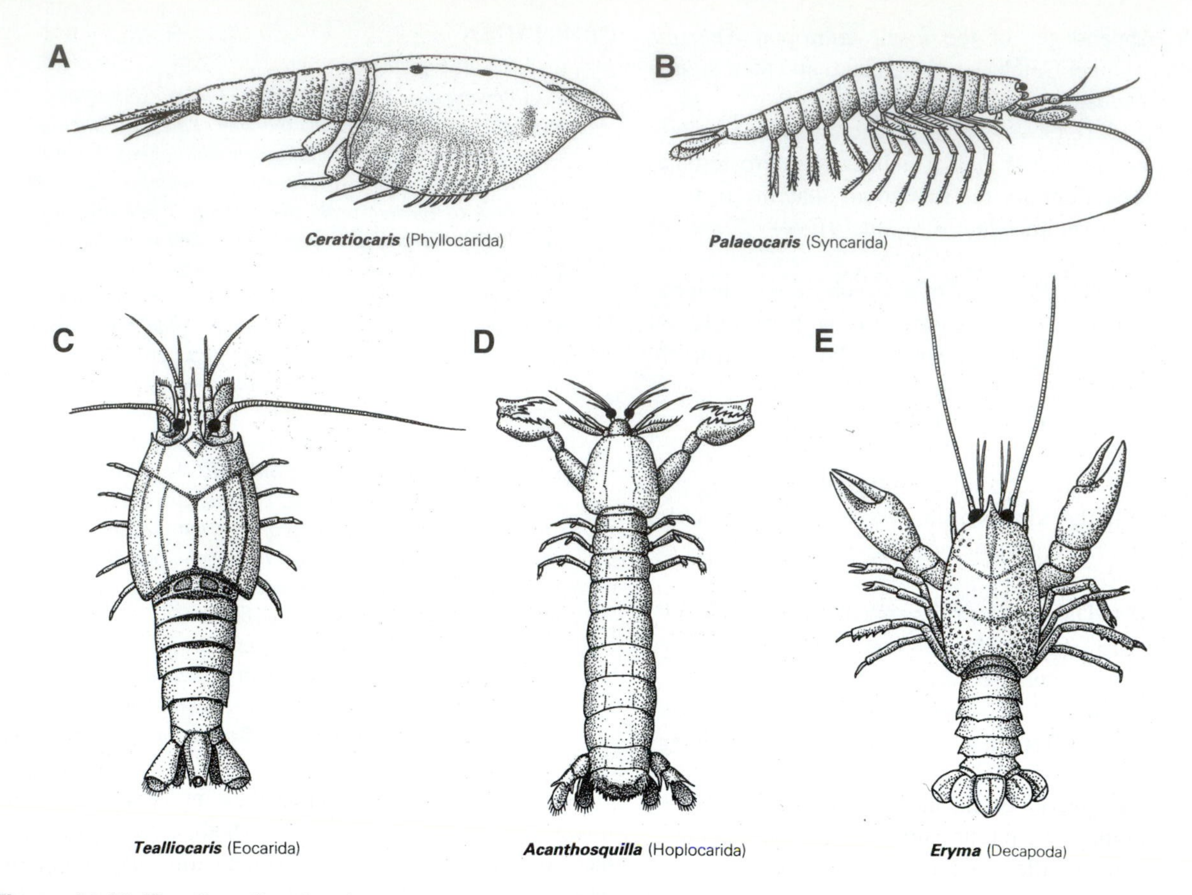

Figure 14.13. The diversity of malacostracan crustaceans is illustrated by these representative taxa. (A) The phyllocarid *Ceratiocaris* (Silurian). (B) The syncarid *Palaeocaris* (Carboniferous) (C) The eocarid *Tealliocaris* (Carboniferous) (D) The hoplocarid *Acanthosquilla* (Recent). (E) The decapod *Eryma* (Jurassic). (From Clarkson, 1993.)

Systematics

There are a wide variety of crustacean classifications, but most of the living taxa have no fossil record, so we need not consider the hundreds of different groups. However, a few taxa of crustaceans are important as fossils. The classification that follows is largely from Brusca and Brusca (1990), with some modifications after McLaughlin (1980) and Schram (1986, 1990). Below the class level, most of the taxa that have no fossil record have been omitted to emphasize only those groups with some paleontological importance.

Class Remipedia: strange worm-like crustaceans that are unspecialized in many features; they are restricted to cave habitats; only one poorly preserved fossil is known.

Class Cephalocarida: tiny benthic crustaceans, with only nine known species; no fossil record.

Class Branchiopoda: fairy shrimps, tadpole shrimps, water "fleas," "clam shrimps"; sparse fossil record.

Class Malacostraca: crabs, shrimps, lobsters, krill, pillbugs, amphipods

Subclass Phyllocarida: phyllocarids (Fig. 14.13A)

Subclass Eumalacostraca

Superorder Syncarida: syncarids (Fig. 14.13B)

Superorder Eocarida: eocarids (Fig. 14.13C)

Superorder Peracarida: many different kinds of shrimp-like forms, including opossum shrimps, "sand fleas," "beach hoppers," and the terrestrial isopods ("pill bugs," "sow bugs," or "wood lice")

Superorder Eucarida

Order Euphausiacea: krill

Order Decapoda: crabs, shrimps, lobsters

Infraorder Brachyura: true crabs (Fig. 14.14)

Infraorder Palinura: spiny lobsters

Infraorder Anomura: hermit crabs, sand crabs, king crabs

Infraorder Astacidea: crayfish, lobsters

Infraorder Thalassinidea: mud and ghost shrimps

(*Plus many other shrimp-like groups with little or no fossil record.*)

Class Maxillopoda: ostracodes, copepods, barnacles, and many other shrimp-like forms

Subclass Cirripedia: barnacles (Fig. 14.15)

Subclass Ostracoda: ostracodes (Fig. 14.16)

Figure 14.14. Beatifully preserved and nearly complete specimen of a crab, *Lobocarcinus pustulosus*, from the Miocene of New Zealand. (From Feldman and Fordyce, 1996; photo courtesy R. Feldmann.)

Class Malacostraca

One of the earliest known groups of crustaceans was the **phyllocarids**, a group with a large bivalved carapace, short antennae, and powerful biting mandibles (Fig. 14.13A). Beneath the carapace were four pairs of biramous, leaf-like appendages used for filter feeding. The abdomen projects from the carapace, and six leaf-like appendages branch out from it. Fossil phyllocarids first occur in the Early Cambrian, although they are best known from the Middle Cambrian Burgess Shale, where they are exquisitely preserved. In some marginal marine and brackish-water facies of the Paleozoic, phyllocarids were extremely common. Some got to be very large as well. The Silurian genus *Ceratiocaris* was up to 75 cm long—a mighty big "shrimp." About 20 species in six genera of phyllocarids still survive today, although most are less than 15 mm long; the largest living forms are about 4 cm long.

Extinct shrimp-like crustaceans known as **eocarids** first occurred in the Late Devonian and underwent a great adaptive radiation in the Carboniferous (Fig.14.13B). They are best known from the great coal swamp deposits of the Carboniferous, suggesting that they inhabited fresh or swampy, brackish water, although this environment may just be more favorable for preservation of their thin cuticle than the normal marine environment. After this great late Paleozoic diversification, the eocarids disappeared during the Permian catastrophe.

Another important Paleozoic group are the **syncarids**, which were small, rather elongate crustaceans (Fig. 14.13C), which lacked a carapace, and have biramous thoracic appendages. Syncarids flourished in the late Paleozoic, but were thought to be extinct at the end of the Permian. In 1892, however, a living syncarid, *Anaspides*, was discovered living in Tasmanian lakes and rivers. Since then, a few other genera have been reported from the same area, true relicts or living fossils of the Paleozoic.

The most familiar of all the crustaceans are the **decapods**, the group that includes most of the crabs, lobsters, crayfish, and an incredible variety of shrimp-like forms (Fig. 14.13E, 14.14). In addition to having 10 walking appendages (hence the name Decapoda, Greek for "ten legs"), they have three pairs of feeding appendages around the mouth (maxillipeds). Their body is composed of 20 segments, of which 6 are fused to form the head, 8 comprise the thorax, and 6 belong to the abdomen. They occupy a tremendous variety of aquatic habitats and every possible depth, from the freshwater crayfish to the great variety of shallow marine crabs, lobsters, and shrimp, to the burrowing mud and ghost shrimps and sand crabs, to the deep-water lobsters and crabs, to the many shrimp-like forms that swim or float in the open ocean. Some species, such as the robber crab, live entirely on the land, returning to the sea only to breed. Decapods also feed in nearly every possible manner, including suspension feeding, predation, herbivory, scavenging, and more.

Decapods may have a partially or completely calcified skeleton, so their fossilization potential can be excellent. Plotnick (1986) counted about 366 fossil genera, although many of these are known from partial specimens. Decapods first appeared in the Devonian, and later diversified to become one of the most successful groups of the Mesozoic. One of their greatest innovations is the development of large claws or pincers that make them very effective predators. Crabs and lobsters use their claws to crush shells, or to crack them and peel away the lip of the aperture, so they can consume the mollusc inside. The increase in predation by crabs and lobsters was probably one of the main factors behind the "Mesozoic marine revolution" (see Chapter 8). These new predators forced molluscs to burrow to escape them, and may have driven the majority of the brachiopods that survived the Permian catastrophe to extinction.

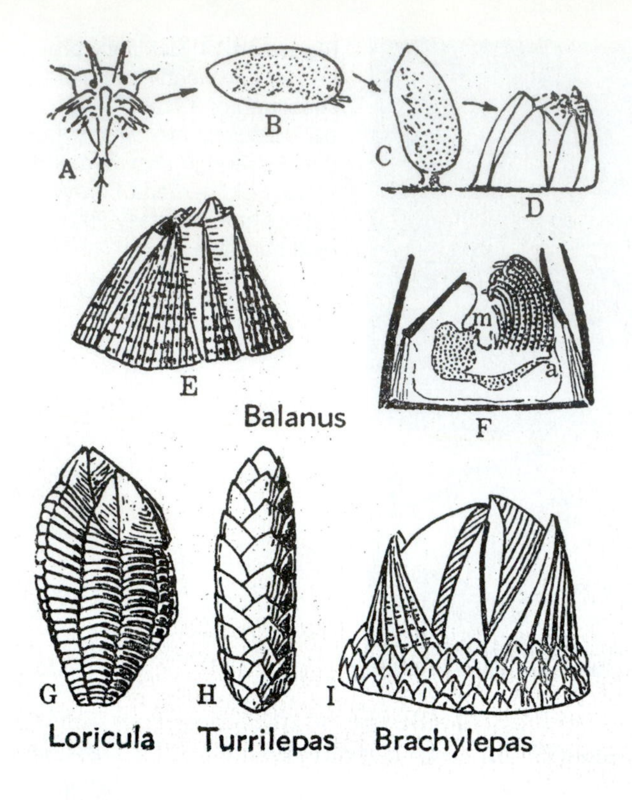

Figure 14.15. (A) Nauplius larva of *Balanus* changes into an ostracode-like cypris larva (B) before attaching by the head (D) and metamorphosing into an adult barnacle (E). In cross section (F), the plates (solid black) are moved by muscles to allow the filter-feeding legs to beat (m, mouth; a, anus). (G) The Cretaceous stalked barnacle *Loricula*. (H) A pine-cone-shaped Silurian form *Turrilepas*. (I) The Cretaceous barnacle *Brachylepas*. (From Moore et al., 1953). (J) These giant barnacles from the Miocene Pancho Rico Formation of central California were almost a foot long and formed dense "reefs." (Photo courtesy K. Whittlesey.)

Of the decapods, the most diverse and successful group is the true crabs, infraorder Brachyura ("short tail" in Greek). They have modified the basic shrimp-like body by greatly expanding and widening the carapace, and reducing the abdomen to a few segments that curl under the back of the carapace (Fig. 14.14). This protects their gill chamber and allows crabs to burrow more effectively, giving them additional niches to exploit. With this wide, flat body, and their ability to scuttle sideways at a rapid rate and hide in crevices or burrows, crabs have become one of the most versatile groups of benthic marine scavengers and predators.

Class Maxillopoda

Subclass Cirripedia—Two groups of maxillopods have calcified skeletons that yield a good fossil record: the barnacles and the ostracodes. Most people are surprised to learn that barnacles are crustaceans, unrelated to colonial filter feeders such as corals. Because of their calcareous shell, barnacles were usually misclassified as molluscs until the 1830s, when Thompson and Burmeister identified the nauplius larvae of barnacles and realized they were crustaceans. When Charles Darwin returned from his *Beagle* voyage in 1836, he found that virtually nothing was known about the barnacles he had collected, so he set about dissecting them and trying to understand them. His children were so used to his daily routine of dissection and description that once when they visited another house, they asked, "Where does he do his barnacles?" Darwin spent over a decade studying barnacles, and not only verified their crustacean affinities, but developed the foundation for our modern understanding of the group. His two-volume barnacle monograph, published in 1851 and 1854, may have postponed his work on *The Origin of Species*, but it also established Darwin's scientific reputation so that his other work would be taken more seriously.

There are about 1000 living species of barnacles. Most are free-living individuals that cluster in large assemblages wherever there is a hard surface on which to attach (including not only wharf pilings, rocks, ship bottoms, but even the skins of whales). One order burrows into the calcareous shells of molluscs and corals. Two other orders are parasitic on corals, echinoderms, and other crustaceans, completely invading the internal body cavity of the host. However, all barnacles begin as a free-swimming nauplius larva once they hatch from eggs (Fig. 14.15A-D). After two or three molts, the nauplius changes to a more shrimp-like animal, with a bivalved shell over its back. Eventually, it finds a hard surface to attach to, and adheres to it with its head, casting off its bivalved shell. Its body undergoes a profound metamorphosis, so that most of the shrimp-like features are unrecognizable, and the upward-turned fringing limbs of the thorax are modified into long, curling filtering devices that rhythmically sweep through the water trapping food. The subclass name, Cirripedia (Greek, *cirri*, "curl," and *ped*, "foot"), refers to this characteristic.

Because they secrete calcareous plates, adult barnacles can be common fossils. They are first known from the Ordovician, and a variety of body forms, including tall, pine-cone-shaped barnacles, are typical of the Paleozoic (Fig. 14.15G-H). The more familiar acorn barnacles and gooseneck barnacles are both known from the Mesozoic, and are common in many Cenozoic localties as well. In some places, they produce large mounds of barnacles

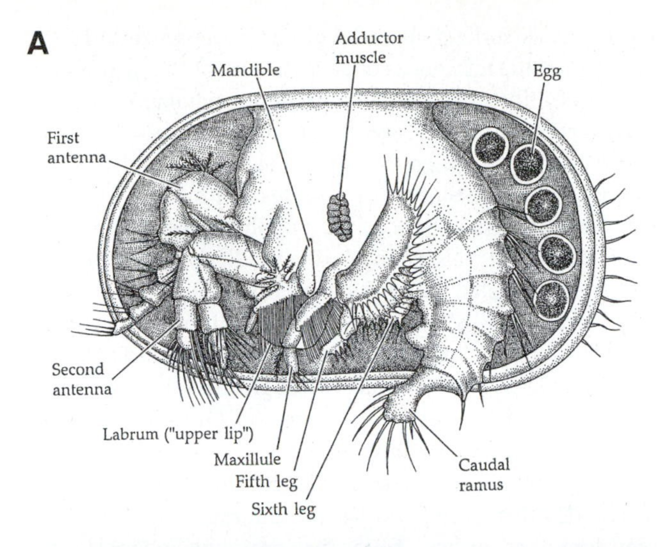

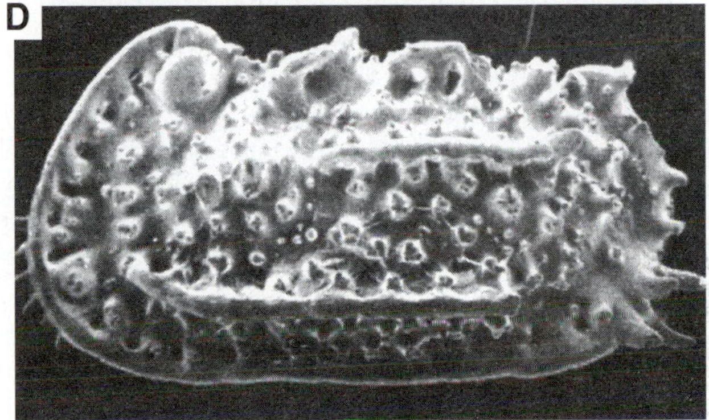

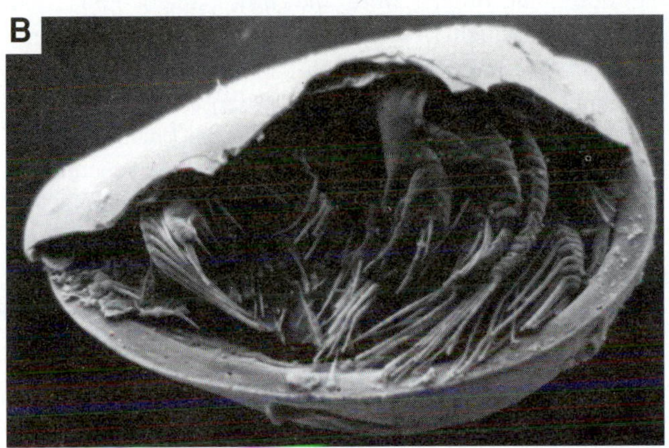

Figure 14.16. (A) Anatomical features of the ostracode *Cytherella,* with the left valve removed. (From Brusca and Brusca, 1990.) (B) This Late Cambrian ostracode *Vestrigothia* is preserved with its appendages intact. (C) Ordovician leperditids were shaped like kidney beans, and reached 3 cm in length. (D) This Pliocene planktonic ostracode *Calinocythereis* was very tiny (magnified 25x). (From Dott and Prothero, 1994.)

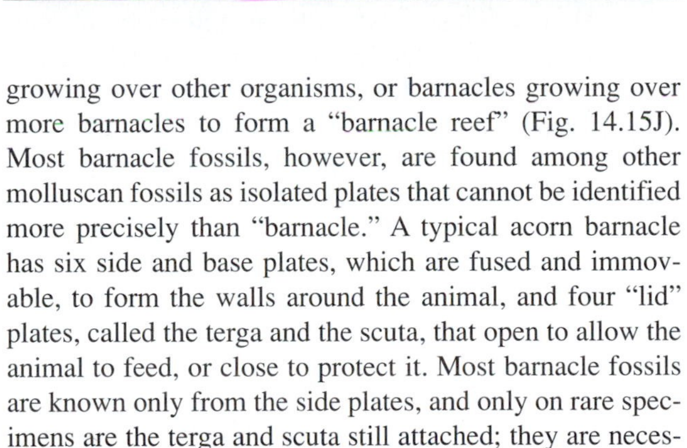

growing over other organisms, or barnacles growing over more barnacles to form a "barnacle reef" (Fig. 14.15J). Most barnacle fossils, however, are found among other molluscan fossils as isolated plates that cannot be identified more precisely than "barnacle." A typical acorn barnacle has six side and base plates, which are fused and immovable, to form the walls around the animal, and four "lid" plates, called the terga and the scuta, that open to allow the animal to feed, or close to protect it. Most barnacle fossils are known only from the side plates, and only on rare specimens are the terga and scuta still attached; they are necessary for precise identification.

Subclass Ostracoda—By far the most commonly fossilized arthropods are the ostracodes, which are crustaceans that secrete a pair of kidney-bean-shaped calcareous shells hinged over their back (Fig. 14.16A). Most are microscopic forms less that 2 mm in length, so they are primarily the domain of micropaleontologists who specialize in ostracodes, rather than arthropod paleobiologists who work with megascopic trilobites and other arthropods. However, a few living species are 20 to 30 mm in length, and some Paleozoic ostracodes were up to 80 mm in length. Most living ostracodes are benthic, and clearly, these large forms were bottom-dwellers. Some ostracodes are small enough to live in the plankton.

For such a small aquatic organism, ostracodes are ecologically very versatile. Most are filter feeders living on tiny diatoms, algae, protistans, and detrital material. They live in fresh, brackish, saline, and even hypersaline waters, and some even live on land in damp moss or supratidal sandy regions. Others are parasitic, or live in deep caves. In the marine environment, they range from the intertidal zone to the abyssal depths (down to 7000 m), and live both on the surface and burrowed beneath it, or cling to the stems of plants. Although they cannot fly, freshwater ostracodes can disperse through the air when their tiny eggs are carried in the mud stuck to the feet of birds, or blown from pond to pond. In many lakes, ostracodes are extremely abundant, and can accumulate in such large quantities that they are the major component of freshwater limestones. Kaesler (1987, p. 241) called them "the accessory minerals of the biosphere" since they are present in the background of nearly every marine setting. A small sample of almost any Phanerozoic biogenic sediment yields ostracodes in abundance.

Because they are so tiny, rapidly evolving, and abundant, ostracodes are studied like other microfossils. Their primary use is in biostratigraphy, especially when there are no foraminifera for this purpose. Ostracodes survive in core samples drilled for oil, and they are common in the deep-sea cores recovered by oceanographic vessels. They are also used in paleoecology and marine biogeography. Because benthic ostracodes are not very mobile, they show strong endemism into biogeographic provinces, and they are also restricted into depth assemblages. Many are associated with specific oceanographic conditions, so they are also valuable tools of paleoceanography. Over 2000 species of ostracodes are alive today, so their biology can be studied in detail. However, many thousands more are known from their fossils, so they have the most complete fossil record of any arthropod group.

Living almost entirely enclosed within their shells, ostracodes are very simple compared to other crustaceans (Fig. 14.16A). They have reduced the number of segments and appendages, and fused the head to the thorax. Only seven pair of appendages remain, five of which are attached to the head. The first pair of antennae (antennules) can also be very long and are used for swimming in planktonic forms. In benthic forms, they sweep the path in front of the ostracode and help maintain balance. The second pair of antennae are also long, and are the principal appendages for locomotion (swimming or walking). The third and fourth pairs of appendages (the mandible and maxilla) are used for feeding. The fifth pair of cephalic appendages has a variety of different functions: most benthic taxa use them for walking while planktonic taxa use them for feeding. The fifth appendages may show strong sexual dimorphism if the males use them for grasping females. The first and second pair of thoracic legs may also be used for walking, or for cleaning other body parts. Ostracodes have no abdominal appendages, and some groups even lack an abdomen altogether.

The rest of the soft anatomy is relatively simplified, probably because of the tiny body size and enclosure in a shell. Ostracodes have a simple straight digestive tract that begins at the mouth and labrum ("upper lip") and exits at the anus in the posterior end of the body. The gills are branched off the appendages, and there is a simple nerve cord along the back, and a simple circulatory system. Planktonic species tend to have large compound eyes, while benthic species have smaller tripartite eyes paired on the side of the cephalothorax, and both may have a median eye in front of the forehead. Bands of adductor muscles, which close the carapace, attach in different ways to the inside, and these muscle scars can be an important taxonomic feature for identification.

These soft parts are seldom preserved in fossils, except in unusual circumstances (Fig. 14.16B). Most ostracodes are known only from their calcified carpaces, so the anatomical features of the shell are critical for identification. A tremendous variety of external ornamentation patterns are known, and these are described in great detail by ostracode taxonomists. The details of the hinge region, the pattern of adductor muscle scars, and the other ridges, grooves, and features of the internal part of the shell are also very useful. As with our treatment of other microfossil groups, however, we will not discuss these details here. If the reader is serious about learning to recognize different ostracodes, the references cited at the end of this section are very useful for ostracode identification and anatomy.

Although ostracodes are widely used in biostratigraphy and paleoecology, there has been much less emphasis and interest in their evolutionary patterns and history. The oldest known ostracodes are members of the order Archaeocopida, which appeared in the Early Cambrian and may have persisted until the Early Ordovician. These ostracodes are typically 0.5 cm in length or more, larger than most later taxa. Several different groups then radiated in the Ordovician and persisted through the Paleozoic. The leperditiacopids, such as the common Ordovician genus *Leperditia* (Fig. 14.16C), were giants (in ostracode terms), reaching several centimeters in length; they are the largest ostracodes that ever lived. In the Silurian and Devonian, the dominant orders were the tiny palaeocopids and podocopids, which rarely exceed 2 mm in length. These trends in size reduction continue to the present (although the living myodocopids are almost as large as the leperditiacopids). There is also a trend through the Phanerozoic in reduction of the muscle-scar pattern, with fewer secondary scars (although there are many reversals and exceptions to this trend). Hinges have tended to become shorter and more curved through time, especially compared to the Paleozoic groups. There are also trends in the changes of surface ornamentation and especially the rib pattern on the shell that can be homologized between groups.

Patterns of evolution in ostracodes have only recently begun to be studied. Many specialists have noticed that ostracode populations tend to be highly variable, with many instances of polymorphism that are poorly understood. Benson (1975) demonstrated that most ostracodes show tremendous morphological stasis in the fossil record. Reyment (1982) examined a lineage of Cretaceous ostracodes of the genus *Oertliella*, and discovered both long-term stability and a high degree of variability and dimorophism just before the descendant species arises. Reyment (1985) conducted a detailed multivariate study of the Eocene ostracode *Echinocythereis*. He found that two speciation events occurred after a long period of stasis in ornament and in shape and size. The lineage underwent a rapid decrease in size, acccompanied by ornamental changes, in about 30,000 years. The transition to the next species in the lineage was more gradual and marked mostly by subtle changes in the ornamentation.

For Further Reading

Bate, R. H., E. Robinson, and L. M. Sheppard, eds. 1982. *Fossil and Recent Ostracods*. British Micropalaeontological Series.

Benson, R. H. 1975. Morphological stability in the Ostracoda. *Bulletins of American Paleontology* 65:13-46.

Brusca, R. C., and G.J . Brusca. 1990. *Invertebrates*. Sinauer Associates, Inc., Sunderland, Mass.

Darwin, C. R. 1851-1854. *A Monograph on the Subclass*

Cirripedia, vols. 1-2. Ray Society, London.

Feldmann, R. M. 1990. Decapod crustacean paleobiogeography: resolving the problem of small sample size, pp. 303-315, *in* Mickulic, D. G., ed. *Arthropod Paleobiology: Short Courses in Paleontology* 3. Paleontological Society and University of Tennessee Press.

Hartmann, G., ed. 1976. Evolution of post-Paleozoic Ostracoda. *Abhandlungen und Verhandlungen des Naturwissenschaften Vereins* 18/19:7-336.

Kaesler, R. L. 1987. Superclass Crustacea, pp. 241-264, *in* Boardman, R. S., A. H. Cheetham, and A. J. Rowell, eds. *Fossil Invertebrates*. Blackwell Scientific Publishers, Cambridge, Mass.

Loffler, H., and D. Danielopol, eds. 1977. *Aspect of Ecology and Zoogeography of Recent and Fossil Ostracoda*. W. Junk, Publishers, The Hague.

McLaughlin, P. 1980. *Comparative Morphology of Recent Crustacea*. Freeman, San Francisco.

Moore, R. C. and C. W. Pitrat, eds. 1961. *Treatise on Invertebrate Paleontology. Part Q. Arthropoda 3.* Crustacea. Geological Society of America and University of Kansas Press, Lawrence, Kansas.

Pokorny, V. 1978. Ostracodes, pp. 109-149, *in* Haq, B.U., and A. Boersma, eds. *Introduction to Marine Micropaleontology*, Elsevier, New York.

Reyment, R. A. 1982. Analysis of trans-specific evolution in Cretaceous ostracods. *Paleobiology* 8:293-306.

Reyment, R. A. 1985. Phenotypic evolution in a lineage of the Eocene ostracod *Echinocythereis*. *Paleobiology* 11:174-194.

Schmitt, W. L. 1965. *Crustaceans*. University of Michigan Press, Ann Arbor.

Schram, F. R. 1986. *Crustacea*. Oxford University Press, Oxford.

Schram, F. R. 1990. Crustacean phylogeny, pp. 285-302, *in* Mickulic, D. G., ed. *Arthropod Paleobiology: Short Courses in Paleontology* 3. Paleontological Society and University of Tennessee Press.

TRACHEATA (= UNIRAMIA)

The most spectacular evolutionary radiation among the uniramians has, of course, been within the insects, which inhabit nearly every conceivable terrestrial and freshwater habitat and, less commonly, even the sea's surface and littoral region. Insects are also found in many unlikely places such as oil swamps and seeps, sulfur springs, glacial streams, and brine ponds. They often live where few other metazoa can exist. Close to a million species of insects have been described, and authorities estimate that 20-50 million species remain undescribed. The beetle family Curculionidae (the weevils) contain about 65,000 described species and is itself larger than any other animal phylum except the Mollusca.

Insects pollinate most of the flowering plants and food crops, and provide us with luxuries such as honey, beeswax, and silk. Honeybees alone pollinate $20 billion worth of crops in the United States and produce $200 million worth of honey annually. Insects are key items in the diets of many terrestrial animals and play a major role as reducer-level organisms. However, they also consume about one-third of our potential annual harvest and transmit major diseases such as malaria and yellow fever. Every year billions of dollars are spent on insect control of various sorts. If we apportioned pages to animal groups based on the number of species, overall abundance, or economic importance, insect chapters could easily fill 90% of this text.

Richard and Gary Brusca, *Invertebrates*, 1990

The last major group of arthropods we will consider is the tracheates, composed of the myriapods (millipedes, centipedes, and several other groups) plus insects. Most tracheates have unbranched appendages (as the old name "uniramians" suggests) and a single pair of antennae. The body is divided into a distinct cephalon, with 4 to 5 pairs of appendages in front of the mouth (antennae, mandibles and two pairs of maxillae).

Class Myriapoda

Myriapods include the multilegged millipedes, centipedes, and their relatives. The name "Myriapoda" means "ten-thousand legs" in Greek, which is a poetic exaggeration, since millipedes ("thousand legs" in Greek) actually have about 100 to 200 pairs of legs, and centipedes ("hundred legs" in Greek) have at most a few dozen pairs. Living millipedes (subclass Diplopoda) are somewhat caterpillar-like, with as many as a hundred segments bearing two pairs of short legs. They have a long, tubular body protected by a calcified cuticle with a waxy covering, and coil up into a tight spiral for protection. Millipedes are mostly scavengers and detritus feeders, living in rotting vegetation; none have the ability to bite or kill prey. Possible burrows in soil horizons from the Upper Ordovician suggest that millipedes may have been the first animals to move onto the land, although there are no body fossils to support this (Retallack and Feakes, 1987). The oldest millipede body fossils are known from the Late Silurian of Scotland, and there are also specimens from the Devonian and Mississippian. Millipedes were extremely abundant in the Pennsylvanian coal swamp deposits, and some reached record lengths of 20 cm (8 inches). They have changed very little since the late Paleozoic, and over 10,000 species are alive today. Another extinct group, the Arthroplerida, were herbivorous and truly gigantic by millipede standards, reaching over 1 m in length.

By contrast, centipedes (subclass Chilopoda) have fewer segments, and each segment bears only one pair of legs. The body is much more flattened, the legs are much longer, and centipedes are very fast compared to millipedes or most other animals their size. Centipedes are also active predators, with their first head appendage modified into poison claws and fangs, and they can bite their prey and paralyze it instantly (which is why people must be careful of them—some have bites that are lethal even to humans). The oldest fossil centipedes are known from the Early Devonian, and their fossils are particularly common in the Pennsylvanian coal swamps. Today, there are about 2500 living species.

Class Insecta (= Hexapoda)

By most criteria, insects are the most important animals on the planet and should be studied in the greatest detail (as suggested by the opening quotation). However, insects have received much less study by paleontologists because they rarely fossilize compared to most other arthropods, or even compared to hard-shelled groups like molluscs. At the generic and specific level, their record is relatively poor: only 6000 fossil genera are known, less than 1% of the hundreds of thousands of living genera, and the millions of genera that must have existed through the Phanerozoic. However, at coarser levels of resolution, the insect fossil record is much better. Labandeira and Sepkoski (1993) point out that 1263 insect families are known from the fossil record, which is about 63% of all the living families. All the 30 or so recognized orders of insects have a fossil record. For comparison, this is more complete than the 825 families of four-legged vertebrates, many of which have a very poor or no fossil record.

The reason for this discrepancy is the nature of preservation of the insect cuticle. Most fossil insects are known from pieces of amber (particularly the Oligocene Baltic amber, and certain Cretaceous ambers), or from *Lagerstätten* with very fine-grained lake deposit preservation, such as the Oligocene Florissant lake beds of Colorado, the Pennsylvanian Mazon Creek nodules of Illinois, and lake bed faunas from the Permian (such as those from Elmo, Kansas, and Chekarda, Russia). With such a discontinuous record, it is possible to see through rare "windows" on insect diversity that show most of the families that were alive at a given time, but the long gaps in between these excellent localities limit the number of genera and species that are fossilized. There are also periods of apparently low insect diversity, such as the Mississippian and the Triassic, that may simply be artifacts of the absence of good terrestrial deposits of that age. Recently, the tremendous fossil record of insects from Russian deposits has become better known to Western scientists, filling in many of the gaps.

Even with a less than perfect fossil record, it is possible to reconstruct the phylogeny of insects based on the great diversity of living taxa (Fig. 14.17). The most primitive insects are the wingless forms, such as the collembolans (springtails), silverfish, and bristletails. As expected, the oldest known insect fossils are springtails from the Middle Devonian Rhynie Chert of Scotland. The next major step in insect evolution was the development of wings (subclass Pterygota), although primitive pterygotes cannot fold their wings against their bodies, but hold them out (as in dragonflies, damselflies, and mayflies). Winged insects first appear during the Early Pennsylvanian in great diversity, with both dragonflies (one of them, *Meganeura*, had a 70-cm-wide wingspan, and was the largest insect known) and mayflies well represented, as well as six other extinct orders. Apparently, insect wings evolved in the Mississippian, but so far there are no insect localities of Devonian or Mississippian age (primarily because of the widespread high sea levels and marine conditions that restricted the number of terrestrial deposits) to document how wings first developed.

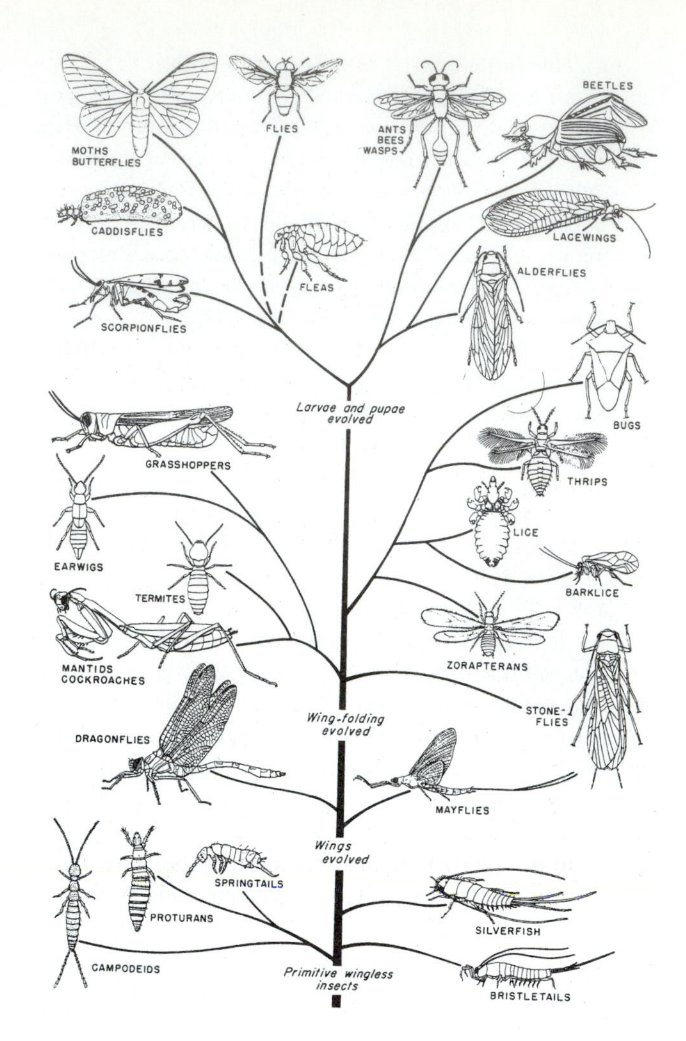

Figure 14.17. Diversification of the insects.

The abundant Pennsylvanian coal swamp deposits and Permian lake deposits record the first appearance of most of the remaining insect orders, suggesting that an enormous adaptive radiation had taken place early in insect evolution. Most of the Neoptera (winged insects that can fold their wings against the body) appeared at this point, or possibly earlier in the Mississippian. The Pennsylvanian coal swamps were dominated not only by huge dragonflies, but also by the first blattodeans (cockroaches up to 10 cm in length), and by orthopterans (grasshoppers, crickets, katydids, and their relatives), as well as extinct groups such as the Palaeodictyoptera, which had piercing, sucking mouthparts for consuming sap and spores. In the Permian, the remaining complement of Neoptera appeared, including the Plecoptera (stoneflies), Psocoptera (booklice), Hemiptera (true bugs) and Homoptera (aphids, cicadas, leafhoppers). All these groups have growth stages (instars or nymphs) that resemble miniature adults. By contrast, the most advanced insects are holometabolous; that is, they undergo a complete metamorphosis from a larva (typically a worm-like form, such as a caterpillar, grub, or maggot) that then pupates and eventually emerges as an adult of completely different body form (Fig. 14.17). Holometabolous groups such as the Coleoptera (beetles), Neuroptera (lacewings and ant lions), Mecoptera (scorpionflies), Diptera (flies), and Trichoptera (caddis flies) all first appeared in the Late Pennsylvanian or Permian, and most of the rest of the insect orders appeared in the Mesozoic. Thus, the great radiation of insects was already underway by the Pennsylvanian and Permian, and continued to accelerate through the Mesozoic and Cenozoic.

The final stage of insect evolution occurred in the Jurassic and Cretaceous, when two major groups, the Lepidoptera (butterflies and moths) and Hymenoptera (ants, bees, wasps), as well as the Isoptera (termites), Mantodea (mantises), and Dermaptera (earwigs) finally appeared. Conventionally, the great radiation of butterflies, moths, and bees, which are important plant pollinators, was thought to be a coevolutionary response to the appearance of flowering plants. However, Labandeira and Sepkoski (1993) point out that most of these groups (as well as many others with the appropriate mouthparts for pollination) had appeared long before flowers, and there is no evidence that their evolution was accelerated by the radiation of flowering plants in the Early Cretaceous.

Only one orders, the Siphonaptera (fleas) first appeared in the Tertiary. This is not surprising, since fleas are parasites of mammals, and would be expected to radiate along with mammals in the early Cenozoic.

For Further Reading

Almond, J. E. 1985. The Silurian-Devonian fossil record of the Myriapoda. *Philosophical Transactions of the Royal Society of London* (B) 309:227-237.

Beall, B. S., and C. C. Labandeira. 1990. Macroevolutionary patterns of the Chelicerata and Tracheata, pp. 257-284, *in* Mickulic, D.G., ed. *Arthropod Paleobiology: Short Courses in Paleontology* 3. Paleontological Society and University of Tennessee Press.

Callahan, P. S. 1972. *The Evolution of Insects*. Holiday House, New York.

Carpenter, F.M. 1992. Hexapoda. *Treatise on Invertebrate Paleontology. Part R. Arthropoda 3-4*. . Geological Society of America and University of Kansas Press, Lawrence.

Carpenter, F. M., and L. Burnham. 1985. The geological record of insects. *Annual Reviews of Earth and Planetary Sciences* 13:297-314.

Gray, J., and W. Shear. 1992. Early life on land. *American Scientist* 80:444-456.

Kaesler, R. L. 1987. Class Insecta, pp. 264-269, *in* Boardman, R.S., A.H. Cheetham, and A. J. Rowell, eds. *Fossil Invertebrates*. Blackwell Scientific Publishers, Cambridge, Mass.

Labandeira, C. C. 1994. A compendium of fossil insect families. *Milwaukee Public Museum Contributions in Biology and Geology* 88.

Labandeira, C. C., B. S. Beall, and F. M. Hueber. 1988. Early insect diversification: evidence from a Lower Devonian bristletail from Quebec. *Science* 242:913-916.

Labandeira, C. C., and J. J. Sepkoski, Jr. 1993. Insect diversity and the fossil record. *Science* 261: 310-315.

Retallack, G. J. and C. R. Feakes. 1987. Trace fossil evidence for Late Ordovician animals on land. *Science* 235:61-63.

Shear, W. A. 1990. Silurian-Devonian terrestrial arthropods. pp. 197-213, *in* Mickulic, D.G., ed. *Arthropod Paleobiology: Short Courses in Paleontology* 3. Paleontological Society and University of Tennessee Press.

Shear, W. A., and J. Kukalova-Peck. 1990. The ecology of Palaeozoic terrestrial arthropods: the fossil evidence. *Canadian Journal of Zoology* 68:1807-1834.

Shear, W. A., P. M. Bonamo, J. D. Grierson, W. D. Rolfe, E. L. Smith, and R. A. Norton. 1984. Early land animals in North America: evidence from Devonian age arthropods from Gilboa, New York. *Science* 224:492-494.

Wootton, R. J. 1981. Palaeozoic insects. *Annual Review of Entomology* 26:319-344.

Figure 15.1. This densely fossiliferous Cretaceous limestone is full of mollusc fossils, including the ammonite *Hoplites* (center), numerous high-spired gastropods (top and bottom), bivalves (top) and tusk-like scaphopods. (Photo by the author.)

Chapter 15

Kingdom of the Seashell

The Molluscs

If there were competitions among invertebrates for size, speed, and intelligence, most of the gold and silver medals would go to the squids and octopuses. But it is not these flashy prizewinners that make the phylum Mollusca the second largest of the animal kingdom, with more than 100,000 described species. That honor has been won for the phylum mostly by the slow and steady snails, with some help from the even slower clams and oysters. The name Mollusca means "soft-bodied," and the tender succulent flesh of molluscs, more than any other invertebrates, is widely enjoyed by humans. But many molluscs are better known for the hard shells that these slow-moving, vulnerable animals secrete as protection against potential predators. Ironically, it is for the beauty and value of these shells that many molluscs are most ardently hunted by humans, in some cases nearly to extinction.

Ralph and Mildred Buchsbaum, *Animals Without Backbones*, 1987

INTRODUCTION

When we think of marine invertebrates, we usually think of seashells, which are the product of the molluscs, including the clams, snails, and many other animals. Molluscs have long had an important role in human society. Many cultures have prized mollusc shells, using them for money, tools, containers, musical devices, fetishes, or decoration. Collecting and displaying seashells is still a big business today, turning many people into amateur malacologists (people who study molluscs or their shells). Humans eat a great variety of "shellfish" (mostly clams, oysters, and scallops), and many people consider abalone, conch, squid (*calamari*), octopus, and land snails (*escargot*) to be great delicacies. Molluscs have long been important in human diets, as shown by the many "shell middens" full of the molluscs left by prehistoric cultures. Today, many molluscs are commercially harvested in large numbers. The annual world squid and octopus fishery exceeds 2 million metric tons per year, and clams, oysters, and scallops are also big business.

Arthropods may be the most diverse and abundant phylum of animals overall, but molluscs are the most common group in the marine realm, with as many as 130,000 living species. They are also a diverse group, occupying an wide range of habitats despite the limitations of their body plan. Molluscs live in every marine environment, from the intertidal zone to the abyssal depths, and in the plankton, as well as in freshwater and terrestrial habitats. From a simple limpet-like ancestor crawling along with a flat cap-shaped shell, molluscs have evolved into many different shapes. The **gastropods** are common not only on the shallow marine seafloor, but also evolved into tiny planktonic pteropods. Land snails and slugs were the only marine invertebrate other than arthropods to invade the terrestrial realm. The **bivalves** have lost their heads, but became very successful with their hinged shells at living on or below the sea bottom, or swimming above it (scallops), and living in brackish and fresh water, too. The **cephalopods**, on the other hand, have well-developed heads with tentacles, and eyes as sophisticated as those of the vertebrates. They are active swimmers and predators, and the octopus is more intelligent than any other invertebrate. Some cephalopods are slow-moving shelled forms like the nautilus, yet some are as fast as the jet-propelled squids. The **polyplacophorans** (chitons) also creep along hard surfaces and graze algae, while the **scaphopods** ("tusk shells") are found in shallow marine habitats. Some molluscs can get very large. The giant squid reportedly reaches about 50 feet (18 m) in length, while the giant clam can be over 1 m long, and the marine snail *Campanile giganteum* has a shell over 1 m long.

For paleontologists, molluscs are even more important. Although not as numerous as arthropods in the living fauna, molluscs are much more likely to fossilize, because most have a hard, calcareous shell (Fig. 15.1). Thus, the fossil record of molluscs is proportionately better than any other living marine invertebrate phylum. Over 60,000 species of fossil molluscs have been described, giving them a representation second only to the foraminifera, and consequently, more invertebrate paleontologists study molluscs than any other phylum. There are more specialists in ammonites, or gastropods, or bivalves, than there are paleontologists specializing in most other groups of invertebrate fossils.

All the major classes of molluscs (except scaphopods) were established in the Cambrian, but they played a relatively minor role at first. In the Ordovician, large nautiloids became the earth's first major predators with preservable shells. With their long tentacles and parrot-like beaks, they could catch and eat anything that moved, and may have changed the ecology of the seafloor forever. In the later Paleozoic, the great radiation of predatory ammonoids was important not only for seafloor ecology, but also because they are one of the best biostratigraphic indicators of the Devonian through Permian. Bivalves and gastropods were also common, but played a subordinate role to the "Paleozoic fauna" that was dominated by brachiopods, bryozoans, crinoids, and tabulate and rugosid corals.

But the real dominance of the molluscs came after the Permian catastrophe wiped out the major groups of the Paleozoic. Ammonoids were nearly extinguished, but two genera survived, and produced a radiation of ammonites that swarmed in the Mesozoic seas. They were so abundant, rapidly evolving, and distinctive that Mesozoic time is marked by ammonites. Bivalves and gastropods both flourished in place of the brachiopods, possibly because these molluscs have anti-predatory adaptations like burrowing or swimming that brachiopods lacked. The Triassic seas were essentially modern in their dominance of bivalves and gastropods. With the extinction of the ammonites at the end of the Cretaceous, the Cenozoic seas became the realm of bivalves and gastropods (at least as far as the fossil record is concerned). When a beachcomber collects seashells today, nearly all are molluscs.

SYSTEMATICS

> *In the great class of molluscs, though we can homologize the parts of one species with those of another and distinct species, we can indicate but few serial homologies; that is, we are seldom enabled to say that one part or organ is homologous with another in the same individual. And we can understand this fact; for in molluscs, even in the lowest members of the class, we do not find nearly so much indefinite repetition of any one part, as we find in the other great classes of the animal and vegetable kingdoms.*
>
> Charles Darwin, 1859, *On the Origin of Species*

The Latin word *molluscus* means "soft," in allusion to their soft bodies. In ancient Rome, a *mollusca* was a kind of soft nut with a thin but hard shell. (Some German and American scientists, and Webster's dictionary, prefer to spell it "mollusk," but the adjective is still spelled "molluscan.") The concept of the Mollusca has gone through many centuries of changes since Aristotle first recognized soft-bodied animals with shells. Linnaeus formalized the term "Mollusca" for a wide spectrum of soft-bodied invertebrates, including not only cephalopods, slugs, and pteropods, but also tunicates, anemones, medusae, echinoderms, brachiopods, and polychaetes. He placed other groups, such as bivalves, shelled gastropods, chitons, nautiloids, barnacles, and serpulid worms in his "Testacea." The modern concept of Mollusca consisting of bivalves, gastropods, and cephalopods essentially dates back to Cuvier (1795). The barnacles were not purged from the molluscs until the 1830s, the tunicates in 1866, and the brachiopods finally at the end of the nineteenth century.

At lower levels, the classification of molluscs has been equally convoluted. The proper name for bivalves has long been controversial, with many authors preferring terms like "Pelecypoda" or "Lamellibranchiata." ("Pelecypoda" is used by many paleontologists, since "Bivalvia" originally included all bivalved animals, including brachiopods. However, it has not been accepted by the most biologists, even though they reject "Coelenterata" and accept "Cnidaria" for similar reasons.) At the generic and specific level, the classification is even more confused. Many species of gastropods and bivalves are burdened with several names (invalid synonyms), and some actually bear over a hundred synonyms. This is largely due to the popularity of shell collecting with amateur malacologists, who frequently name a new species based on minor differences in shell morphology, without looking at variation, or considering the soft anatomy (Brusca and Brusca, 1990, p. 697). To make matters worse, many of these invalid species are based on type specimens which are in private shell collections, making them unavailable to the scientific community, and therefore, hard to determine whether they are valid or not (although these species would not be valid under the Code of Zoological Nomenclature, since their type specimens are not easily available for study.)

Although molluscs have varied the basic body plan in many different ways, there are a number of features that characterize most members of the phylum. Most molluscs have a soft, fleshy, unsegmented body, which is elongate and bilaterally symmetrical, with a distinct head and tail end. The head concentrates the sensory structures, including eyes in many groups, or sensory tentacles in others. Inside the mouth is a ribbon-like string of tiny teeth, known as the **radula**, which is usually made of organic material (although chitons literally have an iron bite, with teeth made of magnetite). The lower part of the body is modified into a distinct foot, which can be used for crawling (as in snails and chitons), burrowing (clams and tusk shells), or as tentacles (cephalopods). Another distinctive feature is the fleshy covering of the body, known as the mantle, which secretes a calcareous shell in many taxa. The shell has a distinctive structure, with a blackish organic layer on the outside (the **periostracum**), and two or more layers of calcite prisms in the core of the shell. There may also be an inner **nacreous** layer of aragonite, or "mother of pearl," which is familiar from the iridescent luster seen inside abalone shells and, of course, pearls. Most molluscs have gills, which occupy a chamber within the mantle (**mantle cavity**), although these are highly modified in many groups.

Given how much we know about the evolution and soft

anatomy of molluscs, it is surprising how recently rigorous phylogenetic hypotheses about the interrelationships of the molluscs have begun to appear. At a Paleontological Society short course on molluscs, Gould (1985) criticized his colleagues for not proposing explicit phylogenetic hypotheses that could be tested, and that could serve as a framework for the many other types of research done on molluscs. "Phylogeny is not only the ultimate framework for relationships; it is also the testing ground upon which almost any other kind of hypothesis about pattern—structural, ecological, and so forth—must be mapped. In this context, my first major unhappiness about the current status of our knowledge arises from our failure to construct a phylogenetic consensus for mollusks either in the large (or relationships among classes, or even their definitions) or, for so many crucial cases, in the small (cladistics of families)." Although there is clearly a tremendous potential for molecular phylogenetic analysis in a group with so many living members, at the time relatively little had been done. Gould (1985, p. 262) put it this way, "I can make no stronger recommendation and express no fonder hope than this: let us join with our biochemical colleagues to resolve the basis of molluscan cladogeny. It can be done. And when we do it, morphology will finally be freed for its proper utility to biologists—not as an imperfect device for viewing cladogeny through a glass darkly, but as embodiments of structural and functional hypotheses about the evolution of forms tested against an independetly established cladogenetic framework."

Since 1980, however, there has been an explosion of research on molluscan phylogenetics. Not only have the traditional anatomical systems been studied in more detail than at any time since the nineteenth century, but many additional systems have been examined, often at the ultrastructural level with an electron microscope. Some of these systems include the the osphradia (taste organs), gills, excretory systems, and sperm. In addition, a wide variety of molecular techniques have now been applied, including gel electrophoresis, cellular DNA content, chromosome numbers, shell matrix proteins, and nucleic acid sequences (see Bieler, 1992, for a review). Although there is still much conflicting data and other problems to be resolved, the cladistic analyses (Götting, 1980; Salvini-Plawen, 1980; Lauterbach, 1983; Wingstrand, 1985; Brusca and Brusca, 1990; Peel, 1992) seem to produce remarkably consistency about the overall branching pattern of the higher molluscan groups. Following the emerging consensus on molluscan phylogeny, the major classes of molluscs are listed below (based on Pojeta et al., 1987).

Phylum Mollusca

As defined by most authors, molluscs are bilaterally symmetrical (but some are secondarily asymmetrical) coelomate animals covered by a thick fleshy mantle that may secrete a calcite shell. Most have a large muscular foot that may be modified for creeping (snails, chitons), digging (bivalves, scaphopods), or into tentacles (cephalopods).

Subphylum Amphineura

Class Caudofoveata

A small, worm-like, shell-less group that live upside-down in burrows on the seafloor; about 70 living species; no fossil record.

Class Aplacophora

Aplacophorans ("bearing no shell" in Greek) or solenogasters are worm-like, shell-less molluscs that live in and around cnidarians in water deeper than 200 m. Although they have no external shell, they have internal calcite spicules. About 250 living species are known, but they have almost no fossil record.

Class Polyplacophora

The chitons are familiar from the rocks of tidepools, where they cling with their broad muscular foot and scrape algae off the rocks with their radulae, which are made of magnetite (iron oxide). Unlike other molluscs, their shells are divided into an armor of eight (the name *Polyplacophora* means "bearers of many shells" in Greek) overlapping plates (Figs. 15.2B, 15.3) surrounded by a broad muscular band of mantle known as the girdle, and also by a ring of nerves surrounding the body. After a chiton dies, its plates tend to break up and disassociate, so their fossil record is meager, with only about 100 species known; about 600 species are alive at present.

Subphylum Cyrtosoma

Class Monoplacophora

Monoplacophora ("single shell bearers" in Greek) were long known only as early Paleozoic fossils until 1952, when a living species (*Neopilina galatheae*) was discovered in the deep waters off Costa Rica (Fig. 15.2A). The discovery of this living fossil was not only surprising, but in addition *Neopilina* showed a faint segmentation of paired gills, "kidneys," "hearts," gonads, and muscles, suggesting that molluscs arose from segmented worms, but then lost their segmentation. Most monoplacophorans have limpet-like, cap-shaped shells, but their segmented soft anatomy is much more primitive than any limpet. Although only 11 species in 3 genera survive today, they were an important group in the early experimental phase of molluscan evolution, with 39 genera in the Cambrian and 18 fossil genera from the Ordovician, Silurian, and Devonian (after which the group was thought to be extinct).

Class Gastropoda

By far the largest class of all the molluscs is the gastropods, consisting of about 40,000 to 100,000 living species of snails and slugs, and many thousands of fos-

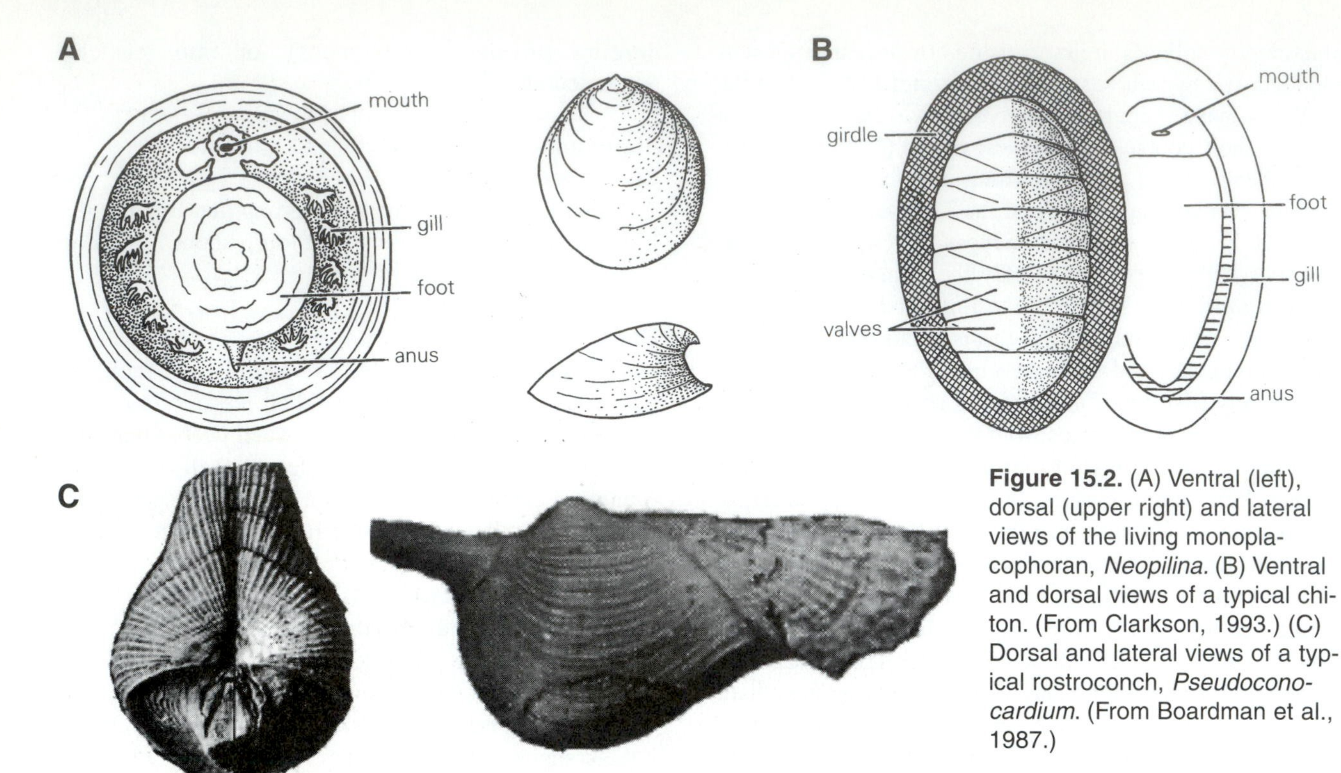

Figure 15.2. (A) Ventral (left), dorsal (upper right) and lateral views of the living monoplacophoran, *Neopilina.* (B) Ventral and dorsal views of a typical chiton. (From Clarkson, 1993.) (C) Dorsal and lateral views of a typical rostroconch, *Pseudoconocardium.* (From Boardman et al., 1987.)

sil species. Gastropods modify their undersides into a long foot (hence the name *gastro*, "stomach," and *podos*, "foot") along which they creep. Most have a head with eyes and tentacles, and their mantle usually secretes a shell. Three subclasses are recognized:

Subclass Prosobranchia ("forward gills" in Greek)—most shelled marine snails

Subclass Opisthobranchia ("backward gills" in Greek)—sea slugs, sea hares, nudibranchs

Subclass Pulmonata ("lung-bearing" in Greek)—land snails and slugs

Class Cephalopoda

The most active and intelligent molluscs are the cephalopods, which have modified their foot into a ring of tentacles around their mouth (hence the name *cephalos*, "head," and *podos*, "foot," in Greek). They have large, sophisticated eyes, a parrot-like beak, and a relatively large brain suited for their active, predatory mode of life. Their mantle cavity is modified for jet propulsion backwards, but some can also swim forward with fin-like extensions of their mantle. Primitively cephalopods had conical shells divided into chambers as they grew, although many groups (especially squids, octopuses, and belemnites) lost their external shell and had only an internal stiffening rod, or no shell at all. Today, only one shelled cephalopod (the chambered nautilus) and the shell-less squids and octopuses survive, comprising about 650 living species, but many thousands of extinct cephalopod species (especially ammonites) are known from the fossil record.

Subphylum Diasoma

Class Rostroconchia

The only extinct class of molluscs is the rostroconchs (Fig. 15.2C), known from about 35 genera from the Cambrian to Late Permian. They have a pseudobivalved shell with an anterior partition called the **pegma**, and some also have a posterior pegma. The pegma connects the two valves and serves as the attachment site for the pedal retractor muscles. Since they are enclosed by two symmetrical valves, rostroconchs look superficially like bivalves, but the lack of a ligament, bivalve-type adductor muscles, and many other features have led paleontologists to place them in their own class, thought to be ancestral to both bivalves and scaphopods.

Class Bivalvia (= Pelecypoda, = Lamellibranchia)

The second most common group of molluscs is the bivalves (clams, oysters, scallops, and their relatives), which have enclosed their bodies in a hinged calcareous shell. Consequently, they have lost their heads and most sense organs, and their foot is modified like an axe-like wedge for burrowing (*pelecypod* means "axe foot" in Greek). They have also expanded and modified their gills for both respiration and filter feeding. About 8000 to 15,000 living species are known, but many more are known from the fossil record (about 7000 from the Paleozoic, 15,000 from the Mesozoic, and 20,000 from the pre-Holocene Cenozoic).

Class Scaphopoda

The tusk shells (Fig. 15.3) live in a long, conical taper-

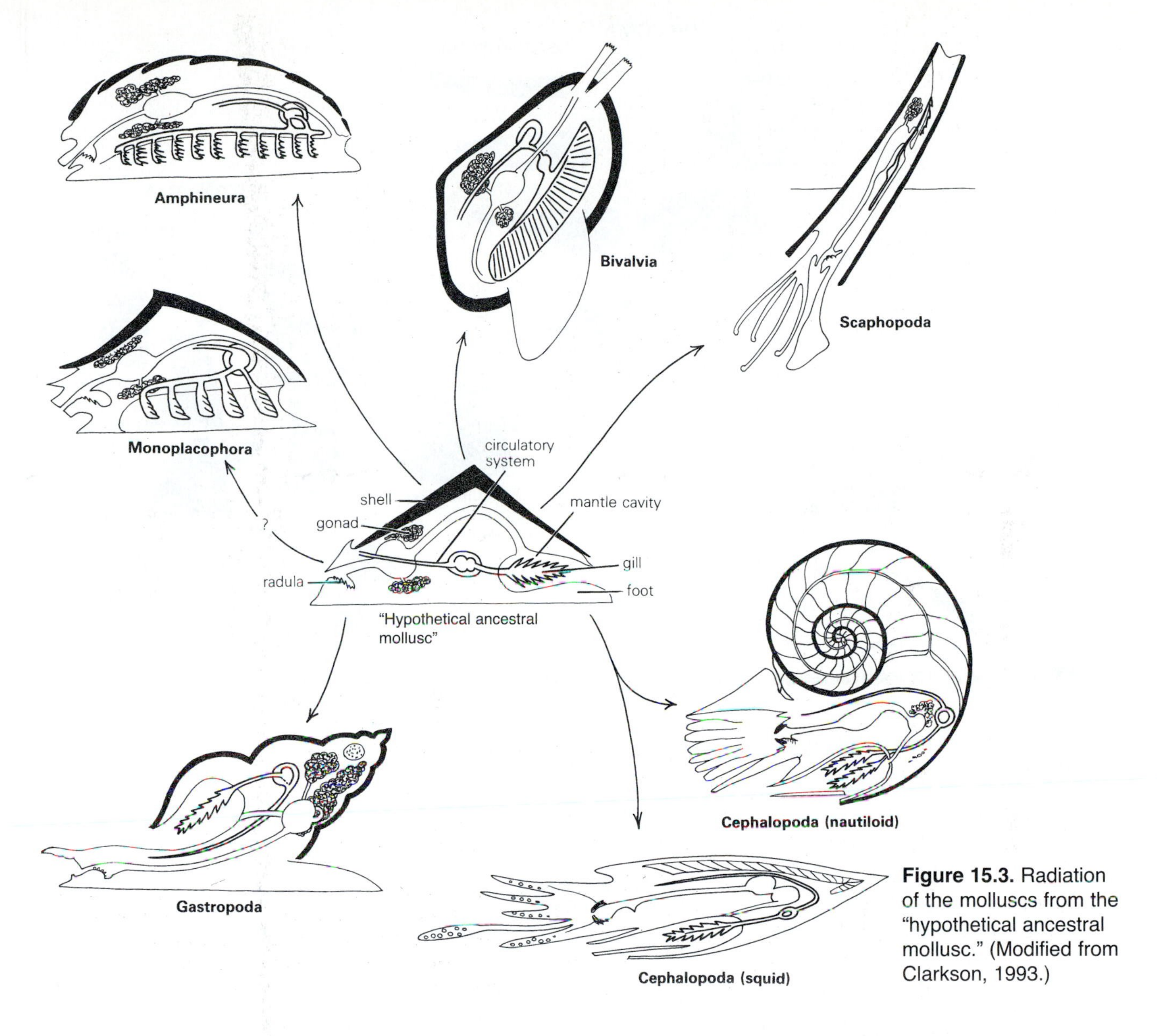

Figure 15.3. Radiation of the molluscs from the "hypothetical ancestral mollusc." (Modified from Clarkson, 1993.)

ing tubular shell half buried in the sand, where they use their foot to dig and their tentacles to catch prey in the sand. Although never very diverse or variable on this simple body form, they are known as early as the Ordovician, and are still common in some modern sandy beaches. About 350 species still survive. Most are only a few centimeters in length, but some Pennsylvanian specimens from Texas were almost 60 cm (2 feet) long.

MOLLUSC ORIGINS AND DIVERSIFICATION

Malacologists have taken a variety of approaches to understanding the origin of molluscs. Some zoologists have visualized molluscs as derived from flatworms with the shell added. However, other zoologists have argued that the segmented worms might be a better ancestral type, because the vestiges of segmentation in monoplacophorans suggests that molluscs had segmentation initially, and then secondarily lost their paired gills and muscles. The traditional model postulates a "**hypothetical ancestral mollusc**" (or HAM) or "hypothetical archimollusc" that has the primitive ancestral morphology capable of producing all the molluscan classes (Fig. 15.3). Most versions of the HAM have a simple, snail-like body covered by a cap-like limpet shell, although HAM differs from limpets (which are gastropods) in that its digestive and respiratory tracts run straight from front to rear, with the gills and anus in the back. As we shall see in the next section, gastropods have twisted these tracts around so their mantle cavity, gills, and anus all exit the front of the shell over the head.

From such a simple archetype, paleontologists and zoologists tried to make a chiton by segmenting the shell into eight plates, a gastropod by twisting the digestive and respiratory tracts around, a bivalve by dividing the shell into two and losing the head, and so on. However, as Pojeta (1980) pointed out, the HAM reconstruction leads to some problems. It requires that monoplacophorans are secondar-

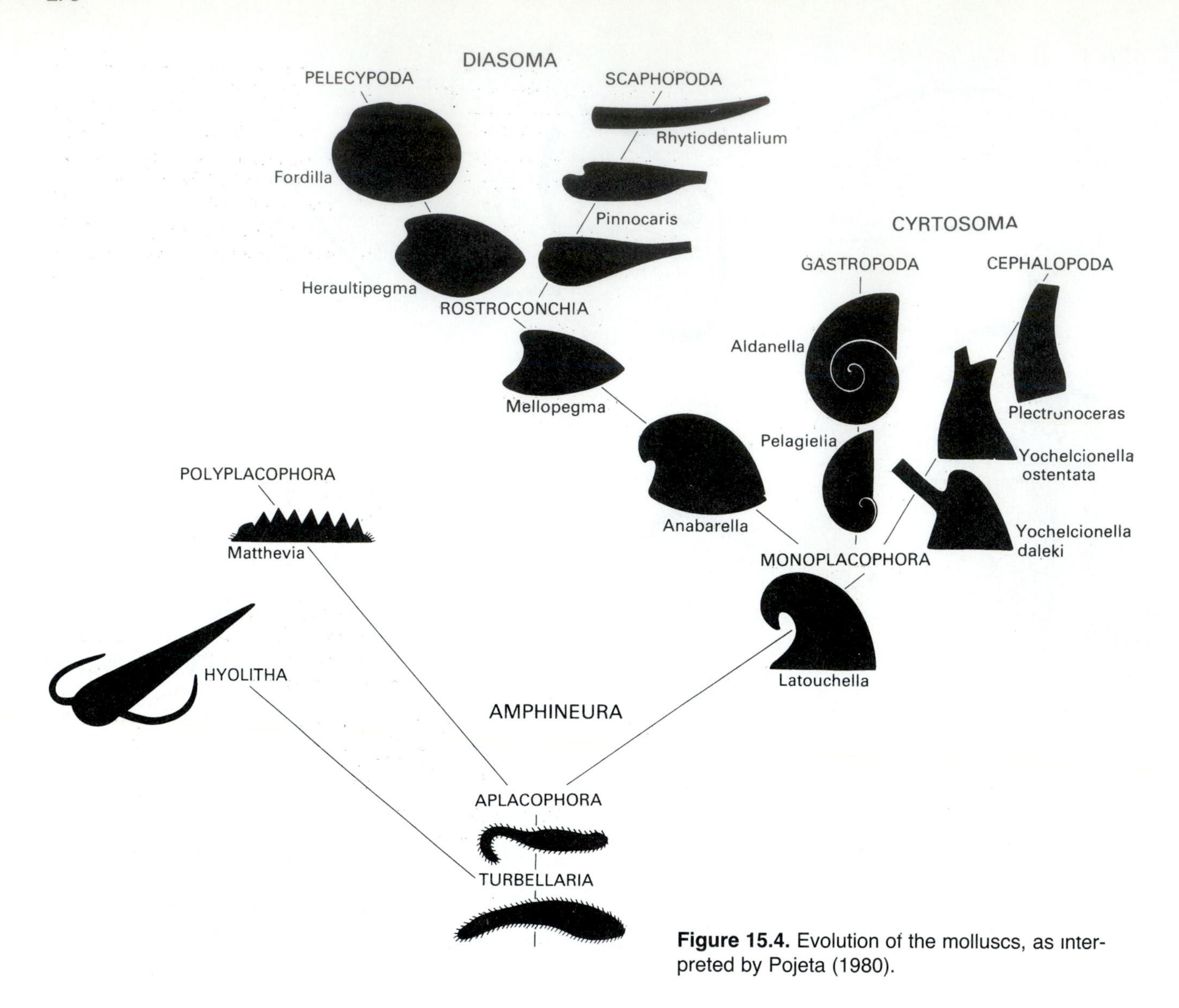

Figure 15.4. Evolution of the molluscs, as interpreted by Pojeta (1980).

ily segmented from an unsegmented ancestor, and that the aplacophorans are degenerate forms that have lost their shells and most of their other anatomical features. Morton (1967) wrote, "The danger is that in mixing genealogical ideas with morphology our archetype may become like an heraldic animal—a lowest common multiple of incompatible origins." The HAM concept may have been useful decades ago when little was known about molluscan origins, but with the discovery of monoplacophorans, aplacophorans, rostroconchs, and many other primitive living and fossil taxa, it is no longer a useful hypothesis. Moreoever, it has no place in cladistic analyses, which use only shared derived characters to determine relationships—the HAM is a composite of almost entirely primitive characters.

Instead of hypothetical mollusc ancestors, some paleontologists have approached the problem by trying to trace the origin of the classes from the simple cap-shaped fossils found in the Cambrian (Pojeta, 1980; Pojeta et al., 1987; Runnegar, 1983, 1985). These hypotheses visualize molluscs as arising from a shell-less aplacophoran ancestor, which then diverged into two branches: the polyplacophorans (chitons) and the group derived from the segmented monoplacophorans, the first group to form a univalved shell (Fig. 15.4). From the simple cap-shaped monoplacophoran shell, the gastropods can be derived by making their shell taller and aperture smaller, so that it develops into an asymmetric coiled form. If a tall, conical monoplacophoran shell is separated into partitions, however, this yields the earliest Cephalopoda. Since both groups are built on the basic body plan of an elongate cone that may be asymmetrically spiraled, the Gastropoda and Cephalopoda are united into a Subphylum Cyrtosoma (Greek, "curved body"). The other major variation on this bauplan is to vary the aperture to produce a laterally compressed symmetrical conical shape. This modification is seen in the rostroconchs, and from this plan paleontologists have derived the scaphopods and bivalves, which together are placed in the Subphylum Diasoma (Greek, "through body").

For Further Reading

Bieler, R. 1992. Gastropod phylogeny and systematics. *Annual Reviews of Ecology and Systematics* 23:311-338.

Bottjer, D. J., C. S Hickman, and P. D. Ward, eds. 1985. *Mollusks, Notes for a Short Course*. University of Tennessee Studies in Geology 13.

Ghiselin, M. T. 1988. The origin of molluscs in the light of molecular evidence. *Oxford Surveys in Evolutionary Biology* 5: 66-95.

Gould, S. J. 1985. Molluscan paleobiology—as we creep toward the millennium: a critique of papers presented at the 1985 Short Course, *in* Bottjer, D. J., C. S. Hickman, and P. D. Ward, eds. *Mollusks, Notes for a Short Course*. University of Tennessee Studies in Geology 13:258-267.

Morton, J.E. 1965. *Molluscs*. Hutchinson, London.

Pojeta, J., Jr. 1980. Molluscan phylogeny. *Tulane Studies in Geology and Paleontology* 16:55-80.

Pojeta, J., Jr., B. Runnegar, J.S. Peel, and M. Gordon, Jr. 1987. Phylum Mollusca, pp. 270-435, *in* Boardman, R.S., A.H. Cheetham, and A. J. Rowell, eds. *Fossil Invertebrates*. Blackwell Scientific Publishers, Cambridge, Mass.

Runnegar, B. 1983. Molluscan phylogeny revisited. *Memoirs of the Association of Australasian Paleontologists* 1:121-144.

Runnegar, B. 1985. Origin and early history of mollusks, *in* Bottjer, D. J., C. S. Hickman, and P. D. Ward, eds. *Mollusks, Notes for a Short Course*. University of Tennessee Studies in Geology 13:17-32.

Runnegar, B. 1996. Early evolution of the Mollusca: the fossil record, pp. 77-87, *in* Taylor, J., ed. *Origin and Evolutionary Radiation of the Mollusca*. Oxford University Press, New York.

Runnegar, B., and P. A. Jell. 1976. Australian Middle Cambrian molluscs and their bearing on early molluscan evolution. *Alcheringa* 1:109-138.

Runnegar, B., and J. Pojeta, Jr. 1974. Molluscan phylogeny: the paleontological viewpoint. *Science* 186:311-317.

Salvini-Plawen, L. V. 1991. Origin, phylogeny, and classification of the phylum Mollusca. *Iberus* 9: 1-33.

Taylor, J., ed. 1996. *Origin and Evolutionary Radiation of the Mollusca*. Oxford University Press, New York.

Trueman, E. R., and M. R. Clarke, eds., Evolution, *in* Wilbur, K. M., ed., *The Mollusca*, vol. 10. Academic Press, Orlando, Florida.

Yochelson, E. L. 1978. An alternative approach to the interpretation of the phylogeny of ancient mollusks. *Malacologia* 17:165-192.

Yonge, C.M., and T. E. Thompson. 1976. *Living Marine Molluscs*. William Collins Sons, London.

GASTROPODS

What is the main thing about a snail? That's right, the snail is slow. He believes in just taking it easy, and he is so slow at it that one gets all tired out just watching him. Following a snail around for any length of time makes me a total wreck. I don't know why I do it. When a snail wants to go anywhere, he travels on the underside of his physique, twitching himself along by wave-like contractions of the muscles. This is not a satisfactory means of locomotion, if you've every tried it, and carrying your house on your back at the same time would hardly improve matters. A snail never hurries to an appointment. He is sure that his date will be a day or two late, anyhow, so what's the use?

Will Cuppy, *How to Attract the Wombat*, 1935

The most diverse group of molluscs, both now and in the past, is the gastropods. There are between 40,000 and 100,000 species of gastropods, making up about 80% of the Mollusca. As the quote from Will Cuppy shows, most people think of snails and slugs as boring, slow-moving creatures, but even with their limitations, the gastropods have been remarkably successful. From the primitive cap-shaped limpets and abalones, gastropods have diversified into the widest variety of habitats of any group of molluscs, despite the fact that their creeping, snail-like body is least modified from the ancestral molluscan body plan. Marine **prosobranch** molluscs have developed in many directions, with a great variety of elaborately coiled shells that are adapted both for surface creeping and burrowing. Another marine group, the **opisthobranchs**, have greatly reduced their shells (the bubble shells) or lost them altogether (sea slugs, sea hares, nudibranchs, and a variety of parasitic shell-less molluscs). The pteropods (a group of opisthobranchs) are so tiny and have such a reduced shell that they live within the plankton. The **pulmonates** have modified their mantle cavity into a type of lung, allowing them to breathe air on land, and exploit a niche unavailable to marine gastropods.

Gastropods have also adapted to a wide variety of diets as well. Many gastropods, such as limpets and abalones cling to rocks and scrape off algae with their radulae, while others cling directly to larger plants on which they feed. Some marine gastropods are scavengers or detritus feeders. Others are active predators, drilling through the shells of their prey (as do whelks and moon snails) or even harpooning their prey with a poisoned barb (as does the cone snail). Worm-shells are uncoiled gastropods that cement themselves to a hard substrate and filter feed; the free-floating planktonic pteropods are also filter feeders. Land snails and slugs usually feed on plants, although many are scavengers, or feed on detritus and dead vegetation. For such sluggish organisms, gastropods are remarkably versatile, diverse, and successful in a wide variety of ecological niches.

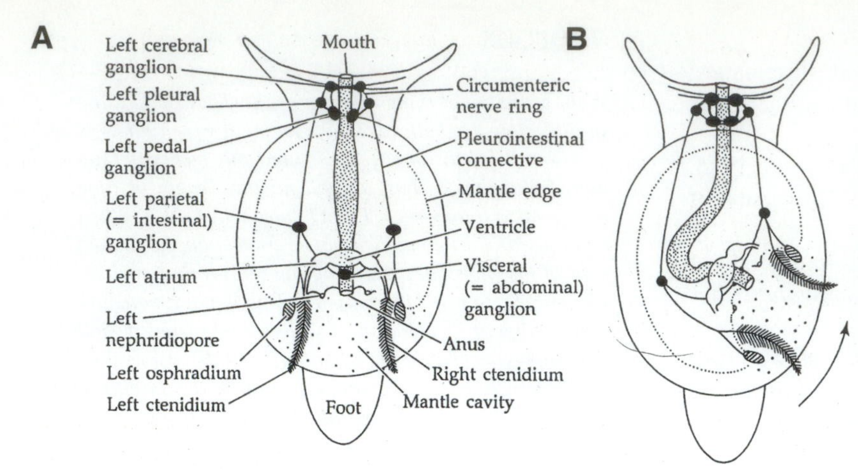

Figure 15.5. (A-D) Dorsal views of an adult gastropod. (A) Hypothetical untorted gastropod. (B-C) Stages of torsion as they might appear in an adult snail. (D) After torsion. The mantle cavity, gills, anus, and nephridiopores are moved from a posterior to anterior position, just above the head. Many structures that were on the right side end up on the left, and vice versa. (E) Lateral views of a veliger larva before and after torsion. After torsion, the head can be withdrawn into the mantle cavity. (F) The principal nerves and ganglia before and after torsion. Note that after torsion, the nerves are twisted into a "figure-8" loop. (From Brusca and Brusca, 1990.)

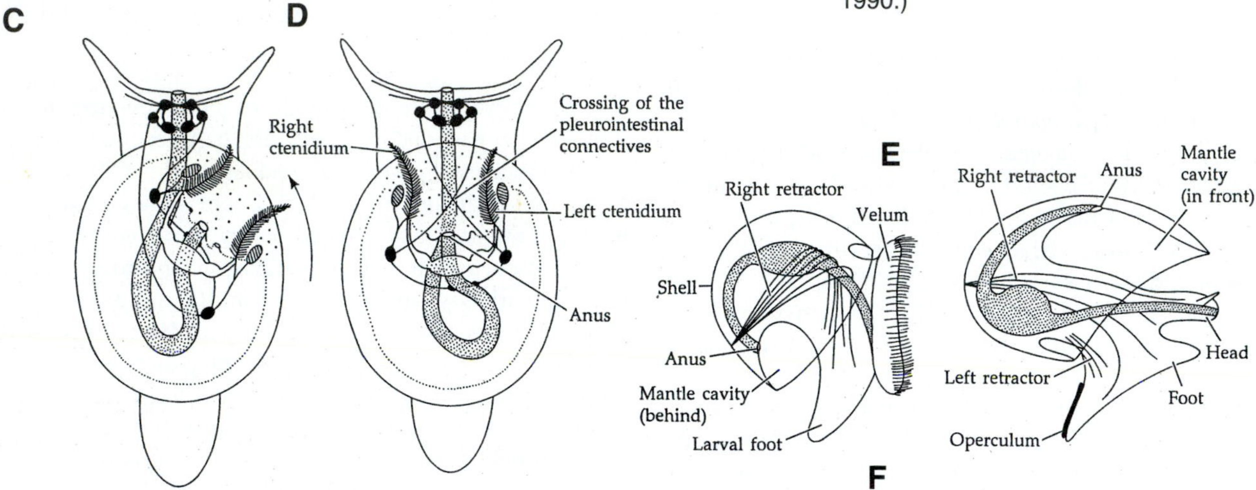

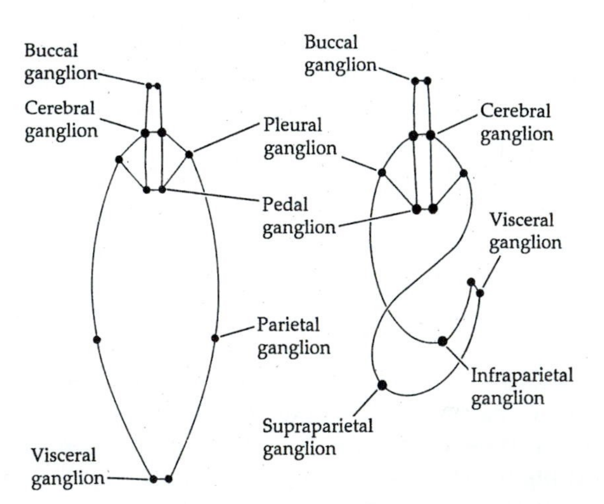

One of the most amazing transformations in the animal kingdom (comparable to the metamorphosis of a butterfly) occurs in the gastropods, and is known as **torsion**. Embryonic gastropods (**veligers**) begin their development with their anus and gills at the rear of the body, the condition that is thought to be primitive for molluscs. As they develop, the right larval retractor muscles retracts asymmetrically, and the entire visceral mass, containing the digestive, nervous, and respiratory tract, is forced to twist around into a U-bend (Fig. 15.5). By the time torsion is completed, the larval snail has its looping nerve cord twisted into a figure eight, its left and right gills reversed in position and located in the mantle cavity right over the head, and its anus discharging right over the head as well. Why would the simple snails go through such contortions? A number of ideas have been suggested over the years (reviewed by Lever, 1979, and Signor, 1985). They include:

1. *Larval retraction hypothesis*: Garstang (1929) suggested that torsion provided space for the larvae to retract the head and foot into the mantle cavity for protection. However, recent studies have shown that this provides no real protection against predators.

2. *Larval settling hypothesis*: Ghiselin (1966) suggested that torsion was advantageous to settling larvae by balancing the shell over the head. However, later research showed that post-torsional larvae of one species were unable to crawl for four days after torsion, so torsion does not convey an advantage to recently settled larvae.

3. *Veliger swimming hypothesis*: Underwood (1972) argued that torsion occurs to better balance the body mass in the swimming veliger larvae. So far, no one has tested this hypothesis with experiments or observations.

4. *The well-adapted adult*: Morton (1958) suggested that torsion is an adult specialization, designed to get the man-

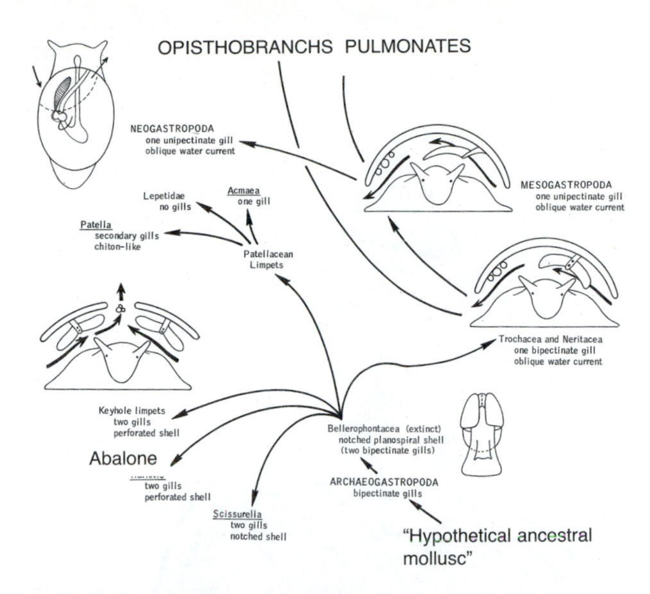

Figure 15.6. Evolution of water circulation and gas exchange in gastropod, showing the development of various asymmetrical arrangements of gills and apertures as water flow is modified. (From Barnes, 1974.)

tle cavity (with the gills, anus, and chemoreceptors) in the front of the animal, so the snail can easily ventilate its gills, wash away wastes, and taste the oncoming water. However, these advantages only apply to the more advanced gastropods; the primitive living gastropods, such as the limpets, do not have their gills in a mantle cavity in the front, yet torsion appeared early in gastropod evolution.

5. *The opercular imperative*: Stanley (1982) argued that torsion allowed the gastropod to retract into its mantle cavity for protection, and draw the lid on the back of its foot (known as the operculum) in to close the aperture. However, the most primitive gastropods do not have a narrow aperture or operculum, so this does not explain why torsion occurs in these gastropods as well.

6. *The helicospiral hypothesis*: Pojeta and Runnegar (1976) argued that torsion arises in response to the helicospiral coiling of the shell, which unbalances the animal. Torsion supposedly rotates the body mass to balance the mass of the spiral shell. Again, this hypothesis does not explain why torsion occurs in primitive limpet-like gastropods, which are symmetrical and balanced (unless one assumes that the most primitive gastropods were coiled, and limpets are secondarily flat and unspiraled. Some Early Cambrian specimens, such as *Aldanella*, do have this morphology).

In short, there is probably no simple, single answer to the puzzle of gastropod torsion. It probably does confer some of the advantages during embryology as suggested previously, because it is a major part of embryonic development. It may also be advantageous to gastropods that have narrow apertures, because they need to have their gills, anus, chemoreceptors, and mantle cavity oriented forward, both to get clean water through them, and maybe also to allow them to retreat into a narrow aperture for protection. In other words, each of the advantages of torsion suggested above may play at least some role in its selective advantage, but no single explanation is sufficient.

In addition to these puzzles is the fact that some gastropods secondarily lose their torsion. Detorted gastropods are most common among the unshelled opisthobranchs, where the lack of a restictive aperture may obviate the need to have the mantle cavity and gills in front. Indeed, some opisthobranchs, such as the colorful sea slugs known as nudibranchs, have their gills prominently displayed as trembling feather-like protuberances on the middle of their backs. They do not need to worry about protecting themselves with a shell, however, because most contain strong toxins that make them distasteful to predators, as their bright warning colors indicate.

Systematics

Gastropods have traditionally been divided into three major groups: the prosobranchs, which include nearly all shelled marine gastropods; the opisthobranchs, most of which have rotated their gill chambers to the back (secondarily detorted) and lost their shells; and the terrestrial and freshwater pulmonates, which have modified the mantle cavity into a air-breathing organ. The unique specializations of opisthobranchs and pulmonates clearly demonstrate that they are natural, monophyletic groups. However, the prosobranchs are by definition a "wastebasket" group that includes all the primitive gastropods that were ancestral to opisthobranchs and pulmonates, as well as many highly specialized marine gastropods.

In addition, the prosobranchs were traditionally subdivided into three groups: the primitive Archaeogastropoda, the more advanced Mesogastropoda (both paraphyletic ancestral groups), and the specialized Neogastropoda (a natural, monophyletic group). The Archaeogastropoda and Mesogastropoda are grades of evolution that are more primitive than the neogastropods, rather than natural groups defined by shared derived characters.

A good way to visualize the differences between the groups can be seen in the evolution of their shell form and configuration of gills and mantle cavity (Fig. 15.6). If we imagine an ancestral gastropod morphotype (not too different from HAM) with a simple cap-shaped shell, and untorted body with the gills and mantle cavity in the rear, then the next step would be the most primitive prosobranchs, the limpets, with their simple cap-shaped shell. Some, such as the keyhole limpets, volcano limpets, and abalones, have symmetrically paired gills, and draw water through the mantle cavity along the sides of their body and then out the holes in the the top of the shell. More specialized limpets (the patellaceans) no longer have a full pair of gills, but reduce them and have an asymmetric flow of water around the body, because they also lack the excurrent holes found in keyhole limpets and abalones.

A number of different molluscan types apparently were derived from the limpet-like body form (Fig. 15.6). The conical Trochacea (top shells) and Neritacea (nerites, or dogtooth periwinkles) are slightly more advanced prosobranchs that have a fully spiraled shell and a narrow aperture. However, they have only a single, double-branched (bipectinate) gill apparatus, and the water currents flow obliquely through their mantle cavity. From such an ancestor, some zoologists visualize the origin of opisthobranchs, pulmonates, and mesogastropods. Some opisthobranchs still have a single bipectinate gill and a shell, although the majority have greatly reduced both of these. Pulmonates are thought to have arisen by loss of the gill entirely and development of the mantle cavity into a kind of lung. The main line of marine gastropod evolution leads to the intermediate Mesogastropoda, which have a single gill with only one main branch (unipectinate), and an oblique flow of water through their mantle cavity. Finally, the highly derived Neogastropoda also have a single unipectinate gill, but they have modified the edge of their mantle into a long tube called a siphon, which allows them to pump water in and out of their mantle cavity while their body is retracted into the shell or while they are burrowing.

Recently, a number of cladistic analyses have been performed on the gastropods, although there is still no consensus on a classification of the major subgroups (see Bieler, 1992; Ponder and Lindberg, 1996). Almost all the analyses agree that the limpets are the most primitive of the living gastropods. Ponder and Lindberg (1996) found that the keyhole limpets, top shells, and slit shells clustered to form a monophyletic group (the "Vetigastropoda"), and almost all the other marine prosobranchs (except the nerites) clustered into another monophyletic group (the "Caenogastropoda"). The opisthobranchs, pulmonates, and two minor groups formed a third large monophyletic cluster, the "Heterobranchia." This hypothesis was supported by sequence analysis of 28S ribosomal RNA (Tillier et al., 1992). If future systematic research on gastropods supports this arrangement, classifications will use names such as "vetigastropods" and "caenogastropods" instead of "archaeogastropods" and "mesogastropods."

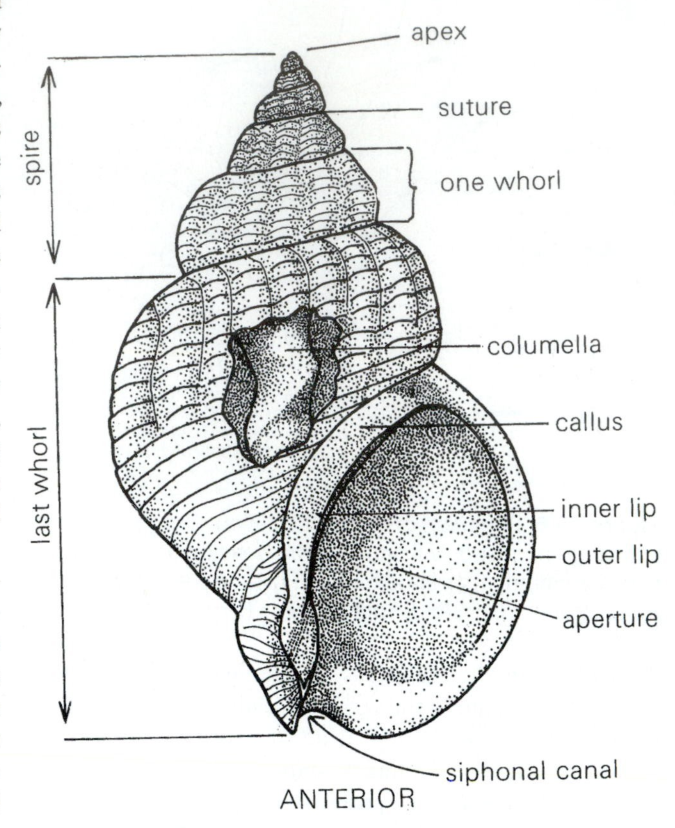

Figure 15.7. Terminology of the gastropod shell. This example is a dextral (right-handed) shell, since the aperture is to the right in this standard orientation. A sinistral (left-handed) shell would be the mirror image of this.

Morphology

Although soft anatomical features are important for understanding gastropod evolution, paleontologists are necessarily restricted to studying the features of the hard shell (Fig. 15.7). A gastropod shell is essentially a tall cone spirally wrapped around an axis in a corkscrew fashion (**trochospiral**). Looking at the shell with the aperture facing you, and the pointed end (**spire**) pointed up, you will find that most gastropods have their aperture on the right (**dextral**). In a few cases, however, the shells spiral in the opposite direction, producing a mirror-image, left-handed (**sinistral**) shell. From the **apex** of the spire, one can count the number of complete **whorls** until the last and largest one, which is the **body whorl** that actually housed most of the snail. Each whorl is joined to the next along a **suture**, and the angle formed by the point of the spire, or the **spiral angle**, is also important in identification. The aperture may have a number of distinctive characteristics used to distinguish taxa. The outer edge of the aperture is known as the **peristome**, while the inner lip (nearest the central axis of the shell) has a smooth callus known as the **inductura**, which rests over the back of the snail. At the base of the shell is a depression in the center of the last whorl known as the **umbilicus**, which may be covered by an expanded callus in some taxa. The advanced Neogastropoda have an elongated trough-shaped protrusion from the aperture known as the **siphonal canal**, which protects their fleshy siphon. The surface of the shell may be ornamented in a number of ways. The growth lines, which run perpendicular to the growth axis of the shell, are often well marked, or there may be protuberances that run perpendicular to the whorls called **ribs**. If they are very prominent, they are known as **varices** (singular, **varix**). Ornamentation that runs parallel to the growth axis of the shell (i.e., down the spiral) is known as **costae**. Some gastropods have large **spines** that protrude from various points along the shell. Finally, if you slice or break open a shell, you will see that the junction of the inner whorls of adjoining spirals produces a fused structure along the axis of the shell known as the **columella**.

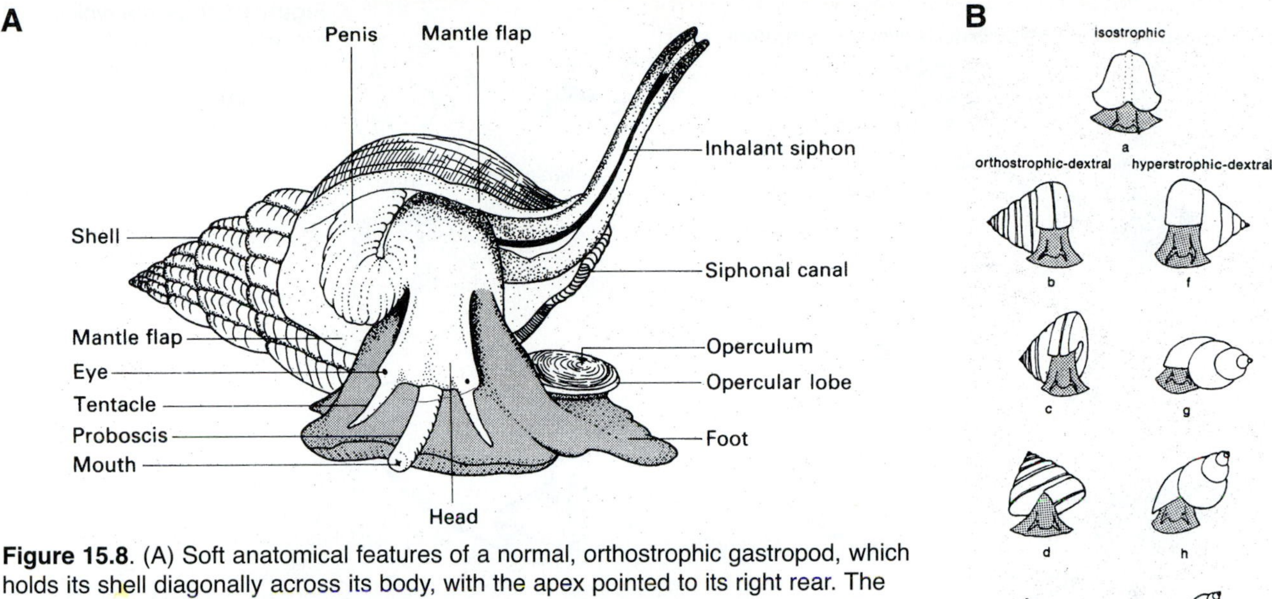

Figure 15.8. (A) Soft anatomical features of a normal, orthostrophic gastropod, which holds its shell diagonally across its body, with the apex pointed to its right rear. The siphon protrudes upward, partially protected by the siphonal notch in the shell. (From Pojeta et al., 1987.) (B) Evolution of orthostrophic and hyperstrophic gastropod shells from a planispiral, isostrophic form. Once the shell becomes asymmetrical, it is unbalanced, so the orthostrophic snail swings the shell slightly backward; the hyperstrophic shell is swung slightly forward. (C) A scenario for the origin of hyperstrophic and orthostrophic gastropods from a flat, uncoiled monoplacophoran ancestor. (B and C from Linsley, 1978.)

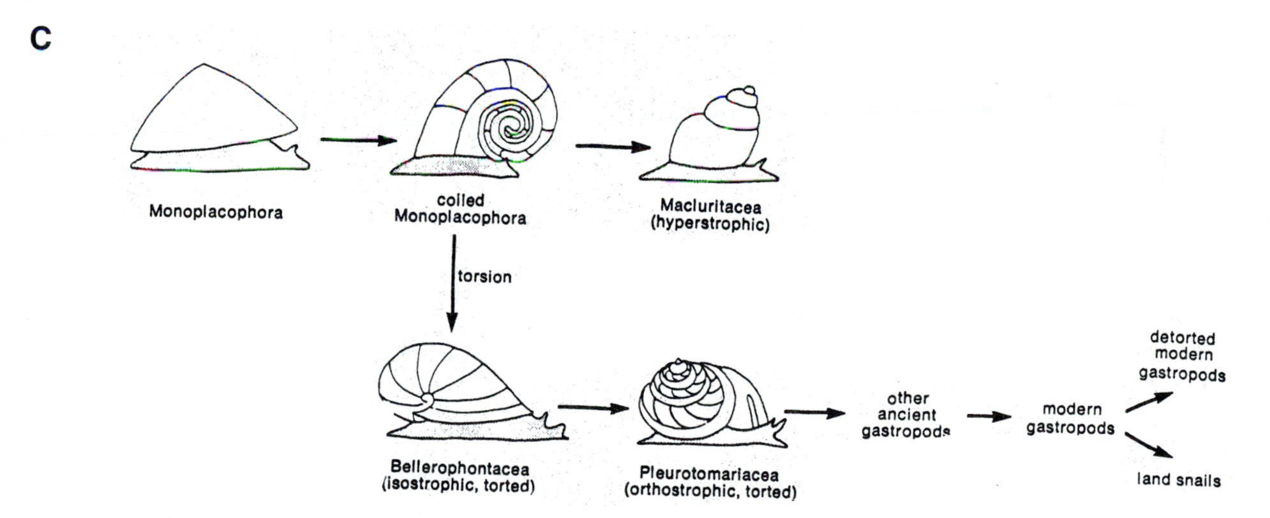

Ecology and Evolution

Evolutionary history—Much is known about the ecology of the various shell forms within the gastropods, because most have living relatives or analogues. For example, we know that limpets cling with their broad foot to hard rocky surfaces in the intertidal zone, pulling their shell down around them for protection. However, the cap-shaped shell is not the only primitive gastropod known. A variety of conical, low-spired fossils are known from the Early Cambrian, and these are usually interpreted as gastropods. Interestingly, three different patterns of spiral shells are known from the early Paleozoic (Fig. 15.8A, B). The first group, the bellerophontids, have a planispiral shell that is symmetrical over the back of the mollusc (**isostrophic** coiling). A number of paleontologists have argued that the bellerophontids were not gastropods at all, but coiled monoplacophorans, because some specimens show what appear to be paired muscle scars (suggesting the segmentation of monoplacophorans). Whatever bellerophontids were, they were a common group in the Paleozoic, but went extinct at the end of the Triassic.

From the symmetrical limpets and bellerophontids, two asymmetrically coiled, trochospiral arrangements are possible (Figs. 15.8, 15.9). A right-handed shell with the whorls displaced to the left side will result in a spiral that is **hyperstrophic**. Apparently this kind of shell was carried diagonally across the back with the apex pointed forward. This unusual arrangement is known primarily from one extinct group, the Macluritacea (Fig. 15.9), which were relatively large snails from the Ordovician through Devonian.

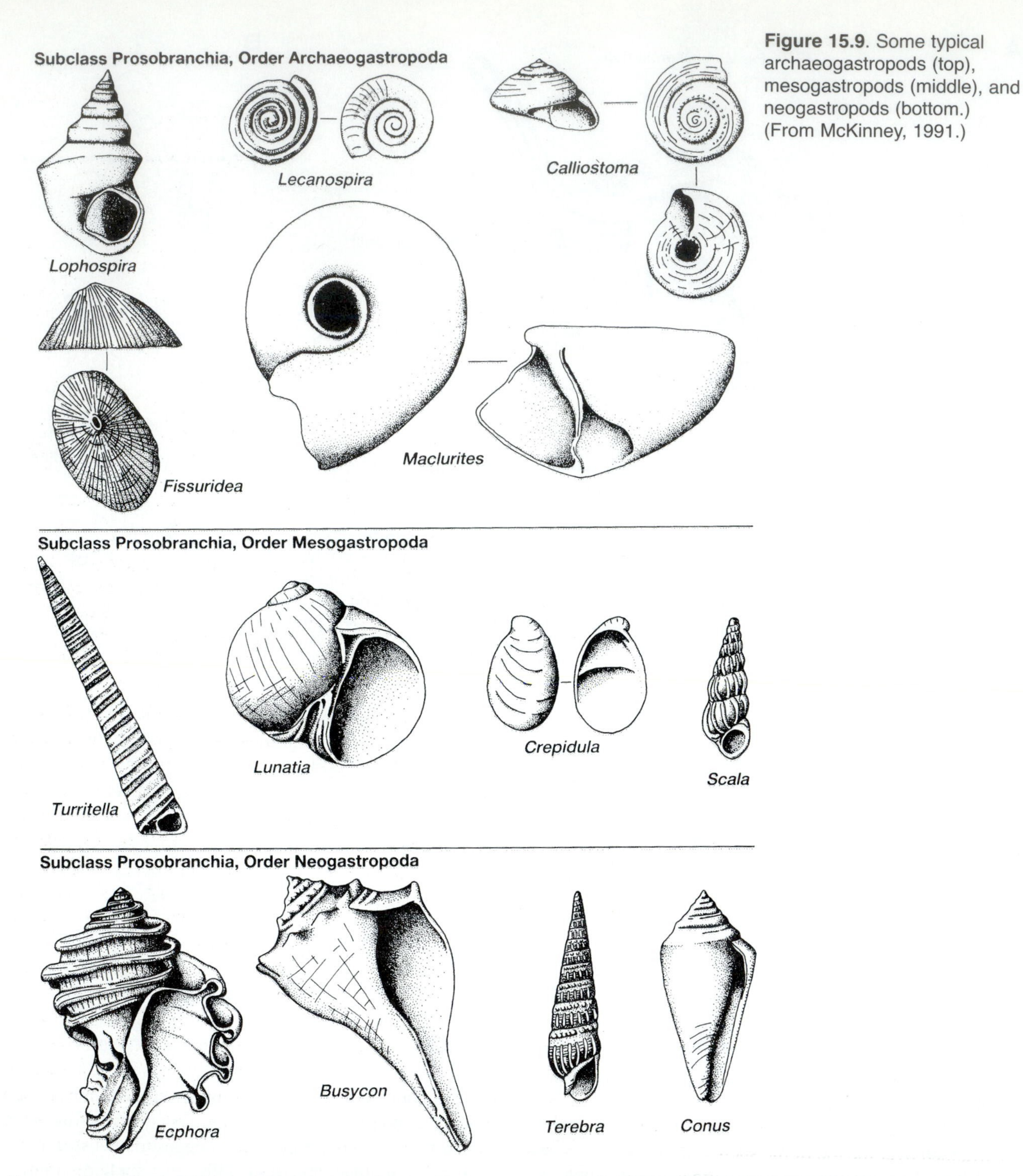

Figure 15.9. Some typical archaeogastropods (top), mesogastropods (middle), and neogastropods (bottom.) (From McKinney, 1991.)

Some paleontologists think that the macluritids were not very mobile, but lay flat on the seafloor and fed by means of cilia.

The majority of trochospiral gastropods, however, have the whorls displaced to the right side in a dextral shell, forming the normal **orthostrophic** shell, which is carried diagonally over the back with the apex pointing backward and to the right side (Fig. 15.8A). During the Paleozoic, a great variety of orthostrophic shells with both high and low spires are known, and most of these are considered prosobranchs. These archaic groups dominated the gastropod faunas of the Paleozoic, but went into decline during the Mesozoic, probably because they lacked the burrowing ability that enabled more advanced gastropods to escape predators during the escalation of the Mesozoic marine revolution. Besides the limpets and nerites, the best known living prosobranchs are the top shells, which are strongly conical. Today, conical and turban-shaped shells with small round apertures are associated with snails that live epifaunally on hard substrates, as the tidepool- and reef-dwelling top shells, tegulas, and nerites do today. Apparently, the stressful intertidal environment provides a refuge from the

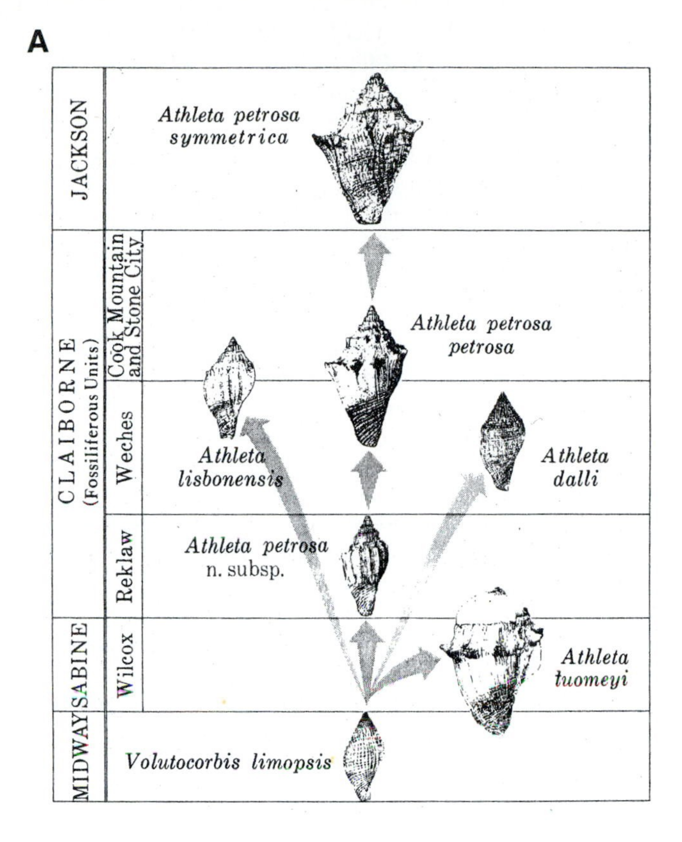

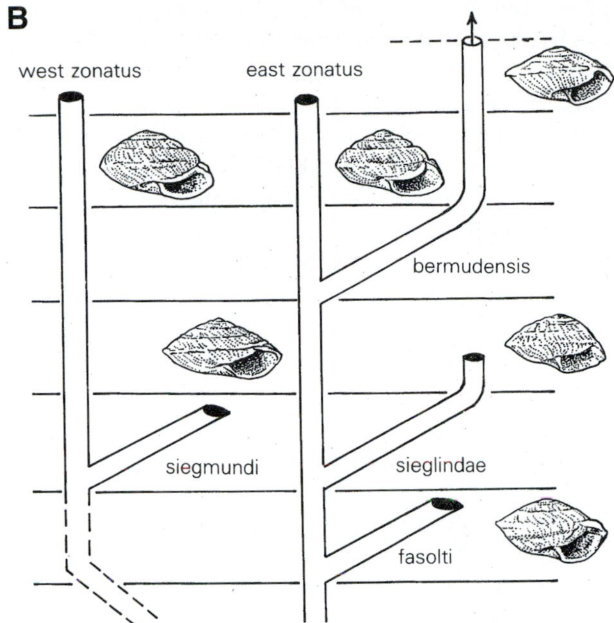

Figure 15.10. (A) Gradualistic interpretation of the evolution of Eocene *Athleta* from the Gulf Coastal Plain (After Rodda and Fisher, 1964). (B) Punctuational interpretaion of the evolution of *Poecilozonites* from the Pleistocene of Bermuda, showing iterative paedomorphosis. (From Gould, 1969.)

heavy predation found in the subtidal depths, allowing both limpets and top-shaped prosobranchs to persist in spite of their lack of anti-predatory devices.

In the late Paleozoic, the mesogastropods, with their single unipectinate gills (Fig. 15.6), evolved. They occasionally show a siphonal notch, indicating that some may have had a small siphon. Their greatest diversity occurred during the Mesozoic and Cenozoic, when the presence of a siphon allowed them to burrow and escape predation. Today, the mesogastropods include a great variety of shell forms, including the high-spired turritellas, ceriths, and wentletraps, as well as the the lower-spired sundial shells, helmet shells, periwinkles, and conchs (Fig. 15.9). Many of these long conical and cylindrical shells act as sand plows, enabling the snail to burrow efficiently. Some mesogastropods have some very peculiar lifestyles. For example, the cowries are distinctive in that their mantle is completely wrapped over the top of the shell, exposing the gills, and leaving the shiny coating of aragonitic nacre on the outside of the shell that makes them so beautiful. The naticids, or moon snails, have a small spherical shell that rides in the middle of a large flat-bodied snail that burrows just beneath the surface of the sand. Moon snails prey on other molluscs by cruising along until they encounter a prey item, which they engulf in their huge foot. Then they rasp a distinctive beveled hole through the shell of the prey (Fig. 8.10E), until they have penetrated the shell and can eat its contents.

In the Jurassic, the most advanced group, the Neogastropoda, originated and quickly diversified to become the dominant living group of marine snails (Figs. 15.6, 15.9). Neogastropods have a long siphon and usually a protracted siphonal notch on the aperture, which enables them to burrow quite deeply. Many of the familiar families of seashells, such as the carnivorous whelks and muricids, volutes, and olives, are neogastropods; so are the tulip and spindle shells, the harp shells, the miters, the dog whelks, the tower shells, and augers. The most unusual members of this group are the cone shells, which have modified some of their radular teeth into poisoned harpoons. With these devices, they are capable of spearing and ingesting a fish, and divers have been known to suffer the poisonous sting of the cone shell.

Because they are not as well fossilized in marine habitats, the fossil record of opisthobranchs and pulmonates is less well known. Shelled opisthobranchs are known as early as the Mississippian, but most taxa lack shells and therefore do not fossilize. Pulmonates apparently crawled out of the ocean before the Pennsylvanian, because they are well represented in the freshwater and terrestrial deposits of the Pennsylvanian coal swamps. Pulmonates can be extraordinarily abundant in some ancient lake and floodplain deposits. This makes them very useful for many different types of evolutionary and geological studies, especially because their ecological tolerances are well known, making them excellent climatic indicators.

Patterns of evolution—Since their shells are often preserved in great abundance, a number of studies have been conducted on detailed evolutionary histories of specific gastropod lineages (Fig. 15.10). For example, the turritellid gastropods are extraordinarily common in many Cenozoic deposits, and the details of the external sculpture of their shells has been used to zone the Cenozoic in many places. Fisher et al. (1964) and Rodda and Fisher (1964) documented apparently gradual trends in the evolution of *Athleta* from the Eocene rocks of the Texas Gulf Coastal Plain, with the shells changing both in shape and number of spines through time. However, the samples are insufficient to document phyletic gradualism by today's standards. Schindel (1982) documented long periods of stasis in the Pennsylvanian gastropod *Glabrocingulum*. Some of the most spectacular gastropod radiations, however, are known from pulmonates that evolved and speciated rapidly in response to changes in their restricted lake or land habitats. For example, the freshwater snails of Lake Turkana in northeast Kenya (Williamson, 1981), or those in the isolated seas of the Paratethys in eastern Europe (Geary, 1990, 1992), show spectacular radiations and extinctions. In most cases, these studies have demonstrated multiple episodes of rapid speciation and long-term stasis, rather than gradual evolution. One of the original examples of punctuated equilibrium was Gould's (1969) studies of the land snail *Poecilizonites* in Bermuda. Since that time, Gould (1984, 1989) has documented a great variety of evolutionary trends in Caribbean land snails, such as the widespread genus *Cerion*.

For Further Reading

Bieler, R. 1992. Gastropod phylogeny and systematics. *Annual Reviews of Ecology and Systematics* 23:311-338.

Haszprunar, G. 1988. On the origin and evolution of major gastropod groups, with special reference to the Streptoneura. *Journal of Molluscan Studies* 54:367-441.

Kohn, A. J. 1985. Gastropod paleoecology, *in* Bottjer, D. J., C. S. Hickman, and P. D. Ward, eds. *Mollusks, Notes for a Short Course*. University of Tennessee Studies in Geology 13:174-189.

Linsley, R. M. 1977. Some "laws" of gastropod shell form. *Paleobiology* 3:196-206.

Linsley, R. M. 1978. Shell form and evolution in the gastropods. *American Scientist* 66: 432-441.

Moore, R. C., and C. W. Pitrat, eds., 1960. *Treatise on Invertebrate Paleontology, Part I, Mollusca 1.* Geological Society of America and University of Kansas Press, Lawrence.

Peel, J. S. 1987. Class Gastropoda, pp. 304-329, *in* Boardman, R. S., A. H. Cheetham, and A. J. Rowell, eds. *Fossil Invertebrates*. Blackwell Scientific Publishers, Cambridge, Mass.

Ponder, W. F., and D. R. Lindberg. 1996. Gastropod phylogeny—challenges for the 1990s, pp. 135-154, *in* Taylor, J., ed. *Origin and Evolutionary Radiation of the Mollusca*. Oxford University Press, New York.

Signor, P. W. 1985. Gastropod evolutionary history, *in* Bottjer, D. J., C. S. Hickman, and P. D. Ward, eds. *Mollusks, Notes for a Short Course*. University of Tennessee Studies in Geology 13:157-173.

Vermeij, G. J. 1977. The Mesozoic marine revolution: gastropods, predators, and grazers. *Paleobiology* 3:245-258.

Vermeij, G. J. 1993. *A Natural History of Shells*. Princeton University Press, Princeton, New Jersey.

BIVALVES

Clams lead quiet, uneventful lives for the most part. Buried in the mud and sand between the tide marks, or farther out in the water, they seldom get around much or hear any important news. Clams don't know what it's all about. They have no heads, so they do not bother with that sort of thing. Clams are very conservative. They voted against having heads in the Ordovician Period and have stuck to it ever since. They never adopt a new idea until it has proven its worth.

Will Cuppy, *How to Attract the Wombat*, 1935

Although not as diverse as gastropods (about 8000 to 15,000 living species), bivalves are frequently the most numerous molluscs in the sea. Many shell deposits are composed entirely of bivalves. When a storm disturbs them and washes them up on the beach, there can be thousands of shells. Bivalves are also the most economically important, with huge volumes of clams, oysters, cockles, scallops, and mussels eaten every year (over 100 million kg of bivalve flesh is consumed in the United States alone in a typical year). The pearl (produced primarily by the pearl oyster, although some are produced by conchs) is the only biologically produced gemstone, and one of the most valuable of all gems. In some cultures, shells were used for currency, and disks cut from shells were known as wampum to the Native Americans. Wampum was the primary currency when Europeans first landed on the East Coast of North America.

To the paleontologist, bivalves are valuable in a very different way. Because their shell form is tightly constrained by the environment in which they live, they are powerful tools for paleoecology. In addition to their general shell shape, bivalve shells have proved useful in recording seasonal changes in their growth lines, and climatic changes in the shell chemistry. Bivalves typically have long stratigraphic ranges, so they are not as commonly used for biostratigraphy, although in certain areas and times (such as the inoceramids of the Western Interior Cretaceous), they have proved useful.

Bivalves are known from every marine habitat, from the great abundance of mussels in the intertidal zone, to the oysters in brackish waters, to the great diversity of bivalves in the surf zone and shallow shelf setting. Some of the largest bivalves are known from the deep volcanic vent settings, where they filter out sulfur-reducing bacteria. About the only marine habitats they have not colonized are those that are anoxic. Like gastropods, bivalves have also invaded the freshwater habitats. Unlike gastropods, however, bivalves require a flow of food- and oxygen-rich water over their gills to survive, so they have never adapted to living on land.

Nevertheless, the success of bivalves is remarkable considering how highly modified they have become. Somehow, their mollusc ancestor, presumably a limpet-like animal, developed a shell hinged over the back, and eventually lost its head completely. In addition to losing their heads, bivalves lack a radula or anterior sense organs, and intelligence or mobility is not their strength (unlike the cephalopods). Instead, they have found a niche as the commonest shelled burrowing animal in marine ecosystems. Where there is reduced predation, they can become the dominant group. In brackish waters, for example, oysters are the most successful and rapidly growing animals in that habitat, and huge "reefs" of oyster-like bivalves occurred repeatedly through the Mesozoic and Cenozoic. As we have already seen, bivalves were relatively unaffected by the great Permian catastrophe (unlike the brachiopods), and took over much of the marine realm in the Mesozoic, when increased predation forced everything below the seafloor. Bivalves also weathered the K/T event with relatively minor losses (although most of the spectacular Cretaceous groups, such as the inoceramids and rudistids, were already gone by the latest Cretaceous), and weathered many other mass extinction events in the Mesozoic and Cenozoic. Bivalves may not be smart or fast or have heads with many sense organs, but they are well adapted for their niches, and have survived many crises that wiped out much faster, more intelligent groups of animals (such as the ammonites).

Morphology

Bivalves look superficially like brachiopods, but the similarity is restricted to the presence of a double-valved shell. As we saw in Chapter 13, the internal anatomy of brachiopods is entirely different from bivalves, and brachiopod shells are usually symmetrical *through* the valves. By contrast, the bivalve shell is symmetrical *between* valves, so that the right valve is usually the mirror image of the left valve (Fig. 15.11). Although bivalves orient themselves in many different ways while they are alive, by convention the shells are always illustrated and described in a standard orientation, with the hinge at the top, and the open edge of the shell (the **commissure**) on the bottom. The foot and mouth are considered anterior (front), while the anus and siphons are posterior (back), because that is the general direction of forward movement in life. If you hold the two articulated valves in your two hands with anterior pointed away from you and the hinge up, then the valve in your right hand is the right valve; the one in your left hand is the left valve.

Unlike a brachiopod, the internal shell cavity of the bivalve does not contain a lophophore, but instead consists of a small headless body and visceral mass. The largest part of the shell volume is occupied by the gills. These serve not only to exchange gases with the seawater, but also as the main filter-feeding apparatus. Food particles trapped by the sticky mucus on the gills are then transported down to the mouth, where they are ingested.

There are other important differences from brachiopods as well. Instead of a pair of oppposed muscles (adductor and diductor) that open and close the shell, bivalves have only **adductor muscles** (usually one in front and one in back). These are capable of tremendous strength in holding the valves shut whenever a threat appears. When the adductors relax, horny **ligaments** adjacent to the hinge area act as a spring, constantly pulling the valves apart. As the bivalve relaxes its adductors, the valves open automatically due to this spring tension. After a bivalve dies, its valves naturally open and frequently become disarticulated. That is why most brachiopods are known from articulated specimens, whereas bivalve fossils are usually found as disarticulated single valves. The next time you are served a pot of steaming bouillabaise in a restaurant, notice how all the clams relaxed and opened once they were put in boiling water.

The soft anatomy of most bivalves is quite simple (Fig. 15.11). The body of the clam lies in the center of the shell interior under the hinge area, with the mouth at one end of a simple digestive tract that includes a stomach and intestine as well as the anus. The body cavity also contains a simple circulatory system and excretory system, and gonads in a sexually mature individual. Part of the body is modified into a long, flexible foot, which protrudes from the front opening of the shell. Filling most of the shell volume is the large area of gills. The inner surface of the shell is lined by the mantle, which secretes the shell. In most bivalves, the posterior portion of the mantle is modified into two chimney-like siphons, one of which is specialized for inflowing water, and the other for expelling waste water.

Much of this anatomy can be read from the bivalve shell (Fig. 15.11). The adductor muscles leave large, oval or circular scars, so in most bivalves there are clear anterior and posterior adductor muscle scars. The edge of the mantle leaves a scar along the inside of the shell called the **pallial line**. In the posterior part of the shell, there is frequently a notch in the pallial line, known as the **pallial sinus**. This is an indication of retractable siphons, and in most cases, the larger the pallial sinus, the larger the siphons (and the deeper the bivalve can burrow).

The pallial sinus gives an indication of the size of siphons and the depth of the burrowing, and in most bivalves, it is pointed toward the surface while the animal is in its burrow. By contrast, the foot is generally pointed down the burrow, because it is responsible for the burrowing in the first place. Burrowing bivalves are remarkably efficient at getting themselves under the sediment, given their lack of large, digging appendages found in most other burrowing organisms. Burrowing starts when the extended foot forms a sharp pointed wedge, which is then pushed down into the sediment. Once it is fully extended, the foot forms a ball-shaped anchor at the tip, and then the entire

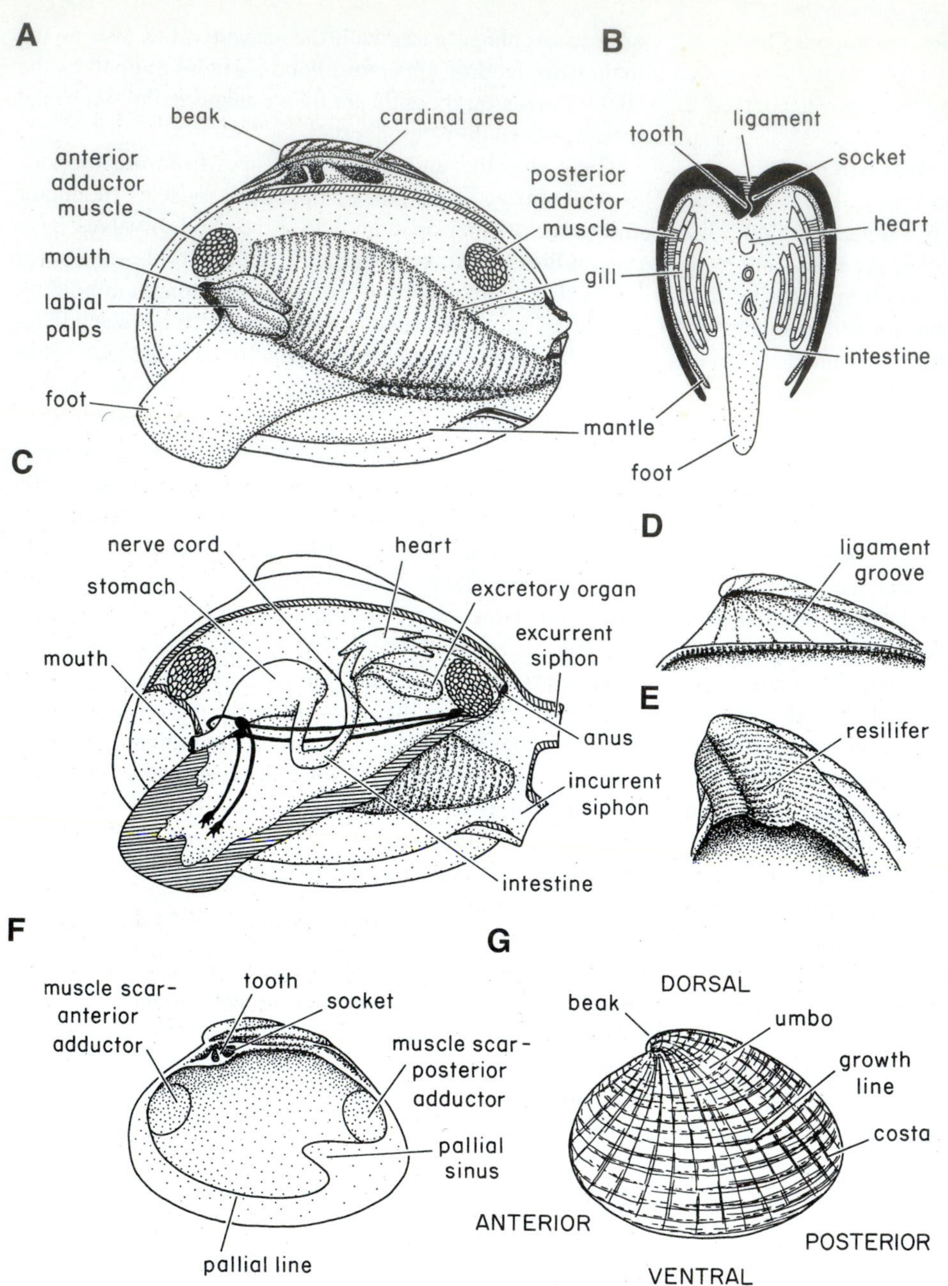

Figure 15.11. (A) Lateral view of a bivalve with the left valve and mantle fold removed. (B) Cross section of bivalve; shell shown in black. (C) Lateral view with left valve, mantle fold, and body wall removed. (D) Cardinal area showing ligamental groove. (E) Cardinal area with resilifer. (F) Lateral view of interior of right valve. (G) Lateral view of exterior of left valve. (From Beerbower, 1968.)

foot muscle contracts, pulling the rest of the animal downward. As it does so, the bivalve forces water out of the mantle cavity, liquefying the sediment around it into a soupy consistency, so it can slide downward easily. Most digging bivalves also rock their shells front to back as they pull downward, which keeps the sediment liquefied and helps shift it around the shell as it sinks. Once the bivalve has dug down to its preferred depth, it protrudes its two long siphons to the surface, so that it can exchange water (and with it, food, oxygen, and waste products) while safely below the surface.

One of the most important anatomical regions for bivalve classification is the gills. The most primitive bivalves have simple, small, leaf-like **protobranch** gills, similar to those in chitons and cephalopods (Fig. 15.12). Many bivalves have large lamellar sheets of gills folded into a W shape; these are known as **filibranch** gills. In the **eulamellibranch** gills, the folded gills are connected by cross-partitions, producing water-filled cavities between them. One specialized group of rock-boring clams has **septibranch** gills, that run transversely across the mantle cavity, enclosing an inner chamber which has only a small opening to the outer cavity.

The hinge area of the shell can be fairly complex. The point of the conical spiral of the shell is called the **beak**; it usually points to the front (**prosogyral**). A few bivalves have beaks and shells that spiral backward (**opisthogyral**). The convex "hump" at the top of the shell is called the **umbo** (plural, **umbones**). In front of the beak is a depression on both valves called the **lunule**, which helps in burrowing. Just behind the umbo is the pocket where the ligament binds the shells together. Some bivalves have a large

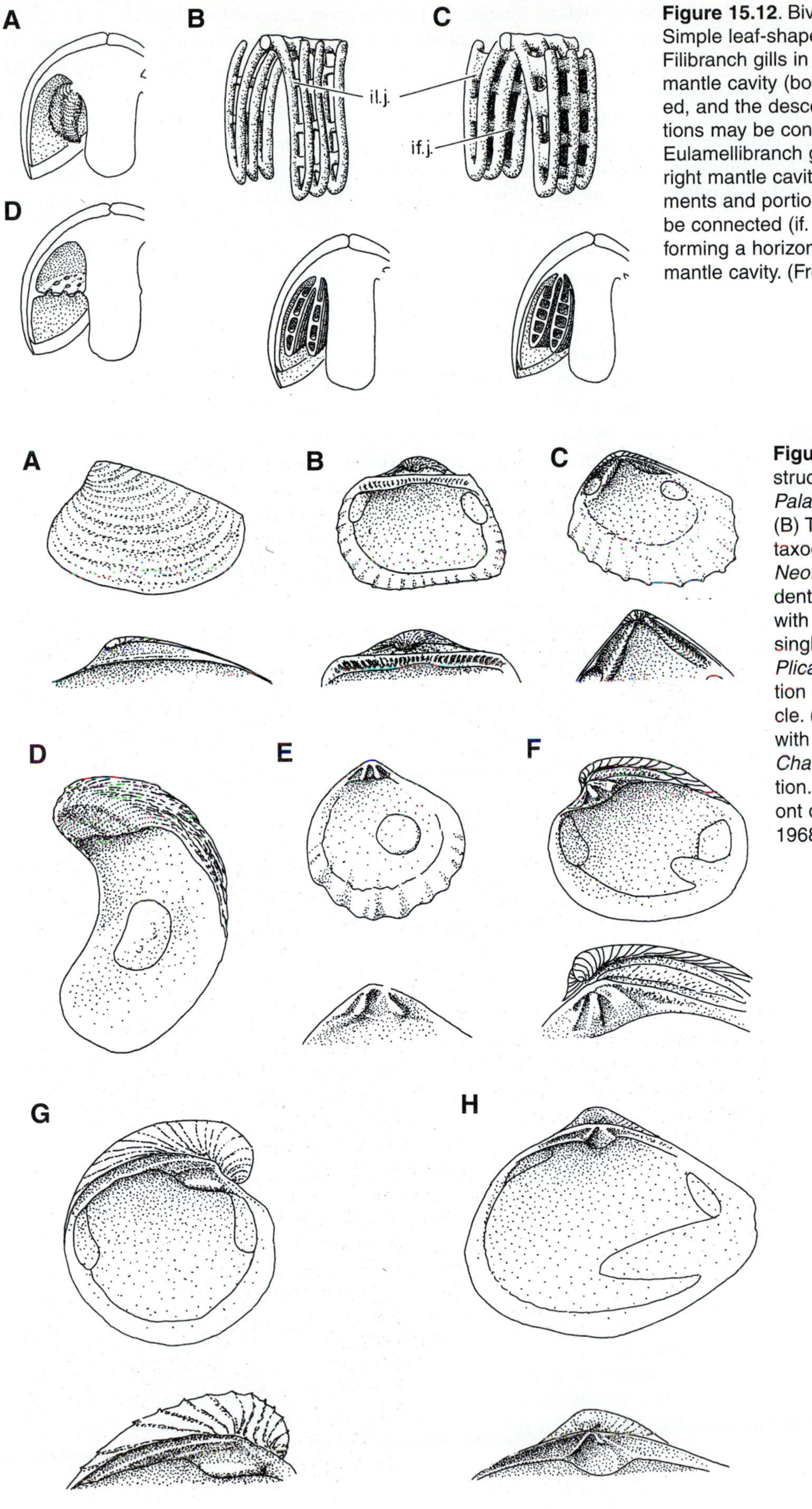

Figure 15.12. Bivalve gill structure. (A) Simple leaf-shaped protobranch gills. (B) Filibranch gills in detail (top) and within right mantle cavity (bottom). Each filament is folded, and the descending and ascending portions may be connected (il. j.). (C) Eulamellibranch gills in detail (top) and in right mantle cavity (bottom). Adjacent filaments and portions of the same filament may be connected (if. j.). (D) Septibranch gills, forming a horizontal perforation in the right mantle cavity. (From Beerbower, 1968.)

Figure 15.13. Bivalve hinge structure and classification. (A) *Palaeoconcha,* no hinge teeth. (B) The ark shell, *Arca,* with a taxodont dentition. (C) *Neotrigonia,* with a schizodont dentition. (D) The oyster, *Ostrea,* with a dysodont dentition and a single adductor muscle. (E) *Plicatula,* with an isodont dentition and a single adductor muscle. (F) The venus clam, *Venus,* with a heterodont dentition. (G) *Chama,* with a pachydont dentition. (H) *Lutraria,* with a desmodont dentition. (From Beerbower, 1968.)

ligamental groove, or even a trough-shaped depression for the ligament called the **resilifer**. Other bivalves have a large projecting internal process with a concavity for the ligament, known as the **chondrophore**. Within the hinge itself (the **cardinal area**) can be a variety of combinations of **teeth** and **sockets** that articulate the two valves. In addition, there may be additional teeth and sockets along the front or back of the hinge area; these are known as **lateral teeth**.

A number of possible configurations of teeth and sockets are known (Fig. 15.13). The most primitive bivalves have a **taxodont** dentition, with rows of numerous subparallel teeth of a similar size. Many bivalves have a **heterodont** dentition, with two or three cardinal teeth below the umbo, as well as elongated anterior and/or posterior lateral teeth. Mussels have a **dysodont** dentition, with just a few small simple teeth at the edge of the valve. The thorny oysters and jingle clams have an **isodont** dentition, with just a few large teeth located on either side of a large pit for a ligament. The trigoniaceans have a unique **schizodont** dentition, in which the teeth are very large, and have a large number of parallel grooves normal to the axis of the tooth. In **desmodont** bivalves, the cardinal teeth are reduced or absent, but accessory lateral ridges along the hinge margin take their place; in some, there is a large chondrophore. Finally, the thick-shelled rudisitid bivalves have huge, heavy, blunt teeth in a **pachydont** dentition.

The exterior of the shell may also have considerable morphological complexity. The **growth lines** form concentric rings away from the beak, because they are formed by the incremental addition of new shell to the edge. There may also be prominent **costae**, or ribs, that radiate out from the beak. The shape and configuration of the ribs and growth lines can be very important in burrowing bivalves, because they help divert the sand as the burrowing proceeds. The shell can be folded into corrugations known as **plications**, commonly seen on shallow-water scallops. In non-burrowing forms, however, a relatively smooth surface is no longer critical, so some have large spines (such as the thorny oysters), or a highly irregular surface texture (like most oysters).

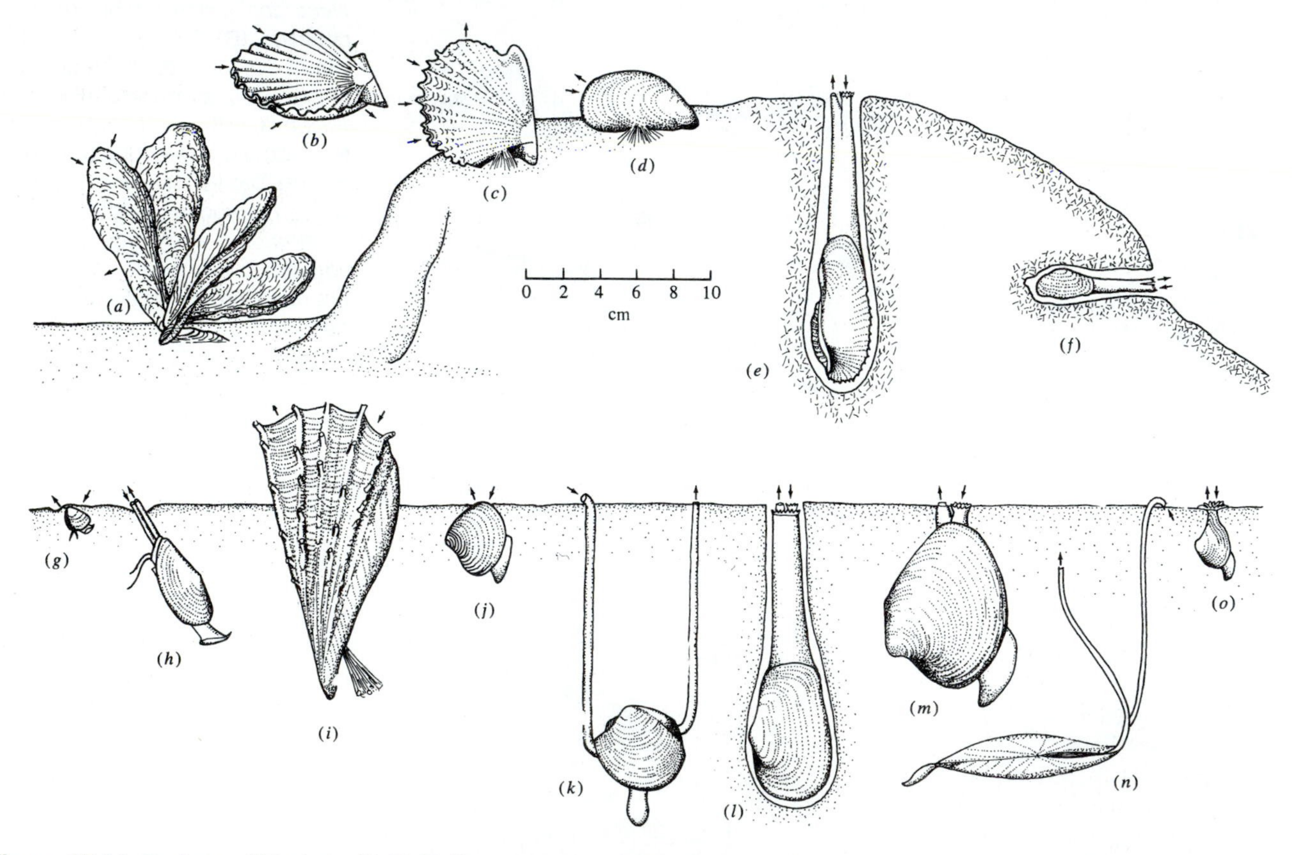

Figure 15.14. Ecology of bivalves. (A-D) Epifaunal suspension feeders. (A) The cemented oyster *Crassostrea.* (B) The swimming scallop *Pecten.* (C) Byssally attached pearl oyster *Pinctada* and (D) the mussel *Mytilus.* (E-O) Infaunal bivalves: (E) the angel wing *Pholas* and (F) the geoduck *Hiatella,* siphonate suspension feeders which bore into rocks. (G) The nut clam *Nucula,* a non-siphonate labial palp deposit feeder. (H) The nut clam *Yoldia,* a siphonate labial palp deposit feeder. (I)The pen shell *Atrina* and (J) the clam *Astarte,* non-siphonate suspension feeders. (K) The lucinid *Phacoides,* an infaunal mucus tube feeder. (L)The soft-shelled clam *Mya* and (M) the venus clam *Mercenaria,* siphonate suspension feeders burrowed into the sediment. (N) The tellin *Tellina,* a siphonate deposit feeder. (O) The septibranch dipper clam *Cuspidaria,* a siphonate carnivore. (After Stanley, 1968.)

Systematics

Bivalve classification schemes based on nearly every organ system, especially the gills and hinge area, have come and gone over the past two centuries. The problem is not with the lower-level taxa, such as the families or superfamilies, which are well-defined natural groups that have remained more or less constant for over a century. For example, scallops (family Pectinidae, superfamily Pectinacea), mussels (family Mytilidae, superfamily Mytilacea), and oysters (family Ostreidae, superfamily Ostreacea) are such well-defined natural taxa and so easy to diagnose that even the casual shopper at a seafood store can tell them apart. The controversy lies in how to combine the superfamilies into larger groups (especially subclasses, superorders, orders, or suborders) within the Class Bivalvia.

Part of the problem lies in the fact that bivalves are relatively simple organisms, with fewer organ systems on which to base a classification scheme. Another factor is the high probability of convergent evolution (especially in shell characters), because bivalves that live in similar habitats and dig in similar ways are constrained to have certain features (especially shell shape and sculpture). In addition, malacologists have long based their classification schemes on soft tissues (such as the gills) that do not fossilize, while paleontologists are necessarily restricted to preservable shell structure. In recent years, paleontologists and malacologists have become more ingenious, looking at a great diversity of organ systems, as well as shell microstructure and molecular characters.

At the present time, the consensus seems to favor the venerable classification developed by Newell (1965, 1969) in the *Treatise of Invertebrate Paleontology*, although some of the higher taxa are defined differently by zoologists, using gill structure, than they are by paleontologists, who focus on hinge features. As further information emerges from new research (especially molecular sequences, which are just now being analyzed), this classification may be highly modified. However, the fundamental features are likely to remain stable.

Class Bivalvia (= Pelecypoda, = Lamellibranchiata)

Subclass Protobranchia

The most primitive groups of clams, with simple protobranch gills, although some extinct forms may have developed filibranch gills. The protobranch gill is a symplesiomorphy, so the group is probably a taxonomic wastebasket for archaic bivalves that might be ancestral to several different lineages. However, two monophyletic orders still survive.

Order Nuculoidea (= Palaeotaxodonta)

The tiny nut clams are rapid shallow burrowers in the surf zone and mudflats (Fig. 15.14G, H). They have a taxodont dentition, and simple aragonitic shell structure. Many have a beak that spirals backward (opisthogyre). The earliest known bivalve fossils closely resemble nut clams in their shell structure. Nut clams are deposit feeders that use labial palps from their mantle to pick food particles out of the sediment.

Order Solemyoidea (= Cryptodonta)

The awning clams have thin, cigar-shaped shells, with a glossy periostracum, and have almost no hinge teeth (dysodont). They are chemosymbiotic, and have a disk-shaped foot with serrated edges, as well as a long siphon, which helps them live in deep, tubular burrows.

Subclass Pteriomorpha (= Filibranchia)

The pteriomorphs are a highly variable group. They all have filibranch or eulamellibranch gills, and most are epifaunal, living attached to a substrate by long, tough **byssal** threads, rather than burrowing beneath the surface. Consequently, they never had siphons, and many have reduced their foot as well.

Order Arcoida

The arc shells (Arcidae) and bittersweet clams (Glycymeridae) are both easy to recognize by their long, taxodont hinges, and their thick box-like shells with heavy ribs. Most live attached to the undersides of rocks, and some have a large gape in their hinge for the byssal threads. Their shells have a cross-lamellar ultrastructure.

Order Mytiloida

The mytiloids include not only the true mussels and the burrowing date mussels (Mytilidae), but also the pen shells (Pinnidae). Mytiloids have two unequal adductor muscles (anisomyarian) and filibranch or eulamellibranch gills, and an almost toothless (dysodont) dentition. Mussels spend their lives attached by byssal threads to rocks in the intertidal environments (Fig. 15.14D), although the date mussels bore into the rocks. Pen shells are large (up to 2 feet long), thin-shelled, fan-shaped shells, which lie partially buried in the sediment with the beak down, and the byssal threads attached to buried rocks or other hard objects as an anchor (Fig. 15.14I).

Order Pterioidia

The Pterioidia include a great diversity of bivalves, such as the scallops (Pectinidae), oysters (Ostreidae), pearl oysters (Pteriidae), thorny oysters (Spondylidae), jingle shells (Anomiidae), file shells (Limiidae), and many extinct groups, such as the giant Cretaceous inoceramids (Fig. 8.10), and many other oyster-like forms, such as the Mesozoic oysters *Gryphaea* and *Exogyra* (Fig. 15.18). Most of these forms are either free-living or byssally attached, although the oysters are usually cemented to their substrate. All pterioids have either unequally developed adductor muscles (anisomyarian), or only a single adductor muscle (monomyarian). Many have a highly reduced foot, or none at all, because they do not burrow.

The scallops are by far the most active of the bivalves. They swim by clapping their valves together like castanets (contracting the large single adductor muscle), which forces water out of the wing-like extensions of their hinge (called the auricles) and gives them jet propulsion (Fig. 15.14B). Scallops may have up to a hundred "eyes" (actually crude light receptors) on the edge of their mantle, which helps them detect the approach of danger and begin swimming. Most oysters (Fig. 15.14A), on the other hand, no longer need to move, so their shells become thick and asymmetrical, with highly modified hinge and muscles.

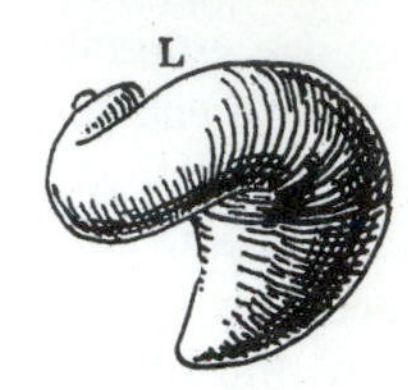

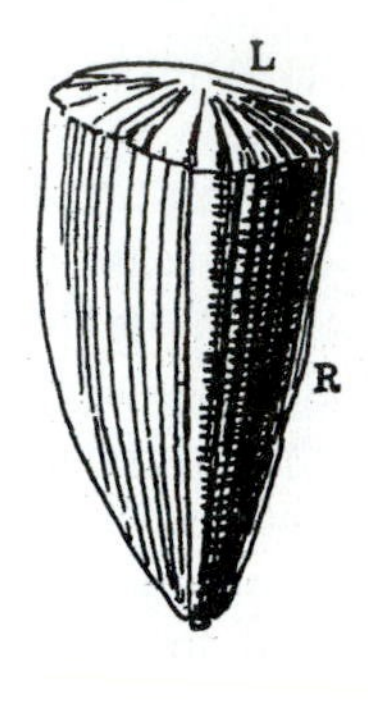

Figure 15.15. The peculiar asymmetrical bivalves with pachydont hinges known as rudistids (order Hippuritoida) were major reef builders of the Cretaceous. The right valve (R) formed a cup or cone attached to the bottom, and the left valve (L) was much smaller, frequently forming a cap over the cone of the right valve. From top to bottom: *Diceras, Caprina, Hippurites.* (From Moore et al., 1953.)

Subclass Heterodonta (= Eulamellibranchia)

Most of the living bivalves are members of the Heterodonta. They are defined by having heterodont dentitions, with well-developed cardinal and lateral teeth, and they also have eulamellibranch gills. Most are burrowing marine forms, with well-developed foot and siphons.

Order Veneroida

Most of the familiar clams are veneroids, including the common venus clams (Veneridae, Fig. 15.14M), the deep-burrowing chemosymbiotic lucines (Lucinidae, Fig.15.14K) and tellins (Tellinidae, Fig. 15.14N), the surf clams (Mactridae, Fig. 15.17), cockles (Cardiidae), giant clams (Tridacnidae), razor clams (Solenidae), and numerous other less familiar families. Almost all have the classic "clam"-shaped shell for active burrowing, and well-developed siphons.

Order Myoida

The myoids are thick-shelled burrowing and boring clams, with asymmetrical valves and degenerate hinges. Most live in deep burrows or rock tunnels, so they have long siphons. They include the soft-shelled clams (Myacidae, Fig. 15.14L), the long-necked geoducks (Hiatellidae, Fig. 15.14F), the ubiquitous corbulas (Corbulidae), the rock-boring angel wings and piddocks (Pholadidae, Fig. 15.14E), and the shipworms (Teredinidae).

Order Hippuritoida

The hippuritoids are an extinct group that includes the highly asymmetrical, colonial rudistids (Fig. 15.15). These bivalves were common reef builders of the Cretaceous. They had one valve that served as a lid over a cone-shaped, attached lower valve. Their shells are extraordinarily thick, and they have thick-toothed, pachydont dentitions.

Subclass Paleoheterodonta

The paleoheterodonts are a paraphyletic assemblage of mostly Paleozoic and Mesozoic groups that some paleontologists believe are ancestral to most of the rest of the more advanced bivalves (especially the Pteriomorpha and Heterodonta).

Order Modiomorpha

A Paleozoic group, thought to be closely related to the heterodonts, because they also have heterodont dentitions.

Order Unionoidia

The dominant freshwater clams are the unionoids, well known from lakes and rivers around the world. They also have heterodont dentitions, and a thick nacreous layer that makes them very popular for the manufacture of buttons and jewelry.

Order Trigonioida

Trigonioids were a distinctive group of Paleozoic-Mesozoic bivalves with a heavily ribbed, triangular shell, and a unique schizodont dentition (Fig. 15.13C). They lived in coarse, shifting, nearshore sands during the Mesozoic, but were long thought

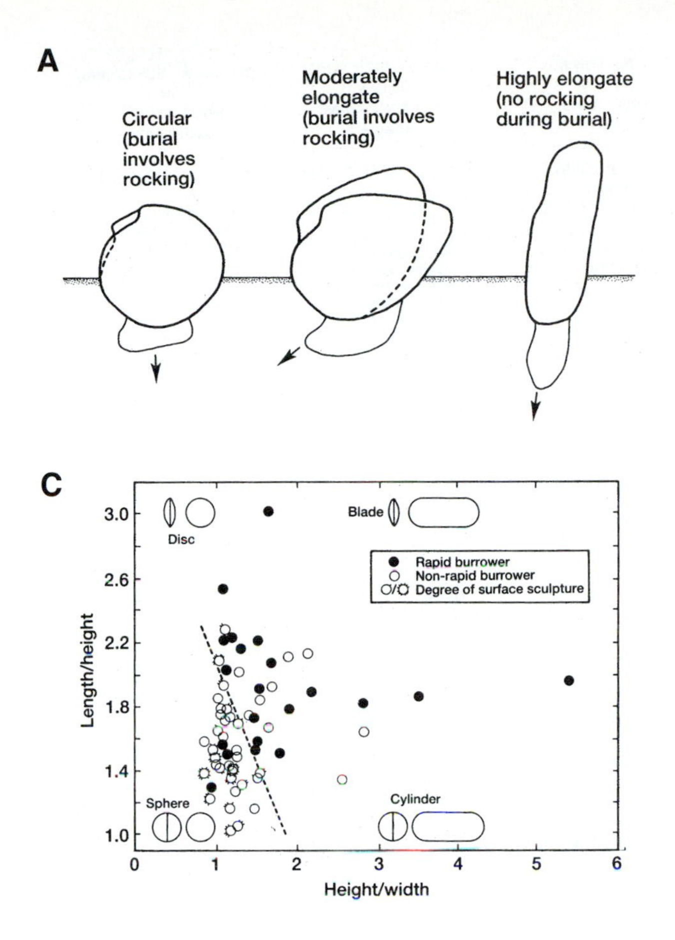

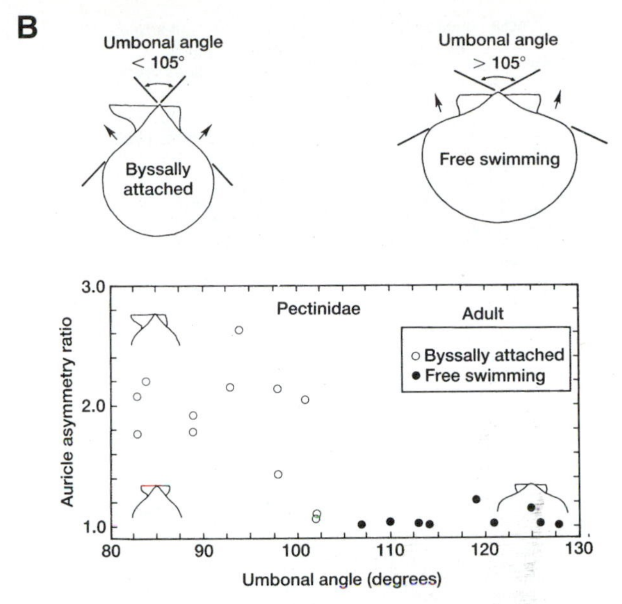

Figure 15.16. (A) Orientation, behavior, and angle of penetration of burrowing bivalves with different degrees of shell elongation. (B) Contrasts between swimming and byssally attached scallops. Swimmers have broader umbonal angles and symmetrical auricles. (C) Relationship between shell shape and burrowing rate in bivalves. To the left of the dashed line, most species are relatively slow burrowers, while the rapid burrowers tend to be right of the dashed line. (After Stanley, 1970.)

to be victims of the K/T event. Then in 1902, specimens of living *Neotrigonia* were found in the southwest Pacific, and this supposed "extinct" group from the Mesozoic was resurrected by a living fossil.

Subclass Anomalodesmata
Order Pholadomyoida

Pholadomyoids are burrowing or boring bivalves (Fig. 15.14O), with highly modified aragonitic shells and desmodont dentition. Their unique feature is their septibranch gills.

Ecology

Despite the constraints of being a slow-moving filter feeder with a two-valved shell, bivalves show a remarkable range of ecological adaptations in marine and fresh waters. In addition, the shape of their shell is strongly constrained by function, so bivalves are excellent organisms for studying functional morphology, and for inferring ancient ecologies. Even with extinct bivalves, there are enough clues from living relatives or analogues to infer the ecology in many cases. No matter what their phylogenetic origin, bivalves follow general ecological and functional rules that govern their shapes.

Although malacologists had documented the ecology of some living bivalves, the landmark study of bivalve ecology was Stanley's (1970) monograph *Relation of shell form to life habits of the Bivalvia (Mollusca)*, published as *Geological Society of America Memoir* 125. In that volume, Stanley studied the actual burrowing habits and location of the shell and siphons in most of the living bivalve groups, using x-rays of the sediment at the bottom of a thin aquarium to photograph the bivalve in life position within its burrow. In addition, Stanley meticulously documented the ecology of many living groups, and inferred some general rules about ecological habits from shell shape. The ecology of a long-extinct form can often be inferred, just by understanding some general rules about shell morphology, position and length of siphons, position and robustness of ribs and costae, and a number of other features.

For example, Stanley (1970) documented some of the features that are characteristic of bivalves that burrow in soft substrates. Their valves are highly symmetrical (equivalved), with two equal adductor muscles (isomyarian) with a distinct pallial line. Those with a circular shell outline tended to rock back and forth over a large angle as they burrow, penetrating the substrate vertically. More elongate species tended to rock asymmetrically, and penetrate the substrate while moving forward, while elongate shells (such as jackknife and razor clams) do not rock at all, but dig straight down (Fig. 15.16A). Clams with highly sculptured and/or thick shells, often with marginal teeth, tend to be shallow burrowers, while deep burrowers are uniformly thin-shelled with a smooth surface. The depth of the pallial sinus is also correlated with depth of burrowing unless the

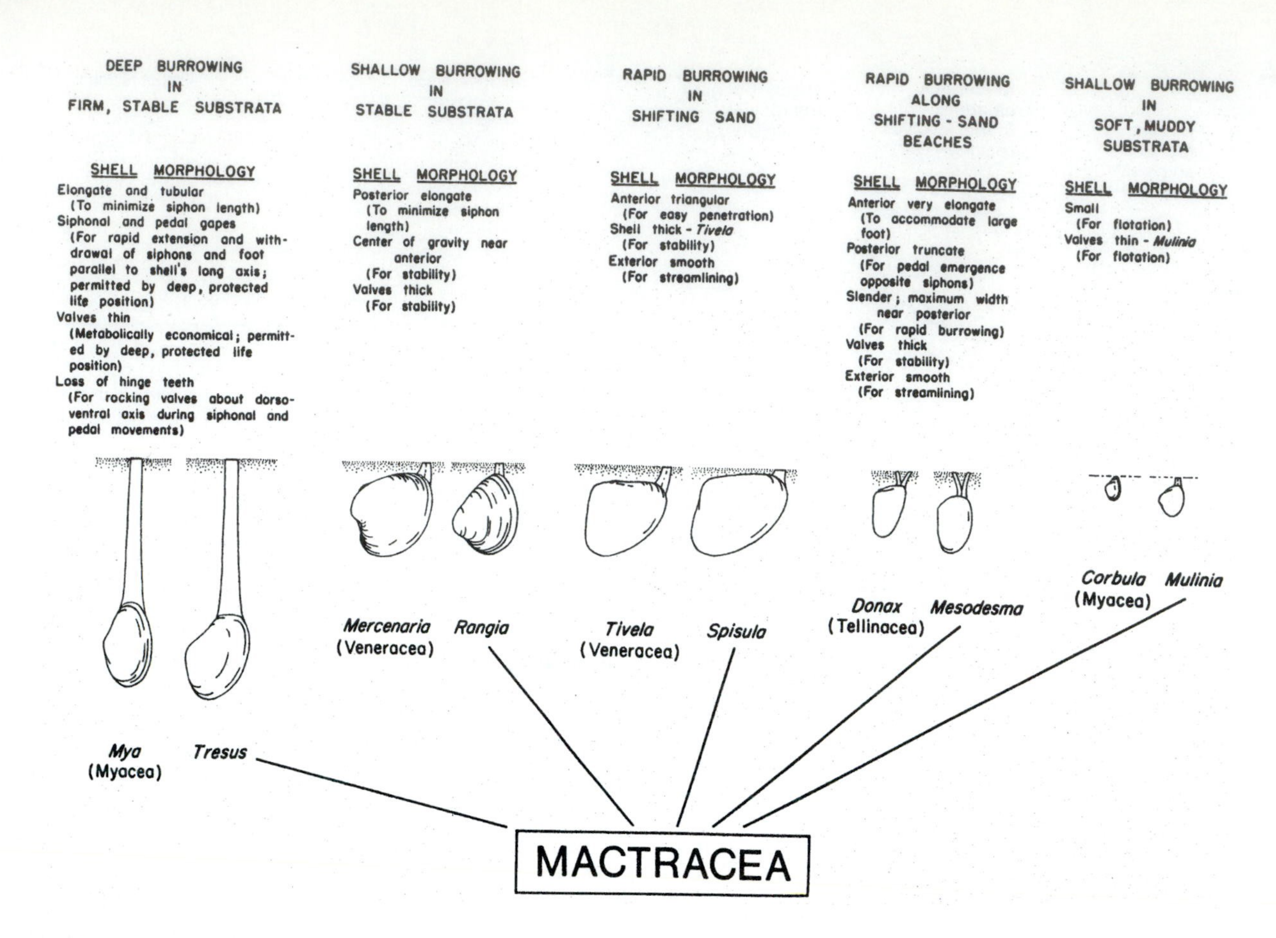

Figure 15.17. Adaptive divergence of the Mactracea (surf clams and their relatives), and their convergence in form and habit with members of other bivalve superfamilies. (After Stanley, 1970.)

species has very slender siphons that don't require much room for retraction. Smaller clams burrow more rapidly than large ones, and those with relatively spherical shells tend to be slower than those with discoidal, bladed, or cylindrical shells (Fig. 15.16B). Highly textured shells tend to produce slow burrowing, except for those with a distinctive divaricate surface pattern, which helps divert the sediment around the shell. Rapidly burrowing animals are often associated with shifting substrates, especially in the surf zone, or with deposit feeding, which means that they must continually burrow to find food.

By contrast, bivalves that bore into hard substrates (partially lithified sediment, wood, or rocks) have a very peculiar cylindrical shape, and tend to have thin shells with asymmetrical adductor muscles (anisomyarian) or a single adductor (monomyarian). Rock-boring specialists, such as the piddocks, have a series of ridges on the back of the shell that are used to cut through the substrate. The substrate is usually something softer than their shell, such as wood or sedimentary rock, but they also excavate limestone (composed of calcite, just like the shell), and some have been known to bore into jade (Mohs hardness 5 to 6)! The most specialized borers of all are the shipworms, which are not worms at all, but bivalves that have long fleshy wormlike bodies and bore into wood (especially in driftwood, ship hulls, and docks). Their shells are reduced to tiny cutting blades at the end of the body, which are rotated like a drill bit to excavate their tunnels.

Swimming bivalves (mostly scallops and their kin) have a very different morphology. Most have very thin shells to reduce weight, with symmetrical wing-like auricles along their hinge for jet propulsion. They tend to have a large umbonal angle (Fig. 15.16C), while byssally attached scallops had a smaller umbonal angle and asymmetrical auricles. Swimmers tend to have a single large adductor muscle (monomyarian) to make the strong contractions needed for swimming stronger and more coordinated than two small adductors could produce. Bivalves that swim with their commissure vertical have both valves equal in curvature, while those that do not swim this way have one valve that is flat and the other convex. The radial ribs and plications are quite marked on very shallow swimmers, but tend to be smoother on those that live below wave base.

Byssally attached bivalves live under a different set of constraints. The anterior adductors are reduced, producing anisomyarian or monomyarian muscles. They have a gape in the shell for the byssus, usually along the ventral commissure, or a byssal sinus between the dorsal and anterior margin of the shell. If they attach with their commissure perpendicular to the substrate (for example, mussels), they tend to have equivalved, elongate shells, with a reduction of the anterior end so that the beak is at the point of the shell. By contrast, if they attach with the commissure horizontal or oblique to the substrate (for example, ark shells or byssally attached scallops), their shells are inequivalved. They also tend to have their hinge lines extended into a single large auricle, possibly to separate inhalant and exhalant currents.

Bivalves that are cemented (such as oysters or rudistids), rather than byssally attached, adhere to different rules. Their valves are highly asymmetrical (inequivalved), and in tropical forms, covered with spines. The attached valve may be very thick, and they have reduced their adductors to a single muscle (monomyarian). Similarly, those that recline on the substrate (like *Gryphaea* or *Exogyra*, Fig. 15.18) are also inequivalved, with the top valve serving as a small lid on the deep, thick-shelled ventral valve.

Bivalve shape is probably more highly constrained than any other invertebrate group we have encountered. For this reason, there is a lot of evolutionary parallelism and convergence in bivalves, with different groups producing the same body form because of similar ecological constraints. Stanley (1970) documented an excellent example of this within the surf clams of the subfamily Mactracea (Fig. 15.17). All mactraceans are easily recognized by the large triangular chondrophore in their hinge area (a phylogenetic character), yet the size and external shape of their shell is largely a reflection of burrowing habits, not phylogenetic origin. Some (e.g., *Mulinia*) are tiny, thin-shelled forms, like the corbulids, and live in shallow burrows in soft muddy substrates. Others (e.g., *Mesodesma*) have slender, thick, smooth shells with long anteriors and truncated posteriors like *Donax* clams. Apparently, these features are for rapid burrowing in the shifting sands of beaches. The mactracean *Spisula* has a thick, smooth shell with a triangular front (like many venus clams, such as *Tivela*) for rapid burrowing in shifting sand, while the mactracean *Rangia* has thick valves with a long posterior and an anteriorly shifted center of gravity (like some other venus clams, such as *Mercenaria*) for rapid burrowing in a stable substrate. Deep burrowers (like the mactracean *Tresus*, or the myacean *Mya*) both have thin, elongate tubular shells, with no hinge teeth and a large gape for the siphon and foot. These examples dramatically demonstrate how characters can be influenced by both phylogeny and ecology. Consequently, phylogenetic analyses and classification schemes in the Bivalvia have been particularly plagued by convergence, and remain controversial

Evolution

Evolutionary history—The earliest known bivalves are tiny, taxodont forms that closely resemble nut clams, including *Pojetaia* from the Early Cambrian of Australia (Jell, 1980; Runnegar and Bentley, 1982), *Tuarangia* from the Middle Cambrian of New Zealand (McKinnon, 1982), and *Fordilla* from the Early Cambrian of several places (Pojeta, 1971, 1975; Pojeta and Runnegar, 1974). According to Runnegar (1985), these early bivalves are probably descended from an ancestor like the Early Cambrian rostroconch *Heraulatipegma*. Surprisingly, there are no bivalves known from the Late Cambrian. By the Early Ordovician, bivalves had undergone a great radiation, with the earliest members of most of the subclasses and orders found in many parts of the world (Pojeta, 1971, 1978, 1985). In addition to the deposit-feeding nut clams, there were shallow-burrowing venus clams with no siphons, byssally attached mytiloids, and deeper-burrowing lucinids, which used a mucus-lined tube to connect to the surface. Even though the major taxa of bivalves were established as early as the Ordovician, they remained much less diverse and abundant through most of the Paleozoic. In many cases, they occupied marginal and nearshore habitats, or specialized in muddy substrates, while brachiopods dominated the shallow subtidal shelf through most of the Paleozoic.

As we have already seen in Chapter 13, the Permian catastrophe completely changed the rules for shelled invertebrates. Not only were the brachiopods decimated, but so were most of the rest of the dominant groups of the "Paleozoic fauna." By contrast, bivalves were relatively unaffected, for reasons that have never been satisfactorily explained. At the beginning of the Mesozoic, the relative diversity balance of brachiopods and bivalves had been completely reset, with many more bivalves surviving than brachiopods (Gould and Calloway, 1980). Brachiopods might have recovered in the Triassic were it not for another important evolutionary change: the advent of many new shell-crushing predators during the Mesozoic marine revolution (Vermeij, 1977). Although archaic bivalves already had an advantage over brachiopods because they were better adapted to burrowing, in the early Mesozoic, bivalves had an additional edge: fusion of the posterior edge of the mantle. This makes it possible not only to develop the mantle into long siphons that allowed them to burrow deeply and efficiently, but in other groups, to seal the mantle for the hydraulic process of burrowing (Stanley, 1968). Although siphonate bivalves may have existed as early as the Devonian, there was a tremendous radiation of siphonate heterodont and desmodont superfamiles of bivalves in the early Mesozoic in response to the pressure of shell-crushing predators. In addition, mantle fusion made it easier for shallow, rapid burrowers to occupy the surf zone (not previously inhabited by any shelled invertebrate), and construct deep burrows and tunnels in the intertidal zone as well. Thus, bivalves expanded both in diversity and also in habitats and tiers, occupying many new niches and depths of penetration of the substrate not previously inhabited by them. Finally, swimming scallops also appeared in the Triassic, exploiting yet another lifestyle that allows them to escape predators.

Since the Triassic, bivalves have continued to diversify and dominate most marine habitats (at least in numbers of fossilizable shells). In the Jurassic and Cretaceous, there were many experiments in large, reclining or attached

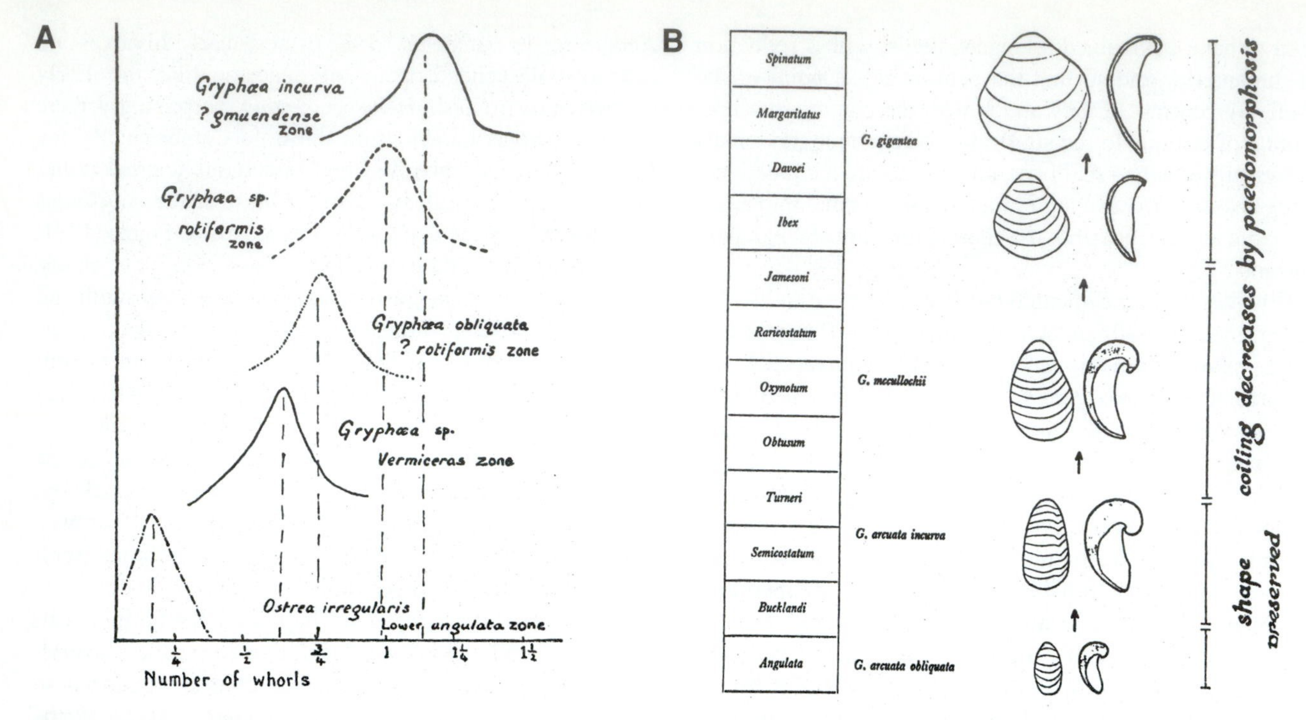

Figure 15.18. (A) Trueman's (1922) original illustration of the supposed gradual trend (shifting histograms) toward increasing coiling in Lower Jurassic gryphaeas. (B) Hallam's (1968) reinterpretation of the evolution of *Gryphaea* through the Lower Jurassic. There was a continual increase in size and initial preservation of adult shape, followed by decreased coiling and shell thickness in higher beds.

bivalves that developed highly assymmetrical oyster-like shells. These include not only the famous Jurassic "devil's toenail" oyster *Gryphaea*, but in the Cretaceous a variety of odd forms, including the highly conical *Exogyra*, the "zig-zag" clams (*Alectryonia*), the "dinner-plate"-shaped inoceramids, and the reef-building rudistids (Fig. 15.15). Rudistids became the dominant reef-forming organisms during the Cretaceous, displacing corals in the tropical region with their massive accumulations of thick shells. Although most of the odd Cretaceous forms did not survive until the end of the Cretaceous, the majority of bivalve groups passed right through the K/T boundary with no ill effects. Their diversification continued through the Cenozoic, with minor extinctions during the late Eocene cooling events, and during local cooling events in the Plio-Pleistocene.

Freshwater bivalves must evolved in the Middle Devonian and by the early Carboniferous they are quite abundant in the lakes and swamps of the coal measures in Europe, and in the Mississippian and Pennsylvanian coal swamps and deltas of North America.

Evolutionary patterns—Stanley (1977) and Miller (1990) reviewed some of the many different examples of evolutionary patterns that can be seen in the Bivalvia. As we have just seen, evolutionary parallelism and convergence are particularly common in this group, and go far beyond the example of the Mactracea in Figure 15.16. For example, Newell and Boyd (1975) documented parallel evolution in the extinct Trigoniaceans. Stanley (1977) also points out that bivalves have a relatively slow but steady rate of diversity increase, apparently because bivalves have a relatively low rate of speciation. This may be because bivalves are relatively eurytopic (occupying huge geographic ranges), which may restrict their ability to fracture into small populations for allopatric speciation. However, bivalves are evolutionarily very conservative by any measure, showing much lower rates not only of speciation, but also lower rates of extinction, and turnover rates in general. Apparently, they occupied a spectrum of well-constrained ecological niches early in their evolution, and the stability of those niches prevents them from making many changes that could be recognized in fossils. About the only major change in the bivalve world was the pressure of new predators in the early Mesozoic, leading to additional diversification of siphonate taxa, deep-burrowing forms, and swimming scallops. Since then, however, bivalves have not occupied many new adaptive zones, or suffered much change in standing diversity.

With their abundant shells, bivalves are also well suited to studying the tempo and mode of evolution. One the earliest such studies was the classic analysis of *Gryphaea* by Trueman (1922). Through the Jurassic beds of England, Trueman claimed to find a gradual increase in the curvature of these bivalves, with the latest forms allegedly so curved in on themselves that they could barely open their lid-like upper valve (Fig. 15.18). However, Trueman's data and conclusions have both been challenged (Hallam, 1968; Gould, 1972; Gould and Eldredge, 1977; Hallam, 1982). Not only did Trueman (1922) ignore major episodes of stasis in earlier Jurassic forms, but he committed numerous errors in sampling, measuring, and analyzing the *Gryphaea* that do show change. Hallam (1968, 1982) and Gould (1972) both concluded that *Gryphaea* actually shows long periods of stasis, and a careful analysis of the specimens (Hallam, 1968) showed that they actually decreased the tightness of their coiling through time (contrary to the widely held misconception). Hallam (1975) found that stasis, with some gradual changes in size, was the norm for most bivalves from the English Jurassic.

Many other bivalve lineages have been examined. Stenzel (1949) studied the Eocene oyster *Cubitostrea* from the Gulf Coastal Plain, and found a steady increase in size and robustness (although the data are insufficient by modern standards to establish phyletic gradualism). Newell (1949) described the size changes in Pennsylvanian-Permian *Myalina*, which showed both size increase and paeodomorphic trends (although not true phyletic gradualism). Kauffman (1978) documented a variety of evolutionary patterns in the bivalves of the Western Interior Cretaceous. Geary (1987) analyzed the Cretaceous cockle *Pleurocardia* from the Western Interior of the High Plains, and found a few short intervals of gradual changes between long intervals of stasis. Miyazaki and Mickevich (1982) studied the evolution of Mio-Pliocene *Chesapecten* scallops, and found good examples of heterochrony, but no gradualism. Stanley (1977, 1979) used the extraordinary species and generic longevities of bivalves, and their low turnover rates, to argue that stasis is the norm thoughout the entire class. However, the most comprehensive study was conducted by Stanley and Yang (1987). They examined the rates of evolution of 19 common lineages of bivalves over the entire Atlantic Coast Neogene, conducting detailed morphometric analyses of every specimen. After statistical analysis of over 43,000 individual measurements, they found there were slight fluctuations in size over the last 17 million years, but that shape has been extremely stable, with no gradual, linear trends that withstood statistical analysis. Their study is extraordinary not only in the size of the data base, but also in the fact that they precisely quantified and analyzed many variables of shape change within the entire unbiased sample of the fauna over a large geographic area and a long span of time. As such, it answers most of the criticisms of previous studies of gradualism versus punctuation, which were plagued by univariate studies of single lineages from a single local area over a short period of time (Gould and Eldredge, 1977; see Chapter 5).

For Further Reading

Allen, J. A. 1985. The Recent Bivalvia: their form and evolution, pp. 337-403, *in* Trueman, E. R., and M. R. Clarke, eds. *The Mollusca, 10: Evolution*. Academic Press, London.

Bottjer, D. J. 1985. Bivalve paleoecology, *in* Bottjer, D. J., C. S. Hickman, and P. D. Ward, eds. *Mollusks, Notes for a Short Course*. University of Tennessee Studies in Geology 13:122-137.

Cope, J. C. W. 1996. The early evolution of the Bivalvia, pp. 361-370, *in* Taylor, J., ed. *Origin and Evolutionary Radiation of the Mollusca*. Oxford University Press, New York.

Cox, L. R., and others. 1969-1971. *Treatise on Invertebrate Paleontology. Part N. Mollusca 6: Bivalvia*. Geological Society of America and University of Kansas Press, Lawrence.

Gould, S. J. 1972. Allometric fallacies and the evolution of *Gryphaea*. *Evolutionary Biology* 6:91-119.

Hallam, A. 1968. Morphology, palaeoecology, and evolution of the genus *Gryphaea* in the British Lias. *Philosophical Transactions of the Royal Society of London* (B) 254:91-128.

Hallam, A. 1975. Evolutionary size increase and longevity in Jurassic bivalves and ammonites. *Nature* 258:493-496.

Hallam, A. 1982. Patterns of speciation in Jurassic *Gryphaea*. *Paleobiology* 8:354-366.

Hallam, A., and A. I. Miller. 1988. Extinction and survival in the Bivalvia, *in* Larwood, G. P., ed. *Extinction and Survival in the Fossil Record. Systematics Association Special Volume* 34: 121-138.

Jell, P. A. 1980. Earliest known pelecypod on earth: a new Early Cambrian genus from South Australia. *Alcheringa* 4:233-239.

Miller, A. I. 1990. Bivalves, pp. 143-161, *in* McNamara, K. J., ed. *Evolutionary Trends*. University of Arizona Press, Tucson.

Miyazaki, J. M., and M. F. Mickevich. 1982. Evolution of *Chesapecten* (Mollusca: Bivalvia, Miocene-Pliocene) and the biogenetic law. *Evolutionary Biology* 15:369-410.

Morton, B. 1996. The evolutionary history of the Bivalvia, pp. 337-359, pp. 135-154, *in* Taylor, J., ed. *Origin and Evolutionary Radiation of the Mollusca*. Oxford University Press, New York.

Newell, N. D. 1965. Classification of the Bivalvia. *American Museum Novitates* 1799:1-13.

Pojeta, J., Jr. 1971. Review of Ordovician pelecypods. *U.S. Geological Survey Professional Paper* 695:1-46.

Pojeta, J., Jr. 1978. The origin and taxonomic diversification of pelecypods. *Philosophical Transactions of the Royal Society of London* (B) 284:225-246.

Pojeta, J., Jr. 1985. Early evolutionary history of diasome molluscs, *in* Bottjer, D. J., C. S. Hickman, and P. D. Ward, eds. *Mollusks, Notes for a Short Course*. University of Tennessee Studies in Geology 13:102-121.

Pojeta, J., Jr. 1987. Class Pelecypoda, pp. 385-435, *in* Boardman, R. S., A. H. Cheetham, and A. J. Rowell, eds. *Fossil Invertebrates*. Blackwell Scientific Publishers, Cambridge, Mass.

Pojeta, J., Jr., and B. Runnegar. 1974. *Fordilla troyensis* and the early history of pelecypod mollusks. *American Scientist* 62:706-711.

Pojeta, J., Jr., and B. Runnegar. 1985. The early evolution

of the diasome molluscs, pp. 295-336, *in* Trueman, E. R., and M. R. Clarke, eds., *The Mollusca, 10: Evolution*. Academic Press, London.

Runnegar, B., and C. Bentley. 1983. Anatomy, ecology, and affinities of the Australian Early Cambrian bivalve *Pojetaia runnegari* Jell. *Journal of Paleontology* 57: 73-92.

Runnegar, B., and J. Pojeta, Jr. 1992. The earliest bivalves and their Ordovician descendants. *American Malacological Bulletin* 9:117-122.

Stanley, S. M. 1968. Post-Paleozoic adaptive radiation of infaunal bivalve molluscs: a conseqence of mantle fusion and siphon formation. *Journal of Paleontology* 42:214-229.

Stanley, S. M. 1970. Relation of shell form to life habits in the Bivalvia (Mollusca). *Geological Society of America Memoir* 125.

Stanley, S. M. 1977. Trends, rates, and patterns of evolution in the Bivalvia, *in* Hallam, A., ed. *Patterns of Evolution as Illustrated by the Fossil Record*. Elsevier, Amsterdam, pp. 209-250.

Stanley, S. M., and X. Yang. 1987. Approximate evolutionary stasis of bivalve morphology over millions of years: a multivariate, multilineage study. *Paleobiology* 13:113-139.

Trueman, A. E. 1922. The use of *Gryphaea* in the correlation of the Lower Lias. *Geological Magazine* 59:256-268.

Vermeij, G. J. 1977. The Mesozoic marine revolution: gastropods, predators, and grazers. *Paleobiology* 3:245-258.

Vermeij, G. J. 1993. *A Natural History of Shells*. Princeton University Press, Princeton, New Jersey.

Yonge, C. M. 1973. Giant clams. *Scientific American* 232:96-105.

Yonge, C. M., and T. E. Thompson, eds. 1978. Evolutonary systematics of bivalve molluscs. *Philosophical Transactions of the Royal Society of London* (B) 284:199-436.

CEPHALOPODS

This is the ship of pearl, which, poets feign
Sails the unshadowed main—
The venturesome bark that flings
On the sweet summer wind its purpled wings
In gulfs enchanted, where the Siren sings,
And the coral reefs lie bare
Where the cold sea-maids rise to sun their streaming hair.

Its webs of living gauze no more unfurl;
Wrecked is the ship of pearl!
And every chambered cell
Where its dim dreaming life was wont to dwell,
As the frail tenant shaped his growing shell
Before thee lies revealed
Its irised ceiling rent, its sunless crypt unsealed!

Year after year beheld the silent toil
That spread his lustrous coil;
Still, as the spiral grew
He left the past year's dwelling for the new,
Stole with soft step its shining archway through,
Built up its idle door,
Stretched in his last-found home, and knew the old no more.

Build thee more stately mansions, O my soul.
As the swift seasons roll!
Leave thy low-vaulted past!
Let each new temple, nobler than the last,
Shut thee from heaven with a dome more vast,
Till thou at length art free,
Leaving thine outgrown shell by life's unresting sea!

Oliver Wendell Holmes, "The Chambered Nautilus," 1858

The largest, most active, and most intelligent of all the molluscs are the class Cephalopoda. Their name means "head foot" in Greek, in reference to the fact that their foot (tentacles) is located around the head. Cephalopods include not only the living squids, octopuses, cuttlefish, and the chambered nautilus (about 650 living species, mostly squids), but also an incredible variety of extinct shelled nautiloids and ammonoids (at least 17,000 described species). All have well-developed heads with eyes as good as those of the vertebrates (but independently evolved). Their nervous system is one of the most sophisticated of all the invertebrates, and octopuses are able to solve quite complex puzzles and mazes. Cephalopods often have elaborate courtship rituals, and squids and octopuses are capable of highly sophisticated, instantly changing color patterns in their skin, which serve not only for camouflage, but also for communicating with others.

Because they are large, active predators, many cephalopods swim rapidly above the seafloor to catch their prey, and to avoid larger predators. By expelling a jet of water out of their mantle cavity, they have a form of "jet propulsion" that propels them backward rapidly. Squids and cuttlefish also use the broad, gently undulating fins along their bodies to move slowly forward, backward, or hover with great precision. Octopuses spend most of their lives crawling along the sea bottom using their tentacles, although they can propel themselves rapidly backward with a jet of water from their mantle cavity, expelling a cloud of ink as they do to confuse the predator. Shelled cephalopods, such as the nautilus (and presumably the extinct forms), use jet propulsion for rapid motions, but suspend themselves above the bottom and use their tentacles to creep along the bottom for slow, delicate foraging.

The earliest cephalopods are known from the Late Cambrian, and by the Ordovician, there were huge straight-shelled nautiloids, with shells as long as 10 m. Presumably, these animals had tentacles that were several meters in length as well. These were the first large predator to ever roam the sea bottom, and may have been responsible for many of the defensive adaptations seen in the Ordovician fauna, especially the changes in the trilobites. Large straight-shelled nautiloids continued to be important predators through the rest of the Paleozoic, although by the Devonian, there were other important new predators: the fish (especially the huge placoderms) and the earliest ammonoids. Ammonoids were one of the most abundant predators of the Mesozoic seas as well.

Cephalopods exhibit a wide range of body size, from tiny forms less then a centimeter in diameter, to ammonites of many different sizes and proportions. The largest ammonite had a diameter of 1.7 m, and may have weighed several hundred pounds. Giant squid are reported to reach 18 m (50 feet) in length, counting their long tentacles. Octopuses range from 5 cm to about 10 m across with their arms spread out. Cephalopods are an exclusively marine group, and never adapted to freshwater or land like some gastropods and bivalves did. However, within the marine realm they occupy almost every part of the ocean. Squids swim in huge numbers in shallow marine waters, forming giant schools that are important food for many other animals. There are also many different types of squid living in intermediate and deep waters, and some poorly known groups, such as the vampire squids, live in waters as deep as 3000 to 5000 m. The giant squid is another animal that has seldom been seen alive, so the behavior and maximum size of these animals can only be estimated from the few specimens washed up on beaches, and the scars that they leave on the skins of their principal predator, the sperm whale. Undoubtedly, some of these huge squids are the source of many of the legends of sea monsters, including the most memorable monster of Jules Verne's *20,000 Leagues under the Sea*.

Cephalopods are intriguing and valuable to humans as well, not just monsters of the deep. Octopus, squid, and cuttlefish are popular as food in many parts of the world, with over 2 million tons of squid caught by fishermen each year. The calcareous stiffening plate of the cuttlefish, known as the cuttlebone, is widely used as a source of calcium for pet birds. Fossil ammonoids have been collected since ancient times, although few knew what they really were. Partially carved ammonites were found in a late Paleolithic cave (about 25,000 years old) in south Germany, showing that they were collected even before recorded history. The Roman historian Pliny the Elder was the first to write about the mysterious "Ammon's horns" (so called because they resembled the coiled horns of the Egyptian ram-god, Ammon). In the Middle Ages, some ammonites were thought to be petrified coiled snakes ("serpent stones"), and mediaeval artisans sometimes carved a snake's head on the last chamber to enhance the snake-like appearance. Yet before 1700 the English microscopist and versatile scientist Robert Hooke recognized the similarity of ammonites to the recently discovered chambered nautilus, and correctly inferred that they were related (Fig. 1.1).

By the early 1800s, the scientific study of cephalopods had advanced greatly, because pioneering biostratigraphers like d'Orbigny, Quenstedt, Oppel, and others found them very useful for their zonation of the Mesozoic. Since that time, the primary impetus for the detailed studies of ammonoids has been biostratigraphic, because they are one of the most useful index fossils for the Devonian through Permian, and the principal means of telling the age of rocks in the Mesozoic. In some parts of the Paleozoic and Mesozoic, the ammonite zones have a resolution of less than 100,000 years, making them much more precise than almost any other fossil group (see Chapter 10).

Systematics and Morphology

Almost all biological and ecological studies of fossil shelled cephalopods must necessarily be based on their only living representative, *Nautilus* (Fig. 15.19). Although shells of *Nautilus* were known to the ancients, the first detailed dissection of their anatomy was undertaken by Richard Owen in 1832. Since that time, a tremendous amount of research has been focused on this one genus (see Ward, 1987, 1988; Saunders and Landman, 1987), because it is the key to understanding ammonoids. Living *Nautilus* is a single genus, with probably six valid species, found entirely in the southwest Pacific (although shells have reached the shores of East Africa and Japan in historic times, either when their range was larger, or simply by postmortem floating). The coiled shell forms an ever-increasing logarithmic spiral, and externally it is ornamented with a variety of irregular brown and white color bands. Inside, the shell is divided into a number of chambers (**camerae**) separated by walls (**septa**), which become increasingly larger along the spiral. The first chamber is called the **protoconch**, and the entire chambered part of the shell is called the **phragmocone**. Each of the chambers is connected by a fleshy stalk extending from the back of the body called the **siphuncle**, which penetrates through a hole in the center of each septum. The unoccupied chambers are filled either with gas or liquid, and the variation of the gas content of these chambers helps maintain near neutral buoyancy.

The soft body of the *Nautilus* occupies only the living chamber, and consists of a compact mass of tissue surrounded by the mantle. A ring of 38 small tentacles surrounds the mouth, with a parrot-like beak in the center of the ring. Behind the tentacles are a pair of well-developed eyes, with excellent vision, especially in the dark. The entire soft body is partially protected by the hard shell, and the rest is covered with a tough, leathery **hood**, which closes over the body when the animal retracts into its shell.

The mouth leads into a crop for holding food, followed by a short U-shaped digestive tract that empties into the mantle cavity. Within the mantle cavity are a pair of feathery gills. The main function of the mantle cavity, however, is jet propulsion. During swimming, *Nautilus* floods the mantle cavity with water, then closes off the entrances and forces the water out the nozzle-like **hyponome** by pressing its body down on the mantle cavity. The rhythmic pumping can cycle very fast, producing relatively rapid spurts of

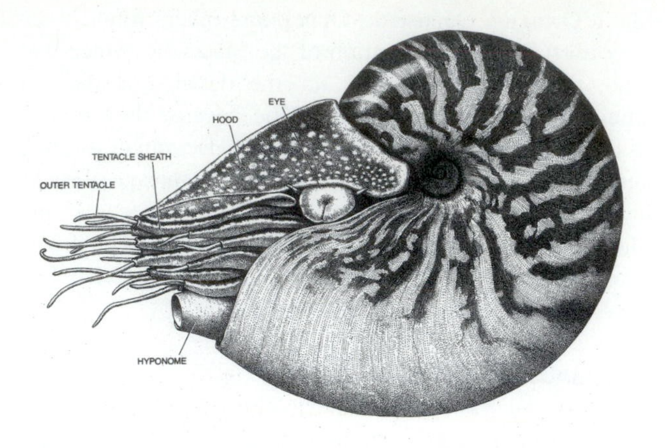

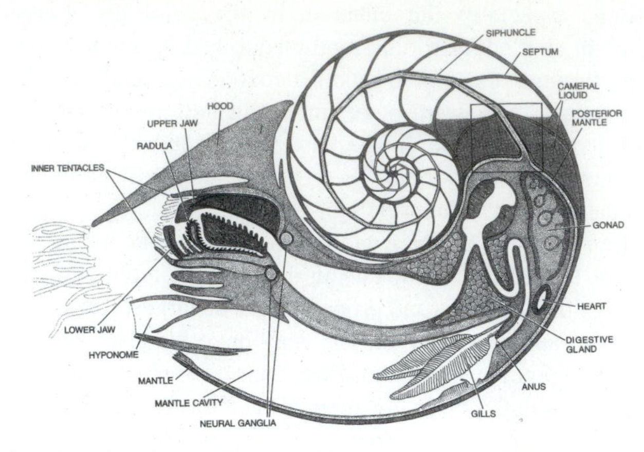

Figure 15.19. External morphology (left) and internal features (right) of the living Nautilus. (From Ward, 1987.)

movement. The center of gravity is a few millimeters below the center of buoyancy, so that the shell is very stable even when this force is applied off-center. Each jet of water does not cause the body to spin, but moves straight backward. Of course, this backward movement is still relatively slow compared to fish or squids, because the wide shell of *Nautilus* is nowhere as streamlined as a squid or a fish. However, backward movement is primarily for short-term escape, and most feeding is done by slowly hovering over the bottom and grabbing prey with the tentacles. *Nautilus* can move in most any direction by changing the direction that its hyponome "nozzle" points.

At one time, it was thought that *Nautilus* actively pumped the water out of each chamber, once it had moved forward and sealed the old chamber up with a new septum behind its body. This misconception is still widely held and is found even in current textbooks. Recent research (Denton and Gilpin-Brown, 1966; Ward, 1987) shows that the gas regulation mechanism works quite differently. The siphuncle controls the fluid content through osmosis by pulling ions out of the cameral fluid until it is less salty than the blood. The cameral fluid will thus tend to flow in the direction of higher salt concentration, which is out of the chamber and into the bloodstream. This process is very slow, with measured rates of only 1 ml per day of liquid removal, so any real change takes weeks. Contrary to popular myth, *Nautilus* cannot rise and sink in the water column in a matter of hours by changing its fluid content. Instead, it evacuates each chamber only once after it is formed, and then keeps its shell at slight negative buoyancy, so it tends to sink slightly when not swimming. To maintain its depth, or to rise to the surface, it must actively propel itself with water from the mantle. Only when it wants to slowly sink to the bottom can it relax.

Nautilus are known to spend their days in deep, dark waters off the reefs of the South Pacific, typically at depths of 150 to 300 m, and seldom above 75 m. In the dark of night, when the competition from fish is decreased, they swim to the surface and forage for crabs, lobsters, and carrion. They grasp their prey with their tentacles and then crush it and eat it with their parrot-like beak. After a night of feeding, *Nautilus* sinks back down to the deep waters where predatory and competitor fish are rare. Experimental studies have shown that shells will implode from the water pressure at depths of 700 to 800 m, and in thin-shelled juveniles, implosion occurred at 300 m. Young *Nautilus* must live near the upper reaches of their range, and as they get older, larger, and more robust, they can hide at greater and greater depths.

The class Cephalopoda is divided into a number of subclasses, some of which are monophyletic, and others serving as paraphyletic ancestral groups. Dzik (1984) recognized only three subclasses (Nautiloidea, Ammonoidea, and Coleoidea), but this makes the Nautiloidea into a huge wastebasket for all primitive cephalopods that are neither ammonoids nor coleoids. Pojeta and Gordon (1987) recognized six, subdividing Dzik's concept of subclass Nautiloidea into four subclasses (the first four below). Pojeta and Gordon's (1987) classification is given below.

Class Cephalopoda

Subclass Nautiloidea

Nautiloids include not only the living genus *Nautilus*, but a wide variety of primitive forms going back as far as the Cambrian (Fig. 15.20). The primitive forms tend to be straight-shelled (**orthoconic**) or slightly curved (**cyrtoconic**), but advanced forms are strongly coiled like the living genus, with each coil completely covering the preceding coils (**involute**) (Fig. 15.21). The siphuncle usually penetrates the center of each septum. The chambers frequently contain calcareous **cameral deposits** to serve as ballast.

Subclass Endoceratoidea

These large to huge orthoconic cephalopods (Fig. 15.20) were the major predators of the Ordovician, and died out by the Silurian (Fig. 15.22). The siphuncle usually passes through the ventral (lower) portion of the septum, and it is filled with conical deposits called **endocones** that are nested within each other like a stack of conical paper cups. These serve as ballast to hold the shell horizontal.

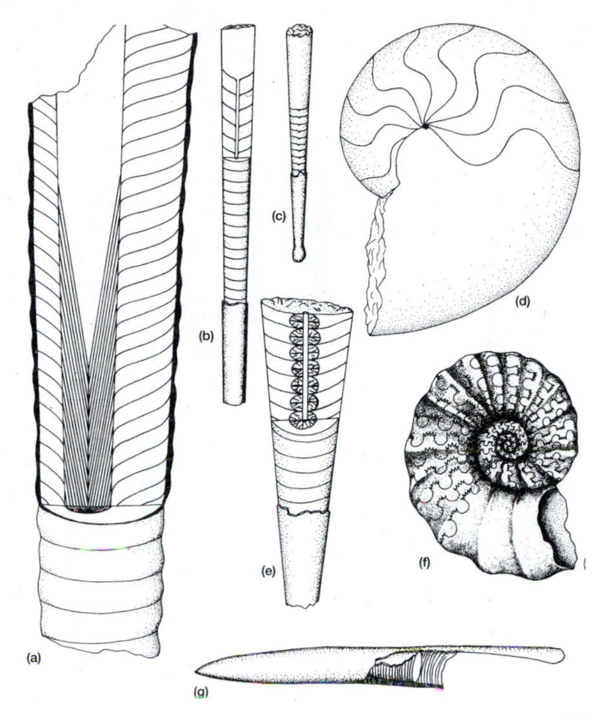

Figure 15.20. Typical examples of the subclasses of the Cephalopoda. (A) Endoceratoidea. (B, D) Nautiloidea. (C) Bactritoidea. (E) Actinoceratoidea. (F) Ammonoidea. (G) Coleoidea. (From McKinney, 1991.)

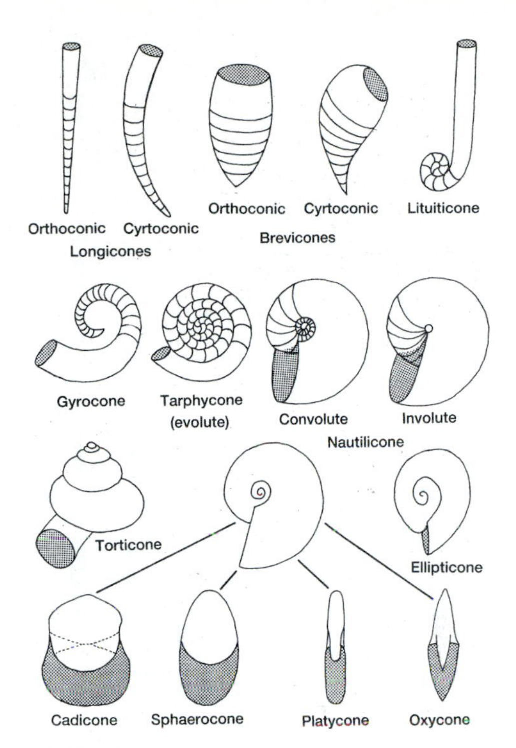

Figure 15.21. Common shapes of cephalopod shells. (From McKinney, 1991.)

Subclass Actinoceratoidea

Another large to huge orthoconic group found only in the Ordovician (Fig. 15.22), actinceratoids can be recognized by the necks of the septa that enclose the siphuncle, which are slightly recurved and enrolled. The connecting rings covering the siphuncle between the septal necks are inflated, sometimes filled with calcareous **siphuncular deposits** (Fig. 15.20). There may also be cameral deposits for additional ballast.

Subclass Bactritoidea

These are small orthoconic forms with a small, globular protoconch that is separated from the succeeding shell by a small constriction (Fig. 15.20). The septa are curved in the more adult chambers, and have a narrow ventral siphuncle with straight septal necks. There are no siphuncular or cameral deposits. Bactritoids appeared in the Ordovician and ranged to the Triassic, and are thought to be the link between the nautiloids and ammonoids (Fig. 15.22).

Subclass Ammonoidea

By far the most successful and diverse subclass of cephalopods were the ammonoids (Fig. 15.22). Ammonoids have shells that are planispiral and usually involute, although some have exposed their inner coils (**evolute**) (Fig. 15.21). In all ammonoids except one group, the siphuncle is ventral, which means that it runs along the outer edge (**venter**) of these planispiral forms. The most important taxonomic feature on ammonoid shells is the **suture**, or the line of contact between the inner shell wall and the septum. If the outer shell is polished or eroded off, this line of contact can be seen and studied (Figs. 15.20, 15.23). Typically, the suture is traced from the curved surface of the shell onto a flat piece of paper, so that it can be described and compared. On a standard suture tracing, a prominent arrow indicates the position of the ventral siphuncle, with the point of the arrow indicating the direction of the head and living chamber. The suture line then trends away from the venter to the center of the coil (**umbilicus**). Each curve on the suture line that is convex (pointed toward the head) is a **saddle**, while the corresponding concavities (pointed away from the head) are called **lobes**. Ammonoids have a standard sequence and terminology of saddles and lobes (Fig. 15.23) that is critical to describing and recognizing specimens.

Originating from the bactritoids in the Devonian, the earliest ammonoids had long, straight suture lines (Fig. 15.24), since the septa were flat or slightly curved (as in modern *Nautilus*). However, most Paleozoic ammonoids had a **goniatitic** suture, where there were as many as eight saddles and lobes with simple curves or zig-zag patterns (Figs. 15.23, 15.24). These curves are a reflection of curves and corrugations where the septum met the inner shell.

Triassic ammonoids tended to have **ceratitic** sutures, which have smooth U-shaped saddles, and lobes with small crenulations (Fig. 15.24). Jurassic and Cretaceous ammonoids had **ammonitic sutures**

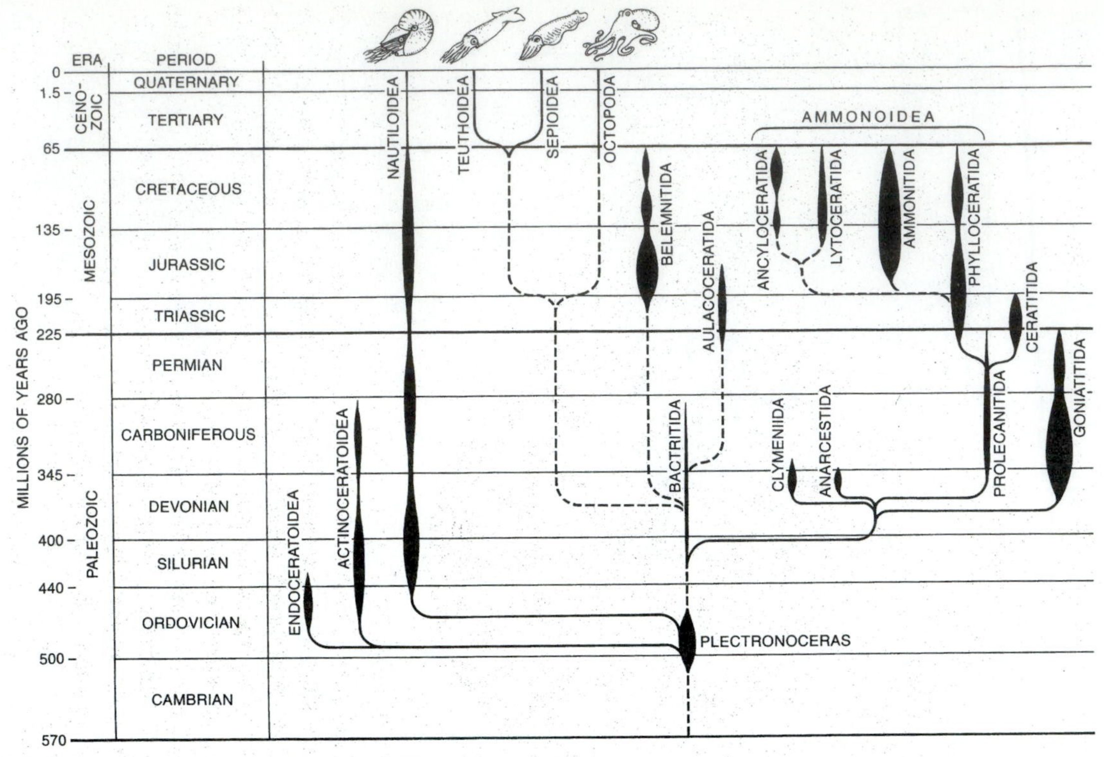

Figure 15.22. Evolution of the cephalopods. (From Ward, 1987.)

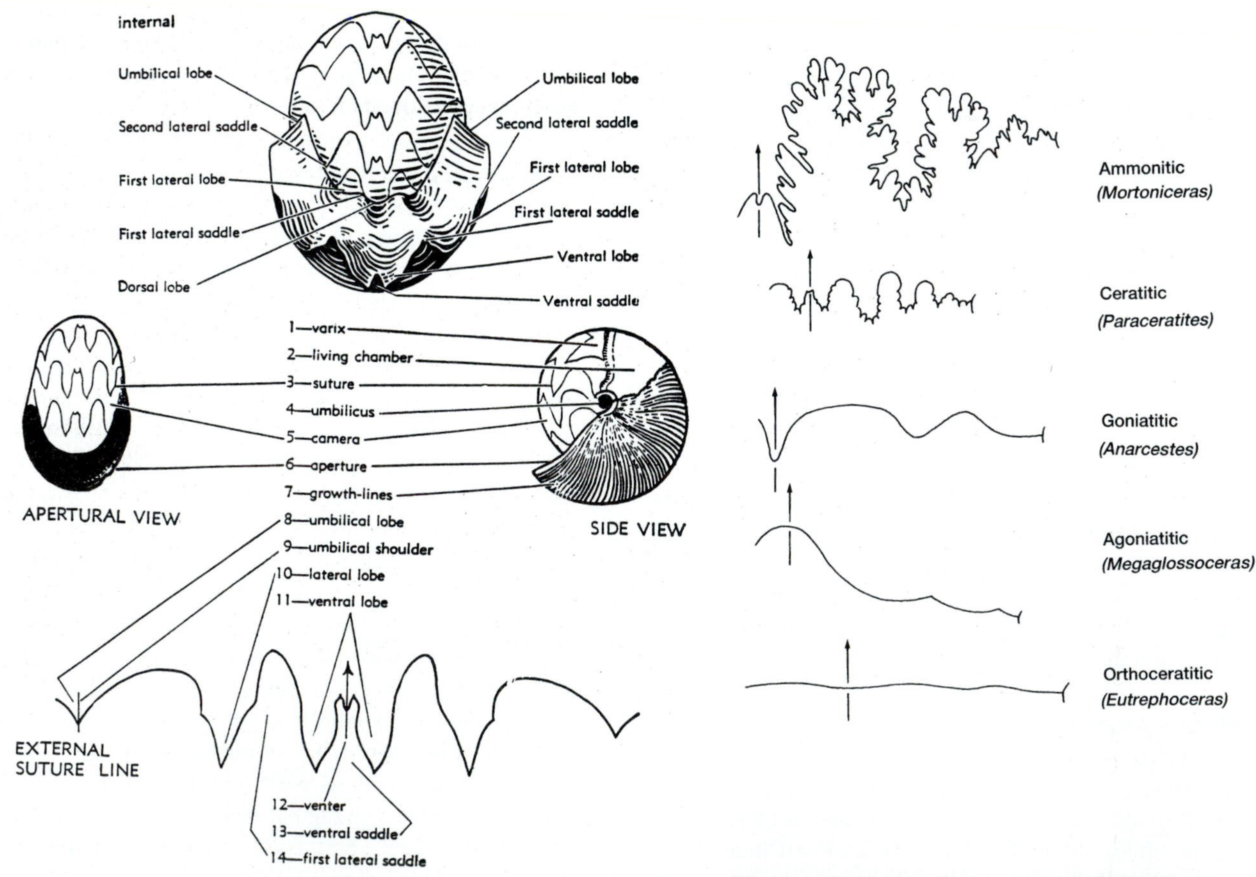

Figure 15.23. Terminology of the ammonoid suture, showing the position of the suture tracing (bottom) with respect to the convolutions on the septum (top) that produces the suture. (From Moore et al., 1953.)

Figure 15.24. The five commonly recognized cephalopod suture types. Nautiloids and early ammonoids have simple sutures (bottom), but ammonoids have increasingly complex sutures (top). (From McKinney, 1991.)

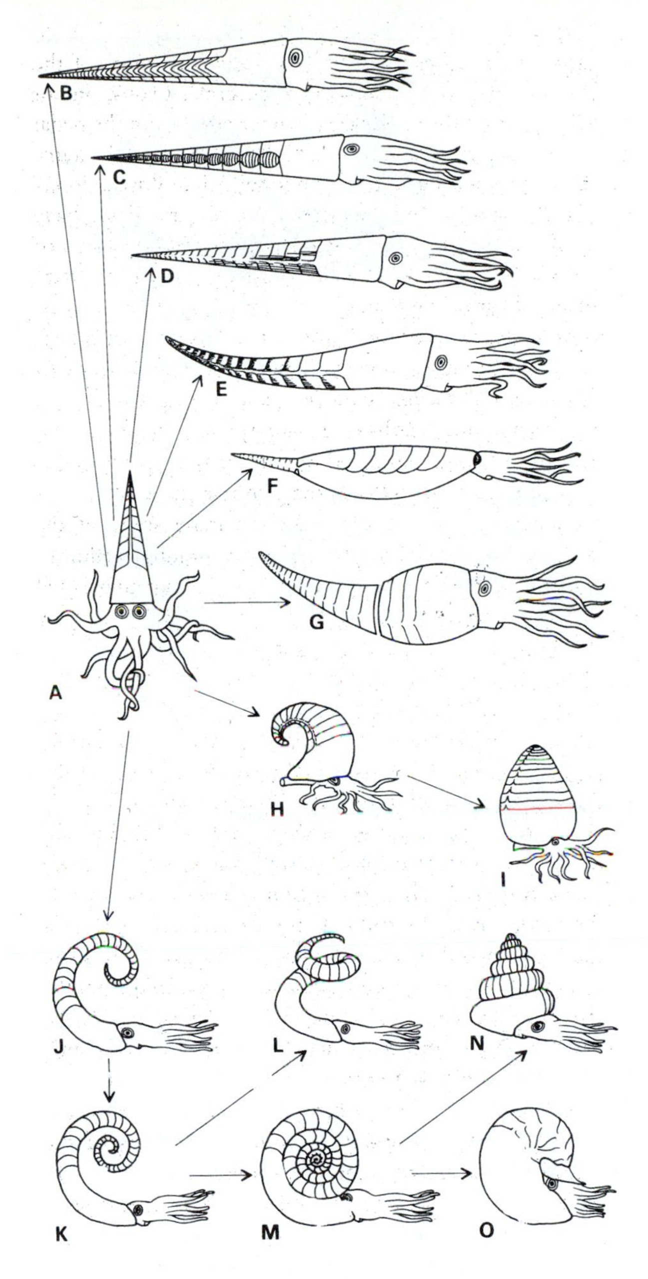

Figure 15.25. Diagram showing the buoyancy control mechanisms and floating positions of shelled cephalopods. From a straight-shelled ancestor with no counterweighting (A), extra shell weighting can be added as endocones (Endoceratoidea, B), complex annular deposits (Actinoceratoidea, C), or cameral deposits (some nautilods, E). Liquid may be retained in the chambers (some nautiloids, D), or the early gas-filled chambers may extend over the body chamber (F) or be shed (G). The chambers can also be coiled over the body (H-O) in a varety of ways. (From Boardman et al., 1987.)

(Fig. 15.24), in which both the lobes and saddles were subdivided by many crenulations, forming incredibly complex, florid patterns. The complexity of these patterns makes it possible to recognize many different taxa in the ammonites, so they are the primary feature used in identification.

Subclass Coleoidea

All the living cephalopods except *Nautilus* are coleoids, including the squids, cuttlefish, and octopus (Fig. 15.22). Most have a highly reduced internal shell (the cuttlebone of cuttlefish, or a thin flexible rod known as a pen that holds the squid's body straight), or none at all (octopus). Consequently, their fossils are not as easy to recognize and study, since they are rarer and have few anatomical features. Although primitive coleoid fossils are known from the Lower Carboniferous, the most common extinct group was the **belemnites**, whose fossils are shaped like large-caliber bullets, and are particularly abundant in the Jurassic and Cretaceous (Fig. 15.20).

Ecology

Because most shelled cephalopods are extinct, we must use the few living analogues for models of their ecology and behavior. Belemnites are thought to belong to an organism much like a modern squid. *Nautilus* is the only living cephalopod with an external shell, so it is used (and perhaps overused) as a starting point for studying ammonoid ecology. In fact, Jacobs and Landman (1991) argue that it is a poor analogue for the wide variety of ammonoids, because it is fundamentally different in many critical ways. For better or worse, however, *Nautilus* serves as a starting point, and a variety of functional and paleoecological studies have given us a richer picture of how most extinct cephalopods lived.

One of the first questions that can be analyzed is how the animal floated when it was alive (Trueman, 1941). The earliest cephalopods from the Late Cambrian were simple conical shells with a few septa, but with no counterweighting of any kind. These relatively small forms must have floated with their heads and tentacles hanging down. This works fine for a small animal, but creates a problem if the shell becomes very long. Consequently, nearly all the Ordovician orthoconic groups (nautiloids, endoceratoids, actinoceratoids) developed one or more methods of ballast in their long shell as a counterbalance. Once the shell was stable and floated horizontally, the cephalopod could catch prey with its tentacles. Endoceratoids accomplished this with endocones (Fig. 15.20A, 15.25B), actinoceratoids with complex siphuncular deposits (Fig. 15.20E, 15.25C), and several nautiloid groups did this with cameral deposits (Fig. 15.25E) or liquid retained in the chambers (Fig. 15.25D). Another solution for animals with an orthoconic shell is to locate the flotation chambers over the living chamber (Fig. 15.25F), or to shed the juvenile gas-filled chambers, leaving a truncated but better balanced shell (Fig. 15.25G). Ultimately, however, the easiest solution is to coil the earlier chambers above the living chamber, so the buoyant part of the shell is above the dead weight of the body. This can be accomplished in many different ways (Fig. 15.25H-O), although the solution preferred by most ammonoids and many nautiloids is the simple planispiral shape (Fig. 15.25K, M, and O).

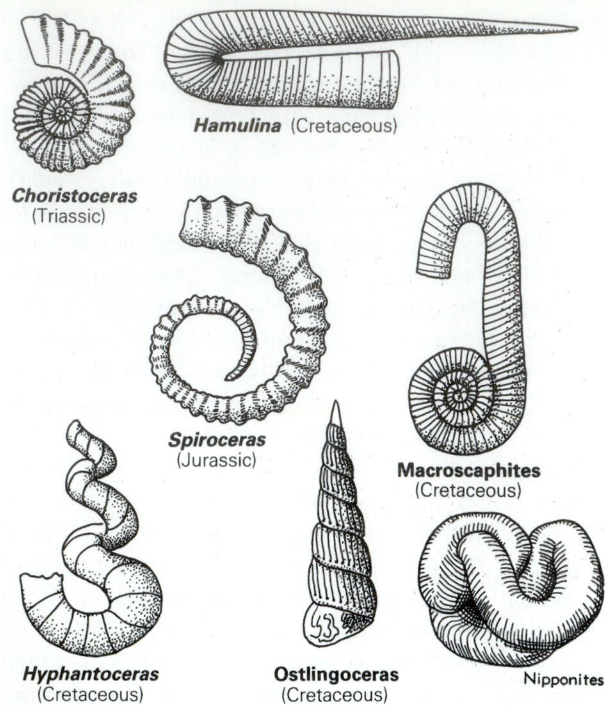

Figure 15.26. From the standard planispiral ammonite, a variety of partially uncoiled or weirdly coiled taxa, known as heteromorphs, evolved. Some were shaped like hairpins, while others were coiled in a question mark shape, or spiraled up an axis, or even tied in a knot (*Nipponites*). (Adapted from Clarkson, 1993.)

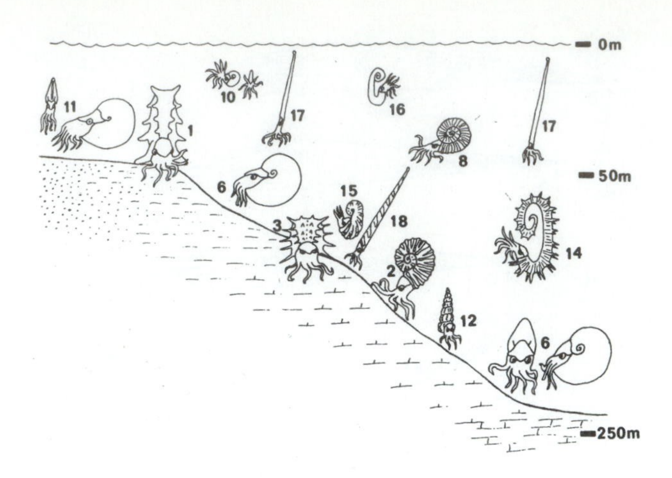

Figure 15.27. Possible habitats of ammonoids from the western interior Cretaceous seaway. 15 are scaphitids, 17 and 18 are baculitids, and 12 are the conispiral turrilitids. (From Batt, 1989.)

Ammonoid shells can be analyzed biomechanically to determine the relative position of the center of buoyancy and the center of gravity. In the most stable orientation, these two centers should lie far apart, or else the shell will be highly unstable and wobble or spin every time force is applied (such as during each pulse of water from the hyponome). Most planispiral ammonoids apparently had a stable orientation with the gas-filled parts of the shell lying over the living chamber as in living *Nautilus*. However, a number of ammonite groups developed uncoiled shapes known as **heteromorphs** (Fig. 15.26). Some were just slightly uncoiled, but others were shaped like hairpins, or question marks, or antelope horns, and some were corkscrew trochospirals like gastropods (although no gastropods ever had sutures or septa). Biomechanical analysis of these shapes suggests that they floated at rather odd angles, and some require that the ammonite's head and tentacles were not in very accessible orientations. These animals probably floated in the open ocean and caught prey that drifted within reach of their tentacles (Fig. 15.27, specimens 16 and 14).

Another approach to understanding the functional morphology and behavior of extinct cephalopods is to place their shells in a tank of flowing water and study the hydrodynamic properties of each shape in a flowing current (summarized in Chamberlain, 1981, and Jacobs, 1992). Not surprisingly, the most streamlined discus-shaped shells (**oxycones**) are the fastest swimmers (Fig. 15.21), while progressively fatter shells (**cadicones**) and globular shells (**sphaerocones**) produce much more drag, and stayed near the bottom. **Serpenticone** shells are thought to be pelagic drifters. Highly ornamented specimens probably didn't attempt to swim fast, but stayed close to the bottom. Trochospiral "corkscrew" heteromorphs were also slow moving, because any strong movement of jet propulsion from their hyponome would have caused the shell to spin on its axis. Chamberlain and Westermann (1976) found that not all shell ornamentation causes drag, however. In some specimens, small bumps and ridges help to break up the boundary layer of turbulence caused by friction of the water as it runs along the shell, and actually help the shell to move through the water faster and more smoothly. The dimples on the golf ball serve the same function. They break up the boundary layer flow so the ball flies through the air with minimum turbulent drag and thus maximum distance.

Combining this information with paleoecological and water depth data, Batt (1989) showed that the Cretaceous ammonites of the Western Interior Seaway actually occupied a wide variety of niches above the seafloor (Fig. 15.27). Some were clearly pelagic floaters, while those with high-drag shells that floated with their heads and tentacles down were probably slow-moving foragers hovering just above the seafloor. The straight-shelled baculitids (Fig. 15.27, specimens 17 and 18) are particularly puzzling. These animals had no counterweighting system, and must have floated slowly with the point of the shell up and head and tentacles down, grabbing whatever prey came within reach. No matter how inefficient this arrangment seems, it must have been very successful, because baculitids evolved very rapidly, and were so common all over the Cretaceous seas that they are the best biostratigraphic indicator for that period.

In addition to inferring the life habit and swimming abilities of extinct cephalopods, another functional problem has long puzzled paleontologists: why do the suture patterns become increasingly complex? The classic explanation dates back to Buckland (1836), who argued that more complex sutures serve as stronger buttresses against the hydrostatic pressure on the shell, comparable to the reinforcing effect seen in corrugated cardboard. However, that argument has been discredited by a number of lines of evidence (Saunders, 1995; Saunders and Work, 1996). For example, Saunders (1995) found that shells with greater sutural strength and complexity did not show a corresponding decrease in thickness that such reinforcement should have provided (but see Jacobs, 1996, for a critique). Although the curvature of the sutures undoubtedly has some function as a support (Jacobs, 1990), the range of complex shapes cannot be fully explained by the buttressing hypothesis alone.

Instead, a number of paleontologists (Mutvei, 1967; Kulicki, 1979; Kulicki and Mutvei, 1988; Ward, 1987; Weitschat and Bandel, 1991; Saunders et al., 1994) have suggested that the complex folding at the edge of the septum is associated with increased surface area for the membrane that helps remove ions from the cameral liquid. In other words, the more complex the suture, the more surface area for the ion-removing membranes, and the faster the chamber can be emptied of fluid. This explanation is corroborated by the recent discovery of specimens with delicately preserved structures that may be remains of this folded membrane, with outpocketings that suggest organs for ion removal (Saunders et al., 1994). Although this doe not explain the entire history of sutural complexity changes, it seems to fit the known evidence much better than the buttressing hypothesis.

Evolution

Evolutionary history—Yochelson et al. (1973) and Runnegar (1985) suggested that cephalopods arose from a monoplacophoran ancestor with a tall, conical, slightly curved shell, such as the Late Cambrian *Knightoconus*. The earliest known cephalopod, *Plectronoceras*, is a tiny Late Cambrian form with a simple conical shell, but it has chambers, septa, and a siphuncle. From such simple origins, the straight-shelled cephalopods (especially endoceratoids, actinoceratoids, and nautiloids) underwent a spectacular Ordovician radiation. Forty genera of endoceratoids are recognized from the Early Ordovician, and 20 genera of actinoceratoids are known from the Middle Ordovician. Not only were they diverse, but they were also the largest predators on earth at the time, occupying a dominant role that they have never since matched. By the Late Ordovician, both endoceratoids and actinoceratoids were in decline, with endoceratoids vanishing in the Silurian, and actinoceratoids straggling on until the Early Carboniferous. Several groups of nautiloids were moderately diverse through the rest of the Paleozoic.

Bactritids arose in the Devonian, but were never very diverse or abundant, with only seven or eight genera at any given time. However, they are important as the ancestors of the ammonoid and coleoid radiations, which also began in the Devonian. From their Early Devonian ancestry, ammonoids diversified rapidly until there were 30 families by the Late Devonian. Their history is then marked by an amazing sequence of repeated evolutionary radiations, followed by extinction crises leaving only a few survivors (Kennedy, 1977; House, 1981; Teichert, 1985; Pojeta and Gordon, 1987). For example, only three genera survived the Frasnian/Famennian extinction event (see Chapter 6), after which there was another radiation of ammonoids in the Famennian. Of about 80 Famennian genera, only two persisted into the Mississippian. This surviving group gave rise to 25 families and over 180 genera of Mississippian ammonoids, all with goniatitic sutures. Ammonoid diversity was cut by the Mississippian/Pennsylvanian extinction, with only nine families surviving, but by the Late Pennsylvanian, there were again 30 families of ammonoids, and 27 are known from the Middle Permian. Their diversity began to decline in the Late Permian, and only two genera survived the Permian catastrophe to become ancestors of the Triassic radiation.

In the Triassic, the ceratitic ammonoids radiated again, reaching a diversity of about 80 families and over 500 genera. Most of the ceratites were wiped out by the end-Triassic extinction event, leaving only a few genera as the ancestors of the final, greatest radiation of ammonitic ammonoids. These again radiate rapidly, with a peak of about 90 families in the Jurassic and 85 families in the Cretaceous. By the latest Cretaceous, however, there were only 11 families, and only five are known by the K/T boundary. Some paleontologists argue that the ammonites declined throughout the latest Cretaceous and may have even been extinct before the K/T impact event, while others attribute this apparent pattern to sampling difficulties in uppermost Cretaceous rocks, and suggest that the ammonites were still thriving right up to the K/T boundary.

Coiled nautiloids were always present in the background of these spectacular radiations and crashes of ammonoids, but were never very diverse, and remained conservative in their body form, with simple sutures. However, nautiloids survived the K/T event, and produced a variety of new forms immediately thereafter (some of which were reminiscent of the ammonoids). Occasionally, they can be found in Tertiary beds that yield abundant bivalves and gastropods, and they had a worldwide distribution through the Cenozoic. Eventually, they became restricted to the southwest Pacific, where only one genus hangs on for survival. It too may be endangered, because their shells are now so valuable that native fishermen have pushed their fragile populations to near extinction (Ward, 1987).

Much less is known about the history of coleoids, because their simple internal shell is less commonly preserved, and reveals less information about their taxonomy than the complex, abundant ammonoid shells. The best specimens of coleoids come from *Lagerstätten* like the Devonian Hunsrückschiefer and Jurassic Solnhofen Limestone (see Chapter 1), where the soft tissues are beautifully preserved around the internal shell. Occasional fossils are known from the Devonian, but they do not become abundantly fossilized until the great radiation of belemnites in the Jurassic and Cretaceous, when their fossils literally covered the seafloor in some places.

Evolutionary patterns—Ammonites have been one of the most popular fossil groups for evolutionary studies, because they have a long fossil record with thousands of well-preserved specimens, showing detailed morphology through millions of years. Almost as soon as the idea of evolution was accepted by paleontologists, ammonites were used as prime examples. In Darwin's lifetime, Alpheus Hyatt described many examples of heterochronic evolution in ammonites, because ammonites preserve their entire ontogeny from the protoconch to the final living chambers, all in a single specimen. Most ammonites were cited as excellent examples of Haeckelian recapitulation (peramorphosis), because the later chambers seemed to show addition of new terminal growth stages past the last living chamber of the ancestor (see Chapter 2). Ammonite specialists were convinced that the entire evolutionary history of a species could be directly observed through the ontogenetic series of its living chambers. However, in the early twentieth century, careful study of some Jurassic ammonites showed that new characters appeared in the early ontogenetic stages and then spread to adult ontogenetic stages in specimens from higher stratigraphic levels. This was the opposite of the recapitulation model, which thought that new features should be added on only in the late ontogenetic stages of the descendant. In other cases, ammonites seem to show paedomorphic progenesis, where the descendant adult shows a juvenile morphology as a result of early sexual maturation (Landman, 1989). (This is in contrast to neoteny, which causes paedomorphosis by slowing down the rate of maturation.) Ironically, the ammonites were long considered the classic case of heterochrony by peramorphic changes like recapitulation, but as cases have been restudied, paedomorphic changes appear to be much more prevalent than peramorphic changes (Swann, 1988; Dommergues, 1990), although Landman (1988) found more peramorphoclines than paedomorphoclines.

At one time, the bizarre appearance and apparent nonfunctionality of some heteromorphs were thought to be driven by mysterious, maladaptive "internal evolutionary forces." These ideas originated with Alpheus Hyatt, and were common at the turn of the century when Darwinian natural selection was unpopular as a mechanism for evolution (see Chapter 5). They were still promoted even after Neo-Darwinism appeared (e.g., Schindewolf, 1936, 1950). To some paleontologists, the appearance of these heteromorphs late in a particular ammonite radiation suggested that the group was experiencing "racial senescence," and had become "degenerate" in its "old age" before final extinction. As Wiedmann (1969) and Kennedy (1977) point out, however, these ideas are nonsense. Heteromorph forms evolved repeatedly in a number of different lineages throughout ammonoid history, suggesting that it was positively selected for. Some of the heteromorphs, especially the baculitids, were extremely diverse and successful and widespread for many millions of years, so they were clearly not "maladaptive." Some heteromorph genera spanned extraordinary periods of time: *Scaphites* lasted over 30 million years and *Baculites* over 20 million years. Finally, heteromorphs do not occur at the end of a radiation but throughout, and are no more common just before the final extinction than earlier in the radiation, so there is no evidence that they are a product of "racial senility."

Another pattern that is typical of ammonoids is **iterative evolution**, or the repeated occurrence of similiar morphologies in different lineages due to convergence. At one time, iterative evolution was thought to be rampant in ammonoids, and specialists such as Frebold (1922) and Arkell (1950) argued that highly specialized ammonitic forms were evolved over and over again from persistently primitive lineage, not descended from slightly less specialized ammonites. In more recent studies, however, paleontologists have shifted away from the idea that there was an enormous amount of convergence, and instead prefer the simpler explanation that only limited iterative evolution occurs. Donovan et al. (1981) abandoned "the theory of Iterative Evolution [because] new discoveries and studies have not substantiated the idea of 'replenishment' of groups by successive homeomorphic waves, especially from the conservative suborders Phylloceratina and Lytoceratina. These suborders now stand in even more isolation than before, clearly distinguished in morphology, and probably in habitat, from the contemporary Ammonitina." In spite of this, however, there are examples of iterative evolution in the ammonoids, although it is much less prevalent than once thought (Landman, 1989).

Ammonite workers, with their dense fossil record of thousands of closely spaced specimens with highly detailed morphology, continue to argue that phyletic gradualism is common in their group (Kennedy, 1977; Kennedy and Wright, 1985; Dommergues, 1990). However, much of the evidence is based on qualitative descriptions and limited comparisons, and has not been analyzed with the kind of statistical rigor suggested by Gould and Eldredge (1977) and exemplified by Stanley and Yang (1987). In some cases, classic examples of gradualism have broken down when examined rigorously. For example, Brinkmann (1929) described apparent gradual evolutionary trends in over 3000 specimens of the Jurassic genus *Kosmoceras*. However, as Callomon (1963) showed, the data do not support Brinkmann's interpretations. First of all, he did not consider the possibility of sexual dimorphism, and many of his "lineages" are actually different male and female morphs. When this is taken into account, a great deal of morphological stasis is actually apparent (Kennedy, 1977). In addition, Brinkmann did not conduct rigorous statistical analyses. He would pick up a specimen, refer it to genus and species, measure it and then discard it. Thus, his data are not reproducible, and because he had already made a taxonomic decision, they could only support his preconceived interpretations. Once these factors are filtered out, about the only gradual trend left in *Kosmoceras* is a gradual increase in size, which is widespread through the ammonoids (Hallam, 1975; Kennedy, 1977; Dommergues, 1990).

For the prevalence of gradualism versus punctuation to be assessed, someone must actually look at all the lineages in a given time and region and rigorously quantify the changes in each to determine if there is true gradualism, and then how common it is. The biostratigraphic focus of ammonite paleontologists may have tended to highlight examples of gradual evolution and ignore cases of stasis. In a few instances, there are examples not only of stasis but of rapid, dramatic change. For example, Okamoto (1988) interpreted the sudden origin of the knot-shaped heteromorph *Nipponites* as due to saltatory evolution, with no functional intermediate stages. Given the rarity of well-documented cases of gradualism in the rest of the Metazoa, it seems likely that the prevalence of gradualism in ammonites is due for a reassessment.

For Further Reading

Arkell, W. J. A. 1950. A classification of Jurassic ammonites. *Journal of Paleontology* 24:354-364.

Arkell, W. J. A., and others. 1957. Mollusca 4. Cephalopoda. Ammonoidea, *in* R. C. Moore, ed. *Treatise on Invertebrate Paleontology*. Part L. Geological Society of America and University of Kansas Press, Lawrence.

Callomon, J. H. 1963. Sexual dimorphism in Jurassic ammonites. *Leicester Literary and Philosophical Society Transactions* 62:21-56.

Denton, E. J., and J. B. Gilpin-Brown. 1966. On the buoyancy of the pearly nautilus. *Journal of the Marine Biological Association of the United Kingdom* 46:723-759.

Dommergues, J.-L. 1990. Ammonoids, pp. 162-187, *in* McNamara, K. J., ed. *Evolutionary Trends*. University of Arizona Press, Tucson.

Dzik, J. 1984. Phylogeny of the Nautiloidea. *Palaeontologica Polonica* 45:1-219.

Hallam, A. 1975. Evolutionary size increase and longevity in Jurassic bivalves and ammonites. *Nature* 258:493-496.

House, M. R., and J. R. Senior, eds. 1981. *The Ammonoidea*. Systematics Association Special Volume 18. Academic Press, New York.

Jacobs, D. K. 1990. Sutural pattern and shell stress in *Baculites* with implications for other cephalopod shell morphologies. *Paleobiology* 16: 336-348.

Jacobs, D. K. 1992. Shape, drag, and power in ammonoid swimming. *Paleobiology* 18:203-220.

Jacobs, D. K. 1996. Chambered cephalopod shells, buoyancy, structure, and decoupling: history and red herrings. *Palaios* 11: 610-614.

Kennedy, W.J. 1977. Ammonite evolution, pp. 251-304, *in* Hallam, A., ed. *Patterns of Evolution as Illustrated by the Fossil Record*. Elsevier, Amsterdam.

Kennedy, W. J., and W. A. Cobban. 1976. Aspects of ammonite biology, biogeography, and biostratigraphy. *Special Papers in Palaeontology* 17:1-94.

Kennedy, W. J., and C. W. Wright. 1985. Evolutionary patterns in Late Cretaceous ammonites. *Special Papers in Palaeontology* 33: 131-143.

Lehmann, U. 1981. *The Ammonites: Their Life and their World*. Cambridge University Press, Cambridge.

Landman, N. H. 1988. Heterochrony in ammonites, pp. 159-182, *in* McKinney, M. L., ed. *Heterochrony in Evolution: A Multidisciplinary Approach*. Plenum, New York.

Landman, N. H. 1989. Iterative progenesis in Upper Cretaceous ammonites. *Paleobiology* 15:95-117.

Landman, N. H., and others, eds. 1996. *Ammonoid Paleobiology*. Plenum Press, New York.

Pojeta, J., Jr., and M. Gordon, Jr. 1987. Class Cephalopoda, *in* Boardman, R.S., A.H. Cheetham, and A. J. Rowell, eds. *Fossil Invertebrates*. Blackwell Scientific Publishers, Cambridge, Mass., pp. 329-358.

Saunders, W. B. 1995. The ammonoid suture problem: relationships between shell and septal thickness and suture complexity in Paleozoic ammonoids. *Paleobiology* 21:343-355.

Saunders, W. B., and N. H. Landman, eds. 1987. *Nautilus, the Biology and Paleobiology of a Living Fossil*. Plenum Press, New York.

Saunders, W. B., and D. M. Work. 1996. Shell morphology and sutural complexity in Upper Carboniferous ammonoids. *Paleobiology* 22:189-218.

Teichert, C. 1985. Crises in cephalopod evolution, *in* Bottjer, D. J., C. S. Hickman, and P. D. Ward, eds. *Mollusks, Notes for a Short Course*. University of Tennessee Studies in Geology 13:202-214.

Teichert, C., and others. 1964. Mollusca 3. Cephalopoda, Nautiloidea, *in* Moore, R. C., ed. *Treatise on Invertebrate Paleontology*. Part L. Geological Society of America and University of Kansas Press, Lawrence.

Trueman, A. E. 1941. The ammonite body chamber, with special reference to the buoyancy and mode of life in the living ammonite. *Quarterly Journal of the Geological Society of London* 96:339-385.

Ward, P. D. 1982. *Nautilus*: have shell, will float. *Natural History* 91:64-69.

Ward, P. D. 1987. *The Natural History of Nautilus*. Allen & Unwin, Boston.

Ward, P. D., and G. E. Westermann. 1985. Cephalopod paleoecology, *in* Bottjer, D. J., C. S. Hickman, and P. D. Ward, eds. *Mollusks, Notes for a Short Course*. University of Tennessee Studies in Geology 13:215-229.

Figure 16.1. Perfectly articulated crinoids from the Mississippian limestones near Crawfordsville, Indiana. (Photo by the author.)

Chapter 16

Spiny Skins
The Echinoderms

The echinoderms may seem, from a human point of view, to be a blind alley of no particular importance. Were we to suppose that life was purposive, that everything was part of a planned progression due to culminate in the appearance of man or some other creature that might rival him in dominating the world, then echinoderms could be dismissed as of no consequence. But such trends are clearer in the minds of men than they are in the rocks. The echinoderms appeared early in the history of life. Their hydrostatic mechanisms proved a serviceable and effective basis for building a variety of bodies, but not susceptible in the end to spectacular development. In the areas that suit them, they are still highly successful. The crown-of-thorns starfish occasionally proliferates to plague proportions and devastates great areas of coral. Crinoids are brought up in trawls from the deep sea several thousand at a time. If it is improbable that any further major dvelopments will come from this stock, it is also unlikely, on the evidence of the last six hundred million years, that the group will disappear as long as life remains possible at all in the seas of the world.

David Attenborough, *Life on Earth*, 1979

For anyone who has ever considered the bizarre panoply that composes the Echinodermata, it is difficult to lose sight of the fact that echinoderms are strange. Their ontogenetic twists and turns of symmetry, absence of certain familiar organ systems, and presence of some less familiar systems might tempt one to imagine an extraterrestrial origin for the Echinodermata. The very strangeness of the Echinodermata is partly responsible for the dedication that specialists in the group feel. We revel in their weirdness.

Rich Mooi, "Evolutionary Dissent," 1989

INTRODUCTION

After the arthropods and molluscs, the echinoderms are the third most commonly skeletonized phylum in the marine realm. Unlike the other two phyla, however, echinoderms have no respiratory system or mechanism for osmotic regulation, so they cannot survive outside marine waters. There are no freshwater or terrestrial echinoderms (a few species are known from brackish water), nor does it appear that they have ever occupied those niches in the past. Nevertheless, within the marine realm, echinoderms can be extraordinarily abundant. As Attenborough points out, sea stars can multiply to plague proportions very quickly, with densities as great as a hundred individuals per square meter. Sea urchins are among the most common denizens of tide pools, and in some soft sea bottoms, sand dollars and heart urchins are extremely abundant. During the later Paleozoic, the shallow seafloor was covered by millions of crinoids (Fig. 16.1) and blastoids, and they left thick deposits of crinoidal limestone that covered entire continents with deposits many meters thick. The volumes of echinoderm debris from just a single formation are impressive. It has been calculated that the total volume of the Lower Mississippian Burlington Limestone is 300 x 10^{10} cubic meters, representing the skeletal remains of approximately 28 x 10^{16} individual crinoid animals! In some habitats, echinoderms are the dominant group. For example, in the deep abyssal plains of the ocean at depths greater than 1000 m, the commonest animals are sea cucumbers and brittle stars (Fig. 16.2), where they make up 90% of the benthic biomass. When deep-sea trawls or cores are brought up, they usually include just these two groups and little else.

The fossil record shows that the five living classes of echinoderms (sea stars, brittle stars, sea urchins, crinoids, and sea cucumbers) are but a tiny remnant of their extinct diversity. Because their hard skeleton of calcite preserves easily, echinoderms have an excellent fossil record. About 6650 species are alive today, but this is about half of their known fossil diversity. Over 3500 genera and 13,000 species are known from the fossil record, distributed into about 20 different classes. Living echinoderms occupy a wide variety of body shapes, from the stalked crinoids to the crawling sea stars to the spiny sea urchins to the worm-like sea cucumbers, but during the past they were even more disparate. Some extinct echinoderms were built on a spiral body plan, whereas others were asymmetrical with a variety of strange appendages. Because of their physiology and other structural constraints, echinoderms do not range as widely in body size as do molluscs or arthropods. The smallest living echinoderms are sea cucumbers and brittle stars less than 1 cm across (and the extinct ctenocystoids and microcrinoids were also a few mm in length), but there

Figure 16.2. In many places, the deep abyssal seafloor is dominated by echinoderms. This photo of the bottom of the San Diego Trough at 1200 m water depth was taken by the bathyscaphe *Trieste.* Hundreds of sea cucumbers (*Scotoplanes*) are walking across a surface covered by thousands of brittle stars (*Amphilepis*). (From Barham et al., 1967; courtesy U.S. Navy.)

are relatively few tiny echinoderms (in contrast to the many microscoic arthropods). At the other end of the size spectrum, there are sea stars over 1 m in diameter, and sea cucumbers that reach 2 m in length. Some extinct crinoids had stems that were tens of meters in length.

Echinoderms also have a wide variety of ecologies and trophic strategies. Crinoids, blastoids, and most other attached echinoderms are filter feeders, trapping food particles with their feathery arms and then conveying them to the mouth. Most sand dollars, heart urchins, and sea cucumbers are detritus feeders, sifting through the sediment for bits of food. Sea urchins graze algae from the rocks in their tidepool habitats. Sea stars are active predators, prying open a clam and extruding their stomachs to eat it inside its own shell, or digesting barnacles or corals right out of their skeletons. Brittle stars show a variety of feeding methods, including deposit feeding, predation, scavenging, and suspension feeding. Some species, such as the basket stars, are suspension feeders which can catch large prey items. But some life habits are not widely used by echinoderms. There are no true planktonic adult echinoderms (although echinoderm larvae are planktonic), in contrast to the many arthropods and pteropod molluscs in the plankton. Nor are there many nektonic, or swimming, echinoderms, although stalkless crinoids can swim for short distances. This is probably because the slow-moving echinoderms have heavy skeletons with no buoyancy mechanism, so efficient swimming is not very easy for animals of their body plan. Finally, there are no parasitic echinoderms, probably because echinoderms seldom reach the tiny body sizes of most parasites.

Despite this great diversity and disparity, all fossil and living echinoderms seem to have a number of unique features in common. Most echinoderms have a skeleton (**test**) whose individual plates (**ossicles**) are made of single crystals of high-magnesium calcite. These are easy to recognize even in partial specimens because they break along the calcite cleavage planes; this also helps in their identification in thin section. Unlike the external skeletons of arthropods, echinoderm skeletons are enclosed in soft tissues, so they are continually remodeled and grow by accretion, rather than by molting, as in arthropods (see Chapter 2). Many echinoderms (especially sea urchins and sea stars) also have spines on their external ossicles, a feature that gave the phylum its name. In Greek, *echinos* means "spiny" (or more literally, "sea urchin" or "hedgehog") and *derma* is "skin."

Of all the features of echinoderms, the most distinctive is their internal **water vascular system**. Although muscles control some movements in the echinoderms, the water vascular system does most of the work. It is analogous to the hydraulic systems on large machinery (or the hydraulic brakes on your car), which gradually but steadily transmit pressure from one area to another via an enclosed cylinder of fluid. Inside the echinoderm test is a closed network of tubes and bulbs filled with water of normal marine salinity (Fig. 16.3). The pressure on the fluid in the tubes is regulated by a plate on the outside called the **madreporite**,

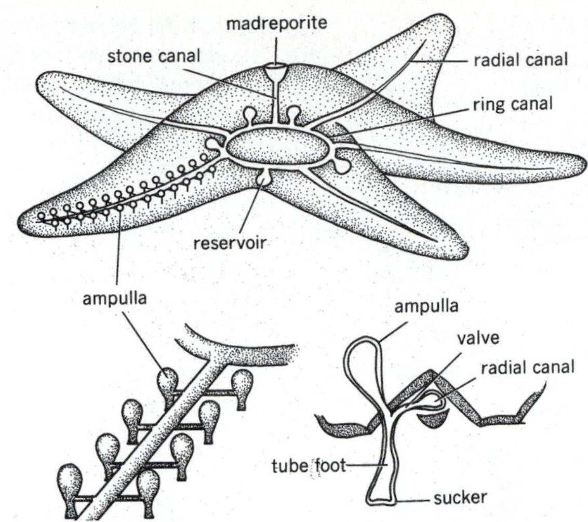

Figure 16.3. All echinoderms have a water vascular system, a sealed network of tubes and canals which operate by hydraulic pressure. Squeezing the bulb (ampulla) in one part of the sea star's arm protrudes its tube feet. (After Stearn and Carroll, 1989.)

which is connected to the main "pipeline" by a flexible **stone canal**, reinforced with rings of calcite to maintain its diameter (like a vacuum cleaner hose, which has rings for support, but flexes between the rings). The water vascular system then radiates around the body of the animal, interconnected by a circular canal around the central part of the body.

At the end of each of the radiating canals are hundreds of tiny **tube feet**, which extend and retract. They are the primary organs for locomotion and many other functions. If you have ever watched a sea star creep along in an aquarium, you have seen their hundreds of tube feet clinging to the glass sides and moving rapidly with extraordinary coordination to cause the sea star to steadily creep along. Each tube foot extends or retracts in response to muscles squeezing on a bulbous **ampulla** at the other end, analogous to squeezing the bulb on an eye-dropper. The strength of hundreds of tiny tube feet, with all those hydraulics powering them, means that a sea star can cling to rocks with extraordinary power, or pull against the adductor muscles of a clam until the bivalve tires and opens up, after which the sea star eats its vulnerable prey. The areas on the test with openings for tube feet in echinoderms are known as **ambulacral areas**; those without such pores are **interambulacral areas** (abbreviated "ambs" and "interambs" in casual conversation among specialists). The ambulacral areas run along the five arms of the sea star or in five discrete rows of pores along the test of a sea urchin.

As larvae, echinoderms are bilaterally symmetrical, but most adults have a secondary pentameral or radial symmetry, with a mouth (often surrounded by arms) at one end, and the anus at the other. Most sea stars, sea urchins, and crinoids no longer have a "head" or "tail," "front" or "back," "right" or "left," and it does not matter which way is front. This radial symmetry is usually divided into five radiating zones (the five arms of a starfish or crinoid, the five ambulacral and interambulacral areas of a sea urchin, and so on), resulting in a **pentameral** symmetry. In the irregular urchins and sand dollars, a secondary bilateral symmetry is superimposed over the radial pentameral symmetry.

Unlike the radial symmetry seen in more primitive invertebrates, such as cnidarians, the radial symmetry of echinoderms is clearly secondary. Not only do the larvae have bilateral symmetry, but it is clear that echinoderms are much more advanced creatures than are cnidarians. Echinoderms are fully coelomate animals, with a mouth and long looping digestive tract ending in an anus, as well as simple reproductive organs and nervous system. Their senses are primitive; they have no eyesight, hearing, taste, or other specialized senses, and respond mainly to touch. Echinoderms also lack separate well-developed circulatory, excretory, or respiratory systems. They accomplish these functions with the circulation of the fluids in their bodies and water vascular system. The lack of excretory and osmoregulatory structures or a discrete respiratory system is largely responsible for the inability of echinoderms to adapt to freshwater or land. Their internal body fluids must be bathed in normal seawater to avoid osmotic imbalances, and to absorb oxygen through their tube feet or other exposed tissues. The lack of respiratory system, the limitations of the water vascular system, and the expense of producing their heavy calcite plates also means that their body size cannot get very large. Consequently, there are no truly giant echinoderms.

Yet these weird "extraterrestrial" animals, with radial symmetry, water vascular system, and tube feet, but no eyes, or circulatory, respiratory, or excretory systems, are our closest relatives among all the invertebrate phyla we have reviewed so far. Chordates (including ourselves and other vertebrates) and echinoderms are sister-taxa, forming a group long known as the **Deuterostomes**. Despite the huge differences in the adults, the evidence for this evolutionary affinity comes from both embryology and molecular biology (Fig. 16.4). Most deuterostomes have a distinctive larval type known as a tornaria, unlike the trochophore larva seen in most invertebrate phyla. This larva forms in a way very different from other invertebrates. As embryonic deuterostome cells divide, they do not split in the spiral pattern found in most invertebrates but in a radial pattern. The cells in most invertebrates are determinate—their fates are decided early in cell division. Deuterostome cells are indeterminate—they can change their fates and become part of a different organ system, or regenerate a new organ, quite late in development. For example, when an early ball of determinate cells is separated, each cell goes on to develop the specific organ it is fated to become. However, if you split up a sea urchin embryo at the 4-cell or 8-cell stage, each cell can go on to develop a complete sea urchin. After the fertilized egg begins to cleave into 2, 4, 8, 16 (and so on) cells, it eventually forms a hollow ball known as a **blastula**. The opening of this ball (the **blastopore**) forms

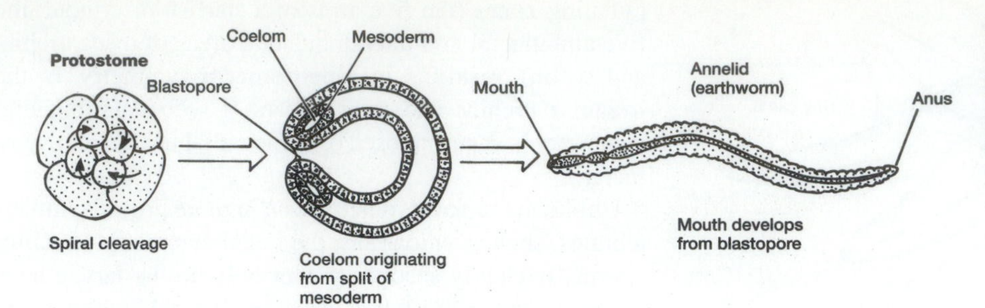

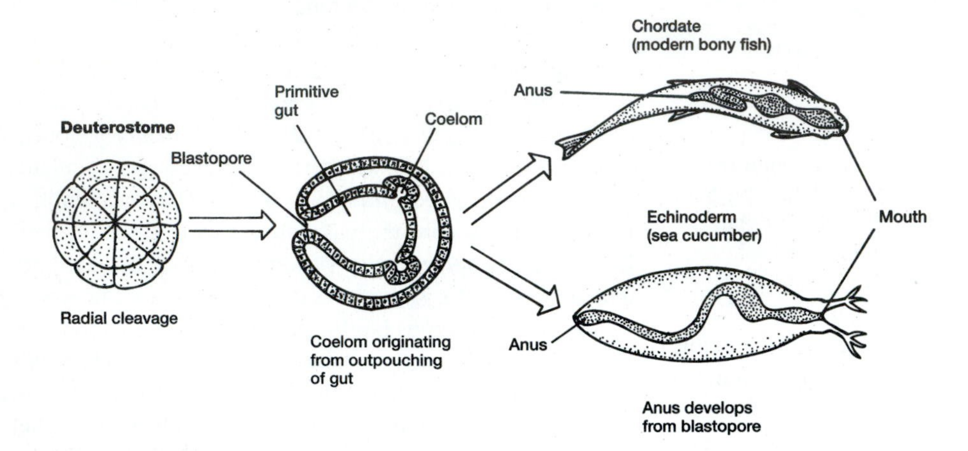

Figure 16.4. All echinoderms are deuterostomes, in that their larvae undergo radial cleavage, and the blastopore becomes the anus, not the mouth. (From Kardong, 1995.)

the mouth in most invertebrates, but in deuterostomes, the polarity is reversed—the blastopore becomes the anus. The mouth forms later in embrology on the opposite side from the blastopore, a feature that gives deuterostomes ("secondary mouth" in Greek) their name. Finally, most invertebrates form their coelom from a split within the the mesodermal wall, while deuterostomes form the coelom from an outpocketing of the endoderm.

All these unique embryological similarities have been corroborated by studies of molecular phylogeny, which always cluster echinoderms and chordates (Field et al., 1988; Lake, 1990; Raff et al., 1994; Halanych et al., 1995). So the next time you watch a sea star creep along the aquarium glass, remember that it's one of your closest relatives among the invertebrates.

SYSTEMATICS

The classification of echinoderms at class-rank and higher has been a topic of considerable dispute for well over a century. It is fair to state that the striking discrepancies among the many rival schemes of classification indicate both a fundamental lack of agreement as to which specific character sets have genealogical significance at high taxonomic rank, and also indicate an acute lack of confidence among echinoderm workers that *__any__* *of the current schemes has a compelling evolutionary basis. In short, there has been a long-held feeling that both paleontologists and neontologists have failed dismally in arriving at a phyletically-valid, high-rank classification of echinoderms.*

Bruce Haugh and Bruce Bell, "Classification Schemes," 1980

The five living classes of echinoderms (Asteroidea, or sea stars; Ophiuroidea, or brittle stars; Echinoidea, or sea urchins and sand dollars; Crinoidea, or sea lilies; and Holothuroidea, or sea cucumbers) are well-defined, monophyletic groups that have been recognized for over a century. The fossil record of echinoderms is far more complex, with many relatively short-lived groups with distinct body plans that have been placed in their own classes, resulting in a scheme with about 20-25 distinct classes. Most of these extinct classes are generally regarded as natural and monophyletic, although there are a few enigmatic extinct forms that appear to be intermediates between two classes. Thus, at the class level and below, echinoderm classification has been fairly stable, and most of the taxa are worth learning because they are clearly natural groups with some utility.

However, when the relationships among the classes are discussed, there is much controversy. For example, Sprinkle (1976, 1980) recognizes five subphylum groupings (Fig. 16.5): the Crinozoa (crinoids and paracrinoids); the Blastozoa (blastoids, cystoids, eocrinoids, and many archaic stalked forms); the Asterozoa (sea stars and brittle stars); the Homalozoa (bilaterally symmetrical forms, such as stylophorans and ctenocystoids); and the Echinozoa (sea urchins, sea cucumbers, plus extinct groups such as the edrioasteroids and helicoplacoids). Beyond these five parallel groupings of subphylum rank, there are few indications of the interrelationships of each subphylum. Paul and Smith (1984) and Smith (1984b) analyzed the echinoderms with cladistic methods (Fig. 16.6), and came up with a twofold arrangement of subphyla: Pelmatozoa (stalked groups, including crinoids and blastoids) and Eleutherozoa (free-

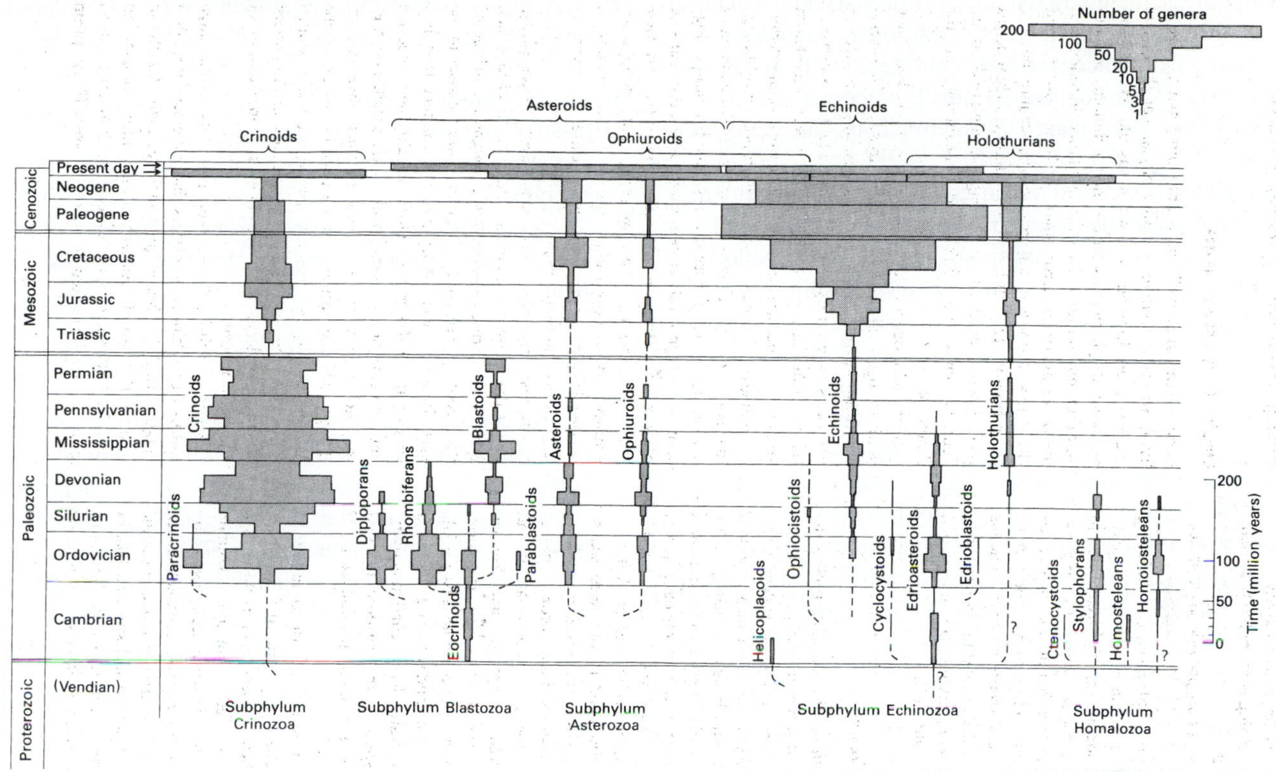

Figure 16.5. The fossil record of echinoderm classes, illustrating their stratigraphic ranges and generic diversity. There were a large number of classes with low diversity in the early Paleozoic, with decreasing numbers of classes through the Paleozoic, and by the late Paleozoic, only a few classes (crinoids and blastoids) were common. After the Permian extinction the echinoids became by far the most diverse class of echinoderms. (After Sprinkle, 1980.)

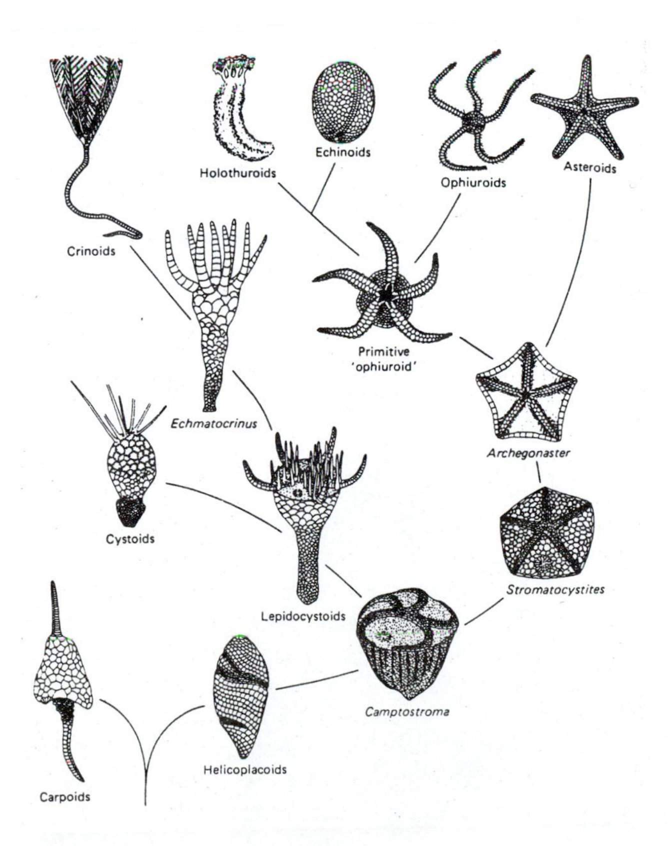

Figure 16.6. Paul and Smith's (1984) suggested phylogeny of the major echinoderm classes.

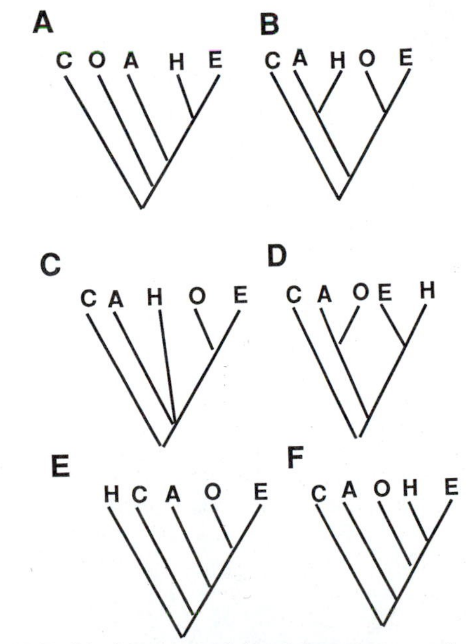

Figure 16.7. Cladograms of the living classes of echinoderms, based on different sources of character data. Taxa: A, Asteroidea; C, Crinoidea; E, Echinodea; H, Holothuroidea; O, Ophiuroidea. Hypothesis A (Raff et al., 1988) based on molecular data; B (Matsumura and Shigel, 1988) based on biochemical data; C (Strathmann, 1988) on ontogenetic data; E (Smiley, 1988) on morphological data; D (Sumrall, 1996) and F (Smith, 1988) based on paleontological data. (Modified from Mooi, 1989.)

living forms, including sea stars, brittle stars, and a clade of echinoids plus sea cucumbers). Odd forms, such as the carpoids and helicoplacoids, were considered to be the remote sister-taxa of the Pelmatozoa plus Eleutherozoa, and some fossil forms were placed in intermediate, ancestral positions.

In 1986, a symposium on echinoderm phylogeny was held in London; it was published two years later (Paul and Smith, 1988). This symposium presented the results of a number of analyses of echinoderm relationships, both molecular and morphological (Fig. 16.7). For example, Raff et al. (1988) reported on the results of molecular analyses. The 18S rRNA sequence data tend to cluster sea urchins and sea cucumbers (consistent with other classification schemes), but progressively more remote sister taxa were sea stars, brittle stars, and crinoids. Unlike other schemes, the sea stars and brittle stars are not grouped by this molecule. A cladistic analysis (Smiley, 1988) of a suite of morphological and embryological characters, polarized differently from the scheme of Smith (1984b), came up with contrasting results. Smiley (1988) grouped sea urchins with brittle stars and placed sea stars, crinoids, and sea cucumbers as progressively more remote sister-taxa. Matsumura and Shigei (1988) examined the amino acid sequences of echinoderm collagen proteins. On this basis, they clustered sea urchins with brittle stars and sea stars with sea cucumbers; crinoids came out as the sister-group of both clades. None of these molecular studies addressed the problem of the extinct groups (since these molecules are not preserved in the fossils), but clearly there is still a lot of controversy about the interrelationships of the living classes. As Raff et al. (1988) put it, "These disagreements derive in large measure from the fact that no tree can be reconciled with all of the available evidence; extensive parallel or convergent evolution must have occurred in the history of the phylum."

Given these problems, we will not use any superclass- or subphylum-level classification below. Instead, we will list the major classes of echinoderms that are widely recognized as natural by most workers. In addition, some of the very rare minor classes that are unlikely to be encountered anywhere but museums are discussed only briefly.

Phylum Echinodermata

Class Crinoidea (Cambrian-Recent; about 1000 genera, including about 165 living genera and 625 living species)

Crinoids are the most numerous and diverse of all the Paleozoic echinoderms, so they are discussed in detail in the next section.

Class Paracrinoidea (Early Ordovician to Early Silurian; 13 to 15 genera)

In contrast to the abundant and diverse Crinoidea, the paracrinoids (Fig. 16.8B) are a minor group known from only a few Ordovician and Silurian deposits (although they may be locally abundant in some Middle Ordovician limestones). They have ambulacrum-bearing arms like crinoids, so Sprinkle (1980) placed them in the subphylum Crinozoa with the crinoids, but now believes they are highly modified blastozoans.

Class Eocrinoidea (Early Cambrian-Late Silurian; about 32 genera)

The earliest known group of stalked, arm-bearing echinoderms (Fig. 16.8A) were the eocrinoids, an "experimental" paraphyletic group that is known chiefly from the Cambrian to Middle Ordovician (Fig. 16.5). They were the most common echinoderms in the Cambrian, but were gone by the Late Silurian (although they may have been ancestral to six other classes: the Rhombifera, Diploporita, Coronoidea, Blastoidea, Parablastoidea, and Paracrinoidea). They have two to five simple ambulacral areas with two rows of brachioles around them, and sutural pores between the plates for respiration. The earliest eocrinoids had a short holdfast and irregularly organized plates, but later forms had a fully developed stalk with columnals and regular rows of plates. Most

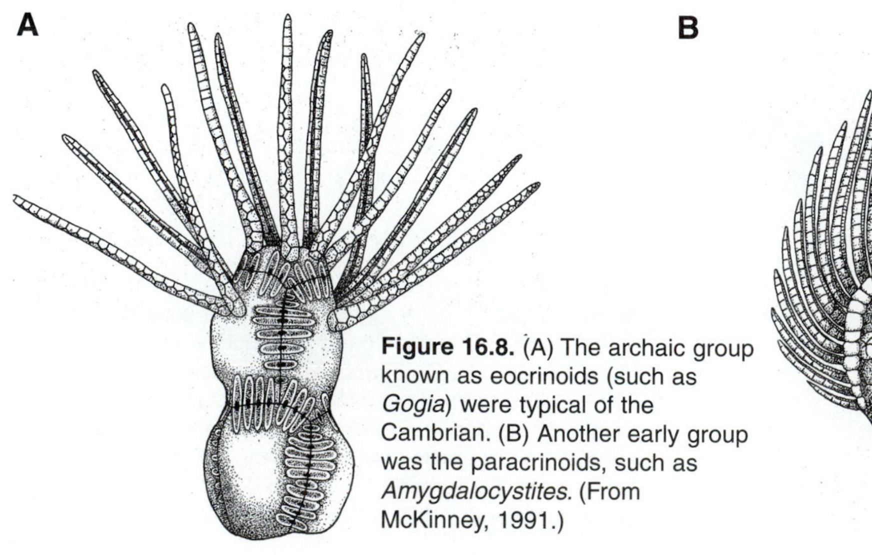

Figure 16.8. (A) The archaic group known as eocrinoids (such as *Gogia*) were typical of the Cambrian. (B) Another early group was the paracrinoids, such as *Amygdalocystites*. (From McKinney, 1991.)

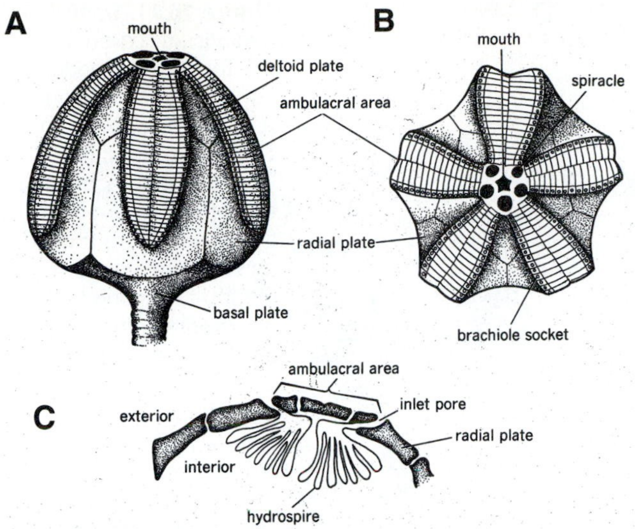

Figure 16.9. Morphology of the common Mississippian blastoid *Pentremites.* (A) Side view of calyx. (B) Top view of calyx. (C) Cross section of ambulacral area. (From Stearn and Carroll, 1989.)

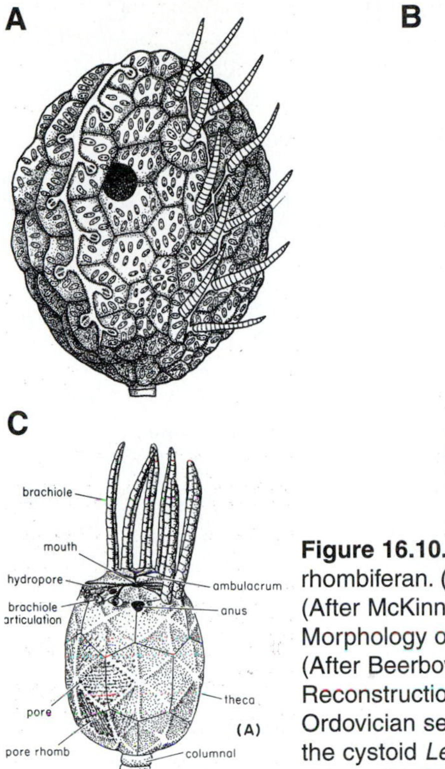

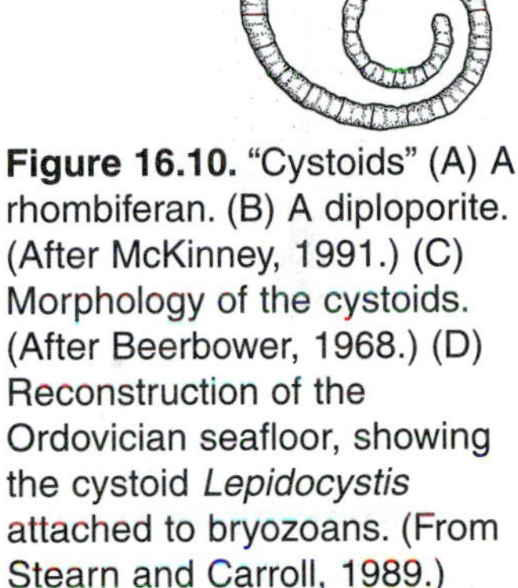

Figure 16.10. "Cystoids" (A) A rhombiferan. (B) A diploporite. (After McKinney, 1991.) (C) Morphology of the cystoids. (After Beerbower, 1968.) (D) Reconstruction of the Ordovician seafloor, showing the cystoid *Lepidocystis* attached to bryozoans. (From Stearn and Carroll, 1989.)

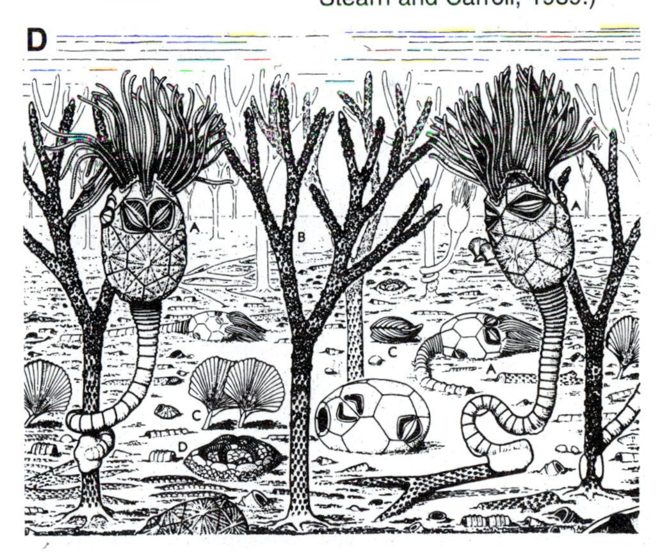

were attached suspension feeders high above the seafloor, but some were apparently unattached and crawled on the sea bottom.

Class Blastoidea (Middle Ordovician-Late Permian; about 95 genera)

Next to crinoids, blastoids are the most common and best known of the Paleozoic stalked echinoderms. Like crinoids, blastoids (Fig. 16.9) had a long stalk and a symmetrical "head" (**theca**) that was typically shaped like a cone, globe, or a flowerbud. Instead of thick arms with a hollow coelomic cavity (as in crinoids), however, the unbranched arms (**brachioles**) of blastoids had no extension of the coelom, and were fragile and slender, so they are usually not preserved. Two rows of brachioles anchored on either side of the five well-defined ambulacral areas, which are covered by plates that closed the pores. Inside the theca behind each ambulacral area was one or more highly folded and pleated structures known as **hydrospires**, which apparently served for respiration. At the top of each ambulacral area were round openings called **spiracles**, which apparently acted as exits for the water currents that entered via inlet pores in the ambulacrum and flowed through the hydrospires. The mouth is a simple opening at the top of the theca, very different from the complex mouth and food grooves of crinoids.

Although blastoids first appeared in the Middle Ordovician, their greatest diversity (Fig. 16.5) was in the Mississippian, when along with the abundant crinoids there were a great many blastoids, especially the flowerbud-shaped *Pentremites* (Fig. 16.9) and the globular *Orbitremites*. They declined again in the late Paleozoic, but suddenly diversified in the Permian, just before their final extinction in the Permian catastrophe.

Classes Rhombifera (Early Ordovician-Late Devonian; about 60 genera) and **Diploporita** (Early Ordovician-Early Devonian; about 42 genera)—"**cystoids**"

Another distinctive group of early Paleozoic echinoderms was the **cystoids**, which are presently divided into two classes, the rhombiferans and the less numerous diploporites. Like crinoids, paracrinoids, and blastoids, they have a "head" (theca) made of calcareous plates, thin arms (brachioles) around the mouth, and a long stalk (Fig. 16.10). Unlike these other groups, however, the stalk was long and flexible, and may have served to attach to a substrate, or may have propelled them when they crawled along the sea bottom. The theca tends to be asymmetrical and irreg-

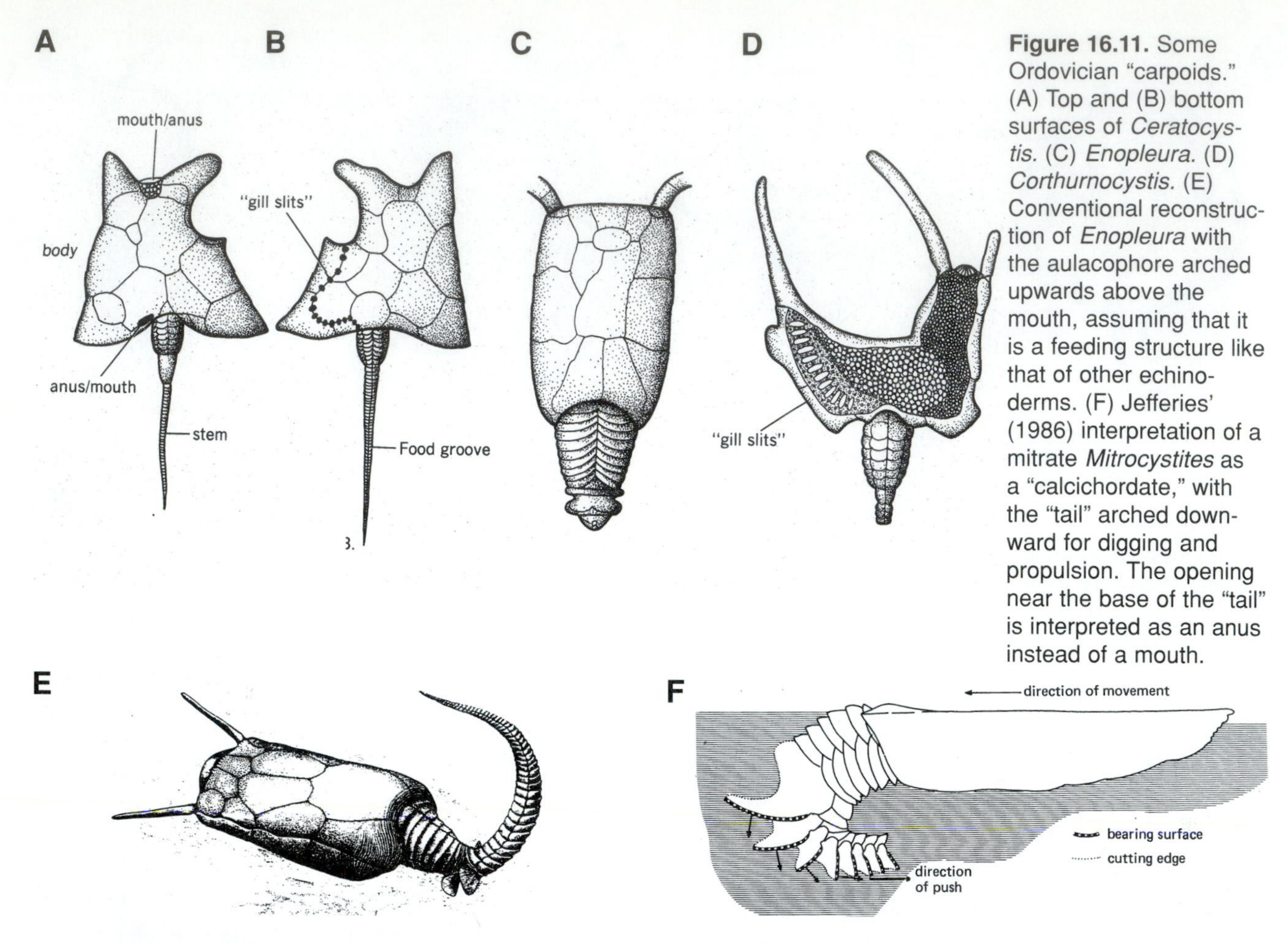

Figure 16.11. Some Ordovician "carpoids." (A) Top and (B) bottom surfaces of *Ceratocystis.* (C) *Enopleura.* (D) *Corthurnocystis.* (E) Conventional reconstruction of *Enopleura* with the aulacophore arched upwards above the mouth, assuming that it is a feeding structure like that of other echinoderms. (F) Jefferies' (1986) interpretation of a mitrate *Mitrocystites* as a "calcichordate," with the "tail" arched downward for digging and propulsion. The opening near the base of the "tail" is interpreted as an anus instead of a mouth.

ular in shape, and the plates are also irregular without the regular cycles seen in crinoids or blastoids. Some cystoids were flattened, with a different arrangement of plates on the bottom (**antanal**) and top (**anal**) side; these were most likely crawling forms.

The most diagnostic feature of rhombiferan cystoids are their distinctive respiratory structures known as **pore rhombs**, which have a variety of different patterns of slits opening to hydrospire-like folds within the theca. Most pore rhombs occupy the space between two different plates. In the diploporites, on the other hand, the plates are perforated by a series of paired pores, giving them their name. These pores apparently allowed coelomic fluids to flow through the thecal plates, making it possible to exchange gases with the surrounding seawater. The mouth is in the center of the upper surface of most cystoids, and food grooves radiate out from it over the plates. These grooves bear erect brachioles, so cystoids apparently fed in a manner similar to that of crinoids and blastoids.

Cystoids were never very abundant, although they are relatively common in some Middle Ordovician rocks. Their diversity dwindled through the Silurian and Devonian (Fig. 16.5), and they went extinct during the Late Devonian (Frasnian-Famennian) extinction event.

Class Stylophora and their relatives (Middle Cambrian-Pennsylvanian; about 50 genera)—homalozoans or "**carpoids**"

Although echinoderms come in an amazing array of shapes for animals of a common body plan, some of the most unusual were the **carpoids** (Fig. 16.11). There is little agreement as to what these animals were, or even how to classify them. Sprinkle (1980) recognizes four carpoid classes within his subphylum Homalozoa: the Stylophora, the Homoiostelea, the Homostelea, and the Ctenocystoidea. Jefferies (1986) places the same carpoids in his subphylum Calcichordata, which is subdivided into classes Cornuta, Mitrata, and Soluta.

The carpoid body is largely asymmetrical, and lacks the typical columns or arms or porous plates or ambulacral areas found in most echinoderms. The flattened, irregular lobed theca has a long, flexible pointed **aulacophore** that did not attach to the substrate like a crinoid stalk. Some echinoderm specialists interpret this structure as a feeding apparatus of some sort. There is an opening near the base of the aulacophore that most interpret as the mouth. At the other end was an opening that was apparently the anus. An irregular row of pores on one side of the animal have been identified as gills slits by some scientists and as ambulacral pores by others.

Carpoids violate so many of the basic rules of the echinoderm body plan that some scientists have questioned whether they are echinoderms at all. The most vocal of these critics is Richard Jefferies (1975, 1986, 1990), who considers them primitive chordates instead of echinoderms. In his view, the aulacophore is actually an armored tail-like structure, and the opening at the base of the tail is the anus; the one at the opposite end is the mouth (Fig. 16.11F). By careful serial sections of well-preserved specimens, Jefferies has been able to look at the detailed internal structures of these enigmatic animals. However, the interpretation of these internal structures is largely colored by favored hypotheses. What Jefferies calls the impressions of notochord and dorsal nerve cord (chordate features) are simply water vascular canals typical of echinoderms in the traditional interpretation (e.g., Ubaghs, 1967). In Jefferies' view, carpoids are a separate subphylum Calcichordata, which show that echinoderms and chordates are more closely related to each other than a variety of other near relatives of chordates (such as the tunicates and lancelets discussed in the next chapter).

At the present time, few scientists agree with Jefferies' interpretations (Ubaghs, 1975; Philip, 1979; Jollie, 1982; Kolata and Jollie, 1982; Sprinkle, 1987; Kolata et al., 1991; Peterson, 1994). Not only are carpoids made of calcite (all chordates use calcium phosphate for their skeletal tissues), but the calcite is monocrystalline and oriented in exactly the same crystallographic direction as in other echinoderms (Carlson and Fisher, 1981; Fisher, 1982), and opposite the direction predicted by Jefferies' interpretations. The most convincing evidence is recently discovered specimens preserved in life position, which shows that Jefferies' interpretation of front and back and top and bottom are backward (Fisher, 1993).

Class Ophiuroidea (Early Ordovician-Recent; about 325 genera, including 2000 living species)

The brittle stars resemble sea stars in many ways, so they have frequently been lumped with the Asteroidea in a higher taxon, typically called the Stelleroidea or Asterozoa. However, as we pointed out above, the molecular and morphological evidence does not always cluster the Asteroidea and Ophiuroidea (Fig. 16.7), so the monophyly of this grouping is questionable. Like sea stars, ophiuroids have five arms radiating from a central area, but their arms are long and slender, and discretely different from a disk-like body. Like asteroids, the mouth is on the center of the bottom of the disk, and the anus and madreporite at the top. However, the long slender arms of brittle stars have no internal coelomic cavity, and are not capable of clinging to rocks or pulling open clams. Instead, brittle stars wriggle their arms in a snake-like fashion and slide rapidly along the sea bottom. In fact, they are the fastest of all echinoderms. When disturbed, they readily shed parts of an arm (which wriggles to distract the predator) and flee; eventually, a new arm regenerates.

A few brittle stars live in shallow water, but they must hide in crevices and under rocks. They are far more abundant in the deep sea, where there are few predators. In a few instances, they reach densities of as many as 2000 individuals per square meter (Fig.16.2). Brittle stars feed on detritus that is brought to their mouth through the ambulacral grooves and tube feet underneath the arms, and can also grasp larger pieces of food with their arms and convey them to the mouth. Although they first appeared in the Ordovician, they are very delicate, and break apart even more easily than sea stars, so their fossil record is rather poor, and complete specimens are mostly restricted to *Lagerstätten*.

Class Asteroidea (Early Ordovician-Recent; about 430 genera, including 1500 living species)

The "starfish" or sea stars are the most familiar echinoderms to most people (Fig. 16.3), but they do not fossilize as well as crinoids or echinoids, so their impact on the fossil record is less. Their poor fossilization potential occurs because their calcite plates are small and not tightly sutured together. This allows them great flexibility, but also results in disarticulation when they die. A number of beautifully preserved, complete specimens are known, but they are usually found in *Lagerstätten*, where the preservation is extraordinary. In some places, they occur in great numbers in well-known "starfish beds."

Sea stars usually have five arms, with rows of tube feet running along the base of each, and a mouth in the center of the bottom of the animal. Inside each arm are rows of digestive glands, and in the center of the body are a mass of gonads, and a protrusible stomach. The anus and madreporite are both found on the top of the central part of the sea star.

As mentioned above, starfish are slow moving but voracious predators that can digest a clam, barnacle, or coral right out of its shell. They are also very tough and cling to rocks very strongly, so they can survive in the crashing breakers of the intertidal zone better than most organisms. The oldest sea stars are known from the Ordovician, and apparently they have been an important predator on the seafloor ever since. Bivalves first developed their complex interlocking commissures to resist sea star predation in the Late Ordovician and Silurian, about the time that the first sea stars with protrusible stomachs apparently arose (Carter, 1967). Some authors (Donovan and Gale, 1990) have suggested that the renewed radiation of sea stars in the Triassic was an important component of the escalation of predators in the Mesozoic. Such escalation may have been partially responsible for failure of brachiopods after the Paleozoic, and for the extinction of spirifers and other epifaunal shellfish that did survive the Permian catastrophe.

Class Holothuroidea (Middle Cambrian-Recent; about 200 genera, including 1150 living species)

Sea cucumbers vary from the standard echinoderm body plan in a number of ways. Instead of a continuous armor of calcite plates, they are largely soft-bodied, and their ossicles are reduced to thousands of tiny sclerites embedded throughout the body. With their elongate, sausage-shaped bodies, they lie on their sides and creep along the seafloor with their tube feet. They trap food particles with tentacle-shaped tube feet around the mouth at one end, then digest it in a long coiled gut, and excrete it through the anus at the other end. Although they retain the radial symmetry and tube feet of the phylum, by lying on their sides sea cucumbers have a front and back end. One genus is adapted for slow swimming; others are adapted for burrowing. They are most abundant at abyssal depths, where they cover the seafloor in great numbers since there are fewer predators, and abundant organic detritus (Fig. 16.2).

Sea cucumbers have a most unusual method of defense. If disturbed, they will spew their intestines, respiratory apparatus, and other internal organs out through their anus, presumably to drive away the predator. This evisceration is apparently not fatal to the sea cucumber, because they will regenerate the lost organs if given enough time.

Since they have no hard skeleton, whole fossil specimens of sea cucumbers are very rare, and known only from a few *Lagerstätten*.The oldest one known is *Redoubtia*, from the Middle Cambrian Burgess Shale. Their microscopic calcareous spicules are very common in the sediment, so they indicate the presence of sea cucumbers through the entire Phanerozoic. However, they have not been studied as much as foraminifera and ostracodes that occur with them in the sediment.

Class Helicoplacoidea (Early Cambrian; 4 genera)

Odd body plans are the norm for the echinoderms, but none is odder than that of the helicoplacoids (Fig. 16.12). These strange organisms are known from just a few localities in the Lower Cambrian of the White Mountains of California and a few other places in western North America. They were first discovered and described only 35 years ago (Durham and Caster, 1963, 1966; Durham, 1967, 1993). Their elongate, spindle-shaped bodies are covered with rows of tiny ossicles, which are arranged in spiral rows down from the apex at one end. The mouth is on the side of the theca, where the ambulacral grooves branch. One ambulacral groove spirals up from the mouth to the top, and two ambulacral grooves spiral down around

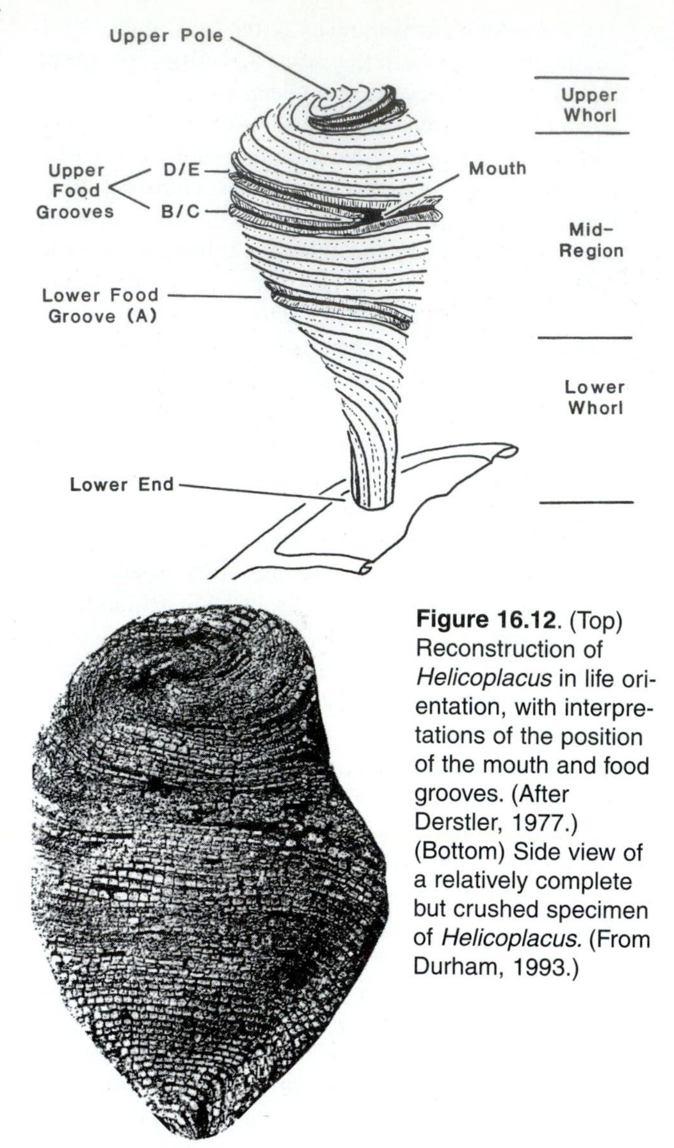

Figure 16.12. (Top) Reconstruction of *Helicoplacus* in life orientation, with interpretations of the position of the mouth and food grooves. (After Derstler, 1977.) (Bottom) Side view of a relatively complete but crushed specimen of *Helicoplacus.* (From Durham, 1993.)

the spindle away from the mouth, and presumably bore tube feet for transporting food up to the mouth, and for respiration. However, there is no attachment structure, stem, madreporite, or anal opening visible on the specimens, so their biology is still quite mysterious.

Even their life habits are controversial. At one time, scientists reconstructed them as creeping slowly across the seafloor on their sides, feeding on detritus as sea cucumbers do. Recently, Durham (1993) described specimens that preserved the broken-off bases of helicoplacoids imbedded in the sediment, evidence that they stood upright with the pointed end buried in the sediment. In this position, they apparently trapped suspended food particles along their spiral ambulacral groove. Since they are known from only a few Early Cambrian specimens, their paleoecology will probably always be controversial. However, their peculiar spiral symmetry suggests that they are a very unusual and early experiment in echinoderm evolution (Figs. 16.5, 16.6).

Because helicoplacoids are made of hundreds of small, loosely articulated plates, they break up easily. However, hundreds of complete (but flattened) specimens have been found that show the plates in articulation. Their distinctive, saddle-shaped ossicles are very abundant in the Lower Cambrian limestones of the Poleta Formation in the White Mountains, and on a few bedding planes, dozens of individuals can be found side-by-side. However helicoplacoids lived, they were extremely common during the Early Cambrian in western North America, and then died out by the late Early Cambrian, leaving no descendants or apparent close relatives.

Class Edrioasteroidea (Early Cambrian-Late Pennsylvanian; about 35 genera).

Another peculiar class of echinoderms are the edrioasteroids (Fig. 16.13), which demonstrate yet another radically different body plan within the basic echinoderm design. Most are small (about 1 cm in diameter) disk-shaped animals that attached to hard surfaces. On the top surface are five curving ambulacral areas that resemble the arms of a sea star. Through the ambulacral pores protruded tube feet, which captured food particles and conveyed them to the mouth in the center. The rest of the top surface is interambulacral area. The anus is covered by a pyramid-shaped arrangement of plates, and located on the top surface, but off-center in the posterior interambulacral area.

Although edrioasteroids were never very diverse, and seldom common in most early Paleozoic faunas, in some localities they occur by the hundreds. In the Upper Ordovician beds of the Cincinnati Arch, they are typically found encrusting the top valve of large strophomenide brachiopods (Fig. 8.9C). Although their diversity declined after the Ordovician, they straggled on until the Pennsylvanian (Fig. 16.5).

Figure 16.13. Reconstruction of the Ordovician edrioasteroid *Carneyella,* showing the tube feet protruding from the ambulacral areas. This specimen is attached to a strophomenide brachiopod. (From Bell, 1980.)

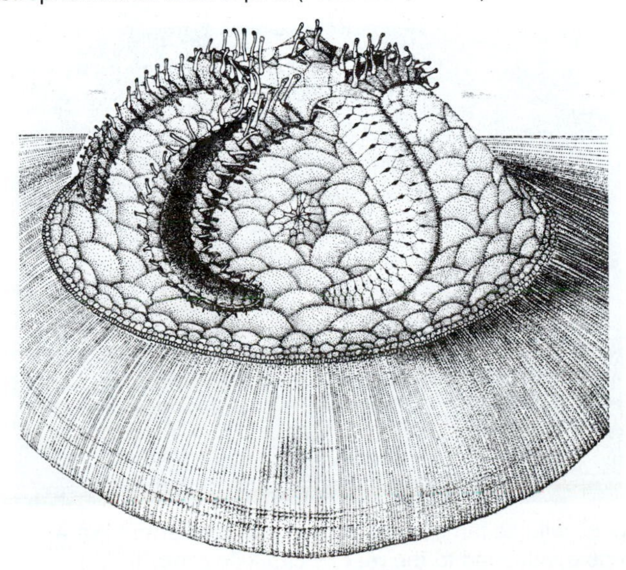

Class Echinoidea (Late Ordovician-Recent; about 765 genera, including 940 living species)

Echinoids are by far the most numerous and best fossilized of all Mesozoic and Cenozoic echinoderms, so they are discussed in greater detail below.

CRINOIDS

Morphology and Systematics

Crinoids are the only surviving group of stalked echinoderm, so our interpretations of the many extinct stalked groups are largely based on them. Today, there are two major groups: the rare stalked crinoids (Fig. 16.15A), which are generally found in low diversity (only 25 living genera) in habitats that have few predators, and the more common (130 living genera) stalkless crinoids ("feather stars"), which are highly mobile and move around much like brittle stars (Fig. 16.15B). In total, there are about 600 to 700 living species. In the Paleozoic and Mesozoic, however, there was a great diversity of stalked crinoids (between 5000 and 6000 species in about 850 genera), and these dominate the fossil record of the class.

The body of the crinoid (Fig. 16.14) consists of the cup-like "head," or **calyx**, from which radiate five or more arms. These arms contain extensions of the coelom inside them and have a nervous system and circulatory system. They also have ambulacral food grooves on the inner side that channel food toward the mouth in the center of the calyx. (Conventionally, the arms and mouth are at the **oral** end of the animal, and the stalk is on the **aboral** end.) The arms are usually branched into multiple rami. In some crinoids, each of these bears unbranched **pinnules** that increase its surface area and filtering capability. Living stalked crinoids (Fig. 16.15A) form a circular fan with their arms curved backward from the calyx (concave part of the fan facing upcurrent), which helps them trap food particles as they pass through the arms. Food is trapped by the sticky tube feet on the arms and pinnules, then mucus carries it down the ambulacral food groove and toward the mouth. Once ingested, the food passes through a simple looping digestive tract that ends in an anus on the oral surface of the calyx near the mouth.

The calyx itself consists of a series of plates, usually arranged in five columnar areas at the base of each arm (Figs. 16.14, 16.16). The most distinctive plates are known as the **radials**, which occur near the base of each arm. In the aboral direction (i.e., toward the stalk) are a series of plates that contact the stalk. If there is only one series (**monocyclic**), these are **basal** plates; some crinoids also have a set of **infrabasal** plates between the basals and the stalk (so they are **dicyclic**). Plates above the radials that were originally part of the arms but are now incorporated into the calyx are known as **fixed brachials**, and the spaces

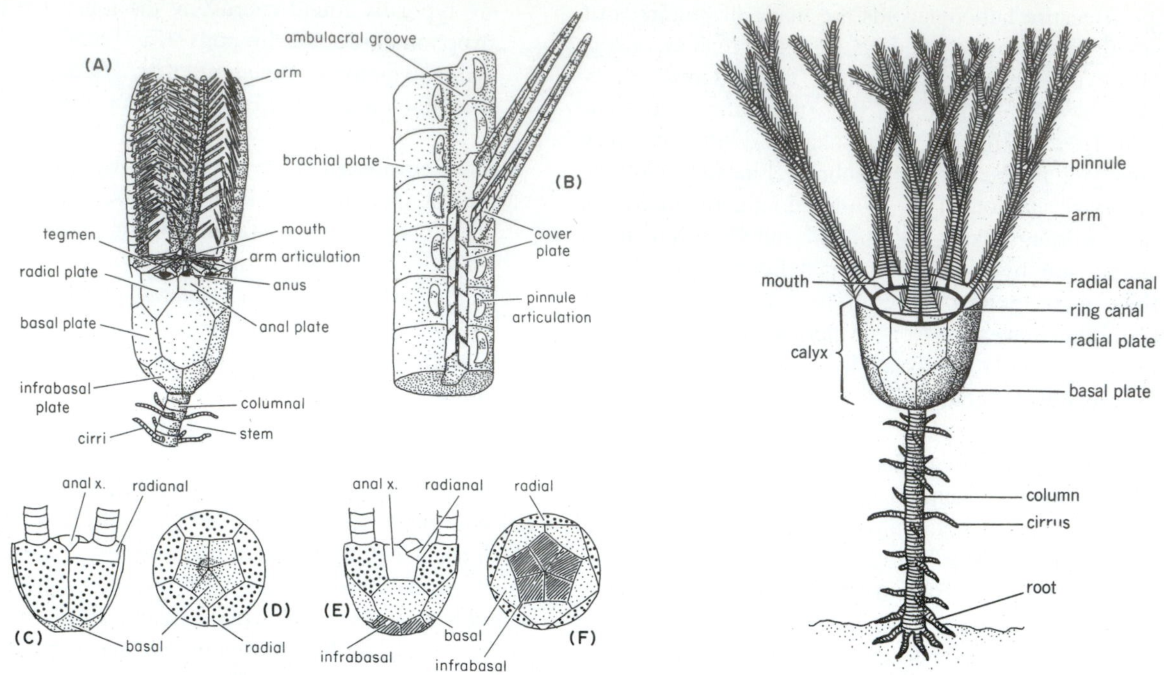

Figure 16.14. Basic anatomical terminology of crinoids. (A) Lateral view of a crinoid calyx with two arms and most of the stem removed. (B) Detail of arm with most pinnules removed. (C-D) Monocyclic crinoid in lateral and aboral view, with a single cycle of plates below the radials (stippled). (E-F) Dicyclic crinoid, with two cycles of plates below radials. (From Beerbower, 1968.) (right) Reconstruction of a complete crinoid. (From Stearn and Carroll, 1989.)

between them may be filled with **interbrachial** plates. Finally, the oral surface inside the arms may be roofed by a **tegmen**, and in some taxa, small tegmen plates may roof over the anus and mouth. Although this calyx plate terminology may seem confusing at first, it is crucial to identifying crinoids, since most diagnostic features are based on the configuration of the plates.

The stalk (or pelma, or **column**) of the crinoid consists of a series of doughnut- or Lifesaver-shaped **columnals** of monocrystalline calcite. In life, these are linked together by a collagenous ligament that connect each columnal, but they break up and disassociate into hundreds of tiny columnals when the crinoid dies. Some columnals are

Figure 16.15. Two examples of living crinoids. (A) A stalked crinoid, with its fan of arms curved into the current like a parachute. (From Clarkson, 1993.) (B) A free-living comatulid crinoid, attached to the reef by cirri. (Courtesy IMSI.)

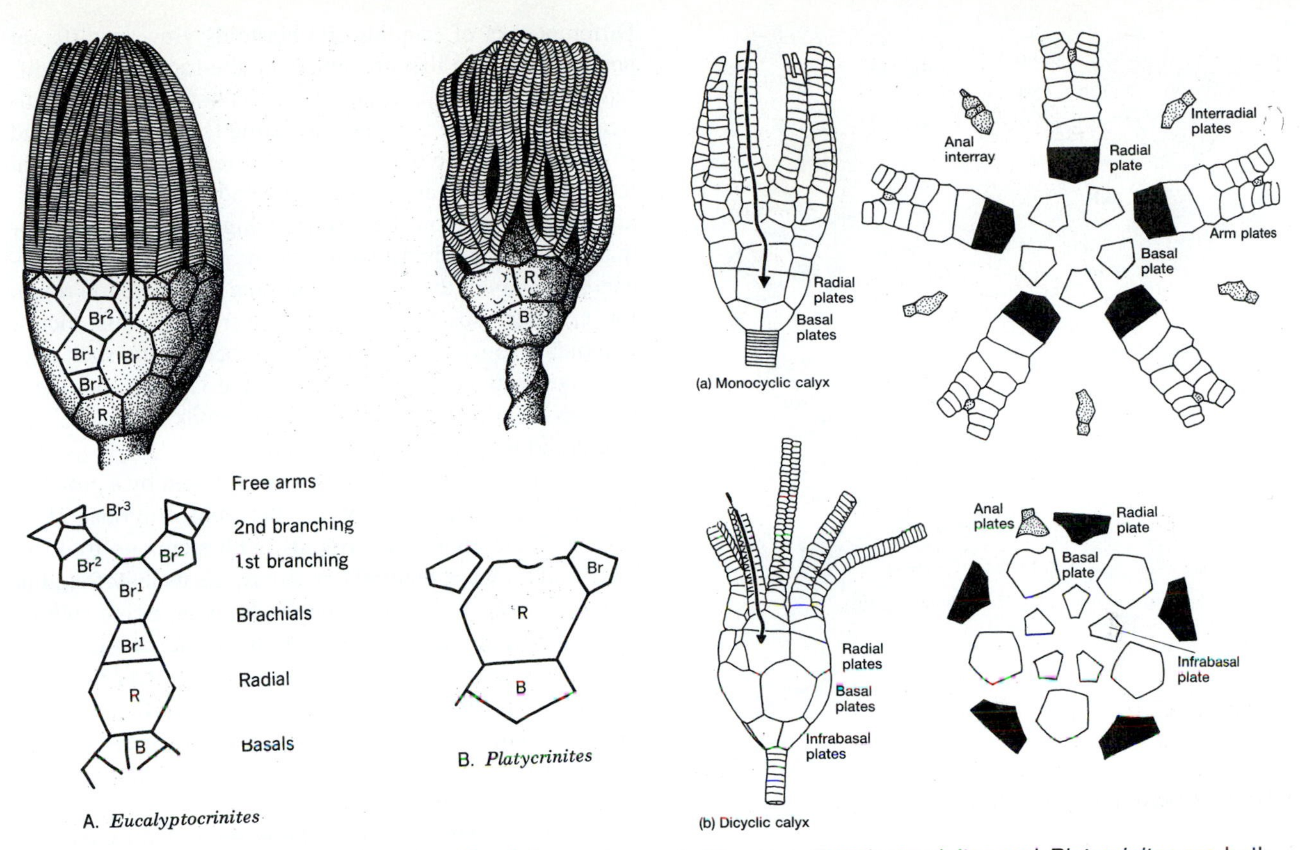

Figure 16.16. (left) Relationship between the plates and arms in the calyx. *Eucalyptocrinites* and *Platycrinites* are both monocyclic camerates, with multiple brachials and interbrachials. (After Stearn and Carroll, 1989.) (right) Arrangement of plates with monocyclic plates below the radials, and dicyclic plates below the radials. (After McKinney, 1991.)

smooth, but many have complex ridges on their articulating surfaces, which enable them to interlock like a stack of poker chips. Branching off the column are the tentacle-like **cirri**, which are useful for grasping seaweed or other objects and clinging firmly. In many complete specimens, the base of the column is a root-like or disk-like structure called a **holdfast**, which looks remarkably like plant roots, for the same functional reason—it is the most effective way for a long-stemmed organism to anchor to soft substrates.

According to some authors, the oldest known possible crinoid is *Echmatocrinus* (Fig. 16.6) from the famous Middle Cambrian *Lagerstätten* of the Burgess Shale in British Columbia. Four main subclasses of crinoids are recognized in the fossil record (Fig. 16.17). The oldest known (Early Ordovician-?Early Triassic) subclass is the paraphyletic **inadunates**, which have rigid calycal plates, arms that are free above the radial plates (i.e., no fixed brachials or interbrachials), and a mouth covered by the tegmen (Fig. 16.17A). The second major Paleozoic group is the **flexibles** (Middle Ordovician-Permian), in which the multiple calycal plates are only loosely united and (as the name suggests) flexible. In their calyces, they usually have brachials, radials, basals, three infrabasals, and a flexible tegmen with an exposed food groove and mouth (Fig. 16.17B, C). Their arms have no pinnules. The third important subclass of the Paleozoic is the **camerates** (Early Ordovician-Permian), which have a rigid calyx and food groove covered by a tegmen, but (unlike inadunates) multiple fixed brachials and interbrachial plates (Figs. 16.16, 16.17D, E). All three of these groups apparently died out during the Permian catastrophe (Fig. 16.5). The only surviving subclass of crinoids is the **articulates** (Early Triassic-Recent), which, as the name implies, have very flexible, articulated arms (Fig.16.17F). The dorsal part of the calyx is greatly reduced, and there are five infrabasals in the ventral cup. Their arms are also unbranched, with extensive pinnules. The articulates consist of five orders (three extinct) of stalked crinoids, and the unstalked order Comatulida, or "feather stars"(Fig. 16.15B).

Ecology

Today, there are so few living crinoids (and so few studies that have been undertaken) that their ecological preferences have to be interpreted with care when making assumptions about extinct groups. The few living stalked crinoids are found today in deeper waters below 100 m and as deep as 5000 m. The more diverse stalkless comatulid crinoids (Fig. 16.15B) can live in shallow water, but they are highly mobile, so they avoid many kinds of predators. Comatulids creep along the seafloor much like a brittle star until they find a good spot on which to attach with their cirri, and then they feed (often at night). Crinoid filtration fans are very effective screening devices. Even with a cur-

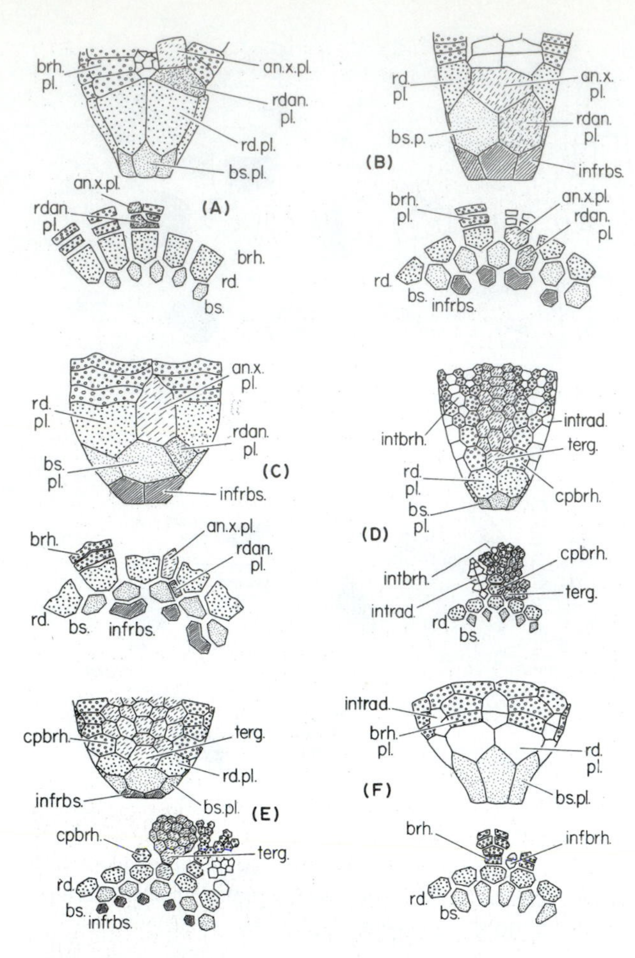

Figure 16.17. Various arrangements of calycal plates in crinoids. (A) Monocyclic cup of the inadunate *Iocrinus.* (B) Dicyclic cup of a flexible. (C) Dicyclic cup of the flexible *Lecanocrinus.* (D) Monocyclic cup of the camerate *Glyptocrinus.* (E) Dicyclic cup of the camerate *Ptychocrinus.* (F) Monocyclic cup of the articulate *Dadocrinus.* Abbreviations: an. x. pl., anal x plate; brh. pl., brachial plate; intbrh, interbrachial; rd. pl., radial plate; rdan. pl., radianal plate; terg., tergal. (From Beerbower, 1968.)

rent speed of only 2 cm/sec, a fan with an area of about 500 cm^2 can filter about 1 liter per second.

It is clear that Paleozoic crinoids and blastoids did not face as many kinds of predators (primarily fish) that evolved during the Mesozoic marine revolution and made the world so difficult for brachiopods, crinoids, bryozoans, and many other groups that dominated the Paleozoic fauna. In contrast to the Mesozoic seas, the shallow waters of the Paleozoic seas were rich in crinoids, and during certain times (such as the Mississippian or Early Carboniferous), they were the dominant group on the shallow seafloor, occurring in the millions. Studies of crinoidal deposits from the late Mississippian of Kentucky (Chestnut and Ettensohn, 1988) showed that crinoids and blastoids have distinctive ecological facies associations. They were particularly common in the shallow areas of the back-barrier lagoon, especially on washover lobes on the landward side of the barrier, and on shoals within the lagoon (Fig. 16.18). Different taxa of crinoids and blastoids grew to different heights, allowing them to subdivide the food found at different levels of currents above the sea bottom. Such stratification of feeding is known as tiering (see Chapter 8) and is analogous to the way trees of different heights occur in the jungle forest canopy. Crinoids had the longest stems, and so could exploit the resources higher off the seafloor than could blastoids. However, the more turbulent, higher-energy waters of the barrier-front sand belt were too rough for crinoids, so their remains are rare in this facies. Crinoids also occurred in lesser numbers in the sheltered, low-energy muddy-bottom areas of the lagoons, but these habitats were better suited for brachiopods and especially fenestellid bryozoans like *Archimedes*.

Yet another ecological niche was exploited by a group of Mesozoic articulates known as the Pentacrinitidae (Fig. 16.19). These enormous crinoids, with stems up to 15 m long, and crowns about 80 cm across, are usually found in deep-water black shales during the Jurassic. Some authors have compared them to kites, with their long stems allowing them to feed high above the bottom, and their huge fan providing lift. However, there are a number of mechanical problems with this interpretation. Currently, most specialists favor the idea that they lived a pseudo-planktonic life, hanging down from pieces of driftwood and trapping surface plankton. This interpretation is supported by the fact that they are frequently found with fossilized driftwood, and in one spectacular slab in a museum in Tübingen, 50 crinoid specimens are preserved close together in black shale with their crowns face down (Fig. 16.19). In addition, their stems are more flexible at the base (probably to cope with storms), but rigid near the crown (where rigidity is needed to control the angle of the filtration fan).

Evolution

Evolutionary history—Although a possible crinoid, *Echmatocrinus*, is known from the Burgess Shale, the major subclasses crinoids do not begin to appear and radiate until the Ordovician (Fig. 16.5). The inadunates appeared first, and by the Early Ordovician gave rise to the other two major Paleozoic subclasses, the flexibles and camerates. Although crinoids were the most diverse group of echinoderms in the Ordovician, they shared the seafloor with almost all the other known orders of echinoderms, including many archaic, experimental groups. Among these were the eocrinoids and edrioasteroids surviving from the Cambrian, as well as paracrinoids, cystoids, rhombiferans, diploporans, asteroids, ophiuroids, and the earliest echinoids; spicules of possible holothurians are known as well. Only the helicoplacoids, ctenocystoids, and homosteleans of the Early and Middle Cambrian were absent during the Ordovician peak of echinoderm diversity.

As the more successful classes of echinoderms (especially crinoids) continued to diversify through the Paleozoic, most of the archaic, less efficient experimental groups were weeded out. The Devonian extinction wiped

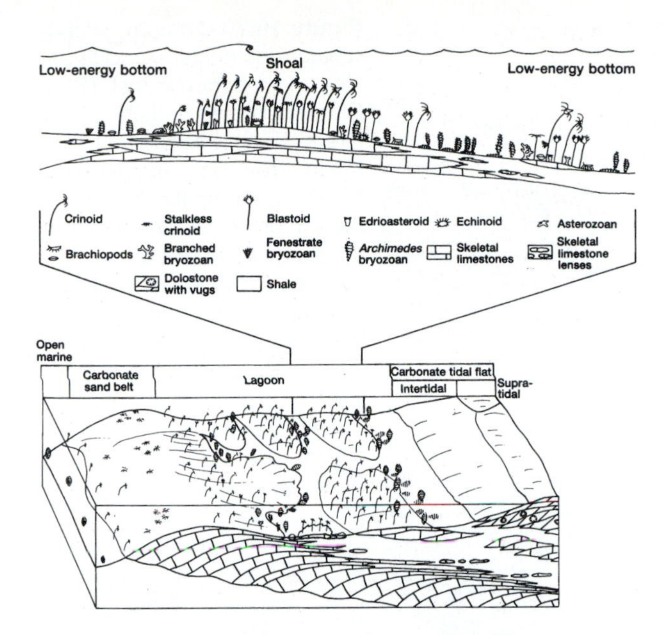

Figure 16.18. Reconstruction of environments in south-central Kentucky during the late Mississippian, showing the sedimentary record produced when facies belts shifted seaward. Dense crinoid gardens inhabited shoals and sandy shallows in the lagoon, as shown by the enlargement of a single shoal at the top of the diagram. (Modified from Chestnut and Ettensohn, 1988.)

out the last of the cystoids and most of the carpoids, leaving only the five living classes plus the blastoids and edrioasteroids for the rest of the Paleozoic. During this time, crinoids make up about 90% of fossilized echinoderm diversity, especially during their heyday in the Mississippian when there were more than 150 genera of inadunates, flexibles, and camerates. Crinoid diversity was still well over 100 genera when the Permian catastrophe hit, wiping out the inadunates, flexibles, and camerates (as well as the abundant blastoids).

When the articulates evolved from the inadunates in the Early Triassic, they inherited a planet that was nearly empty of high-level suspension feeders. Some Lower Triassic deposits contain extensive monospecific crinoid assemblages, which flourished like weeds in the aftermath of the Permian event. By the Jurassic, articulates had begun to diversify, but never reached the level of the Paleozoic heyday of crinoids, possibly because of all the new predators in the Mesozoic oceans. These include the gigantic pseudo-planktonic pentacrinids mentioned above, as well as the stalkless comatulids. Besides the comatulids, there were several independent appearances of stemless forms, all apparently adapted for exploiting new niches with fewer predators. But the great shallow seaways covered with gardens of crinoids never reappeared after the Paleozoic. Modern crinoids are represented by a handful of stalked forms that live in deeper water, and by the mobile comatulids. Apparently, the seas are now too hostile for crinoids, as they are for the brachiopods and other dominant groups of the less dangerous Paleozoic.

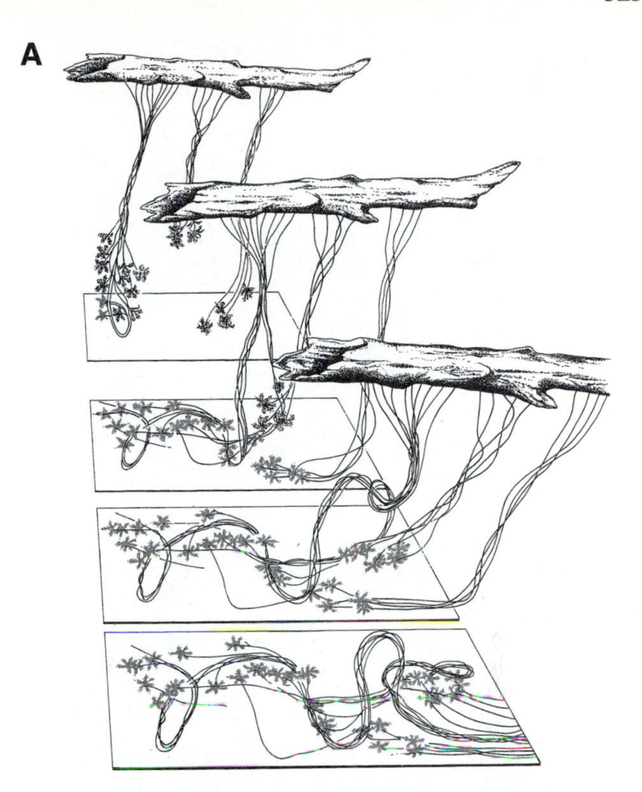

Figure 16.19. Reconstruction of the giant pelagic articulate crinoid *Seirocrinus.* (A) A series of reconstructions showing (from top to bottom) how the crinoids hanging from driftwood became tangled and collected on the bottom. (B) A slab of Jurassic Posidonia shale which preserves many specimens of *Seirocrinus* deposited in this manner. (From Seilacher et al., 1968.)

Evolutionary patterns—The excellent fossil record of crinoids, with their abundant specimens and complex plate morphology, makes them ideal for studying patterns of evolution. Prior to the publication of Eldredge and Gould's (1972) punctuated equilibrium model of evolution, most crinoid lineages were interpreted as evolving gradually. Afterward, most of these lineages were reinterpreted as showing stasis and punctuation (Paul, 1982; Simms, 1990; McNamara, 1990). In spite of their punctuational pattern of evolution, crinoids show a variety of evolutionary trends through their long history.

One of the most common trends is phyletic size increase,

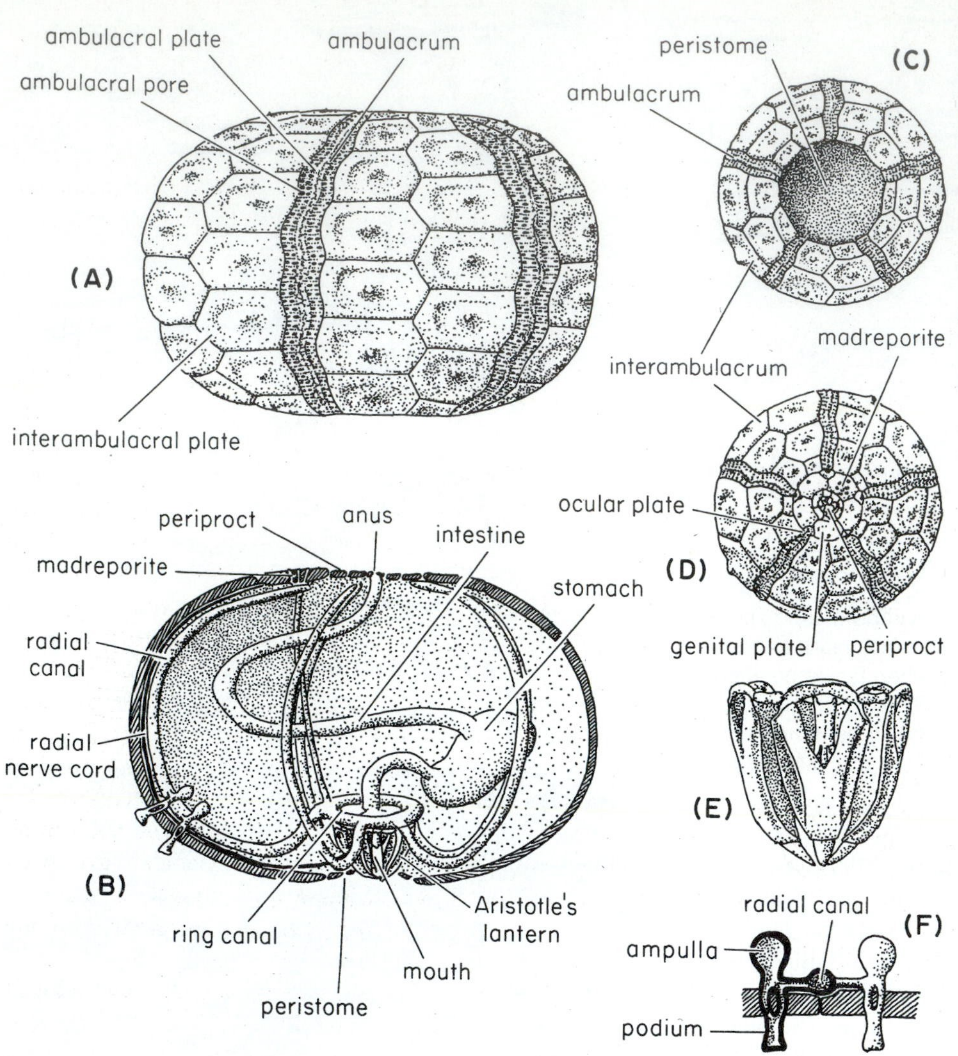

Figure 16.20. Anatomy of the regular echinoids. (A) Lateral view of an echinoid test. (B) Diagram of soft anatomy, in lateral view, with one side of the test removed. In a living echinoid, the space around the viscera shown here is filled by digestive glands and gonads. (C) Oral view. (D) Aboral view. (E) The echinoid jaw structure are five hinged teeth known as "Aristotle's lantern." (F) Cross section of the ambulacrum, showing radial canals and two tube feet. (From Beerbower, 1968.)

although Simms (1990) points out that these trends may be an artifact of changes in variance of the population. As the smaller, ancestral taxa die out, their larger descendants survive and the mean size of the lineage becomes larger. Similarly, the number of longer-stemmed taxa increased through the Paleozoic, so the variance of this character increased without any decrease in short-stemmed crinoids. The calyces of many crinoids change through time, from the complex, rigid conical cups of the earliest inadunates, to the simpler, more flexible bowl-shaped calyces of the flexibles and articulates of the later Paleozoic and Mesozoic. The significance of this trend is unclear, although it may allow crinoids greater ecological flexibility, or may just be a mechanism to produce a skeleton with less energy expended in crystallizing calcite (a difficult biochemical process). Another possible trend is the increase of pinnulate arms with multiple pinnules through time, and the tendency for the apical plates (especially the brachials) to multiply as the arms take over a larger portion of the calyx. A number of other possible trends are reviewed by Simms (1990). It is difficult to generalize too much about evolutionary trends in crinoids, since they occupy so many different possible shapes at once. Most of what has been called a "trend" in the past is simply the increase in variance of the whole population.

ECHINOIDS

Morphology and Ecology

Although fragments of crinoids may be the most numerous echinoderm fossils, the echinoids (sea urchins, sand dollars, heart urchins, sea biscuits, and their kin) have the best fossil record in terms of complete specimens, because they have a hard test that fossilizes very well. Many are burrowing forms, so they are already buried when they die and have a good chance of fossilization. All echinoids (Fig. 16.20) have a test formed of multiple plates of calcite, and most are tightly fused together. The mouth is usually at the bottom of the test, and the anus at the top, and a long, looping digestive tract within the body connects them. The madreporite plate is just off center from the anus on the top of the test, and connects by a stone canal to the water vascular system that runs through the animal. The test usually

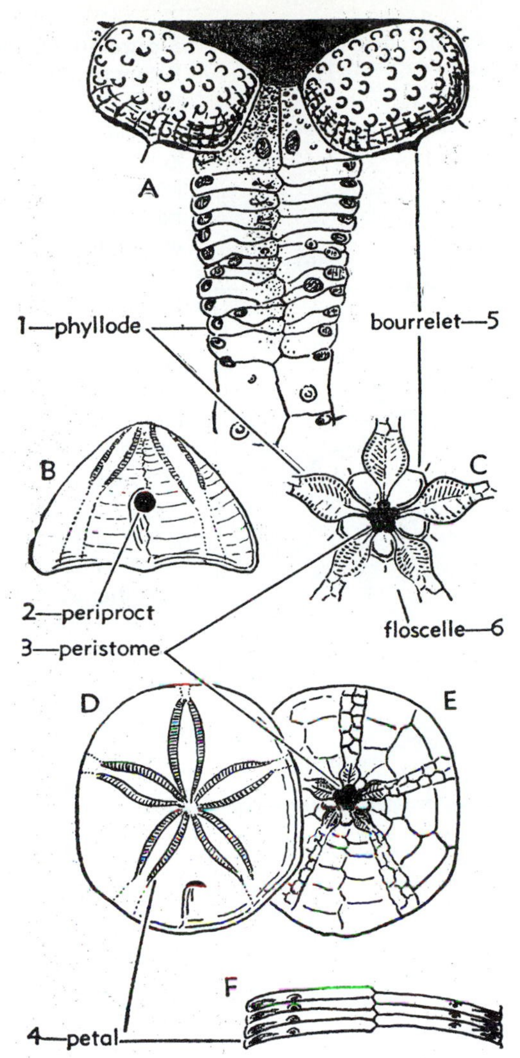

Figure 16.21. Anatomy of the irregular echinoid *Cassidulus* from the Cretaceous. D shows the aboral view, with the ambulacral petals (detailed in F), while E is the oral view, showing the region around the mouth (peristome, C), which has distinctive plates called phyllodes and bourrelets (A). B is a posterior view, showing the region around the anus (periproct). (From Moore et al., 1953.)

has five porous ambulacral areas, from which protrude the tube feet. In most echinoids, the tube feet on the bottom (oral side) are long and tipped with suction cups, so they can provide locomotion. Those on the top (aboral side) are for respiration, or for moving debris off the test. In addition to the tube feet, there are long stalks terminated with pincer-like claws known as **pedicellaria**, which protect the test against smaller organisms, settling larvae, and other debris.

The most striking feature of echinoids, however, is the spines that gave the group its name. Spines come in a variety of shapes and lengths, from the thick spines of slate-pencil urchins, to the thin, sharp spines found in most echinoids. Each spine is made of a single crystal of calcite, and is supported at the base by a ball-and-socket joint, allowing it to pivot and face a wide range of directions. In addition to the large spines that serve for protection, there are smaller spines that protect the muscles at the base of the larger spines.

Primitively, echinoids have a radial symmetry, with no front or back end. Round echinoids ("sea urchins") have long been known as the "**regular**" echinoids, although this is a primitive, paraphyletic group. Most live in the intertidal and shallow subtidal environment, creeping along the rocky surfaces and grazing algae with their jaws, known as **Aristotle's lantern** (Fig. 16.20). They live in crevices for protection, and their spines are very long and point in all directions. Symmetrical regular echinoids have been around since the Ordovician, and have changed relatively little since then (although they are uncommon in the Paleozoic).

Regular echinoids gave rise to bilaterally symmetrical, or **irregular**, echinoids in the Jurassic. Many became specialized for forward movement, and modified their adult radial symmetry to become bilaterally symmetrical, like their larvae (Fig. 16.21). In these groups, the mouth tends to move from the bottom of the test to the front, and the anus from the top of the test to the back. Many of these groups also reduce the size, but increase the number of

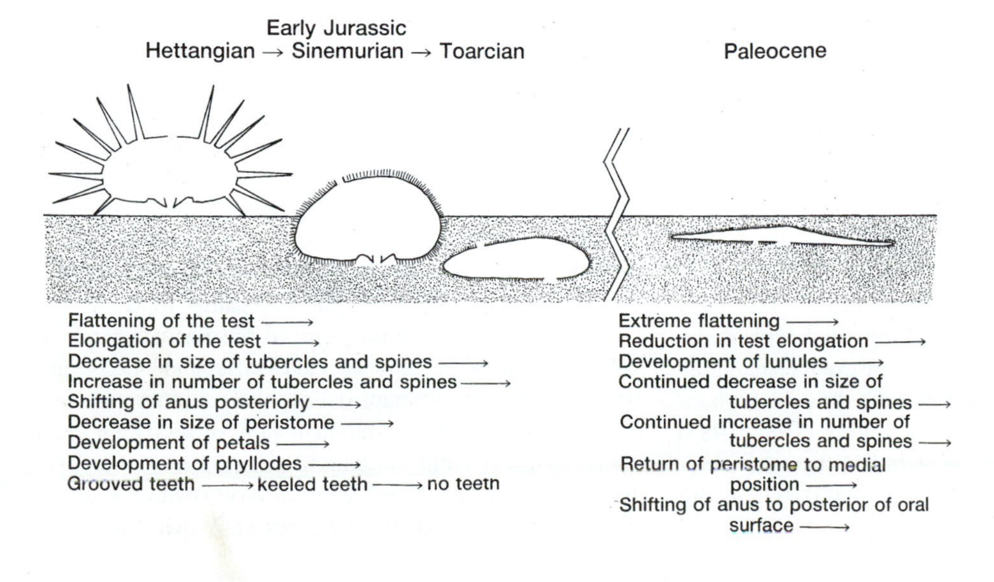

Figure 16.22. General trends in the echinoid test with the transformation from epifaunal regular echinoids to burrowing irregular echinoids, with the sand dollars representing the extreme modification of this trend. (Modified from Kier, 1982.)

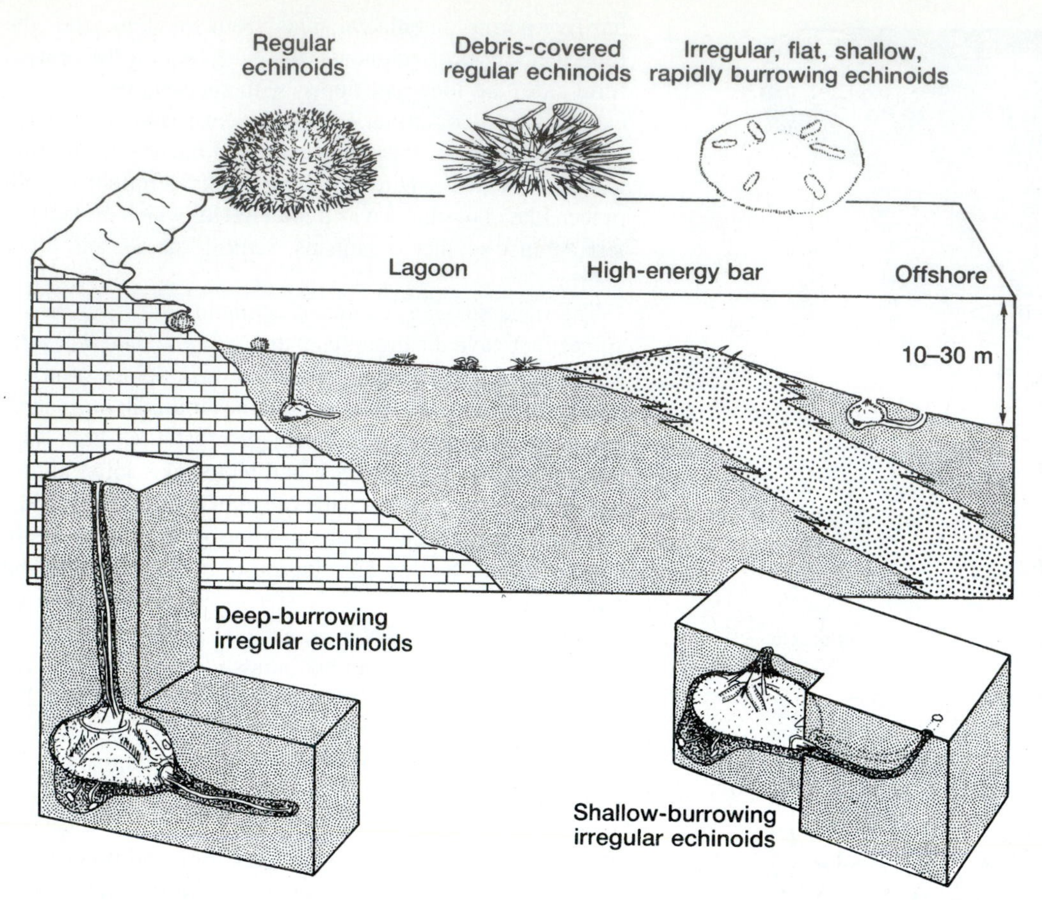

Figure 16.23. Life habits of different types of echinoids. Regular echinoids live in rock crevices in the intertidal and subtidal zone. Irregulars, such as sea biscuits and heart urchins, live in shallow and deep burrows in the subtidal and lagoonal sea bottom. They excavate their burrows and then use elongate tube feet to maintain a vent to the surface, and a waste tunnel behind them. Sand dollars, on the other hand, are found in subtidal sandy bottoms, and are very shallow burrowers. (Modified from Kier, 1982.)

their spines, until they have hundreds of tiny hair-like bristles. Because irregular echinoids have become modified for burrowing, large spines create drag and are no longer an advantage. Once they become burrowing animals, irregular echinoids feed primarily on detritus in the sediment, and may lose their jaw apparatus altogether. There is also no advantage to having tube feet that run from mouth to anus, so most irregular echinoids have reduced the ambulacral areas to five petal-shaped areas on the top of the test (Fig. 16.21).

Irregular echinoids come in a wide variety of shapes, from tall, shallow-burrowing forms (Figs. 16.22, 16.23), to the flatter, deeper-burrowing heart urchins. Deep burrowers use long tube feet to excavate a respiratory canal to the surface, and a sanitary canal behind their anus. The most extremely modified of all echinoids are the sand dollars, which first appeared in the Paleocene, and radiated widely in the Cenozoic (Fig. 16.24). These are adapted for shallow, very rapid burrowing. When they are feeding, they protrude from the sediment at an angle like a shingle buried in the sediment, causing currents to flow over their test and down to their mouth, bringing food with them. Many sand dollars have perforations in their tests called lunules. Traditionally, these features were thought to channel feeding currents to the mouth, but Telford (1982, 1983) has shown that these structures reduce the lift on the wing-like test of sand dollars, and prevent them from being picked up in the currents.

Evolution

Evolutionary history—Complete regular echinoids are rare in the Paleozoic. Although they occur as early as the Middle Ordovician, only a handful of genera are known at any one time, and only a few localities from the Mississippian produce abundant specimens. This is partly because regular echinoid plates are loosely sutured together and break up much more readily than the solid tests of irregular echinoids (Kier, 1977). Early Paleozoic echinoids tended to be small with few plates, although their size and number of plates increased in the later Paleozoic. All the typical Paleozoic regulars became extinct during the Permian catastrophe except for one genus (*Miocidaris*) from a group with reduced and standardized plating known as the cidaroids, which survive today. Apparently, *Miocidaris* was the ancestor of the great Mesozoic radiation of irregular echinoids, as well as all subsequently appearing regulars. Since the beginning of the Mesozoic, regular echinoids have greatly increased the number of ambulacral plates and thus the number of tube feet, which increases their ability to respire and gather food.

In just 10 to 15 million years of the Early Jurassic, the echinoids radiated into a wide variety of body plans derived from the symmetrical regular echinoids. Different lineages became more flattened and shifted their mouth and anus away from the central position on the test (Fig. 16.22). By the Cretaceous, echinoids occupied a wide variety of habitats, from the tidepool rock-dwelling regular

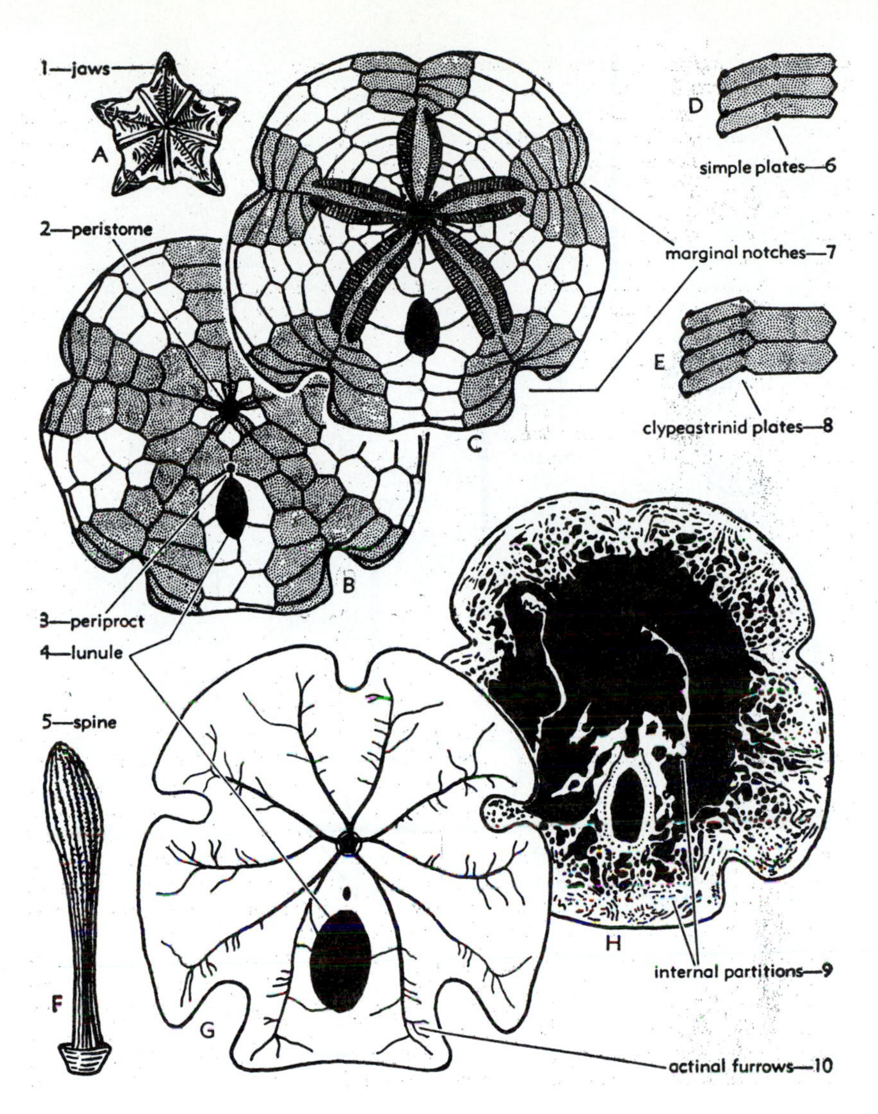

Figure 16.24. Anatomy of *Encope,* a clypeasteroid, or sand dollar. (A) The jaws, or Aristotle's lantern, much reduced from that seen in regular echinoids (Fig. 16.20). (B, C) Oral and aboral views of the test. Ambulacral areas stippled. The mouth and peristome are still in the center of the oral (lower) side, but the anus and periproct has shifted from the top to the front edge of the perforation called a lunule. (D, E). Details of the plate structure in sand dollars. (F). Aboral spine. (G) Oral view showing food-gathering grooves on surface. (H) Section of test, showing internal structures. (From Moore et al., 1953.)

echinoids, to the various flattened and dome-shaped species, to the deep-burrowing heart urchins with their extensive canals and tunnels. Their diversity was severely reduced by the K/T event, but recovered rapidly in the early Cenozoic.

The last major event of echinoid evolution was the evolution of sand dollars in the Paleocene of Africa. With their unique adaptations to shallow burrowing, they exploited new niches, and soon spread around the world in the Eocene. Echinoid diversity took another hit during the Eocene-Oligocene extinctions (McKinney et al., 1992), but recovered again in the Miocene. Today, there are apparently as many genera and species of echinoids as there have ever been at any time in the geological past.

Evolutionary patterns—Because of their excellent fossil record, there have been many studies of evolutionary patterns and trends in echinoids. One of the earliest of these studies was by Rowe (1899), who examined the changes in the heart urchin *Micraster* through the Cretaceous Chalk of southeastern England (Fig. 16.25). Collecting some 2000 specimens through the chalk cliffs, he noticed that the test became broader through time, the mouth moved to the front and the lip below it became more pronounced, and the paired ambulacra lengthened and became straighter. Supposedly, the lineage evolved into three successive species, named *M. corbovis, M. cortestudinarium,* and *M. coranguinum* (translated as "bull's heart," "turtle's heart," and "eel's heart," respectively). However, Rowe's stratigraphic data were inadequate for very fine-scale resolution of the changes through time, and he did not examine the geographic variation of *Micraster*. Instead, following the assumptions of his generation, he plotted the specimens to show a classic gradualistic evolutionary sequence, and this soon became a textbook case for phyletic gradualism (Eldredge and Gould, 1972; Gould and Eldredge, 1977). However, later studies debunked Rowe's (1899) gradualistic sequence (Kermack, 1954; Ernst, 1970; Stokes, 1977; Smith, 1984a). Fine-scale stratigraphic data showed that there was no linear, gradually progressive trend in *Micraster*, but instead, multiple rapid speciation events followed by stasis (Fig. 16.25). In addition, the evolution of these heart urchins did not take place only in England, but in two separate biogeographic provinces, a northern province and the Anglo-Parisian province. Thus, the pattern is further complicated by migration events as well as a complex pattern of speciation.

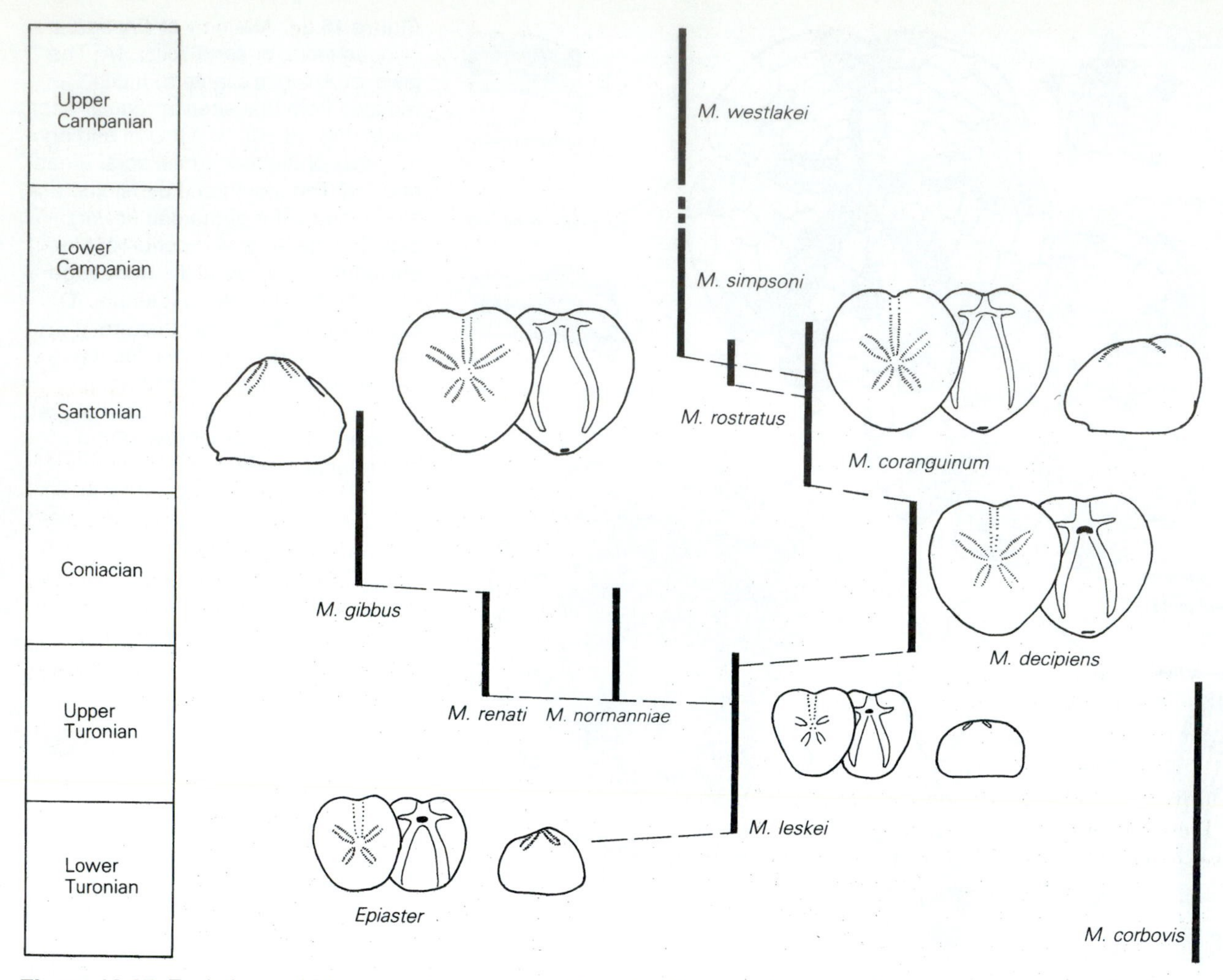

Figure 16.25. Evolutionary history and time ranges of *Epiaster-Micraster* lineage in the Upper Cretaceous Chalk of England and France. (From Clarkson, 1993.)

This is not to say that the progressive changes in morphology noted by Rowe (1899) aren't real—only that they did not evolve by gradual evolution in a single lineage. Rather, these features appeared many times in different taxa, so eventually all later forms are more advanced than the early forms. The ecological meaning of these changes is still disputed. Nichols (1959) argued that they were adaptations for progressively deeper burrowing, but Smith (1984) argued that the features were more likely adaptations for living in the soupy calcareous ooze that became the Chalk.

McNamara (1990) reviewed the evolutionary trends seen in a number of other echinoid lineages. The shape and configuration and number of pores in the ambulcral areas shows a great deal of variation in echinoids, with a number of progressive trends observed in different lineages. The interambulacral areas show increased numbers of plates, and variations for both increased and decreased spines in different lineages. As we have seen above, several different lineages evolve different types of irregular test shapes from the primitive round regular ancestral morphotype, and flattening of the test occurs independently in just about every group of irregular echinoid. Finally, a number of lineages show size increases through time, although in certain parts of the fossil record, size decreases are almost as common.

Many of these evolutionary trends can be attributed to heterochrony (Fig. 2.8B). McNamara (1988, 1990) gives examples of how these shape and anatomical differences can be attributed to either peramorphosis or paedomorphosis; both occur with great frequency in the echinoids. However, other extrinsic factors, such as the pressure of new predators with the ability to drill through echinoid tests, may have also driven many of these evolutionary transformations, especially those that allowed deeper burrowing.

For Further Reading

Broadhead, T.W., and Waters, J. A., eds. 1980. *Echinoderms: Notes for a Short Course*. University of Tennessee Studies in Geology 4.

Jangoux, M. and J. M. Lawrence, eds. 1983. *Echinoderm Studies*. A. A. Balkema, Rotterdam.

Jefferies, R. P. S. 1986. *The Ancestry of the Vertebrates*. British Museum (Natural History), London.

Kermack, K. A. 1959. A biometrical study of *Micraster coranguinum* and *M.* (*Isomicraster*) *senonis*. *Philosophical Transactions of the Royal Society of London* (B) 237:375-428.

Kier, P. M. 1965. Evolutionary trends in Paleozoic echinoids. *Journal of Paleontology* 39:436-465.

Kier, P. M. 1974. Evolutionary trends and their functional significance in the post-Paleozoic echinoids. *Paleontological Society Memoir* 5.

Lawrence, J.M., 1987. *A Functional Biology of Echinoderms*. Croon Helm Ltd., London.

McNamara, K. J., 1990. Echinoids, pp. 205-231, *in* McNamara, K. J., ed. *Evolutionary Trends*. Belhaven, London.

Nichols, D. 1969. *Echinoderms*. Hutchinson, London.

Nichols, D. 1974. *The Uniqueness of the Echinoderms*. Oxford University Press, London.

Paul, C.R.C., 1977. Evolution of primitive echinoderms, pp. 123-158, *in* Hallam, A., ed. *Patterns of Evolution as Illustrated in the Fossil Record*. Elsevier, Amsterdam.

Paul, C.R.C., and A. B. Smith, 1984. The early radiation and phylogeny of the echinoderms. *Biological Reviews* 59:443-481.

Paul, C.R.C., and A. B. Smith, eds. 1988. *Echinoderm Phylogeny and Evolutionary Biology*. Clarendon Press, Oxford.

Rowe, A. W. 1899. An analysis of the genus *Micraster*, as determined by rigid zonal collecting from the zone of *Rhychonella cuvieri* to that of *Micraster coranguinum*. *Quarterly Journal of the Geological Society of London* 55:494-547.

Simms, M. J. 1990. Crinoids, pp. 188-204, *in* McNamara, K. J., ed., *Evolutionary Trends*. Belhaven, London.

Smith, A.D., 1984a. *Echinoid Palaeobiology*. George Allen and Unwin, London.

Smith, A. D., 1984b. Classification of the Echinodermata. *Palaeontology* 27:431-459.

Sprinkle, J., 1973. *Morphology and evolution of blastozoan echinoderms*. Harvard University Museum of Comparative Zoology Special Publication, Cambridge, Mass.

Sprinkle, J. 1983. Patterns and problems in echinoderm evolution. pp. 1-18, *in* Jangoux, M. and J. M. Lawrence, eds. *Echinoderm Studies*. A. A. Balkema, Rotterdam.

Sprinkle, J., and Kier, P. M., 1987. Phylum Echinodermata, pp. 550-611, *in* Boardman, R.S., A.H. Cheetham, and A. J. Rowell, eds. *Fossil Invertebrates*. Blackwell Scientific Publishers, Cambridge, Mass.

Ubaghs, G. 1975. Early Paleozoic echinoderms. *Annual Reviews of Earth and Planetary Sciences* 3:79-98.

Waters, J. A., and C. Maples, eds. 1997. *Echinoderm Paleobiology*. Paleontological Society Short Courses in Paleontology 10.

Figure 17.1. The most complete specimen of a giant sauropod ever discovered is this five-story-tall *Brachiosaurus* from the Upper Jurassic Tendaguru beds of Tanzania, now standing in the main dome of the Museum für Naturkunde in Berlin. (Photo courtesy Museum für Naturkunde, Berlin.)

Chapter 17

Dry Bones

Vertebrates and Their Relatives

I thought that I would like to see
The early world that used to be,
That mastodonic mausoleum,
The Natural History Museum.
At midnight in the vasty hall
The fossils gathered for a ball.
High above notices and bulletins
Loomed up the Mesozoic skeletons.
Aroused by who knows what elixirs,
They ground along like concrete mixers.
They bowed and scraped in reptile pleasure,
And then began to tread the measure.
There were no drums or saxophones,
But just the clatter of their bones,
A rolling, rattling, carefree circus
Of mammoth polkas and mazurkas.
Pterodactyls and brontosauruses
Sang ghostly prehistoric choruses.
Amid the megalosauric wassail
I caught the eye of one small fossil.
Cheer up, old man, he said, and winked—
It's kind of fun to be extinct.

Ogden Nash, "Next!" 1952

INTRODUCTION

In the previous chapters, we have examined the major phyla of invertebrates that have a good fossil record. These are the primary subject matter of most invertebrate paleontology classes, and the study of fossil vertebrates is usually relegated to another, more advanced graduate-level course, or seldom taught in colleges at all. However, many paleontology classes are now beginning to cover both the vertebrates and invertebrates, and the discussion in this book is meant to give a broad overview of chordate evolution. Certainly, the lack of courses on fossil vertebrates is not for lack of interest. Dinosaur paraphernalia is a huge business, and this aspect of paleontology has a high public profile (Fig. 17.1). Many paleontologists first got hooked when they succumbed to dinomania as children, and parlayed that childhood interest into their life's work. Unfortunately, of the millions spent on dinosaur paraphernalia every year, only a tiny fraction of it actually supports the research that makes it all possible.

Nevertheless, an understanding of the broad features of vertebrate evolution is a worthwhile endeavor, even for those who will never take another course in paleontology. For one thing, we are vertebrates, and we want to understand our roots and where we came from. Although vertebrates are not as diverse or numerically abundant as molluscs or arthropods (there are about 45,000 living species of vertebrates, only a few percent of all the animal species on earth), their larger body sizes and sophisticated adaptations give them a dominant role on both the land and sea, especially in the higher levels of the food pyramid. With their incredible ecological diversity, vertebrates occupy the deepest oceanic waters, cover the land, and reign in the air. Today, one species of vertebrate (*Homo sapiens*) has completely changed the face of the planet, wiping out thousands of other species of vertebrates and invertebrates, while causing the proliferation of certain other vertebrate species, such as cattle, pigs, chickens, rats, and pigeons. Humans might not survive much longer on this planet, but some vertebrates, such as rats, will probably persist as long as the cockroaches and cyanobacteria.

For Further Reading

Benton, M. J., 1997. *Vertebrate Palaeontology* (2nd ed.). Chapman & Hall, London.

Carroll, R. M., 1988. *Vertebrate Paleontology and Evolution*. Freeman, New York.

Colbert, E.H., and M. Morales, 1991. *Evolution of the Vertebrates* (4th ed.). Wiley, New York.

Pough, F. H., J. B. Heiser, and W. N. McFarland, 1996. *Vertebrate Life* (4th ed.). Prentice Hall, Upper Saddle River, N. J.

Prothero, D. R., and R. M. Schoch, eds. 1994. *Major Features of Vertebrate Evolution.* Paleontological Society Short Courses in Paleontology 7.

Radinsky, L. B. 1987. *The Evolution of Vertebrate Design.* University of Chicago Press, Chicago.

THE ROAD TO AMPHIOXUS

Gill-slits, tongue bars, synapticulae
Endostyle and notochord: all these you will agree
Mark the protochordate from the fishes in the sea,
And tell alike for them and us their lowly pedigree.
Thyroid, thymus, subnotochordal rod;
These we share with lampreys, the dogfish and the cod—
Relics of the food-trap that served our early meals,
And of tongue-bars that multiplied the primal water-wheels.

Walter Garstang, *Larval Forms,* 1951

What Is a Vertebrate?

Where do vertebrates come from? First, we must define what we mean by "chordate" or "vertebrate." **Vertebrates** are a group of animals that have a number of unique specializations, including bone, red blood cells, a thyroid gland, and a backbone made of numerous bony or cartilaginous segments (the vertebrae). All mammals, birds, reptiles, amphibians, bony fish, sharks, and lampreys are vertebrates. The phylum **Chordata** includes vertebrates and several sister-groups that have certain unique features in common. The name "chordate" refers to the flexible rod of cartilage, the **notochord**, along the back that serves for support (Fig. 17.2). This rod becomes incorporated into the backbone of vertebrae in the vertebrates. However, the primitive relatives of vertebrates have no bony spine, only a notochord, and all vertebrate embryos (including you) had a notochord in the early stages of development before it was replaced by the spinal column. This stiffening rod is important not only to support the elongate body, but also in swimming. When the muscles along the body contract, they pull on the notochord and cause it to flex side-to-side, resulting in an efficient swimming motion. If there were no notochord, the contraction of these muscles would telescope the body and cause it to collapse.

Just above the notochord is a nerve cord in the **dorsal** (along the back) position. By contrast, annelids, arthropods, and molluscs have a nerve cord that runs in a **ventral** (along the belly) position. The front end (**anterior**) of most chordates usually has a concentration of sense organs surrounding a cartilaginous or bony structure, the braincase, that encloses and protects the brain. The mouth is in the front, and passes into a "throat" region (the **pharynx**). In primitive vertebrates, the pharynx is muscular, and serves to pump food particles through the digestive tract, and to pump oxygenated water past the gills. The rest of the digestive tract may have a differentiated stomach, intestines, or other organs, and ends in an anus that is behind the midpoint of the body. Behind the anus, the back end (**posterior**) of the body is usually composed of a muscular tail, composed mostly of V-shaped muscle masses known as **myomeres**, which propel the body in a tadpole-like fashion. The main vessels of the circulatory system run along the belly of the chordate body plan, in contrast to the dorsal heart and circulatory system of annelids, arthropods, and molluscs. The internal coelomic cavity of the chordate contains not only the digestive and respiratory system, but also excretory and reproductive organs, and frequently, many other organ systems as well.

Although most vertebrates are characterized by a hard bony skeleton and have a decent chance of fossilization, the earliest vertebrates and their ancestors were soft-bodied, and so fossilize rarely. For this reason, the primary approach to understanding vertebrate origins has been to look at all the close living relatives of the vertebrates and their embryonic history, and then try to place the few available fossils in this context (Fig. 17.3). Three important groups of living animals give us examples of the steps in the evolution of the basic chordate body plan. They are the **hemichordates** (pterobranchs, "acorn worms," and the extinct graptolites), the **urochordates** (the tunicates or "sea squirts"), and the **cephalochordates** (the amphioxus or lancelets).

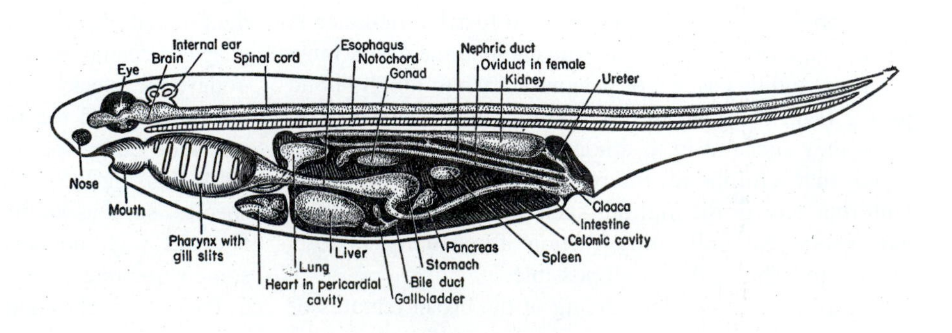

Figure 17.2. Diagrammatic section through the an idealized vertebrate body plan, showing the relative position of the major organs. (From Romer, 1970.)

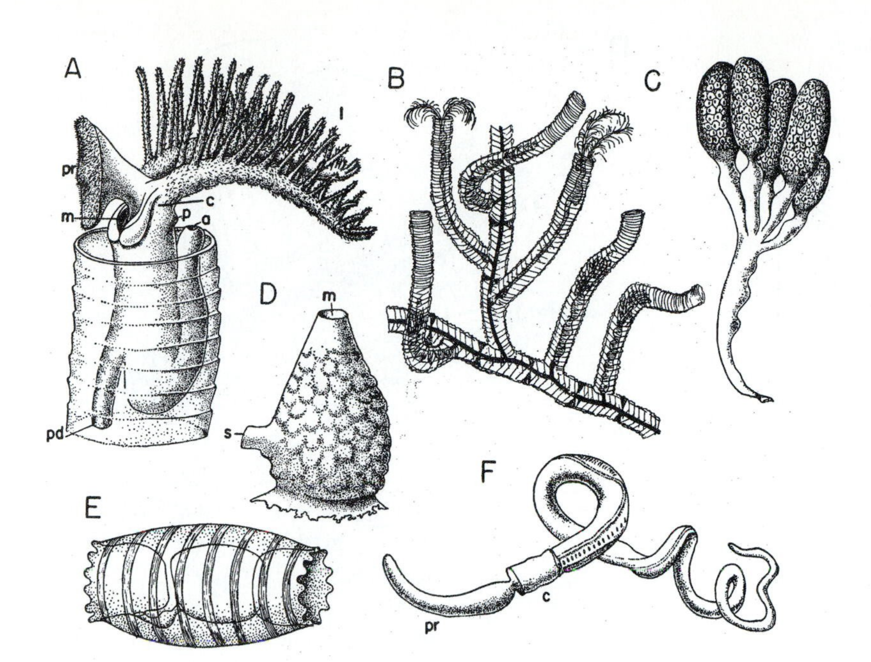

Figure 17.3. Tunicates and hemichordates. (A) A individual from a colony of *Rhabdopleura* projecting from its enclosing tube. (B) A colony of *Rhabdopleura.* (C) A colonial sessile tunicate. (D) A solitary tunicate. (E) A free-floating tunicate, or salp. (F) An acorn worm, *Balanoglossus.* Abbreviations: a, anus; c, collar region; l, lophophore; m, mouth, p. pore or opening from coelom; pd, peduncle (stalk) by which individual is attached to remainder of colony; pr, proboscis; s, siphon to carry off water and body products. (After Romer, 1970.)

Hemichordates

Of all the soft-bodied animals in the sea that we have seen so far, our closest relatives are the phylum Hemichordata ("half-chordates" in Greek). They are usually placed in a different phylum from the chordates, since they do not have the diagnostic notochord. Instead, they have a tubular structure in the position of the notochord that is actually derived from a pouch off the digestive tract, and is thought to be equivalent to the embryonic precursor of the notochord. Hemichordates do have a few derived chordate features, including a pharynx with multiple openings, and a dorsal nerve cord and ventral blood vessel.

Today, the hemichordates are represented by about 90 species in two classes of invertebrates that look as different from vertebrates as could possibly be imagined. One group, class Enteropneusta (about 80 species), is known as the "acorn worms" because they have a worm-like body, with a proboscis on the front that is used for burrowing and for trapping food particles (Fig. 17.3F). Mucus flows along the proboscis, capturing the food and transporting it to a collar-like structure that ingests the good stuff and rejects the rest. Most acorn worms are just a few centimeters long, although some are as long as 2.5 m. Acorn worms live in a U-shaped burrow in shallow marine waters, but they are not a common member of the seafloor community. Most people would never be able to distinguish them from any other marine worm, but to the astute eye, there is one important clue: multiple pharyngeal openings just behind the collar, showing that they are relatives of the chordates. In addition, they have a dorsal nerve cord and ventral blood vessel, another chordate feature. Their elongate, bilaterally symmetrical body is also similar to that of the chordate body plan, and very different from the radial symmetry of the other group of deuterostomes, the echinoderms.

As hard as it is to imagine an acorn worm as our cousin, the other group of hemichordates, the class Pterobranchia (about 10 living species in 3 genera) is even less like vertebrates. As adults (Fig. 17.3A, B), they are tiny colonial filter-feeding animals, with a fan of tentacles for catching food particles, which are then processed by the small proboscis and collar before they enter the mouth. Pterobranchs still possess a pharynx with pharyngeal slits, but almost all other similarities to chordates have been lost. Living pterobranchs like *Rhabdopleura* form large colonies of multiple individuals, each secreting a long segmented tube of organic matter with a very distinctive structure. Each individual animal has a long tube called a **stolon** that connects it with the rest of the animals in the colony.

Mystery Solved: The Graptolites

It is this distinctive segmented tube structure of their colonial skeleton that eventually solved one of the greatest mysteries in all of paleontology: the **graptolites**. Since the late 1700s, people had puzzled over the small markings that looked like tiny hacksaw blades on lower Paleozoic black shales. They were called graptolites ("written on stone" in Greek) because they looked like graphite pencil markings. These fossils were found in many lower Paleozoic marine rocks, but they were practically the only fossils in the deep-water black shales. Even though people had no idea what organism they represented, graptolites eventually became one of the best index fossils for the early Paleozoic. Not only did they evolve rapidly through the Ordovician and Silurian, but they were apparently pelagic animals that floated in surface waters worldwide, so they could be used for global correlation. After they died and sank to the bottom, they were fossilized in both deep-water shales and also shallow-water deposits, so they could be used to correlate formations of very different environments and facies. Graptolites became the basis for Ordovician through Silurian biostratigraphy, and were studied in detail for over a century without a clue as to what kind of organism made

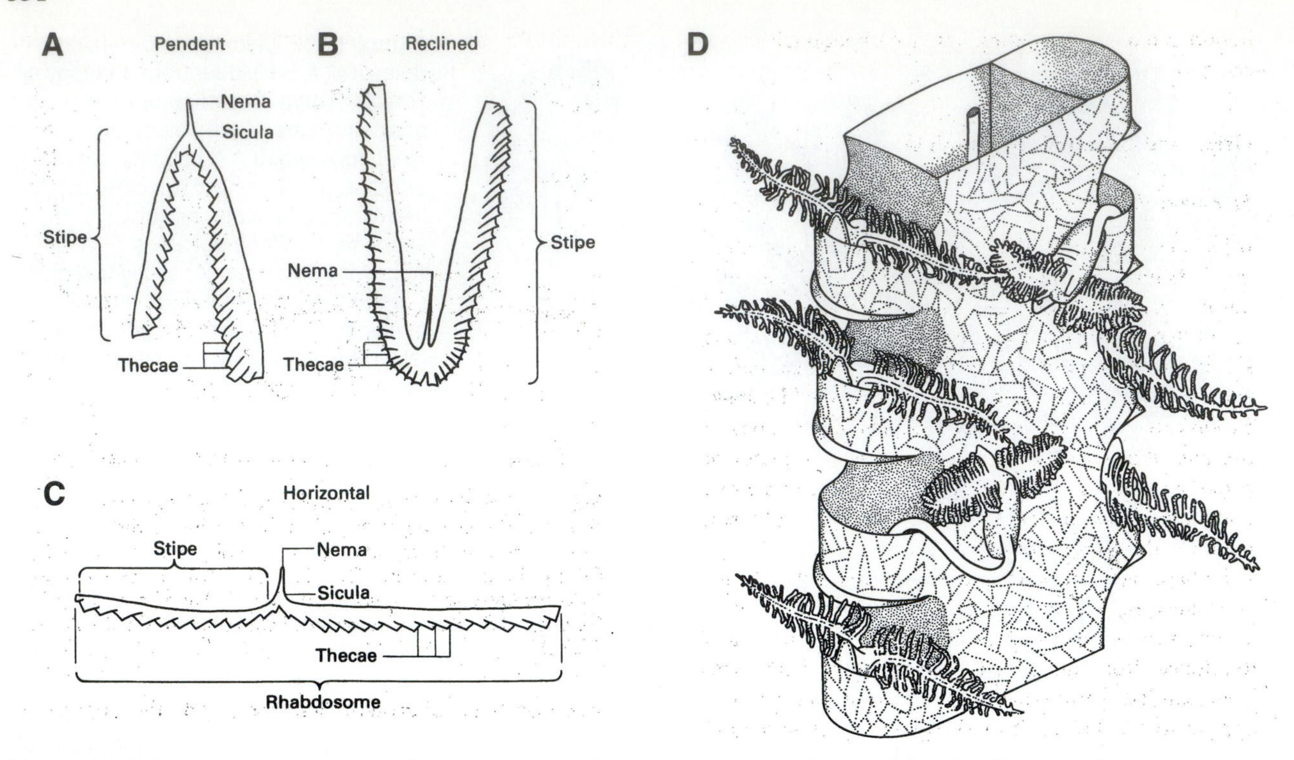

Figure 17.4 Morphology of the graptolite rhabdosome. (A) *Didymograptus* (pendent or "tuning fork" form). (B) *Isograptus* (reclined form). (C) *Didymograptus* (horizontal form). (D) Part of the biserial rhabdosome of *Climacograptus* showing its distinctive "bandage" cortex texture, with a hypothetical reconstruction of the graptoloid zooids. (From Boardman et al., 1987.)

them. Graptolites, like the conodonts discussed below, are classic examples of how biostratigraphy can proceed without any knowledge of the biology of the organism. Biostratigraphers needed only to document objective patterns of graptolite range distributions to construct a zonation; no biology was required.

As long as scientists looked only at the broken, flattened, carbonized remnants of graptolites, there was not much evidence to link them to any living organism. Some paleontologists classed them as bryozoans, others as cnidarians; both are groups with a similar colonial habit. The solution to the problem occurred when Koszlowski (1948)

Figure 17.5. Reconstructions of some floating and drifting graptolites. (From Fenton and Fenton, 1958.)

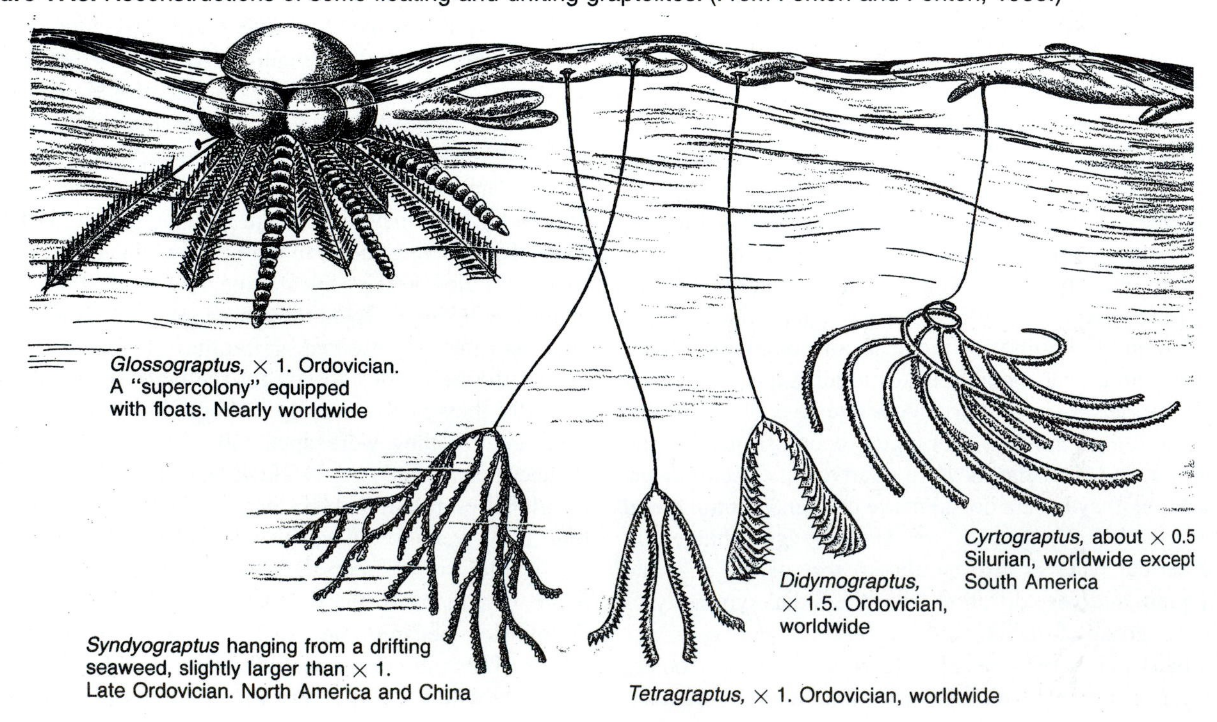

described three-dimensional graptolite specimens dissolved out of cherts, so they were uncrushed and their detailed structures could be examined. (Since then, limestones have yielded excellent three-dimensional specimens as well.) After etching, these delicate specimens were then embedded in paraffin and serially sectioned, allowing Koszlowski to study their structures in microscopic detail. Even though there were no soft tissues preserved, the detailed wall structure of the graptolite tubes (Fig. 17.4) and the presence of stolons are so similar to living hemichordates like *Rhabdopleura* that there is little doubt any longer that graptolites were hemichordates. The tube-like colonies of graptolites have an overlapping half-ring structure that is very similar to that of *Rhabdopleura*, and they grow, branch, and bud in a very similar fashion. The entire graptolite colony (**rhabdosome**) is divided into one or more branches (**stipes**) that support dozens of cups or tubes (**thecae**) that housed the individual organisms, or **zooids** (Fig. 17.4). The details of the size, shape, and spacing of the thecae are very important in identification, as are the spines and other features of the aperture. At the "root" of the rhabdosome is the **sicula**, or the cup of the first zooid,

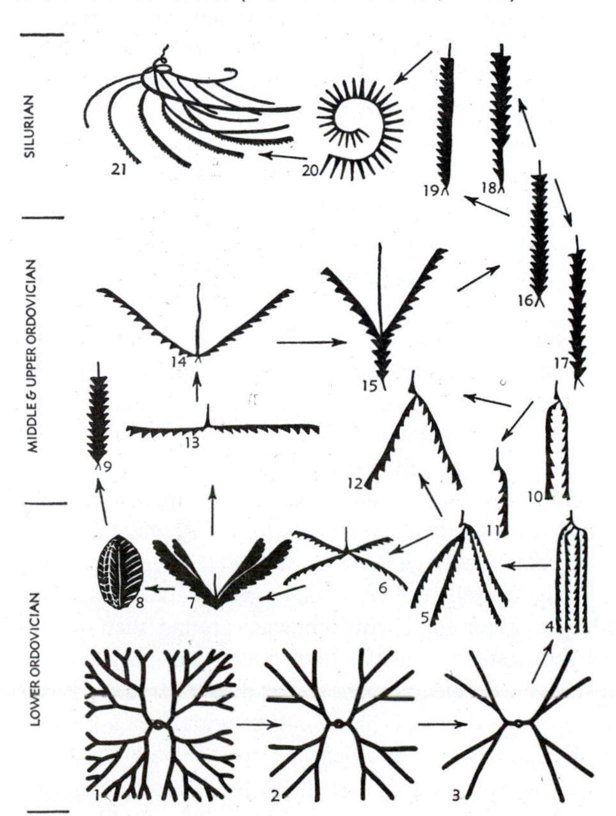

Figure 17.6. Evolution in the shape of graptolite colonies. Trends are as follows: 1-4, reduction in the number of stipes from many to four; 4-8, variations on the four-branched form; 9, loss of two rows of thecae produces biserial forms (*Glossograptus*); 10-11, loss of branches resulting in two-stipe and one-stipe forms; 12-17, elevation of stipes in two-branched forms; 16-18, loss of one row of thecae, producing uniserial *Monograptus;* 19-21, modification of the monograptids, producing coiled unbranched and branched forms. (From Moore et al., 1953.)

that founded the colony, and it may have a distinctive spine projecting from its aperture called the **virgella**. Many graptolites have a long, thread-like "stem" from the sicula known as the **nema**, that attached them to some kind of floating object. A recently described pterobranch, *Cephalodiscus graptoloides*, has a nema almost identical to that of graptolites (Dilly, 1993). The branching patterns of the first few thecae at the base of the colony is a key character to understanding their evolution and ontogeny.

Based on the assumption that the zooids were much like *Rhabdopleura*, the graptolite organism is usually reconstructed as an animal whose fan-like filtering devices protruded from the aperture of the theca, but was connected to all the other members of the colony by a stolon (Fig. 17.4D). The earliest (Late Cambrian) graptolites are dense bushy colonies known as dendroids. In these forms, the entire rhabdosome attached to some other solid substrate, such as rock or floating seaweed, allowing them to filter seawater. In the Ordovician, the multibranched dendroid graptolites evolved to simpler forms with fewer and fewer stipes, until some of them were reduced to as few as four or even two branches (Figs. 17.5, 17.6). These graptolites tended to be lighter, so they could be supported by a smaller floating piece of debris; some apparently had bubble-like floats at the top of the cluster of stipes (Fig. 17.5). By floating at the surface, graptolites were one of the first metazoans to exploit the immense food resources in the surface-water plankton. As mentioned above, most graptolites floated over the oceans worldwide, but they were especially common in deeper waters over the shelf edge and upper slope, apparently above oceanic upwelling currents that would have caused immense plankton blooms. Other graptolites apparently lived at greater depths, allowing them to exploit resources at different levels in the water column. Very different graptolites lived in the shallow nearshore waters, although they are much less common than, nor as cosmopolitan as, their open-ocean counterparts. They are also less useful for biostratigraphy, since they are not as widespread as the planktonic taxa.

Even though the rhabdosome of graptolites was a relatively simple structure, they showed remarkable morphological diversity. Not only were there important details to be found in the ontogeny of the first few thecae, and in the shape and spininess of the apertures, but the colonies themselves varied greatly in shape (Fig. 17.6). Four-branched rhabdosomes may have their stipes draped at many different angles, and come in a variety of shapes, thicknesses, and curvatures. One graptolite (*Phyllograptus*) has four thick leaf-shaped stipes that resemble the vanes on a dart (Fig. 17.6, #8). Many Ordovician graptolites have two rows of thecae on either side of the stipe (**biserial**). Toward the end of the Ordovician, the stipes became further simplified, and the characteristic Silurian-Early Devonian genus *Monograptus* has a single row of thecae (**uniserial**) on a single stipe (Fig. 17.6, #16-18). In the Silurian, a number of graptolites coiled up into elaborate branched spirals and other unusual twisted shapes, which apparently made

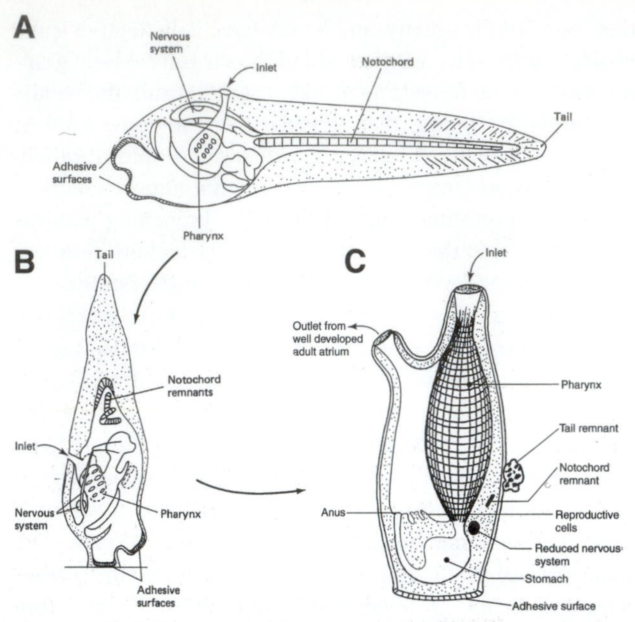

Figure 17.7. The free-swimming tunicate larva (A) attaches and undergoes metamorphosis (B) into a sessile adult (C). (From Pough et al., 1996.)

them slowly spin around in the water as they floated, further enhancing its coverage of seawater for filtering.

During their heyday in the Ordovician and Silurian, graptolites evolved so rapidly that there are 53 successive graptolite zones recognized for this interval. The species *Monograptus uniformis* is used as the indicator of the Silurian/Devonian boundary. Yet despite this rapid evolution, graptolites were not really that diverse compared to other phyla we have studied. Only about 250 genera are recognized worldwide, but the biostratigraphically significant genera are so few and so distinctive that their entire evolution and biostratigraphy can be summarized on a single wall chart (Churkin and Carter, 1972). In the Early Devonian, graptolite diversity was reduced to a few lineages (primarily the rare bushy dendroids and *Monograptus*), and by the Middle Devonian, the planktonic graptolites were extinct and only a few relict dendroids remained to straggle on into the Carboniferous. Nobody knows why planktonic graptolites died out after such a long history of success; there were no other major extinctions at the Lower/Middle Devonian boundary. But several authors have suggested that the radiation of fish in the Devonian may have provided a new predator that these large, vulnerable planktonic organisms could not avoid.

Urochordates

Hemichordates such as the acorn worms, pterobranchs, and graptolites are not considered chordates because they lack a notochord. However, there are organisms in the ocean today that have these features, yet they are still a long way from being vertebrates. These are known as the **tunicates**, ascidians, or "sea squirts," and they are usually placed within the Chordata as the subphylum Urochordata. The name "sea squirt" is an undignified but apt description for our close relatives. As adults, they are a soft-bodied little sac that pumps water in through the "chimney" (**inhalant siphon**) at the top, filters it through a basket-like pharynx, into the surrounding cavity called the **atrium**, and then expels it from an **exhalant siphon** (Figs. 17.3C, D, E, 17.7C). They have a flexible outer body sheath called a **tunic**, whence they get the name "tunicates." Although this simple body plan may not seem impressive, they are a very successful group, with over 2000 living species. They are so small and translucent, however, that most beachcombers and divers never even see them.

How could such a strange little sac of jelly be a chordate? As we have mentioned, the presence of the pharynx in adults is one clue. Better evidence, however, can be seen in their larvae (Fig. 17.7A), which not only has a pharyngeal basket, but also a notochord, dorsal nerve cord, and muscular tail with paired myomeric muscles. This tadpole-like larva swims for a few hours to a few days, trying to find a hard surface. The adhesive papillae at the anterior end attach, and within 5 minutes the tail begins to degenerate and the notochord disappears (Fig. 17.7B). About 18 hours later, the metamorphosis is nearly complete, and the body has reduced to a simple sac filled with a pharyngeal basket.

This remarkable metamorphosis is a classic case of ontogeny providing important phylogenetic clues that are lost in highly specialized adults, a trend that we will see again and again. Garstang (1928) first suggested that neotenic retention of larval characteristics may have been very important in chordate evolution. He visualized a sequence (Fig. 17.8) of steps of evolution from lophophorates (the probable sister-group of deuterostomes) to echinoderms to hemichordates to urochordates, and eventually, to higher chordates. In each step, the retention of larval characteristics allows organisms to continue on this main evolutionary pathway, while the highly specialized adults go off in their own adaptations. Clearly, the odd adult tunicate body form could not have been ancestral to higher chordates, so the next step must have been a neotenic retention of their tadpole-like larva, which is very similar to many othe primitive chordates.

Cephalochordates

The next step in chordate evolution is another tiny, soft-bodied organism known as the amphioxus or lancelet (Fig. 17.9). It is best known from the species *Branchiostoma lanceolatum*, although about 25 species are known from temperate and tropical seas worldwide. This tiny sliver of flesh is usually only a few centimeters in length, and swims like an eel while it is a larva. As adults, lancelets burrow in the sandy seafloor tail first, leaving only their heads protruding to filter feed with tentacles around their mouth. Their elongate, worm-like body has many chordate features: a well-defined notochord and dorsal nerve cord, ventral blood vessel, pharyngeal basket with over a hundred slit-like openings on the sides, and postanal tail. Unlike the larval tunicates, however, they have well-developed

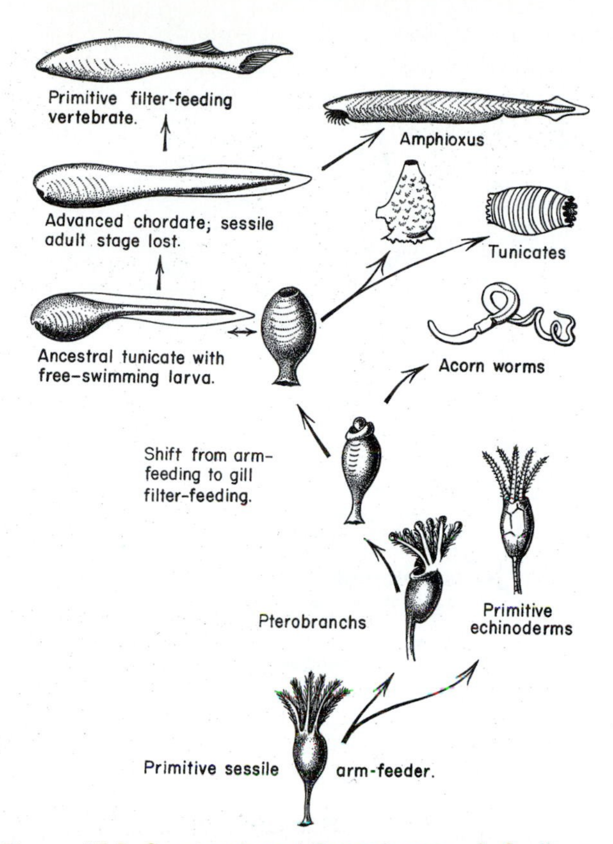

Figure 17.8. Garstang's and Romer's scenario for the evolution of chordates. From a sessile arm feeder came both the echinoderms and the pterobranchs. A shift from arm feeding to gill feeding produces the acorn worms. Tunicates represent the culmination of this gill-feeding stage, but chordates escaped this adult specialization through neotenic retention of their free-swimming larvae. Eventually, the sessile adult stage is lost, producing amphioxus, and finally a filter-feeding jawless vertebrate. (From Romer, 1970.)

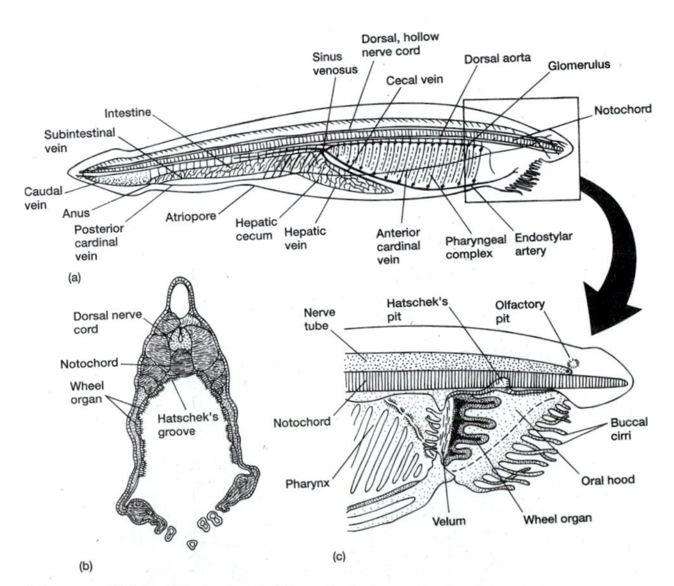

Figure 17.9. The anatomy of the cephalochordate or amphioxus (genus *Branchiostoma*). (A) Lateral view. (B) Cross section through oral region. (C) Enlargement of mouth. (From Kardong, 1995.)

myomeres that run the length of the body, rather than just in part of the tail. They also have other additional advanced chordate features, including a liver and kidney, a more advanced nervous system, and many genetic and molecular features that are unique to lancelets and higher chordates.

Although lancelets definitely have a front end with a mouth, they do not yet have a well-defined head. They have no eyes, but they do have a photosensitive pigment spot in the front that detects light and darkness. Their mouth has no jaws, but instead uses tentacles with many tiny cilia to trap floating food particles. The food then passes through folded, cilia-lined tracts called the **wheel organ** (so-called because the cilia give the impression of a wheel in motion). On the right dorsal side of these tracts is **Hatschek's pit**, which secretes mucus to help collect the food particles, and may be the homologue of the pituitary gland in higher chordates. The **endostyle** produces more mucus to hold the food particles together as they pass through the pharynx.

Such a small, soft-bodied animal would be unlikely to fossilize, but fortunately there are specimens in several *Lagerstätten*. The Middle Cambrian Burgess Shale yields an animal known as *Pikaia* (Fig. 17.10), which has a broad, lancelet-like body with visible notochord and myomeres and a distinct tail fin. Another Middle Cambrian *Lagerstätte* in China has produced a similar animal known as *Yunnanozoon*. Thus, we have a fossil record in the Cambrian not only of hemichordates (graptolites) but also of cephalochordates, and as we shall see below, even vertebrates are known from the Late Cambrian. The evolutionary radiation of the chordates and their primitive relatives must have occurred rapidly entirely within the Early-Middle Cambrian or possibly earlier. Another specimen, *Palaeobranchiostoma*, is known from the Permian of South Africa.

For Further Reading

Barrington, E. T. W. 1965. *The Biology of Hemichordata and Protochordata*. Freeman, San Francisco.

Berry, W. B. N., 1977. Graptolite biostratigraphy: a wedding of classical principles and current concepts, pp. 321-338, *in* Kauffman, E., and Hazel, J.E., eds. *Concepts and Methods of Biostratigraphy*. Dowden,

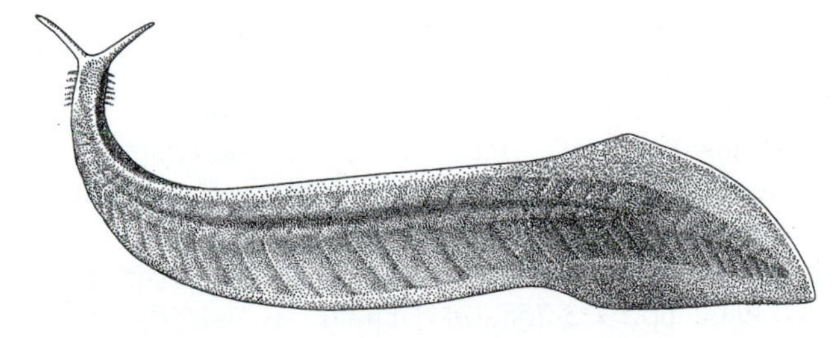

Figure 17.10. Reconstruction of the Middle Cambrian Burgess Shale cephalochordate *Pikaia,* one of the oldest fossils known from the chordate lineage. (From Gould, 1989.)

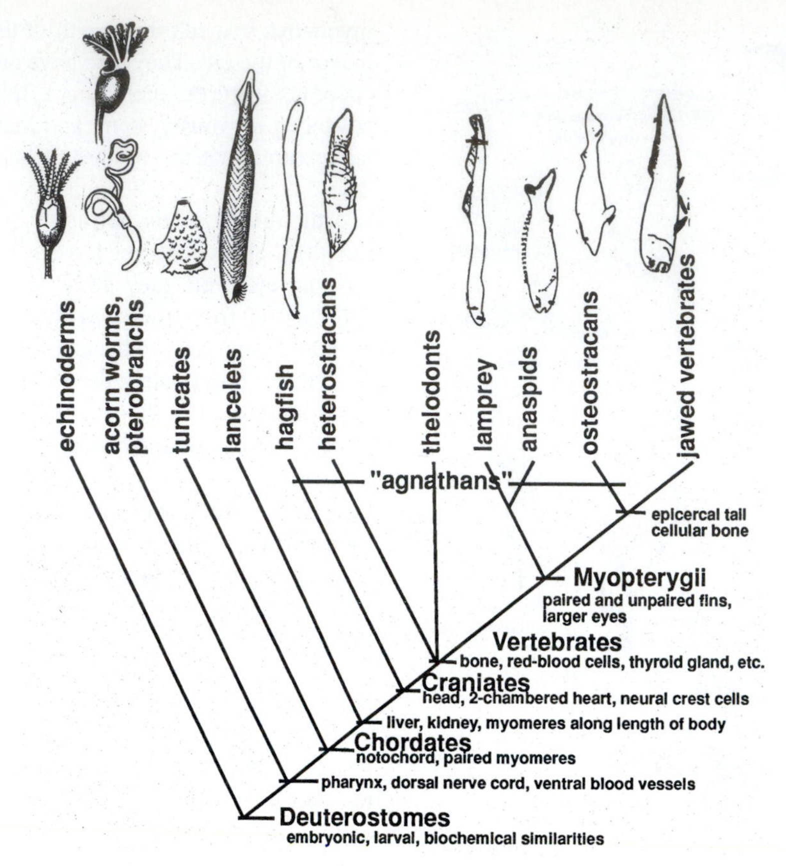

Figure 17.11. Cladogram of the primitive vertebrates and their sister-taxa.

Hutchinson, & Ross, Stroudsburg, Penn.

Berry, W. B. N. 1987. Phylum Hemichordata (including Graptolithina), pp. 612-635, *in* Boardman, R.S., A.H. Cheetham, and A. J. Rowell, eds. *Fossil Invertebrates*. Blackwell Scientific Publishers, Cambridge, Mass.

Bulman, O. M. B., 1970. Graptolithina, *in* Teichert, C., ed. *Treatise on Invertebrate Paleontology, Part V (revised)*. The Geological Society of America and the University of Kansas Press, Lawrence.

Churkin, J., Jr, and Carter, C., 1972. Graptolite identification chart for field determination of geologic age. *U. S. Geological Survey Oil and Gas Investigations Chart* OC 66.

Gee, H. 1997. *Before the Backbone: Views on the Origin of Vertebrates*. Chapman & Hall, New York.

Koszlowski, R., 1948. Les graptolithes et quelques nouveaux groupes d'animaux du Tremadoc de la Pologne. *Palaeontologia Polonica* 3:1-235.

Palmer, D., and Rickards, R. B., eds. 1991. *Graptolites: Writing in the Rocks*. Boydell Press, London.

Rickards, R. B. 1975. Palaeoecology of the Graptolithina, an extinct class of the Phylum Hemichordata. *Biological Reviews* 50:397-436.

Rickards, R. B. 1977. Patterns of evolution in the graptolites, pp. 333-358, *in* Hallam, A., ed. *Patterns of Evolution as Illustrated in the Fossil Record*. Elsevier, Amsterdam.

Schaeffer, B. 1987. Deuterostome monophyly and phylogeny. *Evolutionary Biology* 20: 179-235.

GETTING A HEAD: THE CRANIATES

Among other habits that have endeared them to seafarers, slime eels like to enter dead or dying bodies on the ocean bottom by way of mouth, gills, or anus, and gobble up everything except bones and skin, which remain intact. Fish immobilized in gill nets are particularly susceptible. What's left of the fish is "a bag of bones, literally, like it had been sucked dry by a high-powered vacuum cleaner." Slime eels are often still inside the fish when the bloated gill net spills its contents onto the fisherman's deck. . . the hags ensconced in the victims are typically well-fed and at ease, "smiling, slimy, usually snoring—gently." In one case, a record, a single cod contained 123 slime eels, in a pink, wriggling mass.

Richard Conniff, *Spineless Wonders*, 1996

Slimy Scavengers and Parasites

The next step above the lancelet in complexity (Fig. 17.11) is represented by the **craniates**, or the chordates with a well-defined head region, including a definite brain at the front end of the nerve cord and cranial nerves connected to well-defined sense organs (eyes, nose, ears). Craniates are also the first chordates to have more than just a notochord for support, but a true internal skeleton made of cartilage. Craniates have many other specializations, including a much more sophisticated circulatory system with a two-chambered heart. During their embryology, they have a distinctive region along the spine called the **neural**

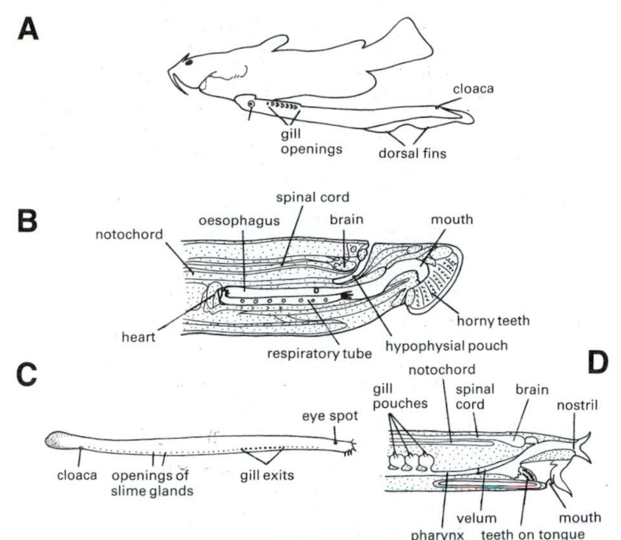

Figure 17.12. Living jawless fishes. (A) Lamprey feeding attached to a fish. (B) Longitudinal section of the head of the lamprey. (C) External lateral view of the Pacific hagfish. (D) Longitudinal section of the head of a hagfish. (From Benton, 1990.)

crest, a feature found in all vertebrate embryos. Neural crest cells are very important in embryology, since they migrate from this region to form the skeleton, skin, nervous system, sense organs, and much of the digesitive and circulatory system of most vertebrates.

The most primitive living craniates are two jawless eel-like animals, the lamprey and the hagfish (Fig. 17.12). You could not imagine two more unappealing relatives if you tried! **Hagfishes** (also known as "slime eels" or "slime hags") are a group of eel-like craniates that are not considered true vertebrates because they lack vertebrae and other derived vertebrate characters. About 60 species in six genera are known, restricted to the shallow continental shelf. They burrow along the seafloor seeking out dead and dying fish and polychaete worms, which they can slurp up faster than a strand of spaghetti. They use tooth-like processes on their muscular tongue to rasp out pieces of living fish, although they prefer to burrow into the body cavity of a dead fish and eat it from the inside. When they are trying to tear a chunk off a fish, they grasp onto a protruding surface (like the gills or anus) and then tie themselves into a knot. They then force the knot against the surface of their victim until they rip a piece loose. This same kind of knotting behavior is useful for wriggling out of the grasp of a predator.

Hagfishes are known as "slime eels" because they produce copious amounts of mucus when they need to evade a predator. Their bodies are lined with 90 slime pores, and it is said that they can fill a two-gallon bucket of water with slime in a matter of minutes. The mucus contains a special kind of protein that unfolds in water and gathers up the water molecules to form a jelly-like substance that is impossible to wash off. Fishermen hate hagfish, because hagfish scavenge the fish caught in gill nets so quickly that the catch may be ruined. However, hagfish are becoming overfished because their tanned skins are used for a popular kind of leather sold to yuppies as "eelskin." Their populations have been decimated in east Asian waters and off the west coast of North America, and are rapidly declining in the Atlantic.

Although they look superficially similar to hagfishes, the **lamprey** (Fig. 17.12A, B) is a more advanced animal, with most of the specializations of vertebrates (Fig. 17.11). Lampreys are parasites that attach to the side of another fish with their suction-cup mouth, armed with hundreds of tiny teeth and a rasping tongue. They clamp on and open a festering wound in the side of the host fish, using an anticoagulant in their saliva to keep the blood flowing, and usually wait until the host dies before they suck out the body fluids. Like hagfish, lampreys live mostly in marine waters, but they swim up rivers to breed in lakes and streams. If they encounter rapids or a waterfall, they use their sucker-like mouth to cling to rocks and slowly creep up the cataracts. They have become established in many bodies of freshwater, such as the Great Lakes. In recent years, humans have built canals that connected the Great Lakes, aiding the spread of lampreys through the lakes and their tributary streams and rivers. Lampreys have become a major economic problem in the region, because they quickly decimate the native fish species. Alarmed officials have tried to use nets across the Saint Lawrence River to keep them out, and poisons and electrical barriers have been tried in other rivers and lakes. Although native fish populations are now beginning to recover, the effort was very expensive, and cannot be relaxed, because once again humans have introduced an exotic species to a region that had no effective predators, and now we must spend the money to control it ourselves.

These two organisms are the only living groups of jawless vertebrates, but they are not very representative of the earliest stages of vertebrate evolution. They are but a tiny remnant of a huge radiation of armored jawless vertebrates that flourished in the early Paleozoic (Fig. 17.13). The earliest known fossils of these jawless fish may be bony plates of *Anatolepis* from the Upper Cambrian Deadwood Sandstone of Wyoming. Although no complete fish are known, the bony plates clearly have all the detailed histological features of vertebrate bone. Isolated plates are also known from the Ordovician Harding Sandstone of Colorado, and the Ordovician of Australia yields a few complete specimens. The best complete specimens are known from the Silurian, and especially from the Devonian, when they reached their peak of diversity.

All these fish had an internal skeleton made mostly of cartilage (as do lampreys and sharks), but had bony plates around the head region, and bony scales covering much of the body. Their jawless mouths were usually simple slits that could be used to feed on detritus, or filter feed by passing water through the pharynx. One group, the **pteraspids** (also called the heterostracans) had a laterally compressed body, with the head and thoracic region covered by solid bony armor (Fig. 17.13A, B). Most were only a few cen-

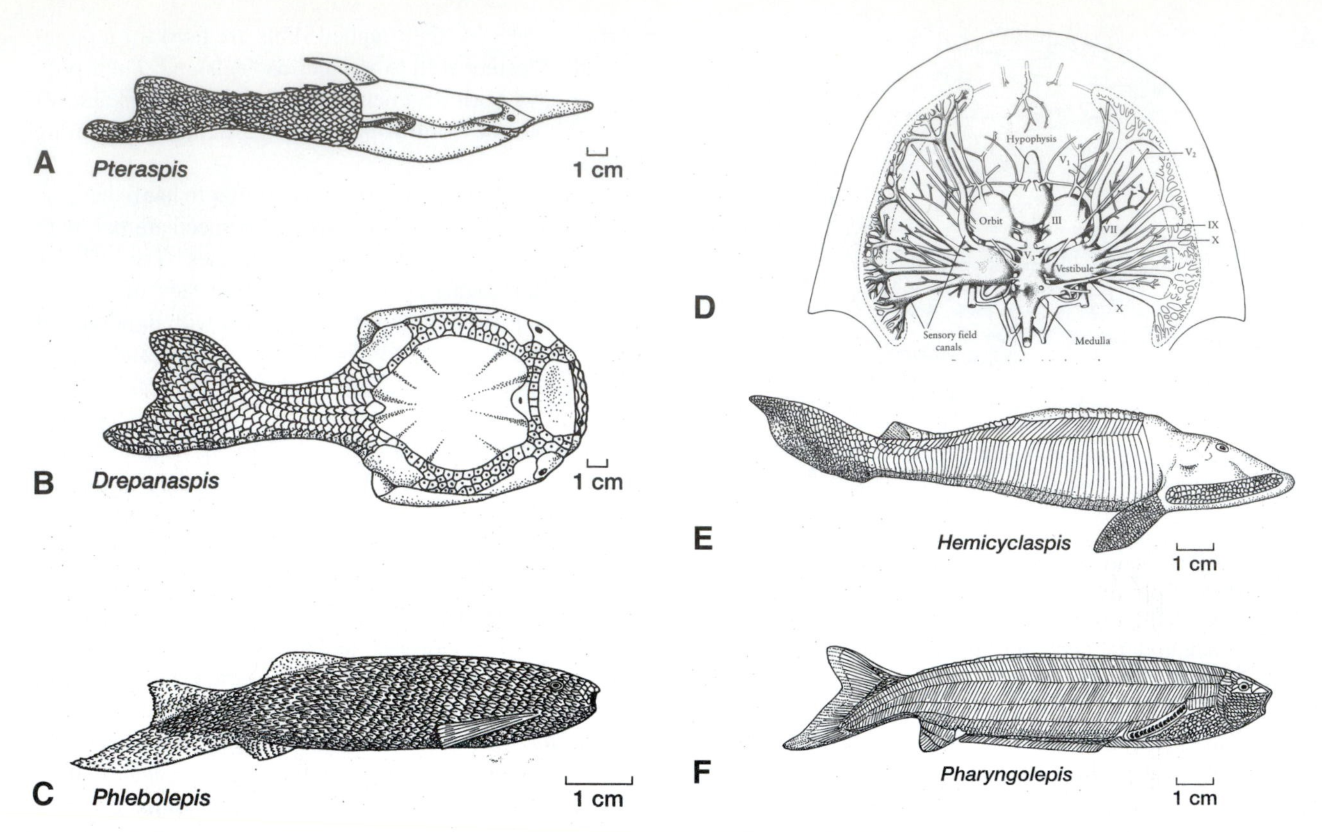

Figure 17.13. Jawless fish came in a variety of shapes in the Silurian and Devonian. (A, B) The pteraspids or heterostracans had large plates around their head and thorax, and a reversed heterocercal tail. (C) The thelodonts were a simple tube of scales with a reversed heterocercal tail. (D) Stensiö's (1963) reconstruction of the internal nerves and organs of the cephalaspid *Kiaeraspis*. (E) The cephalaspids or osteostracans had a flattened body, a well-developed head shield with upward-pointing eyes and pectoral flaps, and a heterocercal tail. (F) The anaspids were characterized by rows of plates and scales and a reversed heterocercal tail. (From Kardong, 1995.)

timeters long, although large ones were up to 2 m long. The rest of the body was covered by small bony scales, and the the tail was asymmetrical with its major lobe pointed downward (a **reversed heterocercal** or **hypocercal** tail). Some had small spines protruding from their head shield, or a dorsal fin spine. However, they apparently had no pectoral or pelvic fins, and thus no stabilizing mechanism for swimming, so they must have swum very erratically. Another group, the **thelodonts**, are very poorly known, since they were apparently not much more than a fish-like bag of scales with a reversed heterocercal tail (Fig. 17.13C).

The best known group is the **cephalaspids**, or osteostracans, which have a large flattened head shield (Fig. 17.13E). The head shield has openings for two upward-directed eyes and a pineal eye (the so-called "third eye" found in many vertebrates) between them, and a broad area around the brim for sensing motion in the water. The internal structure of these head shields has been exhaustively monographed by the Swedish paleontologist Erik Stensiö, who has traced the detailed course of all the nerves and blood vessels and other structures within these exquisitely preserved specimens (Fig. 17.13D). Behind the head shield, cephalaspids had pectoral flaps that would have helped control swimming, although these were not true fins with bony supports that are used for paddling. The rest of the body was covered with fine bony scales, and the lobe of the asymmetrical tail bends upward, as in sharks (a **heterocercal** tail). The flattened bottom of the head shield, with its scoop-like mouth opening, and the generally flattened body with the heterocercal tail, suggests that cephalaspids were bottom-feeders who probably scooped up detritus and sifted out the food with their pharynx and gills.

A fourth group, the **anaspids**, are known from specimens with fine rows of bony scales, a simple slit-like mouth, and paired eyes, and a reversed heterocercal tail (Fig. 17.13F). Of all extinct jawless fish, the anaspids are thought to be the closest relative of lampreys (Fig. 17.11).

A variety of higher taxa have been created to group jawless vertebrates. The term "Agnatha" ("without jaws" in Greek) is the most widely used, but unfortunately it is a paraphyletic wastebasket for craniates without jaws. Likewise, the extinct bony forms were long called "ostracoderms," but that group is also paraphyletic and even excludes the living taxa. The two living jawless craniates were long called "cyclostomes" ("round mouth" in Greek), but this assemblage is polyphyletic, since the lamprey and the hagfish are not even closely related. In recent years, all these obsolete terms have gradually drifted out of the literature, and most scientists refer to monophyletic groups like craniates, vertebrates, or gnathostomes. If you want to talk about "agnaths," for example, "jawless vertebrates" or "jawless craniates" conveys the same information content without the misleading implication that they are a natural group.

Another longstanding controversy has been the issue of whether vertebrates originated in freshwater or in saltwater. When armored cephalaspids were first found in abundance in the freshwater fluvial deposits of the Devonian Old Red Sandstone in Britain, the former environment was favored. As older localities were found, however, it was apparent that all the Cambrian and Ordovician specimens came from normal marine deposits (Darby, 1982). Romer and Grove (1935) argued that the complicated brain and nervous system of the earliest vertebrates suggested that they needed to swim in the turbulent waters of streams. In his provocative little book *From Fish to Philosopher* (1953), physiologist Homer Smith argued that the vertebrate kidney is a structure designed to pump out excess water. Since vertebrate body fluids are saltier than freshwater, if they lived in freshwater initially they would have needed a strong kidney or they would have bloated up with water. Vertebrate body fluids are less concentrated in salts than seawater, so the kidney would not have been as useful in marine settings. In this case, the problem is getting rid of salts or gaining water, not losing it. However, Smith's arguments have since fallen out of favor (Northcutt and Gans, 1983), primarily because the kidney's main function is getting rid of nitrogenous wastes (urine or ammonia), and it only secondarily became used for osmoregulation. Vertebrate kidneys now work equally well in fresh, brackish, and salt water. More importantly, all the sister-taxa of vertebrates are marine, as are the majority of invertebrates, so it is a much less parsimonious hypothesis to assume, against all the evidence, that freshwater origins are required by the kidney or brain structure.

Another Mystery Solved: The Conodonts

If the biological relationships of graptolites were paleontology's longest-running mystery, the affinities of conodonts were a close second. These microscopic tooth-shaped objects (Fig. 17.14) have been a mystery ever since C.H. Pander discovered them in 1856. Walter Sweet (1985) called them "those fascinating little whatzits." As in the case of graptolites, no one had a clue about their biology for over a century, yet this fact did not prevent them from becoming useful for biostratigraphy. In fact, conodonts were so ubiquitous during the Paleozoic and Triassic that practically any marine sedimentary rock (especially organic-rich sandstones and limestones dissolved in acid) of that age yields them. Conodonts evolved so rapidly and they are so abundant that they have become the biostratigraphic index fossil of choice for most of the Paleozoic, supplementing graptolites in the Ordovician and Silurian, and gonitatitic ammonoids in the Devonian through Permian. Conodonts are also of great economic importance. Not only are they the primary tool for correlation of Paleozoic rocks in the search for oil, but also conodonts change color with increasing temperature of burial (Epstein et al., 1977). Those that have never been heated above 80°C are pale yellow, and they get darker brown with increasing temperature, until they turn black at temperatures above 500°C. Such information can be critical in determining whether rocks were heated enough to turn their hydrocarbons into oil, or heated so much that no porosity or hydrocarbons survive.

Like vertebrate bone, conodonts are made of calcium phosphate, so many scientists have guessed they were related to vertebrates (although some groups of worms also produce tooth-like phosphatic structures, and inarticulate brachiopods also use this compound). Their tooth-like structure suggests they might be teeth of some sort of a minute vertebrate, but there are problems: most conodonts show no wear on the crowns of the "teeth," and the growth lines of these structures show that even the tips of the cusps were normally embedded in tissue. For a century, each distinctive conodont shape was described as a new taxon, but eventually specialists began to realize that many different

Figure 17.14. Hypothetical reconstruction of a conodont apparatus (*Ellisonia*, from the Permian of Texas), using associations of conodonts found in rocks. A bilaterally symmetrical element (A) probably lay across the midline of the body, and each of the other elements (B-G) apparently were symmetrically arranged on each side. (From Boardman et al., 1987.)

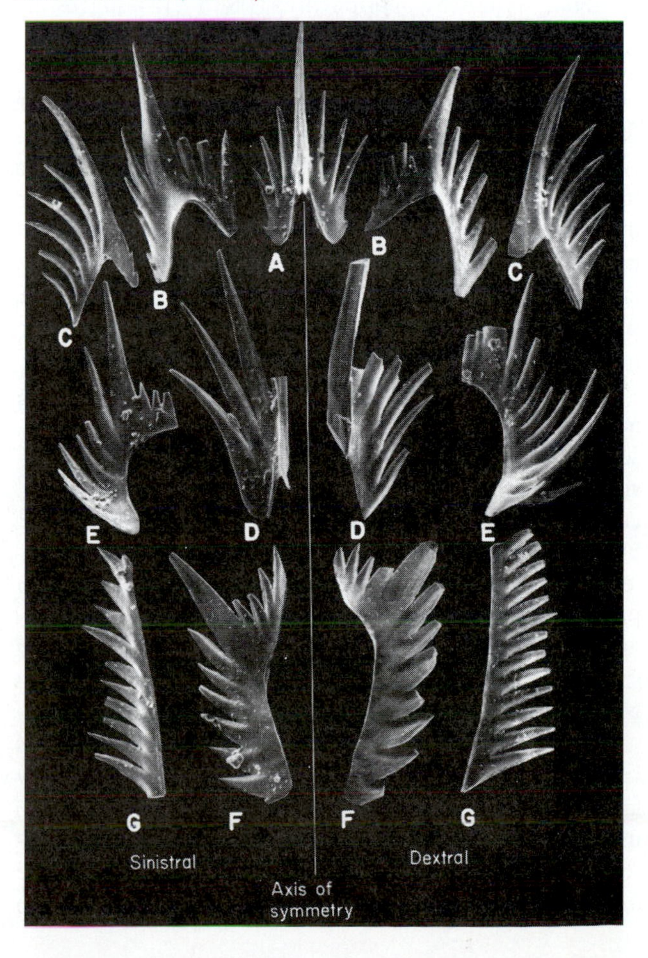

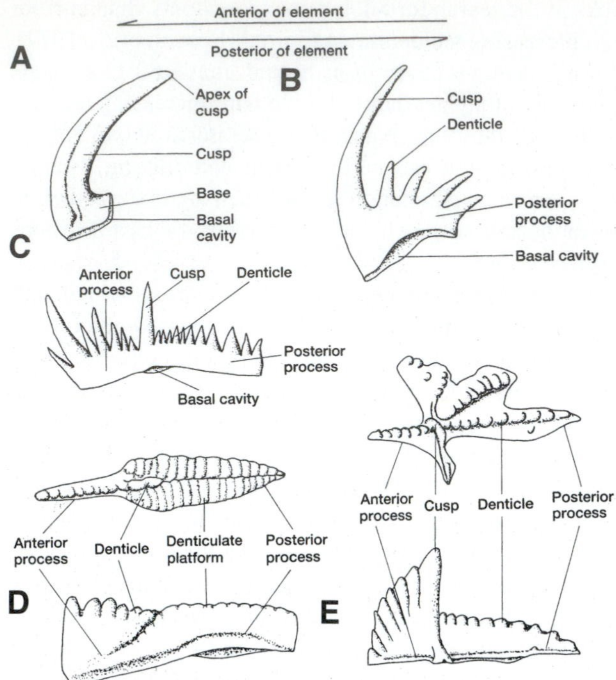

Figure 17.15. The three basic forms of conodont elements and their descriptive terminology. (A) Coniform element. (B-C) Ramiform elements. (D-E) Pectinoform and platform elements. (From McKinney, 1991.)

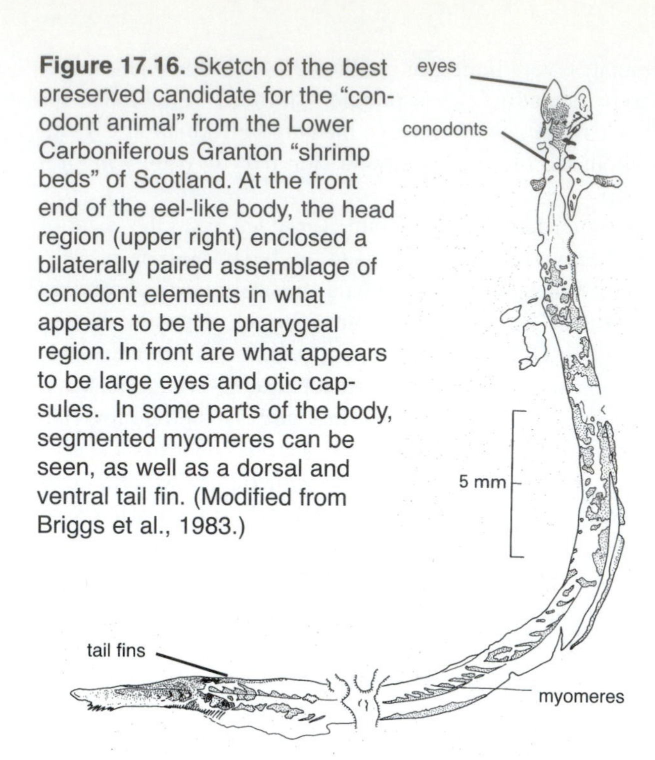

Figure 17.16. Sketch of the best preserved candidate for the "conodont animal" from the Lower Carboniferous Granton "shrimp beds" of Scotland. At the front end of the eel-like body, the head region (upper right) enclosed a bilaterally paired assemblage of conodont elements in what appears to be the pharygeal region. In front are what appears to be large eyes and otic capsules. In some parts of the body, segmented myomeres can be seen, as well as a dorsal and ventral tail fin. (Modified from Briggs et al., 1983.)

conodonts were parts of the same animal. Specimens were found with a variety of different conodonts associated in a bilaterally symmetrical **apparatus** (Fig. 17.14), suggesting that they all supported some kind of structure in a bilaterally symmetrical animal. This created problems for conodont taxonomy, because many different named taxa were found to belong to the same apparatus, and were apparently all synonyms. In the last 30 years, conodont workers have made many revisions and adjustments, so a large number of apparatuses and associations have been documented. Unfortunately, 99% of conodonts are still found as isolated elements in residues of rocks dissolved with acid, so synonymies have still not been determined for the majority of these taxa.

Yet, despite this handicap, conodont paleontologists have made enormous progress describing what is preserved. The earliest conodonts are simple **coniform**, cusp-shaped objects (Fig. 17.15), with a broad base and a long, curving tip. More advanced conodonts are multicusped, blade-like (**ramiform**) elements, and the multicusped, flattened **platform** elements, which have a broad base with a pit underneath, and many cusps and ridges on top. The terminology of each of these elements is relatively simple and straightforward, but nevertheless hundreds of different genera of conodonts have been described.

The earliest conodonts are simple coniform elements known as protoconodonts. These are known from the Early Cambrian, but many authors doubt that they are related to more advanced conodonts, because there are important differences in their chemistry and ultrastructure, and because they were exposed to wear and may have actually been used as teeth. However, two groups of more advanced conodonts, the paraconodonts and euconodonts, are both placed in the Conodontophorida. They radiated in the Early-Middle Cambrian, and by the Middle Ordovician there were at least 60 genera known. Conodont diversity crashed in the Middle Silurian, then recovered in the Devonian when it reached a peak of about 30 genera. After the Late Devonian extinctions, conodonts were much less diverse through the later Paleozoic, but a few genera did survive the Permian catastrophe, and straggle on through the entire Triassic before their final extinction.

So what are the "fascinating little whatzits"? In recent years, a number of specimens have been advanced as the "conodont animal," only to be discredited. Most had conodonts in their stomach area, so they might better be described as conodont predators. In 1983, Briggs and others described fossils from the Granton "Shrimp Bed" of the Lower Carboniferous of Scotland (Fig. 17.16) that are our best candidate for the "conodont animal." The best specimen has a long eel-like body, with a complete conodont apparatus in what appears to be its mouth or throat region (Fig. 17.17). The body appears to have had eyes, ears, a pharynx, fins supported by rays, a dorsal fin, and segmented muscles. Another specimen from the Lower Silurian of Wisconsin is less well preserved, but shows similar structures (Mickulic et al., 1985; Smith et al., 1987). Sansom and others (1992) even reported that conodonts have cellular bone, like vertebrates. Based on this evidence, most specialists are now convinced that the conodont animal was a chordate, and probably a craniate as well (Aldridge et al., 1986, 1993; Peterson, 1994). It is now reconstructed much like a hagfish or lamprey, with a bristling array of conodont elements in its pharygeal region. Although they are not true teeth, some conodont elements may have func-

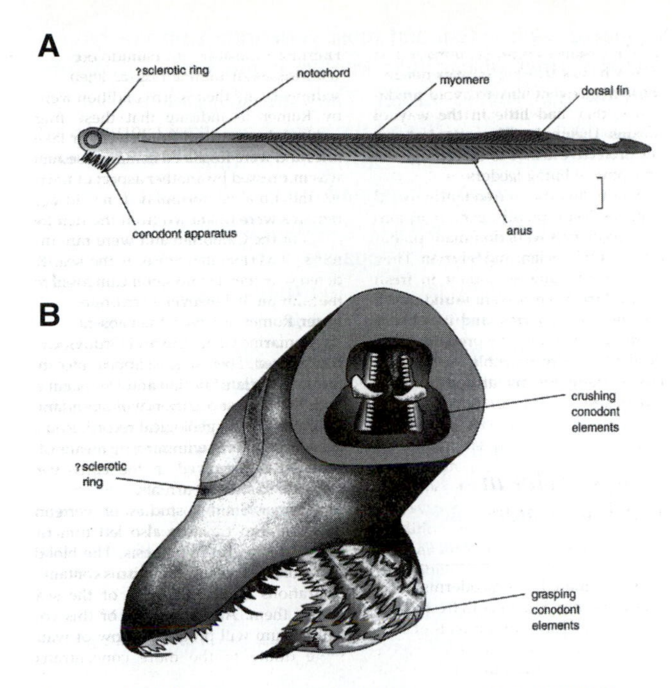

Figure 17.17. Reconstruction of the conodont animal in (A) lateral view and (B) close-up cutaway of the head to show the position of the conodonts. (From Dzik, 1993.)

tioned as grasping organs to hold a prey item once they reached the pharynx (as the hagfish uses its tiny teeth). So conodonts, which had long been in the realm of micropaleontology and studied in biological isolation from nearly all other fossil groups, may turn out to be one of our closer relatives.

For Further Reading

Aldridge, R., ed. 1987. *Palaeobiology of Conodonts*. Ellis Horwood Publishers, Chichester.

Aldridge, R. J., D. E. G. Briggs, E.N.K. Clarkson, and M. P. Smith. 1986. The affinities of conodonts—new evidence from the Carboniferous of Scotland. *Lethaia* 19:279-291.

Aldridge, R. J., D.E. G. Briggs, M. P. Smith, E. N. K. Clarkson, and D. L. Clark. 1993. The anatomy of conodonts. *Philosophical Transactions of the Royal Society of London* (B) 340: 405-421.

Aldridge, R. J., and M. A. Purnell. 1996. The conodont controversies. *Trends in Ecology and Evolution* 111: 463-468.

Briggs, D. E. G. 1992. Conodonts: a major extinct group added to the vertebrates. *Science* 256: 1284-1286.

Briggs, D. E. G., Clarkson, E. N. K., and Aldridge, R. J. 1983. The conodont animal. *Lethaia* 16:1-14.

Clark, D. L. 1984. *Conodont Biofacies and Provincialism*. Geological Society of America Special Paper 196.

Clark, D.L., and others, 1981. Conodonta, *in* Robison, R. A., ed. *Treatise on Invertebrate Paleontology*, Part W. Geological Society of America and the University of Kansas, Lawrence.

Clark, D. L., 1987, Phylum Conodonta, pp. 636-662, *in* Boardman, R. S., A. H. Cheetham, and A. J. Rowell, eds. *Fossil Invertebrates*. Blackwell Scientific Publishers, Cambridge, Mass..

Conniff, R. 1991. The most disgusting fish in the sea. *Audubon* 93:100-118.

Epstein, A. G., J. B. Epstein, and L. D. Harris. 1977. Conodont color alteration—an index to organic metamorphism. *U.S. Geological Survey Professional Paper* 995.

Forey, P. L. 1984. Yet more reflections on agnathan-gnathostome relationships. *Journal of Vertebrate Paleontology* 4:330-343.

Forey, P., and P. Janvier. 1984. Evolution of the earliest vertebrates. *American Scientist* 82:554-565.

Forey, P., and P. Janvier, 1993. Agnathans and the origin of jawed vertebrates. *Nature* 361:129-134.

Hardisty, M. W., and I. C. Potter, eds. 1971. *The Biology of Lampreys*. Academic Press, New York.

Janvier, P. 1981. The phylogeny of the Craniata, with particular reference to the significance of new fossil "agnathans." *Journal of Vertebrate Paleontology* 1:121-159.

Janvier, P. 1984. The relationships of the Osteostraci and Galeaspida. *Journal of Vertebrate Paleontology* 4:344-358.

Jrgensen, J. R., and others, eds. 1997. *The Biology of Hagfishes*. Chapman & Hall, London.

Long, J. A. 1995. *The Rise of Fishes*. John Hopkins University Press, Baltimore.

Maisey, J. G. 1986. Heads and tails: a chordate phylogeny. *Cladistics* 2:201-256.

Maisey, J. G. 1996. *Discovering Fossil Fishes*. Henry Holt, New York.

Moy-Thomas, J., and R. S. Miles, 1971. *Palaeozoic Fishes*. W. B. Saunders, Philadelphia.

Müller, K. J., 1978. Conodonts and other phosphatic microfossils, pp. 277-291, *in* Haq, B. U., and A. Boersma, eds. *Introduction to Marine Micropaleontology*, Elsevier, New York.

Northcutt, R. G., and C. Gans, 1983. The genesis of neural crest and epidermal placodes: a reinterpretation of vertebrate origins. *Quarterly Review of Biology* 58:1-28.

Peterson, K. J. 1994. The origin and early evolution of the craniates, pp. 14-37, *in* Prothero, D.R., and R. M. Schoch, eds. *Major Features of Vertebrate Evolution*. Paleontological Society Short Courses in Paleontology 7.

Repetski, J. E. 1978. A fish from the Upper Cambrian of North America. *Science* 200: 529-531.

Rhodes, F. H. T., ed. 1973. Conodont paleozoology. *Geological Society of America Special Paper* 141.

Smith, H., 1953. *From Fish to Philosopher*. Little, Brown, Boston.

Sweet, W. C. 1985. Conodonts—those fascinating little whatzits. *Journal of Paleontology* 59:485-494.

Sweet, W. C. 1988. *The Conodonta*. Oxford University Press, New York.

JAWS: THE GNATHOSTOMES

It is hard to imagine life without jaws: giant killer sharks, carnivorous dinosaurs, saber-toothed tigers, and that talkative neighbor just would not be the same without them. The acquisition of jaws is perhaps the most profound and radical evolutionary step in craniate history, after the development of the head itself.

John Maisey, *Discovering Fossil Fishes*, 1996

When we hear the word "jaws," we think of the deep, foreboding music and the terrifying giant great white shark of the hit movie based on Peter Benchley's novel. We seldom stop to think how important jaws are to vertebrates, or how limited vertebrates were without them. The jawless craniates could do little more than feed on suspended food or detritus, or in the case of lampreys and hagfish, suck the fluids out of their victims. Jaws are essential for grabbing a food item, and armed with teeth, they allow an animal to chop up the food to edible sizes. The evolution of jaws made it possible for vertebrates to exploit a wide variety of food sources, and with this skill to evolve into many different habitats and body sizes, including the largest fish in the sea and the largest land animals. Jaws are also critical to many other functions, such as manipulation of objects. Vertebrates use their jaws for functions as different as digging holes, carrying pebbles or vegetation to build nests, grasping mates during courtship or copulation, carrying their young around, and making sounds or speech.

Where did such a fundamental innovation come from? The fossil record yields several kinds of jawed fishes in the Silurian and Devonian, but they already have fully developed jaws. The classical notion dates back to the embryological work of Carl Gegenbaur (1872), and is known as the gill arch theory (Fig. 17.18A). According to Gegenbaur, primitive jawless fish and most vertebrate embryos have a series of rings of cartilage known as **gill arches** that support the pharyngeal basket. They are suspended from the cartilaginous braincase and the notochord, and are separated by the gill openings. During embryonic development, the gill arches change shape and migrate. The first gill arch, which originally formed a ring around the mouth of jawless craniates, becomes two separate cartilages: the **palatoquadrate**, or upper jaw, and the **mandible**, or lower jaw (also known as Meckel's cartilage in embryos). The second gill arch (the **hyoid arch**) is made of paired upper and lower cartilages that buttress and support the jaw hinge. The remaining gill arches continue to be supports for the gill apparatus in most fish. Unfortunately, recent research has discredited this long-accepted scenario. More careful research in embryology has shown that the cartilaginous first and second gill arches of lampreys are not homologous with the jaws of gnathostomes. Instead, the jaws are *de novo* features derived from the neural crest cells, and have no homologue in any jawless vertebrate (Maisey, 1989, 1994).

The presence of jaws is such an important innovation that it defines a taxon, the **gnathostomes** (*gnathos* is "jaw" and *stoma* is "mouth" in Greek), which includes all vertebrates except the jawless craniates (Fig. 17.19). However, jaws are not the only evolutionary novelty of the group. Gnathostomes are also defined by having their gills lying on the outside of their gill supports (Fig. 17.18B). By contrast, in jawless vertebrates such as the lamprey, the gills sit inside the gill arches. The gill arches in jawless vertebrates are a complex web of cartilage, but in gnathostomes they are completely separated, free, and segmented. In the ear region, gnathostomes have three semicircular canals that are used to detect motion and keep their balance in three different perpendicular planes; lampreys have only two

A

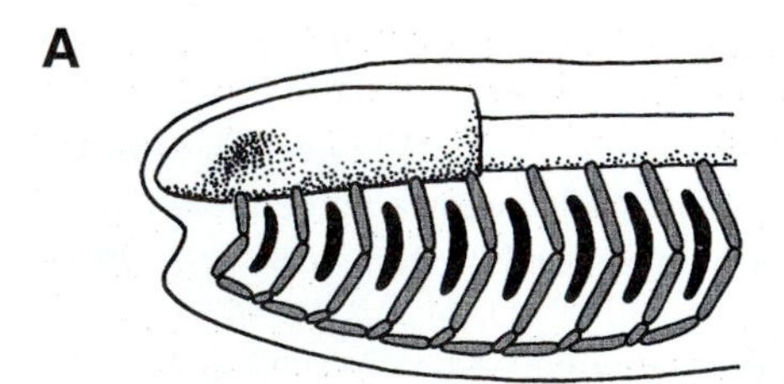

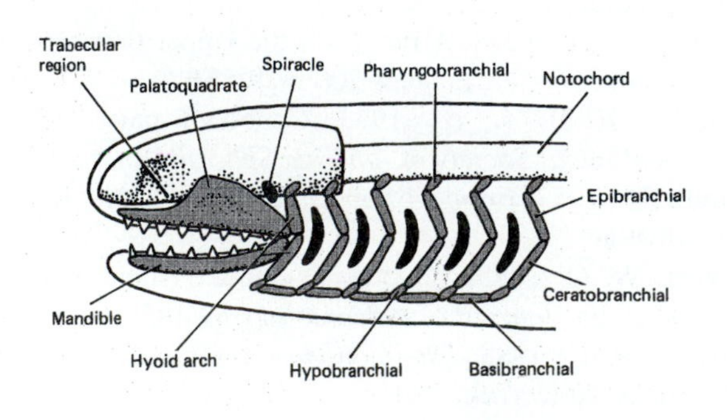

Figure 17.18. (A) Scenario of evolution of the vertebrate jaw from the anterior gill arches. (From Romer, 1970.) (B) Horizontal section through the head of a lamprey (left) and a shark (right) showing the position of the gills. (From Moy-Thomas and Miles, 1971.)

B

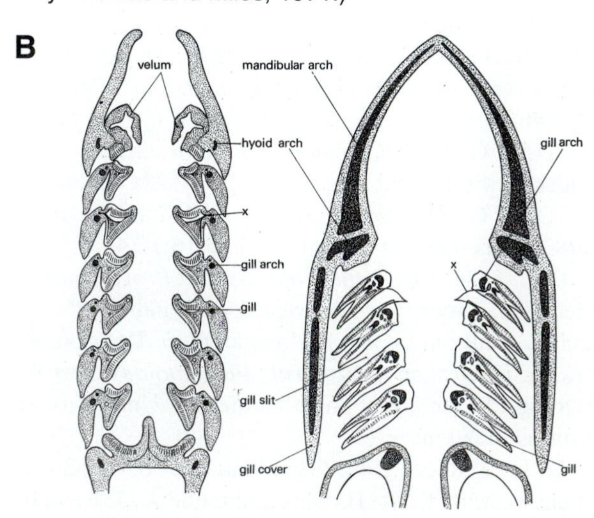

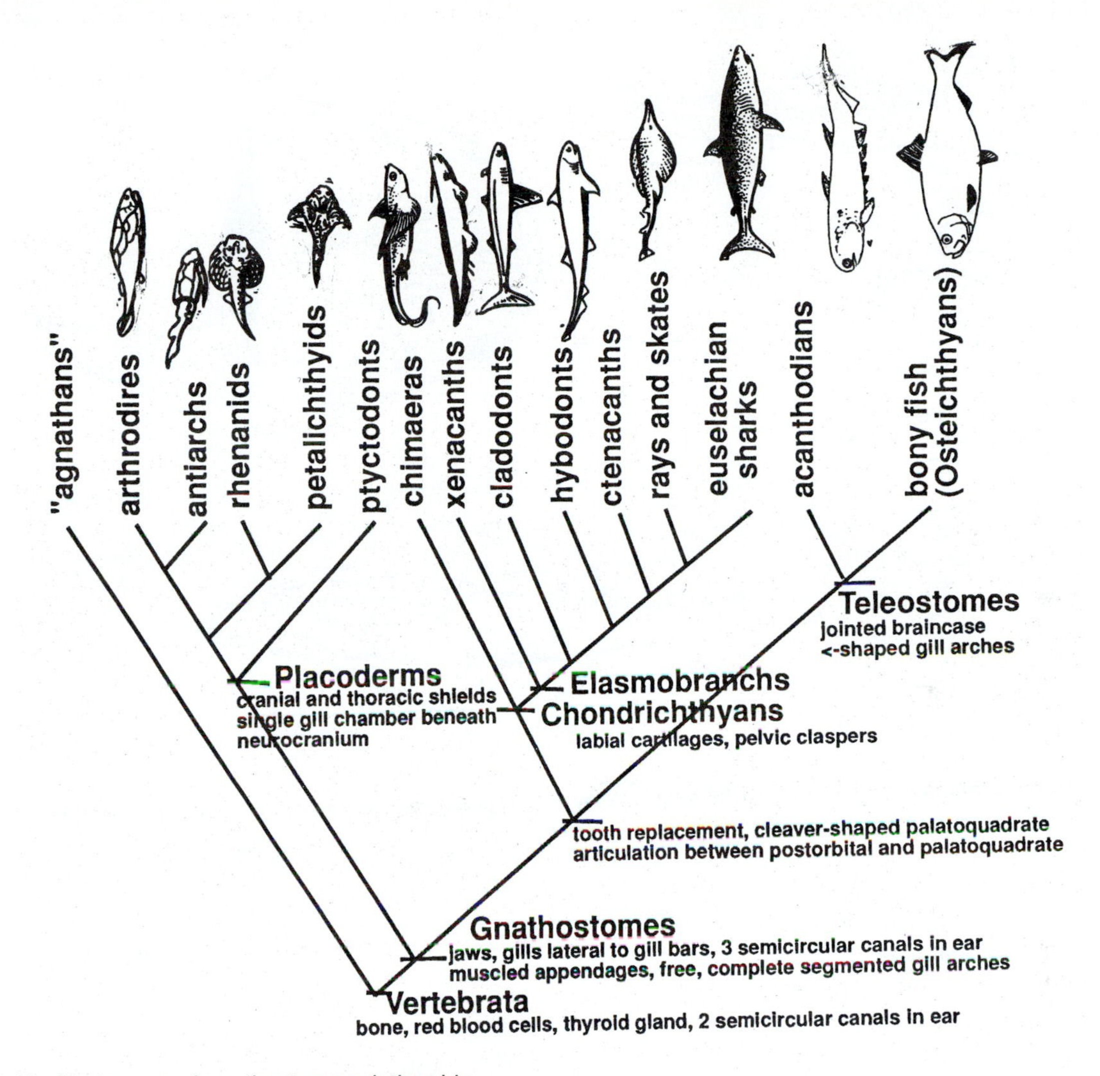

Figure 17.19. Cladogram of gnathostome relationships.

semicircular canals. Finally, gnathostome fins have cartilaginous supports with muscles to allow them to move and flex; these are connected to the pectoral and pelvic girdles (shoulder bones and hip bones in mammals). This is just the beginning of a long list of gnathostome synapomorphies. Maisey (1986) and other authors have pointed out at least 30 more, making the gnathostomes one of the best-supported monophyletic groups known.

Four different groups of jawed vertebrates are known from the Devonian: the extinct **placoderms**, the sharks and their relatives (**chondrichthyans**), the extinct **acanthodians** or "spiny sharks," and the bony fish (**osteichthyans**). Let us examine each in turn.

Placoderms

The Devonian has been nicknamed "The Age of Fish," and the dominant group of Devonian fish was the **placoderms** (Fig. 17.20). As the name implies (*plakos* is "plate" and *derma* is "skin" in Greek), they have armored plates (sometimes as thick as 5 cm) in their skin. Most placoderms have a bony shield covering their head, and another covering their thorax; these are the parts that fossilize best. The rest of the body had no dermal armor, but was supported by a cartilaginous skeleton, as it is in sharks. The placoderm skeleton includes distinct fin supports for a well-developed pectoral fin (improving their control while swimming) and pelvic fin, a long dorsal fin on the back, and a heterocercal tail. In this respect, they are very shark-like. However, placoderms have a number of unique features that tell us they are a separate branch of gnathostome evolution, unrelated to sharks or any other group of fish. For one thing, the bony head and thoracic plates are composed of separate elements that have no known homologue in any other group of vertebrates. The standard terminology of fish skull bones does not work on placoderms, so paleontologists have been forced to develop a completely different set of names for placoderm bones. Clearly, these bones were independently ossified from the skin. Even more fundamental is the relationship of the jaw bones and muscles. In most vertebrates, the jaw muscles lie outside the jaw bones, but in placoderms the jaw muscles are medial to the jaw hinge. Instead of the multiple gill slits of

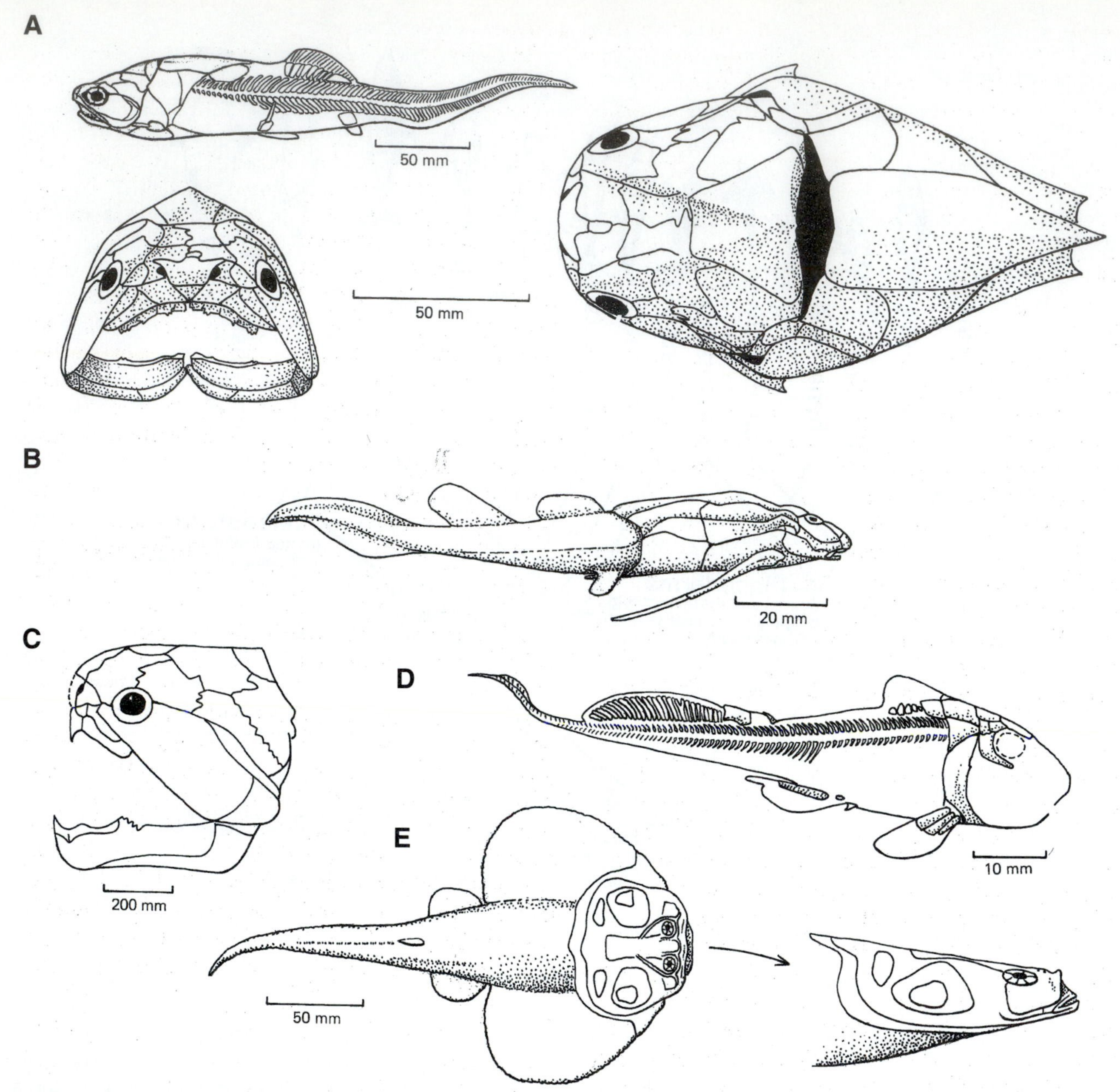

Figure 17.20. A variety of placoderms. (A) Side, front, and top views of a Middle Devonian arthrodire, *Coccosteus.* (B) The peculiar armored antiarch *Bothriolepis,* with a jointed exoskeleton that supported paired appendages. (C) The giant predatory arthrodire *Dunkleosteus.* (D) The chimaera-like ptycdodont *Rhamphodopsis.* (E) The ray-like rhenanid *Gemuendina.* (From Pough et al., 1996.)

sharks and bony fish, placoderms have a single gill opening that lies just beneath the braincase, and exits through an opening between the two hinged parts of the bony armor. Finally, placoderms do not have true teeth, but instead the edges of their head shield and mandibular shield are modified into sharp-edged biting plates that are not homologous with the teeth of other vertebrates.

Within this distinctive body plan, placoderms displayed an amazing variety of adaptations and lifestyles. The most spectacular were the **arthrodires**, the superpredators of the Devonian (Fig. 17.20A). The name "arthrodire" means "jointed neck" in Greek, and this describes their most striking feature: the head shield is connected to the thoracic shield by a ball-and-socket joint, allowing the head to pivot upward and the jaw to drop, forming a very wide gaping bite. Judging from the sharp biting plates, these animals were the most voracious animals of their time. Most were only a few meters in length, but the giant *Dunkleosteus* (also known as *Dinichthys*) was over 10 m (30 feet) long, the largest animal the earth had ever seen until the dinosaurs came along (Fig. 17.20C). Arthrodires must have been nasty, aggressive predators, eating everything that moved in the Devonian seas, including the largest sharks. Some arthrodire plates have puncture wounds from other arthrodires, suggesting that they fought among themselves and may have eaten each other when the opportunity arose.

Early arthrodires had a thoracic shield that went around the entire body, and had only small pectoral fins. During arthrodire evolution, the thoracic shield was reduced on the ventral side, and the pectoral fins became larger, making their bodies lighter, more flexible, and more effective for swimming. More than 200 genera in 20 different families of arthrodires are recognized. Some had modified their tooth plates for other purposes, such as blunt plates for crushing molluscs, or pick-like plates for stabbing. On many specimens, the eye was supported by a ring of bone called the **sclerotic ring,** which provided protection against water pressure for the large eyeballs of these highly visual predators.

Another group of placoderms, the **antiarchs** (Fig. 17.20B), had a completely different lifestyle. These small, mostly freshwater forms are best known from great numbers of specimens from the Devonian Old Red Sandstone in Britain, and the equivalent rocks in Nova Scotia, although they occurred worldwide (even in Antarctica). They were even more heavily armored than arthrodires, with a long thoracic shield that covered almost half their length and a flattened head shield with eyes that pointed upward, suggesting that they were bottom-dwellers. They even had jointed armor on their pectoral fins that resembles the jointed limbs of a crab or lobster. Behind all this armor was an unprotected tail with a pair of long dorsal fins, and a heterocercal tail fin. With all that weight on their bodies, antiarchs were clearly not efficient swimmers, but must have sculled slowly along the bottom, using their jaws to catch and crush molluscs, arthropods, and other soft-bodied invertebrates. Antiarchs are most abundant in the Middle Devonian, and apparently they replaced many of the armored jawless fish from the Early Devonian.

Placoderms evolved into many other body forms as well. One group, the **ptyctodonts** (Fig. 17.20D), had reduced body armor, a long whip-like tail, and thick tooth plates for crushing molluscs (superficially like the modern ratfish or chimaera). Another group, the **rhenanids** (Fig. 17.20E), had a flattened body with broad pectoral fins like the living skates and rays, yet this is clearly convergent evolution; they are placoderms, not chondrichthyans. Yet another group, the **petalichthyids**, had flattened bodies with long bony spines projecting in an arc in front of their pectoral fins. Clearly, the placoderms were highly successful in exploiting nearly every aquatic niche in the sea and in the rivers during the Devonian. The antiarchs radiated in the Middle Devonian, while the most spectacular advanced arthrodires lived in the early Late Devonian. Then the great Frasnian-Famennian extinction hit, and fish diversity plummeted. About 35 of 46 families died out during this great crisis, including the last of the armored jawless fish, virtually all the placoderms, and 10 families of lobe-finned fish. When the world recovered in the latest Devonian and the Carboniferous, placoderms were no longer important, and the seas were ruled by their modern masters: the sharks and bony fish.

Chondrichthyans

The sharks and their relatives (chondrichthyans) also diversified alongside the placoderms in the Late Silurian and Devonian. However, sharks never had large bony plates on their bodies, so they do not fossilize as well. The only bony elements in the shark are the teeth and the sharp objects imbedded in the skin known as denticles, plus the spines found in the fins of many sharks. In some sharks, the cartilage of the backbone may become secondarily calcified, but most sharks have no bone in their internal skeleton. Thus, our fossil record of sharks includes relatively few specimens with a complete body outline, but their isolated teeth, spines, and denticles are very common as fossils. Although the primitive retention of a skeleton made of cartilage (*chondros* is "cartilage" and *ichthys* is "fish" in Greek) is not a synapomorphy, the chondrichthyans have many unique features that show they are a natural group. For example, chondrichthyans have a distinctive set of cartilages around the mouth called labial cartilages, and many distinctive features of the fin structure seen in no other fish. The bony structure of teeth and denticles are also unique to chondrichthyans.

Their most unusual structure is a set of rod-like cartilages that trail behind the pelvic fin and anal opening in male chondrichthyans known as **pelvic claspers**; these allow internal fertilization. During copulation, the male wraps his body around the female and inserts one of the claspers into her urogenital opening, known as the cloaca. The claspers often have hooks and spines on them to prevent them from slipping out. Sperm is then injected into the female, and she then nurses the eggs inside her body until they are ready to be laid in their distinctive egg cases. Some sharks even give birth to live young. This kind of internal fertilization is very different from the external fertilization of most fish, which lay their eggs in the water in clumps attached to a surface; the male then sprays them with sperm. Sharks give birth to only a few well-developed young that have a good chance of survival, whereas fish without the internal fertilization mechanism lay hundreds of eggs, only a few of which survive and grow to maturity.

There are many other unusual features seen in the chondrichthyans. They do not have a swim bladder or any other air-filled sac, as is found in most other fish, so they have trouble keeping neutrally buoyant. Many sharks have an oil-filled liver to reduce their density, but it is not as effective as an air pocket, so sharks must swim continuously to avoid sinking. (Contrary to the popular myth, however, sharks do not need to swim to breathe—their gills have very effective pumping mechanisms that work fine when the fish is stationary.) The body fluids of a shark are saltier than seawater and even more so than freshwater, so they have a special rectal salt gland to get rid of salts, and they also retain urea in their blood to maintain osmotic balance. Although a few sharks (stingrays, sawfish, and bull sharks) are known to live in brackish or freshwater, most sharks have trouble maintaining their osmotic balance in freshwater. Without a swim bladder, sharks also have no backup

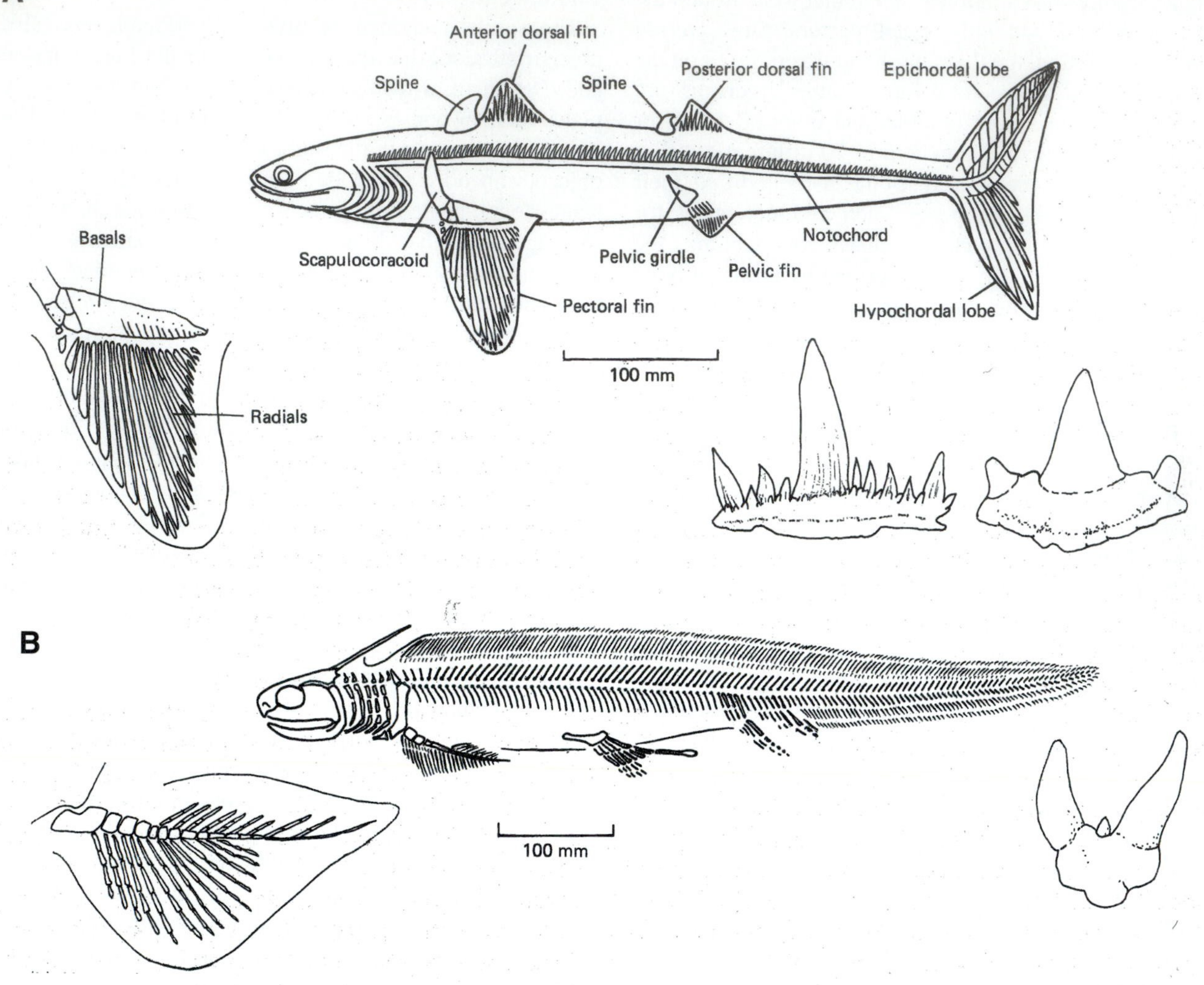

Figure 17.21. Early elasmobranch sharks. (A) *Cladoselache* from the Upper Devonian Cleveland Shale, Ohio, showing the body, details of the broad-based pectoral fin, and the distinctive "cladodont" teeth. (B) The Permian freshwater shark *Xenacanthus,* with the details of its "archipterygial" fin and double-pronged teeth. (From Moy-Thomas and Miles, 1971.)

source of oxygen, so they are at a disadvantage if the oxygen levels in a lake drop too low.

Two main groups of chondrichthyans are alive today, and both are known in the fossil record as well. All the familiar sharks and rays are known as **elasmobranchs** (Fig. 17.19). The earliest elasmobranchs, such as the Late Devonian **cladodont** shark *Cladoselache* (Fig. 17.21A), had a shark-like body form, but there are important differences. They have a very wide, triangular pectoral fin with a broad base that clearly could not be flexed or rotated; this fin must have served as a stiff stabilizer. There were two small dorsal fins, each with a thick horn-like spine in front of it. The heterocercal tail fin had a well-supported lower lobe, making it almost symmetrical. These sharks are best known from the Upper Devonian Cleveland Shale of Ohio, where they reach 2 m (6 feet) in length. However, they were probably prey for their contemporaries, the giant arthrodires like *Dunkleosteus*, which were 5 times their size.

Another important group of late Paleozoic sharks were the **xenacanths**, or pleuracanths (Fig. 17.21B). They are abundantly preserved in the freshwater and deltaic deposits of the Carboniferous coal swamps and the Permian flood-plain redbeds, where their distinctive double-pronged teeth are among the more common fossils. They even survived the Permian catastrophe, although they disappeared for unknown reasons during the Triassic. Some xenacanths were huge, with bodies over 3 m (10 feet) long, and thus were the freshwater fish the world had yet known. They are also one of the most anatomically peculiar sharks. Instead of a heterocercal tail, they are the only sharks with a symmetrical tail more like an eel. The large pectoral and pelvic fins were supported by a central rod of cartilage and branching fin rays, a structure seen in no other shark or fish. The single long dorsal fin ran the length of the body and merged with the tail fin. On top of their head was a long bony spine that projected up and backward. No one has a clue what this was for, although extraneous spines are a common feature in many sharks. The earliest xenacanths look more like normal sharks, without the weird fins or tail. Their peculiar eel-like body shape is thought to have been an adaptation for living in the sluggish waters of swamps and lakes, and lunging at their prey from the black water (rather than the slow steady swimming of marine sharks).

Sharks went through yet another evolutionary radiation beginning in the late Paleozoic and continuing through the Mesozoic. Most early Mesozoic sharks are known as **hybodonts**, a group that is the sister-taxon of the living sharks. Hybodonts had pectoral and pelvic fins with a much more modern support structure, and narrow bases so they could be turned and flexed for good control of swimming. They had two dorsal fins, each with a thick spine in front. Their teeth are highly differentiated, with multi-cusped pointed teeth for piercing up in front, and blunt-cusped teeth for crushing toward the back of the jaw. Hybodonts were common in both marine and freshwaters during the Mesozoic, but reached only about 2 m in length, so they probably didn't prey on the larger dinosaurs.

In the Late Triassic and Jurassic, we find the first evidence of sharks with calcified vertebrae, a feature of the modern group of sharks, or **neoselachians**. Since then, they have radiated into hundreds of species, including not only familiar forms like the dogfish and great white shark and hammerhead (360 living species), but also 456 living species of skates and rays. The largest of the extinct sharks were the giant great white sharks (*Carcharodon megalodon*), which reached almost 12 m (40 feet) in length. Neoselachians have further modified their jaws so that the upper jaw cartilage (the palatoquadrate) is suspended from the hyomandibular cartilage (the upper element of the hyoid arch). When the shark gapes, it not only drops its lower jaw, but it can also protrude its upper jaw, increasing its bite capacity. The next time you see a film of a shark opening its mouth, watch the upper jaw bulge forward. Neoselachians exhibit a wide variety of lifestyles. Most of the familiar forms are voracious open-ocean predators, but the flat-bodied skates and rays swim with their huge pectoral fins and spend much of their time lying buried in the sediment. The largest living chondrichthyans, such as the manta ray, whale shark, and basking shark, however, are plankton-feeders. They open their mouths as they swim and filter plankton, crustaceans, and fish from the water with their gill apparatus. A similar adaptation occurs in the largest marine mammals, the baleen whales.

The other major branch of the chondrichthyans are the **holocephalans**, also known as the chimaeras, ratfish, or rabbit fish (Fig. 17.22). These fish are alive today (30 species in 6 genera), but few people other than deep-water fishermen or ichthyologists ever see them, since they live in waters over 80 m deep. They have large eyes (adapted for the dark waters where they live), a long rat-like tail, very broad pectoral fins, and are very spiny. Their short snout bears large pavement-shaped tooth plates for crushing molluscs, their principal prey. The name "holocephalan" means "whole head" in Greek, because unlike other living sharks, they have fused the upper jaw cartilage to the braincase, so it is no longer mobile or protrusible. This helps reinforce the jaws within the head for the stresses of shell crushing. Instead of mulitple gill slits, holocephalans have a single gill opening. As different as they are from typical elasmobranch sharks, they still have all the key chondrichthyan synapomorphies, including pelvic claspers (which are shaped like a medieval mace). Some authors consider them the sister-taxon of the elasmobranchs (including the cladodonts), while others place them between the cladodonts and the xenacanths within the elasmobranchs. The earliest fossil holocephalans are known from the Carboniferous, although they are much rarer as fossils than are the elasmobranchs.

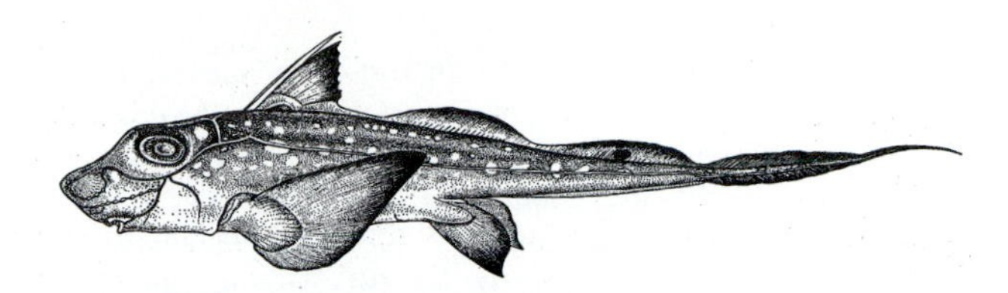

Figure 17.22. The living holocephalan or chimaera. (After Romer, 1970.)

For Further Reading

Compagno, L. J. V., 1977. Phyletic relationships of living sharks and rays. *American Zoologist* 17:303-322.

Gardiner, B. G., 1984. The relationships of placoderms. *Journal of Vertebrate Paleontology* 4:379-395.

Long, J. A. 1995. *The Rise of Fishes*. Johns Hopkins University Press, Baltimore.

Maisey, J. G. 1984. Higher elasmobranch phylogeny and biostratigraphy. *Zoological Journal of the Linnean Society* 82:33-54.

Maisey, J. G. 1984. Chondrichthyan phylogeny: a look at the evidence. *Journal of Vertebrate Paleontology* 4:359-371.

Maisey, J. G. 1994. Gnathostomes (jawed vertebrates), pp. 38-56, *in* Prothero, D. R., and R. M. Schoch, eds. *Major Features of Vertebrate Evolution*. Paleontological Society Short Courses in Paleontology 7.

Maisey, J. G. 1996. *Discovering Fossil Fishes*. Henry Holt and Company, New York.

Moy-Thomas, J., and R. S. Miles, 1971. *Palaeozoic Fishes*. W. B. Saunders, Philadelphia.

Patterson, C. 1965. The phylogeny of the chimaeroids. *Philosophical Transactions of the Royal Society of London* (B) 249: 101-219.

Schaeffer, B., and M. Williams. 1977. Relationships of fossil and living elasmobranchs. *American Zoologist* 17:293-302.

Schultze, H.-P., ed. 1979. *Handbook of Paloichthyology*. Gustav Fischer, Stuttgart.

Young, G. C. 1986. The relationships of placoderm fishes. *Zoological Journal of the Linnean Society of London* 88:1-57.

FISH BONES: THE OSTEICHTHYANS

The term "fish" is of value on restaurant menus, to anglers and aquarists, to stratigraphers and in theological discussions of biblical symbolism. Many systematists use it advisedly and with caution. Fishes are gnathostomes that lack tetrapod characters; they have no unique derived characteristics. We can conceptualize fishes with relative ease because of the great evolutionary gaps between them and their closest living relatives, but that does not mean they comprise a natural group. The only way to make the fishes monophyletic would be to include tetrapods, and to regard the latter merely as a kind of fish. Even then, the term "fish" would be a redundant colloquial equivalent of "gnathostome" (or "craniate," depending upon how far down the phylogenetic ladder one wished to go).

John Maisey, "Gnathostomes," 1994

Acanthodians

After placoderms and chondrichthyans, the third great group of middle Paleozoic fish was the **acanthodians** (Figs. 17.19, 17.23). They are known from isolated spines from the Late Ordovician, so they are by far the earliest known jawed vertebrates. Long misnamed the "spiny sharks," they were thought to be related to sharks based on the superficial similarities of their body, and their early occurrence. However, when detailed work was done on their internal anatomy, gill structure, and braincases, it turned out that acanthodians are actually highly derived animals, closely related to the higher bony fish in a cluster called the **teleostomes** (Figs. 17.19, 17.24). Acanthodians are a classic case of why oldest does not equal most primitive, and why we cannot blindly connect fossils in stratigraphic order to reconstruct their phylogenetic history. Not only did acanthodians first appear about 50 million years before their more primitive relatives, the placoderms and chondrichthyans, but they also persisted until the Early Permian (a total span of over 160 million years), long after the placoderms and archaic sharks were extinct.

The most striking feature of acanthodians is their multiple spines (hence the nickname "spiny sharks"; the Greek word for "thorn" is *akanthos*). All their fins are supported by spines on the front edge, including the two dorsal fins, and in some taxa, the paired rows of six fins that run along the ventral side between the pectoral and pelvic fins. They even have a separate anal fin. These fins were probably not very mobile, however, so they did not make the fish more maneuverable, but probably served as stabilizing mechanisms for such an active swimmer. The bony spines were sunk deep into the body, and their structure is very distinctive. Most acanthodians are known from a few spines, but in some Devonian deposits, there are rare impressions of complete fish. These specimens show that they had large eyes and a short snout, and very advanced jaws with a single row of unreplaced teeth embedded in them (not the multiple teeth of sharks, which are continuously shed). They did have some primitive, shark-like features, such as a heterocercal tail and five gill arches. The majority of acanthodians are known from freshwater deposits, although their spines are also common in marine rocks.

Bony Fish: The Osteichthyans

Acanthodians are the sister-group to two major groups of fish whose skeletons are made of bone (Fig. 17.24). These two groups together are known as the osteichthyans (*osteon* is "bone" and *ichthys* is "fish" in Greek). They are diagnosed by a wide variety of unique specializations, most of which involve bone. They have dermal bone (bone derived from a skin precursor in embryology) in their shoulder girdle, palate, along the margin of the jaw, in the bones that cover the gill flap, and in their hyoid region. The remaining bones of the head and skeleton are derived from a precursor of skeletal cartilage in embryology. In addition, bony fish have some sort of gas-filled chamber (lung or swim bladder) derived from the gut region. Finally, the muscles of the lower jaw attach from the medial side, rather than from the outside as in other fish.

One group of bony fish, the lobe-finned fish or **sarcopterygians** (*sarkos* is "flesh" and *pterygos* is "wing" in Greek) have a series of robust bones and muscles supporting a lobed fin. These include the lungfish, coelacanth, a number of extinct fish called "rhipidistians," and all four-legged land vertebrates (tetrapods). They will be discussed in the next section.

The other group, the ray-finned fish or **actinopterygians** (*aktinos* is "ray" and *pterygos* is "wing" in Greek), support their fins with many thin parallel rods of bone. These ray fins have a unique characteristic in that they articulate with the limb girdles directly, rather than by intermediate bones (as in sharks and lobe-fins). In addition to the ray fins, actinopterygians are defined by many other anatomical features, including a single dorsal fin (most other fish we have seen have two or more). Primitive actinopterygians have a unique type of body armor known as **ganoid** scale, which is composed of a thin layer of protein called ganoine overlying layers of dentin and bone. There are also many soft-

Figure 17.23. Reconstruction of the acanthodian *Euthacanthus*. (From Moy-Thomas and Miles, 1971.)

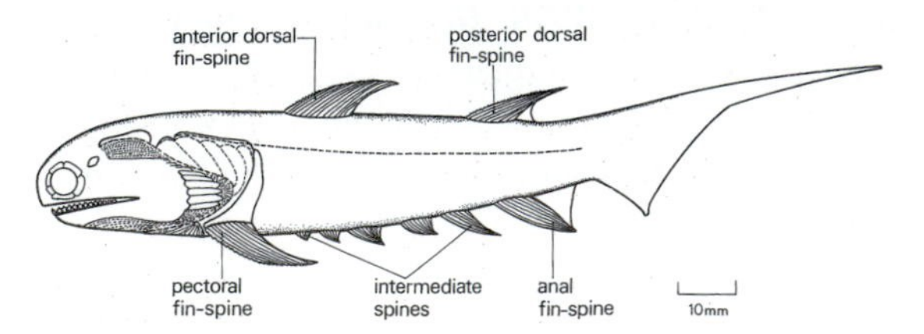

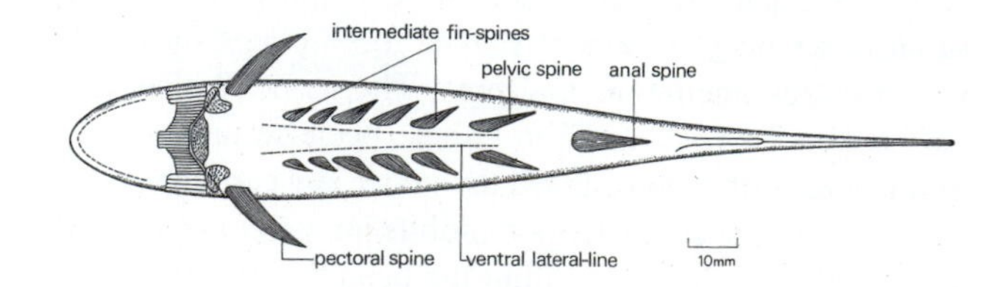

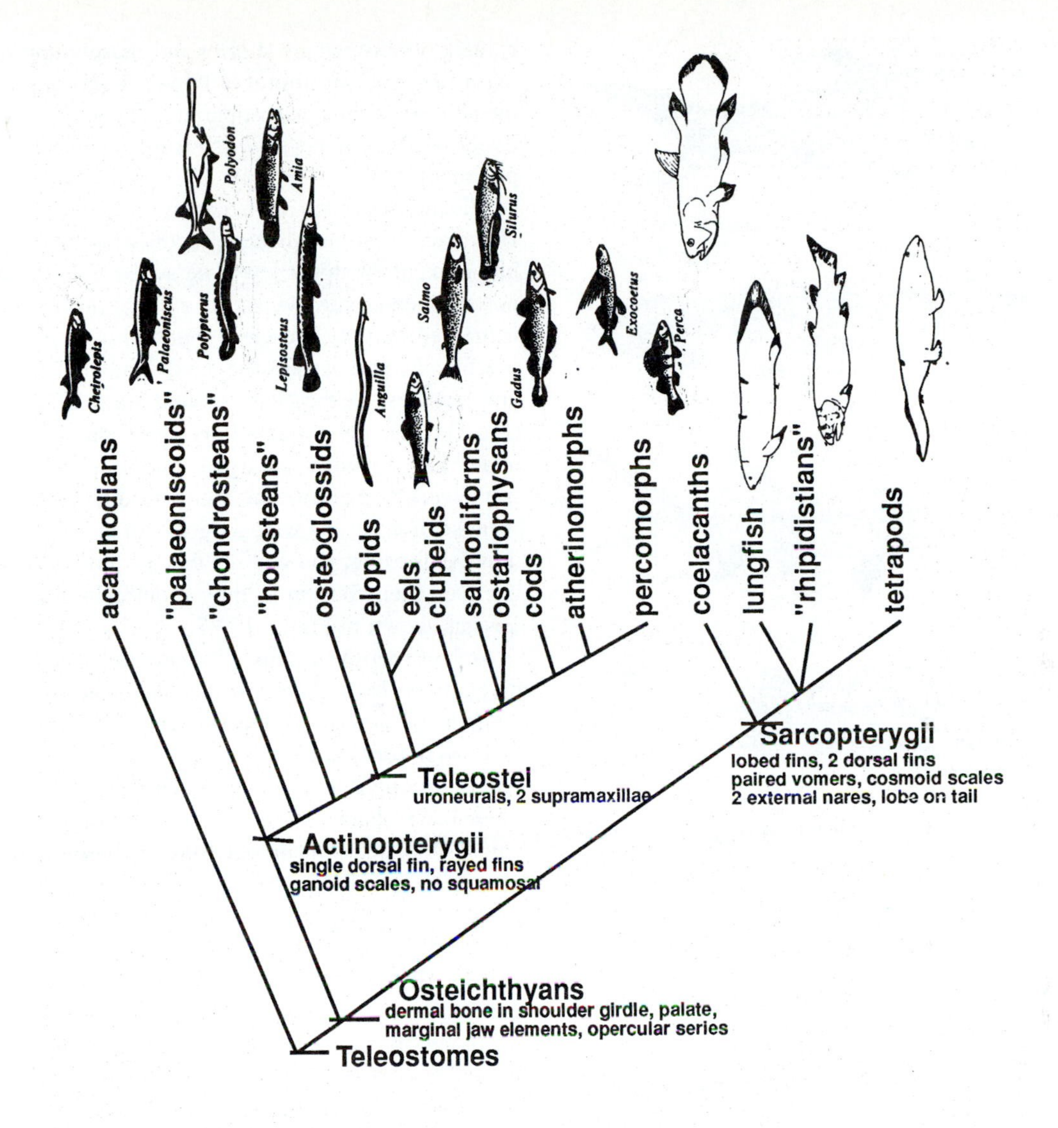

Figure 17.24. Cladogram of the osteichthyans.

tissue and biochemical and molecular characters that are unique to the actipterygians, and found in no other group of vertebrates.

The most primitive actinopterygians used to be placed in a wastebasket group, the "**paleoniscoids**" (Fig. 17.25A). Like the other early fish, they are known from scraps in the Late Silurian, but the paleoniscoids diversified into a number of well-preserved taxa in the Devonian. However, these primitive actinopterygians were relatively rare in the Devonian compared to placoderms or armored jawless fish or sharks, and even rare compared to coelacanths and lungfish (the reverse is true today). After the Late Devonian extinctions, paleoniscoids became the commonest of fossil fish in the later Paleozoic, and straggled on through the early Mesozoic before disappearing in the Cretaceous, replaced by more advanced fish (Fig. 17.26). Compared to more advanced actinopterygians, these fish still had a great variety of primitive features. Their skulls were encased in a continuous roof of dermal bone (modern fish have largely reduced the bone in their skulls, especially the dermal bone). They had very large eyes and a short snout. They had the typical ray fins, but they were triangular in shape and heavy, and included not only pectoral and pelvic fins, but also an anal fin. Unlike advanced bony fish, they still had the primitive heterocercal tail. Although complete paleoniscoid fossils are rare, their thick, distinctively rhomboid scales are common fossils of the late Paleozoic.

Archaic actinopterygians were characteristic of the late Paleozoic and Triassic-Jurassic, and three lineages of these primitive fish still survive today (Fig. 17.25B, C, D). They include the spiny sturgeon (source of caviar), the paddlefish *Polyodon*, and a strange-looking African freshwater fish known as the bichir or reedfish (genus *Polypterus*). Although these fish have lost the heavy dermal bones of the skull found in paleoniscoids, they still have archaic jaw and fin configurations and (in sturgeons and paddlefishes) the heterocercal tail. In the older literature, these archaic fish were lumped into a paraphyletic assemblage known as the "chondrosteans" (Fig. 17.26), because much of their internal skeleton remains cartilaginous. Sturgeons have a most-

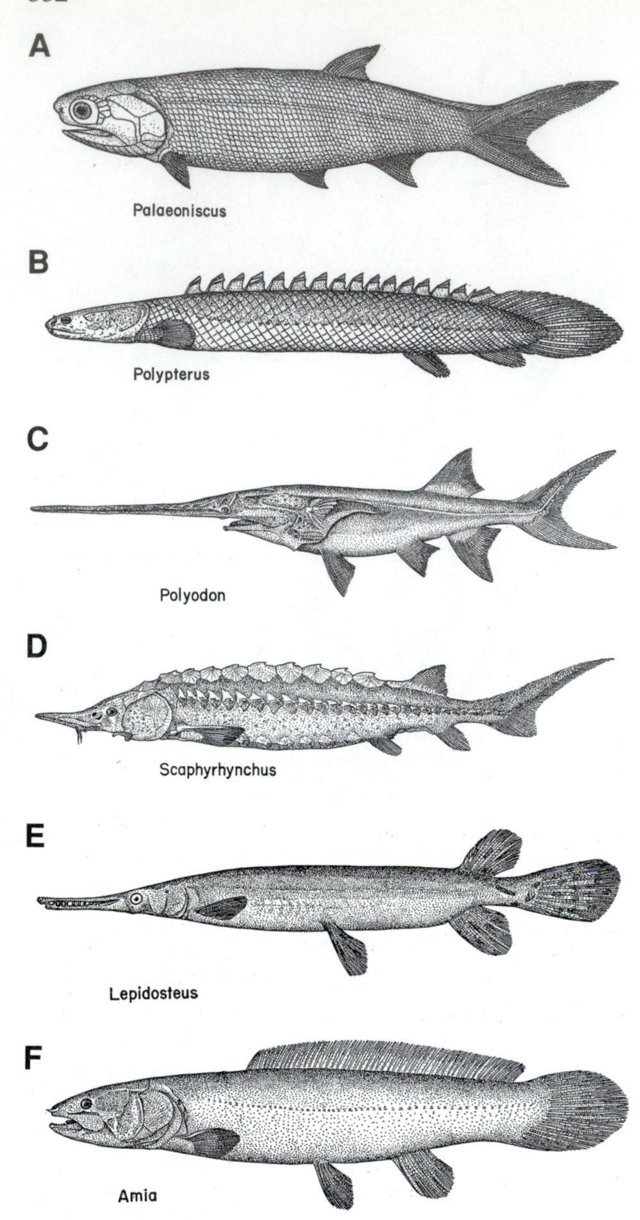

Figure 17.25. Some primitive actinopterygians. (A) The extinct paleoniscoid *Palaeoniscus.* (B) The living bichir. (C) The sturgeon. (D) The paddlefish. (E) The gar. (F) The bowfin. (After Romer, 1970.)

ly cartilaginous skeleton, but many small bony ossicles in their skin. Some sturgeons can reach truly impressive sizes of 6 m (20 feet) in length. Paddlefishes are equally distinctive, with their long flat prow on their nose, and a branchial basket that flares out to filter feed on plankton that have been stirred up out of the mud with their long noses.

From the great radiation of archaic actinopterygians of the late Paleozoic, a second great radiation of fish arose in the Permian, the **neopterygians** (Fig. 17.24). The more primitive neopterygians (often placed in the old paraphyletic taxon, the "holosteans") were the dominant fish of the Mesozoic, especially the Cretaceous (Fig.17.26). Neopterygians show a considerable number of advances over archaic actinopterygians. For one thing, their tail vertebrae no longer curve upward into a shark-like heterocercal tail, but instead are shorter and sharply upturned, forming a symmetrical **homocercal** tail. Even more striking is the change in their jaw apparatus. Paleoniscoids had relatively solid, bony jaws, which were capable only of a simple snapping bite (Fig. 17.27). Their jaw muscles were small and restricted to a narrow slot in the jaw hinge. However, in more advanced actinopterygians, the maxillary bone of the upper jaw is detached from the skull at the posterior end, and swings on a hinge at the anterior end, restrained only by muscles, which allows the jaw to open much wider and permits the development of much stronger jaw muscles. In the final, **teleost** stage of jaw evolution (Fig. 17.27), the premaxillary bone on the front of the upper jaw becomes enlarged, and is part of a pivoting framework that allows the entire mouth to open wide with great suction. If you watch a goldfish or some other aquarium fish feeding, you will see that most of them do not bite, but suck their food in to their mouths by protruding their jaws as shown in Figure 17.27.

Archaic neopterygians ("holosteans") showed a great diversity of body forms in the Mesozoic (Fig. 17.26). In addition to fishes with the standard fusiform body shape, there were torpedo-shaped fast swimmers, eel-shaped fish, fish with long dorsal fins, and deep-bodied fish. Many of these body shapes were also seen among the "paleoniscoids," and are seen again among the living teleost fish, so clearly a lot of convergent evolution must have taken place. Although most of this "holostean" radiation was extinct by the end of the Cretaceous, two lineages (Fig. 17.26E, F) are still found today: the bowfin (*Amia*) and the gar (*Lepisosteus*). Although these archaic fish are far outnumbered by their more diverse and successful teleost relatives, they are remarkably durable and versatile fish. They are found in most bodies of freshwater in the southeastern United States, surviving as "living fossils," although their fossil record shows that both bowfins and gars once lived in many parts of the world. Bass fishermen have repeatedly attempted to eradicate gars, because these fish are voracious predators of the bass, and have spread to most lakes and rivers in the southern United States. With their heavily armored bodies, long needle-nosed snouts, and distinctive heavy scales, gars are very easily recognized, and their fossil scales are common in the dinosaur beds of the west. Some alligator gars are known to reach 4 m (12 feet) in length. Gars even have adaptations for breathing air when the water in their pond becomes too stagnant, so they are true survivors.

Modern Fish: The Teleosts

The final radiation of the most advanced actinopterygians, the **teleosts**, began in the Middle Triassic, exploded in the Cretaceous, and today teleosts make up 99% of the living fish. Except for the fish discussed above, all other living fish are teleosts (Fig. 17.26). With over 20,000 species, teleosts are half of all the living vertebrates. In other words, all the species of birds, mammals, amphibians, reptiles, and sharks put together equals the number of

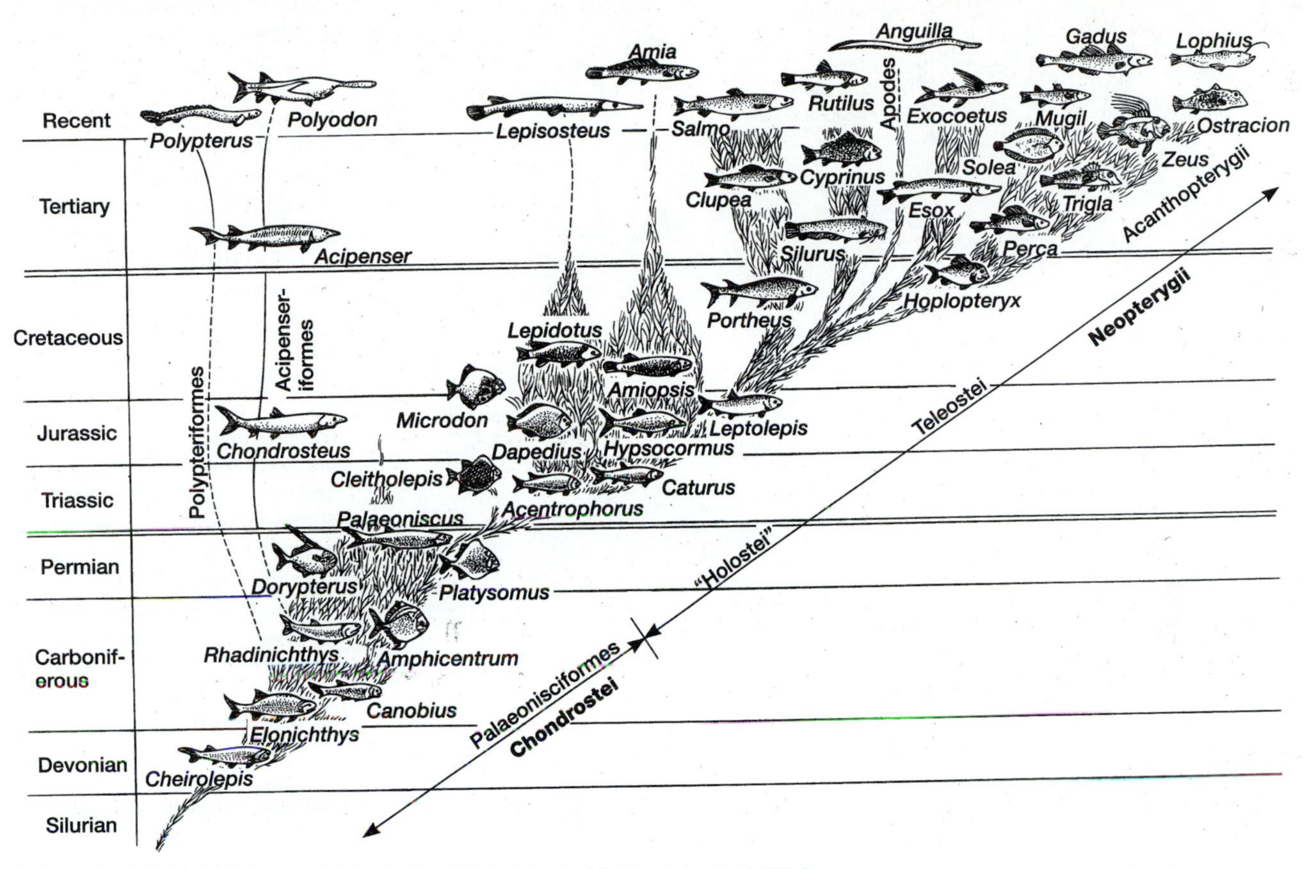

Figure 17.26. Evolutionary radiation of bony fishes. (After Kardong, 1996.)

teleosts alive today. As tetrapod chauvinists, we like to think of the Mesozoic as the "Age of Dinosaurs" and the Cenozoic as the "Age of Mammals," but if sheer numbers of individuals or species or morphological diversity mean anything, among vertebrates the Cretaceous and Cenozoic have always been the "Age of Teleosts." (Of course, insects and crustaceans have all vertebrates beat in terms of either numbers of individuals or taxonomic diversity.)

As we saw above (Fig. 17.27), teleosts have many special adaptations in their jaw mechanism that allows the mouth to protrude and suck up their food, rather than to bite hard. Among the unique bones in this mechanism are two supramaxillary bones, which are found in no other group of animals. The rest of the skull is also lightened and the bony components reduced, making it a highly mobile, kinetic structure of bony struts that can flex and stretch much more easily than the heavy skulls of archaic fish.

Most teleosts have a **swim bladder**, a gas-filled chamber in their bodies that makes them neutrally buoyant and much more adept at swimming and hovering in the water. For this reason, their pectoral and pelvic fins are no longer as critical for providing forward thrust, but instead can be used for fine steering, hovering, and even backward swimming. Consequently, the pectoral fin shifts up to the middle of the body, just behind the gill opening, and the pelvic fins shift forward, so the fish can turn on a dime.

Even more diagnostic is the anatomy of their tail (Fig. 17.27). The bones of the symmetrical homocercal tail of neopterygians have been further modified to a distinctive series of radiating elements known as **uroneurals**, originating from the upturned end of the spinal column. The detailed configuration of the tail is one of the key features used in teleost systematics.

Despite their incredible diversity and importance, it is impossible in a brief chapter like this to give the teleosts coverage commensurate with their numbers. Nevertheless, we will outline a few of the important features of teleost evolution. Prior to 1966, teleost classification was, to use Patterson's (1994, p. 74) words, ". . . chaotic. The systems then in use consisted of a long string of orders, arranged in a kind of *scala naturae*, from primitive to advanced, but with no hierarchical structure above the rank of order." Modern teleost systematics essentially began with the landmark study by Greenwood and others (1966). Even though that paper was pre-cladistic, it cut through over a century of confusion about teleost relationships, and allowed scientists to understand their evolution in a rigorous fashion for the first time. As Patterson (1977) points out, this late start in understanding such a fundamental problem is comparable to mammalogists not recognizing until recently the differences between marsupials and placentals (a distinction that has been recognized for over a century).

Many extinct groups of teleosts are known from the Cretaceous, including such paraphyletic assemblages as the pholidophorids and leptolepids. One of the most spectacular group of Cretaceous teleosts were the ichthyodecti-

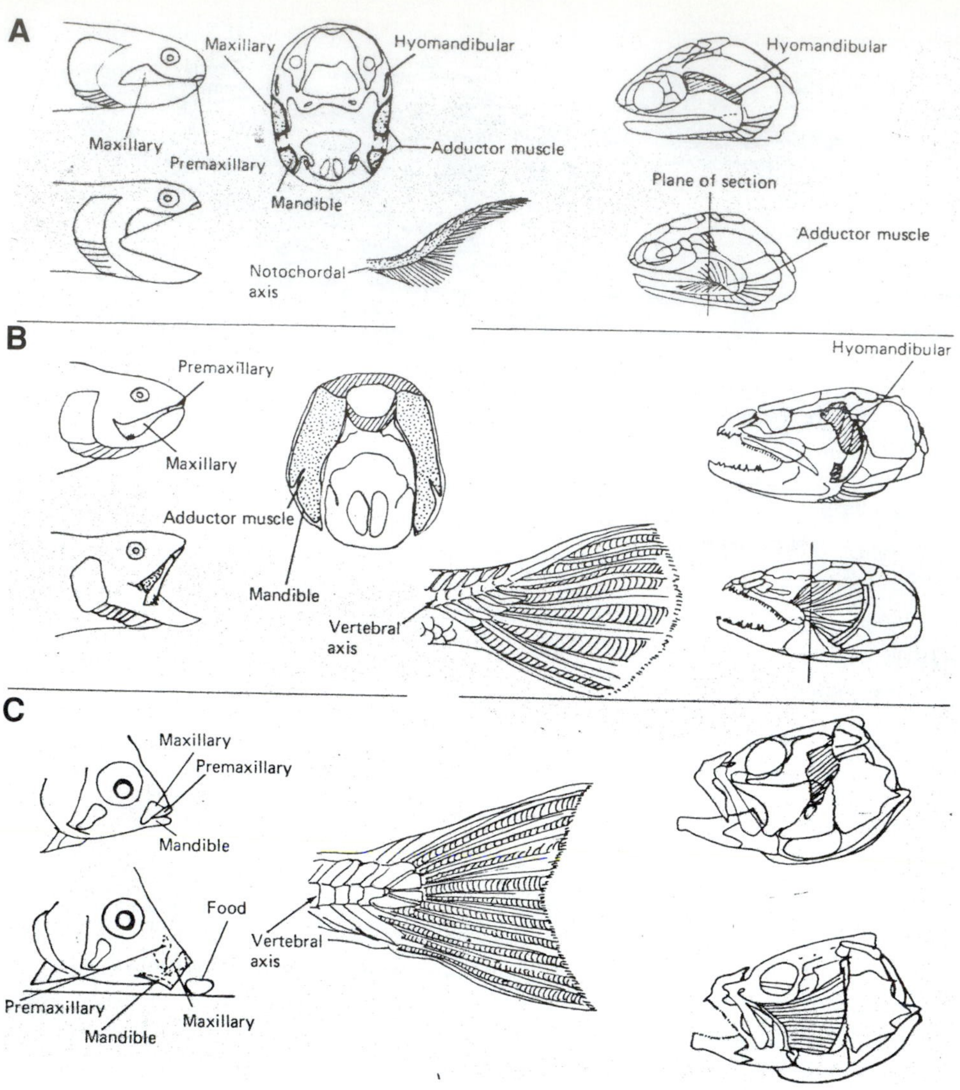

Figure 17.27. Morphology of the three grades of bony fish. (A) "Paleoniscoids" had a simple "snap-trap" jaw with the maxillary fused to the face (left). The jaw adductor muscles (right) fit into a pocket with the bone of the jaw joint, limiting their size and mobility. The tail (center) was heterocercal (as in a sturgeon.) (B) "Holosteans" have some mobility in the maxilla (left), and the back of the skull is considerably less bony, allowing much larger adductor muscles (center and right). Although the tail support still curves upward, it is very shortened. (C) Teleost jaws (left) have a higly mobile maxilla and premaxilla so they can protrude their mouthparts. The skull (right) is mostly thin bony struts, allowing great flexibility and mobility and room for larger muscles. The tail (center) is homocercal, with symmetrical vanes even though the final bones of the spine (the uroneurals) which support the tail are flexed slightly upward. (From Pough et al., 1996.)

formes, including the giant *Xiphactinus* (Fig. 17.28). Among the living teleosts, the most primitive group (Fig. 17.24) are the **osteoglossomorphs** (Fig. 17.29A). These fish were common in the Cretaceous, but today they are restricted to about 217 species found mostly in tropical freshwaters. Osteoglossomorphs include *Arapaima*, a huge (4.5 m, or 14 foot) freshwater predator from the Amazon; the meter-long Amazonian predator *Osteoglossum*, along with the mudskippers, knifefish, elephant fish, mooneyes, and featherbacks, and some electric fish as well. The next branch point in the teleosts are the **elopomorphs** (Figs. 17.24, 17.29B), a group of about 350 species including tarpons, eels, true electric eels, and many of the unusual fish of the deep ocean. All elopomorphs have a distinctive, ribbon-like larva known as the leptocephalus. After osteoglossomorphs and elopomorphs, the third major branch point up the cladogram of the teleosts is the **clupeomorphs** (Fig. 17.29C), or the herrings, shad, sardines, anchovies, and their relatives. Of the 350 living species, most are specialized for living in large schools and feeding on plankton in the open ocean with their specially adapted gills. Clupeomorphs are of tremendous economic importance, since herring, sardines, and anchovies are among the most common food fish in many parts of the world.

After these groups, the remaining higher teleosts are all grouped into a clade known as the Euteleostei. The most primitive euteleosts are a huge group known as the **ostariophysans** (Fig. 17.29D, E, F). About 6500 species of ostariophysans are living today, and they are the dominant

Figure 17.28. The giant Cretaceous ichthyodectiform fish *Xiphactinus* from the Cretaceous chalk beds of western Kansas. (Photo courtesy Sternberg Memorial Museum.)

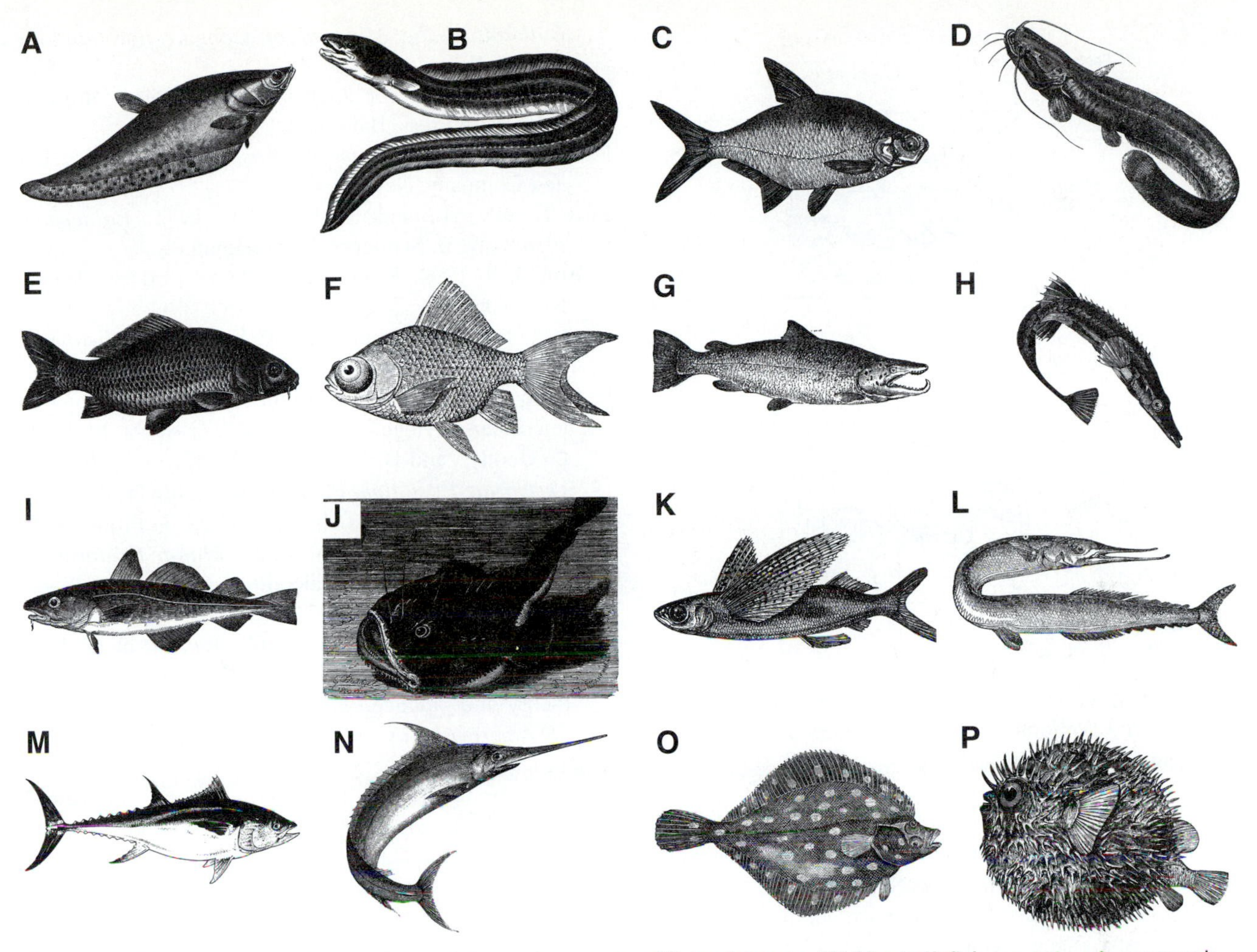

Figure 17.29. Representative examples of the major groups of living teleosts. (A) The knifefish, an osteoglossomorph. (B) An eel, an elopomorph. (C) A herring, a clupeomorph. (D-F) Ostariophysans include (D) catfish, (E) carp, and (F) goldfish. (G-H) Salmoniformes include (G) salmon and (H) pike. (I-J) Paracanthopterygians include (I) the cod and (J) anglerfish. (K-L) The atherinomorphs include (K) the flying fish and (L) the half-beak. (M-P). Percomorphs include the (M) tuna, (N) swordfish, (O) flatfish such as this flounder and (P) pufferfish. (Images courtesy IMSI.)

group of freshwater fish on this planet. Most freshwater fish in your fish tank, backyard pond, or in the aquarium store are ostariophysans, including the goldfish, carp, minnows, suckers, loaches, tetras, piranhas, and catfish. Ostariophysans have a unique series of bones attached to the front of the swim bladder known as the **Weberian apparatus** (Fig. 17.30). These bones connect the swim bladder to the inner ear, allowing them to hear the sounds amplified by the hollow chamber of the swim bladder, and thus greatly enhancing their sensitivity to sound.

Also included in the euteleosts are the **salmoniformes** (Fig. 17.29G, H), including the salmon, trout, pike, pickerel, muskellunge, and lantern fish of the deep sea. Some of these fish are fast-swimming predators of the freshwaters and lakes, and are very important to both sportsmen and commercial fishermen. Next up the cladogram are the **paracanthopterygii**, with about 1160 living species, including the cod and anglerfish and their relatives (Fig. 17.29I-J).

Finally, the largest group of advanced teleosts are the **acanthomorphs,** or "spiny teleosts" (Fig. 17.29K-P), including about 15,000 living species in 300 families. Acanthomorphs bear hundreds of erectible spines on their bodies, so that when they are threatened, they bristle like a porcupine and become harder to swallow. Among the many familiar groups of acanthomorphs (Fig. 17.29K, L) are the **atherinomorphs** (silversides, grunions, half-beaks, killifish, seahorse, barracuda, flying fish, guppies, mollies, swordtails, kissing fish). The atherinomorphs are distinguished by their extraordinarily protrusible mouthparts, which is why many of them can "kiss" or feed with a similar motion. Finally, the largest of all the acanthomorph groups are the **percomorphs** (about 9000 living species), including the perches, bass, cichlids, tuna, marlin, swordfish, angelfish, remoras, scorpionfish, stonefish, puffers, porcupine fish, sunfish, flounders, sole, halibut, and many other freshwater and marine forms (Fig. 17.29M-P). As the list of species indicates, percomorphs are incredibly diverse, not only in body shape and habitat, but also in their incredible variety of adaptations. Many of them, such as the perch, tuna, sunfish, and flatfish like the flounder, sole, and halibut, are also important to recreational or commer-

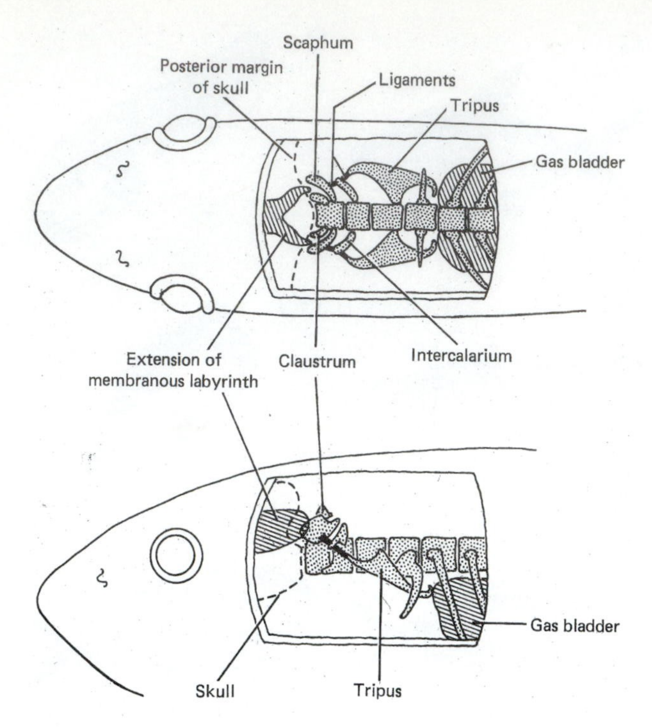

Figure 17.30. Position of the ear in the ostariophysans in relation to the Weberian ossicles (the scaphum, claustrum, intercalarium, and tripus). (From Pough et al., 1996.)

cial fishermen as well. The explosive adaptive radiation of percomorphs occurred in the early Cenozoic, about the same time as the radiation of mammals, but there are about twice as many living species of percomorphs as there are of mammals. Percomorphs are such a large and complex group that they are still relatively poorly understood, in contrast to the systematics of the smaller groups, which have been studied since 1966.

The subject of ichthyology has come a long way since Izaak Walton's "The Compleat Angler." We cannot do it justice in an introductory textbook such as this one, which focuses mainly on fossil groups. Some of the references at the end of this section are highly recommended for those with an interest in understanding the amazing diversification of bony fish.

For Further Reading

Alexander, R. McN. 1970. *Functional Design in Fishes.* Hutchinson, London.

Bone, Q., and N. B. Marshall. 1982. *Biology of Fishes.* Blackie & Sons, Glasgow.

Greenwood, P.H., R. S. Miles, and C. Patterson, eds. 1973. *Interrelationships of Fishes.* Academic Press, London.

Greenwood, P. H., D. E. Rosen, S. H. Weitzman, and G. S. Myers. 1966. Phyletic studies of teleostean fishes, with a provisional classification of living forms. *Bulletin of the American Museum of Natural History* 131:339-456.

Lauder, G. V., and K. F. Liem. 1983. The evolution and interrelationships of the actinopterygian fishes. *Bulletin of the Museum of Comparative Zoology* 150:95-197.

Long, J. A. 1995. *The Rise of Fishes.* Johns Hopkins University Press, Baltimore.

Maisey, J. G. 1996. *Discovering Fossil Fishes.* Henry Holt and Company, New York.

Moy-Thomas, J., and R. S. Miles. 1971. *Palaeozoic Fishes.* W. B. Saunders, Philadelphia.

Nelson, J. S. 1994. *Fishes of the World* (3rd ed.) Wiley, New York.

Norman, J. R., and Greenwood, P. H. 1975. *A History of Fishes*, 3rd ed. Ernest Benn, London.

Patterson, C. 1977. The contribution of paleontology to teleostean phylogeny, pp. 579-643, *in* Hecht, M. K., P. C. Goody, and B. M. Hecht, eds. *Major Patterns of Vertebrate Evolution.* Plenum Press, New York.

Patterson, C. 1994. Bony fishes, pp. 57-84, *in* Prothero, D. R., and R. M. Schoch, eds. *Major Features of Vertebrate Evolution.* Paleontological Society Short Courses in Paleontology 7.

Patterson, C., and D. E. Rosen. 1977. Review of ichthyodectiform and other Mesozoic teleost fishes and the theory and practice of classifying fossils. *Bulletin of the American Museum of Natural History* 158:81-172.

Rosen, D. E. 1982. Teleostean interrelationships, morphological function, and evolutionary inference. *American Zoologist* 22:261-273.

Schaeffer, B., and D. E. Rosen. 1961. Major adaptive levels in the evolution of the actinopterygian feeding mechanism. *American Zoologist* 1:187-204.

LOBE-FINS: THE SARCOPTERYGIANS

> *I went onto the deck of the trawler* Nerine *and there I found a pile of small sharks, spiny dogfish, rays, starfish, and rat tail fish. I said to the old gentleman, "These all look much the same, perhaps I won't bother with these today." Then, as I moved them, I saw a blue fin and pushing off the fish, the most beautiful fish I have ever seen was revealed. It was 5 feet long and pale mauvy blue with iridescent silver markings. "What is this?" I asked the old gentleman. "Well, lass," said he, "this fish snapped at the Captain's fingers as he looked at it in the trawler net. It was trawled with a ton and a half of fish plus all these dogfish and others." "Oh," I said, "this I will definitely take to the museum."*
>
> M. Courtenay-Latimer, *My Story of the First Coelacanth*, 1949

Although actinopterygians make up over 99% of the fish living today, from our tetrapod perspective, the most important development among the bony fish was the evolution of our ancestors, the lobe-finned fish, or sarcopterygians. When the Osteichthyes split into the Actinopterygii and Sarcopterygii in the Late Silurian (Figs. 17.24, 17.31), the two branches of bony fish were nearly equal in diversi-

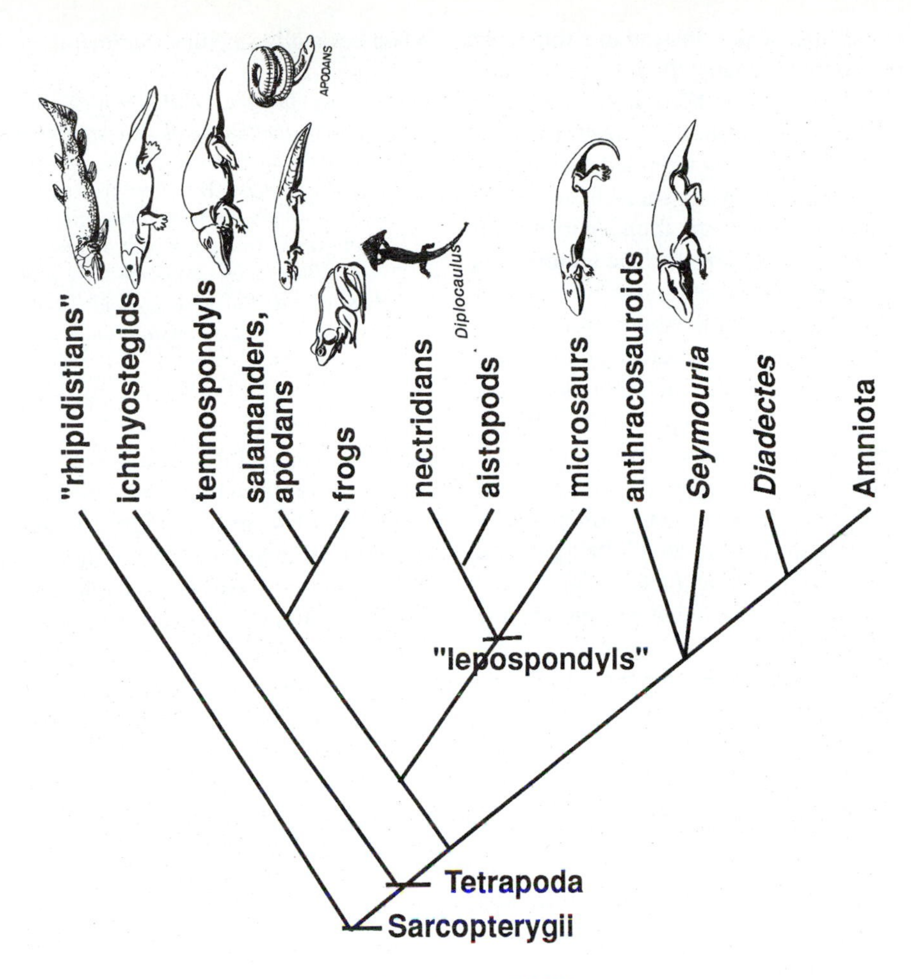

Figure 17.31. Cladogram of the sarcopterygians and primitive tetrapods.

ty. Sarcopterygians then became very abundant and diverse in the Devonian, with up to 19 families of lungfish and 3 families of coelacanths (equal to all the Devonian families of chondrichthyans, acanthodians, and actinopterygians *combined*). By the Late Devonian, they had given rise to the first tetrapods on land. Much of this sarcopterygian diversity became extinct during the Permian catastrophe, although lungfish remained common in freshwater habitats in the Mesozoic, and coelacanths were quite common in the Cretaceous marine habitats before apparently becoming extinct.

Today, there are only four living species of non-tetrapod sarcopterygians: the sole surviving coelacanth (*Latimeria chalumnae*, from the western Indian Ocean), and three species of lungfish: the African *Protopterus*, the South American *Lepidosiren*, and the Australian *Neoceratodus* (Fig. 17.33). All are true living fossils, tiny remnants of their former diversity clinging to survival, which give us clues about the step from fish to tetrapods. All sarcopterygians are characterized by a number of shared derived characters, but the most obvious is the muscular lobed fin that gives the group its name. Unlike the delicate rays that support the actinopterygian fin, the sarcopterygian fin is supported by robust bony elements (homologous with the tetrapod arm and leg bones), which in turn support the fin rays (Fig. 17.34). Another distinctive feature of sarcoptery-

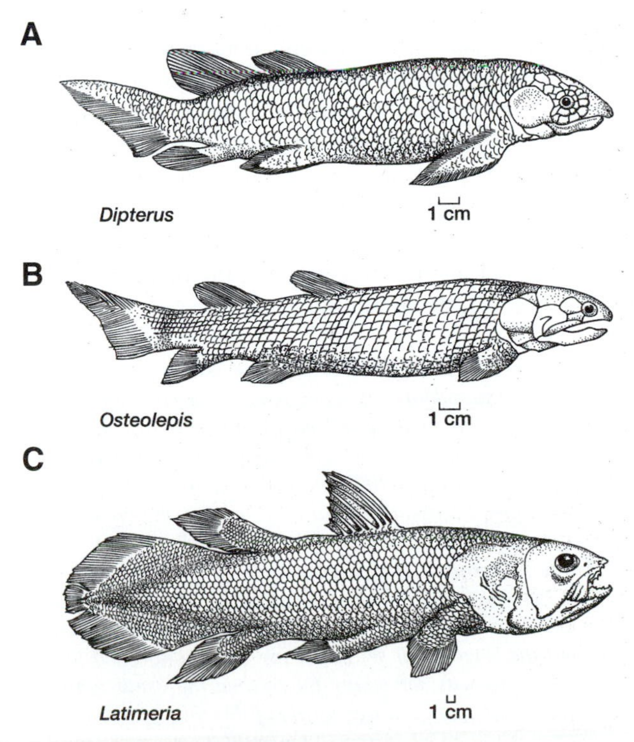

Figure 17.32. The three major groups of sarcopterygians. (A) lungfish or dipnoans, (B) "rhipidistians" and (C) coelacanths. (From Kardong, 1995).

gians is the histology of their scales. Instead of a thin layer of ganoine found in actinopterygians, the lobe-fin scale is covered by a thick layer of porous bone called **cosmine**; such scales are called cosmoid scales. The teeth are covered by cosmine, and have a uniquely wrinkled enamel surface, which looks like a labyrinth when sliced in cross section. These **labyrinthodont** teeth are found not only in primitive sarcopterygians, but in most archaic tetrapods as well (some of which were once placed in the wastebasket taxon "Labyrinthodontia" based on this tooth structure). In addition, there are many unique characters of the jaws, jaw-support mechanisms, gill arches, and shoulder girdle that are not found in any other group of fish (Maisey, 1986). In general, the sarcopterygians have reduced the teeth on the edge of the jaw (typical of actinopterygians) and emphasized the teeth on the palate for holding prey as they swallow. Sarcopterygians have a sac off their gut used for respiration (the lung in lungfish and tetrapods).

The more primitive of the two living groups are the **coelacanths**, or Actinistia (Figs. 17.31, 17.32C). They are easily recognized by the abundance of their lobed fins, including paired pectorals, pelvics, two tandem dorsal fins, an anal fin, and a symmetrical tail with a lobe in the center. The other distinctive feature is the triangular shape of the operculum, or the bone covering the gills. Coelacanths were as common as the paleoniscoids in both marine and freshwater habitats in the late Paleozoic and Triassic (Schaeffer, 1952), but their fossils suggested that they became extinct at the end of the Cretaceous. For almost a century, there was no evidence to suggest that they survived into the Cenozoic. Then, in 1938 a remarkable discovery was made at the mouth of the Chalumna River in South Africa. Fishermen trawled up a huge, shiny blue fish that looked like nothing they had ever seen before. The local museum curator, Marjorie Courtenay-Latimer, was summoned and immediately recognized its importance. Although the fish weighed 127 pounds, and was already dead and rotting rapidly, she did her best to preserve it, but eventually had to discard everything but the skin. A few weeks later, her letter and sketches of the fish reached the ichthyologist J. L. B. Smith. As he wrote later,

> *Then I turned the page and saw the sketch, at which I stared and stared, at first in puzzlement, for I did not know any fish of our own or indeed of any seas like that; it looked more like a lizard. And then a bomb seemed to burst in my brain, and beyond that sketch and the paper of the letter I was looking at a series of fishy creatures flashed up as on a screen, fishes no longer here, fishes that had lived in dim past ages gone, and of which often only fragmentary remains in rocks are known. I told myself sternly not to be a fool, but there was something about the sketch that seized on my imagination and told me that this was something far beyond the usual run of fishes in our seas . . . I was afraid of this thing, for I could see something of what it would mean if it were true, and I also realized only too well what it would mean if I said it was it was not* (Smith, 1956, p. 62).

When he finally saw the specimen,

> *Coelacanth—yes, God! Although I had come prepared, that first sight hit me like a white-hot blast and made me feel shaky and queer, my body tingled. I stood as if stricken to stone. Yes, there was not a shadow of doubt, scale by scale, bone by bone, fin by fin, it was a true Coelacanth. It could have been one of those creatures of 200 million years ago come alive again. I forgot everything else and just looked and looked, and then almost fearfully went close up and touched and stroked* (Smith, 1956, p. 73).

Smith named the specimen *Latimeria* (after the discoverer) *chalumnae* (after the place it was found), and it caused a sensation in the scientific world in 1939. However, 13 years of diligent searching by Smith and many others failed to turn up another specimen in South African waters. They even put out "wanted" posters with a reward of £100 for another specimen, and circulated them all over the African coast. Finally, in 1952 another specimen was found in the Comoros Islands north of Madagascar, and it was preserved soon enough that all of its internal organs remained intact. Scientists soon realized that the original South African specimen had strayed far from home; the Comoros was the last refuge of the coelacanth. In the next 20 years, 83 specimens were hauled out of the deep waters off these steep volcanic islands. In fact, the coelacanths are now becoming so rare and valuable that humans are driving them to final extinction, and there may be more specimens in museums now than there are in the waters of the Comoros.

Even though *Latimeria* is an undoubted coelacanth, with the distinctive fins and triangular operculum, it has many peculiarities of its own. Fossil coelacanths have some bone, but *Latimeria* is mostly cartilaginous, and even retains a notochord in its spinal column. *Latimeria* has a peculiar spiral intestine, shaped like a long straight cylinder with a helical divider running down its length. This type of intestine is found elsewhere only in sharks. Instead of an air-filled lung, *Latimeria* has a swim bladder filled with fatty tissue that gives it some neutral buoyancy. When preserved specimens were dissected, completely developed embryos were found inside, showing that *Latimeria* gives birth to live young, rather than laying eggs. *Latimeria* has phosphorescent eyes and a deep blue color consistent with the depths of 200 to 500 m where it lives. It comes near the surface only at night, so it is hard to catch, and is almost never seen alive in its habitat. Using a submarine, Hans Fricke and his colleagues were able to observe living specimens, and found that it has a very peculiar "dance" when it swims, flexing its fins in surprising ways, and even standing on its head (Fricke et al., 1987). However, it does not seem to use its lobed fins to "walk" in a way that anticipates walking in tetrapods.

Although not as strange as coelacanths, lungfishes (the dipnoans) have many of their own peculiarities (Fig.

17.33). Late Paleozoic and Triassic fossil lungfishes (Fig. 17.32A) were common in both marine and freshwaters, but the three surviving genera are all restricted to freshwater lakes on the Gondwana continents of Africa, Australia, and South America (probably a relict vicariant distribution). They use their lungs to supplement their respiration in stagnant lakes and ponds. When the lake dries up completely, they estivate by encasing themselves in a coccoon of dried mud and mucus, with only a breathing hole at one end, and protect themselves from drying up. When the rains come again, the coccoon breaks down and the lungfish swims free. Unlike the highly kinetic, mobile skulls of actinopterygian fish, lungfish have fused the upper jaw (palatoquadrate) to the braincase, so their bite can crush their prey. They have no teeth on the edge of the mouth, but instead their palates are armed with peculiar ridged toothplates, which are their most commonly and distinctively fossilized remains. The dermal bones of the skull are also peculiar, with a completely different pattern from that found in other groups of bony fish. The three living species have a symmetrical tail (similar to that of the pleuracanth sharks—Fig. 17.21B), but this is clearly a secondary feature, since the earliest lungfish have a heterocercal tail (Fig. 17.32A).

In addition to the lungfishes and coelacanths, there are extinct sarcopterygians that are placed in the wastebasket taxon "rhipidistia" (Figs. 17.31, 17.32B). These fish have most of the primitive features of lobe-finned fish, instead of the highly specialized characteristics of the living lungfish and coelacanths. Traditionally, one group of "rhipidistians," the osteolepiformes (Fig. 17.32B) are considered the best candidate for the ancestry of tetrapods. However, Rosen and others (1981) challenged this traditional belief and argued that lungfish are closer to tetrapods than are "rhipidistians." They based this argument on the presence of certain derived features (such as the lungs and internal nostrils, or choanae, in lungfish and tetrapods) not found in osteolepiformes, and also showed that much of the case for osteolepiform-tetrapod relationships was based on ancestor worship, not on shared derived characters. Since this unorthodox view was proposed, a number of scientists have disputed their conclusion, reinterpreting the same characters and using comparable cladistic arguments (Jarvik, 1981; Thomson, 1981; Holmes, 1985; Maisey, 1986; Janvier, 1986; Panchen and Smithson, 1988; Schultze, 1987, 1991; Chang, 1991), although Forey (1987) briefly tried to salvage the lungfish-tetrapod hypothesis.

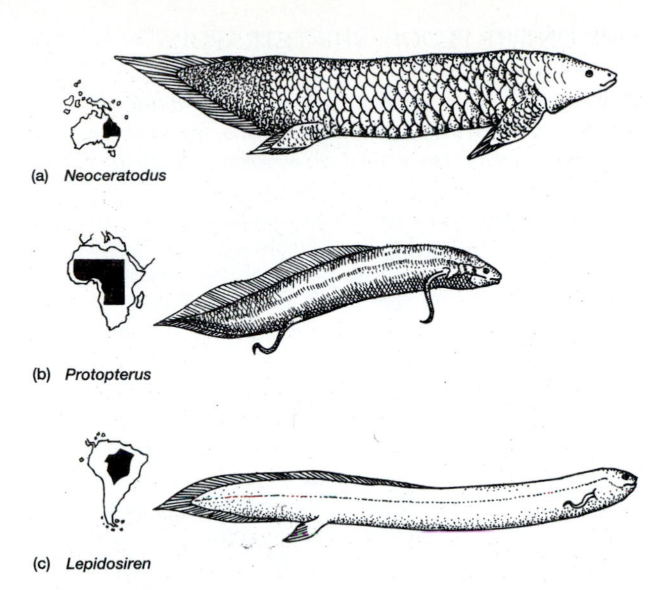

Figure 17.33. The three living lungfish, found in three different southern continents. (From Kardong, 1995.)

For Further Reading

Bemis, W. E., Burggren, W. W., and Kemp, N. E., eds. 1987. *The Biology and Evolution of Lungfishes*. Liss, New York.

Forey, P. L. 1997. *History of the Coelacanth Fishes*. Chapman & Hall, London

Fricke, H. 1988. Coelacanths, the fish that time forgot. *National Geographic* 173:824-838.

Fricke, H., O. Reinecke, H. Hofer, and W. Nachtigall. 1987. Locomotion of the coelacanth *Latimeria chalumnae* in its natural environment. *Nature* 329:331-333.

Holmes, E. B., 1985. Are lungfishes the sister group of tetrapods? *Biological Journal of the Linnaean Society* 25(4): 379-397.

Long, J. A. 1995. *The Rise of Fishes*. John Hopkins University Press, Baltimore.

Maisey, J. G. 1996. *Discovering Fossil Fishes*. Henry Holt and Company, New York.

McCosker, J. E., and M. D. Lagios, eds. 1979. The biology and physiology of the living coelacanth. *California Academy of Sciences Occasional Papers* 134:1-175.

Moy-Thomas, J., and R. S. Miles. 1971. *Palaeozoic Fishes*. W. B. Saunders, Philadelphia.

Rosen, D. E., P. Forey, B. G. Gardiner, and C. Patterson. 1981. Lungfishes, tetrapods, paleontology and plesiomorphy. *Bulletin of the American Museum of Natural History* 141:357-474.

Schaeffer, B. 1952. The Triassic coelacanth fish *Diplurus*, with observations on the evolution of the Coelacanthini. *Bulletin of the American Museum of Natural History* 99:25-78.

Smith, J. L. B. 1956. *Old Fourlegs: The Story of the Coelacanth*. Longman, Green, London.

Thomson, K. S. 1969. The biology of lobe-finned fishes. *Biological Review* 44:91-154.

Thomson, K. S. 1981. A radical look at fish-tetrapod relationships. *Paleobiology* 7:153-156.

Thomson, K. S. 1991. *Living Fossil*. W. W. Norton, New York.

Ward, P. D. 1992. *On Methuselah's Trail: Living Fossils and the Great Extinctions*. Freeman, New York.

FOUR ON THE FLOOR: THE TETRAPODS

In my day it was believed that the place for a fish was in the water. A perfectly sound idea, too. If we wanted fish, for one reason or another, we knew where to find it. And not up a tree.

For many of us, fish are still associated quite definitely with water. Speaking for myself, they always will be, though certain fish seem to feel differently about it. Indeed, we hear so much these days about the climbing perch, the walking goby, and the galloping eel that a word in season appears to be needed.

Times change, of course—and I only wish I could say for the better. I know all that, but you will never convince me that fish that is out on a limb, or strolling around in vacant lots, or hiking across the country, is getting a sane, normal view of life. I would go so far as to venture that such a fish is not a fish in its right mind.

Will Cuppy, *How to Become Extinct*, 1941

One of the most dramatic steps ever taken in the history of life was the invasion of land by animals and plants. As we saw in previous chapters, a variety of arthropods (millipedes, scorpions, spiders, and eventually, insects) and vascular plants accomplished this in the Silurian (and possibly Late Ordovician). Vertebrates did not crawl out onto land until the Late Devonian, almost 100 million years after arthropods and vascular plants had colonized the land surface. Yet as radical as this step seems, it is not as difficult to accomplish as it first appears. A variety of teleosts, including the mudskippers, walking catfishes, and a variety of nearshore fish, are adept at crawling using their ray fins, and adapting their gills for breathing air, or gulping air with their swim bladders. Some can live on land for days as long as the conditions are humid, and are able to crawl for considerable distances. Mudskippers spend most of their lives out of the water, and walking catfish are so mobile that they have infested many of the lakes in the Southeast, despite attempts to eradicate them.

Nevertheless, permanent life on land requires a number of adaptations that no teleost has completely developed. They fall in several categories: respiration, locomotion, lack of buoyancy, sense organs, and desiccation:

Respiration—As we have just seen, the lungfish and probably the "rhipidistians" already had an air-breathing lung in place, which helps them survive when their ponds dry up. (This may have been one of the important factors that drove the fish onto the land, since the drying of ponds forces fish to crawl out or die.) An aquatic fish can gulp air and then force it back into the lungs by diving downward. But on land, a tetrapod needs a pumping mechanism to force the air backward. Salamanders and some other amphibians do this by expanding and contracting the rib cage, while frogs pump air with the sac on the base of the throat. The earliest tetrapods have broad flanges on their ribs, apparently to help with pumping the air into the lungs. They also still had gills as well, so they had not completely given up on breathing in water (Coates and Clack, 1991).

Locomotion—Once again, sarcopterygian fins already have robust bony supports for their fin rays, so all that is needed for the lobed fin to become a tetrapod limb is further development of these robust supporting bones into the bones of the tetrapod arm or leg, and the replacement of the fin rays with toes. Although fossils of these specimens are rare, the most advanced osteolepiform fins and the earliest tetrapod limbs are not very different (Fig. 17.34). Even more surprising is the recent discovery (Coates and Clack, 1990) that the earliest tetrapods had a wide range of number of toes on their hands and feet—eight on some specimens, seven or six on others. The standard count of five toes and fingers is a later development that stabilized after the earliest tetrapods had crawled out onto land (see Gould's essay, "Eight little piggies," 1993).

But strong limbs with toes are not enough. To support the body on land without the buoyancy of water, the limbs must have bony connections to the main axis of the spinal column. This is comparable to the engineering problem of a supporting the span of a bridge. To hold the long horizontal span of the bridge stable, two strong vertical support columns need to be firmly attached at either end. For this reason, the earliest known tetrapods have already considerably enlarged the pelvic elements and fused them to the spinal column (in fish, the pelvic bones float in the body wall connected only by muscles). Likewise, the shoulder girdle, which was originally part of the dermal bones of the skull roof and gill covering elements, have become detached from the head and fused to the spine, allowing the tetrapod to move its head independently of its front limbs. Clearly, a freely turning head (which no fish has) allows a land predator to see better, and to turn its head quickly to catch prey.

Lack of buoyancy—The vertebral column of fish is only loosely joined together, since the water supports their body weight. However, when the spine no longer has the support of water, it must become a much stronger, more directly articulated structure. Consequently, the vertebrae of tetrapods developed a variety of bony spines and articulations between them that not only join the vertebrae together directly while permitting movement between them, but the enlarged spines are also attachment points for much stronger bundles of muscles and tendons that run along and reinforce the spinal column. The details of how the various bony elements of tetrapod vertebrae are put together have long been used in the systematics of different primitive tetrapod groups.

Desiccation—Most fish have thin, soft scales so their skin is effective in allowing gases to pass through, and many respire through their skin more than they do through their gills. However, a permeable skin is a disadvantage in the dry air, so most tetrapods have developed an impermeable protein called **keratin** in their scales that protects them from drying up. (Keratin is the same protein found in your hair, fingernails, and in bird feathers.) Keratinous scales are relatively rare in primitive tetrapods, but are found in all

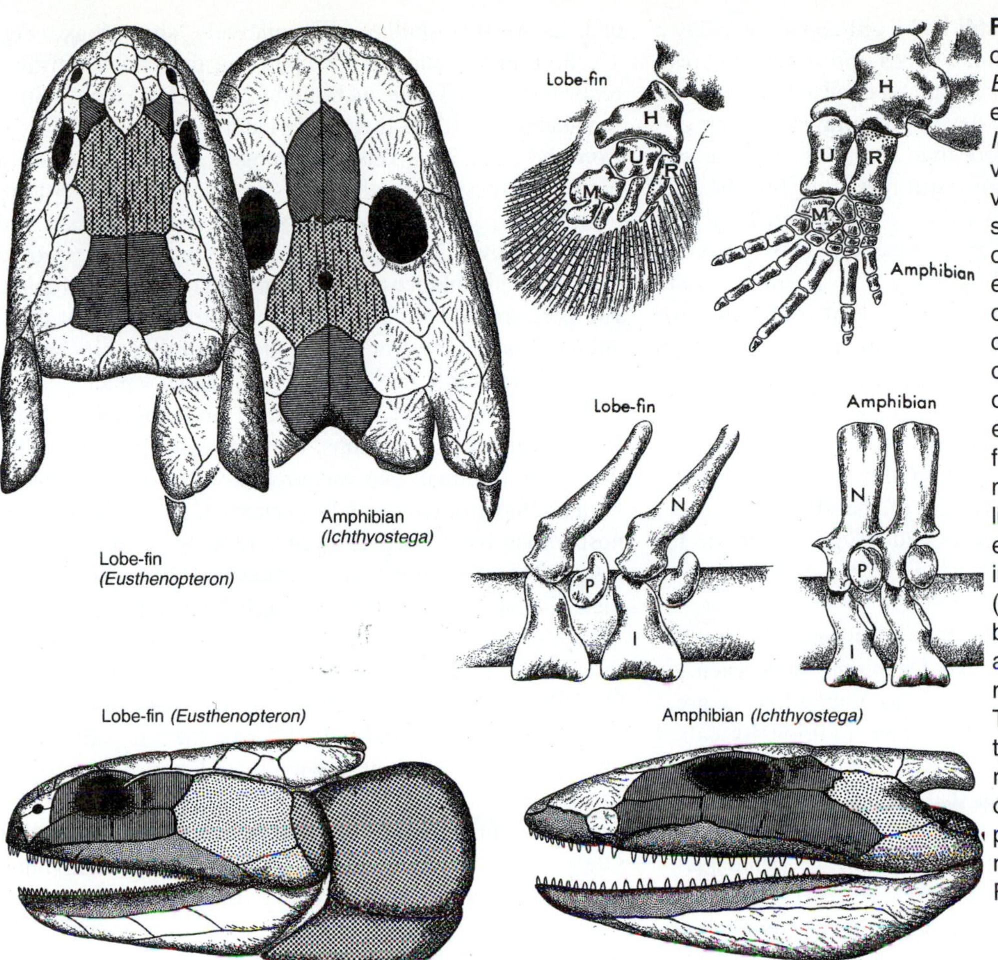

Figure 17.34. Comparison of the "rhipidistian" fish *Eusthenopteron* and the earliest known tetrapod *Ichthyostega.* In the top view (upper left) and side view (lower left) of the skull, note how the bones of the snout region are expanded at the expense of the bones of the gill covers. *Ichthyostega* has only a tiny remnant of the operculum that once covered the gills. The lobed fin (upper right) has been modified into a terrestrial limb, with the robust bony elements of the fin becoming the upper arm bone (humerus, H), lower arm bones (radius and ulna, R and U), and wrist elements (metacarpals, M). The vertebrae (right center) are locked together for rigidity, with gradual fusion of the intercentrum (I), pleurocentrum (P) and neural arch (N). (From Fenton and Fenton, 1958.)

amniotes, which live permanently out of the water. Among living amphibians (frogs and salamanders), the skin is permeable and helps with respiration, since they have such small body volume and seldom stray far from water.

Another area of water loss is the mouth and nasal cavity. Every time you breathe, you lose a certain amount of moisture (as you notice on a subfreezing day). Fish constantly open their mouths and swallow water, which then passes through the pharynx and out the gills. Tetrapods have restricted the flow of external fluids by breathing through a passage (the **choana**) that connects the nasal opening to the mouth cavity. In many tetrapods, this restricted nasal passage is lined with an area of folded tissues that further trap moisture before exhaling. Even without these, however, the choana conserves most of the moisture in the body of the animal while allowing it to breathe normally.

Sense organs—The senses needed in an aquatic life are very different from those used on land. In water, the light is distorted by the odd refraction patterns (think of the distorted images in a fish tank) and most bodies of water are too muddy for light penetration, so sight is not as important. However, sight is a very important sense in the air, and most early tetrapods had large eyes, with well-developed eyelids to protect the eyeball from drying out. Most fish have a series of pits along the side of the head and body called the **lateral line system**, which aids in detecting changes in pressure in the water (especially the waves caused by nearby obstacles or moving animals). This system is obsolete on land, since air is so much less dense than water that the lateral line is useless in detecting changing air pressures. The earliest tetrapods (which were still aquatic) still have traces of the lateral line, but later forms lose it completely. Instead, the best way to detect changes in air pressure is a large membrane for hearing. Most fish cannot hear directly, although the ostariophysan teleosts pick up sound in their swim bladders using the Weberian ossicles mentioned above. In the earliest tetrapods, however, the bones of the gill covers are attached to the bones of the hyoid arch (including the **hyomandibular**), which in turn is attached to the braincase and inner ear region. When the the gill covers were reduced, they left the hyomandibular behind, supporting a skin membrane that became the eardrum of tetrapods. Living amphibians have a large eardrum, whose vibrations are transmitted by the hyomandibular directly to the inner ear region. In your ear, the hyomandibular cartilage that you had as an embryo became the "stirrup" bone (**stapes**) of your inner ear, transmitting sound to the sensory region.

Different stages of evolution of all these features can be seen in the earliest tetrapods, such as *Ichthyostega* (Fig. 17.34) and *Acanthostega*, and other specimens known primarily from the Upper Devonian of Spitsbergen and

Greenland. These animals still had a gill apparatus (shown by the gill-covering opercular bones and other features), but the flanges on the ribs show that they also had lungs for breathing. They have fully developed limbs and limb girdles, although they had more than five toes in specimens that have preserved feet. They still had a tail fin, showing that locomotion in the water was also important. Their vertebrae have all the necessary connections and spines for suspension and support of the body. The lateral lines are still visible on the skull, although there is also a notch at the back of the skull for the eardrum, the **otic notch**. Compared to osteolepiformes, the eyes are large, and placed farther back on the skull, because the snout has lengthened considerably (presumably to aid in catching prey). In later tetrapods, there was a continuation of the trend to lengthen the snout and shorten the region of the skull behind the eyes, especially as the bones around the gill covers disappeared.

The Radiation of Tetrapods ("Amphibians")

From the transitional Late Devonian taxa such as *Ichthyostega*, the Carboniferous was marked by a great radiation of tetrapods that have been traditionally called "amphibians." However, the traditional use of the word "amphibians" as all tetrapods that are not amniotes (i.e., do not lay a land egg) is clearly paraphyletic (Fig. 17.31). Many of these fossil "amphibians" (e.g., the paraphyletic anthracosaurs) are closer to amniotes. However, the groups that are clearly sister-groups of the living amphibians (frogs and salamanders) can be referred to the Amphibia as a monophyletic group. Let us examine each of these groups in turn.

Temnospondyls—The late Paleozoic terrestrial and freshwater habitats were dominated by the **temnospondyls** (formerly called "labyrinthodonts"), which had long bodies and large, flat skulls, and relatively short legs (Fig. 17.35A). Some, such as *Eryops* from the Lower Permian redbeds of west Texas, were up to 2 m (6 feet) long, with a flat skull one-fifth of their body length. The skull of *Edops* from the same beds was even longer, suggesting an even bigger body size. In the Triassic, the last of the temnospondyls, or the metoposaurs, were extremely flat, and some were over 3 m in length. A few taxa, such as armored *Cacops*, were smaller (only 40 cm long) and less flattened, but these are exceptions to the general trend. Most of these animals must have lurked in the water like crocodiles, using their huge mouths to gulp down smaller prey. Although they were the largest land animals in the Pennsylvanian coal swamps, they were not the top predators of the Permian; that was probably the role of finbacks such as *Dimetrodon*. Over 170 genera in 30 familes of temnospondyls have been recorded, ranging from the Early Carboniferous to the Late Cretaceous, although most of their diversity occurred in the late Paleozoic and Triassic.

Lepospondyls—A second possibly monophyletic group of late Paleozoic tetrapods are known as the **lepospondyls** (17.35B), which are recognized by the distinctive spool-like shape of their vertebrae (although there is some question as to whether they are monophyletic). Two distinct groups of tetrapods have been clustered into the lepospondyls. One group, the microsaurs, look much like salamanders or lizards, with a long, cylindrical body, and deep, sturdy skull, but relatively tiny limbs. The other group, the nectrideans, is best know from animals like *Diplocaulus*, whose skull flared into a bizarre "boomerang" shape. There have been many speculations about the function of these strange "horns," but the best hypothesis suggests that they were like an airfoil that helped these animals swim. A third group, the aïstopods, h lost their limbs and had a very snake-like body form.

A

C

B

Figure 17.35. Representative examples of the three major group of extinct amphibians, all from the Permian of Texas. (A) *Eryops,* a large temnospondyl. (B) *Diplocaulus,* the "boomerang-head" lepospondyl. (C) *Seymouria,* a very amniote-like anthracosaur. (From Fenton and Fenton, 1958.)

"Anthracosaurs"—The tetrapods that are the sister-group to reptiles and synapsids are lumped into the paraphyletic group, the anthracosaurs (Fig. 17.35C). Like their descendants, they had a deep skull with a short snout, large eyes, long strong legs, and a relatively short body and tail. They have many anatomical features of the reptiles as well, so it is hard to decide when the "anthracosaurs" end and when the reptiles begin. This problem is discussed further in the next section.

Lissamphibians—The three groups of living amphibians (frogs, salamanders, and the legless caecilians) are now represented by about 4000 living species. The oldest known frogs occur in the Early Triassic, and the oldest salamanders in the Late Jurassic, and they already show essentially modern body forms. The origin of living amphibians has long been a puzzle, but currently the best evidence suggests that they are descended from a group of temnospondyls like *Cacops* and *Doleserpeton*. These animals have teeth with a distinctive base (**pedicellate** teeth), a feature found elsewhere only in the lissamphibians.

For Further Reading

Ahlberg, P. E. and A. R. Milner. 1994. The origin and early diversification of the tetrapods. *Nature* 368:507-514.

Bolt, J. R. 1977. Dissorophid relationships and ontogeny, and the origin of Lissamphibia. *Journal of Paleontology* 51:235-249.

Carroll, R. L. 1977. Patterns of amphibian evolution: an extended example of the incompleteness of the fossil record, pp. 405-437, *in* Hallam, A., ed. *Patterns of Evolution*. Elsevier, Amsterdam.

Carroll, R. L. 1992. The primary radiation of terrestrial vertebrates. *Annual Reviews of Earth and Planetary Sciences* 20:45-84.

Carroll, R. L., and P. Gaskill. 1978. The order Microsauria. *Memoirs of the American Philosophical Society* 126:1-211.

Coates, M. I., and J. A. Clack. 1990. Polydactyly in the earliest known tetrapod limbs. *Nature* 347:66-69.

Fracasso, M. A. 1994. Amphibia: disparity and diversification of early tetrapods, pp. 108-128, *in* Prothero, D. R., and R. M. Schoch, eds. *Major Features of Vertebrate Evolution*. Paleontological Society Short Course 7.

Milner, A. R. 1988. The relationships of the living amphibians, pp. 59-102, *in* Benton, M. J., ed. *The Phylogeny and Classification of the Tetrapods, vol. 1: Amphibians, Reptiles, Birds*. Clarendon Press, Oxford.

Milner, A. R. 1990. The radiations of temnospondyl amphibians, pp. 321-349, *in* Taylor, P. D., and G. P. Larwood, eds. *Major Evolutionary Radiations*. Clarendon Press, Oxford.

Panchen, A. L., ed. 1980. *The Terrestrial Environment and the Origin of Land Vertebrates*. Academic Press, London.

Panchen, A. L., and T. R. Smithson. 1987. Character diagnosis, fossils, and the origin of tetrapods. *Biological Reviews* 62:341-438.

Panchen, A. L., and T. R. Smithson. 1988. The relationships of the earliest tetrapods, pp. 1-32, *in* Benton, M. J., ed. *The Phylogeny and Classification of the Tetrapods, Vol. 1: Amphibians, Reptiles, Birds*. Clarendon Press, Oxford.

Schultze, H.-P., and L. Trueb, eds. 1991. *Origins of the Higher Groups of Tetrapods: Controversy and Consensus*. Cornell University Press, Ithaca, New York.

Thomson, K. S. 1993. The origin of the amphibia. *Quarterly Review of Biology* 37:189-241.

Thomson, K. S. 1994. The origin of the tetrapods, pp. 85-107, *in* Prothero, D. R., and R. M. Schoch, eds. *Major Features of Vertebrate Evolution*. Paleontological Society Short Course 7.

LAND EGGS: THE AMNIOTES

The most fundamental innovation is the evolution of another internal fluid-filled sac, the amnion, in which the embryo floats. Amniotic fluid has roughly the same composition as seawater, so that in a very real sense, the amnion is the continuation of the original fish or amphiban eggs together with its microenvironment, just as a space suit contains an astronaut and a fluid that mimics the Earth's atmosphere. All of the rest of the amniote egg is add-on technology that is also required for life in an alien environment, and in that sense it corresponds to the rest of the space station with its food storage, fuel supply, gas exchangers, and sanitary disposal systems.

Richard Cowen, *History of Life*, 1990

Amphibians have successfully invaded the land in many respects, but in one important feature they remained tied to the water. Although some can even live in deserts, all amphibians must return to the water to reproduce. Most do so by laying huge masses of tiny eggs that are vulnerable to being eaten or drying up, but enough survive and hatch to form larvae (the tadpole stage of frogs) and eventually adults. Some tree frogs push this type of reproduction to the limit by laying their eggs in damp areas glued to leaves on branches, but ultimately their reproduction is tied to a moist environment. The ability to lay an egg that does not require such restrictive conditions would give some tetrapods an advantage in exploiting new habitats.

Such an egg is known as the **amniotic** (or **cleidoic**) **egg**, and it is laid by all reptiles, birds, and even by the egg-laying mammals such as the platypus. This is one of many shared specializations that defines a group known as the **amniotes**. Instead of the hundreds of tiny, thin-shelled eggs laid by amphibians, amniotes lay fewer, larger eggs with a shell (either leathery or calcareous) that protects the egg and resists water loss while allowing limited exchange of gases through its pores. Inside the egg are numerous specialized systems (Fig. 17.36). The embryo is surrounded by

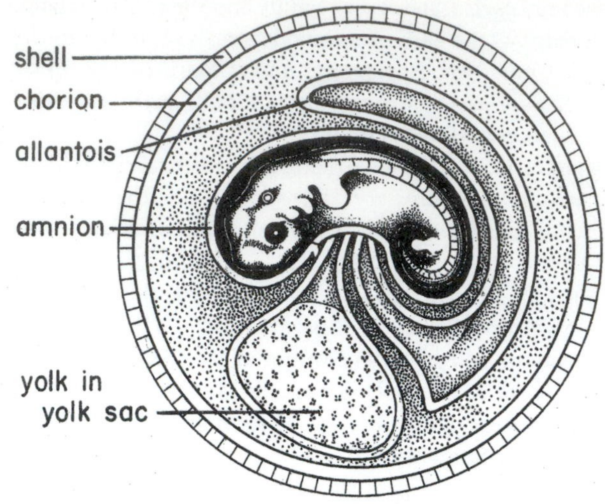

Figure 17.36. Basic structure of the amniotic egg. The entire egg is surrounded by the shell and the chorionic membrane. The embryo itself is surrounded by the amniotic membrane, and it is also supported by two other sacs: the yolk sac, which provides food, and the allantois, which stores wastes. As the embryo grows, the yolk sac shrinks, and the allantois expands. (From Romer, 1933.)

a membrane called the **amnion**, which is also filled with amniotic fluid that buffers the embryo against shock, temperature change, and other rapid, stressful fluctuations in the environment. Attached to the gut of the embryo but outside the amnion is the **yolk sac**, which provides food for the developing embryo so that it can emerge from the egg relatively self-sufficient, and not a partially developed tadpole found in the amphibians. A third sac off the hindgut of the embryo is the **allantois**, which serves for waste storage and also for respiration. As the embryo develops, the yolk sac dwindles, and the amnion and allantois get larger as their contents expand. Amphibian embryos produce waste in the form of urea, which easily dissolves away in their aquatic habitats, but amniote embryos have a limited water supply, so they secrete their nitrogenous wastes in the form of crystallized uric acid, so that it can be stored without wasting water. Finally, the amnion, yolk sac, and allantois are surrounded by the fluid albumin ("egg white") that fills the rest of the egg cavity, which in turn is completely surrounded by a membrane, the **chorion**, that lies just beneath the eggshell.

In addition to protecting the young from drying up and predation, and allowing them to hatch far from water, the amniotic egg has other implications. Each egg is more costly to produce, so fewer can be laid, and each embryo must develop much further before it hatches to ensure that some survive. In addition, the eggs cannot be fertilized by a male who simply swims nearby and sprays the floating egg mass with sperm. The amniotic egg can be produced only by **internal fertilization**. Males and females must copulate so that the female can carry the sperm to the eggs inside her, so they can begin development. Of course, this is not the first time that internal fertilization appears in the animals. Most land-living arthropods have independently developed it, as have the sharks (which are also capable of giving birth to live young).

Clearly, the amniotic egg is an important innovation, a breakthrough that allowed one group of tetrapods to exploit an entire range of new habitats. Unfortunately, it is not a feature that fossilizes very readily, and even more rarely in association with the organism that laid it. Although amniote skeletal fossils are known from the early Carboniferous, the oldest known fossil amniotic egg is Permian in age, and some paleontologists question whether it is even an egg. It is possible that some of the "anthracosaurs" (discussed above) may have laid an amniotic egg. If they are not true amniotes, they still have many of the derived features of amniotes, and so they are the closest sister-groups. For example, many of these "anthracosaurs" have lost the otic notch in the back of the skull for the eardrum (an amphibian feature), and instead have a solid skull roof in the back, with a tubular canal around the stapes for hearing. Most of these "anthracosaurs" have modified the limb girdles for a more upright, efficient form of locomotion; they were not nearly as sprawling as the huge, flat-bodied temnospondyls. They also had high-domed, deeper skulls, with a narrow snout and short region behind the eyes, in contrast to the flat-skulled temnospondyls and lepospondyls. There are numerous specializations in the wrist and ankle bones that accompany their more active locomotion. Their neck vertebrate become specialized into an atlas (the first neck vertebra, which supports the skull) and axis (the second vertebra, along which the head pivots and turns). This allowed them to swivel their heads rapidly to catch prey. Finally, the muscles and bones of the palate region were modified so that they had a much stronger bite force, rather than the relatively weak "snapping" motion of temnospondyls. Many even have fangs on the palate to secure their prey.

Various combinations of these characters have been used to recognize early amniotes, although the distinction is not very clear-cut. For example, *Solenodonsaurus* still has the otic notch, but also has the advanced jaw muscles. *Gephyrostegus* has an otic notch and primitive vertebrae and limb girdles, but has an advanced ankle region. *Limnoscelis* and *Seymouria* (Fig. 17.35C) were much more

Figure 17.37. *Westlothiana lizziae*, the earliest known amniote, from the Early Carboniferous of Scotland. (From Smithson et al., 1994.)

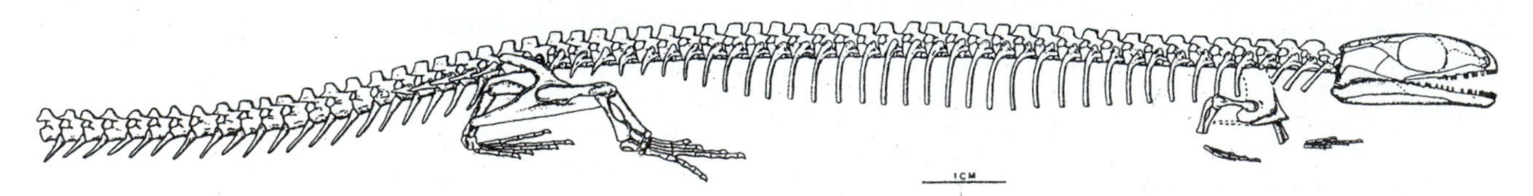

reptilian with no otic notch, but do not have the advanced ankle bones. *Diadectes*, a pig-sized herbivorous animal from the Permian, had an atlas-axis complex, strong limbs with an advanced ankle, but still had the otic notch. Most paleontologists now regard these "anthracosaurs," which are known mostly from the late Carboniferous or Permian, as extinct side branches of tetrapod evolution that show various combinations of amniote characters long after the amniotes had split off. This is because tiny, lizard-like animals such as *Westlothiana* (Smithson et al., 1994) are now known from much earlier deposits (lower Carboniferous of Scotland), so the amniote-"anthracosaur" split must have occurred long before *Solenodonsaurus, Gephyrostegus, Limnoscelis*, or *Diadectes* lived.

Westlothiana from the early Carboniferous (Fig. 17.37), and *Hylonomus* and *Paleothyris* from the middle Carboniferous, are now considered as representative of the earliest amniotes. These animals were built much like slender lizards, with long gracile limbs and toes, and a very long trunk and tail. They had relatively small heads in proportion to the their bodies, but their deep skulls had very effective jaw muscles, and so they were very effective in catching insects. Their relatively large eyes also suggest that they may have been active predators, possibly at night as well as in the day. The best specimens of *Hylonomus* were found inside the hollow trunks of giant club moss trees (lycopods) from the middle Carboniferous beds of Joggins, Nova Scotia. Traditionally, it was thought that these animals had been trapped inside half-buried rotting logs, but more recent analysis suggests that they probably lived and hunted in and around these hollow trees.

Holes in their Heads

From a relatively primitive creature like *Westlothiana*, the amniotes radiated into many distinct lineages. The traditional way of subdividing the amniotes has focused on one character, the openings on the side of the skull known as the **temporal fenestrae** (Fig. 17.38). These openings serve not only to increase the area for the attachment of the temporal muscles (which pull the jaw shut), but also allow room for the expansion of bulging jaw muscles as amniotes developed larger and stronger jaws. Four conditions of temporal fenestrae are recognized:

1) Anapsids have no temporal fenestrae, so the skull roof

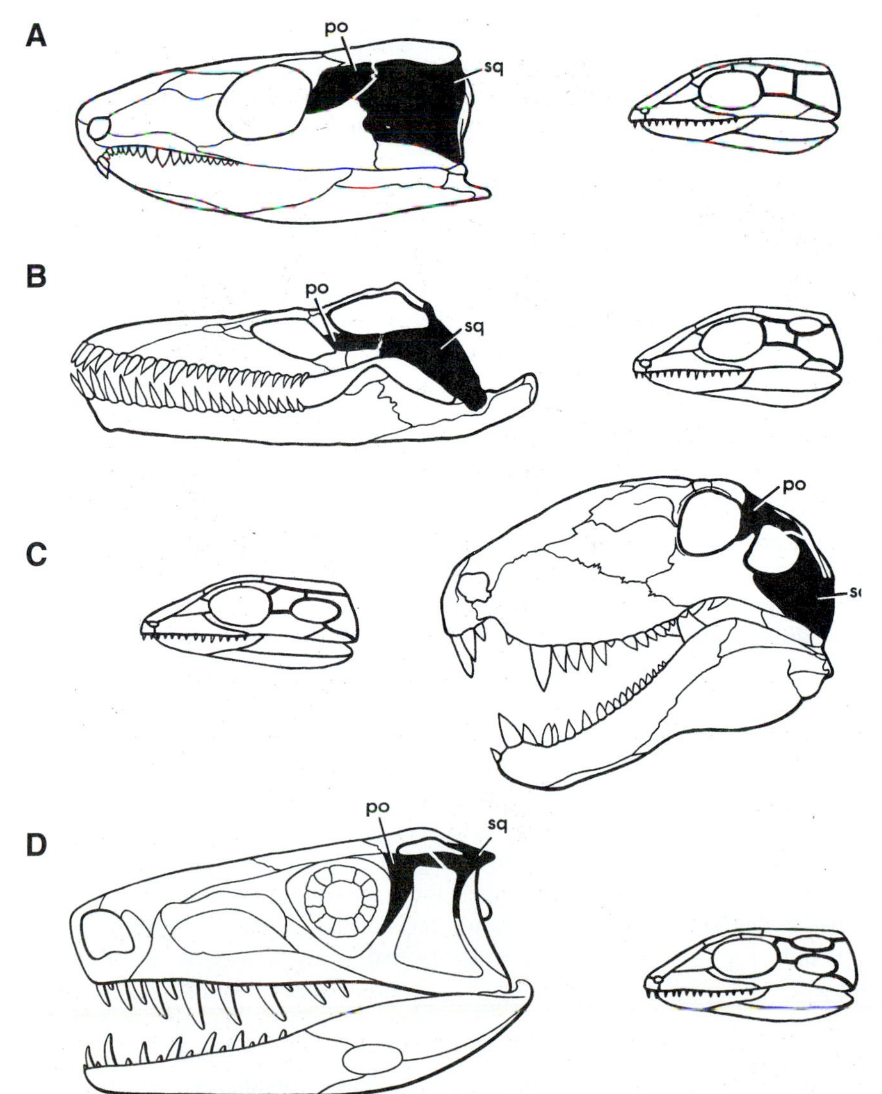

Figure 17.38. Examples of the four main patterns of temporal fenestrae in the skull roofs of amniotes. (A) The primitive anapsid condition has no temporal opening, as seen in this specimen of *Captorhinus.* It is found in most primitive amniotes and in turtles. (B) The euryapsid condition has a temporal fenestra above the postorbital (po) and squamosal (sq) bones. (These bones are shown by the bold outline in the schematic views of the skulls.) It is found in ichthyosaurs and plesiosaurs, such as this *Muraenosaurus.* (C) The synapsid condition has a temporal fenestra below the postorbital-squamosal contact. It is found in the "mammal-like" reptiles, such as this *Dimetrodon.* (D) The diapsid condition has two temporal fenestrae, one above and one below the postorbital-squamosal contact. It is found in most reptiles, as well as dinosaurs, pterosaurs, and birds, and in this primitive archosaur, *Euparkeria.* (From Colbert and Morales, 1991.)

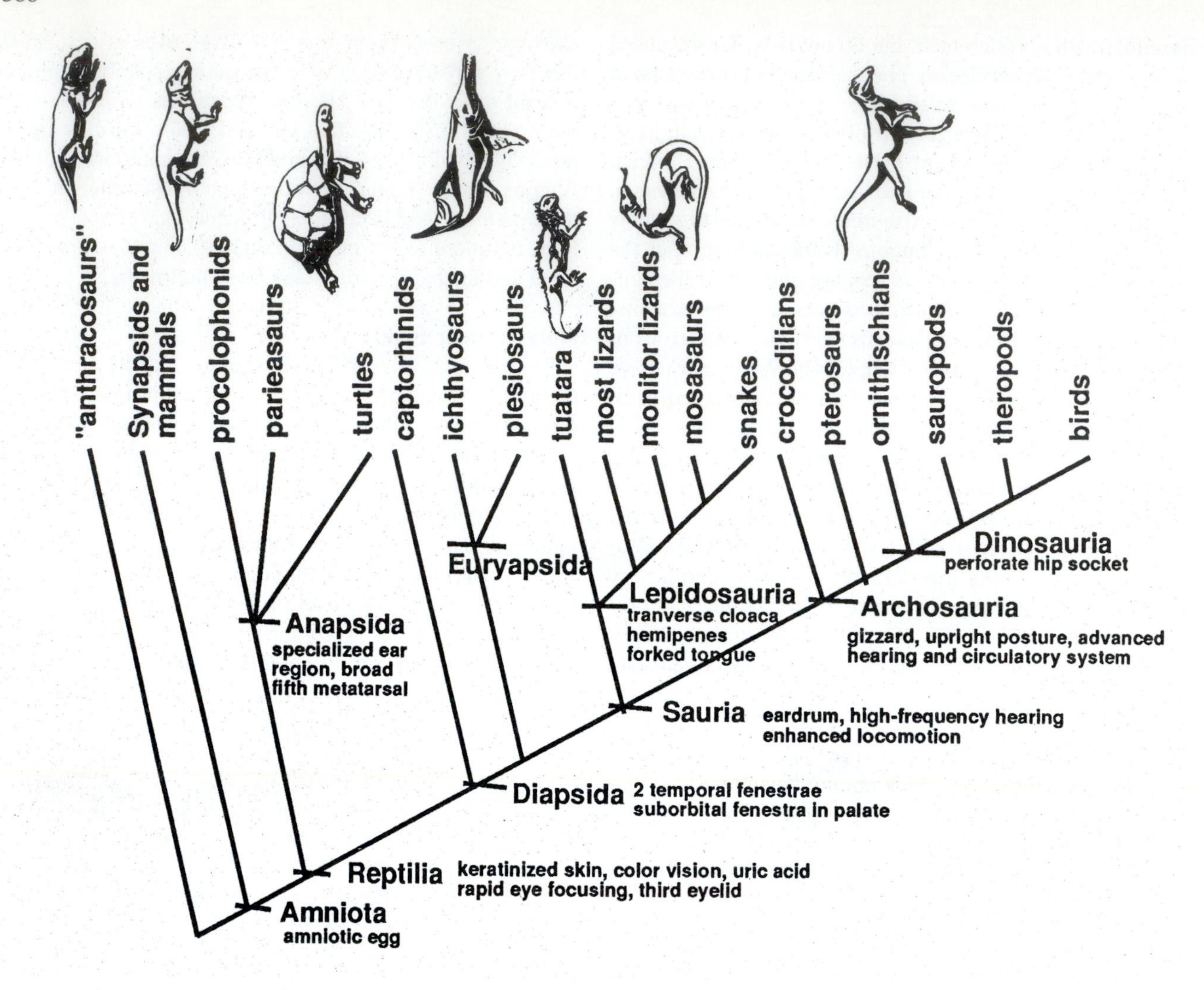

Figure 17.39. Relationships of the amniotes.

is solid (Fig. 17.38A). This condition is found in most of the primitive amniotes, such as *Westlothiana, Hylonomus*, and *Paleothyris*, as well as more advanced animals such as the aquatic *Mesosaurus*, and many other extinct amniotes of the Permian. The only living anapsids are the turtles. Clearly, the lack of a temporal fenestra is a primitive features, so anapsids (defined this way) are a primitive, paraphyletic group.

2) Synapsids have a single temporal fenestra, located relatively low on the back of the skull, below the postorbital-squamosal bones (Fig. 17.38C). This condition is unique to the "mammal-like reptiles" and their descendants, the mammals (see below).

3) Euryapsids had a single temporal fenestra, but high on the skull, above the postorbital-squamosal bones (Fig. 17.38B). This condition is restricted to the marine reptiles, primarily the ichthyosaurs and plesiosaurs. (In older books, you might read that the ichthyosaurs supposedly had yet another condition, known as a "parapsid" skull, with the temporal fenestra very high on the skull within the parietal and temporal bones. However, this interpretation was later shown to be based on the mistaken identification of bones on a badly crushed ichthyosaur skull. When less distorted ichthyosaurs were described, it became apparent that they had the same euryapsid condition as found in plesiosaurs).

4) Diapsids have two temporal fenestrae, one above and one below the postorbital-squamosal bones (Fig. 17.38D). This condition applies to most reptiles, including the lizards and snakes (lepidosaurs), the crocodiles, pterosaurs, dinosaurs, and birds (archosaurs), and a number of other extinct taxa.

The temporal fenestrae seem to define three monophyletic groups (plus the primitive, paraphyletic "anapsids"), so they have been used in amniote classification for almost a century. Relying only on these characters, however, has left open some interesting questions: were diapsids derived from synapsids that gained an upper temporal opening, or euryapsids that gained a lower temporal fenestra? In recent years, a detailed analysis of many other characters besides the temporal fenestrae has shown that the classification was too dependent upon this single character state (Gauthier et al., 1988a, 1988b; De Queiroz and

Gauthier, 1992; Gauthier, 1994; Laurin and Reisz, 1994, 1996). For example, a comprehensive cladistic analysis (Fig. 17.39) of all the character states shows that the euryapsids are a clade within the diapsids (apparently they have secondarily lost their lower temporal fenestra), and that many of the supposed "anapsids" (such as *Hylonomus, Paleothyris*, and *Protorothyris*) also fall within the diapsids.

These analyses suggest that the emphasis on this single character of the temporal region, and the attempt to derive each group from a primitive "anapsid" ancestor completely ignored the wide variety of lines of evidence that give a completely different phylogenetic geometry from that traditionally recognized. In the modern, cladistic scheme of relationships, the fundamental split within the amniotes is between the synapsids and the reptiles. Both the earliest synapsids (*Protoclepsydrops* from the early Carboniferous and *Archaeothyris* from the middle Carboniferous) and the earliest reptiles (*Westlothiana, Hylonomus*, and *Paleothyris*) are known from deposits of about the same age, suggesting that synapsids and reptiles are separate, contemporaneous branches, not ancestors and descendants. (The earliest animal widely regarded as a diapsid, *Petrolacosaurus*, is known from the late Carboniferous.) This pattern of relationships is supported by a wide variety of other characters. Synapsids are *not* "mammal-like reptiles," but a separate clade that split off very early from the branch that led to the groups we recognize as reptiles (turtles, lizards, snakes, crocodiles, and their extinct relatives). Synapsids never had anything to do with reptiles in any sense of the word, and should never be called "mammal-like reptiles." We will discuss synapsids below.

Once we move past the synapsid branch point (Fig. 17.39), we find that the reptiles can be defined as a natural monophyletic group, with many shared derived characters (Gauthier, 1994; Laurin and Reisz, 1996), including scaly, keratinized skin, color vision, rapid eye focusing, a third eyelid, and nitrogenous wastes secreted as uric acid, not urea. The reptiles next split into two major branches: the anapsids and diapsids. As defined by Gauthier and others (1988a, 1988b; Gauthier, 1994), the anapsids are defined not by their primitive lack of temporal fenestrae, but by their highly specialized ear region, with the ear drum supported by the squamosal bone, rather than by the quadrate as found in other reptiles. The anapsid foot also has a number of unique features, including a broadly sloping fifth metatarsal (the foot bone supporting smallest toe on your foot).

The extinct anapsid groups included the huge, sprawling **pareiasaurs**, which were the size of a small hippo (Fig. 17.40A), and were covered with bony armored knobs all over their bodies. Their massive, warty skulls were marked by bony protuberances that pointed downward from the cheek bones and lower jaw. These monsters were among the largest herbivores of the Late Permian, especially in Russia, China, and South Africa, where they faced bear-sized predatory synapsids. The most familiar anapsids are the turtles, which first appeared in the Late Triassic with all the features we associate with turtles. The earliest turtle, *Proganochelys*, already has most of the specializations of the skull and skeleton, including ribs that are completely fused into the shell, and the toothless beak, the deeply curved quadrate to support the eardrum, and other turtle specializations in the skull. Turtles have never developed temporal fenestrae like other reptiles, so they evolved a deep notch in the back of the skull roof through which their jaw muscles pass, allowing them to achieve a strong bite force in a completely different fashion from other reptiles.

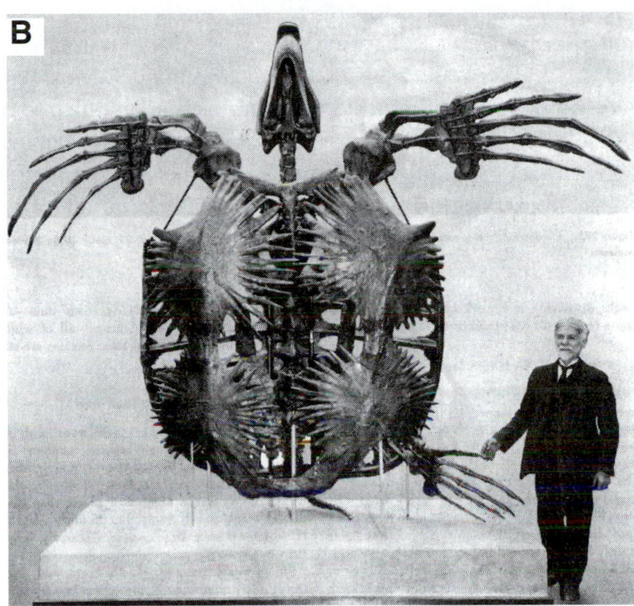

Figure 17.40. Some extinct anapsids. (A) The huge (2.4 m, or 8 feet long) pariesaur *Bradysaurus* from the Middle Permian of South Africa. (From Fenton and Fenton, 1958.) (B) The giant sea turtle *Archelon* from the Cretaceous chalk beds of Kansas. (Photo courtesy Yale Peabody Museum.)

In the Jurassic, two major lineages of turtles appeared. Most of the familiar groups of turtles, including most of the aquatic turtles and land tortoises of the northern hemisphere, are **cryptodires** ("hidden neck" in Greek). They have a number of specializations, but the most distinctive is a neck that can fold up in a vertical plane within the shell to allow the head to be pulled inside. The giant tortoises of the Galápagos Islands and of the islands of the Indian

Ocean are highly specialized cryptodires, as was *Meiolania*, a giant turtle from the Pleistocene islands north of New Zealand, which had a spike-covered head and spiky tail club. The largest cryptodires, however, are the sea turtles, which today can reach up to 2 m in length and weigh almost a ton. In the Cretaceous seas, there were even larger sea turtles, such as the giant *Archelon*, which was over 3 m long and 2 m across (Fig. 17.40B).

The other major group, the **pleurodires** ("side neck" in Greek), or "side-necked turtles," fold their necks in a horizontal S-shaped curve so their head is pulled sideways under the lip of their shell for protection. The three living genera are now restricted to the southern continents (Africa, Madagascar, Australia, and South America), apparently a relict Gondwana distribution. During the Mesozoic, they had a worldwide distribution, although they were never as diverse as cryptodires. The most spectacular pleurodire, *Stupendemys*, from the Pliocene of Venezuela, was almost 3 m long and may have weighed several tons.

Two Holes: The Diapsids

The diapsids are also an easy group to recognize. Not only do most of them have two temporal fenestrae in their skulls, but they also have an additional opening, the suborbital fenestra, in the palate region just below the eye. In addition, early diapsids have relatively long and slender hands and feet, with the bases of the wrist and toe bones overlapping extensively, a feature not seen in anapsids nor synapsids. For this reason, some of the earliest reptiles that had an "anapsid" skull (such as the captorhinids, and *Hylonomus, Paleothyris*, and *Protorothyris*) are now considered diapsids (Heaton and Reisz, 1986; de Queiroz and Gauthier, 1992; Gauthier, 1994).

Diapsids split into two major clades: the euryapsids and the saurians (Fig. 17.39). Euryapsids are first known from *Claudiosaurus* from the Permian of Madagascar, and during the Mesozoic, they radiated into two great groups of marine reptiles: the dolphin-like ichthyosaurs, and the paddling plesiosaurs. The ichthyosaurs (Figs. 1.9B) are one of the most amazing examples of evolutionary convergence ever found in the fossil record. Their bodies are highly streamlined like that of a fish or whale, except that ichthyosaurs were clearly reptiles that had secondarily gone back to the sea. Their tail is supported by a vertebral column that bends downward, unlike the tail of fish or sharks, or the horizontal flukes of whales and dolphins. The ichthyosaur head merged smoothly with the body with no neck, and had a long slender snout with many conical teeth for catching fish or squid, and most had large eyes for seeing in dark, murky waters. They had a prominent dorsal fin, and their front and hind limbs were modified into flippers, with the finger bones multiplying into nearly a hundred tiny elements. Most ichthyosaurs were a few meters in length, although the whale-sized *Shonisaurus* from the Triassic of Nevada was over 15 m long. Extraordinary specimens from the Jurassic Holzmaden Shales of Germany preserve their body outlines as a thin black film. One extraordinary specimen shows a young ichthyosaur emerging from the birth canal, the same mode of live birth employed by whales and dolphins (Fig. 8.10A). Like whales and dolphins, they were not capable of dragging their bodies onto the land to lay eggs, so live birth was a necessity in such completely aquatic amniotes.

The other major group of euryapsids was the plesiosaurs, which became specialized for marine life in a completely different manner from ichthyosaurs. Instead of the fusiform, fish-like body for fast swimming, plesiosaurs have a broad, heavy body with large paddles. Although they were not fast swimmers, they are thought to have been steady "subaqueous flyers" which could cruise underwater for long distances. Their lack of speed in some in some taxa was compensated for by a long, serpentine neck (with as many as 76 vertebrae) that allowed them to quickly reach in almost any direction and catch fish and squid with their long, protruding teeth. Some of these long-necked plesiosaurs, such as *Elasmosaurus*, reached 10 m in length, almost half of which was neck. Another group, the pliosaurs, had relatively short necks but long beaks, so they had to catch prey by swimming up to it and then snapping with their long tooth-filled snouts. The giant of all the short-necked plesiosaurs was the Australian *Kronosaurus*, which had a skull almost 4 m long, and a body over 12 m long. The most primitive plesiosaurs are the Triassic nothosaurs, an excellent transitional form to the highly specialized, long-necked animals of the Jurassic and Cretaceous. The alleged "Loch Ness monster" is usually thought to be a plesiosaur that has survived since the end of the Cretaceous, but there are many problems with this idea. For one thing, Loch Ness is a glacial fjord that was completely under ice until about 10,000 years ago, so "Nessie" could not have lived there since the Mesozoic. In addition, there would have to be a population of "Nessies" for these animals to persist for even thousands of years. If there is truly a population of air-breathing monsters, we should have had at least one reliable sighting by now. Recently, most of the best photographs of "Nessie" have been revealed to be hoaxes, so the credibility of this "fish tale" has vanished to zero.

Lepidosaurs: The "Scaly Reptiles"

Besides the euryapsids, the other major branch of the diapsids is the Sauria, which is divided into two large, well-defined groups: the **lepidosaurs** and the **archosaurs**. The lepidosaurs include the lizards, snakes, and the tuatara (Fig. 17.39). There are now over 6000 species of lizards, and over 2500 species of snakes (compared to only about 250 species of turtles and two dozen species of crocodilians), so most of the living reptiles (except birds) are lepidosaurs. At least 55 synapomorphies (Gauthier, 1994) define the lepidosaurs, including scales made out of layers of two proteins (α and ϕ keratin). In lepidosaurs, the skin is shed all at once, rather than piecemeal as in crocodilians or turtles, and the urogenital opening (the cloaca) is a slit oriented transversely across the body. During copulation, the male

Figure 17.41. The tuatara *Sphenodon,* a lepidosaurian reptile now found only in New Zealand. Although it looks superficially like a lizard, it is a living fossil. (After Lydekker.)

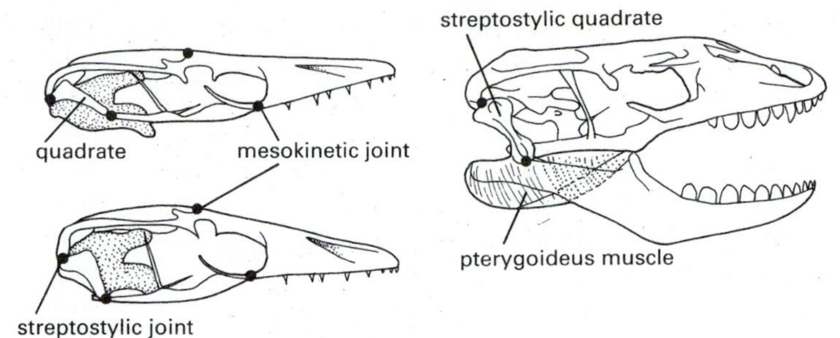

Figure 17.42. Squamate jaw mechanics. Skull of the monitor lizard (upper left) showing the motion of the streptostylic quadrate. In the upper right is a lizard skull with the jaws open, so the pterygoideus muscles have maximum leverage. (From Benton, 1990.)

snake or lizard inserts one of a pair of hemipenes into the female's cloaca to transfer sperm. When not in use, the hemipenes shrivel and turn inside out. One of the most distinctive features found in the lepidosaurs is the forked tongue, which carries smells to the vomeronasal organ on the roof of the mouth. In the low-budget dinosaur movies, they use close-ups of lizards for dinosaurs, or reconstruct dinosaur models with forked tongues, but no dinosaurs (which are archosaurs) ever had a forked tongue.

The most archaic of the living lepidosaurs is the tuatara, *Sphenodon*, a lizard-like "living fossil" now restricted to a few islands off the coast of New Zealand (Fig. 17.41). Today, the tuatara hides in its remote island habitat, catching invertebrates and crushing them with rows of palatal teeth. Both fossil and living sphenodontids are easy to recognize by their hooked beak on the front of the skull, and by their triangular marginal teeth that are fused to the edge of the jaw. Sphenodontids were quite diverse in the Triassic (at least nine genera in Britain alone), and were found widely throughout the world during the Mesozoic. Today, the sole surviving genus straggles on, its niche on most continents occupied by the lizards.

Once the sphendontids are excluded (Fig. 17.39), the rest of the lepidosaurs consist of a large monophyletic group, the **squamates** (lizards and snakes). Their most distinctive feature is loss of the bony bar across the base of the lower temporal opening (Fig. 17.42), allowing the quadrate bone in the jaw hinge to flex back and forth (**streptostyly**). This jaw hinge flexibility allows them to open their mouth very wide to swallow large prey. In many squamates, there are additional hinges in other parts of the skull, making the skull bones highly moveable, or kinetic. In the snakes, the bones of the skull are reduced to thin struts, and the sutures between the bones are joined only by tendons, so the entire skull is capable of tremendous flexion and expansion. If you have ever seen a snake stretching its entire head around a much larger prey item, you have seen evidence of the enormous flexibility of some squamate skulls. Squamates are also distinctive in that their teeth are attached to the inside of the jaw (**pleurodont**), rather than emerging from sockets on the edge of the jaw (thecodont, as in archosaurs).

Some paleontologists have attributed lizard-like Permian fossils, such as *Paliguana* or *Palaeagama*, to the lizards, but they are probably sister-taxa to the lepidosaurs rather than squamates. The oldest-known unquestioned squamate fossils come from the Late Jurassic, long after they must have diverged from their common ancestor with the Triassic sphendontids. When they first appear in the Late Jurassic, squamates have already differentiated into six main lines, including the four main lineages of living lizards (the geckos; the iguanids and chameleons; the skinks and lacertids; and the monitor lizards and their relatives), plus two other groups: the amphisbaenids (blind, burrowing, limbless lizard-like reptiles) and the snakes. A transitional fossil between snakes and lizards has just been described from middle Cretaceous marine beds of the Middle East (Caldwell and Lee, 1997). Although it is fully snake-like in its body and skull, *Pachyrhachis* still had well-developed hindlimbs (but no forelimbs), and apparently evolved its serpentine body for swimming in the ocean. This specimen also corroborated the long-held hypothesis that snakes are just highly modified varanid lizards, related to the monitor lizards, goannas, and Komodo dragons of Australia and southeast Asia. One extinct monitor lizard from the Pleistocene of Australia, *Megalania*, was truly gigantic, reaching 6 m in length (compared to 3 m for the largest Komodo dragon).

The most spectacularly modified of all the monitor lizards, however, was a group that returned to the sea: the **mosasaurs**. These "sea serpents" had a typically varanid skull (complete with a hinge in the lower jaw, and streptostylic jaw hinge), yet their bodies were almost completely modified for marine life. The feet are paddles, and their tail has expanded into a vertical fin. Over 20 genera are known from the Cretaceous seaways, some of which were over 10 m in length. Most had rows of sharp, conical teeth for catching fish and squid, and numerous ammonite specimens have been reported with a V-shaped row of circular bite marks, almost certainly inflicted by a mosasaur (see Fig. 8.10C). The seas of the Cretaceous had a wide variety of large marine reptiles, all derived from different roots within the amniotes. Not only were there mosasaurs (modified varanid lizards), but also the plesiosaurs and

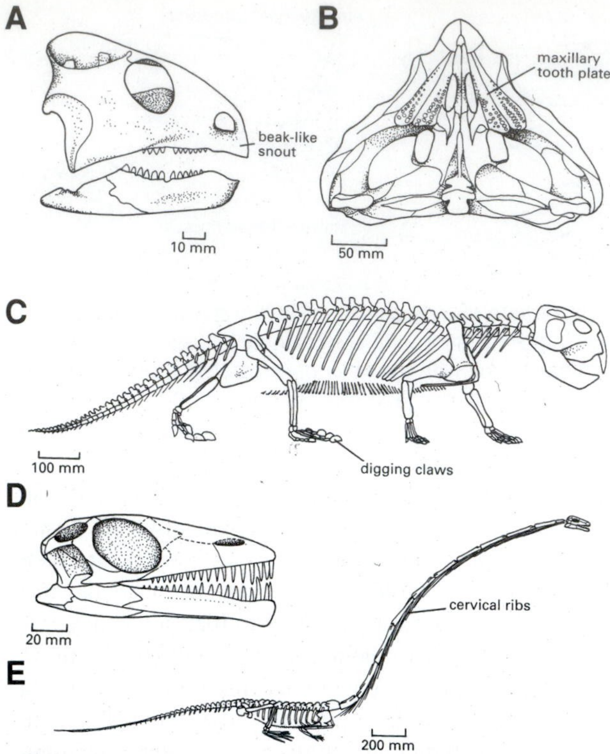

Figure 17.43. Some primitive Triassic archosauromorphs (formerly called "thecodonts".) (A) *Trilophosaurus,* with its peculiar teeth with cross-crests. (B-C) The pig-sized beaked rhynchosaur *Hyperodapedon.* (D-E). Skull and skeleton of the long-necked *Tanystropheus.* (From Benton, 1990.)

ichthyosaurs (both euryapsids), marine turtles, and even fully marine crocodiles (which are archosaurs).

Archosaurs: The "Ruling Reptiles"

The other major branch of the Sauria (besides the lepidosaurs) is the Archosauria, or "ruling reptiles" (Fig. 17.39). Today, they include the crocodilians and the birds (with over 9000 species, they are still a major group), but in the past they also included the dinosaurs, the "flying reptiles" or pterosaurs, and many other extinct groups. Archosaurs (especially dinosaurs and their primitive Triassic sister-taxa) completely "ruled" the world in the Mesozoic. Except for bats, all flying vertebrates (birds and pterosaurs) are archosaurs. Like the lepidosaurs, the archosaurs are a well-defined group with over 75 syapomorphies (Gauthier et al., 1988; Gauthier, 1994). Their most obvious features are related to their more upright posture and more active lifestyles. For example, their limbs are partly or completely under the axis of the body and move in a fore-and-aft plane, and they have a stiffened spine so that they run more efficiently without the side-to-side wiggles of lizards or snakes. Archosaurs have relatively large lungs that are pumped by a diaphragm muscle, rather than the inefficient rib pumping of other amniotes. Archosaurs have completely divided ventricles in the heart, a pulmonary artery with three semilunar valves, and muscular lateral valves to the right of the auriculoventricular orifice. They are also unique in having a muscular gizzard filled with rocks or sand for grinding up their food (in the absence of grinding teeth and chewing abilities). Since many of these features do not fossilize, extinct archosaurs are recognized by a variety of skeletal features, including an extra hole in the skull in front of the eye (the **antorbital fenestra**), another in the jaw (**mandibular fenestra**), and an eye opening shaped like an inverted triangle. Archosaur teeth are inserted in sockets along the rim of the jaw (**thecodont** teeth), and their thigh bone (femur) has an extra knob, called the fourth trochanter, for additional leg muscles that help hold them upright. There are also many unique specializations in the foot, particularly in the upper row of ankle bones.

The earliest archosauromorphs, the **protorosaurs**, are known from the Upper Permian of Europe, and persisted into the Triassic. The most bizarre of these primitive archosaurs was the long-necked *Tanystropheus*, which had a neck almost 2 m long and the rest of the body less than 1 m in length (Fig. 17.43D-E). Its neck was composed of a few elongate vertebrae, so it was not very flexible or sinuous, but must have been more like a stiff fishing rod. Nobody knows what these strange animals did with their odd necks, but they apparently didn't have the mobility to snap them sideways to catch mobile prey—yet their teeth are simple pegs for catching prey, not for eating immobile plants or other hard food.

In the Triassic, there was a great variety of other primitive archosaurs roaming the landscape around the world in addition to the protorosaurs. Most of these archaic archosaurs were once lumped in a "wastebasket" taxon known as the "thecodonts," because they all share the primitive archosaurian feature of thecodont teeth. However, this wastebasket group has outlived its usefulness, and in the last few years, it has been broken up as the relationships of archosaurs are becoming clear (Fig. 17.43B-C). The **rhynchosaurs**, with fat bodies and short limbs, and a hooked beak on a short, flattened snout, were the commonest herbivorous animals of the Triassic, and must have been prey for many different large carnivorous synapsids and archosaurs. After the divergence of protorosaurs and rhynchosaurs, the remaining archosaurs can be divided into two natural groups (Fig. 17.39): the crocodiles and their relatives (**Pseudosuchia**), and the pterosaurs, dinosaurs, birds, and their relatives (**Ornithosuchia**).

The pseudosuchians include an interesting variety of Triassic archosaurs that once dominated the landscape (Fig. 17.44). Several of these were among the largest predators on land before the dinosaurs arrived. One group, the **phytosaurs** (their name means "plant lizard," a misnomer in view of their crocodile-like teeth), were built much like crocodiles long before crocodiles had evolved (Fig. 17.44A). However, phytosaurs have many skeletal differences from crocodilians, including a nostril on the top of their head (rather than at the tip of the snout, as in crocodiles), so this similarity is mostly due to convergence.

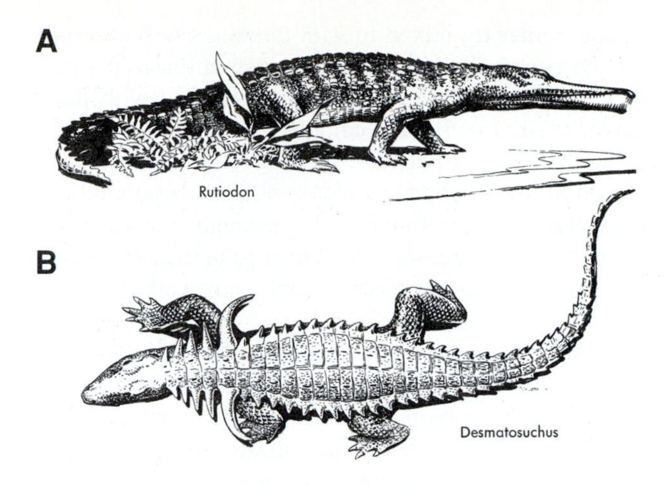

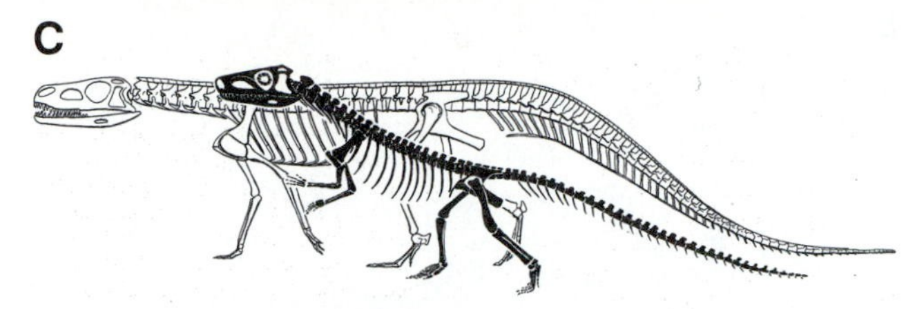

Figure 17.44. Primitive archosaurs ("thecodonts") of the Triassic. (A) The crocodile-like phytosaur *Rutiodon.* (B) The armored aetosaur *Desmatosuchus.* (C) The large predatory erythrosuchid *Ticinosuchus* (background) and the bipedal ornithosuchian *Euparkeria* (foreground). (From Colbert and Morales, 1991.)

Another group of predators was the long-bodied, deep-skulled animals known as **erythrosuchids** ("bloody crocodiles"), which reached up to 5 m in length (Fig. 17.44C). Not all pseudosuchians were predators. The armored **aetosaurs** had tiny leaf-shaped teeth for eating plants, and some had long spikes protruding from the edge of their armor plating along their back, especially in the shoulder region (Fig. 17.44B).

Crocodilians were also present in the Late Triassic among all these other pseudosuchians, but the earliest crocodilians (such as *Protosuchus*) were delicately built animals with relatively long, slender legs and short snouts—the niche for the large-bodied, short-limbed body form we recognize as crocodilian today was still occupied by phytosaurs in the Triassic. By the Jurassic, phytosaurs were extinct, and crocodilians began to diversify in earnest. More than 150 fossil genera are known, and in the Mesozoic, they ranged from fully marine forms with a tail fin and webbed feet (the geosaurs and metriorhynchids), to forms with tall skulls and narrow, laterally compressed snouts and teeth (*Sebecus*), to the gigantic Cretaceous crocodile *Deinosuchus* ("terror crocodile"), which reached 15 m in length, and had a skull 2 m long! It was large enough to eat dinosaurs. Crocodilians continued to be among the largest predators on land during the early Cenozoic, with some known as far north as the Arctic Circle in the early Eocene. However, they are very sensitive to cold climates, and with the cooling since the Oligocene, crocodilians have become restricted to tropical to warm temperate latitudes.

The other major branch of the archosaurs is the ornithosuchians (Fig. 17.39). In contrast to the mostly quadrupedal, sprawling pseudosuchians, most of the early ornithosuchians were bipedal, and had an upright posture even when they secondarily returned to quadrupedal locomotion (as happened with the gigantic dinosaurs). The earliest Triassic ornithosuchians include agile, bipedal forms such as *Euparkeria* (Fig. 17.44C), which looked like a small predatory dinosaur. Others, such as *Lagosuchus* ("rabbit crocodile") had very long slender hindlimbs that suggested they might have hopped. These Middle Triassic forms eventually were followed in the Late Triassic by two major ornithosuchian clades, the pterosaurs and the dinosaurs. Both of these advanced groups are united by a very distinctive feature in their ankles, known as the **mesotarsal joint**. Instead of the usual hinge between the shin bone (tibia) and the ankle bones, the pterosaurs and dinosaurs have their foot hinged between the first and second row of ankle bones (Fig. 17.46). As a consequence, the first row of ankle bones is usually fused to the end of the tibia, and no longer functions as a separate series of ankle elements. The next time you eat a chicken or turkey drumstick, notice the caps of cartilage on the end of the bone. These are the remnants of the first row of ankle bones, now fused to the tibia ("drumstick") as a cap of cartilage.

Primitive pterosaurs such as *Eudimorphodon* first are known in the Late Triassic with fully developed wings for flying (Fig. 17.45). Unlike other flying animals, pterosaurs supported their wing membranes with a single elongate finger bone (the fourth, or "ring" finger), along with stiffened fibrous rods that held the wing membrane out. By contrast, bats use all five fingers to support their wings, and birds have fused all their fingers together, shaping and supporting their wing with feather shafts instead. Pterosaurs had many other specializations for flight as well. Their bones were very light and hollow to minimize their weight. The breastbone was broadly expanded for the attachment of flight muscles, and the pelvis was fused with the ribs in the hip region of the spine into a single bone called the **synsacrum** (the same thing occurred in a different fashion in birds). As we saw in Chapter 8, there is considerable debate about the relative flying and soaring abilities of pterosaurs, and whether their wing membranes connected to their hind feet, but there is no longer any doubt that they had true powered flight.

In the Jurassic, there were a wide variety of pterosaurs, including the many small bird-sized forms like *Pterodactylus* and *Rhamphorhynchus* found in the Upper Jurassic Solnhofen Limestone (Figs. 1.14A, 17.45A). In the Cretaceous, pterosaurs reached the pinnacle of their diversity and evolution. *Pterodaustro* (from the Upper Cretaceous of Argentina—Fig. 17.45F) had a long curved beak with hundreds of tiny strainer teeth, like the baleen of whales. *Pteranodon* had 7-m wingspan, and a long crest on

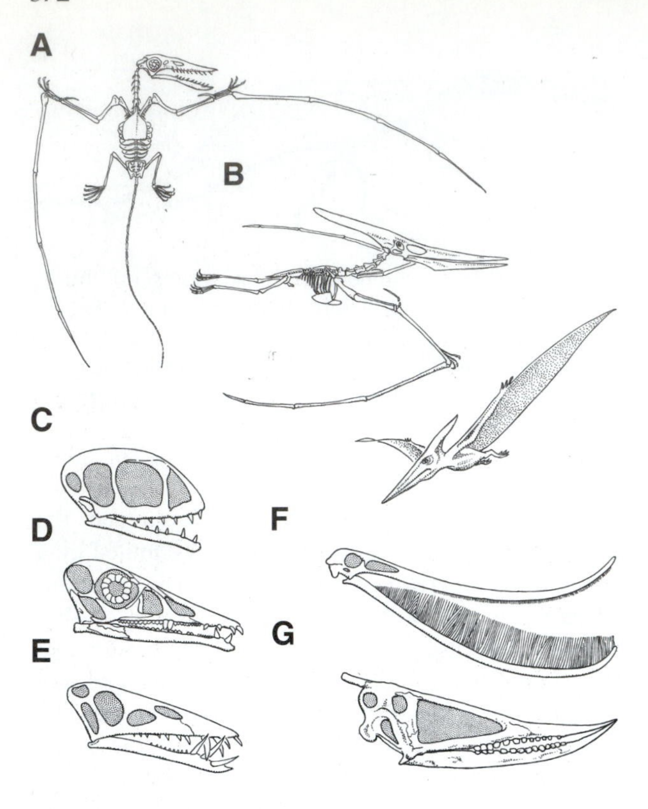

Figure 17.45. Pterosaurs were highly diverse in shape and size. (A) *Rhamphorhynchus* from the Upper Jurassic Solnhofen limestone was sparrow sized, and had a long tail with a vane at the tip. (B) *Pteranodon* from the Cretaceous chalk of Kansas had no tail, but a long crest. (C) *Anurognathus* may have been insectivorous. (D) The most primitive pterosaur was *Eudimorphodon.* (E) *Dorygnathus* had teeth for eating fish. (F) *Pterodaustro* had a comb-like array of teeth for sieving plankton. (G) *Dsungaripterus* may have pulled molluscs from the rock with its can-opener beak and then crushed them with its molariform teeth. (From Pough et al., 1996.)

the back of its head (Fig. 17.45B). It is best known from the Cretaceous marine rocks of Kansas, so it presumably soared over the sea, plucking fish from the surface waters. The largest of all, however, was the giant Texas pterosaur, *Quetzalcoatlus*, with an 11- to 15-m wingspan; it was as large as small airplane. More than 90 species of pterosaurs are known from the Mesozoic, and considering how delicate their bones are, and how rarely they fossilize, this suggests that they were an extremely successful group of flying animals. If they had not vanished during the K/T event, they might still be flying overhead and competing with the birds today.

Dinosaurs arose in the Late Triassic, and quickly came to dominate the terrestrial realm, driving the more primitive archosaurs (such as the phytosaurs and erythrosuchids) and the large, carnivorous synapsids into extinction. Most people are familiar with many of the common dinosaurs thanks to popular culture, but a true scientific definition is not so simple. Pterosaurs were not dinosaurs, nor were synapsids such as the finback *Dimetrodon* (even though they are frequently mixed in with dinosaurs in books and toys). Dinosaurs are defined by a number of shared derived characters. Most of these features have to do with their posture. All dinosaurs have completely vertical limbs supporting their bodies (whether bipedal or quadrupedal), so the head of the thigh bone had a right angle bend, and the hip socket is perforated right through the middle. Primitively, dinosaurs walked on the tips of their toes (**digitigrade**), although the massive sauropods walked on the soles of their feet (plantigrade). There are other specializatons of the ankle not seen in any other group of vertebrates as well. In short, dinosaurs are not diagnosed by their large size, or inclusion in popular books, movies, and merchandise, but by a suite of shared specializations.

Two main groups of dinosaurs are recognized. One group, the Saurischia, or "lizard-hipped" dinosaurs, had the primitive condition of the pelvis, with the pubic bone pointing forward (Fig. 17.46). Although this primitive character state is not enough to define the group, there are many derived characters that unite the Saurischia, exclusive of the pelvis. The most obvious of these is the elongate, S-shaped neck, and the asymmetrical hand with a distinct thumb, although there are at least nine other synapomorphies in the hand, skull, and postcranial skeleton (Gauthier, 1996). Saurischians include the huge, long-necked sauropods (familiar from animals such as "*Brontosaurus*" or *Apatosaurus*, *Brachiosaurus*, and *Diplodocus*), and all the predatory dinosaurs, or theropods (including the ostrich-mimic dinosaurs, *Allosaurus, Velociraptor, Deinonychus*, and *Tyrannosaurus*, among others).

The other main group, the Ornithischia, or "bird-hipped" dinosaurs, include a wide variety of herbivorous forms: the duckbills, iguanodonts, stegosaurs, armored ankylosaurs, bone-headed pachycephalosaurs, and the horned ceratopsians. They are called "bird-hipped" because the pubic bone of the pelvis has a backward extension that runs along the base of the ischium, a condition similar to that seen in birds. However, this is misleading, since it only superficially resembles the avian condition, and besides, birds are saurischian dinosaurs. Ornithischians are diagnosed by many other characters, including an extra bone on the front of the lower jaw, the predentary bone (an old name for ornithischians is "predentates"), and cheek teeth that are inset into the jaw, suggesting that they had fleshy cheeks to hold their food in while they chewed. (Most saurisichians had a wide gape with no evidence of cheeks.)

Dinosaurs have fascinated people since they were first discovered in the early 1800s, but in recent years the interest in dinosaurs has reached a fever pitch, both in the popular culture and in professional circles. Part of this excitement has been new theoretical ideas about the dinosaurs that arose in the 1960s, especially the idea that dinosaurs might have been "warm-blooded." This notion was first suggested by John Ostrom, but zealously promoted by his former student Robert Bakker, who has made a career out of being a "dinosaur heretic." When Bakker first presented

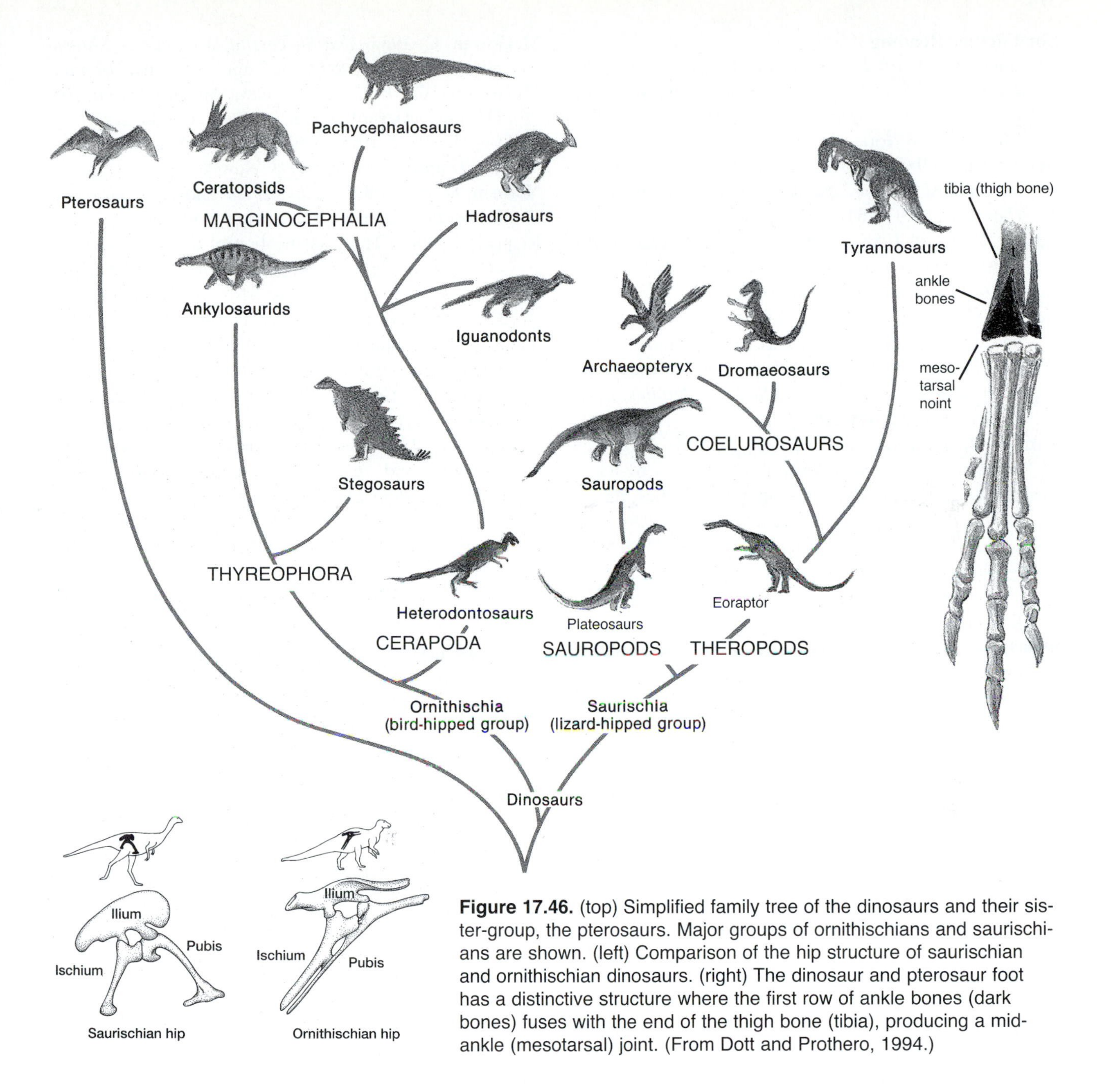

Figure 17.46. (top) Simplified family tree of the dinosaurs and their sister-group, the pterosaurs. Major groups of ornithischians and saurischians are shown. (left) Comparison of the hip structure of saurischian and ornithischian dinosaurs. (right) The dinosaur and pterosaur foot has a distinctive structure where the first row of ankle bones (dark bones) fuses with the end of the thigh bone (tibia), producing a mid-ankle (mesotarsal) joint. (From Dott and Prothero, 1994.)

the evidence for dinosaur **endothermy** (body heat generated by internal metabolism, not from the environment), it seemed impressive, but since the late 1970s, most of the evidence has been shown to be ambiguous. To make a long story short, it seems likely that the smaller dinosaurs and pterosaurs were very active animals in order to run and fly, and probably were endothermic. These animals have a large surface area relative to their small volume, and they would lose heat at a very rapid rate unless they had some kind of insulating covering, such as fur or feathers. By contrast, the largest dinosaurs have the opposite problem. As body size increases, the surface area increases only as a square, but the volume increases as a cube, so large animals have too little surface area for their volume. Consequently, large terrestrial animals (such as elephants) have a problem getting rid of heat. Elephants manage by flapping their ears (which are fully of blood vessels, and are primarily for dumping heat) and by bathing frequently. Camels can allow their body temperature to fluctuate through the day, so that they are cold in the morning, and take a long time to warm up in the desert, and then slowly cool down at night. This kind of thermal inertia means that large dinosaurs must have had a constant body temperature by virtue of their body size alone (**inertial homeothermy**), and would not have needed any special regulatory mechanism or internal source of heat in the warm climates of the Mesozoic. Indeed, if they had been endothermic, they would have had problems getting rid of their excess body heat.

For Further Reading

Alexander, R. M. 1989. *Dynamics of Dinosaurs and other Extinct Giants*. Columbia University Press, New York.

Bakker, R. T. 1986. *The Dinosaur Heresies*. William Morrow, New York.

Benton, M. J. 1985. Classification and phylogeny of diapsid reptiles. *Zoological Journal of the Linnean Society (London)*. 84:97-164.

Benton, M. J., ed. 1988. *The Phylogeny and Classification of the Tetrapods. vol. 1: Amphibians, Reptiles, Birds*. Oxford Clarendon Press, Oxford.

Benton, M. J., and J. Clark. 1988. Archosaur phylogeny and the relationships of the Crocodylia, pp. 295-338, *in* M. J. Benton, ed. *The Phylogeny and Classification of the Tetrapods, vol. 1: Amphibians, Reptiles, Birds*. Clarendon Press, Oxford.

Carroll, R. L. 1964. The earliest reptiles. *Zoological Journal of the Linnean Society (London)* 45:61-83.

Carroll, R. L. 1982. Early evolution of the reptiles. *Annual Reviews of Ecology and Systematics* 13:87-109.

Estes, R., and G. K. Pregill, eds. 1988. *The Phylogenetic Relationships of the Lizard Families: Essays Commemorating Charles L. Camp*. Stanford University Press, Palo Alto, Calif.

Evans, S. E. 1988. The early history and relationships of the Diapsida, pp. 221-260, *in* Benton, M. J., ed. *The Phylogeny and Classification of the Tetrapods, vol. 1: Amphibians, Reptiles, Birds*. Clarendon Press, Oxford.

Fastovsky, D. E., and D. B. Weishampel. 1996. *The Evolution and Extinction of the Dinosaurs*. Cambridge University Press, Cambridge.

Gaffney, E. S. 1975. A phylogeny and classification of the higher categories of turtles. *Bulletin of the American Museum of Natural History* 155:387-436.

Gauthier, J. 1986. Saurischian monophyly and the origin of birds. *Memoirs of the California Academy of Sciences* 8: 1-55.

Gauthier, J. 1994. The diversification of the amniotes, pp. 129-159, *in* Prothero, D. R., and R. M. Schoch, eds. *Major Features of Vertebrate Evolution*. Paleontological Society Short Course 7.

Gauthier, J., A. G. Kluge, and T. Rowe. 1988a. Amniote phylogeny and the importance of fossils. *Cladistics* 4:105-209.

Gauthier, J., A. G. Kluge, and T. Rowe. 1988b. The early evolution of the Amniota, pp. 103-155, *in* Benton, M. J., ed. *The Phylogeny and Classification of the Tetrapods, vol. 1: Amphibians, Reptiles, Birds*. Clarendon Press, Oxford.

Heaton, M. J., and R. R. Reisz. 1986. Phylogenetic relationships of captorhinomorph reptiles. *Canadian Journal of Earth Sciences* 23:402-418.

Laurin, M. and R. R. Reisz. 1996. A reevaluation of early amniote phylogeny. *Zoological Journal of the Linnean Society* 113:165-223.

Lucas, S. G. 1997. *Dinosaurs, The Textbook* (2nd ed.). W. C. Brown, Dubuque, Iowa.

McGowan, C. 1983. *The Successful Dragons: A Natural History of Extinct Reptiles*. Samuel Stevens, Toronto.

McGowan, C. 1991. *Dinosaurs, Spitfires, and Sea Dragons*. Harvard University Press, Cambridge.

McLoughlin, J. C. 1979. *Archosauria: A New Look at the Old Dinosaur*. Viking Press, New York.

Norman, D. B. 1985. *The Illustrated Encyclopedia of the Dinosaurs*. Crescent Books, New York.

Rieppel, O. 1988. The classification of the Squamata, pp. 261-294. *in* Benton, M. J., ed. *The Phylogeny and Classification of the Tetrapods, vol. 1: Amphibians, Reptiles, Birds*. Clarendon Press, Oxford.

Schultze, H.-P., and L. Trueb, eds. 1991. *Origins of the Higher Groups of Tetrapods: Controversy and Consensus*. Cornell University Press, Ithaca, New York.

Sereno, P. C. 1991. Basal archosaurs: phylogenetic relationships and functional implications. *Journal of Vertebrate Paleontology Memoir* 2:1-53.

Sereno, P. C. 1997. The origin and evolution of dinosaurs. *Annual Reviews of Earth and Planetary Sciences* 25: 435-490.

Smithson, T. R., R. L. Carroll, A. L. Panchen, and S. M. Andrews. 1994. *Westlothiana lizziae* from the Visean of East Kirkton, West Lothian, Scotland, and the amniote stem. *Transactions of the Royal Society of Edinburgh* 84: 383-412.

Thomas, R. D. K., and E. C. Olson, eds. 1980. *A Cold Look at the Warm-Blooded Dinosaurs*. AAAS Selected Symposium 28. Westview Press, Colorado.

Weishampel, D. B., P. Dodson, and H. Osmolska, eds. 1990. *The Dinosauria*. University of California Press, Berkeley.

Wellnhofer, P. 1991. *The Illustrated Encyclopedia of Pterosaurs*. Crescent Books, New York.

FEATHERED DINOSAURS: THE BIRDS

The dinosaurs fly past in herds
Singing their song without words
They're small ones, it's true,
Warm and feathery, too,
But they're here—and we call them birds.

Richard Cowen, *History of Life*, 1990

Two years after the publication of *On the Origin of Species*, the scientific world was startled by the discovery of a fossil that had enormous implications for evolution. Originally, Darwin had been quite apologetic about the lack of transitional fossils between major taxa, but in 1861 the first fossils of *Archaeopteryx* were discovered and described from the Upper Jurassic Solnhofen Limestone. Every skeletal feature was dinosaurian, except for two important bird-like characters: the fused collarbones forming a wishbone, and the presence of feathers. Darwinians

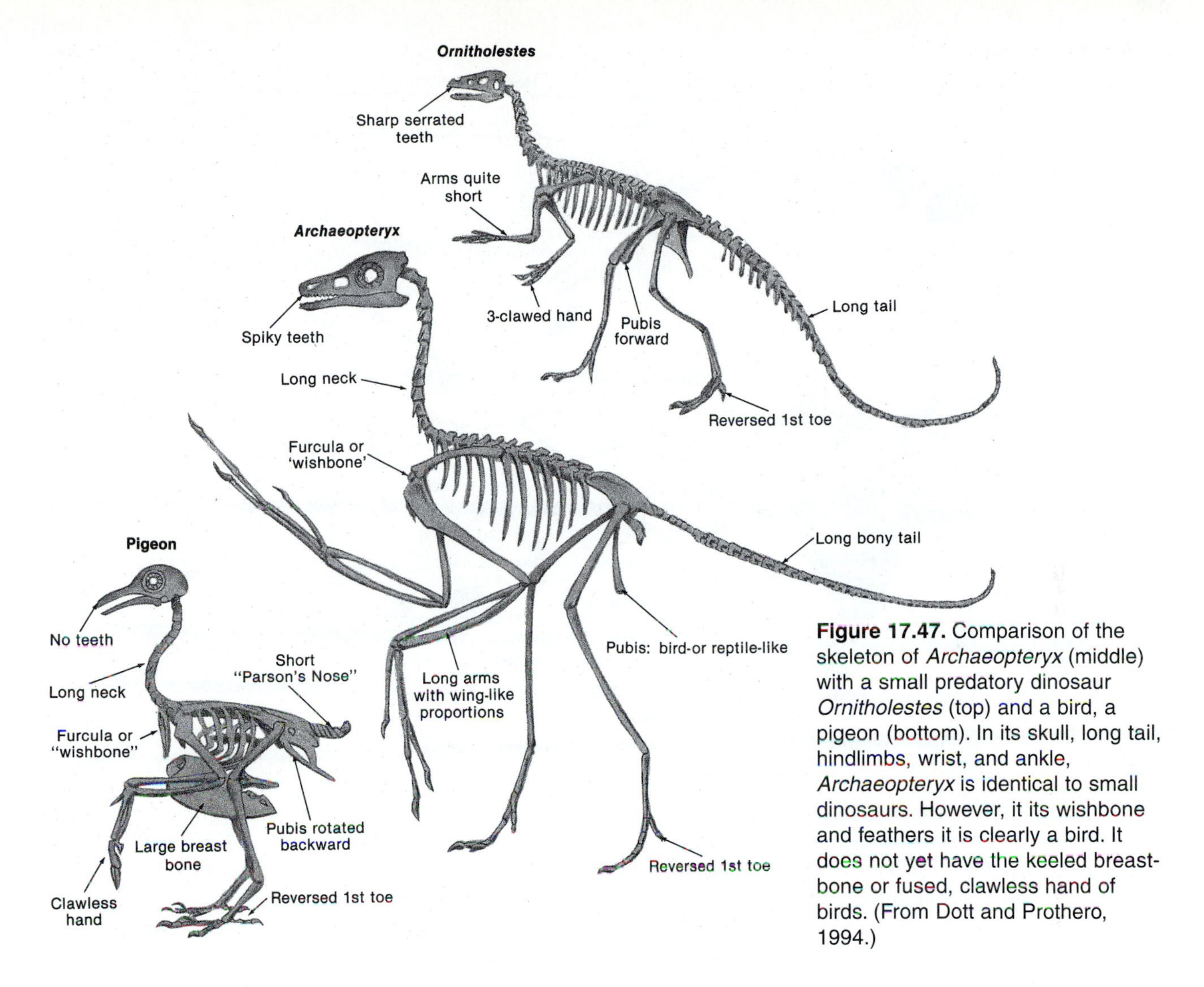

Figure 17.47. Comparison of the skeleton of *Archaeopteryx* (middle) with a small predatory dinosaur *Ornitholestes* (top) and a bird, a pigeon (bottom). In its skull, long tail, hindlimbs, wrist, and ankle, *Archaeopteryx* is identical to small dinosaurs. However, it its wishbone and feathers it is clearly a bird. It does not yet have the keeled breastbone or fused, clawless hand of birds. (From Dott and Prothero, 1994.)

quickly pointed out that this fossil was a perfect transitional form between birds and reptiles, since it still had many reptilian features, including thecodont teeth, a long bony tail, and a dinosaurian hand with long fingers and claws, and a dinosaurian foot as well (Fig. 17.47). In the following century and a half, six more specimens were found that added further details to the story. The "Berlin" specimen (Fig. 4.1), found in 1877, was the most complete and well articulated of the seven known specimens, and Feduccia (1996, p. 29) calls it "the most important natural history specimen in existence, comparable perhaps in scientific and even monetary value to the Rosetta stone. Beyond doubt, it is the most widely known and illustrated fossil animal—a perfectly preserved Darwinian intermediate, a bird that has anatomical features of a reptile, feathers, and a long, lizard-like tail." Other specimens have been found that show the broad breastbone for the attachment of flight muscles, and one specimen was long misidentified as the small dinosaur *Compsognathus* before its faint feather impressions were discovered. These specimens, and a careful analysis of Mesozoic reptiles, have given us a startling perspective: birds are most closely related to certain types of theropod dinosaurs. The fact that one specimen of *Archaeopteryx* could be misidentified as a dinosaur shows this, although it is not a new idea—Thomas Henry Huxley first suggested it soon after *Archaeopteryx* was described. Dozens of synapomorphies (Gauthier and Padian, 1985; Gauthier, 1986) support the close relationship of birds and certain theropods, such as *Deinonychus*. One of the most striking is in the wrist structure, which has a fused assemblage of wrist bones shaped like a half-moon (the **semilunate carpal**). This wrist bone allows the hand to rotate and extend in a characteristic grasping fashion once used by predatory dinosaurs, but also characteristic of the hand motion during the extensional wing stroke of birds. Today, there is little doubt among most paleontologists that birds are descended from a subgroup of the theropod dinosaurs (Fig. 17.46), the coelurosaurs (familiar from such animals as *Deinonychus* and *Velociraptor*). Cladistically speaking, birds *are* dinosaurs (Ostrom, 1973, 1975, 1976; Bakker and Galton, 1974; Gauthier and Padian, 1985; Gauthier, 1986). In a strictly cladistic classification, the "Class Aves" would be demoted to a taxon within the Dinosauria, Saurischia, Theropoda, and Dromaeosauria, although more conservative scientists persist in placing birds in their own class simply due to their evolutionary divergence, diversity, and

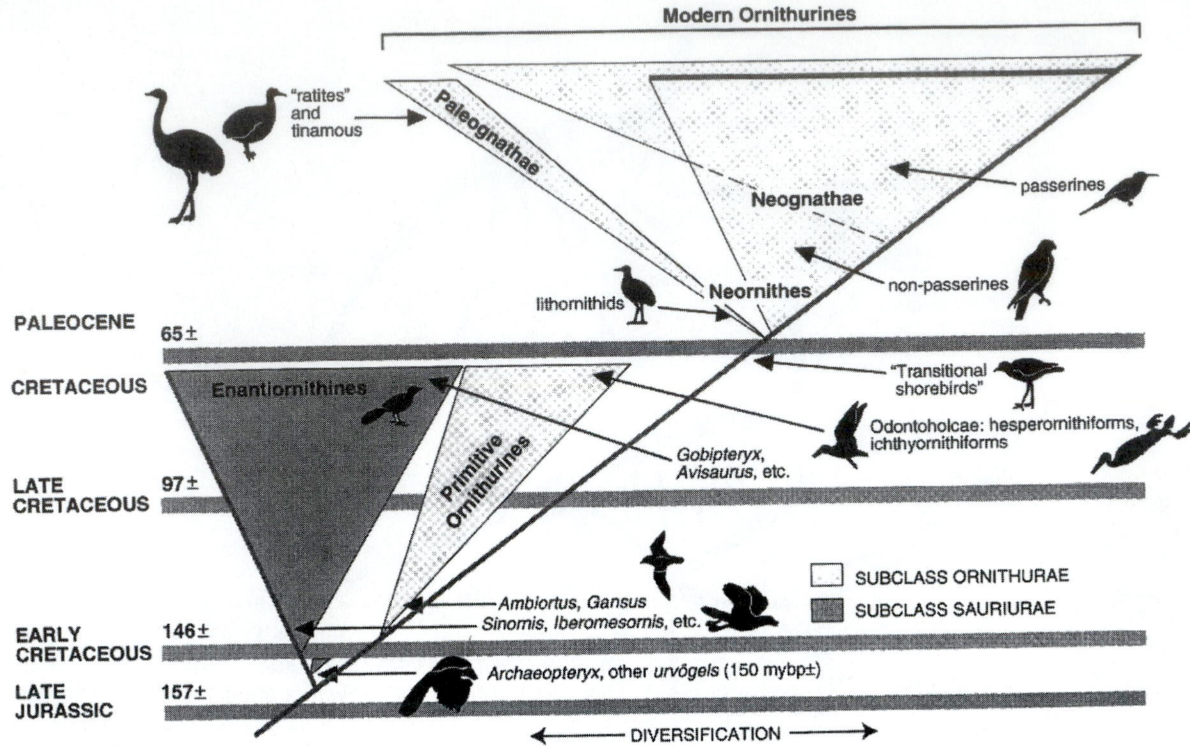

Figure 17.48. Diagrammatic summary of bird evolution. (From Feduccia, 1994.)

success. As we saw in Chapter 4, this argument mixes evolutionary relationships with ecological factors such as disparity and diversity. Some taxonomists find this acceptable, but cladists feel that mixing phylogeny and ecology makes such a classification confusing and hard to decipher.

A few dissenters (e.g., Martin, 1991; Feduccia, 1996) persist in arguing that birds are not related to dinosaurs, but their arguments mostly consist of nitpicking at the characters that support dinosaurian origin, or just disagreeing with cladistics altogether. None have been able to provide a convincing case, based on as many syapomorphies, for some taxon other than theropod dinosaurs that is closer to birds. Their case has been further weakened by as yet unpublished discoveries of a Cretaceous dinosaur with feathers, and additional dinosaur specimens with forearms and claws that further strengthen the bird-dinosaur link (Gibbons, 1996).

Archaeopteryx is known only from the Late Jurassic, so the bird/dromaeosaur divergence must have occurred before then. (An alleged Late Triassic bird dubbed "*Protoavis*" has been described, but most paleontologists are unconvinced that it is more than a collection of unassociated scraps of different Triassic archosaurs.) For a long time after the discovery of *Archaeopteryx*, only a few other Mesozoic birds were known, since bird bones are delicate and rarely fossilize (47 of the 155 living families of birds have no fossil record, and most of these 47 living families are known only from the Pleistocene). In the last decade, however, the number of described Cretaceous birds had multiplied, and a new picture of Mesozoic bird evolution is emerging (Fig. 17.48). Besides *Archaeopteryx*, the only other known Jurassic bird is *Confuciusornis* from China (Hou et al., 1995), which already had a toothless beak, but still had long fingers and claws. By the Early Cretaceous, however, bird fossils imply that a tremendous diversification was occurring. Sereno and Rao (1992) described *Sinornis* from the Lower Cretaceous of China, which also had a hand made of long, unfused, clawed fingers, and a toothed beak, but a wrist joint that allows the wings to fold against the body, feet with an opposable first toe for perching, and all the tail bones were fused into a single, reduced element, the **pygostyle**. *Cathayornis* from the Lower Cretaceous of Jiufotang, China (Zhou et al., 1992), was like *Sinornis* in retaining teeth in its beak, and a primitive pelvis and hindlimb, but was a powerful flier. Sanz et al. (1988) described *Iberomesornis* from the Lower Cretaceous of Las Hoyas, Spain, a bird with a pygostyle, larger wings, and a well-developed breastbone, but still retaining the 13 or 14 back vertebrae (rather than the 5 to 6 found in modern birds), an unfused hand and unfused ankle. *Concornis* from the same deposits (Sanz et al., 1995) was a larger bird with more advanced features, including fusion of the thigh bone with the upper row of ankle bones (**tibiotarsus**). Most of these early Cretaceous birds are now placed in an extinct group, the **enantiornithes**, or "opposite birds," because their foot bones (metatarsals) fuse from the ankle down to the toes, rather than from the toes up, as in all living birds. The enantiornithes (first recognized by Cyril Walker in 1981) diversified into a huge radiation of Cretaceous birds, but none survived the K/T event (Fig. 17.48).

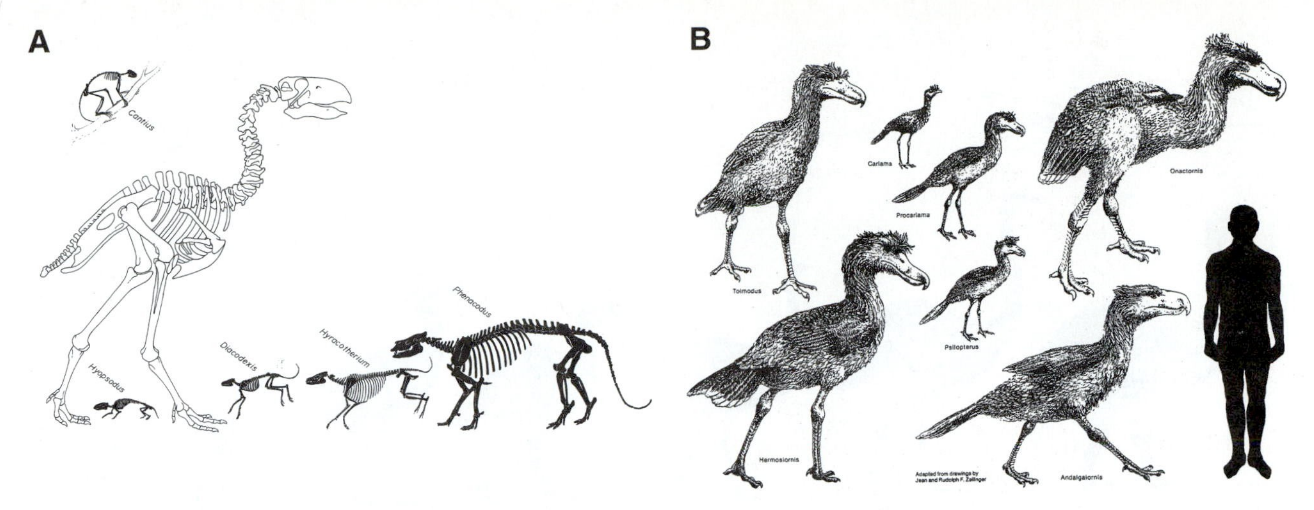

Figure 17.49. (A) Comparison of *Diatryma* with some of the Eocene mammals that might have been its prey. (From Witmer and Rose, 1991.) (B) The South American phorusrhacids shown with their living relative, the cariama, and a six-foot man for scale. (From Marshall, 1978.)

A second major clade of Mesozoic birds was the **ornithurines**, which are Cretaceous sister taxa to the living bird radiation (Fig. 17.48). They are represented by *Ambiortus* from the Lower Cretaceous of Mongolia (Kurochkin, 1985), with a fully modern wishbone, fused hand bones and fingers, a keel on its breastbone for the flight muscles, and even feather impressions. *Gansus* from the Lower Cretaceous of Gansu Province, China, has a hind limb that suggests that it was an early shorebird (Hou and Liu, 1984). *Chaoyangia* from the Lower Cretaceous *Cathayornis* locality (Hou et al., 1993) is known primarily from a pelvis, but it is modern in many aspects, as are the vertebrae and ribs. By the Late Cretaceous, the ornithurines were represented by many good specimens of marine birds known from the chalk beds of Kansas, Texas, and Alabama. These include the loon-like *Hesperornis*, and the tern-like *Ichthyornis*, which were decribed by O.C. Marsh over a century ago in his famous monograph on toothed birds, *Odontornithes* (1880). This monograph was originally published by the U.S. Geological Survey, and caused a scandal on the floor of Congress, when a fundamentalist Congressman was outraged that taxpayer dollars were spent on studying birds with teeth. To his mind, these were a biblical impossibility.

Although the bird fossil record is scrappy so that it is difficult to say how abruptly these archaic birds died out at the end of the Cretaceous, the fact remains that few survived into the Cenozoic. Instead, there was a renewed radiation of modern bird families from a few survivors related to the ornithurines, so by the Eocene there was a huge diversity of birds, many from modern families. In this respect, the Paleocene bird radiation event resembles the enormous evolutionary radiation of mammals in the Paleocene, after the extinction of the non-avian dinosaurs cleared the way for large terrestrial vertebrates. Ironically, the avian dinosaurs also underwent a huge evolutionary radiation after their close relatives died out. In some areas, the avian dinosaurs did not completely yield the role of large terrestrial predators. Whenever there were no large mammalian predators during the Cenozoic, it seems that birds often occupied that role. In the early-middle Eocene of North America, there was a gigantic predatory bird known as *Diatryma*, which stood over 2 m tall, and had a head with a huge, sharp beak over 0.5 m (1.5 feet) long, and may have weighed as much as 175 kg (385 pounds) (Fig. 17.49A). In the Oliogocene and Miocene of South America, there was an entire family of giant predatory birds, the phorusrhachids, which also had large sharp beaks and stood 2 to 3 m tall (Fig. 17.49B). These birds apparently originated in Europe during the Eocene, but crossed to South America in the Oligocene at the same time that New World monkeys and caviomorph rodents did. They remained confined to South America until the Panamanian land bridge arose in the Pliocene, when gigantic forms such as the 3-m (10-foot) tall *Titanis* migrated north, where it is known from the late Pliocene of Florida, and recently other specimens were recovered from the Pleistocene of Texas.

Both of these groups were large, walking birds that were too large to fly. The largest flightless birds of all, however, occur within a group known as the **ratites** (Fig. 17.50), and include the ostrich of Africa, the rhea of South America, the emu and cassowary of Australia, and the kiwi of New Zealand. In addition to the living ratites, there were extinct giants, such as the moas of New Zealand, which were up to 3.7 m (12 feet) tall, and were hunted to extinction by the Maoris by about 400 years ago. The most massive bird of all, however, is the famous "elephant bird," *Aepyornis*, from the Pleistocene of Madagascar, which weighed close to 454 kg (1000 pounds), stood almost 3 m tall, and laid an egg almost 1 foot long with a 2-gallon capacity. They, too, were hunted to extinction when humans first arrived on Madagascar. Ratites are members of the most primitive group of living birds, the Palaeognathae, because they have a number of unique features besides flightlessness. The **palaeognathous palate** (Fig. 17.51) is more primitive than the palate seen in most birds (**neognathous**), but it does

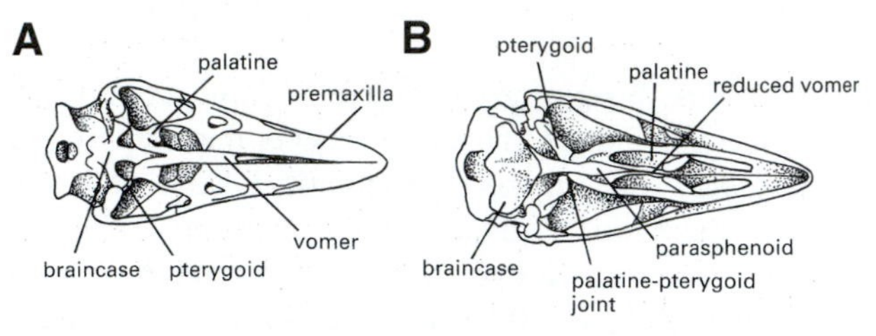

Figure 17.50. (top) Comparison of various flightless birds, including the ratites, the predatory phorusrhachids, and *Diatryma.* Note that the predators have large, robust beaks and skulls. Scale is in meters. (From Witmer and Rose, 1991.)

Figure 17.51. (left) Comparison of the palates of a palaeognathous bird (A) and neognathous bird (B). (From Benton, 1990.)

have a number of specializations that argue for the monophyly of palaeognathous birds (Cracraft, 1988).

The striking Gondwana distribution of the living ratites has led some authors to suggest that their distribution is a vicariant relict of Gondwanaland, implying that their divergence began before these continents rifted apart in the middle Cretaceous (Cracraft, 1973, 1974). However, Houde (1988) pointed out that their close relatives, the lithornithids, are known from the Paleocene of Europe and Asia, and there are fossil ostriches from the Eocene of Germany and the Miocene of central Europe and Asia, and fossil ratites from the Paleocene of France. Apparently, ratites had a worldwide distribution during the early Cenozoic, so that only the surviving relicts are confined to Gondwana continents.

All the remaining living birds have the derived neognathous palate, and comprise an enormous radiation of some 9000 species, all of which originated and diversified during the Cenozoic. The evolution and relatioships of these living orders of birds is still highly controversial. Cracraft (1988) published the first cladistic analysis of the living birds, but it has been strongly criticized by more traditional taxonomists (see Feduccia, 1996). Sibley and Ahlquist (1990) produced a completely different geometry of bird relationships using DNA hybridization (see Chapter 4), but these results have also been extensively criticized (see Cracraft, 1987, and Feduccia, 1996, p. 176, for a critique). Clearly, the story of the huge radiation of Cenozoic birds is too large for a short chapter such as this. The reader is referred to Feduccia (1996) for an up-to-date account.

For Further Reading

Chiappe, L. M. 1995. The first 85 million years of bird evolution. *Nature* 378:349-355.

Cracraft, J. 1973. Continental drift, paleoclimatology, and the evolution and biogeography of birds. *Journal of Zoology (London)* 169:455-545.

Cracraft, J. 1974. Phylogeny and evolution of the ratite birds. *Ibis* 116:494-521.

Cracraft, J. 1985. The origin and early diversification of birds. *Paleobiology* 12:383-389.

Cracraft, J. 1987. DNA hybridization and avian phylogenetics. *Evolutionary Biology* 21:47-96.

Cracraft, J. 1988. The major clades of birds, pp. 339-361, *in* Benton, M. J., ed. *The Phylogeny and Classification of the Tetrapods, vol. 1: Amphibians, Reptiles, Birds.* Clarendon Press, Oxford.

Feduccia, A. 1994. Tertiary bird history: notes and comments, pp. 178-188, *in* Prothero, D. R., and R. M. Schoch, eds. *Major Features of Vertebrate Evolution.* Paleontological Society Short Course 7.

Feduccia, A. 1995. Explosive radiation of Tertiary birds and mammals. *Science* 267:637-638.

Feduccia, A. 1996. *The Origin and Evolution of Birds.* Yale University Press, New Haven.

Gauthier, J. 1986. Saurischian monophyly and the origin of birds. *Memoirs of the California Academy of Sciences* 8:1-55.

Gauthier, J. A., and Padian, K. 1985. Phylogenetic, functional, and aerodynamic analyses of the origin of birds, pp. 185-197, *in* Hecht, M. K., J. H. Ostrom, G.

Viohl, and P. Wellnhofer, eds. *The Beginnings of Birds*. Freunde des Juras-Museums, Eichstätt.

Hecht, M. K., J. H. Ostrom, G. Viohl, and P. Wellnhofer, eds. *The Beginnings of Birds*. Freunde des Juras-Museums, Eichstätt.

Hou, L., Z. Zhou, L. D. Martin, and A. Feduccia. 1995. A beaked bird from the Jurassic of China. *Nature* 377: 616-618.

Houde, P. 1988. Palaeognathous birds from the early Tertiary of the Northern Hemisphere. *Publications of the Nuttall Ornithological Club* 22:1-148.

Marsh, O. C. 1880. Odontornithes: a monograph on the extinct toothed birds of North America. *Report on the Geological Exploration of the Fortieth Parallel* 7:1-201.

Marshall, L. G. 1994. The terror birds of South America. *Scientific American* 270 (2):90-95.

Martin, L.D. 1991. Mesozoic birds and the origin of birds, pp. 485-540, *in* Schultze, H.-P., and L. Trueb, eds. *Origins of the Higher Groups of Tetrapods*. Comstock Publishing Company, Ithaca, New York.

Martin, L. D. 1995. The Enantiornithes: terrestrial birds of the Cretaceous in avian evolution. *Courier Forschunginstitut Senckenberg* 181:23-36.

Olson, S. L. 1985. The fossil record of birds. *Avian Biology* 8:79-252.

Ostrom, J. H. 1973. The ancestry of birds. *Nature* 242:136.

Ostrom, J. H. 1975. The origin of birds. *Annual Reviews of Earth and Planetary Sciences* 3:55-77.

Ostrom, J. H. 1976. *Archaeopteryx* and the origin of birds. *Biological Journal of the Linnean Society (London)* 8:91-182.

Ostrom, J. H. 1979. Bird flight: how did it begin? *American Scientist* 67:46-56.

Ostrom, J. H. 1991. The question of the origin of birds, pp. 467-484, *in* Schultze, H.-P., and L. Trueb, eds. *Origins of the Higher Groups of Tetrapods*. Comstock Publishing Company, Ithaca, New York.

Ostrom, J. H. 1994. On the origin of birds and of avian flight, pp. 160-177, *in* Prothero, D. R., and R. M. Schoch, eds. *Major Features of Vertebrate Evolution*. Paleontological Society Short Course 7.

Sanz, J. L., J. F. Bonaparte, and A. Lacasa. 1988. Unusual Early Cretaceous birds from Spain. *Nature* 331: 433-435.

Sanz, J. L., and A. D. Buscalioni. 1992. A new bird from the early Cretaceous of Las Hoyas, Spain, and the early radiation of birds. *Palaeontology* 35:829-845.

Sanz, J. L., J. M. Chiappe, and A. D. Buscalioni. 1995. The osteology of *Concornis lacustris* (Aves: Enantiornithes) from the Lower Cretaceous of Spain and a re-examination of its phylogenetic significance. *American Museum Novitates* 3133:1-23.

Sereno, P. C., and Rao Chenggang. 1992. Early evolution of avian flight and perching: new evidence from the Lower Cretaceous of China. *Nature* 255:845-848.

Sibley, C. G., and J. E. Ahlquist. 1990. *Phylogeny and Classification of Birds*. Yale University Press, New Haven.

Walker, C. A. 1981. New subclass of birds from the Cretaceous of South America. *Nature* 292:51-52.

Witmer, L. M. 1991. Perspectives on avian origins, pp. 427-466, *in* Schultze, H.-P., and L. Trueb, eds. *Origins of the Higher Groups of Tetrapods*. Comstock Publishing Company, Ithaca, New York.

Witmer, L. M., and K. D. Rose. 1991. Biomechanics of the jaw apparatus of the gigantic Eocene bird *Diatryma*: implications for diet and mode of life. *Paleobiology* 17:95-120.

Zhou, Z., F. Jin, and J. Zhang. 1992. Preliminary report on a Mesozoic bird from Liaoning, China. *Chinese Science Bulletin* 37:1365-1368.

FURRY FOLK: SYNAPSIDS AND MAMMALS

> *Of all the great transitions between major structural grades within vertebrates, the transition from basal amniotes to basal mammals is represented by the most complete and continuous fossil record, extending from the Middle Pennsylvanian to the Late Triassic and spanning some 75 to 100 million years.*
>
> James Hopson, "Synapsid Evolution and the Radiation of Non-Eutherian Mammals," 1994

At the very beginning of their history, amniotes split into two lineages, the **synapsids** and the reptiles. Traditionally, the earliest synapsids have been called the "mammal-like reptiles" but this is a misnomer. The earliest synapsids had nothing to do with reptiles as the term is normally used (referring to the living reptiles and their extinct relatives). Early synapsids are "reptilian" only in the sense that they initially retained a lot of primitive amniote characters. Part of the reason for the persistence of this archaic usage is the pre-cladistic view that the synapsids are descended from "anapsid" reptiles, so they are also reptiles. In fact, a lot of the "anapsids" of the Carboniferous, such as *Hylonomus*, which once had been postulated as ancestral to synapsids, are actually derived members of the diapsids (Gauthier, 1994). Furthermore, the earliest reptiles (*Westlothiana* from the early Carboniferous) and the earliest synapsids (*Protoclepsydrops* from the Early Carboniferous and *Archaeothyris* from the Middle Carboniferous) are equally ancient, showing that their lineages diverged at the beginning of the Carboniferous, rather than synapsids evolving from the "anapsids." For all these reasons, it is no longer appropriate to use the term "mammal-like reptiles." If one must use a non-taxonomic term, "protomammals" is a alternative with no misleading phylogenetic implications.

From their origin in the Early Carboniferous, an amazing array of synapsid fossils shows the transition from early amniote to mammal in remarkable detail. Yet though the fossil record is excellent, many features that distinguish

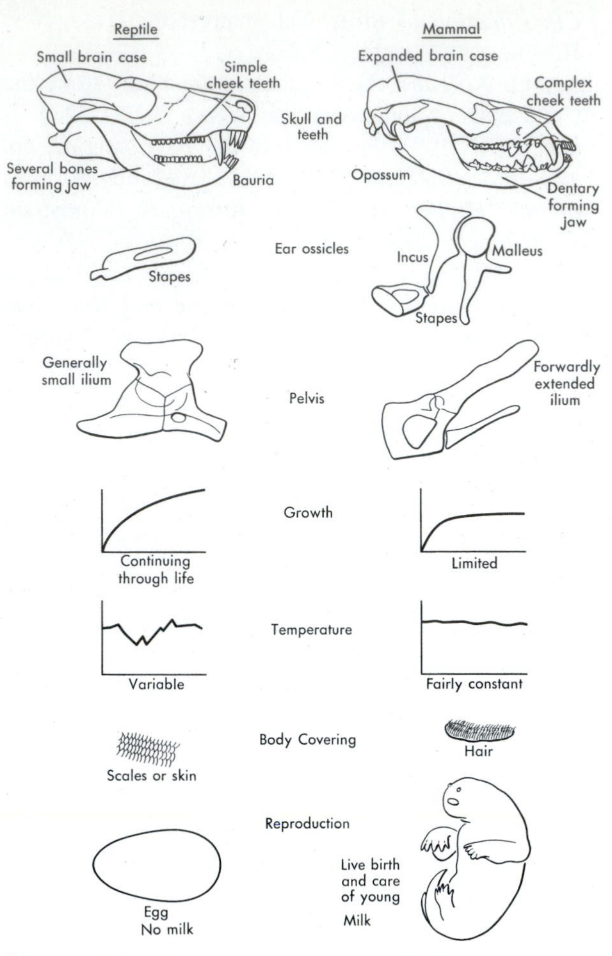

Figure 17.52. Some important anatomical and physiological differences between primitive amniotes and mammals. (From Colbert and Morales, 1991.)

mammals from reptiles do not fossilize (Fig. 17.52). Distinguishing features include:

1. Physiological characters—Mammals are usually defined as homeothermic (having constant body temperature) endothermic amniotes with hair. They also have other features related to their high metabolism and active lifestyles, such as a four-chambered heart, a diaphragm for actively pumping air in and out of the lungs, and a sophisticated brain with an enlarged neocortex. Most of these characters do not preserve in the skeleton. The internal molds of the brain cavity are known from many synapsids, so it is possible to determine when the enlargement of the neocortex occurs.

2. Reproductive characters—Another distinctive characteristic of mammals is their mode of reproduction. Most mammals (except the egg-laying platypus and echidna) give birth to live young, which the females then nurse with milk from their mammary glands. Instead of laying eggs and then abandoning them, mammals invest a lot of parental care into each offspring, so that fewer are born, and they are born more helpless than hatchling reptiles or amphibians. Young mammals grow rapidly after birth, but their growth slows down to a terminal, adult growth stage (in contrast to most other animals, which grow continuously throughout their lives). The best way to detect a pattern of terminal growth is by the presence of cartilaginous caps on the ends of the long bones of juveniles, indicating that the animals underwent rapid growth as a juvenile, and then stopped growing when these caps fused to the shaft of the bone. Unfortunately, the other reproductive features have a very low fossilization potential, although there are indirect means of detecting some of them.

For paleontologists, the transformation to mammals must be detected in skeletal features that have at least some fossilization potential. Most of these give indirect evidence for mammalian physiology and reproduction. For example, there are many modifications of the skull and jaws for chewing and eating food more rapidly and efficiently, which is required for an animal with high metabolism. The teeth in early synapsids are simple cones or pegs for catching and puncturing prey, but later in synapsid evolution, the teeth become differentiated into nipping incisors in front, a large stabbing canine to catch and hold the prey, and multicusped cheek teeth (premolars and molars) for chewing up the food. Reptiles replace their teeth continuously throughout their lives, but mammals replace their deciduous teeth ("baby teeth") only once, and the molars are never replaced. Many primitive amniotes have teeth on the palate and in the throat region for holding a struggling prey item, but mammals have teeth only on the margin of their jaws.

In reptiles, the nasal passage opens into the front of the mouth cavity, so that when a lizard is slowly swallowing a large prey item, it must hold its breath while there is food in its mouth. Clearly, the high metabolism of mammals would not allow them to hold their breath for long while eating or chewing. For this reason, the bones of the upper jaw grow toward the midline and form a **secondary palate** roofing over the original amniote palate, so that the internal nasal passage is enclosed and opens in the throat. (If you feel the roof of your mouth with your tongue, you can detect the suture along the midline of your palate. Some babies have a birth defect called cleft palate, where the two halves of the secondary palate fail to grow together, making it difficult for them to eat and breathe.)

The primitive synapsid jaw was a simple snap-trap mechanism, with a strong temporal muscle pulling up on the jaw and inserting on the top of the skull behind the eyes. Numerous bones made up this primitive jaw: the dentary in front, which bore the teeth; the articular, which formed the jaw hinge with the quadrate bone of the skull; the coronoid, forming a ridge on the top of the back of the jaw; the angular and surangular, on the back corner of the jaw, and several others (Fig. 17.53). Such a jaw was suitable for grabbing and crushing prey, but not for chewing. A single-element jaw is mechanically much stronger against the pressures and torque of the chewing motion than one with numerous elements that are sutured together; the sutures are lines of weakness under stress. Through synapsid evolution, the post-dentary elements of the jaw become smaller and smaller as the dentary becomes the

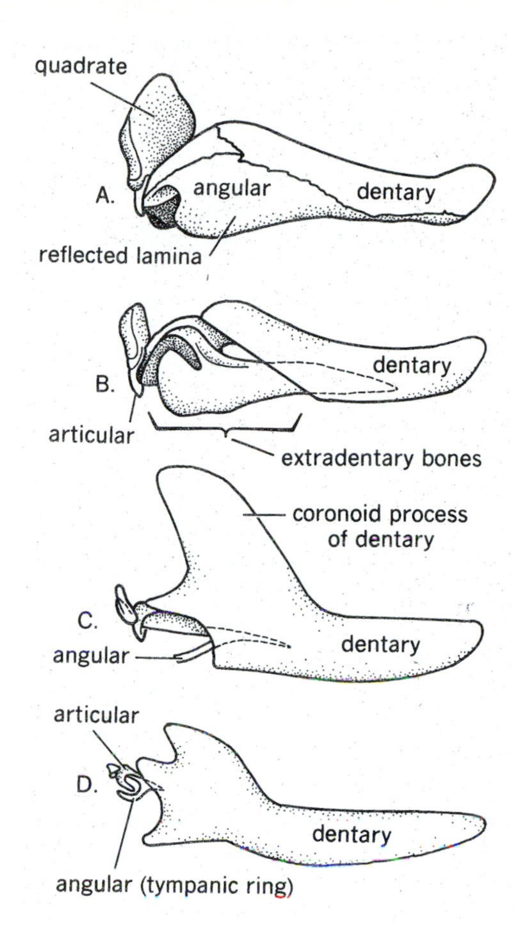

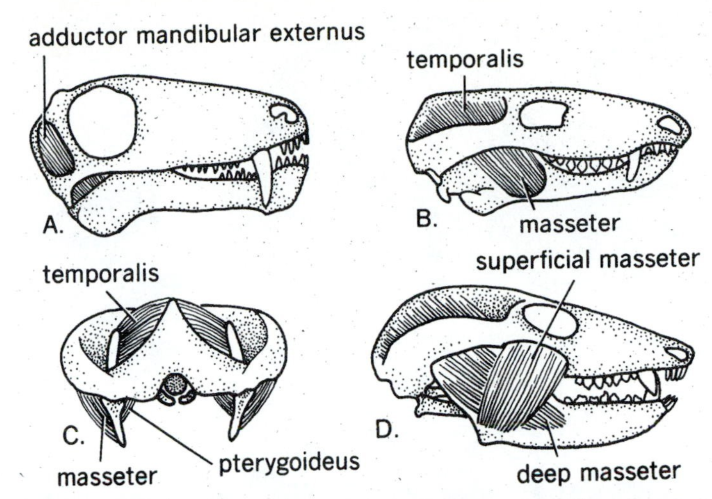

Figure 17.53. (left) Transformation of the jaw and skull region during synapsid evolution. Primitive synapsids such as the finback *Dimetrodon* (A) had a jaw with many different bones, hinged from the quadrate bone of the skull. In more advanced synapsids, the dentary expands until it becomes the only jaw bone, and the non-dentary bones shrink and are eventually lost, except for the quadrate and articular bones of the jaw joint, which go to the ear. (From Stearn and Carroll, 1989.)

Figure 17.54. (above) Primitive synapsids such as *Biarmosuchus* (A) had only a single jaw adductor muscle. *Thrinaxodon* (B-C) had both temporal and masseter muscles, and advanced cynodonts (D) divided the masseter into two branches for complex chewing motions. (From Stearn and Carroll, 1989.)

primary, and eventually the only, bone of the jaw. As the post-dentary elements reduced in size and most of them disappeared, the dentary extended back and took their place as the main area of muscle attachment. Eventually, the dentary developed a tall coronoid process to which the temporal muscles attached, allowing them to have even greater bite strength. In addition, a pair of new muscles, the masseters, arose between the outer edge of the cheek bones and the outer side of the jaw, allowing front-back and side-to-side motion in chewing (Fig. 17.54).

Eventually, the non-dentary bones of the jaw were lost completely (although some persisted even in the earliest mammals) as the dentary expanded backward and took their place. In advanced synapsids, the dentary reaches far enough back to touch the squamosal bone of the skull and develop a dentary/squamosal jaw joint, replacing the old reptilian quadrate/articular jaw joint. In some specimens, such as *Diarthrognathus* ("two jaw joint" in Greek), both jaw joints operated side-by-side on each side of the head. Eventually, however, the dentary/squamosal joint took over completely, and then the quadrate and articular no longer functioned as a jaw joint. Instead of vanishing, however, they took over a new function. In reptiles, they served not only as a jaw hinge, but are also able to transmit sound to the ear, since most reptiles can hear with their lower jaws (Fig. 17.55). (The snake charmer's flute is for the spectators, not the cobra, since snakes cannot hear well when their jaw is up off the ground in a threat posture.) Once the quadrate and articular became detached from the jaw hinge, they took up a new role as bones of the middle ear. The quadrate became the incus, or "anvil" bone, and the articular became the malleus, or "hammer" bone. (The "stirrup" bone, or stapes, has been in the ear since the early tetrapods.) When sound vibrates your eardrum, the chain of bones—"hammer," "anvil," and "stirrup," or malleus, incus, and stapes—that transmits this vibration to the inner ear is actually a remnant of your reptilian jaw apparatus. This amazing story is apparent not only in synapsid fossils, but also in mammalian embryology. When you began your development, your ear bones started out as part of your jaw, but were transferred entirely to your ear later in ontogeny.

Other skeletal modifications are apparent as synapsids became mammals. The early amniotes had a sprawling posture, with the legs held out from the side of the body, but midway through synapsid evolution, the body adopted an erect posture, with the limbs held under the body and moving rapidly fore and aft. These skeletal changes are particularly evident in the shoulder blade, which flares out into a broad triangle with a ridge down the middle for more complex muscle insertions. The hips became long and narrow for greater flexibility, with forward expansions of the ilium bone for stronger leg muscles, and eventually the three bones of the pelvis fused into a single bone (not the multiple bones of the primitive amniote hip). The free ribs of the chest are linked together with a breastbone, forming

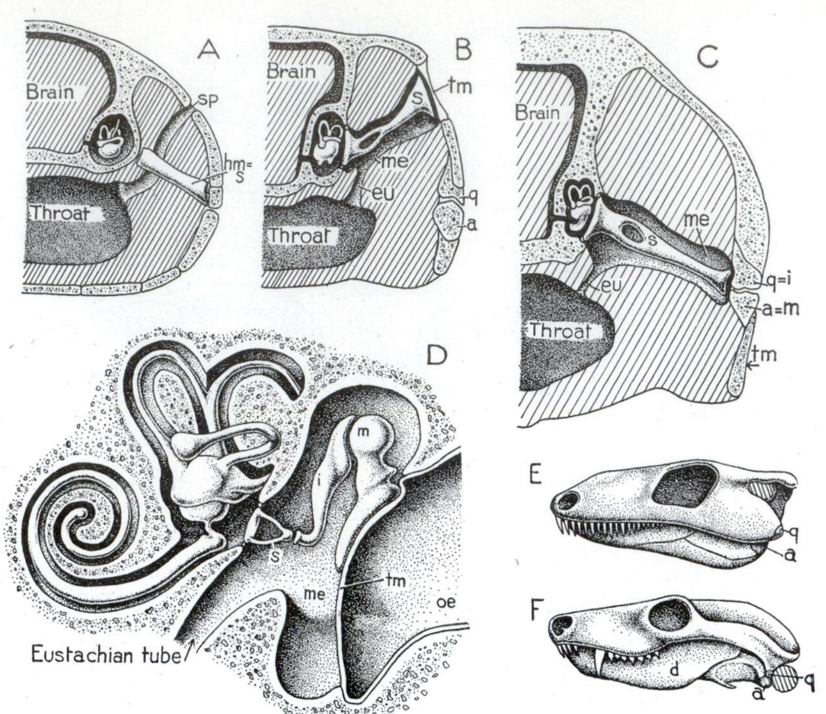

Figure 17.55. Evolution of the mammalian ear. In the fish (A), the hyomandibular bone (hm) acts as a bony strut, transmitting sound from the gill arches to the inner ear. In amphibians (B) it becomes the stapes (s) which vibrates with the eardrum (tm). In primitive amniotes (C) the stapes transmits sound from the jaw joint bones, the quadrate (q) and articular (a). These bones then become the incus (i) and malleus (m) of the middle ear of mammals (D). As these bones shift to the ear, the dentary (d) bone of the jaw contacts the squamosal bone of the skull to establish a new jaw joint (E-F). Other abbreviations: eu, eustachian tube; me, middle ear cavity; oe, outer ear cavity; sp, spiracle. (From Romer, 1970.)

a solid rib cage. This means that advanced synapsids could not breathe by flexing their ribs, but must have had a diaphragm to pump their lungs within the rigid rib cage. The ribs of the lower back, on the other hand, were lost, allowing the trunk to become more flexible. The small lower temporal opening of primitive synapsids became larger and larger as the jaw muscles expanded, until only a thin cheek bone, the zygomatic arch, remained. In many advanced synapsids and in most mammals, the temporal opening is so large that the bony bar between it and the eye is lost. The single ball joint that connects the skull to the vertebral column (the occipital condyle) in reptiles split into two small ball joints on either side of the spinal column, allowing much greater strength and stability in moving the head to catch and hold prey.

Most of these skeletal features can be traced through the course of synapsid evolution (Figs. 17.56, 17.57). For example, the earliest synapsids (mostly from the Pennsylvanian-Early Permian) are known as the "**pelycosaurs**" (a paraphyletic group unless it includes the rest of the synapsids), and include such familiar forms as the predatory finback *Dimetrodon* (often mistaken for a dinosaur). This animal, and the herbivorous finback *Edaphosaurus*, had large "sails" along their backs supported by long spines extending from their vertebrae. Many ingenious ideas have been proposed for the function of these fins, but the most plausible is that they served as heat gathering and dumping devices for thermoregulation. They have almost exactly the amount of surface area for an animal of that body volume to allow them to dump heat when the sail is out of the sun, and pick up heat when it is exposed broadside to the sun. This suggests that the earliest synapsids were not yet endotherms, but used sunning behavior to regulate their body temperature (as do most living reptiles). The "pelycosaurs" were primitive synapsids in many other aspects. They had a sprawling posture with a simple shoulder blade, small iliac blade on the pelvis, and simple thigh bone. Their teeth were simple pegs (although those in the canine position were a bit larger) replaced multiple times, and they had no secondary palate; instead, they had many teeth on their original, reptilian palate and in the throat region. They had a single ball joint in the back of the skull, a small brain, as well as a jaw composed of a small dentary and many accessory jaw bones. Indeed, the primary feature that

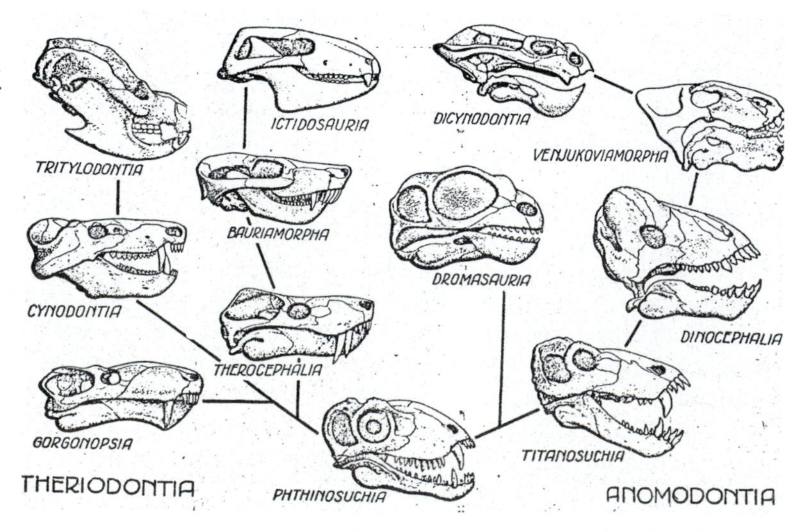

Figure 17.56. Variation in the skulls of therapsids and cynodonts. (From Romer, 1966.)

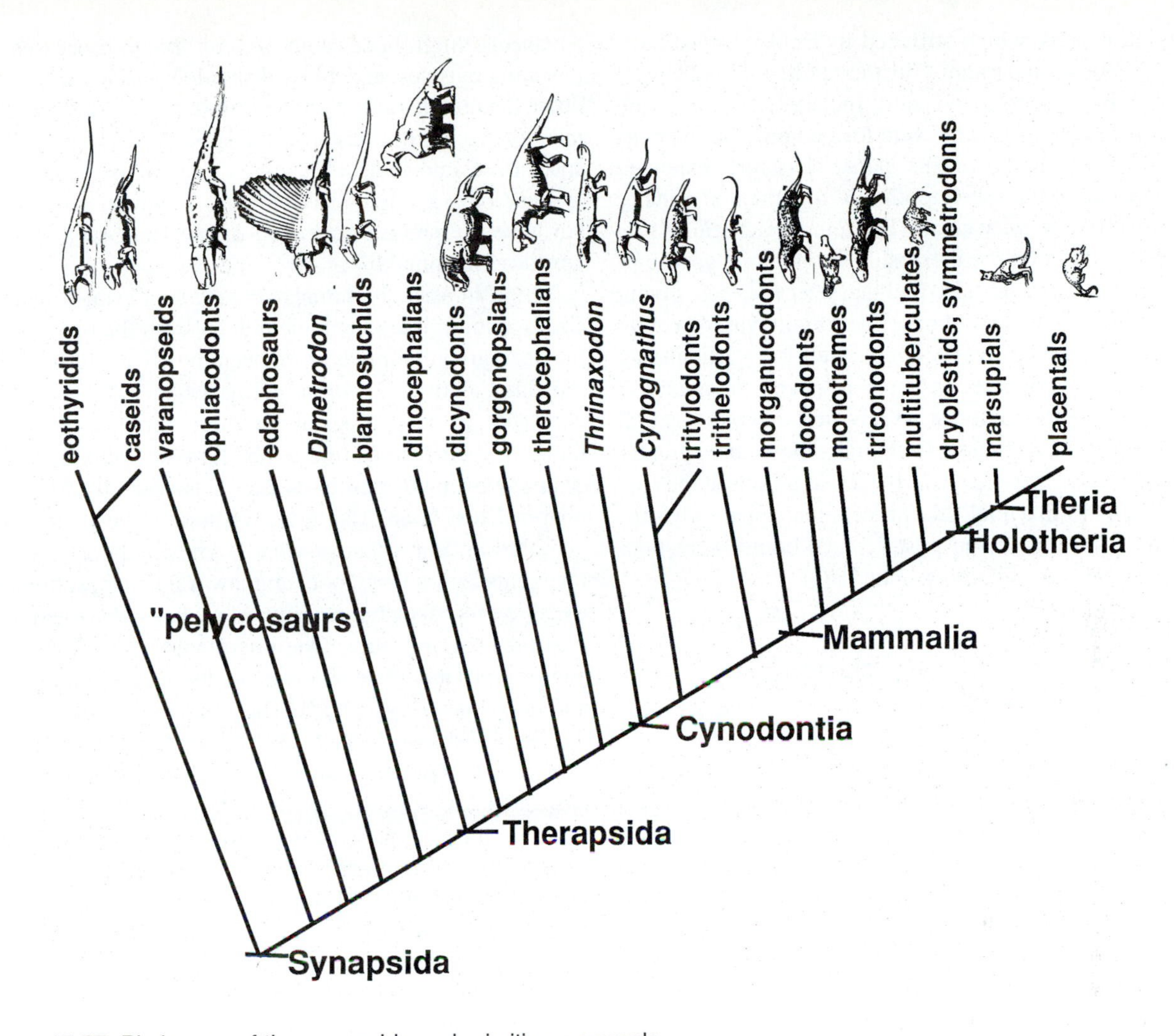

Figure 17.57. Phylogeny of the synapsids and primitive mammals.

earmarks the "pelycosaurs" as synapsids is the presence of the lower temporal opening, although it is small, indicating relatively small jaw muscles.

By the Late Permian, the first radiation of "pelycosaurs" became extinct, and a "second wave" of synapsids, the "**therapsids**" (also a paraphyletic group unless it includes the higher synapsids and mammals) came to dominate the landscape (Fig. 17.57). Many of these "therapsids" (such as the biarmosuchids, gorgonopsians, and therocephalians) were wolf-sized or bear-sized predators, with huge canine teeth and much stronger jaws (as shown by the enlarged dentary and temporal openings). They also had a more upright posture, and the more advanced forms even had a short secondary palate. These predators could have killed and eaten not only more primitive reptiles of the Late Permian (such as the parieasaurs discussed above), but also two groups of herbivorous "therapsids," the dicynodonts and the dinocephalians. The dicynodonts ("two dog teeth" in Greek) had an almost toothless beak (except for large canines) for munching plants, and a sliding jaw joint that gave them a chewing motion. With over 70 genera, they were the dominant herbivores of the Late Permian of Russia and South Africa, and some, such as the cow-sized *Kannemeyeria*, survived until the Late Triassic. The dinocephalians ("terrible head" in Greek) were huge (some up to 5 m long and weighing over a ton) sprawling, hippo-like "therapsids," with a massive ribcage and thick, stocky limbs. The bone over the braincase was very thick, suggesting to some paleontologists that they engaged in head-to-head butting (Barghusen, 1975). Some genera, such as *Estemmenosuchus*, had bizarre crests and flanges sticking out from their head, probably for display to other members of their species. About 40 genera of dinocephalians are known from the Late Permian of Russia and South Africa, but they did not survive until the end of the Permian.

The Permian catastrophe decimated most of the "therapsid" radiation except the dicynodonts and the lineage that led to third radiation of synapsids in the Triassic, the "**cynodonts**" (again, a paraphyletic group unless it contains the mammals). The "cynodonts" were mostly weasel- to dog-sized predators with many advances toward the mammalian condition (although *Cynognathus* was bear-sized). As exemplified by *Thrinaxodon*, they had an upright posture, with a more advanced shoulder blade and pelvis, and an additional muscle attachment point, the greater trochanter, on the thigh bone. *Thrinaxodon* had broad flanges on its ribs, which stiffened its trunk; it must have breathed with a diaphragm. In the skull, the teeth were almost fully differentiated into incisors, canines, and molariform cheek teeth, and were located only on the margin of the jaw. The secondary palate was now extended near to the end of the cheek tooth row, and the non-dentary jaw elements were very small. The temporal opening was already quite large in the therapsids, but there is good evidence that these animals also had masseter muscles, giving them complex chewing motions. Their brain was also quite a bit larger, and there was a double ball joint articulating with the neck vertebrae. In most respects, "cynodonts" were very close to mammals, except that they still had all the accessory jaw bones (although these are greatly reduced). In some "cynodonts," the angular bone at the corner of the jaw form a hook-like process that apparently supported an eardrum, so the post-dentary bones were both supporting the jaw and aiding in hearing.

Hearing with Their Jaw Bones: The Mammals

With malleus
Aforethought
Mammals
Got an earful
of their ancestors'
Jaw.

John Burns, *Biograffiti*, 1975

The synapsids had been the dominant and most diverse land animals through most of the Permian and Triassic, but by the Late Triassic, they were largely replaced by the rapidly evolving archosaurs (especially the dinosaurs). Most of the "cynodont" lineages (along with the dicynodont stragglers from the Permian) were extinct. Instead, there were a number of very small-bodied, mammal-like "cynodont" lineages that approach mammals in most features, yet most paleontologists are unwilling to call them mammals. Some, like the tritylodonts, had a very rodent-like skull with long incisors, no canines, rows of molars with multiple cusps for grinding; they had a long body shaped like a weasel. Others, such as the ictidosaurs or trithelodonts, had very advanced jaws—*Diarthrognathus* (mentioned above) had both a dentary/squamosal and quadrate/articular jaw joint operating side-by-side. However, most paleontologists do not regard a fossil as mammalian until it had a robust dentary/ squamosal jaw joint; others use the presence of an incus and malleus in the middle ear as their criterion for a mammal. This condition first appears in the Latest Triassic and Early Jurassic with tiny, rat-sized animals such as *Morganucodon* and *Sinoconodon* (Fig. 17.58). Although they have a robust dentary/ squamosal jaw joint, they retain vestiges of some of the other non-dentary jaw bones on the inside and back of the jaw. They had an upright mammalian posture, with a long blade on the iliac portion of the fused pelvis, an advanced thigh bone with several bony processes for attaching muscles, and a broad shoulder blade with a spine down the middle (although the primitive amniote interclavicle bone was still present in the shoulder). Their teeth are specialized into incisors, canines, premolars, and molars, and had only a single replacement. However, they did not yet have the precise occlusion of the teeth seen in more advanced mammals.

In the Jurassic, mammals became small (rat- to shrew-sized) animals that may have hidden from the dinosaurs in the undergrowth, or may have been mostly nocturnal. They remained as tiny, nocturnal animals under the feet of the dinosaurs, or in the trees above them, through about two-thirds of their history (the entire Jurassic and Cretaceous, spanning over 120 million years). Consequently, Mesozoic mammal fossils are also tiny, and tend to be fragmentary and hard to find. Most of what is known about Mesozoic mammals comes from tooth and jaw fragments, although in recent years, skeletal remains have been discovered for many the major groups. One important lineage is the morganucodonts and their relatives, including the triconodonts and docodonts (Fig. 17.57). These mammals tend to have the main cusps of their cheek teeth oriented in a line. For example, a typical triconodont tooth has a single main cusp, and lower cusps in front and in back of it. These cusps occluded directly *on top of*, or *just in front of*, the corresponding cusps on the opposite tooth in the jaw. The other main lineage of Mesozoic mammals had a completely different occlusal pattern. The cusps of their teeth were arranged in a triangular fashion, with the triangles of the upper molars fitting into the V-shaped valleys *between* (rather than on top of) the reversed triangles of the lower molars. This "reversed triangle" condition is a derived feature found in all the higher mammals, variously known as the "Theria" or the "Holotheria" (Hopson, 1994).

The most primitive of these holotherians was *Kuehneotherium* from the Upper Triassic of England. The best known of the holotheres were common in the Upper Jurassic dinosaur beds of North America and Europe, and were once known by the paraphyletic wastebasket taxon "pantotheres." These primitive therians are sister-taxa to the living mammals, and include the dryolestids, the symmetrodonts, and the paurodonts (Prothero, 1981). Some dryolestids and symmetrodonts still had tiny vestiges of the non-dentary jaw bones, but in other respects they are fully mammalian. Their high-crowned triangular teeth were very narrow in the front-to-back direction, since there were as many as eight or nine molars on each side of the jaw (most living mammals only have three or four).

In addition to the triconodont lineage and the dryolestid-symmetrodont (therian) lineage, there were at least two other lines of mammalian evolution whose systematic position has been controversial. One group was the **multituberculates** (Fig. 17.59). They looked somewhat squirrel-like with their chisel-shaped incisors in front, and the gap between the incisors and cheek teeth. In some multituberculates, the skeleton was also quite squirrel-like, with a

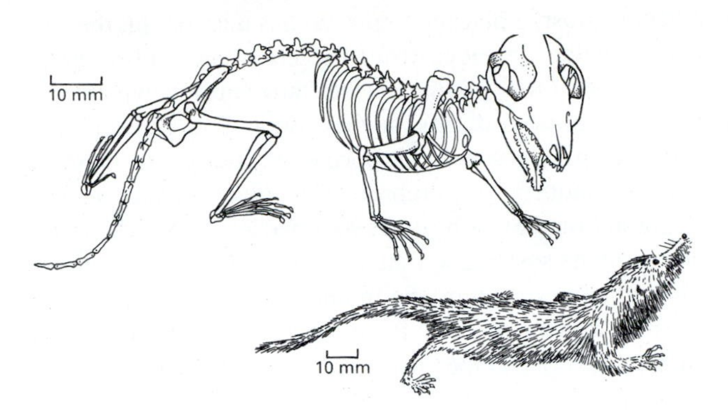

Figure 17.58. One of the earliest mammals was the shrew-sized *Morganucodon*. (From Benton, 1990.)

long, prehensile tail, and an ankle that could twist backward, allowing them to climb down trees headfirst. Multituberculates get their name from their complex molars, which are broad grinding mills with two or three rows of cusps ("tubercles") for processing nuts and seeds. Most advanced multituberculates also had an enlarged, blade-like tooth in front of the lower cheek teeth, which may have been used for slicing open hard nuts, seeds, and fruits. Despite their rodent-like appearance, multituberculates were still primitive mammals in many features, so their rodent features are due to evolutionary convergence. The earliest multituberculates appeared in the Late Jurassic, long before rodents evolved, and the group persisted until the late Eocene, a span of over 180 million years. By this criterion, they survived longer than any other order of mammals, living or extinct, and could be considered one of the most successful groups of mammals ever. Apparently, multituberculates finally met their doom in the Eocene due to competition from the rodents, which apparently were more successful in occupying the same niche.

The position of the multituberculates within the mammals has long been controversial, with some arguing that they are prototherians, and others arguing that they are related to primitive therians. The latest consensus places them as a sister-group to the holotherians, between the triconodonts and dryolestids (Fig. 17.57). Another line of mammalian evolution has also been a mystery, although a few fossils were left behind. This group is the **monotremes**, or the egg-laying mammals, such as the duckbilled platypus, and the "spiny anteater" or echidna. Today, they are confined to Australia and New Guinea, but fossil monotremes are also known from the Paleocene of Argentina. Monotremes lay a pea-sized, soft-shelled egg, which is sticky and carried in a shallow slit (not quite a pouch) on the female's belly. Once it hatches, the young lap up milk extruded from mammary glands without nipples. Monotremes have a much more primitive physiology that other living mammals. For example, their body temperature is not very well regulated, but fluctuates with the environmental temperature (although not as much as does a true ectotherm). Their urogenital tract is essentially a primitive amniote **cloaca** (Latin for "sewer"), into which both the reproductive openings and the waste openings from the kidney (the urethra) empty.

So what are monotremes related to? The monotreme skeleton still has many primitive synapsid features, including archaic bones in shoulder and hip that are found in no other living mammal. These elements would suggest that monotremes are more primitive than dryolestids, and possibly related to the "prototherians." Since Mesozoic mammals are known primarily from teeth, we need tooth fossils to make proper comparisons, but only juvenile monotremes have teeth, and they are so peculiar that for years no one could make sense of them. In 1975, a Miocene monotreme was described that gave a slightly better insight into the origins of their teeth, and in 1985 a Cretaceous monotreme preserved in opal was found in an opal mine and rescued just before it was made into jewelry. These specimens have teeth that are slightly more advanced than dryolestids, although they are not fully tribosphenic either. This evidence seemed to suggest that monotremes are in fact therian mammals that diverged sometime in the Middle Jurassic, possibly before dryolestids but after symmetrodonts, and have persisted with little change since then. However, more recent analyses of the ear region and skeleton suggest that monotremes are more primitive than triconodonts, and at the very base of the mammalian radiation. The most important implication of the relationships of monotremes is that they give us a glimpse of the biology and physiology of their Mesozoic counterparts. Based on monotremes, it is likely that some other Mesozoic mammals laid eggs, but had mammary glands without nipples, and very poor thermoregulation.

By the Early Cretaceous, mammals with more advanced teeth are known, living alongside the surviving archaic groups, such as the triconodonts, symmetrodonts, and dryolestids. These more advanced mammals have added a new

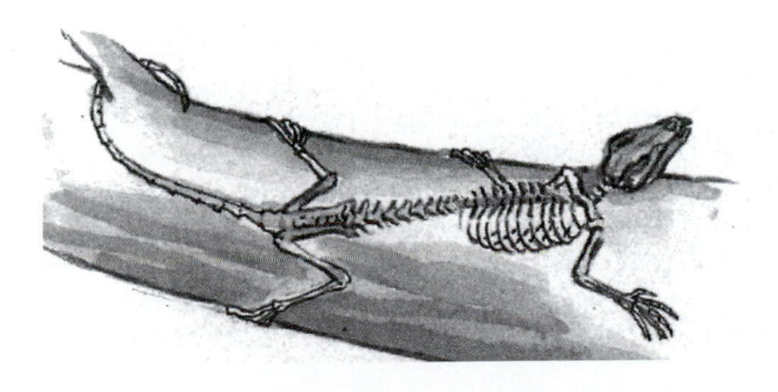

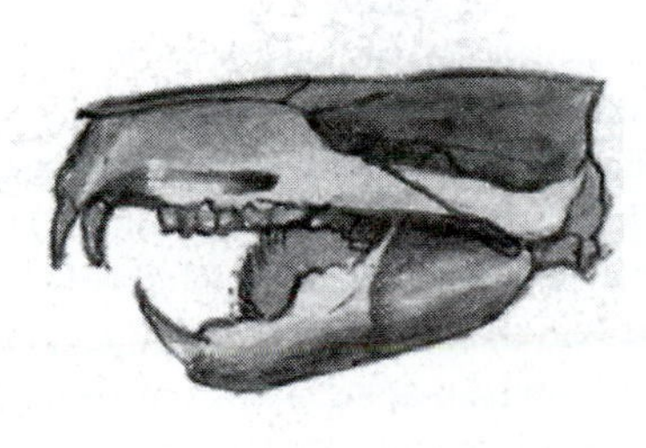

Figure 17.59. The squirrel-like multituberculates were the longest-lived group of mammals, ranging from the Jurassic to the Eocene. (From Dott and Prothero, 1994.)

cusp to the upper tooth (the protocone), making it essentially a modern mammalian molar (known as the **tribosphenic** tooth). This basic tribosphenic prototype will be highly modified in later mammals, but the position and homologies of the primary cusps are the same, no matter what the tooth is used for. By the late Early Cretaceous, the tribosphenic therian mammalian lineage had split into the two major living groups, the marsupials (Metatheria) and the placentals (Eutheria).

By contrast, a placental embryo has a an organ, the **placenta**, which develops from the chorionic and amniotic membranes (also known as the "afterbirth"), which protects the embryo during development. The placenta also serves to pass gases, food, hormones, and waste products between mother and embryo. The placenta has another important function—it serves as a barrier against the mother's immune system, so that when the embryo develops its own immune signature, the mother's system will not reject it as a "foreign object." By contrast, marsupials have no such protection, so the young are born before they can suffer immune rejection. In addition, the membrane surrounding the embryo (the **trophoblast**) is nowhere near as efficient at supplying food and gases to the embryo as is a placenta, so the marsupial embryo must be born prematurely so that it can nurse.

Marsupial reproduction allows multiple generations of young to be raised at once. A marsupial mother can carry one baby in the pouch, an embryo in the uterus, and take care of a third generation still staying in her vicinity, so the generational turnover can be quite rapid. If a marsupial mother is in great danger from a predator or starvation, she can drop the babies in her pouch at minimal risk to herself, and live to breed again. By contrast, a placental mother cannot abort her fetus without great risk to herself, so she is obliged to carry it to term, even if it means death for her. In other words, marsupial mothers make less parental investment in each young, but suffer less risk as a result. The main disadvantages of marsupial reproduction are that the young are born with a smaller neocortex in the brain, due to their abbreviated development, and require a longer time to mature and be weaned from the mother. By contrast, some mammals (such as rabbits or rodents) can shorten their generation time until the babies are weaned, and thus can produce offspring faster than most marsupials.

Today, marsupials comprise most of the native fauna of Australia, and are restricted to that continent (except for the opossums and their South American relatives). In the Cretaceous, however, marsupials were widespread and found on most of the continents (they were the most common mammals in North America during the reign of *Tyrannosaurus*). After the K/T event, however, the balance shifted to placentals on the northern continents, and only opossums persisted through much of the Cenozoic in Europe or North America. By contrast, the marsupials did very well on the southern continents of South America and Australia, where there was little placental competition (and also Antarctica, before it froze over). In South America, there were no large carnivorous placental mammals during the early Cenozoic, so marsupials occupied that niche. Some of them (the borhyaenids) were shaped much like wolves or hyaenas, while another (*Thylacosmilus*) was a sabertooth that closely resembles the placental sabertooth cat. Most of these marsupial predators disappeared as placental carnivorans came from North America when the Panamanian land bridge opened in the Pliocene. However, South America still supports a large diversity of opossum-

Pouched Life: The Marsupials

> *The opossum is a marsupial and marsupials are animals who carry their young around in an abdominal pouch or marsupium. As they have done this for millions and millions of years, they are not likely to stop, no matter how you and I feel about it. Baby opossums are born in a rudimentary or unfinished state, from four to twenty at once. They are only half an inch long and smaller around than a honey bee. This seems hardly worth while, but it suits the mother opossum, and she is the one directly involved. She thinks the other animals are crazy for having such enormous babies. If one of the children comes out before his time, she hisses, "You get right back in the marsupium." (The opossum language consists of faint hisses, growls and grunts. It is perfectly intelligible to insiders).*
>
> Will Cuppy, *How to Attract the Wombat,* 1935

Most people are familiar with the opossum, kangaroo, koala bear, and thanks to Looney Tunes, the Tasmanian devil. All these animals are **marsupials**, or pouched mammals, so called because they carry their young in a pouch. Marsupials are the dominant group of mammals only on the island continent of Australia, where there are few native placental mammals. In many instances where marsupials have been forced to compete with placentals, they have lost, and this has led people to think that they are more primitive and inferior to placentals. However, marsupials are not inferior, just very different. The most obvious difference is in their reproduction, which works very differently from that of a placental (Fig. 17.60). A female marsupial has a pair of uteri (unlike the single uterus of a placental) that open into a vagina with three different branches, a central medial vagina (the birth canal), and two lateral vaginas that lead into it. The penis of many male marsupials is forked, so that it can deliver sperm to both lateral vaginas. Once fertilization occurs, the embryo develops for only a few weeks, after which the young is born essentially prematurely, with only its forelimbs functional. These limbs are important, because the embryo must crawl up the mother's belly fur and find the opening of the pouch. Once it reaches the pouch, it crawls in and clamps onto a nipple, where it completes its development.

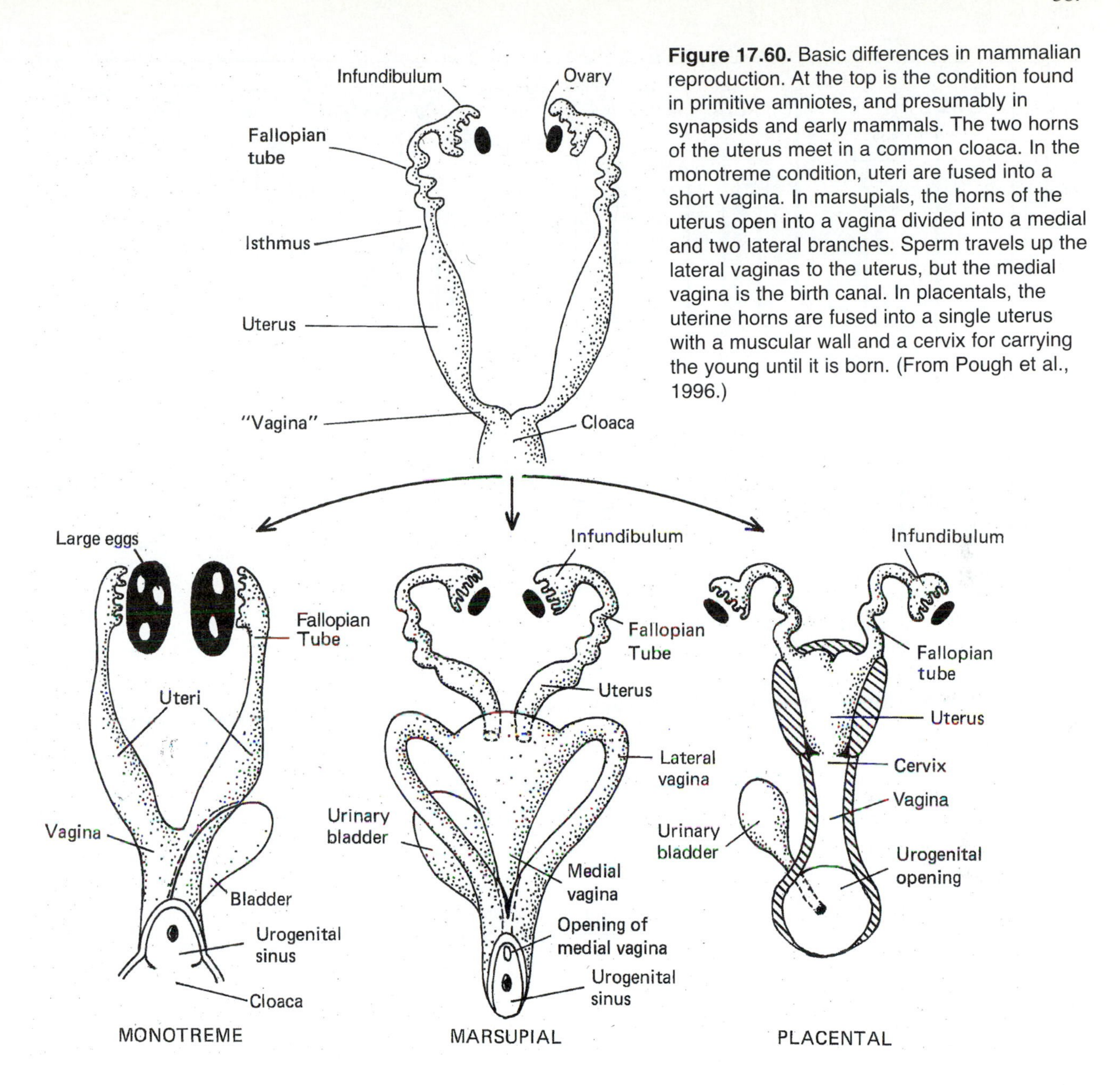

Figure 17.60. Basic differences in mammalian reproduction. At the top is the condition found in primitive amniotes, and presumably in synapsids and early mammals. The two horns of the uterus meet in a common cloaca. In the monotreme condition, uteri are fused into a short vagina. In marsupials, the horns of the uterus open into a vagina divided into a medial and two lateral branches. Sperm travels up the lateral vaginas to the uterus, but the medial vagina is the birth canal. In placentals, the uterine horns are fused into a single uterus with a muscular wall and a cervix for carrying the young until it is born. (From Pough et al., 1996.)

like marsupials.

In Australia, the situation was even simpler. Only one possible placental fossil is known from that continent before humans arrived with their animals in the Pleistocene, so Australia was apparently completely isolated from placentals during most of the Cenozoic. In the absence of such placental competition, marsupials evolved into a great variety of body forms to fill the niches occupied by placentals on other continents (Fig. 9.1). There were marsupial equivalents of moles, mice, cats, flying squirrels, wolves, groundhogs, anteaters, and many other body forms. In addition, there are many body forms that placentals never invented. Kangaroos are the main herbivorous marsupials, but they get along by hopping, an innovation that hoofed placental mammals never discovered. In the Australian Pleistocene, there were giant wombats the size of rhinos, and kangaroos almost twice the size of any living species. There was even a marsupial "lion," *Thylacoleo*, that had a peculiarly short skull with long cutting blades in its jaws instead of multiple shearing teeth, like a placental carnivore.

Most of these giant Pleistocene marsupials vanished as the Ice Ages ended and climate changed. However, the most dangerous change for marsupials was the invasion of aborigines to Australia about 40,000 years ago, and with them their placental dogs (dingoes). When Europeans came about two centuries ago, they brought other destructive placentals, such as goats, rats, and rabbits. Today, many of the native Australian marsupials are endangered as their habitats disappear and placental mammals continue to take over. Australia has long been a "living museum" of unique animals that evolved in isolation through over 70 million years, but that "museum" may vanish within another century.

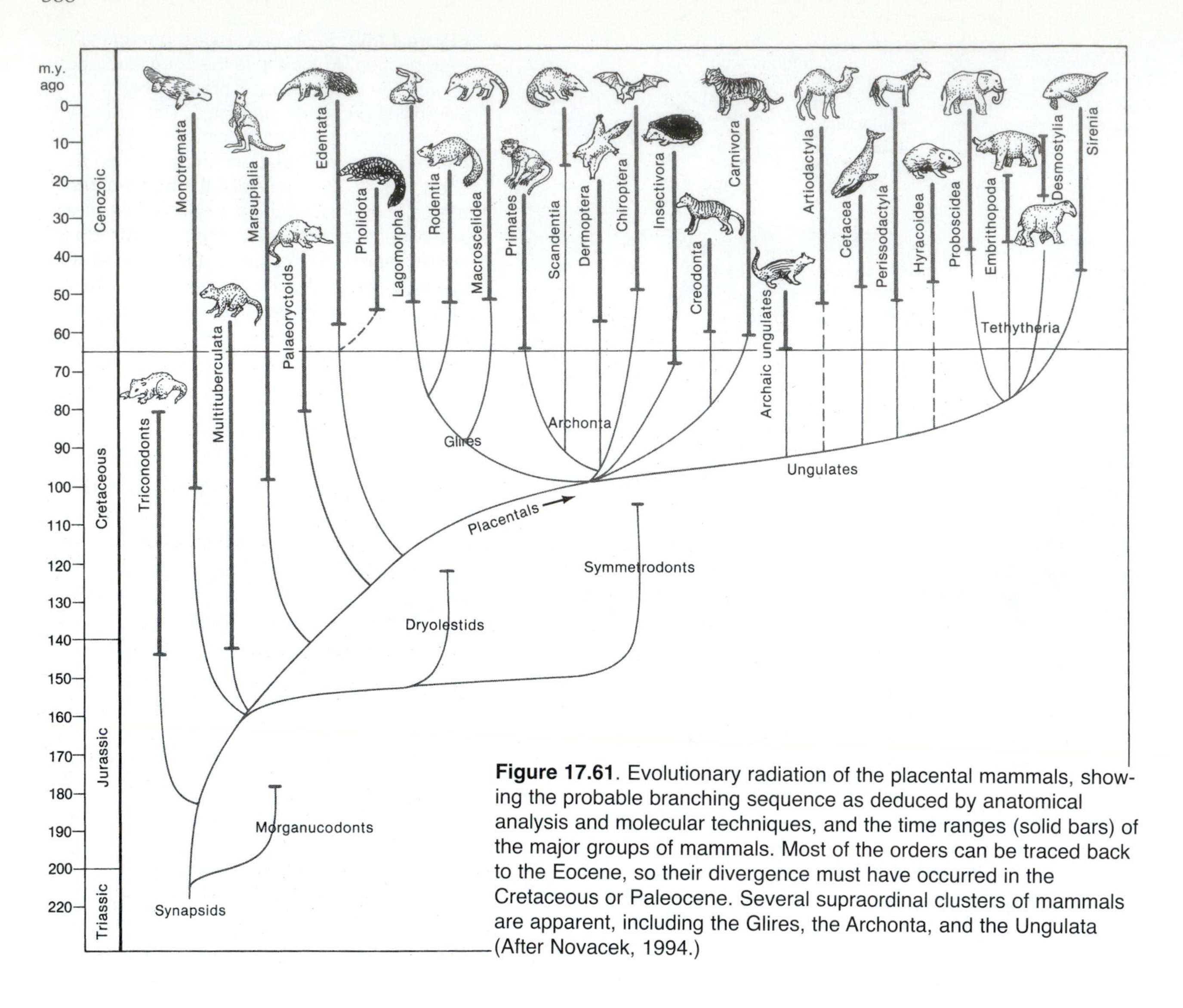

Figure 17.61. Evolutionary radiation of the placental mammals, showing the probable branching sequence as deduced by anatomical analysis and molecular techniques, and the time ranges (solid bars) of the major groups of mammals. Most of the orders can be traced back to the Eocene, so their divergence must have occurred in the Cretaceous or Paleocene. Several supraordinal clusters of mammals are apparent, including the Glires, the Archonta, and the Ungulata (After Novacek, 1994.)

The Placental Explosion

The placental or eutherian mammals comprise about twenty living orders and several extinct ones. The morphological and adaptive range of this group is extraordinary; diversification has produced lineages as varied as human and their primate relatives, flying bats, swimming whales, ant-eating anteaters, pangolins, and aardvarks, a baroque extravagance of horned, antlered, and trunk-nosed herbivores (ungulates), as well as the supremely diverse rats, mice, beaver and porcupines of the order Rodentia. Such adaptive diversity, and the emergence of thousands of living and fossil species, apparently resulted from a radiation beginning in the late Mesozoic between 65 and 80 million years ago. This explosive radiation is one of the more intriguing chapters in vertebrate history.

Michael J. Novacek, "The Radiation of Placental Mammals," 1994

Placentals make up about 95% of the fossil and living mammals. In Simpson's (1945) classification of mammals, there were over 2600 placental genera, compared to a few hundred marsupials, and a few dozen multituberculates, monotremes, and other Mesozoic forms. The number of described taxa has greatly increased in the last 50 years. Teeth that are recognizably placental are known from the late Early Cretaceous (about 110 Ma), and by the early Late Cretaceous (about 85 Ma), some of the main branches (such as the earliest hoofed mammals, or ungulates) had already differentiated. A diverse fauna of placentals is found in the uppermost Cretaceous (65 Ma) beds which entombed *Tyrannosaurus* and *Triceratops*, the Hell Creek Formation of Montana and the Lance Formation of Wyoming, as well as in the Upper Cretaceous beds of Mongolia. Although most of these animals are rat- to cat-sized insectivorous forms, it is already possible to recognize the earliest primate-like fossils (*Purgatorius*), several kinds of hoofed mammals (*Protungulatum*), insectivorans (*Batodon*), fossils that have been linked to the carnivorous mammals (*Cimolestes*), and others.

Once the non-avian dinosaurs were gone, however, placental mammals underwent an explosive adaptive radiation, so that by the early Eocene, nearly all of the 20 or so living orders, and numerous extinct ones, had appeared (Fig. 17.61). This includes not only true carnivorans, insectivorans, rodents, primates, and several orders of hoofed mammals, but animals as different as bats and whales. Evolutionary biologists have long regarded this as one of the most spectacular adaptive radiations ever documented, although as we have seen, the same thing may have happened with early Cenozoic birds, and three times in the ammonites.

For over a century, paleontologists have tried to piece together the origin and early history of each of the orders of mammals, primarily by studying the scrappy teeth and jaws collected from the Cretaceous and Paleocene. Despite all this effort, however, little progress was made from the time of William King Gregory's (1910) massive monograph, *The Orders of Mammals*, until the late 1970s. This was due to several problems. For one thing, a lot of the important evidence is available from anatomy other than the teeth and jaws (especially from the braincase, ear region, and other parts of the skull and skeleton), and yet mammalian paleontologists persisted in trying to trace ancestral-descendant sequences of teeth back through the rocks. Another problem was that the most studied collections were primarily from North America and Europe, so paleontologists tended to try to link together fossils found in the same area, neglecting the possibility of immigration from other continents. When the excellent fossil record of the Paleocene of China was finally opened for study by international scientists in the late 1970s and 1980s, many of the important "missing links" turned up in Asia, not in North America or Europe.

The breakthrough came in the 1970s due to two developments: cladistic analysis and molecular biology. Starting with McKenna's (1975) landmark paper on mammal phylogeny, paleontologists began to apply cladistic analysis to the problem of eutherian interrelationships, and began to use the neglected data base of shared derived characters in the skull, ear region, skeleton, and non-skeletal anatomy. In addition, most of the orders of mammals have living representatives, so it is possible to analyze their molecular similarities as well. Together, these parallel research programs have made tremendous progress in understanding placental evolution, and debunked many of the longstanding myths that persisted over the years. Some of the critical papers were published in volumes edited by Benton (1988) and by Szalay, Novacek, and McKenna (1993), and much of the research has been summarized by Novacek (1986, 1990, 1992, 1994) and Novacek and Wyss (1986). Our current perspective on placental evolution looks very different from that published even in the more recent textbooks (e.g., Carroll, 1988; Colbert and Morales, 1991).

Edentates—One of the first myths to be debunked was the notion that insectivores were the ancestors of all other mammals. It is true that most Mesozoic mammals were small, insectivorous creatures, since they could not grow large and eat other kinds of food as long as the dinosaurs reigned (the multituberculates were the only important herbivorous mammals in the Mesozoic). But the well-defined monophyletic group known as the order Insectivora (composed of shrews, moles, and hedgehogs) was improperly expanded to include a whole zoo full of unrelated beasts, including tree shrews, elephant shrews, and many extinct Mesozoic and early Cenozoic mammals sometimes thrown in the wastebasket order "Proteutheria." Bats were supposedly derived from this amorphous cloud of animals because some have an insectivorous diet; it turns out they are more closely related to primates. At one time or another, all of the rest of the mammalian orders were also traced to one or more "insectivores" of the Late Cretaceous.

When McKenna (1975) applied cladistic analysis to the problem, however, a surprising result emerged. Of the living placentals, the most primitive group was not the Insectivora (shrews, moles, and hedgehogs), but the **edentates** (anteaters, sloths, and armadillos, known as the order Xenarthra). Although the name "edentate" implies that they are toothless, only anteaters fit that description, since sloths and armadillos have simple teeth with no coating of enamel. Because they don't have an abundant fossil record of teeth, edentates were long neglected in the analysis. But their remaining anatomical features show that they are very primitive placentals, lacking many of the specializations found in all other eutherians (the Epitheria of McKenna, 1975). For example, edentate females have a uterus simplex, which is divided by a septum and has no cervix. Edentate metabolism tends to be much slower and less well regulated than that of other placentals. Edentates still retain a few primitive amniote bones that all other placentals have lost, and their brain and neural development is also much less advanced. One of the most consistent characters is the presence of a stirrup-shaped stapes in epitherian placental mammals. Edentates have the primitive amniote rod-like stapes with no hole at the base for the stapedial artery. In addition, edentates have many unique synapomorphies of their own, including extra articulations between the vertebrae (the name Xenarthra refers to these "strange joints"), odd fusions of the hip region with the vertebrae of the back and tail, and many other unusual features in the shoulder, ankle, and skull.

Part of the reason for the neglect of edentates is that most of their evolution took place in isolation in South America, so they were seldom studied by northern hemisphere paleontologists. Their Cretaceous ancestors were among the earliest mammals to evolve on that continent while it was mostly separate from the rest of the world, and consequently, through the Cenozoic, South America hosted a wide variety of edentates, including the huge ground sloths (*Megatherium* towered over 6 m tall and weighed 3 tons) and giant relatives of the armadillos, the glyptodonts (which were over 2 m long, weighed 2 tons including 400 kg of bony armor, and had a spiked club at the tip of their tail). However, edentates were not always confined to

South America. An anteater, *Eurotamandua*, is known from the middle Eocene Messel *Lagerstätten* of Germany, and a strange edentate-like animal, *Ernanadon*, has been described from the Paleocene of China. Edentates were among the few South American natives to successfully march north across the Panamanian land bridge in the Plio-Pleistocene against to the tide of North American mammals heading south. Ground sloths, armadillos, glyptodonts were all common in the Pleistocene of North America. The "scaly anteaters," or pangolins (order Pholidota) are presently restricted to tropical Africa and southeast Asia, but they are known from Eocene and Oligocene fossils in Europe, North America, and China. Although considered a separate order from edentates, some molecular data and anatomical analyses place them closer to the edentates than to anything else.

After the edentates branched off (sometime in the early Late Cretaceous before 85 Ma), the remaining placentals (Epitheria) split up into into five supraordinal groups (Fig. 17.61): the **Insectivora** (moles, shrews, and hedgehogs); the carnivorous mammals (the true **Carnivora**, plus the extinct creodonts); the **Glires** (rodents, rabbits, elephant shrews, and their extinct relatives); the **Archonta** (primates, bats, tree shrews, colugos, and their extinct relatives); and the **ungulates** (the hoofed mammals, including whales). How these five supraordinal groups are related is still controversial. Some molecular and morphological data, for example, tend to support a relationships between ungulates and carnivores, but other information clusters insectivores, carnivores, and archontans. For example, the presence of a bony support in the male penis (the baculum bone) is known from Primates, Rodentia, Insectivora, Carnivora, and Chiroptera (= bats), suggesting a connection between archontans (Primates, Chiroptera), Insectivora, Carnivora, and Glires (Rodentia). However, this character could be primitive for the Epitheria, and then lost in the ungulates, which are the only epitheres lacking a baculum. For the present, we will treat these five taxa as equal groups of supraordinal rank. Let us consider each in turn.

Insectivora—Today, there are three main living groups of insectivorous mammals, the shrews, moles, and hedgehogs, that form a well-defined natural group, the order Insectivora (sometimes called the Lipotyphla). This taxon is easy to recognize by a number of features besides the sharp, high-crowned "reversed triangle" teeth used for their insectivorous diet. They lose several bones in their zygomatic arch, and have several other modifications of the front of the skull. In their braincase, they have an unusual circulatory arrangement, and there are many other features about their anatomy (such as the lack of a caecum, the blind digestive pouch branching off the intestine found in most mammals). Shrews are the most primitive of all the insectivorans, yet there are some 245 living species, and they have an extensive fossil record. Despite their primitive body form, shrews have a number of specializations, including pigmented teeth. Shrews are fast-moving, active, voracious predators, eating almost continuously to feed their tiny bodies, which lose heat at a high rate due to their tiny size. They can use echo location to find their prey, and they will attack animals much larger than they are. These include not only insects and other arthropods, but also worms and small reptiles, birds, and mammals. Some have poison in their saliva that helps immobilize their prey.

Hedgehogs are mostly known from Eurasia and Africa today, but they once were common in North America as well. In Europe, these spiny little animals are well known for hunting insects and worms in the undergrowth, and rolling up into a spiny ball when threatened. Not all hedgehogs are small, however. In the Miocene, some Mediterranean islands supported a giant hedgehog as big as a medium-sized dog, which had a robust skull with large incisors for killing sizeable prey.

Moles are familiar for their tunnels, which serve not only for protection, but primarily to trap prey. When a worm drops into the tunnel and thrashes around, the mole runs to kill and eat it. Due to their subterranean habitat, moles have greatly reduced their eyes, and many are completely blind. They rely on their smell, hearing, and some have special organs on their nose to feel their way around in the darkness. Moles are also highly modified for digging, with robust, muscular forearms and large claws on their hands for pushing earth aside.

In addition to these groups, the lipotyphlan insectivorans include several other living oddities, including the golden moles and the otter shrews of Africa, and the tenrecs, a radiation of insectivorans unique to Madagascar. Another unique "living fossil," the shrew-like solenodon, was once found on the islands of Cuba and Hispaniola, apparently isolated on those islands since they first rose out of the sea. Most are now thought to be extinct since the 1930s thanks to humans and their accompanying rats.

Carnivorans and creodonts—The earliest placentals (and most living insectivores) had molar teeth with a strongly "reversed triangle" pattern for chopping up the hard cuticle of insects and other arthropods. Cutting up flesh is another matter. It requires teeth that are modified into long, sharp points and blades. The upper blades must shear against the edges of the lower blades in a scissor-like fashion to cut up muscles, tendons, and even break bones. This kind of dentition is so stereotyped for any fully carnivorous animal that several different groups have independently developed it (such as the carnivorous marsupials, and even some hoofed mammals), and once it is developed, it is highly constrained and changes very little. Thus, we must look to other parts of the anatomy (especially the braincase) to decipher the relationships of carnivores, and get past the convergent evolution of their teeth.

Once this is done, two main groups of carnivorous mammals are recognized: the extinct creodonts (an early experiment in carnivory that became extinct in the Pliocene), and the members of the living order Carnivora, or the carnivorans (Fig. 17.62). (Note that "carnivores" describes their diet, but "carnivorans" is a taxonomic term referring to

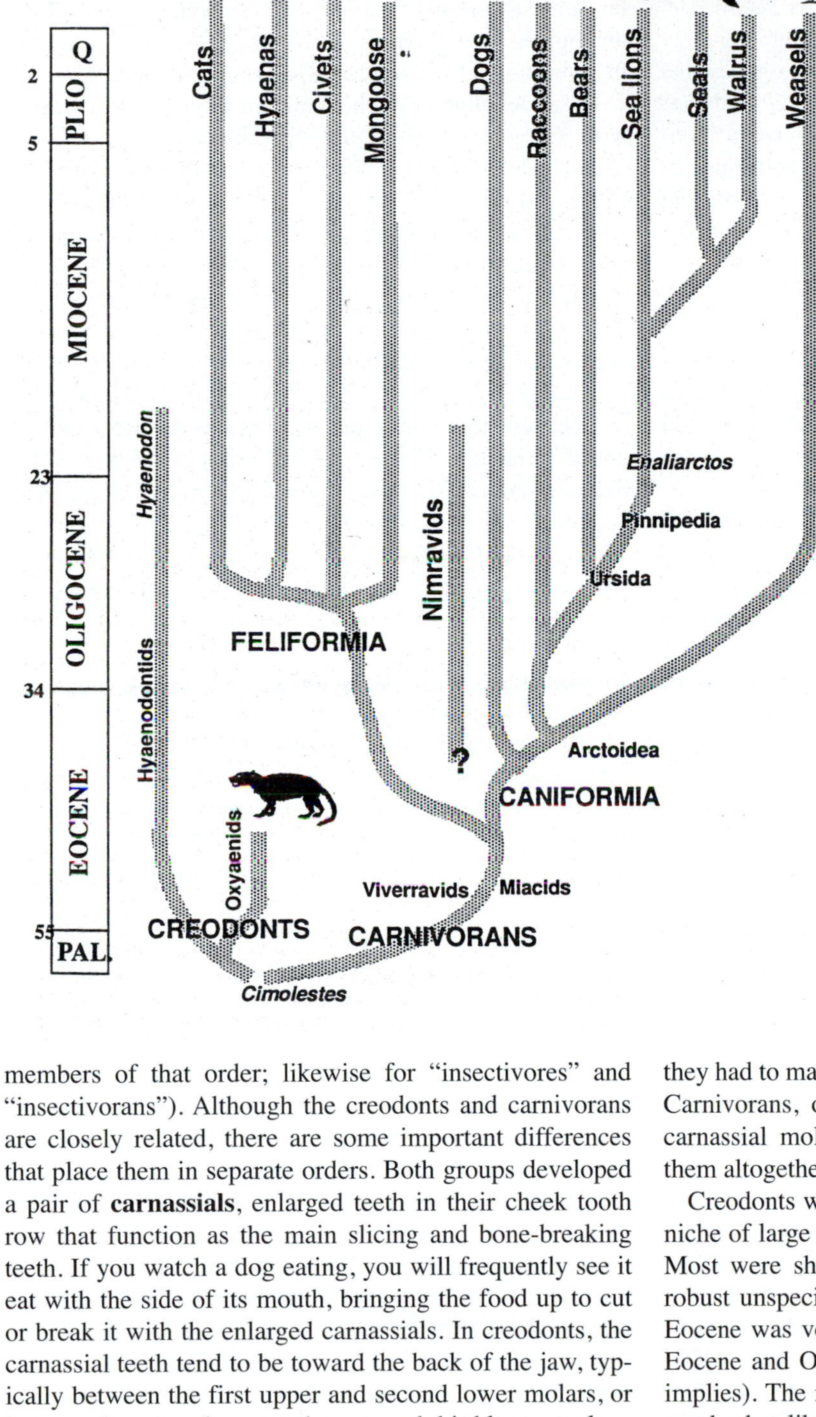

Figure 17.62. Evolution of the carnivorans and creodonts. (From Prothero, 1994.)

members of that order; likewise for "insectivores" and "insectivorans"). Although the creodonts and carnivorans are closely related, there are some important differences that place them in separate orders. Both groups developed a pair of **carnassials**, enlarged teeth in their cheek tooth row that function as the main slicing and bone-breaking teeth. If you watch a dog eating, you will frequently see it eat with the side of its mouth, bringing the food up to cut or break it with the enlarged carnassials. In creodonts, the carnassial teeth tend to be toward the back of the jaw, typically between the first upper and second lower molars, or in some, between the second upper and third lower molars. Carnivorans, on the other hand, always use their last upper premolar and first lower molar as the carnassial pair. This distinction is important for a number of reasons. Not only is it a diagnostic feature that can be found consistently, no matter how carnivores modify their diets (for example, bears and raccoons are omnivores, and pandas eat bamboo), but it is important for functional reasons as well. Creodonts had much less evolutionary flexibility, because the location of carnassials so far back in the jaw meant that they had to maintain a very standard, stereotyped dentition. Carnivorans, on the other hand, can modify their postcarnassial molars in many different ways, or even lose them altogether, permitting many dietary specializations.

Creodonts were among the first mammals to occupy the niche of large predators during the Paleocene and Eocene. Most were shaped like dogs or weasels, with relatively robust unspecialized limbs, although *Patriofelis* from the Eocene was very bear-like, and *Hyaenodon* from the late Eocene and Oligocene was very hyena-like (as the name implies). The middle Eocene creodont *Apataelurus* developed saber-like canines, just as two groups of carnivorans and the marsupial *Thylacosmilus* did (see Fig. 7.15). *Sarkastodon* from the late Eocene of Mongolia was a huge predator larger than any bear, with enormous blunt teeth for bone crushing. Most of the creodonts were extinct by the late Eocene, but the hyaenodont lineage persisted well into the Pliocene in Eurasia and Africa. In Africa during the Miocene, there was a huge hyenodont named *Megistotherium*, which had a skull twice as long as a tiger, with huge canines and enormous jaw muscles.

The oldest known carnivorans are weasel-like and raccoon-like animals from the Paleocene and early Eocene which have been loosely lumped into the "wastebasket" group called the "miacids." As the creodonts declined in the late Eocene, the order Carnivora radiated into a variety of body forms. and differentiated into two branches: the Caniformia (dogs, bears, raccoons, weasels, seals, and their relatives) and the Feliformia (cats, hyenas, civets, mongooses, and their relatives) (Fig. 17.62). The caniforms differentiated very early, with the dogs (Canidae) appearing by the end of the middle Eocene, and diversifying on many continents throughout the Cenozoic. In addition to the typical body forms of wolves, foxes, and coyotes, dogs have also come in weasel-like shapes, and one group, the borophagines, were huge, bone-crushing hyena-like predators in North America in the Miocene (since North America never had true hyenas). Bears, raccoons, and the weasel clan (otters, minks, wolverines, badgers, skunks, and so on) appeared somewhat later, although they too were very diverse throughout the later Cenozoic on the northern continents.

Seals, sea lions, and walruses have long been placed in their own order "Pinnipedia," but recent analyses show that they are closely related to primitive bears, and should be a subgroup of the Carnivora. The transition from bears to seals can be demonstrated by the early Miocene *Enaliarctos*, which has many primitive bear-like features in the skull and teeth, even though it had flippers and other aquatic features like seals (and it is found in marine rocks).

The feliform branch of the Carnivora can probably be traced to a group of Paleocene-Eocene "miacids" called the viverravids (Fig 17.62). However, the first true cats are not known until *Proailurus* of the early Miocene. From these roots, dozens of cat genera are known, including at least four different genera of saber-toothed cats, found worldwide during the Plio-Pleistocene. The mongoose and civet lineages go back at least to the Oligocene, and the earliest hyena fossils are known from the Miocene.

In addition to these familiar living families, there was also an extinct family of carnivorans known as the nimravids. These animals were extremely cat-like in appearance, occupying the normal cat and sabertooth ecological niches during the late Eocene and Oligocene. They have long been called "paleofelids" or "false cats," but this is misleading. All their cat-like features (especially the teeth and jaws) are due to evolutionary convergence. The details of their skull, braincase, ear region, and skeleton show that they are not cats at all. Some paleontologists argue that in fact they are closer to caniforms, while others place them as a distant relative of the feliforms. However, in nearly every textbook and illustration of the famous Badlands nimravids such as *Dinictis* and sabertooth *Hoplophoneus*, they are still mislabeled as "cats."

Archontans—Many people are surprised (and sometimes uncomfortable) to learn that our own order, the Primates, is most closely related to the bats, tree shrews, and colugos. Yet that is the conclusion supported by a variety of anatomical characters (particularly in the braincase and foot region) that unite a supraordinal group called the Archonta by Gregory (1910). In addition, a great variety of molecular analyses seem to support the Archonta as well. Most archontans have features for living in the trees, and from there became gliders (colugos) or flying mammals (bats).

The oldest known archontans are the primates, starting with *Purgatorius* from the uppermost Cretaceous Hell Creek beds of Montana. During the early Cenozoic, lemur-like and squirrel-like primates were among the most common and diverse mammals in North America and Europe, because these regions were still covered by tropical jungle vegetation, even up to the Arctic Circle. As climates cooled and the forests retreated in the late Eocene and Oligocene, primates disappeared from most of their former habitats, eventually becoming restricted to Africa by the early Oligocene. From this origin, they again spread around the world. In the early Oligocene, the ancestors of New World monkeys apparently rafted from Africa to the island continent of South America, where they radiated into spider monkeys, howler monkeys, marmosets, and all the other New World monkeys (the family Cebidae). Meanwhile the Old World monkeys (baboons, rhesus monkeys, macaques, and their kin) continued to diversify in Africa through the Oligocene and Miocene. By the early Miocene, they competed with a great radiation of apes as well, and for much of the Miocene, apes were more common and diverse in Africa than were monkeys. By the Pliocene, one group of apes, the hominids, split off from the rest, and they were our ancestors.

The bat fossil record is not known until the middle Eocene, with extraordinary specimens of complete artciulated bats from the Green River Shale of Wyoming, which entombed *Icaronycteris*, and the Messel *Lagerstätten* of Germany. These earliest known bats already had wings supported by the elongated bones of all five fingers, and the size and proportions of a modern bat. However, they still retained many primitive features lost in later bats, including a full placental dentition (modern bats have reduced dentitions), an unfused breastbone with no keel for the wing muscles, claws on the fingers, and a long tail that was free of the wing membrane. From this origin, bats quickly diversified into the tiny Microchiroptera (the insect-eating bats, which live in caves and hunt flying prey at night by echolocation) and the much bigger Megachiroptera (the fruit bats, which live in trees, and fly during the day to seek fruit). By the Oligocene, most of the families of bats had differentiated. Today, bats are the second most diverse order of mammals alive, after rodents. There are over 780 species, 140 genera, and 17 families of Microchiroptera living today, and a smaller number of fruit bats.

The tree shrews (order Scandentia), as the name implies, were long lumped with the true shrews in the Insectivora, but in recent years their similarity to ancestral primates has been noted again and again. Tree shrews lack most of the derived characters of the true lipotyphlan insectivorans, but instead have a number of derived archontan features. Even though they live much like true shrews, they have become the models for what the ancestral archontan and early primates must have looked like.

The colugos, or "flying lemurs," of the order Dermoptera are represented by a single living genus, *Cynocephalus*, which is found only in southeast Asia. The name "flying lemurs" is misleading, since they neither fly nor are they lemurs. Instead, they resemble primitive primates or insectivores, but have developed a gliding membrane between the front and hindlimbs, which they use to sail from branch to branch like a flying squirrel (which does not fly, but also glides). The colugos were long an evolutionary mystery, a single genus in their own isolated order, until they were connected to some fairly common fossils from the Paleocene and Eocene of North America called plagiomenids. This group was unusually common in the Canadian Arctic in the early Eocene, when that region was warm and forested, and survived in North America until the late Oligocene. After that, there is no further fossil record of dermopterans until their sole living representative.

Glires—By far the most common and successful group of placentals is the order Rodentia. They are incredibly diverse (over 40% of the living mammals, or at least 350 genera and 1700 species, are rodents), disparate (occupying body forms from the pig-sized capybara, to aquatic beavers and muskrats, gliding, tree-climbing, and burrowing squirrels, spiny porcupines, subterranean gophers and naked mole rats, and hundreds of different kinds of rats and mice), and they are also incredibly abundant. One only needs to think about the ability of rats or mice or hamsters to multiply to realize why they are by far the most common mammals on the planet. If it were not for their predators, the earth would be a planet of rodents. Rodents are usually the dominant group in the small-body-size niche, but occasionally they become huge. The largest living rodent, the capybara, weighs about 40 kg, but the Miocene capybara *Telicomys* was the size of a rhinoceros, and the the Pleistocene beaver *Castoroides* weighed about 200 kg and reach 2.5 m in length, as large as a bear.

Rodents have a number of unique features, but their most obvious is their pair of chisel-like (gliriform) upper and lower incisors, which are used to gnaw their hard-shelled food and vegetation, and in some groups, to cut down trees or dig tunnels or burrows. These incisors are constantly growing, with open roots, and must be continuously worn down into a sharp point by abrading them together (Fig. 17.63). If there is a problem with occlusion so the incisors are not sharpened down, they will continue growing in a curve until they curl around and puncture the top of the skull. There is a toothless gap (diastema) behind the incisors, and then a row of premolars and molars that are adapted for grinding their diet of seeds, nuts, and vegetation.

This small-bodied, seeds/nuts/vegetation-gnawing diet and lifestyle was very successful, as demonstrated by the fact that multituberculates occupied this niche for most of the Mesozoic, and several groups of primates also were built like rodents in the Paleocene. When rodents spread from Asia to North America and Europe in the early Eocene, however, they began to displace the earlier occupants, so that by the Oligocene, multituberculates and rodent-like primates were extinct. The early rodents had very primitive **protrogomorph** skulls (Fig. 17.63), with the masseteric muscles attached only to a limited area along the base of the zygomatic arch (as in other mammals). By the late Eocene, they had diversified into three main lineages. The **sciuromorphs** are only slightly more specialized than the ancestral protrogomorphs, with the masseter muscles extending up along the front of the zyomatic arch to the side of the snout. Sciuromorphs include the squirrels and all their relatives, including chipmunks, woodchucks and marmots, and the beavers. In the second condition, known as **hystricomorph**, the masseter muscle passes up through the zygomatic arch and onto the snout through a hole for the passage of nerves called the infraorbital foramen. Hystricomorphs include not only the porcupines (both North American and African), and most other African rodents, but also the incredible radiation of

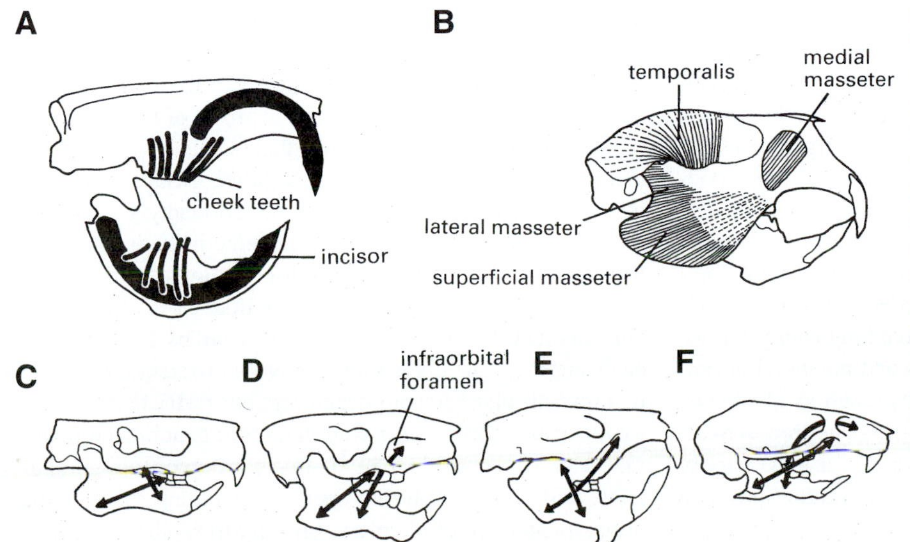

Figure 17.63. Rodents have many important specializations of the teeth and jaws. (A) Their ever-growing chisel-like incisors have deep roots, and must be constantly sharpened. (B) Configuration of the jaw muscles in a porcupine. A branch of the medial masseter passes through the infraorbital foramen on the front of the skull. The positon of the masseter muscles is shown in the (C) protrogomorph, (D) hystricomorph, (E) sciuromorph, and (F) myomorph conditions of the rodent skull. (After Benton, 1990.)

native South American rodents, the caviomorphs (including the guinea pigs, capybaras, chinchillas, agoutis, and many less familiar animals). The caviomorphs first arrived in South America in the Oligocene, presumably from African hystricomorph ancestors that rafted there across the Atlantic. (The same scenario applies to the New World monkeys as well.) The most specialized condition is known as **myomorph**, and it combines a strand of the masseter passing along the front of the zygomatic arch with another passing through the infraorbital foramen. The vast majority of rodents, including the rats, mice, hamsters, voles, lemmings, and their kin, exhibit this condition.

Most people are surprised to learn that rabbits are not rodents, but they have always been placed in their own order Lagomorpha, along with the hamster-like pikas. Lagomorphs have two pairs of chisel-like incisors, in contrast to the single pair in rodents, and a number of other unique specializations. Although rabbits were originally classed with rodents, for most of this century opinion swung away from this hypothesis, and attributed their similarities to parallelism. However, the last decade has seen opinion shift back again to the grouping of lagomorphs and rodents, called the Glires by Gregory (1910). Not only is there a lot of anatomical evidence to support it, but also much of the recent molecular data suggest a close relationship as well. In addition, recent studies of Chinese Paleocene eurymylids show that both rabbits and rodents probably originated from a eurymylid ancestor, and then emigrated to other regions in the Eocene.

The third group assigned to the Glires is the elephant shrews, or Macroscelidea, which resemble hopping shrews with a long snout. They were long placed with the Insectivora because of shared primitive characters and their insectivorous diet. However, they have a number of anatomical specializations shared with rabbits and rodents, and recent molecular data also support their inclusion in the Glires.

Ungulates—After rodents and bats, the third largest group of placentals is the hoofed mammals, or ungulates. Hoofed mammals make up about 33% of the living and extinct mammalian genera, and nearly all the large-bodied herbivores are ungulates. They include the even-toed **artiodactyls** (pigs, hippos, camels, deer, antelopes, giraffes, cattle, sheep, and goats), the odd-toed **perissodactyls** (horses, rhinos, tapirs, and their extinct kin), the **tethytheres** (elephants, manatees, and their extinct relatives), the woodchuck-like **hyraxes** or conies, and, surprisingly, the whales. Ungulates have dominated not only the large herbivore niche through most of the Cenozoic, but also are the dominant aquatic predators and filter feeders, and some were even carnivorous. Some ungulates have long slender limbs for fast running (especially antelopes and horses), but others are large-bodied with robust limbs (such as elephants, rhinos, hippos, and many extinct group). The tree hyraxes even climb trees. Ungulates have occupied a wide variety of ecological niches given the constraints of their body size and diet.

Until recently, the interrelationships of the major ungulate groups was obscured by a paraphyletic ancestral "wastebasket" group, the order "Condylarthra" (Fig. 17.64). "Condylarths" had nothing in common except that they were primitive ungulates that were not members of any of the living orders. As long as this "wastebasket" group covered up the evidence, there was no possibility that ungulate relationships could be deciphered. However, when a cladistic analysis was applied to the group (Cifelli, 1983; Prothero et al., 1988), there was a clear pattern of branching among the ungulate groups that has withstood repeated testing from additional morphological and molecular analyses (Prothero, 1993; Thewissen and Domning, 1992; Court, 1990; Archibald, 1998). It turned out that throwing taxa into the "Condylarthra" hid a phylogenetic pattern for over a century, but a focus on shared derived characters (plus the great increase in numbers of taxa and characters) was able tease out that pattern.

The earliest ungulates are known from the early Late Cretaceous (about 85 Ma) of Uzbekistan (Archibald, 1996), and show that the major placental divergences must have come quite early. Better specimens of ungulates are known from the latest Cretaceous, where *Protungulatum* is among the more common taxa. Although these Mesozoic ungulates are known mostly from isolated teeth and bones, they still have diagnostic features. Their molars are square and lower-crowned, with rounder cusps, for eating vegetation rather than insects, and they already have distinctive features of the ankle that are recognizably ungulate.

In the Paleocene, the ungulates split into a number of distinct clades. Some of these archaic ungulates (such as the arctocyonids, hyopsodonts, and periptychids) have long been lumped into the order "Condylarthra," but each is distinctive and related to a different part of the ungulate radiation. The Paleocene arctocyonids were the most primitive of the ungulates, about the size and shape of a raccoon, and probably with a similarly omnivorous diet. The hyopsodonts, on the other hand, were most common in the Eocene, and were among the last of the surviving "condylarths." They were shaped somewhat like dachshunds, except that their multicusped teeth were clearly adapted for grinding vegetation. Another group of "condylarths," the phenacodonts, are not closely related to the other archaic ungulates, but are actually the sister-group of the clade that includes perissodactyls and tethytheres.

Surprisingly, one of the first ungulate groups to branch off was the even-toed ungulates, or artiodactyls. They are so called because the axis of symmetry in their hand and foot runs between the third and fourth digits, so they usually have either two or four toes. Artiodactyls also have a very distinctive ankle bone that has a pulley-like facet on each surface. This gives their feet very efficient movement in a fore-aft plane for rapid running, but restricts their ability to rotate their feet in a way that more generalized mammals can. Artiodactyls are the largest group of living ungulates, with over 190 living species, including most of the domesticated hoofed mammals (cattle, sheep, goats,

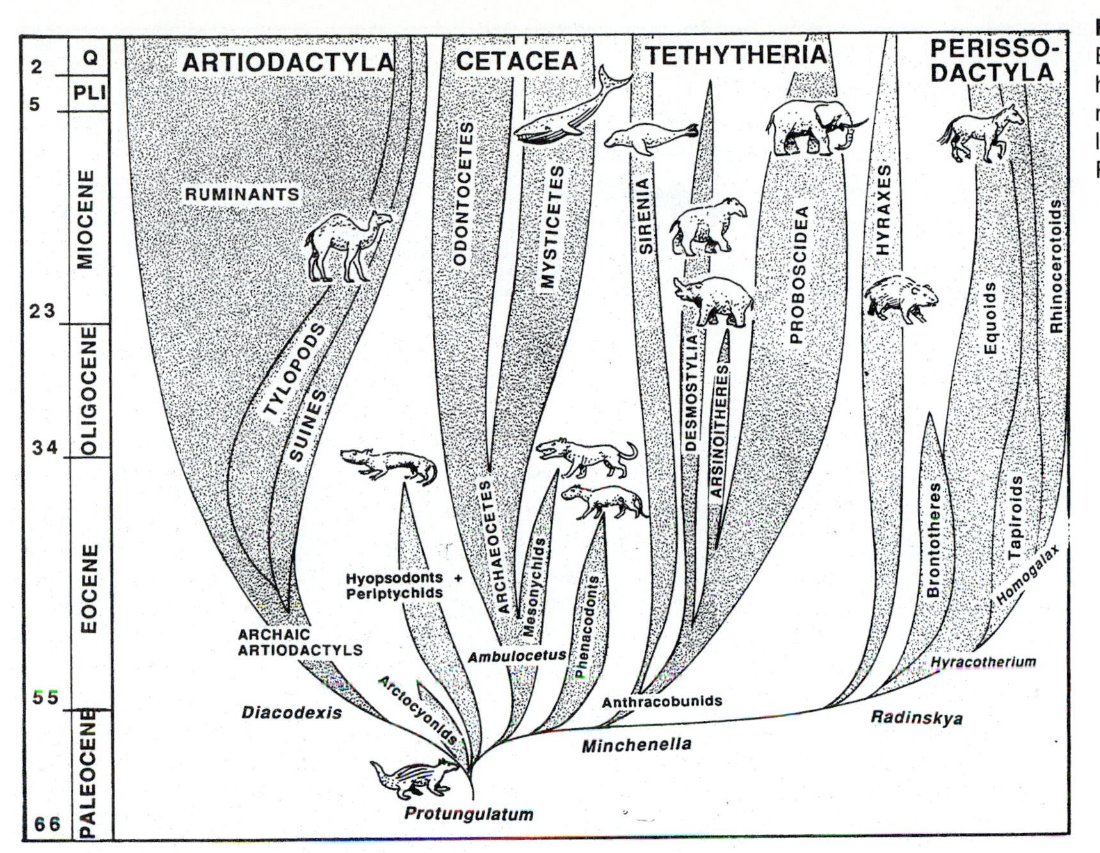

Figure 17.64. Evolution of the hoofed mammals, or ungulates. (From Prothero, 1994.)

camels, pigs) and they are the source of most of our meat, milk, and wool.

The earliest artiodactyls are known from the lower Eocene rocks of Pakistan, and shortly thereafter they spread to the rest of Eurasia and North America. These early forms were very delicately built, resembling a small hornless antelope, and some had such long legs that they may have hopped. During the Eocene, these archaic artiodactyls quickly diversified into a great variety of lineages—the heavy-bodied, omnivorous pigs and their American relatives, the peccaries or javelinas; the aquatic hippos; and the early camels, which did not yet have humps, but were built more like deer. Camels were once a strictly North American group, playing the roles on this continent that were occupied by other groups elsewhere. For example, in the Miocene there were long-necked, long-legged "giraffe-camels," delicate "gazelle-camels," and others that paralleled the shapes of many African antelopes (since North America never hosted true antelopes). In the Pliocene, camels migrated to South America across the Panamanian land bridge, giving rise to the llamas, alpacas, guanacos, and vicuñas still living there today. In the Pleistocene, they also crossed the Bering Strait to the Old World, where they evolved into dromedaries and Bactrian camels, the only groups with a hump. Then about 10,000 years ago they became extinct in their North American homeland.

In the Oligocene, another great evolutionary breakthrough occurred when a group of artiodactyls, the **ruminants**, developed a four-chambered stomach system. Ruminants first swallow their food and then let it ferment in the first stomach chamber, the rumen, where cellulose-digesting bacteria help break up the plant matter. When they have a chance, ruminants regurgitate food from the rumen and "chew their cud," which helps break it down even further. (Their appearance of thoughtfulness during this time has let to the term "rumination" for someone "chewing over" an idea.) By the time it is swallowed again, most of the nutrients can be absorbed by the intestines, so ruminants get the maximum nutrition out of each bite of vegetation. By contrast, most other herbivorous mammals (horses, rhinos, elephants, rabbits) are hindgut fermenters, and have no specialized foregut fermentation chamber, so they can get only a limited amount of nutrition out of the relatively indigestible cellulose in the food as it passes rapidly through their intestine and caecum. Consequently, hindgut fermenters must eat much larger quantities of food than ruminants, and are not as efficient or versatile. (Rabbits get around this by eating their feces, so the food goes through their digestive tract twice). With this great innovation, the ruminants (especially the deer, giraffes, cattle, antelopes, goats, and sheep) eventually became the dominant hoofed mammals of the later Cenozoic, and pushed out many other groups, such as the horses.

One of the most amazing stories in evolutionary biology is the origin of whales from land mammals (Fig. 17.65). By the middle Eocene, there were archaic fossil whales with a fully whale-like body, including a horizontal tail fluke, forelimbs modified into flippers, and no hindlimbs. However, their distinctive triangular teeth gave a clue as to

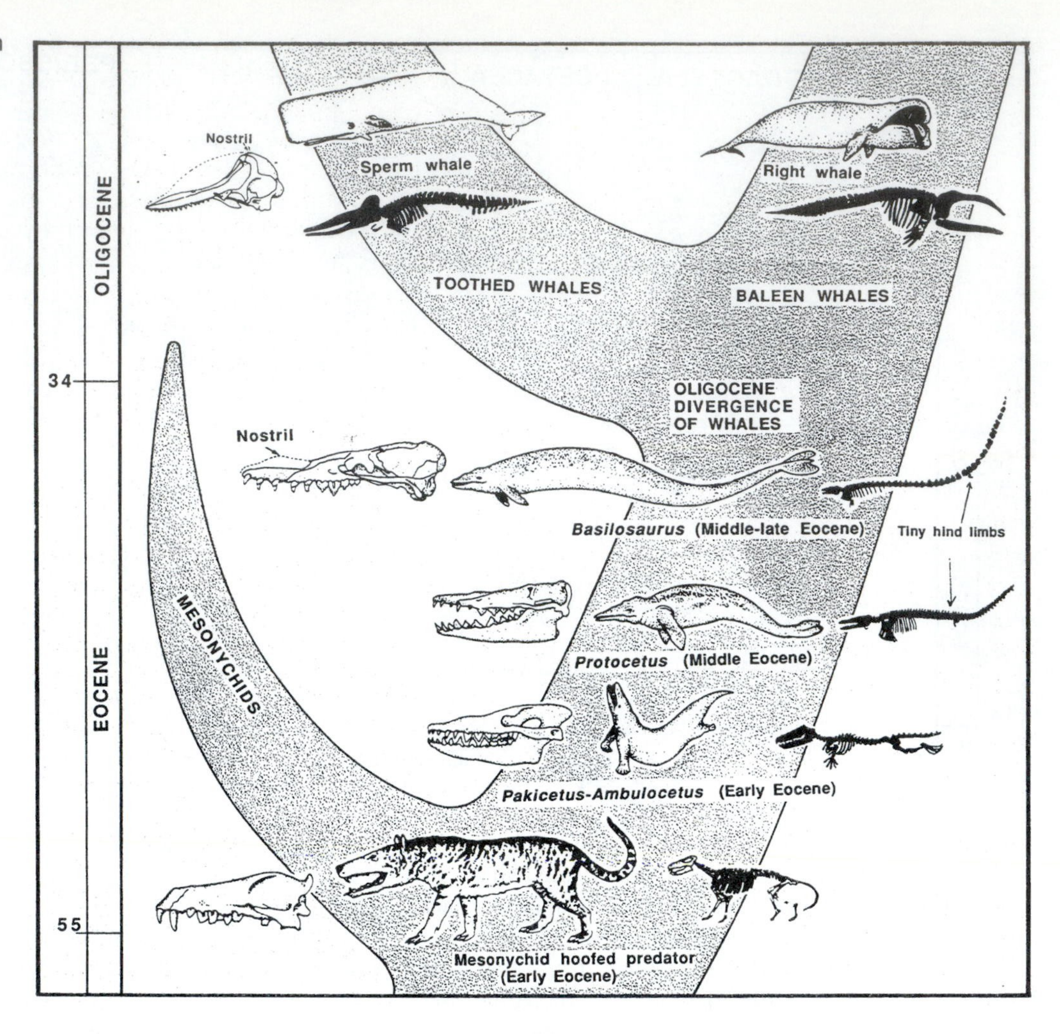

Figure 17.65. Evolution of the whales from the hooved carnivorous mesonychids. (From Prothero, 1994.)

their origins among a group of hoofed mammals known as the mesonychids.

Mesonychids were the first group of mammals to become specialized meat-eaters, appearing in the middle Paleocene before more specialized carnivorous mammals (the creodonts, and eventually the true carnivorans) occupied that niche. Most mesonychids were the size and shapes of large wolves or bears (although some were as small as foxes). They had a heavy robust skull armed with sharp canine teeth, and huge round-cusped molar teeth suitable not only for eating meat, but also for bone crushing. Their body was also very wolf-like, with a long tail and limbs. Like many modern carnivores, they walked on the tips of their long toes, rather than flat-footed. Despite all these carnivorous adaptations, however, mesonychids were derived from hoofed mammals. The proof is in their toes, which had hooves rather than claws.

By the early Eocene, mesonychids had reached their peak of diversity, with wolf-sized beasts such as *Mesonyx* or *Harpagolestes* reigning as the largest carnivorous mammals of their time. However, they had to share their world with two other groups of carnivorous mammals: the creodonts (which soon surpassed them in size and diversity); and the true carnivorans (which were still weasel-sized, and did not become large dog-sized or cat-sized predators until the Oligocene). By the middle Eocene, the mesonychids rapidly declined in North America and Eurasia, where they had once dominated. The reasons for this decline are unclear. It was a time of major climatic change, with global cooling and drying that destroyed the dense forests that mesonychids had once ruled. With the coming of open habitats, the prey species became faster and more agile. Large, clumsy predators like mesonychids might have had difficulty finding cover to ambush their prey. In addition, some paleontologists speculate that mesonychids were not as efficient at eating meat as the creodonts or carnivorans, since the blunt, rounded cusps of mesonychid molars never developed the specialized, scissor-like carnassial shear found in more specialized carnivorous mammals. For whatever reasons, the mesonychids were very rare in the late middle Eocene, and they disappeared from North America at the end of the middle Eocene, and from Asia in the late Eocene. The last of the Asian mesonychids, however, was a truly spectacular beast known as *Andrewsarchus*. Only one specimen of this animal is known, but it is a skull almost a meter long, more than twice the size of any bear that has ever lived! If the rest of the animal were also bear-like, it would have been about 4 m long, and 2 m high at the shoulder, and weighed almost four times as much as the largest known bear.

As the last of the mesonychids died out in the late Eocene, their close relatives, the whales, were already established in the oceans of the world. The earliest whales have many features of the braincase and skull, and especially their distinctive, triangular bladed teeth, that are very similar to the condition found in mesonychids. For years the oldest known whales of the early middle Eocene were fully aquatic animals without hindlimbs, and very different from mesonychids. Recently, however, numerous transitional forms between whales and mesonychids have been found from the early Eocene of Africa and Asia. The most impressive of these is *Ambulocetus* from the early Eocene of Pakistan. Although it still has a mesonychid skull and teeth, its front and hind feet are both adapted for swimming, yet it does not yet have a tail fluke. Other fossil whales have even more specialized front flippers, and have reduced their hindlimbs to tiny vestiges, and a tail with a horizontal fluke. The transformation from a carnivorous hoofed mammal to a fully aquatic whale is now one of the best documented major evolutionary transitions in the fossil record.

By the Oligocene, the archaic archaeocete whales were extinct, and were replaced by a radiation of the two modern groups of cetaceans, the odontocetes (toothed whales, including sperm whales, killer whales, dolphins, and porpoises) and the mysticetes (baleen whales, including the blue whale, right whale, humpback whale, gray whale, and many others). The more familar toothed whales are predators, feeding on fish and squid with their many conical teeth. The baleen whales, on the other hand, are toothless, and their mouth is filled with screens of horny tissue called baleen, which is used to filter out small fish and plankton. Baleen whales such as the blue whale swallow a large mouthful of seawater, and as they close their mouths, they force out the water through the filter, leaving all the food trapped in their mouths.

After the branch points for the artiodactyls, the hyopsodonts and periptychids, and the whales plus mesonychids, the remaining ungulates are a natural monophyletic groups now known as the Altungulata (Prothero and Schoch, 1989). These include the sheep-like "condylarths" known as phenacodonts (Fig. 17.64), and two major clades, the tethytheres and the perissodactyls. Tethytheres were not recognized as a group until McKenna named them in 1975, but they have a great variety of shared derived characters that unite them (Domning et al., 1986; Tassy and Shoshani, 1988; Fischer and Tassy, 1993; Shoshani, 1993). These include a single pair of teats on the breasts (like humans), eyes that are shifted far forward on the skull, cheekbones that contain a broadly expanded portion of the rear skull bones, and teeth that do not erupt from below, as in most mammals, but from the back, pushing the old teeth out the front of the jaw.

The most familiar tethytheres are the elephants and their kin, the order Proboscidea (Fig. 17.66). The two living species of elephants are but a tiny remnant of a long, distinguished history of mammoths, mastodonts, and other unusual animals. The earliest proboscideans are known from the late Paleocene of Africa, and by the Eocene they were shaped like small hippos, without trunks or tusks. In the Oligocene they diverged into numerous lineages: the deinotheres, with their downward-deflected lower tusks; the true mastodonts; and the gomphotheres, with small upper and lower tusks. Some evolved into beasts with enormous broad tusks shaped like shovels, while others had various combinations of two and four tusks with different lengths and curvatures. In the late Pleistocene, only the true mastodonts and the diversity of mammoths remained, and most of these were driven to extinction at the end of the Pleistocene.

Closely related to the Proboscidea are several other groups that had long been zoological mysteries, placed in their own isolated orders. The order Sirenia includes the manatees and dugongs, or "sea cows." These animals are completely aquatic, losing their hindlimbs and developing a circular fluke on the tail, and paddles for forelimbs. They are restricted to freshwater lakes, rivers, and estuaries, browsing the water plants, but they are so slow and docile that they are now on the endangered species list as a result of hunting and injuries from speedboats hitting them.

Another longstanding mystery was a peculiar rhino-like group from the African Oligocene known as arsinoitheres, the order Embrithopoda. These elephant-sized animals had a pair of huge, sharp, recurved bony horns on their noses, and no one had a clue as to what they were related to. However, when more archaic Eocene arsinoitheres were found in Mongolia and Turkey, McKenna and Manning (1977) suggested that they were tethytheres, and this has since been borne out (Court, 1990).

Yet another paleontological puzzle were the peculiar Pacific Miocene marine mammals known as desmostylians. These walrus-sized animals had hoofed feet rather than flippers, with a broad shovel-like tusked jaw containing bizarre molars that look more like a bundle of barrels than anything else. They were long placed in their own order Desmostylia, with no apparent relationships to anything else, until Domning et al. (1986) described an unusually primitive specimen known as *Behemotops*, and showed that desmostylians were actually tethytheres, distantly related to sirenians and proboscideans.

Still controversial are the woodchuck-like hoofed mammals known as hyraxes or conies, the order Hyracoidea. These little animals are today restricted to rocky outcrops in east African and the Middle East, but during the early Cenozoic, they were among the most common hoofed mammals in Africa, evolving into beasts with hippo-like bodies, and many other shapes as well. Traditionally, they were thought to be related to tethytheres, and that hypothesis is still supported by some (Novacek and Wyss, 1986; Novacek et al., 1988; Shoshani, 1993), while others place them with perissodactyls (Prothero et al., 1988; Fisher, 1989; Fischer and Tassy, 1993).

The perissodactyls are the order of herbivorous "odd-toed" hoofed mammals that includes the living horses,

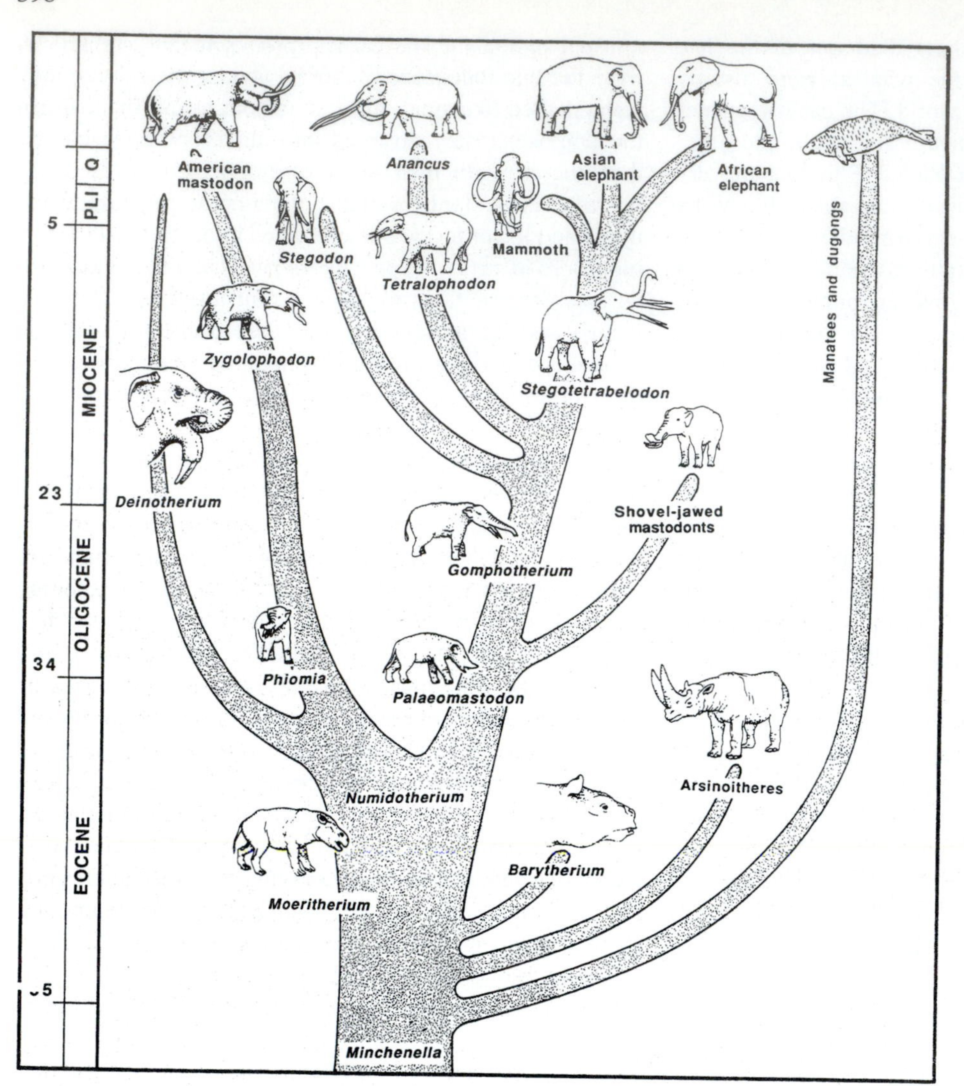

Figure 17.66. Evolution of the tethytheres (elephants, manatees and their extinct relatives). (From Prothero, 1994.)

zebras, asses, tapirs, rhinoceroses, and their extinct relatives (Fig. 17.67). They are recognized by a number of unique specializations, but their most diagnostic feature is their feet. Most perissodactyls have either one or three toes on each foot, and the axis of symmetry of the foot runs through the middle digit. They are divided into three groups: the Hippomorpha (horses and their extinct relatives); the Titanotheriomorpha (the extinct brontotheres); and the Moropomorpha (tapirs, rhinoceroses, and their extinct relatives).

Perissodactyls were once thought to have evolved in Central America from the phenacodonts, an extinct group of archaic hoofed mammals placed in the invalid taxon "Condylarthra." However, in 1989, a specimen recovered from upper Paleocene deposits in China was described and named *Radinskya*. This specimen shows that perissodactyls originated in Asia around 57 million years ago, and were unrelated to North American phenacodonts. *Radinskya* is very similar to the earliest relatives of the tethytheres. This agrees with other evidence that perissodactyls are more closely related to tethytheres than they are to any other group of mammals.

By the early Eocene, the major groups of perissodactyls had differentiated, and migrated from Asia to Europe and North America. Before the Oligocene, the brontotheres and the archaic tapirs were the largest and most abundant hoofed mammals in Eurasia and North America. After these groups became extinct, horses and rhinoceroses were the most common perissodactyls, with a great diversity of species and body forms. Both groups were decimated during another mass extinction about 5 million years ago, and today only 5 species of rhinoceros, four species of tapir, and a few species of horses, zebras, and asses cling to survival in the wild. The niches of large hoofed herbivores have been taken over by the ruminant artiodactyls, such as cattle, antelopes, deer, and their relatives.

From their Asian origin, the hippomorphs spread all over the northern continents. In Europe, the horse-like palaeotheres substituted for true horses. North America became the center of evolution of true horses, which occasionally migrated to other continents. *Protorohippus* (once called *Hyracotherium* or "*Eohippus*") was a terrier-sized horse with four toes on the front feet that lived in th early Eocene. Its descendants evolved into many different lin-

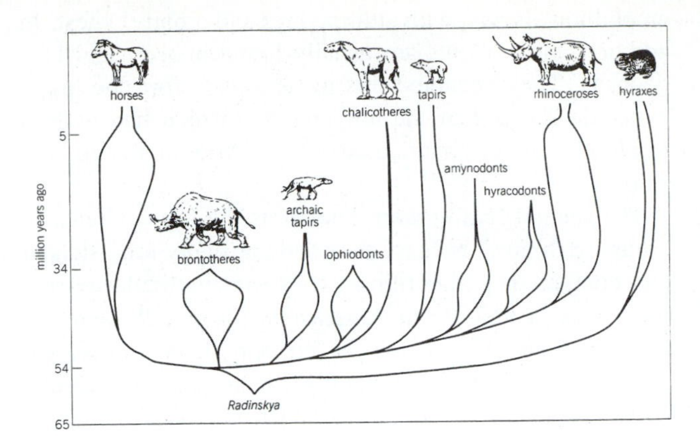

Figure 17.67. Evolution of the perissodactyls. (From Prothero, 1994.)

eages living side-by-side. The late Eocene-early Oligocene collie-sized three-toed horses *Mesohippus* and *Miohippus* were once believed to be sequential segments on the unbranched trunk of the horse evolutionary tree. However, they coexisted for millions of years, with five different species of the two genera living at the same time and place. From *Miohippus*-like ancestors, horses diversified into many different ecological niches. One major lineage, the anchitherines, retained low-crowned teeth, presumably for browsing soft leaves in the forests. Some anchitherines, such as *Megahippus*, were almost as large as the modern horse. *Anchitherium* migrated from North America to Europe in the late early Miocene, the first true horse to reach Europe.

In the middle Miocene, there were at least 12 different lineages of three-toed horses in North America, each with slightly different ecological specializations. This situation is analogous to the diversity of modern antelopes in East Africa. The ancestors of this great radiation of horses are a group of three-toed, pony-sized beasts that have long been lumped into the "wastebasket" genus "*Merychippus*." However, recent analyses have shown that the species of "*Merychippus*" are ancestral to many different lineages of horses. True *Merychippus* was a member of the hipparion lineage, a group of three-toed horses that developed highly specialized teeth, and had a distinctive concavity in the bone on the front of the face. Hipparions were a highly diverse and successful group of horses, with seven or eight different genera spread not only across North America, but also migrating to Eurasia. Merychippines were also ancestral to lineages such as *Calippus* (a tiny dwarf horse), *Protohippus*, and *Astrohippus*.

On two different occasions (*Pliohippus* and *Dinohippus*) three-toed horses evolved into lineages with a single toe on each foot. In the early Pliocene, most of these three-toed and one-toed horse lineages became extinct, leaving only *Dinohippus* to evolve into the modern horse *Equus*. The main lineage of horses that survived the latest Miocene extinctions were known as the equines. The living genus *Equus* first appeared in the Pliocene, and was widespread throughout the northern hemisphere. When the Isthmus of Panama rose about 2.5 million years ago, horses also spread to South America. There they evolved into distinctive horses with a short proboscis known as the hippidions. At the end of the last Ice Age (about 10,000 years ago), horses became extinct in the New World. They were reintroduced to their ancestral homeland by Columbus in 1493. Wild horses that have escaped from domesticated stock are known as mustangs.

Brontotheres or titanotheres began as pig-sized, hornless animals about 53 million years ago, and quickly evolved into multiple lineages of cow-sized animals with long skulls and no horns. In the late middle Eocene (between 40 and 47 million years ago), there were six different lineages of brontotheres. Some had long skulls, while others had short snouts and broad skulls. Still others had a pair of tiny blunt horns on the tip of their noses. Between 37 and 34 million years ago, their evolution culminated with huge, elephant-sized beasts bearing paired blunt horns on their noses. Throughout their history, brontotheres were the largest animals in North America. They also appeared in Asia in the late Eocene, where beasts such as *Embolotherium*, with a huge single "battering-ram" horn evolved.

Recent research has shown that the extinction of brontotheres about 34 million years ago was due to a global climatic change (triggered by the first Antarctic glaciers) that caused worldwide cooling and drying of climates. This climatic change decimated the forests of the temperate regions and eliminated most of the soft, leafy vegetation on which brontotheres fed.

The earliest moropomorphs, such as *Homogalax,* occur in lower Eocene strata. They are virtually indistinguishable from the earliest horses, such as *Protorohippus*. From this unspecialized ancestry, a variety of archaic tapir-like animals diverged. Most retained the simple leaf-cutting teeth characteristic of tapirs, and like brontotheres, they died out at the end of the Eocene when their forest habitats shrank. Only the modern tapirs, with their distinctive long proboscis, still survive in the jungles of Central and South America (three species), and southeast Asia (one species). All are stocky, pig-like beasts with short stout legs and oval hooves, and a short tail. They have no natural defenses against large predators (such as jaguars or tigers) except fleeing through dense brush and swimming to make their escape.

The horse-like clawed chalicotheres are closely related to some of these archaic tapirs. When chalicotheres were first discovered, paleontologists refused to believe that the claws belonged to a hoofed mammal related to horses and rhinos. However, many specimens have clearly shown that chalicotheres are an example of a hoofed mammal that has secondarily regained its claws. There has been much speculation as to what chalicothere claws were used for. Traditionally, they were considered useful for digging up roots and tubers, except that the fossilized claws show no sign of the characteristic scratches due to digging. Instead, chalicotheres apparently used their claws to haul down

Figure 17.68. Life-size restoration of the gigantic hyracodont rhinoceros *Paraceratherium,* the largest land mammal that ever lived. Its close relative, the running rhino *Hyracodon,* stands just below it to the right. The African elephants standing further to the right give a sense of scale. (Photo courtesy University of Nebraska State Museum.)

limbs and branches to eat leaves (much as ground sloths did), rather than for digging. *Chalicotherium* had such long forelimbs and short hindlimbs that it apparently knuckle-walked like a gorilla, with its claws curled inward. Chalicotheres were always rare throughout their history in North America and Eurasia, but nevertheless survived until the Ice Ages in Africa.

Rhinoceroses have been highly diverse and successful throughout the past 50 million years. They have occupied nearly every niche available to a large herbivore, from dog-sized running animals, to several hippo-like forms, to the largest land mammal that ever lived, *Paraceratherium.* Most rhinoceroses were hornless. Unlike the horns of cattle, sheep, and goats, rhino horns are made of cemented hair fibers, and have no bony core, so they rarely fossilize. The presence and size of the horn must be inferred from the roughened area on the top of the skull where it once attached.

The earliest rhinos, known as *Hyrachyus,* were widespread over Eurasia and North America in the early middle Eocene, and are even known from the Canadian Arctic. They apparently crossed back and forth between Europe and North America using a land bridge across the North Atlantic (before that ocean opened to its present width). From *Hyrachyus,* three different families of rhino diverged. One family, the amynodonts, were hippo-like amphibious forms, with stumpy legs and a barrel chest. In addition, amynodonts are usually found in river and lake deposits. They occupied this niche long before the hippo evolved. The last of the amynodonts, which had a short trunk like an elephant, died out in Asia in the middle Miocene.

The second family were known as the hyracodonts, or "running rhinos," because they had unusually long slender legs compared to other rhinos. They were particularly common in Asia and North America in the middle and late Eocene. The last of the North American forms was *Hyracodon,* which was about the size and proportions of a Great Dane, and survived until the late Oligocene. The second group of hyracodonts were the gigantic indricotheres (Fig. 17.68), which were the largest mammals in Asia during the late Eocene and Oligocene (about 40 to 30 million years ago). The biggest of all was *Paraceratherium* (once called *Baluchitherium* or *Indricotherium*), which was 18 feet (6 meters) tall at the shoulder and weighed 44,000 lb (20,000 kg). It was so tall that it must have browsed leaves from the tops of trees, as giraffes do today. Despite its huge bulk, it did not have the massive limbs and short, compressed toes of most giant land animals, such as sauropod dinosaurs, brontotheres, or elephants. Instead, it reveals its heritage as a running rhino by retaining its long slender toes—even though it was much too large to run. Indricotheres were also the last of the hyracodonts, vanishing from Asia in the middle Miocene.

The third family is the true rhinoceroses, or family Rhinocerotidae. They first appeared in Asia and North America in the late middle Eocene, and lived side-by-side with the hyracodonts and amynodonts on both continents. Up until this point, all the rhinoceroses we have mentioned were hornless. Rhinos with horns first appeared in the early late Oligocene; two different lineages independently evolved paired horns on the tip of the nose. Both of these groups became extinct in the late early Miocene, when two new subfamilies immigrated to North America from Asia: the browsing (leaf-eating) aceratherines, and hippo-like grazing teleoceratines. In the middle and late Miocene, browzer-grazer pairs of rhinos were found all over the grasslands of Eurasia, Africa, and North America. The teleoceratine *Teleoceras* was remarkably similar to hippos in its short limbs, massive barrel-shaped body, and high-crowned teeth for eating gritty grasses.

A mass extinction event that occurred about 5 million years ago wiped out North American rhinos, and decimated most of the archaic rhino lineages (especially the teleoceratines and aceratherines) in the Old World. The surviving lineages diversified in Eurasia and Africa, and even thrived during the Ice Ages. For example, the woolly rhinoceros was widespread in the glaciated regions of Eurasia, although it never crossed into North America (unlike the woolly mammoth or bison, which did). The five living species of rhinoceros, are all on the brink of extinction due to heavy poaching for their horns.

Chapter 18

Fossilized Behavior

Trace Fossils

There is no branch of detective science so important and so much neglected as the art of tracing footsteps.

Sherlock Holmes, in Sir Arthur Conan Doyle's *A Study in Scarlet*, 1891

In a sense, the field of ichnology is both old and new. Its guiding principles were known to a few workers many years ago, and these principles are now being rediscovered by scores of current workers. As is true in the development of any science, ichnologists have indeed gotten some occasional pebbles mixed in with their snowball; but they have also exposed many misconceptions and have made numerous positive gains. Ichnology today is rapidly approaching that plateau at which the subdiscipline will settle comfortably into the ever-growing accumulation of "standard" but highly useful methods or procedures in geology. Ichnology is not a new "magic wand," to render sister subdisciplines obsolete, but neither can it be glibly ignored by anyone seriously interested in ancient life or environmental reconstructions.

Robert W. Frey, *The Study of Trace Fossils*, 1975

INTRODUCTION

Paleontologists traditionally have focused on the skeletal and shelly remains of organisms ("body fossils"), but in recent years, they have come to appreciate the importance of **trace fossils**: sedimentary structures formed by organisms, such as trackways, trails, borings, and burrows. Trace fossils are also known as **ichnofossils** (Greek, *ichnos*, "trace") or **biogenic sedimentary structures**. The study of trace fossils has many important applications. Sedimentary geologists use them as powerful tools for interpreting sedimentary environments and paleobathymetry (see Pemberton et al., 1992; Prothero and Schwab, 1996). Structural geologists can use them to tell which way was originally the top of the bed in highly deformed rocks. Trace fossils give valuable clues to the properties of the sediment before it was lithified (its water and/or air content, compaction, coherence, and stability), so they are important indicators of diagenetic history. In fact, the burrowing action of organisms (**bioturbation**) is in itself an important sedimentary process, often responsible for reworking sediments and destroying bedding features to produce massive, structureless beds.

Trace fossils are even more important to paleontologists. Not only do they give clues to paleoecology and paleoenvironments, but they also represent the fossilized behavior of organisms. As such, they are often the only direct evidence we have for the activities of ancient life. Interpreted properly, they can tell us a tremendous amount about extinct animals that we would otherwise never know.

Trace fossils also have their limitations. Although some are relatively easy to interpret, many are mysterious, with no clues as to the tracemaker. Rarely do the tracemaker and the organism occur together (Fig. 18.1), so most trace fossils cannot be associated with a known organism. Some kinds of trace fossils, particularly vertical tubular burrows in shallow marine sands and meandering feeding traces on deep, muddy bottoms, are probably made by several different kinds of organisms. Similarly, some organisms make a variety of different traces, depending upon which behavior is occurring. There is relatively little one-to-one correspondence between traces and their tracemakers.

PRESERVATION

Although their limitations can be severe, trace fossils also have many advantages over body fossils:

1. They are often preserved in rocks where body fossils may be rare (especially shallow marine sandstones and deep marine shales). In many places they are far more abundant than body fossils, or they may be the *only* fossils preserved. Often there is a negative association between trace fossils and body fossils; one seems to exclude the other. This may be because many environments that enhance the preservation of trace fossils are inhospitable to a high diversity of organisms, so few body fossils are left. In favorable environments, a high diversity of organisms tends to bioturbate the sediment so much that few trace fossils are preserved.

2. Diagenesis may destroy or distort body fossils, but has little effect on trace fossils, and in some cases may even enhance them. For example, when iron oxides percolate through the bedrock, they often precipitate along trace fossil surfaces and make them more visible.

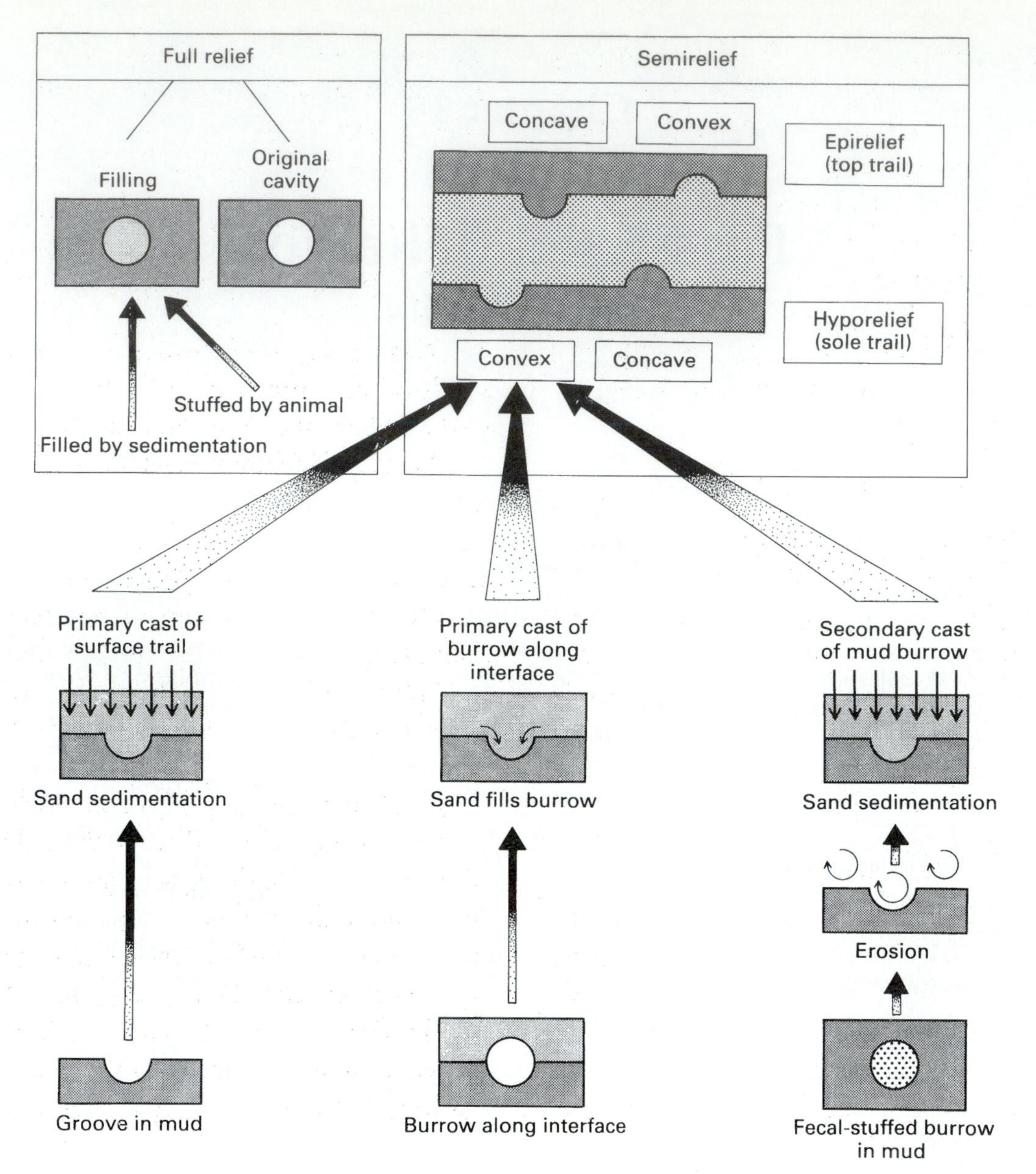

Figure 18.2. Types of preservation of trace fossils. The dark stippling indicates mud; the light stippling indicates silt or fine sand. (After Seilacher, 1964.)

3. Because trace fossils are a part of the sedimentary rock, they are not subject to post-mortem transport and are automatically *in situ*. As we saw in Chapter 1, this is not always true of body fossils, which can be transported to form unnatural death assemblages (thanatocoenoses). Trace fossils are much better indicators of the local sedimentary environment that transported body fossils.

Trace fossils can be preserved several ways, depending upon the local sedimentary and diagenetic conditions. Often the most difficult aspects of interpreting trace fossils is determining their three-dimensional appearance from a limited exposure or cross sections. The ichnologist has to be very specific about the nature of the preservation, and its three-dimensional aspects, so that the actual shape and orientation can be described correctly.

The entire trace fossil may be preserved in three-dimensional **full relief** (Fig. 18.2), with all sides of the trace exposed. In some cases, the trace may have been filled by later sediment, or by the material left behind by the animal, and then breaks out into a full cast of the trace when the surrounding matrix is shattered. In other cases, the original trace exists in full relief with the original cavity unfilled. Such traces may be recovered from the inside of a rock by pouring a molding latex into it and pulling out the latex impression after it has dried.

More often, the tracemaker moved in and out of the depositional interface, or did not leave behind enough material to fully fill the trace. In other instances, an originally cylindrical burrow collapsed, or is partially eroded, and then formed a trough or depresssion (Fig. 18.2). Such traces are known as **semireliefs**. They have the shape of concave troughs or depressions on the upper surface of a bed (**concave epireliefs**), which may also be preserved cast as a sole mark on the overlying bed (**convex hyporeliefs**).

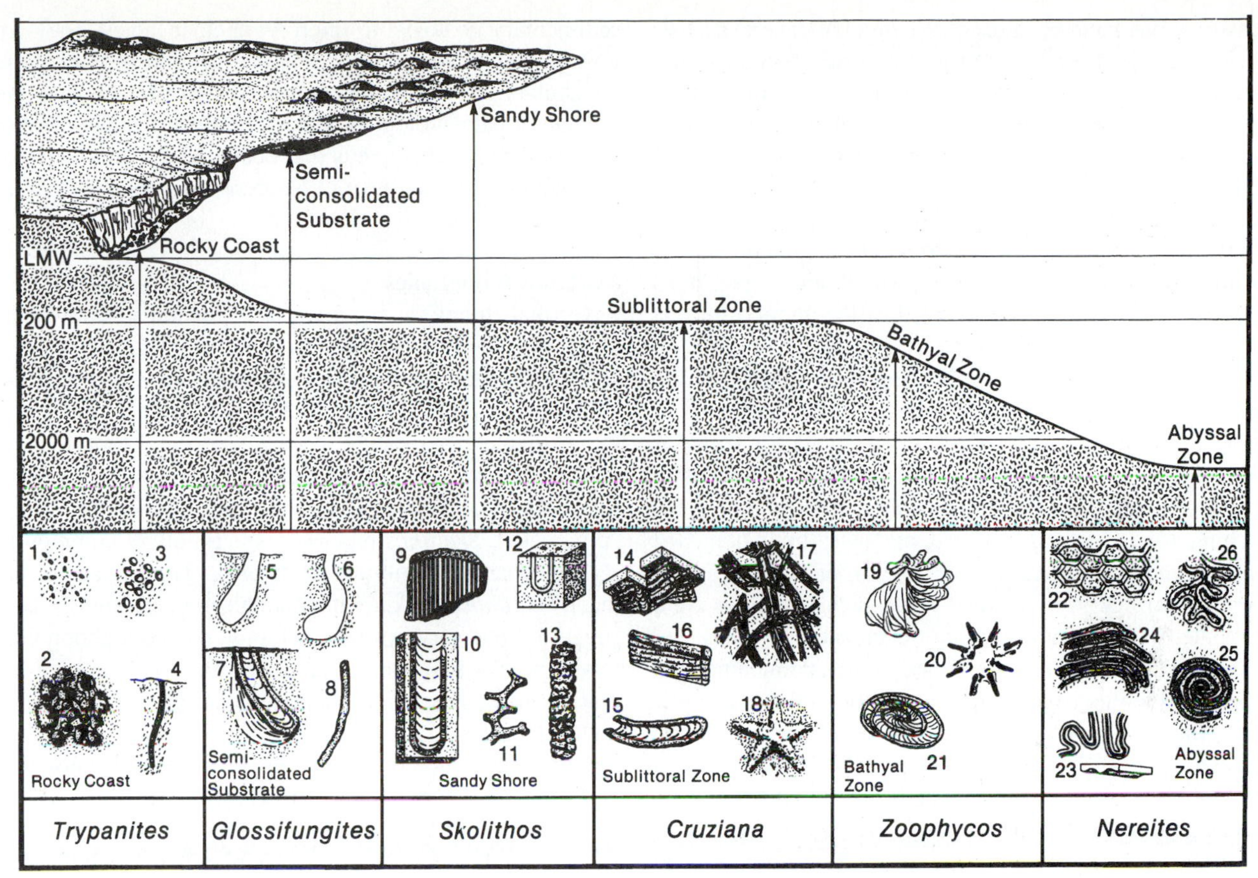

Figure 18.3. Some of the common ichnofacies and examples of the trace fossils that occur in them. 1, *Koupichnium* (horseshoe crab tracks); 2, *Isopodichnus;* 3, borings of *Polydora,* a polychaete annelid; 4, *Entobia,* borings of a clionid sponge; 5, echinoid borings; 6, algal borings; 7, pholadid bivalve borings; 8, *Diplocraterion*; 9, unlined crab burrow; 10, *Skolithos*; 11, *Thalassinoides*; 12, *Diplocraterion*; 13, *Ophiomorpha*; 14, *Arenicolites*; 15, *Phycodes*; 16, *Rhizocorallium*; 17, *Teichichnus*; 18, *Diplichnites* (trilobite tracks); 19, *Cruziana*; 20, *Rusophycus*; 21, *Asteriacites*; 22, *Zoophycos*; 23, *Lorenzinia*; 24, *Paleodictyon*; 25, *Taphrhelminthopsis*; 26, *Helminthoida*; 27, *Spiroraphe*; 28, *Cosmoraphe*. (Modified from Ekdale et al., 1984.)

Some tracemakers produce ridges or hills of sediment, which are **convex epireliefs** on the top of the original sedimentary bed, or also can be preserved as a **concave hyporelief** on the underside of the overyling bed. These are much rarer than concave epireliefs and convex hyporeliefs, however.

Trace fossils preserved as semireliefs are usually more valuable to ichnologists because they are often preserved better. In addition, they are easier to study because they are found on the top surface or the sole of a bedding plane. The three-dimensional geometry of full relief burrows is often much harder to understand because they are imbedded in the rock, and may require serial sections to get a complete image of their shape.

CLASSIFICATION

Many different schemes of classifying trace fossils have been developed over the years. At first, most trace fossils were regarded as fossil plants, so ichnologists adopted a system of giving each trace fossil a Linnaean taxonomic name, complete with genus and species. Although this is convenient and allows us to talk about trace fossils in a consistent manner (as the Linnaean system does so well for body fossils), strictly speaking it is not comparable to the taxonomy of body fossils. Trace fossils are not organisms, but fossilized behavior. Because one organism can produce several different types of traces, and several different organisms may leave the same trace, the Linnaean name gives the misleading connotation that a trace fossil is a single biological entity. It is analogous to giving a different species name to footprints produced by the same individual wearing different shoes, or the same individual walking, running, and hopping. Nevertheless, the practice of giving Linnaean names to trace fossils is so well established that it persists for lack of a better alternative. Many (but not all) trace fossil genera have characteristic suffixes, such as *-ichnus, -ichnites, -craterion, -opus*, and others, to indicate that they are not true biological genera, but ichnogenera.

The analogy with Linnaean classification breaks down since ichnogenera cannot be clustered into ichnofamilies, ichnorders, ichnoclasses and ichnophyla. Because traces are not connected by the hierarchical structure of evolutionary ancestry and descent (unlike body fossils, which come from one kind of organism), there is no natural scheme of higher categories analogous to the Linnean families, orders, classes, and phyla. Instead, other schemes of classifications have been used. When ichnogenera are listed sequentially, they tend not to be clustered in any categories, and instead may simply be listed in alphabetical order. Since trace fossils are really different kinds of fossilized behavior, one scheme of classification (originated by Seilacher, 1953, and formalized by Simpson, 1975; Ekdale et al., 1984; and Frey and Pemberton, 1985) clusters them into behavioral categories. These include:

Cubichnia, or resting traces, which are impressions caused when the animal interrupted its locomotion for rest and refuge. These usually occur as shallow depressions on the bedding surface, often reflecting the anatomy of the undersurface of the trackmaker. **Domichnia**, on the other hand, are dwelling traces, such as deep elongate burrows or deep excavations that served as a longer-term residence of the animal. They may also serve as a trap or tunnel for catching food. Most cylindrical burrows (both branched and unbranched, vertical or horizontal), or U-shaped burrows, or borings in rocks, and other such structures were permanently inhabited by the tracemaker, so they are domichnia. Unlike most other trace fossils, domichnia tend to be found within beds rather than on top of them. In some cases, the domichnia bear signs of forced escape by the organism inside; these are known as **fugichnia**. This is common in trace fossils that were buried in waves of new sediment, forcing the animal to dig out and escape, or to dig its burrow upward to continue its flow of water and function.

When the animal is not resting, but moving, it can make a wide variety of traces. **Repichnia** are made by normal crawling motions, and they are usually continuous, elongate trails, often with delicate marks from the leg motions of the tracemaker. These tend to be found on the tops of bedding surface. **Paschichnia** are grazing traces, usually horizontal geometric patterns that show the organism was systematically combing the surface of the sediment ("grazing") for food. **Agrichnia** translates literally as "farming traces." Like paschichnia, they are regularly patterned burrow systems, often on the top of the bedding plane (although they are usually preserved on the sole of the bed above). However, they reflect a network of permanent dwelling and feeding behavior in a local area, rather than continuous movement through the sediment without returning. **Fodichnia** are deposit feeding traces, formed when organisms made three-dimensional burrows, eating the sediment to digest out all of the food within it. These are usually horizontal within beds, but they can be at any angle, and they can be simple or complexly branched.

ICHNOFACIES

Despite all these other systems of classification by their shape or name or the types of behavior they represent, the most practical scheme of clustering trace fossils is by their environmental associations. It has long been known that certain types of trace fossils were found in certain types of sedimentary environments and depth conditions (Fig. 18.3). These **ichnofacies** have become standard tools of the sedimentary geologist to interpret ancient facies. They are now so widely encountered that they should be part of the vocabulary of any competent geoscientist, not just the domain of paleontologists. Each ichnofacies is named after a characteristic ichnogenus that occurs there, although that ichnogenus does not have to be present for the ichnofacies to be recognized.

Skolithos Ichnofacies

Vertical tubelike burrows are formally known as *Skolithos*. They are believed to have been formed by tube-dwelling organisms that lived in rapidly moving water and shifting sands (Fig. 18.4A, B). Most of the tubes are 1 to 5 mm in diameter and can be as long as 30 cm. In some cases, they are densely clustered together to form thick layers of sandstone that resemble organ pipes (hence the name "piperock"). *Skolithos* piperock is particularly common in shallow marine Cambrian sandstones. The organism that made *Skolithos* is unknown, although some geologists have suggested phoronids (a burrowing wormlike lophophorate related to brachiopods) or tube worms. It is also possible that the tracemaker is extinct, since *Skolithos* is rare after the Cretaceous.

Another common burrow in this ichnofacies is known as *Ophiomorpha* (Fig. 18.4A, C). These vertical cylindrical burrows are similar to *Skolithos*, except that they are slightly larger in diameter (0.5 to 3 cm), branch extensively, and have a bumpy outer surface caused by the pellets that lined the burrow used by the burrower as "bricks" to reinforce the walls. Typically, they are also less densely clustered than *Skolithos*, and may have short horizontal connecting burrows between the vertical tubes. In cross section, they appear as a circular or ovoid structure, often with a dark ring of organic matter from the pellet lining. In contrast to *Skolithos*, we know what produces *Ophiomorpha* today (they are known as early as the Permian). The modern tracemakers are burrowing ghost shrimps belonging to the Family Callianassidae (Fig. 18.4D).

A third common shallow marine ichnofossil is *Diplocraterion* (Fig. 18.4A, E, F). *Diplocraterion yoyo* is a very evocative (but now invalid) name for a trace that tells of a very specific behavior. It is a curved burrow found between the arms of a vertical U-shaped tube that presumably housed a burrowing organism. When the openings were buried by sediment, the organism moved up in its burrow; when the upper part of the burrow was eroded away, the animal dug in deeper. The sequence of U-shaped burrow traces thus responded like a yo-yo to the rise and fall of the sediment/water interface.

All of these burrows suggest a rapidly shifting substrate that required animals to dig deep vertical burrows, and rebuild them repeatedly when waves eroded them or buried them in more sediment. Most of these animals were apparently filter feeders that used the sediment strictly for shelter, not as a source of food, so they are mostly domichnia or fugichnia. Additional sedimentological evidence indicates that this ichnofacies is very shallow marine, and this

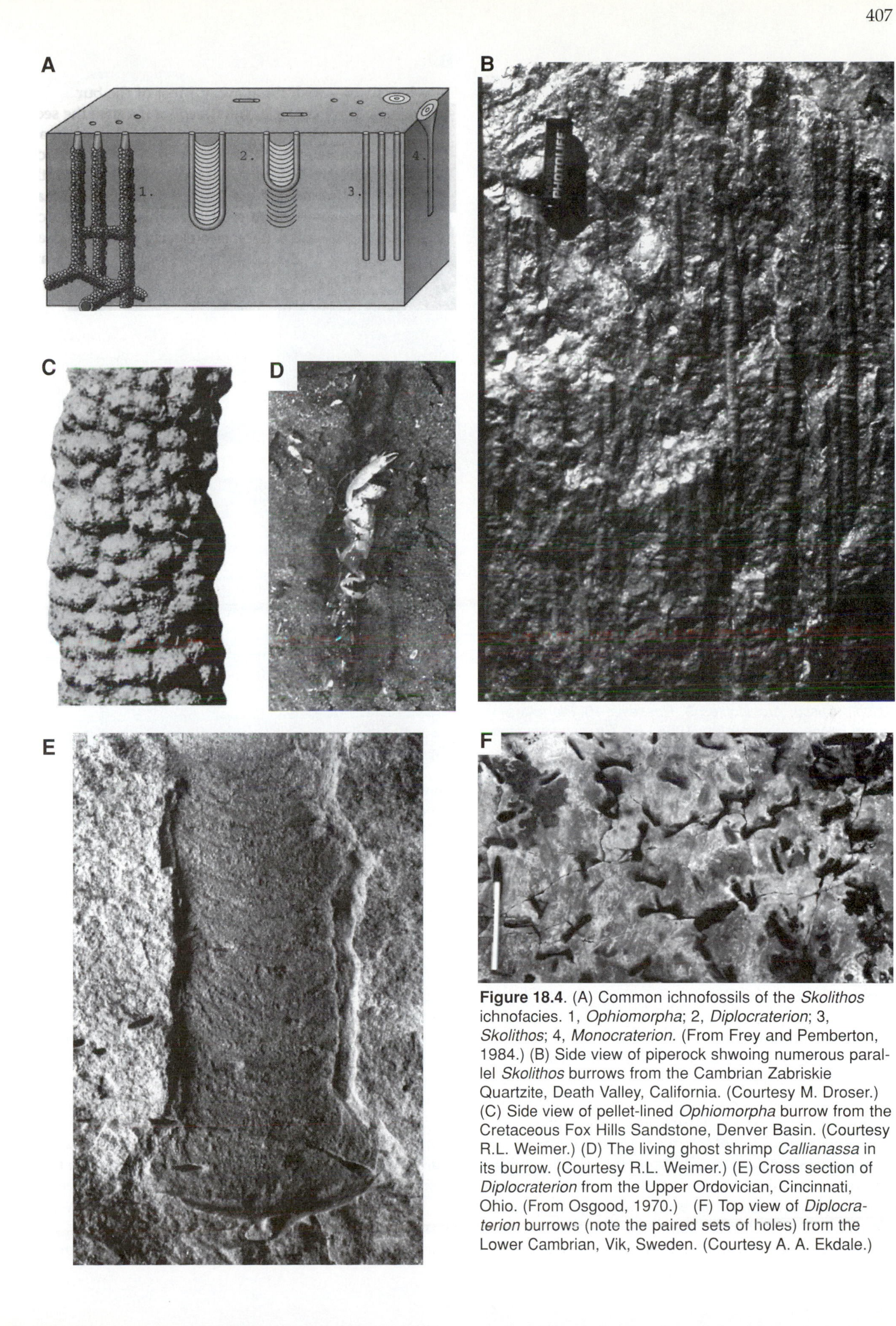

Figure 18.4. (A) Common ichnofossils of the *Skolithos* ichnofacies. 1, *Ophiomorpha*; 2, *Diplocraterion*; 3, *Skolithos*; 4, *Monocraterion.* (From Frey and Pemberton, 1984.) (B) Side view of piperock shwoing numerous parallel *Skolithos* burrows from the Cambrian Zabriskie Quartzite, Death Valley, California. (Courtesy M. Droser.) (C) Side view of pellet-lined *Ophiomorpha* burrow from the Cretaceous Fox Hills Sandstone, Denver Basin. (Courtesy R.L. Weimer.) (D) The living ghost shrimp *Callianassa* in its burrow. (Courtesy R.L. Weimer.) (E) Cross section of *Diplocraterion* from the Upper Ordovician, Cincinnati, Ohio. (From Osgood, 1970.) (F) Top view of *Diplocraterion* burrows (note the paired sets of holes) from the Lower Cambrian, Vik, Sweden. (Courtesy A. A. Ekdale.)

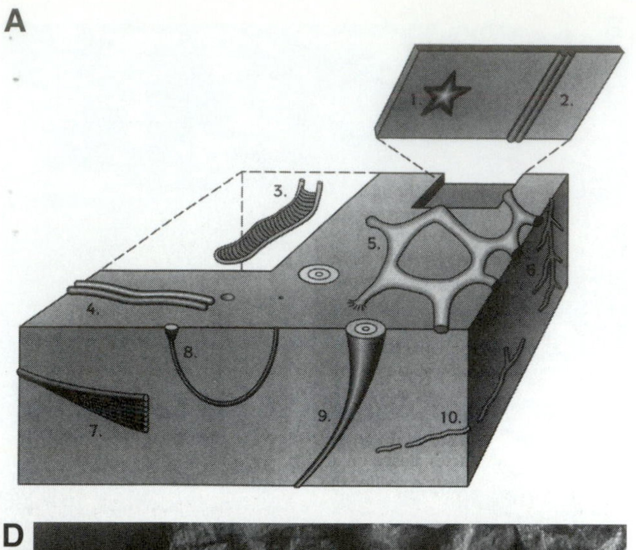

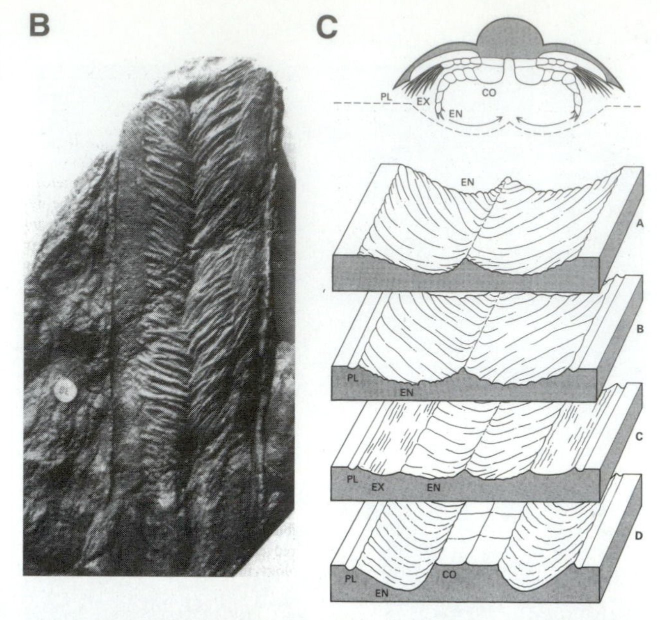

Figure 18.5. (A) Common trace fossils of the *Cruziana* facies. 1, *Asteriacites*; 2, *Cruziana*; 3, *Rhizocorallium*; 4, *Aulichnites*; 5, *Thalassinoides*; 6, *Chondrites*; 7, *Teichichnus*; 8, *Arenicolites*; 9, *Rossellia*; 10, *Planolites*. (From Frey and Pemberton, 1984.) (B) *Cruziana* from the Upper Cambrian of Wales. (Courtesy T.P. Crimes.) (C) Mode of formation of *Cruziana*. The endopodite (EN) portion of the walking leg is responsible for most of the excavation, but evidence of the coxae (CO), exopodites (EX), and pleurae (PL) may be present. (From Crimes et al., 1970.) (D) *Thalassinoides* burrows are a complex three-dimension web, which is usually compacted into a jack-straw-like web of burrows when viewed in a two-dimensional bedding plane. (Photo by the author.) (E) *Asteriacites* (Upper Ordovican, northern Kentucky) preserved as a convex hyporelief. The scratches of the tube feet of this sart fish can be seen. (From Osgood, 1970.)

is consistent with the known environmental preferences of callianassid shrimps. In summary, the *Skolithos* ichnofacies usually represents organisms inhabiting clean, well-sorted nearshore sands with high levels of wave and current energy.

Cruziana Ichnofacies

Horizontal U-shaped troughs with many intermediate, rib-like striations are known as *Cruziana*, and these occur in moderate- to low-energy sands and silts of the shallow shelf (Figs. 18.3, 18.5). *Cruziana* are often preserved as the convex hyporelief cast of the trough-shaped burrow, rather than as the original concave burrow (Fig. 18.5B, C). Most *Cruziana* are believed to represent the crawling traces of trilobites, since they are long troughs that appear to bear the scratch traces of trilobites legs, formed as they burrowed through the shallow sediment. Their occurrence in rocks of Cambrian to Permian age (the same stratigraphic range as the trilobites) further reinforces this interpretation.

Another common ichnofossil in this facies is *Thalassinoides* (Fig. 18.5D). This is a general name for a complex three-dimensional network of cylindrical burrows that form an irregular web of crisscrossing tubes about 1-7 cm in diameter. Apparently, this burrower was mining shallow marine silts or muds for their nutrients as well as seeking protection in the complex web of burrows (they are both fodichnia and domichnia). The branching network of *Thalassinoides* burrows are very similar to those of *Ophiomorpha*, except that the former are not lined with organic pellets. The organisms that produced *Thalassinoides* in the fossil record probably were mostly decapod crustaceans (i.e., crabs, shrimps, and lobsters), as a wide variety of decapods create such burrows today.

In addition to these two typical ichnogenera, there are a

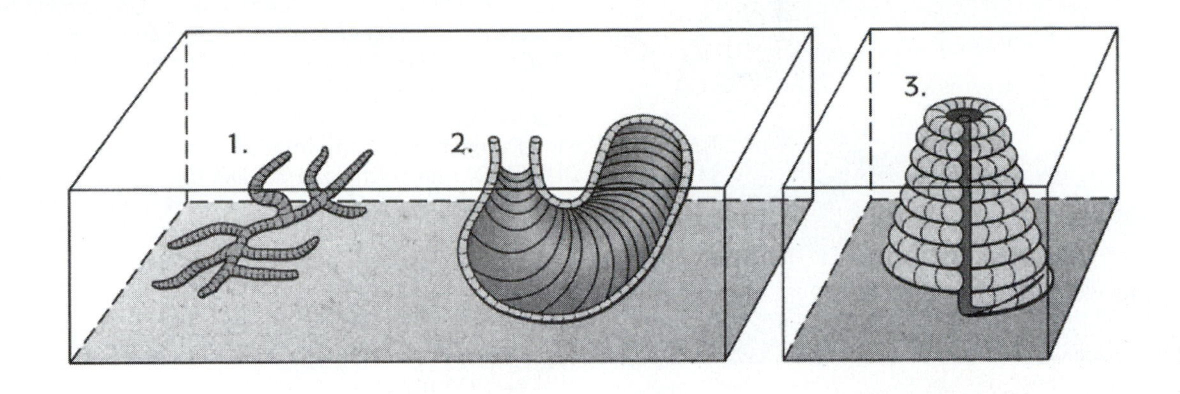

Figure 18.6. (A) Typical trace fossils of the *Zoophycos* ichnofacies. 1, *Phycosiphon*; 2, *Zoophycos*; 3, *Spirophyton.* (From Frey and Pemberton, 1984.) (B, C) Typical *Zoophycos* traces, showing a pattern of complex arcuate feeding traces in three dimensions. (B) is from the Oligocene Amuri Limestone, Vulcan Gorge, Canterbury, New Zealand. (C) is is from the Eocene Saraceno Formation, Satanasso Valley, Italy. (Photos courtesy A. A. Ekdale.)

number of additional less common trace fossils that are characteristic of this ichnofacies. They include (see Fig. 18.5A) the star-shaped *Asteriacites* (the cubichnia of sea stars—Fig. 18.5E), the U-shaped *Rhizocorallium* (like a horizontal *Diplocraterion*), the U-shaped *Arenicolites*, the conical *Rossellia*, and the horizontal burrow known as *Planolites*. Most are traces of organisms that used the substrate both as shelter and also mined the sediment for food particles (fodichnia and domichnia). The *Cruziana* ichnofacies is also the most diverse of all ichnofossil communities, and it is commonly associated with finer-grained sediments than those which produce *Skolithos*. Based on all these lines of evidence, most ichnologists consider the *Cruziana* ichnofacies to indicate shallow marine waters between the low tide line and storm wave base, typically from the middle to outer shelf (Fig. 18.3). In fact, the top surfaces of storm deposits are often overprinted by *Cruziana* ichnofacies activity that occurred on the fresh sea bottom right after a major storm.

Zoophycos Ichnofacies

Broad, looping infaunal feeding traces known as *Zoophycos* occur in low-energy muds and muddy sands (Fig. 18.6). Traditionally, they were considered indicators of deep waters along the continental slope below storm wave base, but above the continental rise where turbidites accumulate. In the standard ichnofacies scheme, this places *Zoophycos* between the *Cruziana* and *Nereites* ichnofacies (Fig. 18.3). However, further study has shown that *Zoophycos* can be found in a great variety of depths (Frey and Seilacher, 1980; Bottjer et al., 1988). *Zoophycos* appear to represent a highly versatile, opportunistic trace-maker, since they occasionally occur in the *Cruziana* and *Nereites* ichnofacies. Instead of being primary depth indicators, *Zoophycos* are more closely associated with lowered oxygen levels and abundant organic material in the sediment in quiet water settings. These conditions are common on the outer shelf and continental slope, but they also occur in shallower water of Paleozoic and Mesozoic epeir-

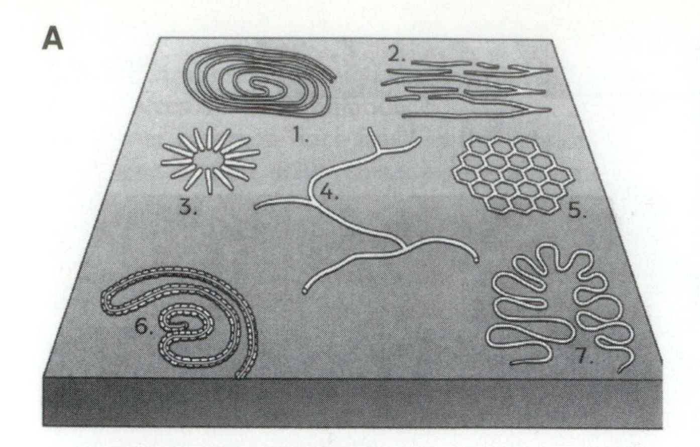

Figure 18.7. (A) Typical deep-water trace fossils of the *Nereites* ichnofacies. 1, *Spiroraphe*; 2, *Urohelminthoida*; 3, *Lorenzinia*; 4, *Megagrapton*; 5, *Paleodictyon*; 6, *Nereites*; 7, *Cosmoraphe*. (From Frey and Pemberton, 1984.) (B) Two different meandering feeding traces, *Spirophycus* (larger burrows) and *Phycosiphon* (smaller burrows), from the Permian Oquirrh Formation, Wasatch Mountains, Utah. (C) *Paleodictyon,* a net-like trace, from the Middle Jurassic of the Ziz Valley, Morocco. (Photos courtesy A. A. Ekdale.)

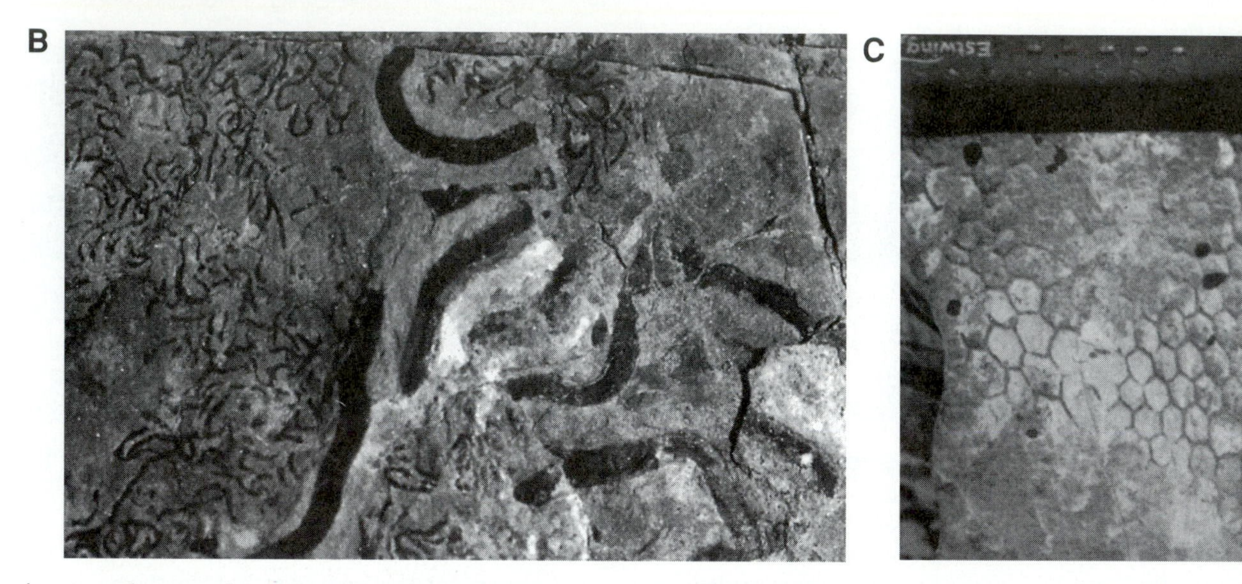

ic seas wherever the water was quiet enough but low in oxygen content.

Besides *Zoophycos*, relatively few other trace fossils are known from this community. The horizontal branched feeding trace known as *Phycosiphon* and the helically spiraling burrow known as *Spirophyton* are among the few commonly found with *Zoophycos*. The lack of diversity in the *Zoophycos* ichnofacies also suggests that it may commonly represent a relatively hostile, oxygen-stressed environment where only a few low-oxygen-tolerant burrowers can thrive.

Nereites Ichnofacies

In contrast to the *Zoophycos* ichnofacies, the interpretation of the *Nereites* ichnofacies is relatively straightforward. Meandering feeding traces on bedding plane surfaces are called *Nereites* and are usually found in the abyssal plains, often associated with turbidites and deep pelagic muds (Fig. 18.7). Almost all the ichnogenera in this facies are horizontal burrows in the top few centimeters of the muddy bottom. They all display some sort of pattern of meandering or zigzagging across the bottom, showing that the animal was systematically mining the seafloor for food and detritus (pascichnia or fodichnia).

Glossifungites, *Trypanites*, and *Teredolites* Ichnofacies

The four ichnofacies just discussed primarily reflect differences in water depth and physical energy. Three other ichnofacies are known from a variety of depths, since they are controlled by the nature of the substrate. The *Glossifungites* ichnofacies (Fig. 18.8A) is typical of firm but unlithified substrates, such as dewatered muds. Such dewatering usually occurs after burial, so the substrate is only available to be colonized by the tracemaker if it is later exposed by erosion. This can occur in shallow coastal waters due to wave erosion, or in deep submarine channels when a turbidity current scours away previously deposited sediments. Naturally, such surfaces are likely to be disconformities.

The traces in this ichnofacies include vertical, cylindrical, U-shaped, or teardrop-shaped borings, sparsely to densely branched burrows, or mixtures of borings and burrows, as well as fan-shaped *Rhizocorallium* and *Diplocraterion*. This ichnofacies can include a variety of traces that look like *Skolithos, Thalassinoides*, and *Diplocraterion*, and simple U-shaped burrows (*Arenicolites*). These burrows also tend to avoid rocky obstructions, staying only in the firm mud.

Fully lithified substrates, on the other hand, are colonized by rock-boring organisms of the *Trypanites* ichnofacies (Fig. 18.8B). Such hard substrates include reefs, rocky coastlines, beachrock, and hardground surfaces caused when a drop in sea level lithified the sediment. They may even bore into igneous substrates. Almost all the traces of this ichnofacies are tubular or vase-shaped domichnia of animals that can bore into rocks for shelter, such as pholadid bivalves. In addition, there may be excavations caused by the raspings and gnawings of algal grazers, such as limpets, chitons, and echinoids. Unlike *Glossofungites* traces, those of the *Trypanites* ichnofacies cut through everything in their path, and do not avoid the more lithified obstacles.

A third substrate-dependent ichnofacies occurs with organisms that burrow into wood, known as the *Teredolites* ichnofacies (*Teredo* is the wood-boring bivalve known as a "shipworm") (Fig. 18.8C). This ichnofacies can occur wherever wood is exposed to water, including logjams in rivers and lakes, driftwood in the ocean, as well as human-created substrates, such as docks, wharf pilings, and ship hulls. Because wood can float, these traces can move far from their point of origin, so only *in situ* borings that have not been transported clearly belong to the *Teredolites* ichnofacies. Most traces in this setting are club-shaped borings, stumpy to elongate cylindrical borings, and other impressions that reflect the texture of the wood.

Terrestrial Trackways and Traces

Above the normal high tide level is a region of backshore beach deposits, coastal sand dunes, washover fans, and supratidal flats known as the *Psilonichnus* ichnofacies (Fig. 18.9). This region is colonized by only a few types of organisms that tolerate relatively dry conditions except when the area is flooded by unusually high spring tides, major storms, and hurricanes. The major inhabitant of this ichnofacies today are ghost crabs, which excavate J-, Y-, or U-shaped burrows with a bulbous basal chamber. The Y-shaped burrows in the fossil record are called *Psiloichnus*. There may also be the vertical shafts of burrowing spiders, horizontal tunnels formed by foraging insects and tetrapods, and ephemeral surface tracks and trails of many different animals which wander across the sand. This facies also exhibits abundant plant root traces, as well as plant fossils themselves. The *Psilonichnus* ichnofacies falls on the boundary between terrestrial and marine ichnofacies, so it is often found adjacent in space and in section to the *Skolithos* ichnofacies, or transitional between the *Skolithos* ichnofacies and nonmarine ichnofacies.

Trace fossils found in continental red beds of lakes, rivers, and floodplains have traditionally been called the *Scoyenia* ichnofacies. They include a variety of small, horizontal, lined back-filled feeding burrows (such as the ichnogenus *Scoyenia*), curved unlined feeding burrows, sinuous crawling traces, vertical cylindrical shafts, and tracks and trails of many different kinds of vertebrates and invertebrates.

The trace fossils of terrestrial sand dunes are also distinctive (Fig. 18.10). Among them are the branching tun-

A

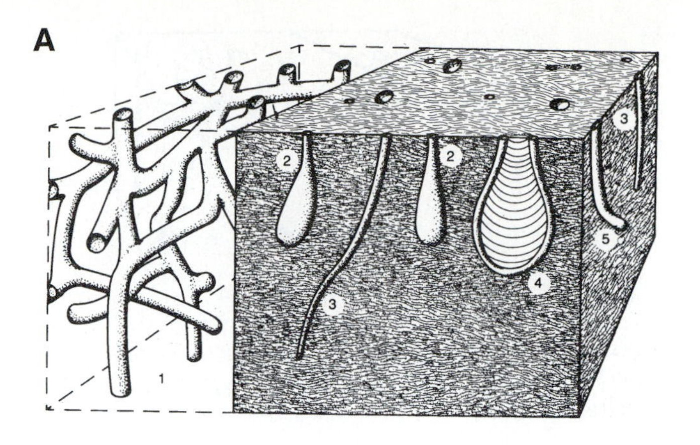

B

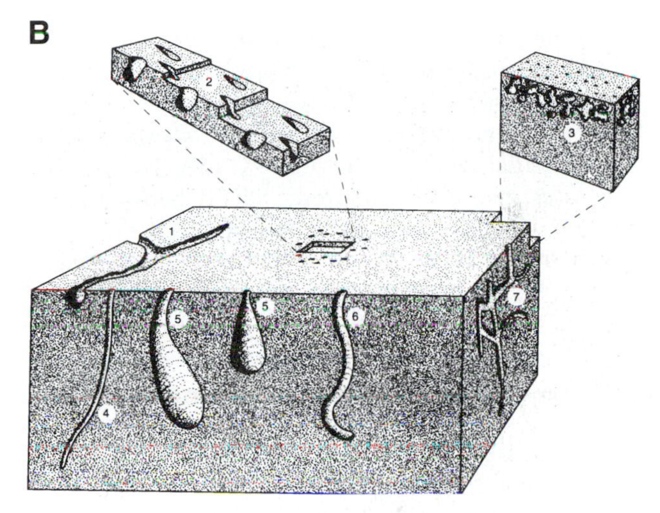

C

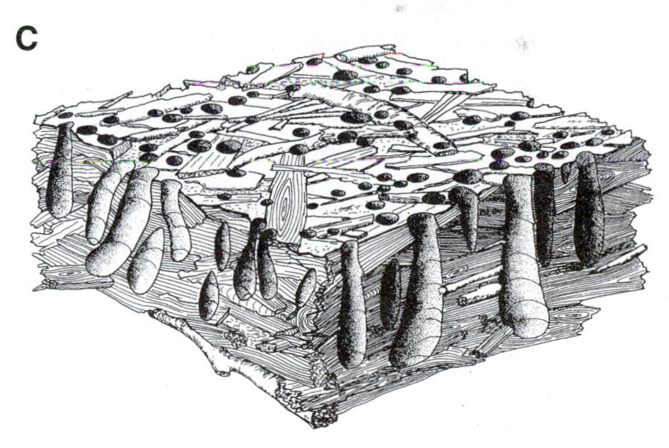

Figure 18.8. (A) Trace fossils of the *Glossifungites* facies. 1, *Thalassinoides*; 2, *Gastrochaenolites*; 3, *Skolithos* or *Trypanites*-like structures; 4, *Diplocraterion*; 5, *Psiloichnus*; 6, *Arenicolites*; 7, *Rhizocorallium.* (B)Trace fossils of the *Trypanites* ichnofacies. 1, unnamed echinoid grooves; 2, *Rogeralla*; 3, *Entobia*; 4, *Trypanites*; 5, *Gastrochaenolites*; 6, *Trypanites*; 7, unnamed polychaete borings. (C) Trace fossils of the *Teredolites* ichnofacies. (From Frey and Pemberton, 1984.)

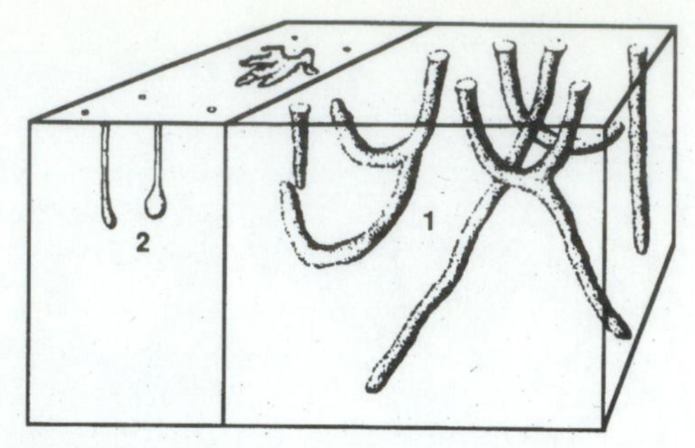

Figure 18.9. Trace fossils of the *Psiloichnus* ichnofacies. 1, *Psiloichnus*; 2, *Macanopsis.* (From Frey and Pemberton, 1984.)

nels of ants, as well as burrows known to be produced today by organisms such as crickets, wolf spiders, beetles, crane flies, sand wasps, toads, gastropods, and sweat bees. Boucot (1990, pp. 328-345) documents a wide spectrum of other terrestrial and freshwater trace fossils, such as caddisfly cases, mayfly larval burrows, bivalve burrows, lungfish burrows, mud wasp nests, butterfly cocoons, leaves eaten by a variety of insects, wasp burrows, earthworm cocoons and burrows, spider tracks, and many other remarkable fossils. Most of these are very rare in the fossil record, but some of these terrestrial traces are common in the certain formations. For example, in the dune deposits of the Oligocene-Miocene Arikaree Group in western Nebraska, the burrows of sweat bees are one of the most common sedimentary structures.

The study of marine trace fossils is still a very young field, but the study of terrestrial trackways is much older, dating back to the discovery of the first dinosaur trackways in 1802. These were originally described by Pliny Moody as the tracks of "Noah's raven," but by 1828, these tracks and similar ones in England were correctly interpreted as produced by the newly discovered giant fossils that were later called dinosaurs. Since that time, there have been numerous detailed studies of terrestrial trackways (summarized by Sarjeant, 1975; Gillette and Lockley, 1989; Lockley, 1991; Lockley and Hunt, 1995). Terrestrial trackways are valuable for more than the fact that they give some evidence of the non-marine nature of the beds, and some indication of the cohesiveness and water content of the sediment. They are also powerful tools for understanding the locomotion and behavior of the trackmaker. Such traces may lead to quite precise estimates of speed as well as insights about posture, activity levels and metabolism, and even social behavior.

The biomechanical analysis of the trackway begins with measurements of standard parameters (Fig. 18.11). In addition to standard measurements of length and width of the tracks themselves, it is valuable to measure the **step length** (distance from corresponding right foot to left foot impression), and the **stride length** (distance from one foot impression to the next one left by the same foot). The relative stride length can be calculated by dividing the stride length by animal's height at the hip, giving a proportional estimate of how long the stride really is (Thulborn, 1989). When the relative stride length is greater than 2.0, Alexander (1976) found that most terrestrial tetrapods change from a walking gait to a trotting gait; when the relative stride length exceeds 2.9, then the animal employs a running gait. Most dinosaurs, for example, leave trackways of walking gaits, although a few are known with fully running gaits, and relative stride lengths as high as 5.0. Using these methods, Alexander (1976) was able to estimate the

Figure 18.10. Characteristic organic traces in modern sand dune deposits. (From Frey and Pemberton, 1984.)

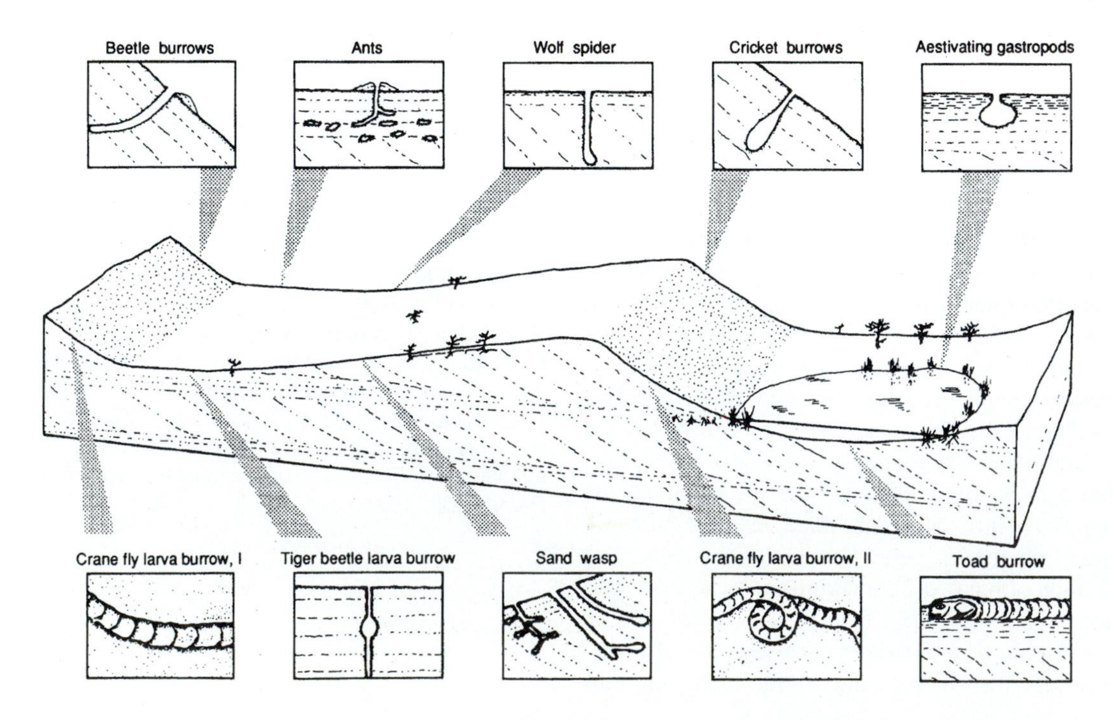

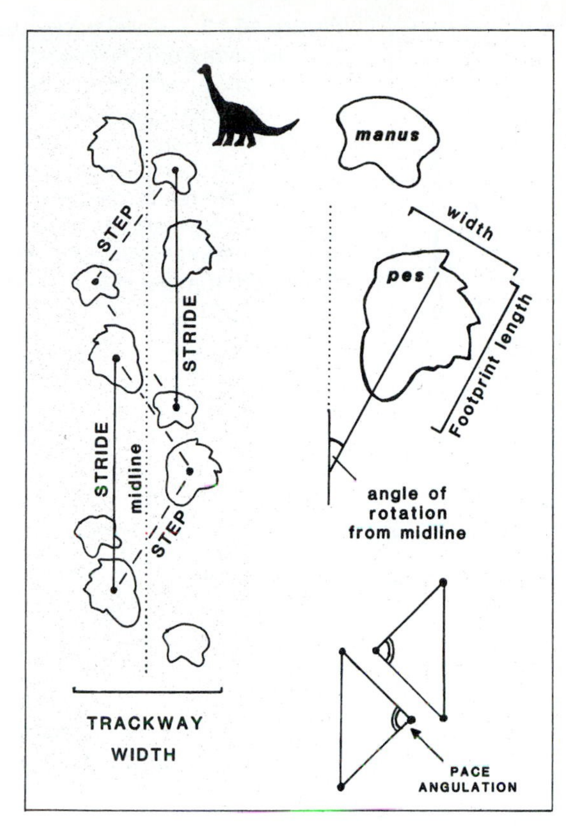

Figure 18.11. Trackway of a quadrupedal dinosaur, showing standard features that help trackers interpret locomotion. Manus = hand; pes = foot. (From Lockley, 1991.)

speeds of several dinosaurs. Most were moving at a slow walk (less than 5 km/hr), but this does not tell us their maximum speed. In addition to these estimates, the **pace angulation** (angle between step line segments) is valuable in determining the body width of the animal, and to what degree an animal sprawled or walked upright (Fig. 18.11).

Different gaits employed by vertebrates moving at the same speed can also be detected. For example, in tetrapods there are two forms of rapid locomotion: the **trot** and the **pace**. We are all familiar with how a horse trots, although we may have never analyzed how the limbs actually move. Careful study (aided by frame-by-frame analysis of movie footage) shows that in a trot, the horse moves the diagonally opposite limbs (left forefoot and right hindfoot, right forefoot and left hindfoot) in unison (Fig. 18.12). In their trackways, the impressions on the same side of the body are closely paired. By contrast, camels move by pacing, or moving both limbs on the same side of the body in unison. This is why they have their peculiar swaying motion that makes some riders seasick on the "ship of the desert." On pacing trackways, the steps of each stride are very evenly spaced, with no pairing as in trotting trackways.

Three-toed horse trackways from several localities show that the horse trot is very ancient. A series of middle Miocene camel trackways from the Barstow Formation in the Mojave Desert bore extinct protolabine camel tracks with clear evidence of a pacing motion even back in the Miocene (Webb, 1972). Other trackways from the same

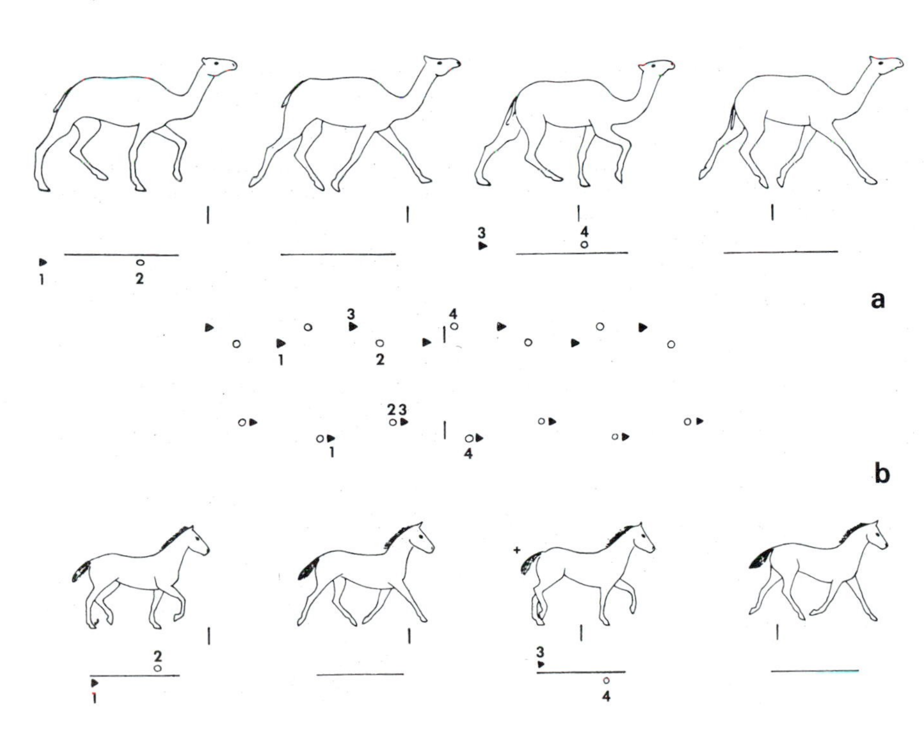

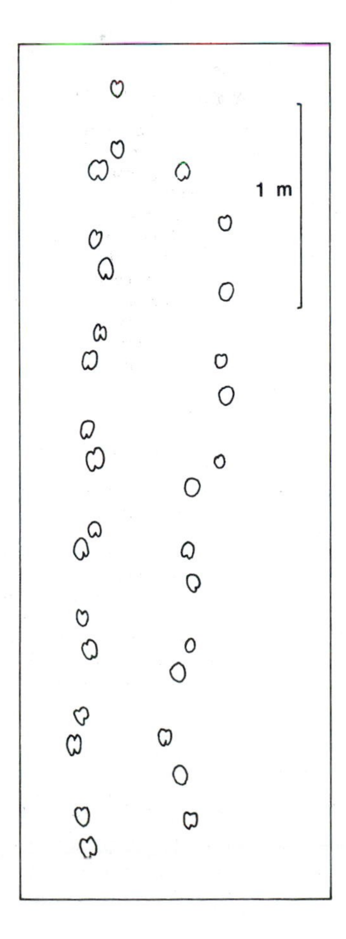

Figure 18.12. (above) Comparison of the pace of a camel, where the limbs on the same side move in unison (above) with the trot of a horse, where the diagonal limbs move in unison (below). (From Boucot, 1990.) (right) Tracks of a Miocene camel (*Procamelus* or *Protolabis*) from the Barstow Formation, California. (From Lockley and Hunt, 1995.)

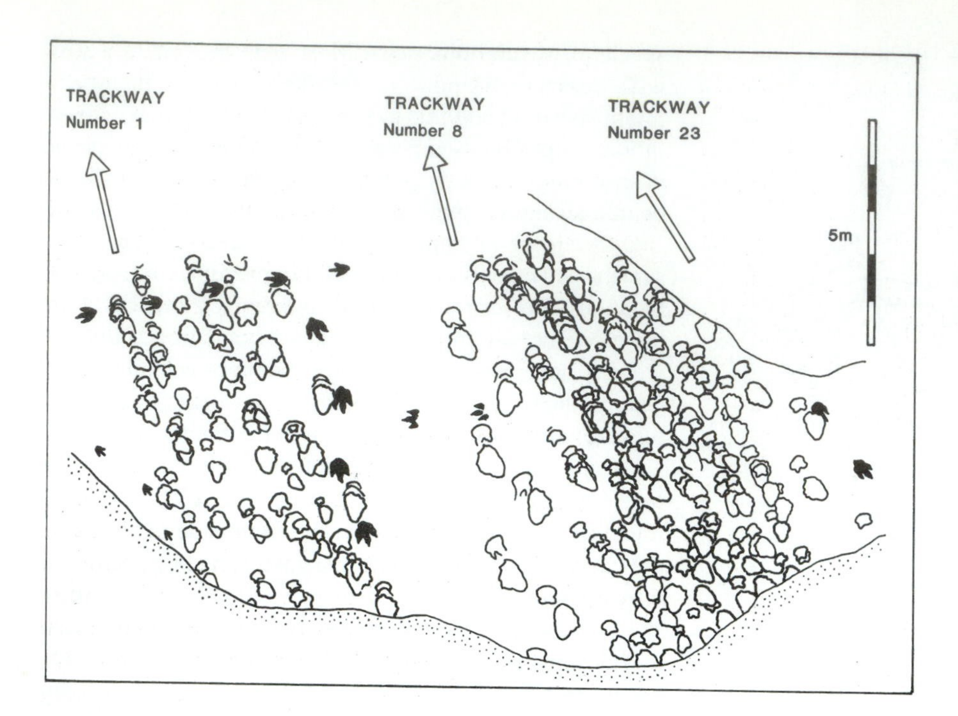

Figure 18.13. Parallel trackways of sauropods from Davenport Ranch, Texas, which show that a herd of at least 23 animals all moved in the same direction. About 5 theropod dinosaurs also crossed the area in various directions, possibly stalking the herd of prey. Sauropod trackways are numbered from left to right. (From Lockley, 1991.)

area showed the running motion of extinct pronghorns, and one even shows a predatory bear-dog (amphicyonid) chasing a pronghorn.

In addition to posture, gait, and speed, large numbers of trackways give valuable evidence of many other aspects of behavior. Lockley (1991) summarizes the insights into dinosaurian behavior gained by trackways. In several places, the trackways of large sauropod dinosaurs are moving in the same direction in clusters, indicating that they moved in herds (Fig. 18.13). Some paleontologists argue that trackways show that the juveniles were in the center of the moving herd for protection by larger adults on the perimeter, although Lockley (1991) is skeptical of these claims. Other trackways are famous for apparent stalking of dinosaurs by their predators. These show large three-toed theropod trackways moving parallel and overprinting their prey tracks, or crossing the area soon after the prey species moved by. In one tracksite, the sauropod trackways actually crushed a bed of freshwater bivalves.

ICHNOFABRIC

The total record of sedimentary rocks fabric resulting from bioturbation is termed **ichnofabric**. This includes discrete identifiable trace fossils as well as unidentifiable bioturbation structures. One important aspect of ichnofabric is that it can be used to estimate the degree to which the burrowing has reworked the sediment. Not all sedimentary rocks show equal degrees of bioturbation, and the relative amounts of bioturbation may yield important clues about the environment. Paleontologists have found it useful to quantify this degree of bioturbation, based on how physically disturbed the sediment appears, just as sedimentologists quantify the grain size, bedding, and sedimentary structures of a bed in their descriptions. This degree of bioturbation is known as an **ichnofabric index**. Droser and Bottjer (1986; see also Bottjer and Droser, 1991) created a standard series of five images of different degrees of bioturbation, or ichnofabric indices, for quick standardized description and comparison of any outcrop or hand sample (Fig. 18.14). Ichnofabric index 1 is essentially undisturbed, with all original sedimentary structures present. Index 2 has discrete isolated trace fossils cutting across the original bedding, so that up to 10% of the original bedding is disturbed. Index 3 shows 10 to 40% of the bedding disturbed, with isolated burrows (although there may be some overlap). In index 4, the last vestiges of the bedding are still discernible, so the rock is about 40 to 60% disturbed, but the burrows overlap and are not very well defined. Index 5 has completely lost the original sedimentary bedding, but the burrows are still discrete in places and the fabric is not mixed. Ichnofabric index 6 is completely homogenized sediment with no primary bedding and few distinct trace fossils.

TRACE FOSSILS THROUGH TIME

Trace fossils also give us valuable clues about the history of life that body fossils alone cannot. Most of the wormlike burrowers that leave trace fossils have no body fossil record, so we can detect their presence only from their traces. In addition, some of the earliest metazoan animals were soft-bodied that left traces but no body fossils.

The oldest possible trace fossils are tubelike structures of questionable biologic origin from the Medicine Peak Quartzite of Wyoming, which is 2.0 to 2.5 billion years old. If these are really biological features, then they may indicate that some kind of wormlike metazoan was around early in the Proterozoic, which is much earlier than most paleontologists would accept. The oldest uncontested trace fossils are simple burrows found with the famous soft-bodied late Proterozoic (about 600 Ma) Ediacara fauna,

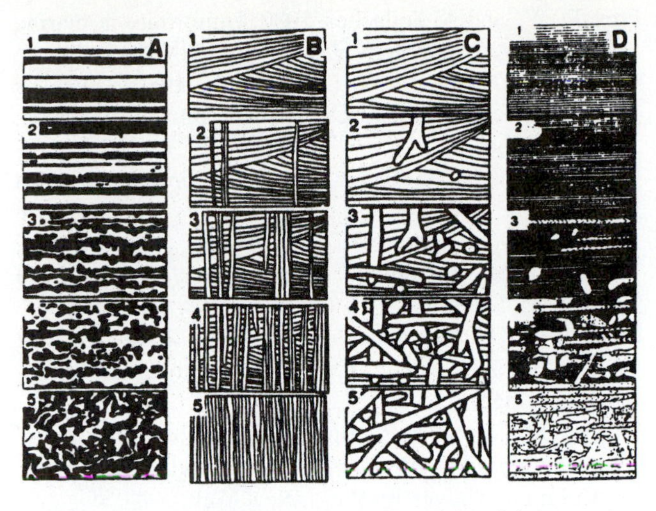

Figure 18.14. Visual images of various ichnofabric indices. Schematic diagrams of ichnofabric indices 1 through 5 for strata deposited in (A) shelf environments; (B) high-energy nearshore sandy environments dominated by *Skolithos*; (C) high-energy nearshore environments dominated by *Ophiomorpha*; (D) deep-sea deposits. Ichnofabric indices are defined as follows: 1, no bioturbation recorded (all original sedimentary structures intact); 2, discrete isolated trace fossils (up to 10% of original bedding disturbed); 3, approximately 10-40% of original bedding disturbed, burrows are generally isolated, but locally overlap; 4, last vestiges of bedding discernible, approimately 40-60% disturbed, burrows overlap and are not always well defined; 5, bedding is completely disturbed, but burrows are still discrete in places and the fabric is not mixed. Ichnofabric index 6 (not shown) is totally homogenized sediment. (From Bottjer and Droser, 1994.)

which is known entirely from impressions of a variety of large but unskeletonized animals.

In the earliest Cambrian (540 Ma) Nemakit-Daldynian Stage, there is a greater variety of surface traces, including *Phycodes* and other presumed arthropod traces, showing that there were soft-bodied arthropods long before the appearance of skeletonized trilobites. In the next stage of the Cambrian (Tommotian Stage), the trace fossil assemblage is augmented by a variety of deep vertical burrows (especially *Skolithos*), showing that the seafloor was now being burrowed by metazoans ("worms") with tubular bodies that were held rigid by hydrostatic pressure of the coelomic fluid inside (no flatworm can burrow very deeply). Body fossils of trilobites and many other groups do not appear until the third stage of the Early Cambrian, the Atdabanian. In addition to the increase in diversity of trace fossils through the first three stages of the Cambrian, there is also an increase the average degree of bioturbation (as measured by the ichnofabric indices just discussed). An additional explosion in trace fossil diversity and increase in ichnofabric indices occurred in the Late Ordovician, in conjunction with the great Ordovician radiation of metazoans (Bottjer and Droser, 1994).

Trace fossil diversity remains high through the rest of the Paleozoic. One would expect that the Permian catastrophe, which wiped out 95% of species in the marine realm, would have also diminished the diversity of trace fossils. Bottjer et al. (1988) studied one deep-water Permo-Triassic section in China, but detected no net change in diversity or in ichnofabric indices. However, this may be because the deep-water environment is always relatively low in diversity and stable against major changes. Studies of shallow-water sections, which have high Permian trace fossil diversity, are necessary to answer this question conclusively.

Since the Permian, trace fossil diversity has held remarkably constant, with the exception of the deep-water *Nereites* facies, which has increased in diversity beginning in the Cretaceous. It is unclear why this is so. One suggestion is that the rapid radiation of angiosperms in the Early Cretaceous produced much more detrital organic matter that eventually drifted to the deep ocean and increased its food supply. Another possibility is that there was more deep-water deposition since the breakup of Pangea in the Jurassic, yielding a greater volume of deep-water sediments to be bioturbated.

Another approach to changing trace fossil diversity has been to look for onshore-offshore trends in origination of trace fossil genera, analogous to the onshore-offshore trends in origination of major groups of body fossils (see Fig. 8.21). Bottjer et al. (1988) found that the earliest *Zoophycos* (known from the Early Ordovician) occur on the inner shelf, but by the Early Silurian they had spread to the slope and deep basin facies. *Zoophycos* remained common in the nearshore facies until the Cretaceous, after which it became restricted to deep and or low-oxygen waters. Such a pattern is consistent with the onshore origination of many Paleozoic body fossil groups as well.

Bottjer et al. (1988) also examined the occurrences of *Ophiomorpha*, which are typical of the Mesozoic and Cenozoic. The oldest known *Ophiomorpha* are Lower Permian and are found in nearshore sediments. By the Late Jurassic, *Ophiomorpha* was found in the inner shelf environment, and by the middle Cretaceous in slope and deep basin environments. Since the Mesozoic, it has been known to occur in almost all environments, although it is still more common in the nearshore settings than farther offshore. Once again, it appears that the organisms which made particular trace fossil taxa originated first in the nearshore environment, and then gradually colonized more offshore settings, although in this case the tracemaker was still primarily a nearshore organism.

There is much information to be gleaned about not only the sedimentary history of a region, but also the history of life itself if the geologist has a clear understanding and an eye for trace fossils.

For FurtherReading

Alexander, R. M. 1976. Estimates of the speeds of dinosaurs. *Nature* 261:129-130.

Bottjer, D. J., and M. L. Droser. 1991. Ichnofabric and basin analysis. *Palaios* 6:199-205.

Bottjer, D. J., and M. L. Droser. 1994. The history of Phanerozoic bioturbation, pp. 155-176, *in* Donovan, S. K., ed. 1994. *The Palaeobiology of Trace Fossils*. Wiley, Chichester.

Bottjer, D. J., M. L. Droser, and D. Jablonski. 1988. Palaeoenvironmental trends in the history of trace fossils. *Nature* 333:252-255.

Bromley, R. G. 1996. *Trace Fossils, Biology, Taphonomy, and Applications*. Chapman & Hall, London.

Crimes, T. P. 1975. The stratigraphical significance of trace fossils, pp. 109-130, *in* Frey, R. W., ed. *The Study of Trace Fossils*. Springer-Verlag, New York.

Crimes, T. P., and J. C. Harper, eds. 1977. Trace fossils 2. *Geological Journal Special Issue* 9.

Curran, H. A., ed. 1985. *Biogenic Structures: Their use in Interpreting Depositional Environments*. SEPM Special Publication 35.

Donovan, S. K., ed. 1994. *The Palaeobiology of Trace Fossils*. Wiley, Chichester.

Droser, M. L., and D. J. Bottjer. 1986. A semiquantitative field classification of ichnofabric. *Journal of Sedimentary Petrology* 56:558-559.

Droser, M. L., and D. J. Bottjer. 1993. Trends and patterns of Phanerozoic ichnofabrics. *Annual Reviews of Earth and Planetary Sciences* 21:205-225.

Ekdale, A. A. 1988. Pitfalls of paleobathymetric intepretations based on trace fossil assemblages. *Palaios* 3:464-472.

Ekdale, A. A., R. G. Bromley, and S. G. Pemberton, eds. 1984. *Ichnology: The Use of Trace Fossils in Sedimentology and Stratigraphy*. SEPM Short Course Notes 15.

Frey, R. W., ed. 1975. *The Study of Trace Fossils*. Springer-Verlag, New York.

Frey, R. W., and S. G. Pemberton. 1985. Biogenic structures in outcrops and cores. I. Approaches to ichnology. *Bulletin of Canadian Petroleum Geology* 333:72-115.

Frey, R. W., and A. Seilacher. 1980. Uniformity in marine invertebrate ichnology. *Lethaia* 13:183-208.

Gillette, D. D., and M. G. Lockley, eds. 1989. *Dinosaur Tracks and Traces*. Cambridge University Press, Cambridge.

Häntzschel, W. 1975. Trace fossils and problematica, in Teichert, C., ed. *Treatise on Invertebrate Paleontology, part W. Miscellanea, Supplement I*. Geological Society of America and University of Kansas, Lawrence.

Lockley, M. G. 1991. *Tracking Dinosaurs: A New Look at the Ancient World*. Cambridge University Press, Cambridge.

Lockley, M. G., and A. P. Hunt. 1995. *Dinosaur Tracks and Other Fossil Footprints of the Western United States*. Columbia University Press, New York.

Osgood, R. G., Jr. 1987. Trace Fossils, pp. 663-674, *in* Boardman, R. S., A. H. Cheetham, and A. J. Rowell, eds. *Fossil Invertebrates*. Blackwell Scientific Publishers, Cambridge, Mass.

Pemberton, S. G., J. A. MacEachern, and R. W. Frey. 1992. Trace fossil facies models: environmental and allostratigraphic significance, pp. 47-72, *in* Walker, R. G., and N. P. James, eds. *Facies Models: Response to Sea Level Change*. Geological Association of Canada, Toronto.

Sarjeant, W. A. S. 1975. Fossil traces and impressions of vertebrates, pp. 283-324, *in* Frey, R. W., ed. *The Study of Trace Fossils*. Springer-Verlag, New York.

Seilacher, A. 1953. Studien der Palichnologie. I. Über die Methoden der Palichnologie. *Neues Jahrbuch der Geologie und Paläontologie, Abhandlungen* 96:421-452.

Seilacher, A. 1967. Bathymetry of trace fossils. *Marine Geology* 5: 413-428.

Simpson, S. 1975. Classification of trace fossils, pp. 39-54, *in* Frey, R. W., ed. *The Study of Trace Fossils*. Springer-Verlag, New York.

Thulborn, R. A. 1989. The gaits of dinosaurs, pp. 39-50, *in* Gillette, D. D., and M. G. Lockley, eds. 1989. *Dinosaur Tracks and Traces*. Cambridge University Press, Cambridge.

Glossary

Abdomen. Posterior portion of the body in most arthropods.
Aboral. Direction opposite the mouth in an echinoderm.
Abundance zone. Biostratigraphic zone based on sudden increases in the abundance of a certain fossil (=peak zone, acme zone).
Acanthopore. Small tube adjacent and parallel to zooecial walls of a bryozoan, marked on surface with a projecting spine.
Acceleration. Peramorphosis by adding more development stages in a shorter time.
Accretionary growth. Growth by gradual addition of increments of tissue.
Accuracy. The closeness to the truth of a data estimate.
Acme zone (see Abundance zone).
Adaptationist programme. The idea that every feature can be explained as an adaptation under the influence of natural selection (see Panselectionism).
Adaptive landscape. Hypothetical surface with peaks of optimal adaptation, and valleys of poor adaptation. Theoretically, organisms are constantly striving to maintain themselves on peaks, and avoid entering valleys, which would lead to their death.
Additional growth. Growth by addition of discrete new parts.
Adductor muscles. Muscles attached to the interior surfaces of both valves that pull them shut. Opposed by diductor muscles in a brachiopod, or the ligament in a bivalve.
Aerobic. Waters that have at least 1.0 ml dissolved oxygen per liter.
Agglutinated. Foraminiferal test composed of grains of foreign material (usually sand) cemented together.
Agrichnia. Traces caused by an animal continuously "farming" the same area for food.
Ahermatypic corals. Corals that do not use symbiotic algae in their tissues.
Allantois. Waste storage chamber in the amniotic egg.
Allele. One of several alternative states of the same gene.
Allen's rule. Animals in colder climates tend to have shorter, stubbier limbs, ears, and other appendages to help conserve heat.
Allometric growth. Growth in which at least one dimension changes at a much faster rate than another.
Allopatric. Populations living in different areas, so they are genetically isolated.
Allopatric speciation model. Species tend to form in genetically isolated peripheral populations found in different areas (allopatric).
Ambulacrum. Narrow tract or groove extending radially from the mouth of an echinoderm, and bearing the pores for the tube feet.
Ammonitic suture. Cephalopod suture characterized by highly complex fluting of the septum, and a convoluted tracing on the exterior of the shell.
Amnion. Membrane that surrounds the embryo in the amniotic egg.
Amniotic egg. Land egg, which contains not only an embryo, but also several membranes for protection. (= **cleidoic** egg).
Amoeboid cell. Sponge cell with irregular, changing shape, lying between epidermal cells and collar cells in body wall of sponge.
Ampulla. Bulb-like structure attached to internal end of tube foot, which contracts and forces water into the tube foot to extend it.
Anaerobic. Waters that have less than 0.1 ml dissolved oxygen per liter.
Anagenesis. Phyletic evolution within a single lineage.
Analogous. Having a similar structure and function but derived from different evolutionary roots.
Anapsid. Having no temporal fenestrae.
Ancestrula. Initial individual of a bryozoan colony.
Anterior. Toward the front of an animal.
Antorbital fenestra. Additional skull opening in front of the eye.
Aperture. Opening in shell through which the molluscan body is extended or withdrawn.
Apex. The small end or point of the cone or spire on a molluscan shell.
Apparatus. Bilaterally symmetrical assemblage of paired conodonts found associated in a rock.
Aristotle's lantern. Complex system of calcareous elements that form the jaws in some echinoids.
Assemblage zone. A class of biostratigraphic range zones based on the association of three or more taxa.
Assemblage. A group of fossils that occur at the same stratigraphic level or interval.
Autecology. The ecology of individual organisms, and their relationship to the environment.
Autotrophs. Organisms that produce their own food (i.e., plants via photosynthesis).
Avicularium. Specialized bryozoan individual bearing a beak-like structure worked by strong muscles.
Background extinction. Normal, average rates of extinction.
Bathyal. Living at extreme ocean depths.
Bauplan. The basic body plan or design of an organism.
Beached Viking funeral ship. A landmass that moves from one continent to another, carrying its fossils to a new location.
Beak. Point of bivalve shell at the beginning of the spiral.
Benthic. Bottom dwelling
Bergmann's rule. Animals in colder climates tend to have larger body sizes to help conserve heat.

Bernoulli's principle. In a moving fluid, the sum of the pressure and velocity are constant. If one increases, then the other must decrease.

Binomen. Two-part name (genus plus species in the Linnaean system).

Biocenosis. Original living community of organisms.

Biochronology. Sequence of fossils worked out partially by biostratigraphy, but also by non-stratigraphic methods (such as stage of evolution), so it is not entirely based on local biostratigraphic range zones.

Biodiversity. The study of the diversity of life on earth.

Biogeography. The study of the geographic distribution of organisms.

Biological species concept. An array of populations that are actually or potentially interbreeding, and that are reproductively isolated from other such arrays under natural conditions.

Biomass. The total weight of living organisms in a given area.

Biome. An ecological community that characterized a natural region, particularly a plant community.

Biosphere. The entire realm of living organisms on earth.

Biostratigraphy. Stratigraphy based on the fossil content of the rocks.

Biostratinomy. Processes that occur when organic remains are incorporated into the rock.

Biramous. Branched arthropod appendage.

Biserial. Foraminiferal test with two rows of chambers.

Blastopore. Initial opening of the gastrula.

Body whorl. The last and largest loop of the spiral gastropod shell, containing most of the soft tissues and terminating in the aperture.

Bourrelets. Elevated areas around the mouth of echinoids, located in the interambulacral areas.

Brachial valve (= dorsal valve.) Brachiopod valve to which the brachidium is attached; typically the smaller valve with no beak and no pedicle opening.

Brachidium. Calcareous support of the lophophore.

Brachiophore. Short, stout processes projecting from the hinge line of the brachial valve into the interior of the valve.

Brackish. Having between 0.5 to 30 parts per mil dissolved salts, i.e., a mixture of fresh and normal marine waters, typically found in deltas, lagoons, and estuaries.

Brine. A solution with greater than 80 parts per mil dissolved salts.

Byssus. Threadlike material used by bivalves to attach themselves to the substrate.

Cadicone. Ammonoid shell with a fat, globular shape.

Calice. Oral end of the corallite on which the basal disk of the polyp rests.

Calyx. Head of crinoid surrounding the viscera, but not including the arms.

Camera. Chamber in a cephalopod shell closed off from the living chamber.

Canal. Tube leading from external pore of sponge to cloaca, serving to conduct water flow.

Carbon isotopes. Different atomic weights of the element carbon. Carbon-12 is the common isotope, but the rare stable isotope carbon-13 is also important in biological systems, and the radioactive isotope carbon-14 is important for radiocarbon dating.

Carbonate compensation depth (CCD). Depth in ocean below which the rate of dissolution of carbonate compensates for the rate of supply from above, typically about 4000 to 5000 m in depth.

Carbonization. Process of fossilization wherein the original organic material in a fossil has been reduced to a film of carbon.

Cardinal septum. One of the original septa in a coral, it lies in the plane of bilateral symmetry of the corallite.

Cardinal teeth. Teeth that fit into sockets in the hinge area of the bivalve.

Carnassial teeth. Enlarged shearing/crushing teeth found in carnivorous mammals.

Central dogma of genetics. Information can only flow from genotype to phenotype, but phenotypic information can never get back into the genotype (= Weismannism).

Cephalon. Front portion of a trilobite (the head) consisting of several fused segments and bearing the eyes and mouth.

Cheek. Portion of the outer cephalon lateral and anterior to the glabella.

Chelicerae. First set of oral appendages in the chelicerate arthropods, usually modified into mouthparts.

Chemosynthesis. Producing basic nutrients by the breakdown of chemicals such as methane or hydrogen sulfide, rather than by photosynthesis involving light and chlorophyll.

Chilidial plate. Plate at the side of the opening (notothyrium) in the brachial valve for the pedicle.

Chitin. Organic polysaccharide found in the exoskeleton of arthropods.

Cladistics. Classification based solely on phylogeny, as defined by shared evolutionary novelties, or shared derived characters (= phylogenetic systematics).

Cladogenesis. Branching evolution into two or more species.

Cladogram. A branching diagram of relationships between three or more taxa, with no implication of time sequence, or of ancestry and descent.

Climax community. The final, mature stage of ecological succession, characterized by a community of organisms that is stable over long periods of time.

Cline. A character gradient in a geographically variable population.

Cloaca. Central cavity of sponge (= spongocoel).

Coelom. Internal fluid-filled body cavity of most animals, commonly containing the viscera.

Coenosteum. Skeletal tissue connecting the corallites in a colonial coral or between individual bryozoan zooecia.

Columella. Central pillar in a spiral shell formed by the fusion of the inner walls of the whorls in a gastropod, or longitudinal rod in the axis of a corallite.

Columnals. Circular or polygonal plates that form the stem of a stalked echinoderm.

Commissure. Line of juncture between the two valves of a brachiopod or bivalve.

Competition. Interaction of two organisms striving for the same resource (food, space, mates, light, etc.).

Competitive exclusion (see Gause's competitive exclusion principle).

Connecting ring. Calcareous ring forming the walls of the siphuncle between the septa.

Convergence. Independent evolution to a very similar body form in distantly related lineages (see Parallelism).

Convolute. Coiled shell in which part of the outer whorl extends in toward the center and completely covers inner whorls.

Corallite. Skeleton formed by individual coral polyp, consisting of wall, septa, and accessory structures such as tabulae and dissepiments.

Corridor. A geographic connection between two faunal regions that allows most organisms to pass.

Costa. Ridge on surface of shell, typically radiating out from the apex of a gastropod, or beak of a bivalve.

Counter septum. One of the original septa in a coral, it lies directly opposite the cardinal septum in the plane of bilateral symmetry.

Cranidium. Central part of the cephalon including the axial lobe (glabella) and bounded by the facial suture.

Crura. Basal portions of brachidium.

Cubichnia. Resting traces, caused when an animal interrupts it locomotion for rest or refuge.

Cystiphragm. Calcareous plate extending from bryozoan zooecial wall partway across the tube. The surface is domed, convex upward or inward.

Delthyrium. Opening in pedicle valve of brachiopod adjacent to hinge line, allowing exit of pedicle.

Deltidial plate. Plates on either side of the pedicle opening (delthyrium) in the pedicle valve. They constrict the opening or even close it off completely.

Derived. Anatomical character specialized beyond the primitive condition.

Desma. Sponge spicule with irregular, knotty growth form.

Desmodont. Bivalve hinge with a large ligament inside the hinge area.

Dextral. Right-handed coiling, with the aperture on the right side when the apex is pointed upward and the aperture faces the viewer.

Diaphragm. Calcareous plate extending transversely across with of zooecial tube in a bryozoan. Surface is flat or gently curved.

Diapsid. Having two temporal fenestrae, one above and one below the postorbital-squamosal bar.

Dicyclic. Crinoid cup characterized by a double circlet of plates aboral to the radial plates.

Diductor muscles. Muscles that open the valves of a brachiopod.

Discrete inheritance. The idea that phenotypic characters are not blended out by continued outcrossing, but can maintain their integrity and reappear in later generations.

Disparity. A measure of the total range of variation of morphological shapes, and differences between those shapes.

Dispersalism. The idea that dispersal from a point of origin accounts for most geographic distributions of organisms.

Dissepiment. Small curved plate forming a vesicle within a corallite.

Diversity. A measure of the total number of taxa at a given time and place.

Domichnia. Dwelling traces formed as a long-term residence for an animal.

Dominant allele. An allele that is expressed, even when paired with a recessive allele in a heterozygote.

Dorsal. Along the back of an animal.

Doublure. Portion of the exoskeleton that bends under to form a border around the underside of the cephalon.

Dysaerobic. Waters that have between 1.0 and 0.1 ml dissolved oxygen per liter.

Dysodont. Bivalve hinge with reduced hinge teeth and a large ligament inside the hinge area.

Ecological biogeography. Explanations of the distribution of living organism based largely on present-day processes and conditions, or those of the recent past.

Ecophenotypic variation. Normal variation within a population of genetically similar individuals due to differences in their environment and differences in their growth.

Ecosystem. The sum of all the physical and biological characteristics of a given area.

Ectotherm. An organism that relies on external heat sources for its body heat.

Elvis taxa. Fossils that have converged in their morphology on some extinct taxon, giving the false appearance that the earlier fossil's range has been extended.

Endocones. Conical calcareous deposits inside a siphuncle.

Endotherm. An organism that produces body heat internally through metabolic processes.

Epifaunal. Living on the surface.

Epirelief. Sedimentary structure protruding from the top of a sedimentary bed.

Epitheca. Sheath of skeletal material that forms the wall of the corallite.

Equivalve. Right and left valves are symmetrical across the plane of the commissure.

Escalation. An "arms race" between a species that evolves a new adaptation, and others that must evolve new defenses in response to that adaptation.

Escalator counterflow. The possibility that animals can persist on a mid-oceanic ridge island, or a chain of hotspot islands, which are continually sinking under the ocean as new islands arise. The history of that animal on that island is older than the island itself, since the animal can go "up the down escalator."

Etymology. The origin and meaning of a word.

Eukaryotes. Organisms whose DNA is enclosed within a nucleus.

Eulamellibranch. Bivalve gill structure where the folds of the gills are interconnected in many complex ways.

Euryapsid. Having a single temporal fenestra above the postorbital-squamosal connection.

Euryhaline. Tolerating a broad range of salinities.

Eurytopic. Occupying a wide range of habitats, or geographically widespread.

Evolute. Coiled shell in which the outer whorls do not cover any of the inner whorls.

Evolutionary species concept. Lineage evolving separately from others and with its own unitary evolutionary role and tendencies.

Evolutionary synthesis. The mid-twentieth-century combination of genetics, systematics, and paleontology to form the dominant school of evolutionary thought (see Neo-Darwinism).

Evolutionary taxonomy. Traditional system of classification, which mixes phylogenetic classification with overall phenetic differences.

Evolutionary tree. A branching diagram of relationships that also shows a time axis and may show ancestors and descendants.

Exoskeleton. External skeleton of an animal (typical of arthropods).

Exotic terranes. Rock bodies that are not native to a region, but have been brought there from far away by plate motions.

Extrapolationism. The idea that processes operating at one level (e.g., microevolution) can be extrapolated to a different level (e.g., macroevolution).

Facial suture. Line along which the exoskelton of the cephalon splits when a trilobite molts. It may be limited to the margin of the cephalon (marginal suture) or it may pass across the cephalon in front of the genal angle (proparian) or behind it on the posterior border of the cephalon (opisthoparian).

Facies fossils. Fossils that tend to be highly environmentally restricted, so their first or last appearance in a given area is more a reflection of the appearance of their favored environment than of their evolutionary first or last occurrence.

Filibranch. Bivalve gill structure where the folds of the gills are only slightly interconnected.

Filter bridge. A geographic connection between two faunal regions that allows some organisms to pass, but not others.

First appearance datum (FAD). The horizon where a given fossil first appears in a section.

Fodichnia. Traces caused by deposit feeding, where the animal ingests the sediment and digests all of the food out, and leaves behind a trail of digested castings.

Fold. Elevation of a brachiopod valve along the midline, accompanied by a depression (sulcus) on the other valve.

Food chain. The sequence of organisms in which each is food for a higher member of the sequence.

Food web. The complex of feeding relationships between organisms in an ecosystem.

Fossula. Unusually wide space between septa caused by a failure of one or more septa to develop as rapidly as the others. If it is due to an aborted cardinal septum, it is a cardinal fossula.

Founder principle. A type of random genetic drift wherein the gene pool of a new population differs from the parent population simply by sampling accident due to a small founder population.

Fusiform. Spindle shaped.

Gastrula. First ball of cells formed from an embryo.

Gause's competitive exclusion principle. Whenever two or more organisms occupy the same niche, one will exclude the other, or else they will subdivide the niche into subniches to minimize competition.

Genal angle. Posterior lateral corner of the cephalon. Typically terminates in a spine, or may be rounded.

Genal spine. Spine extending posteriorly from the genal angle of the cephalon.

Gene flow. The transfer or exchange of genetic material from one population to another by interbreeding.

Gene pool. The sum total of all the genes in a given population.

Generalized tracks. Connections of the geographic ranges of many different organisms and their nearest relatives, showing a widespread pattern of geographic similarity.

Genetic assimilation. Apparent inheritance of acquired characteristics caused by environmental stresses on developing embryos.

Genotype. The genetic or hereditary information of an individual.

Germ line. Line of genetic inheritance from parent to offspring.

Gill arches. Rods of cartilage that support the gill apparatus in fishes.

Glabella. Elevated axial portion of cephalon; the anterior part of the axial lobe on a trilobite.

Glossopetrae. "Tongue stones," the medieval name for fossilized shark teeth.

Gonitatitic suture. Type of cephalopod suture characterized by simple fluting, with a single series of zig-zag lobes and saddles.

Graphic correlation. Quantitative correlation method that plots the distributional data of one stratigraphic section against another to produce a composite standard section.

Growth lines. Series of fine to coarse ridges on the outer surface of a shell. They tend to be concentric around the beak of a brachiopod or bivalve, or spiral down the shell of a cephalopod or gastropod.

Growth series. An arrangement of specimens thought to represent the ontogeny of a species.

Gyre. A circular motion of water in each of the world's major ocean basins.

Habitat. The actual physical environment in which a given organism lives.

Hardground. Emergent erosional surface that has been cemented during its exposure to nonmarine diagenesis.

Hermatypic corals. Corals that harbor symbiotic algae in their tissues.

Heterocercal tail. Asymmetric tail with support in upper lobe caused by upward flexion of spine.

Heterochrony. 1. Evolution by changes in developmental timing; 2. Asychrony of biostratigraphic events, or biostratigraphic assemblages, due to time-transgressive first or last occurrences.

Heterodont. Bivalve hinge structure characterized by teeth and sockets in both the cardinal area, and also lateral to it.

Heteromorph. Aberrant cephalopod shell shape that no longer follows planispiral growth patterns.

Heterotroph. Organism that gets its food by consuming other organisms (e.g., animals and fungi).

Heterozygote advantage. A mechanism whereby selection favors the heterozygote condition, maintaining a recessive gene that might otherwise be selected against.

Heterozygous. Having two different alleles at the same locus.

Hierarchical. Having one group clustered within a larger or higher group.

Hinge line. Line of articulation between the two valves of a brachiopod or bivalve shell.

Historical biogeography. Explanations of the distribution of organisms (including fossils) based largely on long-term changes in the earth's continents and climate.

Holaspid. Final growth stage of a trilobite, recognized when its full adult complement of thoracic segments is present.

Holochroal eye. Trilobite compound eye made of hundreds of tiny, closely packed calcite lenses.

Holotype (see Type specimen).

Homeotherm. An organism that maintains a constant body temperature.

Homeotic genes. Regulatory genes that cause one organ system in an organism to be replaced by another.

Homocercal tail. Symmetrical tail, with reduced support from the vertebral column.

Homologous. Having a common evolutionary origin.

Homonym. Two taxonomic names that are spelled differently but sound the same.

Homozygous. Having two identical alleles at the same locus.

Hood. Tough fleshy structure that covers the head of *Nautilus*, and covers the aperture when the head is withdrawn.

Hopeful monster. Unusual individual produced by an abrupt change in genotype.

Hydrospire. Folded sheet of calcite in the interior of a blastoid, lying just beneath the ambulacral pores.

Hyomandibular. Major bony component of the hyoid, or second gill arch.

Hypermorphosis. Peramorphosis by shutting off growth at a later stage.

Hypersaline. Having between 40 to 80 parts per mil dissolved salts.

Hyperstrophic. A rare molluscan shell type in which the whorls form an inverted cone, so that the apex pointed forward and down, rather than up and back, as in most gastropods.

Hypodigm (see Syntype).

Hyporelief. Sedimentary structure protruding from the base of a sedimentary bed (= sole mark).

Hypostome. Plate on the front underside of the cephalon directly in front of the mouth.

Ichnofabric. Degree of reworking and bioturbation, as measured by the disruption of primary sedimentary structures.

Ichnofacies. Association of trace fossils in a given sedimentary environment.

Ichnofossils (see Trace fossils).

Index fossils. Biostratigraphic indicator fossils that ideally are distinctive, widespread, abundant, independent of facies, rapidly changing, and short ranging.

Industrial melanism. Selection favoring a dark or melanic variety of individual due to darkening of the background by industrial pollution.

Inequivalve. Shell form in which one valve is different in size or shape from the other.

Inertial homeothermy. The tendency of animals of large body size to retain their body heat simply by the fact that large masses lose or gain heat very slowly.

Infaunal. Burrowing beneath the surface.

Interambulacral. Area between the rays of the ambulacra.

Interarea. Planar or curved surface between the beak and the hinge line on either valve of a brachiopod.

Intertidal. Between low and high tides (= littoral).

Interval zone. 1. A class of range zones based on two first or last occurrences; 2. An overlapping range zone based on the interval between two successive first occurrences, or two successive last occurrences.

Involute. Coiled shell in which part of outer whorl extends in toward the center and covers part of the inner whorl.

Isodont. Bivalve hinge structure characterized by two large subequal teeth on one side, and corresponding sockets on the other.

Isometric growth. Growth wherein dimensions change at the same relative rate.

Iterative evolution. Repeated evolution of the same (or highly similar) external body form by convergence.

Jaccard coefficient. A measure of faunal similarity, where the number of taxa in common between the two samples (C) is divided by the sum of the number of taxa unique to fauna A and to fauna B, or C/(A + B - C).

Jugum. Connection between the left and right half of the arms of the brachidium.

Keel. Ridge along outer margin of test or shell.

Keratin. Protein found in most vertebrate dermal tissues; the basic component of hair, nails, claws, horns, feathers, and scales.

Lagerstätten. Fossil deposit with extraordinary preservation.

Lamarckian inheritance. The idea that characters acquired during an individual's lifetime can be passed on to the offspring.

Last appearance datum (LAD). The horizon where a given fossil last appears in a section.

Lateral line system. Series of canals along the body of some fishes that help sense changes in water pressure, and thus detect movement of other fishes and foreign objects and obstacles.

Lateral teeth. Teeth and sockets found along hinge line in front or in back of cardinal area.

Latitudinal diversity gradient. The observation that diversity is highest in the tropics, and decreases toward the poles.

Law of correlation of parts. The anatomical features of an organism tend to be highly interrelated and correlated, so that the knowledge of some features can usually predict the rest.

Lazarus taxa. Fossils that were apparently extinct at a given horizon, but turn up in much later horizons, showing that they survived the extinction event.

Lectotype. A type specimen designated from the original hypodigm by a later author, since the original author failed to designate a holotype.

Ligament. Elastic tissue attaching the bivalve shells along the hinge line, and serving to open the valves automatically.

Lineage zone. A range zone based on the successive evolutionary first occurrences within a single evolving lineage.

Littoral. Intertidal.

Lobe. Flexure of cephalopod suture line concave toward aperture.

Loop. Brachidium made up of a pair of simply curved or doubly bent longitudinal "arms" and a relatively simple jugum connecting the anterior ends of these arms.

Lophophore. Feather-like or circular structure bearing tentacles for filter feeding, found in both bryozoans and brachiopods.

Lunule. In bivalves, a depressional long the commissure and in front of the beak. In sand dollars, a large perforation of the shell.

Macroevolution. Large-scale evolutionary change, typically between major body plans or different organ systems, and usually marking changes between higher level taxa (e.g., orders, classes).

Macromutation. Mutation that causes an abrupt change in phenotype.

Madreporite. Sieve-like plate covering the external opening to the water vascular system of echinoderms.

Mandible. Lower jaw.

Mandibular fenestra. Fenestra in the lower jaw.

Mantle cavity. Chamber between the lateral and posterior folds of the mantle and the sides of the body and foot, often containing the gills.

Mantle. Dorsal and lateral portion of molluscan body wall, covering the visceral hump, and folding around the body and foot.

Marsupials. Pouched mammals.

Mass extinction. An extinction event that wipes out a significant percentage of the earth's living organisms.

Meckel's cartilage. Embryonic cartilage that precedes the lower jaw.

Median septum. Calcareous ridge built along the midline of the brachiopod valve.

Medusa. A free-living, jellyfish-like stage of the cnidarian life cycle.

Meraspid. Intermediate growth stages of a trilobite.

Mesoglea. Layers of cells and gelatinous connective material between the endoderm and ectoderm of a cnidarian.

Mesopore. Zooecium of unusually small size set between larger zooecia, and characterized by numerous transverse partitions (diaphragms).

Microevolution. Small-scale evolutionary change, usually at the species level.

Molting. The shedding of the skin or exoskeleton of an organism (= ecdysis).

Monocyclic. Crinoid cup characterized by a single circlet of plates aboral to the radial plates.

Monographic bursts. Artificial inflation of diversity statistics when the appearance of a major publication or monograph greatly increases the published diversity of some taxon.

Monophyletic group. A taxon that includes all descendants of common ancestor (= natural group, holophyletic group).

Morphological species concept. A diagnosable cluster of individuals within which there is a pattern of ancestry and descent, and beyond which there is not.

Morphospace. Theoretical multidimensional space whose axes are different morphological features.

Mutation. A spontaneous, heritable change in genotype.

Myomeres. V-shaped segmented muscle masses in chordates.

Nacreous layer. Iridescent layer of aragonite found in the inner part of some molluscan shells.

Nannoplankton. Submicroscopic (less than 100 microns) planktonic organisms.

Natural selection. Organisms that inherit favorable variations will survive and reproduce those variations in the next generation.

Necrolysis. Post-mortem decay.

Nektonic. Free swimming.

Nema. Delicate tube to which the base of a graptolite colony is attached.

Neo-Darwinism. Twentieth-century revival of the emphasis on Darwinian natural selection, combined with modern genetics.

Neo-Lamarckism. Late-twentieth-century revival of the idea that there is some sort of inheritance of acquired characters.

Neoteny. Paedomorphosis by slowing down development, so that sexual maturity occurs earlier.

Neotype. A new type specimen from the original hypodigm designated by a later reviser if the original holotype specimen is lost.

Neritic. Subtidal.

Neutralism. The idea that some features are selectively neutral, or invisible to natural selection.

Niche. The sum of all the physical, chemical, and biological limits on an organism, including its way of life, its habitat, and the role it plays in the ecosystem.

Noah's ark. A landmass that moves from one continent to another, carrying its fauna with it to colonize the new area.

Nomen dubium. A taxonomic name of doubtful validity, since the original material is inadequate for diagnosis.

Nomen nudum. A taxonomic name that is invalid because the original namer failed to provide some important part of its definition.

Notochord. Cartilaginous rod that supports the body of chordates, and precedes the bony backbone in the embryology of most vertebrates.

Notothyrium. Opening in the brachial valve adjacent to and outside the hinge line, forming part of the opening for the pedicle.

Numerical taxonomy. Classification scheme based on overall similarity, as quantified by computer (=phenetics).

Objective synonym. Two names given to the same specimen.

Ontogeny. The life history of an individual.

Operational taxonomic unit (OTU). Each specimen or taxon in a data matrix, the basic unit of numerical taxonomic analysis.

Opisthoma. Abdomen in chelicerate arthropods.

Oppel zone. An assemblage zone with more precise definitions of beginning and end based on a single first or last occurrence of a fossil.

Oral. Direction toward the mouth in an echinoderm.

Orthocone. Cephalopod shell made of a simple, straight cone.

Orthostrophic. The common gastropod shell type in which the whorls coil to form a cone with the apex pointed up and backward as it sits on the back of the snail.

Osculum. Large opening from the internal cavity of a sponge to the exterior.

Outgroup. A taxon that is outside those included in the original cladistic analysis, and is used for determination of the primitive character state.

Overlapping range zones. Biostratigraphic range zones defined by the overlapping ranges of two or more species (=concurrent range zone).

Oxycone. Ammonoid shells with a streamlined, discus-like shape.

Oxygen isotopes. Different atomic weights of the element oxygen. Oxygen-16 is the common isotope, but the heavy isotope oxygen-18 is also important, although very rare.

Oxygen minimum zone. A zone of low oxygen levels in the ocean, typically between 600 to 10,000 m water depth.

Pachydont. Bivalve hinge structure characterized by one or more very large teeth.

Paedomorphocline. A series of specimens that appears to show paedomorphic changes through time.

Paedomorphosis. Retention of juvenile features into sexual maturity (see Progenesis, Neoteny).

Palatoquadrate. Cartilage that functions as the upper jaw.

Paleobiogeography. The study of ancient distributions of organisms.

Pallial line. Line found on the inner surface of the bivalve shell, parallel to and just inside the edge, where the mantle contact terminated.

Pallial sinus. Embayment on the posterior portion of the pallial line where the siphons protruded.

Panselectionism. The idea that all parts of the phenotype of an organism are continually under scrutiny from natural selection, and that no feature is selectively neutral.

Paradigm method of functional morphology. The process of generating functional explanations by comparing the organism to an idealized model, or paradigm.

Parallelism. Independent evolution to a very similar body form in closely related lineages (see Convergence).

Paraphyletic group. A taxon that includes some but not all descendants of a common ancestor.

Paratype. Additional specimens chosen as reference standards for a taxon, supplementing the holotype.

Paschichnia. Traces caused by an animal "grazing" through the sediment in search of food.

PDB ("Pee Dee Belemnite"). The laboratory standard by which mass spectrometers are calibrated for the measurement of stable isotopes of carbon and oxygen.

Pedicle opening. Opening that permits the pedicle to exit shell.

Pedicle valve (= ventral valve). Valve to which the pedicle is attached.

Pedicle. Muscular stalk that attaches to the inner surface of the pedicle valve, and attaches the brachiopod to the substrate.

Pedipalps. Second set of oral appendages in the chelicerate arthropods, sometimes modified into claws.

Pelagic. Living in the open ocean.

Pelvic claspers. Rods of cartilage in the pelvic fins of male sharks that help with copulation and internal fertilization.

Pentameral symmetry. Five-fold symmetry.

Peramorphocline. A series of specimens that appears to show peramorphic changes through time.

Peramorphosis. Addition of ontogenetic stages beyond the normal adult reproductive stage.

Perforate. Foraminiferal test with many small openings in chamber walls.

Periostracum. Dark organic material that covers the outside of some molluscan shells.

Permineralization. Process of fossilization wherein the original hard parts of an organism have additional mineral material precipitated in the pores.

Petal. Petal-shaped portion of ambulacrum on aboral surface of irregular echinoids.

Phenetics (see Numerical taxonomy).

Phenotype. The physical characteristics of an individual, as dictated by its genotype and developmental history.

Photic zone. Ocean depths where light can penetrate, typically less than 50 m, and never more than 200 m.

Phragmocone. Portion of the shell containing the camerae and septa.

Phyletic gradualism. Gradual anagenetic evolutionary change.

Phyllode. Depressed petal-shaped portion of ambulacrum adjacent to mouth in echinoids.

Phylogenetic systematics (see Cladistics).

Phylogeny. Evolutionary history.

Pinnule. Thin structures that branch off the primary arms in crinoids.

Pioneer community. The earliest stage of succession, characterized by organisms that establish the community and prepare it for further successional change.

Placenta. A membrane that surrounds the developing embryo in eutherian mammals.

Planispiral. Shell that spirals in a plane.

Planktonic (also, **plankic**). Free floating.

Pleiotropy. A single gene having multiple effects.

Pleural lobes. Portion of thoracic segments lateral to axial lobes.

Pleural spine. Spine protruding from a thoracic segment.

Pleurodont teeth. Teeth that are attached to the inside margin of the jaw.

Plica. Fold involving the full thickness of the shell that extends radially from the beak to the commissure.

Poikilotherm. An organism that allows its body temperature to fluctuate over a wide range.

Polymorphism. Maintaining a wide phenotypic variability in a single population.

Polyp. An attached, hydra-like stage of the cnidarian life cycle.

Polyphyletic group. A taxon that includes groups from two unrelated lineages.

Populational species concept. A broader definition of species, based on the idea that natural populations are highly variable (see Typological species concept).

Posterior. Toward the rear of an animal.

Precision. The reliability (as measured by the reproducibility) of a data estimate.

Priority. The taxonomic principle that the first name given to a taxon is usually the valid name.

Progenesis. Paedomorphosis by attaining sexual maturity at an early stage of development.

Prokaryotes. Organisms whose nucleic acids are not enclosed in a nucleus.

Proloculus. Initial chamber of foraminiferal test.

Prosoma. Head segment in chelicerate arthropods.

Protaspid. Earliest calcified juvenile stage of a trilobite.

Protobranch. Simple leaf-like gills in bivalves.

Protoconch. Initial chamber of a cephalopod shell, the first to be formed by the embryonic cephalopod.

Pseudoextinction. Apparent "extinction" of species because its descendant is given a different species name.

Pseudopod. Lobate or thread-like extension of protoplasm in an amoeba, foraminiferan, or radiolarian.

Pseudopunctate. Shell microstructure characterized by structureless rods of calcite penetrating the shell normal to its surface.

Pull of the Recent. Artificial distortion of diversity of more recent organisms by the fact that if they have one fossil occurrence and are alive today, they are given credit for having existed through periods of time when they have no fossil record.

Punctate. Shell microstructure characterized by small canals extending perpendicularly from the inner wall to the outer wall of the shell.

Punctuated equilibrium. The idea that species change rapidly during episodes of speciation, then remain stable for long periods of time between speciation events.

Pygidium. Posterior portion of trilobite (tail) consisting of one or more fused segments.

Radial plate. One of a circlet of cup plates to which the arms attach in a crinoid.

Radula. Strip of horny material bearing file-like teeth, found in the mouth of most molluscs.

Rain shadow effect. Desert created behind a mountain range, which tends to take most of the moisture out of the clouds coming from the ocean.

Range zone. All the strata that contain a given fossil.

Rarefaction. Statistical technique that assesses how large a sample is required to obtain different degrees of sampling completeness.

Realms. Biogeographic regions on the continental scale.

Recessive allele. An allele that is not expressed when the dominant allele is present, but only when homozygous with another recessive allele.

Recrystallization. Process of fossilization wherein the original mineral grains of a skeleton have been dissolved and reprecipitated into new, usually larger crystals.

Red Queen hypothesis. Species must constantly improve to avoid extinction, since the environment constantly changes and competitors constantly evolve new adaptations.

Reductionism. The idea that all processes reduce to the fundamental level, e.g., all macroevolution reduces to microevolution, or that organisms are nothing more than their genes.

Regulatory genes. Genes that determine which structural genes are expressed, and which ones are ignored.

Repichnia. Traces formed by normal walking or crawling motions of an animal.

Replacement. Process of fossilization where the original material has been replaced by a new mineral.

Resilifer. Depression inside hinge line that holds an internal ligament (resilium).

Resolution. The ability to distinguish two or more events that are closely spaced in time.

Retrovirus. A virus that can copy its genetic code back into the genes of its host organism.

Reversed heterocercal tail. Asymmetric tail with support in lower lobe caused by downward flexion of spine.

Rhabdosome. Entire graptolite colony.

Ring canal. Hollow tube forming a closed ring around the inside of the mouth in an echinoderm. It interconnects all the canals in the water vascular system.

Ring species. A circular cline, where the end members overlap geographically but do not interbreed.

Saddle. Flexure of cephalopod suture line convex toward aperture.

Saltationism. Rapid, discontinous evolutionary change.

Scenario. An elaborate explanatory story, usually only partly supported by data.

Schizochroal eye. Trilobite compound eye made of a few dozen relatively large calcite lenses.

Schizodont. Bivalve hinge structure characterized by teeth diverging sharply from beneath the beak.

Sclerotic ring. Ring of bones surrouding and protecting the eyeball.

Secondary palate. Additional palate that grows over the original reptilian palate, enclosing the internal nasal chamber in a tube that exits in the throat.

Septibranch. Bivalve gill type that has a perforate diaphragm extending horizontally across the mantle cavity, dividing it into upper and lower portions.

Septum. 1. One of several longitudinal plates arranged radially between the axis and wall of the corallite; 2. Transverse partitions in a cephalopod shell separating camerae.

Serpenticone. Ammonoid shell that looks like a flat-coiled snake.

Sexual dimorphism. Phenotypic differences between males and females of the same species.

Sibling species. Species that cannot be distinguished based on external morphology, but have differences of behavior or genetics that allows them to recognize each other and not interbreed.

Sicula. Tube formed by initial individual of a graptolite colony.

Signor-Lipps effect. Problems in sampling biostratigraphic ranges will mean that a sudden, abrupt extinction event will have the appearance of a gradual extinction. Few organisms that were alive at a mass extinction will leave fossils right up to the final horizon.

Simpson coefficient. A measurement of faunal similarity, C/N x 100, where C is the number of taxa found in both samples, and N is the number of taxa found in the smaller sample.

Sinistral. Left-handed coiling, with the aperture on the right side when the apex is pointed upward and the aperture faces the viewer.

Siphon. Tubular extension of mantle border that carries water to and from the mantle cavity.

Siphonal notch. Notch at the edge of the aperture that protects the siphon.

Siphuncle. Tube extending back from the living chamber through septa to protoconch of a cephalopod.

Siphuncular deposits. Calcareous deposits within the siphuncle of a cephalopod.

Sister group. A taxon that is the nearest relative of another taxon.

Socket. Depression along hinge line that receives projecting tooth from the opposite valve.

Soma. Phenotype of an organism.

Speciation. Process of formation of new species.

Species sorting. The idea that species are hierarchical entities above the level of individuals or populations, and have properties as species that are not properties of their constituent populations.

Sphaerocone. Ammonoid shell with a nearly spherical shape.

Spicule. Needle-like, rod-like, or star-like skeletal element of a sponge.

Spiralium. One of a pair of spirally coiled calcareous ribbons that forms the brachidium in spirifer brachiopods.

Spondylium. Curved plate in midline of peak of pedicle valve.

Stage. A chronostratigraphic unit composed of several biostratigraphic zones, and equivalent to an "age" in time stratigraphy.

Stapes. Innermost middle ear bone that attaches to the window in the inner ear; the "stirrup" bone.

Steinkern. Internal mold of fossil ("stone cast" in German).

Stenohaline. Tolerating a narrow range of salinities.

Stenotopic. Occupying a narrow range of habitat, or geographically restricted.

Step. Distance between corresponding left and right foot impressions.

Stipe. Branch of graptolite colony.

Stolon. Dense tube extending through the thecae of a graptolite, and connecting them in life.

Stone canal. Short canal leading from the ring canal to the external opening of the water vascular system in an echinoderm.

Streptostyly. The ability to flex the jaw joint backward and forward in squamates, since the ventral bar beneath the lower temporal fenestra has been lost.

Stride. Distance between the track of one foot and the track caused by the next footfall of the same foot.

Structural constraints. Properties of a phenotype that are dictated by fundamental engineering and design constraints, and do not necessarily require a special adaptationist explanation.

Structural genes. Genes that produce the basic phenotypic building blocks of organisms.

Subjective synonym. Two different specimens or species that were originally given different names, but are probably the same thing.

Subtidal. Below low tide level (= neritic).

Succession. The regular changes in the components of an ecological community.

Supratidal. Above mean high tide.

Survivorship curve. A plot of the number of taxa surviving against the duration of their survival.

Suture. Line of intersection between septum and inner surface of the shell wall in a cephalopod.

Sweepstakes dispersal. Low-probability dispersal to distant oceanic islands through rafting, storms blowing animals long distances, and other such chance methods.

Swim bladder. Gas-filled chamber in bony fish that helps make them buoyant.

Sympatric. Populations living in the same area, so they can be genetically interbreeding.

Symplesiomorphy. Shared primitive characteristic.

Synapomorphy. Shared evolutionary specialization, or shared derived character.

Synapsid. Having a single temporal fenestra below the postorbital-squamosal connection.

Synecology. The ecology of communities of organisms, and their relationships to the environment.

Synonym. Two different taxonomic names that represent the same thing.

Syntype. A series of specimens chosen as a reference standard for a taxon (= hypodigm).

Systematics. The scientific study of the kinds and diversity of organisms and of any and all relationships between them.

Tabulae. Transverse partitions in a corallite.

Taphonomy. The study of the processes of fossilization.

Taxodont. Bivalve hinge structure characterized by numerous small subequal teeth arranged in a row along the hinge line.

Taxon. A collective group of species or of populations, usually recognized and delimited by morphological features.

Taxon range zone. A range zone based on the first or last occurrence of a single taxon.

Taxonomy. The science of classification.

Tegmen. Oral surface of crinoid body.

Teilzone. Partial or local range zone.

Telson. Spine protruding from the pygidium of a trilobite, usually pointing posteriorly along the midline.

Temporal fenestrae. Openings on the side of the skull for increased jaw muscles.

Test. A shell, especially in microfossils.

Thanatocenosis. "Death assemblage" of fossils that have been brought together after death by sedimentary processes.

Theca. Individual tube in a graptolite colony, or plates enclosing the viscera in stalked echinoderms, or skeletal deposit enclosing the corallite.

Thecodont teeth. Teeth that are inserted in sockets.

Thorax. Middle section of an arthropod between the head and tail.

Tiering. The number of different levels above and below the sediment-water interface occupied by organisms.

Torsion. Rotation of the viscera so that the anus and mantle cavity are brought from the back of the animal to a position over the head.

Trace fossils. Sedimentary structures formed by organisms, such as trackways, trails, borings, and burrows (= ichnofossils, *Lebensspuren*).

Trochospiral. Spiraling up an axis.

Trophic pyramid. Pyramid of relative biomass, wherein the largest biomass of producers (= plants) is the base of the pyramid, and each trophic level above has less biomass.

Trophic. Referring to feeding relationships.

Type specimen. Specimen that is chosen as the reference standard for a taxon, usually of a species (=holotype).

Typological species concept. An older species concept based on the idea that each individual in a species was an imperfect representation of an ideal "type" (originally thought to be in the mind of God). Typological species tend to be very narrowly defined and oversplit, with each slight variation used as justification for a different species, since they do not take into account the natural variation of populations (see Populational species concept).

Umbilicus. Depression along the central axis of the spiral where the inner walls of the whorl fail to meet.

Umbo. Elevated, convex portion of the valve, forming the "hump" above the bivalve beak.

Uniramous. Unbranched arthropod appendage.

Uniserial. Foraminiferal test composed of a single row of chambers.

Uroneurals. Bones of the teleostean fish tail that radiate from the upturned end of the spinal column.

Van't Hoff's rule. For every 10°C of temperature increase up to an optimal level, the biological reaction rates increase by a factor of 1 to 6, depending upon species.

Venter. Portion of cephalopod whorl farthest from axis of coiling.

Ventral. Along the belly of an animal.

Vicariance. The idea that fragmentation (by mountains, opening oceans, and so on) of an originally large continuous geographic range accounts for the geographic distribution of organisms.

Water mass. A large body of water within an ocean that is similar in temperature and salinity.

Weberian apparatus. Set of bones in front of the swim bladder in ostariophysan teleost fishes that conducts sound from the swim bladder amplification chamber.

Whorl. Single complete turn of a spiral shell.

Zoarium. Skeleton of entire bryozoan colony.

Zombie effect. Apparent range extension of an extinct organism due to reworking of its fossils into younger strata by sedimentary processes.

Zooecium. The skeleton of an individual bryozoan animal.

Zooid. The individual bryozoan, consisting both of its skeleton (zooecium) and the soft parts.

Zooxanthellae. Symbiotic algae living within the tissues of an animal.

Bibliography

Ager, D. V. 1964. The British Mesozoic Committee. *Nature* 203:1059.

Ager, D. V. 1973. *The Nature of the Stratigraphical Record* (1st ed.). Macmillan, London.

Ager, D. V. 1981. *The Nature of the Stratigraphical Record* (2nd ed.). Wiley, New York.

Ahlberg, P. E., and A. R. Milner. 1994. The origin and early diversification of the tetrapods. *Nature* 368:507-514.

Akersten, W. A. 1985. Canine function in *Smilodon* (Mammalia: Felidae: Machairodontidae). *Contributions in Science of the Natural History Museum of Los Angeles County* 356:1-22.

Aldridge, R. J., ed. 1987. *Palaeobiology of Conodonts*. Ellis Horwood Publishers, Chichester.

Aldridge, R. J., D. E. G. Briggs, E. N. K. Clarkson, and M. P. Smith. 1986. The affinities of conodonts—new evidence from the Carboniferous of Scotland. *Lethaia* 19:279-291.

Aldridge, R. J., D. E. G. Briggs, M. P. Smith, E. N. K. Clarkson, and D. L. Clark. 1993. The anatomy of conodonts. *Philosophical Transactions of the Royal Society of London* (B) 340: 405-421.

Aldridge, R. J., and M. A. Purnell. 1996. The conodont controversies. *Trends in Ecology and Evolution* 111: 463-468.

Alexander, R. M. 1970. *Functional Design in Fishes*. Hutchinson, London.

Alexander, R. M. 1976. Estimates of the speeds of dinosaurs. *Nature* 261:129-130.

Alexander, R. M. 1989. *Dynamics of Dinosaurs and other Extinct Giants*. Columbia University Press, New York.

Alexander, R. R. 1974. Morphologic adaptations of the bivalve *Anadara* from the Pliocene of the Kettleman Hills, California. *Journal of Paleontology* 48:633-651.

Allen, J. A. 1985. The Recent Bivalvia: their form and evolution, pp. 337-403, *in* Trueman, E. R., and M. R. Clarke, eds., *The Mollusca, 10: Evolution*. Academic Press, London.

Almond, J. E. 1985. The Silurian-Devonian fossil record of the Myriapoda. *Philosophical Transactions of the Royal Society of London* (B) 309:227-237.

Alvarez, L. W., W. Alvarez, F. Asaro, and H. V. Michel. 1980. Extra-terrestrial cause for the Cretaceous-Tertiary extinction. *Science* 208: 1095–1108.

Alvarez, W., F. Asaro, H. V. Michel, and L. W. Alvarez. 1982. Iridium anomaly approximately synchronous with terminal Eocene extinctions. *Science* 216: 886–888.

Anderson, L. C. 1992. Naticid gastropod predation on corbulid bivalves: effects of physical factors, morphological features, and statistical artifacts. *Palaios* 7:602-620.

Anstey, R. S. 1990. Bryozoans, pp. 232-252, *in* McNamara, K. J., ed., *Evolutionary Trends*. University of Arizona Press, Tucson.

Archibald, J. D. 1996a. *Dinosaur Extinction and the End of an Era: What the Fossils Say*. Columbia University Press, New York.

Archibald, J. D. 1996b. Fossil evidence for a Late Cretaceous origin of "hoofed" mammals. *Science* 272:1150-1153.

Archibald, J. D. 1998. Archaic ungulates, *in* Janis, C., K. M. Scott, and L. L. Jacobs, eds., *Evolution of Tertiary Mammals of North America*. Cambridge University Press, Cambridge.

Arkell, W. J. A. 1950. A classification of Jurassic ammonites. *Journal of Paleontology* 24:354-364.

Arkell, W. J. A., and others. 1957. Mollusca 4. Cephalopoda. Ammonoidea, *in* R. C. Moore, ed. *Treatise on Invertebrate Paleontology*. Part L. Geological Society of America and University of Kansas Press, Lawrence.

Asaro, F., L. W. Alvarez, W. Alvarez, and H. V. Michel. 1982. Geochemical anomalies near the Eocene/Oligocene and Permian/ Triassic boundaries. *Geological Society of America Special Paper* 190: 517–528.

Askin, R. A., D. H. Elliott, S. R. Jacobson, F. T. Kyte, X. Li, and W. J. Zinsmeister. 1994. Seymour Island: a southern high-latitude record across the KT boundary. New developments regarding the KT event and other catastrophes in earth history. *Lunar and Planetary Science Institute Contribution* 825:7-8.

Attenborough, D. 1979. *Life on Earth*. Little, Brown, Boston.

Aubry, M. P., F. M. Gradstein, and L. F. Jansa. 1990. The late Early Eocene Montagnais meteorite: No impact on biotic diversity. *Micropaleontology* 36: 164–172.

Ausich, W. I., and D. J. Bottjer. 1982. Tiering in suspension-feeding communities on soft substrata throughout the Phanerozoic. *Science* 216:173-174.

Ausich, W. I., and D. J. Bottjer. 1985. Phanerozoic tiering in suspension-feeding communities on soft substrata: implications for diversity, pp. 255-274, *in* Valentine, J. W., ed., *Phanerozoic Diversity Patterns: Profiles in Macroevolution*. Princeton University Press, Princeton, New Jersey.

Babcock, L. 1991. The enigma of conulariid affinities, *in* Simonetta, A. M., and S. Conway Morris, eds., *The Early Evolution of Metazoa and the Significance of Problematic Taxa*. Cambridge University Press, Cambridge, pp. 133-143.

Bakker, R. T. 1972. Anatomical and ecological evidence for endothermy in dinosaurs. *Nature* 238:81-85.

Bakker, R. T. 1975. Dinosaur renaissance. *Scientific American* 232(4): 58-78.

Bakker, R. T. 1986. *The Dinosaur Heresies*. William Morrow, New York.

Bakker, R. T., and P. M. Galton. 1974. Dinosaur monophyly and a new class of vertebrates. *Nature* 248:168-172.

Ball, I. R. 1976. Nature and formulation of biogeographic

hypotheses. *Systematic Zoology* 24:407-430.

Balsam, W. L., and S. Vogel. 1973. Water movement in archaeocyathids: evidence and implications of passive flow in models. *Journal of Paleontology* 47:979-984.

Bambach, R. K. 1973. Tectonic deformation of composite-mold fossil Bivalvia (Mollusca). *American Journal of Science* 273A:409-430.

Bambach, R. K. 1977. Species richness in marine benthic environments through the Phanerozoic. *Paleobiology* 3:152-167.

Barrick, R. E., and W. J. Showers. 1994. Thermophysiology of *Tyrannosaurus rex*: evidence from oxygen isotopes. *Science* 265:222-224.

Barrington, E. T. W. 1965. *The Biology of Hemichordata and Protochordata*. W. H. Freeman, San Francisco.

Barry, J. C. 1995. Faunal turnover and diversity in the terrestrial Neogene of Pakistan, pp. 115-134, *in* Vrba, E. S., G. H. Denton, T. C. Partridge, and L. H. Burckle, eds., *Paleoclimate and Evolution, with Emphasis on Human Origins*. Yale University Press, New Haven.

Bassett, M. G. 1985. Towards a common language in stratigraphy. *Episodes* 8(2):87-92.

Bate, R. H., E. Robinson, and L. M. Sheppard, eds. 1982. *Fossil and Recent Ostracods*. British Micropalaeontological Series.

Batt, R. J. 1989. Ammonite shell morphotype distribution in the Western Interior Greenhorn Sea and some paleoecological implications. *Palaios* 4:32-42.

Batten, R. L., and R. Schweikert. 1981. Discussion of Nur and Ben-Avraham, 1981, pp. 359-366, *in* Nelson, G., and D. E. Rosen, eds., *Vicariance Biogeography: A Critique*. Columbia University Press, New York.

Bayer, F. M., et al. 1956. Coelenterata, *in* Moore, R. C., ed., *Treatise on Invertebrate Paleontology*. Part F. Geological Society of America and University of Kansas Press, Lawrence, Kansas.

Beall, B. S., and C. C. Labandeira. 1990. Macroevolutionary patterns of the Chelicerata and Tracheata, pp. 257-284, *in* Mickulic, D. G., ed., *Arthropod Paleobiology: Short Courses in Paleontology* 3. Paleontological Society and University of Tennessee Press.

Beck, M. W. 1996. On discerning the causes of late Pleistocene megafaunal extinctions. *Paleobiology* 22:91-103.

Behrensmeyer, A. K. 1975. The taphonomy and paleoecology of Plio-Pleistocene vertebrate assemblages east of Lake Rudolf, Kenya. *Bulletin of the Museum of Comparative Zoology, Harvard University* 146:473-578.

Behrensmeyer, A. K., and S. M. Kidwell. 1985. Taphonomy's contribution to paleobiology. *Paleobiology* 11:105-119.

Behrensmeyer, A. K., and S. M. Kidwell, eds. 1993. *Taphonomic Approaches to Time Resolution in Fossil Assemblages. Short Courses in Paleontology* 6. Paleontological Society and University of Tennessee Press.

Béland, P., and D. A. Russell. 1980. Dinosaur metabolism and predator/prey ratios in the fossil record, pp. 85-102, *in* Thomas, R. D. K., and E. C. Olson, eds., *A Cold Look at the Warm-Blooded Dinosaurs*. Westview Press, Boulder, Colo.

Bemis, W. E., Burggren, W. W., and Kemp, N. E., eds. 1987. *The Biology and Evolution of Lungfishes*. Liss, New York.

Benson, R. H. 1975. Morphological stability in the Ostracoda. *Bulletins of American Paleontology* 65:13-46.

Benton, M. J. 1983. Dinosaur success in the Triassic: a non-competitive ecological model. *Quarterly Review of Biology* 58:29-55.

Benton, M. J. 1985. Classification and phylogeny of diapsid reptiles. *Zoological Journal of the Linnean Society (London)* 84:97-164.

Benton, M. J. 1986. More than one event in the late Triassic mass extinctions. *Nature* 321:857-861.

Benton, M. J. 1987. Progress and competition in macroevolution. *Biological Reviews* 62:305-338.

Benton, M. J., ed. 1988. *The Phylogeny and Classification of the Tetrapods. Vol. 1: Amphibians, Reptiles, Birds*. Clarendon Press, Oxford.

Benton, M. J. 1991. What really happened in the Late Triassic? *Historical Biology* 5:263-278.

Benton, M. J. 1993. Late Triassic extinctions and the origin of the dinosaurs. *Science* 260:760-770.

Benton, M. J. 1994. Late Triassic to Middle Jurassic extinctions among continental tetrapods: testing the pattern, *in* Fraser, N. C., and H. -D. Sues, eds., *In the Shadow of the Dinosaurs*. Cambridge University Press, Cambridge, pp. 366-397.

Benton, M. J. 1997. *Vertebrate Palaeontology* (2nd ed.). Chapman & Hall, London.

Benton, M. J., and J. Clark. 1988. Archosaur phylogeny and the relationships of the Crocodylia, pp. 295-338, *in* M. J. Benton, ed., *The Phylogeny and Classification of the Tetrapods, vol. 1: Amphibians, Reptiles, Birds*. Clarendon Press, Oxford.

Berggren, W. A. and D. R. Prothero. 1992. Eocene-Oligocene climatic and biotic evolution: An overview, pp. 1–28, *in* D. R. Prothero and W. A. Berggren, eds., *Eocene-Oligocene Climatic and Biotic Evolution*. Princeton University Press, Princeton, New Jersey.

Berggren, W. A., and J. A. Van Couvering. 1978. Biochronology. *American Association of Petroleum Geologists Memoir* 6:39-55.

Berggren, W. A., and K. G. Miller. 1988. Paleogene tropical planktonic foraminiferal biostratigraphy and magnetobiochronology. *Micropaleontology* 34:362-380.

Berggren, W. A., Kent, D. V., Swisher, C. C., III, and Aubry, M. -P., 1995, A revised Cenozoic geochronology and chronostratigraphy. *SEPM Special Publication* 54:129-212.

Bergström, J. 1973. Organization, life, and systematics of trilobites. *Fossils and Strata* 2:1-69.

Bergström, J. 1975. Functional morphology and evolution of the xiphosurids. *Fossils and Strata* 4:291-305.

Berguist, P. R. 1978. *Sponges*. University of California Press, Berkeley.

Berry, W. B. N. 1977. Graptolite biostratigraphy: a wedding of classical principles and current concepts, pp. 321-338, *in* Kauffman, E., and Hazel, J. E., eds., *Concepts and Methods of Biostratigraphy*. Dowden, Hutchinson, & Ross, Stroudsburg, Pennsylvania.

Berry, W. B. N. 1987a. *Growth of a Prehistoric Time Scale Based on Organic Evolution* (2nd ed.). Blackwell Scientific Publishing, Palo Alto, California.

Berry, W. B. N. 1987b. Phylum Hemichordata (including Graptolithina), pp. 612-635, *in* Boardman, R. S., A. H. Cheetham, and A. J. Rowell, eds., *Fossil Invertebrates*. Blackwell Scientific Publishers, Cambridge, Mass.

Bice, D. M., C. R. Newton, S. McCauley, P. W. Reiners, and C. A. McRoberts. 1992. Shocked quartz at the Triassic/Jurassic boundary in Italy. *Science* 255: 443-446.

Bieler, R. 1992. Gastropod phylogeny and systematics. *Annual Reviews of Ecology and Systematics* 23:311-338.

Boardman, R. S., and A. H. Cheetham. 1987. Phylum Bryozoa, pp. 497-549, *in* Boardman, R. S., A. H. Cheetham, and A. J. Rowell, eds., *Fossil Invertebrates*. Blackwell Scientific

Publishers, Cambridge, Mass.
Boardman, R. S., et al., 1983. *Bryozoa*, Vol. 1, Part G, in Robinson, R. A., ed., *Treatise on Invertebrate Paleontology*. Geological Society of America and University of Kansas Press, Lawrence.
Boaz, N. T., and A. K. Behrensmeyer. 1976. Hominid taphonomy: transport of human skeletal parts in an artificial fluviatile environment. *American Journal of Physical Anthropology* 45:53-60.
Bohlin, B. 1940. Food habits of the machaerodonts, with special regard to *Smilodon*. *Bulletin of the Geological Institute of Uppsala* 28:156-174.
Bolt, J. R. 1977. Dissorophid relationships and ontogeny, and the origin of Lissamphibia. *Journal of Paleontology* 51:235-249.
Bone, Q., and Marshall, N. B. 1982. *Biology of Fishes*. Blackie & Sons, Glasgow.
Bottjer, D. J. 1985. Bivalve paleoecology, pp. 122-137, *in* Bottjer, D. W., C. S. Hickman, and P. D. Ward, eds., *Mollusks, Notes for a Short Course*. University of Tennessee Department of Geological Sciences Studies in Geology 13.
Bottjer, D. J., and W. I. Ausich. 1986. Phanerozoic development of tiering in soft substrata suspension-feeding communities. *Paleobiology* 12:400-420.
Bottjer, D. J., and M. L. Droser. 1991. Ichnofabric and basin analysis. *Palaios* 6:199-205.
Bottjer, D. J., and M. L. Droser. 1994. The history of Phanerozoic bioturbation, pp. 155-176, *in* Donovan, S. K., ed. 1994. *The Palaeobiology of Trace Fossils*. Wiley, Chichester.
Bottjer, D. J., M. L. Droser, and D. Jablonski. 1988. Palaeoenvironmental trends in the history of trace fossils. *Nature* 333:252-255.
Bottjer, D. J., C. S. Hickman, and P. D. Ward, eds. 1985. *Mollusks, Notes for a Short Course*. University of Tennessee Department of Geological Sciences Studies in Geology 13.
Bottjer, D. J., and D. Jablonski. 1988. Paleoenvironmental patterns in the evolution of the post-Paleozoic benthic marine invertebrates. *Palaios* 3:540-560.
Bottomley, R., and D. York. 1988. Age measurements of the submarine Montagnais impact crater and the periodicity question. *Geophysical Research Letters* 14(12): 1409–1412.
Boucot, A. J. 1975. *Evolution and Extinction Rate Controls*. Elsevier, Amsterdam.
Boucot, A. J. 1990. *Evolutionary Paleobiology of Behavior and Coevolution*. Elsevier, Amsterdam.
Boucot, A. J., W. Brace, and R. DeMar. 1958. Distribution of brachiopod and pelecypod shells by currents. *Journal of Sedimentary Petrology* 28:321-332.
Bowes, D. R., and C. D. Waterston, eds. 1985. Fossil arthropods as living animals. *Transactions of the Royal Society of Edinburgh* 76:101-399.
Bralower, T. J., R. M. Leckie, W. V. Sliter, and H. R. Thierstein. 1995. An intergrated Cretaceous microfossil biostratigraphy. *SEPM Special Publication* 54:65-80.
Bramwell, C. D. 1971. Aerodynamics of *Pteranodon*. *Biological Journal of the Linnean Society* 3:313-328.
Bramwell, C. D., and G. R. Whitfield. 1974. Biomechanics of *Pteranodon*. *Philosophical Transactions of the Royal Society of London* (B) 267:503-581.
Branch, G. M. 1984. Competition between marine organisms: ecological and evolutionary implications. *Annual Review of Oceanography and Marine Biology* 22:429-493.
Brasier, M. D. 1980. *Microfossils*. Allen & Unwin, London.
Bretsky, P. W. 1973. Evolutionary patterns in the Paleozoic Bivalvia: documentation and some theoretical considerations. *Geological Society of America Bulletin* 84:2079-2096.
Bretsky, P. W., and S. S. Bretsky. 1975. Succession and repetition of Late Ordovician fossil assemblages from the Nicolet River Valley, Quebec. *Paleobiology* 1:225-237.
Briggs, D. E. G. 1992. Conodonts: a major extinct group added to the vertebrates. *Science* 256: 1284-1286.
Briggs, D. E. G., D. L. Bruton, and H. B. Whittington. 1979. Appendages of the fossil arthropod *Aglaspis spinifer* (Upper Cambrian, Wisconsin) and their significance. *Palaeontology* 22:167-180.
Briggs, D. E. G., E. N. K. Clarkson, and R. J. Aldridge. 1983. The conodont animal. *Lethaia* 16:1-14.
Briggs, D. E. G., and R. A. Fortey. 1989. The early radiation and relationships of the major arthropod groups. *Science* 256:241-243.
Briggs, D. E. G., R. A. Fortey, and M. A. Wills. 1992. Morphological disparity in the Cambrian. *Science* 256:1670-1673.
Briggs, D. E. G., and P. D. Lane, eds. 1983. Trilobites and other arthropods. *Special Papers in Palaeontology 30*. Academic Press, London.
Brinkmann, R. 1929. Statistischbiostratigraphische Untersuchungen an mitteljurassischen Ammoniten über Artbegriff und Stammensentwicklung. *Abhandlungen Ges. Wissenschaften Göttingen, Math. -Phys. Kl., N. F.,* 13:1-249.
Broadhead, T. W., and Waters, J. A., eds., 1980. *Echinoderms: Notes for a Short Course*. University of Tennessee Department of Geological Sciences Studies in Geology 4.
Brodal, A., and R. Fänge, eds. 1963. *The Biology of Myxine*. Universitetsforlaget, Oslo.
Bromley, R. G. 1990. *Trace Fossils, Biology and Taphonomy*. Special Topics in Palaeontology, Unwin & Hyman, London.
Brower, J. C. 1980. Pterosaurs: how they flew. *Episodes* 1980(4):21-24.
Brower, J. C. 1983. The aerodynamics of *Pteranodon* and *Nyctosaurus*, two large pterosaurs from the Upper Cretaceous of Kansas. *Journal of Vertebrate Paleontology* 3:84-124.
Brundin, L. 1966. Transantarctic relationships and their significance. *Kungliga Svenska Vetenskapsakademiens Hanglingar, series* 4, 11:1-472.
Brusca, R. C., and G. J. Brusca. 1990. *Invertebrates*. Sinauer Associates, Sunderland, Mass.
Buchsbaum, R., and M. Buchsbaum. 1987. *Animals without Backbones* (3rd ed.). University of Chicago Press, Chicago.
Bukry, D. 1973. Low-latitude coccolith biostratigraphic zonation. *Initial Reports of the Deep-Sea Drilling Project* 15:685-703.
Bulman, O. M. B. 1970. Graptolithina, *in* Teichert, C., ed., *Treatise on Invertebrate Paleontology, Part V (revised)*. The Geological Society of America and the University of Kansas Press, Lawrence, Kansas.
Burns, J. 1975. *Biograffiti*. Harvard University Press, Cambridge, Mass.
Cain, A. J., and P. M. Sheppard. 1954. Natural selection in *Cepaea*. *Genetics* 39:89-116.
Caldwell, M. W., and M. S. Y. Lee. 1997. A snake with legs from the marine Cretaceous of the Middle East. *Nature* 386:705-709.
Callahan, P. S. 1972. *The Evolution of Insects*. Holiday House, New York.
Callaway, J., and E. L. Nicholls, eds. 1997. *Ancient Marine Reptiles*. Academic Press, San Diego.
Callomon, J. H. 1963. Sexual dimorphism is Jurassic ammonites. *Leicester Literary and Philosophical Society Transactions*

62:21-56.

Campbell, K. A., and D. J. Bottjer. 1993. Fossil cold-seeps (Jurassic-Pliocene) along the convergent margin of western North America. *National Geographic Research and Exploration* 9:326-343.

Campbell, K. E., Jr. 1995. A new date for the beginning of the Great American Faunal Interchange. *Journal of Vertebrate Paleontology* 15(3):21A.

Campbell, K. E., Jr., and C. D. Frailey. 1996. The Great American Faunal Interchange: rewriting the script. *Paleontological Society Special Publication* 8:63.

Carlson, S. J. 1990. The articulate brachiopod hinge mechanism: morphological and functional variation. *Paleobiology* 15:364-386.

Carlson, S. J. 1991. A phylogenetic perspective on articulate brachiopod diversity and the Permo-Triassic extinctions, pp. 119-142, *in* Dudley, E. C., ed., *The Unity of Evolutionary Biology* 1. Dioscorides Press, Portland, Oregon.

Carlson, S. J. 1992. Evolutionary trends in the articulate brachiopod hinge mechanism. *Paleobiology* 18:344-366.

Carlson, S. J. 1993. Phylogeny and evolution of 'pentameride' brachiopods. *Palaeontology* 36:807-837.

Carlson, S. J. 1995. Phylogenetic relationships among extant brachiopods. *Cladistics* 11:131-197.

Carlson, S. J., and D. C. Fisher. 1981. Microstructural and morphologic analysis of a carpoid aulacophore. *Abstracts with Programs of the Geological Society of America* 13(7): 422.

Carpenter, F. M., and L. Burnham. 1985. The geological record of insects. *Annual Reviews of Earth and Planetary Sciences* 13:297-314.

Carpenter, F. M. 1992. Hexapoda. *Treatise on Invertebrate Paleontology. Part R. Arthropoda 3-4*. . Geological Society of America and University of Kansas Press, Lawrence.

Carroll, R. L. 1964. The earliest reptiles. *Zoological Journal of the Linnean Society (London)* 45:61-83.

Carroll, R. L. 1977. Patterns of amphibian evolution: an extended example of the incompleteness of the fossil record, pp. 405-437, *in* Hallam, A., ed., *Patterns of Evolution*. Elsevier, Amsterdam.

Carroll, R. L. 1982. Early evolution of the reptiles. *Annual Reviews of Ecology and Systematics* 13:87-109.

Carroll, R. L. 1988. *Vertebrate Paleontology and Evolution*. W. H. Freeman, New York.

Carroll, R. L. 1992. The primary radiation of terrestrial vertebrates. *Annual Reviews of Earth and Planetary Sciences* 20:45-84.

Carroll, R. L., and P. Gaskill. 1978. The order Microsauria. *Memoirs of the American Philosophical Society* 126:1-211.

Carruthers, R. G. 1910. On the evolution of *Zaphrentis delanouei* in the Lower Carboniferous times. *Quarterly Journal of the Geological Society of London* 66:523-536.

Carter, R. M. 1967. On the biology and palaeontology of some predators of bivalved molluscs. *Palaeogeography, Palaeoclimatology, Palaeoecology* 4:29-65.

Cavanaugh, C. M. 1985. Symbiosis of chemoautotrophic bacteria and marine invertebrates from hydrothermal vents and reducing sediments. *Bulletin of the Biological Society of Washington* 6:373-388.

Cerling, T. E. 1984. The stable isotopic composition of modern soil carbonate and its relation to climate. *Earth and Planetary Science Letters* 71:229-240.

Cerling, T. E. 1992. Development of grasslands and savannas in East Africa during the Neogene. *Palaeogeography, Palaeoclimatology, Palaeoecology* 97:241-247.

Cerling, T. E., J. Quade, Y. Wang, and J. R. Bowman. 1989. Carbon isotopes in soils and paleosols as ecology and paleoecology indicators. *Nature* 341:138-139.

Cerling, T. E., Y. Wang, and J. Quade. 1993. Global ecological change in the late Miocene: expansion of C4 ecosystems. *Nature* 361:344-345.

Chamberlain, J. A., Jr. 1981. Hydromechanical design of fossils cephalopods, pp. 289-336, *in* House, M. R., and J. R. Senior, eds. *The Ammonoidea*. Systematics Association Special Volume 18. Academic Press, New York.

Chamberlain, J. A., Jr., and G. E. G. Westermann. 1976. Hydrodynamic properties of cephalopod shell ornament. *Paleobiology* 2:168-172.

Chang, M. -M. 1991. "Rhipidistians," dipnoans, and tetrapods, pp. 3-28, *in* Schultze, H. -P., and L. Trueb, eds., *Origins of the Higher Groups of Tetrapods: Controversy and Consensus*. Cornell University Press, Ithaca, New York.

Charig, A. J. 1976. "Dinosaur monophyly and a new class of vertebrates": a critical review, *in* Bellairs, A. A., and C. B. Cox, eds., *Morphology and Biology of Reptiles*. Linnean Society of London Symposium 3:65-104.

Chave, K. E. 1964. Skeletal durability and preservation, pp. 377-387, *in* Imbrie, J., and N. D. Newell, eds., *Approaches to Paleoecology*. Wiley, New York.

Cheetham, A. H. 1986. Tempo and mode of evolution in a Neogene bryozoan: rates of morphologic change with and across species boundaries. *Paleobiology* 12:190-202.

Cheetham, A. H. 1987. Tempo and mode of evolution in a Neogene bryozoan: are trends in a single morphologic character misleading? *Paleobiology* 13:286-296.

Cheetham, A. H., and J. E. Hazel. 1969. Binary (presence-absence) similarity coefficients. *Journal of Paleontology* 43:1130-1136.

Chestnut, D. R., Jr., and F. R. Ettensohn. 1988. Hombergian (Chesterian) echinoderm paleontology and paleoecology, south-central Kentucky. *Bulletins of American Paleontology* 95:1-102.

Churkin, J., Jr, and C. Carter. 1972. Graptolite identification chart for field determination of geologic age. *U. S. Geological Survey Oil and Gas Investigations Chart* OC 66.

Cifelli, R. L. 1981. Patterns of evolution among Artiodactyla and Perissodactyla (Mammalia). *Evolution* 35:433-440.

Cifelli, R. L. 1982. The petrosal structure of *Hyopsodus* with respect to that of some other ungulates, and its phylogenetic implications. *Journal of Paleontology* 56:795-805.

Cifelli, R. L. 1993. The phylogeny of the native South American ungulates, pp. 195-216, *in* Szalay, F. S., M. J. Novacek, and M. C. McKenna, eds., *Mammal Phylogeny, vol. 2: Placentals*. Springer-Verlag, New York.

Clack, J. A. 1997. Devonian tetrapod trackways and trackmakers: a review of the fossils and footprints. *Palaeogeography, Palaeoecology, Palaeoclimatology* 130:227-250.

Clark, D. L. 1984. Conodont biofacies and provincialism. *Geological Society of America Special Paper* 196.

Clark, D. L. 1987. Phylum Conodonta, pp. 636-662, *in* Boardman, R. S., A. H. Cheetham, and A. J. Rowell, eds., *Fossil Invertebrates*. Blackwell Scientific Publishers, Cambridge, Mass.

Clark, D. L., and others, 1981. Conodonta, *in* Robison, R. A., ed., *Treatise on Invertebrate Paleontology*, Part W. Geological Society of America and the University of Kansas, Lawrence, Kansas.

Clarke, A. 1993. Temperature and extinction in the sea: a physiologist's view. *Paleobiology* 19:499-519.

Clarke, K. U. 1973. *The Biology of Arthropods*. Edward Arnold, London.

Clarkson, E. N. K. 1979. The visual system of trilobites. *Palaeontology* 22:1-22.

Clyde, W. C., and D. C. Fisher. 1997. Comparing the fit of stratigraphic and morphologic data in phylogenetic analysis. *Paleobiology* 23:1-19.

Coates, A. G., and J. B. C. Jackson. 1987. Clonal growth, algal symbiosis, and reef formation by corals. *Paleobiology* 13:363-378.

Coates, M. I., and J. A. Clack. 1990. Polydactyly in the earliest known tetrapod limbs. *Nature* 347:66-69.

Colbert, E. H. 1964. Climatic zonation andterrestrial faunas, pp. 617-639, *in* Nairn, A. E. M., ed., *Problems in Paleoclimatology*. Wiley-Interscience, New York.

Colbert, E. H., and M. Morales, 1991. *Evolution of the Vertebrates* (4th ed.). Wiley, New York.

Compagno, L. J. V. 1977. Phyletic relationships of living sharks and rays. *American Zoologist* 17:303-322.

Conniff, R. 1991. The most disgusting fish in the sea. *Audubon* 93:100-118.

Conniff, R. 1996. *Spineless Wonders*. Henry Holt, New York.

Cope, J. C. W. 1996. The early evolution of the Bivalvia, pp. 361-370, *in* Taylor, J., ed., *Origin and Evolutionary Radiation of the Mollusca*. Oxford University Press, New York.

Copper, P. 1988. Ecological succession in Phanerozoic reef ecosystems: it is real? *Palaios* 3:136-151.

Copper, P., ed. 1997. *Proceedings of the Third International Brachiopod Congress* (in press).

Corliss, J. B., and others. 1979. Submarine thermal springs on the Galapagos rift. *Science* 203:1073-1083.

Court, N. 1990. Periotic anatomy of *Arsinoitherium* (Mammalia, Embrithopoda) and its phylogenetic implications. *Journal of Vertebrate Paleontology* 10:170-182.

Cowen, R. 1975. "Flapping valves" in brachiopods. *Lethaia* 8:23-29.

Cowen, R. 1990. *History of Life* (2nd ed.). Blackwell Scientific Publishers, New York.

Cowie, J. A., W. Ziegler, and J. Remane. 1989. Stratigraphic commission accelerates progress, 1984 to 1989. *Episodes* 12(2):79-83.

Cox, L. R., and others. 1969-1971. *Treatise on Invertebrate Paleontology. Part N. Mollusca 6: Bivalvia*. Geological Society of America and University of Kansas Press, Lawrence.

Cracraft, J. 1973. Continental drift, paleoclimatology, and the evolution and biogeography of birds. *Journal of Zoology (London)* 169:455-545.

Cracraft, J. 1974. Phylogeny and evolution of the ratite birds. *Ibis* 116:494-521.

Cracraft, J. 1985. The origin and early diversification of birds. *Paleobiology* 12:383-389.

Cracraft, J. 1987. DNA hybridization and avian phylogenetics. *Evolutionary Biology* 21:47-96.

Cracraft, J. 1988. The major clades of birds, pp. 339-361. *in* Benton, M. J., ed., *The Phylogeny and Classification of the Tetrapods, Vol. 1: Amphibians, Reptiles, Birds*. Clarendon Press, Oxford.

Crimes, T. P. 1975. The stratigraphical significance of trace fossils, pp. 109-130, *in* Frey, R. W., ed., *The Study of Trace Fossils*. Springer-Verlag, New York.

Crimes, T. P., and J. C. Harper, eds. 1977. Trace fossils 2. *Geological Journal Special Issue* 9.

Croizat, L. 1952. *Manual of Phytogeography*. W. Junk, The Hague.

Croizat, L. 1958. *Panbiogeography*. Published by the author, Caracas.

Croizat, L. 1964. *Space, Time, Form: The Biological Synthesis*. Published by the author, Caracas.

Croizat, L. 1981. Biogeography: past, present, and future, pp. 501-514, *in* Nelson, G., and D. E. Rosen, eds., *Vicariance Biogeography: A Critique*. Columbia University Press, New York.

Croizat, L. 1982. Vicariance/vicariism, panbiogeography, "vicariance biogeography," etc. : a clarification. *Systematic Zoology* 31:291-303.

Cubitt, J. M., and R. A. Reyment, eds. 1982. *Quantitative Stratigraphic Correlation*. Wiley, New York.

Cuppy, W. 1931. *How to Tell your Friends from the Apes*. Liveright, New York.

Cuppy, W. 1941. *How to Become Extinct*. University of Chicago Press, Chicago.

Cuppy, W. 1949. *How to Attract the Wombat*. University of Chicago Press, Chicago.

Curran, H. A., ed. 1985. Biogenic structures: their use in interpreting depositional environments. *SEPM Special Publication* 35.

Darby, D. G. 1982. The early vertebrate *Astraspis*, habitat based on lithologic association. *Journal of Paleontology* 56:1187-1196.

Darlington, P. D. 1957. *Zoogeography*. Wiley, New York.

Darwin, C. R. 1851-1854. *A Monograph on the Subclass Cirripedia*, vols. 1-2. Ray Society, London.

Davis, M., P. Hut, and R. A. Muller. 1984. Extinction of species by periodic comet showers. *Nature* 308: 715–717.

Dawkins, R. 1976. *The Selfish Gene*. Oxford University Press, Oxford.

De Muizon, C., and L. D. Marshall. 1992. *Alcideorbignya inopinata* (Mammalia, Pantodonta) from the early Paleocene of Bolivia: phylogenetic and paleobiogeographic implications. *Journal of Paleontology* 66:499-520.

De Queiroz, K., and J. Gauthier. 1992. Phylogenetic taxonomy. *Annual Reviews of Ecology and Systematics* 23:449-480.

Debrenne, F., and J. Vacelet. 1984. Archaeocyatha: is the sponge model consistent with their structural organization? *Paleontographica Americana* 54:358-369.

Denton, E. J., and J. B. Gilpin-Brown. 1966. On the buoyancy of the pearly nautilus. *Journal of the Marine Biological Association of the United Kingdom* 46:723-759.

Dobzhansky, T. 1937. *Genetics and the Origin of Species*. Columbia University Press, New York.

Dodd, J. R., and R. J. Stanton, Jr. 1975. Paleosalinities with a Pliocene bay, Kettleman Hills, California: a study of the resolving power of isotopic and faunal techniques. *Geological Society of America Bulletin* 86:51-64.

Dommergues, J. -L. 1990. Ammonoids, pp. 162-187, *in* McNamara, K. J., ed., *Evolutionary Trends*. University of Arizona Press, Tucson.

Domning, D. P., C. E. Ray, and M. C. McKenna. 1986. Two new Oligocene desmostylians and a discussion of tethytherian systematics. *Smithsonian Contributions to Paleobiology* 59:1-56.

Donovan, D. T., J. H. Callomon, and M. K. Howarth. 1981. Classification of the Jurassic Ammonitina, pp. 101-155, *in* House, M. R., and J. R. Senior, eds., *The Ammonoidea*. Systematics Association Special Volume 18. Academic Press,

New York.

Donovan, S. K., ed. 1994. *The Palaeobiology of Trace Fossils*. Wiley, Chichester.

Donovan, S. K., and A. S. Gale. 1990. Predatory asteroids and the decline of the articulate brachiopods. *Lethaia* 23:77-86.

Douglass, R. C. 1977. The development of fusulinid biostratigraphy, pp. 463-482, *in* Kauffman, E., and Hazel, J. E., eds., *Concepts and Methods of Biostratigraphy*. Dowden, Hutchinson, & Ross, Stroudsburg, Penn.

Driscoll, E. G. 1967. Experimental field study of shell abrasion. *Journal of Sedimentary Petrology* 37:1117-1123.

Driscoll, E. G. 1970. Selective bivalve shell destruction in marine environments, a field study. *Journal of Sedimentary Petrology* 40:898-905.

Driscoll, E. G., and T. P. Weltin. 1973. Sedimentary parameters as factors in abrasive shell reduction. *Palaeogeography, Palaeoclimatology, Palaeoecology* 13:275-288.

Droser, M. L., and D. J. Bottjer. 1986. A semiquantitative field classification of ichnofabric. *Journal of Sedimentary Petrology* 56:558-559.

Droser, M. L., and D. J. Bottjer. 1993. Trends and patterns of Phanerozoic ichnofabrics. *Annual Reviews of Earth and Planetary Sciences* 21:205-225.

Dullo, W. C. 1983. Fossildiagenese im miozenen Leitha-Kalk der Paratethys von Österreich: Ein Beispiel für Faunenverschiebungen durch Diageneseunterschiede. *Facies* 8: 1-112.

Durham, J. W. 1967a. Notes on the Helicoplacoidea and the early echinoderms. *Journal of Paleontology* 41:97-102.

Durham, J. W. 1967b. The incompleteness of our knowledge of the fossil record. *Journal of Paleontology* 41:559-565.

Durham, J. W. 1993. Observations on the Early Cambrian helicoplacoid echinoderms. *Journal of Paleontology* 67:590-603.

Durham, J. W., and K. E. Caster. 1963. Helicoplacoidea: a new class of echinoderms. *Science* 140:820-822.

Durham, J. W., and K. E. Caster. 1966. Helicoplacoids, p. U131-136, in Moore, R. C., ed., *Treatise on Invertebrate Paleontology, Pt. U. Echinodermata* 3(1). Geological Society of America and University of Kansas Press, Lawrence.

DuToit, A. 1937. *Our Wandering Continents*. Oliver & Boyd, Edinburgh.

Dutro, J. T., and R. S. Boardman. 1981. *Lophophorates: Notes for a Short Course*. University of Tennessee Department of Geological Sciences Studies in Geology 5.

Dzik, J. 1984. Phylogeny of the Nautiloidea. *Palaeontologica Polonica* 45:1-219.

Eaton, J. G., J. I. Kirkland, and K. Doi. 1989. Evidence of reworked Cretaceous fossils and their bearing on the existence of Tertiary dinosaurs. *Palaios* 4:281-286.

Edwards, L. E. 1982. Quantitative biostratigraphy: the methods should suit the data, pp. 45-60, *in* Cubitt, J. M., and R. A. Reyment, eds., *Quantitative Stratigraphic Correlation*. Wiley, New York.

Eisma, D. 1965. Shell-characteristics of *Cardium edule* L. as indicators of salinity. *Netherlands Journal of Sea Research* 2:493-540.

Ekdale, A. A. 1988. Pitfalls of paleobathymetric intepretations based on trace fossil assemblages. *Palaios* 3:464-472.

Ekdale, A. A., R. G. Bromley, and S. G. Pemberton, eds. 1984. *Ichnology: The Use of Trace Fossils in Sedimentology and Stratigraphy*. SEPM Short Course Notes 15.

Eldredge, N. 1970. Observations on the burrowing behavior in *Limulus polyphemus* (Chelicerata, Merostomata) with implications for the functional anatomy of trilobites. *American Museum of Natural History Novitates* 2436:1-17.

Eldredge, N. 1972. Systematics and evolution of *Phacops rana* (Green 1832) and *Phacops iowensis* (Delo 1935) from the Middle Devonian of North America. *Bulletin of the American Museum of Natural History* 147:49-113.

Eldredge, N. 1974. Revision of the suborder Synziphosurina (Chelicerata, Merostomata) with remarks on merostome phylogeny. *American Museum of Natural History Novitates* 2543.

Eldredge, N. 1977. Trilobites and evolutionary patterns, pp. 305-332, *in* Hallam, A., ed. *Patterns of Evolution as Illustrated in the Fossil Record*. Elsevier, Amsterdam.

Eldredge, N. 1989. *Macroevolutionary Dynamics: Species, Niches, and Adaptive Peaks*. McGraw-Hill, New York.

Ellis, R. 1975. *The Book of Sharks*. Grosset & Dunlap, New York.

Elton, C. 1947. *Animal Ecology*. Macmillan, New York.

Emerson, S. B., and L. Radinsky. 1980. Functional analysis of sabertooth cranial morphology. *Paleobiology* 6:295-313.

Epstein, A. G., Epstein, J. B., and L. D. Harris. 1977. Conodont color alteration—an index to organic metamorphism. *U. S. Geological Survey Professional Paper* 995.

Ernst, G. 1970. Zur Stammgeschicte und stratigraphischen Bedeutung der Echiniden-Gattung *Micraster* in der nordwest deutschen Oberkreide. *Mitteilungen der Geologischer-Paläontologische Institut der Universität Hamburg* 39:117-135.

Erwin, D. H., and M. L. Droser. 1993. Elvis taxa. *Palaios* 8:623-624.

Estes, R. and J. H. Hutchinson. 1980. Eocene lower vertebrates from Ellesmere Island, Canadian Arctic Archipelago. *Palaeogeography, Palaeoclimatology, Palaeoecology* 30: 325–347.

Estes, R., and G. K. Pregill, eds. 1988. *The Phylogenetic Relationships of the Lizard Families: Essays Commemorating Charles L. Camp*. Stanford University Press, Palo Alto.

Evans, S. E. 1988. The early history and relationships of the Diapsida, pp. 221-260, *in* M. J. Benton, ed., *The Phylogeny and Classification of the Tetrapods, vol. 1: Amphibians, Reptiles, Birds*. Clarendon Press, Oxford.

Evernden, J. F., D. E. Savage, G. H. Curtis, and G. T. James. 1964. Potassium-argon dates and the Cenozoic mammalian chronology of North America. *American Journal of Science* 262:145-198.

Eyles, J. M. 1985. William Smith, Sir Joseph Banks, and the French geologists, pp. 37-50, *in* Wheeler, A., and J. H. Price, eds., *From Linnaeus to Darwin*. Society for the History of Natural History, London.

Fallaw, W. C. 1979. A test of the Simpson Coefficient and other binary coefficients of faunal similarity. *Journal of Paleontology* 53:1029-1034.

Farlow, J. O. 1980. Predator/prey biomass ratios, community food webs, and dinosaur physiology, pp. 55-84, *in* Thomas, R. D. K., and E. C. Olson, eds., *A Cold Look at the Warm-Blooded Dinosaurs*. Westview Press, Boulder, Colo.

Fastovsky, D. E., and D. B. Weishampel. 1996. *The Evolution and Extinction of the Dinosaurs*. Cambridge University Press, Cambridge.

Faul, H., and C. Faul. 1983. *It Began with a Stone*. Wiley, New York.

Feduccia, A. 1994. Tertiary bird history: notes and comments, pp. 178-188, *in* Prothero, D. R., and R. M. Schoch, eds., *Major Features of Vertebrate Evolution. Short Courses in Paleontology* 7. Paleontological Society and University of Tennessee Press.

Feduccia, A. 1995. Explosive radiation of Tertiary birds and mammals. *Science* 267:637-638.

Feduccia, A. 1996. *The Origin and Evolution of Birds*. Yale University Press, New Haven.

Feldmann, R. M. 1990. Decapod crustacean paleobiogeography: resolving the problem of small sample size, pp. 303-315, *in* Mickulic, D. G., ed., *Arthropod Paleobiology: Short Courses in Paleontology* 3. Paleontological Society and University of Tennessee Press.

Ferris, V. R. 1980. A science in search of a paradigm? Review of the symposium, "Vicariance biogeography: a critique. " *Systematic Zoology* 29:67-76.

Field, K. G., and others. 1988. Molecular phylogeny of the animal kingdom. *Science* 239:748-753.

Fischer, M., and P. Tassy. 1993. The interrelation between Proboscidea, Sirenia, Hyracoidea, and Mesaxonia: the morphological evidence, pp. 217-234, *in* Szalay, F. S., M. J. Novacek, and M. C. McKenna, eds., *Mammal Phylogeny. Vol. II: Placentals*. Springer-Verlag, New York.

Fisher, C. R. 1990. Chemoautotrophic and methanotrophic symbioses in marine invertebrates. *Reviews in Aquatic Sciences* 2:399-436.

Fisher, D. C. 1975. Swimming and burrowing in *Limulus* and *Mesolimulus*. *Fossils and Strata* 4:281-290.

Fisher, D. C. 1977. Functional significance of spines in the Pennsylvanian horseshoe crab *Euproops danae*. *Paleobiology* 3:175-195.

Fisher, D. C. 1982a. Phylogenetic and macroevolutionary patterns within the Xiphosurida. *Proceedings of the Third North American Paleontological Convention* 1:175-180.

Fisher, D. C. 1982b. Stylophoran skeletal crytallography: testing the calcichordate theory of vertebrate origins. *Abstracts with Programs of the Geological Society of America* 14(7): 488.

Fisher, D. C. 1984. The Xiphosurida: archetypes of bradytely? pp. 196-213, *in* Eldredge, N., and S. M. Stanley, eds., *Living Fossils*. Springer Verlag, New York.

Fisher, D. C. 1993. Life orientation of mitrate stylophorans and its implication for the calcichordate theory of vertebrate origins. *Abstracts with Programs of the Geological Society of America* 25(6): A105.

Fisher, D. C. 1994. Stratocladistics: morphological and temporal patterns and their relation to phylogenetic process, pp. 133-171, *in* Grande, L., and O. Rieppel, eds., *Interpreting the Hierarchy of Nature*. Academic Press, San Diego.

Fisher, W. L., P. U. Rodda, and J. W. Dietrich. 1964. Evolution of the *Athleta petrosa* stock (Eocene, Gastropoda) of Texas. *University of Texas Publications* 6413:1-117.

Flessa, K. W., S. G. Barnett, D. B. Cornue, M. A. Lomaga, N. Lombardi, J. M. Miyazaki, and A. S. Murer. 1979. Geologic implications of the relationship between mammalian faunal similarity and geographic distance. *Geology* 7:15-18.

Flessa, K. W., and R. H. Thomas. 1985. Modelling the biogeographic regulation of evolutionary rate, pp. 355-376, *in* Valentine, J. W., ed., *Phanerozoic Diversity Patterns: Profiles in Macroevolution*. Princeton University Press, Princeton, New Jersey.

Flynn, J. J. 1986. Faunal provinces and the Simpson coefficient. *University of Wyoming Contributions to Geology, Special Paper* 3:317-338.

Flynn, J. J., B. J. MacFadden, and M. C. McKenna. 1984. Land mammal ages, faunal heterochrony, and temporal resolution in Cenozoic terrestrial sequences. *Journal of Geology* 92:687-705.

Flynn, J. J., N. A. Neff, and R. H. Tedford, 1988. Phylogeny of the Carnivora, pp. 73-116, *in* Benton, M. J., ed., *The Phylogeny and Classification of the Tetrapods, vol. 2: Mammals*. Clarendon Press, Oxford.

Foote, M. 1992. Paleozoic record of morphological diversity in blastozoan echinoderms. *Proceedings of the National Academy of Sciences* 89:7325-7329.

Foote, M. 1993a. Discordance and concordance between morphological and taxonomic diversity. *Paleobiology* 19:185-204.

Foote, M. 1993b. Contributions of individual taxa to overall morphological disparity. *Paleobiology* 19:185-204.

Foote, M. 1995. Morphological diversification of Paleozoic crinoids. *Paleobiology* 21: 273-299.

Foote, M., and S. J. Gould. 1992. Cambrian and Recent morphological disparity. *Science* 258:1816.

Forey, P. L. 1984. Yet more reflections on agnathan-gnathostome relationships. *Journal of Vertebrate Paleontology* 4:330-343.

Forey, P. L. 1987. Relationships of lungfishes, pp. 75-91, *in* Bemis, W. E., W. W. Burggren, and N. E. Kemp, eds., *The Biology and Evolution of Lungfishes*. Liss, New York.

Forey, P. L., and P. Janvier, 1993. Agnathans and the origin of jawed vertebrates. *Nature* 361:129-134.

Forey, P. L., and P. Janvier. 1984. Evolution of the earliest vertebrates. *American Scientist* 82:554-565.

Fortey, R. A. 1990. Trilobite evolution and systematics, pp. 44-65, *in* Mickulic, D. G., ed., *Arthropod Paleobiology: Short Courses in Paleontology* 3. Paleontological Society and University of Tennessee Press.

Fortey, R. A., and R. P. S. Jefferies. 1982. Fossils and phylogeny—a compromise approach. *Systematics Association Special Volume* 21:197-234.

Fortey, R. A., and R. A. Owens. 1990. Trilobites, pp. 121-142, *in* McNamara, K. J., ed., *Evolutionary Trends*. University of Arizona Press, Tucson.

Fortey, R. A., and H. B. Whittington. 1989. The Trilobita as a natural group. *Historical Biology* 2:125-138.

Fracasso, M. A. 1994. Amphibia: disparity and diversification of early tetrapods, pp. 108-128, *in* Prothero, D. R., and R. M. Schoch, eds., *Major Features of Vertebrate Evolution. Short Courses in Paleontology* 7. Paleontological Society and University of Tennessee Press.

Frebold, H. 1922. Phylogenie und Biostratigraphie der Amaltheen im mittleren Lias von Nordwestdeutschland. *Jahrber. Niedersächs. Geol. Ver.* 15:1-26.

Frey, R. W., ed. 1975. *The Study of Trace Fossils*. Springer-Verlag, New York.

Frey, R. W., and A. Seilacher. 1980. Uniformity in marine invertebrate ichnology. *Lethaia* 13:183-208.

Frey, R. W., and S. G. Pemberton. 1985. Biogenic structures in outcrops and cores. I. Approaches to ichnology. *Bulletin of Canadian Petroleum Geology* 333:72-115.

Fricke, H., 1988. Coelacanths, the fish that time forgot. *National Geographic* 173:824-838.

Fricke, H., O. Reinecke, H. Hofer, and W. Nachtigall. 1987. Locomotion of the coelacanth *Latimeria chalumnae* in its natural environment. *Nature* 329:331-333.

Gaffney, E. S. 1975. A phylogeny and classification of the higher categories of turtles. *Bulletin of the American Museum of Natural History* 155:387-436.

Gardiner, B. G. 1984. The relationships of placoderms. *Journal of Vertebrate Paleontology* 4:379-395.

Garstang, W. 1928. The morphology of the Tunicata. *Quarterly Journal of Microscopic Science* 72:51-87.

Garstang, W. 1929. The origin and evolution of larval forms. *Report of the British Association, Glasgow*, pp. 77-98.

Garstang, W. 1951. *Larval Forms with other Zoological Verses*. Basil Blackwell, Oxford.

Gauthier, J. 1986. Saurischian monophyly and the origin of birds. *Memoirs of the California Academy of Sciences* 8:1-55.

Gauthier, J. 1994. The diversification of the amniotes, pp. 129-159, *in* Prothero, D. R., and R. M. Schoch, eds., *Major Features of Vertebrate Evolution. Short Courses in Paleontology* 7. Paleontological Society and University of Tennessee Press.

Gauthier, J., A. G. Kluge, and T. Rowe. 1988a. Amniote phylogeny and the importance of fossils. *Cladistics* 4:105-209.

Gauthier, J., A. G. Kluge, and T. Rowe. 1988b. The early evolution of the Amniota. pp. 103-155, *in* M. J. Benton, ed., *The Phylogeny and Classification of the Tetrapods, Vol. 1: Amphibians, Reptiles, Birds*. Clarendon Press, Oxford.

Gauthier, J. A., and K. Padian. 1985. Phylogenetic, functional, and aerodynamic analyses of the origin of birds, pp. 185-197, *in* Hecht, M. K., J. H. Ostrom, G. Viohl, and P. Wellnhofer, eds., *The Beginnings of Birds*. Freunde des Juras-Museums, Eichstätt.

Geary, D. 1987. Evolutionary tempo and mode in a sequence of the Upper Cretaceous bivalve *Pleuriocardia*. *Paleobiology* 13:140-151.

Geary, D. 1990. Patterns of evolutionary tempo and mode in the radiation of *Melanopsis* (Gastropoda: Melanopsidae). *Paleobiology* 16:492-511.

Geary, D. 1992. An unusual pattern of divergence between two fossil gastropods: ecophenotypy, dimorphism, or hybridization? *Paleobiology* 18: 93-109.

Gee, H. 1997. *Before the Backbone: Views on the Origin of Vertebrates*. Chapman & Hall, New York.

Ghiselin, M. T. 1966. The adaptive significance of gastropod torsion. *Evolution* 20:337-348.

Ghiselin, M. T. 1988. The origin of molluscs in the light of molecular evidence. *Oxford Surveys in Evolutionary Biology* 5: 66-95.

Gibbons, A. 1996. New feathered fossil brings dinosaurs and birds closer. *Science* 274:720-721.

Gillette, D. D., and M. G. Lockley, eds. 1989. *Dinosaur Tracks and Traces*. Cambridge University Press, Cambridge.

Goldschmidt, R. 1940. *The Material Basis of Evolution*. Yale University Press, New Haven, CT.

Goodman, M., M. M. Miyamoto, and J. Czelusniak. 1987. Pattern and process in vertebrate phylogeny revealed by coevolution of molecules and morphologies, pp. 141-176, *in* Patterson, C., ed., *Molecules and Morphology in Evolution: Conflict or Compromise?* Cambridge University Press, Cambridge.

Gorczynski, R. M., and E. J. Steele. 1980. Inheritance of acquired immunological tolerance to foreign histocompability antigens in mice. *Proceedings of the National Academy of Sciences* 77:2871-2875.

Goreau, T. F., N. I. Goreau, and T. J. Goreau. 1979. Corals and coral reefs. *Scientific American* 245(5): 110-121.

Götting, K. -J. 1974. *Malakozoologie*. G. Fisher, Stuttgart.

Gould, S. J. 1969. An evolutionary microcosm: Pleistocene and Recent history of the land snail *P.* (*Poecilozonites*) in Bermuda. *Bulletin of the Museum of Comparative Zoology, Harvard University* 138:407-532.

Gould, S. J. 1972. Allometric fallacies and the evolution of *Gryphaea*. *Evolutionary Biology* 6:91 119.

Gould, S. J. 1977. Size and shape, pp. 171-178, *in* Gould, S. J., *Ever Since Darwin*. W. W. Norton, New York.

Gould, S. J. 1980. Crazy Old Randolph Kirkpatrick, pp. 227-235, *in* Gould, S. J., *The Panda's Thumb*. W. W. Norton, New York.

Gould, S. J. 1980. Might we fit inside a sponge's cell, pp. 245-258, *in* Gould, S. J., *The Panda's Thumb*. W. W. Norton, New York.

Gould, S. J. 1980. The promise of paleobiology as a nomothetic, evolutionary discipline. *Paleobiology* 5:96-118.

Gould, S. J. 1984. Morphological channeling by structural constraint: convergence in styles of dwarfing and gigantism in *Cerion*, with a description of two new fossil species and a report on the discovery of the largest *Cerion*. *Paleobiology* 10:172-194.

Gould, S. J. 1984. *The Flamingo's Smile*. W. W. Norton, New York.

Gould, S. J. 1985. Molluscan paleobiology—as we creep toward the millennium: a critique of papers presented at the 1985 Short Course, pp. 258-267, *in* Bottjer, D. W., C. S. Hickman, and P. D. Ward, eds., *Mollusks, Notes for a Short Course*. University of Tennessee Department of Geological Sciences Studies in Geology 13.

Gould, S. J. 1989. *Wonderful Life: The Burgess Shale and the Nature of History*. W. W. Norton, New York.

Gould, S. J. 1991. Eight (or fewer) little piggies. *Natural History* 1991: 22-29.

Gould, S. J. 1991. The disparity of the Burgess Shale arthropod fauna and the limits of cladistic analysis: why we must strive to quantify morphospace. *Paleobiology* 17:411-423.

Gould, S. J. 1993. How to analyze Burgess Shale disparity—a reply to Ridley. *Paleobiology* 19: 522-524.

Gould, S. J., and C. B. Calloway. 1980. Clams and brachiopods—ships that pass in the night. *Paleobiology* 6:383-396.

Gould, S. J., and D. S. Woodruff. 1986. Evolution and systematics of *Cerion* (Mollusca: Pulmonata) on New Providence Island: a radical revision. *Bulletin of the American Museum of Natural History* 182:389-490.

Gould, S. J., and N. Eldredge. 1977. Punctuated equilibria: the tempo and mode of evolution reconsidered. *Paleobiology* 3:115-151.

Gradstein, F. M., J. P. Agterberg, J. C. Brower, and W. J. Schwazacher, eds., 1985. *Quantitative Stratigraphy*. D. Reidel, Dordrecht, Netherlands.

Gradstein, F. M., F. P. Agterberg, J. G. Ogg, J. Hardenbol, P. van Veen, J. Thierry, and Z. Huang. 1995. A Triassic, Jurassic, and Cretaceous time scale. *SEPM Special Publication* 54: 95-126.

Graham, R. W., and E. L. Lundelius, Jr. 1984. Coevolutionary disequilibrium and Pleistocene extinctions, pp. 223-249, *in* Martin, P. S., and R. G. Klein, eds., *Quaternary Extinctions: A Prehistoric Revolution*. The University of Arizona Press, Tucson.

Grant, R. E. 1972. The lophophore and feeding mechanisms of the Productidina (Brachiopoda). *Journal of Paleontology* 46:213-248.

Grant, R. E. 1975. Methods and conclusions in functional analysis: a reply. *Lethaia* 8:31-33.

Grassé, P. P. 1977. *Evolution of Living Organisms*. Academic Press, New York.

Gray, J., and W. Shear. 1992. Early life on land. *American Scientist* 80:444-456.

Greene, H. W. 1986. Diet and arboreality in the emerald monitor, *Varanus prasinus*, with comments on the study of adaptation. *Fieldiana* 1370:1-12.

Greenwood, P. H., Rosen, D. E., Weitzman, S. H., and Myers, G.

S. 1966. Phyletic studies of teleostean fishes, with a provisional classification of living forms. *Bulletin of the American Museum of Natural History* 131:339-456.

Greenwood, P. H., Miles, R. S., and Patterson, C., eds. 1973. *Interrelationships of Fishes*. Academic Press, London.

Gregory, W. K. 1910. The orders of mammals. *Bulletin of the American Museum of Natural History* 27:1-524.

Gupta, A. P., ed. 1979. *Arthropod Phylogeny*. Van Nostrand Reinhold, New York.

Haeckel, E. 1874. *Anthropogenie*. W. Engelman, Leipzig.

Hagmeier, E. M. 1966. A numerical analysis of the distributional patterns of North American mammals, II. Re-evaluation of the provinces. *Systematic Zoology* 15: 279-299.

Hagmeier, E. M., and C. D. Stults. 1964. A numerical analysis of the distributional patterns of North American mammals. *Systematic Zoology* 13:125-155.

Halanych, K. M., J. D. Bacheller, A. Aguinaldo, S. M. Liva, and D. M. Hillis. 1995. Evidence from 18S DNA that lophophorates are protostome animals. *Science* 267:1641.

Hallam, A. 1968. Morphology, palaeoecology, and evolution of the genus *Gryphaea* in the British Lias. *Philosophical Transactions of the Royal Society of London* (B) 254:91-128.

Hallam, A. 1975. Evolutionary size increase and longevity in Jurassic bivalves and ammonites. *Nature* 258:493-496.

Hallam, A. 1978. How rare is phyletic gradualism, and what is its evolutionary significance? Evidence from Jurassic bivalves. *Paleobiology* 4:16-25.

Hallam, A. 1981. The end-Triassic bivalve extinction event. *Palaeogeography, Palaeoclimatology, Palaeoecology* 35:1-44.

Hallam, A. 1982. Patterns of speciation in Jurassic *Gryphaea*. *Paleobiology* 8:354-366.

Hallam, A. 1986. Evidence of displaced terranes from Permian to Jurassic faunas around the Pacific margins. *Journal of the Geological Society of London* 143:209-216.

Hallam, A. 1989. The case for sea-level change as a dominant causal factor in mass extinctions of marine invertebrates. *Philosophical Transactions of the Royal Society of London* (B) 325:437-455.

Hallam, A. 1990. The end-Triassic mass extinction event. *Geological Society of America Special Paper* 585:577-583.

Hallam, A., and A. I. Miller. 1988. Extinction and survival in the Bivalvia, pp. 121-138, *in* Larwood, G. P., ed., *Extinction and Survival in the Fossil Record*. Systematics Association Special Volume 34.

Hallock, P., and W. Schlager. 1986. Nutrient excess and the demise of reefs and carbonate platforms. *Palaios* 1:389-398.

Hamilton, W. 1969. Mesozoic California and the underflow of the Pacific mantle. *Geological Society of America Bulletin* 80:2409-2430.

Hancock, J. M. 1977. The historic development of biostratigraphic correlation, pp. 3-22, *in* Kauffman, E., and Hazel, J. E., eds., *Concepts and Methods of Biostratigraphy*. Dowden, Hutchinson, & Ross, Stroudsburg, Penn.

Hankin, E. H., and D. M. S. Watson. 1914. On the flight of pterodactyls. *The Aeronautical Journal* 72:1-12.

Hansen, T. A. 1980. Influence of larval dispersal and geographic distribution on species longevity in neogastropods. *Paleobiology* 6:193-207.

Hansen, T. A. 1987. Extinction of late Eocene to Oligocene molluscs: Relationship to shelf area, temperature changes, and impact events. *Palaios* 2: 69–75.

Hansen, T. A. 1988. Early Tertiary radiation of marine molluscs and the long-term effects of the Cretaceous-Tertiary extinction. *Paleobiology* 14: 37–51.

Hansen, T. A., and P. H. Kelley. 1995. Spatial variation of naticid gastropod predation in the Eocene of North America. *Palaios* 10:268-278.

Hanson, C. B. 1980. Fluvial taphonomic processes: models and experiments, pp. 156-181, *in* Behrensmeyer, A. K., and A. P. Hill, eds., Fossils in the Making. University of Chicago Press, Chicago.

Häntzschel, W. 1975. Trace fossils and problematica, *in* Teichert, C., ed. *Treatise on Invertebrate Paleontology, part W. Miscellanea, Supplement I*. Geological Society of America and University of Kansas, Lawrence.

Haq, B. U., T. R. Worsley, L. H. Burckle, K. G. Douglas, L. D. Keigwin, Jr., N. D. Opdyke, S. M. Savin, M. H. Sommer, E. Vincent, and F. Woodruff. 1980. Late Miocene marine carbon-isotope shift and synchroneity of some phytoplanktonic biostratigraphic events. *Geology* 8:427-431.

Hardisty, M. W., and I. C. Potter, eds. 1971. *The Biology of Lampreys*. Academic Press, New York.

Harper, C. W., Jr. 1987. Might Occam's canon explode the Death Star? A moving average model of biotic extinctions. *Palaios* 2: 600–604.

Hartman, W. D., and T. E. Goreau. 1970. Jamaican coralline sponges: their morphology, ecology, and fossil representatives. *Zoological Society of London Symposium* 25:205-243.

Hartman, W. D., J. W. Wendt, and F. Wiedenmayer. 1980. *Living and Fossil Sponges*. University of Miami Rosenstiel School of Marine and Atmospheric Sciences, Sedimenta VIII.

Hartmann, G., ed. 1976. Evolution of post-Paleozoic Ostracoda. *Abhandlungen und Verhandlungen des Naturwissenschaften Vereins* 18/19:7-336.

Haszprunar, G. 1988. On the origin and evolution of major gastropod groups, with special reference to the Streptoneura. *Journal of Molluscan Studies* 54:367-441.

Haugh, B., and B. Bell. 1980. Classification schemes, pp. 94-105, *in* Broadhead, T. W., and Waters, J. A., eds., *Echinoderms: Notes for a Short Course*. University of Tennessee Department of Geological Sciences Studies in Geology 4.

Hays, J. D., and N. J. Shackleton. 1976. Globally synchronous extinction of the radiolarian *Stylatractus universus*. *Geology* 4:649-652.

Hazel, J. E. 1970. Atlantic continental shelf and slope of the United States—ostracode zoogeography in the southern Nova Scotian and northern Virginian faunal provinces. *U. S. Geological Survey Professional Paper* 529E:1-21.

Hazlehurst, G. A., and J. M. V. Rayner. 1992. Flight characteristics of Triassic and Jurassic Pterosauria: an appraisal based on wing shape. *Paleobiology* 18:447-463.

Heaton, M. J., and R. R. Reisz. 1986. Phylogenetic relationships of captorhinomorph reptiles. *Canadian Journal of Earth Sciences* 23:402-418.

Hecht, M. K., and A. Hoffman. 1986. Why not Neodarwinism? A critique of paleobiological challenges. *Oxford Surveys in Evolutionary Biology* 3:1-47.

Hecht, M. K., J. H. Ostrom, G. Viohl, and P. Wellnhofer, eds., *The Beginnings of Birds*. Freunde des Juras-Museums, Eichstätt.

Heck, K. J., Jr. 1980. Competitive delusion? *Paleobiology* 6:241-243.

Heckel, P. H. 1974. Carbonate buildups in the geological record: a review. *SEPM Special Publication* 18:90-154.

Hedberg, H. D. 1976. *International Stratigraphic Guide. A Guide to Stratigraphical Classification, Terminology, and Procedure*. Wiley, New York.

Heissig, K. 1986. No effect of the Ries impact event on the local mammal fauna. *Modern Geology* 10: 171–179.

Heptonstall, W. B. 1971. An analysis of the flight in the Cretaceous pterodactyl *Pteranodon ingens*. *Scottish Journal of Geology* 7:61-78.

Hickman, C. S. 1988. Analysis of form and function in fossils. *American Zoologist* 28:775-793.

Hill, A. 1981. Why study palaeoecology? *Nature* 293:340.

Hill, D. 1972. Archaeocyatha, pp. 2-158, *in* Teichert, C., ed., *Treatise on Invertebrate Paleontology*. Part E. Geological Society of America and University of Kansas Press, Lawrence.

Hill, D. 1981. Coelenterata, Supplement 1, Rugosa and Tabulata, in Teichert, C., ed., *Treatise on Invertebrate Paleontology*. Part F. Geological Society of America and University of Kansas Press, Lawrence.

Hillis, D. M., and C. Moritz, eds. 1990. *Molecular Systematics*. Sinauer Associates, Sunderland, Mass.

Hillis, D. M., A. Larson, S. K. Davis, and E. A. Zimmer. 1990. Nucleic acids III: Sequencing, pp. 318-370, *in* Hillis, D. M., and C. Moritz, eds. *Molecular Systematics*. Sinauer Associates, Sunderland, Mass.

Hitchin, R., and M. J. Benton. 1997. Congruence between parsimony and stratigraphy: comparisons of three indices. *Paleobiology* 23:20-32.

Hoedemaker, P. H., and others. 1993. Ammonite zonation for the Lower Cretaceous of the Mediterranean region, basis for stratigraphic correlations within IGCP Project 262. *Revista Española Paleontologia* 8:117-120.

Hoffman, A. 1979. Community paleoecology as an ephiphenomenal science. *Paleobiology* 5: 347-379.

Hoffman, A. 1982. Punctuated versus gradual mode of evolution: a reconsideration. *Evolutionary Biology* 15:411-436.

Hoffman, A. 1983. Paleobiology at the crossroads: a critique of some modern paleobiological research programs, pp. 241-272, *in* Grene, M., ed., *Dimensions of Darwinism: Themes and Counterthemes in Twentieth-Century Evolutionary Biology*. Cambridge University Press, Cambridge.

Hoffman, A. 1989. *Arguments on Evolution*. Oxford University Press, New York.

Hoffman, A. 1992. Twenty years later: punctuated equilibrium in retrospect, pp. 121-138, *in* Somit, A. and S. A. Peterson, eds., *The Dynamics of Evolution: The Punctuated Equilibrium Debate in the Natural and Social Sciences*. Cornell University Press, Ithaca, New York.

Hoffman, A., and J. Ghiold. 1986. Randomness in the pattern of 'mass extinctions' and 'waves of originations.' *Geological Magazine* 122:1–4.

Holmes, E. B. 1985. Are lungfishes the sister-group of tetrapods? *Biological Journal of the Linnean Society* 24:379-397.

Hooker, J. J. 1989. Character polarities in early perissodactyls and their significance for *Hyracotherium* and infraordinal relationships, pp. 79-101, *in* Prothero, D. R., and R. M. Schoch, eds., *The Evolution of Perissodactyls*. Oxford University Press, New York.

Hopson, J. A. 1994. Synapsid evolution and the radiation of non-eutherian mammals, pp. 190-219, *in* Prothero, D. R., and R. M. Schoch, eds., *Major Features of Vertebrate Evolution. Short Courses in Paleontology* 7. Paleontological Society and University of Tennessee Press.

Horner, J., and J. Gorman. 1988. *Digging Dinosaurs*. Workman Publishing Company, New York.

Hotton, N., P. D. MacLean, J. J. Roth, and E. C. Roth, eds., 1986. *The Ecology and Biology of Mammal-Like Reptiles*. Smithsonian Institution Press, Washington, D. C.

Hou, L., and Z. Liu. 1984. A new fossil bird from the Lower Cretaceous of Gansu and early evolution of birds. *Scientia Sinica* 27:1296-1302.

Hou, L., and J. Zhang. 1993. A few fossil bird from the Lower Cretaceous of Liaoning. *Vertebrata PalAsiatica* 7:217-224.

Hou, L., Z. Zhou, L. D. Martin, and A. Feduccia. 1995. A beaked bird from the Jurassic of China. *Nature* 377: 616-618.

Houde, P. 1988. Palaeognathous birds from the early Tertiary of the Northern Hemisphere. *Publications of the Nuttall Ornithological Club* 22:1-148.

Hough, J. 1949. The habits and adaptations of the Oligocene saber tooth carnivore *Hoplophoneus*. *U.S. Geological Survey Professional Paper* 221H:123-137.

House, M. R. 1985. The ammonoid time-scale and ammonoid evolution. *Geological Society of London Memoir* 10:273-283.

House, M. R., and J. R. Senior, eds. 1981. *The Ammonoidea*. Systematics Association Special Volume 18. Academic Press, New York.

Huelsenbeck, J. P. 1994. Comparing the stratigraphic record to estimates of phylogeny. *Paleobiology* 20:470-483.

Huelsenbeck, J. P., and B. Rannala. 1997. Maximum likelihood estimation of phylogeny using stratigraphic data. *Paleobiology* 23:174-180.

Hull, D. 1981. The principles of biological classification: the use and abuse of philosophy. *Philosophy of Science Association 1978* 2:130-153.

Hurst, J. M. 1975. *Resserella sabrinae* Bassett, in the Wenlock of Wales and the Welsh Borderland. *Journal of Paleontology* 49:316-328.

Hut, P., W. Alvarez, W. P. Elder, T. Hansen, E. G. Kauffman, G. Keller, E. M. Shoemaker, and P. Weismann. 1987. Comet showers as a cause of mass extinctions. *Nature* 329:118–126.

Hutchison, J. H. 1982. Turtle, crocodilian and champsosaur diversity changes in the Cenozoic of the north-central region of the western United States. *Palaeogeography, Palaeoclimatology, Palaeoecology* 37:149–164.

Huxley, T. H. 1862. The Anniversary Address. *The Quarterly Journal of the Geological Society of London* 18:xl-liv.

Jablonski, D. 1980. Apparent verse real biotic effects of transgression and regression. *Paleobiology* 6:398-407.

Jablonski, D. 1982. Evolutionary rates and models in Late Cretaceous gastropods. *Proceedings of the Third North American Paleontological Convention* 1:257-262.

Jablonski, D. 1985. Marine regressions and mass extinctions: a test using the Recent biota, pp. 335-354, *in* Valentine, J. W., ed., *Phanerozoic Diversity Patterns: Profiles in Macroevolution*. Princeton University Press, Princeton, New Jersey.

Jablonski, D. 1986a. Causes and consequences of mass extinctions, *in* Elliott, D. K., ed., *Dynamics of Extinction*. Wiley, New York, p. 183-229.

Jablonski, D. 1986b. Background and mass extinctions: the alternation of macroevolutionary regimes. *Science* 231:129-133.

Jablonski, D. 1993. The tropics as a source of evolutionary novelty through time. *Nature* 364:142-144.

Jabonski, D., and D. J. Bottjer. 1983. Soft-bottom epifaunal suspension-feeding assemblages in the Late Cretaceous: implications for the evolution of benthic paleocommunities, pp. 747-812, *in* Tevesz, M. J. S., and P. L. McCall, eds., *Biotic Interactions in Recent and Fossil Benthic Communities*. Plenum Press, New York.

Jablonski, D., K. W. Flessa, and J. W. Valentine. 1985. Biogeography and paleobiology. *Paleobiology* 11:75-90.

Jablonski, D., J. J. Sepkoski, Jr., D. J. Bottjer, and P. M. Sheehan. 1983. Onshore-offshore patterns in the evolution of Phanerozoic shelf communities. *Science* 222:1123-1125.

Jackson, J. B. C. 1974. Biogeographic consequences of eurytopy and stenotopy among marine bivalves and their evolutionary significance. *American Naturalist* 108:541-560.

Jackson, J. B. C., and A. H. Cheetham. 1995. Phylogeny reconstruction and the tempo of speciation in cheilostome Bryozoa. *Paleobiology* 20:407-423.

Jacobs, D. K. 1990. Sutural pattern and shell stress in *Baculites* with implications for other cephalopod shell morphologies. *Paleobiology* 16:336-348.

Jacobs, D. K. 1992. Shape, drag, and power in ammonoid swimming. *Paleobiology* 18:203-220.

Jacobs, D. K. 1996. Chambered cephalopod shells, buoyancy, structure, and decoupling: history and red herrings. *Palaios* 11: 610-614.

Jaeger, J. -J., and J. -L. Hartenberger. 1975. Pour utilization systématique de niveaux-repéres en biochronologies mammalienne. *Troisieme Réunion Annuel Sciences de Terre* 201.

Janis, C. 1993. Tertiary mammal evolution in the context of changing climates, vegetation, and tectonic events. *Annual Reviews of Ecology and Systematics* 24:467-500.

Janis, C., K. M. Scott, and L. L. Jacobs, eds. 1998. *Evolution of Tertiary Mammals of North America*. Cambridge University Press, Cambridge.

Janvier, P. 1981. The phylogeny of the Craniata, with particular reference to the significance of new fossil "agnathans. " *Journal of Vertebrate Paleontology* 1:121-159.

Janvier, P. 1984. The relationships of the Osteostraci and Galeaspida. *Journal of Vertebrate Paleontology* 4:344-358.

Jarvik, E. 1981. Review of "Lungfishes, Tetrapods, Paleontology, and Plesiomorphy." *Systematic Zoology* 30:378-384.

Jefferies, R. P. S. 1979. The origin of chordates—a methodological essay, pp. 443-477, *in* House, M. R., ed., *The Origin of Major Invertebrate Groups*. Academic Press, New York.

Jefferies, R. P. S. 1986. *The Ancestry of the Vertebrates*. British Museum (Natural History), London.

Jefferies, R. P. S. 1990. The solute *Dendrocystoides scoticus* from the Upper Ordovician of Scotland and the ancestry of chordates and echinoderms. *Palaeontology* 33:631-679.

Jell, P. A. 1980. Earliest known pelecypod on earth: a new Early Cambrian genus from South Australia. *Alcheringa* 4:233-239.

Johnson, A. L. A., and M. J. Simms. 1989. The timing and cause of Late Triassic marine invertebrate extinctions: evidence from scallops and crinoids, pp. 174-194, *in* Donovan, S. K., ed., *Mass Extinctions: Processes and Evidence*. Columbia University Press, New York.

Johnson, J. G. 1975. Allopatric speciation in fossil brachiopods. *Journal of Paleontology* 49:646-661.

Johnson, R. G. 1964. The community approach to paleoecology, pp. 107-134, *in* Imbrie, J., and N. D. Newell, eds., *Approaches to Paleoecology*. Wiley, New York.

Jollie, M. 1982. What are the 'Calcichordata'? and the larger question of the origin of the chordates. *Zoological Journal of the Linnean Society* 75:167-188.

Jones, D. L., N. J. Silberling, and J. Hillhouse. 1977. Wrangellia—a displaced terrane in northwestern North America. *Canadian Journal of Earth Sciences* 14:2565-2577.

Jones, D. S., and I. R. Quitmyer. Marking time with bivalve shells: oxygen isotopes and season of annual increment formation. *Palaios* 11:340-346.

Kaesler, R. L. 1982. Paleoecology and paleoenvironments. *Journal of Geological Education* 30:204-214.

Kaesler, R. L. 1987a. Superclass Crustacea, pp. 241-264, *in* Boardman, R. S., A. H. Cheetham, and A. J. Rowell, eds., *Fossil Invertebrates*. Blackwell Scientific Publishers, Cambridge, Mass.

Kaesler, R. L. 1987b. Class Insecta, pp. 264-269, *in* Boardman, R. S., A. H. Cheetham, and A. J. Rowell, eds., *Fossil Invertebrates*. Blackwell Scientific Publishers, Cambridge, Mass.

Kauffman, E. G. 1978. Evolutionary rates and patterns among Cretaceous Bivalvia. *Philosophical Transactions of the Royal Society* (B) 284:277-304.

Kauffman, E., and J. E. Hazel, eds. 1977. *Concepts and Methods of Biostratigraphy*. Dowden, Hutchinson, & Ross, Stroudsburg, Penn.

Kaufmann, R. 1933. Variations-statistische Untersuchungen über die Artabwandlung und Atrumbildung an der oberkambrischen Trilobitengattung *Olenus* Dalm. *Abhandlung du Geologisches-Paläontologisches Institut der Univerisität Griefswald* 10:1-54.

Keller, G. 1993. The Cretaceous/Tertiary boundary transition in the Antarctic Ocean and its global implications. *Marine Micropaleontology* 21:1-45.

Kelley, P. H. 1988. Predation by Miocene gastropods of the Chesapeake Group: stereotyped and predictable. *Palaios* 3:436-448.

Kelley, P. H. 1989. Evolutionary trends within bivalve prey of Chesapeake Group naticid gastropods. *Historical Biology* 2:139-156.

Kelley, P. H. 1991. Cannibalism by Chesapeake Group naticid gastropods: a predictable result of stereotyped predation. *Journal of Paleontology* 65:75-79.

Kelley, P. H., and T. A. Hansen. 1993. Evolution of the naticid gastropod predator-prey system: an evaluation of the hypothesis of escalation. *Palaios* 8:358-375.

Kemp, T. S. 1982. *Mammal-Like Reptiles and the Origin of Mammals*. Academic Press, London.

Kemp, T. S. 1988. Interrelationships of the Synapsida, pp. 1-22, *in* Benton, M. J., ed., *The Phylogeny and Classification of the Tetrapods, vol. 2: Mammals*. Clarendon Press, Oxford.

Kennedy, W. J. 1977. Ammonite evolution, pp. 251-304, *in* Hallam, A., ed., *Patterns of Evolution as Illustrated by the Fossil Record*. Elsevier, Amsterdam.

Kennedy, W. J., and W. A. Cobban. 1976. Aspects of ammonite biology, biogeography, and biostratigraphy. *Special Papers in Palaeontology* 17:1-94.

Kennedy, W. J., W. A. Cobban, and G. R. Scott. 1992. Ammonite correlation of the uppermost Campanian of Western Europe, the U.S. Gulf Coast, Atlantic Seaboard and Western Interior, and the numerical age of the base of the Maastrichtian. *Geological Magazine* 129:497-500.

Kennedy, W. J., and C. W. Wright. 1985. Evolutionary patterns in Late Cretaceous ammonites. *Special Papers in Palaeontology* 33:131-143.

Kennett, J. P. 1976. Phenotypic variation in some Recent and Late Cenozoic planktonic foraminifera, pp. 111-170, *in* Hedley, R. H., and C. G. Adams, eds., *Foraminifera*, vol. 2. Academic Press, New York.

Kennett, J. P. and L. D. Stott. 1990. Proteus and Proto-Oceanus: Ancestral Paleogene oceans as revealed from Antarctic stable isotopic results, ODP Leg 113. *Proceedings of the Ocean Drilling Program, Scientific Results* 113:865–880.

Kennett, J. P. and L. D. Stott. 1991. Abrupt deep-sea warming, paleoceanographic changes and benthic extinctions at the end of the Palaeocene. *Nature* 353: 225–229.

Kermack, K. A. 1959. A biometrical study of *Micraster coranguinum* and *M.* (*Isomicraster*) *senonis*. *Philosophical Transactions of the Royal Society of London* (B) 237:375-428.

Kidwell, S. M., and K. W. Flessa. 1995. The quality of the fossil record: populations, species, and communities. *Annual Reviews of Earth and Planetary Sciences* 26:269-299.

Kier, P. M. 1965. Evolutionary trends in Paleozoic echinoids. *Journal of Paleontology* 39:436-465.

Kier, P. M. 1974. Evolutionary trends and their functional significance in the post-Paleozoic echinoids. *Paleontological Society Memoir* 5.

Kier, P. M. 1977. The poor fossil record of the regular echinoid. *Paleobiology* 3:168-174.

King, G. 1990. *The Dicynodonts: A Study in Paleobiology*. Chapman & Hall, London.

Kitchell, J. A. 1985. Evolutionary paleoecology: recent contributions to evolutionary theory. *Paleobiology* 11:91-104.

Kitchell, J. A. 1986. The evolution of predator-prey behavior: naticid gastropods and their molluscan prey, pp. 88-110, *in* Nitecki, M., and J. A. Kitchell, eds., *Evolution of Animal Behavior: Paleontological and Field Approaches*. Oxford University Press, Oxford.

Kitchell, J. A., C. H. Boggs, J. F. Kitchell, and J. A. Rice. 1981. Prey selection by naticid gastropods: experimental tests and application to the fossil record. *Paleobiology* 7:533-552.

Kitchell, J. A., and G. Estabrook. 1986. Was there a 26-Myr periodicity of extinctions? *Nature* 321: 534–535.

Kitchell, J. A., and D. Pena. 1984. Periodicity of extinctions in the geologic past: Deterministic versus stochastic explanations. *Science* 226: 689–692.

Kjellesvig-Waering, E. N. 1961. The Silurian eurypterids of the Welsh Borderland. *Journal of Paleontology* 35:784-835.

Kjellesvig-Waering, E. N. 1986. A restudy of the fossil Scorpionida of the world. *Palaeontographica Americana* 55:1-287.

Klein, R. G. 1984. Mammalian extinctions and Stone Age people in Africa, pp. 553-573, *in* Martin, P. S., and R. G. Klein, eds., *Quaternary Extinctions: A Prehistoric Revolution*. The University of Arizona Press, Tucson.

Kleinpell, R. M. 1938. *Miocene Stratigraphy of California*. American Association of Petroleum Geologists, Tulsa, Oklahoma.

Kleinpell, R. M., and D. W. Weaver. 1963. Oligocene biostratigraphy of the Santa Barbara embayment. *University of California Publications in Geological Sciences* 43:1-77.

Knoll, A. H. 1987. Protists and Phanerozoic evolution in the ocean, *in* Lipps, J. H., ed., *Fossil Prokaryotes and Protists*. University of Tennessee Studies in Geology 18:248-264.

Knoll, A. H., R. K. Bambach, D. E. Canfield, and J. P. Grotzinger. 1996. Comparative earth history and Late Permian mass extinction. *Science* 273:452-457.

Koch, P. L., J. C. Zachos, and P. D. Gingerich. 1992. Correlation between isotope records in marine and continental carbon reservoirs near the Palaeocene/Eocene boundary. *Nature* 358:310-322.

Kohn, A. J. 1985. Gastropod paleoecology, pp. 174-189, *in* Bottjer, D. W., C. S. Hickman, and P. D. Ward, eds., *Mollusks, Notes for a Short Course*. University of Tennessee Department of Geological Sciences Studies in Geology 13.

Kolata, D. R., and M. Jollie. 1982. Anomalocystoid mitrates (Stylophora, Echinodermata) from the Champlainian (Middle Ordovician) Guttenberg Formation of the upper Mississippi Valley region. *Journal of Paleontology* 56:631-653.

Kolata, D. R., T. J. Frest, and R. H. Mapes. 1991. The youngest carpoid: occurrence, affinities, and life mode of a Pennsylvanian (Morrowan) mitrate from Oklahoma. *Journal of Paleontology* 65:844-855.

Korth, W. W. 1979. Taphonomy of microvertebrate fossil assemblages. *Annals of the Carnegie Museum* 48:235-285.

Koszlowski, R., 1948. Les graptolithes et quelques nouveaux groupes d'animaux du Tremadoc de la Pologne. *Palaeontologia Polonica* 3:1-235.

Krause, D. W. 1986. Competitive exclusion and taxonomic displacement in the fossil record: the case of rodents and multituberculates in North America. *University of Wyoming Contributions to Geology, Special Paper* 3: 95-117.

Krause, D. W. and M. C. Maas. 1990. The biogeographic origins of late Paleocene–early Eocene mammalian immigrants to the Western Interior of North America. *Geological Society of America Special Paper* 243: 71–105.

Kripp, D. 1943. Ein Lebensbild von *Pteranodon ingens* auf flugtechnischer Grundlage. *Nova Acta Leopoldina Abhandlungen der Kaiserlich Leopoldinisch-Carolinisch Deutschen Akademie der Naturforscher* 12:217-246.

Krishtalka, L. 1989. The naming of the shrew, pp. 28-37, *in* Krishtalka, L., *Dinosaur Plots*. William Morrow, New York.

Kulicki, C. 1979. The ammonite shell: its structure, development, and biological significance. *Palaeontologica Polonica* 39:79-142.

Kulicki, C., and H. Mutvei. 1988. Functional interpretation of ammonite septa, pp. 713-718, *in* Wiedmann, J., and J. Kullman, eds., *Cephalopods Present and Past*. Schweizerbartsche, Stuttgart.

Kurochkin, E. N. 1985. A true carinate bird from Lower Cretaceous deposits in Mongolia and other evidence of early Cretaceous birds in Asia. *Cretaceous Research* 6:271-278.

Kurtén, B. 1952. The Chinese *Hipparion* fauna. *Societas Scientiarum Fennica, Commentationes Biologicae* 13:1-82.

Kurtén, B. 1968. *Pleistocene Mammals of Europe*. Columbia University Press, New York.

Kurtén, B. 1988. *Before the Indians*. Columbia University Press, New York.

Kurtén, B., and E. Anderson. 1980. *Pleistocene Mammals of North America*. Columbia University Press, New York.

Labandeira, C. C., and B. S. Beall. 1988. Early insect diversification: evidence from a Lower Devonian bristletail from Quebec. *Science* 242:913-916.

Labandeira, C. C., and B. S. Beall. 1990. Arthropod terrestriality, pp. 214-256, *in* Mickulic, D. G., ed., *Arthropod Paleobiology: Short Courses in Paleontology* 3. Paleontological Society and University of Tennessee Press.

Labandeira, C. C., and J. J. Sepkoski. 1993. Insect diversity and the fossil record. *Science* 261: 310-315.

Lake, J. A. 1990. Origin of the Metazoa. *Proceedings of the National Academy of Sciences* 87:763-766.

Landman, N. H. 1988. Heterochrony in ammonites, pp. 159-182, *in* McKinney, M. L., ed., *Heterochrony in Evolution: A Multidisciplinary Approach*. Plenum Press, New York.

Landman, N. H. 1989. Iterative progenesis in Upper Cretaceous ammonites. *Paleobiology* 15:95-117.

Landman, N. H., and others, eds. 1996. *Ammonoid Biology*. Plenum Press, New York.

Larwood, G. P., ed. 1973. *Living and Fossil Bryozoa*. Academic Press, London.

Larwood, G. P., and C. Nielsen, eds. 1981. *Recent and Fossil*

Bryozoa. Olsen & Olsen, Fredensborg, Denmark.

Lauder, G. V. 1995. On the inference of function from structure, pp. 1-18, *in* Thomason, J. J., ed., *Functional Morphology in Vertebrate Paleontology*. Cambridge University Press, New York.

Lauder, G. V., and K. F. Liem. 1983. The evolution and interrelationships of the actinopterygian fishes. *Bulletin of the Museum of Comparative Zoology* 150:95-197.

Lauder, G. V., and K. F. Liem. 1991. The role of historical factors in the evolution of complex organic functions, pp. 63-78, *in* Wake, D. B., and G. Roth, eds., *Complex Organismal Functions: Integration and Evolution in Vertebrates*. Wiley, Chichester.

Laurin, M. and R. R. Reisz. 1995. A re-evaluation of early amniote phylogeny. *Zoological Journal of the Linnean Society* 113:165-223.

Lauterbach, K. -E. von. 1983. Erörterungen zur Stammesgeschichte der Mollusca, insbesondere der Conchifera. *Zeitschrift für Zoologische Systematik Evolutionforschung* 21:201-216.

Lawrence, J. M. 1987. *A Functional Biology of Echinoderms*. Croon Helm Ltd., London.

Lazarus, D. B. 1983. Speciation in pelagic Protista and its study in the planktonic microfossil record: a review. *Paleobiology*12:175-189.

Lazarus, D. B., and D. R. Prothero. 1984. The role of stratigraphic and morphologic data in phylogeny reconstruction. *Journal of Paleontology* 58:163-172.

Lehmann, U. 1981. *The Ammonites: Their Life and their World*. Cambridge University Press, Cambridge.

Lever, J. 1979. On torsion in gastropods, p. 5-23, *in* Van der Spoel, S., A. C. Van Bruggen, and J. Lever, eds., *Pathways in Malacology*. Scheltema and Hokelma, Utrecht.

Levi-Setti, R. 1993. *Trilobites* (2nd ed.). University of Chicago Press, Chicago.

Lewin, R. 1987. *Bones of Contention: Controversies in the Search for Human Origins*. Simon & Schuster, New York.

Lewontin, R. C. 1974. *The Genetic Basis of Evolutionary Change*. Columbia University Press, New York.

Lewontin, R. C. 1982. Keeping it clean. *Nature* 300: 113-114.

Lewontin, R. C., and J. L. Hubby. 1966. A molecular approach to the study of genic heterozygosity in natural populations. II. Amount of variation and degree of heterozygosity in natural populations of *Drosophila pseudoobscura*. *Genetics* 54:595-605.

Li, C. K., R. W. Wilson, and M. R. Dawson. 1987. The origin of rodents and lagomorphs. *Current Mammalogy* 1:97-108.

Lidgard, F. K. McKinney, and P. D. Taylor. 1993. Competition, clade replacement, and a history of cyclostome and cheilostome bryozoan diversity. *Paleobiology* 19:352-371.

Liebermann, B. S., C. E. Brett, and N. Eldredge. 1995. A study of stasis and change in two species lineages from the Middle Devonian of New York State. *Paleobiology* 21: 15-27.

Lillegraven, J. A. 1974. Biological considerations of the marsupial-placental dichotomy. *Evolution* 29:707-722.

Lillegraven, J. A., Z. Kielan-Jaworowska, and W. A. Clemens, eds. 1979. *Mesozoic Mammals: The First Two-Thirds of Mammalian History*. University of California Press, Berkeley.

Linsley, R. M. 1977. Some "laws" of gastropod shell form. *Paleobiology* 3:196-206.

Linsley, R. M. 1978. Shell form and evolution in the gastropods. *American Scientist* 66: 432-441.

Lockley, M. G. 1991. *Tracking Dinosaurs: A New Look at the Ancient World*. Cambridge University Press, Cambridge.

Lockley, M. G., and A. P. Hunt. 1995. *Dinosaur Tracks and other Fossil Footprints of the Western United States*. Columbia University Press, New York.

Loffler, H., and D. Danielopol, eds., 1977. *Aspect of Ecology and Zoogeography of Recent and Fossil Ostracoda*. W. Junk, The Hague.

Long, J. A. 1995. *The Rise of Fishes*. Johns Hopkins University Press, Baltimore.

Loper, D. E., and K. McCartney. 1986. Mantle plumes and the periodicity of magnetic field reversals. *Geophysical Research Letters* 13:1525–1528.

Loper, D. E., K. McCartney, and G. Buzyna. 1988. A model of correlated episodicity in magnetic-field reversals, climate, and mass extinctions. *Journal of Geology* 96:1–15.

Lucas, S. G. 1997. *Dinosaurs, The Textbook* (2nd ed.). W. C. Brown, Dubuque, Iowa.

Luckett, W. P., and J. -L. Hartenberger, eds. 1985. *Evolutionary Relationships among Rodents*. Plenum Press, New York.

Lundelius, E. L., Jr. 1983. Climatic implications of late Pleistocene and Holocene faunal associations in Australia. *Alcheringa* 7:125-149.

Luria, S. E., S. J. Gould, and S. Singer. 1981. *A View of Life*. Benjamin/Cummings, Menlo Park, CA.

MacArthur, R. H., and E. O. Wilson. 1967. *The Theory of Island Biogeography*. Princeton University Press, Princeton, New Jersey.

MacFadden, B. J. 1992. *Fossil Horses: Systematics, Paleobiology, and Evolution of the Family Equidae*. Cambridge University Press, Cambridge.

MacKinnon, D. I. 1985. New Zealand late Middle Cambrian molluscs and the origin of the Rostroconchia and Bivalvia. *Alcheringa* 9:65-81.

MacKinnon, D. I., D. E. Lee, and J. D. Campbell, eds. 1991. *Brachiopods through Time: Proceedings of the Second International Brachiopod Congress, Dunedin, New Zealand*. A. A. Balkema, Rotterdam.

MacReady, P. 1985. The great pterodactyl project. *Engineering and Science* 49:18-24.

Maglio, V. J. and H. B. S. Cooke, eds. 1978. *Evolution of African Mammals*. Harvard University Press, Cambridge.

Maisey, J. G. 1984a. Chondrichthyan phylogeny: a look at the evidence. *Journal of Vertebrate Paleontology* 4:359-371.

Maisey, J. G. 1984b. Higher elasmobranch phylogeny and biostratigraphy. *Zoological Journal of the Linnean Society* 82:33-54.

Maisey, J. G. 1986. Heads and tails: a chordate phylogeny. *Cladistics* 2:201-256.

Maisey, J. G. 1994. Gnathostomes (jawed vertebrates), pp. 38-56, *in* Prothero, D. R., and Schoch, R. M., eds., *Major Features of Vertebrate Evolution*. *Short Courses in Paleontology* 7. Paleontological Society and University of Tennessee Press.

Maisey, J. G. 1996. *Discovering Fossil Fishes*. Henry Holt and Company, New York.

Makowski, H. 1963. Problems of sexual dimorphism in ammonites. *Acta Palaeontologica Polonica* 12:1-92.

Makurath, J. H., and E. J. Anderson. 1973. Inter- and intraspecific variation in gypidulid brachiopods. *Evolution* 27: 303-310.

Manton, S. M. 1973. Arthropod phylogeny—a modern synthesis. *Journal of Zoology* 171:11-130.

Manton, S. M. 1977. *The Arthropoda*. Oxford University Press, Oxford.

Margulis, L., and K. V. Schwartz. 1982. *Five Kingdoms: An Illustrated Guide to Life on Earth*. W. H. Freeman, New York.

Marincovich, L., Jr. 1993. Delayed extinction of Mesozoic marine mollusks in the Paleocene Arctic Ocean. *Geological Society of America Abstracts with Programs* 25(6):295.

Marsh, O. C. 1880. Odontornithes: a monograph on the extinct toothed birds of North America. *Report on the Geological Exploration of the Fortieth Parallel* 7:1-201.

Marshall, C. R. 1990. Confidence intervals on stratigraphic ranges. *Paleobiology* 16:1-10.

Marshall, C. R. 1991. Estimation of taxonomic ranges from the fossil record, pp. 19-38, *in* Gilinsky, N., and P. W. Signor, eds., *Analytical Paleobiology. Short Courses in Paleontology* 4. Paleontological Society and University of Tennessee Press.

Marshall, C. R. 1994. Confidence intervals on stratigraphic ranges: partial relaxation of the assumption of randomly distributed fossil horizons. *Paleobiology* 20:459-469.

Marshall, C. R. 1995. Distinguishing between sudden and gradual extinctions in the fossil record: predicting the position of the Cretaceous-Tertiary iridium anomaly using the ammonite fossil record on Seymour Island, Antarctica. *Geology* 23:731-734.

Marshall, C. R. 1997. Confidence intervals on stratigraphic ranges with nonrandom distribution of fossil horizons. *Paleobiology* 23:165-173.

Marshall, L. G. 1984. Who killed Cock Robin? An investigation of the extinction controversy, pp. 785-806, *in* Martin, P. S., and R. G. Klein, eds., *Quaternary Extinctions: A Prehistoric Revolution*. The University of Arizona Press, Tucson.

Marshall, L. G. 1994. The terror birds of South America. *Scientific American* 270 (2): 90-95.

Martin, L. D. 1991. Mesozoic birds and the origin of birds, pp. 485-540, *in* Schultze, H. -P., and L. Trueb, eds., *Origins of the Higher Groups of Tetrapods*. Comstock Publishing Company, Ithaca, New York

Martin, L. D. 1995. The Enantiornithes: terrestrial birds of the Cretaceous in avian evolution. *Courier Forschunginstitut Senckenberg* 181:23-36.

Martin, P. S. 1984. Prehistoric overkill: the model, pp. 354-403, *in* Martin, P. S., and R. G. Klein, eds., *Quaternary Extinctions: A Prehistoric Revolution*. University of Arizona Press, Tucson.

Martin-Kaye, P. 1951. Sorting of lamellibranch valves on beaches in Trinidad, B. W. I. *Geological Magazine* 88: 432-434.

Martini, E., 1971, Standard Tertiary and Quaternary calcareous nannoplankton zonation, pp. 739-785, *in* Farinacci, A., ed., *Proceedings of the Second Planktonic Conference:* Roma, Tecnoscienza.

Martinsson, A., ed. 1975. Evolution and morphology of the Trilobita, Trilobitoidea, and Merostomata. *Fossils and Strata* 4:1-467.

Matsumara, T., and M. Shigei. 1988. Collagen biochemistry and the phylogeny of echinoderms, pp. 43-52, *in* Paul, C. R. C., and A. B. Smith, eds. *Echinoderm Phylogeny and Evolutionary Biology*. Clarendon Press, Oxford.

Matthew, W. D. 1910. The phylogeny of the Felidae. *Bulletin of the American Museum of Natural History* 26:289-316.

Matthew, W. D. 1915. Climate and evolution. *Annals of the New York Academy of Sciences* 24:171-318.

Maxson, L. R., and R. D. Maxson. 1990. Proteins II: Immunological techniques, pp. 127-155, *in* Hillis, D. M., and C. Moritz, eds. *Molecular Systematics*. Sinauer Associates, Sunderland, Mass.

Maynard Smith, J. 1958. *The Theory of Evolution*. Penguin, New York.

Mayr, E. 1942. *Systematics and the Origin of Species*. Columbia University Press, New York.

Mayr, E. 1992. Speciational evolution or punctuated equilibria, pp. 21-53, *in* Somit, A. and S. A. Peterson, eds., *The Dynamics of Evolution: The Punctuated Equilibrium Debate in the Natural and Social Sciences*. Cornell University Press, Ithaca, New York.

McCall, P. L. and M. J. S. Tevesz. 1983. Soft-bottom succession and the fossil record, pp. 157-194, *in* Tevesz, M. J. S., and P. L. McCall, eds., *Biotic Interactions in Recent and Fossil Benthic Communities*. Plenum Press, New York.

McCarthy, B. 1977. Selective preservation of mollusc shells in a Permian beach environment, Sydney Basin, Australia. *Neues Jahrbuch Geologie und Paläontologie Monatsheft* 8:466-474.

McCosker, J. E., and M. D. Lagios, eds. 1979. The biology and physiology of the living coelacanth. *California Academy of Sciences Occasional Papers* 134:1-175.

McDougall, K. 1980. Paleoecological evaluation of late Eocene biostratigraphic zonations of the Pacific Coast of North America. *Paleontological Society Monograph* 2.

McGhee, G. R., Jr. 1980. Shell form in the biconvex articulate Brachiopoda: a geometric analysis. *Paleobiology* 6:57-76.

McGhee, G. R., Jr. 1989. The Frasnian-Famennian extinction event, pp. 133-151, *in* Donovan, S. K., ed., *Mass Extinctions: Processes and Evidence*. Columbia University Press, New York.

McGhee, G. R., Jr. 1990. The Frasnian-Famennian mass extinction record in the eastern United States, pp. 161-168, *in* Walliser, O. H., and E. G. Kauffman, eds., *Extinction Events in Earth History*. Springer-Verlag, Berlin.

McGhee, G. R., Jr. 1995. Late Devonian bioevents in the Appalachian Sea: immigrations, extinction, and species replacements, *in* Brett, C. E., and G. C. Baird, eds., *Paleontological Events: Stratigraphic, Ecologic, and Evolutionary Implications*. Columbia University Press, New York.

McGhee, G. R., Jr. 1996. *The Late Devonian Mass Extinction: The Frasnian/Famennian Crisis*. Columbia University Press, New York.

McGowan, C. 1983. *The Successful Dragons: A Natural History of Extinct Reptiles*. Samuel Stevens, Toronto.

McGowan, C. 1991. *Dinosaurs, Spitfires, and Sea Dragons*. Harvard University Press, Cambridge.

McIntosh, R. P. 1980. The background and some current problems of theoretical ecology. *Synthèse* 43:195-255.

McKenna, M. C. 1973. Sweepstakes, filters, corridors, Noah's arks, and beached Viking funeral ships in paleogeography, pp. 295-308, *in* Tarling, D. H., and S. K. Runcorn, eds., *Implications of Continental Drift to the Earth Sciences*, vol. 1. Academic Press, London.

McKenna, M. C. 1975. Toward a phylogenetic classification of the Mammalia, pp. 21-46, *in* Luckett, W. P., and F. S. Szalay, eds. *Phylogeny of the Primates*. Plenum Press, New York.

McKenna, M. C. 1981. Discussion of Hallam, 1981, pp. 335-338, *in* Nelson, G., and D. E. Rosen, eds., *Vicariance Biogeography: A Critique*. Columbia University Press, New York.

McKenna, M. C. 1981. Early history and biogeography of South America's extinct land mammals, pp. 43-77, *in* Ciochon, R. L., and Chiarelli, A. B., eds., *Evolutionary Biology of New World Monkeys and Continental Drift*. Plenum Publishing, New York.

McKenna, M. C. 1983. Holarctic landmass rearrangement, cosmic events, and Cenozoic terrestrial organisms. *Annals of the Missouri Botanical Garden* 70: 459–489.

McKenna, M. C., and E. Manning. 1977. Affinities and palaeobiogeographic significance of the Mongolian Paleogene genus

Phenacolophus. *Géobios, Memoire Special* 1:61-85.

McKinney, F. K. 1993. A faster-paced world? Contrasts in biovolume and life-process rates in cyclostome (Class Stenolaemata) and chelostome (Class Gymnolaemata) bryozoans. *Paleobiology* 19:335-351.

McKinney, F. K., and J. B. C. Jackson. 1989. *Bryozoan Evolution*. Unwin Hyman, Boston.

McKinney, M. L., B. D. Carter, K. J. McNamara, and S. K. Donovan. 1992. Evolution of Paleogene echinoids: a global and regional view, pp. 348-367, *in* Prothero, D. R., and W. A. Berggren, eds., *Eocene-Oligocene Climatic and Biotic Evolution*. Princeton University Press, Princeton, New Jersey.

McLaughlin, P. 1980. *Comparative Morphology of Recent Crustacea*. W. H. Freeman, San Francisco.

McLoughlin, J. C. 1979. *Archosauria: A New Look at the Old Dinosaur*. Viking Press, New York.

McLoughlin, J. C. 1980. *Synapsida: A New Look into the Origin of Mammals*. Viking Press, New York.

McMasters, J. H. 1984. Reflections of a palaeoaerodynamicist. *Perspectives in Biology and Medicine* 29:331-384.

McNamara, K. J. 1988. Heterochrony and the evolution of echinoids, pp. 149-163, *in* Paul, C. R. C., and A. B. Smith, eds. *Echinoderm Phylogeny and Evolutionary Biology*. Clarendon Press, Oxford.

McNamara, K. J., 1990. Echinoids, pp. 205-231, *in* McNamara, K. J., ed., *Evolutionary Trends*. Belhaven, London.

McShea, D. W. 1993. Arguments, test, and the Burgess Shale—a commentary on the debate. *Paleobiology* 19:399-402.

Mickulic, D. G., D. E. G. Briggs, and J. Kluessendorf. 1985. A Silurian soft-bodied biota. *Science* 228:715-717.

Mickulic, D. G., ed. 1990. *Arthropod Paleobiology: Short Courses in Paleontology* 3. Paleontological Society and University of Tennessee Press.

Miller, A. I. 1990. Bivalves, pp. 143-161, *in* McNamara, K. J., ed., *Evolutionary Trends*. University of Arizona Press, Tucson.

Miller, F. X. 1977. The graphic correlation method in biostratigraphy, pp. 165-186, *in* Kauffman, E., and Hazel, J. E., eds., *Concepts and Methods of Biostratigraphy*. Dowden, Hutchinson, & Ross, Stroudsburg, Penn.

Miller, K. G., W. A. Berggren, J. Zhang and A. A. Palmer-Julson. 1991. Biostratigraphy and isotope stratigraphy of upper Eocene microtektites at Site 612: How many impacts? *Palaios* 6: 17–38.

Milner, A. R. 1988. The relationships of the living amphibians, pp. 59-102, *in* Benton, M. J., ed., *The Phylogeny and Classification of the Tetrapods, Vol. 1: Amphibians, Reptiles, Birds*. Clarendon Press, Oxford.

Milner, A. R. 1990. The radiations of temnospondyl amphibians, pp. 321-349, *in* Taylor, P. D., and G. P. Larwood, eds., *Major Evolutionary Radiations*. Clarendon Press, Oxford.

Miyazaki, J. M., and M. F. Mickevich. 1982. Evolution of *Chesapecten* (Mollusca: Bivalvia, Miocene-Pliocene) and the biogenetic law. *Evolutionary Biology* 15:369-410.

Monger, J. W. H., and C. A. Ross. 1971. Distribution of the fusulinaceans in the western Canadian Cordillera. *Canadian Journal of Earth Sciences* 8:259-278.

Mooi, R. 1989. Evolutionary dissent: a review of "Echinoderm Phylogeny and Evolutionary Biology," edited by C. R. C. Paul and A. B. Smith. *Paleobiology* 15:437-444.

Moore, R. C. and C. W. Pitrat, eds. 1961. *Treatise on Invertebrate Paleontology. Part Q. Arthropoda 3*. Crustacea. Geological Society of America and University of Kansas Press, Lawrence, Kansas.

Moore, R. C., and C. W. Pitrat, eds., 1960. *Treatise on Invertebrate Paleontology, Part I, Mollusca 1*. Geological Society of America and University of Kansas Press, Lawrence, Kansas.

Moore, R. C., ed. 1959. *Treatise on Invertebrate Paleontology Part O, Arthropoda 1*. Geological Society of America and University of Kansas Press, Lawrence, Kansas.

Morgan, E. 1982. *The Aquatic Ape*. Stein & Day, New York.

Morris, D. 1967. *The Naked Ape*. McGraw-Hill, New York.

Morton, B. 1996. The evolutionary history of the Bivalvia, pp. 337-359, *in* Taylor, J.,, ed., *Origin and Evolutionary Radiation of the Mollusca*. Oxford University Press, New York.

Morton, J. E. 1958. Torsion and the adult snail. *Proceedings of the Malacological Society of London* 33:2-10.

Morton, J. E. 1967. *Molluscs*. Hutchinson, London.

Moss, W. W. 1983. Taxa, taxonomists, and taxonomy, pp. 72-75, *in* Felsenstein, J., ed., *Numerical Taxonomy*. Springer-Verlag, New York.

Moy-Thomas, J., and R. S. Miles, 1971. *Palaeozoic Fishes*. W. B. Saunders, Philadelphia.

Müller, K. J., 1978. Conodonts and other phosphatic microfossils, pp. 277-291, *in* Haq, B. U., and A. Boersma, eds., *Introduction to Marine Micropaleontology*, Elsevier, New York.

Murphy, M. A. 1977. On chronostratigraphic units. *Journal of Paleontology* 52:123-129.

Muscatine, L., and H. M. Lenhoff. 1974. *Coelenterate Biology*. Academic Press, New York.

Mutvei, H. 1967. On the microscopic shell structure in some Jurassic ammonoids. *Neues Jahrbuch für Geologie und Paläontologie Abhandlungen* 129:157-166.

Nelson, J. S. 1994. *Fishes of the World* (3rd ed.). Wiley, New York.

Newell, N. D. 1949. Phyletic size increase, an important trend illustrated by fossil invertebrates. *Evolution* 3:103-124.

Newell, N. D. 1965. Classification of the Bivalvia. *American Museum Novitates* 1799:1-13.

Newell, N. D., and D. W. Boyd. 1975. Parallel evolution in early Trigoniacean bivalves. *Bulletin of the American Museum of Natural History* 154:53-162.

Newell, N. D., and D. W. Boyd. 1985. Permian scallops of the pectinacean family Streblochondriidae. *American Museum Novitates* 2831.

Newman, W. A., and S. M. Stanley. 1981. Competition wins out overall: reply to Paine. *Paleobiology* 7:561-569.

Nichols, D. 1959. Changes in the chalk heart-urchin *Micraster* interpreted in relation to living forms. *Philosophical Transactions of the Royal Society of London* (B), 242:347-437.

Nichols, D. 1969. *Echinoderms*. Hutchinson, London.

Nichols, D. 1974. *The Uniqueness of the Echinoderms*. Oxford Biology Readers 35, Oxford University Press, London.

Noma, E. and A. L. Glass. 1987. Mass extinction pattern: Result of chance. *Geological Magazine* 124: 319–322.

Norman, D. B. 1985. *The Illustrated Encyclopedia of the Dinosaurs*. Crescent Books, New York.

Norman, J. R., and P. H. Greenwood. 1975. *A History of Fishes* (3rd ed.). Ernest Benn, London.

Norris, R. D. 1986. Taphonomic gradients in shelf fossil assemblage: Pliocene Purisima Formation, California. *Palaios* 1:256-270.

North American Commission on Stratigraphic Nomenclature. 1983. North American Stratigraphic Code. *American Association of Petroleum Geologists Bulletin* 62:912-931.

Northcutt, R. G., and C. Gans. 1983. The genesis of neural crest and epidermal placodes: a reinterpretation of vertebrate origins. *Quarterly Review of Biology* 58:1-28.

Novacek, M. J. 1986. The skull of leptictid insectivorans and a higher-level classification of eutherian mammals. *Bulletin of the American Museum of Natural History* 183:1-111.

Novacek, M. J. 1990. Morphology, paleontology, and the higher clades of mammals. *Current Mammalogy* 2:507-543.

Novacek, M. J. 1992. Mammalian phylogeny: shaking the tree. *Nature* 356:121-125.

Novacek, M. J. 1994. The radiation of placental mammals, pp. 220-237, *in* Prothero, D. R., and R. M. Schoch, eds., *Major Features of Vertebrate Evolution. Short Courses in Paleontology* 7. Paleontological Society and University of Tennessee Press.

Novacek, M. J., and A. R. Wyss. 1986. Higher-level relationships of the Recent eutherian orders: morphological evidence. *Cladistics* 2:257-287.

Novacek, M. J., A. R. Wyss, and M. C. McKenna. 1988. The major groups of eutherian mammals, pp. 31-73, *in* Benton, M. J., ed., *The Phylogeny and Classification of the Tetrapods, Vol. 2: Mammals*. Clarendon Press, Oxford.

Nur, A., and Z. Ben-Avraham. 1981. Lost Pacifica continent: a mobilistic speculation, pp. 341-358, *in* Nelson, G., and D. E. Rosen, eds., *Vicariance Biogeography: A Critique*. Columbia University Press, New York.

Okada, H., and Bukry, D., 1980, Supplementary modification and introduction of code numbers to the low-latitude coccolith biostratigraphic zonation (Bukry, 1973, 1975). *Marine Micropaleontology* 5:321-325.

Okamoto, T. 1988. Developmental regulation and morphological saltation in the heteromorph ammonite *Nipponites*. *Paleobiology* 14:272-286.

Oliver, W. A., Jr. 1980. The relationship of scleractinian corals to the rugose corals. *Paleobiology* 6:146-160.

Oliver, W. A., Jr., and A. G. Coates. 1987. Phylum Cnidaria, pp. 140-193, *in* Boardman, R. S., A. H. Cheetham, and A. J. Rowell, eds., *Fossil Invertebrates*. Blackwell Scientific Publishers, Cambridge, Mass.

Olsen, P. E., N. H. Shubin, and M. H. Anders. 1987. New Early Jurassic tetrapod assemblages constrain Triassic-Jurassic extinction event. *Science* 237:1025-1029.

Olson, E. C. 1960. Morphology, paleontology, and evolution, pp. 523-546, *in* Tax, S., ed., *Evolution after Darwin, vol. I: The Evolution of Life*. University of Chicago Press, Chicago.

Olson, S. L. 1985. The fossil record of birds. *Avian Biology* 8:79-252.

Orth, C. 1989. Geochemistry of bio-event horizons, pp. 37-72, *in* Donovan, S. K., ed., *Mass Extinctions: Processes and Evidence*. Columbia University Press, New York.

Osborn, H. F., and W. D. Matthew. 1909. Cenozoic mammal horizons of western North America. *U.S. Geological Survey Bulletin* 361:1-138.

Osgood, R. G., Jr. 1987. Trace Fossils, pp. 663-674, *in* Boardman, R. S., A. H. Cheetham, and A. J. Rowell, eds., *Fossil Invertebrates*. Blackwell Scientific Publishers, Cambridge, Mass.

Ostrom, J. H. 1973. The ancestry of birds. *Nature* 242:136.

Ostrom, J. H. 1975. The origin of birds. *Annual Reviews of Earth and Planetary Sciences* 3:55-77.

Ostrom, J. H. 1976. *Archaeopteryx* and the origin of birds. *Biological Journal of the Linnean Society* 8:91-182.

Ostrom, J. H. 1979. Bird flight: how did it begin? *American Scientist* 67:46-56.

Ostrom, J. H. 1991. The question of the origin of birds, pp. 467-484, *in* Schultze, H. -P., and L. Trueb, eds., *Origins of the Higher Groups of Tetrapods*. Comstock Publishing Company, Ithaca, New York.

Ostrom, J. H. 1994. On the origin of birds and of avian flight, pp. 160-177, *in* Prothero, D. R., and R. M. Schoch, eds., *Major Features of Vertebrate Evolution. Short Courses in Paleontology* 7. Paleontological Society and University of Tennessee Press.

Owen-Smith, N. 1987. Pleistocene extinctions: the pivotal role of megaherbivores. *Paleobiology* 13:351-362.

Padian, K. 1979. The wings of pterosaurs: a new look. *Discovery* 14:21-29.

Padian, K. 1983. A functional analysis of flying and walking in pterosaurs. *Paleobiology* 9:218-239.

Padian, K. 1985. The origins and aerodynamics of flight in extinct vertebrates. *Palaeontology* 28:413-433.

Padian, K. 1987. The case of the bat-winged pterosaur: typological taxonomy and the influence of pictorial representation, pp. 65-81, *in* Czerkas, S. J., and E. C. Olson, eds., *Dinosaurs Past and Present,* Vol. 2. Natural History Museum of Los Angeles County, Los Angeles, and University of Washington Press, Seattle.

Padian, K. 1995. Form versus function: the evolution of a dialectic, pp. 264-277, *in* Thomason, J. J., ed., *Functional Morphology in Vertebrate Paleontology*. Cambridge University Press, New York.

Padian, K., and J. M. V. Rayner. 1993. The wings of pterosaurs. *American Journal of Science* 293A:91-166.

Paine, R. T. 1966. Food web complexity and species diversity. *American Naturalist* 100:65-76.

Paine, R. T. 1981. The forgotten roles of disturbance and predation. *Paleobiology* 7:553-560.

Paine, R. T. 1983. On paleoecology: an attempt to impose order on chaos. *Paleobiology* 9:86-90.

Palmer, A. R. 1982. Predation and parallel evolution: recurrent parietal plate reduction in balanomorph barnacles. *Paleobiology* 8:31-44.

Palmer, D., and R. B. Rickards, eds. 1991. *Graptolites: Writing in the Rocks*. Boydell Press, London.

Panchen, A. L., ed. 1980. *The Terrestrial Environment and the Origin of Land Vertebrates*. Academic Press, London.

Panchen, A. L., and T. R. Smithson. 1987. Character diagnosis, fossils, and the origin of tetrapods. *Biological Reviews* 62:341-438.

Panchen, A. L., and T. R. Smithson. 1988. The relationships of the earliest tetrapods, pp. 1-32. *in* Benton, M. J., ed., *The Phylogeny and Classification of the Tetrapods, Vol. 1: Amphibians, Reptiles, Birds*. Clarendon Press, Oxford.

Patterson, C. 1965. The phylogeny of the chimaeroids. *Philosophical Transactions of the Royal Society of London* (B) 249: 101-219.

Patterson, C. 1977. The contribution of paleontology to teleostean phylogeny, pp. 579-643, *in* Hecht, M. K., Goody, P. C., and Hecht, B. M, eds., *Major Patterns of Vertebrate Evolution*. Plenum Press, New York.

Patterson, C. 1994. Bony fishes, pp. 57-84, *in* Prothero, D. R., and Schoch, R. M., eds., *Major Features of Vertebrate Evolution. Short Courses in Paleontology* 7. Paleontological Society and University of Tennessee Press.

Patterson, C., and Rosen, D. E. 1977. Review of ichthyodectiform and other Mesozoic teleost fishes and the theory and practice of classifying fossils. *Bulletin of the American Museum of Natural*

History 158:81-172.

Patterson, C., and A. B. Smith. 1987. Is the periodicity of extinctions a taxonomic artefact? *Nature* 330: 248–251.

Paul, C. R. C. 1975. A reappraisal of the paradigm method of functional analysis of fossils. *Lethaia* 8:15-21.

Paul, C. R. C. 1977. Evolution of primitive echinoderms, pp. 123-158, *in* Hallam, A., ed. *Patterns of Evolution as Illustrated in the Fossil Record*. Elsevier, Amsterdam.

Paul, C. R. C. 1982. The adequacy of the fossil record, pp. 75-117, *in* Joysey, K., and A. Friday, eds., *Problems of Phylogenetic Reconstruction*. Academic Press, London.

Paul, C. R. C., and A. B. Smith, eds. 1988. *Echinoderm Phylogeny and Evolutionary Biology*. Clarendon Press, Oxford.

Paul, C. R. C., and A. B. Smith. 1984. The early radiation and phylogeny of the echinoderms. *Biological Reviews* 59:443-481.

Paulay, G. 1990. Effects of late Cenozoic sea-level fluctuations on the bivalve faunas of tropical oceanic islands. *Paleobiology* 16:415-434.

Peel, J. S. 1987. Class Gastropoda, pp. 304-329, *in* Boardman, R. S., A. H. Cheetham, and A. J. Rowell, eds., *Fossil Invertebrates*. Blackwell Scientific Publishers, Cambridge, Mass.

Pemberton, S. G., J. A. MacEachern, and R. W. Frey. 1992. Trace fossil facies models: environmental and allostratigraphic significance, pp. 47-72, *in* Walker, R. G., and N. P. James, eds., *Facies Models: Response to Sea Level Change*. Geological Association of Canada, Toronto.

Pennycuick, C. J. 1988. On the reconstruction of pterosaurs and their manner of flight, with notes on vortex wakes. *Biological Reviews* 63:299-331.

Peters, R. H. 1991. *A Critique for Ecology*. Cambridge University Press, Cambridge.

Peterson, K. J. 1994. The origin and early evolution of the craniates, pp. 14-37, *in* Prothero, D. R., and Schoch, R. M., eds., *Major Features of Vertebrate Evolution. Short Courses in Paleontology* 7. Paleontological Society and University of Tennessee Press.

Petrunkevich, A. 1952. Macroevolution and the fossil record of the Arachnida. *American Scientist* 40:99-122.

Philip, G. M. 1979. Carpoids—echinoderms or chordates? *Biological Reviews of the Cambridge Philosophical Society* 54:439-471.

Pianka, E. 1983. *Evolutionary Ecology*. Harper & Row, New York.

Platnick, N. D. 1976. Drifting spiders or continents? Vicariance biogeography of the spider family Laroniinae (Araneae: Gnaphosidae). *Systematic Zoology* 24:101-109.

Plotnick, R. E. 1985. Life-based mechanisms for swimming in eurypterids and portunid crabs. *Transactions of the Royal Society of Edinburgh, Earth Sciences* 76:325-337.

Plotnick, R. E. 1986. Taphonomy of a modern shrimp: implications for the arthropod fossil record. *Palaios* 1:286-293.

Pojeta, J., Jr. 1971. Review of Ordovician pelecypods. *U. S. Geological Survey Professional Paper* 695:1-46.

Pojeta, J., Jr. 1975. *Fordilla troyensis* Barrande and early pelecyopod phylogeny. *Bulletins of American Paleontology* 67:363-384.

Pojeta, J., Jr. 1978. The origin and taxonomic diversification of pelecypods. *Philosophical Transactions of the Royal Society of London* (B) 284:225-246.

Pojeta, J., Jr. 1980. Molluscan phylogeny. *Tulane Studies in Geology and Paleontology* 16:55-80.

Pojeta, J., Jr. 1985. Early evolutionary history of diasome molluscs, pp. 102-121, *in* Bottjer, D. W., C. S. Hickman, and P. D. Ward, eds., *Mollusks, Notes for a Short Course*. University of Tennessee Department of Geological Sciences Studies in Geology 13.

Pojeta, J., Jr. 1987. Class Pelecypoda, pp. 385-435, *in* Boardman, R. S., A. H. Cheetham, and A. J. Rowell, eds., *Fossil Invertebrates*. Blackwell Scientific Publishers, Cambridge, Mass.

Pojeta, J., Jr., and M. Gordon, Jr. 1987. Class Cephalopoda, pp. 329-358, *in* Boardman, R. S., A. H. Cheetham, and A. J. Rowell, eds., *Fossil Invertebrates*. Blackwell Scientific Publishers, Cambridge, Mass.

Pojeta, J., Jr., and B. Runnegar. 1974. *Fordilla troyensis* and the early history of pelecypod mollusks. *American Scientist* 62:706-711.

Pojeta, J., Jr., and B. Runnegar. 1976. The paleontology of rostroconch mollusks and the early history of the Phylum Mollusca. *U. S. Geological Survey Professional Paper* 968:1-88.

Pojeta, J., Jr., and B. Runnegar. 1985. The early evolution of the diasome molluscs, pp. 295-336, *in* Trueman, E. R., and M. R. Clarke, eds., *The Mollusca, 10: Evolution*. Academic Press, London.

Pojeta, J., Jr., B. Runnegar, J. S. Peel, and M. Gordon, Jr. 1987. Phylum Mollusca, pp. 270-435, *in* Boardman, R. S., A. H. Cheetham, and A. J. Rowell, eds., *Fossil Invertebrates*. Blackwell Scientific Publishers, Cambridge, Mass.

Pokorny, V. 1978. Ostracodes, pp. 109-149, *in* Haq, B. U., and A. Boersma, eds., *Introduction to Marine Micropaleontology*. Elsevier, New York.

Ponder, W. F., and D. R. Lindberg. 1996. Gastropod phylogeny—challenges for the 1990s, pp. 135-154, *in* Taylor, J., ed., *Origin and Evolutionary Radiation of the Mollusca*. Oxford University Press, New York.

Pough, F. H., J. B. Heiser, and W. N. McFarland, 1996. *Vertebrate Life* (4th ed.). Prentice-Hall, Upper Saddle River, New Jersey.

Prothero, D. R. 1981. New Jurassic mammals from Como Bluff, Wyoming, and the interrelationships of the non-tribosphenic Theria. *Bulletin of the American Museum of Natural History* 167:277-326.

Prothero, D. R. 1982. How isochronous are mammalian biostratigraphic events? *Proceedings of the Third North American Paleontological Convention* 2:405-409.

Prothero, D. R. 1993. Ungulate phylogeny: morphological versus molecular evidence, pp. 173-181, *in* Szalay, F. S., M. J. Novacek, and M. C. McKenna, eds. *Mammal Phylogeny. Vol. II: Placentals*. Springer-Verlag, New York.

Prothero, D. R. 1994a. The late Eocene-Oligocene extinctions. *Annual Reviews of Earth and Planetary Sciences* 22:145-165.

Prothero, D. R. 1994b. *The Eocene-Oligocene Transition, Paradise Lost*. Columbia University Press, New York.

Prothero, D. R. 1994c. Mammalian evolution, pp. 238-270, *in* Prothero, D. R., and R. M. Schoch, eds., *Major Features of Vertebrate Evolution. Short Courses in Paleontology* 7. Paleontological Society and University of Tennessee Press.

Prothero, D. R. 1996. Camelidae, pp. 609-651, *in* Prothero, D. R., and R. J. Emry, eds., *The Terrestrial Eocene-Oligocene Transition in North America*. Cambridge University Press, New York.

Prothero, D. R., and J. M. Armentrout. 1985. Magnetostratigraphic correlation of the Lincoln Creek Formation, Washington: implications for the age of the Eocene/Oligocene boundary. *Geology*, 13:208-211.

Prothero, D. R., and D. B. Lazarus. 1980. Planktonic microfossils and the recognition of ancestors. *Systematic Zoology* 29:119-129.

Prothero, D. R., E. M. Manning, and M. Fischer. 1988. The phylogeny of the ungulates, pp. 201-235, *in* Benton, M. J., ed., *The Phylogeny and Classification of the Tetrapods, Vol. 2: Mammals*. Clarendon Press, Oxford.

Prothero, D. R., and R. M. Schoch, eds. 1989. *The Evolution of Perissodactyls*. Oxford University Press, New York.

Prothero, D. R., and R. M. Schoch, eds. 1994, *Major Features of Vertebrate Evolution. Short Courses in Paleontology* 7. Paleontological Society and University of Tennessee Press.

Prothero, D. R., and R. M. Schoch. 1998. *Horns, Hooves, and Flippers: The Evolution of Hoofed Mammals and their Relatives*. Johns Hopkins University Press, Baltimore.

Prothero, D. R., and F. Schwab. 1996. *Sedimentary Geology*. W. H. Freeman, New York.

Prothero, D. R., and J. Sutton. 1998. Magnetic stratigraphy of the Narizian-Zemorrian (Middle Eocene-Late Oligocene) San Lorenzo Formation, Santa Cruz County, California. *Pacific Section SEPM Volume* (in press).

Prothero, D. R., and M. R. Thompson. 1998. Magnetic stratigraphy of the type Refugian stage (late Eocene-Oligocene), western Santa Ynez Range, Santa Barbara County, California. *Pacific Section SEPM Volume* (in press).

Prothero, D. R., and K. E. Whittlesey. 1998. Magnetostratigraphy and biostratigraphy of the Orellan and Whitneyan land mammal "ages" in the White River Group. *Geological Society of America Special Paper* (in press).

Purnell, M. A. 1995. Microwear in conodont elements and macrophagy in the first vertebrates. *Nature* 374:798-800.

Quade, J., and T. E. Cerling. 1994. Expansion of C4 grasses in the late Miocene of northern Pakistan: evidence from stable isotopes and paleosols. *Palaeogeography, Palaeoclimatology, Palaeoecology* 115:91-116.

Quade, J., N. Solounias, and T. E. Cerling. 1994. Stable isotopic evidence from paleosol carbonates and fossil teeth in Greece for forest or woodlands over the past 11 Ma. *Palaeogeography, Palaeoclimatology, Palaeoecology* 108:41-53.

Quinn, J. F. 1987. On the statistical detection of cycles in extinctions in the marine fossil record. *Paleobiology* 13: 456–478.

Raasch, G. O. 1939. Cambrian Merostomata. *Geological Society of America Special Paper* 19.

Radinsky, L. B. 1987. *The Evolution of Vertebrate Design*. University of Chicago Press, Chicago.

Raff, R. A., and others. 1988. Molecular analysis of distant phylogenetic relationships in the echinoderms, pp. 29-42, *in* Paul, C. R. C., and A. B. Smith, eds. *Echinoderm Phylogeny and Evolutionary Biology*. Clarendon Press, Oxford.

Raff, R. A., C. R. Marshall, and J. M. Turbeville. 1994. Using DNA sequences to unravel the Cambrian radiation of animal phyla. *Annual Reviews of Ecology and Systematics* 25:351.

Rampino, M. R. and R. B. Stothers. 1984. Terrestrial mass extinctions, cometary impacts, and the Sun's motion perpendicular to the galactic plane. *Nature* 308:709–712.

Rampino, M. R. and R. B. Stothers. 1988. Flood basalt volcanism during the past 250 million years. *Science* 241: 663–668.

Ramsbottom, W. H. C. 1979. Rates of transgression and regression in the Carboniferous of northwest Europe. *Journal of the Geological Society of London* 136:147-153.

Rau, W. W. 1958. Stratigraphy and foraminiferal zonation in some of the Tertiary rocks of southwestern Washington. *U.S. Geological Survey Oil and Gas Investigations Chart* OC 57.

Rau, W. W. 1966. Stratigraphy and foraminifera of the Satsop River area, southern Olympic Peninsula, Washington. *Washington Division of Mines and Geology Bulletin* 53:1-66.

Raup, D. M. 1967. Geometric analysis of shell coiling: coiling in ammonoids. *Journal of Paleontology* 41:43-65.

Raup, D. M. 1972. Taxonomic diversity during the Phanerozoic. *Science* 177: 1065-1071.

Raup, D. M. 1976a. Species richness in the Phanerozoic: a tabulation. *Paleobiology* 2:279-288.

Raup, D. M. 1976b. Species richness in the Phanerozoic: an interpretation. *Paleobiology* 2:289-297.

Raup, D. M. 1982. Biogeographic extinction: a feasibility test. *Geological Society of America Special Paper* 190:277-281.

Raup, D. M., and A. Michelson. 1965. Theoretical morphology of the coiled shell. *Science* 147:1294-1295.

Raymond, A. 1988. The paleoecology of a coal-ball deposit from the Middle Pennsylvanian of Iowa dominated by cordaitalean gymnosperms. *Reviews of Paleobotany and Palynology* 53:233-250.

Rea, D. K., J. C. Zachos, R. M. Owen, and P. D. Gingerich. 1990. Global change at the Paleocene-Eocene boundary: climatic and evolutionary consequences of tectonic events. *Palaeogeography, Palaeoclimatology, Palaeoecology* 79:117–128.

Remane, A. 1971. *Die Grundlagen des Natürlichen Systems der vergleichenden Anatomie und der Phylogenetik*. Koeltz, Königstein-Taunus.

Repenning, C. A. 1967. Palearctic-Nearctic mammalian dispersal in the late Cenozoic, pp. 288-311, *in* Hopkins, D. M., ed., *The Bering Land Bridge*. Stanford University Press, Stanford, California.

Repetski, J. E. 1978. A fish from the Upper Cambrian of North America. *Science* 200:529-531.

Retallack, G. J. 1983. Late Eocene and Oligocene fossil paleosols from Badlands National Park, South Dakota. *Geological Society of America Special Paper* 193.

Retallack, G. J., and C. R. Feakes. 1987. Trace fossil evidence for Late Ordovician animals on land. *Science* 235:61-63.

Reyment, R. A. 1982. Analysis of trans-specific evolution in Cretaceous ostracods. *Paleobiology* 8:293-306.

Reyment, R. A. 1985. Phenotypic evolution in a lineage of the Eocene ostracod *Echinocythereis*. *Paleobiology* 11:174-194.

Rhodes, F. H. T., ed. 1973. Conodont paleozoology. *Geological Society of America Special Paper* 141.

Rickards, R. B. 1977. Patterns of evolution in the graptolites, pp. 333-358, *in* Hallam, A., ed. *Patterns of Evolution as Illustrated in the Fossil Record*. Elsevier, Amsterdam.

Rickards, R. B. 1975. Palaeoecology of the Graptolithina, an extinct class of the Phylum Hemichordata. *Biological Reviews* 50:397-436.

Ricklefs, R. E. 1973. *Ecology* (2nd ed.). Chiron Press, New York.

Ridley, M. 1990. Dreadful beasts. *London Review of Books* 28 (6).

Ridley, M. 1993. Analysis of the Burgess Shale. *Paleobiology* 19:519-522.

Rieppel, O. 1988. The classification of the Squamata, pp. 261-294. *in* Benton, M. J., ed., *The Phylogeny and Classification of the Tetrapods, Vol. 1: Amphibians, Reptiles, Birds*. Clarendon Press, Oxford.

Rigby, J. K. 1971. Sponges and reef and related facies through time, *in* Rigby, J. K., ed., *Reefs through Time. Proceedings of the North American Paleontological Convention*, J:1374-1388. Allen Press, Lawrence, Kansas.

Rigby, J. K. 1987. Phylum Porifera, pp. 116-139, *in* Boardman, R. S., A. H. Cheetham, and A. J. Rowell, eds., *Fossil*

Invertebrates. Blackwell Scientific Publishers, Cambridge, Mass.

Rigby, J. K., and C. W. Stearn, eds. 1983. *Sponges and Spongiomorphs: Notes for a Short Course*. University of Tennessee Department of Geological Sciences Studies in Geology 7.

Rigby, J. K., and R. W. Gangloff. 1987. Phylum Archaeocyatha, pp. 107-115, *in* Boardman, R. S., A. H. Cheetham, and A. J. Rowell, eds., *Fossil Invertebrates*. Blackwell Scientific Publishers, Cambridge, Mass.

Rigby, J. K., Jr., K. R. Newman, J. Smit, S. Van der Kaars, R. E. Sloan, and J. K. Rigby. 1987. Dinosaurs from the Paleocene part of the Hell Creek Formation, McCone County, Montana. *Palaios* 2:296-302.

Rio, M., M. Roux, M. Renard, and E. Schein. 1992. Chemical and isotopic features of present day bivalve shells from hydrothermal vents or cold seeps. *Palaios* 7:351-360.

Robison, R. A. 1967. Ontogeny of *Bathyuriscus fimbriatus* and its bearing on the affinities of corynexochid trilobites. *Journal of Paleontology* 43:213-221.

Robison, R. L., and R. L. Kaesler. 1987. Phylum Arthropoda, pp. 205-269, *in* Boardman, R. S., A. H. Cheetham, and A. J. Rowell, eds., *Fossil Invertebrates*. Blackwell Scientific Publishers, Cambridge, Mass.

Rodda, P. U., and W. L. Fisher. 1964. Evolutionary features of *Athleta* (Eocene, Gastropoda) from the Gulf Coastal Plain. *Evolution* 18:235-244.

Romer, A. S., and B. H. Grove 1934. Environment of the earliest vertebrates. *American Midland Naturalist* 16:805-856.

Rosen, D. E., P. Forey, B. G. Gardiner, and C. Patterson. 1981. Lungfishes, tetrapods, paleontology and plesiomorphy. *Bulletin of the American Museum of Natural History* 141:357-474.

Rosen, D. E. 1982. Teleostean interrelationships, morphological function, and evolutionary inference. *American Zoologist* 22:261-273.

Rowe, A. W. 1899. An analysis of the genus *Micraster*, as determined by rigid zonal collecting from the zone of *Rhychonella cuvieri* to that of *Micraster coranguinum*. *Quarterly Journal of the Geological Society of London* 55:494-547.

Rowell, A. J., and R. E. Grant. 1987. Phylum Brachiopoda, pp. 445-496, *in* Boardman, R. S., A. H. Cheetham, and A. J. Rowell, eds., *Fossil Invertebrates*. Blackwell Scientific Publishers, Cambridge, Mass.

Rozanov, A. Y. 1974. Homological variability of the archaeocyathans. *Geological Magazine* 111: 107-120.

Rudwick, M. J. S. 1961. The feeding mechanism of the Permian brachiopod *Prorichthofenia*. *Palaeontology* 3:450-471.

Rudwick, M. J. S. 1970. *Living and Fossil Brachiopods*. Hutchinson and Co., Ltd., London.

Rudwick, M. J. S. The inference of function from structure in fossils. *British Journal for the Philosophy of Science* 15:27-40.

Runnegar, B. 1983. Molluscan phylogeny revisited. *Memoirs of the Association of Australasian Paleontologists* 1:121-144.

Runnegar, B. 1985. Origin and early history of mollusks, pp. 17-32, *in* Bottjer, D. W., C. S. Hickman, and P. D. Ward, eds., *Mollusks, Notes for a Short Course*. University of Tennessee Department of Geological Sciences Studies in Geology 13.

Runnegar, B. 1996. Early evolution of the Mollusca: the fossil record, pp. 77-87, *in* Taylor, J.,, ed., *Origin and Evolutionary Radiation of the Mollusca*. Oxford University Press, New York.

Runnegar, B., and C. Bentley. 1983. Anatomy, ecology, and affinities of the Australian Early Cambrian bivalve *Pojetaia runnegari* Jell. *Journal of Paleontology* 57:73-92.

Runnegar, B., and J. Pojeta, Jr. 1974. Molluscan phylogeny: the paleontological viewpoint. *Science* 186:311-317.

Runnegar, B., and J. Pojeta, Jr. 1992. The earliest bivalves and their Ordovician descendants. *American Malacological Bulletin* 9:117-122.

Runnegar, B., and P. A. Jell. 1976. Australian Middle Cambrian molluscs and their bearing on early molluscan evolution. *Alcheringa* 1:109-138.

Rye, D. M., and M. A. Sommer II. 1980. Reconstructing paleotemperatures and paleosalinity regimes with oxygen isotopes, pp. 169-202, *in* Rhoads, D. C., ed., *Skeletal Growth of Aquatic Organisms: Biological Records of Environmental Change*. Plenum Press, New York.

Salvador, A. 1994. *International Stratigraphic Guide* (2nd ed.). International Union of Geological Sciences and the Geological Society of America, Boulder, Colorado.

Salvini-Plawen, L. V. 1980. Proposed classification of the Mollusca. *Malacologia* 19:249-278.

Salvini-Plawen, L. V. 1991. Origin, phylogeny, and classification of the phylum Mollusca. *Iberus* 9:1-33.

Sansom, I. J., and others. 1992. Presence of the earliest vertebrate hard tissues in conodonts. *Science* 256:1308-1311.

Sanz, J. L., J. F. Bonaparte, and A. Lacasa. 1988. Unusual Early Cretaceous birds from Spain. *Nature* 331: 433-435.

Sanz, J. L., and A. D. Buscalioni. 1992. A new bird from the early Cretaceous of Las Hoyas, Spain, and the early radiation of birds. *Palaeontology* 35:829-845.

Sanz, J. L., J. M. Chiappe, and A. D. Buscalioni. 1995. The osteology of *Concornis lacustris* (Aves: Enantiornithes) from the Lower Cretaceous of Spain and a re-examination of its phylogenetic significance. *American Museum Novitates* 3133:1-23.

Sarjeant, W. A. S. 1975. Fossil traces and impressions of vertebrates, pp. 283-324, *in* Frey, R. W., ed., *The Study of Trace Fossils*. Springer-Verlag, New York.

Saunders, W. B. 1995. The ammonoid suture problem: relationships between shell and septal thickness and suture complexity in Paleozoic ammonoids. *Paleobiology* 21:343-355.

Saunders, W. B., and N. H. Landman, eds. 1987. *Nautilus, the Biology and Paleobiology of a Living Fossil*. Plenum Press, New York.

Saunders, W. B., and R. H. Swan. 1984. Morphology and morphologic diversity of mid-Carboniferous (Namurian) ammonites in time and space. *Paleobiology* 10:195-228.

Saunders, W. B., P. D. Ward, and T. L. Daniel. 1994. Cameral liquid transport: resolution of the ammonoid suture problem? *Geological Society of America Abstracts with Programs* 26(7): A374.

Saunders, W. B., and D. M. Work. 1996. Shell morphology and sutural complexity in Upper Carboniferous ammonoids. *Paleobiology* 22:189-218.

Savage, D. E. 1977. Aspects of vertebrate paleontological stratigraphy and geochronology, pp. 427-442, *in* Kauffman, E., and Hazel, J. E., eds., *Concepts and Methods of Biostratigraphy*. Dowden, Hutchinson, & Ross, Stroudsburg, Penn.

Savage, D. E., and D. E. Russell. 1983. *Mammalian Paleofaunas of the World*. Addison Wesley, Reading, Mass.

Savage, R. J. G., and M. R. Long. 1986. *Mammal Evolution: An Illustrated Guide*. Facts-on-File Publications, New York.

Savarese, M. 1992. Functional analysis of archaeocyathan skeletal morphology and its paleobiological implications. *Paleobiology* 18:465-480.

Schaeffer, B. 1952. The Triassic coelacanth fish *Diplurus*, with observations on the evolution of the Coelacanthini. *Bulletin of the American Museum of Natural History* 99:25-78.

Schaeffer, B. 1987. Deuterostome monophyly and phylogeny. *Evolutionary Biology* 20:179-235.

Schaeffer, B., and D. E. Rosen. 1961. Major adaptive levels in the evolution of the actinopterygian feeding mechanism. *American Zoologist* 1:187-204.

Schaeffer, B., and M. Williams. 1977. Relationships of fossil and living elasmobranchs. *American Zoologist* 17:293-302.

Scheltema, R. S. 1977. Dispersal of marine invertebrate organisms: paleobiogeographic and biostratigraphic implications, pp. 73-108, *in* Kauffman, E. G., and J. E. Hazel, eds., *Concepts and Methods of Biostratigraphy*. Dowden, Hutchinson & Ross, Stroudsburg, Penn.

Scheltema, R. S. 1978. On the relationships between the dispersal of pelagic veliger larvae and the evolution of marine prosobranch gastropods, pp. 303-322, *in* Battaglia, B., and J. A. Beardmore, eds., *Marine Organisms: Genetics, Ecology, and Evolution*. Plenum Press, New York.

Schindel, D. 1980. Microstratigraphic sampling and the limits of paleontological resolution. *Paleobiology* 6:408-426.

Schindel, D. 1982. Punctutations in the Pennsylvanian evolutionary history of *Glabrocingulum* (Mollusca: Archaeogastropoda). *Geological Society of America Bulletin* 93:400-408.

Schindel, D. E. 1990. Unoccupied morphospace and the coiled geometry of gastropods: architectural constraint or geometric covariation? pp. 270-304, *in* Ross, R. M., and W. D. Allmon, eds., *Causes of Evolution, A Paleontological Perspective*. University of Chicago Press, Chicago.

Schindewolf, O. H. 1936. *Paläontologie, Entwicklungslehre und Genetik: Kritik und Synthese*. Bornträger, Berlin.

Schindewolf, O. H. 1950. *Grundfragen der Paläontologie*. E. Schweitzbartsche Verlagbuchhandlung, Stuttgart.

Schmitt, W. L. 1965. *Crustaceans*. University of Michigan Press, Ann Arbor.

Schopf, T. J. M. 1974. Permo-Triassic extinction: relation to seafloor spreading. *Journal of Geology* 82:129-143.

Schopf, T. J. M. 1978. Fossilization potential of an intertidal fauna: Friday Harbor, Washington. *Paleobiology* 4:261-270.

Schopf, T. J. M. 1979. The role of biogeographic provinces in regulating marine faunal diversity through geologic time, pp. 449-457, *in* Gray, J., and A. J. Boucot, eds., *Historical Biogeography, Plate Tectonics, and the Changing Environment*. Oregon State University Press, Corvallis, Oregon.

Schopf, T. J. M. 1977. Patterns and themes of evolution among the Bryozoa, pp. 159-207, *in* Hallam, A., ed. *Patterns of Evolution as Illustrated in the Fossil Record*. Elsevier, Amsterdam.

Schram, F. R. 1986. *Crustacea*. Oxford University Press, Oxford.

Schram, F. R. 1990. Crustacean phylogeny, pp. 285-302, *in* Mickulic, D. G., ed., *Arthropod Paleobiology: Short Courses in Paleontology* 3. Paleontological Society and University of Tennessee Press.

Schrock, R. R., and W. H. Twenhofel. 1953. *Principles of Invertebrate Paleontology*. McGraw-Hill, New York.

Schultz, C. B., and C. M. Falkenbach. 1956. Miniochoerinae and Oreonetinae, two new subfamilies of oreodonts. *Bulletin of the American Museum of Natural History* 109:373-482.

Schultz, C. B., and C. M. Falkenbach. 1968. The phylogeny of the oreodonts. *Bulletin of the American Museum of Natural History* 139:1-498.

Schultze, H. P., ed. 1979. *Handbook of Paloichthyology*. Gustav Fischer, Stuttgart.

Schultze, H. -P. 1987. Dipnoans as sarcopterygians, pp. 39-74, *in* Bemis, W. E., Burggren, W. W., and Kemp, N. E., eds., 1987. *The Biology and Evolution of Lungfishes*. Liss, New York.

Schultze, H. -P. 1991. A comparison of controversial hypotheses on the origin of tetrapods, pp. 29-67, *in* Schultze, H. -P., and L. Trueb, eds., *Origins of the Higher Groups of Tetrapods: Controversy and Consensus*. Cornell University Press, Ithaca, New York.

Schwartz, R. D., and P. B. James. 1984. Periodic mass extinctions and the Sun's oscillation around the galactic plane. *Nature* 308: 712–713.

Secord, J. A. 1986. *Controversy in Victorian Geology: The Cambrian-Silurian Dispute*. Princeton University Press, Princeton, New Jersey.

Seilacher, A. 1953. Studien der Palichnologie. I. Über die Methoden der Palichnologie. *Neues Jahrbuch für Geologie und Paläontologie Abhandlungen* 96:421-452.

Seilacher, A. 1967. Bathymetry of trace fossils. *Marine Geology* 5:413-428.

Seilacher, A. 1970. Arbeitskonzept zur Konstruktions-Morphologie. *Lethaia* 3:393-396.

Seilacher, A. 1972. Divaricate patterns in pelecypod shells. *Lethaia* 5:325-343.

Selden, P. A. 1983. Autecology of Silurian eurypterids. *Special Papers in Palaeontology* 32:39-54.

Selden, P. A. 1988. The arachnid fossil record. *British Journal of Entomology and Natural History* 1:15-18.

Selden, P. A., and D. J. Siveter. 1988. The origin of the limuloids. *Lethaia* 20:383-392.

Sepkoski, J. J., Jr. 1978. A kinetic model of Phanerozoic taxonomic diversity, I. Analysis of marine orders. *Paleobiology* 4:223-251.

Sepkoski, J. J., Jr. 1979. A kinetic model of Phanerozoic taxonomic diversity, II. Early Phanerozoic families and multiple equilibria. *Paleobiology* 5: 222-251.

Sepkoski, J. J., Jr. 1981. A factor analytic description of the Phanerozoic marine record. *Paleobiology* 7:36-53.

Sepkoski, J. J., Jr. 1982. A compendium of fossil marine families. *Milwaukee Public Museum Contributions to Biology and Geology* 51:1-125.

Sepkoski, J. J., Jr. 1989. Periodicity in extinction and the problem of catastrophism in the history of life. *Journal of the Geological Society of London* 146:7-19.

Sepkoski, J. J., Jr. 1991. A model for onshore-offshore changes in faunal diversity. *Paleobiology* 17:58-77.

Sepkoski, J. J., Jr., R. K. Bambach, D. M. Raup, and J. W. Valentine. 1981. Phanerozoic marine diversity and the fossil record. *Nature* 293:435-437.

Sepkoski, J. J., Jr., and A. I. Miller. 1985. Evolutionary faunas and the distribution of Paleozoic benthic communities in space and time, pp. 153-190, *in* Valentine, J. W., ed., *Phanerozoic Diversity Patterns: Profiles in Macroevolution*. Princeton University Press, Princeton, New Jersey.

Sereno, P. C. 1991. Basal archosaurs: phylogenetic relationships and functional implications. *Journal of Vertebrate Paleontology Memoir* 2:1-53.

Sereno, P. C. 1997. The origin and evolution of dinosaurs. *Annual Reviews of Earth and Planetary Sciences* 25: 435-490.

Sereno, P. C., and C. Rao. 1992. Early evolution of avian flight and perching: new evidence from the Lower Cretaceous of China. *Nature* 255:845-848.

Sessions, S. K. 1990. Chromosomes: molecular cytogenetics, pp. 156-203, *in* Hillis, D. M., and C. Moritz, eds. *Molecular*

Systematics. Sinauer Associates, Sunderland, Mass.

Shaw, A. B. 1964. *Time in Stratigraphy*. McGraw-Hill, New York.

Shear, W. A. 1990. Silurian-Devonian terrestrial arthropods, pp. 197-213, *in* Mickulic, D. G., ed., *Arthropod Paleobiology: Short Courses in Paleontology* 3. Paleontological Society and University of Tennessee Press.

Shear, W. A., P. M. Bonamo, J. D. Grierson, W. D. Rolfe, E. L. Smith, and R. A. Norton. 1984. Early land animals in North America: evidence from Devonian age arthropods from Gilboa, New York. *Science* 224:492-494.

Shear, W. A., and J. Kukalova-Peck. 1990. The ecology of Palaeozoic terrestrial arthropods: the fossil evidence. *Canadian Journal of Zoology* 68:1807-1834.

Sheehan, P. M. 1977. Species diversity in the Phanerozoic: a reflection of labor by systematists? *Paleobiology* 3:325-328.

Sheehan, P. M. 1982. Brachiopod macroevolution at the Ordovician-Silurian boundary. *Proceedings of the Third North American Paleontological Convention* 2:477-481.

Sheldon, P. R. 1987. Parallel gradualistic evolution of Ordovician trilobites. *Nature* 330:561-563.

Sheldon, P. R. 1997. Plus ça change—a model for stasis and environmental evolution in different environments. *Palaeogeography, Palaeoclimatology, Palaeoecology* 127:209-227.

Shimer, H. W., and R. R. Schrock. 1944. *Index Fossils of North America*. M. I. T. Press, Cambridge, Mass.

Shipman, P. 1981. *Life History of a Fossil*. Harvard University Press, Cambridge.

Shoemaker, E. M., and R. F. Wolfe. 1986. Mass extinctions, crater ages, and comet showers, pp. 338–386, *in* Smoluchowski, R. S., J. N. Bahcall, and M. S. Matthews, eds., *The Galaxy and the Solar System*. University of Arizona Press, Tucson.

Shoshani, J., 1993. Hyracoidea-Tethytheria affinity based on myological data, pp. 235-256, *in* Szalay, F. S., M. J. Novacek, and M. C. McKenna, eds., *Mammal Phylogeny. Vol. II: Placentals*. Springer-Verlag, New York.

Sibley, C. G., and J. Ahlquist. 1983. The phylogeny and classification of birds based on DNA-DNA hybridization, pp. 245-292, *in* Johnson, R. F., ed., *Current Ornithology* 1. Plenum Press, New York.

Sibley, C. G., and J. E. Ahlquist. 1990. *Phylogeny and Classification of Birds*. Yale University Press, New Haven.

Signor, P. W. 1982. A critical re-evaluation of the paradigm method of functional inference. *Neues Jahrbuch für Geologie und Paläontologie Abhandlungen* 164:59-63.

Signor, P. W. 1985. Gastropod evolutionary history, pp157-173, *in* Bottjer, D. W., C. S. Hickman, and P. D. Ward, eds., *Mollusks, Notes for a Short Course*. University of Tennessee Department of Geological Sciences Studies in Geology 13.

Signor, P. W., and J. H. Lipps. 1982. Sampling bias, gradual extinction patterns, and catastrophes in the fossil record. *Geological Society of America Special Paper* 190:291-296.

Simberloff, D. 1970. Taxonomic diversity of island biotas. *Evolution* 24:23-47.

Simberloff, D. 1974a. Equilibrium theory of island biogeography. *Annual Reviews of Ecology and Systematics* 5:151-182.

Simberloff, D. 1974b. Permo-Triassic extinctions: effects of area on biotic equilibrium. *Journal of Geology* 82:267-274.

Simberloff, D., K. L. Heck, E. D. McCoy, and E. F. Connor. 1981. There have been no statistical tests of cladistic biogeographical hypotheses, pp. 40-93, *in* Nelson, G., and D. E. Rosen, eds., *Vicariance Biogeography: A Critique*. Columbia University Press, New York.

Simms, M. J. 1990. Crinoids, pp. 188-204, *in* McNamara, K. J., ed., *Evolutionary Trends*. Belhaven, London.

Simms, M. J., and A. H. Ruffell. 1989. Synchroneity of climatic change and extinction in the late Triassic. *Geology* 17:265-268.

Simms, M. J., and A. H. Ruffell. 1990. Climatic and biotic change in the late Triassic. *Journal of the Geological Society of London* 147:321-327.

Simpson, G. G. 1934. *Attending Marvels*. University of Chicago Press, Chicago.

Simpson, G. G. 1936. Data on the relationships of local and continental mammalian faunas. *Journal of Paleontology* 10:410-414.

Simpson, G. G. 1940. Mammals and land bridges. *Journal of the Washington Academy of Sciences* 30:137-163.

Simpson, G. G. 1941. The function of saber-like canines in carnivorous mammals. *American Museum Novitates* 1130.

Simpson, G. G. 1944. *Tempo and Mode in Evolution*. Columbia University Press, New York.

Simpson, G. G. 1945. The principles of classification and a classification of the mammals. *Bulletin of the American Museum of Natural History* 85:1-350.

Simpson, G. G. 1947. Evolution, interchange, and resemblance of North American and Eurasian Cenozoic mammalian faunas. *Evolution* 1:218-220.

Simpson, G. G. 1953. *Major Features of Evolution*. Columbia University Press, New York.

Simpson, G. G. 1980. *Splendid Isolation: The Curious History of South American Mammals*. Yale University Press, New Haven.

Simpson, S. 1975. Classification of trace fossils, pp. 39-54, *in* Frey, R. W., ed., *The Study of Trace Fossils*. Springer-Verlag, New York.

Smiley, S. 1988. The phylogenetic relationships of holothurians: a cladistic analysis of the extant echinoderm classes, pp. 69-84, *in* Paul, C. R. C., and A. B. Smith, eds. *Echinoderm Phylogeny and Evolutionary Biology*. Clarendon Press, Oxford.

Smith, A. B. 1984a. *Echinoid Palaeobiology*. Allen & Unwin, London.

Smith, A. B. 1984b. Classification of the Echinodermata. *Palaeontology* 27:431-459.

Smith, A. B. 1994. *Systematics and the Fossil Record: Documenting Evolutionary Patterns*. Blackwell Scientific, London.

Smith, A. B., and C. Patterson. 1988. The influence of taxonomic method on the perceptions of patterns of evolution. *Evolutionary Biology* 23: 127–216.

Smith, H. 1953. *From Fish to Philosopher*. Little, Brown, Boston.

Smith, J. L. B. 1956. *Old Fourlegs: The Story of the Coelacanth*. Longman, Green, London.

Smith, M. P., D. E. G. Briggs, and R. J. Aldridge. 1987. A conodont animal from the lower Silurian of Wisconsin, U. S. A., and the apparatus architecture of panderodontid conodonts, pp. 91-104, *in* Aldridge, R. J., ed., *Palaeobiology of Conodonts*. Ellis Horwood, Chichester.

Smithson, T. R., R. L. Carroll, A. L. Panchen, and S. M. Andrews. 1994. *Westlothiana lizziae* from the Visean of East Kirkton, West Lothian, Scotland, and the amniote stem. *Transactions of the Royal Society of Edinburgh* 84: 383-412.

Sprinkle, J. 1973. *Morphology and evolution of blastozoan echinoderms. Harvard University Museum of Comparative Zoology Special Publication*, Cambridge, Mass.

Sprinkle, J. 1976. Classification and phylogeny of pelmatozoan echinoderms. *Systematic Zoology* 25: 83-91.

Sprinkle, J. 1980. An overview of the fossil record, pp. 15-26, *in* Broadhead, T. W., and Waters, J. A., eds., *Echinoderms: Notes for a Short Course*. University of Tennessee Department of Geological Sciences Studies in Geology 4.

Sprinkle, J., and Kier, P. M., 1987. Phylum Echinodermata, pp. 550-611, *in* Boardman, R. S., A. H. Cheetham, and A. J. Rowell, eds., *Fossil Invertebrates*. Blackwell Scientific Publishers, Cambridge, Mass.

Srinivasan, M. S., and J. P. Kennett. 1981. A review of Neogene planktic foraminiferal biostratigraphy: applications in the equatorial and South Pacific. *SEPM Special Publication* 32:395-432.

Stanley, G. D., Jr. 1981. Early history of scleractinian corals and its geological consequences. *Geology* 9:507-511.

Stanley, G. D., Jr., ed. 1996. *Paleobiology and Biology of Corals: Notes for a Short Course*. University of Tennessee Department of Geological Sciences Studies in Geology 10.

Stanley, S. M. 1968. Post-Paleozoic adaptive radiation of infaunal bivalve molluscs: a consequence of mantle fusion and siphon formation. *Journal of Paleontology* 42:214-229.

Stanley, S. M. 1970. Relation of shell form to life habits in the Bivalvia (Mollusca). *Geological Society of America Memoir* 125.

Stanley, S. M. 1975. Why clams have the shape they have: an experimental analysis of burrowing. *Paleobiology* 1:48-58.

Stanley, S. M. 1977. Trends, rates, and patterns of evolution in the Bivalvia, pp. 209-250, *in* Hallam, A., ed., *Patterns of Evolution as Illustrated by the Fossil Record*. Elsevier, Amsterdam.

Stanley, S. M. 1979. *Macroevolution: Pattern and Process*. W. H. Freeman, New York.

Stanley, S. M. 1982. Gastropod torsion: predation and the opercular imperative. *Neues Jahrbuch für Geologie und Paläontologie Abhandlungen* 164:95-107.

Stanley, S. M. 1984. Mass extinctions in the ocean. *Scientific American* 250(6):46-54.

Stanley, S. M. 1987. *Extinction*. Scientific American Library, New York.

Stanley, S. M. 1990. Delayed recovery and the spacing of major extinctions. *Paleobiology* 16:401-414.

Stanley, S. M., and L. D. Campbell. 1981. Neogene mass extinction of western North Atlantic molluscs. *Nature* 293:457-459.

Stanley, S. M., and W. A. Newman. 1980. Competitive exclusion in evolutionary time: the case of the acorn barnacles. *Paleobiology* 6:173-183.

Stanley, S. M., and X. Yang. 1987. Approximate evolutionary stasis of bivalve morphology over millions of years: a multivariate, multilineage study. *Paleobiology* 13:113-139.

Stanton, R. J., Jr. 1976. The relationship of fossil communities to the original communities of living organisms, pp. 107-142, *in* Scott, R. W., and R. R. West, eds., *Structure and Classification of Paleocommunities*. Dowden, Hutchinson, & Ross, Stroudsburg, Penn.

Stanton, R. J., Jr., and J. R. Dodd. 1970. Paleoecologic techniques: comparison of faunal and geochemical analyses of Pliocene paleoenvironments, Kettleman Hills, California. *Journal of Paleontology* 44:1092-1121.

Stanton, R. J., Jr., and J. R. Dodd. 1981. *Paleoecology: Concepts and Applications* (2nd ed., 1990). Wiley, New York.

Stearn, C. W. 1975. The stromatoporoid animal. *Lethaia* 8:89-100.

Steele, E. J. 1981. *Somatic Selection and Adaptive Evolution*. University of Chicago Press, Chicago.

Stehli, F. G. 1968. Taxonomic diversity gradients in pole location: the Recent model, pp. 163-227, *in* Drake, E. T., ed., *Evolution and Environment*. Yale University Press, New Haven.

Stehli, F. G., R. G. Douglas, and N. D. Newell. 1969. Generation and maintenance of gradients in taxonomic diversity. *Science* 164:947-949.

Stehli, F. G., and S. D. Webb, eds. 1985. *The Great American Biotic Interchange*. Plenum Press, New York.

Stehli, F. G., and J. W. Wells. 1971. Diversity and age patterns in hermatypic corals. *Systematic Zoology* 20:115-126.

Stein, R. S. 1975. Dynamic analysis of *Pteranodon ingens*: a reptilian adaptation for flight. *Journal of Paleontology* 49:534-548.

Stenzel, H. B. 1949. Successional speciation in paleontology: the case of the oysters of the *sellaeformis* stock. *Evolution* 3:34-50.

Stets, J., A. -R. Ashraf, H. K. Erben, G. Hahn, U. Hambach, K. Krumsiek, J. Thein, and P. Wurster. 1995. The Cretaceous-Tertiary boundary in the Nanxiong Basin (continental facies, southeast China), *in* MacLeod, N., and G. Keller, eds., *The Cretaceous-Tertiary Mass Extinction: Biotic and Environmental Effects*. W. W. Norton, New York.

Stevens, M. S., and J. B. Stevens. 1996. Merycoidodontinae and Miniochoerinae, pp. 498-573, *in* Prothero, D. R., and R. J. Emry, eds., *The Terrestrial Eocene-Oligocene Transition in North America*. Cambridge University Press, New York.

Stigler, S. M., and M. J. Wagner. 1987. A substantial bias in nonparametric tests for periodicity in geophysical data. *Science* 238:940–945.

Stokes, R. B. 1977. The echinoids *Micraster* and *Epiaster* from the Turonian and Senonian of southern England. *Palaeontology* 20:805-821.

Størmer, L. 1952. Phylogeny and taxonomy of fossil horseshoe crabs. *Journal of Paleontology* 26:630-639.

Størmer, L. 1955. Arthropoda 2: Merostomata, pp. 4-41, *in* Moore, R. C., ed., *Treatise on Invertebrate Paleontology, Part P*. Geological Society of America and University of Kansas Press, Lawrence.

Stott, L. D. 1992. Higher temperatures and lower oceanic pCO_2: A climate enigma at the end of the Paleocene Epoch. *Paleoceanography* 7:395–404.

Stucky, R. K. 1990. Evolution of land mammal diversity in North America during the Cenozoic. *Current Mammalogy* 2:375-432.

Surlyk, F. 1979. Review of "The Ecology of Fossils," edited by W. S. McKerrow. *Paleobiology* 4:444-446.

Swan, R. H. 1988. Heterochronic trends in Namurian ammonoid evolution. *Palaeontology* 31:1033-1051.

Swart, P. K., and G. D. Stanley, Jr. 1989. Intraskeletal variations in the carbon and oxygen isotopic composition of a Late Triassic coral *Toechastrea major*: implications for the development of symbiotic associations in scleractinian corals. *Geological Society of America Abstracts with Programs* 21(6):111.

Sweet, W. C. 1985. Conodonts—those fascinating little whatzits. *Journal of Paleontology* 59:485-494.

Sweet, W. C. 1988. *The Conodonta*. Oxford University Press, New York.

Sweet, W. C., and S. M. Bergstrom, eds. 1971. Symposium on conodont biostratigraphy. *Geological Society of America Memoir* 127.

Szalay, F. S., M. J. Novacek, and M. C. McKenna, eds. 1993. *Mammal Phylogeny*. Springer-Verlag, Berlin.

Talbot, M. R. 1990. A review of the palaeohydrological interpretation of carbon and oxygen isotopic ratios in primary lacustrine carbonates. *Chemical Geology: Isotope Geosciences Section* 80.261-279.

Tassy, P., and J. Shoshani. 1988. The Tethytheria: elephants and their relatives, pp. 283-316, *in* Benton, M. J., ed., *The Phylogeny and Classification of the Tetrapods, Vol. 2: Mammals*. Clarendon Press, Oxford.

Taylor, J., ed. 1996. *Origin and Evolutionary Radiation of the Mollusca*. Oxford University Press, New York.

Taylor, M. E. 1977. Late Cambrian of western North America: trilobite biofacies, environmental significance, and biostratigraphic implications, pp. 397-426, *in* Kauffman, E., and Hazel, J. E., eds., *Concepts and Methods of Biostratigraphy*. Dowden, Hutchinson, & Ross, Stroudsburg, Penn.

Tedford, R. H. 1970. Principles and practices of mammalian geochronology in North America. *Proceedings of the North American Paleontological Convention* F:666-703.

Teichert, C. 1985. Crises in cephalopod evolution, pp. 202-214, *in* Bottjer, D. W., C. S. Hickman, and P. D. Ward, eds., *Mollusks, Notes for a Short Course*. University of Tennessee Department of Geological Sciences Studies in Geology 13.

Teichert, C., and others. 1964. Mollusca 3. Cephalopoda, Nautiloidea, *in* R. C. Moore, ed. *Treatise on Invertebrate Paleontology. Part L.* Geological Society of America and University of Kansas Press, Lawrence.

Telford, M. 1983. An experimental analysis of lunule function in the sand dollar *Mellitta quiquiesperforata*. *Marine Biology* 75:125-134.

Thaler, L. 1972. Datation, zonation, et mamméferes. *Bureau de Recherches Géologie et Minières Mémoires* 77:411-424.

Thayer, C. W. 1981. Ecology of living brachiopods, pp. 110-126, *in* Dutro, J. T., and R. S. Boardman, eds., *Lophophorates: Notes for a Short Course*. University of Tennessee Department of Geological Sciences Studies in Geology 5.

Thewissen, J. G. M., and D. P. Domning. 1992. The role of phenacodontids in the origin of the modern orders of ungulate mammals. *Journal of Vertebrate Paleontology* 12:494-504.

Thomas, R. D. K., and E. C. Olson, eds. 1980. *A Cold Look at the Warm-Blooded Dinosaurs*. AAAS Selected Symposium 28. Westview Press, Colorado.

Thompson, M. L., H. E. Wheeler, and W. R. Danner. 1950. Middle and Upper Perman fusulinids of Washington and British Columbia. *Cushman Foundation for Foraminiferal Research, Contribution* 1, Parts 3 and 4, no. 8, pp. 46-63.

Thomson, K. S. 1969. The biology of lobe-finned fishes. *Biological Review* 44:91-154.

Thomson, K. S. 1981. A radical look at fish-tetrapod relationships. *Paleobiology* 7: 153-156.

Thomson, K. S. 1991. *Living Fossil*. W. W. Norton, New York.

Thomson, K. S. 1993. The origin of the amphibia. *Quarterly Review of Biology* 37:189-241.

Thomson, K. S. 1994. The origin of the tetrapods, pp. 85-107, *in* Prothero, D. R., and R. M. Schoch, eds., *Major Features of Vertebrate Evolution. Short Courses in Paleontology* 7. Paleontological Society and University of Tennessee Press.

Thulborn, R. A. 1973. Thermoregulation in dinosaurs. *Nature* 245:51-52.

Thulborn, R. A. 1989. The gaits of dinosaurs, pp. 39-50, *in* Gillette, D. D., and M. G. Lockley, eds. 1989. *Dinosaur Tracks and Traces*. Cambridge University Press, Cambridge.

Tillier, S., M. Masselot, P. Hervé, and A. Tillier. 1992. Phylogénie moléculaire des Gastropoda (Mollusca) fondée sure le séquençage partiel de l'ARN ribosomique 28 S. *Comptes Rendus Academie de Sciences (Paris), Series 3*, 134:79-85.

Tipper, H. W. 1981. Offset of an upper Pliensbachian geographic zonation in the North American Cordillera by transcurrent movement. *Canadian Journal of Earth Sciences* 18:1788-1792.

Tozer, E. T. 1982. Marine Triassic faunas of North America: their significance for assessing plate and terrane movements. *Geologische Rundschau* 71:1077-1104.

Tracy, C. R. 1976. Tyrannosaurs: evidence for endothermy? *American Naturalist* 110:1105-1106.

Tremaine, S. D. 1986. Is there evidence of a solar companion star? pp. 409–416, *in* Smoluchowski, R. S., J. N. Bahcall, and M. S. Matthews, eds., *The Galaxy and the Solar System*. University of Arizona Press, Tucson.

Trueman, A. E. 1922. The use of *Gryphaea* in the correlation of the Lower Lias. *Geological Magazine* 59:256-268.

Trueman, A. E. 1941. The ammonite body chamber, with special reference to the buoyancy and mode of life in the living ammonite. *Quarterly Journal of the Geological Society of London* 96:339-385.

Trueman, E. R., and M. R. Clarke, eds. 1985. Evolution, *in* Wilbur, K. M., ed., *The Mollusca*, vol. 10. Academic Press, Orlando, Florida.

Tunnicliffe, V. 1992. The nature and origin of the modern hydrothermal vent fauna. *Palaios* 7:338-350.

Turbeville, J. M., J. R. Schutz, and R. A. Raff. 1994. Deuterostome phylogeny and the sister group of the Chordates: evidence from molecules and morphology. *Molecular Biology and Evolution* 11:648.

Ubaghs, G. 1967. Stylophora, p. S495-S565, *in* R. C. Moore, ed., *Treatise on Invertebrate Paleontology. Part S, Echinodermata 1*. Geological Society of America and University of Kansas Press, Lawrence.

Ubaghs, G. 1975. Early Paleozoic echinoderms. *Annual Reviews of Earth and Planetary Sciences* 3:79-98.

Underwood, A. J. 1972. Spawning, larval development and settlement behavior of *Gibbula cineraria* (Gastropoda: Prosobranchia) with a reappraisal of torsion in gastropods. *Marine Biology* 17:341-349.

Valentine, J. W. 1966. Numerical analysis of marine molluscan ranges in the extratropical northwest Pacific shelf. *Limnology and Oceanography* 11:198-211.

Valentine, J. W. 1967. The influence of climatic fluctuations on species diversity within the Tethyan provincial system. *Systematics Association Publication* 7:153-166.

Valentine, J. W. 1969. Patterns of taxonomic and ecologic structure in the shelf benthos during Phanerozoic times. *Palaeontology* 12:684-709.

Valentine, J. W. 1970. How many marine invertebrate fossil species? A new approximation. *Journal of Paleontology* 44:410-415.

Valentine, J. W. 1973a. *Evolutionary Paleoecology of the Marine Biosphere*. Prentice-Hall, Englewood Cliffs, New Jersey.

Valentine, J. W. 1973b. Phanerozoic taxonomic diversity: a test of alternate models. *Science* 180: 1078-1079.

Valentine, J. W., T. C. Foin, and D. Peart. 1978. A provincial model of Phanerozoic marine diversity. *Paleobiology* 4:55-66.

Valentine, J. W., and E. M. Moores. 1970. Plate tectonic regulation of faunal diversity and sea level: a model. *Nature* 228:657-659.

Van Couvering, J. A. H. 1980. Community evolution in East Africa during the late Cenozoic, pp. 272-298, *in* Behrensmeyer, A. K., and A. P. Hill, eds., *Taphonomy and Vertebrate Paleoecology, with Special Reference to the Late Cenozoic of Sub-Saharan Africa*. University of Chicago Press, Chicago.

Van Valen, L. 1973. A new evolutionary law. *Evolutionary Theory* 1:1-30.

Van Valkenburgh, B. 1985. Locomotor diversity within past and present guilds of large predatory mammals. *Paleobiology* 11:406-428.

Van Valkenburgh, B. 1988. Trophic diversity in past and present guilds of large predatory mammals. *Paleobiology* 14:155-173.

Vermeij, G. J. 1977. The Mesozoic marine revolution: gastropods, predators, and grazers. *Paleobiology* 3:245-258.

Vermeij, G. J. 1987. *Evolution and Escalation: A Natural History of Life*. Princeton University Press, Princeton, New Jersey.

Vermeij, G. J. 1993. *A Natural History of Shells*. Princeton University Press, Princeton, New Jersey.

Visscher, H., H. Brinkhuis, D. L. Dilcher, W. C. Elsik, Y. Eshet, C. V. Looy, M. R. Rampino, and A. Traverse. 1996. The terminal Paleozoic fungal event: evidence of terrestrial ecosystem destabilization and collapse. *Proceedings of the National Academy of Sciences* 93:2155-2158.

Voigt, E. 1979. The preservation of slightly or non-calcified fossil Bryozoa (Ctenostomata and Cheilostomata) by bioimmuration, pp. 541-564, *in* Larwood, G. P., and M. B. Abbott, eds., *Advances in Bryozoology*. Academic Press, London.

Voorhies, M. R. 1969. Taphonomy and population dynamics of an early Pliocene vertebrate fauna, Knox County, Nebraska. *University of Wyoming Contributions in Geology, Special Paper* 1: 1-69.

Voorhies, M. R. 1981. Ancient ashfall creates Pompeii of prehistoric rhinos. *National Geographic* 159 (1):66-75.

Vrba, E. S. 1985. Environment and evolution: alternative causes of the temporal distribution of evolutionary events. *South African Journal of Science* 82:229-236.

Waddington, C. H. 1957. *The Strategy of the Genes*. Allen & Unwin, London.

Wagner, P. J. 1995. Stratigraphic tests of cladistic hypotheses. *Paleobiology* 21:153-178.

Walker, A., and P. Shipman. 1996. *The Wisdom of the Bones*. Alfred Knopf, New York.

Walker, C. A. 1981. New subclass of birds from the Cretaceous of South America. *Nature* 292:51-52.

Walker, K. R., and L. P. Alberstadt. 1975. Ecological succession as an aspect of structure in fossil communities. *Paleobiology* 1:238-257.

Wallace, A. R. 1876. *The Geographical Distribution of Animals*. Macmillan, London.

Wang, Y., T. E. Cerling, and B. J. MacFadden. 1994. Fossil horses and carbon isotopes: new evidence for Cenozoic dietary, habitat, and ecosystem changes in North America. *Palaeogeography, Palaeoclimatology, Palaeoecology* 107:269-279.

Ward, P. D. 1980. Comparative shell shape distributions in Jurassic-Cretaceous ammonoids and Jurassic-Tertiary nautiloids. *Paleobiology* 6:32-43.

Ward, P. D. 1982. *Nautilus*: have shell, will float. *Natural History* 91:64-69.

Ward, P. D. 1987. *The Natural History of Nautilus*. Allen & Unwin, Boston.

Ward, P. D., 1992. *On Methuselah's Trail: Living Fossils and the Great Extinctions*. W. H. Freeman, New York.

Ward, P. D., and G. E. Westermann. 1985. Cephalopod paleoecology, pp. 215-229, *in* Bottjer, D. W., C. S. Hickman, and P. D. Ward, eds., *Mollusks, Notes for a Short Course*. University of Tennessee Department of Geological Sciences Studies in Geology 13.

Warme, J. E., A. A. Ekdale, and S. F. Ekdale. 1976. Raw material of the fossil record, pp. 143-169, *in* Scott, R. W., and R. R. West, eds., *Structure and Classification of Paleocommunities*. Dowden, Hutchinson, & Ross, Stroudsburg, Penn.

Waterston, C. D. 1957. The Scottish Carboniferous Eurypterida. *Transactions of the Royal Society of Edinburgh* 66:265-288.

Waterston, C. D. 1979. Problems of functional morphology and classification of stylonurid eurypterids (Chelicerata, Merostomata), with observations on Scottish Silurian Stylonuroidea. *Transactions of the Royal Society of Edinburgh* 70:251-322.

Webb, S. D. 1972. Locomotor evolution in camels. *Forma et Functio* 5:99-112.

Webb, S. D. 1977. A history of savanna vertebrates in the New World. Part I: North America. *Annual Reviews of Ecology and Systematics* 8:355–380.

Webb, S. D. 1991. Ecogeography and the Great American Interchange. *Paleobiology* 17:266-280.

Wegener, A. 1915. *Die Entstehung der Kontinente und Ozeane*. Sammlung Vieweg, Braunschweig, Vieweg.

Weishampel, D. B., P. Dodson, and H. Osmolska, eds. 1990. *The Dinosauria*. University of California Press, Berkeley.

Weitschat, W., and K. Bandel. 1991. Organic components in phragmocones of Boreal Triassic ammonoids: implications for ammonoid biology. *Paläontologische Zeitschrift* 65:269-303.

Wellnhofer, P. 1987. Die Flughaut von *Pterodactylus* (Reptilia, Pterosauria) am Beispiel des Wiener Exemplares von *Pterodactylus kochi* (Wagner). *Annalen des Naturhistorisches Museums Wien* 88A: 149-162.

Wellnhofer, P. 1988. Terrestrial locomotion in pterosaurs. *Historical Biology* 1:3-16.

Wellnhofer, P. 1991. *The Illustrated Encyclopedia of Pterosaurs*. Crescent Books, New York.

Werman, S. D., M. S. Springer, and R. J. Britten. 1990. Nucleic acids I: DNA-DNA hybridization, pp. 204-249, *in* Hillis, D. M., and C. Moritz, eds. *Molecular Systematics*. Sinauer Associates, Sunderland, Mass.

Whitmire, D. P., and A. A. Jackson IV. 1984. Are periodic mass extinctions driven by a distant solar companion? *Nature* 308: 713–715.

Whitmire, D. P,. and A. A. Jackson IV. 1985. Periodic comet showers and Planet X. *Nature* 313: 36–38.

Whittaker, R. H. 1969. New concepts on the kingdoms of organisms. *Science* 163:150-160.

Whittington, H. B. 1957. The ontogeny of trilobites. *Biological Reviews* 32:421-469.

Whittington, H. B. 1966. Phylogeny and distribution of Ordovician trilobites. *Journal of Paleontology* 40:696-737.

Whittington, H. B. 1981. Paedomorphosis and cryptogenesis in trilobites. *Geological Magazine* 11:591-602.

Wiedmann, J. 1969. The heteromorphs and ammonoid extinction. *Biological Reviews* 48:159-194.

Wilde, P., and W. B. N. Berry. 1984. Destabilisation of the oceanic density structure and its significance to marine "extinction" events. *Palaeogeography, Palaeoclimatology, Palaeoecology* 48:143-162.

Williams, A. 1968. Evolution of the shell structure in the articulate brachiopods. *Special Papers in Palaeontology* 2:1-55.

Williams, A., and A. J. Rowell. 1965. Brachiopoda, *in* Moore, R. C., ed., *Treatise on Invertebrate Paleontology*. Part H. Geological Society of America and University of Kansas Press, Lawrence.

Williams, A., and J. M. Hurst. 1977. Brachiopod evolution, pp. 79-121, *in* Hallam, A., ed. *Patterns of Evolution as Illustrated in the Fossil Record*. Elsevier, Amsterdam.

Williams, H. S. 1901. The discrimination of time values in geology. *Journal of Geology* 9:570-585.

Williamson, P. G. 1981. Paleontological documentation of speciation in Cenozoic molluscs from Turkana Basin. *Nature* 293: 437-443.

Willis, M. A., D. E. G. Briggs, and R. A. Fortey. 1994. Disparity as an evolutionary index: a comparison between Cambrian and Recent arthropods. *Paleobiology* 20:93-130.

Wilson, E. O. 1996. Little things that rule the world, pp. 139-146, *in* Wilson, E. O. *In Search of Nature*. Island Press, Washington, D. C.

Wilson, E. O., and W. H. Bossert. 1971. *A Primer of Population Biology*. Sinauer Associates, Stamford, Conn.

Wingstrand, K. G. 1985. On the anatomy and relationships of recent Monoplacophora. *Galathea Report* 16:7-94.

Witmer, L. M. 1991. Perspectives on avian origins, pp. 427-466, *in* Schultze, H. -P., and L. Trueb, eds., *Origins of the Higher Groups of Tetrapods*. Comstock Publishing Company, Ithaca, New York.

Witmer, L. M., and K. D. Rose. 1991. Biomechanics of the jaw apparatus of the gigantic Eocene bird *Diatryma*: implications for diet and mode of life. *Paleobiology* 17:95-120.

Woese, C. R. 1987. Bacterial evolution. *Microbiology Reviews* 51:221-271.

Wolfe, J. A. 1978. A paleobotanical interpretation of Tertiary climates in the Northern Hemisphere. *American Scientist* 66: 694–703.

Wolfe, J. A. 1990. Estimates of Pliocene precipitation and temperature based on multivariate analysis of leaf physiognomy. *U.S. Geological Survey Open-File Report* 90-94:39–42.

Wolfe, J. A. 1994. Tertiary climatic change at the middle latitudes of North America. *Palaeogeography, Palaeoclimatology, Palaeoecology* 108:195-205.

Wood, H. E., II, R. W. Chaney, J. Clark, E. H. Colbert, G. L. Jepsen, J. B. Reeside, and C. Stock. 1941. Nomenclature and correlation of the North American continental Tertiary. *Geological Society of America Bulletin* 52:1-48.

Wood, R. A. 1993. Nutrients, predators, and the history of reefs. *Palaios* 8:526-543.

Wood, R. A. 1995. The changing biology of reef-building. *Palaios* 10:517-529.

Wood, R. A., A. Y. Zhuravlev, and F. Debrenne. 1992. Functional biology and ecology of the Archaeocyatha. *Palaios* 7:131-156.

Woodburne, M. O. 1977. Definition and characterization in mammalian chronostratigraphy. *Journal of Paleontology* 51:220-234.

Woodburne, M. O., ed. 1987. *Cenozoic Mammals of North America: Geochronology and Biostratigraphy*. University of California Press, Berkeley.

Woodburne, M. O. 1989. Hipparion horses: a pattern of world-wide dispersal and endemic evolution, pp. 197-233, *in* Prothero, D. R., and R. M. Schoch, eds., *The Evolution of Perissodactyls*. Oxford University Press, New York.

Woodburne, M. O. 1996. Reappraisal of *Cormohipparion* from the Valentine Formation, Nebraska. *American Museum Novitates* 3163:1-56.

Woodburne, M. O., and C. C. Swisher III. 1995. Land mammal high-resolution geochronology, intercontinental overland dispersals, sea level, climate, and vicariance. *SEPM Special Publication* 54:335-364.

Wootton, R. J. 1981. Palaeozoic insects. *Annual Review of Entomology* 26:319-344.

Wray, J. L. 1971. Ecology and geologic distribution, pp. 5. 1-5. 6, *in* Ginsburg, R., R. Rezak, and J. L. Wray, eds., *Geology of Calcareous Algae: Notes for a Short Course*. Comparative Sedimentology Laboratory, University of Miami.

Wright, S. 1932. The roles of mutation, inbreeding, cross-breeding, and selection in evolution. *Proceedings Fourth International Congress of Genetics* 1:356-366.

Yochelson, E. L. 1978. An alternative approach to the interpretation of the phylogeny of ancient mollusks. *Malacologia* 17:165-192.

Yochelson, E. L., R. H. Flower, and G. F. Webers. 1973. The bearing of a new later Cambrian monoplacophoran genus *Knightoconus* upon the origin of the Cephalopoda. *Lethaia* 6:275-310.

Yonge, C. M. 1973. Giant clams. *Scientific American* 232:96-105.

Yonge, C. M., and T. E. Thompson. 1976. *Living Marine Molluscs*. William Collins, London.

Yonge, C. M., and T. E. Thompson, eds. 1978. Evolutonary systematics of bivalve molluscs. *Philosophical Transactions of the Royal Society of London* (B) 284:199-436.

Young, G. C., 1986. The relationships of placoderm fishes. *Zoological Journal of the Linnean Society of London* 88:1-57.

Zhou, Z., F. Jin, and J. Zhang. 1992. Preliminary report on a Mesozoic bird from Liaoning, China. *Chinese Science Bulletin* 37:1365-1368.

Ziegler, A. M. 1966. The Silurian brachiopod *Eocoelia hemisphaerica* (J. de C. Sowerby) and related species. *Palaeontology* 9:523-543.

Ziegler, A., L. R. M. Cocks, and R. K. Bambach. 1968. The composition and structure of Lower Silurian marine communities. *Lethaia* 1:1-27.

Zinsmeister, W. J., R. M. Feldmann, M. O. Woodburne, and D. H. Elliott. 1989. Latest Cretaceous/ earliest Tertiary transition on Seymour Island, Antarctica. *Journal of Paleontology* 63:731-738.

Index

Credits

Figure 1.11. Reprinted by permission of the publisher from *Life History of a Fossil*, by Pat Shipman, Cambridge, Mass.: Harvard University Press. Copyright © 1981 by the President and Fellows of Harvard College.

Figure 1.14. From *Principles of Paleontology* 2/E by Raup and Stanley © 1978 by W. H. Freeman and Company. Used with permission.

Figure 1.14D. Photograph by Dr. J. L. Franzen, Forschunginstitut Senckenberg.

Figure 1.5B. Courtesy of the George C. Page Museum.

Figure 1.6A. Reprinted with permission from A. J. Boucot, 1990, *Evolutionary Paleobiology of Behavior and Coevolution*, Fig. 79, p. 90, with kind permission from Elsevier Science-NL, Sara Burgerhartstraat 25, 1055 KV Amsterdam, The Netherlands.

Figure 2.5A from *Paleontology: The Record of Life*, by C. Stearn and R. L. Carroll, copyright © 1989 John Wiley & Sons, Inc. Reprinted by permission of John Wiley & Sons, Inc.

Figure 2.14 (bottom) from *The Wisdom of the Bones* by Alan Walker and Pat Shipman. Copyright © 1996 by Alan Walker and Pat Shipman. Reprinted with permission of Alfred A. Knopf, Inc.

Figure 2.14 (top). Copyright © 1982 by Coloring Concepts Inc. Reprinted by permission of Harper Collins, Inc.

Figure 2.15C. From R. K. Bambach, 1973, *American Journal of Science*, Fig. 6, p. 419. Reprinted with Permission of the American Journal of Science.

Figure 3.1 was obtained from IMSI's Master Photos Collection, 1895 Francisco Blvd. East, San Rafael, CA 94901-5506, USA.

Figure 3.2. Reprinted by permission of Blackwell Science, Inc.

Figure 3.3. from *Systematics and the Origin of Species* by Ernst Mayr. Copyright © 1982 by Columbia University Press. Reprinted with permission of the publisher.

Figure 3.4. from *Systematics and the Origin of Species* by Ernst Mayr. Copyright © 1982 by Columbia University Press. Reprinted with permission of the publisher.

Figure 4.7. Reprinted with permission from Fitch and Margoliash, 1967, *Science*, v. 155, p. 279-284, Fig. 1. Copyright © 1967 American Association for the Advancement of Science.

Figure 5.3. Reprinted by permission of Blackwell Science, Inc.

Figure 6.3. Reprinted with permission from Raup and Sepkoski, 1982, *Science*, v. 215, p. 1501, fig. 1. Copyright © 1982 American Association for the Advancement of Science.

Figure 6.5. Reprinted with permission from Raup and Sepkoski, 1986, *Science*, v. 231, p. 84, fig. 1. Copyright © 1986 American Association for the Advancement of Science.

Figure 7.1. Reprinted with permission from Raup and Michelson, 1965, *Science*, v. 147, p. 1294-1295, Fig. 1. Copyright © 1965 American Association for the Advancement of Science.

Figure 7.15C. © 1985 Mark Hallett Illustrations.

Figure 8.10C. Reprinted with permission from A. J. Boucot, 1990, *Evolutionary Paleobiology of Behavior and Coevolution*, Fig. 167, p. 185, with kind permission from Elsevier Science-NL, Sara Burgerhartstraat 25, 1055 KV Amsterdam, The Netherlands.

Figure 8.10E. Reprinted with permission from A. J. Boucot, 1990, *Evolutionary Paleobiology of Behavior and Coevolution*, Fig. 147, p. 165, with kind permission from Elsevier Science-NL, Sara Burgerhartstraat 25, 1055 KV Amsterdam, The Netherlands.

Figure 8.12. From Miller et al., 1987, *Paleoceanography*, v. 2, p. 4, Fig. 2, copyright by the American Geophysical Union.

Figure 8.18. From J. A. Valentine, *Phanerozoic Diversity Patterns*, copyright 1985 by Princeton University Press. Reprinted by permission of Princeton University Press.

Figure 8.21. From J. A. Valentine, *Phanerozoic Diversity Patterns*, copyright 1985 by Princeton University Press. Reprinted by permission of Princeton University Press.

Figure 8.5. From J. A. Valentine, *Phanerozoic Diversity Patterns*, copyright 1985 by Princeton University Press. Reprinted by permission of Princeton University Press.

Figure 8.7. From J. A. Valentine, *Phanerozoic Diversity Patterns*, copyright 1985 by Princeton University Press. Reprinted by permission of Princeton University Press.

Figure 8.9A. Reprinted with permission from A. J. Boucot, 1990, *Evolutionary Paleobiology of Behavior and Coevolution*, Fig. 45, p. 55, with kind permission from Elsevier Science-NL, Sara Burgerhartstraat 25, 1055 KV Amsterdam, The Netherlands.

Figure 8.9D. Reprinted with permission from A. J. Boucot, 1990, *Evolutionary Paleobiology of Behavior and Coevolution*, Fig. 321, p. 390, with kind permission from Elsevier Science-NL, Sara Burgerhartstraat 25, 1055 KV Amsterdam, The Netherlands.

Figure 8.11 from *Dynamic Stratigraphy: An Introduction to*

Sedimentation, by R. Matthews, ©1984. Reprinted by permission of Prentice-Hall, Inc., Upper Saddle River, NJ.

Figure 9.1. from *Life: An Introduction to Biology*, Second Edition, by George G. Simpson and William S. Beck, copyright © 1965 by Harcourt Brace & Company, and renewed 1993 by William S. Beck, Elizabeth Simpson Wurr, Helen S. Vishniac, and Joan S. Burns, reproduced by permission of the publisher.

Figure 9.2 from *Paleontology: The Record of Life*, by C. Stearn and R. L. Carroll, copyright © 1989 John Wiley & Sons, Inc. Reprinted by permission of John Wiley & Sons, Inc.

Figure 9.10A. From Wilson, E. O., and R. A. Macarthur, *Theory of Island Biogeography*, copyright © 1967 by Princeton University Press. Reprinted by permission of Princeton University Press.

Figure 9.14A. From *Geological Society of America Special Paper* 241, p. 71, by D. Krause and M. Maas. Reproduced with permission of the publisher, the Geological Society of America, Boulder, Colorado, USA. Copyright © 1978 Geological Society of America.

Figure 9.15. from *Vicariance Biogeography* by Gareth Nelson and Donn Rosen. Copyright © 1981 by Columbia University Press. Reprinted with permission of the publisher.

Figure 9.16. from *Vicariance Biogeography* by Gareth Nelson and Donn Rosen. Copyright © 1981 by Columbia University Press. Reprinted with permission of the publisher.

Figure 9.3. From A. N. Strahler, 1971. *The Earth Sciences*, 2e.

Figure 9.7. By permission of the Society of Systematic Biologists.

Figure 10.4 from *Geologic Time* (2e) by Don L. Eicher, © 1976. Reprinted by permission of Prentice-Hall, Inc., Upper Saddle River, NJ.

Figure 10.5 from *Geologic Time* (2e) by Don L. Eicher, © 1976. Reprinted by permission of Prentice-Hall, Inc., Upper Saddle River, NJ.

Figure 10.7 from *Geologic Time* (2e) by Don L. Eicher, © 1976. Reprinted by permission of Prentice-Hall, Inc., Upper Saddle River, NJ.

Figure 10.9 from *Geologic Time* (2e) by Don L. Eicher, © 1976. Reprinted by permission of Prentice-Hall, Inc., Upper Saddle River, NJ.

Figure 10.18. From *Sedimentary Geology: An Introduction to Sedimentary Rocks and Stratigraphy*, by D. R. Prothero and F. Schwab, © 1996 by W. H. Freeman & Co. Used with permission.

Figure 10.19. From *Sedimentary Geology: An Introduction to Sedimentary Rocks and Stratigraphy*, by D. R. Prothero and F. Schwab, © 1996 by W. H. Freeman & Co. Used with permission.

Figure 10.8. From *Interpreting the Stratigraphic Record*, by D. R. Prother, © 1990 by W. H. Freeman & Co. Used with permission.

Figure 11.1. Courtesy Scripps Institution of Oceanography/UCSD.

Figure 11.16. Reprinted by permission of Blackwell Science, Inc.

Figure 11.9. By permission of the Society of Systematic Biologists.

Figure 12.1 was obtained from IMSI's Master Photos Collection, 1895 Francisco Blvd. East, San Rafael, CA 94901-5506, USA.

Figure 14.9. From P. R. Sheldon, *Nature*, 1987, v. 330, p. 561-563, Fig. 1. Reprinted with permission from *Nature*, copyright © 1987. Macmillan Magazines Ltd.

Figures 15.11, 15.12, 15.13 from *Search for the Past* (2e) by J. R. Beerbower, ©1968. Reprinted by permission of Prentice-Hall, Inc., Upper Saddle River, NJ.

Figures 15.14, 15.16, 15.17. From *Geological Society of America Memoir* 125, by S. M. Stanley. Reproduced with permission of the publisher, the Geological Society of America, Boulder, Colorado, USA. Copyright © 1970 Geological Society of America.

Figure 16.2. Reprinted from *Deep Sea Research*, v.14, p. 798, fig. 6, © 1978. with kind permission of Elsevier Science, Ltd., The Boulevard, Langford Lane, Kidlington OX5 1GB, U.K.

Figure 16.9 from *Paleontology: The Record of Life*, by C. Stearn and R. L. Carroll, copyright © 1989 John Wiley & Sons, Inc. Reprinted by permission of John Wiley & Sons, Inc.

Figure 16.10A-C from *Search for the Past* (2e) by J. R. Beerbower, ©1968. Reprinted by permission of Prentice-Hall, Inc., Upper Saddle River, NJ.

Figure 16.10D from *Paleontology: The Record of Life*, by C. Stearn and R. L. Carroll, copyright © 1989 John Wiley & Sons, Inc. Reprinted by permission of John Wiley & Sons, Inc.

Figure16.11 from *Paleontology: The Record of Life*, by C. Stearn and R. L. Carroll, copyright © 1989 John Wiley & Sons, Inc. Reprinted by permission of John Wiley & Sons, Inc.

Figure 16.14 from *Search for the Past* (2e) by J. R. Beerbower, ©1968. Reprinted by permission of Prentice-Hall, Inc., Upper Saddle River, NJ.

Figure 16.16 from *Paleontology: The Record of Life*, by C. Stearn and R. L. Carroll, copyright © 1989 John Wiley & Sons, Inc. Reprinted by permission of John Wiley & Sons, Inc.

Figure 16.17 from *Search for the Past* (2e) by J. R. Beerbower, ©1968. Reprinted by permission of Prentice-Hall, Inc., Upper Saddle River, NJ.

Figure 16.20 from *Search for the Past* (2e) by J. R. Beerbower, ©1968. Reprinted by permission of Prentice-Hall, Inc., Upper Saddle River, NJ.

Figure 17.7 from *Vertebrate Life* (4e) by Pough et al., 1996. Reprinted by Permission of Prentice-Hall, Inc., Upper Saddle River, NJ.

Figure 17.10 from *Wonderful Life: The Burgess Shale and the Nature of History*, by Stephen Jay Gould. Copyright © 1989 by Stephen Jay Gould. Reprinted by permission of W. W. Norton & Company, Inc.

Figure 17.20 from *Vertebrate Life* (4e) by Pough et al., 1996. Reprinted by Permission of Prentice-Hall, Inc., Upper Saddle River, NJ.

Figure 17.27 from *Vertebrate Life* (4e) by Pough et al., 1996. Reprinted by Permission of Prentice-Hall, Inc., Upper Saddle River, NJ.

Figure 17.28. By permission of Sternberg Museum of Natural History, Fort Hays State University, Hays, KS.

Figure 17.29 was obtained from IMSI's Master Photos Collection, 1895 Francisco Blvd. East, San Rafael, CA

94901-5506, USA.

Figure 17.30 from *Vertebrate Life* (4e) by Pough et al., 1996. Reprinted by Permission of Prentice-Hall, Inc., Upper Saddle River, NJ.

Figure 17.38 from E. H. Colbert and M. Morales, *Evolution of the Vertebrates* (4e), copyright © 1991 Wiley-Liss, Inc. Reprinted by permission of Wiley-Liss, Inc., a division of John Wiley & Sons, Inc.

Figure 17.44 from E. H. Colbert and M. Morales, *Evolution of the Vertebrates* (4e), copyright © 1991 Wiley-Liss, Inc. Reprinted by permission of Wiley-Liss, Inc., a division of John Wiley & Sons, Inc.

Figure 17.52 from E. H. Colbert and M. Morales, *Evolution of the Vertebrates* (4e), copyright © 1991 Wiley-Liss, Inc. Reprinted by permission of Wiley-Liss, Inc., a division of John Wiley & Sons, Inc.

Figure 17.53 from *Paleontology: The Record of Life,* by C. Stearn and R. L. Carroll, copyright © 1989 John Wiley & Sons, Inc. Reprinted by permission of John Wiley & Sons, Inc.

Figure 17.54 from *Paleontology: The Record of Life,* by C. Stearn and R. L. Carroll, copyright © 1989 John Wiley & Sons, Inc. Reprinted by permission of John Wiley & Sons, Inc.

Figure 17.66. Courtesy University of Nebraska State Museum.

Figure 18.1. Reprinted with permission from A. J. Boucot, 1990, *Evolutionary Paleobiology of Behavior and Coevolution,* Fig. 255, p. 314, with kind permission from Elsevier Science-NL, Sara Burgerhartstraat 25, 1055 KV Amsterdam, The Netherlands.

Figure 18.11. From Martin Lockley, 1991, *Tracking Dinosaurs*, Fig. 2.2, p. 15. Reprinted with permission of Cambridge University Press.

Figure 18.12. Reprinted with permission from A. J. Boucot, 1990, *Evolutionary Paleobiology of Behavior and Coevolution,* Fig. 275, p. 341, with kind permission from Elsevier Science-NL, Sara Burgerhartstraat 25, 1055 KV Amsterdam, The Netherlands.

Figure 18.13. From Martin Lockley, 1991, *Tracking Dinosaurs*, Fig. 7.1, p. 73. Reprinted with permission of Cambridge University Press.

Quotation on p. 213 from *Life on Earth* by David Attenborough. Copyright © 1979 by David Attenborough Productions, Ltd. by permission of Little, Brown and Company.

Quotation on p. 309 from *Life on Earth* by David Attenborough. Copyright © 1979 by David Attenborough Productions, Ltd. by permission of Little, Brown and Company.

Poem on p. 331, "Next!" by Ogden Nash, from *Verses from 1929 On* by Ogden Nash. Copyright 1953 by Ogden Nash. By Permission of Little, Brown, and Company.

Quotation on p. 338 from *Spineless Wonders* by Richard Conniff © 1996 by Richard Conniff. Reprinted by permission of Henry Holt and Company, Inc.

Quotation on p. 344 from *Discovering Fossil Fishes* by John Maisey © 1996 by John Maisey. Reprinted by permission of Henry Holt and Company, Inc.